Intertidal Invertebrates of California

TEXT CONTRIBUTORS

Donald P. Abbott

Zach M. Arnold

William C. Austin

Gerald J. Bakus

J. Laurens Barnard

Robert D. Beeman

Darl E. Bowers

Joe H. Brumbaugh

Roy L. Caldwell

Fenner A. Chace, Jr.

J. Wyatt Durham

William G. Evans

Howard M. Feder

W. Gordon Fields

John S. Garth

William B. Gladfelter

Eugene C. Haderlie

Michael G. Hadfield

Janet Haig

Cadet Hand

Joel W. Hedgpeth

F. G. Hochberg, Jr.

Welton L. Lee

Milton A. Miller

Andrew Todd Newberry

William A. Newman

Donald J. Reish

Mary E. Rice

Dorothy F. Soule

John D. Soule

Carol D. Wagner

Gary C. Williams

Russel L. Zimmer

Intertidal Invertebrates of California

Robert H. Morris, Donald P. Abbott, and Eugene C. Haderlie

WITH 31 TEXT CONTRIBUTORS

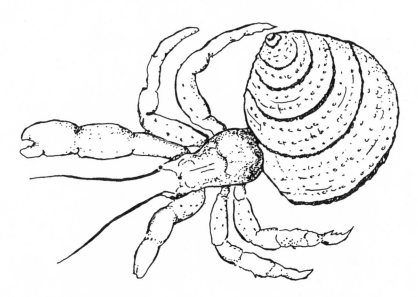

Stanford University Press, Stanford, California

Publication of this work was made possible by support

from The David and Lucile Packard Foundation

Stanford University Press

Stanford, California

Printed in the United States of America

ISBN 0-8047-1045-7

Original edition 1980

Last figure below indicates year of this printing:

90 89 88 87 86 85 84 83

To Blandine, Izzie, and Aileen

Preface

This is a book for those who find the California seashore at low tide a place of beauty, wonder, and interest, and who want information on the common animals that live there. We have not tried to cover all the animals, but have considered many that are conspicuous enough to attract the eye and distinctive enough to be identified from a clear photograph and a short description of distinguishing characters. For groups like the crabs, where the species are relatively well known, the coverage of intertidal forms is virtually complete. Groups in which the species are not readily identifiable from pictures are less well represented here, though we have included samplings of more distinctive forms. The smallest animals, which require microscopic examination to identify or even to find in the first place, have generally been omitted unless, as in the case of the Foraminifera and the copepod *Tigriopus*, they sometimes occur in such large numbers as to be conspicuous despite their small size.

The book deals with the diversity of animal life, in a tradition that is both old and new. Biologists' efforts to describe and name the different kinds of animals go back for centuries. In earlier days the work was a sort of cataloging, but even before Darwin's time the best classifications, based mainly on comparative anatomy, began to resemble those in use today. As the concept of evolution spread, those dealing with biological diversity attempted to work out genealogical relationships among existing forms and to portray them by means of evolutionary trees and improved systems of classification.

Even as this went on, however, the emphasis in biological research was shifting, and in the first half of the twentieth century the main thrust became the study of how living things function. Physiology and biochemistry made enormous strides. Researchers in these fields uncovered increasing evidence of the fundamental unity of life, and learned that at the molecular, genetic, and cellular level many basic processes are similar in all living things. The breakthroughs since the gross structure of DNA was established in 1953 have literally revolutionized biology.

This thrust continues unabated, but there is a renewed wave of interest in biological diversity. For one thing, the new tools now available, such as the transmission and scanning electron microscopes, have enabled researchers to see things that were previously invisible. The results have been extraordinarily exciting and revealing, and biologists have been looking at the whole world of morphology, histology, and cytology all over again. The comparative approach, always a powerful one, has led to results and understanding not obtainable from research using only a single species. Elegant comparative studies of such varied structures as photoreceptor organs, teeth, mollusk shells, cilia, and the whole gamut of intracellular organelles have yielded fresh insights for all fields of biology.

For another thing, it has become apparent that certain species or types of organisms offer unique potentials for particular kinds of research; they provide "systems" much better suited to pursuing certain inquiries than the traditional frogs and white rats. Thus, the squids and the polychaete worm *Myxicola* provide neurobiologists with the largest nerve fibers in the animal kingdom, and the sand crab *Emerita* supplies some of the largest sensory neurons yet described. The barnacles *Balanus nubilus* and *B. aquila*, with extraordinarily large muscle cells, are ideal for certain types of muscle research. The tiny flatworm *Polychoerus* contains the largest mitotic spindles and the largest centrioles yet found in any metazoan. The opisthobranch mollusks, with their large ganglionic cells and simple nervous systems, and the octopus, with its vastly more complex brain and behavior, are helping us begin to understand the relationship between nervous system structure and function. Thus more and more invertebrates are joining the fruit flies, hydras, crayfishes, earthworms, sea urchins, and the echiuran *Urechis* as important laboratory animals.

It has become clear, too, that despite their underlying similarities, organisms show some remarkable differences even at the biochemical level, differences that intrigue the chemists who study the variety of compounds ("natural products") formed by different living creatures. Some researchers are challenged by the ecological and evolutionary aspects of chemical diversity, others by the problems raised with regard to molecular synthesis. Those with a more practical bent point to the array of actually or potentially useful specialized substances that have been discovered in invertebrates, for example the antibacterial, antiviral, and antileukemia materials variously reported in sponges, cnidarians, tunicates, and other groups; the prostaglandins in certain sea whips; the commercial pesticide based on the toxins of a marine annelid (discovered when fishermen noted that flies died after alighting on certain bait worms); and a luminescent protein from a medusa that is useful for monitoring calcium levels in medical research.

Along another track, ecologists and their intellectual forebears living close to the land have long known that a diversity of living things is essential for the proper functioning of natural biological communities. The concept that animals are dependent ultimately upon plants ("All flesh is grass") dates from at least Biblical times. Today we recognize ever more clearly that the biogeochemical processes that recycle the materials essential to all living creatures—including ourselves—depend on a huge variety of species residing in different places and doing different things. Natural communities generally embrace greater numbers of species than comparable communities disturbed by man, and on the whole, communities with many species are probably considerably more stable than those with few species. Reduction in species diversity is generally detrimental, resulting in a more precarious balance of nature. This is probably just as true of the sea as it is of the land. A great many species of no direct use to man (and doubtless some classed as harmful) are nevertheless linked to us in the great ecological cycles. For an unknown but probably very large number of these, our own welfare may depend on their welfare.

Population biologists with a more theoretical bent are also taking a strong interest in diversity. The blending of the concepts of ecology with those of behavior and genetics has led to fresh and more quantitative investigations of the varied strategies that plants and animals employ in satisfying their needs for energy and materials, in coping with the local hazards to existence, and in reproducing their kind. Studies of such shore dwellers as limpets, hermit crabs, sea anemones, and sea stars have yielded new insights in these areas, and offer splendid opportunities for further work.

As well, there is an interest in the diversity of life by those who look along the edges of the sea for food. A variety of tasty marine invertebrates and algae have joined fishes in tickling the palates and filling the bellies of coastal peoples since prehistoric times. The kitchen middens left by California's coastal Indian tribes show that oysters, mussels, clams, abalones, limpets, chitons, shrimps, lobsters, and crabs provided important sources of food in the past, and they are still esteemed today. Conservation measures recently adopted by the State will help preserve some natural populations along the beaches and rocky shores. Techniques for mass culture and accelerated growth of edible fishes and shellfishes are being investigated in many areas, and mariculture of an increasing number of species will surely become economically feasible in a hungry world.

In view of this renewed interest in animal diversity, and with the ever-increasing popularity of the seashore as a place for both natural history recreation and serious scientific study, we feel there is a present need for a volume that summarizes a good deal of what we know about our common larger shore invertebrates, illustrates the species in color, and tells the reader where to go for more details.

This is a book for the interested layman, the professional biologist and teacher, and students at levels ranging from high school to graduate school. For some, the information contained may be all that is needed; for those with more exacting needs it is a starting point, and the references cited provide stepping-stones into the detailed scientific literature. Teachers will find that the book contains both information and an assortment of leads to studies that could be carried out by students, in many cases with little more equipment than eyes and imagination. The book may also prove useful to the researcher looking for suitable animals for particular laboratory or field studies or to the researcher who is simply on the prowl, looking for new possibilities.

The book had its origin more than a decade ago, when Robert H. Morris, an artist and photographer, then with the California State Parks and Recreation Department, approached Stanford University

Press with more than a thousand photographs of marine animals, made systematically at localities from one end of the state to the other; all animals were tentatively identified, and the work was backed by a preserved collection of the specimens photographed. Impressed with the possibilities, William W. Carver, executive editor, contacted Professor Donald P. Abbott, a resident faculty member at Stanford University's Hopkins Marine Station in Pacific Grove, California. The original plan was for a book centered on the illustrations, each picture captioned to provide the name, habitat, and geographical distribution of the species, along with such descriptive notes as would be needed to ensure proper identification. Scientists with special knowledge of the various groups were contacted and asked to provide identifications and the basic accompanying information, to select additional species to be illustrated to make coverage of the group representative, and, where deemed appropriate, to add some natural history notes and references for those who wished to know more. While work got under way on the manuscript, Morris undertook additional trips to sites recommended by specialists, to find and photograph additional species.

As the manuscript accumulated, the parts showing the greatest disparity between chapters were those dealing with "natural history." Moreover, it became increasingly evident that most of the interesting information on California species in the scientific literature had never been brought together in relatively non-technical form, though short selections were available in such widely used volumes as Johnson and Snook's *Seashore animals of the Pacific coast* (New York: Macmillan, 1935); MacGinitie and MacGinitie's *Natural history of marine animals* (second edition; New York: McGraw-Hill, 1968); Ricketts and Calvin's *Between Pacific tides* (fourth edition, revised by J. Hedgpeth; Stanford, Calif.: Stanford University Press, 1968); and Smith and Carlton's *Light's manual: Intertidal invertebrates of the central California coast* (third edition; Berkeley and Los Angeles: University of California Press, 1975). Discussions with colleagues and students led to the decision to provide a large selection of this biological information, in a form understandable to the intelligent reader willing to use a desk dictionary now and then, and to provide references for those desiring further details. Gathering the desired information greatly lengthened both the task of writing and the book itself and led us to seek a third collaborator, Professor Eugene C. Haderlie of the Naval Postgraduate School, in Monterey, California.

In sharing the work with their numerous scientific collaborators, Abbott and Haderlie are responsible for the text and references. Morris has provided most of the illustrations and had major responsibility for securing and reproducing others not supplied by the chapter authors. He has also prepared the legends for them and the small sketches in the text, reviewed the text to locate spots where scientific jargon might be reduced without reducing the information content, and compiled the index. As contributors and editors, Abbott and Haderlie made efforts to see that content and format were reasonably consistent from one chapter to the next. Our contributors have been generous in accepting revisions where space and consistency required deletions or additions.

Many people have contributed to this book, and in a variety of ways. First and foremost we wish to thank those who have written chapters to the book. Next, we extend thanks to the colleagues and students who have contributed in various other ways, providing factual details, critical acumen, pertinent references, and unpublished details of work under way. Among these, we are most deeply indebted to Myra Keen, professor emeritus of geology at Stanford University, for extensive help with the mollusks. And it is with deep sorrow that we record the death of our friend Allyn G. Smith, who contributed substantially to the chapter on chitons. Others who have provided much needed help include Einar Anderson, Alan Baldridge, Charles Baxter, Susan Baxter, David W. Behrens, Ralph Buchsbaum, Nancy Burnett, Robin Burnett, Robert Carlson, Jay C. Carroll, Faylla A. Chapman, Jules Crane, Earl Ebert, Douglas Eernisse, David Epel, Rimmon C. Fay, Daniel W. Gotshall, Janine Haderlie, Norine Haven, Galen Hilgard, Anson Hines, Charles Jacoby, Roy Johnson, Jr., Samuel Johnson, Robert King, Christopher Kitting, Charles Lambert, Gretchen Lambert, James McLean, Mitsuaki Nakauchi, Sharon Nugent, Chris Patton, Dorothy Paul, John Pearse, Vicki Pearse, David Phillips, Stephen C. Piper, Virginia Scofield, Robert Sellers, Ralph I. Smith, Ronald C. Tipper, Brian Tissot, Victor Vacquier, James Watanabe, Steven Webster, Lani West, Mary Wicksten, and perhaps others we have unfortunately overlooked. Our thanks to all of them.

We would like to give special thanks, as well, to all of those who have generously contributed their photographs for use in this book. All photos except those by Morris are individually credited in the picture legends, and a list of contributors appears on page 665.

It is a pleasure to thank those at Stanford University Press who have helped to make this a book. Most of all we extend thanks to Jean McIntosh, our editor. A biologist herself, she made a thousand constructive suggestions, maintained a stern attention to format and details of style, and with a sensitivity for clarity and coherence time and again used her green pencil to improve the text. Our thanks go, as well, to William W. Carver, who started us off and who has shown unflagging interest, read many sections and provided valuable comments, and made judicious use of both the carrot and the stick, as needed, when things were lagging. We are also indebted to Martha Johnston for her imaginative design.

Two wives, Dr. Isabella A. Abbott and Mrs. Aileen E. Haderlie, deserve thanks in ways extending from critical reading of manuscript and proof to putting up with many a weekend of neglect. Without their love and support the book would never have been finished.

Blandine L. Morris also deserves her share of thanks. Although her contribution was that of a proud and supportive mother, her encouragement was needed and appreciated by her most wayward son. Regrettably, her time ran out before the book was completed, but she would have been very pleased with the end result.

Finally, it is with heartfelt thanks that we acknowledge the very generous gift of The David and Lucile Packard Foundation, which has supported the cost of publication to a degree that allows a book expensive to produce to be sold at a price that many biologists, teachers, and students can afford.

D.P.A.

Contents

Intertidal Invertebrates of California

Introduction

The Pacific coast of North America supports a wealth of marine life matched by few other shorelines in the world. One important factor contributing to this is the fertilization of inshore waters by the minerals essential for plant growth. Some of the nutrients arrive in runoff from the land, but more are brought in by the seasonal upwelling (February to July) of deep ocean waters from regions beyond the narrow continental shelf. Together with sunshine, the nutrients support a rich phytoplankton and a lush plant growth on rocky shores and in offshore kelp beds. This in turn provides food for many animals and microorganisms. Another factor favoring diversity of seashore life is the variety of marine habitats available, sometimes within quite limited regions. Wave-swept rocks and beaches alternate with more peaceful bays, and each habitat bears a characteristic but different biota. The surface currents of the sea may also contribute to the diversity of life seen, carrying inshore at different seasons pelagic organisms from farther north or farther south and distributing the larvae of benthic forms up and down the coast. To these favorable factors we can add, for California at least, a generally temperate coastal climate, without great seasonal fluctuations.

The results of these and other factors can be seen by visiting the rocky shore at low tide. Living things appear in such abundance in the low intertidal zone that the main limiting factor is simply standing room—space for organisms to attach, cling, crawl, hide, and burrow.

The unaided eye does not see all the life present. Some forms are concealed below the surface, but many more remain unseen because of their small size. Among them are the bacteria, protozoans, microscopic algae and fungi, and many small species of such multicellular animals as nematode worms, creeping copepods, amphipods, mites, and flatworms. Important as some of these small forms are in the ecology of the shore, special methods are often needed to find and identify them. For the most part, we shall look past them and focus on the larger invertebrates, which, together with the larger plants and a few species of vertebrates, are the conspicuous inhabitants of the intertidal zone.

Biologists have been investigating California shore invertebrates for nearly a century. In addition, many California species, or their close relatives, occur in other parts of the world and are studied there. Information relevant to California forms already fills a small library, and new studies are being published at an ever-accelerating pace. The mood of researchers is reflected in a sign posted in the library of Stanford University's Hopkins Marine Station: "The problem of keeping up with the literature has been replaced by the problem of how to fall behind most strategically."

Despite this information deluge, most of the possible questions about our commonest species remain unanswered. It is a delightful but humbling experience for the professional marine biologist to visit the seashore at low tide with an excited and articulate child. Eagerly the questions come: "What is it? Can I touch it? What does it eat? Why is it colored that way? What makes it stick to the rocks? How old is it? Where are the babies?" Within mere seconds the expert is made to realize how little we know in detail about many abundant forms.

The questions the professional biologist brings to his marine colleagues are even more varied. We are often asked, "What sorts of studies have been made on this beast? Has anyone looked into ——?" The blank represents practically anything, for example, food habits, sperm morphology, range of temperatures tolerated, body sterol content, territorial behavior, osmotic regulation, muscle physiology, breeding season, chromosome number, sex hormones, trace metal content, early embryology, life span, number of eggs produced, effect of DDT on larval activity, and so on, endlessly.

Even where some information is already available, it is usually not enough to provide the kind of understanding that permits prediction. This statement applies not only to individual species but to whole biological communities along our shores. For example, during and after the underwater oil-well blowout at Santa Barbara, California, in 1969, studies were made to determine, among other things, the extent of damage caused to the organisms of sea and shore. The most casual observer could see that the beach was a sticky mess and that oiled birds suffered a high mortality, but beyond that it proved difficult to make firm quantitative statements about damage. The impact of the spill could not be truly measured because data on the biota before the spill were inadequate to provide standards for comparison. This would have been the case regardless of where the spill occurred on the coast.

Accidents like this, a need to better assess the effects of industrial waste disposal in coastal waters, and a growing public interest and sense of responsibility have led to the establishment and conduct of ecological baseline surveys at several points along the coast. Some have been conducted by federal or state agencies, others by commercial firms or student groups. More and better ecological survey studies are needed in *most* areas of California. They represent projects that can be carried out, year by year on a continuing basis, by advanced biology classes; adequately planned and supervised, and carefully executed, such studies can be of great educational value to participants and can provide very useful information to local communities. We hope this book will stimulate and facilitate such studies.

Organization of the Text

The particular order in which the various phyla are presented in books on invertebrates is largely arbitrary. We follow tradition in starting with the morphologically simpler forms (protozoans, sponges, etc.). We depart from tradition in culminating with the arthropods rather than the chordates. The tradition of ending with chordates stems from man's view that he and his kin sit at the top of the evolutionary tree; were we intelligent cephalopods, no doubt our invertebrate texts would culminate with a chapter on mollusks. Although not lacking in human bias, the authors prefer to end this book on *in*vertebrates with that largest of all invertebrate phyla, the arthropods.

Within each chapter, the accounts of the individual species covered include the following: the name of the animal; brief notes on its habitat and the geographical range of the species; an abbreviated description intended to augment the photograph in confirming identification; a summary of selected information available on the biology of the species; and a list of references to articles or books where further details may be found. These categories of information are discussed in more detail below.

The name of the animal, and its place in the classification of animals

Zoologists have described and named roughly a million species of multicellular animals (Metazoa) so far. The work is far from being completed and may never get done, for at the present rate of change in the world's natural habitats, undescribed species are probably becoming rare or extinct as fast as the remaining ones are being described. One need not visit exotic spots to find "new" species. On almost any trip to the California seashore one encounters among the smaller forms numerous species that have never been described or named by anyone. The same can be said of our terrestrial fields and forests.

Notwithstanding, the metazoan animals already known fall into approximately 30 large groups called phyla, each of which differs sufficiently from the others that its evolutionary relationships with other phyla are matters of conjecture and sometimes controversy. All but three or four of these phyla are represented in California shore waters. However, most of the common, macroscopic, free-living marine animals fall into the 15 metazoan phyla covered by this book.

Within each phylum, taxonomists arrange the species in a hierarchy of categories reflecting a judgment of their degrees of similarity or difference. Phyla are divided into classes, the classes into orders, the orders into families, the families into genera, and the genera into species. The levels of classification from phylum to family are often called "higher categories," "higher taxa," or "major groups"; the genera and species represent "lower categories." In taxonomic groups embracing large numbers of species, intermediate categories are often used (e.g., subclasses, suborders, superfamilies, sections). In this book the higher categories deemed important are given directly above the heading in each species account.

The scientific name of a species normally consists of two parts, the name of the genus (always capitalized) and the name of the species (not capitalized). In this book, scientific names are printed in italics or boldface type. Following the species name at the head of each species account is the surname of the person who first described and named the species, and the year the name was published. Where the author's name and the date appear in parentheses, this indicates that the species is no longer placed in the genus to which the author of the species originally assigned it. Occasionally one sees a scientific name with three or four parts, for example, *Uca* (*Leptuca*) *crenulata crenulata* (Lockington, 1877). Here, the names, in order from left to right, represent *Genus* (*Subgenus*) *species subspecies* (Author, year).

If a species is well known under another scientific name in recent scientific publications, either through incorrect identification or because the two scientific names are synonyms applied to the same species, we have placed the other name in parentheses, following the author and date of the current scientific name. Ordinarily, synonyms are not given.

Common names are given for some species. Most of the species included here have no widely accepted common names, though the California Department of Fish and Game has designated common names for some species frequently mentioned in its publications. Decisions on use of scientific and common names have been left to the authors of individual chapters.

Habitat, geographical distribution, and depth distribution of the species

Information in these categories is given insofar as it is available, but always with the following qualifications. With respect to habitat, the statements made are probably broadly reliable, but within a particular local area a given species may be absent from the sort of habitat described, or it may occur in a type of habitat not specifically mentioned. The limits given for maximum depths, and for northern and southern geographical limits along the coast, simply represent the greatest ranges that have been published or otherwise recorded so far for each species. In no sense whatever are they limits within which an animal is "supposed" to be or stay, or limits

beyond which it cannot survive. For most of the species, no one has ever made a determined attempt to establish the exact distributional limits. San Diego, California, is given as the southern limit for many species, not because the Mexican border represents a barrier to marine animals but because it represents a barrier to people; for many forms no one has bothered to look south of San Diego. Even these limitations need qualification. Northern species, less tolerant of warm conditions, may extend their distributions southward much farther in the cooler subtidal waters than in the warmer intertidal zone; thus numerous species common intertidally in Washington and British Columbia may be found in California only by scuba diving or dredging. For southern forms whose northern limits may be set by their ability to tolerate low winter temperatures, specimens are often found thriving (at least temporarily) in the warmer waters of protected bays far north of their usual northern limits on the colder open coast. At different seasons of the year, pelagic larvae of marine invertebrates can be carried either southward along the coast by the California Current, or northward by surface or subsurface countercurrents, bringing about somewhat anomalous distributions.

It is no exaggeration to say that most of the ranges given here in fairly specific terms are "wrong" in some sense. The extreme ranges set by the collection of isolated individuals are often deceptive; the vast bulk of the species population may have a more limited geographical spread. Ranges can also be expected to fluctuate with minor climatic changes; in oceanographically "warm" years (such as were experienced along the west coast in 1957–59) many warm-water species extend their ranges northward, while cold-water species tend to withdraw toward the north. At any rate, the ranges presented here should be thought of as rough approximations of a fluctuating parameter, not necessarily reflecting accurately the distribution of the main species population, and subject to change with shifts in climate and increases in knowledge. Almost any trip to the shore would result in extensions of previously published ranges were one able to identify all of the animals present.

General aspects of west coast zoogeographical regions are discussed by Briggs (1974). He noted that northern California is in the Oregon(ian) Province, which extends from Point Conception north to Queen Charlotte Islands and the southern boundary of Alaska (Dixon Entrance). Southern California falls in his so-called San

Diego Province, which extends from Point Conception (Santa Barbara Co.) south to Bahía Magdalena (Baja California). The southern extent of Briggs's San Diego Province is roughly equivalent to that of the so-called Californian Province of other authors, and therefore the latter name is used here.

Newell (1948), Valentine (1966), and Newman (1979) recognize Point Conception not as a boundary of abrupt biotic change, but rather as the center of a Transition Zone, or region of faunal change between the Oregonian and Californian Provinces, extending from approximately Monterey Bay in the north to San Diego in the south. Most but not all species have their northern or southern limits in the Transition Zone, however, as was recently noted in idoteid isopods (Brusca & Wallerstein, 1979). Furthermore, the Channel Islands of Southern California lie within the Transition Zone and their special relationship has been discussed by Seapy and Littler (1980).

In a few groups of marine invertebrates, such as mollusks and barnacles, the knowledge of both the systematics and distribution of species has advanced to the point where data presently available are useful not only in designating the biogeographical provinces, but also in demonstrating that the Transition Zone contains species not found in the provinces to the north or south. These species are known as short-range endemics. Further discussion of the marine provinces and the short-range endemics of the Californian Transition Zone will be found in Chapter 20 (Cirripedia: The Barnacles).

Species are limited not only in their horizontal (latitudinal) distribution along the coast but also in their vertical distribution on the shore at any one locality. We are concerned mainly with distribution in the intertidal zone, though many forms present there occur as well in deeper waters. On our shores the tides rise and fall twice each day in a cycle averaging 24 hours and 50 minutes in length. The peaks and lows of the tide include, in order, higher high water, lower low water, lower high water, and higher low water. The height of each peak and low in the cycle changes every day; tidal predictions, taken from tide tables published by the U.S. Coast and Geodetic Survey, are printed daily in most coastal newspapers. The baseline for tidal measurements on this coast is the mean level of the lower low waters, taken over a prolonged period; mean lower low water (MLLW) is the 0.0 level of the tide tables.

In general, only a few species inhabit the high intertidal zone and the splash zone just above it, where the residents are exposed most of the time to terrestrial conditions that are harsh for aquatic organisms. More species live in the middle intertidal zone, and a great variety occur in the low intertidal zone and adjacent subtidal regions where they are seldom or never exposed to air. To observe the greatest diversity of life, visit the shore during periods of lower low water, especially when the predicted tides are below the zero level.

A good discussion of the vertical zonation of marine organisms in the intertidal zone is available in Ricketts & Calvin (1968). More extensive general treatments are found in Connell (1972), Doty (1957), Lewis (1964), and Stephenson & Stephenson (1972). Examples of investigations of more specialized situations on the west coast are Dayton (1975), Glynn (1965), and Hodgson (1979). The functional aspects of life in the intertidal zone are very well reviewed by Newell (1970) and Yonge (1949).

Photographs, descriptions, and the identification of species

Each species illustrated is assigned a unique decimal serial number, which appears in the heading for the text account, in the photograph legend, and in the Index entry for that species. We hope the photographs will provide most of the details necessary for identification. However, do not expect all specimens of a species to look exactly alike. If you were to select *one* individual photograph by which to represent "mankind" in a natural history guide to the planet earth (prepared for extraterrestrial visitors), whose picture would you select? Try to think of each picture as a single example of a species population in which there are differences between individuals even of the same age and sex, as well as differences with age. The pictures, together with the written descriptions, should suffice for identifications in most cases. The list of references given after each species usually includes one or two sources helpful in confirming identification. When you have difficulty, remember that you might have in hand an animal not covered in the book.

If you find that the picture, descriptive notes, and references cited leave you in doubt about the genus and species of a specimen you have found, use the scientific or common name of the lowest of the categories above the level of genus (e.g., family, order, class, phylum) to which you are sure the animal belongs. Don't hesitate to call an animal a bryozoan, a copepod, an amphipod, a hermit crab, a brittle star (ophiuroid), a sea star (asteroid), a sponge, an ascidian, or a polyclad flatworm, if you are sure of yourself that far but no farther.

The importance of correct identification at the species level varies with circumstances. Researchers preparing to publish their results should always ask specialists to check dubious identifications, for errors here can lead to confusion and reduce the value of otherwise good work. In marine biology classes the need to know genus and species is usually less critical. Students here should know by name some of the commonest forms in their areas, and learn to use keys and other aids to identification, but they should also learn to be cautious in applying species names when identification is uncertain, and to substitute the names of higher taxa in doubtful cases.

Such is our preoccupation with names that even the proper identification of a species has its hazards. As one of us once wrote, in another connection, "Names. . . . contain such satisfying magic that we are often deluded into thinking that to label something properly is to know all about it. 'That is *Arbacia*, a sea urchin,' we say, and tramp on, satisfied that we have dealt with the beast appropriately and now understand its niche in the cosmos." Beware of this illusion.

Biological or natural history notes

The accounts summarizing information on the biology or natural history of the different species vary greatly in length. Some animals have been studied extensively in field and laboratory, and have provided experimentalists with material for diverse inquiries. The tunicate *Ciona intestinalis*, the sea urchin *Strongylocentrotus purpuratus*, the sand dollar *Dendraster excentricus*, the edible crab *Cancer magister*, and the mussel *Mytilus edulis* are each the subject of hundreds of published scientific papers, and book-length accounts have been written for some of them. Other species are virtually unknown. For some whole animal groups (e.g., crabs, sea stars, sea anemones) there is much to say about numerous species; for others (e.g., foraminifera, sponges, bryozoans, beach insects), the literature offers little specific biological information on California species.

Where the information is available, the accounts stress such things as behavior, food, reproduction, larval stages, predators, symbionts of various sorts, and physiological abilities to cope with fluctuations in environmental factors. Where the organisms are uniquely suited for certain types of research, or have made impor-

tant contributions to knowledge as laboratory animals, we have pointed this out; but we have not tried to summarize (for example) a host of papers dealing with chemical embryology, biochemistry, cytology, and the like. Use of technical terms, here as elsewhere, has been curtailed but not avoided. Most of the terms used either are defined in the chapter introductions or are available in an ordinary desk dictionary. For others, a reader may consult any introductory text in invertebrate zoology (see the References section at the end of this chapter).

For many California species, additional information is given in accounts of other species. As noted earlier, each species illustrated is assigned a unique serial number; for example, the number 20.24 refers to *Balanus glandula*, the twenty-fourth species appearing in Chapter 20. When this species is mentioned in other accounts in the text, the name *B. glandula* is followed by its number (20.24). A given species may receive casual mention in the accounts of several other species, perhaps serving here as food, there as an occasional predator or host. All such mentions and cross-references are included in the Index.

In the discussions of some animals, related species that are not themselves illustrated are sometimes mentioned briefly. Names of California species appear in boldface type when first used; those from other parts of the world appear only in italic type.

In geographic scope the accounts are somewhat provincial, stressing work done on California or other west coast shores, but where information from this coast is meager or lacking, accounts of these species or their close relatives from other parts of the world are sometimes included, and indicated as such. We have tried to be reasonably comprehensive but not encyclopedic.

The reader will soon find that *most* animals along our shores are poorly known in *most* respects. Even for very common species, information is often fragmentary and anecdotal, its reliability uncertain. Sooner or later, some of it will doubtless turn out to be misinformation.

Don't regard this as a drawback, but as a challenge. The opportunities for further research on the biology and ecology of marine invertebrates along our shores are virtually unlimited. The field is wide open, and information on such things as patterns of activity, food habits, reproductive seasons, natality, interactions with other species, etc., can often be gathered without very elaborate equipment. Almost any invertebrate, observed alive for a time, yields information to the observer that appears in no scientific publications.

Readers are encouraged to consider what they might do to help fill some of the gaps.

Treatment of literature citations

The references cited are intended to document our sources, aid in identification, provoke thought, stimulate further investigation, and provide portals of entry into the scientific literature. Complete titles are given in the Literature Cited section at the end of each chapter, to aid in locating sources of particular facts, to indicate the sorts of investigations that have been made, and to permit browsing. General references and review papers are usually cited in chapter introductions; references cited at the end of an individual species account generally apply to that species or to one of its close relatives. Our references favor more recent papers, which in turn cite the older classic contributions. The references also favor works in English, but good papers and books in other languages are included, too.

Some of the works referred to are readily available; others are in technical journals available only in college and university libraries. A few are unpublished master's or doctoral theses, which are available at the library of the university that granted the degree and/or through University Microfilms, Ann Arbor, Michigan. Occasional unpublished student reports have been cited, but only when a copy of the paper is on permanent file at a recognized library (specified in the reference) and a photocopy may be purchased. Not to have cited such papers would have denied the reader some information not otherwise available. Although not all readers will have direct access to university libraries, most such libraries provide photocopies or microfilms of articles at reasonable cost, as do the libraries of Congress and of the Department of Agriculture in Washington, D.C.

Three works dealing with the marine biota of California should prove especially useful to the reader: Abbott & Hollenberg (1976), on the marine algae of California; Ricketts & Calvin (1968), on seashore life in relation to habitats; and Smith & Carlton (1975), a more comprehensive guide to the identification of California invertebrates than we have attempted here. See the References section at the end of this chapter for more information on these titles and for additional sources that may be useful in providing background information on the animal phyla and on invertebrate structure, function, development, and evolution.

Collecting

California's laws and regulations governing the collecting of marine organisms were tightened and made more restrictive in 1972. Collecting today requires a fishing license and for some items a scientific collector's permit. Those planning to collect should familiarize themselves with the regulations governing what may be taken and designating the areas where collecting is prohibited. Printed copies of regulations, revised annually, are available at offices of the Department of Fish and Game.

Those searching for animals along the shore are advised to follow two simple rules.

First, when hunting for organisms, try to disturb the environment as little as possible. Particularly, if you turn over a rock to see what grows or lurks below, replace it right side up when you are done, as nearly as possible in the position in which you found it. A long-undisturbed rock in the low intertidal zone often bears a rich growth of plants on top and a fine growth of sedentary animals below. Lifting it and peeking below causes relatively little damage, but if the rock is overturned and abandoned, both plants and animals quickly die and are replaced by a few fast-growing algae, notably the sea lettuce *Ulva*. Even in the absence of further disturbance it takes a year or more to reestablish the normal "climax" population of organisms of the sort that the rock supported before it was overturned. Rocks that are frequently turned, by people or wave action, bear very little life either above or below. If people hunting abalones or other animals realized that failure to restore rocks to their original positions creates havoc and is a sure way of reducing the future crop, perhaps they would be more careful. Teachers taking classes of students to the shore should make certain the students realize this.

Second, don't bring animals home from the seashore if it is not really necessary. Most species won't survive long except in recirculating saltwater aquaria, and are difficult to preserve in a lifelike condition. Bring your camera, with a close-up lens, to the shore, instead. The animals can be photographed in natural settings or in a shallow pan of seawater, and left behind when you are done. Pictures keep their color better than dried or pickled animals, and are a lot easier to store. These are conservation measures that we as individuals can take, measures that will help give our children's children a chance to see the organisms portrayed in this book, and many more.

References

Three works may be especially useful to the reader:

Abbott, I. A., and G. J. Hollenberg. 1976. Marine algae of California. Stanford, Calif.: Stanford University Press. 827 pp. The welfare of many marine animals is intimately related to that of the marine plants whose company they keep. The plants are often laden with attached animal life, and many marine animals are themselves adorned with attached plants. Directly or indirectly, the algae provide food, shelter, and safe breeding places for many invertebrates, and they provide a variety of products used directly by man. This book is comprehensive, up to date, and beautifully illustrated; it is *the* reference to consult for the identification of algae.

Ricketts, E. F., and J. Calvin. 1968. Between Pacific tides. 4th ed. Revised by J. Hedgpeth. Stanford, Calif.: Stanford University Press. 614 pp. This excellent and highly readable volume considers California seashore life in relation to the different major habitats: open coast, protected outer coast, and bays and estuaries. The approach is ecological, and within that framework much valuable natural history information is interestingly presented.

Smith, R. I., and J. T. Carlton, eds. 1975. Light's manual: Intertidal invertebrates of the central California coast. 3rd ed. Berkeley and Los Angeles: University of California Press. 716 pp. This book contains taxonomic keys and useful illustrations for a much larger number of California invertebrates than we have included in this volume. It covers many of the forms omitted here because of their small size. If you have difficulty in identifying some forms, this book is a fine source of first aid.

At an introductory level, one of the most readable accounts of the invertebrates is still the following:

Buchsbaum, R. 1976 (reissue). Animals without backbones. Chicago, Ill.: University of Chicago Press. 392 pp. Complete revision in process.

At an intermediate level, the following are recommended:

Barnes, R. D. 1968. Invertebrate zoology. 2nd ed. Philadelphia: Saunders. 743 pp.

Barrington, E. J. W. 1979. Invertebrate structure and function. 2nd ed. New York: Wiley. 765 pp.

Fretter, V., and A. Graham. 1976. A functional anatomy of invertebrates. New York: Academic Press. 589 pp.

Gardiner, M. S. 1972. The biology of invertebrates. New York: McGraw-Hill. 954 pp.

Kaestner, A. 1967–70. Invertebrate zoology. Translated and adapted from 2nd German edition by H. W. Levi and L. Levi. New York: Wiley-Interscience. Vol. 1, 1967, 597 pp.; Vol. 2, 1968, 472 pp.; Vol. 3, 1970, 523 pp.

Kumé, M., and K. Dan. 1968. Invertebrate embryology. Translated from the Japanese by J. C. Dan. Published for the U.S. Department of Health, Education, and Welfare and the National Science Foundation (NOLIT Publishing House, Belgrade, Yugoslavia). 605 pp. (Available as TT-67-58050, Clearinghouse for Federal Scientific and Technical Information, Springfield, Va.)

Meglitsch, P. A. 1972. Invertebrate zoology. 2nd ed. New York: Oxford University Press. 834 pp.

Nicol, J. A. C. 1960. The biology of marine animals. New York: Wiley-Interscience. 707 pp.

Russell-Hunter, W. D. 1979. A life of invertebrates. New York: Macmillan. 650 pp.

At a more advanced level, among the most useful references are:

Florkin, M., and B. T. Scheer, eds. 1967– . Chemical zoology. Several vols. New York: Academic Press.

Giese, A. C., and J. S. Pearse, eds. 1974– . Reproduction of marine invertebrates. Several vols. New York: Academic Press.

Grassé, P.-P., ed. 1948– . Traité de zoologie. Vols. 1–11 especially useful. Paris: Masson.

Hyman, L. H. 1940–67. The invertebrates. Vols. 1–6. New York: McGraw-Hill.

Moore, R. C., ed. 1953– . Treatise on invertebrate paleontology. Parts A–W, some parts involving several volumes. Still incomplete, but an invaluable source to the invertebrate zoologist. New York: Geol. Soc. Amer.; Lawrence: University of Kansas Press.

The following cited sources contain useful discussions of biogeographical provinces on the west coast of North America:

Briggs, J. C. 1974. Marine zoogeography. New York: McGraw-Hill. 475 pp.

Brusca, R. C., and B. R. Wallerstein. 1979. Zoogeographic patterns of idoteid isopods in the northeast Pacific, with a review of shallow water zoogeography of the area. Bull. Biol. Soc. Washington 3: 67–105.

Hall, C. A. 1964. Shallow-water marine climates and molluscan provinces. Ecology 45: 226–34.

Newell, I. M. 1948. Marine molluscan provinces of western North America: A critique and a new analysis. Proc. Amer. Philos. Soc. 92: 155–66.

Newman, W. A. 1979. The Californian Transition Zone: Significance of short-range endemics, pp. 399–416, *in* J. Gray and A. Boucot, eds., Historical biogeography, plate tectonics, and the changing environment. 37th Ann. Biol. Colloq. Corvallis: Oregon State University Press.

Seapy, R. R., and M. M. Littler. 1980. Biogeography of rocky intertidal macroinvertebrates of the southern California islands, pp. 307–23, *in* D. M. Power, ed., The California islands: Proceedings of a multidisciplinary symposium. Santa Barbara, Calif.: Santa Barbara Natural History Museum.

Valentine, J. W. 1966. Numerical analysis of marine molluscan ranges on the extratropical northeastern Pacific shelf. Limnol. Oceanogr. 11: 198–211.

For discussions of matters relating to the vertical zonation of marine organisms in the intertidal zone, see these cited sources:

Connell, J. H. 1972. Community interactions on marine rocky intertidal shores. Ann. Rev. Ecol. Syst. 3: 169–92.

Dayton, P. K. 1975. Experimental evaluation of ecological dominance in a rocky intertidal algal community. Ecol. Monogr. 45: 137–59.

Doty, M. S. 1957. Rocky intertidal surfaces, pp. 535–85, *in* J. W. Hedgpeth, ed., Treatise on marine ecology and paleoecology, vol. 1, Ecology. Geol. Soc. Amer. Mem. 67. Baltimore, Md.: Waverly Press. 1,296 pp.

Glynn, P. W. 1965. Community composition, structure, and interrelationships in the marine intertidal *Endocladia muricata—Balanus glandula* association in Monterey Bay, California. Beaufortia 12(148): 1–198.

Hodgson, L. M. 1979. Ecology of a low intertidal red alga, *Gastroclonium coulteri* (Harvey) Kylin. Doctoral thesis, Biological Sciences, Stanford University, Stanford, Calif. 129 pp.

Lewis, J. R. 1964. The ecology of rocky shores. London: English Universities Press. 323 pp.

Newell, R. C. 1970. Biology of intertidal animals. New York: American Elsevier. 555 pp.

Ricketts & Calvin (see above).

Stephenson, T. A., and A. Stephenson. 1972. Life between tidemarks on rocky shores. San Francisco: Freeman. 425 pp.

Yonge, C. M. 1949. The sea shore. London: New Naturalist series, Collins. 311 pp.

For subtidal invertebrates, see the following:

Gotshall, D. W., and L. L. Laurent. 1979. Pacific coast subtidal marine invertebrates: A fishwatcher's guide. Los Osos, Calif.: Sea Challengers. 112 pp.

Chapter 1

Foraminifera: *Shelled Protozoans*

Zach M. Arnold

Foraminifera (=Foraminiferida), the "oil bugs" of the geologist, are amoeboidal protozoans in which the soft protoplasmic body is protected by a hard outer shell, or test. They are distributed worldwide in marine and brackish waters. The largest west coast forms (0.5–1.5 mm in diameter) may be conspicuous, especially when abundant. They are not always easily recognized as protozoans, however. In some the test is a transparent organic sphere; others have calcareous tests with one chamber (unilocular) or with several (multilocular), the boundaries between chambers being marked externally by grooves or sutures. In some the test is spirally coiled, with a central umbilicus, externally resembling the shell of a snail or chambered nautilus. Other foraminiferan tests superficially resemble the tubes of worms or other invertebrates, but are usually much smaller. The test may be ornately sculptured with spines, ridges (costae), or tubercles, or may bear a marginal flange (keel). In still other species the test is formed of sediment particles and debris cemented together (agglutinated); such coverings range from nondescript hovels of sand grains and mica flakes to masterpieces of microscopic masonry in which every particle appears to have been selected carefully, fitted artistically, and cemented skillfully into place. The name "Foraminifera" ("pore bearers") refers to the internal passageways (foramina) through which protoplasm passes from chamber to chamber, not to the pores abounding in the test of so many of the calcareous species.

The order Foraminifera consists of approximately 16,000 fossil and 4,000 living species. About 200 species are found either living in, or washed into, the shallow coastal waters of California.

Foraminifera are easily collected. Their empty tests accumulate in fine sediments, in well-winnowed strandline deposits, and around the bases of marine plants. Living forms occur on the surfaces of plants, attached arborescent invertebrates (hydroids and bryozoans particularly), and shells and rocks. To find the smaller forms one not only must examine sediments washed or brushed from plants, animals, and rocks, but must carefully scrutinize under a stereoscopic microscope "dirty" hydroid or bryozoan colonies, algal fronds, macerated algal holdfasts, and the tiny crevices in rock fragments. In the rigorous surf- and current-swept rocky intertidal areas, living foraminifera may find shelter wherever tiny accumulations of fine sand or silt occur. Among the smaller foraminifera, *Patellina* (biologically one of the best known; see Berthold, 1971; Grell, 1959; Le Calvez, 1938; Myers, 1935; and Zech, 1964) and *Spirillina* (see Myers, 1936), neither easily photographed, occur in a variety of California habitats, including folds of broad-frond green and red algae. Seasonally abundant crops of *Glabratella* (e.g., 1.12) and *Rosalina* (e.g., 1.14) nestle within the forests of large diatoms (*Isthmia nervosa*) that flourish in spring and summer on blades of the eelgrass *Zostera*. Although many foraminifera are equipped for life in intertidal and shallow subtidal waters, by virtue of their small size and their ability to insinuate themselves into microscopic nooks and crannies, to attach securely to organic and inorganic surfaces, to bury themselves industriously with food and debris, and to remain long inactive within resistant membranes, the group as a whole is best represented in the offshore waters of the continental shelf and slope, regions beyond the scope of this book.

Because of their long and rich fossil record (extending from early Paleozoic times) and the ease with which their remains are recov-

Zach M. Arnold is Professor Emeritus of Paleontology at the University of California, Berkeley.

ered and studied, foraminifera are of great value to geologists in correlating sedimentary strata and in interpreting past environmental conditions, particularly in petroleum exploration. For this reason they are among the best known of fossil marine groups. The living organism, however, has received far less attention, and few species are well understood biologically.

The terms "dorsal" and "ventral" are inappropriate for the tests of many of the symmetrical or simple foraminifera. However, they are applied to many multilocular forms with asymmetrically coiled tests, where dorsal (usually the upper or exposed surface) and ventral (usually the lower surface) may differ in chamber arrangement, pore patterns, and ornamentation. The aperture is usually ventral, but in *Cibicides lobatulus* (1.17) it becomes dorsal.

The "body" inside the test may contain one or many nuclei. Pseudopodia, which may be simple and threadlike or anastomosed to form a reticulum, extend either from the main aperture of the test or from a sheet of protoplasm that emerges from the aperture and spreads to cover (and help secrete) the test. Food, consisting of diatoms, other microorganisms, and fine organic detritus, adheres to these pseudopodia and is carried into the test by protoplasmic streaming, as though on an endless belt. Sand, mineral particles, and other inorganic debris used in the construction of the tests of some species are also collected and transported to the building site by the pseudopodia. Undigested matter may be retained inside the body of the animal or haphazardly removed by the pseudopodia, but in a few species it is ejected en masse just prior to reproduction.

Reproduction may be either asexual (by multiple fission) or sexual (involving the fusion of amoeboidal or flagellated gametes). In some species the two processes occur alternately in the life cycle (alternation of generations), and the two generations differ in other significant ways: the sexually reproducing form is initially haploid and possesses a single nucleus, and the initial chamber of its test is relatively large compared to later chambers (the test is megalospheric); the asexually reproducing form is diploid and possesses numerous nuclei, and the initial chamber of its test is relatively small compared to later chambers (the test is microspheric). However, the microspheric test is often larger overall than its megalospheric counterpart.

Meiosis occurs at the end of the asexual generation, as a prelude to asexual reproduction (multiple fission). In some species multiple fission occurs inside the test, the young then escaping from the pa-

rental test, but in others it takes place when a mass of protoplasm has been ejected from the parental test into a protective reproductive cyst. The sexual forms, which are the products of multiple fission, are haploid. They secrete their megalospheric tests and ultimately undergo mitosis to produce gametes that look alike (isogametes). Fertilization occurs when isogametes from two different individuals meet and fuse, restoring the diploid condition. In some species, gametes are liberated freely into the sea, but in others two (or more) parents may form closely attached pairs (or groups) prior to the release of gametes, so that fertilization occurs within the associated megalospheric tests. Paired tests of some species (*Glabratella ornatissima*, 1.12, for example) are often encountered in intertidal collections.

For more detailed information on these organisms one must consult the technical literature. Cushman's well-illustrated book (1948), though biologically outdated, has long been the leading reference work in America. Brady's classic (1884) is indispensable to the specialist; his magnificent plates, together with updated taxonomic notes, are available at a reasonable price (Barker, 1960). The most extensive taxonomic treatment and bibliography are found in the Loeblich & Tappan treatise (1964); these authors have also summarized recent taxonomic developments (1974). The massive catalogue of type descriptions compiled in Ellis & Messina (1940–) must be accessible to the taxonomist. A modern view of the systematic position of the foraminifera and their near allies is given in Honigberg et al. (1964). Myers (1943a,b), Le Calvez (1953, and Grell (1973) have summarized the biology of foraminifera, while Phleger (1961), Murray (1972), and Lee (1974) have assembled much useful biogeographical and ecological information on the group; Bandy (1960) has summarized the general aspects of the correlation between test form and water depth. Methods for collecting and culturing living foraminifera are described in Myers (1937), Lee (1974, 1975), and Arnold (1974). An open-ended series of books edited by Hedley and Adams (1974, 1977) contains review articles on various aspects of foraminiferan research.

Studies of west coast foraminiferan faunas are available in Arnal (1958), Bandy (1953), W. Cooper (1961), Cushman (1910–17, 1925, 1927), Cushman & Moyer (1930), Cushman & Valentine (1930), Hedman (1975), Lankford (1962), Lankford & Phleger (1973), Natland (1933), Phleger & Bradshaw (1966), Reiter (1959), Sliter (1970, 1971), Steinker (1969), Stinemyer & Reiter (1958), Uchio (1960), Walton (1955), and Watkins (1961).

For most of the benthic foraminifera found in inshore waters, a lack of basic knowledge of the life histories and the ranges and limits of variation for individual species has confused the taxonomy, and, in turn, has greatly complicated the assimilation and synthesis of a wealth of detailed biogeographical data into a clear worldwide picture. Much more work on living animals is needed.

Phylum Protozoa / Class Sarcodina / Order Foraminifera / Family Lagynidae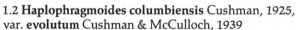

1.1 Iridia serialis Le Calvez, 1935

Abundant on rocks, particularly in silt-laden water, often firmly attached to the bases of marine plants and sessile invertebrates, middle to low intertidal zones; known from Sonoma Co. and Monterey Bay, probably widespread in California; common in Mediterranean Sea.

Test usually 1–5 mm long, formed of small particles cemented together, firmly attached to substratum; test with one to several chambers.

Like related species having fairly nondescript agglutinated tests, this one, though often common, is easily overlooked or mistaken for small accumulations of fine sand or debris. The California species is less well known than the Mediterranean one; a single-chambered form is more often abundant in California than forms with two or more chambers, and the multilocular individuals on our shores seldom develop the progressively larger chambers exhibited by *I. serialis* in the Mediterranean (Le Calvez, 1936, 1938). When placed in laboratory culture, specimens of *Iridia* actively develop pseudopodia, and their bodies soon become green with ingested algae.

Iridia lucida Le Calvez, which produces a transparent, membranous, hemispherical test unadorned with agglutinated materials, is a common associate of *I. serialis*, as is **I. diaphana** Le Calvez, a single-chambered form, which falls between the other two species in its agglutinating habits. Though *I. lucida* is easily overlooked because of the clarity of its test, it is often the most abundant of the three in intertidal and shallow subtidal areas.

Asexual reproduction occurs in all three species. The young produced by *I. diaphana* and *I. serialis* pass through a plank-tonic phase that lasts for several hours to a day or more, after which they settle to the bottom of the culture container and proceed with their development. Both *I. lucida* and *I. diaphana* are able to leave their test, move about in the naked state for a while, and then settle down to construct a new test. In the naked state they may be easily confused with various other amoeboid forms, including the foraminiferan *Allogromia* and its allies.

See Cushman (1920), Le Calvez (1936, 1938), Marszalek (1969), and Marszalek, Wright & Hay (1969).

Phylum Protozoa / Class Sarcodina / Order Foraminifera / Family Lituolidae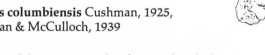

1.2 Haplophragmoides columbiensis Cushman, 1925, var. evolutum Cushman & McCulloch, 1939

A hardy species widely encountered in fine sand and silt washed from plant holdfasts, coralline algae, and bryozoans; common intertidally, also readily dredged at depths up to 40 m or more; Canada to Mexico.

Test 0.7–1 mm in diameter; var. *evolutum* differing from typical form in its more open and depressed umbilical area and more globular chambers; test wall relatively simple, composed of sand particles (principally quartz and feldspar) cemented to organic base or matrix containing iron compounds.

Living specimens are difficult to distinguish from empty tests without observing pseudopodia or crushing to reveal protoplasm. Thin sections of a few tests from Moss Beach (San Mateo Co.) reveal a simple wall consisting of a homogeneous matrix with some embedded sand grains, but some from southern California with a more complex test wall have been referred to the genus *Alveolophragmium*. Sections are essential for accurate differentiation between the two genera. Forms identified as *A. columbiensis* (Cushman) have occasionally been observed to be more numerous in the vicinity of sewage outfalls than elsewhere in waters of comparable depth.

See Bandy, Ingle & Resig (1965a,b), Haderlie (1969), Towe (1967), and Watkins (1961).

Phylum Protozoa / Class Sarcodina / Order Foraminifera
Family Trochamminidae

1.3 Trochammina pacifica Cushman, 1925

Present intertidally, but more abundant subtidally to 50 m in Monterey Bay; widely distributed from Canada to Mexico.

Test 0.3–0.5 mm in diameter, with several whorls forming a low conical spire on one side, the outer whorl with four or five chambers; sutures depressed but distinct; periphery rounded; embedded sand grains generally well matched and uniformly cemented, imparting smooth finish to test; aperture slitlike, ventral, at base of final chamber; easily distinguished from *Haplophragmoides* (e.g., 1.2), in which test is flatter and coiled in a single plane.

This common species of the open continental shelf appears to flourish in the vicinity of sewage outfalls. *T. quadriloba* Höglund, a related species from the northwestern Gulf coast of Florida, under laboratory conditions shows little selectivity in its choice of the mineral particles from which it constructs its test; it can be maintained as a reproducing population for protracted periods in the laboratory, even when deprived of particulate mineral matter.

See Bandy, Ingle & Resig (1965a,b), Hedley, Hurdle & Burdett (1964), Salami (1974), Uchio (1960), and Watkins (1961).

Phylum Protozoa / Class Sarcodina / Order Foraminifera / Family Textulariidae

1.4 Textularia schencki Cushman & Valentine, 1930

Common subtidally, occasionally found intertidally, particularly as displaced tests washed inshore; Canada to Mexico.

Test to 0.9 mm long, 0.5–0.6 mm broad.

In subtidal areas of Monterey Bay this species is often attached to the agglutinated foraminiferan *Rhabdammina abyssorum* Sars, which may reach 12–15 mm in length and is easily mistaken for a worm tube unless pseudopodia are observed. Study of the structure and composition of the test wall of this and related species of agglutinated foraminifera, though challenging, is difficult. Carefully oriented sections prepared by

grinding are essential in disclosing the pores that characterize the wall of *Textilina* but are lacking in *Textularia* (Nørvang, 1966).

See La Croix (1931, 1933) and Nørvang (1966).

Phylum Protozoa / Class Sarcodina / Order Foraminifera / Family Fischerinidae

1.5 Cyclogyra lajollaensis (Uchio, 1960)
(=Cornuspira lajollaensis)

On surfaces and in minute depressions and crevices on algal fronds, various invertebrates, rocks, and debris, often in silty accumulations in folds of the green alga *Ulva*, easily transported by waves and currents and often found in good condition rafted on floating algal fragments, low intertidal and subtidal waters; widely distributed on west coast.

Test about 0.4 mm in diameter, superficially resembling shell of tube worm *Spirorbis*, though smaller and not flattened on attached side.

Cyclogyra lajollaensis is possibly synonymous with *C. involvens* (Reuss) and/or *C. planorbis* (Schultze).

This species has been maintained in small-volume laboratory cultures for several months. No sexual reproduction or dimorphism of the test has been noted, but asexual reproduction occurs. The asexual generation lasts approximately a month under favorable laboratory conditions. At the time of reproduction the parental protoplasm is extruded from the test. The young are differentiated from this multinucleate mass and produce their own tests, the average brood size ranging from 15 to 20.

The population density for representatives of this species on test panels in Monterey Bay is greatest during the months of August, September, and October, and more specimens are found attached to floating panels than to panels at fixed depths in the intertidal or subtidal zones.

See Gougé (1971) and Haderlie (1968, 1969).

Phylum Protozoa / Class Sarcodina / Order Foraminifera / Family Miliolidae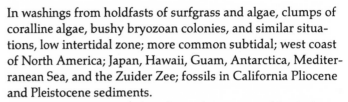

1.6 **Massilina pulchra** Cushman & Gray, 1946

Empty tests and occasional living animals present in coarse washings from algal holdfasts, middle and low intertidal zones; more commonly subtidal; Washington to Baja California.

Test to 2.5 mm in diameter, 0.4 mm thick, markedly compressed; chambers with acute periphery; costae (ridges) oblique, slightly curved, providing a ready means of identification.

This species was originally described from fossils in Pliocene deposits of southern California, but is now known to live in protected areas along California shores, including the Farallon Islands (San Francisco Co.) and Los Coronados (Baja California). A shortening of the terminal chamber of the test appears to characterize the later evolutionary stages of this species, and the periphery of the test may in some cases be sharper than it is in the photograph.

See Lankford (1962) and Lankford & Phleger (1973).

Phylum Protozoa / Class Sarcodina / Order Foraminifera / Family Miliolidae

1.7 **Quinqueloculina angulostriata** Cushman & Valentine, 1930

In washings from algal holdfasts and in clumps of bryozoans and coralline algae, low intertidal zone; more common subtidal on relatively coarse bottoms; San Mateo Co. to Mexico.

Test to 1.5 mm long, 0.6 mm broad; chambers usually quadrate in transverse section, with blunt edges or keels at corner, occasionally broadly rounded, the walls usually with fine longitudinal ridges (costae); aperture with a bifid tooth, a simple tooth, or none.

This species is possibly synonymous with *Q. inca* d'Orbigny.

See Cushman & Valentine (1930) and Lankford (1962).

Phylum Protozoa / Class Sarcodina / Order Foraminifera / Family Miliolidae

1.8 **Quinqueloculina vulgaris** d'Orbigny, 1826

In washings from holdfasts of surfgrass and algae, clumps of coralline algae, bushy bryozoan colonies, and similar situations, low intertidal zone; more common subtidal; west coast of North America; Japan, Hawaii, Guam, Antarctica, Mediterranean Sea, and the Zuider Zee; fossils in California Pliocene and Pleistocene sediments.

Test 1 mm or more in length, easily recognized by its lenticular form, smooth surface, rounded chambers, and prominent tooth in the aperture; variation great, possibly overlapping that of *Q. seminulum* (Linnaeus).

This species may constitute about 7 percent of the foraminifera found among roots of the surfgrass *Phyllospadix*.

See Cushman (1917, pt. 6) and Steinker (1969).

Phylum Protozoa / Class Sarcodina / Order Foraminifera / Family Elphidiidae

1.9 **Elphidium crispum** (Linnaeus, 1758)
(=*Polystomella crispa*)

Common in coarser washings from algal holdfasts, clumps of bryozoans, and coralline algae; most commonly subtidal, but many survive and reproduce intertidally; Canada to Mexico; North Pacific, Japan; British Isles, northern Europe.

Test about 1 mm in diameter; species distinguished from related species of *Elphidium* by number of chambers in outer whorl (10–15), and from **Elphidiella hannai** (Cushman & Grant), a common associate in washings and in strandline deposits, by its retral processes (bridges connecting adjacent chambers, the bridges expressed externally as ridges; see especially last three chambers in the photograph, top) and by the absence of the two rows of pores adjacent to the septa that characterize *Elphidiella*.

Though *Elphidium crispum* was the first species in which the correlation between an alternation of generations (sexual and asexual) and test dimorphism (megalospheric and microspheric) was demonstrated (see the discussion of the life cycle in the introduction to this chapter), much basic biological information, such as the position of meiosis in the cycle and the number of chromosomes in the nucleus, is still lacking.

Nicol's (1944) subdivision of this species along the west coast into three new species and three new subspecies was questioned by Burma (1948), but it is possible the population shows clinal variation. A study of the sort conducted by Chang & Kaesler (1974) on east coast populations of *Ammonia beccarii* (1.11) would be desirable.

See Angell (1975), Burma (1948), Jepps (1942), Lister (1895), Murray (1963), Myers (1942, 1943b), Nicol (1944), and Rosset-Moulinier (1961).

Phylum Protozoa / Class Sarcodina / Order Foraminifera / Family Nonionidae

1.10 **Nonionella basispinata** (Cushman & Moyer, 1930)
(? = Nonion pizarrensis)

Subtidal to at least 100 m, occasionally carried inshore and recovered in washings from algal holdfasts; Canada to Mexico, more abundant in southern part of range.

Test about 0.8 mm long; spines along inner margins of chambers of outer whorl distinctive but not always present, making differentiation of this species from the possibly synonymous *Nonion pizarrensis* (Berry) difficult.

This species, part of the *N. scapha basispinata* group, appears to be potentially useful as an indicator of marine pollution; it is scarcest on the continental shelf in areas affected by sewage effluent.

See Bandy, Ingle & Resig (1965a,b), Cushman (1939), and Cushman & Moyer (1930).

Phylum Protozoa / Class Sarcodina / Order Foraminifera / Family Rotaliidae

1.11 **Ammonia beccarii** (Linnaeus, 1758)
(= Streblus beccarii, Rotalia beccarii)

On muddy and fine sandy bottoms and on surfaces of marine plants in brackish California bays, lagoons, and estuaries, intertidal zone and subtidal waters, occasionally abundant in strandline deposits; common in Salton Sea (Imperial and Riverside Cos.), San Francisco Bay, and Lake Merritt (Oakland, Alameda Co.); widely distributed elsewhere, probably with several geographic subspecies.

Test 0.6–0.9 mm or more in diameter, with a low convex spire on dorsal side; chambers in outer whorl 9–12; characteristic umbilical openings on ventral (concave) surface.

This is often one of the most conspicuous members of brackish-water foraminiferan faunas, and much is known of its distribution and varieties. Test structure and morphological variation have been carefully studied by geologically oriented observers, but the current lack of cytological information on the species makes it hard to discriminate between environmental and hereditary influences on test morphology. Some studies have been made on the effects of temperature, salinity, and food on growth and size, but little is known of the organism's fundamental biology.

See Banner & Williams (1973), Bradshaw (1957, 1961), Brooks (1967), Chang & Kaesler (1974), Cifelli (1962), Hansen & Reiss (1971), Hofker (1930), Schnitker (1974), and Wood, Haynes & Adams (1963).

Phylum Protozoa / Class Sarcodina / Order Foraminifera / Family Glabratellidae

1.12 **Glabratella ornatissima** (Cushman, 1925)
(= Discorbis ornatissima, ?Eponides columbiensis)

Common, associated with *Eponides columbiensis* (1.13) on attached plants and animals, intertidal zone on sheltered rocky coasts; often abundant in strandline deposits and tidepool sediments, where test size may reflect natural sedimentary sorting processes; Canada to central California.

Test to 0.8 mm or more in diameter, 0.3 mm thick, the periphery broadly rounded; dorsal surface smooth, finely perforated, convex; ventral surface with irregular rows of coarse papillae; paired tests, consisting of two adults temporarily united for sexual reproduction, common in the late winter and spring.

Lankford (1962) suggested that *G. ornatissima* represents the megalospheric, sexually reproducing generation of a species for which *Eponides columbiensis* is the microspheric, asexually reproducing generation. Cytological proof of such a relationship is still lacking, but the circumstantial evidence for it is strong (Steinker, 1969, 1973).

See Lankford (1962), Lipps & Erskian (1969), Myers (1940), Seiglie & Bermudez (1965), and Steinker (1969, 1973).

Phylum Protozoa / Class Sarcodina / Order Foraminifera / Family Glabratellidae

1.13 **Eponides columbiensis** (Cushman, 1925)
(=?Glabratella ornatissima, ?Discorbis ornatissima)

Abundant in holdfasts and in clumps of attached invertebrates and algae, low intertidal zone on rocky shores, but more commonly subtidal; often conspicuous in beach deposits of northern California; California north of Point Conception (Santa Barbara Co.).

Test to 1.5 mm in diameter, 0.7 mm wide; dorsal surface similar to that of other species of *Eponides*, but ventral surface (silvered for contrast in the photograph) distinctive.

Eponides columbiensis is easily dislodged by current and wave action, its biconvex test making it particularly vulnerable to displacement. It appears less well adapted to life in turbulent waters than such encrusting or attaching types as *Iridia serialis* (1.1) and *Rosalina columbiensis* (1.14). Though it is one of the most conspicuous and abundant species in some intertidal waters of northern California, little is known of its basic biology or natural history. Circumstantial evidence strongly suggests that it is the asexually reproducing (microspheric) form of a life cycle in which *Glabratella* (=*Discorbis*) *ornatissima* (1.12) is the sexual (megalospheric) form.

See Lipps & Erskian (1969), Myers (1940), Seiglie & Bermudez (1965), and Steinker (1973).

Phylum Protozoa / Class Sarcodina / Order Foraminifera / Family Discorbidae

1.14 **Rosalina columbiensis** (Cushman, 1925)

Widely encountered and often abundant, attached to intertidal plants or to animals such as bryozoans and hydroids, or secluded in minute rock crevices; distribution probably widespread, but taxonomic clarification needed.

Test to 0.5 mm in diameter, 0.15 mm or more thick; early (smaller) chambers characteristically a rich wine-red color; final (largest) chamber with inturned lobe on ventral surface; umbilical region broadly depressed on dorsal side.

This species is highly variable, and possibly synonymous with *R. globularis* (d'Orbigny). The planktonic form *Tretomphalus bulloides* (d'Orbigny) may possibly be the sexual phase of the *Rosalina* life cycle, but proof is lacking.

These foraminifera are attached so securely to the substratum that they may resist vigorous brushing or washing and not be recovered in quantity by crude collecting procedures. The animals live and reproduce successfully in minute nooks and crannies in the most turbulent of rocky intertidal environments. They can be maintained indefinitely in laboratory culture, where they feed and move about actively and reproduce freely, apparently by asexual processes only. In natural situations, the animals tend to be highly motile while young, but they settle later in life and, if conditions are favorable, remain attached until reproduction empties the test of living material.

The functional consequences of the attachment of rosalinids to various marine invertebrates and algae is sometimes unclear. In some cases the foraminifera are obviously using the invertebrates or the algae merely as a substratum that affords security against displacement by wave or current action or burial by accumulating sediment. In others, particularly when the foraminifera lies in a pit within the host tissue (see, for example, Todd, 1965), the possibility of actual damage by the protozoan must be considered (see the discussion of *Gromia oviformis*, 1.18).

Considerable information is available on the structure, ultrastructure, and morphogenesis of the test of *R. floridana* (Cushman) (see Angell, 1967a,b). Similar studies on *R. columbiensis* would enhance the understanding of the organism's biology and might contribute to the resolution of some of the present taxonomic difficulties with the species.

See especially Steinker (1969); see also Angell (1967a,b), Douglas and Sliter (1965), Haderlie (1968, 1969), Hedley & Wakefield (1962), Lankford & Phleger (1973), Myers (1943a), Sliter (1965), and Todd (1965, 1971).

Phylum Protozoa / Class Sarcodina / Order Foraminifera / Family Cassidulinidae

1.15 **Cassidulina subglobosa** Brady, 1881

Occasionally intertidal, generally subtidal, from shallow bays to ocean deeps; Point Conception (Santa Barbara Co.), possibly farther north, to Baja California; widely distributed in all ocean basins.

Test 0.7 mm in greatest dimension; species easily recognized by its relatively large size, irregular contour, subglobular shape, and arrangement and reduced number of chambers.

See Nørvang (1959).

Phylum Protozoa / Class Sarcodina / Order Foraminifera / Family Cassidulinidae

1.16 **Cassidulina limbata** Cushman & Hughes, 1925

Subtidal, but occasionally found in washings from intertidal plant holdfasts, clumps of bryozoans, and coralline algae; in sediments along entire California coast, more abundant in northern half of state; common in Pliocene and Pleistocene deposits in southern California.

Test 0.4–0.6 mm or more in greatest diameter; sutures, broad, the central (umbonal) thickening prominent.

See Nørvang (1966).

Phylum Protozoa / Class Sarcodina / Order Foraminifera / Family Cibicididae

1.17 **Cibicides lobatulus** (Walker & Jacob, 1798)

Firmly attached to plant holdfasts, algal fronds, bryozoans, hydroids, rocks, and shells, intertidal zone and subtidal waters to deep water; west coast north of Point Conception (Santa Barbara Co.); widely distributed in temperate seas.

Test to 0.8 mm or more in length; dorsal surface (the surface of attachment in this species) flattened; sutures broad; pores prominent; periphery weakly lobed.

Like many attached foraminifera, this one is highly variable, and the limits of its variability poorly defined. The life cycle, too, is incompletely understood; Nyholm (1961) has suggested it might encompass forms that have heretofore been assigned to at least five different genera having calcareous tests with perforated walls, and even one genus of agglutinating forms, all of them often occurring together in the same habitat. These forms are widely distributed and easily collected, but the overriding need now is for a study of carefully controlled laboratory populations and a detailed cytological analysis of the developmental stages of all the suspect organisms.

See especially Steinker (1969); see also Cooper (1965), Dupeuble (1962), Nyholm (1961), and Schnitker (1969).

Phylum Protozoa / Class Sarcodina / Order Gromida / Family Gromiidae

1.18 **Gromia oviformis** Dujardin, 1835
(Not illustrated as Fig. 1.18; see Fig. 25.9b.)

Common on holdfasts and bases of surfgrass and algae, on various invertebrates, including sponges (especially *Leucandra heathi*, see under 2.4), hydroids, bryozoans, and crabs, and in any sheltered nook or cranny, middle to low intertidal zone and subtidal waters; Canada to Mexico; North Atlantic coasts of Europe and America, Mediterranean Sea, New Zealand, and Antarctica.

Commonly 1–3 mm in diameter, occasionally to 5 mm; shell spherical, transparent, glassy, exposing light-brown interior protoplasm; organism distinguished from fecal pellets, eggs, and spores, with which it is often confused, by its protoplasmic organization, pseudopodia, and shell, particularly the complex apertural region of the shell.

This protozoan differs from the very similar true foraminifer *Allogromia* by the microscopically non-granular appearance of its pseudopodia at all stages of life, and by the exceptional size of its adults. Various allogromiid foraminifera, including members of the genus *Allogromia*, occur on the west coast, but they are far less conspicuous than the gigantic *G. oviformis*.

Gromia oviformis has often been considered to be a foraminifer, but it is set apart by such fundamental differences as the absence of granules from its pseudopodia, the close association between meiosis and gametogenesis (which occur in separate generations in true foraminifera), and the greater complexity in nuclear cycles and structure of the organic shell. Its protective shell and efficient attachment organelles appear to suit it well for life in rocky intertidal areas. Sexual reproduction appears to be restricted to subtidal waters, the paired tests of "mating" animals occurring only rarely intertidally.

During the spring when the Monterey pine (*Pinus radiata*) is shedding pollen on the Monterey Peninsula, *G. oviformis* often concentrates so many pollen grains inside its protoplasmic body as to acquire a distinctly yellowish color often visible to the unaided eye. Gamete-filled adults are milky white instead of the usual gray to brown of vegetative adults. When brought into the laboratory, mature adults often liberate gametes within a few days, after which their partly empty shells contain only a residue—but often a voluminous one—of indigestible or undigested waste matter.

Gromia oviformis living on kelp holdfasts actually appear to be making use of the tissues of the host. The protoplasm of *Gromia*, clearly discernible through the shell, becomes discolored by the wine-red tissues of the ruptured kelp hapteron, which provides the protozoan with both protection and sustenance.

See Arnold (1972), Hedley (1962), Hedley & Bertaud (1962), Hedley & Wakefield (1962), and Jepps (1926).

Literature Cited

Angell, R. W. 1967a. The test structure and composition of the foraminifer *Rosalina floridana*. J. Protozool. 14: 299–307.

_____. 1967b. The process of chamber formation in the foraminifer *Rosalina floridana* (Cushman). J. Protozool. 14: 568–74.

_____. 1975. The test structure of *Elphidiella hannai*. J. Foram. Res. 5: 85–89.

Arnal, R. E. 1958. Rhizopoda from the Salton Sea, California. Contr. Cushman Found. Foram. Res. 9: 36–45.

Arnold, Z. M. 1972. Observations on the biology of the protozoan *Gromia oviformis* Dujardin. Univ. Calif. Publ. Zool. 100: 1–168.

_____. 1974. Field and laboratory techniques for the study of living Foraminifera, pp. 153–206, *in* Hedley & Adams (1974).

Bandy, O. L. 1953. Ecology and paleoecology of some California Foraminifera. 1. The frequency distribution of recent Foraminifera off California. J. Paleontol. 27: 161–82.

_____. 1960. General correlation of foraminiferal structure with environment. Rep. 21st Session, Internat. Geol. Congress Proc., Internat. Paleontol. Union, Part 22: 7–19.

Bandy, O. L., J. C. Ingle, Jr., and J. M. Resig. 1965a. Foraminiferal trends, Hyperion outfall, California. Limnol. Oceanogr. 10: 314–32.

_____. 1965b. Modification of foraminiferal distribution by the Orange County Outfall, California. Ocean Science and Ocean Engineer., Trans. Joint Conf., June 1965: 54–76.

Banner, F. T., and E. Williams. 1973. The test structure and extrathalamous cytoplasm of *Ammonia* Brünnich. J. Foram. Res. 3: 49–69.

Barker, R. W. 1960. Taxonomic notes on the species figured by H. B. Brady in his report on the Foraminifera dredged by H.M.S. *Challenger* during the years 1873–1876. Soc. Econ. Paleontol. Mineral., Spec. Publ. 9: 1–23, 115 pls.

Berthold, W.-U. 1971. Untersuchungen über die sexuelle Differenzierung der Foraminifere *Patellina corrugata* Williamson mit einem Beitrag zum Entwicklungsgang und Schalenbau. Arch. Protistenk. 113: 147–84.

Brady, H. B. 1884. Report on the Foraminifera dredged by H.M.S. *Challenger* during the years 1873–1876. Rep. Voy. *Challenger*, Zool. 9: 1–814, 115 pls.

Bradshaw, J. S. 1957. Laboratory studies on the rate of growth of the foraminifer "*Streblus beccarii* (Linné) var. *tepida* (Cushman)". J. Paleontol. 31: 1138–47.

_____. 1961. Laboratory experiments on the ecology of Foraminifera. Contr. Cushman Found. Foram. Res. 12: 87–106.

Brooks, A. L. 1967. Standing crop, vertical distribution, and morphometrics of *Ammonia beccarii* (Linné). Limnol. Oceanogr. 12: 667–84.

Burma, B. H. 1948. Studies in quantitative paleontology: I. Some aspects of the theory and practice of quantitative invertebrate paleontology. J. Paleontol. 22: 725–61.

Chang, Y.-M., and R. L. Kaesler. 1974. Morphological variation of the foraminifer *Ammonia beccarii* (Linné) from the Atlantic coast of the United States. Kansas Univ. Paleontol. Contr. Pap. 69: 1–23.

Cifelli, R. 1962. The morphology and structure of *Ammonia beccarii* (Linné). Contr. Cushman Found. Foram. Res. 13: 119–26.

Cooper, S. C. 1965. A new morphologic variation of the foraminifer *Cibicides lobatulus*. Contr. Cushman Found. Foram. Res. 16: 137–40.

Cooper, W. C. 1961. Intertidal Foraminifera of the California and Oregon coast. Contr. Cushman Found. Foram. Res. 12: 47–63.

Cushman, J. A. 1910–17. A monograph of the Foraminifera of the North Pacific Ocean. U.S. Nat. Mus., Bull. 71 (pts. 1–6): 1–596.

———. 1920. Observations on living specimen of *Iridia diaphana*, a species of Foraminifera. Proc. U.S. Nat. Mus. 57: 153–58.

———. 1925. Recent Foraminifera from British Columbia. Contr. Cushman Lab. Foram. Res. 1: 38–45.

———. 1927. Recent Foraminifera from off the west coast of America. Bull. Scripps Inst. Oceanogr., Tech. Ser. 1: 119–88.

———. 1939. A monograph of the foraminiferal family Nonionidae. U.S. Geol. Surv. Prof. Paper 191: 1–69.

———. 1948. Foraminifera, their classification and economic use. Cambridge, Mass.: Harvard University Press. 605 pp.

Cushman, J. A., and D. A. Moyer. 1930. Some recent Foraminifera from off San Pedro, California. Contr. Cushman Lab. Foram. Res. 6: 49–62.

Cushman, J. A., and W. W. Valentine. 1930. Shallow water Foraminifera from the Channel Islands of southern California. Stanford Univ. Dept. Geol., Contr. 1: 1–31.

Douglas, R., and W. V. Sliter. 1965. Taxonomic revision of certain Discorbacea and Orbitoidacea (Foraminiferida). Tulane Stud. Geol. 3: 149–64.

Dupeuble, P. A. 1962. Polymorphisme chez les Cibicidinae actuels de la région de Roscoff (Finistère). Rev. Micropaléontol. 4: 197–202.

Ellis, B. F., and A. R. Messina. 1940– . A catalogue of Foraminifera. Amer. Mus. Natur. Hist., Spec. Publ.

Gougé, J. D. 1971. Biological observations on the foraminifer *Cornuspira lajollaensis* Uchio. Master's thesis, Paleontology, University of California, Berkeley. 94 pp.

Grell, K. G. 1959. Untersuchungen über die Fortpflanzung und Sexualität der Foraminiferen. IV. *Patellina corrugata*. Arch. Protistenk. 104: 211–35.

———. 1973. Protozoology. New York: Springer. 554 pp.

Haderlie, E. C. 1968. Marine fouling organisms in Monterey harbor. Veliger 10: 327–41.

———. 1969. Marine fouling and boring organisms in Monterey harbor. II. Second year of investigation. Veliger 12: 182–92.

Hansen, H. J., and Z. Reiss. 1971. Electron microscopy of rotaliacean wall structures. Bull. Geol. Soc. Denmark 20: 329–46.

Hedley, R. H. 1962. *Gromia oviformis* (Rhizopoda) of New Zealand, with comments on the fossil Chitinozoa. New Zeal. J. Sci. 5: 121–36.

Hedley, R. H., and C. G. Adams, eds. 1974. Foraminifera. Vol. 1. New York: Academic Press. 276 pp.

———. 1977. Forminifera. Vol. 2. New York: Academic Press. 260 pp.

Hedley, R. H., and W. S. Bertaud. 1962. Electron-microscopic observations of *Gromia oviformis* (Sarcodina). J. Protozool. 9: 79–87.

Hedley, R. H., C. M. Hurdle, and I. D. Burdett. 1964. *Trochammina squamata* Jones and Parker (Foraminifera) with observations on some closely related species. New Zeal. J. Sci. 7: 417–26.

Hedley, R. H., and J. Wakefield. 1962. Fine structure of *Gromia oviformis* Rhizopoda: Protozoa. Bull. Brit. Mus. (Natur. Hist.) 18: 69–89.

Hedman, C. M. 1975. Bolinas Lagoon Forminifera: Student laboratory and field guide. College of Marin, Kentfield, Calif. 77 pp.

Hofker, J. 1930. Der Generationswechsel von *Rotalia beccarii* var. *flevensis* nov. var. Z. Zellforsch. Mikr. Anat. 10: 756–68.

Honigberg. B. M., W. Balamuth, E. C. Bovee, et al. 1964. A revised classification of the Phylum Protozoa. J. Protozool. 11: 7–20.

Jepps, M. W. 1926. Contribution to the study of *Gromia oviformis*. Quart. J. Microscop. Sci. 70: 701–19.

———. 1942. Studies on *Polystomella* Lamarck (Foraminifera). J. Mar. Biol. Assoc. U.K. 25: 607–66.

La Croix, E. 1931. Microtexture du test des Textularidae. Bull. Inst. Océanogr. Monaco 582: 1–18.

———. 1933. Le pseudomorphisme chez les Textularidae. Bull. Inst. Océanogr. Monaco 622: 1–12.

Lankford, R. R. 1962. Recent Foraminifera from the near-shore turbulent zone, western United States and northwest Mexico. Doctoral thesis, University of California, San Diego. 235 pp.

Lankford, R. R., and F. B. Phleger. 1973. Foraminifera from the nearshore turbulent zone, western North America. J. Foram. Res. 3: 101–32.

Le Calvez, J. 1936. Observations sur le genre *Iridia*. Arch. Zool. Expér. Gén. 78: 115–31.

———. 1938. Recherches sur les foraminifères. 1. Développement et reproduction. Arch. Zool. Expér. Gén. 80: 163–333.

———. 1953. Ordre des Foraminifères, pp. 149–265, in P.-P. Grassé, ed., Traité de zoologie, vol. 1(2). Paris: Masson. 1,160 pp.

Lee, J. J. 1974. Toward understanding the niche of the Foraminifera, pp. 207–60, in Hedley & Adams (1974).

_____. 1975. Culture of salt marsh microorganisms and micrometazoa, pp. 87–107, *in* W. L. Smith and M. H. Chanley, eds., Culture of marine invertebrate animals. New York: Plenum Press. 338 pp.

Lipps, J. H., and M. G. Erskian. 1969. Plastogamy in Foraminifera: *Glabratella ornatissima.* J. Protozool. 16: 422–25.

Lister, J. J. 1895. Contributions to the life-history of the Foraminifera. Phil. Trans. Roy. Soc. London 186: 401–53.

Loeblich, A. R., Jr., and H. Tappan. 1964. Sarcodina, chiefly "Thecamoebians" and Foraminiferida, pp. 1–900, *in* R. C. Moore, ed., Treatise on invertebrate paleontology, Part C, Protista 2, 2 vols. New York: Geological Society.

_____. 1974. Recent advances in the classification of the Foraminiferida, pp. 1–53, *in* Hedley & Adams (1974).

Marszalek, D. S. 1969. Observations on *Iridia diaphana*, a marine foraminifer. J. Protozool. 16: 599–611.

Marszalek, D. S., R. C. Wright, and W. W. Hay. 1969. Function of the test in Foraminifera. Trans. Gulf Coast Assoc. Geol. Socs. 19: 341–52.

Murray, J. W. 1963. Ecological experiments on Foraminiferida. J. Mar. Biol. Assoc. U.K. 43: 631–42.

_____. 1972. Distribution and ecology of living benthic foraminiferids. New York: Crane, Russak. 274 pp.

Myers. E. H. 1935: Morphogenesis of the test and the biological significance of dimorphism in the foraminifer *Patellina corrugata* Williamson. Bull. Scripps Inst. Oceanogr., Tech. Ser. 3: 393–404.

_____. 1936. The life-cycle of *Spirillina vivipara* Ehrenberg, with notes on morphogenesis, systematics and distribution of the Foraminifera. J. Roy. Microscop. Soc. 56: 120–46.

_____. 1937. Culture methods for marine Foraminifera of the littoral zone, pp. 93–96, *in* P. S. Galtsoff, F. E. Lutz, P. S. Welch, and J. G. Needham, Culture methods for invertebrate animals. Ithaca, N.Y.: Comstock. 590 pp. (Reprinted without change, Dover, N.Y., 1959.)

_____. 1940. Observations on the origin and fate of flagellated gametes in multiple tests of *Discorbis* (Foraminifera). J. Mar. Biol. Assoc. U.K. 24: 201–26.

_____. 1942. A quantitative study of the productivity of the Foraminifera in the sea. Proc. Amer. Philos. Soc. 85: 325–42.

_____. 1943a. Biology, ecology, and morphogenesis of a pelagic foraminifer. Stanford Univ. Publ., Biol. Sci. 9: 5–30.

_____. 1943b. Life activities of Foraminifera in relation to marine ecology. Proc. Amer. Philos. Soc. 86: 439–58.

Natland, M. L. 1933. The temperature and depth distribution of some recent and fossil Foraminifera in the southern California region. Bull. Scripps Inst. Oceanogr., Tech. Ser. 3: 225–30.

Nicol, D. 1944. New west American species of the foraminiferal genus *Elphidium.* J. Paleontol. 18: 172–85.

Nørvang, A. 1959. *Islandiella* n.g. and *Cassidulina* d'Orbigny. Vidensk. Medd. Dansk. Naturhist. Foren. 120: 25–41.

_____. 1966. *Textilina* nov. gen., *Textularia* Defrance and *Spiroplectammina* Cushman (Foraminifera). Biol. Skrif. Kongel. Dansk. Vidensk. Selsk. 15: 1–16.

Nyholm, K.-G. 1961. Morphogenesis and biology of the foraminifer *Cibicides lobatulus.* Zool. Bidrag Uppsala 33: 157–96.

Phleger, R. B. 1961. Ecology and distribution of recent Foraminifera. Baltimore: Johns Hopkins University Press. 297 pp.

Phleger, R. B., and J. S. Bradshaw. 1966. Sedimentary environments in a marine marsh. Science 154: 1551–53.

Reiter, M. 1959. Seasonal variations in intertidal Foraminifera of Santa Monica Bay, California. J. Paleontol. 33: 606–30.

Rosset-Moulinier, M. 1961. Étude systématique et écologique des Elphidiidae et des Nonionidae (Foraminifères) du littoral Breton. I. Les *Elphidium* du groupe *crispum* (Linné). Rev. Micropaléontol. 14: 76–81.

Salami, M. B. 1974. Biology, morphology, and phylogenetic relationships of *Trochammina* cf. *T. quadriloba* Hoglund, an agglutinating foraminifer. Master's thesis, Paleontology, University of California, Berkeley. 51 pp.

Schnitker, D. 1969. *Cibicides, Caribeanella* and the polyphyletic origin of *Planorbulina.* Contr. Cushman Found. Foram. Res. 20: 67–69.

_____. 1974. Ecotypic variation in *Ammonia beccarii* (Linné). J. Foram. Res. 4: 216–23.

Seiglie, G. A., and P. J. Bermudez. 1965. Monografia de la familia de foraminiferos Glabratellidae. Geoscience 12: 15–65.

Sliter, W. V. 1965. Laboratory experiments on the life cycle and ecologic controls of *Rosalina globularis* d'Orbigny. J. Protozool. 12: 210–15.

_____. 1970. *Bolivina doniezi* Cushman and Wickenden in clone culture. J. Protozool. 21: 87–99.

_____. 1971. Predation on benthic foraminifers. J. Foram. Res. 1: 20–29.

Steinker, D. C. 1969. Foraminifera of the rocky intertidal zone of the central California coast with emphasis on the biology of *Rosalina columbiensis* (Cushman). Doctoral thesis, Paleontology, University of California, Berkeley. 266 pp.

_____. 1973. Test dimorphism in *Glabratella ornatissima* (Cushman). Compass of Sigma Gamma Epsilon. 50: 10–21.

Stinemeyer, E. H., and M. Reiter. 1958. Ecology of the Foraminifera of Monterey Bay, California. Exploration Misc. Rep. 1621 (unpublished). Shell Oil Company, Bakersfield, Calif. 168 pp. and appendixes.

Todd, R. 1965. A new *Rosalina* (Foraminifera) parasitic on a bivalve. Deep-Sea Res. 12: 831–37.

_____. 1971. *Tretomphalus* (Foraminifera) from Midway. J. Foram Res. 1: 162–69.

Towe, K. M. 1967. Wall structure and cementation in *Haplophrag-moides canariensis*. Contr. Cushman Found. Res. 18: 147–51.

Uchio, T. 1960. Ecology of the living benthonic Foraminifera from the San Diego, California, area. Cushman Found. Foram. Res., Spec. Publ. 5. 72 pp.

Walton, W. R. 1955. Ecology of living benthonic Foraminifera, Todos Santos Bay, Baja California. J. Paleontol. 29: 952–1018.

Watkins, J. G. 1961. Foraminiferal ecology around the Orange County, California, ocean sewer outfall. Micropaleontol. 7: 199–206.

Wood, A., J. Haynes, and T. D. Adams. 1963. The structure of *Ammonia beccarii* (Linné). Contr. Cushman Found. Foram. Res. 14: 156–57.

Zech, L. 1964. Zytochemische Messungen an den Zellkernen der Foraminiferen *Patellina corrugata* and *Rotaliella heterocaryotica*. Arch. Protistenk. 107: 295–330.

Chapter 2

Porifera: *The Sponges*

Gerald J. Bakus and Donald P. Abbott

Sponges are aquatic animals. Two families occur exclusively in fresh water (Spongillidae and Potamolepidae; see Penney, 1960); but most species are marine, and in the sea they occur at all depths and latitudes. To find most of the kinds illustrated here, one must visit the shore during the lowest tides and look low down, under rocks and ledges, in caves, on surfaces shaded by seaweeds, or on floats or pier pilings. Some sponges are shaped like urns or globes, and others have branching tubular bodies; most intertidal sponges, however, occur as irregular encrusting masses, often brightly colored and varying greatly in area and thickness. A few sponges are soft-bodied, but the great majority are firm or even hard and woody to the touch. This firmness is the result of an internal supporting skeleton consisting of large numbers of tiny calcareous or siliceous spicules, the latter often accompanied by a network of tough spongin fibers. Forms in which the spicules project from the body and form a fur of tiny sharp bristles should be handled carefully, for the spicules can easily penetrate the skin and cause irritation.

Sponges are unique among animals in possessing bodies that are perforated throughout by a network of canals that open to the surrounding water. Water enters the canals through minute pores in the body surface, moves along passageways to chambers lined with cells bearing a single long flagellum (choanocytes or collar cells; see Rasmont, 1959), passes through further corridors, and eventually emerges through one or more large openings (oscules or oscula) at

Gerald J. Bakus is Associate Professor of Biological Sciences at the University of Southern California. Donald P. Abbott is Professor, Department of Biological Sciences and Hopkins Marine Station, Stanford University.

the surface. The flow through the sponge body is driven mainly by the flagella of the collar cells, but sponges exposed to external currents experience an additional "passive flow," as moving water sweeps past the oscule tips and reduces pressure there (Vogel, 1974, 1977, 1978). One sponge may circulate many liters of water through its porous body in a day (e.g., Bidder, 1923; Parker, 1914; Reiswig, 1974, 1975b). The current brings in dissolved oxygen, food particles (bacteria, fine detritus), and dissolved organic matter, and carries away body wastes (Frost, 1976; Reiswig, 1971, 1975a). Excessive sediments in the water can cause clogging and inhibit sponge growth (Bakus, 1968, 1969a).

The sponge body provides a good habitat for other living things, and a large sponge may contain several thousand small organisms of 20–30 different kinds living in its canal system, with additional species growing on its outer surface (see, for example, Bakus, 1966; Pearse, 1934, 1949; Rützler, 1970). Penetration of the sponge body by some commensals and parasites arouses a characteristic defensive tissue response in the sponges (see, for example, Connes, Paris & Sube, 1971). The living tissues of many sponges contain unicellular photosynthetic algae (see, for example, Sarà, 1964, 1965; Sarà & Vacelet, 1973; Vacelet, 1970) and bacteria (Vacelet & Donadey, 1977).

Sponge tissues consist of epithelial and connective elements and contain a complex array of cell types (e.g., Bergquist, 1978; Borojević, 1966; Lévi, 1964, 1970; Simpson, 1963, 1968a,b), some of which are capable of amoeboid locomotion within the sponge body. No organs, in the usual sense, exist, and the only systems worthy of the name in most sponges are the canal-and-chamber system and the skeletal system. A nervous system is lacking (e.g., Bullock, 1965; Jones, 1962), though sponges do display some coordinated behavior (e.g., Pavans de Ceccatty, 1974). Such body functions as digestion, excretion, secretion of spicules, and production of gametes are carried out by individual cells.

Sexual reproduction in sponges shows many features of interest (Bergquist, 1978; Brien, 1973; Fell, 1974b; Fry, 1970b). Eggs arise from amoebocytes, and sperm develop from amoebocytes or transformed collar cells. The sperm are shed to the sea, and sometimes the eggs as well, but more commonly the eggs are retained within the parent and fertilized there. Eggs develop into flagellated larvae, which may be solid (parenchymula) or hollow (amphiblastula). These are released, swim briefly (usually less than 3 days), then settle, metamorphose into tiny sponges, and begin to feed and grow (Burton, 1949a). The growth pattern in some sponges is complex and may include extensive reorganization of the existing canal system. The sponge resulting from such growth is considered an individual by some and a colony by others (e.g., Hartman & Reiswig, 1973; Simpson, 1973).

Many sponges also reproduce asexually, by fragmentation, budding, the formation of swimming pseudolarvae, or other methods.

Sponge embryology is curiously different from that of most multicellular animals (see, for example, Bergquist, 1978; Brien, 1943, 1973; Fell, 1974b). This, together with the unique sponge body plan and the age of the group (dating from the Precambrian), suggests that sponges very early formed a separate branch of the animal kingdom, the Parazoa. Whether they diverged early from other metazoan stocks or arose separately from the Protozoa is unsettled (see, for example, Tuzet, 1963, 1973).

Sponges are important sources of food for some marine animals, including certain nudibranchs, chitons, sea stars, turtles, and rasping fishes; the fishes are especially important predators in shallow tropical waters (e.g., Bakus, 1964, 1969b; Randall & Hartman, 1968). Some sponge species that survive grazing best have been shown to produce distasteful or toxic compounds (e.g., Bakus & Green, 1974; Bakus & Thun, 1979). Sponges are also of biochemical interest for the variety of pigments, sterols, and antibiotic substances they produce (e.g., Bergmann, 1962; Cecil et al., 1976; Florkin & Scheer, 1968; Jakowska & Nigrelli, 1960). Bath sponges are still fished and marketed commercially (e.g., Ferry, 1967), but diseases of uncertain nature and competition from synthetic sponges have brought about a marked decline in world sponge trade.

Classification of sponges is based in good part on the skeleton. About 5,000 species of sponges have been described and named. These fall into four classes, two of which are represented by species in the California intertidal zone. Members of the class Calcarea are usually small (not over 20 cm in greatest dimension, and usually much less), and not conspicuously colored. They bear calcareous spicules, which are needlelike or have three or four rays radiating from a common center. More common, by far, are members of the class Demospongiae, which make up 80 percent of the known species of sponges. Here the spicules, when present, are siliceous, and enormously varied in both size and shape. Some species have spongin fibers as well as spicules, some lack spicules and contain only spongin fibers, and a very few have neither spicules nor spongin. Species of the class Hyalospongiae, or Hexactinellida, do not occur intertidally on California shores, and most forms live at depths greater than 200 m. Spicules are siliceous, commonly have six rays, and may be fused into a complex gridwork. A new group, the class Sclerospongiae, was recently established for a group of sponges, living at subtidal depths in tropical seas around the world, that bear a striking resemblance to the fossil organisms called stromatoporoids (Hartman & Goreau, 1970). Their skeletal parts include a massive calcareous base and usually siliceous spicules and organic fibers.

The California sponges are imperfectly known. Enough of the common species are included here to give one an idea of the variety that occurs, but a collector will very likely find others that have been omitted. Certain species are distinctive enough to be recognized in the field, but some are not, and even a specialist may need to examine the spicules (easily liberated from the tissues with ordinary laundry bleach) under the microscope to identify species with certainty. Hartman (1975) presents a key to some of the common California intertidal sponges, along with a very helpful introduction to "the fearsome terminology of sponge spicules." Other useful papers on west coast sponges include Bakus (1966), De Laubenfels (1927, 1932, 1948, 1961), Dickinson (1945), Koltun (1959, 1966), and Lambe (1893, 1894, 1895). Further details on spicule nomenclature are covered in Bakus & Green (1977), De Laubenfels (1955), O'Connell (1919), and Reid (1970). The terminology of soft parts is discussed in Hyman (1940) and Borojević et al. (1968).

For those wishing further details on sponges as organisms, good general accounts of the phylum are available in Bergquist (1978), Brien (1968), Grassé (1973), Hartman (1960), and Hyman (1940, 1959). A bibliography on sponges for the years 1551 to 1913 is presented in Bidder & Vosmaer-Röell (1928). A discussion of the sponge genera of the world is given in De Laubenfels (1936). Current research on many aspects of sponge biology is well presented in Florkin & Scheer (1968), Fry (1970b), and Harrison & Cowden (1976). Considerable new information on sponges will soon be published from the Second International Symposium on Sponges, held in Paris, December 18–22, 1978.

Phylum Porifera / Class Calcarea / Order Clathrinida

2.1 Clathrina blanca (Miklucho-Maclay, 1868)
(=Leucosolenia macleayi, Guancha blanca)

Abundant in some areas on shaded sides and bottoms of rocks, low intertidal zone; subtidal to 820 m; Palos Verdes Peninsula (Los Angeles Co.) to La Jolla (San Diego Co.), and Cabo San Quintín (Baja California).

Animal vase-shaped and stalked; body flattened in one plane, to 2 cm high and 2 cm in greatest diameter, the stalk to 5 mm long; color white to tan; spicules calcareous, triradiate.

Burton (1963) considered *Clathrina blanca* a synonym of *C. coriacea* De Laubenfels, but Johnson (1976, 1978a,b) has shown clearly that southern California populations of the two are separate species (see also De Laubenfels, 1932, as *Leucosolenia macleayi*).

Closely related forms from Atlantic and Mediterranean waters have been studied from the standpoint of oogenesis, polarity of the egg, fertilization, embryology, larval structure, and method of secretion of the spicules; see Borojević (1969), Fry (1970b, Index), Hartman (1958b), Jones (1970), Korotkova & Gelihovskaia (1963), Lévi (1963), Minchin (1898), Sarà, Liaci & Melone (1966a,b), Tuzet (1947, 1948, 1970), and bibliographies of these papers.

Phylum Porifera / Class Calcarea / Order Leucosoleniida

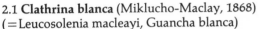

2.2 Leucosolenia eleanor Urban, 1905

On shaded sides and open undersurfaces of rocks, low intertidal zone; subtidal on rocks and boat hulls to 26 m; British Columbia to California; Japan.

Colony encrusting or attached by a narrow base, forming a three-dimensional lattice of slender branching and anastomosing tubes 2 mm or less in diameter; entire mass sometimes reaching 15 cm in greatest diameter, but more commonly 3–6 cm; color white to tan; spicules calcareous, consisting of triradiates, quadriradiates, and small to large rods pointed at both ends.

Burton (1963) synonymized this species with *L. botryoides*, but this decision is questioned by other sponge specialists.

The colonies of *L. eleanor* provide shelter for a diversified assemblage of protozoans, small worms, and small crustaceans. Preparations showing choanocyte activity are easily made by slicing tubes lengthwise, placing them on a glass slide with the inner surface upward, adding a drop of seawater and a coverslip, and examining under high magnification. Panels submerged in Monterey Bay showed larval settlement and growth of small colonies in all months from February to October; colonies 5 cm or more in diameter were present on panels that had been in the water for 3 months or more.

More than 100 species of *Leucosolenia* have been described for the world. They are almost the only adult sponges to exhibit what spongologists call the ascon grade of structure, in which collar cells are distributed along the entire lining of the branching tubes that make up the sponge body. This is the simplest, and perhaps most primitive, type of body structure known in sponges. European species have been studied in detail with respect to development and polarity of the egg, method of fertilization, development of the amphiblastula larva, formation of tissues, growth rate of body and spicules, regeneration, and the effects of cutting, wounding, and application of carcinogens. In some species, fragments as small as 1 mm in diameter can develop into whole sponges. Carcinogens and multiple wounds may cause tubes to transform into stolonlike outgrowths that change polarity and give rise to numerous buds.

For *L. eleanor*, see Laubenfels (1932), Haderlie (1968a,b, 1969, 1971), Hartman (1958b, 1975), Hewatt (1937), McLean (1962), Thompson & Chow (1955), Turner, Ebert & Given (1969), and Urban (1905). For other species, see Burton (1963), Hartman (1958b, 1964), Jones (1954, 1959, 1964, 1970), and Tuzet (1948, 1970).

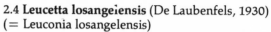

Phylum Porifera / Class Calcarea / Order Sycettida

2.3 **Leucilla nuttingi** (Urban, 1905)
(=Rhabdodermella nuttingi)

Common in some areas under boulders, in caves, and on sides of deeply shaded rocks, low intertidal zone on exposed rocky shores; subtidal on rock walls to 46 m; Sansum Narrows (British Columbia) to Cabo San Quintín (Baja California).

Animal slender, vase-shaped, stalked, and usually in clusters; stalk to 2 cm in length, the body to 3 cm long and 5 mm or more in diameter; color white to tan; spicules calcareous, consisting of small and large triradiates and quadriradiates, small rods with pointed ends, and in some specimens large rods with sharp tips forming an oscular fringe.

This is the simplest sponge of our shores that exhibits the leucon grade of structure, in which the collar cells (choanocytes) are restricted to flagellated chambers that are both tiny and very numerous. Living collar cells can be seen in preparations made by cutting open a freshly collected sponge, scraping the interior with a needle, mounting the resulting debris on a microscope slide in a drop of seawater, adding a coverslip, and examining at high magnification. Amphiblastula larvae are encountered occasionally in such preparations.

At La Jolla (San Diego Co.), sexual reproduction occurs in the warmer months; larval settlement was found on panels submerged from June to September. Farther north, in Monterey Bay, some settlement of larvae was noted in the fall and winter (October to March), and one individual on a panel exposed for a year reached a total length of 5 cm (Haderlie,

1968a). Partial chemical analyses have been made of major body constituents and spicules.

See (usually under either *Leucilla* or *Rhabdodermella*) Burton (1963), Coe (1932, as *Grantia*), Coe & Allen (1937), De Laubenfels (1932), Giese (1966), Haderlie (1968a, 1969, 1971, 1974), Hartman (1958b), Kittridge et al. (1962), McLean (1962), and Thompson & Chow (1955).

Phylum Porifera / Class Calcarea / Order Leucettida

2.4 **Leucetta losangeiensis** (De Laubenfels, 1930)
(= Leuconia losangelensis)

Common to abundant in many areas on sides of rocks, in crevices, and on pilings and floats, low intertidal zone to 111 m depth; southern California to Cabo San Quintín (Baja California) and Gulf of California.

Colony encrusting, to about 3 cm thick and 15 cm in greatest diameter; body texture crusty; surface convoluted and bearing many oscules to 15 mm in diameter; spicules calcareous, consisting of large and small rods pointed at both ends, triradiates, and (in some specimens) quadriradiates. (See Coe, 1932, as "a white encrusting sponge"; Coe & Allen, 1937; and De Laubenfels, 1932.)

Burton (1963) synonymized *Leucetta losangelensis* and other species under *Leuconia barbata* (Duchassaing & Michelotti), but this decision is questioned by other sponge specialists. Coe and Allen (1937) reported a colony growing to a diameter of 15 cm or more from September to the following July.

A related species, **Leucandra heathi** Urban, common in the low intertidal zone and at depths of a few meters on rocks and pilings in central California, is strikingly different in superficial appearance. The body is pear-shaped, attached by the larger end or by one side, and typically bears a single oscule at the narrowed free tip. The body bristles with projecting spicules with pointed tips, which also form a fringe at the osculum. A sphincter just inside the oscule constricts when the animal is disturbed (Dr. Harold Heath, for whom the species is named, once noted that tobacco smoke blown at the orifice causes fairly rapid closure). Individuals attain a height of 4 cm in 4 to 7 months in Monterey Bay. Water flow through the body of a related form, *Leucandra aspera*, was studied by Bidder (1923).

See Haderlie (1968a, 1969, 1971, 1974), Hartman (1958b, 1975), McLean (1962), Ricketts & Calvin (1968), Smith & Haderlie (1969), Tuzet (1970), and Urban (1905).

Phylum Porifera / Class Demospongiae / Order Dendroceratida

2.5 Aplysilla glacialis (Merejkowsky, 1878)
Keratose Sponge

On shaded sides and open undersurfaces of rocks, low intertidal zone; subtidal to 84 m; Pacific coast; northern Europe, Australia, nearly cosmopolitan in temperate seas.

Body encrusting, to 6 mm thick and roughly 6 cm in diameter; color often rose, occasionally yellow; spicules absent; skeleton consisting of dendritic spongin fibers, whose projecting ends form little cones to 5 mm high on exposed surface of body.

Aplysilla glacialis is preyed upon by nudibranchs of the genus *Hypselodoris*. Embryonic development has been studied in a related species, *A. sulfurea*.

See Bloom (1976), Delage (1892), De Laubenfels (1932, 1948), and Turner, Ebert & Given (1969).

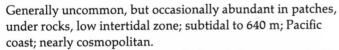

Phylum Porifera / Class Demospongiae / Order Dictyoceratida

2.6 Dysidea fragilis (Montagu, 1814)
Keratose Sponge

Generally uncommon, but occasionally abundant in patches, under rocks, low intertidal zone; subtidal to 640 m; Pacific coast; nearly cosmopolitan.

Body forming a crust to 3 cm thick and about 15 cm in diameter; surface fibrous, bearing many small projections; color often light brown, sometimes gray or lavender, rarely black; spicules absent; principal spongin fibers forming skeleton

nearly transparent but packed with sand grains, the secondary or thinner fibers transparent and often clean of sand.

This species is similar to, and possibly synonymous with, *D. amblia* De Laubenfels (see, as *D. amblia*, De Laubenfels, 1932, 1948; Dickinson, 1945). In southern California it is sometimes cast ashore in large quantities during storms.

Studies of distribution, body minerals, lipids, carbohydrates, and antibiotic substances have been made on this or related species of sponges.

See Bergmann (1949), Bloom (1976), Bowen & Sutton (1951), Jakowska & Nigrelli (1960), Koltun (1959), MacLennan (1970), and Turner, Ebert & Given (1968).

Phylum Porifera / Class Demospongiae / Order Dictyoceratida

2.7 Aplysina fistularis (Pallas, 1766)
(= Verongia aurea, V. thiona, Spongia fulva)
Yellow Sponge, Keratose Sponge

Sometimes common under rock ledges, low intertidal zone; subtidal to 36 m; Pacific coast; nearly cosmopolitan.

Body forming an irregular crust to 4 cm thick and about 38 cm in diameter; color lemon-yellow to yellow-brown, occasionally tinged with green or lavender, dark violet when dried or preserved; spicules absent; skeleton composed of reticulated spongin fibers.

The limpetlike opisthobranch mollusk *Tylodina fungina* (14.13) consumes this sponge, and its body color matches that of the host. Studies of ultrastructure, biochemistry, and ecology have been made of related species.

Wiedenmayer (1977) considers *Verongia aurea* to be an ecophenotype of this common Caribbean species.

See, as *V. thiona*, De Laubenfels (1932, 1948), Dickinson (1945), and Given & Lees (1967). For other species, see Bergmann (1962), Bergquist & Hartman (1969), Cecil et al. (1976), Pavans de Ceccatty, Thiney & Garrone (1970), Reiswig (1970, 1971, 1973, 1974), and Vacelet (1970).

Phylum Porifera / Class Demospongiae / Order Poecilosclerida

2.8 Acarnus erithacus De Laubenfels, 1927

Sometimes common on shaded sides and open undersurfaces of rocks, low intertidal zone; subtidal to 700 m; British Columbia to Cabo San Quintín (Baja California) and Gulf of California.

Body low and encrusting or massive, to about 9 cm in diameter; intertidal specimens scarlet in color (subtidal forms bronze); larger siliceous spicules of several types, including rods capped by small clusters of recurved thorns, and rods with the main shaft either smooth or spiny.

In central California and the Gulf of California this species reaches the low intertidal zone, but in the San Juan Islands (Washington), it is confined to depths between 68 and 192 m, possibly because the great volume of freshwater runoff makes the shallower waters too brackish. *A. erithacus* is eaten by the nudibranchs *Rostanga pulchra* (14.26) and *Doriopsilla albopunctata* (14.34).

Materials exhibiting significant antimicrobial and antiviral activity have been found in extracts of *A. erithacus* from the Gulf of California.

See Bakus (1966), Bloom (1976), Carter & Rinehart (1978), De Laubenfels (1927, 1932), Dickinson (1945), and Turner, Ebert & Given (1968).

Phylum Porifera / Class Demospongiae / Order Poecilosclerida

2.9 Axocielita originalis (De Laubenfels, 1930)
(= Esperiopsis originalis)

Sometimes common on undersides of rocks, low intertidal zone; subtidal to 75 m; British Columbia to Cabo San Quintín (Baja California).

Body forming a thin crust to 4 mm in thickness and roughly 7 cm in diameter; juveniles pale yellow, adults orange-red; siliceous spicules occurring in tracts associated with spongin, the largest spicules including thick and thin rods, all sharply pointed at one end, rounded at the other.

Five other species of intertidal sponges in California are practically identical with this one in gross appearance, and difficult to distinguish in the field: *Microciona microjoanna* (2.12), *Ophlitaspongia pennata* var. *californiana* (2.14), **Isociona lithophoenix** De Laubenfels, **Plocamia karykina** (De Laubenfels), and **Plocamilla illgi** Bakus.

Axocielita originalis is eaten by the nudibranch *Rostanga pulchra* (14.26).

See especially Hartman (1975, as *Axocielita originalis*); see also Bakus (1966), Bloom (1976), De Laubenfels (1932), and Turner, Ebert & Given (1968), all as *Esperiopsis originalis*, and Simpson (1966, as *Axocielita hartmani*).

Phylum Porifera / Class Demospongiae / Order Poecilosclerida

2.10 Hymenamphiastra cyanocrypta De Laubenfels, 1930

On undersurfaces of low intertidal rocks in deep shade; subtidal to at least 37 m, especially under ledges and in caves; Pacific Grove (Monterey Co.) to Point Loma (San Diego Co.).

Body forming an encrusting film to 1 mm thick and to nearly 1 m in diameter, but most often only a few centimeters across; color usually deep cobalt blue, sometimes light orange; spicules siliceous.

This strikingly colored species is readily identified in the field and can scarcely be confused with anything else. (See also Fig. 3.42a.) With regard to the color, De Laubenfels (1932) said, "Throughout the sponge, but especially in the ectosomal [= outer] regions, there are abundant dark blue spiral microbes, 4 μ or 5 μ long and about ½ μ thick, very probably symbionts, and certainly responsible for the blue color of the sponge." P. Castro has described symbiosis between a blue-green alga and *H. cyanocrypta* at the Second International Symposium on Sponges.

See De Laubenfels (1932), Given & Lees (1967), McLean (1962), and Turner, Ebert & Given (1968).

Phylum Porifera / Class Demospongiae / Order Poecilosclerida

2.11 **Lissodendoryx topsenti** (De Laubenfels, 1930)
(=Tedania topsenti)

On shaded sides and open undersurfaces of low intertidal rocks; subtidal on walls and flat surfaces shaded by the brown alga *Pterygophora*; known only in Monterey Co., from Pacific Grove to Granite Creek, south of Carmel; probably more widespread.

Body encrusting to massive, to 3 cm thick and about 5 cm or more in diameter, commonly with large oscules (to 5 mm in diameter); color red-orange to salmon-pink; largest siliceous spicules all rodlike, with one or both ends rounded or knobbed, the shaft occasionally spiny; very tiny spicules (microscleres) absent.

Generic placement of this sponge is uncertain; see Hartman (1975, as *Lissodendoryx topsenti*) and Bakus (1966, as *Kirkpatrickia topsenti*); see also De Laubenfels (1932), McLean (1962), and Turner, Ebert & Given (1968), all as *Tedania topsenti*.

A rather similar California species, **L. firma** Lambe (=*L. noxiosa* De Laubenfels), can be distinguished by its yellowish or tan color, the presence of numerous microscleres, and the possession of a strong odor. Aqueous extracts of the sponge cause symptoms in mice similar to those of histamine poisoning.

See De Laubenfels (1932), Hartman (1975), and Sheikh & Djerassi (1974).

Phylum Porifera / Class Demospongiae / Order Poecilosclerida

2.12 **Microciona microjoanna** De Laubenfels, 1930

Fairly common on hard substrata in harbors and bays, and on shaded sides and open undersurfaces of rocks in calmer waters, low intertidal zone; subtidal to 18 m; British Columbia south at least to Point Vicente (Los Angeles Co.).

Body rather flattened and encrusting, to 2 cm thick and about 7 cm in diameter; color red; largest siliceous spicules

including thin and thick rods, both pointed at one end and blunt at the other, some with spiny shafts.

Five other species of California sponges look virtually identical to this one in the field (see under *Axocielita originalis*, 2.9).

Microciona microjoanna reproduces (apparently by sexual processes) in late spring and early summer; older animals may then die off, for larger specimens are sometimes not seen again until late in the fall. The species is especially suitable for laboratory culture and experimentation.

See especially Fell (1967); see also Bakus (1966), De Laubenfels (1932), Humphreys (1970), and references listed under *M. prolifera* (2.13).

Phylum Porifera / Class Demospongiae / Order Poecilosclerida

2.13 **Microciona prolifera** (Ellis & Solander, 1786)
Redbeard Sponge

Sometimes abundant on oyster shells, and present on rocks in mud and on pilings in calm waters, very low intertidal zone; subtidal to at least 26 m; known from Willapa Bay (Washington) and San Francisco Bay; common on Atlantic coast from Nova Scotia to Gulf of Mexico; possibly cosmopolitan.

Young colonies encrusting, larger colonies bushlike, with numerous erect, fingerlike, anastomosing lobes forming a mass to 20 cm in height and 25 cm in greatest diameter; color red to orange-brown; siliceous spicules of several types, very numerous, making up 45 percent of dry weight of the body.

This species is excellent for experimental work and has been the subject of many studies, mostly on the Atlantic coast. Seasonal changes in east coast populations are correlated with temperature cycles. In North Carolina, growth is most rapid in the colder months, least rapid in the summer. In shallow Connecticut waters the reverse is true; colonies in shallow water regress in winter, lose some cell types, and become inactive. In spring new flagellated chambers and canals form, and ova develop; sexual reproduction begins in July, and larvae are liberated until September or October.

The larvae are flagellated, and swim actively at the surface for 20–30 hours, then go to the bottom and creep for a similar period before finally attaching. They contain most of the cell types found in adult animals (and sometimes some somatic cells of the parent as well), and some spicules. The larvae attach by a small adhesive area free of flagella, the animal spreads out on the surface, and the flagellated cells migrate internally to form flagellated chambers. Adults contain nine types of cells, of which five or six types are needed to produce adult tissue. Explants cut from adult colonies and tied to glass slides readily attach and grow new tissue. Sponge cells dissociated by squeezing bits of sponge against bolting silk or by treating sponge fragments with calcium-free and magnesium-free seawater reaggregate into clumps that develop into small functional sponges. Reaggregation of dissociated cells is common in sponges, but this is one of the few species in which reaggregates form fully functional new individuals. Reaggregated sponges grown in still water bear oscules on chimneys that are perpendicular to the substratum (regardless of orientation), whereas those raised in currents have the chimneys bent downstream.

Food particles can be taken in by the epidermis, and some dissolved organic matter (amino acids) can be absorbed directly from the water; ingestion of food by collar cells has not been observed. Tissue extracts contain antibiotic materials with a fairly broad spectrum of antimicrobial activity. Numerous chemical studies have been made on this and related species in the genus.

Several species of *Microciona* are preyed upon by dorid nudibranchs.

See especially Simpson (1963, 1968a,b); see also Afzelius (1961), Bagby (1963, 1966, 1970), Bergmann (1949, 1962), Bergquist & Hartman (1969), Bloom (1976), Curtis (1970), Fell (1967, 1976), Fry (1970b, Index), Harrison & Cowden (1976, Index), Lévi (1960b), Litchfield & Morales (1976), Reed, Greenberg & Pierce (1976), Reiswig (1975b), Stephens & Schinske (1961), Warburton (1960, 1966), Wells, Wells & Gray (1964), Wilson (1937), and Wilson & Penney (1930).

Phylum Porifera / Class Demospongiae / Order Poecilosclerida

2.14 **Ophlitaspongia pennata** (Lambe, 1895) var. **californiana** De Laubenfels, 1932

Often common on rocks and pebbles and on walls of shaded crevices and caves, lower middle intertidal zone to 2 m below low water; Sooke (Vancouver Island, British Columbia) to near Puertocitos (Gulf of California).

Body forming a firm crust to 5 mm thick and from several cm to nearly 1 m in diameter; surface often with a stellate pattern formed by radiating excurrent grooves around the shallow depressions of the oscules; color orange to scarlet; siliceous spicules consisting of thin and thick rods, both pointed at one end and slightly knobbed at the other, and of much smaller rods bent like an archer's bow.

Five other species of California sponges closely resemble this one in the field (see under *Axocielita originalis*, 2.9).

The photograph shows the red nudibranch *Rostanga pulchra* (14.26) on the sponge surface. This sea slug frequently consumes the sponge and appears to incorporate the red pigment, though the color of the slug does not change when its diet is altered. *O. pennata* is also eaten by the nudibranchs *Aldisa sanguinea* (14.27) and *Hypselodoris*.

Little is known of our variety of *Ophlitaspongia*, but studies of histology and ultrastructure, larval behavior, spicule variation, and carbohydrate constitutents have been carried out on other species in the genus.

See Bakus (1966), Bergquist & Sinclair (1968), Bergquist, Sinclair & Hogg (1970), Bloom (1976), Borojević & Lévi (1964), De Laubenfels (1927, 1932), Fell (1976), Fry (1970a), Given & Lees (1967), Hewatt (1937), Lévi (1960a,b), and MacLennan (1970).

Phylum Porifera / Class Demospongiae / Order Halichondrida

2.15 **Halichondria panicea** (Pallas, 1766) Crumb-of-Bread Sponge

On rocks, around kelp stipes, in mussel beds, and on pier pilings, middle and low intertidal zones; subtidal to about 100 m; cosmopolitan; abundant in some areas, especially north of Point Conception (Santa Barbara Co.).

Shape variable: on rocks often forming encrusting sheets to several centimeters thick, with short oscules irregularly distributed; on protected pilings forming larger masses (to 17 cm in diameter) with elongate oscular tubes; color orange, yellow, or green; siliceous spicules consisting of rods pointed at both ends, in disarray; odor often unpleasant.

This is a widely distributed and widely studied sponge. Preliminary studies by C. Smecher (pers. comm.) suggest that *H. panicea* may comprise a complex of species, distinguishable in part by histological and histochemical techniques.

On the Pacific coast, sexual reproduction occurs in late summer and early fall. Larval settlement occurred on a panel submerged in Monterey Bay for 6 months, from October to March. Observations of colonies over many months in the field in England show that they may move slowly, subdivide, or coalesce with adjacent colonies. The probable lifetime of a colony is 2–3 years.

Halichondria panicea is one of several sponges fed upon by the sea star *Pteraster tesselatus*; it is also eaten by the nudibranchs *Archidoris montereyensis* (14.30), *Archidoris odhneri* (14.29), *Anisodoris nobilis* (14.31), *Cadlina luteomarginata* (14.19), *Discodoris heathi* (14.33), *Diaulula sandiegensis* (14.32), and others. Wounds inflicted in the sponge surface by the predators or other agents are readily repaired, and the sponge produces its own antibiotic materials. Body fragments only 1–2 mm in diameter can reorganize themselves into small functional sponges and reattach themselves to a substratum. Dissociated cells of this species have been used in studies of cell aggregation and cell-to-cell adhesion.

Work on related species includes studies of body amino acids and sterols, experiments with dissociated cells, and observations of reproduction, development, and larval behavior. The larva of *H. moorei* Bergquist can swim with a spiral motion while rotating on its axis, but it spends most of its free larval life (20–60 hours) creeping on the bottom. It tends to creep upward on any vertical face, though it appears indifferent to light. The larva spins slowly while it attaches by the anterior pole. After attachment the larval body spreads out on the surface, and within 72 hours the canal system is functioning.

See Bergmann (1949, 1962), Bergquist & Sinclair (1968), Bergquist, Sinclair & Hogg (1970), Bloom (1976), Burton (1949a), Curtis (1970), Fell (1967, 1976), Fry (1970b, Index), Haderlie (1969), Hartman (1958a, 1975), Humphreys (1970), Jakowska & Nigrelli (1960), Koltun (1959), Korotkova (1970), Litchfield & Morales (1976), MacGinitie & MacGinitie (1968), MacLennan (1970), Mauzey, Birkeland & Dayton (1968), Meewis (1941), Miller (1968), Reiswig (1975b), and Ricketts & Calvin (1968).

Phylum Porifera / Class Demospongiae / Order Haplosclerida

2.16 Haliclona sp.
(? = H. permollis (Bowerbank, 1866))

Abundant in some areas on sides of rocks, in caves, under ledges, in crevices, and on pilings, mostly in shaded areas, middle and low intertidal zones; subtidal to about 50 m; all along California rocky shores and widely distributed around the world.

Body encrusting, forming sheets to 4 cm thick and from a few cm to more than 1 m in diameter; oscules rather regularly spaced, elevated on tubes like small volcanoes or nearly level with surface; color violet to rose, occasionally pale orange or yellow; siliceous spicules consisting of rods pointed at both ends, arranged in a triangular reticulum and connected by spongin.

In California it is eaten by the nudibranch *Diaulula sandiegensis* (14.32).

Sexual reproduction occurs during the spring and early summer. The larva is a parenchymula. Gemmules are not known for *H. permollis*, though they occur in some related species. Dissociated cells placed in seawater may reaggregate but do not normally reconstitute small sponges.

Another California species, usually called **H. ecbasis** De Laubenfels, characteristically grows on mussels on floats in harbors. A central California population is presently known from Bodega Bay (Sonoma Co.), various parts of San Francisco Bay, and Carmel (Monterey Co.). A southern California population, differing in some respects (Hartman, 1975) is known from San Pedro (Los Angeles Co.) to San Diego. The shape of this species is highly variable. In central California, sexual reproduction shows peaks in the spring and fall, but some ripe sponges are found in other seasons. Oogenesis and development of the embryo have been studied in detail by Fell (1969).

A New Zealand species of *Haliclona* (see Bergquist & Sinclair, 1968; Bergquist, Sinclair & Hogg, 1970) has a solid, oval parenchymula larva, which is ciliated all over but bears a ring of longer cilia near the posterior end. A dense cluster of rod-like spicules with pointed ends lies within the larva at the posterior end. The larva swims rapidly, rotating the body, often following a corkscrew course and showing a strong positive response to light. Nine or 10 hours after release the larva becomes photonegative, settles to the bottom, and creeps around for about 2 hours before settling. Attachment is by the anterior end, and within 24 hours the larval body is reorganized into a small sponge with one oscule.

Numerous studies have been made on other species of *Haliclona.* Certain *Haliclona* species contain materials toxic not only to bacteria but also to higher invertebrates and vertebrates. At least one brackish-water species, *H. loosanoffi,* tolerates salinities 25–50 percent that of normal seawater. Different species of *Haliclona* show different degrees of specificity when their dissociated cells are mixed with those of other species. Some *Haliclona* even form multicellular aggregates with dissociated cells of sea anemones. The great variety of sterols present suggests that the genus may include forms not closely related.

The species in the genus *Haliclona* are frequently difficult or impossible to distinguish by microscopic analysis of the spicules and skeleton, even though spicules compose 7–44 percent of the dry weight in species studied. The forms treated in this chapter are most easily distinguished by examining living specimens, but even this may not be reliable. This complex genus, containing more than 200 species, needs extensive revision based on histology, histochemistry, and reproduction. For further information, see especially Griessinger (1971) and Hartman (1975).

For *H. permollis,* see Bergmann (1949, 1962), Bergmann & Feeney (1949), Bloom (1976), De Laubenfels (1932, as *H. cinerea*), Elvin (1976), Giese (1966), Given & Lees (1967, as *H. cinerea*), Hartman (1975), MacGinitie (1935, as *H. cinerea*), Ricketts & Calvin (1968), Sheikh & Djerassi (1974), Thompson & Chow (1955), and Turner,

Ebert & Given (1969). For other species of *Haliclona,* see Bakus (1969b), Bergmann (1949, 1962), Bergmann & Feeney (1949), Bergquist & Hartman (1969), Bergquist & Sinclair (1968), Bergquist, Sinclair & Hogg (1970), Cecil *et al.* (1976), Fell (1967, 1969, 1970, 1974a,b, 1975, 1976), Fry (1970b, Index), Gasic & Galanti (1966), Griessinger (1971), Hartman (1958a), Humphreys (1970), Jakowska & Nigrelli (1960), Lévi (1956), Litchfield & Morales (1976), MacGinitie (1935, as *H. cinerea*), Meewis (1939), Pavans de Ceccatty, Thiney & Garrone (1970), Reiswig (1975a,b), Sarà, Liaci & Melone (1966a,b), Simpson & Fell (1974), and Wells, Wells & Gray (1964).

Phylum Porifera / Class Demospongiae / Order Haplosclerida

2.17 **Haliclona** sp.

Moderately common under rocks and on coralline algae, low intertidal zone on rocky shores; subtidal to 32 m; Pacific Grove (Monterey Co.) to Point Loma (San Diego Co.).

Body encrusting, hemispherical, to 2 cm thick and 5 cm in greatest diameter; oscules to 4 mm in diameter, often elevated on short tubes; color drab light brown, lavender, or pale rosegray; siliceous spicules consisting of rods pointed at both ends, mostly incorporated into spongin fibers to 0.2 mm in diameter.

The generic and even the family affinities of this species have been questioned. Further study is needed.

See De Laubenfels (1932), Hartman (1975), and references under *Haliclona* sp. (2.16).

Phylum Porifera / Class Demospongiae / Order Haplosclerida

2.18 **Haliclona** sp.

On shaded sides and undersurfaces of rocks, low intertidal zone; Pacific Grove (Monterey Co.).

Body a firm crust to 1 cm thick and 5 cm or more in diameter; surface with numerous, nearly uniformly spaced and slightly elevated oscules; color light brown; siliceous spicules consisting of rods pointed at both ends.

See discussion and references under *Haliclona* sp. (2.16).

Phylum Porifera / Class Demospongiae / Order Tetractinellida

2.19 **Geodia mesotriaena** Lendenfeld, 1910

Common in rock crevices and protected caves in some areas, low intertidal zone; subtidal to about 370 m; southeastern Alaska to Gulf of California.

Body subspherical to massive and irregular, to 15 cm thick and 28 cm in greatest diameter, hard and crusty in texture; color white or gray to pale rose; surface with a "fur" of projecting siliceous spicules; large and small spicules of many types present.

This species should be handled with care, as projecting spicules may penetrate the skin and adhere to one's fingers, causing irritation. Specimens with broken spicules may lack the surface fur. Some species of *Geodia* in cold waters attain a diameter of half a meter. Chemical extractions made from a *Geodia* collected in southern California were lethal to mosquito fish (*Gambusia affinis*) in concentrations as low as 8 μg of toxin per ml of tap water (G. Bakus, unpubl. ms.).

See Bergmann (1949, 1962), Bergquist & Hartman (1969), Burton (1949b), De Laubenfels (1932), Fry (1970b, Index), Given & Lees (1967), Harrison & Cowden (1976, Index), Koltun (1966), and Reiswig (1970).

Phylum Porifera / Class Demospongiae / Order Hadromerida

2.20 **Cliona celata** Grant, 1826, var. **californiana** De Laubenfels, 1932
Boring Sponge, Sulfur Sponge

Boring in calcareous materials such as shells of the red abalone, large barnacles, some clams, moon snails, slipper shells, oysters, etc., low intertidal zone; subtidal to at least 120 m; often found in tropical corals, widely distributed on Pacific coast and elsewhere in world.

Exposed portions of body forming lemon-yellow papillae protruding from holes 1–3 mm in diameter in calcareous materials; remainder of sponge hidden in tunnels and galleries below surface; siliceous spicules consisting of rods knobbed at one end and pointed at the other.

The remarkable boring activities of this species and its many relatives in the genus result in the removal of small oval chips from the calcium carbonate substratum. Spicules are not used as flaking tools; instead, tiny cell processes reach out and in a highly localized manner apparently secrete materials that loosen the chips. Contractility of cells may be responsible for the removal of the chips. Burrowing activity is greater in areas of strong currents and high daytime light intensity. For reviews of boring, see Goreau & Hartman (1963) and Rützler & Rieger (1973).

Unlike most sponges, *Cliona* sheds its eggs into the sea, and the larvae develop there. They swim for 20–30 hours, creep about for a similar period, and settle. Swimming larvae contain some amoeboid cells derived from the parental tissues, and therefore are genetic mosaics; whether the parental cells persist long in the offspring or merely serve as food is uncertain.

Cliona celata is reportedly eaten by the nudibranch *Doriopsilla albopunctata* (14.34).

Various *Cliona* species have been studied from the standpoint of anatomy, development, ecology, boring mechanism, biochemistry of sterols and carbohydrates, aggregating properties of dissociated cells, absorption of dissolved organic matter, and presence of antibiotic materials.

See Bakus (1975), Bergmann (1949, 1962), Bloom (1976), De Laubenfels (1932), Driscoll (1967), Fry (1970b, Index), Goreau & Hartman (1963), Hartman (1957, 1958a, 1975), Jakowska & Nigrelli (1960), MacGinitie (1935), MacGinitie & MacGinitie (1968), MacLennan (1970), Neumann (1966), Rützler (1965, 1975), Rützler & Rieger (1973), Simpson & Fell (1974), Stephens & Schinske (1961), Warburton (1958, 1961, 1966), and Wells, Wells & Gray (1964).

Phylum Porifera / Class Demospongiae / Order Hadromerida

2.21 **Spheciospongia confoederata** De Laubenfels, 1930

On shaded rock surfaces and pilings, very low intertidal zone; subtidal to 26 m; Queen Charlotte Islands (British Columbia) to Cabo San Quintín (Baja California).

Body encrusting to massive and irregular, to at least 14 cm thick and 70 cm or more in greatest diameter; surface brown to creamy gray, leathery in texture, the interior yellow, firm in consistency, cavernous; siliceous spicules consisting of rods knobbed at one end and pointed at the other.

The genus *Spheciospongia* contains some of the world's largest sponges. Ricketts & Calvin (1968) reports that *S. confoederata* forms "great, slaty clusters on which several people could be seated." Such colonies are still found occasionally at Point Pinos (Pacific Grove, Monterey Co.; Roy Johnson, Jr., pers. comm.). *S. vesparia,* the Caribbean loggerhead sponge, attains a diameter of nearly 3 m. One specimen, with a volume of about 185,000 cm³, contained in its passages 17,128 other animals of 22 species (mainly shrimps); two 50,000-cm³ specimens contained, respectively, 13,504 and 6,282 other animals. The central areas of very large *S. vesparia* tend to die off and disintegrate, leaving doughnut-shaped specimens. This species is capable of forcibly ejecting soft detritus accidentally taken in, by washing it back out along the same channels by which it entered. The mechanism by which this back washing is accomplished is not clear. Biochemical studies of sterols, cell-membrane lipids, and free amino acids have been carried out on some species of the genus.

See Bergmann (1949, 1962), Bergquist & Hartman (1969), De Laubenfels (1932), Litchfield & Morales (1976), Pearse (1934, 1949), Ricketts & Calvin (1968), and Storr (1976a,b).

Phylum Porifera / Class Demospongiae / Order Hadromerida

2.22 **Suberites ficus** (Johnson, 1842)

On shaded rocks, very low intertidal zone to at least 36 m depth on rocks and gravel in California waters; Bering Sea and Gulf of Alaska to off San Elijo Lagoon (San Diego Co.); Japan, Canadian Arctic, east coast south to Virginia, Greenland, Iceland, Norway to France, Mediterranean Sea, western Africa; nearly cosmopolitan in northern seas.

Body shape variable, encrusting to massive, to 3 cm thick and 36 cm or more in greatest diameter, usually smaller; texture firm and leathery; color orange, yellow-orange, or yellow-brown; siliceous spicules consisting of small to large rods pointed at one end and knobbed at the other.

A widespread species, this was probably one of the first sponges reported from North America. In the Pacific Northwest it frequently grows over snail shells that are carried about by hermit crabs in the subtidal zone. In California it is preyed upon by the nudibranch *Archidoris montereyensis* (14.30).

Small gemmules, lacking spicules, occur in the sponge tissues.

Studies on this and related species have been concerned with systematics and distribution, morphology, biochemistry, comparative serology, and larval ultrastructure, and with the adhesive properties of suspensions of dissociated cells.

See Bergmann (1949, 1962), Bloom (1976), Curtis (1970), De Laubenfels (1932, as *Ficulina suberea*), Diaz, Connes & Paris (1975), Hammen & Florkin (1968), Hartman (1958a), Koltun (1966, as *Suberites domuncula ficus*), Litchfield & Morales (1976), MacLennan (1970), Paris (1961), Simpson & Fell (1974), Tuzet (1963), and Tuzet, Bessiére & Paris (1961).

Phylum Porifera / Class Demospongiae / Order Hadromerida

2.23 **Tethya aurantia** (Pallas, 1766)

On shaded sides and open undersurfaces of rocks, very low intertidal zone; more common in crevices of rock walls in the scuba zone, and extending to 440 m depth; British Columbia to southern California and Gulf of California; widely distributed elsewhere, for example, Europe, Iceland, Russian Arctic, New Zealand.

Body subspherical or hemispherical, to at least 8 cm in greatest diameter; surface of cortex typically covered with low warts, regularly arranged, lacking easily visible pores or ostia; color orange, yellow, or greenish (southern California specimens often overgrown with green algae and diatoms); interior of sponge (medulla) yellow, firm, with conspicuous radial arrangement of larger spicules.

This is one of the best known of the noncommercial demosponge genera, and species have been studied from several points of view. *Tethya aurantia* reproduces both sexually and asexually. Eggs arise from granular amoebocytes; they are fertilized internally and liberated shortly thereafter. Cleavage is complete and equal, giving rise first to an embryo with approximately equal blastomeres and later to a solid, swimming parenchymula larva covered with cells bearing one flagellum each. Oogenesis and larval development are illustrated in Lévi (1956). Development takes only 2–3 days. The larvae settle in irregularities on surfaces and usually aggregate in clusters that fuse during metamorphosis. A single larva can give rise to a small sponge, but more commonly numerous settled larvae in close contact fuse and reorganize to produce a single sponge that is a genetic mosaic.

Asexual reproduction occurs by several means in *T. aurantia* and related species. Buds may arise on stalks at the surface, detach, drift away, and settle singly, each bud yielding a small sponge, or several buds may fuse in a cluster to produce one larger sponge. Stolons may grow out over the substratum from the base of a large sponge, then expand distally to produce new individuals still attached to the parent; whole clones may arise from an original sponge in this manner and occupy a large rock surface. Gemmules, which are found in the interiors of some individuals, vary in size (0.1–2 mm in greatest diameter), but are always homogeneous clusters of cells, tightly packed together inside membranes of spongin.

Experiments with other *Tethya* species show (1) that individuals sliced in two heal together in a day if the two halves are kept pressed together; (2) that the halves of two different conspecific sponges, similarly sliced and held together, heal equally quickly to form bilateral mosaics, which apparently function normally; and (3) that *Tethya* cells, isolated and suspended in water, reaggregate with cells of other genera of sponges, and even reaggregate with suspended cells of sea anemones (*Anemonia*) to form cell clusters viable for at least 72 hours. *Tethya* with bacterial infections make use of spicules in isolating and shedding infected parts. *Tethya* also react protectively to the burrowing activities of commensal crustaceans. Chemical studies include work on amino acids, sterols, and cell-surface materials, as well as spicule composition and comparative serology.

Tethya aurantia is a hardy species; living individuals have been maintained in running seawater aquaria at the Hopkins Marine Station (Pacific Grove, Monterey Co.), without special care for periods of up to 2 years.

See Bergmann (1949, 1962), Bergquist & Hartman (1969), Bergquist, Sinclair & Hogg (1970), Bloom (1976), Burton (1949a), Connes (1966, 1967, 1968), De Laubenfels (1932), Edmondson (1946), Egami & Ishii (1956), Given & Lees (1967), Hammen & Florkin (1968), Koltun (1959, 1970), Lévi (1956), McLean (1962), Paris (1961), Reiswig (1971, 1973, 1974), Sarà, Liaci & Melone (1966a,b), Sheikh & Djerassi (1974), Thompson & Chow (1955), Turner, Ebert & Given (1968), and Tuzet, Bessiére & Paris (1961).

Phylum Porifera / Class Demospongiae / Order Axinellida

2.24 **Higginsia** sp.

On undersides of rocks, low intertidal zone; Pacific Grove (Monterey Co.).

Body encrusting, rubbery in texture to 3 mm thick and at least 5 cm in greatest diameter; color green with yellow blotches; largest siliceous spicules rodlike, pointed at one or both ends; some small spicules rodlike with knob near middle of shaft, others forming spiny rods pointed at both ends.

This is closely related to *H. higginissima*, which has been taken at a depth of 48 m off Isla del Espíritu Santo (Gulf of California).

Higginsia is reportedly eaten by the nudibranch *Cadlina luteomarginata* (14.90).

See Bloom (1976) and Dickinson (1945).

Literature Cited

Afzelius, B. A. 1961. Flimmer-flagellum of the sponge. Nature 191: 1318–19.

Bagby, R. M. 1963. Fine structure of the dermal membrane of the sponge *Microciona prolifera*. Amer. Zool. 3: 515–16 (abstract).

———. 1966. The fine structure of myocytes in the sponges *Microciona prolifera* (Ellis & Solander) and *Tedania ignis* (Duchassaing & Michelotti). J. Morphol. 118: 167–82.

———. 1970. The fine structure of pinacocytes in the marine sponge *Microciona prolifera* (Ellis & Solander). Z. Zellforsch. 105: 579–94.

Bakus, G. J. 1964. The effects of fish-grazing on invertebrate evolution in shallow tropical waters. Allan Hancock Found. Occas. Pap. 27: 1–29.

————. 1966. Marine poecioscleridan sponges of the San Juan Archipelago, Washington. J. Zool., London 149: 415–531.

————. 1968. Sedimentation and benthic invertebrates of Fanning Island, central Pacific. Marine Geol. 6: 45–51.

————. 1969a. Some effects of sedimentation on benthic invertebrates of atoll lagoons, pp. 503–4, *in* A. A. Casteñares, and F. B. Phleger, eds., Coastal lagoons, a symposium, Mem. Internat. Symp. Coastal lagoons, UNAM-UNESCO, Mexico, Nov. 28–30, 1967.

————. 1969b. Energetics and feeding in shallow marine waters. Internat. Rev. Gen. Exper. Zool. 4: 275–369.

————. 1975. Ecology of excavating sponges. Panama Canal Company, Canal Zone. 12 pp.

Bakus, G. J., and G. Green. 1974. Toxicity in sponges and holothurians: A geographic pattern. Science 185: 951–53.

Bakus, G. J., and K. Green. 1977. Porifera of the Pacific coast of North America. Southern California Coastal Water Research Project, Taxonomic Standardization Program: Sponge Workshop. (photocopy) 22 pp.

Bakus, G. J., and M. Thun. 1979. Bioassays on the toxicity of Caribbean sponges. Second International Symposium on Sponges, Paris, December 18–22, 1978. (manuscript) 6 pp.

Bergmann, W. 1949. Comparative biochemical studies on the lipids of marine invertebrates, with special reference to the sterols. J. Mar. Res. 8: 137–76.

————. 1962. Sterols: Their structure and distribution, pp. 103–62, *in* M. Florkin, and H. S. Mason, eds., Comparative biochemistry; vol. 3 part A. New York: Academic Press. 959 pp.

Bergmann, W., and R. J. Feeney, 1949. Contributions to the study of marine products. 23. Sterols from sponges of the family Haliclonidae. J. Org. Chem. 14: 1078–84.

Bergquist, P. R. 1978. Sponges. Berkeley and Los Angeles: University of California Press. 268 pp.

Bergquist, P. R., and W. D. Hartman. 1969. Free amino acid patterns and the classification of the Demospongiae. Mar. Biol. 3: 247–68.

Bergquist, P. R., and M. E. Sinclair. 1968. The morphology and behavior of larvae of some intertidal sponges. New Zeal. J. Mar. Freshwater Res. 2: 426–37.

Bergquist, P. R., M. E. Sinclair, and J. J. Hogg. 1970. Adaptation to intertidal existence: Reproductive cycles and larval behaviour in Demospongiae, pp. 247–71, *in* Fry (1970b).

Bidder, G. P. 1923. The relation of the form of a sponge to its currents. Quart. J. Microscop. Sci. 67: 293–323.

Bidder, G. P., and C. S. Vosmaer-Röell, eds. 1928. Bibliography of sponges, 1551–1913. Cambridge, Eng.: Cambridge University Press. 234 pp.

Bloom, S. A. 1976. Morphological correlations between dorid nudibranch predators and sponge prey. Veliger 18: 289–301.

Borojević, R. 1966. Étude expérimentale de la différenciation des cellules de l'éponge au cours de son développement. Develop. Biol. 14: 130–53.

————. 1969. Étude du développement et de la différenciation cellulaire d'éponges calcaires calcinéennes (genre *Clathrina et Ascandra*). Ann. Embryol. Morphogen. 2: 15–36.

Borojević, R., and C. Lévi. 1964. Étude au microscope electronique des cellules de l'éponge: *Ophlitaspongia seriata* (Grant), au cours de la réorganisation aprés dissociation. Z. Zellforsch. 64: 708–25.

Borojević, R., W. G. Fry, W. C. Jones, C. Lévi, R. Rasmont, M. Sarà, and J. Vacelet. 1968. A reassessment of the terminology for sponges. Bull. Mus. Nat. Hist. Natur. (2) 39: 1224–35.

Bowen, V. T., and D. Sutton. 1951. Comparative studies of mineral constituents of marine sponges. I. The genera *Dysidea, Chondrilla, Terpios*. J. Mar. Res. 10: 153–67.

Brien, P. 1943. L'embryologie des éponges. Bull. Mus. Hist. Natur. Belg. 19: 1–20.

————. 1968. The sponges, or Porifera, pp. 1–30, *in* Florkin & Scheer (1968).

————. 1973. Les demosponges. Morphologie et reproduction, pp. 133–461, *in* Grassé (1973).

Bullock, T. H. 1965. Protozoa, Mesozoa, and Porifera, pp. 433–57, *in* T. H. Bullock, and G. A. Horridge, Structure and function in the nervous systems of invertebrates, vol. 1. San Francisco: W. H. Freeman. 798 pp.

Burton, M. 1949a. Observations on littoral sponges, including supposed swarming of larvae, movement and coalescence in mature individuals, longevity and death. Proc. Zool. Soc. London 118: 893–915.

————. 1949b. Non-sexual reproduction in sponges with special reference to a collection of young *Geodia*. Proc. Linn. Soc. London 160: 163–78.

————. 1963. A revision of the classification of the calcareous sponges, London: Brit. Mus. (Natur. Hist.). 693 pp.

Carter, G. T., and K. L. Rinehart, Jr. 1978. Acarnidines, novel antiviral and antimicrobial compounds from the sponge *Acarnus erithacus* (de Laubenfels). J. Amer. Chem. Soc. 100: 4302–4.

Cecil, J. T., M. F. Stempien, Jr., G. D. Ruggiere, and R. F. Nigrelli. 1976. Cytological abnormalities in a variety of normal and transformed cell lines treated with extracts from sponges, pp. 171–81, *in* Harrison & Cowden (1976).

Coe, W. R. 1932. Season of attachment and rate of growth of sedentary marine organisms at the pier of the Scripps Institution of Oceanography, La Jolla, California. Bull. Scripps Inst. Oceanogr., Tech. Ser. 3: 37–86.

Coe, W. R., and W. E. Allen. 1937. Growth of sedentary marine organisms on experimental blocks and plates for nine successive years at the pier of the Scripps Institution of Oceanography. Bull. Scripps Inst. Oceanogr., Tech. Ser. 4: 101–36.

Connes, R. 1966. Aspects morphologiques de la régénération de *Tethya lyncurium* Lamarck. Bull. Soc. Zool. France 91: 43–53.

———. 1967. Structure et développement des bourgeons chez l'éponge siliceuse *Tethya lyncurium* Lamarck. Recherches expérimentales et cytologiques. Arch. Zool. Expér. Gén. 108: 157–95.

———. 1968. Réactions de defense de l'éponge *Tethya lyncurium* Lamarck, vis-à-vis des micro-organismes et de l'amphipode *Leucothoe spinicarpa*. Vie et Milieu 18: 281–98.

Connes, R., J. Paris, and J. Sube. 1971. Réactions tissulaires de quelques démosponges vis-à-vis de leurs commensaux et parasites. Naturaliste Canad. 98: 923–35.

Curtis, A. S. G. 1970. Problems and some solutions in the study of cellular aggregation, pp. 335–52, in Fry (1970b).

Delage, Y. 1892. Embryogénie des éponges. Développement post-larvaire des éponges siliceuses et fibreuses marines et d'eau douce. Arch. Zool. Expér. Gén. (2) 10: 345–498.

De Laubenfels, M. W. 1927. The red sponges of Monterey Peninsula, California. Ann. Mag. Natur. Hist. (9) 19: 258–66.

———. 1932. The marine and fresh-water sponges of California. Proc. U.S. Nat. Mus. 81: 1–140.

———. 1936. A discussion of the sponge fauna of the Dry Tortugas in particular and West Indies in general, with material for a revision of the families and orders of the Porifera. Carnegie Inst. Washington, Pap. Tortugas Lab. 30: 1–225.

———. 1948. The order Keratosa of the phylum Porifera—a monographic study. Allan Hancock Found. Occas. Pap. 3: 1–217.

———. 1955. Porifera, pp. E22–E112, in R. C. Moore, ed., Treatise on invertebrate paleontology, Part E, Archaeocyatha and Porifera. Lawrence: University of Kansas Press. 122 pp.

———. 1961. Porifera of Friday Harbor and vicinity. Pacific Sci. 15: 192–202.

Diaz, J.-P., R. Connes, and J. Paris. 1975. Étude ultrastructurale de l'ovogénese d'une demosponge: *Suberites massa* Nardo. J. Microscopie 24: 105–16.

Dickinson, M. G. 1945. Sponges of the Gulf of California. Allan Hancock Pacific Exped. 11: 1–252.

Driscoll, E. G. 1967. Attached epifauna–substrate relations. Limnol. Oceanogr. 12: 633–41.

Edmondson, C. H. 1946. Reproduction in *Donatia deformis* (Thiele). Occas. Pap. Bernice P. Bishop Mus. 18; 271–82.

Egami, N. and S. Ishii. 1956. Differentiation of sex cells in united heterosexual halves of the sponge, *Tethya serica*. Annot. Zool. Japon. 29: 199–201.

Elvin, D. W. 1976. Feeding of a dorid nudibranch, *Diaulula sandiegensis*, on the sponge *Haliclona permollis*. Veliger 19: 194–96.

Fell, P. E. 1967. Sponges, pp. 265–76, in F. H. Wilt and N. K. Wessells, eds. Methods in developmental biology. New York: T. Y. Crowell. 813 pp.

———. 1969. The involvement of nurse cells in oogenesis and embryonic development in the marine sponge, *Haliclona ecbasis*. J. Morphol. 127: 133–50.

———. 1970. The natural history of *Haliclona ecbasis* De Laubenfels, a siliceous sponge of California. Pacific Sci. 24: 381–86.

———. 1974a. Diapause in the gemmules of the marine sponge, *Haliclona loosanoffi*, with a note on the gemmules of *Haliclona oculata*. Biol. Bull. 147: 333–51.

———. 1974b. Porifera, pp. 51–132, in A. C. Giese and J. S. Pearse, eds., Reproduction of marine invertebrates, vol. 1. New York: Academic Press. 546 pp.

———. 1975. Salinity tolerance and desiccation resistance of the gemmules of the brackish-water sponge, *Haliclona loosanoffi*. J. Exper. Zool. 194: 409–12.

———. 1976. Analysis of reproduction in sponge populations: An overview with specific information on the reproduction of *Haliclona loosanoffi*, pp. 51–67, in Harrison & Cowden (1976).

Ferry, M. 1967. The sponge industry of Tarpon Springs. Com. Fish. Rev. 29: 82–86.

Florkin, M. and B. T. Scheer, eds. 1968. Chemical zoology. Z. Porifera, Coelenterata, and Platyhelminthes. New York: Academic Press. 639 pp.

Frost, T. M. 1976. Sponge feeding: A review with a discussion of some continuing research, pp. 283–98, in Harrison & Cowden (1976).

Fry, W. G. 1970a. The sponge as a population: A biometric approach. pp. 135–62, in Fry (1970b).

———, ed. 1970b. The biology of the Porifera. Symp. Zool. Soc. London 25. London: Academic Press. 512 pp.

Gasic, G. J., and N. L. Galanti. 1966. Proteins and disulfide groups in the aggregation of dissociated cells of sea sponges. Science 151: 203–5.

Giese, A. C. 1966. Lipids in the economy of marine invertebrates. Physiol. Rev. 46: 244–98.

Given, R. R., and D. C. Lees. 1967. Santa Catalina Island Biological Survey. Survey Rep. 1. Allan Hancock Found., University of Southern California. (mimeo) 126 pp.

Goreau, T. F., and W. D. Hartman. 1963. Boring sponges as controlling factors in the formation and maintenance of coral reefs, pp. 25–54, in R. F. Sognnaes, ed., Mechanisms of hard tissue destruction. Washington, D.C.: Amer. Assoc. Adv. Sci., Publ. 75. 764 pp.

Grassé, P.-P., ed. 1973. Traité de zoologie. 3. Spongaires. Paris: Masson. 716 pp.

Griessinger, J.-M. 1971. Étude des réniérides de Méditerranée (démosponges haplosclérides). Bull. Mus. Nat. Hist. Natur. Paris (Zool.) (3) 3: 97–181.

Haderlie, E. C. 1968a. Marine fouling and boring organisms in Monterey harbor. Veliger 10: 327–41.

————. 1968b. Marine boring and fouling organisms in open water of Monterey Bay, California, pp. 658–79, *in* A. H. Walters and J. S. Elphick, eds., Biodeterioration of materials: Microbiological and allied aspects. Barking, Eng.: Elsevier Publ. 740 pp.

————. 1969. Marine fouling and boring organisms in Monterey harbor. II. Second year of investigation. Veliger 12: 182–92.

————. 1971. Marine fouling and boring organisms at 100 feet depth in open water of Monterey Bay. Veliger 13: 249–60.

————. 1974. Growth rates, depth preference, and ecological succession of some sessile marine invertebrates in Monterey harbor. Veliger 17 (Suppl.): 1–35.

Hammen, C. S., and M. Florkin. 1968. Chemical composition and intermediary metabolism—Porifera, pp. 53–64, *in* Florkin & Scheer (1968).

Harrison, F. W., and R. R. Cowden, eds. 1976. Aspects of sponge biology. New York: Academic Press. 354 pp.

Hartman, W. D. 1957. Ecological niche differentiation in the boring sponges. Evolution 11: 294–97.

————. 1958a. Natural history of the marine sponges of southern New England. Bull. Peabody Mus. Natur. Hist. 12: 1–155.

————. 1958b. A re-examination of Bidder's classification of the Calcarea. Syst. Zool. 7: 97–110.

————. 1960. Porifera, pp. 513–19, *in* McGraw-Hill encyclopedia of science and technology, vol. 10. New York: McGraw-Hill.

————. 1964. Taxonomy of calcareous sponges. Science 144: 711–12.

————. 1975. Phylum Porifera, pp. 32–64, *in* R. I. Smith, and J. T. Carlton, eds., Light's manual: Intertidal invertebrates of the central California coast. 3rd ed. Berkeley and Los Angeles: University of California Press. 716 pp.

Hartman, W. D., and T. F. Goreau. 1970. Jamaican coralline sponges: Their morphology, ecology and fossil relatives, pp. 205–43, *in* Fry (1970b).

Hartman, W. D., and H. M. Reiswig. 1973. The individuality of sponges, pp. 567–84, *in* R. S. Bordman, A. H. Cheetham, and W. A. Oliver, Jr., eds., Animal colonies, development and function through time. Stroudsburg, Pa.: Dowden, Hutchinson, and Ross. 603 pp.

Hewatt, W. G. 1937. Ecological studies on selected marine intertidal communities of Monterey Bay, California. Amer. Midl. Natur. 18: 161–206.

Humphreys, T. 1970. Biochemical analysis of sponge cell aggregation, pp. 325–34, *in* Fry (1970b).

Hyman, L. H. 1940. The invertebrates: Protozoa through Ctenophora. Vol. 1. New York: McGraw-Hill. 726 pp.

————. 1959. The invertebrates: Smaller coelomate groups (Retrospect—Phylum Porifera, pp. 715–18). Vol. 5. New York: McGraw-Hill. 783 pp.

Jakowska, S., and R. F. Nigrelli. 1960. Antimicrobial substances from sponges, pp. 913–16, *in* R. F. Nigrelli, ed., Biochemistry and pharmacology of compounds derived from marine organisms. Ann. N.Y. Acad. Sci. 90: 615–950.

Johnson, M. F. 1976. Conspecificity in calcareous sponges: *Clathrina coriacea* (Montagu, 1818) and *Clathrina blanca* (Miklucho-Maclay, 1868). Doctoral thesis, Biological Sciences, University of Southern California. 442 pp.

————. 1978a. Significance of life history studies of calcareous sponges for species determination. Bull. Mar. Sci. 28: 570–74.

————. 1978b. A comparative study of the external form and skeleton of the calcareous sponges *Clathrina coriacea* and *Clathrina blanca* from Santa Catalina Island, California. Canad. J. Zool. 56: 1669–77.

Jones, W. C. 1954. Spicule form in *Leucosolenia complicata*. Quart. J. Microscop. Sci. 95: 191–203.

————. 1959. Spicule growth rates in *Leucosolenia variabilis*. Quart. J. Microscop. Sci. 100: 557–70.

————. 1962. Is there a nervous system in sponges? Biol. Rev. 37: 1–50.

————. 1964. Photographic records of living oscular tubes of *Leucosolenia variabilis*. II. Spicule growth, form and displacement. J. Mar. Biol. Assoc. U.K. 44: 311–31.

————. 1970. The composition, development, form and orientation of calcareous sponge spicules, pp. 91–123, *in* Fry (1970b).

Kittredge, J. S., D. G. Simonsen, E. Roberts, and B. Jelinek. 1962. Free amino acids of marine invertebrates, pp. 176–86, *in* J. T. Holden, ed., Amino-acid pools: Distribution, formation and function of free amino acids. Amsterdam: Elsevier Publ. 815 pp.

Koltun, V. M. 1959. Siliceous horny sponges of the northern and far-eastern seas of the U.S.S.R. [In Russian.] Opred. Faune S.S.S.R. 67: 1–236.

————. 1966. Tetraxonid sponges of northern and far-eastern seas of the U.S.S.R. [In Russian.] Opred. Faune S.S.S.R. 90: 1–112.

————. 1970. Sponges of the Arctic and Antarctic: A faunistic review, pp. 285–97, *in* Fry (1970b).

Korotkova, G. P. 1970. Regeneration and somatic embryogenesis in sponges, pp. 423–36, *in* Fry (1970b).

Korotkova, G. P., and M. A. Gelihovskaia. 1963. Recherches expérimentales sur le phénomène de polarité chez les éponges calcaires du type ascon. Cah. Biol. Mar. 4: 47–59.

Lambe, L. M. 1893. On some sponges from the Pacific coast of Canada and Behring Sea. Proc. Trans. Roy. Soc. Canad. 10 (1892): 67–78.

————. 1894. Sponges from the Pacific coast of Canada. Proc. Trans. Roy. Soc. Canad. 11(1893): 25–43.

_____. 1895. Sponges from the western coast of North America. Proc. Trans. Roy. Soc. Canad. 12(1894): 113–138.

Lévi, C. 1956. Étude des *Halisarca* de Roscoff. Embryologie et systématique des démosponges. Arch. Zool. Exper. 93: 1–181.

_____. 1960a. Reconstitution du squelette de l'éponge *Ophlitaspongia seriata* (Grant) à partir de suspensions cellulaires. Cah. Biol. Mar. 1: 353–58.

_____. 1960b. Les démosponges des côtes de France. I. Les Clathriidae. Cah. Biol. Mar. 1: 47–87.

_____. 1963. Gastrulation and larval phylogeny in sponges, pp. 375–82, *in* E. C. Dougherty, ed., The lower Metazoa, comparative biology and phylogeny. Berkeley and Los Angeles: University of California Press. 478 pp.

_____. 1964. Ultrastructure de la larve parenchymella de démosponge. I. *Mycale contarenii* (Martens). Cah. Biol. Mar. 5: 97–104, pls. 1–8.

_____. 1970. Les cellules des éponges, pp. 353–64, *in* Fry (1970b).

Litchfield, C., and R. W. Morales. 1976. Are Demospongiae cell membranes unique among living organisms?, pp. 183–200, *in* Harrison & Cowden (1976).

MacGinitie, G. E. 1935. Ecological aspects of a California marine estuary. Amer. Midl. Natur. 16: 629–765.

MacGinitie, G. E., and N. MacGinitie. 1968. Natural history of marine animals. 2nd ed. New York: McGraw-Hill. 523 pp.

MacLennan, A. P. 1970. Polysaccharides from sponges and their possible significance in cellular aggregation, pp. 299–324, *in* Fry (1970b).

Mauzey, K. P., and C. Birkeland, and P. K. Dayton. 1968. Feeding behavior of asteroids and escape responses of their prey in the Puget Sound region. Ecology 49: 603–19.

McLean, J. H. 1962. Sublittoral ecology of kelp beds of the open coast area near Carmel, California. Biol. Bull. 122: 95–114.

Meewis, H. 1939. Contribution à l'étude de l'embryogénèse des Chalinidae: *Haliclona limbata* (Mont.). Ann. Soc. Roy. Belg. 70: 201–43.

_____. 1941. Contribution à l'étude de l'embryogénèse des éponges siliceuses. Développement de l'oeuf chez *Adocia cinerea* (Grant) et *Halichondria coalita* (Bowerbank). Ann. Soc. Zool. Belg. 72: 126–49.

Miller, R. L. 1968. Some observations on the ecology and behavior of *Lucapinella callomarginata*. Veliger 11: 130–34, pl. 16.

Minchin, E. A. 1898. Materials for a monograph of the ascons. I. On the origin and growth of the triradiate and quadriradiate spicules in the family Clathrinidae. Quart. J. Microscop. Sci. 40: 469–587, pls. 38–42.

Neumann, A. C. 1966. Observations on coastal erosion in Bermuda and measurements of the boring rate of the sponge, *Cliona lampa*. Limnol. Oceanogr. 11: 92–108.

O'Connell, M. 1919. The Schrammen collection of Cretaceous siliceous sponges in the American Museum of Natural History. Bull. Amer. Mus. Natur. Hist. 41: 1–261 (spicule nomenclature pp. 34–44).

Paris, J. 1961. Greffes et sérologie chez les éponges siliceuses. Vie et Milieu 11 (Suppl.): 1–82.

Parker, G. H. 1914. On the strength and the volume of the water currents produced by sponges. J. Exper. Zool. 16: 443–46.

Pavans de Ceccatty, M. 1974. Coordinatiion in sponges. The foundations of integration. Amer. Zool. 14: 895–903.

Pavans de Ceccatty, M., Y. Thiney, and R. Garrone. 1970. Les bases ultrastructurales des communications intercellulaires dans les oscules de quelques éponges, pp. 449–66, *in* Fry (1970b).

Pearse, A. S. 1934. Inhabitants of certain sponges at Dry Tortugas. Pap. Tortugas Lab. 28: 117–24.

_____. 1949. Notes on the inhabitants of certain sponges at Bimini. Ecology 31: 149–51.

Penney, J. T. 1960. Distribution and bibliography (1892–1957) of the freshwater sponges. Univ. So. Carolina Publ. Biol. 3: 1–97.

Randall, J. E., and W. D. Hartman. 1968. Sponge-feeding fishes of the West Indies. Mar. Biol. 1: 216–25.

Rasmont, R. 1959. L'ultrastructure des choanocytes d'éponges. Ann. Sci. Natur., Zool. (12) 1: 253–62, 2 pls.

Reed, C., M. J. Greenberg, and S. K. Pierce, Jr. 1976. The effects of the cytochalasins on sponge cell reaggregation: New insights through the scanning electron microscope, pp. 153–69, *in* Harrison & Cowden (1976).

Reid, R. E. H. 1970. Tetraxons and demosponge phylogeny, pp. 63–89, *in* Fry (1970b).

Reiswig, H. M. 1970. Porifera: Sudden sperm release by tropical Demospongiae. Science 170: 538–39.

_____. 1971. Particle feeding in natural populations of three marine demosponges. Biol. Bull. 141: 568–91.

_____. 1973. Population dynamics of three Jamaican Demospongiae. Bull. Mar. Sci. 23: 191–226.

_____. 1974. Water transport, respiration and energetics of three tropical marine sponges. J. Exper. Mar. Biol. Ecol. 14: 231–49.

_____. 1975a. Bacteria as food for temperate-water marine sponges. Canad. J. Zool. 53: 582–89.

_____. 1975b. The aquiferous systems of three marine Demospongiae. J. Morphol. 145: 493–502.

Ricketts, E. F., and J. Calvin. 1968. Between Pacific tides. 4th ed. Revised by J. W. Hedgpeth. Stanford, Calif.: Stanford University Press. 614 pp.

Rützler, K. 1965. Substratstabilität in marinen Benthos als ökologischer Faktor, dargestellt am Biespiel adriatischer Porifera. Internat. Rev. Ges. Hydrobiol. 50: 281–92.

_____. 1970. Spatial competition among Porifera: Solution by epizoism. Oecologia 5: 85–95.

————. 1975. The role of burrowing sponges in bioerosion. Oecologia 19: 203–16.

Rützler, K., and G. Rieger. 1973. Sponge burrowing: Fine structure of *Cliona lampa* penetrating calcareous substrata. Mar. Biol. 21: 144–62.

Sarà, M. 1964. Associazioni di demonspongie con zooxantelle e cianelle. Boll. Zool. 31:359–65.

————. 1965. Associations entre éponges et algues unicellulaires dans la Méditerraneé. Rap. Proc. Verb. Réunions C.I.E.S.M.M. 28: 125–27.

Sarà, M., L. Liaci, and N. Melone. 1966a. Bispecific cell aggregation in sponges. Nature 210: 1167–68.

————. 1966b. Mixed cell aggregation between sponges and the anthozoan *Anemonia sulcata*. Nature 210: 1168–69.

Sarà, M., and J. Vacelet. 1973. Écologie des démosponges, pp. 462–576, *in* Grassé (1973).

Sheikh, Y. M., and C. Djerassi. 1974. Steroids from sponges. Tetrahedron 30: 4095–4103.

Simpson, T. L. 1963. The biology of the marine sponge *Microciona prolifera* (Ellis & Solander). I. A study of cellular function and differentiation. J. Exper. Zool. 154: 135–52.

————. 1966. A new species of clathriid sponge from the San Juan Archipelago. Postilla 103: 1–7.

————. 1968a. The biology of the marine sponge *Microciona prolifera* (Ellis & Solander). II. Temperature-related, annual change in functional and reproductive elements with a description of larval metamorphosis. J. Exper. Mar. Biol. Ecol. 2: 252–77.

————. 1968b. The structure and function of sponge cells: New criteria for the taxonomy of poecilosclerid sponges (Demospongiae). Bull. Peabody Mus., Yale Univ., 25: 1–142.

————. 1973. Coloniality among the Porifera, pp. 549–65, *in* R. S. Boardman, A. H. Cheetham, and W. A. Oliver, Jr., eds., Animal colonies: development and function through time. Stroudsburg, Pa.: Dowden, Hutchinson, and Ross. 603 pp.

Simpson, T. L., and P. E. Fell. 1974. Dormancy among the Porifera: Gemmule formation and germination in fresh-water and marine sponges. Trans. Amer. Microscop. Soc. 93: 544–77.

Smith, S. V., and E. C. Haderlie. 1969. Growth and longevity of some calcareous fouling organisms, Monterey Bay, California. Pacific Sci. 23: 447–51.

Stephens, G. C., and R. A. Schinske. 1961. Uptake of amino acids by marine invertebrates. Limnol. Oceanogr. 6: 175–81.

Storr, J. F. 1976a. Ecological factors controlling sponge distribution in the Gulf of Mexico and the resulting zonation, pp. 261–76, *in* Harrison & Cowden (1976).

————. 1976b. Field observations of sponge reactions as related to their ecology, pp. 277–82, *in* Harrison & Cowden (1976).

Thompson, T. G., and T. J. Chow. 1955. The strontium-calcium atom ratio in carbonate-secreting marine organisms. Pap. Mar. Biol. Oceanogr., Deep-Sea Res. 3 (Suppl): 20–39.

Turner, C. H., E. E. Ebert, and R. R. Given. 1968. The marine environment offshore from Point Loma, San Diego County. Calif. Dept. Fish & Game, Fish Bull. 140: 1–85.

————. 1969. Man-made reef ecology. Calif. Dept. Fish & Game, Fish Bull. 146: 1–221.

Tuzet, O. 1947. L'ovogénèse et la fecondation de l'éponge calcaire *Leucosolenia coriacea* et de l'éponge siliceuse *Reniera elegans*. Arch. Zool. Expér. Gén. 85: 127–48.

————. 1948. Les premiers stades du développement de *Leucosolenia botryoides* Ellis & Solander et de *Clathrina* (*Leucosolenia*) *coriacea* Mont. Ann. Sci. Natur. Paris Zool. (11) 10: 103–14.

————. 1963. The phylogeny of sponges according to embryological, histological, and serological data, and their affinities with the Protozoa and the Cnidaria, pp. 129–48, *in* E. C. Dougherty, ed., The lower Metazoa, comparative biology and phylogeny. Berkeley and Los Angeles: University of California Press. 478 pp.

————. 1970. La polarité de l'oeuf et la symétrie de la larve des éponges calcaires, pp. 437–48, *in* Fry (1970b).

————. 1973. Introduction et place des spongaires dans la classification. Éponges calcaires, pp. 1–132, *in* Grassé (1973).

Tuzet, O., C. L. Bessiére, and J. Paris. 1961. Recherches sérologiques sur les spongiaires et les cnidaires. Bull. Biol. France Belg. 95: 150–54.

Urban, F. 1905. Kalifornische Kalkschwämme. Arch. Naturgesch. 72: 33–76, pls. 6–9.

Vacelet, J. 1970. Description de cellules a bactéries intranucléaires chez des éponges *Verongia*. J. Microscop. 9: 333–46.

Vacelet, J., and C. Donadey. 1977. Electron microscope study of the association between some sponges and bacteria. J. Exper. Mar. Biol. Ecol. 30: 301–14.

Vogel, S. 1974. Current-induced flow through the sponge, *Halichondria*. Biol. Bull. 147: 443–56.

————. 1977. Current-induced flow through living sponges in nature. Proc. Nat. Acad. Sci. 74: 2069–71.

————. 1978. Organisms that capture currents. Sci. Amer. 239: 129–39.

Warburton, F. E. 1958. The manner in which the sponge *Cliona* bores in calcareous objects. Canad. J. Zool. 36: 555–62.

————. 1960. Influence on currents on form of sponges. Science 132: 89.

————. 1961. Inclusion of parental somatic cells in sponge larvae. Nature 191: 1317.

————. 1966. The behavior of sponge larvae. Ecology 47: 672–74.

Wells, H. W., M. J. Wells, and I. E. Gray. 1964. Ecology of sponges in Hatteras Harbor, North Carolina. Ecology 45: 752–67.

Wiedenmayer, F. 1977. Shallow-water sponges of the western Bahamas. Basel and Stuttgart: Birkhäuser Verlag. 287 pp.

Wilson, H. V. 1937. Notes on the cultivation and growth of sponges from reduction bodies, dissociated cells, and larvae, pp. 137–39, *in* P. S. Galtsoff, F. E. Lutz, P. S. Welch, and J. G. Needham (chairman), eds., Culture methods for invertebrate animals. Ithaca, N.Y.: Comstock. Reprinted by Dover, N.Y., 1959. 590 pp.

Wilson, H. V., and J. T. Penney. 1930. The regeneration of sponges (*Microciona*) from dissociated cells. J. Exper. Zool. 56: 73–134.

Chapter 3

Cnidaria (Coelenterata): *The Sea Anemones and Allies*

Eugene C. Haderlie, Cadet Hand, and William B. Gladfelter

The cnidarians, or coelenterates, are a diverse group, including the hydroids, sea anemones, corals, jellyfishes, and their relatives. About 5,500 species are known worldwide, and they are found throughout the seas from the intertidal zone to the greatest depths. A few species occur in fresh water. The group is an ancient one and many fossils of cnidarians occur in Cambrian rocks.

Many cnidarians are large and brightly colored, and some occur in dense populations in the intertidal zone. They are radially symmetrical animals, lacking a head but usually bearing a crown of tentacles around the mouth, which provides the only major opening to the gut. Tiny stinging capsules (nematocysts) are present in all cnidarians, especially in the tentacles. Basically, they are used to capture food, but they provide a powerful deterrent to predators, as well. Suitably stimulated, they explode, in most cases everting a hollow thread that penetrates the predator or prey and injects toxic materials. Present by the millions, the stinging capsules may even cause injury to people.

Cnidarians are carnivores and use their nematocysts to sting and often kill a variety of small invertebrate animals and fishes. The prey is then transferred to the mouth by the tentacles. Many gastropod mollusks, fishes, and pelagic turtles feed on cnidarians and are apparently immune to the stinging cells.

The cnidarian body tends to be modified for either an attached or a swimming way of life. Sedentary forms, called polyps, are gener-

Eugene C. Haderlie is Professor of Oceanography at the Naval Postgraduate School, Monterey. Cadet Hand is Director of the Bodega Marine Laboratory, and Professor of Zoology, at the University of California, Berkeley. William B. Gladfelter is Assistant Professor of Biology at the West Indies Laboratory (St. Croix, U.S. Virgin Islands) of Fairleigh Dickinson University.

ally tubular or cylindrical in shape, attached by their basal ends to the substratum or to one another, and usually bear a ring of tentacles at the free end. Medusae, in contrast, are generally adapted for swimming and are cup- or bowl-shaped. The life cycles of some species include both polyp and medusa stages.

Representatives of all three major classes of the phylum Cnidaria (Hydrozoa, Anthozoa, and Scyphozoa) are found along our shores. The smallest common forms are the hydroids and their allies (class Hydrozoa). Hydroids typically form colonies consisting of delicate branching tubes bearing tiny feeding polyps, which may be quite evident or so small they are best seen with a hand lens. Each colony arises from a single fertilized egg; the egg first develops into a cylindrical ciliated planula larva, which usually swims, and eventually settles to form a single polyp. This initial polyp sends out tubular branches from which additional polyps arise. The branching pattern varies greatly; some species produce upright colonies resembling tiny bushes, trees, or feathers, whereas in others polyps arise from prostrate stolons that adhere to rocks, algae, or the exposed hard parts of animals. The stems, branches, or stolons interconnecting the polyps are tubular and lined with cilia; thus, what one polyp eats may be distributed through the colony. The stolons and the branches of the colony are covered by a thin but tough exoskeleton, which in some forms extends as a protective cup (theca) around each feeding polyp and certain reproductive structures.

The polyp phase of the life cycle is usually considered a juvenile stage, in which the only reproduction is by budding. After a period of feeding and growth, the colony produces medusae, also by budding. In some hydroids, medusa buds arise on the feeding polyps themselves; in other species, medusa buds are produced only by polyps specialized for that function. Medusa buds may develop into tiny medusae, which are released and swim away to feed and grow, or they may remain attached and develop into sessile medusoids, which are arrested developmental stages of medusae. In either case, the medusa or medusoid develops gonads and produces eggs or sperm (rarely both). Sperm are liberated into the sea, but the eggs are often retained, especially by sessile medusoids, and are fertilized in the gonad. In this case they generally develop into planula larvae before they escape to swim and settle. Hydroids are common in the low intertidal zone in all the major habitats along the California coast, from wave-swept rocky shores to the floats, pilings, and eelgrass beds of protected harbors and bays.

Members of the class Anthozoa are polyps that lack a medusa stage altogether. The sea anemones, members of the subclass Zoantharia (=Hexacorallia), constitute the most abundant and conspicuous group in California waters. Individual anemone polyps are not connected to one another. Some attach to rocks or seaweeds by the base; others burrow into sand or mud. At the center of the free end is the mouth, surrounded by an oral disk and one or more rings or cycles of tentacles. The main part of the body between the tentacles and the base, called the column, may be long or short, and smooth or covered by numerous tubercles. The mouth leads to a tubular pharynx, which in turn opens into a capacious gut filled with seawater and partly divided into compartments by radially arranged septa (mesenteries). The water in the gut, under slight pressure when the mouth is closed, provides internal support. The body can expand greatly, by inhaling more water along ciliated tracts in the pharynx, or contract, expelling the water by muscular action. Submerged animals usually have the tentacles extended, but anemones exposed at low tide are often contracted to rounded humps; each has its tentacles and oral disk pulled in for protection and its distal end closed by a powerful sphincter.

Related to the true sea anemones are other anemone-like forms. The best known of these are the true corals, which secrete a limestone skeleton. The class Anthozoa also includes the soft corals, sea pens, and their relatives (subclass Alcyonaria=Octocorallia). The colonies they form vary greatly in size and shape, but the feeding polyps are all very similar and always have eight tentacles, each branched pinnately like a feather.

Several kinds of large jellyfishes (class Scyphozoa) may be found washed up on beaches or floating in harbors and bays. In these forms, the medusa is the conspicuous phase of the life cycle. Most jellyfishes are budded from tiny attached polyps (scyphistoma stages), become free as small juveniles (ephyrae), and grow to adulthood as free-swimming forms.

The excellent classical account of cnidarians in Hyman (1940, 1959) includes an extensive bibliography of older literature. Good modern reviews of many aspects of cnidarian biology and of areas where cnidarian research is most active are to be found in Bullock & Horridge (1965), Campbell (1974), D. Chapman (1974), Crowell (1965), Florkin & Scheer (1968), Mackie (1976), Miller & Wyttenbach (1974), Muscatine & Lenhoff (1974), W. Rees (1966a), and Tokioka & Nishimura (1973). *Hydra* and other cnidarians that can be cultured in the laboratory are making remarkable contributions to an understanding of general problems in development and behavior (Lenhoff & Loomis, 1961; Lentz, 1966). For laboratory culture of hydroids, see Crowell (1967), Mackie (1966), and Worthman (1974).

For general works, especially on the taxonomy and evolution of cnidarians, see D. Chapman (1966), Hand (1966), Kramp (1961), Naumov (1969), W. Rees (1957, 1966b), Russell (1953, 1970), and H. Thiel (1966), which have good bibliographies. Hand (1954, 1955a,b) monographed the central California sea anemones. For the identification of these and other common cnidarians of the central California coast, useful keys are provided by Hand (1975) and J. Rees & Hand (1975).

Much work still remains to be done on the west coast cnidarians, especially the hydroids. The only comprehensive taxonomic work on Pacific coast hydroids is Fraser (1937), which has many shortcomings and is badly in need of revision. Many of our hydroids are poorly known from the standpoint of taxonomy and natural history, and many medusae are taken in California waters for which the hydroid phase is uncertain. Life history studies are especially needed.

Phylum Cnidaria / Class Hydrozoa / Order Hydroida / Suborder Anthomedusae
Family Corynidae

3.1 **Sarsia** sp.
(=Syncoryne sp.)

Common, on rocks and wharf pilings, low intertidal zone in coastal areas protected from strong wave action; also in quiet bays on harbor floats and the eelgrass *Zostera*; genus present on Pacific coast from British Columbia to Chile, and on both coasts of North Atlantic.

Hydroid stage (usually called *Syncoryne*) forming branched colony to 5 cm tall; feeding polyps distinct in bearing scattered capitate (knob-tipped) tentacles; colony orange or pink.

Individual polyps feed readily in the laboratory and may become greatly distended. Polyps bend toward copepods that swim nearby, and also toward a vibrating needle held a few millimeters away. Tiny medusae are budded from the bases of the feeding polyps and released to the sea in spring and summer in California. The medusae, long known under the name *Sarsia*, feed, grow, and reproduce sexually as pelagic forms. Larvae settle mainly during the winter months (November to February) in Monterey Bay.

The medusae have been used in electrophysiological studies. Experiments show that swimming contractions of the bell are controlled and coordinated by pacemaker centers located in bulbs at the bases of the four tentacles.

See N. Berrill (1953b), Fraser (1937, 1946), Gladfelter (1973), Haderlie (1968, 1969, 1971), Horridge (1966), Josephson (1961), Mackie (1971), Mackie & Passano (1968), Passano (1966), and Russell (1953).

Phylum Cnidaria / Class Hydrozoa / Order Hydroida / Suborder Anthomedusae
Family Tubulariidae

3.2 **Tubularia marina** (Torrey, 1902)

Fairly common on exposed and semiprotected rocky shores, low intertidal zone; subtidal to 15 m; Eureka (Humboldt Co.) to Monterey Bay.

Polyps to 5 cm long, solitary or in groups of a few well-spaced individuals; feeding polyps single, pink, occurring at the free ends of long, thin stalks, each polyp with two whorls of tentacles, one basal and one apical, around the mouth.

Single stalks bearing feeding polyps live well in laboratory aquaria. Polyps starved for 1–2 days feed voraciously on active copepods, catching their prey with either whorl of tentacles; well-fed polyps, however, give no response to active food. Medusa buds arise in clusters attached to the feeding polyp between the two whorls of tentacles. The medusoids never become free; they develop three to five tentacles, but remain small and attached as gametes develop. Some polyps bearing sexually mature medusoids have been found in Monterey Bay in February, June, and July. Sperm are shed, but eggs are retained in the sessile medusoids and are not released until they have developed into tentacled actinula larvae, which resemble miniature unstalked feeding polyps.

See Fraser (1937, 1946) and Ricketts & Calvin (1968); see also the references under *T. crocea* (3.3).

Phylum Cnidaria / Class Hydrozoa / Order Hydroida / Suborder Anthomedusae
Family Tubulariidae

3.3 **Tubularia crocea** (Agassiz, 1862)

Common on pilings and floats in harbors and bays, low intertidal zone and subtidal to 40 m; Gulf of Alaska to San Diego; both coasts of North Atlantic.

Colony forming large tangled tuft to 15 cm high; stalks branched, thin, each terminating in one pink feeding polyp with two whorls of about 20 tentacles each.

As in *T. marina* (3.2), medusa buds arise on feeding polyps between the two whorls of tentacles. Male and female medusoids occur on separate animals and remain attached; when the gametes mature in summer, sperm are released to the sea. However, the ripe eggs are retained, and the female reproductive structures release substances that attract swimming sperm. The eggs are fertilized inside the medusoid, and actinula larvae are released. From June through August in San Francisco Bay the actinulae settle in concentrations up to 200 individuals per cm^2, the tangled colonies eventually forming a dominant part of the fouling growth on pilings.

Tubularia crocea can be easily cultured in the laboratory. It catches and consumes a wide variety of animal plankton, especially copepods and small chaetognaths. The long stalks of individual polyps of species of *Tubularia* make these animals favorable for studies of regeneration. Studies have also been made on the ultrastructure of the sperm, embryology, DNA synthesis, and antigenic composition of stems and feeding polyps. *Tubularia* has been used extensively in electrophysiological and pharmacological studies of pacemaker operation and epithelial conduction.

See Akin & Akin (1967), N. Berrill (1952b), Campbell (1967), Campbell & Campbell (1968), Costello & Henley (1971), Crowell (1967), Davidson & Berrill (1948), Day & Harris (1978), Fraser (1937, 1946), Graham & Gay (1945), Hinsch (1974), Josephson (1965, 1974a,b), Josephson & Uhrich (1969), Liu & Berrill (1948), MacGinitie (1935), Mackie (1966), Miller (1966a,b), Miller & Brokaw (1970), Miller & Tseng (1974), Morrill (1959), Morrill & Norris (1963), Parmentier & Case (1973), Pyefinch & Downing (1949), Ricketts & Calvin (1968), Rose (1974), Tardent (1962), Turner, Ebert & Given (1969), and Tweedell (1974).

Phylum Cnidaria / Class Hydrozoa / Order Hydroida / Suborder Anthomedusae Family Corymorphidae

3.4 Corymorpha palma Torrey, 1902

On mudflats, low intertidal zone; subtidal on mud and sand; Newport Bay (Orange Co.) to San Diego.

Individual polyps separate, solitary, to 14 cm tall; numerous short tentacles in a cluster around the mouth, up to 30 longer tentacles around periphery of feeding polyp; medusoids borne on short stalks between tentacular whorls; color creamy white.

Corymorpha palma is abundant at times (to 1 polyp per 25 cm²) on low intertidal mudflats of Newport Bay, and when the tide recedes the polyps collapse and lie on the mud until the sea returns. Under water the animals stand upright, with the feeding polyp facing downstream if there is a current. The base of the stalk bears numerous fine filaments (not shown in the photograph), which help anchor it in the mud. The tip of each filament contains a statocyst consisting of a hollow space

between the two terminal cells housing a small hard statolith. Although the surrounding cells contain no cilia or microvilli, these gravity-sensitive statocysts direct the growth of the filaments down into the mud. Elongation of the filaments downward is due to cell stretching, not cell proliferation.

Corymorpha exhibits fairly complex behavior compared with other hydroids. During periods of quiet water the polyps stand upright, but every 3–8 minutes show a unique bowing behavior. The long tentacles flex inward toward the mouth, the stalk contracts, and the polyp bends over until the tentacles touch the substratum. The electrical activity associated with this behavior has been studied, and five interacting pulse systems located in various parts of the body are involved. The large size of the polyp, the well-developed epitheliomuscular cells, and the large axial endodermal cells in the stalk have made *Corymorpha* favorable for electrophysiological and regeneration studies. Dissociated cells of the stem will aggregate, and some aggregates give rise to new individuals.

Medusoids arise by budding from the body wall between the two whorls of tentacles on the feeding polyp; they remain small and never detach to become free-swimming. The gonads become pale yellow-green as they develop, owing to the presence of unicellular algae. The algal cells persist only until the gonads are ripe, then disappear. Mature eggs, to 0.3 mm in diameter, are fertilized and undergo cleavage in the gonad. Eventually they break out as planula-like larvae that lack cilia and move about on the parent polyp and substratum in an amoeboid manner. In the laboratory, *Corymorpha* produces eggs during most months of the year. Reports of large populations in December and January and again in June, with fewer in between, suggest that most animals live only half a year.

Corymorpha is preyed upon by the nudibranch *Antiopella aureocincta*, which also lays its eggs about the base of the hydroid.

See Ball (1973), Ball & Case (1973), Campbell (1968, 1969), Child (1928), Fraser (1937, 1946), MacGinitie (1938), MacGinitie & MacGinitie (1968), Pardy & Rahat (1972), Parker (1917), Ricketts & Calvin (1968), Torrey (1904, 1905, 1907, 1910a,b), Turner, Ebert & Given (1969), and Wyman (1965).

Phylum Cnidaria / Class Hydrozoa / Order Hydroida / Suborder Anthomedusae
Family Bougainvilliidae

3.5 **Hydractinia milleri** Torrey, 1902

Fairly common in protected situations on sides and undersurfaces of rocks, low intertidal zone; Vancouver Island (British Columbia) to Carmel (Monterey Co.).

Expanded and extended polyps to 0.5 cm tall, the individual polyps closely spaced, arising from a basal mat of stolons; colony forming a pink fuzzy patch, soft to the touch.

Hydractinia colonies live well in laboratory aquaria and provide excellent material for behavioral studies under a dissecting microscope. The several types of polyps making up the colony include large and small feeding polyps, dactylozooids that themselves resemble elongate tentacles, and slender feeding polyps that bear medusoids. The medusoids are reduced to little more than gonads and the tissue supporting them. Colonies bear medusoids of only one sex. Sperm (whitish) and eggs (lavender) are both released. Recently settled colonies have been observed in Monterey Bay in August.

On the east coast, development and differentiation have been studied in related species of *Hydractinia* and the fertilized eggs used in embryology courses. The release of gametes is light-dependent; after a colony has been in the dark for several hours, brief exposure to light results in shedding of eggs and sperm. Sperm become active immediately in the presence of eggs. Sperm ultrastructure is known.

Colonies of east coast *Hydractinia* often overgrow snail shells occupied by hermit crabs. Although the relationship is not obligatory, it appears mutually beneficial. The hermit crab utilizes some food taken by the hydroid. Planktonic forms, such as barnacle larvae, are too small to be captured directly by the hermit crab, but are caught by the feeding polyps of the hydroid living on the shell. Some of the plankton is scraped off the tentacles of the hydroid by the hermit crab, which in this way uses a food source that is otherwise unavailable to it. The hydroid colony also presumably benefits from the association by picking up small fragments of food as the crab feeds while scavenging the bottom. The hydroid is a nonselective carnivore; crustacean larvae, nematodes, and a variety of small benthic animals are consumed if available.

When more than one colony of *Hydractinia* occurs on a single shell, the colonies appear to be incompatible and remain separated by 1–2 mm. On the Texas coast *Hydractinia* lives on shells occupied by *Pagurus* sp. and the hermit crabs appear to be unaffected by the hydroid's nematocysts. However, a larger and more aggressive hermit crab (*Clibanarius vittatus*) is sensitive to the sting of the nematocysts and is thus prevented from using shells covered with the hydroid.

See N. Berrill (1953a), Browning & Lytle (1967), A. Burnett, Sindelar & Diehl (1967), Christensen (1967), Crowell (1967), Fraser (1937, 1946), Haderlie (1971), Hinsch (1974), Ivker (1967, 1972a,b), Miller (1974), Rappaport (1966), and Wright (1973).

Phylum Cnidaria / Class Hydrozoa / Order Hydroida / Suborder Anthomedusae
Family Bougainvilliidae

3.6 **Garveia annulata** Nutting, 1901

Common seasonally in shaded areas on open or semiprotected rocky shores, usually growing on sponges or the branched coralline algae *Bossiella* and *Calliarthron*, low intertidal zone; subtidal to depths of over 120 m; Sitka (Alaska) to Santa Catalina Island (Channel Islands).

Colony to 5 cm tall, consisting of 20–30 polyps; stems, feeding polyps, and globular reproductive structures (gonophores) borne on stems and stolons; color deep to pale orange or yellowish.

Its bright color makes this hydroid conspicuous in late winter and spring on the open coast of California; the species is less abundant in summer and fall. It is sometimes mistaken for a member of the suborder Leptomedusae, because the exoskeleton forms a cup (theca) about the lower parts of feeding polyps and may completely cover the young (but not fully mature) gonophores. The gonophores are sessile medusoid structures, which on a given colony are either all male or all female. Male medusoids shed their sperm to the sea. The female gonophores retain the several large eggs they produce and liberate planula larvae. Little else is known of the species.

See Fraser (1937, 1946) and Ricketts & Calvin (1968).

Phylum Cnidaria / Class Hydrozoa / Order Hydroida / Suborder Anthomedusae
Family Eudendriidae

3.7 **Eudendrium californicum** Torrey, 1902

On rocks and under ledges protected from heavy wave action, low intertidal zone on rocky shores; subtidal to over 120 m; British Columbia to Monterey Bay.

Colony bushy, to 15 cm tall; main stalks erect, stout, arising from common platelike base; branches short, extending out in all planes; exoskeleton on stalk and branches, stiff, brown, annulated; feeding polyps large, pink, with about 16–20 white tentacles.

Eudendrium colonies in central California are best developed in spring and early summer. Older colonies are often overgrown with smaller hydroids, and the exoskeleton shows marks of damage and repair. Freshly collected animals are especially suitable for feeding experiments; as in many other hydroids, touching the mouth gently with animal food causes it to open widely like a bell. Not only are the nematocysts located on the feeding polyp tentacles, but they also form conspicuous belts around the base of the polyp and at the lip of the mouth.

Colonies produce either male or female medusoids, which, when present, usually in the winter months, are found on the sides of the feeding polyps below the whorl of tentacles. Sperm are shed to the sea, but the bright-orange eggs are retained and develop to the planula larva stage before release.

Relatively little is known of the biology of *Eudendrium*. The ultrastructure of the sperm of one species has been investigated. Another species, *E. ramosum* (Linnaeus), is preyed upon by nudibranchs such as *Flabellinopsis iodinea* (14.61).

See N. Berrill (1952a), Fraser (1937, 1946), McLean (1962), Ricketts & Calvin (1968), and Summers (1972).

Phylum Cnidaria / Class Hydrozoa / Order Hydroida / Suborder Anthomedusae
Family Polyorchidae

3.8 **Polyorchis montereyensis** Skogsberg, 1948

Found in coastal and bay plankton, often in harbors with sandy or muddy bottoms; reported only from Monterey Bay; related species in San Francisco Bay, Bodega Bay (Sonoma Co.), and similar localities from British Columbia to San Diego.

Bell globular, to 4 cm tall, with four distinct radial canals bearing lateral branches; tentacles to 90, unbranched, in a single whorl around margin of bell, each with a dark eyespot ringed with red at the base; gonads numerous, pendent, filamentous, suspended below radial canals within bell.

The genus *Polyorchis* includes some of the largest hydromedusae. The medusae occur sporadically; common in some years, especially April through November, they may then be absent for prolonged periods.

Polyorchis alternately swims and drifts, as is typical of hydromedusae. From time to time the animal rests upright on sand in the laboratory (and probably in nature as well). Then, with tentacles shortened, it swims upward. At or near the surface it ceases its swimming movements and drifts slowly downward, allowing the drag of the water to extend its relaxed tentacles into a large halo around the bell. When planktonic food, such as a copepod, is captured, the capturing tentacle contracts, bringing the food to the bell margin. The mouth, on its flexible stalk, then swings over and removes the prey from the tentacle. Sedentary crustaceans such as caprellids are taken as *Polyorchis* swims through subtidal beds of the eelgrass *Zostera*, and medusae resting on the bottom may feed on benthic forms. In an aquarium they can be fed brine shrimp or the copepod *Tigriopus californicus* (26.1); carotenoids of the prey color the entire canal system orange.

Swimming mechanics and behavior have received detailed study. The contraction of the subumbrellar muscles narrows the bell by folding it along predetermined "joints"; expansion of the bell is caused by elasticity of the mesoglea or jelly layer. Animals can be induced to swim by various stimuli, including the turning off of a bright light. Swimming velocities may reach 7 cm per second. The velum, a shelflike membrane extending inward around the inner bell margin, serves to restrict the orifice for water outflow from the bell, thereby increasing the water velocity and the amount of thrust. Turning is accomplished by the asymmetrical contraction of the radial muscles in the velum, which displaces the orifice to one side and causes the animal to turn in the same direction.

The sexes are separate, and in the laboratory mature animals spawn when subjected to strong light after being kept in darkness for several hours. Early development has been studied in a *Polyorchis* from Japan. The first cleavage occurs within 1 hour after fertilization, and the 32-cell stage is reached in 4 hours at a water temperature of 14–15°C. By 20 hours a free-swimming, club-shaped planula larva has developed. No polyp stages have been seen, but young medusae 2 mm high with four tentacles appear in the plankton in Japanese waters in April. These medusae grow rapidly, spawn in June and July, and die off in August.

Only a single published account of a hydroid or polyp stage of *Polyorchis* exists. The report was based on a single hydroid colony consisting of small polyps 1-1.5 mm tall growing on, or partly embedded in, a sponge on the upper surface of the rock scallop *Hinnites giganteus* (15.14). The scallop was collected from a depth of 15 m in Departure Bay (Vancouver Island, British Columbia). Each polyp, which arose singly from a network of prostrate stolons, possessed two cycles of about five capitate tentacles. One to five medusa buds occurred in the middle of the hydranths below the tentacles. The identity of the hydroid as that of *P. penicillatus* (Eschscholtz) was based on the comparability of the exumbrellar nematocyst patches of the newly released medusae with those of young *P. penicillatus* taken from the plankton.

Spermatogenesis occurs in the epidermis of the long filamentous testes and has been studied with the electron microscope. The ultrastructure of the eyespots has also been investigated, revealing complex light-sensitive retinal cells bearing highly modified cilia; no lens is present.

See Anderson & Mackie (1977), Brinkmann-Voss (1977), Eakin & Westfall (1962, 1964), Gladfelter (1970, 1972, 1973), Little (1914), Lyke (1972), Nagao (1963, 1970), Skogsberg (1948), Spencer (1978), and Tamasige & Yamaguchi (1967).

Phylum Cnidaria / Class Hydrozoa / Order Hydroida / Suborder Leptomedusae
Family Campanulariidae

3.9 **Clytia bakeri** Torrey, 1904

Often common on shells of snails (*Nassarius fossatus*, 13.92; *Olivella biplicata*, 13.99) and bivalves (*Tivela stultorum*, 15.32; *Donax gouldii*, 15.54), low intertidal and adjacent subtidal zones along sandy beaches of open coast; San Francisco to Baja California.

Stalks to 12 cm tall, arising in clusters from stolon network on mollusk shell; branches on two sides of stalk arising alternately, each branch terminating in a feeding polyp; gonangia arising at junctions of stalks and branches, sessile, each containing many medusa buds; colony buff or gray in color.

Clytia is the only hydroid found on sandy beaches in California. The colonies never occur on fixed objects such as stones or discarded shells. The hydroids on *Tivela* (see the photograph) always grow at the posterior margin of the shell, adjacent to the clam siphons; when the clam is in the proper position for feeding and respiration, the hydroids are exposed above the sand. The sand level constantly fluctuates on beaches, and the only hard substrata that are constantly repositioned at the surface as the beach level changes are the exposed hard parts of living animals. At irregular periods during the year, enormous numbers of small free medusae (for a long time known as *Phialidium*) are produced and released, forming a significant part of the plankton.

See Fraser (1937, 1946), Hale (1973), Ricketts & Calvin (1968), Roosen-Runge (1970), and Russell (1953).

Phylum Cnidaria / Class Hydrozoa / Order Hydroida / Suborder Leptomedusae
Family Campanulariidae

3.10 **Obelia** sp.

Common throughout the year on rocks, pilings, seaweeds, shells, and other fixed objects along the open coast and in bays and harbors, low intertidal zone to 50 m; numerous species, some poorly known, reported from the Pacific coast; genus distributed worldwide.

Bushy colony to 25 cm tall; stalks delicate, branching in all planes; exoskeleton thin, transparent, often annulated, covering stalk and branches and forming protective cups (thecae) around polyps; feeding polyps occurring at all branch tips; polyps producing medusae developing along main stalks and branches; medusae small, free, flat, with 16 or more tentacles.

Obelia is the hydroid most often studied in zoology classes.

The life cycle includes a small, ovoid, swimming larva (planula), which settles, attaches to the substratum, and rapidly develops into a feeding polyp. This in turn grows and buds off other feeding individuals, which remain connected to the main stalk and form a delicate treelike colony. Mature colonies develop specialized polyps here and there that do not feed, but instead bud off tiny medusae that detach, swim away, grow, and develop gonads. Eggs and sperm are shed into the sea, and each fertilized egg forms a ciliated planula larva. This type of life cycle is often treated in elementary texts as typical of the class Hydrozoa; however, in many hydroids the medusa stage is reduced, remains permanently attached to the colony, and may be represented by little more than gonadal tissue.

Obelia has been used in many studies on growth and differentiation. The colony grows exponentially (as measured by increase in number of feeding polyps) provided the food supply is adequate, and the rate of growth is directly correlated with temperature between 8°C and 20°C. Linear growth in the terminal portion of the colony is also temperature-dependent but independent of food supply. A starved colony continues to increase in height, but polyps lower down on the stem are resorbed. East coast species elongate at rates up to 20 mm per 24 hours at 20°C, but are slower at lower temperatures. At 20°C the medusae can develop from tiny buds to liberation in 24 hours, and new feeding polyps can be formed in a similar period. On mooring lines in Monterey Bay, in water to 50 m deep and at an average temperature below 8°C, *Obelia* colonies can grow to a height of 5 cm in less than a month.

In Bodega Bay harbor (Sonoma Co.), **Obelia dichotoma** (Linnaeus) settles during the winter, spring, and early summer when water temperatures are low. Two or three periods of major settlement are interspersed with periods of no settlement during that time. This species settles on barnacles, the tunicate *Ascidia ceratodes* (12.26), mussels, and substrata free of other animals. Colonies commonly grow to 12 cm in 2–3 months before regressing.

In California, *Obelia* medusae are found in the plankton through most of the year, but under laboratory conditions, one common species (*O. longissima*) buds and releases medusae only at temperatures below 12°C. Medusae probably survive and reproduce for an extended period, for planula larvae settle through most of the year in Monterey Bay. At times, especially in late summer and fall, *Obelia* colonies may dominate fouling communities on pilings at Monterey and La Jolla.

Two east coast species, *O. articulata* and *O. geniculata*, grow and reproduce in the normal way at temperatures below 20°C, but at higher temperature presumptive gonangia continue to grow to relatively massive bulbs, which ultimately break off the parent colony and drift to new settlement sites where each bulb can establish a new colony.

Both the polyps and the medusae of *Obelia* have been used in studies of ultrastructure and physiology. The medusae possess special types of muscle cells (myocytes) in the subumbrella, where striated myofibrils run in two layers at right angles to one another and extend outward from the walls of the myocytes as minute projections making contact with the myofibrils of adjacent cells. The fine structure of the nematocysts is known. Many *Obelia* colonies luminesce when stimulated electrically, or when stimulated chemically with potassium chloride. The light comes from endodermal cells in the stolons and is weak and flickering.

In California, *Obelia* polyps are preyed upon by *Hermissenda crassicornis* (14.66), *Dendronotus frondosus*, and probably other nudibranchs. Grazing by these predators can reduce the lush autumn hydroid colonies to short stubby stalks by spring.

Cornelius (1975) revised the species of the genus *Obelia* and concluded that only three species (**O. bidentata** Clarke, *O. dichotoma*, and **O. geniculata** (Linnaeus) are valid. All three occur in California. In the foregoing text, *O. longissima* (Pallas) is probably referable to *O. bidentata*, as is *O. articulata* (Agassiz).

See Aizu (1967), Beloussov (1973), N. Berrill (1948, 1949b,c), Boyd (1971), D. Chapman (1968), Coe (1932), Cornelius (1975), Costello & Henley (1971), Crowell (1953), Darby (1965), Fraser (1937, 1946), Haderlie (1968, 1969, 1971), Kemp & Berrill (1974), Knight (1968), MacGinitie (1935), Morin & Cook (1971a,b,c), Morin & Reynolds (1974), Standing (1976), Torrey (1909), and Westfall (1965a, 1966).

Phylum Cnidaria / Class Hydrozoa / Order Hydroida / Suborder Leptomedusae
Family Sertulariidae

3.11 **Sertularia furcata** Trask, 1857

Common on bases of blades of surfgrass (*Phyllospadix*) and red algae, low intertidal zone on exposed rocky shores; subtidal to 50 m; British Columbia to San Diego.

Colony consisting of many upright stalks, each about 1 cm high, closely spaced, arising from basal network of stolons; feeding polyps encased in individual translucent cups (thecae), arising opposite one another on stalk; mature colony with conspicuous oval capsules (gonangia) on stolons; colony tan or golden.

As in many other hydroids, the medusoids are not released but produce gametes while they remain sessile within the gonangia. Sperm are liberated into the sea, but the eggs are retained and fertilized in place, developing into ciliated planula larvae, which swim off and settle to start new colonies. The frequency with which colonies occur on *Phyllospadix* suggests that swimming planulae may be chemically attracted by the plants; the matter deserves investigation.

See Fraser (1937, 1946) and Ricketts & Calvin (1968).

Phylum Cnidaria / Class Hydrozoa / Order Hydroida / Suborder Leptomedusae
Family Sertulariidae

3.12 **Sertularella turgida** (Trask, 1857)

Common under rocks and ledges, low intertidal zone on exposed rocky shores; subtidal to 160 m; British Columbia to San Diego.

Colony with stout, robust stalks 2–5 cm tall; feeding polyps, in yellowish translucent cups, arising alternately on opposite sides of stalk, giving stalks a zigzag appearance (a characteristic separating the genus from *Sertularia*, e.g., 3.11); gonangia conspicuous, oval, containing small, sessile medusoids arising on lower parts of stalks between feeding polyps.

This species is one of the most common and widely distributed hydroids on the Pacific coast. Like *Sertularia* it releases

sperm but retains the eggs; fertilization and development into ciliated planula larvae occurs in the gonangia. Little else is known of the biology of the animal.

See Fraser (1937, 1946), McLean (1962), and Nishihira (1967a,b).

Phylum Cnidaria / Class Hydrozoa / Order Hydroida / Suborder Leptomedusae
Family Sertulariidae

3.13 **Abietinaria** sp.

Common under ledges and on surfaces of boulders shaded from sunlight, low intertidal zone on wave-swept rocky shores; numerous species, all in need of further study, reported from Pacific shores, Alaska to San Diego.

Colony 2–5 cm tall; each main stalk with numerous side branches, all borne essentially in one plane, giving the colony a fernlike appearance; feeding polyps in flask-shaped cups, alternate on opposite sides of stalks and branches; gonangia containing sessile medusoids borne on main stalk and lower portions of branches; overall color buff to orange.

Reproduction in *Abietinaria* is similar to that in *Sertularia*. Little is known of the biology of these hydroids.

See Fraser (1937, 1946), McLean (1962), and Ricketts & Calvin (1968).

Phylum Cnidaria / Class Hydrozoa / Order Hydroida / Suborder Leptomedusae
Family Plumulariidae

3.14 **Aglaophenia struthionides** (Murray, 1860)
Ostrich-Plume Hydroid

Low intertidal zone on rocky, wave-beaten shores; subtidal to 160 m; southern Alaska to San Diego.

Plumes arranged in clusters, yellowish to light red, to 13 cm tall, each plume consisting of a central flexible stalk and numerous side branches in one plane.

The nudibranch *Dendronotus subramosus* (14.52) often occurs on, and closely resembles, this hydroid in appearance. See the account of *A. latirostris* (3.15) for a discussion of the life cycle.

See Fraser (1937, 1946), Ricketts & Calvin (1968), and Torrey & Martin (1906).

Phylum Cnidaria / Class Hydrozoa / Order Hydroida / Suborder Leptomedusae Family Plumulariidae

3.15 Aglaophenia latirostris Nutting, 1900
Ostrich-Plume Hydroid

Common on rocks and larger red and brown algae, low intertidal zone along semiprotected rocky shores; subtidal to 35 m; southern Alaska to Santa Barbara.

Plumes 5–8 cm high, usually arranged in clusters, the colony tuftlike; each plume consisting of a simple stalk with numerous, closely set lateral branches (hydrocladia) arising alternately on opposite sides, giving the plume a featherlike appearance; feeding polyps small, protected by exoskeletal cups (hydrothecae) arranged in a closely set row on each branch; some lateral branches in mature plumes bearing inflated regions (corbulae) that contain the reduced medusoids; colony brown to orange or tan.

The ostrich-plume hydroids are among the largest and most common west coast hydroids, conspicuous not only growing on rocks and algae but also frequently cast ashore on beaches after storms. Individual feeding polyps are small, but their arrangement in the plume appears to provide maximum filtration of food from waters passing through the colony, while minimizing duplication of effort between polyps of the plume. Since all parts of the colony are connected by a continuous tubular gut, food ingested by one polyp is potentially available to other parts of the colony. Each feeding polyp is closely associated with three much smaller mouthless polyps (dactylozooids or nematophores), each of which resembles a solitary tentacle tipped by a cluster of nematocysts.

In *Aglaophenia*, free medusae are lacking. The corbulae contain the reduced medusoids; the sexes are separate, and male and female corbulae differ slightly in form. Sperm are shed, but eggs are retained until they develop into large wormlike orange planulae, which eventually become free. Mature colonies kept in seawater aquaria or placed in a dark refrigerator for a few hours and subsequently exposed to sunlight often

release larvae. The larvae crawl slowly for a period, then attach by the anterior end, and eventually produce new colonies.

The sea spider *Tanystylum californicum* commonly inhabits *A. latirostris* in the Monterey area, as do small skeleton shrimps (caprellids).

West coast species of *Aglaophenia* are in need of good comparative biological studies.

See Fraser (1937, 1946), McLean (1962), Ricketts & Calvin (1968), Tardent (1965), and Torrey & Martin (1906).

Phylum Cnidaria / Class Hydrozoa / Order Hydroida / Suborder Leptomedusae Family Plumulariidae

3.16 Plumularia sp.

Common on protected rocks and holdfasts of algae, low intertidal zone along rocky shores; several species along the Pacific coast.

Colony to about 2 cm tall, consisting of upright plumes resembling delicate, sparsely branched feathers; polyps widely spaced on branches; gonangia oval or elongate, arising near bases of branches or on main stalk; colony buff or yellow.

As in *Aglaophenia* (e.g., 3.14, 3.15), each feeding polyp is closely associated with three mouthless stinging polyps (dactylozooids or nematophores), which resemble oversized tentacles. Free medusae are lacking. Sessile medusoids release sperm, but eggs are fertilized and develop in the gonangia. Ciliated planula larvae are eventually released to settle and form new colonies. Members of this genus provide excellent material for the study of polyp behavior. Planulae, if present, are easily released by opening up the gonangia with needles.

See Coe (1932), Fraser (1937, 1946), and Ricketts & Calvin (1968).

3.17 **Allopora porphyra** (Fisher, 1931)
(=Stylantheca porphyra)

Uncommon, low intertidal zone on exposed, rocky coasts; on shaded rock faces protected from direct wave impact; British Columbia to central California.

Colonies ranging in size from mere specks to encrusting sheets covering several hundred square centimeters; vivid purple color, pitted surface, and encrusting habit distinguish this species from the related **A. californica** (Verrill), which is red or pink and forms erect branching colonies subtidally.

Superficially similar to the stony anthozoan corals in their possession of a calcareous skeleton, the hydrocorals are really much more like hydroids in structure and function. Each scalloped pit in the surface of *A. porphyra* contains 1–12 feeding polyps surrounded by a ring of stinging zooids (dactylozooids), which probably play a major role in capturing food. Tiny, reduced medusae, borne in chambers below the surface, are never released; they produce gametes and brood the developing planulae, which finally escape and start new colonies. In a related species from the San Juan Islands (Washington), *A. petrograpta,* planulae leave the parent colony from mid-June to mid-August. These larvae, 2 mm long, orange with white anterior ends, are already equipped with nematocysts. The planulae attach to the substratum, flatten into disks, and within a day begin depositing calcareous skeletal material. Within 2 weeks polyps are formed and begin to feed.

Allopora californica is common subtidally in the Carmel area (Monterey Co.). In November this species produces planulae that settle near adult colonies. Few new colonies survive as long as a year. Those on flat surfaces are usually overgrown by algae or smothered by sediments. Those on sloped or vertical surfaces survive better, and some colonies live many years.

See Boschma (1953), Broch (1942), Fisher (1931, 1938), Fritchman (1974), Moseley (1878), Ostarello (1973), and Thompson & Chow (1955).

3.18 **Velella velella** (Linnaeus, 1758)
(=V. lata)   By-the-Wind-Sailor

Pelagic, floating at the surface in offshore waters, sometimes cast ashore in huge numbers; distributed worldwide in temperate and tropical seas.

Length to 8 cm; intact animal easily recognized by blue color and angular projecting sail.

Velella was once thought to be related to siphonophores such as the Portuguese man-of-war, *Physalia*, but careful study has shown that it is best compared to the polyp of a hydroid such as *Tubularia* (e.g., 3.2, 3.3) or *Corymorpha* (e.g., 3.4), which has taken to a pelagic life, hangs mouth down, and bears a float and sail in place of a stalk. During development, an invagination occurs on the side of the larva opposite the mouth (where other hydroids develop a stalk); the inturned layer then secretes a chitinous exoskeleton forming the gas-filled float and transparent sail.

The sail is angled to the left or right of the main body axis such that in a breeze the animals drift some 45° to the left or right of the true wind direction. Right-handed animals prevail off the California coast, and our prevailing northerly winds tend to hold the population offshore; prolonged southerly or westerly winds bring them onto the beaches.

The lower surface of a mature *Velella* has a large central mouth surrounded by a host of smaller polyps that both feed and bud off small medusae. These polyps exhibit a rhythmic behavior involving contraction and flexing toward the mouth, similar to that observed in the sessile hydroids *Tubularia* and *Corymorpha*. The nervous system contains two distinct nerve networks, but the conduction of impulses also occurs effectively in nerve-free epithelia. The small medusae (formerly known as *Chrysomitra*) become free and eventually sink to depths of up to 2,000 meters. They develop gonads, but sexual reproduction and early development are little known. Tiny juveniles develop a float and begin to secrete gas within it, then rise to the surface of the sea, where they feed on animal plankton and grow. Floating *Velella* and young medusae bear symbiotic algae in their tissues, and may gain some nourishment from them. In turn *Velella* are preyed upon by pelagic gastropods, especially by the nudibranch *Glaucus* and the bubble-rafting snail *Ianthina*. The bare chitinous floats of dead *Velella* may serve as places of attachment for the pelagic goose-necked barnacles *Lepas anatifera* (20.4) and *L. fascicularis* (20.6).

See Bieri (1959, 1961, 1966), Bieri & Krinsley (1958), Brinckmann (1964), Carl (1948), Edwards (1966), Fields & Mackie (1971), Hardy (1956), Herring (1967), Knudsen (1963), LeLoup (1929), Mackie (1959, 1960, 1961), Ricketts & Calvin (1968), and Savilov (1961).

Phylum Cnidaria / Class Scyphozoa / Order Stauromedusae
Family Halicyathidae

3.19 **Manania** sp.

Uncommon on coralline and other red algae, low intertidal pools along rocky shores; on subtidal red algae to 10 m depth; Moss Beach (San Mateo Co.) to Santa Rosa Island (Channel Islands).

Body 1–2 cm long, the unattached end with eight marginal clusters of knobs; stalk red or brown, the knobs white.

Seven species of stalked scyphozoans (Stauromedusae) are known from the Pacific coast, three from California shores. *Manania* and the more common *Haliclystus* show an interesting combination of characteristics of both larval and adult jellyfishes. In terms of behavior they have more in common with sea anemones than with jellyfishes: they cannot swim and remain relatively motionless, attached to algae by a basal attachment disk. The stalk can contract somewhat, the knob-studded umbrella (calyx) can fold, and the animal can slowly move about by gliding on the disk. If it becomes detached, it clings to the substratum with the knobby tentacles until the basal disk has reattached.

Little is known of the biology of most California stauromedusans. Related species on the east coast are known to feed on small crustaceans, such as gammarid amphipods and caprellids, and on small gastropods that come in contact with the nematocysts of the marginal knobs.

See M. Berrill (1962, 1963), Child (1933), Gwilliam (1956, 1960), Ricketts & Calvin (1968), H. Thiel (1966), and Uchida (1973).

Phylum Cnidaria / Class Scyphozoa / Order Semaeostomeae / Family Pelagiidae

3.20 **Chrysaora melanaster** Brandt, 1838

Pelagic in open coastal waters, often washed up on beaches; North Pacific to southern California.

Bell to 30 cm in diameter, with radial yellow or brownish lines and 24 reddish, marginal tentacles in single whorl; oral arms fringed and prominent.

Little is known of the biology of this species. However, anatomical, physiological, developmental, and pharmacological studies have been made on related species from the Atlantic. For example, the sea nettle, *C. quinquecirrha*, found in Chesapeake Bay in dense swarms during the summer, has become a nuisance by stinging bathers and clogging water intakes on ships, and has therefore stimulated much research.

Along the California coast *Chrysaora* is preyed upon by the blue rockfish *Sebastes* (formerly *Sebastodes*) *mystinus*, which stalk the jellyfish, and, avoiding the marginal tentacles, bite off pieces of the oral arms.

See Alexander (1964), J. Burnett (1971), Burnett & Goldner (1969, 1970a,b), Burnett & Sutton (1969), Calder (1971), Cargo & Schultz (1966, 1967), G. Chapman (1953), Gladfelter (1970, 1973), Gotshall, Smith & Holbart (1965), and Rice & Powell (1970).

Phylum Cnidaria / Class Scyphozoa / Order Semaeostomeae / Family Pelagiidae

3.21 **Pelagia colorata** Russell, 1964

Usually in continental shelf waters, sometimes common inshore in open bays and harbors (such as Monterey), trapped in tidepools, or cast ashore; known only from Pacific coastal waters of the United States; genus worldwide in distribution.

Bell to more than 80 cm in diameter, dome-shaped; marginal tentacles eight, long, in single ring; oral arms prominent, fringed; color pale silver, with deep purple radial bands and interradial blotches.

This spectacular jellyfish has long been known under the name of its widely distributed but much smaller relative, **P. noctiluca** Forsskål (= *P. panopyra*), whose bell diameter together with its oral arm length rarely reaches 10 cm. In contrast, in large undamaged specimens of *P. colorata* from offshore waters, the oral arms reach several meters in length. The nematocysts sting fiercely.

In species of *Pelagia*, the fertilized egg develops to a planula larva, which in turn develops directly into a free-swimming ephyra stage (really a juvenile jellyfish) without intervention of a sessile, asexually reproducing polyp stage such as occurs in most other jellyfishes. Active young crabs, particularly *Cancer gracilis* (25.18), are often found clinging to *P. colorata*; healthy specimens are even found in the gut cavity of the medusa. The association is widely known but needs serious study. *P. colorata* is reportedly fed upon by the ocean sunfish *Mola mola*, as well as the blue rockfish *Sebastes mystinus*.

Pelagia noctiluca from the Atlantic and Mediterranean luminesces brilliantly; sometimes the entire bell glows, at other times only localized regions of the bell.

See G. Chapman (1959), Gladfelter (1970, 1973), Gotshall, Smith & Holbart (1965), Heymans & Moore (1924), and Russell (1964, 1970).

Phylum Cnidaria / Class Scyphozoa / Order Semaeostomeae / Family Ulmaridae

3.22 **Aurelia aurita** (Linnaeus, 1758)
Moon Jelly

Medusae found in coastal waters and washed up on beaches; polyps (scyphistomae) found sometimes in abundance on pilings and floats and under rocks and shells, low intertidal and shallow subtidal zones in bays and harbors; worldwide distribution.

Medusa bell dish-shaped, to 40 cm in diameter, scalloped into eight lobes, with numerous, small, short marginal tentacles and four moderately fringed oral arms; gonads four in number, conspicuous, horseshoe-shaped, bright purple in male, opaque white or yellow in female, visible through translucent gray or blue bell; sedentary polyps to 1.5 cm long when extended.

The male *Aurelia* medusa sheds sperm into the sea, which are taken up by the female; fertilization is internal, and developing embryos are retained in grooves on the oral arms. The planulae are released and swim free to settle and develop into small polyps (scyphistomae), which bud actively and are often found locally in dense colonies, as under the floating docks in Bodega Bay harbor (Sonoma Co.). Tiny jellyfish (ephyrae) are budded off from these polyps by serial transverse fission (strobilation). The polyps feed like small anemones. In contrast, the medusae trap plankton such as copepods in mucus on the surface of the bell or oral arms, and transport it to the mouth via ciliary tracts. Apparently nematocysts play little role in the capture of food.

Aurelia, because of its size, hardiness, wide distribution, and availability, has been used in many investigations, including studies of swimming rhythm. The nerve net in the subumbrella has been mapped. The swimming beat originates in one of the eight marginal sense organs (rhopalia), which act as pacemakers. Medusae from more northern seas have a more rapid beat at a given temperature than those from southern areas.

When deprived of food *Aurelia* medusae undergo remarkable "degrowth," in which an animal the size of a dinner plate may shrink to the size of a small coin, yet be a perfect miniature. When supplied with food, the animal can again grow to normal size. *Aurelia* does well in laboratory aquaria and will feed on brine-shrimp larvae.

On the California coast at Bodega Bay harbor (Sonoma Co.) and Tomales Bay (Marin Co.), young polyps are first found in February; they strobilate to produce young medusae (ephyrae) through March. The medusae grow synchronously and very rapidly, reaching maximum size and reproduction rate by June. By July most have died. A few survive and live another full year, reproducing continuously. Much the same cycle has been noted on the east coast, where cleavage, early embryology, and ultrastructure of the sperm have been studied.

The polyps also live well in the laboratory and have provided material for a number of studies. Strobilating polyps synthesize thyroxine, and iodine stimulates release of ephyrae. Strobilation is also temperature-dependent, and the polyps can osmoregulate to a degree when exposed to seawater of varying salt concentrations.

Along the California coast *Aurelia* medusae are often found in dense swarms, and at such times are fed upon by the blue rockfish *Sebastes mystinus.*

See N. Berrill (1949a), Calder (1971), D. Chapman (1965), G. Chapman (1953), Costello & Henley (1971), Crowell (1967), Custance (1966), Dillon (1977), Hamner & Jenson (1974), Hinsch (1974), Horridge (1954, 1956), Kuwabara, Sato & Noguchi (1969), Luch & Martin (1967), Mayer (1910), Romanes (1885), Schwab (1972), Shick (1973, 1974), Silverstone & Cutress (1974), Southward (1955), Spangenberg (1965, 1967, 1968, 1971, 1974), M. Thiel (1959), and Yasuda (1968, 1969).

Phylum Cnidaria / Class Anthozoa / Subclass Alcyonaria / Order Stolonifera Family Clavulariidae

3.23 **Clavularia** sp.

On shaded sides and undersurfaces of rocks and in crevices, low intertidal zone, and in subtidal areas; occasionally on decorator crabs (*Loxorhynchus*, e.g., 25.10, 25.11); Sonoma Co. to southern Baja California.

Colony forming a thin encrusting patch to 10 cm or more across; polyps well spaced, the retracted ones forming low mounds on the colony, the fully expanded ones showing eight pinnately branched tentacles; colony cream to pink or orange.

Clavularia is one of the very few "soft corals" to invade the California rocky intertidal zone. Nothing is known of its biology.

See Hickson (1915) and Ricketts & Calvin (1968).

Phylum Cnidaria / Class Anthozoa / Subclass Alcyonaria / Order Gorgonacea Family Muriceidae

3.24 **Muricea appressa** (Verrill, 1864)
Sea Fan

In pools, very low intertidal zone; subtidal to 30 m on rocky bottoms; Newport Bay (Orange Co.) to Zorritos (Peru).

Colony to 1 m high, but usually 15–20 cm in Newport Bay, consisting of an upright bush with slender branchlets to 15 cm long and 3–4 mm in diameter, the surface prickly, owing to projecting thorny spicules, the color burnt orange or reddish brown, rarely yellowish brown; calyces (cups bearing polyps) projecting, with strong lower lip upturned and more or less appressed to the branches; extended polyps minute, white in color.

Muricea belongs to the gorgonians, an order that includes the sea whips and sea fans common in some tropical waters. The colony is supported by an internal axial skeleton of gorgonin, a hard dark protein resembling horn. The soft tissues are impregnated with calcareous spicules. In general, sea fans orient at right angles to the prevailing current; water filters through, and the polyps capture from it small organisms and particles.

Studies by Grigg in southern California on *Muricea fruticosa* and *M. californica* (which may prove to be synonymous with *M. appressa*) indicate that as the colonies grow they form annual growth rings in the skeleton. In *M. californica* the sexes are separate, with a sex ratio of 1:1. Reproduction occurs during months when the mean temperature is above 14°C. The larvae swim for about 30 days, then settle and produce slowly growing colonies that reach sexual maturity in 5–10 years.

See especially Grigg (1970, 1972, 1974, 1975, 1976, 1977); see also Block (1974), Ricketts & Calvin (1968), Verrill (1869), and Wainwright & Dillon (1969).

Phylum Cnidaria / Class Anthozoa / Subclass Alcyonaria / Order Pennatulacea
Family Virgularidae

3.25 **Stylatula elongata** (Gabb, 1863)
Sea Pen

Common on mud flats, middle and low intertidal zones (Newport Bay, Orange Co.); subtidal to 70 m (Monterey Bay); Tomales Bay (Marin Co.) to San Diego.

To 60 cm tall, slender, rough to the touch; central axis stiff and brittle, the upper half with semicircular leaflike flanges, about as wide as axis; overall color buff, gray, or greenish.

The sea pen body is divided functionally into two regions, an upper rachis and a lower peduncle. The animals live in vertical burrows in muddy or sandy bottoms and move up and down by peristalsis of the peduncle. *Stylatula* can disappear completely below the surface rather quickly, then reemerge after the disturbance or receding tide has passed. Each of the leaflike structures borne on the rachis consists of a single row of tiny feeding polyps, laterally fused to one another and each equipped with eight pinnately branched tentacles. The whole animal forms an effective device for trapping small living and particulate matter from the water. Between adjacent leaves of feeding polyps the rachis bears clusters of tiny polyps (siphonozooids), whose function is to circulate water through the colony. Entire animals, or sections of the rachis 2–5 cm long (excellent for study under the dissecting scope), live indefinitely in seawater aquaria. The polyps contract when handled, but expand rapidly and beautifully in a few minutes. Like other sea pens, *Stylatula* luminesces when disturbed, especially at night or after prolonged exposure to darkness. After spawning and fertilization, the eggs develop within a day into planula larvae, which settle and develop tentacles by the fourth day.

See Fager (1968), MacGinitie (1938), Morin (1976), and Ricketts & Calvin (1968).

Phylum Cnidaria / Class Anthozoa / Subclass Alcyonaria / Order Pennatulacea
Family Renillidae

3.26 **Renilla köllikeri** Pfeffer, 1886
Sea Pansy

Common on sand flats, low intertidal and adjacent subtidal areas; Coal Oil Point (Santa Barbara Co.) to Isla Cedros (Baja California).

Colony to 8 cm in diameter, purple, consisting of a flat, heart-shaped rachis with a fleshy peduncle attached at indentation and extending downward into sand; feeding polyps with eight white tentacles extending from upper surface of expanded colony.

Renilla rests with the rachis flat on the surface of the sand and the peduncle sunk vertically in the bottom, its tip expanded as an anchor. The upper surface is usually dusted with a light layer of sand, the polyps rising above it. Water is pumped into the colony through tiny specialized polyps (lateral siphonozooids) and passes out through a centrally located axial polyp. An expanded colony may have a diameter more than twice that of a contracted one.

In southern California, planula larvae are released in July, and young colonies from 0.7 to 5 cm in diameter can be found from August through December. In subtidal aggregations on sandy bottoms the sea pansies move about a great deal. In creeping, muscular waves of contraction pass along the edges of the rachis and the peduncle trails behind. Rates of movement of 2 mm per minute have been recorded. After settling out subtidally as juveniles, many *Renilla* migrate inshore to populate sand flats in the low intertidal zone.

The food of *Renilla* appears to be microzooplankton; the polyps have difficulty trapping and engulfing even such small planktonic animals as the nauplius larvae of barnacles. The colonies are preyed upon by the nudibranch *Armina californica* (14.57) and the sea star *Astropecten armatus* (8.1). Adult colonies exposed to nudibranch attack generally withdraw their polyps. *Astropecten* consumes entire large colonies. The amount of predation in an area generally determines the size of the *Renilla* population.

When attacked by a predator, or stimulated electrically or mechanically, the colony luminesces brilliantly, the wave of light spreading over the rachis at a rate of 5–10 cm per second. Studies on related species show that, as in most luminescent animals, a blue light (488 nm) is generated by the oxidation of the substance luciferin in the presence of the enzyme luciferase. In *Renilla* this stimulates a fluorescent reaction, resulting in emission of the green light (508 nm) characteristically seen in the living animal.

Renilla species have also been investigated to determine the toxins in the nematocysts, the ultrastructure of the feeding zooids, and the mechanism whereby 1-glutamic acid suppresses bioluminescence and muscular activity. The animals contain a free-amino-acid pool rich in taurine. Calcium carbonate spicules are abundant; many endoderm cells contain small oval-shaped aragonite crystals, and the mesoglea harbors elongate spicules of calcite.

In California the polyps of *Renilla* are often parasitized by minute copepods of the genus *Lamippe*.

See Anderson & Case (1975), Bertsch (1968), Buck (1973), Buck & Hanson (1967), Case & Morin (1966), Coleman & Teague (1973), Dunkelberger & Watabe (1972), Fager (1968), Huang & Mir (1972), Human (1973), Ivester & Dunkelberger (1971), Kastendiek (1975), Kittredge et al. (1962), Lyke (1965), MacGinitie (1938), MacGinitie & MacGinitie (1968), Morin (1974, 1976), Nicol (1955a,b), Nutting (1909), Parker (1920a,b), Ricketts & Calvin (1968), and Shimomura & Johnson (1975).

Phylum Cnidaria / Class Anthozoa / Subclass Zoantharia / Order Actiniaria Family Halcampidae

3.27 Halcampa decemtentaculata Hand, 1954

Fairly common, buried in sand, gravel, or accumulated debris in holdfasts of algae or surfgrass roots or among masses of colonial ascidians, low intertidal zone and adjacent subtidal waters along rocky shores; Puget Sound (Washington) to Carmel (Monterey Co.).

Expanded animal to nearly 8 cm long, but usually about 2 cm in length and 0.3 cm in diameter; contracted animal nearly spherical; basal part of body consisting of inflatable digging organ (physa); column white, cream, purple, or brown, often with longitudinal rows of white spots; tentacles ten, short,

conical, pink, white, brown, or purple, often marked with brownish V-shaped bands; eggs gray-green, about 0.3 mm in diameter.

A related species, **H. crypta** Siebert & Hand (known from Marin Co. and San Juan Island, Puget Sound, Washington), has 12 tentacles and produces olive-green eggs about 0.5 mm in diameter.

Few details are available on the biology of these primitive burrowing species, but spawning and development have been studied in *Halcampa duodecimcirrata* Carlgren of the North Atlantic. The adults of this species live in vertical burrows in gravel substrata, with their tentacles flush with the surface, and presumably feed on detritus. In late fall the anemones become sexually mature and spawn. The male stretches up out of its burrow and releases sperm. The female then rises and may stretch toward a discharging male before releasing eggs. The ripe eggs, 0.3 mm in diameter, undergo meiosis before fertilization. Some eggs are fertilized after discharge; others are fertilized while still within the mother's coelenteron. The fertilized egg cleaves superficially, leaving a central mass of yolk. The planula hatching from the egg case does not become planktonic, but burrows immediately into the substratum and feeds by means of cilia around the mouth until the tentacles appear at about 6 months of age.

See Hand (1954), Nyholm (1949), Siebert & Hand (1974), Spaulding (1974), and Strathmann (1969).

Phylum Cnidaria / Class Anthozoa / Subclass Zoantharia / Order Actiniaria Family Halcampidae

3.28 Cactosoma arenaria Carlgren, 1931

In holdfasts of large brown algae in large permanent tidepools, and among tunicate colonies consisting of sand-encrusted lobes, low intertidal zone on rocky shores; more commonly subtidal; Monterey Bay to San Pedro (Los Angeles Co.).

Column about 2–4 cm in height, brownish or orange with a cuticle in which foreign material collects; tentacles in adult specimens 24, short, conical, arranged in three whorls, brown or orange marked with V-shaped bands.

Cactosoma, like *Halcampa* (e.g., 3.27) and *Harenactis* (e.g., 3.40), lacks a true base and basilar muscles, but the lower terminal end (physa) is usually flattened into a disk and used for attachment. Little is known of the biology of the species.

See Carlgren (1931) and Hand (1954).

Phylum Cnidaria / Class Anthozoa / Subclass Zoantharia / Order Actiniaria Family Actiniidae

3.29 Epiactis prolifera Verrill, 1869
Proliferating Anemone

Common on and under rocks and on algae and eelgrass, middle and low intertidal zones and in subtidal waters on outer rocky coasts and in bays; southern Alaska to southern California.

Body low, squat, the tentacular crown to 5 cm in diameter, usually less; base broad, about twice diameter of column; tentacles 48–96, short, conical, each with a terminal pore; overall color highly variable, green, brown, orange, blue, or gray, either solid color or blotched with another color, the base usually with short, white radiating lines, the column with brownish-red stripes or, on green specimens, dark-green stripes.

From 25 to 50 percent of all *Epiactis* observed in the field have tiny juvenile anemones attached to the column around the base. The young are not formed by budding but are produced sexually. Recent studies show that *E. prolifera* has a sex life unique among animal species but akin to that of some orchid plants (gynodioecy). Young adults (with a pedal disk 8–15 mm in diameter), which make up the bulk of the population seen in the field, are nearly all functional females. As these become mothers they continue to grow in size, and testes begin to develop along with the ovaries; those animals that live long enough to reach full maturity (with a pedal disk at least 20.5 mm in diameter) are active simultaneous hermaphrodites. Since the sperm come only from mature hermaphrodites, all eggs from all females, young and old, are fertilized by animals that have survived a long time and thereby demonstrated their fitness to father the next generation. The majority of young are produced by crossing (outbreeding), but mature hermaphrodites may be self-fertile; thus the oldest and most successful individuals in the population may be perpetuating their desirable traits by a bit of inbreeding.

In large individuals both eggs and sperm can be found in all stages of development on a single mesentery, with little or no separation of formative tissues. There is no well-marked breeding season. The eggs expelled by adults are pink or pinkish orange, spherical, 0.4 mm in diameter, and usually embedded in mucus. In early development many nuclear divisions occur before the cytoplasm cleaves, the nuclei arranging themselves in two concentric layers, which ultimately develop into ectoderm and entoderm. The embryos are not buoyant and lack cilia; after discharge from the mother's mouth, maternal cilia move them across the oral disk and down the column, where they become attached near the parent's base in a shallow depression. Both mucus and specialized large nematocysts localized in the basal region seem to play a role in anchoring the young to their mother. Very soon after settlement, the larvae grow tentacle buds and become juvenile anemones. Large adults have been observed brooding 30 or more juveniles, all apparently derived from the brooding parent. The young probably receive no nourishment directly from the adult, but survive on the large store of yolk until they can begin feeding themselves. The young remain on the parent's column until they are at least 3 months old and 4 mm in diameter, then glide off to take up an independent existence. As the young often continue to live for some time near the parent, anemone concentrations of 20 or more per m^2 are not uncommon.

Adult *Epiactis* probably feed mainly on small crustaceans; detailed dietary studies are lacking. Starved *Epiactis* discharge nematocysts much more readily than well-fed animals; they may also ingest young anemones that have become detached from the parent's base, but these are normally regurgitated unharmed even after several hours in the gut. *Epiactis* is preyed upon by the nudibranch *Aeolidia papillosa* (14.67) and the sea star *Dermasterias imbricata* (8.3).

See Dunn (1972, 1975a,b, 1977a,b), Hand (1955a), Hewatt (1937), Lenhoff (1965), Mauzey, Birkeland & Dayton (1968), McLean (1962), Mariscal (1973), Mariscal, Bigger & McLean (1976), Mariscal & McLean (1976), Mariscal, McLean & Hand (1977), Solomon (1975), Spaulding (1974), and Waters (1975).

Phylum Cnidaria / Class Anthozoa / Subclass Zoantharia / Order Actiniaria Family Actiniidae

3.30 Anthopleura xanthogrammica (Brandt, 1835)
(= Cribrina xanthogrammica)

On rocks in tidepools and deep channels on exposed rocky shores, and on concrete pilings in open bays and harbors, low intertidal and subtidal zones; Alaska to Panama.

Column to 17 cm in diameter and 30 cm in height, covered with irregular, compound, adhesive tubercles; tentacular crown to 25 cm in diameter; base of somewhat greater diameter than column, very firmly adherent to rocks or pilings; contracted animal forming a hemispherical mound; tentacles numerous, short, conical, pointed or blunt, in six or more circles in narrow band around margin; column and base usually green to dark greenish brown, lighter in shaded locations, the tentacles greenish, bluish, or white, not marked or banded, never pink at tips, the oral disk flat, gray-blue, green, or greenish blue.

The giant green anemone is one of the most spectacular animals in tidepools along the Pacific coast. Although basically a solitary form, in favorable localities it occurs in numbers up to 14 per m². Animals living exposed to bright sunlight are brilliantly green. The color is due in part to green pigment in the anemone epidermis and in part to symbiotic algae living in tissues lining the gut. The latter may be unicellular green algae (zoochlorellae) alone or in combination with dinoflagellates (zooxanthellae). Experiments have shown that the zooxanthellae contribute to the nutrition of the host by providing organic materials synthesized by the plants. The possible role of the zoochlorellae in the nutrition of the anemone is not clear. In anemones living in caves or other deeply shaded areas, both types of symbionts are reduced in number and may be totally absent.

Anthopleura xanthogrammica does not reproduce by fission, but releases sperm and brownish eggs in late spring and summer. Development has not been followed, but the larvae apparently swim or float freely for a period of time and become widely dispersed. Small animals are sometimes found in mussel beds, but the largest adults occur only in permanent pools. The adults are often crowded and have tentacular crowns in contact, yet *A. xanthogrammica* does not display the aggressive behavior seen in a smaller relative, *A. elegantissima* (3.31).

The hermit crab *Pagurus samuelis* (24.10) often walks up and down the column of the anemone, and even walks through and strokes the tentacles and probes the mouth opening, all without being stung. Possibly the hermit crab becomes so coated with mucus from the anemone that the anemone responds to the crab as though it were its own tissue. Hermit crabs not previously associated with anemones may be eaten or merely taken into the gastrovascular cavity and later released.

In many tropical regions anemones commonly have small damselfishes living in and among the tentacles in a symbiotic relationship. Although *A. xanthogrammica* is not normally associated with such fishes, experiments show that anemone fish from the tropics become gradually acclimated when put in a tank with *Anthopleura* as the only available anemone. By making light contact with the anemone's tentacles over a period of several hours, the fish develop a coating that keeps them from being stung by the anemone's nematocysts; the fish can then move with impunity through the tentacles of *Anthopleura*. If the coating of a fish is wiped off and the animal is then placed among the anemone's tentacles, it is stung and often killed.

The major foods of these large anemones are detached mussels, crabs, sea urchins, and small fishes. Predators include the nudibranch *Aeolidia papillosa* (14.67) and the snail *Epitonium tinctum* (13.50), both of which feed on the tentacles, and the snails *Opalia chacei* (13.51) and *O. funiculata* (13.52) and the sea spider *Pycnogonum stearnsi* (27.1), which feed on the column. In Puget Sound (Washington), small anemones are preyed upon by the sea star *Dermasterias imbricata* (8.3).

Anthopleura xanthogrammica has been used in several chemical and pharmacological studies. Its tissues are the source of a new vertebrate heart stimulant and of a pheromone mediating an alarm response in both *A. xanthogrammica* and *A. elegantissima*. The animal has also been found to flex or retract its tentacles when subjected to ultraviolet radiation at different wave lengths.

See Belcik (1968), Broughton (1975), Clark & Kimeldorf (1970), Dayton (1973), Francis (1973b), Gee (1913), Giese (1966), Hand (1955a), Hewatt (1937), Howe (1976), Kittredge et al. (1962), Koehl (1976), Lindberg (1976), MacGinitie (1935), Mariscal (1965, 1969a, 1970a,b, 1973), Martin (1963), Mauzey, Birkeland & Dayton (1968), McLean (1962), Muscatine (1971, 1974), Salo (1977), Strain, Manning & Hardin (1944), Tanaka et al. (1977), Torrey (1906), and Waters (1975).

Phylum Cnidaria / Class Anthozoa / Subclass Zoantharia / Order Actiniaria Family Actiniidae

3.31 Anthopleura elegantissima (Brandt, 1835)
(=Cribrina elegantissima, Bunodactis elegantissima)
Aggregating Anemone

Abundant on rock faces or boulders, in tidepools or crevices, on wharf pilings, singly or in dense aggregations; characteristic of middle intertidal zone of semiprotected rocky shores of both bays and outer coast; Alaska to Baja California.

Aggregating individuals to 6 cm in column diameter and 8 cm across tentacular crown; solitary individuals often larger, to 25 cm across tentacular crown; column light green to white, twice as long as wide when extended, with longitudinal rows of adhesive tubercles (verrucae) often bearing attached debris; verrucae usually branched only on upper sixth of column; tentacles numerous, short, in five or more cycles, variously colored but often with pink, blue, or lavender tips; aggregating forms easily recognized by the clones they produce (see below); solitary forms usually distinguishable from *A. xanthogrammica* (3.30) by their branched verrucae, general coloration, and softer, thinner body walls.

The wartlike tubercles (verrucae) on the column are adhesive structures to which pebbles, shell fragments, and bits of seaweed adhere through the action of nematocyst-like structures (spirocysts). The adherent material reduces desiccation and gives protection from solar radiation during exposure at low tide. When the animals are contracted, their bodies form low hemispherical mounds, and a bed of aggregated anemones may resemble a bed of coarse sand or shell gravel.

The green color of these animals is due to the presence of a fluorescent green pigment in the epidermis and to symbiotic, unicellular green algae (zoochlorellae) or dinoflagellates (zooxanthellae), or both, in the tissues lining the gut. Studies using radioactive carbon indicate that some organic matter synthesized by the zooxanthellae is transferred to the host anemone, thereby contributing to its nutrition. Anemones kept in darkness for an extended period gradually lose all their algae, and then become indifferent to light and shade. Anemones possessing the symbionts, however, under laboratory conditions move about, either toward or away from light, until some optimum intensity is reached. Anemones experimentally exposed to oxygen gradients move toward the side of higher oxygen tension; thus it appears that the oxygen produced by the symbiotic algae may help control phototaxis. In addition, anemones devoid of symbionts do not expand or contract under different light conditions, whereas those with zooxanthellae expand in moderate light and contract in strong light.

Anthopleura elegantissima reproduces sexually, and near San Francisco spawning has been observed in September. In anemones at Morro Bay (San Luis Obispo Co.), ova are present as early as February and grow steadily until their release in July; the ovarian tissue is then resorbed and new eggs do not appear until the next February. Sperm are released through the summer.

Anthopleura elegantissima also reproduces asexually by longitudinal fission. This process results in aggregations or clones of anemones pressed together in concentrations of several hundred per square meter. All of the anemones within a clone are genetically identical, and thus have the same color pattern and are of the same sex. Anemones on the periphery of clones exhibit aggressive behavior toward non-clonemates. When genetically different animals make tentacular contact, knob-like swellings (acrorhagi), located on a collar just outside the tentacular ring and packed with large nematocysts, become inflated and are pushed out toward the other anemone. If contact is made, the tissue touched by the acrorhagi is stung and damaged; it ultimately dies and is sloughed off. One or both animals on a border between two different clones may be injured; they tend to draw away from one another so that adjacent clones are usually separated by an anemone-free zone on the rock face. The bare corridors between clones may be up to 5 cm wide. Clones periodically observed for several years showed little change, though occasional individuals moved. The anemones on the clonal edges of corridors ("warriors") differ in several respects from their clonemates: they are usually smaller than mid-clone anemones, have larger acrorhagi, and are often without gonads at a time when the other anemones are ripe.

In coastal areas where thermal outfalls from power plants increase the water temperature by as much as 10°C, *A. elegantissima* populations survive, but store less lipid and show a somewhat reduced reproductive ability, possibly owing to increased stress and metabolic demands imposed by the warmer water.

This species has been investigated for biologically active materials, and crude algal-free extracts of its tissues show activity against ascitic tumors in mice. Its free amino acids differ from those found in *A. xanthogrammica*. Biochemical studies show that its feeding behavior is controlled by asparagine, which causes tentacular bending when food is brought to the mouth, and by reduced glutathione, which controls ingestion of food once it reaches the mouth. When one member of a clone of *A. elegantissima* is injured, it releases a pheromone (anthopleurine), and anemones downstream from the injured animal react by rapidly flexing the tentacles toward the column, then retracting the tentacles and constricting the marginal sphincter.

Anthopleura elegantissima feeds on copepods, isopods, amphipods, and other small animals that contact the tentacles. It is preyed upon by the nudibranch *Aeolidia papillosa* (14.67), which usually attacks the column, by the snail *Epitonium tinctum* (13.50) which attacks the tips of the tentacles, and by sea stars such as *Dermasterias imbricata* (8.3) that can engulf an entire small anemone. In some anemones, small pink amphipods (*Allogaussia recondita*) make a home in the gastrovascular cavity.

See Belcik (1968), Broughton (1975), Buchsbaum (1968), Carlisle, Turner & Ebert (1964), Childress (1970), Davis (1962), Dunn, Kashiwagi & Norton (1975), Ford (1964), Francis (1973a,b, 1976), Fredericks (1976), Hand (1955a), Hewatt (1937), Howe (1976), Howe & Sheikh (1975), Jennison (1975), Kittredge et al. (1962), Lindstedt (1971), Martin (1963, 1968), Mauzey, Birkeland & Dayton (1968), Muscatine (1961, 1971), Muscatine & Hand (1958), Muscatine & Lenhoff (1965), Pearse (1974a,b), Phillips (1961), Salo (1977), Stasek (1958), Trench (1971), and Waters (1975).

Phylum Cnidaria / Class Anthozoa / Subclass Zoantharia / Order Actiniaria Family Actiniidae

3.32 **Anthopleura artemisia** (Pickering in Dana, 1848)

Typically attached to rocks buried in sand, or in abandoned holes made by rock borers, occasionally on pilings or floats, low intertidal and subtidal waters along open coast and in protected bays: Alaska to southern California.

Column to 5 cm in diameter, capable of elongation to over five times its diameter, black, gray, or brown grading to white or pink near base, with tubercles well developed on upper third; tentacular crown to 7 cm in diameter, the tentacles numerous, tapering, somewhat transparent, in five cycles, colored variously red, white, orange, black, or blue, either solid color or in patterns, sometimes with each cycle differing in color; oral disk also variously colored.

This species is easily distinguished from other members of the genus by its distinctive habitat and the bright color patterns of its tentacles and disk. Normally only the tentacular crown is exposed, and when disturbed or at low tide, the anemone retracts and withdraws some distance below the sand surface.

Unicellular symbiotic algae occur in the tissues, but are restricted to the upper portions of the body, which are normally exposed to light. Like *A. xanthogrammica* (3.30), this species occurs as solitary individuals. But like *A. elegantissima* (3.31), it reproduces by longitudinal fission; and in the laboratory, when isolated individuals are brought into close contact, they exhibit agonistic and aggressive behavior similar to that described for *A. elegantissima*.

A related species from Hawaii, *A. nigrescens*, exhibits interesting "escape" behavior. On being attacked by predatory nudibranchs, the anemone responds by shedding its column armor of pebbles and shell fragments and detaching the pedal disk from the rock, whereupon it is moved about by wave action, often rolling along the bottom. This usually separates prey and predator, and the anemone then reattaches in a new position.

See Francis (1973b), Hand (1955a), Powell (1964), Rosin (1969), and Waters (1975).

Phylum Cnidaria / Class Anthozoa / Subclass Zoantharia / Order Actiniaria Family Actiniidae

3.33 **Tealia crassicornis** (Müller, 1776)

Common on the sides and undersurfaces of rocks, low intertidal zone and subtidal waters along exposed rocky shores; Alaska to south of Carmel (Monterey Co.); both coasts of North Atlantic.

Column to 8 cm in diameter, 8–10 cm in height, smooth, lacking adhering foreign material, variable in color, red or green, with irregular patches of green, yellow, or red; collar below tentacles well developed; tentacles short, stout, blunt, semitransparent when extended, arranged in four or five cycles, with varying patterns or pale shades of cream, pink, blue, or green, the tips usually white; oral disk same color as tentacles.

In east coast populations the eggs are retained and fertilized internally; the young are brooded between the mesenteries of the gastrovascular cavity and released as small, well-formed, young anemones. In California, fertilization of this anemone is internal, but brooding of the young has yet to be observed.

In Puget Sound (Washington), *T. crassicornis* releases both eggs and sperm into the sea, with spawning occurring in the spring. The eggs are yolky, to 0.7 mm in diameter, and covered with minute spines or microvilli.

Following fertilization, superficial cleavage results in a solid, ciliated blastula. Absorption of the central yolk produces a gastrovascular cavity, and by the sixth day following

fertilization a cone-shaped planula is produced, which swims near the bottom. If tubes of the annelid worms *Sabellaria cementarium* (18.31) or *Phyllochaetopterus prolifica* (18.22) are available, the planulae settle on them on about the tenth or eleventh day. If not, the larvae settle on small stones on about the twenty-seventh day. The settled planulae rapidly develop into miniature anemones 0.6 mm in diameter and 0.8 mm in height. The first four tentacles appear 7 days after settlement, the second four after another 5 days, and at this stage the young anemones begin to feed. Growth is slow, and 2 months after settlement the young are only 0.8 mm in diameter; at this stage they possess 12 tentacles. After one year anemones provided with adequate food are 10 mm in diameter and have up to 35 tentacles. When starved, young animals remain alive for long periods but do not change size. Size therefore is dependent on the amount of food available, not age. In the field, anemones 10–15 mm in diameter are often sexually mature, and these are presumed to be at least one year old. In the laboratory, apparently well-fed animals 14 months old and 12 mm in diameter still showed no gonadal development.

When *T. crassicornis* is kept in an aquarium with the shrimp *Lebbus grandimanus*, the shrimps aggregate in groups on the column of the anemone just below the tentacles. The shrimps then slowly move up through the tentacles and across the oral disk and appear to feed on the egested wastes and dead tissues of the anemone.

See Chia & Spaulding (1972), Hand (1955a), Hoffman (1967), McLean (1962), Spaulding (1974), and Waters (1975).

Phylum Cnidaria / Class Anthozoa / Subclass Zoantharia / Order Actiniaria Family Actiniidae

3.34 **Tealia coriacea** (Cuvier, 1798)

Fairly common, low intertidal zones on rocky outer coast or in bays, typically lying buried in patches of sand, gravel, or shell between large rocks, with only tentacular crown exposed; also subtidal on rock walls and in crevices to 15 m; Alaska to south of Carmel (Monterey Co.); Atlantic coast of Europe.

Column to 10 cm in diameter, 14 cm in height, with heavy

thick tubercles (verrucae) to which sand grains and other foreign material adhere strongly, dull brownish red to bright brick red, never variegated with green; tentacles short, stout, blunt, in four cycles, colored green, pink, or blue, with one or more bands of white or pink; oral disk gray, blue, red, or pink.

Little is known of the biology of this species. It is preyed upon in Puget Sound (Washington) by the sea star *Dermasterias imbricata* (8.3).

See Belcik (1968), Hand (1955a), Martin (1963), McLean (1962), and Mauzey, Birkeland & Dayton (1968).

Phylum Cnidaria / Class Anthozoa / Subclass Zoantharia / Order Actiniaria Family Actiniidae

3.35 **Tealia lofotensis** (Danielssen, 1890)

Common on rocks and walls of surge channels, low intertidal zone and subtidal waters to 15 m depth on exposed outer coast; on concrete piles and on marina floats in Monterey harbor; northern Washington to San Diego; both sides of North Atlantic.

Column to 10 cm in diameter, 15 cm in height, smooth, free of adherent material, scarlet or crimson with white spots arranged in longitudinal rows; tentacles slender, somewhat elongate, scarlet to crimson, lacking bands or marks; oral disk reddish, less bright than tentacles.

Little is known of the biology of this beautiful and distinctive animal. It survives well in aquaria and should prove favorable for physiological, pharmacological, and behavioral studies.

See Hand (1955a) Martin (1963), and McLean (1962).

Phylum Cnidaria / Class Anthozoa / Subclass Zoantharia / Order Actiniaria Family Isanthidae

3.36 **Zaolutus actius** Hand, 1955

Often common, found buried up to the tentacular crown in sand or mud, low intertidal zone in bays and estuaries, subtidal on sandy bottoms on open coast; Monterey Bay to San Diego.

Body elongate, usually to 1.5 cm in diameter, 17 cm in length, but occasionally reaching 50 cm in length; column cylindrical when expanded, sometimes narrower in basal half and expanding toward tentacles, smooth, gray, gray-green, or cream; tentacles 100 to 192, in up to six cycles, long, pointed, transparent, with opaque gray tips, sometimes with white bands.

This burrowing form is attached by the base to small stones, worm tubes, or shells buried some distance below the surface, and can withdraw below the sand when disturbed. In shallow subtidal depths off La Jolla *Zaolutus* occurs in densities to 10 per m², and the population is limited in part by the scarcity of suitable buried attachment sites. Along with the tube-building polychaete *Owenia fusiformis*, *Zaolutus* stabilizes the substratum by reducing sand shift due to surge. Areas thus stabilized may be populated by other benthic animals. *Zaolutus* normally moves only up and down in the sand, but it can also move laterally. It feeds on minute zooplankton or detritus particles, which it traps in a mucus sheet on the tentacles and disk and carries to the mouth by ciliary action. Larger items of animal food appear to be taken by some individuals, rejected by others.

In southern California, recently settled young have been observed from March to July, and the life span appears to be about one year. The wentletrap snail *Epitonium tinctum* (13.50) is a specific predator and may be found with its proboscis inserted deeply into the column of the anemone. *Zaolutus* is also eaten by the flatfish *Pleuronichthys ritteri*.

See Fager (1964, 1968), Hand (1955a), and MacGinitie (1935).

Phylum Cnidaria / Class Anthozoa / Subclass Zoantharia / Order Actiniaria
Family Metridiidae

3.37 **Metridium senile** (Linnaeus, 1767)
Plumose Anemone

On pilings, floats, breakwaters, and jetties in bays and harbors, low intertidal zone and subtidal waters; southern Alaska to southern California; both sides of North Atlantic.

Typically 2–5 cm in tentacular crown diameter, but subtidal animals reaching 25 cm in crown diameter and 50 cm in height; column when expanded much longer than wide, overhung by broad disk, colored white, cream, tan, orange, or brown; tentacular crown round in young animals, lobed or pleated in larger ones, the tentacles very numerous, slender, short, tapering distally, covering most of oral disk, transparent when expanded, same color as column when contracted.

When seen on wharf piles, *Metridium* characteristically occurs in dense aggregations of small individuals, all in contact, in the low intertidal zone; much larger solitary individuals occur lower down (subtidally).

The animals generally appear motionless, but time-lapse movies show slow rhythms of expansion and contraction. The body can assume a wide range of shapes. As in other anemones, the fluid in the gut, under slight positive pressure from the muscles of the body wall, acts as a hydraulic skeleton, providing internal support. Additonal support, along with elasticity and extensibility, is provided by the mesoglea or middle layer of the body wall, which contains an amorphous polymer network of collagen. An anemone crawls slowly along the substratum by muscular waves of its base.

When the animal is handled roughly, long threads (acontia), which are attached to the mesenteries in the gut and are rich in nematocysts, are forced out through the mouth and through weak spots in the body wall (cinclides). When placed in contact with the green aggregating anemone *Anthopleura elegantissima* (3.31), *Metridium* also uses its acontia as weapons. *Metridium* reproduces asexually as it moves about, by breaking off and leaving behind small fragments of the pedal disk; the process is called pedal laceration. The isolated fragments round up and develop into new, tiny anemones, which feed and grow. In this manner dense clones of individuals, all the same color and genetically identical, are produced.

In dense clones of the smaller *Metridium*, animals on the border of the clone often develop "catch" tentacles. These are derived from among the regular tentacles arising closest to the mouth. In a single anemone, as many as 19 of these may be present. The catch tentacles have a complement of nematocysts unlike that of regular feeding tentacles. Capable of great expansion, the tentacles can stretch out for 12 cm or more and explore their surroundings, extending and retracting rhythmically. The catch tentacles respond neither to food nor to contact with the tentacles or columns of clonemates. However, if the regular feeding tentacles of an anemone make contact with a *Metridium* of another clone, one or both of the anemones expand their catch tentacles. If the tip of a catch tentacle touches a non-clonemate it sticks and stings, and the stinging tip breaks off. The attacked individual may bend away, contract, or move away, and after some time tissue damage is seen at the site of the clinging tentacle tip. Adjacent distinct clones of *Metridium* tend to remain separated from one another by a space several centimeters wide.

Anemones from the center of a clone usually bear no catch tentacles. They develop them in a period of about 9 weeks if moved out to the border of the clone or otherwise placed in contact with non-clonemates. When a border *Metridium*, armed with catch tentacles, encounters the green anemone, *Anthopleura elegantissima*, it may lash it repeatedly, causing it to contract or move away. The catch tentacles are clearly used for aggression against non-clonemates and anemones of other species.

Metridium reproduces by sexual as well as asexual means. Males release sperm with wedge-shaped heads into the sea. The presence of sperm stimulates the females to release their eggs, which are pinkish and about 0.1 mm in diameter. The eggs undergo maturation divisions before release, and fertilization is external. The zygote divides to produce a hollow ciliated blastula, a gastrula, and eventually a planula larva that swims in the plankton. In Monterey harbor, planulae settle and metamorphose into young anemones in July and August, and from October to December.

Both large and small *Metridium* feed mainly on small zooplankton, and feeding appears to be non-selective. Small benthic organisms of the pilings, such as polychaetes, and even scraps of fish and squid, may occasionally be taken as food, especially by the smaller anemones.

Metridium can acclimate to temperature changes, and its metabolic rate is often positively correlated with temperature. It also tolerates brackish conditions and is found in areas of San Francisco Bay having a salinity as low as 68 percent that of pure seawater. Attempts to establish conditioned reflexes have yielded negative results. The nematocysts have been investigated for fine structure and for pharmacological effects. The nematocyst toxin contains a protein fraction and dialyzable material that includes aromatic amines.

The nudibranch *Aeolidia papillosa* (14.67) feeds on small *Metridium* (but not larger animals). In Puget Sound (Washington), the sea star *Dermasterias imbricata* (8.3) feeds on fairly large *Metridium*.

See especially Batham & Pantin (1950a,b,c,d, 1951, 1954) and Purcell (1977a,b); see also Alexander (1962), Batham (1960), Batham Pantin & Robson (1960), Carlisle, Turner & Ebert (1964), Chao (1975), Clark & Dewel (1974), Day & Harris (1978), Elder (1973), Felice (1958), Gemmill (1920), Giese (1966), Goodwin & Telford (1971), Gosline (1971a,b), Grimstone et al. (1958), Haderlie (1969), Hand (1955b), Harris (1971), Hinsch (1974), Lindberg (1976), Mariscal, Bigger & McLean (1976), Mariscal & McLean (1976), Mariscal, McLean & Hand (1977), Martin (1963), Mauzey, Birkeland & Dayton (1968), Pantin (1965), Parker (1896), Phillips (1956), Phillips & Abbott (1957), Robson (1957, 1961), Robson & Josephson (1969), Ross (1964), Sassaman & Mangum (1970), Spaulding (1974), Torrey (1898), Waters (1975), and Westfall (1965a).

Phylum Cnidaria / Class Anthozoa / Subclass Zoantharia / Order Actiniaria Family Metridiidae

3.38 **Metridium exilis** Hand, 1955

In semiprotected situations under rocks or ledges, and in crevices on sides of surge channels, low intertidal zone and subtidal waters along rocky coasts; Mendocino Co. to Carmel Bay (Monterey Co.).

Crown to 1.4 cm in diameter, to 1.8 cm in height; column smooth when extended, pale pink to bright orange; tentacles to 100, tapered and pointed, translucent, sometimes with white or orange bands.

This small anemone, like *M. senile* (3.37), responds to rough handling by protruding slender pale threads (acontia), loaded with nematocysts, from the mouth and from pores on the column. Little is known of its biology.

See Haderlie et al. (1974) and Hand (1955b).

Phylum Cnidaria / Class Anthozoa / Subclass Zoantharia / Order Actiniaria Family Haliplanellidae

3.39 **Haliplanella luciae** (Verrill, 1898)

Uncommon, on undersides of stones in well-protected pools, middle intertidal zone on outer rocky coast; sometimes in bays and estuaries on pilings and floating docks; Washington to southern California; Japan and both sides of North Atlantic.

Tentacular crown to 3.5 cm in diameter, 3 cm in height; column cylindrical, smooth, green, gray, or brown, with or without vertical orange or white stripes; tentacles to 100, slender, tapering, fully retractile, usually transparent, sometimes gray or light green flecked with white.

European studies show that in some individuals the innermost cycles of tentacles near the mouth include 1–18 "catch" tentacles (see the discussion under *Metridium senile*, 3.37). When contracted, these tentacles are shorter, blunter, and more opaque than ordinary tentacles; when fully extended, they are both larger in diameter than normal tentacles and three to four times as long. The catch tentacles are not involved in feeding and are never brought to the mouth. Instead they reach out and appear to search the surrounding area. When the tip of one touches another anemone, of the same or a different species, it stings and becomes stuck fast to the victim. The tentacle then retracts slowly, and the tip is broken off, remaining behind. The stung anemone contracts forcefully and may die as a result of the attack; at the least, severe tissue damage occurs at the site of the sting. An anemone surviving an attack may move away from further contact with the aggressor, and a main function of long catch tentacles may be to maintain separation between anemones. If an animal is starved for a period of 3 months, the catch tentacles become transformed into ordinary tentacles.

Although this species reproduces sexually, it also reproduces asexually, both by a process called pedal laceration (see the discussion of *Metridium senile*) and by longitudinal fission. Many individuals bear a pale line down one side of the column, marking the site of regeneration following the last fission. Atlantic coast populations are usually composed of one or a few asexual clones. Mass mortality may result when the colony is exposed to extremes in temperature or salinity. On the east coast, *H. luciae* may also respond to temperature increase by contracting and undergoing encystment, with a much lower metabolic rate.

The species is preyed upon by wentletrap snails (*Epitonium* spp.).

See Atoda (1973), Hand (1955b), Hausman (1919), Johnson & Shick (1977), Robertson (1963), Sassaman & Mangum (1970), Shick (1976), Shick & Lamb (1977), and Williams (1975).

Phylum Cnidaria / Class Anthozoa / Subclass Zoantharia / Order Actiniaria Family Ilyanthiidae

3.40 **Harenactis attenuata** Torrey, 1902

Buried in mud or sand in protected bays, low intertidal zone; subtidal on gravel bottoms along open coast to 15 m depth; Humboldt Bay to San Diego Bay.

Body elongate, normally 8–12 cm long but sometimes extending to 25 cm or more, the tentacular spread to 3 cm; base often swollen into anchorlike bulb; tentacles 24, short; column wrinkled, yellowish white to gray.

Off La Jolla (San Diego Co.), *Harenactis* is the most abundant invertebrate in sand epifauna at depths of 5–10 m. Like *Halcampa* (e.g., 3.27), this species lacks a true base at the aboral end and is an active burrower. It extends its tentacular crown above the surface of the sand when the tide is in and actively feeds on animal detritus, small *Dendraster excentricus* (11.7), polychaetes, and small crustaceans. It has not been observed feeding on active prey. When disturbed, *Harenactis* re-

tracts into the substratum and anchors itself securely by its swollen basal bulb (which may be broken off when the animals are collected). Observations indicate that this anemone lives for 6 years or more. No predators on *Harenactis* have been seen.

See Fager (1969), MacGinitie (1935), McLean (1962), and Ricketts & Calvin (1968).

Phylum Cnidaria / Class Anthozoa / Subclass Zoantharia Order Corallimorpharia / Family Corallimorphidae

3.41 **Corynactis californica** Carlgren, 1936

Abundant on shaded rocks and ledges, low intertidal zone to 30 m depth on rocky shores; also on concrete wharf pilings and plastic foam floats (Monterey harbor), and on legs of offshore oil towers (southern California); Sonoma Co. to San Diego.

Average column height and diameter about 1 cm; expanded tentacular crown to 2.5 cm in diameter, column smooth, trumpet-shaped or flaring to the tentacular crown, colored most commonly red, crimson, or pink, but also purple, brown, orange, yellow, buff, or nearly white, each animal of one solid color; tentacles numerous, club-tipped, not fully retractile, arranged in radial rows.

Corynactis and other genera in the order Corallimorpharia more closely resemble the true stony corals than they do the sea anemones, but lack a calcareous exoskeleton. The very large nematocysts found in the tentacle tips are excellent for classroom study. Nematocyst discharge can be demonstrated easily by flattening a tentacle between a glass slide and coverslip, placing it under the microscope, and adding a drop of 10 percent acetic acid so that it is drawn under the coverslip. Investigations on the nematocysts of *Corynactis* species show that undischarged nematocysts have osmotic pressures up to 140 atmospheres, and they also reveal how the thread "unfolds" as it turns inside out.

When attacked by the anemone *Anthopleura elegantissima* (3.31) in experimental conditions, *Corynactis* has been observed to extend its mesenteries through the mouth; the filaments at the mesentery edges possess nematocysts and perhaps help in repelling attack. Electron microscope studies of the epidermis of *Corynactis*, *Balanophyllia elegans* (3.42), *Metridium senile* (3.37), and other zoantharians reveal organelles (ciliary cones) that may be receptors of mechanical stimuli, and other structures, each consisting of a cilium surrounded by short microvilli that possibly detect chemical stimuli.

Biochemical investigations have revealed that the free-amino-acid pool of *Corynactis* is high in glutamic acid, alanine, valine, and the leucines, but low in taurine.

Corynactis californica reproduces asexually by longitudinal fission and produces clones all of one color, some covering a square meter or more of substratum. The animals feed actively on copepods, nauplius larvae, and other small animals.

See Carlisle, Turner & Ebert (1964), Chao (1975), Hand (1954), Kittredge et al. (1962), Mariscal (1974a,b,c), Mariscal, Bigger & McLean (1976), Mariscal & McLean (1976), Mariscal, McLean & Hand (1977), Martin (1963), McLean (1962), Picken (1953), Picken & Skaer (1966), and Robson (1953).

Phylum Cnidaria / Class Anthozoa / Subclass Zoantharia / Order Madreporaria
Family Dendrophylliidae

3.42 **Balanophyllia elegans** Verrill, 1864
Orange Cup Coral

Common on and under shaded rocks, on sides of surge channels, and under ledges, low intertidal zone and subtidal to 10 m depth along open coast; Oregon to southern California.

Skeleton in large animals to 1 cm in diameter and height; polyp nearly completely retractile into stony cuplike skeleton, colored bright orange to yellow, with tentacles bearing wartlike batteries of nematocysts.

This solitary coral is the only true coral (order Madreporaria) that occurs intertidally on the California coast, though other genera occur at scuba depths. Animals carefully removed from the rocks live well in laboratory aquaria, feed-

ing on living and dead animal matter. Food is caught primarily by the tentacles, but in addition the mouth may open widely, permitting mesenteries with their nematocyst-laden margins to trap food. Recent studies of the tentacles of *Balanophyllia elegans* (3.42), *Corynactis californica* (3.41), *Metridium senile* (3.37), and related forms show that organelles called spirocysts, which resemble nematocysts and occur along with them, evert to produce tangles of sticky tubules. Spirocysts may be important in capturing prey, in attaching coelenterates to the substratum, or both.

Unlike many tropical reef corals, *Balanophyllia* bears no symbiotic algae in its tissues. It can, however, absorb dissolved organic carbon exudates liberated by brown algae such as *Macrocystis*.

In reproduction, the eggs are released into the parent's gastrovascular cavity, where they are fertilized and undergo development to the planula stage. In central California, wormlike orange planula larvae are liberated year-round but mainly in winter (Y. Fadlallah, pers. comm.). These settle and metamorphose into tiny polyps, which then secrete a skeleton; the whole process may be watched in the laboratory.

See Andrews (1945), Durham (1949), Fankboner (1976), Mariscal (1974a,b,c), Mariscal, Bigger & McLean (1976), Mariscal & McLean (1976), Mariscal, McLean & Hand (1977), McLean (1962), and Yonge (1932).

Phylum Cnidaria / Class Anthozoa / Subclass Zoantharia / Order Ceriantharia
Family Cerianthidae

3.43 **Pachycerianthus fimbriatus** McMurrich, 1910

Fairly common in soft muddy bottoms of bays and harbors and on protected sandy bottoms of the outer coast, low intertidal zone and subtidal waters in southern California; only from subtidal waters in northern California; distribution poorly known.

Large individuals to 35 cm long when expanded, inhabiting a tough, slippery, black, secreted tube projecting above mud surface, into which animal withdraws very rapidly when disturbed; tentacular crown with two sets of translucent whitish to golden-brown tentacles, the inner circle usually held over the mouth, the outer circle projecting upward or outward.

Several species of the order Ceriantharia occur along the Pacific coast. Superficially they resemble sea anemones, but they show important anatomical differences, including possession of an anal pore at the aboral end of the body. Little is known of the biology of this particular species, but related forms have been examined with regard to the nature of the nematocysts, the muscular system, the nervous system, and the response mechanisms. Their responses, and the rates at which major nerve tracts conduct impulses (up to 1.3 m per second), are unusually rapid for a coelenterate. In many species the tentacles fluoresce.

In Monterey Bay, the tentacles of *Pachycerianthus* are fed upon by the nudibranch *Dendronotus iris* (14.53). The cerianthid usually reacts by withdrawing into its tube; the nudibranch may be drawn into the tube as well, where it may complete its feeding on the tentacles before emerging. The cerianthids do not seem to be seriously injured by such predation and may live for 10 years or more.

Examination of the process of tube building in one ceriantharid reveals that the tube is formed largely by eversible organelles similar to spirocysts and nematocysts but differing from both; they have been named ptychocysts.

See Arai (1965a,b, 1971, 1972), Horridge (1958), Mariscal, Conklin & Bigger (1977), Moore (1927), Ross & Horridge (1957), Torrey & Kleeberger (1909), and Wobber (1970).

Literature Cited

Aizu, S. 1967. Transformation of ectoderm to endoderm in *Obelia*. Sci. Rep. Tôhôku Univ. 33: 375–82.

Akin, G. C., and J. R. Akin. 1967. The effects of trypsin in regeneration inhibitors in *Tubularia*. Biol. Bull. 133: 82–89.

Alexander, R. M. 1962. Visco-elastic properties of the body wall of sea anemones. J. Exper. Biol. 39: 373–86.

———. 1964. Visco-elastic properties of the mesoglea of jellyfish. J. Exper. Biol. 41: 363–69.

Anderson, P. A. V., and J. F. Case. 1975. Electrical activity associated with luminescence and other colonial behavior in the pennatulid *Renilla köllikeri*. Biol. Bull. 149: 80–95.

Anderson, P. A. V., and G. O. Mackie. 1977. Electrically coupled, photosensitive neurons control swimming in jellyfish. Science 197: 186–88.

Andrews, H. L. 1945. The kelp beds of the Monterey region. Ecology 26: 24–37.

Arai, M. N. 1965a. The ceriantharian nervous system. Amer. Zool. 5: 424–29.

———. 1965b. Contractile properties of a preparation of the column of *Pachycerianthus torreyi* (Anthozoa). Comp. Biochem. Physiol. 14: 323–37.

———. 1971. *Pachycerianthus* (Ceriantharia) from British Columbia and Washington. J. Fish. Res. Bd. Canada 28: 1677–80.

———. 1972. The muscular system of *Pachycerianthus fimbriatus*. Canad. J. Zool. 50: 311–17.

Atoda, K. 1973. Pedal laceration of the sea anemone, *Haliplanella luciae*. Publ. Seto Mar. Biol. Lab. 20: 299–313.

Ball, E. E. 1973. Electrical behavior in the solitary hydroid *Corymorpha palma*. I. Spontaneous activity in whole animals and in isolated parts. Biol. Bull. 145: 223–42.

Ball, E. E., and J. F. Case. 1973. Electrical activity and behavior in the solitary hydroid *Corymorpha palma*. II. Conducting systems. Biol. Bull. 145: 243–64.

Batham, E. J. 1960. The fine structure of epithelium and mesoglea in a sea anemone. Quart. J. Microscop. Sci. 101: 481–85.

Batham, E. J., and C. F. A. Pantin. 1950a. Inherent activity in the sea anemone, *Metridium senile*. J. Exper. Biol. 27: 290–301.

———. 1950b. Muscular and hydrostatic action in the sea-anemone, *Metridium senile*. J. Exper. Biol. 27: 264–89.

———. 1950c. Phases of activity in the sea-anemone, *Metridium senile*. J. Exper. Biol. 27: 377–99.

———. 1951. The organization of the muscular system of *Metridium senile*. Quart. J. Microscop. Sci. 92: 27–54.

———. 1954. Slow contraction and its relation to spontaneous activity in the sea-anemone *Metridium senile*. J. Exper. Biol. 31: 84–103.

Batham, E. J., C. F. A. Pantin, and E. A. Robson. 1960. The nerve-net of the sea-anemone *Metridium senile*: The mesenteries and the column. Quart. J. Microscop. Sci. 101: 487–510.

Belcik, F. P. 1968. Metabolic rate in certain sea anemones. Turtox News 46: 178–81.

Beloussov, L. V. 1973. Growth and morphogenesis of some marine Hydrozoa according to histological data and time-lapse studies. Publ. Seto Mar. Biol. Lab. 20: 315–66.

Berrill, M. 1962. The biology of three New England Stauromedusae, with a description of a new species. Canad. J. Zool. 40: 1249–62.

———. 1963. Comparative functional morphology of the Stauromedusae. Canad. J. Zool. 41: 741–52.

Berrill, N. J. 1948. A new method of reproduction in *Obelia*. Biol. Bull. 95: 94–99.

———. 1949a. Developmental analysis of Scyphomedusae. Biol. Rev. 24: 393–410.

———. 1949b. Growth and form of calyptoblastic hydroids. I. Comparison of a campanulid, campanularian, sertularian and plumularian. J. Morphol. 85: 297–336.

———. 1949c. The polymorphic transformations of *Obelia*. Quart. J. Microscop. Sci. 90: 235–64.

———. 1952a. Growth and form in gymnoblastic hydroids. IV. Relative growth in *Eudendrium*. J. Morphol. 90: 20–32.

———. 1952b. Growth and form in gymnoblastic hydroids. V. Growth cycle in *Tubularia*. J. Morphol. 90: 583–602.

———. 1953a. Growth and form in gymnoblastic hydroids. VI. Polymorphism within the hydractinids. J. Morphol. 92: 241–72.

———. 1953b. Growth and form in gymnoblastic hydroids. VII. Growth and reproduction in *Syncoryne* and *Coryne*. J. Morphol. 92: 273–302.

Bertsch, H. 1968. Effect of feeding by *Armina californica* on the bioluminescence of *Renilla köllikeri*. Veliger 10: 440–41.

Bieri, R. 1959. Dimorphism and size distribution in *Velella* and *Physalia*. Nature 184: 1333.

———. 1961. Post-larval food of the pelagic coelenterate *Velella lata*. Pacific Sci. 15: 553–56.

———. 1966. Feeding preferences and rates of the snail, *Ianthina prolongata*, the barnacle *Lepas anserifera*, the nudibranchs *Glaucus atlanticus* and *Fiona pinnata*, and the food web in the marine neuston. Publ. Seto Mar. Biol. Lab. 14: 161–70.

Bieri, R., and D. H. Krinsley. 1958. Trace elements in the pelagic coelenterate, *Velella lata*. J. Mar. Res. 16: 246–54.

Block, J. H. 1974. Marine sterols from some gorgonians. Steroids 23: 421–24.

Boschma, H. 1953. The Stylasterina of the Pacific. Zool. Meded. 32: 165–84.

Boyd, M. J. 1971. Fouling community structure and development in Bodega Harbor, California. Doctoral thesis, Zoology, University of California, Davis. 191 pp.

Brinckmann, A. 1964. Observations on the structure and development of the medusa of *Velella velella* (Linné, 1758). Vidensk. Medd. Naturh. Foren. 126: 327–36.

Brinkmann-Voss, A. 1977. The hydroid of *Polyorchis penicillatus* (Eschscholtz) (Polyorchidae, Hydrozoa, Cnidaria). Canad. J. Zool. 55: 93–96.

Broch, H. 1942. Investigations of the Stylasteridae (Hydrocorals). Skrift. Norske. Videns-Akad. Oslo, I. Mat. naturv; Kl. ann. 1942: 1–113.

Broughton, P. S. 1975. The symbiotic relationship between the hermit crab *Pagurus samuelis* (Arthropoda: Decapoda) and the sea anemones *Anthopleura elegantissima* and *A. xanthogrammica* (Cnidaria: Actiniaria). Research paper, Biol. 175h, library, Hopkins Marine Station of Stanford University, Pacific Grove, Calif. 20 pp.

Browning, R. F., and C. F. Lytle. 1967. *Hydractinia* as a food collecting device for hermit crabs. Amer. Zool. 7: 770 (abstract).

Buchsbaum, V. M. 1968. Behavioral and physiological responses to light by the sea anemone *Anthopleura elegantissima* as related to its algal symbiotes. Doctoral thesis, Biological Sciences, Stanford University, Stanford, Calif. 123 pp.

Buck, J. 1973. Bioluminescent behavior in *Renilla*. I. Colonial responses. Biol. Bull. 144: 19–42.

Buck, J., and F. E. Hanson, Jr. 1967. Zooid response in *Renilla*. Biol. Bull. 133: 459.

Bullock, T. H. 1965. Coelenterata and Ctenophora, pp. 459–534, *in* T. H. Bullock and G. A. Horridge, Structure and function in the nervous system of invertebrates, vol. I. San Francisco: Freeman. 798 pp.

Burnett, A. L., W. Sindelar, and N. Diehl. 1967. An examination of polymorphism in the hydroid *Hydractinia echinata*. J. Mar. Biol. Assoc. U.K. 47: 645–58.

Burnett, J. W. 1971. An ultrastructural study of the nematocytes of the polyp of *Chrysaora quinquecirrha*. Chesapeake Sci. 12: 225–30.

Burnett, J. W., and R. Goldner. 1969. The effects of *Chrysaora quinquecirrha* (sea nettle) toxin on the rat cardiovascular system. Proc. Soc. Exper. Biol. Med. 132: 353–56.

———. 1970a. Effect of *Chrysaora quinquecirrha* (sea nettle) toxin on rat nerve and muscle. Toxicon 8: 179–81.

———. 1970b. Partial purification of sea nettle (*Chrysaora quinquecirrha*) nematocyst toxin. Proc. Soc. Exper. Biol. Med. 133: 978–81.

Burnett, J. W., and J. S. Sutton. 1969. The fine structural organization of the sea nettle fishing tentacles. J. Exper. Zool. 172: 335–48.

Calder, D. R. 1971. Nematocysts of *Aurelia*, *Chrysaora* and *Cyanea* and their utility in identification. Trans. Amer. Microscop. Soc. 90: 269–74.

Campbell, R. D. 1967. Cell proliferation and morphological patterns in the hydroids *Tubularia* and *Hydractinia*. J. Embryol. Exper. Morphol. 17: 607–16.

———. 1968. Holdfast movement in the hydroid *Corymorpha palma*: Mechanism of elongation. Biol. Bull. 134: 26–34.

———. 1969. A statocyst lacking cilia in the hydroid polyp *Corymorpha palma*. Amer. Zool. 9: 1140.

———. 1974. Cnidaria, pp. 133–99, *in* A. C. Giese and J. S. Pearse, eds., Reproduction of marine invertebrates, vol. 1. New York: Academic Press. 546 pp.

Campbell, R. D., and F. Campbell. 1968. *Tubularia* regeneration: Radial organization of tentacles, gonophores and endoderm. Biol. Bull. 134: 245–51.

Cargo, D. G., and L. P. Schultz. 1966. Notes on the biology of the sea nettle, *Chrysaora quinquecirrhia*, in Chesapeake Bay. Chesapeake Sci. 7: 95–100.

————. 1967. Further observations on the biology of the sea nettle and jellyfishes in Chesapeake Bay. Chesapeake Sci. 8: 209–20.

Carl, G. C. 1948. An unusual abundance of *Velella velella* Linné (Coelenterata: Siphonophorae) in inshore waters. Canad. Field Natur. 62: 158–59.

Carlgren, O. 1931. Zur Kenntnis der Actiniaria *Abasilaria*. Ark. Zool. 23A: 1–48.

Carlisle, J. G., Jr., C. H. Turner, and E. E. Ebert. 1964. Artificial habitat in the marine environment. Calif. Dept. Fish & Game, Fish Bull. 124: 1–93.

Case, J., and J. Morin. 1966. Glutamate suppression and neuroeffector processes in a coelenterate. Amer. Zool. 6: 525 (abstract).

Chao, C. 1975. Inter-specific aggression between three sympatric anemones, *Anthopleura elegantissima*, *Metridium senile*, and *Corynactis californica*, at the Monterey wharf. Research paper, Biol. 175h, library, Hopkins Marine Station of Stanford University, Pacific Grove, Calif. 13 pp.

Chapman, D. M. 1965. Coordination in a scyphistoma. Amer. Zool. 5: 455–64.

————. 1966. Evolution of the scyphistoma, pp. 51–75, *in* Rees (1966a).

————. 1968. A new type of muscle cell from the subumbrella of *Obelia*. J. Mar. Biol. Assoc. U.K. 48: 667–88.

————. 1974. Cnidarian histology, pp. 1–92, *in* Muscatine & Lenhoff (1974).

Chapman, G. 1953. Studies of the mesoglea of coelenterates. I. Histology and chemical properties. Quart J. Microscop. Sci. 94: 155–76.

————. 1959. The mesoglea of *Pelagia noctiluca*. Quart. J. Microscop. Sci. 100: 599–610.

Chia, F.-S. and J. G. Spaulding. 1972. Development and juvenile growth of the sea anemone, *Tealia crassicornis*. Biol. Bull. 142: 206–18.

Child, C. M. 1928. Axial development in aggregates of dissociated cells from *Corymorpha palma*. Physiol. Zool. 1: 419–61.

————. 1933. Reconstitution in *Haliclystus auricula* Clark. Sci. Rep. Tôhôku Imp. Univ., (4), Biology 8: 75–106.

Childress, L. 1970. Intraspecific aggression and its relation to the distribution pattern of the clonal sea anemone, *Anthopleura elegantissima*. Doctoral thesis, Biological Sciences, Stanford University, Stanford, Calif. 123 pp.

Christensen, H. E. 1967. Ecology of *Hydractinia echinata* (Fleming) (Hydroidea, Athecata). I. Feeding biology. Ophelia 4: 245–75.

Clark, E. D., and D. J. Kimeldorf. 1970. Tentacle response of the sea anemone *Anthopleura xanthogrammica* to ultraviolet and visible radiations. Nature 227: 856–57.

Clark, W. H., and W. C. Dewel. 1974. The structure of the gonads, gametogenesis, and sperm-egg interaction in the Anthozoa. Amer. Zool. 14: 495–510.

Coe, W. R. 1932. Season of attachment and rate of growth of sedentary marine organisms at the pier of the Scripps Institute of Oceanography, La Jolla, California. Bull. Scripps Inst. Oceanogr., Tech. Ser. 3: 37–86.

Coleman, D. E., and C. Teague. 1973. Sea pansies. Pacific Discov. 26: 28–29.

Cornelius, P. F. S. 1975. The hydroid species of *Obelia* (Coelenterata, Hydrozoa: Campanulariidae), with notes on the medusa stage. Bull. Brit. Mus. (Natur. Hist.), Zoology 28: 249–93.

Costello, D. P., and C. Henley. 1971. Methods for obtaining and handling marine eggs and embryos. 2nd ed. Marine Biological Laboratory, Woods Hole, Mass. 247 pp.

Crowell, S. 1953. The regression-replacement cycle of hydranths of *Obelia* and *Campanularia*. Physiol. Zool. 26: 319–27.

————. ed. 1965. Behavioral physiology of coelenterates. Amer. Zool. 5: 335–589.

————. 1967. Coelenterates, pp. 257–64, *in* F. H. Wilt and N. K. Wessells, eds., Methods in developmental biology. New York: Crowell. 813 pp.

Custance, D. R. N. 1966. The effect of a sudden rise in temperature on the strobilae of *Aurelia aurita*. Experientia 22: 588–89.

Darby, R. L. 1965. A laboratory study of growth and timing of sexuality in the colonial marine hydroid *Obelia longissima*. Amer. Zool. 5: 616 (abstract).

Davidson, M. E., and N. J. Berrill. 1948. Regeneration of primordia and developing hydranths of *Tubularia*. J. Exper. Zool. 107: 465–78.

Davis, C. 1962. One equals two: A sea anemone achieves its plurality in singular fashion. Natur. Hist. 71: 61–63.

Day, R. M., and L. G. Harris. 1978. Selection and turnover of coelenterate nematocysts in some aeolid nudibranchs. Veliger 21: 104–9.

Dayton, P. K. 1973. Two cases of resource partitioning in an intertidal community: Making the right prediction for the wrong reason. Amer. Natur. 107: 662–70.

Dillon, T. M. 1977. Effects of acute change in temperature and salinity on pulsation rates in ephyrae of the scyphozoan *Aurelia aurita*. Mar. Biol. 42: 31–35.

Dunkelberger, D., and N. Watabe. 1972. Electron microscope study of the coenenchyme in the pennatulid colony *Renilla reniformis*, with special emphasis on spicule formation. Amer. Zool. 12: 716–17.

Dunn, D. F. 1972. Natural history of the sea anemone *Epiactis prolifera* Verrill, 1869, with special reference to its reproductive biology. Doctoral thesis, Zoology, University of California, Berkeley. 187 pp.

————. 1975a. Gynodioecy in an animal. Nature 253: 528–29.

————. 1975b. Reproduction of the externally brooding sea anemone *Epiactis prolifera* Verrill, 1869. Biol. Bull. 148: 199–218.

_____. 1977a. Dynamics of external brooding in the sea anemone *Epiactis prolifera*. Mar. Biol. 39: 41–47.

_____. 1977b. Locomotion by *Epiactis prolifera* (Coelenterata: Actiniaria). Mar. Biol. 39: 67–70.

Dunn, D. F., M. Kashiwagi, and T. R. Norton. 1975. The origin of antitumor activity in sea anemones. Comp. Biochem. Physiol. 50C: 133–35.

Durham, J. W. 1949. Ontogenetic stages of some simple corals. Univ. Calif. Publ. Geol. 28: 137–72.

Eakin, R. M., and J. A. Westfall. 1962. Fine structure of photoreceptors in the hydromedusan, *Polyorchis penicillatus*. Proc. Nat. Acad. Sci. 48: 826–33.

_____. 1964. Further observations on the fine structure of some invertebrate eyes. Z. Zellforsch. 62: 310–32.

Edwards, C. 1966. *Velella velella* (L.): The distribution of its dimorphic forms in the Atlantic Ocean and the Mediterranean, with comments on its nature and affinities, pp. 283–96, *in* H. Barnes, ed., Some contemporary studies in marine science. London: Allen & Unwin. 716 pp.

Elder, H. Y. 1973. Distribution and functions of elastic fibers in the invertebrates. Biol. Bull. 144: 43–63.

Fager, E. W. 1964. Marine sediments: Effects of a tube-building polychaete. Science 143: 356–59.

_____. 1968. A sand-bottom epifaunal community of invertebrates in shallow water. Limnol. Oceanogr. 13: 448–64.

Fankboner, P. V. 1976. Accumulation of dissolved carbon by the solitary coral *Balanophyllia elegans*—an alternative nutritional pathway, pp. 111–16, *in* Mackie (1976).

Fields, W. G., and G. O. Mackie. 1971. Evolution of the Chondrophora: Evidence from behavioural studies on *Velella*. J. Fish. Res. Bd. Canada 28: 1595–1602.

Filice, F. P. 1958. Invertebrates from the estuarine portion of San Francisco Bay and some factors influencing their distributions. Wasmann J. Biol. 16: 159–211.

Fisher, W. K. 1931. California hydrocorals. Ann. Mag. Natur. Hist. (10) 8: 391–99.

_____. 1938. Hydrocorals of the North Pacific Ocean. Proc. U.S. Nat. Mus. 84: 493–554.

Florkin, M., and B. T. Scheer. 1968. Chemical Zoology. 2. Porifera, Coelenterata and Platyhelminthes. New York: Academic Press. 639 pp.

Ford, C. E. 1964. Reproduction in the aggregating sea anemone, *Anthopleura elegantissima*. Pacific Sci. 18: 138–45.

Francis, L. 1973a. Clone specific segregation in the sea anemone *Anthopleura elegantissima*. Biol. Bull. 144: 64–72.

_____. 1973b. Intraspecific aggression and its effect on the distribution of *Anthopleura elegantissima* and some related sea anemones. Biol. Bull. 144: 73–92.

_____. 1976. Social organization within clones of the sea anemone *Anthopleura elegantissima*. Biol. Bull. 150: 361–76.

Fraser, C. M. 1937. Hydroids of the Pacific coast of Canada and the United States. University of Toronto Press. 207 pp.

_____. 1946. Distribution and relationship in American hydroids. University of Toronto Press. 464 pp.

Fredericks, C. 1976. Oxygen as a limiting factor in phototaxis and in interclonal spacing of *Anthopleura elegantissima*. Mar. Biol. 38: 25–28.

Fritchman, H. K., II. 1974. The planula of the stylasterine hydrocoral *Allopora petrograpta* Fisher: Its structure, metamorphosis and development of the primary cyclosystem. Proc. 2nd Internat. Coral Reef Symp. 2: 245–58.

Gee, W. 1913. Modifiability in the behavior of the California shore-anemone, *Cribrina xanthogrammica* Brandt. J. Anim. Behav. 3: 305–28.

Gemmill, J. F. 1920. The development of the sea anemones *Metridium dianthus* (Ellis) and *Adamsia palliata* (Bohad.). Phil. Trans. Roy. Soc. London B209: 351–75.

Giese, A. C. 1966. Lipids in the economy of marine invertebrates. Physiol. Rev. 46: 244–98.

Gladfelter, W. B. 1970. Swimming in medusae (Cnidaria: Hydrozoa and Scyphozoa): A study in functional morphology and behavior. Doctoral thesis, Biological Sciences, Stanford University, Stanford, Calif. 235 pp.

_____. 1972. Structure and function of the locomotor system of *Polyorchis montereyensis*. Helgoländer Wiss. Meeresunters. 23: 38–79.

_____. 1973. A comparative analysis of the locomotor system of medusoid Cnidaria. Helgoländer Wiss. Meeresunters. 25: 228–72.

Goodwin, M. H., and M. Telford. 1971. The nematocyst toxin of *Metridium*. Biol. Bull. 140: 389–99.

Gosline, J. M. 1971a. Connective tissue mechanics of *Metridium senile*. I. Structural and compositional aspects. J. Exper. Biol. 55: 763–74.

_____. 1971b. Connective tissue mechanics of *Metridium senile*. II. Visco-elastic properties and macromolecular model. J. Exper. Biol. 55: 775–95.

Gotshall, D. W., J. G. Smith, and A. Holbart. 1965. Food of the blue rockfish *Sebastodes mystinus*. Calif. Fish & Game 51: 147–62.

Graham, H. W., and H. Gay. 1945. Season of attachment and growth of sedentary marine organisms at Oakland, California. Ecology 26: 375–86.

Grigg, R. W. 1970. Ecology and population dynamics of the gorgonians *Muricea californica* and *Muricea fruticosa*. Doctoral thesis, University of California, San Diego. 261 pp.

_____. 1972. Orientation and growth form of the sea fans. Limnol. Oceanogr. 17: 185–92.

_____. 1974. Growth rings: Annual periodicity in two gorgonian corals. Ecology 55: 876–81.

————. 1976. Age structure of a longevous coral: A relative index of habitat suitability and stability. Amer. Natur. 109: 647–57.

————. 1977. Population dynamics of two gorgonian corals. Ecology 58: 278–90.

Grimstone, A. V., R. W. Horne, C. F. A. Pantin, and E. A. Robson. 1958. The fine structure of the mesenteries of the sea-anemone *Metridium senile*. Quart. J. Microscop. Sci. 99: 523–40.

Gwilliam, G. F. 1956. Studies on west coast Stauromedusae. Doctoral thesis, Zoology, University of California, Berkeley. 191 pp.

————. 1960. Neuromuscular physiology of a sessile scyphozoan. Biol. Bull. 119: 454–73.

Haderlie, E. C. 1968. Marine fouling organisms in Monterey harbor. Veliger 10: 327–41.

————. 1969. Marine fouling and boring organisms in Monterey harbor. 2. Second year of investigation. Veliger 12: 182–92.

————. 1971. Marine fouling and boring organisms at 100 feet depth in open water of Monterey Bay. Veliger 13: 249–60.

Haderlie, E. C., J. C. Mellor, C. S. Minter III, and G. C. Booth. 1974. The sublittoral benthic fauna and flora off Del Monte Beach, Monterey, California. Veliger 17: 185–204.

Hale, L. J. 1973. The pattern of growth of *Clytia johnstoni*. J. Embryol. Exper. Morphol. 29: 283–309.

Hamner, W. M., and R. M. Jenson. 1974. Growth, degrowth, and irreversible cell differentiation in *Aurelia aurita*. Amer. Zool. 14: 833–49.

Hand, C. H. 1954. The sea anemones of central California. Part 1. The corallimorpharian and athenarian anemones. Wasmann J. Biol. 12: 345–75.

————. 1955a. The sea anemones of central California. Part 2. The endomyarian and mesomyarian anemones. Wasmann J. Biol. 13: 37–99.

————. 1955b. The sea anemones of central California. Part 3. The acontiarian anemones. Wasmann J. Biol. 13: 189–251.

————. 1966. On the evolution of the Actiniaria, pp. 135–46, *in* Rees (1966a).

————. 1975. Class Anthozoa, pp. 85–94, *in* R. I. Smith and J. T. Carlton, eds., Light's manual: Intertidal invertebrates of the central California coast. 3rd ed. Berkeley and Los Angeles: University of California Press. 716 pp.

Hardy, A. C. 1956. The open sea, its natural history: The world of plankton. Boston, Mass.: Houghton Mifflin. 335 pp.

Harris, L. G. 1971. Ecological observations on a New England nudibranch-anemone association. Amer. Zool. 11: 699 (abstract).

Hausman, L. A. 1919. The orange striped anemone (*Sagartia luciae* Verrill): An ecological study. Biol. Bull. 37: 363–71.

Herring, P. J. 1967. The pigments of plankton at the sea surface. Symp. Zool. Soc. London 19: 215–35.

Hewatt, W. G. 1937. Ecological studies on selected marine intertidal communities of Monterey Bay, California. Amer. Midl. Natur. 18: 161–206.

Heymans, C., and A. R. Moore. 1924. Luminescence in *Pelagia noctiluca*. J. Gen. Physiol. 6: 273–80.

Hickson, S. J. 1915. Some Alcyonaria and a *Stylaster* from the west coast of North America. Proc. Zool. Soc. London 1915: 541–57.

Hinsch, G. W. 1974. Comparative ultrastructure of cnidarian sperm. Amer. Zool. 14: 457–65.

Hoffman, D. L. 1967. Symbiosis between shrimp and anemones. Amer. Zool. 7: 205 (abstract).

Horridge, A. 1954. Observations on the nerve fibers of *Aurelia aurita*. Quart. J. Microscop. Sci. 95: 85–92.

————. 1956. The nervous system of the ephyra larva of *Aurelia aurita*. Quart. J. Microscop. Sci. 97: 59–73.

————. 1958. The co-ordination of the responses of *Cerianthus* (Coelenterata). J. Exper. Biol. 35: 369–82.

————. 1966. Some recently discovered underwater vibration receptors in invertebrates, pp. 395–405, *in* H. Barnes, ed., Some contemporary studies in marine science. London: Allen & Unwin. 716 pp.

Howe, N. R. 1976a. Behavior evoked by an alarm pheromone in the sea anemone *Anthopleura elegantissima*. Doctoral thesis, Biological Sciences, Stanford University, Stanford, Calif. 99 pp.

————. 1976b. Behavior of sea anemones evoked by the alarm pheromone anthopleurine. J. Compar. Physiol. 107: 67–76.

Howe, N. R., and Y. M. Sheikh. 1975. Anthopleurine: A sea anemone alarm pheromone. Science 189: 386–88.

Huang, C. L., and G. N. Mir. 1972. Toxicological and pharmacological properties of sea pansy *Renilla mulleri*. J. Pharmacol. Sci. 60: 1620–22.

Human, V. L. 1973. Albinism in three species of marine invertebrates from southern California. Calif. Fish & Game 59: 89–92.

Hyman, L. 1940. The invertebrates: Protozoa through Ctenophora. Vol. 1. New York: McGraw-Hill. 726 pp.

————. 1959. The invertebrates: Smaller coelomate groups. Vol. 5. New York: McGraw-Hill. 783 pp.

Ivester, S., and D. Dunkelberger. 1971. Ultrastructure of the autozooid of the sea pansy, *Renilla reniformis* (Pallas). Amer. Zool. 11: 695–96.

Ivker, F. 1967. Localization of tissue incompatibility of the overgrowth reaction in *Hydractinia echinata*. Biol. Bull. 143: 162–74.

————. 1972a. Hierarchy of histo-incompatibility in *Hydractinia echinata*. Biol. Bull. 143: 162–74.

————. 1972b. The role of histo-incompatibility in establishing a zone of sterility between two colonies of *Hydractinia echinata* on a single shell. Amer. Zool. 12: 705 (abstract).

Jennison, B. L. 1975. The effect of increased temperature on reproduction in the sea anemone *Anthopleura elegantissima*. Amer. Zool. 15: 787.

Johnson, L. L., and J. M. Shick. 1977. Effects of fluctuating temperature and immersion on asexual reproduction in the intertidal sea anemone *Haliplanella luciae* (Verrill) in laboratory culture. J. Exper. Mar. Biol. Ecol. 28: 141–49.

Josephson, R. K. 1961. The response of a hydroid to weak water-borne disturbances. J. Exper. Biol. 38: 17–27.

———. 1965. Pacemakers and conducting systems in the hydroid *Tubularia*. Amer. Zool. 5: 483–90.

———. 1974a. Factors affecting muscle activation in the hydroid *Tubularia*. Biol. Bull. 147: 594–607.

———. 1974b. The strategies of behavioral control in a coelenterate. Amer. Zool. 14: 905–15.

Josephson, R. K., and J. Uhrich. 1969. Inhibition of pacemaker systems in the hydroid *Tubularia*. J. Exper. Biol. 50: 1–14.

Kastendiek, J. 1975. The behavior, distribution, and predatory-prey interactions of *Renilla köllikeri*. Doctoral thesis, Zoology, University of California, Los Angeles. 235 pp.

Kemp, N. E., and N. J. Berrill. 1974. Fine structure of epidermis and perisarc in growing stolons of *Obelia* and *Campanularia*. Amer. Zool. 14: 1301 (abstract).

Kittredge, J. S., D. G. Simonson, E. Roberts, and B. Jelinek. 1962. Free amino acids of marine invertebrates, pp. 176–86, *in* J. T. Holden, ed., Amino acid pools: Distribution, formation and function of free amino acids. Amsterdam: Elsevier. 815 pp.

Kittredge, J. S., A. F. Isbell, and R. R. Hughes. 1967. Isolation and characterization of the N-methyl derivatives of 2-aminoethylphosphoric acid from the sea anemone, *Anthopleura xanthogrammica*. Biochemistry 6: 289–95.

Knight, D. P. 1968. Cellular basis for quinone tanning of the perisarc in the thecate hydroid *Campanularia* (*Obelia*) *flexuosa*. Nature 218: 585–86.

Knudsen, J. W. 1963. Notes on the barnacle *Lepas fascicularis* found attached to the jellyfish *Velella*. Bull. So. Calif. Acad. Sci. 62: 130–31.

Koehl, M. A. R. 1976. Mechanical design in sea anemones, pp. 23–31, *in* Mackie (1976).

Kramp, P. L. 1961. Synopsis of medusae of the world. J. Mar. Biol. Assoc. U.K. 40: 1–469.

Kuwabara, R., S. Sato, and N. Noguchi. 1969. Ecological studies on the medusa *Aurelia aurita*. 1. Distribution of *Aurelia* patches in the northeast region of Tokyo Bay in summer 1966 and 1967. Bull. Jap. Soc. Sci. Fish. 35: 156–62.

LeLoup, E. 1929. Recherches sur l'anatomie et le développement de *Velella spirans* Forsk. Arch. Biol., Paris, 39: 397–478.

Lenhoff, H. M. 1965. Mechanical stimulation of feeding in *Epiactis prolifera*. Nature 207: 1003.

Lenhoff, H. M., and W. F. Loomis. 1961. The biology of *Hydra* and of some other coelenterates. Coral Gables, Fla.: University of Miami Press. 467 pp.

Lentz, T. L. 1966. The cell biology of *Hydra*. New York: Wiley. 199 pp.

Lindberg, W. J. 1976. Starvation behavior of the sea anemones, *Anthopleura xanthogrammica* and *Metridium senile*. Biologist 58: 81–88.

Linstedt, K. H. 1971. Biphasic feeding response in a sea anemone: Control by asparagine and glutathione. Science 173: 333–34.

Little, E. V. 1914. The structure of the ocelli of *Polyorchis penicillata*. Univ. Calif. Publ. Zool. 11: 307–28.

Liu, C. K., and N. J. Berrill. 1948. Gonophore formation and germ cell origin in *Tubularia*. J. Morphol. 83: 39–60.

Luch, J. M., and E. W. Martin. 1967. Evidence for osmoregulatory activity by scyphistoma of *Aurelia aurita*. Amer. Zool. 7: 734 (abstract).

Lyke, E. B. 1965. The histology of the sea pansies, *Renilla reniformis* (Pallas) and *Renilla köllikeri* (Pfeffer), with a note on the fine structure of the latter species. Doctoral thesis, Zoology, University of Wisconsin, Madison. 247 pp.

———. 1972. An ultrastructural examination of spermatogenesis in the hydromedusa, *Polyorchis penicillatus*. Amer. Zool. 12: 704 (abstract).

MacGinitie, G. E. 1935. Ecological aspects of a California marine estuary. Amer. Midl. Natur. 16: 629–725.

———. 1938. Notes on the natural history of some marine animals. Amer. Midl. Natur. 19: 207–19.

MacGinitie, G. E., and N. MacGinitie. 1968. Natural history of marine animals. 2nd ed. New York: McGraw-Hill. 523 pp.

Mackie, G. O. 1959. The evolution of the Chondrophora (Siphonophora: Disconanthae): New evidence from behavioral studies. Trans. Roy. Soc. Canada 53: 7–20.

———. 1960. The structure of the nervous system in *Velella*. Quart. J. Microscop. Sci. 101: 119–33.

———. 1961. Factors affecting the distribution of *Velella velella* (Chondrophora). Internat. Rev. Ges. Hydrobiol. 47: 26–32.

———. 1966. Growth of the hydroid *Tubularia* in culture, pp. 397–410, *in* Rees (1966a).

———. 1971. Neurological complexity in medusae: A report of central nervous organization in *Sarsia*, pp. 269–80, *in* Actas del I simposio internacional de zoofilogenia. University of Salamanca.

———. ed. 1976. Coelenterate ecology and behavior. New York: Plenum. 744 pp.

Mackie, G. O., and L. M. Passano. 1968. Epithelial conduction in hydromedusae. J. Gen. Physiol. 52: 600–621.

Mariscal, R. N. 1965. Observations on acclimation behavior in the symbiosis of anemone fish and sea anemones. Amer. Zool. 5: 694 (abstract).

———. 1969a. An experimental analysis of the protection of *Amphiprion xanthurus* Cuvier and Valenciennes and some other anemone fishes from sea anemones. J. Exper. Mar. Biol. Ecol. 4: 134–49.

———. 1969b. The protection of the anemone fish *Amphiprion xanthurus*, from the anemone, *Stoichactis kenti*. Experientia 25: 1114.

———. 1970a. A field and laboratory study of the symbiotic behavior of fishes and sea anemones from the tropical Indo-Pacific. Univ. Calif. Publ. Zool. 91: 1–33.

———. 1970b. The nature of the symbiosis between Indo-Pacific anemone fishes and sea anemones. Mar. Biol. 6: 58–65.

———. 1973. The control of nematocyst discharge during feeding by sea anemones. Publ. Seto Mar. Biol. Lab. 20: 695–702.

———. 1974a. Nematocysts, pp. 129–78, *in* Muscatine & Lenhoff (1974).

———. 1974b. Scanning electron microscopy of the sensory epithelia and nematocysts of corals and a corallimorpharian sea anemone. Proc. 2nd Internat. Coral Reef Symp. 1: 519–32.

———. 1974c. Scanning electron microscopy of the sensory surface of the tentacles of sea anemones and corals. Z. Zellforsch. 147: 149–56.

Mariscal, R. N., C. H. Bigger, and R. B. McLean. 1976. The form and function of cnidarian spirocysts. 1. Ultrastructure of the capsule exterior and relationship to the tentacle sensory surface. Cell Tissue Res. 168: 465–74.

Mariscal, R. N., E. J. Conklin, and C. H. Bigger. 1977. The ptychocyst, a major new category of cnida used in tube construction by a cerianthid anemone. Biol. Bull. 152: 392–405.

Mariscal, R. N., and R. B. McLean. 1976. The form and function of cnidarian spirocysts. 2. Ultrastructure of the capsule tip and wall and mechanism of discharge. Cell Tissue Res. 169: 313–21.

Mariscal, R. N., R. B. McLean, and C. Hand. 1977. The form and function of cnidarian spirocysts. 3. Ultrastructure of the thread and the function of spirocysts. Cell Tissue Res. 178: 427–33.

Martin, E. J. 1963. Toxicity of dialyzed extracts of some California anemones (Coelenterata). Pacific Sci. 17: 302–4.

———. 1968. Specific antigens released into sea water by contracting anemones (Coelenterata). Comp. Biochem. Physiol. 25: 169–76.

Mauzey, K. P., C. Birkeland, and P. Dayton. 1968. Feeding behavior of asteroids and escape responses of their prey in the Puget Sound region. Ecology 49: 603–19.

Mayer, A. G. 1910. Medusae of the world, 3 vols. Washington, D.C.: Carnegie Institution. 735 pp.

McLean J. H. 1962. Sublittoral ecology of the kelp beds of the open coast area near Carmel, California. Biol. Bull. 122: 95–114.

Miller, R. L. 1966a. Chemotaxis during fertilization in the hydroids *Tubularia* and *Gonothyrea*. Amer. Zool. 6: 509–10.

———. 1966b. Gel filtration of the chemotactants of the hydroids *Campanularia*, *Tubularia*, and *Gonothyrea*. Amer. Zool. 6: 611–12.

———. 1974. Sperm behavior close to *Hydractinia* and *Ciona* eggs. Amer. Zool. 14: 1250 (abstract).

Miller, R. L., and C. J. Brokaw. 1970. Chemotactic turning behavior of *Tubularia* spermatozoa. J. Exper. Biol. 52: 699–706.

Miller, R. L., and C. Y. Tseng. 1974. Properties and partial purification of the sperm attractant of *Tubularia*. Amer. Zool. 14: 467–86.

Miller, R. L., and C. R. Wyttenbach, eds. 1974. The developmental biology of the Cnidaria. Amer. Zool. 14: 440–866.

Moore, M. M. 1927. The reaction of *Cerianthus* to light. J. Gen. Physiol. 8: 509–18.

Morin, J. G. 1974. Coelenterate bioluminescence, pp. 397–438, *in* Muscatine & Lenhoff (1974).

———. 1976. Probable functions of bioluminescence in the Pennatulacea (Cnidaria, Anthozoa), pp. 629–38, *in* Mackie (1976).

Morin, J. G., and I. M. Cooke. 1971a. Behavioral physiology of the colonial hydroid *Obelia*. I. Spontaneous movements and correlated electrical activity. J. Exper. Biol. 54: 689–706.

———. 1971b. Behavioral physiology of the colonial hydroid *Obelia*. II. Stimulus-initiated electrical activity and bioluminescence. J. Exper. Biol. 54: 707–21.

———. 1971c. Behavioral physiology of the colonial hydroid *Obelia*. III. Characteristics of the bioluminescent system. J. Exper. Biol. 54: 723–35.

Morin, J. G., and G. T. Reynolds. 1974. The cellular origin of bioluminescence in the colonial hydroid *Obelia*. Biol. Bull. 147: 397–410.

Morrill, J. B. 1959. Antigenic differences between stem and hydranth in *Tubularia*. Biol. bull. 117: 319–26.

Morrill, J. B., and E. Norris. 1963. Multiple forms of hydrolytic enzymes in *Tubularia crocea*. Amer. Zool. 3: 551–52.

Moseley, H. N. 1878. On the structure of the Stylasteridae, a family of the hydroid stony corals. Phil. Trans. Roy. Soc. London 169: 425–503.

Muscatine, L. 1961. Symbiosis in marine and fresh water coelenterates, pp. 255–68, *in* Lenhoff & Loomis (1961).

———. 1971. Experiments on green algae coexistent with zooxanthellae in sea anemones. Pacific Sci. 25: 13–21.

———. 1974. Endosymbiosis of cnidarians and algae, pp. 359–95, *in* Muscatine & Lenhoff (1974).

Muscatine, L., and C. Hand. 1958. Direct evidence for the transfer of materials from symbiotic algae to the tissues of a coelenterate. Proc. Nat. Acad. Sci. 44: 1259–63.

Muscatine, L., and H. M. Lenhoff. 1965. Symbiosis of *Hydra* and algae. II. Effects of limited food and starvation on growth of symbiotic and aposymbiotic *Hydra*. Biol. Bull. 129: 316–28.

———, eds. 1974. Coelenterate biology: Reviews and new perspectives. New York: Academic Press. 501 pp.

Nagao, Z. 1963. The early development of the anthomedusa, *Polyorchis karafutoensis* Kishinouye. Annot. Zool. Japon. 36: 187–93.

———. 1970. The metamorphosis of the anthomedusa *Polyorchis karafutoensis* Kishinouye. Publ. Seto Mar. Biol. Lab. 28: 21–35.

Naumov, D. V. 1969. Hydroids and hydromedusae of the U.S.S.R. Zool. Inst. Akad. Sci. U.S.S.R. English trans. Jerusalem: Israel

Program of Scientific Translation (I.P.S.T. Press), available through U.S. Dept. Commerce, Washington, D.C. 660 pp.

Nicol, J. A. C. 1955a. Nervous regulation of luminescence in the sea pansy *Renilla köllikeri*. J. Exper. Biol. 32: 619–35.

———. 1955b. Observations on luminescence in *Renilla* (Pennatulacea). J. Exper. Biol. 32: 299–320.

Nishihira, M. 1967a. Dispersal of the larvae of the hydroid, *Sertularella muirensis*. Bull. Biol. Sta. Asamushi 13: 48–56.

———. 1967b. Observations on the selection of algal substrata by hydrozoan larvae, *Sertularella muirensis*, in nature. Bull. Biol. Sta. Asamushi 13: 35–48.

Nutting, C. C. 1909. Alcyonaria of the California coast. Proc. U.S. Nat. Mus. Washington, D.C. 35: 681–727.

Nyholm, K. G. 1949. On the development and dispersal of athenaria actinia with special reference to *Halcampa duodecimcirrata* M. Sars. Zool. Bidrag, Uppsala, 27: 466–509.

Ostarello, G. L. 1973. Natural history of the hydrocoral *Allopora californica* Verrill (1866). Biol. Bull. 145: 548–64.

Pantin, C. F. A. 1965. Capabilities of the coelenterate behavior machine. Amer. Zool. 5: 581–89.

Pardy, R. L., and M. Rahat. 1972. Algae associated with the hydroid *Corymorpha*. Amer. Zool. 12: 719.

Parker, G. H. 1896. The reactions of *Metridium* to food and other substances. Bull. Mus. Compar. Zool., Harvard, 29: 107–19.

———. 1917. The activities of *Corymorpha*. J. Exper. Zool. 24: 303–31.

———. 1920a. Activities of colonial animals. 1. Circulation of water in *Renilla*. J. Exper. Zool. 31: 343–67.

———. 1920b. Activities of colonial animals. Z. Neuromuscular movements and phosphoresence of *Renilla*. J. Exper. Zool. 31: 475–515.

Parmentier, J., and J. Case. 1973. Pharmacological studies of coupling between electrical activity and behaviour in the hydroid *Tubularia crocea* (Agassiz). Comp. Gen. Pharmacol. 4: 11–15.

Passano, L. M. 1966. Dual pacemaker control of swimming in the jellyfish *Sarsia*. Amer. Zool. 6: 350–51.

Pearse, V. B. 1974a. Modification of sea anemone behavior by symbiotic zooxanthellae: Expansion and contraction. Biol. Bull. 147: 641–51.

———. 1974b. Modification of sea anemone behavior by symbiotic zooxanthellae: Phototaxis. Biol. Bull. 147: 630–40.

Phillips, J. H. 1956. Isolation of active nematocysts of *Metridium senile* and their chemical composition. Nature 178: 932.

———. 1961. Isolation and maintenance in tissue culture of coelenterate cell lines, pp. 245–54, *in* Lenhoff & Loomis (1961).

Phillips, J. H., and D. P. Abbott. 1957. Isolation and assay of the nematocyst toxin of *Metridium senile fimbriatum*. Biol. Bull. 113: 296–301.

Picken, L. E. R. 1953. A note on the nematocysts of *Corynactis viridis*. Quart. J. Microscop. Sci. 94: 203–27.

Picken, L. E. R., and R. J. Skaer. 1966. A review of researches on nematocysts, pp. 19–50, *in* Rees (1966a).

Powell, D. C. 1964. Fluorescence in the sea anemone *Anthopleura artemisia*. Bull. Amer. Littoral Soc. 2: 17.

Purcell, J. E. 1977a. Aggressive function and induced development of catch tentacles in the sea anemone *Metridium senile* (Coelenterata, Actiniaria). Biol. Bull. 153: 355–68.

———.1977b. The diet of large and small individuals of the sea anemone *Metridium senile*. Bull. So. Calif. Acad. Sci. 76: 168–72.

Pyefinch, K. A., and F. S. Downing. 1949. Notes on the general biology of *Tubularia larynx* Ellis and Solander. J. Mar. Biol. Assoc. U.K. 28: 21–43.

Quinn, R. J., M. Kashiwagi, T. R. Norton, S. Shibata, M. Kuchii, and R. E. Moore. 1974. Antitumor activity and cardiac stimulatory effects of constituents of *Anthopleura elegantissima*. J. Pharmacol. Sci. 63: 1798–1800.

Rappaport, R. 1966. Experiments concerning the cleavage furrow in invertebrate eggs. J. Exper. Zool. 161: 1–8.

Rees, J. T., and C. Hand. 1975. Class Hydrozoa, pp. 65–85, *in* R. I. Smith and J. T. Carlton, eds., Light's manual: Intertidal invertebrates of the central California coast. 3rd ed. Berkeley and Los Angeles: University of California Press. 716 pp.

Rees, W. J. 1957. Evolutionary trends in the classification of capitate hydroids and medusae. Bull. Brit. Mus. (Natur. Hist.) 4: 455–534.

———. ed. 1966a. The Cnidaria and their evolution. Symp. Zool. Soc. London 16. New York: Academic Press. 449 pp.

———. 1966b. The evolution of the Hydrozoa, pp. 199–222, *in* Rees (1966a).

Rice, N. E., and W. A. Powell. 1970. Observations on three species of jellyfishes from Chesapeake Bay with special reference to their toxins. I. *Chrysaora quinquecirrha*. Biol. Bull. 139: 180–87.

Ricketts, E. F., and J. Calvin. 1968. Between Pacific tides. 4th ed. Revised by J. Hedgpeth. Stanford, Calif.: Stanford University Press. 614 pp.

Robertson, R. 1963. Wentletraps (Epitoniidae) feeding on sea anemones and corals. Proc. Malacol. Soc. London 35: 51–63.

Robson, E. A. 1953. Nematocysts of *Corynactis*; the activity of the filaments during discharge. Quart. J. Microscop. Sci. 94: 229–35.

———. 1957. Structure and hydromechanics of the musculoepithelium in *Metridium*. Quart J. Microscop. Sci. 98: 265–78.

———. 1961. A comparison of the nervous system of two sea-anemones, *Calliactis parasitica* and *Metridium senile*. Quart. J. Microscop. Sci. 102: 319–26.

Robson, E. A., and R. K. Josephson. 1969. Neuromuscular properties of mesenteries from the sea anemone *Metridium*. J. Exper. Biol. 50: 51–68.

Romanes, G. J. 1885. Jellyfish, starfish, and sea urchins. New York: Appleton. 323 pp.

Roosen-Runge, E. C. 1970. Life cycle of the hydromedusan *Philalidium gregarium* (A. Agassiz) in the laboratory. Biol. Bull. 139: 203–21.

Rose, S. M. 1974. Bioelectric control of regeneration in *Tubularia*. Amer. Zool. 14: 797–803.

Rosin, R. 1969. Escape response of the sea anemone *Anthopleura nigrescens* (Verrill) to its predatory eolid nudibranch *Harviella* Baba spec. nov. Veliger 12: 74–77.

Ross, D. M. 1964. The behaviour of some sessile coelenterates in relation to some conditioning experiments, pp. 43–54, *in* W. H. Thorp and D. Davenport, eds., Learning and associated phenomena in invertebrates. Anim. Behav., Suppl. No. 1. London: Bailliere, Tindall & Cassell. 100 pp.

Ross, D. M., and G. A. Horridge. 1957. Responses of *Cerianthus* (Coelenterata). Nature 180: 1368–70.

Russell, F. S. 1953. The medusae of the British Isles. Cambridge, Eng.: Cambridge University Press. 530 pp.

———. 1964. On scyphomedusae of the genus *Pelagia*. J. Mar. Biol. Assoc. U.K. 44: 133–36.

———. 1970. The medusae of the British Isles. 2. Pelagic Scyphozoa with a supplement to the first volume on Hydromedusae. Cambridge, Eng.: Cambridge University Press. 284 pp.

Salo, S. 1977. Observations on feeding, chemoreception and toxins in two species of *Epitonium*. Veliger 20: 168–72.

Sassaman, C., and C. P. Mangum. 1970. Patterns of temperature adaption in North American Atlantic coastal actinians. Mar. Biol. 7: 123–30.

Savilov, A. I. 1961. The distribution of different ecological forms of *Velella lata* Ch. and Eys. and *Physalia utriculus* (La Martinière) Esch. in the North Pacific. [In Russian, with English summary.] Trudy Inst. Okeanol. 45: 223–39.

Schwab, W. E. 1972. Some effects of ionic variation on the swimming rhythm of *Aurelia aurita*. Amer. Zool. 12: 693 (abstract).

Shibata, S., D. F. Dunn, M. Kuchii, M. Kashiwagi, and T. R. Norton. 1974. Cardiac stimulant action of extracts of coelenterates on rat atria. J. Pharmacol. Sci. 63: 1332–33.

Shick, J. M. 1973. Effects of salinity and starvation on the uptake and utilization of dissolved glycine by *Aurelia aurita* polyps. Biol. Bull. 144: 172–79.

———. 1974. Temperature effects on uptake and on kinetics of uptake of dissolved glycine by *Aurelia aurita* polyps. Amer. Zool. 14: 1249 (abstract).

———. 1976. Ecological physiology and genetics of the colonizing actinian *Haliplanella luciae*, pp. 137–46, *in* Mackie (1976).

Shick, J. M., and A. N. Lamb. 1977. Asexual reproduction and genetic population structure in the colonizing sea anemone *Haliplanella luciae*. Biol. Bull. 153: 604–17.

Shimomura, O., and F. H. Johnson. 1975. Chemical nature of bioluminescence systems in coelenterates. Proc. Nat. Acad. Sci. 72: 1546–49.

Siebert, A. E., Jr., and C. Hand. 1974. A description of the sea anemone *Halcampa crypta*, new species. Wasmann J. Biol. 32: 327–36.

Silverstone, M. P., and C. E. Cutress. 1974. Evidence hormones initiate strobiliation in *Aurelia*. Amer. Zool. 14: 1255 (abstract).

Skogsberg, T. 1948. A systematic study of the family Polyorchidae (Hydromedusae). Proc. Calif. Acad. Sci. (4)26: 101–24.

Solomon, D. 1975. Spawning and relations between parent and young in *Epiactis prolifera*. Research paper, Biol. 175h, library, Hopkins Marine Station of Stanford University, Pacific Grove, Calif. 14 pp.

Southward, A. J. 1955. Observations on the ciliary currents of the jellyfish *Aurelia aurita* (L.). J. Mar. Biol. Assoc. U.K. 34: 201–16.

Spangenberg, D. B. 1965. Cultivation of the life stages of *Aurelia aurita* under controlled conditions. J. Exper. Zool. 159: 303–18.

———. 1967. Iodine induction of metamorphosis in *Aurelia*. J. Exper. Zool. 165: 441–50.

———. 1968. Statolith differentiation in *Aurelia aurita*. J. Exper. Zool. 169: 487–500.

———. 1971. Thyroxine induced metamorphosis in *Aurelia*. J. Exper. Zool. 178: 183–94.

———. 1974. Thyroxine in early strobilization in *Aurelia aurita*. Amer. Zool. 14: 825–31.

Spaulding, J. G. 1974. Embryonic and larval development in sea anemones (Anthozoa: Actiniaria). Amer. Zool. 14: 511–20.

Spencer, A. N. 1978. Neurobiology of *Polyorchis*. I. Function of effector systems. J. Neurobiol. 9: 143–57.

Standing, J. D. 1976. Fouling community structure: Effects of the hydroid, *Obelia dichotoma*, on larval recruitment, pp. 155–64, *in* Mackie (1976).

Stasek, C. R. 1958. A new species of *Allogaussia* (Amphipoda, Lysiannassidae) found living within the gastrovascular cavity of the sea anemone *Anthopleura elegantissima*. J. Washington Acad. Sci. 48: 119–26.

Strain, H. H., W. M. Manning, and G. Hardin. 1944. Xanthophylls and carotenes of diatoms, brown algae, dinoflagellates, and sea anemones. Biol. Bull. 86: 169–91.

Strathmann, M., ed. 1974. Methods in developmental biology; technique for the study of eggs, embryos, and larvae of marine invertebrates at Friday Harbor Laboratories. (Mimeo.) Friday Harbor Laboratories, University of Washington.

Summers, R. G. 1972. An ultrastructural study of the spermatozoon of *Eudendrium racemosum*. Z. Zellforsch. 132: 147–66.

Tamasige, M., and T. Yamaguchi. 1967. Equilibrium orientation controlled by ocelli in an anthomedusa, *Polyorchis karafutoensis*. [In Japanese, with English summary.] Zool. Mag., Tokyo, 76: 35–36.

Tanaka, M., M. Haniu, K. T. Yasunobu, and T. R. Norton. 1977. Amino acid sequence of the *Anthopleura xanthogrammica* heart stimulant, anthopleurin A. Biochemistry 16: 204–8.

Tardent, P. 1962. Morphogenetic phenomena in the hydrocaulus of *Tubularia*. Pubbl. Staz. Zool. Napoli 33: 50–63.

———. 1965. Ecological aspects of the morphodynamics of some Hydrozoa. Amer. Zool. 5: 525–29.

Thiel, H. 1966. The evolution of the Scyphozoa: A review, pp. 77–117, *in* Rees (1966a).

Thiel, M. E. 1959. Beiträge zur Kenntnis der Wachstrums und fortpflanzungsverhältnisse von *Aurelia aurita*. Abh. Verh. naturw. Ver. Hamburg 3: 13–26.

Thomson, T. G., and T. J. Chow. 1955. The strontium-calcium atom ratio in carbonate secreting organisms. Pap. Mar. Biol. Oceanogr., Deep-Sea Res. 3 (Suppl.): 20–39.

Tokioka, T., and S. Nishimura, eds. 1973. The proceedings of the Second International Symposium on Cnidaria. Publ. Seto Mar. Biol. Lab. 20: 1–793.

Torrey, H. B. 1898. Observations on monogenesis in *Metridium*. Proc. Calif. Acad. Sci. 1: 345–60.

———. 1904. Biological studies on *Corymorpha*. 1. *C. palma* and environment. J. Exper. Zool. 1: 395–422.

———. 1905. The behavior of *Corymorpha*. Univ. Calif. Publ. Zool. 2: 333–40.

———. 1906. The California shore anemone, *Bundoactis xanthogrammica*. Univ. Calif. Publ. Zool. 3: 41–45.

———. 1907. Biological studies on *Corymorpha*. 2. The development of *Corymorpha palma* from the egg. Univ. Calif. Publ. Zool. 4: 253–98.

———. 1909. The Leptomedusae of the San Diego region. Univ. Calif. Publ. Zool. 6: 11–31.

———. 1910a. Biological studies on *Corymorpha*. 3. Regeneration of hydranth and holdfast. Univ. Calif. Publ. Zool. 6: 205–21.

———. 1910b. Note on geotropism in *Corymorpha*. Univ. Calif. Publ. Zool. 6: 223–24.

Torrey, H. B., and F. L. Kleeberger. 1909. Three species of *Cerianthus* from southern California. Univ. Calif. Publ. Zool. 6: 115–25.

Torrey, H. B., and A. Martin. 1906. Sexual dimorphism in *Aglaophenia*. Univ. Calif. Publ. Zool. 3: 47–52.

Trench, R. K. 1971. The physiology and biochemistry of zooxanthellae symbiotic with marine coelenterates. Proc. Roy. Soc. London B177: 225–64.

Turner, C. H., E. E. Ebert, and R. R. Given. 1969. Man-made reef ecology. Calif. Dept. Fish & Game, Fish Bull. 146: 1–221.

Tweedell, K. S. 1974. DNA synthesis and cell movement during regeneration in *Tubularia*. Amer. Zool. 14: 805–20.

Uchida, T. 1973. The systematic position of the Stauromedusae. Publ. Seto Mar. Biol. Lab. 20: 133–39.

Verrill, A. E. 1869. Review of the corals and polyps of the west coast of America. Trans. Conn. Acad. Arts Sci. 1: 377–558.

Wainwright, S. A., and J. R. Dillon. 1969. On the orientation of sea fans. Biol. Bull. 136: 130–39.

Waters, V. L. 1975. Food preference of the nudibranch *Aeolidia papillosa*, and the effect of the defenses of the prey on predation. Veliger 15: 174–92.

Westfall, J. A., 1965a. Development of nematocysts in Cnidaria. Amer. Zool. 5: 665 (abstract).

———. 1965b. Nematocysts of the sea anemone *Metridium*. Amer. Zool. 5: 377–93.

———. 1966. Electron microscopy of the basitrich and its associated structures in *Obelia*. Amer. Zool. 6: 554–55.

Williams, R. B. 1975. Catch-tentacles in sea anemones: Occurrence in *Haliplanella luciae* (Verrill) and a review of current knowledge. J. Natur. Hist. 9: 241–48.

Wobber, D. R. 1970. A report on the feeding of *Dendronotus iris* on the anthozoan *Cerianthus* sp., from Monterey Bay, California. Veliger 12: 383–87.

Worthman, S. G. 1974. A method of raising clones of the hydroid *Phialidium gregarium* (A. Agassiz, 1862) in the laboratory. Amer. Zool. 14: 821–24.

Wright, H. O. 1973. Effect of commensal hydroids on hermit crab competition in the littoral zone of Texas. Nature 241: 139–40.

Wyman, R. 1965. Notes on the behavior of the hydroid, *Corymorpha palma*. Amer. Zool. 5: 491–97.

Yasuda, T. 1968. Ecological studies on the jellyfish, *Aurelia aurita* in Urazoko Bay, Fukui Prefecture. II. Occurrence pattern of the ephyra. Bull. Jap. Soc. Sci. Fish. 34: 983–87.

———. 1969. Ecological studies on the jellyfish, *Aurelia aurita* in Urazoko Bay, Fukui Prefecture. I. Occurrence pattern of the medusae. Bull. Jap. Soc. Sci. Fish. 35: 1–6.

Yonge, C. M. 1932. A note on *Balanophyllia regia*, the only eupsammid coral in the British fauna. J. Mar. Biol. Assoc. U.K. 18: 219–24.

Chapter 4

Platyhelminthes: *The Flatworms*

Eugene C. Haderlie

The flatworms are divided taxonomically into three classes: Turbellaria (mainly free-living, predaceous forms), Trematoda (flukes), and Cestoda (tapeworms). As adults, the flukes and tapeworms are parasites, mainly in vertebrate animals, although many flukes use intertidal gastropods as intermediate hosts for their larval stages. Our main concern here is with the free-living Turbellaria.

Turbellarians are an ancient group but have left few fossils. There are about 3,000 species of Turbellaria known worldwide. Most of these are marine and live on the bottom from the intertidal zone to deep water. A few swim as pelagic animals in the sea. A small number of species of turbellarians live in fresh water or in damp places on land. Free-living flatworms are the most primitive animals to show true bilateral symmetry and a beginning of cephalization, that is, the concentration of nervous coordinating centers and sensory organs at the anterior end of the body. They are distinguished from all other animals in having thin, leaflike bodies that are usually ciliated, and most move with a gliding locomotion that carries them smoothly over the substratum. Their color varies from drab shades of brown and tan to brilliant hues and patterns; some pelagic free-swimming turbellarians are practically transparent. Most species of the intertidal area live on the undersides of rocks or boulders, but they also move about on algae, over the bottoms of tidepools, and in the interstitial water of sand and gravel.

Of the several orders of Turbellaria represented in California intertidal waters, only two (Acoela and Polycladida) are treated here. Small turbellarians belonging to other orders are abundant on sand, gravel, rocky crevices, and seaweeds. Zoologists have been aware of their presence, but have largely ignored them.

Most acoels are inconspicuous (see Kozloff, 1965a), but one bright-red species, *Polychoerus carmelensis* (4.1), is often abundant in tidepools of the Monterey Peninsula. Acoels have a mouth on the ventral surface but usually lack a hollow gut. They feed on diatoms and small animals such as copepods, which they digest in interior cells.

Polyclads, by contrast, are conspicuous and fairly common. Their flattened bodies are up to 15 cm or more long and have highly branched digestive systems. The mouth is located near the middle of the body on the ventral surface; through it the large skirtlike pharynx can be everted to envelop food. These flatworms are carnivores and feed on many kinds of small animals, including copepods, amphipods, isopods, and encrusting bryozoans. All polyclads are hermaphroditic and have complicated reproductive systems, but self-fertilization does not occur. Following copulation, the eggs are fertilized internally, then pass to the outside as part of a gelatinous strand that adheres to the substratum. Hatching occurs several days or weeks after spawning. Some polyclads develop directly into juvenile worms; others pass through a free-swimming larval stage.

The identification of polyclads at the species level is often difficult, and positive determination usually must be made by an expert, who often has to make sagittal sections of the reproductive organs for microscopic study. In some species the color pattern may be distinctive. Other anatomical features useful in identification are the presence or absence of tentacles (either nuchal or neck

Eugene C. Haderlie is Professor of Oceanography at the Naval Postgraduate School, Monterey.

tentacles, or tentacles on the anterior margin of the body), and the distribution of eyes, or ocelli.

A good general account of acoels, polyclads, and other turbellarians is that of Hyman (1951). Karling (1974) discusses the division of the Turbellaria into orders. The works of Hyman (1953, 1955, 1959) are most useful in identification of turbellarians. Distribution of most species of free-living flatworms along the coast is very poorly known. A key to the species from the central California coast is found in Haderlie (1975). Continuing studies on the Pacific coast indicate that a rich variety of species of turbellarians, many of which are new to science, occur along our shores (e.g., Holleman, 1972; Holleman & Hand, 1962; Holmquist & Karling, 1972; Hyman, 1954; Karling, 1962a,b, 1964, 1966a, 1967, 1977; Karling & Schockaert, 1977; Kozloff, 1965a,b; Schockaert & Karling, 1970).

Freshwater turbellarians have been used extensively in experimental studies of many sorts. Marine forms, too, have provided research materials for studies on fine structure, digestion, regeneration, development, behavior and learning, and other aspects of biology (e.g., Antonius, 1970; Bullock, 1965; Clark, 1964; Clark & Cowey, 1958; Florkin & Scheer, 1968; Hendelberg, 1974; Hyman, 1951; Jennings, 1957; Karling, 1961, 1966b; Koopowitz, 1970; MacRae, 1966; Moore, 1933, 1945; Olmstead, 1922; Riser & Morse, 1974; Thomas, 1970), but much more work is needed.

Most turbellarians are free-living, but many species of polyclads have symbiotic associations with marine invertebrates that appear to border on parasitism. Some of the predatory forms, such as the oyster leech of the east coast, enter the shells of oysters and feed on the soft parts and thus can be of economic significance. A few turbellarians appear to be truly parasitic, such as the rhabdocoel **Syndesmis franciscana** (Lehman) (= *Syndisyrinx franciscana*), which occurs in the gut of California sea urchins (Giese, 1958; Lehman, 1946).

Length to 6 mm; anterior end rounded or pointed, posterior end blunt and deeply notched with a contractile caudal filament arising medially from the notch; color usually deep orange, but varying from ivory to red-orange.

Polychoerus shuns light less than most other flatworms, and during early-morning low tides may be so abundant as to color the bottoms of pools orange. Any disturbance of the water causes the worms to move downward, and they retreat below the gravel as water enters the pool on a rising tide. They feed mainly on tiny crustaceans, and tolerate salinity conditions of 75–125 percent seawater indefinitely.

The animals are hermaphroditic, but copulation occurs and impregnation may be mutual. Fertilization is internal. Developing spermatids have two flagella; these are incorporated by the maturing sperm as two undulating membranes, which become active only after copulation. Meiosis, which occurs after fertilization, results in two polar bodies, which remain within the egg. Development proceeds to the metaphase of the first cleavage, but halts at this stage until the eggs are released by a rupture of the body wall (which heals rapidly). The eggs are deposited in small jelly masses on stones and in folds of *Ulva* from late June through August. They are relatively large (to 0.3 mm in diameter), and the asters and spindle formed during cleavage are the largest known in the animal kingdom. Behavior of the extraordinarily large centrioles (to 5 μm long) has led to important conclusions regarding the cause of spiral cleavage. In *Polychoerus* the spiral cleavage starts with the two-cell stage.

See Armitage (1961), D. Costello (1946, 1960a,b,c, 1961a,b, 1970, 1973a,b), D. Costello & Henley (1971), D. Costello, Henley & Ault (1969), D. Costello & H. Costello (1968), H. Costello & D. Costello (1938a,b, 1939), Dörjes & Karling (1975), Henley, Costello & Ault (1968), Keil (1929), Kozloff (1965a), and Schwab (1967).

Phylum Platyhelminthes / Class Turbellaria / Order Acoela
Family Convolutidae

4.1 Polychoerus carmelensis Costello & Costello, 1938

Often abundant on stones, shells, and fronds of the green alga *Ulva*, in pools, high and upper middle intertidal zones; Monterey Bay area.

Phylum Platyhelminthes / Class Turbellaria / Order Polycladida
Suborder Acotylea / Family Stylochidae

4.2 Stylochus tripartitus Hyman, 1953

Fairly common, under rocks, low intertidal zone, and on kelp holdfasts; Oregon to Baja California.

Length to 50 mm; body oval in outline; nuchal tentacles with eyes; marginal eyes limited to anterior half of body; color buff with numerous, tiny, elongate, brown spots giving the worm an overall brownish hue.

This species, along with **Notoplana inquieta** (Heath & McGregor), is a major predator on the barnacle *Balanus pacificus* (20.27). The flatworms reach the barnacles as planktonic larvae and enter through the opercular aperture, then proceed to eat the living barnacles at rates up to 0.05 g barnacle tissue per gram of flatworm per day.

See Haderlie & Donat (1978), Hurley (1975, 1976), and Hyman (1953).

Phylum Platyhelminthes / Class Turbellaria / Order Polycladida
Suborder Acotylea / Family Stylochidae

4.3 **Kaburakia excelsa** Bock, 1925

Uncommon, under rocks, low intertidal zone; among mussels and other fouling organisms on boat bottoms; Alaska to southern California.

To 10 cm long and 7 cm wide when extended and active; nuchal tentacles large, retractile, bearing eyes inside and at the base; marginal eyes completely encircling the body; color tan, heavily marked with uniformly distributed dark-brown dashes, giving body an overall brownish hue.

This is one of the largest flatworms of the coast. A boat brought to dry dock in Monterey yielded dozens of the animals creeping among the mussels and other fouling growth.

See Hyman (1953, 1955) and MacGinitie & MacGinitie (1968).

Phylum Platyhelminthes / Class Turbellaria / Order Polycladida
Suborder Acotylea / Family Leptoplanidae

4.4 **Notoplana acticola** (Boone, 1929)

One of the most common polyclads of rocky shores, especially abundant under boulders, high and middle intertidal zones; throughout California.

Adults 25–60 mm long when extended and moving, the body generally widest in front, tapering to the rear; tentacular eyes in rounded clusters with scattered eyes lying anterior, posterior, and sometimes lateral to them; cerebral eyes about 25, in an elongate band; color tan or pale gray with darker markings along midline.

Notoplana is a carnivore capable of ingesting other animals up to half its own bulk. It has been observed feeding on limpets (*Collisella digitalis*, 13.12) and small acorn barnacles. Captive worms have been seen to feed on the red nudibranch *Rostanga pulchra* (14.26) and become pink in color. Experiments on feeding behavior have shown that both local and centrally controlled reflexes are involved. Following experimental severing of neural pathways, the nerves rapidly rejoin and pathways are reestablished. Studies of the eyes show that each one consists of a single, cup-shaped pigment cell covering six to ten retinal cells.

On the California coast near Dillon Beach (Marin Co.), Thum (1974) found that through the year the majority of mature *N. acticola* were functional hermaphrodites, with both gonads active and mature eggs and sperm present. Small numbers of worms bearing only ovaries were observed during the spring months, but animals with only testes were found throughout the year in numbers from 10–50 percent of the population. Up to two-thirds of all worms examined through the year had sperm in their seminal receptacles, evidence of copulation. Egg laying apparently occurs from late spring to early fall.

See Boone (1929), Ewer (1965), Glynn (1965), Haderlie (1971), Haderlie & Donat (1978), Hyman (1953, 1955), Keenan, Koopowitz & Bernardo (1979), Koopowitz (1974, 1975), Koopowitz, Bernardo & Keenan (1979), Koopowitz, Keenan & Bernardo (1979), MacRae (1966), and Thum (1974).

Phylum Platyhelminthes / Class Turbellaria / Order Polycladida
Suborder Acotylea / Family Leptoplanidae

4.5 **Phylloplana viridis** (Freeman, 1933)

Often common, on the broad eelgrass *Zostera* in sheltered bays, shallow subtidal waters; Puget Sound (Washington) to Humboldt Bay.

Body elongate, to 25 mm long and 8 mm wide, thin and soft; color a uniform light green.

See Hyman (1953).

Phylum Platyhelminthes / Order Polycladida / Suborder Acotylea Family Hoploplanidae

4.6 **Hoploplana californica** Hyman, 1953

Common, low intertidal zone on wharf pilings in association with the encrusting bryozoan *Celleporaria brunnea* (6.12) in Monterey harbor; subtidal to 16 m in Newport Harbor (Orange Co.); Monterey Bay and Newport Bay.

Body to 12 mm long, oval in shape; dorsal surface papillate, the papillae pointed, longer and thicker on central part of body, yellowish to orange; nuchal tentacles conspicuous, pointed, conical, arising from light area on dorsum; tentacular eyes around bases of tentacles; cerebral eyes in two oval clusters medial and anterior to tentacles; body orange to reddish gray, with small dark spots between bases on papillae.

The presence of dorsal papillae distinguishes this species from other common flatworms.

In Monterey Bay, *H. californica* has been found only in association with *Celleporaria brunnea* (as shown in Fig. 6.12). The flatworm's color pattern and texture so closely mimic the red and orange tentacles and dark sessile avicularia of *Celleporaria* that the animals usually pass unnoticed, and only the light spot at the base of each of the worm's tentacles reveals its presence. When touched, the worms are relatively stiff and move sluggishly; they cling tightly and move only when sharply prodded. When they are overturned, the highly branched, white digestive tract can be seen, often with an extruded, ruffled pharynx. Below the worm the bryozoan colony is bleached white and only the skeletal parts remain, indicating that *Hoploplana* feeds on the bryozoan.

See Haderlie (1968, 1969), Haderlie & Donat (1978), and Hyman (1953).

Phylum Platyhelminthes / Class Turbellaria / Order Polycladida Suborder Acotylea / Family Planoceridae

4.7 **Alloioplana californica** (Heath & McGregor, 1912) (=Planocera californica)

Sometimes relatively abundant, most commonly under boulders that rest on damp sand or gravel, upper middle intertidal zone; throughout California.

Body to 24 mm long and 14 mm wide, elliptical, firm in consistency; nuchal tentacles nipplelike, contractile, bearing numerous eyes; tentacular eyes borne in a diagonal series anterior and posterior to tentacle bases; cerebral eyes also present; ground color light, transparent olive with a highly branched digestive tract of chocolate brown or dark green forming zigzag lines radiating from the central main intestine to the periphery.

Near Monterey, this worm is often found with fragments of small snails in the gut. Encrusting egg masses to 6 mm in diameter are found under boulders and in crevices through most of the year.

See Heath & McGregor (1912), Hyman (1953), and Koopowitz (1970).

Phylum Platyhelminthes / Class Turbellaria / Order Polycladida Suborder Acotylea / Family Pseudoceridae

4.8 **Thysanozoon sandiegense** Hyman, 1953

Uncommon, under rocks, low intertidal zone; Dana Point (Orange Co.) to La Jolla (San Diego Co.).

Body to 30 mm long, with a distinct notch at the anterior end; dorsal surface black, bearing lighter-colored papillae.

Known from a limited area, and never properly described, this animal is reported to be a rapid swimmer that escapes readily when the rocks under which it hides are moved.

Another species occurring in southern California, **T. californicum** Hyman, has been used in neurophysiological studies related to the retractile nature of the papillae. A relative from South Africa, *T. brocchii*, has had the fine structure of the epidermis determined.

See Bedini & Papi (1974), Hyman (1953, p. 363), Johnson & Snook (1927), and Koopowitz (1974).

Phylum Platyhelminthes / Class Turbellaria / Order Polycladida
Suborder Cotylea / Family Pseudoceridae

4.9 Pseudoceros luteus (Plehn, 1898)

Fairly common under stones and sometimes on the surface film, low intertidal pools; reported from Monterey Peninsula and Corona del Mar (Orange Co.).

Body thin and delicate, to 50 mm long, elongate with a deeply ruffled margin; color whitish, densely so along the margin and speckling the body, with a narrow, middorsal black stripe forking anteriorly between the marginal tentacular flaps.

This worm swims beautifully, with undulating movements of the ruffled margin.

See Hyman (1953) and MacGinitie & MacGinitie (1968).

Phylum Platyhelminthes / Class Turbellaria / Order Polycladida
Suborder Cotylea / Family Pseudoceridae

4.10 Pseudoceros montereyensis Hyman, 1953

Uncommon, under stones, middle intertidal zone on rocky shores; reported only from Pacific Grove (Monterey Co.).

Body to 90 mm long, but more typically about 40 mm long and 20 mm wide, oval, with gracefully ruffled margin; color basically whitish with darker markings; one black stripe tinged with red extending middorsally, another encircling the body just in from the margin but absent anteriorly between the tentacles; dorsal surface with elongate black spots, and a few red spots and white spots; tentacles banded with black; cerebral eyes forming an inverted V-shaped cluster between the tentacles and the anterior end of the middorsal stripe.

See Hyman (1953, p. 370) and MacGinitie & MacGinitie (1968).

Phylum Platyhelminthes / Class Turbellaria / Order Polycladida
Suborder Cotylea / Family Euryleptidae

4.11 Eurylepta aurantiaca (Heath & McGregor, 1912)

Uncommon, under stones and creeping sluggishly on bottom rocks and wharf piles, shallow subtidal water; Vancouver Island (British Columbia) to San Diego.

Body to 28 mm long and 9 mm wide; eyes distributed on tentacles and forming two well-defined oval clusters of about 50 eyes each over the brain; ventral surface with a sucker, permitting the animal to cling tenaciously to rocks; body yellowish-pink, salmon, or reddish, peppered with minute white spots, with a bright-pink middorsal streak extending back from the cerebral eyes.

See Haderlie & Donat (1978), Heath & McGregor (1912), Hyman (1953), and Johnson & Snook (1927).

Phylum Platyhelminthes / Class Turbellaria / Order Polycladida
Suborder Cotylea / Family Euryleptidae

4.12 Eurylepta californica Hyman, 1959

Uncommon, under stones and on algae, low intertidal zone of rocky shores; central California.

Body to at least 30 mm long and 12 mm wide; flaring tentacles each with a heavy black and red mark at base and a small median black mark; cerebral eyes numerous, very small, massed to form an inverted V; color pattern very distinctive, the body grayish white with a white margin, a narrow white middorsal stripe, and irregularly distributed white spots, this pattern crisscrossed with narrow black lines, some terminating in red tips near the border.

See Hyman (1959).

Phylum Platyhelminthes / Class Turbellaria / Order Polycladida
Suborder Cotylea / Family Euryleptidae

4.13 **Eurylepta** sp.

Known only from exposed, but deeply shaded, surfaces of seaweed; low intertidal zone; Pacific Grove (Monterey Co.).

Body oval in outline, about 28 mm long and 9 mm wide, with a ruffled margin; anterior tentacles flaring, with internal eyes; cerebral eyes forming a distinct cluster; background color light yellow; dorsal surface with numerous, concentric, white lines and darker yellow margin.

An as yet unnamed form, clearly in the genus *Eurylepta*.

Phylum Platyhelminthes / Class Turbellaria / Order Polycladida
Suborder Cotylea / Family Euryleptidae

4.14 **Prostheceraeus bellostriatus** Hyman, 1953

Uncommon, under stones, low intertidal zone on rocky shores; more commonly subtidal on wharf pilings; Monterey Bay to southern California.

Body to 35 mm long and 25 mm wide, strikingly colored, with alternating black and white longitudinal stripes, a mid-dorsal orange stripe, and an orange border, the white stripes sometimes tinged with red, the black stripes often bearing orange spots; ventral surface white except for orange margin; anterior tentacles dense black and flaring; tentacular eyes at tentacle bases and on the anterior margin between tentacles; cerebral eyes in one or two small clusters.

On pilings in Monterey harbor this species lives among mussels and sea anemones.

See Hyman (1953, 1959) and Johnson & Snook (1927).

Phylum Platyhelminthes / Class Turbellaria / Order Polycladida
Suborder Cotylea / Family Euryleptidae

4.15 **Enchiridium punctatum** Hyman, 1953

Uncommon, low intertidal zone along rocky shores; subtidal to 15 m; southern California.

Body about 40 mm long and 10 mm wide; cerebral eyes in two groups; marginal eyes completely encircling body; sucker ventral, near posterior end; color white, cream, or greenish, with brown or black spots.

See Hyman (1953).

Literature Cited

Antonius, A. 1970. Sense organs in marine Acoela. Amer. Zool. 10: 550 (abstract).

Armitage, K. B. 1961. Studies on the biology of *Polychoerus carmelensis* (Turbellaria, Acoela). Pacific Sci. 15: 203–10.

Bedini, C., and F. Papi. 1974. Fine structure of the turbellarian epidermis, pp. 108–47, *in* Riser & Morse (1974).

Boone, E. S. 1929. Five new polyclads from the California coast. Ann. Mag. Natur. Hist. 10: 33–46.

Bullock, T. H. 1965. Platyhelminthes, pp. 535–77, *in* T. H. Bullock and G. A. Horridge, Structure and function in the nervous systems of invertebrates, vol. 1. San Francisco: Freeman. 798 pp.

Clark, R. B. 1964. Dynamics in metazoan evolution. The origin of the coelom and segments. Oxford: Clarendon. 313 pp.

Clark, R. B., and J. B. Cowey. 1958. Factors controlling the change of shape of certain nemerteans and turbellarian worms. J. Exper. Biol. 35: 731–48.

Costello, D. P. 1946. The giant cleavage spindle of the egg of *Polychoerus carmelensis*. Anat. Rec. 96: 146 (abstract).

———. 1960a. The "polar suns" (centrospheres) of the egg of *Polychoerus* (Turbellaria: Acoela). Anat. Rec. 137: 346 (abstract).

———. 1960b. The internal polocytes of the egg of *Polychoerus carmelensis* (Turbellaria: Acoela). Anat. Rec. 137: 346–47 (abstract).

———. 1960c. The giant cleavage spindle of the egg of *Polychoerus carmelensis*. Biol. Bull. 119: 285 (abstract).

———. 1961a. On the orientation of centrioles in dividing cells and its significance: A new contribution to spindle mechanics. Biol. Bull. 120: 285–312.

————. 1961b. The orientation of centrioles in dividing cells and its significance. Biol. Bull. 121: 368 (abstract).

————. 1970. Identical linear order of chromosomes in both gametes of the acoel turbellarian *Polychoerus carmelensis*: A preliminary note. Proc. Nat. Acad. Sci. 67: 1951–58.

————. 1973a. A new theory on the mechanics of ciliary and flagellar motility. I. Supporting observations. Biol. Bull. 145: 279–91.

————. 1973b. A new theory on the mechanics of ciliary and flagellar motility. II. Theoretical considerations. Biol. Bull. 145: 292–309.

Costello, D. P., and H. M. Costello. 1968. Immotility and motility of acoel turbellarian spermatozoa, with special reference to *Polychoerus carmelensis*. Biol. Bull. 135: 417.

Costello, D. P., and C. Henley. 1971. Methods for obtaining and handling marine eggs and embryos. 2nd ed. Marine Biological Laboratory, Woods Hole, Mass. 247 pp.

Costello, D. P., C. Henley, and C. R. Ault. 1969. Microtubules in spermatozoa of *Childia* (Turbellaria, Acoela) revealed by negative staining. Science 163: 678–79.

Costello, H. M., and D. P. Costello. 1938a. A new species of *Polychoerus* from the Pacific coast. Ann. Mag. Natur. Hist. (11) 1: 148–55.

————. 1938b. Copulation in the acoelous turbellarian *Polychoerus carmelensis*. Biol. Bull. 75: 85–89.

————. 1939. Egg laying in the acoelous turbellarian *Polychoerus carmelensis*. Biol. Bull. 76: 80–89.

Dörjes, J., and T. G. Karling. 1975. Species of Turbellaria Acoela in the Swedish Museum of Natural History, with remarks on their anatomy, taxonomy and distribution. Zool. Scripta 4: 175–89.

Ewer, D. W. 1965. Networks and spontaneous activity in echinoderms and platyhelminthes. Amer. Zool. 5: 563–72.

Florkin, M., and B. T. Scheer, eds. 1968. Chemical zoology. 2. Porifera, Coelenterata, and Platyhelminthes. New York: Academic Press. 639 pp.

Giese, A. C. 1958. Incidence of *Syndesmis* in the gut of two species of sea urchins. Anat. Rec. 132: 441–42.

Glynn, P. W. 1965. Community composition, structure, and interrelationships in the marine intertidal *Endocladia muricata*–*Balanus glandula* association in Monterey Bay, California. Beaufortia 12: 1–198.

Haderlie, E. C. 1968. Marine fouling organisms in Monterey harbor. Veliger 10: 327–41.

————. 1969. Marine fouling and boring organisms in Monterey harbor. II. Second year of investigation. Veliger 12: 182–92.

————. 1971. Marine fouling and boring organisms at 100 ft depth in open water of Monterey Bay. Veliger 13: 249–60.

————. 1975. Phylum Platyhelminthes, pp. 100–111, *in* R. I. Smith and J. T. Carlton, eds., Light's manual: Intertidal invertebrates of the central California coast. Berkeley and Los Angeles: University of California Press. 716 pp.

Haderlie, E. C., and W. Donat III. 1978. Wharf piling fauna and flora in Monterey harbor, California. Veliger 21: 45–69.

Heath, H., and E. A. McGregor. 1912. New polyclads from Monterey Bay, California. Proc. Acad. Natur. Sci. Philadelphia 64: 455–88.

Hendelberg, J. 1974. Spermiogenesis, sperm morphology, and biology of fertilization in the Turbellaria, pp. 148–64, *in* Riser & Morse (1974).

Henley, C., D. P. Costello, and C. R. Ault. 1968. Microtubules in axial filament complexes of acoel turbellarian spermatozoa, as revealed by negative staining. Biol. Bull. 135: 422–23.

Holleman, J. J. 1972. Marine turbellarians of the Pacific coast. Proc. Biol. Soc. Washington 85: 405–12.

Holleman, J. J., and C. Hand. 1962. A new species, genus, and family of marine flatworms (Turbellaria: Tricladida, Maricola) commensal with mollusks. Veliger 5: 20–22.

Holmquist, C., and T. G. Karling. 1972. Two new species of interstitial marine triclads from the North American Pacific coast, with comments on evolutionary trends and systematics in Tricladida (Turbellaria). Zool. Scripta 1: 175–84.

Hurley, A. C. 1975. The establishment of populations of *Balanus pacificus* Pilsbry (Cirripedia) and their elimination by predatory Turbellaria. J. Anim. Ecol. 44: 521–32.

————. 1976. The polyclad flatworm *Stylochus tripartitus* Hyman as a barnacle predator. Crustaceana 31: 110–11.

Hyman, L. H. 1951. The invertebrates: Platyhelminthes and Rhynchocoela. Vol. 2. New York: McGraw-Hill. 550 pp.

————. 1953. The polyclad flatworms of the Pacific coast of North America. Bull. Amer. Mus. Natur. Hist. 100: 269–392.

————. 1954. A new marine triclad from the coast of California. Amer. Mus. Novitates 1679: 1–5.

————. 1955. The polyclad flatworms of the Pacific coast of North America: Additions and corrections. Amer. Mus. Novitates 1704: 1–11

————. 1959. Some Turbellaria from the coast of California. Amer. Mus. Novitates 1943: 1–17.

Jennings, J. B. 1957. Studies on the feeding, digestion and food storage in free-living flatworms (Platyhelminthes: Turbellaria). Biol. Bull. 112: 63–80.

Johnson, M. E., and H. J. Snook. 1927. Seashore animals of the Pacific coast. New York: Macmillan. 659 pp.

Karling, T. G. 1961. Zur Morphologie, Entstehungsweise und Funktion des Spaltrüssels der Turbellaria Schizorhynchia. Ark. Zool. 13: 253–86.

————. 1962a. Marine Turbellaria from the Pacific coast of North America. I. Plagiostomidae. Ark. Zool. 15: 113–41.

————. 1962b. Marine Turbellaria from the Pacific coast of North America. II. Pseudostomidae and Cylindrostomidae. Ark. Zool. 15: 181–209.

_____. 1964. Marine Turbellaria from the Pacific coast of North America. III. Otoplanidae. Ark. Zool. 16: 527–41.

_____. 1966a. Marine Turbellaria from the Pacific coast of North America. IV. Coelogynoporidae and Monocelididae. Ark. Zool. 18: 493–528.

_____. 1966b. On nematocysts and similar structures in turbellarians. Acta Zool. Fennica 116: 1–28.

_____. 1967. On the genus *Promesostoma* (Turbellaria) with descriptions of four new species from Scandinavia and California. Sarsia 29: 257–68.

_____. 1974. On the anatomy and affinities of the turbellarian orders, pp. 1–16, *in* Riser & Morse (1974).

_____. 1977. Taxonomy, phylogeny and biogeography of the genus *Austrorhynchus* Karling (Turbellaria; Polycystididae). Mikrofauna Meeresboden 61: 153–65.

Karling, T. G., and E. R. Schockaert. 1977. Anatomy and systematics of some Polycystididae (Turbellaria, Kalyptorhynchia) from the Pacific and S. Atlantic. Zool. Scripta 6: 5–19.

Keenan, L., H. Koopowitz, and K. Bernardo. 1979. Primitive nervous systems: Action of aminergic drugs and blocking agents on activity in the ventral nerve cord of the flatworm *Notoplana acticola*. J. Neurobiol. 10: 397–407.

Keil, E. M. 1929. Regeneration in *Polychoerus caudatus* Mark. Biol. Bull. 57: 225–44.

Koopowitz, H. 1970. Feeding behavior and the role of the brain in the polyclad flatworm, *Planocera gilchristi*. Anim. Behav. 18: 31–35.

_____. 1974. Some aspects of the physiology and organization of the nerve plexus in polyclad flatworms, pp. 198–212, *in* Riser & Morse (1974).

_____. 1975. Recovery of function and feeding behavior in polyclad flatworms. Amer. Zool. 15: 778 (abstract).

Koopowitz, H., K. Bernardo, and L. Keenan. 1979. Primitive nervous systems: Electrical activity in ventral nerve cords of the flatworm, *Notoplana acticola*. J. Neurobiol. 10: 367–81.

Koopowitz, H., L. Keenan, and K. Bernardo. 1979. Primitive nervous systems: Electrophysiology of inhibitory events in flatworm nerve cords. J. Neurobiol. 10: 383–95.

Kozloff, E. N. 1965a. New species of acoel turbellarians from the Pacific coast. Biol. Bull. 129: 151–66.

_____. 1965b. *Desmote inops* sp. n. and *Fallacohospes inchoatus* gen. and sp. n., umagillid rhabdocoels from the intestine of the crinoid *Florometra serratissima* (A. H. Clark). J. Parasitol. 51: 305–12.

Lehman, H. E. 1946. A histological study of *Syndisyrinx franciscanus* gen. et sp. nov., an endoparasitic rhabdocoel of the sea urchin *Strongylocentrotus franciscanus*. Biol. Bull. 91: 295–311.

MacGinitie, G. E., and N. MacGinitie. 1968. Natural history of marine animals. 2nd ed. New York: McGraw-Hill. 523 pp.

MacRae, E. K. 1966. The fine structure of photoreceptors in a marine flatworm. Z. Zellforsch. 75: 469–84.

Moore, A. R. 1933. On the role of the brain and cephalic nerves in the swimming and righting movements of the polyclad worm, *Planocera reticulata*. Sci. Rep. Tôhôku Imp. Univ. 8: 193–200.

_____. 1945. The individual in simpler forms. Univ. Oregon Monogr., Stud. Psychol. 2: 1–143.

Olmstead, J. D. 1922. The role of the nervous system in the locomotion of certain marine polyclads. J. Exper. Zool. 36: 57.

Riser, N. W., and M. P. Morse, eds. 1974. Biology of the Turbellaria. New York: McGraw-Hill. 530 pp.

Schockaert, E. R., and T. G. Karling. 1970. Three new anatomically remarkable Turbellaria Eukalyptorhynchia from the North American Pacific coast. Ark. Zool. (2) 23: 237–53.

Schwab, R. G. 1967. Overt responses of *Polychoerus carmelensis* (Turbellaria: Acoela) to abrupt changes in ambient water temperature. Pacific Sci. 21: 86–90.

Thomas, M. B. 1970. Transition between helical and protofibrilar configurations in doublet and singlet microtubules in spermatozoa of *Stylochus zebra* (Turbellaria, Polycladida). Biol. Bull. 138: 219–34.

Thum, A. B. 1974. Reproductive ecology of the polyclad turbellarian *Notoplana acticola* (Boone, 1929) on the central California coast, pp. 431–45, *in* Riser & Morse (1974).

Chapter 5

Nemertea: *The Ribbon Worms*

Eugene C. Haderlie

Nemertea (=Rhynchocoela, often called ribbon worms or proboscis worms) are about as common as polyclad flatworms along the California coast. There are about 800 species of nemerteans known worldwide. They are mainly marine animals and occur at all depths in the sea, but a few species live in fresh water and damp places on land, and a few live as commensals.

Nemerteans constitute an ancient group and are clearly derived from the flatworms, but they are more advanced in structure, as evidenced by a circulatory system and a digestive system terminating in an anus. They are distinct from other worms in having soft, elongate, narrow, non-segmented bodies that are highly contractile. The body is covered with cilia and often flattened, at least in part, but in some species it may be rounded in cross section. A cephalic lobe may or may not be distinguishable at the anterior end. The nemertean body is exceedingly extensible: several intertidal nemerteans that measure about 8 cm long when contracted are capable of extending to at least 45 cm; and one species living in mud at Morro Bay (San Luis Obispo Co.) is reported to reach a fully extended length of 19 m (MacGinitie & MacGinitie, 1968).

A unique anatomical feature of the nemerteans is an eversible proboscis, which can be shot out anteriorly to capture prey. It is not connected with the digestive system proper, and when not in use it lies in a cavity dorsal to the gut. The proboscis is sometimes as long as, or longer than, the body, and when retracted may be somewhat coiled within its sheath. The extended proboscis coils about the prey, such as a small annelid worm, and secretes a very sticky and

Eugene C. Haderlie is Professor of Oceanography at the Naval Postgraduate School, Monterey.

sometimes toxic mucus that aids in holding it captive. A living nemertean can often be made to evert the proboscis by placing the worm in fresh water. In some species (order Hoplonemertea) the proboscis is armed near its tip with a sharply pointed stylet mounted on a firm base. The stylet may be used to stab the prey repeatedly. Lost or broken stylets are replaced by spares, which are secreted and held ready in accessory stylet pouches near the stylet base. Nemerteans are carnivorous, feeding mainly at night on annelids, mollusks, crustaceans, and even small fishes. They in turn are fed upon by crabs and birds.

Many nemerteans are strikingly colored. Some are white or yellow, but most have the dorsal surface pigmented in various shades of orange, red, brown, or green. Some are plain, but others are patterned with longitudinal or circular stripes or bands in a contrasting color. A few pelagic forms are nearly transparent.

Most nemerteans live on the bottom, but some burrow into mud or sand and may produce delicate membranous or parchmentlike tubes in which they live; others reside in the crevices of algal holdfasts, among the blades of densely branched seaweeds, or under rocks. Some nemerteans are pelagic and live at depths of several hundred meters in the open sea. It is interesting to note that of the 29 species of bathypelagic nemerteans reported from the entire Pacific Ocean, 16 have been found in the Monterey Submarine Canyon (Coe, 1954a,b).

The majority of nemerteans are free-living, but along our coast one species, *Malacobdella grossa* (5.11), lives as a commensal in the mantle cavities of clams and snails and feeds on plankton brought in by the ciliary currents of the mollusk.

The sexes are separate in most nemerteans, and sexual dimorphism is uncommon (Coe, 1920). Development is generally more or less direct, but some forms give rise to a pilidium larval stage, which is planktonic. The eggs of several species have been used for embryological studies, and the worms themselves have served in experimental studies on regeneration and muscular physiology. Some can withstand prolonged starvation for months or even a year or more, while shrinking in size and ultimately undergoing some dedifferentiation. One Japanese species is luminescent along all of the body except the anterior tip.

Nemerteans are classified into two subgroups (often considered subclasses), the Anopla and the Enopla, primarily on the basis of the position of the central nervous system in relation to the musculature and the presence or absence of a stylet on the proboscis. Each of these two groups in turn is divided into two orders.

Field identification of nemerteans is often difficult unless the coloration is unique. In addition to size and color, such details as shape of head, location of ocelli (eyes), and presence of grooves either around the "neck" (nuchal grooves) or along the side of the head (cephalic grooves or furrows) are of importance in recognizing species.

A good general account and bibliography of the group appears in Hyman (1951). The classic monograph by Bürger (1895) is still a very valuable reference for anatomy and histology. Willmer (1970) contains much information on the cytology of nemerteans, and an expanded account of the old but still unorthodox idea that vertebrates originated from nemerteanlike forms. What is known of the nervous system is summarized in Bullock (1965). Recent studies of nemerteans include work on cytology (Hinsch & Clark, 1970), fine structure (Eakin & Westfall, 1968), feeding and digestion (Jennings, 1960, 1962, 1968, 1969), and locomotion and shape changes (Clark, 1964; Clark & Cowey, 1958). The best general reference on the natural history of nemerteans is Coe (1943), and other publications by Coe (1901, 1905a,b, 1940) are useful for the identification of species found on the California coast. Roe (1970, 1971) has made the first studies on nemertean populations. A key to central California species is available in Haderlie (1975).

Phylum Nemertea / Class Anopla / Order Paleonemertea / Family Tubulanidae

5.1 **Tubulanus frenatus** (Coe, 1904)
(=Carinella frenata)

Uncommon, among algae and mussels on rocky shores, among fouling organisms on pilings, and in mud in sheltered bays, low intertidal zone; southern California.

Length to 50 cm; anterior third of body yellow or ochre, the remainder sage green; dorsal surface with up to 100 dark-brown transverse bands (the anterior three broad, the remainder narrow) and with three dark-brown or black longitudinal stripes (the middorsal one widest and extending to the anterior end, the lateral stripes narrower and ending anteriorly at the transverse neck band).

See Coe (1904; 1905a, as *Carinella frenata*; 1940).

Phylum Nemertea / Class Anopla / Order Paleonemertea / Family Tubulanidae

5.2 **Tubulanus polymorphus** Renier, 1804
(=Carinella speciosa, C. rubra)

Uncommon, under stones, especially heavy boulders embedded in gravel, among mussels and other growth, and in mud, low intertidal and adjacent subtidal zones; Aleutian Islands and southern Alaska to Piedras Blancas Point (San Luis Obispo Co.).

Body soft, pliable, capable of great changes in proportions, to 3 m long when fully extended; cephalic lobe broad and rounded; color uniform, bright red, orange-yellow, or deep vermilion, often very conspicuous in the field.

This worm is an excellent source of gametes for fertilization and embryological studies, because large females produce many eggs during the summer. Coe (1940) noted that eggs can be teased from fragments of the body, washed in clean seawater, and fertilized by the addition of sperm. Development is rapid and direct.

See Coe (1901, as *Carinella speciosa*; 1905a, as *C. rubra*; 1940).

Phylum Nemertea / Class Anopla / Order Paleonemertea / Family Tubulanidae

5.3 **Tubulanus sexlineatus** (Griffin, 1898)
(=Carinella sexlineata, C. dinema)

Relatively uncommon, inhabiting transparent tubes among algae, mussels, and other organisms on rocks and pilings, low intertidal and subtidal regions; Sitka (Alaska) to southern California.

Averaging about 20 cm in length, though extended specimens may stretch more than 1 m; dorsal surface deep brown and bearing five narrow, longitudinal, white stripes, with indications of a sixth stripe midventrally, the stripes sometimes broken, consisting of longitudinal rows of white dots; body encircled by up to about 150 white bands, the bands in the midbody region often as wide as the brown areas between bands.

See Coe (1901, as *Carinella dinema*; 1905a, as *C. sexlineata*; 1940).

Phylum Nemertea / Class Anopla / Order Paleonemertea / Family Carinomidae

5.4 **Carinoma mutabilis** Griffin, 1898
(=C. griffini)

Uncommon, typically in low intertidal zone on soft bottoms, buried in sand, sandy mud, or clay, sometimes to depth of 45 cm; subtidal on wharf pilings; Vancouver Island (British Columbia) to Bahía de los Angeles (Gulf of California).

Averaging about 20 cm in length, but ranging to 50 cm; body slender, rounded in cross section anteriorly, but considerably flattened posteriorly; cephalic lobe wider than neck, variable in shape; color pure white anteriorly, cream in intestinal region, lacking conspicuous markings.

Spawning occurs in the summer months. Females collected and placed in clean seawater may spawn spontaneously, or the eggs can be obtained by cutting the animals longitudinally. After fertilization, development is rapid and direct.

See Coe (1901, as *C. griffini*; 1905a, 1940).

Phylum Nemertea / Class Anopla / Order Heteronemertea / Family Baseodiscidae

5.5 **Baseodiscus punnetti** (Coe, 1904)
(=Taeniosoma punnetti)

Uncommon, under rocks among red algae, low intertidal zone; subtidal on bottom detritus to 100 m or more; Cabrillo Point near Fort Bragg (Mendocino Co.) to Gulf of California.

Body flabby, when expanded to 60 cm long and 1 cm wide, when wholly contracted very short, fat, subcylindrical; color brownish red, mahogany, or dark red above, cream or gray below; head white at anterior tip, bearing many minute ocelli in irregular clusters on each side.

This species is often snagged on hooks trailing the bottom in Monterey Bay. The worms are sexually mature in August.

See Coe (1904, 1905a, as *Taeniosoma punnetti*; 1940).

Phylum Nemertea / Class Anopla / Order Heteronemertea / Family Lineidae

5.6 **Micrura verrilli** Coe, 1901

Sometimes common, under rocks and in beds of algae or eelgrass, low intertidal zone on rocky shores; subtidal in kelp holdfasts; Prince William Sound (Alaska) to Monterey Bay.

Length to 50 cm; dorsal surface mostly covered by a broad, deep-purple longitudinal stripe, interrupted by 30–40 narrow, white transverse bands; sides and ventral surface pure ivory-white; cephalic lobe with a conspicuous triangular dorsal marking of bright orange or vermilion; small caudal cirrus sometimes present (often lost).

Micrura verrilli held in aquaria build mucus tubes and may thrive for months. Eggs and sperm mature in the spring and early summer in southern California, and fertilized eggs rapidly reach the pilidium larval stage. Regenerative capacity is marked; young worms that have body regions posterior to the brain removed regenerate them, but posterior portions do not regenerate new anterior ends.

See Coe (1901, 1905a, 1940).

Phylum Nemertea / Class Anopla / Order Heteronemertea / Family Lineidae

5.7 Cerebratulus californiensis Coe, 1905

Fairly common, in soft sediments in protected bays and harbors, low intertidal zone; subtidal to 50 m or more; Puget Sound (Washington) to Bahía Tenacatita (Jalisco, Mexico).

Length to 15 cm or more; body rounded in cross section in anterior esophageal region, flattened and to 4–5 mm wide posteriorly; head normally broader than neck, bearing a deep cephalic groove laterally on each side; color variable, especially in immature specimens, often a rosy flesh or cream anteriorly, the intestinal region less rosy, often buff, the lateral nerves marked by conspicuous reddish lines.

Cerebratulus californiensis is fragile and sensitive to disturbance, and often breaks into many fragments when collected or handled. A close relative, *C. herculeus* Coe, also fragments readily, but grows much larger; an individual taken by G. E. MacGinitie at Morro Bay (San Luis Obispo Co.) was estimated to reach 19 m when extended. *C. californiensis* breeds during May and June in southern California, and during July in Monterey Bay. Development is indirect and includes a pilidium larval stage. Excellent descriptive and experimental embryological studies have been carried out on a related east coast form, *C. lacteus*, whose anatomy is also known in detail. All species of *Cerebratulus* can swim by rapid dorsoventral undulations of the flattened body.

See Coe (1895, 1905a, 1940), Costello & Henley (1971, containing numerous embryological references), Freeman (1974), MacGinitie & MacGinitie (1968), and Wilson (1900).

Phylum Nemertea / Class Enopla / Order Hoplonemertea
Family Emplectonematidae

5.8 Emplectonema gracile (Johnson, 1837)

Very common among barnacles and mussels on pier pilings, less common under stones in muddy areas, and in algal holdfasts and empty barnacle shells in moist shaded areas, upper middle intertidal zone on rocky shores; Unalaska (Aleutian Islands, Alaska) to Ensenada (Baja California); Chile; northern Europe, Mediterranean Sea, Madeira.

Body to 50 cm long, but more commonly 5–15 cm; slender, yellowish green to dark green dorsally, pale greenish yellow to white ventrally; head pale or colorless at anterior tip, bearing ocelli in two groups on each side; stylets smooth, slender, slightly curved.

At low tide, individuals of this hardy species are often found huddled gregariously in tangled clusters. They move out when wetted again by the sea and are most active at night. They tolerate brackish conditions and live well in laboratory aquaria (though they may crawl out of the water). Mounted in seawater and pressed between large glass slides, they make excellent whole mounts for student examination under the compound microscope; inverted proboscis, stylet, accessory style pouches, ocelli, brain, lateral blood vessels, gut, gonads, and other structures are clearly visible in the living animal.

The worms appear to be rather unselective carnivores; in the field they have been seen to feed on injured barnacles, eggs of the snail *Nucella emarginata* (13.83), and polychaete worms.

The gonads mature as early as March in southern California, and small juveniles are abundant there in April and May. Further north, reproduction occurs in the early summer. Development is direct, without a pilidium larva.

See Bürger (1895, containing a detailed account of the anatomy), Coe (1901, 1905a, 1940), Delsman (1915), Glynn (1965), and Haderlie (1968).

Phylum Nemertea / Class Enopla / Order Hoplonemertea
Family Emplectonematidae

5.9 Paranemertes peregrina Coe, 1901

Common on rocky shores under rocks and among mussels and coralline algae, also in muddy bays, middle and low intertidal zones; often out and crawling on damp exposed rocks or mud during early-morning low tides or at night; Kamchatka (U.S.S.R.) and Aleutian Islands (Alaska) to Ensenada (Baja California); Japan.

Body sometimes exceeding 25 cm in length, but more commonly 6–15 cm; dorsal and lateral body surfaces purplish brown, dark brown, or orange-brown, the ventral surface and sides of head deep yellow to white; neck region dorsally marked with a narrow, pale, inverted V-shaped mark; surfaces of stylets ridged, giving them a braided appearance.

One of the most common nemerteans of rocky shores, this hardy species thrives in the laboratory; living specimens make excellent whole mounts for student examination under the microscope (see also *Emplectonema gracile*, 5.8). Exposure to fresh water or very dilute acetic acid sometimes causes spectacular eversion of the proboscis. In the laboratory, hungry individuals sometimes capture and ingest nereid worms placed in front of them.

Roe (1970, 1971) made an extensive study of the life history and predator-prey interactions of populations of *Paranemertes* on the open rocky coast of Washington and in muddy habitats in Puget Sound. In females, which in many populations significantly outnumber males, the eggs require up to 6 months to mature, but males produce sperm more rapidly. Spawning occurs primarily in the spring and summer, but may continue through the fall and winter months, any one population spawning for a period of about 1 month. The eggs are deposited singly or in gelatinous clusters. Approximately 30 hours after fertilization, blastulae are formed; the eggs hatch on the third day. Development is rapid, and within 2 weeks stylets and a rudimentary proboscis are formed. Individual worms live for as long as 1.75 years and may spawn up to three times.

Paranemertes remains hidden when covered by the tide, but when exposed during especially low water it emerges and actively seeks out polychaetes, which it captures and ingests. In Washington, the main prey is the nereid *Platynereis bicanaliculata* (18.12). However, many species of nereid polychaetes show strong escape responses to *Paranemertes*. The nemertean can ingest prey slightly greater in diameter than itself. It must make physical contact with the prey, for it cannot detect it at a distance. On contact with prey, the head of *Paranemertes* recoils, and the proboscis everts and wraps around the prey, which is paralyzed or killed by the wounds inflicted by the central stylet, and possibly by secretions containing toxic ma-

terials pumped from the posterior proboscis glands into the victim's body. The mouth then takes in the annelid. Digestion involves both extra- and intracellular processes, and defecation occurs from 12 to 33 hours after feeding.

See Coe (1901, 1905a, 1940), Corrêa (1964), Gibson (1970), and Roe (1970, 1971).

Phylum Nemertea / Class Enopla / Order Hoplonemertea / Family Amphiporidae

5.10 **Amphiporus bimaculatus** Coe, 1901

Relatively common, under rocks, in beds of algae or mussels, in rock crevices, and under sheets of encrusting coralline algae, low intertidal zone on rocky shores; Sitka (Alaska) to Ensenada (Baja California).

Length 4–15 cm, width 2–6 mm, the body dorsoventrally flattened, broadest in the midbody region and gradually tapering toward both ends; head somewhat pointed and set off from the body by conspicuous nuchal grooves; body deep red, or brownish red to yellow-brown dorsally and pale to flesh-colored ventrally; head pale or white, with two large, symmetrical, triangular or oval dark spots dorsally.

This species lives well in laboratory aquaria and secretes copious amounts of mucus. When disturbed it may swim actively by powerful dorsoventral undulations of the body. Spawning has been observed in mid-July in Monterey Bay.

The genus *Amphiporus* is the most common nemertean genus on the Pacific coast of North America, and is represented by at least 17 species. One of the most common species (though not the most conspicuous) is the small, uniformly white or pale pink **A. imparispinosus** Griffin, which sometimes abounds in beds of the red alga *Corallina vancouveriensis* and in other intertidal situations; it occurs from Siberia and the Bering Sea to Ensenada (Baja California). Living *A. imparispinosus*, mounted between two glass slides, are excellent for study under the compound microscope.

The stylets of this genus have recently been investigated. Each consists of an outer, inorganic cortex, apparently composed of a crystalline form of calcium and strontium phosphates, and an inner, rod-shaped organic portion. As in the case of sponge spicules, the stylets of these nemerteans are formed within a single cell in the accessory stylet sac.

See Coe (1901, 1905a, 1940) and Wourms (1976).

Phylum Nemertea / Class Enopla / Order Bdellonemertea
Family Malacobdellidae

5.11 **Malacobdella grossa** (O. F. Müller, 1776)

Fairly common, in the mantle cavities of different species of clams, low intertidal and subtidal areas; British Columbia to central California; east coasts of Canada and New England; northern Europe and Mediterranean Sea.

Body to about 3.5 cm long and 1 cm wide, usually smaller, short, flattened, leechlike, with a distinct anterior notch and a rounded muscular sucker posteriorly; color pinkish to pale cream or almost white.

This curious nemertean occurs in at least 22 species of Atlantic bivalves. On the Pacific coast it is found in *Macoma secta* (15.51), *M. nasuta* (15.50), *Siliqua patula* (15.65), occasionally *Tresus nuttallii* (15.48), and doubtless other clams as well. Usually there is no more than one worm in a given host. The nemerteans are food robbers rather than true parasites. They feed on phytoplankton and small zooplankton organisms that the clams have filtered from the water and trapped in mucus. The living tissues of the bivalves are not eaten or otherwise harmed, and the nemertean produces no measurable effect upon its host. In England, where it occurs regularly in the clams *Zirfaea crispata* and *Hiatella arctica* (15.69), up to 90 percent of the clam population may be infested. There the worm breeds throughout the year with early spring and midsummer peaks, which coincide with periods of increased food availability. The ratio of male to female worms is 1:1.

Malacobdella itself is host to various parasites, including protozoa (ciliates, gregarines, and other sporozoa) and trematodes. The sporozoan *Haplosporidium malacobdellae* can cause parasitic castration of the nemertean, or may even cause death through rupture of the body wall.

See especially Gibson (1967, 1968, 1970), Gibson & Jennings (1969); see also Coe (1940, 1945), Corrêa (1964), Gering (1911), Guberlet (1925), Jennings (1968), Maclaren (1901), Quayle (1960), and Riepen (1933).

Literature Cited

Bullock, T. H. 1965. Nemertinea, pp. 580–95, *in* T. H. Bullock and A. G. Horridge, Structure and function in the nervous systems of invertebrates, vol. 1. San Francisco: Freeman. 798 pp.

Bürger, O. 1895. Die Nemertinen des Golfes von Neapel. Fauna und Flora des Golfes von Neapel. Monogr. 22: 1–743.

Clark, R. B. 1964. Dynamics in metazoan evolution. The origin of the coelom and segments. Oxford: Clarendon. 313 pp.

Clark, R. B., and J. B. Cowey. 1958. Factors controlling the change of shape of certain nemerteans and turbellarian worms. J. Exper. Biol. 35: 731–48.

Coe, W. R. 1895. On the anatomy of a species of nemertean (*Cerebratulus lacteus* Verrill), with remarks on certain other species. Trans. Conn. Acad. Arts Sci. 9: 479–514.

———. 1901. Papers from the Harriman Alaska Expedition. XX. The Nemerteans. Proc. Washington Acad. Sci. 111: 1–110. (Reprinted without change in 1904 as Nemerteans of the Pacific coast of North America. Part I. The nemerteans of the expedition. Harriman Alaska Expedition 11: 1–110.)

———. 1904. Nemerteans of the Pacific coast of North America. Part II. Harriman Alaska Expedition 11: 111–220.

———. 1905a. Nemerteans of the west and northwest coast of North America. Bull. Mus. Compar. Zool., Harvard, 47: 1–319.

———. 1905b. Synopsis of the North American invertebrates. Nemerteans I. Amer. Natur. 39: 425–47.

———. 1920. Sexual dimorphism in nemerteans. Anat. Rec. 17: 352 (abstract).

———. 1940. Revision of the nemertean fauna of the Pacific coasts of North, Central and northern South America. Allan Hancock Pacific Exped. 2: 247–323.

———. 1943. Biology of the nemerteans of the Atlantic coast of North America. Trans. Conn. Acad. Arts Sci. 35: 129–328.

———. 1945. *Malcobdella minuta*, a new commensal nemertean. J. Washington Acad. Sci. 35: 65–67.

———. 1954a. Bathypelagic nemerteans of the Pacific Ocean. Bull. Scripps Inst. Oceanogr. 6: 225–86.

———. 1954b. Geographic distribution and means of dispersal of the bathypelagic nemerteans found in the great submarine canyon at Monterey Bay, California. J. Washington Acad. Sci. 44: 324–26.

Corrêa, D. D. 1964. Nemerteans from California and Oregon. Proc. Calif. Acad. Sci. 31: 515–58.

Costello, D. P., and C. Henley. 1971. Methods of obtaining and handling marine eggs and embryos. 2nd ed. Marine Biological Laboratory, Woods Hole, Mass. 247 pp.

Delsman, H. 1915. Eifurchung und Gastrulation bei *Emplectonema*. Helder Tidjschr. Nederland. Dierk. Vereen., (2) 14: 68–114.

Eakin, R. M., and J. A. Westfall. 1968. Fine strucutre of nemertean ocelli. Amer. Zool. 8: 803 (abstract).

Freeman, G. 1974. The role of asters in setting up localizations of developmental potential during embryogenesis in the nemertine *Cerebratulus lacteus*. Amer. Zool. 14: 1299 (abstract).

Gering, G. 1911. Beiträge zur Kenntniss von *Malacobdella grossa* (Müll.) Z. Wiss. Zool. 97: 673–720.

Gibson, R. 1967. Occurrence of the entocommensal rhynchocoelan, *Malacobdella grossa,* in the oval piddock, *Zirfaea crispata,* on the Yorkshire coast. J. Mar. Biol. Assoc. U.K. 47: 301–17.

———. 1968. Studies on the biology of the entocommensal rhynchocoelan *Malacobdella grossa*. J. Mar. Biol. Assoc. U.K. 48: 637–56.

———. 1970. The nutrition of *Paranemertes peregrina* (Rhynchocoela: Hoplonemertea). II. Observations on the structure of the gut and proboscis, site and sequence of digestion, and food reserves. Biol. Bull. 139: 92–106.

Gibson, R., and J. B. Jennings. 1969. Observations on the diet, feeding mechanisms, digestion and food reserves of the entocommensal rhynchocoelan *Malacobdella grossa*. J. Mar. Biol. Assoc. U.K. 49: 17–32.

Glynn, P. W. 1965. Community composition, structure, and interrelationships in the marine intertidal *Endocladia muricata–Balanus glandula* association in Monterey Bay, California. Beaufortia 12: 1–198.

Guberlet, J. E. 1925. *Malacobdella grossa* from the Pacific coast of North America. Publ. Puget Sound Biol. Sta. 5: 1–13.

Haderlie, E. C. 1968. Marine fouling organisms in Monterey harbor. Veliger 10: 327–41.

———. 1975. Phylum Nemertea (Rhynchocoela), pp. 112–20, *in* R. I. Smith and J. T. Carlton, eds., Light's manual: Intertidal invertebrates of the central California coast. Berkeley and Los Angeles: University of California Press. 716 pp.

Hinsch, G., and W. H. Clark. 1970. Centriolar satellites. Amer. Zool. 10: 523 (abstract).

Hyman, L. H. 1951. The invertebrates: Platyhelminthes and Rhynchocoela. Vol. 2. New York: McGraw-Hill. 550 pp.

Jennings, J. B. 1960. Observations on the nutrition of the rhynchocoelan *Lineus ruber* (O. F. Müller). Biol. Bull. 119: 189–96.

———. 1962. A histochemcial study of digestion and digestive enzymes in the rhynchocoelan *Lineus ruber* (O. F. Müller). Biol. Bull. 122: 63–72.

———. 1968. A new astomatous ciliate from the entocommensal rhynchocoelan *Malacobdella grossa* (O. F. Müller). Arch. Protistenk. 110: 422–25.

———. 1969. Ultrastructual observations on the phagocytic uptake of food materials by the ciliated cells of the rhynchocoelan intestine. Biol. Bull. 137: 476–85.

MacGinitie, G. E., and N. MacGinitie. 1968. Natural history of marine animals. 2nd ed. New York: McGraw-Hill. 523 pp.

Maclaren, N. H. W. 1901. On the blood vascular system of *M. grossa*. Zool. Anz. 24: 126–29.

Quayle, D. B. 1960. The intertidal beivalves of British Columbia. British Columbia Prov. Mus. Handbook 17. 104 pp.

Riepen, O. 1933. Anatomie und Histologie von *Malacobdella grossa* (Müll.). Z. Wiss. Zool. 143: 323–424.

Roe, P. 1970. The nutrition of *Paranemertes peregrina* (Rhynchocoela: Hoplonemertea). Studies on food and feeding behavior. Biol. Bull. 139: 80–91.

———. 1971. Life history and predator-prey interactions of the nemertean *Paranemertes peregrina* Coe. Doctoral thesis, Zoology, University of Washington. 129 pp.

Willmer, E. N. 1970. Cytology and evolution. 2nd ed. New York: Academic Press. 649 pp.

Wilson, C. B. 1900. The habits and early development of *Cerebratulus lacteus* (Verrill). A contribution to physiological morphology. Quart. J. Microscop. Sci. 43: 97-198.

Wourms, J. P. 1976. Structure, composition, and unicellular origin of nemertean stylets. Amer. Zool. 16: 213 (abstract).

Chapter 6

Bryozoa and Entoprocta: *The Moss Animals*

John D. Soule, Dorothy F. Soule, and Donald P. Abbott

The groups Ectoprocta and Entoprocta, formerly united in the phylum Bryozoa (=Polyzoa), differ in many respects and are now generally treated as separate phyla. The modern tendency is to retain the term Bryozoa for the Ectoprocta only (e.g., Mayr, 1968; J. Soule & D. Soule, 1968; Ryland, 1970; Larwood, 1973), and this treatment is followed here. The phyla Bryozoa and Entoprocta are treated separately below.

The Bryozoa, or Ectoprocta, are virtually all colonial forms, with each colony (zoarium) composed of many small attached individuals (zooids). A colony originates from a single, sexually produced individual (the ancestrula) and increases by asexual budding of new individuals. Bryozoans are widely distributed in the sea, and many are found on rocky shores that are exposed at only the lowest tides, or on harbor pilings or ships' hulls. At first glance some colonies may be mistaken for bushy types of hydroids, branching corals, or marine algae.

The individual zooids in a bryozoan colony are usually less than 1 mm long, and each is encased in a secreted outer cuticle or exoskeleton, which stiffens the colony and provides support and protection for the enclosed soft parts. The exoskeleton surrounding a single individual is termed the zooecium. It is provided with an opening (the aperture) through which the zooid may extend its tentacles to feed, and into which it can withdraw rapidly when disturbed. In some groups the aperture is covered by a hinged lid (the operculum).

The individual zooids are composed of two main structural parts, the cystid and the polypide. The cystid consists of the boxlike zooecium and the living layers of the body wall that line the zooecium and surround the body cavity or perivisceral coelom. The polypide includes the lophophore, which bears the circlet of tentacles that surround the mouth, the tentacle sheath that encases the lophophore when retracted, and the viscera and associated muscles. The polypide moves conspicuously when the individual extends to feed or withdraws inside the zooecium.

Bryozoans feed on bacteria, phytoplankton, other small organisms, and tiny particles of organic detritus, which are swept to the mouth in water currents created by the ciliated tentacles. The mouth lies in the center of the circle formed by the bases of the tentacles. The gut is U-shaped, and the anus empties on the wall of the membranous tentacle sheath, outside the circlet of tentacles, hence the name "ecto-procta."

Feeding individuals (autozooids) usually form the major portion of a bryozoan colony. However, many bryozoan colonies contain additional zooids (collectively called heterozooids) that are markedly modified in connection with particular functions, such as protection, reproduction, brooding of embryos, anchorage to the substratum, or provision of joints in the colony (see Silén, 1977). Among the specialized heterozooids, the protective avicularia are usually smaller than the autozooids, but some are large and conspicuous. They may replace entire autozooids, or may originate directly from

John D. Soule is Professor of Histology and Assistant Dean at the School of Dentistry, and Professor of Biological Sciences and Curator of Bryozoa at the Allan Hancock Foundation, of the University of Southern California. Dorothy F. Soule is Director of Harbors Environmental Projects of the Institute for Marine and Coastal Studies and Curator of Bryozoa, Allan Hancock Foundation, and Adjunct Professor of Environmental Engineering, of the University of Southern California. Donald P. Abbott is Professor, Department of Biological Sciences and Hopkins Marine Station, Stanford University.

the individual zooecia. The avicularian polypide is greatly reduced or absent, and the avicularium consists mainly of a small cystid equipped with an oversized operculum and the muscles to operate it. Stalked avicularia resemble the heads of birds, the operculum operating as a lower jaw that can be opened wide or snapped shut on small intruding animals. Sessile avicularia look like small boxes or knobs, each bearing on its free surface an enlarged operculum that can be opened or closed like a movable jaw. The avicularia can handle small crustaceans (see *Bugula californica*, 6.4) but are of little use against such bryozoan predators as browsing nudibranchs and fishes.

Living ectoprocts are most commonly subdivided into two classes: the Gymnolaemata, a large group of species nearly all marine in habitat, and the Phylactolaemata, a smaller group of freshwater species not considered here, although they are widespread. The Gymnolaemata in this chapter fall into three orders, two of which (orders Ctenostomata and Cyclostomata) are relatively small, whereas the third (order Cheilostomata) is large and includes more than half the species considered.

A variety of reproductive modes is found in bryozoans. Sexual reproduction usually results in the production of new colonies, whereas asexual reproduction, by budding, increases the number of zooids within an existing colony. Most bryozoans of the order Cheilostomata are considered to be hermaphroditic, but cross-fertilization between autozooids or colonies occurs. In some species the male or female gonads mature at different times (sequentially), preventing self-fertilization. Reproduction in most bryozoan species has not been studied live or histologically, but a few cheilostomes are known to have morphologically different individual male and female zooecia (e.g., *Hippothoa*) in the same colony (Silén, 1977). Sperm release through tentacle tips has been observed in some species (Silén, 1966; Bullivant, 1967).

The Ctenostomata and some cheilostomes, like the members of the freshwater class Phylactolaemata, have only a membranous sac for the maturation of ova, but most cheilostomes produce external structures called ovicells within which ova are brooded (Ström, 1977). The ovicell is a very specialized structure that lacks ovaries, testes, and polypide. It may originate from a single zooecium or may have a complex origin. In cases that have been studied, the ovicell chamber is formed by contributions from (1) the maternal zooecium that furnishes the egg, (2) the transverse walls of the ma-

ternal and next distal (anterior) zooecium, and (3) in some instances the frontal (ventral) surfaces of adjacent zooecia (Ryland, 1968; D. Soule, 1973; Woollacott & Zimmer, 1972b). Ovicells may be hidden (immersed) in some species, but in many species they form a distinctive bulge or hood anterior to the aperture, which may or may not be closed by the operculum. Ovicells, when present, offer good characters for the identification of species.

Female reproductive structures in the order Cyclostomata are quite different from the ovicells of cheilostomes. In the family Crisiidae they are single whole individuals (gonozooids) modified for egg production and brooding. However, in most cyclostomes the gonozooid expands and merges with adjacent autozooids or extrazooidal spaces to form a composite structure better termed an ooecium. Cyclostomes may be sequentially hermaphroditic, or the individuals in the colony may be of separate sexes (Borg, 1926; Silén, 1977).

There are several types of larvae among the bryozoans. In some species eggs and sperm are shed into the sea, where fertilization and development of a planktonic larval stage occur. In most species, however, the eggs are retained in ovicells, gonozooids, or ooecia, and the embryos are brooded for a time. In some ctenostomes and some cheilostomes (e.g., *Membranipora tuberculata*, 6.1), the planktonic larvae are known as cyphonautes, tiny, free-swimming forms with a bivalve shell. Although the cyphonautes in some species feed on plankton (planktotrophic), others lack a complete digestive tract and are nourished by stored yolk (lecithotrophic). Other non-feeding cheilostome larvae, called coronate larvae, lack shells and have ciliary girdles of various types for swimming. Some coronate larvae are non-feeding spheroids, covered with cilia, that do not move about extensively.

In the order Cyclostomata the early embryonic stages undergo an asexual multiplication (polyembryony) within the gonozooid or ooecium. A single fertilized egg undergoes cleavage to form a primary embryo. This divides into secondary and sometimes tertiary embryos, each of which goes on to form a larva. The larvae known are highly modified lecithotrophic spheroids (Borg, 1926; Ström, 1977).

In any case, the bryozoan larva typically swims or moves about for a time, after which it becomes negatively phototactic, tests the substratum, and settles. The larva attaches temporarily by sticky secretions; subsequent eversion of the metasomal sac accomplishes permanent attachment. Metamorphosis follows rapidly and the adult ancestrula of the colony results (see Zimmer & Woollacott (1977) on metamorphosis; J. Soule (1973) and J. Soule & D. Soule (1977) on bioadhesives).

The bryozoans are an ancient group, with a fossil record extending from the early Paleozoic. Between 3,000 and 4,000 living species of bryozoans are known in the world's oceans. Nearly 250 are recorded from California, and though many are known only from deeper water, the shore collector will still find a bewildering variety much greater than that portrayed in this chapter. Identification of some forms is difficult, but the non-specialist can learn to recognize the more distinctive common species.

Colony growth form is a good starting point in identification. In general, colonies tend to be either (1) recumbent (adnate), with the whole colony closely adhering to the substratum, or (2) erect, with the colony attached over a smaller area and sending branches or processes composed of many individuals into the water above. Some caution must be exercised, for certain species may form erect growths in quiet waters and produce flattened encrustations in turbulent areas. Recumbent bryozoans include (1) soft-bodied species, which either form fleshy, gelatinous clumps or send creeping stolons along the substratum, (2) encrusting species with a mineralized exoskeleton that may be lightly calcified (flexible) or heavily calcified (rigid), and (3) tubular species. Erect bryozoans include those that are (1) branching, and often quite bushy, (2) foliaceous or flustraform (with flattened leaflike or bladelike processes), (3) fenestrate or reteporid (with upright stiffened blades fused into a latticelike or lacelike structure), and (4) tubular.

Other characters used in bryozoan taxonomy relate to the size, shape, ornamentation (spines, pits, ridges, and pores), and degree of calcification of individual zooecia, as well as the form, number, and placement of heterozooids such as avicularia. A hand lens ($10\times$ to $15\times$) is essential for field recognition of most common genera, and it is folly to attempt more serious study of the bryozoans without a good stereoscopic dissecting microscope providing magnifications to at least $70\times$.

Excellent general accounts of the Bryozoa are found in Brien (1960a), Hyman (1959), and Ryland (1970). An outstanding recent volume (Woollacott & Zimmer, 1977) contains reviews of numerous aspects of bryozoan biology, including sexual reproduction and development (Franzén, 1977; Ström, 1977; Zimmer & Woollacott, 1977a,b), polymorphism (Silén, 1977), feeding (Winston, 1977), life history strategies (Farmer, 1977; J. Soule & D. Soule, 1977), aging (Gordon, 1977), detailed structure of various parts and systems (Bobin, 1977; Lutaud, 1977; Sandberg, 1977), behavior (Ryland, 1977), and population genetics (Schopf, 1977). Reviews of recent research are presented in Annoscia (1968), Boardman, Cheetham & Oliver (1973), Larwood (1973), McCammon & Reynolds (1977), Pouyet (1975), and Ryland (1967, 1974, 1976). Relationships of Bryozoa with other phyla are considered by Brien (1960a,b), Farmer (1977), Hyman (1959), Nielsen (1971, 1977a,b), and Zimmer (1973). The best comprehensive systematic study for the Pacific coast of the Americas is the three-volume work of Osburn (1950, 1952, 1953). Keys useful for identification of California shore species appear in J. Soule, D. Soule & Pinter (1975) and Ross (1970). Bryozoans of the Gulf of California and Baja California are covered in J. Soule (1959, 1961, 1963) and D. Soule & J. Soule (1964).

The phylum Entoprocta includes small animals in which the body consists of a slender stalk topped by an expanded flowerlike calyx. The calyx contains most of the viscera and bears distally a ring of ciliated tentacles used to capture microscopic, particulate food. Superficially, entoproct individuals resemble hydroids, from which they are easily distinguished by their behavior. When disturbed, the zooids roll their tentacles inward and the stalk bends, bringing the calyx down against the substratum, hence the common name "nodding heads," often applied to the group. The entoprocts most commonly seen form colonies.

About 120 species of entoprocts are known for the world. In California, entoprocts are relatively common on the undersurfaces of intertidal rocks, in the fouling community on floats and pilings in bays, and as commensals on the bodies of other animals, but their small size makes them inconspicuous. For more detailed accounts see Brien (1959), Hyman (1951), Nielsen (1971), Mariscal (1975a), and D. Soule & J. Soule (1965); Mariscal (1975b) provides a key to the California species.

Phylum Ectoprocta / Class Gymnolaemata / Order Cheilostomata
Suborder Anasca / Family Membraniporidae

6.1 **Membranipora tuberculata** (Bosc, 1802)

Commonly encrusted on floating brown algae, especially the kelps *Macrocystis* and *Cystoseira*, also on smaller plants inshore, especially the red alga *Gelidium*, occasionally on shells or wood, low intertidal zone to shallow subtidal depths; prevalent in warm temperate and tropical waters of Atlantic, Pacific, and Indian Oceans, wherever floating *Sargassum* or *Fucus* is found; Pleistocene fossils known from Santa Monica (Los Angeles Co.) and Newport Beach (Orange Co.).

Colony forming a white crust to several centimeters in diameter, consisting of a single layer of zooids and having a fine reticulate honeycomb appearance; individuals 0.5–0.8 mm long, rectangular, covered by a lightly calcified membrane, and with a heavily calcified rim, the rim bearing calcified tubercles at the distal corners, as well as tiny spines projecting inward from side of rim toward center of individual.

The membraniporids are the only members of the order Cheilostomata known to have a planktotrophic cyphonautes larva, a free-swimming, plankton-feeding stage with a triangular bivalve shell. This settles and metamorphoses to an ancestrula (often double or twinned), which buds to form a flat, encrusting colony. The relatively transparent living colonies are excellent for observation under the microscope.

Membranipora tuberculata is one of three species in the genus commonly encrusting the kelp *Macrocystis*; under favorable conditions colonies grow rapidly and may coalesce to cover whole kelp blades in 3–4 weeks. Overgrown blades often bear 0.5–1 kg of bryozoans per m² of blade surface, and occasionally perhaps ten times as much.

Although *Membranipora* species are among the best-known marine bryozoans, the species have often been confused. **Membranipora membranacea** (Linnaeus) is primarily an Atlantic species; many Pacific identifications of that species may actually be **M. villosa** Hincks, which, in turn, may be a synonym of *M. isabelleana* (d'Orbigny). Usually *M. tuberculata*

is much more heavily calcified than the other species; however, young colonies may be difficult to differentiate.

For details of systematics, structure, reproduction, and ecology, see Atkins (1955a,b), Banta (1969), Bobin (1977), Franzén (1956), Haderlie (1968a), Lutaud (1959, 1961, 1977), O'Donoghue (1927), Osburn (1950), Pinter (1969), Robertson (1908, as *M. tehuelca*), Rucker (1968), Ryland (1977), Silén (1944a,b,c, 1966), D. Soule & J. Soule (1967), J. Soule & Duff (1957), J. Soule, D. Soule & Pinter (1975, keys and figures of species), Strathmann (1973), Turner & Strachan (1969), Winston (1977, 1978), Woollacott & North (1971), and Woollacott & Zimmer (1977, index).

Phylum Ectoprocta / Class Gymnolaemata / Order Cheilostomata
Suborder Anasca / Family Thalamoporellidae

6.2 **Thalamoporella californica** (Levinsen, 1909)

An abundant fouling organism on pilings, wharves, ship hulls, rocks, and some algae in shallow warm waters, low intertidal zone to 145 m; northern Channel Islands and Point Conception (Santa Barbara Co.) to Colombia and Galápagos Islands; fossils in Pleistocene deposits at Newport Beach (Orange Co.).

Colony consisting of a basal crust with numerous, coarse, upright projections; projections dichotomously branched and consisting of many zooids; individuals 0.5–0.6 mm long and about 0.3 mm wide, arched distally above the aperture; cryptocyst (calcareous plate underlying transparent front wall of zooecium) porous, and with a large hole on either side, near operculum, through which muscles pass to the membranous frontal wall; overall, individual resembling a flask with two handles at the neck.

Many ovicells, in the form of bilobed calcareous hoods that rise above the individuals, are easily seen when the colony is fertile. Up to four large pink ova can be seen through the translucent centers of the ovicells. Members of the genus *Thalamoporella* possess unique internal spicules, which can be seen only under high magnification by crushing a portion of a dried specimen in a drop of water on a microscope slide. The spicules of *T. californica* are all shaped like curved calipers.

This species is one of the most important bryozoans encrusting the giant kelp *Macrocystis*; where best developed, under dense kelp canopies, it averages 284 gm per m² of blade surface and constitutes nearly half the total animal biomass present. Inshore plants most commonly bearing colonies in southern California are the red algae *Gelidium, Lithothrix,Pterocladia,* and *Gigartina,* and the brown alga *Egregia.*

The anatomy and embryology of a related species (*T. evelinae*), which has separate male and female zooids borne on the same colony, are beautifully shown in Marcus (1941). See also Osburn (1950), Pinter (1969), Powell & Cook (1966), Robertson (1908, as *T. rozieri*), Rucker (1968), Silén (1938, 1977), D. Soule & J. Soule (1964), J. Soule & Duff (1957), J. Soule & D. Soule (1970), J. Soule, D. Soule & Pinter (1975), Turner & Strachan (1969), Winston (1978), and Woollacott & North (1971).

Phylum Ectoprocta / Class Gymnolaemata / Order Cheilostomata Suborder Anasca / Family Scrupocellariidae

6.3 Tricellaria occidentalis (Trask, 1857)

On rocks and algae, low intertidal zone to 30 m or more; also on offshore kelp; Queen Charlotte Islands (British Columbia) to Isla Cedros (Baja California); Japan; abundant in San Francisco Bay area.

Colony forming white bushy tuft not over 2 cm high; branches arising alternately, the node joints yellow-brown, the internodes usually formed by three zooecia; individual zooecia about 0.4 mm long, with a frontal membranous covering bordered by five to seven long spines; a specialized spine (scutum) projecting from border out over frontal membrane below midline area; avicularia present; ovicells large, globose, and perforate.

A nominal subspecies, **T. occidentalis catalinensis** (Robertson), generally having a more elaborate scutum, is abundant in warmer waters off southern California.

See Mawatari (1951b), Osburn (1950), Pinter (1969), Robertson (1905, as *Menipea occidentalis*), J. Soule, D. Soule & Pinter (1975), and Woollacott & North (1971).

Phylum Ectoprocta / Class Gymnolaemata / Order Cheilostomata Suborder Anasca / Family Bicellariellidae

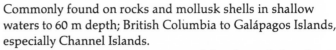

6.4 Bugula californica (Robertson, 1905)

Commonly found on rocks and mollusk shells in shallow waters to 60 m depth; British Columbia to Galápagos Islands, especially Channel Islands.

Colony composed of one to several fronds, each frond to 5 cm or more in height, formed by spiral whorls of branches, giving the colony the appearance of having been trimmed evenly at several levels; color usually whitish, but sometimes tan, or (during sexual reproduction) somewhat orange; branches formed of two rows of individuals, widening to four or five zooids at bifurcation points; individuals about 0.5–0.6 mm long and 0.2–0.25 mm wide; examination with hand lens showing three spines at end of each zooid, globular ovicells that sit straight across end of zooid, and large "bird's head" avicularia attached at middle of side walls.

The *Bugula californica* reported as a fouling organism from ports such as San Francisco Bay and Los Angeles harbor has recently been recognized as **B. stolonifera** Ryland. Although very similar to *B. californica, B. stolonifera* is grayish and lacks the distinctive, whorled colony pattern; it is found on both sides of the Atlantic Ocean.

Under a dissecting microscope the activities of the *B. californica* avicularia are readily seen; the ability of avicularia to catch and hold small objects can be demonstrated easily by probing them with a hair, or by introducing copepods, or the nauplius larvae of barnacles or brine shrimp, into the dish. Soule and Soule have seen this species use avicularia of several zooecia to pull apart small crustaceans (caprellid amphipods); the tentacles then directed the fragments into the polypide mouths. Experiments with other species suggest that the avicularia play a role in defending the colony against small disruptive invaders.

The coronate larvae of *B. californica* settle in Los Angeles harbor year around, in warm years. In Monterey Bay they settle in all but the winter months; colonies growing on shallow suspended panels reached a height of 30 mm in 4 months.

See Calvet (1900), Haderlie (1968b, 1969, 1974), Kaufman (1968), Maturo (1959), Mawatari (1946), Osburn (1950), Reish (1963), Robertson (1905), Rucker (1968), Schneider (1963), S. Smith & Haderlie (1969), D. Soule, J. Soule & Henry (1979), J. Soule, D. Soule & Pinter (1975), Thompson & Chow (1955), and Woollacott & North (1971).

Phylum Ectoprocta / Class Gymnolaemata / Order Cheilostomata
Suborder Anasca / Family Bicellariellidae

6.5 Bugula neritina (Linnaeus, 1758)

A common fouling organism on harbor installations and boat hulls, low intertidal zone to 80 m depth; Monterey Bay to Canal Zone (Panama) and Galápagos Islands; Gulf of California as far north as Isla Ángel de la Guarda; in warm waters around the world.

Colony bushy, 3–10 cm in height, easily distinguished from similar species by its reddish-brown or purplish-brown color; individuals 0.6–0.8 mm long, 0.2–0.3 mm wide; spines and avicularia absent; ovicell set diagonally at tip of zooecium.

This cosmopolitan species has been studied extensively in many areas because of its abundance and general importance as a fouling organism. The animals may be cultured in the laboratory on a diet of the flagellate *Monochrysis lutheri*.

Individual zooids are hermaphroditic, and sperm are probably released through pores at the tips of the tentacles. Developing coronate larvae are brooded one at a time in large ovicells, each of which consists of a small heterozooid that lacks a differentiated polypide and that combines with parts of both the maternal zooid and the zooid just distal to it. The embryo, surrounded by maternal tissues, is nourished by yolk and other materials from the mother. A placenta-like system aids in the transport of extraembryonic nutrients.

Released larvae swim rapidly by means of cilia; at first they swim toward light, but later some become photonegative. Two pigmented spots, one on each side, appear to be eye-spots of a unique type; each contains one cell, probably a photoreceptor, provided with a cluster of unmodified cilia held at right angles to the incoming light. The larvae, still laden with yolk, lack mouths and indeed any trace of a gut. They usually swim for less than 10 hours, then settle, and attach by inversion of an internal sac that, in metamorphosis, gives rise to the ancestrula epidermis. The body, still nearly solid, elongates and forms a cuticle. Larval organs degenerate almost completely, and tissues of the first zooids derive largely or wholly from larval ectoderm. Many environmental factors, such as temperature, light, materials in the water, and character of the substratum, affect attachment and metamorphosis. At Monterey the release and settlement of larvae occur through most of the year; they virtually cease in midwinter (December to February) and reach a peak in June and July. At Los Angeles and San Diego, settlement is most abundant from April through October, but larvae can be obtained in all months.

Growth is rapid. At La Jolla (San Diego Co.), 3-week-old colonies were 8 mm high, with 8 branches and over 50 zooids; at 6 weeks they were 12 mm high, with 32–64 branches and about 300 zooids. Colonies in Monterey harbor on shallow submerged panels reached a height of 120 mm in 4–6 months, and some colonies persisted for at least 14 months. Since sexual reproduction begins 6–8 weeks after settlement (La Jolla), three or four generations a year are possible.

Bugula neritina rarely settles on plants other than the giant kelp *Macrocystis*, and it is not abundant there. It provides a favorite perch for the skeleton shrimp *Caprella californica* (22.8).

See especially Woollacott & Zimmer (1971, 1977, 1978); see also Bullivant (1967, 1968b), Calvet (1900), Coe (1932), Gooch & Schopf (1970), Grave (1930), Haderlie (1968b, 1969, 1974), Hammen & Osborne (1959), Keith (1971), Loeb & Walker (1977), Lynch (1947, 1961), Maturo (1959), Mawatari (1951a), Miller (1946), Osburn (1950), Pinter (1969), Reish (1963), Rucker (1968), Ryland (1960, 1976, 1977), S. Smith & Haderlie (1969), J. Soule, D. Soule & Pinter (1975), Ström (1977), Winston (1978), Woollacott & North (1971), Woollacott & Zimmer (1972a,b,c, 1975), and Zimmer & Woollacott (1977a,b).

Phylum Ectoprocta / Class Gymnolaemata / Order Cheilostomata
Suborder Anasca / Family Bicellariellidae

6.6 **Dendrobeania laxa** (Robertson, 1905)

In shaded areas protected from strong surf, low intertidal zone on rocky coasts; subtidal to 90 m; British Columbia to San Pedro (Los Angeles Co.).

Colony extensive, recumbent, the fronds numerous, curved, petal-like, of variable size and shape; rootlike attachment processes (radicles) intertwining with fronds to form a tangled mass; individuals about 0.7–0.9 mm long, with an uncalcified frontal membrane surrounded by heavy spines; avicularia lacking; ovicell round, prominent, and faintly striated.

See Calvet (1900), Osburn (1950), and J. Soule, D. Soule & Pinter (1975).

Phylum Ectoprocta / Class Gymnolaemata / Order Cheilostomata
Suborder Ascophora / Family Schizoporellidae

6.7 **Schizoporella unicornis** (Johnston, 1847)

Common on rocks and shells along open coast and on wharf pilings in bays and harbors, low intertidal zone to 60 m depth; Strait of Georgia (British Columbia) to southern California; Hawaii, Japan; western Europe, Mediterranean Sea.

Colony encrusting, the overall color brown, golden yellow, or reddish orange; zooids hexagonal in shape, each 0.5–0.6 mm long, 0.3–0.4 mm wide, in district longitudinal rows separated by definite grooves; aperture wider than long, with a shallow, U-shaped proximal sinus, and with a tiny tooth inside the aperture at each corner of proximal lip; umbonate frontal wall with many small pores; avicularia, when present, single or paired; ovicell globose, finely perforate, with radiating flutings on surface.

This species is often confused with **S. errata** (Waters) on both Atlantic and Pacific coasts. Ross & McCain (1976) provides evidence of the presence of *S. unicornis* in Spanish Bay (Washington) in 1927. Powell (1970) suggests that *S. unicornis* was introduced to California from Japan in 1932 with Pacific

oysters planted at Morro Bay (San Luis Obispo Co.), and that *S. errata* was introduced from the Atlantic coast along with the oyster *Crassostrea virginica.*

Atlantic coast "*S. unicornis*" (=*S. errata*) has been the subject of recent studies on population genetics and calcification of the shell. The exoskeleton contains calcium carbonate in the forms of both calcite and aragonite.

Studies of *S. unicornis* in southern California have shown (1) that sperm are released through pores at the tentacle tips, (2) that the population suffered a high mortality during a "red tide," and (3) that the larvae tend to settle on substrata other than algae.

See Bullivant (1967), Gooch & Schopf (1970), Haderlie (1974), Maturo (1959), Pelluet & Hayes (1936), Pinter (1969), Powell (1970), Reish (1963), Ross & McCain (1976), Rucker (1968), Sandberg (1977), Schopf (1977), Schopf & Allen (1970), Schopf & Gooch (1971), D. Soule & J. Soule (1968), J. Soule, D. Soule & Pinter (1975), Strathmann (1977), Sutherland (1977, 1978), and Winston (1977, 1978).

Phylum Ectoprocta / Class Gymnolaemata / Order Cheilostomata
Suborder Ascophora / Family Schizoporellidae

6.8 **Hippodiplosia insculpta** (Hincks, 1882)

Abundant in lower middle and low intertidal zones on rocky shores, encrusting stones, mollusk shells, hydroids, algae, and other bryozoans; common on *Phyllochaetopterus prolifica* tubes (18.22) on piles in Monterey harbor; subtidal to 234 m; Alaska to Gulf of California and Isla del Coco (Costa Rica); Pleistocene fossils at San Pedro (Los Angeles Co.).

Colony encrusting, to 5 cm or more in diameter, but also rising in double-layered frills or fanlike folds; color light yellow or pale tan, light orange when ova are present; individuals 0.5–0.7 mm long, 0.3–0.4 mm wide, rectangular to hexagonal in shape; distinguishable from other species of the genus on this coast by the presence of large globular ovicells that are imperforate but bear radiating striations and by the absence of avicularia.

This species is excellent for observation under the dissecting microscope. Individuals readily extend the lophophore in freshly collected colonies, and the embryos or larvae can be liberated by rupturing the ovicells carefully with a sharp needle. A comparison of colonies collected at different latitudes shows that individual zooid size may decrease as water temperature increases.

See Calvet (1900), Osburn (1952), J. Soule & Duff (1957), J. Soule, D. Soule & Pinter (1975), Thompson & Chow (1955), and Woollacott & North (1971).

Phylum Ectoprocta / Class Gymnolaemata / Order Cheilostomata
Suborder Ascophora / Family Eurystomellidae

6.9 **Eurystomella bilabiata** (Hincks, 1884)

On stones and mollusk shells, low intertidal zone on rocky shores; subtidal to 64 m; Alaska to Bahía de Tenacatita (Mexico); common in central California (Monterey area), but most abundant in waters north of California; Pleistocene fossils from San Pedro (Los Angeles Co.) and Japan.

Colony flat, encrusting in a single-layered sheet to 5 cm or more in diameter; color rose-red, red-orange, or brown; individuals averaging 0.6–0.7 mm long and 0.5 mm wide, smooth and shiny, arranged in a distinct regular pattern in the colony; apertures of zooids large and bell-shaped; frontal wall calcified, imperforate, with a raised umbo in the center; spines and avicularia absent (see also Fig. 12.31).

The species name *bilabiata* of this beautiful and conspicuous form calls attention to the unusual pair of lips or folds that operate with the operculum to close the aperture when the tentacles are withdrawn.

Colonies from central California often bear on their surfaces numerous, small, black folliculinid protozoans. These sessile ciliates inhabit tubular shells from which they extend two ciliated lobes to feed.

Eurystomella bilabiata is preyed upon by the nudibranch *Hopkinsia rosacea* (14.37). The color of the nudibranch closely matches that of rose-red colonies of *E. bilabiata*, and the pigment responsible (hopkinsiaxanthin) appears to be identical in the two species.

See McDonald & Nybakken (1978), Osburn (1952), Robertson (1908, as *Lepralia bilabiata*), J. Soule & Duff (1957), and J. Soule, D. Soule & Pinter (1975).

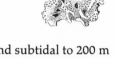

Phylum Ectoprocta / Class Gymnolaemata / Order Cheilostomata
Suborder Ascophora / Family Reteporidae

6.10 **Phidolopora pacifica** (Robertson, 1908)

On rocks in deep, sheltered tidepools and subtidal to 200 m along open coast and in bays; British Columbia to Peru; most abundant in shallow subtidal waters from Oregon to southern California, and common on breakwaters of Los Angeles harbor and at scuba depths around the Channel Islands.

Colony fenestrate, forming an erect, lacy, convoluted mass to 110 mm across and 65 mm high, orange to orange-pink; zooecia to about 0.6 mm long, 0.3 mm wide, frontal avicularia and ovicells often prominent.

The zooids all open on one side of the stiff, lacelike sheets that make up the colony. Water emerging from between the tentacles of feeding zooids passes through the holes in the lace without disruption of incoming feeding currents. The fenestrated form not only provides for more efficient feeding, but it renders the colony more resistant to currents than if it were a solid sheet of zooids.

In Puget Sound (Washington) the sea star *Dermasterias imbricata* (8.3) occasionally feeds on *Phidolopora*.

See Haderlie (1969), Mauzey, Birkeland & Dayton (1968), Osburn (1952), Robertson (1908, as *Retepora pacifica*), J. Soule & Duff (1957), J. Soule, D. Soule & Pinter (1975), Thompson & Chow (1955), Turner, Ebert & Given (1969).

Phylum Ectoprocta / Class Gymnolaemata / Order Cheilostomata
Suborder Ascophora / Family Celliporidae

6.11 **Costazia robertsoniae** Canu & Bassler, 1923 (=Celliporina robertsoniae)

Common on rocks and algae, low intertidal zone to 100 m depth along rocky shores; British Columbia to northern Mexico, but most common from Dillon Beach (Marin Co.) to La

Jolla (San Diego Co.); Pleistocene fossils at Santa Monica and San Pedro (both Los Angeles Co.).

Colony rough in appearance, encrusting and forming small nodules or rising in irregularly forked branches to 50 mm or more in height; individuals about 0.6–0.7 mm long and 0.4 mm wide, heavily calcified, with few pores on frontal wall but many pores at edges between adjacent zooids, aperture rounded but with a V-shaped notch; raised avicularia small, lying on each side of the aperture and often above it where an ovicell is lacking; ovicells exposed or embedded, each with a frontal area bearing triangular pores around the edge.

This is one of the bryozoans that most commonly encrust blades of the giant kelp *Macrocystis* from Monterey Bay to Baja California.

See Osburn (1952), Rucker (1968), J. Soule & Duff (1957), J. Soule, D. Soule & Pinter (1975), and Woollacott & North (1971).

Phylum Ectoprocta / Class Gymnolaemata / Order Cheilostomata
Suborder Ascophora / Family Celliporidae

6.12 Celleporaria brunnea (Hincks, 1884)
(= Holoporella brunnea)

Common on rocks, shells, worm tubes, or algae, on rocky shores from low intertidal zone to depths of 185 m; on wharf piles in harbors; British Columbia to Colombia and Galápagos Islands; extremely common in Gulf of California; Pleistocene fossils at San Pedro (Los Angeles Co.).

Colony encrusting, to several centimeters in diameter, building a heavy multilayered crust to 10 mm or more in thickness, composed of more or less randomly arranged individuals heaped on one another, often produced into coarse folds and branches; color gray or brown; individuals about 0.6–0.8 mm long, 0.4 mm wide, each with a suboral spike giving the colony a bristly appearance; avicularian mandibles large, distinctive, black or brown; extended tentacles conspicuously blood-red in color.

This is the most common encrusting bryozoan in Monterey harbor. *C. brunnea* larvae settled on submerged panels there in all months of the year; the rate was highest from March through September, lowest in December and January. (In Los Angeles–Long Beach harbor results were similar; larvae did not settle during the winter months on panels at 3 m depth.) At Monterey very little settlement was noted at 15 m depths on panels 215 m from shore. Growth was fastest in colonies just below the level of the lowest tides—one colony here reached a diameter of 6 cm in 7 months. The growth rate declined with increasing depth. Old colonies on pilings become thickened and form globular crustose masses up to 6 cm in diameter.

Living specimens are excellent for study under a dissecting microscope, for the zooids extend readily. Many of the zooids in larger colonies have asymmetrical lophophores, the tentacles on one side being longer than those on the other. In this species, and in *C. albirostris*, localized clusters of zooids appear to behave in a coordinated manner in feeding and in retracting. Retracting zooids often show a curious response involving a writhing of the tentacles against the surface of the colony. These features of asymmetry and behavior are not unique, but appear in some other genera and species as well.

In Monterey harbor, *C. brunnea* is preyed upon by the flatworm *Hoploplana californica* (4.6), which mimics the bryozoan in color and texture. The flatworm eats mainly the older zooids at the centers of the colonies, and thus has little effect on marginal growth. (Both species are shown in Fig. 4.6.)

See especially Haderlie (1968a,b, 1969, 1974); see also Bullivant (1967), Marcus (1938), Osburn (1952), Pinter (1969), Reish (1963), S. Smith & Haderlie (1969), D. Soule, J. Soule & Henry (1979), J. Soule & Duff (1957), J. Soule, D. Soule & Pinter (1975), and Winston (1978).

Phylum Ectoprocta / Class Gymnolaemata / Order Cyclostomata
Suborder Tubuliporina / Family Diastoporidae

6.13 Diaperoecia californica (d'Orbigny, 1852)
(= Idmonea californica)

On rocks, mollusk shells, and other hard substrata, from just below the level of the lowest tides to depths over 185 m along rocky shores; British Columbia to Costa Rica; common in

Gulf of California and along Pacific coast of Mexico, and dredged in quantity off Los Angeles harbor and Santa Catalina Island (Channel Islands); Pleistocene fossils from Santa Barbara, and from Santa Monica and San Pedro (both Los Angeles Co.).

Colony coral-like in appearance, the branches erect or spreading, 25 mm or more in height, often anastomosing to form large reticular masses 20–25 cm in diameter and 10 cm high, the branches formed by successive rows of four or five zooids fused together for all or part of their length; individuals tubular, larger than most cyclostomes, averaging 0.22 mm in diameter, curved, bearing circular terminal apertures; reproductive chamber a large individual (gonozooid) usually occurring below bifurcating branches and appearing as a perforate swollen area surrounding the bases of a number of tubules.

One of the most common ectoprocts at scuba-diving depths in southern California, particularly near the tops of natural reefs. The interstices of the branches provide a habitat reminiscent of the crevices between branches on a coral head. Branches of a colony that touch become fused, and growing branches that reach rock surfaces become secondarily attached. Colonies taken in calm, deep water have thin, more erect branches; in those from more-exposed locations the branches are thicker and more flattened. Small colonies are widely distributed, but usually not abundant, on the giant kelp *Macrocystis*.

Anatomy and biology of related forms are covered in Borg (1926), Harmelin (1975), and Nielsen (1970). See also Osburn (1953), Robertson (1910, as *Idmonea californica*), J. Soule & Duff (1957), J. Soule, D. Soule & Pinter (1975), Turner, Ebert & Given (1969), and Woollacott & North (1971).

Phylum Ectoprocta / Class Gymnolaemata / Order Cyclostomata
Suborder Articulata / Family Crisiidae

6.14 **Crisulipora occidentalis** Robertson, 1910

On rocks along open shores, on dock pilings, in clusters of the mollusk *Mytilus* (15.9, 15.10) in sheltered waters, low intertidal zone to 86 m depth; Monterey Bay to Peru; Japan;

Brazil; common from San Pedro (Los Angeles Co.) to San Diego, Channel Islands, and Gulf of California.

Colony to 30 mm high, white to pale tan, fragile, erect, branching, bushy, with jointed rootlets; autozooids tubular, variable in length, fused at their bases but flaring out separately along branches, averaging 0.12 mm in diameter at tips; apertures located at free tips of zooids and lacking opercula; avicularia absent; female reproductive structures consisting of separate individuals (gonozooids) with inflated bodies much larger than those of autozooids; surface of both autozooids and gonozooids bearing minute perforations.

In gross appearance this species resembles *Bugula* (i.e., 6.4, 6.5) and other related erect bryozoans. However, examination with a hand lens clearly reveals the characteristic features of cyclostome ("round mouth") bryozoans: the tubular shape of the zooids, the round terminal apertures, and the absence of opercula.

The sexes are separate in *C. occidentalis*; only the female colonies possess gonozooids. As often happens in cyclostomes, the embryo subdivides to give rise to numerous larvae; gonozooids may bear many offspring in various stages of development. In Monterey Bay, swimming larvae have been observed to settle and start colonies in all months of the year, but the main settlement is from May through September. Faster-growing colonies reach 50 mm in height in 6–7 months.

For general cyclostome biology, see Borg (1926), Harmelin (1975), Nielsen (1970), Ross (1977), Winston (1977, 1978), and Zimmer & Woollacott (1977a). For *C. occidentalis*, see Coe (1932), Haderlie (1968a,b, 1969, 1974), Nielsen (1970), Osburn (1953), Robertson (1903, 1910), S. Smith & Haderlie (1969), and J. Soule, D. Soule & Pinter (1975).

Phylum Ectoprocta / Class Gymnolaemata / Order Cyclostomata
Suborder Articulata / Family Crisiidae

6.15 **Filicrisia franciscana** (Robertson, 1910)

On rocks and algae on outer coast, on pilings and floats in harbors, upper middle to low intertidal zone and subtidal to 100 m depth; Orca (Alaska) to southern California; Japan.

Colony bushy, 15–25 mm high; branches straight; zooecia

tubular, 0.6–0.9 mm long, three to five per internode; gono-zooids swollen and recumbent; colony white, with joints on branches often black and conspicuous.

This small and delicate form occurs higher in the intertidal zone than any other California ectoproct; it attaches to the bases of such red algae as *Endocladia*, *Gelidium*, and *Gastroclonium*, and in higher pools is associated with strands of the green algae *Enteromorpha* and *Chaetomorpha*. Colonies often grow on blades of giant kelp, as well.

Colonies are probably either male or female. Origin of the germ cells and embryonic development have been followed in some detail. At about the 20–24-cell stage, the primary embryo develops multicellular projections from which secondary embryos arise. These, in turn, may develop tertiary embryos. Eventually all embryos, including the primary, develop into larvae that are liberated to swim and settle. In Monterey harbor, settlement occurs from May to October.

See Borg (1926), Coe (1932), Glynn (1965), Haderlie (1968a, 1969), Nielsen (1970), Osburn (1953), Pinter (1969), Robertson (1903, as *Crisia occidentalis*; 1910, as *C. franciscana*), J. Soule, D. Soule & Pinter (1975), Woollacott & North (1971), and Zimmer & Woollacott (1977a).

Phylum Ectoprocta / Class Gymnolaemata / Order Ctenostomata Suborder Carnosa / Family Flustrellidridae

6.16 Flustrellidra corniculata (Smitt, 1871)
(=Flustrella corniculata)

A cold-water species characteristic of rocky shores, often growing on calcareous algae and stipes of the large brown alga *Laminaria*, low intertidal zone; subtidal to 75 m or more; Alaska to Point Buchon (San Luis Obispo Co.); northern Europe.

Colony pale tan to dark brown, often 4–6 cm long, forming cylindrical sheath on algal stipes or branching into small cylinders or leafy shapes depending on nature and shape of substratum; feeding individuals about 0.7–0.8 mm long and about 0.4–0.6 mm wide; spines, produced by numerous special zooids, tall, chitinous, branched, giving colony a coarse fuzzy appearance.

This species takes its specific name *corniculata* ("provided with little horns") from the conspicuous spines borne on special zooids. Post-larval development of the polypide and muscles has been followed. A related Atlantic species, *F. hispida*, produces a shell-bearing larva that is nourished by stored yolk and does not feed on plankton. Its attachment and metamorphosis have been studied.

See especially J. Soule (1954, as *Flustrella corniculata*); see also Barrois (1877), Loeb & Walker (1977), Osburn (1953, as *Flustrella corniculata*), Pace (1906), Prouho (1890), Ryland (1960), L. Smith (1973), J. Soule, D. Soule & Pinter (1975), Ström (1977), Winston (1977), and Zimmer & Woollacott (1977a,b).

Phylum Ectoprocta / Class Gymnolaemata / Order Ctenostomata Suborder Stolonifera / Family Vesiculariidae

6.17 Zoobotryon verticillatum (Delle Chiaje, 1828)

A common fouling organism in warmer harbors and at lowered salinities, often seen floating in tangled spaghetti-like masses from docks, floats, and ship hulls; San Diego to Gulf of California and Central America, and, in recent years, in harbors as far north as Long Beach (Los Angeles Co.) and Corona del Mar (Orange Co.); circumtropical, extending into warm Mediterranean waters.

Colony stringy, to 1–2 m long, consisting of tangles of soft flaccid stolons 0.4.–0.7 mm in diameter; individuals barrel-like, ovoid, 0.36–0.5 mm long, arising in groups from the stolons.

The biology of this species has been studied at Newport harbor (Orange Co.). The zooids have small mouths and feed only on particles 45 μm or less in diameter. Small flagellates are readily taken as food. Newly budded zooids are sexually precocious, and mature sperm develop in the body cavity before the immature animal first extends its tentacles. The sperm emerge from pores on the tentacle tips, and most are emitted during the first 6 hours of active life. Fertilization is internal, and single white larvae are found in zooids undergoing regression. Larval release occurs mainly in the morning in the laboratory; larvae are attracted to light, and attach to the walls of tanks just below the water surface. At metamorphosis a tubular ancestrula is produced, with a polypide that is fully functional (unusual in ctenostome ancestrulae). A stolon arises from the base of the ancestrula, branches repeatedly, and produces buds that become autozooids. These function for about 6 days, then degenerate. Colonies fed and kept at 25–26°C grew and produced larvae; those fed and held at 20°C did not.

See especially Bullivant (1967, 1968a,b,c); see also Osburn (1953), Ries (1936), D. Soule & J. Soule (1967), J. Soule (1954), Winston (1977, 1978), and Zimmer & Woollacott (1977a,b).

Phylum Ectoprocta / Class Gymnolaemata / Order Ctenostomata
Suborder Stolonifera / Family Vesiculariidae

6.18 **Bowerbankia gracilis** (O'Donoghue, 1926)

Common on low intertidal rocks and pilings in quieter waters; Puget Sound (Washington) to Isla del Espíritu Santo (Baja California); Atlantic coast from Greenland to Brazil; cosmopolitan, if all identifications are correct.

Colony consisting of stolons from which arise clusters of erect, tubular zooecia about 1–1.5 mm long, usually densely packed to form a brown encrusting fuzz.

Bowerbankia, like many ctenostomes, possesses, at the base of its esophagus, a gizzard, equipped with a set of chitinous teeth, each tooth secreted by a single cell. It takes approximately 1 hour for ingested particles to traverse the gut from mouth to anus.

In sexual reproduction, the zooids produce only one egg at a time. Enlargement of an egg is accompanied by regression of the polypide. The egg is retained after fertilization and is brooded in the vestibule of the parent. A non-feeding coronate larva develops there, escapes by the aperture, swims, settles, metamorphoses, and buds to form a colony. Development of the polypide and muscles has been followed, for both the zooid structure and the order of development of the muscle sets are important in separating bryozoans of the order Ctenostomata into two major subgroups.

Bowerbankia gracilis is abundant in Los Angeles and Monterey harbors. At Monterey, larvae settle from January or February to October, with the greatest settlement from April to September. Colonies then grow rapidly and often cover settled barnacles and ascidians (e.g., *Ascidia ceratodes*, 12.25).

See Bobin (1977), Eiben (1976), Franzén (1956), Haderlie (1968a,b, 1969, 1974), Osburn (1953), Ryland (1960, 1970), J. Soule (1954), J. Soule, D. Soule & Pinter (1975), Ström (1977), Winston (1977, 1978), and Zimmer & Woollacott (1977a,b).

Phylum Ectoprocta / Class Gymnolaemata / Order Ctenostomata
Suborder Stolonifera / Family Triticellidae

6.19 **Triticella elongata** (Osburn, 1912)

On carapace and appendages and inside gill chamber of the pea crab *Scleroplax granulata* (25.41), in tunnels made by burrowing organisms in mud flats of protected bays; Elkhorn Slough (Monterey Bay); Atlantic coast from Massachusetts to North Carolina.

Colony consisting of creeping stolons; individual zooids 0.9–1.8 mm long and 0.18–0.24 mm wide, arising erect, in pairs or clusters, from stolonial internodes, each zooid with an elongated narrow stalk (pedicel), the stalk expanding into an inflated ovoid area containing tentacles and alimentary system; aperture terminal.

This species is excellent for observations of bryozoan behavior. Living on the body of a tunnel-dwelling crab, the zooids are accustomed to being knocked about; when disturbed, they contract, only to reemerge and commence feeding almost immediately.

Reproduction has been observed in a related European species, *T. koreni*, which lives on the burrowing crustacean *Calocaris*. The eggs are released, but having sticky external membranes, they adhere for a day to the exterior of the bryozoan. A ciliated gastrula hatches and swims away, to complete its development in the plankton. Rudiments of a larval gut develop, but feeding probably does not occur. Larval development takes only a week, but the larvae may spend months in the pelagic realm before settling.

See Franzén (1956, 1977), Osburn (1953), J. Soule (1954), J. Soule, D. Soule & Pinter (1975), Ström (1977), and Zimmer & Woollacott (1977).

Phylum Entoprocta / Family Pedicellinidae

6.20 **Barentsia benedeni** (Foettinger, 1887)

On pilings, floats, worm tubes, barnacle and bivalve shells, and other hard substrata in harbor and estuarine areas, low intertidal and adjacent subtidal zones; San Francisco Bay; Europe.

Colony to several centimeters in diameter, consisting of prostrate stolons bearing upright individuals often 1–4 mm long, each with an enlarged muscular base, a stalk, and a distal calyx; calyces bowl-shaped, 0.1–0.3 mm deep, carrying a fringe of tentacles; stalk in young zooids often plain and unjointed, in older zooids usually with 1–12 (average 4) joints marked by muscular swellings.

This is one of several *Barentsia* species reported from California. It is often confused with **B. gracilis** (M. Sars), a cosmopolitan species that differs from *B. benedeni* in lacking joints marked by muscular enlargements along the stalk.

Barentsia benedeni colonies in San Francisco Bay (Palo Alto) reproduce sexually and brood larvae from February to July or August. Autumn colonies lack larger zooids, and many stalks lack calyces. Regeneration of calyces increases in January. The brooded developing larvae are attached to the upper surface of the parental calyx, where they feed by diverting food from the adjacent parental food grooves with their ciliary tracts.

For adult and larval structure and behavior of *B. benedeni*, see Mariscal (1965, as *B. gracilis*). For other California species of *Barentsia*, see Mariscal (1975b) and Osburn (1953). For extensive reviews and bibliographies on entoprocts, see Hyman (1951), Mariscal (1975a), and Nielsen (1971). See also Atkins (1932), Brien (1956–57, 1959, 1960a), Brien & Papyn (1954), Franzén (1956), Haderlie (1968b, 1969), Nielsen (1977a,b), Prenant & Bobin (1956), D. Soule & J. Soule (1965), and Woollacott & Eakin (1973).

Literature Cited

Annoscia, E., ed. 1968. Proceedings of the First International Conference of Bryozoa. Atti Soc. Ital. Sci. Natur. Mus. Civ. Storia Nat. Milano 108: 1–377.

Atkins, D. 1932. The ciliary feeding mechanism of the entoproct Polyzoa, and a comparison with that of the ectoproct Polyzoa. Quart. J. Microscop. Sci. 75: 393–423.

———. 1955a. The cyphonautes larvae of the Plymouth area and the metamorphosis of *Membranipora membranacea* (L.). J. Mar. Biol. Assoc. U.K. 34: 441–49.

———. 1955b. The ciliary feeding mechanism of the cyphonautes larva (Polyzoa Ectoprocta). J. Mar. Biol. Assoc. U.K. 34: 451–66.

Banta, W. C. 1969. The body wall of the cheilostome Bryozoa. II. Interzoidal communication organs. J. Morphol. 129: 149–70.

Barrois, G. 1877. Mémoire sur l'embryologie des bryozoaires. Trav. Inst. Zool. Wimereux 1: 1–305.

Boardman, R. S., A. H. Cheetham, and W. A. Oliver, Jr., eds. 1973. Animal colonies: Development and function through time. Stroudsburg, Pa.: Dowden, Hutchinson, & Ross. 603 pp.

Bobin, G. 1977. Interzooecial communications and the funicular system, pp. 397–33, *in* Woollacott & Zimmer (1977).

Borg, F. 1926. Studies on recent cyclostomatous Bryozoa. Zool. Bidrag Uppsala 10: 181–507.

Brien, P. 1956–57. Le bourgeonnement des endoproctes et leur phylogénèse. Ann. Soc. Roy. Zool. Belg. 87: 27–43.

———. 1959. Classe des endoproctes ou kamptozoaires, pp. 927–1007, *in* P.-P. Grassé, ed., Traité de zoologie, vol. 5 (1). Paris: Masson. 1,053 pp. + 63-page addendum.

———. 1960a. Le bourgeonnement et la phylogénèse des endoproctes et des ectoproctes: Réflexions sur les processus de l'évolution animale. Bull. Acad. Roy. Belg. (Sci.) (5) 46: 748–66.

———. 1960b. Classe de bryozoaires, pp. 1053–1335, *in* P.-P. Grassé, ed., Traité de zoologie, vol. 5 (2). Paris: Masson. 2,219 pp.

Brien, P., and L. Papyn. 1954. Les endoproctes et la class de bryozoaires. Ann. Soc. Roy. Zool. Belg. 85: 59–87.

Bullivant, J. S. 1967. Release of sperm by Bryozoa. Ophelia 4: 139–42.

———. 1968a. Attachment and growth of the stoloniferous ctenostome bryozoan *Zoobotryon verticillatum*. Bull. So. Calif. Acad. Sci. 67: 199–202.

———. 1968b. The rate of feeding of the bryozoan *Zoobotryon verticillatum*. New Zeal. J. Mar. Freshwater Res. 2: 111–34.

———. 1968c. The method of feeding of lophophorates (Bryozoa, Phoronida, Brachiopoda). New Zeal. J. Mar. Freshwater Res. 2: 135–46.

Calvet, L. 1900. Contributions a l'histoire naturelle des bryozoaires ectoproctes marins. Thèses, Fac. Sci. Paris, Montpellier. 488 pp., 13 pls.

Coe, W. R. 1932. Season of attachment and rate of growth of sedentary marine organisms at the pier of the Scripps Institution of Oceanography, La Jolla, California. Bull. Scripps Inst. Oceanogr., Tech. Ser. 3: 37–86.

Eiben, R. 1976. Einfluss von Benetzungsspannung und Ionen auf die Substratbesiedlung und das Einsetzen der Metamorphose bei Bryozoenlarven (*Bowerbankia gracilis*). Mar. Biol. 37: 249–54.

Farmer, J. D. 1977. An adaptive model for the evolution of the ectoproct life cycle, pp. 487–517, *in* Woollacott & Zimmer (1977).

Franzén, A. 1956. On spermiogenesis, morphology of the spermatozoan, and biology of fertilization among invertebrates. Zool. Bidrag Uppsala 31: 355–480, 6 pls.

————. 1977. Gametogenesis of bryozoans, pp. 1–22, *in* Woollacott & Zimmer (1977).

Glynn, P. W. 1965. Community composition, structure, and interrelationships in the marine intertidal *Endocladia muricata–Balanus glandula* association in Monterey Bay, California. Beaufortia 12: 1–198.

Gooch, J. L., and T. J. M. Schopf. 1970. Population genetics of marine species of the phylum Ectoprocta. Biol. Bull. 138: 138–56.

Gordon, D. P. 1977. The aging process in bryozoans, pp. 335–76, *in* Woollacott & Zimmer (1977).

Grave, B. H. 1930. The natural history of *Bugula flabellata* at Woods Hole, Massachusetts, including the behavior and attachment of the larvae. J. Morphol. 49: 355–84.

Haderlie, E. C. 1968a. Marine boring and fouling organisms in open water of Monterey Bay, California, pp. 658–79, *in* A. H. Walters and J. S. Elphick, eds., Biodeterioration of materials. Proc. 1st Internat. Biodeter. Symp. Barking, Essex, Eng.: Elsevier. 740 pp.

————. 1968b. Marine fouling and boring organisms in Monterey harbor. Veliger 10: 327–41.

————. 1969. Marine fouling and boring organisms in Monterey harbor. II. Second year of investigation. Veliger 12: 182–92.

————. 1974. Growth rates, depth preference and ecological succession of some sessile marine invertebrates in Monterey harbor. Veliger 17 (Suppl.): 1–35.

Hammen, C. S., and P. J. Osborne. 1959. Carbon dioxide fixation in marine invertebrates: A survey of major phyla. Science 130: 1490–10.

Harmelin, J. G. 1975. Relations entre la forme zoariale et l'habitat chez les bryozoaires cyclostomes. Conséquences taxonomiques, pp. 369–84, *in* Pouyet (1975).

Hyman, L. H. 1951. The invertebrates: Acanthocephala, Aschelminthes, and Entoprocta. Vol. 3. New York: McGraw-Hill. 572 pp.

————. 1959. The invertebrates: Smaller coelomate groups. Vol. 5. New York: McGraw-Hill. 784 pp.

Kaufmann, K. W., Jr. 1968. The biological role of *Bugula*-type avicularia (Bryozoa), preliminary report, pp. 54–60, *in* Annoscia (1968).

Keith, D. E. 1971. Substrate selection in caprellid amphipods of southern California, with emphasis on *Caprella californica* Stimpson and *Caprella equilibra* Say (Amphipoda). Pacific Sci. 25: 387–94.

Larwood, G. P., ed. 1973. Living and fossil Bryozoa—recent advances in research. Proc. 2nd Conf. Internat. Bryozool. Assoc. London: Academic Press. 634 pp.

Loeb, M. J., and G. Walker. 1977. Origin, composition, and function of secretions from pyriform organs and internal sacs of four settling cheilo-ctenostome bryozoan larvae. Mar. Biol. 42: 37–46.

Lutaud, G. 1959. Étude cinématographique du bourgeonnement chez *Membranipora membranacea* (L.) bryozoaire chilostome. I. Le developpement du polypide. Bull. Soc. Zool. France 84: 167–73.

————. 1961. Contribution à l'étude du bourgeonnement et la croissance des colonies chez *Membranipora membranacea* (Linné), bryozoaire chilostome. Ann. Soc. Roy. Zool. Belg. 91: 157–300.

————. 1977. The bryozoan nervous system, pp. 377–410, *in* Woollacott & Zimmer (1977).

Lynch, W. F. 1947. The behavior and metamorphosis of the larva of *Bugula neritina* (Linnaeus): Experimental modification of the length of the free-swimming period and the responses of the larvae to light and gravity. Biol. Bull. 92: 115–50.

————. 1961. Extrinsic factors influencing metamorphosis in bryozoan and ascidian larvae. Amer. Zool. 1: 59–66.

Marcus, E. 1938. Bryozoarios marinhos brasileiros—II. Bol. Fac. Phil. Sci. Letr. Univ. S. Paulo 4, Zoologia 2: 1–196, 29 pls.

————. 1941. Sôbre os Bryozoa do Brasil. Bol. Fac. Phil. Sci. Letr. Univ. S. Paulo 22, Zoologia 5: 3–208, 18 pls. [English summary, pp. 139–54.]

Mariscal, R. N. 1965. The adult and larval morphology and life history of the entoproct *Barentsia gracilis* (M. Sars, 1835). J. Morphol. 116: 311–38.

————. 1975a. Entoprocta, pp. 1–41, *in* A. C. Giese, and J. S. Pearse, eds., Reproduction of marine invertebrates. 2. Entoprocts and lesser coelomates. New York: Academic Press. 344 pp.

————. 1975b. Phylum Entoprocta, pp. 609–13, *in* R. I. Smith and J. Carlton, eds., Light's manual: Intertidal invertebrates of the central California coast. 3rd ed. Berkeley and Los Angeles: University of California Press. 716 pp.

Maturo, F. J. S., Jr. 1959. Seasonal distribution and settling rates of estuarine Bryozoa. Ecology 40: 116–27.

Mauzey, K. P., C. Birkeland, and P. K. Dayton. 1968. Feeding behavior of asteroids and escape responses of their prey in the Puget Sound region. Ecology 49: 603–19.

Mawatari, S. 1946. On the metamorphosis of *Bugula californica* Robertson. [In Japanese.] Misc. Rep. Res. Inst. Nat. Resources, Tokyo, 9: 21–28

————. 1951a. Natural history of a common fouling bryozoan *Bugula neritina* (Linnaeus). Misc. Rep. Res. Inst. Nat. Resources, Tokyo, 19–21: 47–54.

————. 1951b. On *Tricellaria occidentalis* (Trask), one of the fouling bryozoans in Japan. Misc. Rep. Res. Inst. Nat. Resources, Tokyo, 22: 9–16.

Mayr, E. 1968. Bryozoa versus Ectoprocta. Syst. Zool. 17: 213–16.

McCammon, H. M., and W. A. Reynolds, organizers. 1977. Biology of lophophorates. Amer. Zool. 17: 1–150.

McDonald, G. R., and J. W. Nybakken. 1978. Additional notes on the food of some California nudibranchs with summary of known food habits of California species. Veliger 21: 110–19.

Miller, M. A. 1946. Toxic effects of copper on attachment and growth of *Bugula neritina*. Biol. Bull. 90: 122–40.

Nielsen, C. 1970. On metamorphosis and ancestrula formation in cyclostomatous bryozoans. Ophelia 7: 217–56.

———. 1971. Entoproct life-cycles and the entoproct/ectoproct relationship. Ophelia 9: 209–341.

———. 1977a. Phylogenetic considerations: The protostomian relationships, pp. 519–34, *in* Woollacott & Zimmer (1977).

———. 1977b. The relationship of Entoprocta, Ectoprocta and Phoronida. Amer. Zool. 17: 149–50.

O'Donoghue, C. H. 1927. Observations on the early development of *Membranipora villosa* Hincks. Contr. Canad. Biol. (n.s.) 3: 249–63.

Osburn, R. C. 1950. Bryozoa of the Pacific coast of America. Cheilostomata Anasca. Allan Hancock Pacific Exped. 14: 1–269.

———.1952. Bryozoa of the Pacific coast of America. Cheilostomata Ascophora. Allan Hancock Pacific Exped. 14: 271–611.

———. 1953. Bryozoa of the Pacific coast of America. Cyclostomata, Ctenostomata, Entoprocta, and addenda. Allan Hancock Pacific Exped. 14: 613–841.

Pace, R. M. 1906. On the early stages in the development of *Flustrella hispida* (Fabricius) and on the existence of a "yolk nucleus" in the egg in this form. Quart. J. Microscop. Sci. 50: 435–78, pls. 22–25.

Pelluet, D., and F. R. Hayes. 1936. *Schizoporella unicornis*—a bryozoan new to Canada. Proc. N.S. Inst. Sci. 19: 157–59.

Pinter, P. 1969. Bryozoan-algal associations in southern California waters. Bull. So. Calif. Acad. Sci. 68: 199–218.

Pouyet, S., ed. 1975. Bryozoa 1974. Proc. 3rd Conf. Internat. Bryozool. Assoc. Docum. Lab. Géol. Fac. Sci. Lyon 3: 1–690.

Powell, N. A. 1970. *Schizoporella unicornis*—an alien bryozoan introduced into the Strait of Georgia. J. Fish. Res. Bd. Canada 27: 1847–53.

Powell, N. A., and P. L. Cook. 1966. Conditions inducing polymorphism in *Thalamoporella rozieri* (Audouin) (Polyzoa, Anasca). Cah. Biol. Mar. 7: 53–59.

Prenant, M., and G. Bobin. 1956. Bryozoaires. Faune de France 60. Fed. Franc. Soc. Sci. Natur., Paris. 398 pp.

Prouho, H. 1890. Recherches sur la larve de *Flustrella hispida* (Gray). Arch. Zool. Expér. Gén. (2) 8: 409–59, pls. 22–24.

Reish, D. J. 1963. Mass mortality of marine organisms attributed to the "red tide" in southern California. Calif. Fish & Game 49: 265–70.

Ries, E. 1936. Fütterungsversuche bei *Zoobotryon* (Bryozoa). Z. Vergl. Physiol. 23: 64–99.

Robertson, A. 1903. Embryology and embryonic fission in the genus *Crisia*. Univ. Calif. Publ. Zool. 1: 115–56, pls. 12–15.

———. 1905. Non-incrusting chilostomatous Bryozoa of the west coast of North America. Univ. Calif. Publ. Zool. 2: 235–322, pls. 4–16.

———. 1908. The incrusting chilostomatous Bryozoa of the west coast of North America. Univ. Calif. Publ. Zool. 4: 253–344, pls. 14–24.

———. 1910. The cyclostomatous Bryozoa of the west coast of North America. Univ. Calif. Publ. Zool. 6: 225–84, pls. 18–25.

Ross, J. R. P. 1970. Keys to the Recent Cyclostomata Ectoprocta of marine waters of northwest Washington State. Northwest Sci. 44: 154–69.

———. 1977. Microarchitecture of body wall of extant cyclostome ectoprocts. Amer. Zool. 17: 93–105.

Ross, J. R. P., and K. W. McCain. 1976. *Schizoporella unicornis* (Ectoprocta) in coastal waters of the northwestern United States and Canada. Northwest Sci. 50: 160–71.

Rucker, J. B. 1968. Skeletal mineralogy of cheilostome Bryozoa, pp. 101–10, *in* Annoscia (1968).

Ryland, J. S. 1960. Experiments on the influence of light on the behavior of polyzoan larvae. J. Exper. Biol. 37: 783–800.

———. 1967. Polyzoa. Oceanogr. Mar. Biol. Ann. Rev. 5: 343–69.

———. 1968. Terminological problems in Bryozoa, pp. 225–36, *in* Annoscia (1968).

———. 1970. Bryozoans. London: Hutchinson University Library. 175 pp.

———. 1974 (publ. 1976). Behaviour, settlement and metamorphosis of bryozoan larvae: A review. Thalassia Jugoslavica 10: 239–62.

———. 1976. Physiology and ecology of marine bryozoans. Adv. Mar. Biol. 14: 285–443.

———. 1977. Taxes and tropisms of bryozoans, pp. 411–36, *in* Woollacott & Zimmer (1977).

Sandberg, P. A. 1977. Ultrastructure, minerology, and development of bryozoan skeletons, pp. 143–81, *in* Woollacott & Zimmer (1977).

Schneider, D. 1963. Normal and phototropic growth reactions in the marine bryozoan *Bugula avicularia*, pp. 357–71, *in* E. C. Dougherty, Z. N. Brown, E. D. Hanson, and W. D. Hartman, eds., The lower Metazoa: Comparative biology and phylogeny. Berkeley and Los Angeles: University of California Press. 478 pp.

Schopf, T. J. M. 1977. Population genetics of bryozoans, pp. 459–86, *in* Woollacott & Zimmer (1977).

Schopf, T. J. M., and J. R. Allen. 1970. Phylum Ectoprocta, order Cheilostomata: Microprobe analysis of calcium, magnesium, strontium, and phosphorus in skeletons. Science 169: 280–82.

Schopf, T. J. M., and J. L. Gooch. 1971. Gene frequencies in a marine ectoproct: A cline in natural populations related to sea temperature. Evolution 25: 286–89.

Silén, L. 1938. Zur Kenntnis des Polymorphismus der Bryozoen. Die Avicularien der Cheilostomata Anasca. Zool. Bidrag Uppsala 17: 149–366, 18 pls.

_____. 1944a. On the division and movements of the alimentary canal of the Bryozoa. Ark. Zool. 35A(12): 1–40.

_____. 1944b. The main features of the development of the ovum, embryo and ooecium in the ooeciferous Bryozoa Gymnolaemata. Ark. Zool. 35A(17): 1–34.

_____. 1944c. Origin and development of the Cheilo-Ctenostomatous stem of Bryozoa. Zool. Bidrag Uppsala 22: 1–59.

_____. 1966. On the fertilization problem in the gymnolaematous Bryozoa. Ophelia 3: 113–40.

_____. 1977. Polymorphism, pp. 183–231, *in* Woollacott & Zimmer (1977).

Smith, L. W. 1973. Ultrastructure of the tentacles of *Flustrellidra hispida* (Fabricius), pp. 335–42, *in* Larwood (1973).

Smith, S. V., and E. C. Haderlie. 1969. Growth and longevity of some calcareous fouling organisms, Monterey Bay, California. Pacific Sci. 23: 447–51.

Soule, D. F. 1973. Morphogenesis of giant avicularia and ovicells in some Pacific Smittinidae, pp. 485–95, *in* Larwood (1973).

Soule, D. F., and J. D. Soule. 1964a. The Ectoprocta (Bryozoa) of Scammon's Lagoon, Baja California, Mexico. Amer. Mus. Novitates 2199: 1–56.

_____. 1964b. Clarification of the family Thalamoporellidae (Ectoprocta). Bull. So. Calif. Acad. Sci. 63: 193–200.

_____. 1965. Two new species of Loxosomella, Entoprocta, epizoic on Crustacea. Allan Hancock Found. Publ. Occas. Pap. 29: 1–19.

_____. 1967. Faunal affinities of some Hawaiian Bryozoa (Ectoprocta). Proc. Calif. Acad. Sci. 13: 265–272.

_____. 1968. Bryozoan fouling organisms from Oahu, Hawaii with a new species of *Watersipora*. Bull. So. Calif. Acad. Sci. 67: 203–18.

Soule, D. F., J. D. Soule, and C. A. Henry. 1979. The influence of environmental factors on the distribution of estuarine bryozoans, as determined by multivariate analysis. 4th Internat. Conf. Internat. Bryozool. Assoc. London: Academic Press. (in press).

Soule, J. D. 1954. Post-larval development in relation to the classification of the Bryozoa Ctenostomata. Bull. So. Calif. Acad. Sci. 53: 13–34.

_____. 1959. Results of the Puritan-American Museum of Natural History Expedition to western Mexico. 6. Anascan Cheilostomata (Bryozoa) of the Gulf of California. Amer. Mus. Novitates 1969: 1–54.

_____. 1961. Results of the Puritan-American Museum of Natural History Expedition to western Mexico. 13: Ascophoran Cheilostomata (Bryozoa) of the Gulf of California. Amer. Mus. Novitates 2053: 1–66.

_____. 1963. Results of the Puritan-American Museum of Natural History Expedition to western Mexico. 18. Cyclostomata, Ctenostomata (Ectoprocta) and Entoprocta of the Gulf of California. Amer. Mus. Novitates 2144: 1–34.

_____. 1973. Histological and histochemical studies on the bryozoan-substrate interface, pp. 343–47, *in* Larwood (1973).

Soule, J. D., and M. M. Duff. 1957. Fossil Bryozoa from the Pleistocene of southern California. Proc. Calif. Acad. Sci. (4) 29: 87–146.

Soule, J. D., and D. F. Soule. 1968. Perspectives on the Bryozoa-Ectoprocta question. Syst. Zool. 17: 468–70.

_____. 1970. New species of *Thalamoporella* (Ectoprocta) from Hawaii examined by scanning electron microscopy. Amer. Mus. Novitates 2417: 1–18.

_____. 1977. Fouling and bioadhesion: Life strategies of bryozoans, pp. 437–57, *in* Woollacott & Zimmer (1977).

Soule, J. D., D. F. Soule, and P. A. Pinter. 1975. Phylum Ectoprocta (Bryozoa), pp. 579–608, *in* R. I. Smith and J. T. Carlton, eds., Light's manual: Intertidal invertebrates of the central California coast. 3rd ed. Berkeley and Los Angeles: University of California Press. 716 pp.

Strathmann, R. 1973. Function of lateral cilia in suspension feeding of lophophorates (Brachiopoda, Phoronida, Ectoprocta). Mar. Biol. 23: 129–36.

Ström, R. 1977. Brooding patterns of bryozoans, pp. 23–55, *in* Woollacott & Zimmer (1977).

Sutherland, J. P. 1977. Effect of *Schizoporella* (Ectoprocta) removal on the fouling community at Beaufort, North Carolina, U.S.A., pp. 155–76, *in* B. C. Coull, ed., Ecology of marine benthos. Belle W. Baruch Library in Marine Science No. 6. Columbia: University of South Carolina Press. 467 pp.

_____. 1978. Functional roles of *Schizoporella* and *Styela* in the fouling community at Beaufort, North Carolina. Ecology 59: 257–64.

Thompson, T. G., and T. J. Chow. 1955. The strontium-calcium atom ratio in carbonate-secreting marine organisms. Pap. Mar. Biol. Oceanogr., Deep Sea Res. 3 (Suppl.): 20–39.

Turner, C. H., E. E. Ebert, and R. R. Given. 1969. Man-made reef ecology. Calif. Dept. Fish & Game, Fish Bull. 146. 221 pp.

Turner, C. H., and A. R. Strachan. 1969. The marine environment in the vicinity of the San Gabriel River mouth. Calif. Fish & Game 55: 53–68.

Winston, J. E. 1977. Feeding in marine bryozoans, pp. 233–71, *in* Woollacott & Zimmer (1977).

_____. 1978. Polypide morphology and feeding behavior in marine ectoprocts. Bull Mar. Sci. 28: 1–31.

Woollacott, R. M., and R. M. Eakin. 1973. Ultrastructure of a potential photoreceptoral organ in the larva of an entoproct. J. Ultrastruc. Res. 43: 412–25.

Woollacott, R. M., and W. J. North. 1971. Bryozoans of California and northern Mexico kelp beds, pp. 455–79, *in* W. J. North, ed., The biology of giant kelp beds (*Macrocystis*) in California. Nova Hedwigia 32: 1–600.

Woollacott, R. M., and R. L. Zimmer. 1971. Attachment and metamorphosis of the cheilo-ctenostome bryozoan *Bugula neritina* (Linné). J. Morphol. 134: 351–82.

_____. 1972a. Fine structure of a potential photoreceptor organ in the larva of *Bugula neritina* (Bryozoa). Z. Zellforsch. 123: 458–69.

_____. 1972b. Origin and structure of the brood chamber in *Bugula neritina* (Bryozoa). Mar. Biol. 16: 165–70.

_____. 1972c. A simplified placenta-like brooding system in *Bugula neritina* (Bryozoa), pp. 130–31, *in* C. J. Arceneux, ed., 30th Ann. Proc. Electron Microscopy Soc. Amer., Los Angeles.

_____. 1975. A simplified placenta-like system for the transport of extraembryonic nutrients during embryogenesis of *Bugula neritina* (Bryozoa). J. Morphol. 147: 355–77.

_____, eds. 1977. Biology of bryozoans. New York: Academic Press. 566 pp.

_____. 1978. Metamorphosis of cellularoid bryozoans, pp. 49–63, *in* F.-S. Chia and M. E. Rice, eds., Settlement and metamorphosis of marine invertebrate larvae. New York: Elsevier/North Holland Biomedical. 290 pp.

Zimmer, R. L. 1973. Morphological and developmental affinities of the lophophorates, pp. 593–99, *in* Larwood (1973).

Zimmer, R. L., and R. M. Woollacott. 1977a. Metamorphosis, ancestrulae, and coloniality in bryozoan life cycles, pp. 91–142, *in* Woollacott & Zimmer (1977).

_____. 1977b. Structure and classification of gymnolaemate larvae, pp. 57–89, *in* Woollacott & Zimmer (1977).

Chapter 7

Brachiopoda and Phoronida

Russel L. Zimmer and Eugene C. Haderlie

The Brachiopoda and Phoronida constitute separate phyla usually considered to be related to each other and to the Ectoprocta as well. Differing views on phylogenetic relationships are summarized in Farmer (1977), Hyman (1959), Nielsen (1977a,b), and Zimmer (1973).

Brachiopods have a compact body that is largely encased between two shells, one dorsal and one ventral. Of modest size, brachiopods seldom exceed a few centimeters in greatest shell dimension. In some (class Articulata), the two valves are dissimilar and provided with a toothed hinge that restricts shell movement to an opening and closing action. These animals have a short, slender attachment stalk (pedicel) that secures them permanently to shell or rock. The stalk is flexible and the animal can move enough to dislodge sediment or to orient the shell to local currents. Other brachiopods (class Inarticulata) lack a hinge, and the two shells are more mobile. The more familiar inarticulate brachiopods have a long stout pedicel that anchors the animal within a vertical burrow in sand or mud. Worldwide in distribution, brachiopods are exclusively marine. They are of limited occurrence in the intertidal zone, but more common on the continental shelf, and extend to depths of 5,000 m. Fewer than 300 living species are known, but the fossil species described number more than 30,000, and they are found as far back as Cambrian time.

Phoronids have a slender body 4–25 cm long, encased in a longer membranous tube that in most species is encrusted with sand

Russel L. Zimmer is Associate Professor of Biological Sciences at the University of Southern California. Eugene C. Haderlie is Professor of Oceanography at the Naval Postgraduate School, Monterey.

grains. Phoronid tubes are usually buried vertically in sand, mud, or fine gravel, but the animals may burrow in limestone or cement their tubes to shell, stones, or even wooden floats. All phoronids are marine, and most of the 12 to 16 recognized species (assigned to two genera) are from the temperate zone. Phoronids occur from the intertidal zone, where they are sometimes very abundant, down to depths of a hundred meters or more. The phoronids probably have existed as a group for a long time, but their soft bodies have not been preserved as fossils; many paleontologists, however, interpret the fossil *Skolithus* to represent the burrows or tubes of phoronids.

In both phyla the body cavity is a coelom, since it is lined with epithelium, and the body itself has two main divisions, the lophophore and the trunk. The lophophoral region is marked externally by a ring of tentacles that surrounds the mouth and is often thrown into complicated loops or spirals. In phoronids the tentacles form a delicate tuft at the distal end of the body, whereas in brachiopods they occupy much of the cavity enclosed by the shells. Along the tentacles are cilia that produce a current of water important in both respiration and feeding. In attached brachiopods the animals use external water movements to augment their ciliary feeding currents by orienting the gape of the shell at right angles to the prevailing ambient flow (M. C. LaBarbera in Vogel, 1978). The diet consists largely of unicellular flagellates and diatoms; colloidal particles and dissolved organic matter may also be used (Chuang, 1956; Emig & Thouveny, 1976; McCammon, 1969; McCammon & Reynolds, 1976; but see Cowen, 1971). Non-living detritus particles rich in organic matter are also ingested. The major portion of the body, the trunk, contains the digestive tract (see Morton, 1960), the reproductive organs, the ciliated funnels that serve both as excretory organs and genital ducts, and most of the musculature. Virtually the entire length of a phoronid is formed of this region; in brachiopods, however, the trunk occupies the back portion of the shell cavity and includes the pedicel and the two body-wall flaps that secrete the shell and line its inner surface.

Phoronids have a well-developed circulatory system, with two main vessels in the trunk and capillaries into each tentacle. Under slight magnification, some of these vessels may be seen, since the blood is bright red, owing to the presence of the respiratory pigment hemoglobin in nucleated cells. Blood in the rudimentary circulatory system of the Brachiopoda either is colorless or contains the red pigment hemerythrin.

As in other sedentary animals, the nervous systems are relatively simple, and, with the exception of statocysts in some brachiopods, complex sensory organs are lacking.

All brachiopods and phoronids reproduce sexually. Each phylum has characteristic larvae that are free-swimming for at least a short time; since the adults are sedentary, these larval stages afford the only opportunity for geographical spread of the species. Asexual reproduction occurs in some phoronids, but not in brachiopods.

Neither phylum is of direct economic importance, but brachiopods, with an impressive fossil record, are of considerable use to paleontologists in dating sedimentary rocks. Both phyla intrigue zoologists, for in many characters they are intermediate between the annelid-mollusk and echinoderm-chordate assemblages that represent two major lines of animal evolution (see Hyman, 1959; Siewing, 1974; Zimmer, 1967).

Brachiopods and bivalved mollusks superficially resemble one another in possessing two shells. However, a brachiopod may always be recognized externally by the bilateral symmetry of each of its two shells and (with the exception of adult *Crania* and *Craniscus*) by the presence of a pedicel. Phoronids are readily confused with tube-dwelling polychaetes, but in the latter the tentacles are branched or threadlike, and the body is divided into segments bearing setae; in the phoronids, by contrast, the tentacles are straight and finger-shaped and the body is unsegmented and lacks setae. The identification of brachiopods to species is based almost exclusively on the architecture of the two shells. The lack of hard parts makes phoronids more difficult to identify; the size and color of the animal and the shape and structure of the tube provide valu-

able clues for field identification, but most of the systematically important characters can be determined only on microscopically thin sections.

Excellent accounts of brachiopods as a group are those of Hyman (1959), Rudwick (1970), and Williams & Rowell (1965). The Cenozoic brachiopods are covered in the monograph of Hertlein & Grant (1944), and Bernard (1972) has a useful key to living west coast (British Columbia) brachiopods. Selected aspects of brachiopod biology are considered in Reynolds & McCammon (1977) and accompanying papers in the symposium on Biology of Lophophorates. For further information on phoronids, see especially Hyman (1959), also Emig (1975, 1977), Marsden (1959), Strathmann (1973), and the papers of Zimmer, including his 1975 key to west coast intertidal phoronids.

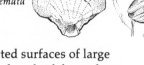

Phylum Brachiopoda / Class Articulata / Order Telotremata

7.1 Terebratalia transversa Sowerby, 1846

Scarce, attached to undersides or protected surfaces of large rocks, low intertidal zone; more commonly subtidal to at least 1,800 m, in clean, quiet water; Alaska to Baja California.

Valves to 56 mm wide, broader than long, smooth or with prominent radial ribbing, the concentric sculpturing conspicuous or modest; pedicel valve with slightly elevated median ridge and rounded notch at margin opposite pedicel; shell color gray, yellow, or reddish.

Terebratalia transversa has existed as a distinct species for over 20 million years and is represented by fossils in sedimentary rock of Miocene age. The animal is solitary and does not occur in clusters as do some brachiopods.

Along the Pacific coast, *T. transversa* breeds from November through February, and in Washington the larvae settle and metamorphose during these months. The fertilized eggs develop into a characteristic (but nameless) larva that does not feed during its short planktonic life. The formation of the coeloms in this larva is reminiscent of that in echinoderms and hemichordates, but the larval setae resemble those of polychaetes even in their fine structure. After attachment of the larva, there is a considerable period of differentiation before the typical features of the adult are evident.

Growth rates of *T. transversa* have been monitored in Washington. Individuals at 12 months of age average 10 mm in width, at 24 months 25 mm, and at 48 months 40 mm; growth rates then decline and the animals reach maximum size (50–56 mm wide) at 9–10 years. Sexual maturity occurs during the second year, when the animals are about 12 mm wide. The often-prominent growth lines on the shell are not good indicators of age.

Individuals of *T. transversa* live well in laboratory aquaria and have been used in studies on feeding and on the growth, differentiation, and structure of the lophophore, including cytology of the myoepithelial cells of the tentacles.

The main predators on these brachiopods, at least in low intertidal and shallow subtidal waters, appear to be crabs, which chip off part of the margin of the shell to get at the soft parts inside. *T. transversa* may survive such attacks, and many show areas of asymmetrical shell growth where repairs have been made.

Two other species of brachiopods are occasionally encountered along California shores. **Platidia (Morrissia) hornii** Gabb (class Articulata) is attached to stones or shells in the low intertidal zone along semiprotected rocky shores and in subtidal waters at least to 400 m, from central California to central Mexico. Its valves are nearly circular, only a few millimeters in greatest dimension, and the aperture for the pedicel is shared by both valves, rather than restricted to one. **Glottidia albida** Hinds (class Inarticulata) is found burrowed in sand or sandy mud in quiet water in the low intertidal zone and in subtidal waters to 160 m, from Tomales Bay (Marin Co.) to Acapulco (Mexico). This species has an elongated shell to 35 mm in length; the valves are similar, with parallel sides, a rounded anterior end, and a pointed posterior end. The valves are smooth and white with red-brown or brownish markings.

For *T. transversa*, see Atkins (1959), Bernard (1972), Flammer (1963), Gustus & Cloney (1972), Long (1963, 1964), Mattox (1955), McCammon & Reynolds (1976), Paine (1969), Reed & Cloney (1975, 1977), Reynolds & McCammon (1977), Smith & Gordon (1948), and Strathmann (1973). For other species, see Bernard (1972), Chuang (1956, 1960), Cowen (1971), Hertlein & Grant (1944), Jones & Barnard (1963), Mattox (1955), Paine (1963), and Smith & Gordon (1948).

Phylum Phoronida

7.2 **Phoronis vancouverenisis** Pixell, 1912 (? = P. hippocrepia)

Often common, in clumps or sheets of intertwined individuals on pilings and floats in harbors or on protected rocks in bays, or burrowing into shells or calcareous rocks, low intertidal zone and adjacent subtidal waters; British Columbia to southern California.

Individual worms to 40 mm long, 1–2 mm wide; color whitish but red blood often visible through body wall; tubes membranous, flexible, not sand-encrusted, typically occurring in clumps averaging about 70 mm in diameter.

This is the only Pacific coast phoronid forming clusters of intertwined individuals. On concrete pilings in Monterey harbor and elsewhere, these phoronids may dominate the fouling growth at 1.5–6 m below zero tide level. The clusters harbor a variety of polychaetes, flatworms, nemerteans, and crustaceans, as well as a rich surface film of diatoms. Deep within larger clusters anaerobic conditions often prevail, and sulfur bacteria release hydrogen sulfide.

Phoronis vancouverensis is hermaphroditic. When reproducing, adults possess two pairs of lophophoral organs (accessory sex glands) in the space formed by the indentation of the tentacular ring. One pair facilitates the brooding of the embryos, which in early stages are white and spherical but in later stages possess a purple pigment and are of complex shape. Individual broods contain embryos at various stages of development. The second pair of lophophoral organs functions in spermatophore elaboration. Sperm released into the body cavity accumulate in the nephridia (excretory organs) and pass to the male accessory gland on the lophophore, where they are formed into a packet and surrounded by one or more membranes before being released into the sea. These sperm packets are probably taken up by adjacent individuals so that egg fertilization is internal, although this has not been demonstrated (see the account of *Phoronopsis viridis*, **7.3**).

The brooded embryos develop into pelagic actinotroch larvae, which feed on small plankton and eventually settle, undergo metamorphosis, and grow. Larval tentacles, used in feeding, are retained in juvenile phoronids of this species.

Colonies of *P. vancouverensis* can form rapidly, leading to speculation that the individuals may reproduce asexually. A few cases of spontaneous fission have been observed, in which the body divided in the reproductive region, then regenerated the missing parts to form two complete individuals. In the laboratory, when phoronids are experimentally sectioned through the reproductive region, each half regenerates the lost parts. In nature, the tentacular crown may be periodically cast off and regenerated. Laboratory experiments show, however, that this last is not a means of reproduction, for the cast-off crown cannot itself regenerate a new body.

Three other species of the genus *Phoronis* are also found along the California coast. All are small and occupy slender, sand-encrusted tubes that never form the intertwined clusters characteristic of *P. vancouverensis*.

Phoronis architecta Andrews lives in the low intertidal zone or subtidally to 30 m or more, from Puget Sound (Washington) to southern California. Typically, individuals have straight tubes up to 40 mm long and 1 mm wide that are oriented vertically in the muddy sand of bays or in clean sand along the open coast; occasionally, individuals occur on the undersides of rocks on unprotected shores, and these have irregular tubes. The lophophore of this phoronid is usually flesh-colored, rarely violet or reddish, with a white band encircling the tentacles and with white crescentic flecks just above and below the tentacle bases on opposite sides.

Phoronis psammophila Cori is similar to *P. architecta* but lacks the white pattern of pigmentation on the lophophore. The tube of the third species, **P. pallida** (Schneider) is 12–20 mm long, 0.3–0.5 mm wide, and distinctly flexed one-third of the way toward the base. The elongate trunk is differentiated into several distinct zones. This species, earlier known only from dredge hauls, has been found on intertidal mud flats in Bodega Bay harbor (Sonoma Co.) in the mucus-cemented walls of the burrows of the blue mud shrimp, *Upogebia pugettensis* (24.1). There is circumstantial evidence that *P. pallida* is an obligatory commensal in the burrows of this and other thalassinid shrimp. Like *P. vancouverensis*, this phoronid is hermaphroditic and possesses spermatophoral glands, but unlike *P. vancouverensis*, it sheds its eggs directly into the sea in the summer and fall.

For *P. vancouverensis*, see Bullivant (1968), Donat (1975), Haderlie & Donat (1978), Marsden (1957, 1959), Strathmann (1973), and Zimmer (1964, 1967, 1973). For related species, see Brooks & Cowles (1905), Marsden (1959), Silén (1952, 1954a,b), and Thompson (1972).

Phylum Phoronida

7.3 **Phoronopsis viridis** Hilton, 1930 (? = P. harmeri)

Often common in sand or sandy mud, low intertidal zone in bays and estuaries; Oregon to central California.

Tubes vertical, usually to 180 mm long, 2 mm or less wide, nearly straight, stiff, sand-encrusted, closed at basal end; lophophore bright green to gray-green, often flecked with white.

Along our Pacific coast are populations of *Phoronopsis* that may represent several species; if so, they are so similar as to defy easy separation. The populations at Bodega Bay harbor (Sonoma Co.), Tomales Bay (Marin Co.), Elkhorn Slough (Monterey Co.), and Morro Bay (San Luis Obispo Co.) have usually been identified as *P. viridis*, for which the white-flecked green lophophore is diagnostic. Those from Oregon and Washington lack the green pigmentation and have been identified as *P. harmeri* Pixell.

The immense beds of *P. viridis* found in central California, with as many as 1,200 individuals per m², are unique in the world. Phoronids in most areas are rare animals, and most zoologists know them only from textbooks.

The open upper ends of the tubes of *P. viridis* are generally about 6 cm or less below the surface of the substratum, but can be as deep as 20 cm. But the animals, usually shorter than the tubes, keep a passage open through the mud up to the water above, so that the lophophore may be extended above the substratum for feeding and respiration. When removed completely from the tube, an individual quickly secretes a soft transparent covering. In the field as many as half the animals lack lophophores, and many others have lophophores partially regenerated. It is not known whether the tentacular crowns in nature are cast off spontaneously (as in *Phoronis van-*

couverensis, 7.2) or removed by predators, but individuals placed in laboratory tanks frequently shed and then regenerate their lophophores. *P. viridis* can survive for 6 weeks or more when totally covered with anaerobic mud or sealed in a container filled with black sulfurous sediment.

The animals are not hermaphrodites. During the spring and early summer the males have large spermatophoral organs within the lophophoral concavity.

Spermatophores formed by the males are released, drift off in the seawater, and make chance contact with the lophophores of adjacent individuals in the population. After loss of the spermatophoral membranes, the sperm mass lyses a passage through the tentacle wall. Eventually sperm reach the trunk cavity that contains coelomic oocytes in ripe females. The actual event of fertilization has not been observed, but is presumed to be internal; the eggs of isolated females develop as soon as released into the seawater and occasionally will develop before release.

The development of *P. viridis* is easily observed and has been followed in detail. Superficially, the cleavage pattern appears spiral, but this is probably due to the pressure exerted by the extraembryonic membranes, since most features associated with genuine spiral cleavage (as it occurs in flatworms, annelids, and mollusks) are lacking. If a dozen animals are collected during the reproductive season (spring and summer) and kept in shallow bowls of water, the eggs, about 60 μm in diameter, are usually spawned and settle to the bottom of the dish within a few hours. The eggs are presumably already fertilized, and if pipetted into dishes of fresh seawater, development up to the characteristic actinotroch larval stage can be followed. Development beyond this stage, to metamorphosis, requires special planktonic food and is difficult to achieve in the laboratory. The literature contains one unconfirmed observation of fission, but the frequency of this means of asexual reproduction is unknown.

Phoronopsis viridis possesses giant nerve fibers 25–35 μm in diameter; they mediate a quick withdrawal response and have received some neurophysiological study. In a large population of these phoronids at Bodega Bay, the genetic variability at 39 loci has been investigated using electrophoresis. Antiseptic compounds (bromophenols) have been detected in the body tissues.

A related species, **P. californica** Hilton, inhabits vertical tubes in sediments in the low intertidal zone and in subtidal waters to at least 35 m, from Point Conception (Santa Barbara Co.) to Newport Bay (Orange Co.). The tubes of this species are about 25 cm long and 2–7 mm wide. The lophophore is bright orange to flesh color with white flecks, and the tentacles are arranged in four to nine whorls on each side. Giant nerve fibers extending the length of the body permit the animal to retract quickly into the tube when disturbed or attacked by predators.

See Ayala et al. (1975), Connell (1963), Emig (1974, 1975, 1977), Johnson (1959, 1967), Marsden (1959), Rattenbury (1953, 1954), Robin (1964), Sheikh & Djerassi (1975), Wilson & Bullock (1958), and Zimmer (1964, 1967, 1972).

Literature Cited

Atkins, D. 1959. The growth stages of the lophophore and loop of the brachiopod *Terebratalia transversa* (Sowerby). J. Morphol. 105: 401–26.

Ayala, F. J., J. W. Valentine, L. G. Barr, and G. S. Zumwalt. 1975. Genetic variability in a temperate intertidal phoronid, *Phoronopsis viridis*. Biochem. Genet. 11: 413–27.

Bernard, F. R. 1972. The living brachiopods of British Columbia. Syesis 5: 73–82.

Brooks, W. K., and R. P. Cowles. 1905. *Phoronis architecta*: Its life history, anatomy, and breeding habits. Mem. Nat. Acad. Sci. Washington 10: 69–111.

Bullivant, J. S. 1968. The method of feeding of lophophorates (Bryozoa, Phoronida, Brachiopoda). New Zeal. J. Mar. Freshwater Res. 2: 135–46.

Chuang, S. H. 1956. The ciliary feeding mechanism of *Lingula unguis* (L.) (Brachiopoda). Proc. Zool. Soc. London 127: 167–89.

———. 1960. An anatomical, histological, and histochemical study of the gut of the brachiopod, *Crania anomala*. Quart. J. Microscop. Sci. 101: 9–18.

Connell, J. H. 1963. Territorial behavior and dispersion in some marine invertebrates. Res. Pop. Ecol. 5: 87–101.

Cowen, R. 1971. The food of articulate brachiopods—a discussion. J. Paleontol. 45: 137–39.

Donat, W. 1975. Subtidal concrete piling fauna in Monterey harbor, California. Master's thesis, Oceanography, Naval Postgraduate School, Monterey, Calif. 85 pp.

Emig, C. C. 1974. Phoronidiens récoltés lors de la campagne "Bi-cores" du N/O "Jean Charcot." Tethys 4: 423–28.

———. 1975. The systematics and evolution of the phylum Phoronida. Z. Zool. Syst. Evol. 12: 128–51.

———. 1977. Embryology of Phoronida. Amer. Zool. 17: 21–37.

Emig, C. C., and Y. Thouveny. 1976. Absorption directe d'un acide amine par *Phoronis psammophila*. Mar. Biol. 37: 69–73.

Farmer, J. D. 1977. An adaptive model for the evolution of the ectoproct life cycle, pp. 487–517, *in* R. M. Woollacott and R. L. Zimmer, eds., Biology of bryozoans. New York: Academic Press. 566 pp.

Flammer, L. I. 1963. Embryology of the brachiopod *Terebratalia transversa*. Master's thesis, Zoology, University of Washington, Seattle. 71 pp.

Gustus, R. M., and R. A. Cloney. 1972. Ultrastructural similarities between setae of brachiopods and polychaetes. Acta Zool. 53: 229–33.

Haderlie, E. C., and W. Donat III. 1978. Wharf piling fauna and flora in Monterey harbor, California. Veliger 21: 45–69.

Hertlein, L. G., and U. S. Grant. 1944. The Cenozoic Brachiopoda of western North America. Publ. Math. Phys. Sci., Univ. Calif. Los Angeles 3: 1–236.

Hyman, L. H. 1959. The invertebrates: Smaller coelomate groups. Vol. 5. New York: McGraw-Hill. 783 pp.

Johnson, R. G. 1959. Spatial distribution of *Phoronopsis viridis* Hilton. Science 129: 1221.

———. 1967. The vertical distribution of the infauna of a sand flat. Ecology 48: 571–78.

Jones, G. F., and J. L. Barnard. 1963. The distribution and abundance of the inarticulate brachiopod *Glottidia albida* (Hinds) on the mainland shelf of southern California. Pacific Natur. 4: 27–52.

Long, J. A. 1963. Cortical granules in brachiopod development. Amer. Zool. 3: 504 (abstract).

———. 1964. The embryology of three species representing three superfamilies of articulate brachiopods. Doctoral thesis, Zoology, University of Washington, Seattle. 185 pp.

Marsden, J. R. 1957. Regeneration in *Phoronis vancouverensis*. J. Morphol. 101: 307–23.

———. 1959. Phoronidea from the Pacific coast of North America. Canad. J. Zool. 37: 87–111.

Mattox, N. T. 1955. Observations on the brachiopod communities near Santa Catalina Island, pp. 73–86, *in* Essays in the natural sciences in honor of Captain Allan Hancock, on the occasion of his birthday, July 26, 1955. Los Angeles: University of Southern California Press. 345 pp.

McCammon, H. M. 1969. The food of articulate brachipods. J. Paleontol. 43: 976–85.

McCammon, H. M., and W. A. Reynolds. 1976. Experimental evidence for direct nutrient accumulation by the lophophore of articulate brachiopods. Mar. Biol. 34: 41–51.

Morton, J. E. 1960. The functions of the gut in ciliary feeders. Biol. Rev. 35: 92–140.

Nielsen, C. 1977a. Phylogenetic considerations: The protostomian relationships, pp. 519–34, *in* R. M. Woollacott and R. L. Zimmer, eds., Biology of bryozoans. New York: Academic Press. 566 pp.

———. 1977b. The relationships of Entoprocta, Ectoprocta and Phoronida. Amer. Zool. 17: 149–50.

Paine, R. T. 1963. Ecology of the brachiopod *Glottidia pyramidata*. Ecol. Monogr. 33: 255–80.

———. 1969. Growth and size distribution of the brachiopod *Terebratalia transversa* Sowerby. Pacific Sci. 23: 337–43.

Rattenbury, J. C. 1953. Reproduction in *Phoronopsis viridis*: The annual cycle in the gonads, maturation and fertilization of the ovum. Biol. Bull. 104: 182–96.

———. 1954. The embryology of *Phoronopsis viridis*. J. Morphol. 95: 289–349.

Reed, C. G., and R. A. Cloney. 1975. Myoepithelial cells in the tentacles of an articulate brachiopod. Amer. Zool. 15: 782 (abstract).

———. 1977. Brachiopod tentacles: Ultrastructure and functional significance of the connective tissue and myoepithelial cells in *Terebratalia*. Cell Tissue Res. 185: 17–42.

Reynolds, W. A., and H. M. McCammon. 1977. Aspects of the functional morphology of the lophophore in articulate brachiopods. Amer. Zool. 17: 121–32.

Robin, Y. 1964. Biological distribution of guanidines and phosphagens in marine Annelida and related phyla from California, with a note on pluriphosphagens. Comp. Biochem. Physiol. 12: 347–67.

Rudwick, M. J. S. 1970. Living and fossil brachiopods. London: Hutchinson University Library. 200 pp.

Sheikh, Y. M., and C. Djerassi. 1975. 2,6-Dibromophenol and 2,4,6-tribromophenols—antiseptic secondary metabolites of *Phoronopsis viridis*. Experientia 31: 265.

Siewing, R. 1974. Morphologische Untersuchungen zum Archicoelomatenproblem. 2. Die Körpergliederung bei *Phoronis muelleri* de Selys-Longchamps (Phoronida). Ontogenese—Larvae—Metamorphose—Adultus. Zool. Jahrb. Anat. 92: 275–318.

Silén, L. 1952. Research on Phoronidea of the Gullmar Fjord area (west coast of Sweden). Ark. Zool. 4: 95–140.

———. 1954a. Developmental biology of the Phoronidea of the Gullmar Fjord area of the west coast of Sweden. Acta Zool. 35: 215–57.

———. 1954b. On the nervous system of *Phoronis*. Ark. Zool. 6: 1–40.

Smith, A. G., and M. Gordon, Jr. 1948. The marine mollusks and brachiopods of Monterey Bay, California, and vicinity. Proc. Calif. Acad. Sci. (4) 26: 147–245.

Strathmann, R. 1973. Function of lateral cilia in suspension feeding of lophophorates (Brachiopoda, Phoronida, Ectoprocta). Mar. Biol. 23: 129–36.

Thompson, R. K. 1972. Functional morphology of the hind-gut gland of *Upogebia pugettensis* (Crustacea, Thalassinidea) and its role in burrow construction. Doctoral thesis, Zoology, University of California, Berkeley. 202 pp.

Vogel, S. 1978. Organisms that capture currents. Sci. Amer. 239: 128–39.

Williams, A., and A. J. Rowell. 1965. Brachiopod anatomy, pp. H6–57, *in* R. C. Moore, ed., Treatise on invertebrate paleontology, Part H, Brachiopoda, vol. 1. New York: Geol. Soc. Amer.; Lawrence: University of Kansas Press.

Wilson, D. M., and T. H. Bullock. 1958. Electrical recording from giant fiber and muscle in phoronids. Anat. Rec. 132: 518–19 (abstract).

Zimmer, R. L. 1964. Reproductive biology and development of Phoronida. Doctoral thesis, Zoology, University of Washington, Seattle. 416 pp.

_____. 1967. The morphology and function of accessory reproductive glands in the lophophores of *Phoronis vancouverensis* and *Phoronopsis harmeri*. J. Morphol. 121: 159–78.

_____. 1972. Structure and transfer of spermatozoa in *Phoronopsis viridis*, pp. 108–9, *in* C. J. Arceneaux, ed., 30th Ann. Proc. Electron Microscopy Soc. Amer., Los Angeles.

_____. 1973. Morphological and developmental affinities of the lophophorates, pp. 593–99, *in* G. P. Larwood, ed., Living and fossil Bryozoa—recent advances in research. 2nd Conf. Internat. Bryozool. Assoc. New York: Academic Press. 652 pp.

_____. 1975. Phylum Phoronida, pp. 613–16, *in* R. I. Smith and J. T. Carlton, eds., Light's manual: Intertidal invertebrates of the central California coast. 3rd ed. Berkeley and Los Angeles: University of California Press. 716 pp.

_____. 1978. The comparative structure of the preoral hood coelom in Phoronida and the fate of this cavity during and after metamorphosis, pp. 23–40, *in* F.-S. Chia and M. E. Rice, eds., Settlement and metamorphosis of marine invertebrate larvae. New York: Elsevier/North Holland Biomedical. 290 pp.

Echinodermata: *Introduction*

Donald P. Abbott and Eugene C. Haderlie

The echinoderms are a distinctive group strangely unlike most other multicellular animals. The great American zoologist Libbie Hyman (1955) once referred to the phylum as "a noble group especially designed to puzzle zoologists." The name ("echinos"= "hedgehog" or "urchin," "derma" = "skin") stresses the superficially hard and spiny nature of many species. The spines are part of a unique endoskeleton composed of a large number of hard parts, each—whatever its shape—a single crystal of calcite embedded in living tissue. The uniqueness of echinoderm skeletal features suggests that long ago in Precambrian seas a group of bilaterally symmetrical animals commenced to secrete hard parts in the form of spongy crystals of calcium carbonate (calcite). Perhaps these proved of some value in deterring predators, and selection favored more of the same. Ultimately, development of a weighty armor must have enforced a more sedentary benthic life, and later, in this connection, the group evolved a secondary radial symmetry. Of the earliest stages in the evolution of the group we have no traces, but some clues to ancestry are seen in the curious pattern of embryonic and later development. In ontogeny up to the pelagic larval stage, most echinoderms show bilateral symmetry. A radical metamorphosis then converts the larva into a sedentary animal that is usually well endowed with hard parts, and converts the original bilateral symmetry into a secondary five-parted radial symmetry. Perhaps this sequence of developmental steps reflects the gross events of evolution.

Donald P. Abbott is Professor, Department of Biological Sciences and Hopkins Marine Station, Stanford University. Eugene C. Haderlie is Professor of Oceanography at the Naval Postgraduate School, Monterey.

In their pattern of development, and also in their biochemistry, the echinoderms share some features with the Chordata and the Hemichordata (vertebrates and their relatives). The nature of the evolutionary relationship that these resemblances reflect is still disputed, some authorities arguing that ancient echinoderms are direct ancestors of the chordates (see, for example, Gislén, 1930; Jefferies, 1975), others regarding them as a collateral branch stemming from still earlier common ancestors (see Bone, 1972, for a summary of several views).

In any event, the fossil record shows a great flowering of echinoderm groups in the early Paleozoic, when 20 or more classes existed. Some still retained traces of bilateral symmetry in the adult stage. Only five classes of echinoderms have living representatives today. One, the sea lilies (Crinoidea), occurs only in deeper water off California. The other four classes are conspicuous in both intertidal and subtidal habitats on California shores: sea stars (Asteroidea), sea cucumbers (Holothuroidea), brittle stars (Ophiuroidea), and sea urchins (Echinoidea). Most of the world's 6,000 species of living echinoderms are exclusively marine, but a few tolerate estuarine conditions.

Echinoderms possess a unique system of water-filled tubes, the water-vascular system, derived from the coelom. In most groups the system is open to the sea at one or more conspicuous plates (sieve plates, or madreporites). The water-vascular system uses muscular action and hydraulic pressure to operate the podia (often called tube feet) of various sorts; these appendages may aid in locomotion, clinging, food capture, and even respiration. The areas where the podia protrude from the body are called ambulacra ("ambulare" = "to walk"). The body bears no clear head in the adult stage, and the nervous system remains primitively organized. Yet these animals have long been highly successful, and in places the sea bottom is virtually paved with echinoderms, especially brittle stars.

Most echinoderms have separate sexes (although males and females generally look alike externally) and reproduce by shedding their gametes into the sea, where external fertilization occurs. Some species produce large numbers of small eggs, which develop into free-swimming, plankton-feeding larvae that bear little resemblance to the adults. Others produce yolk-filled larvae that swim about but are independent of planktonic food. Still others produce a few yolky eggs and brood the young to a juvenile stage, bypassing a pelagic phase altogether.

Adult echinoderms are nearly all benthic forms. In their food habits, some, like the starfishes, are mainly carnivores, whereas others, like the sea urchins, are predominantly herbivores. Many echinoderms are filter feeders and have diverse mechanisms for trapping small organisms and food particles from the water. Others, living on muddy bottoms, ingest the sediments and digest the small organisms and organic detritus contained therein.

For excellent general treatments of the echinoderms, see the concise accounts by Clark (1962) and Nichols (1969), the longer works by Cuénot (1948), Dawydoff (1948), and especially Hyman (1955), and the echinoderm volumes (parts S and U) of Moore (1966–67). Physiology, biochemistry, and some aspects of ecology and behavior are covered in Binyon (1972), Nichols (1964), Pentreath & Cobb (1972), and the numerous papers in Boolootian (1966), Florkin & Scheer (1969), and Millott (1967). Those wishing to rear embryos may consult Costello & Henley (1971). For larval biology, see Strathmann (1971, 1974).

Literature Cited

Binyon, J. 1972. Physiology of echinoderms. New York: Pergamon. 264 pp.

Bone, Q. 1972. The origin of chordates. Oxford Biol. Readers 18. London: Oxford University Press. 16 pp.

Boolootian, R. A., ed. 1966. Physiology of Echinodermata. New York: Interscience. 822 pp.

Clark, A. M. 1962. Starfishes and their relations. London: British Museum (Natur. Hist.). 119 pp.

Costello, D. P., and C. Henley. 1971. Methods for obtaining and handling marine eggs and embryos. 2nd ed. Marine Biological Laboratory, Woods Hole, Mass. 247 pp.

Cuénot, L. 1948. Anatomie, éthologie et systématique des échinodermes, pp. 3–275, *in* P. Grassé, ed., Traité de zoologie, vol. 11. Paris: Masson. 1,077 pp.

Dawydoff, C. 1948. Embryologie des échinodermes, pp. 277–363, *in* P. Grassé, ed., Traité de zoologie, vol. 11. Paris: Masson. 1,077 pp.

Florkin, M., and B. T. Scheer, eds. 1969. Chemical zoology. 3. Echinodermata, Nematoda and Acanthocephala. New York: Academic Press. 625 pp.

Gislén, T. 1930. Affinities between the Echinodermata, Enteropneusta, and Chordonia. Zool. Bidrag Uppsala 12: 199–304.

Hyman, L. H. 1955. The invertebrates: Echinodermata. Vol. 4. New York: McGraw-Hill. 763 pp.

Jefferies, R. P. S. 1975. Fossil evidence concerning the origin of the chordates, pp. 253–318, *in* E. J. W. Barrington and R. P. S. Jefferies, eds., Protochordates. Symp. Zool. Soc. London 36. London: Academic Press. 361 pp.

Millott, N., ed. 1967. Echinoderm biology. Symp. Zool. Soc. London 20. London: Academic Press. 240 pp.

Moore, R. C., ed. 1966–67. Treatise on invertebrate paleontology. Part S, Echinodermata 1, 650 pp. Part U, Echinodermata 3, 695 pp. New York: Geol. Soc. Amer.; Lawrence: University of Kansas Press.

Nichols, D. 1964. Echinoderms: Experimental and ecological. Oceanogr. Mar. Biol. Ann. Rev. 2: 393–423.

———. 1969. Echinoderms. 4th ed. London: Hutchinson University Library. 192 pp.

Pentreath, V. W., and J. L. S. Cobb. 1972. Neurobiology of Echinodermata. Biol. Rev. 47: 363–92.

Strathmann, R. R. 1971. The feeding behavior of planktotrophic echinoderm larvae: Mechanisms, regulation, and rates of suspension feeding. J. Exper. Mar. Biol. Ecol. 6: 109–60.

———. 1974 (publ. 1976). Introduction to function and adaptation in echinoderm larvae. Thalassia Jugoslavica 10: 321–39.

Chapter 8

Asteroidea: *The Sea Stars*

Howard M. Feder

Sea stars, or starfishes, are typically star-shaped, with a central area, the disk, gradually merging with the symmetrically placed arms. Although most species have five arms, individual animals with arms numbering anywhere from 4 to 24 are not uncommon along the California coast, and isolated living arms of *Linckia columbiae* (8.2) are occasionally found. Characteristically, the starfish body is somewhat flattened. The oral surface, bearing the mouth at its center and the ambulacral grooves with their rows of tube feet along the arms, is held against the substratum, while the anus and the conspicuous sieve plate (madreporite) lie exposed on the aboral surface of the disk.

The calcareous skeleton in sea stars is composed of many separate ossicles or plates, bound together with connective tissue into a framework that can be held rigid or may be moved about in a surprisingly flexible manner by the circular and radial muscles of the arms and disk. Tubercles and spines are often conspicuous at the body surface. Many sea stars also possess, among the spines, numerous small, pincerlike pedicellariae, which may protect the body surface from the settlement of small organisms or may capture food (see Chia & Amerongen, 1975; Hyman, 1955). However, pedicellariae are lacking in some forms, including *Patiria miniata* (8.4) and *Dermasterias imbricata* (8.3). Spaces between the ossicles and the spines bear soft, saclike extensions, the dermal branchiae or papules, where fluids of the body cavity are separated by only a thin layer of tissue from the surrounding sea. The papules serve for respiration and excretion; extended under water, they may give the sea star a furry appearance, but they are usually retracted when the animal is removed from water or disturbed.

The tube feet, or podia, which extend from the grooves below the arms, are used for attachment to rocks or prey, or for walking on rocks or sand. They even serve for burrowing in softer sediments. In most sea stars the podia terminate in adhesive disks, which attach by means of suction (J. Smith, 1946, 1947) or mucus (Chaet & Philpott, 1960; Defretin, 1952a,b) or both (Nichols, 1966). Sea stars, especially those of rocky shores, move about quite slowly. *Pisaster ochraceus* (8.13) may remain essentially immobile for weeks, and even when active creeps only about 3-4 cm per minute; *Pycnopodia helianthoides* (8.16), perhaps the fastest of the sea stars, moves at about 75–115 cm per minute.

Asteroids are found in all the seas of the world, living on the bottom from the intertidal zone down to great depths. A few species tolerate brackish conditions but none live in fresh water. More than 2,000 living species have been named, with the greatest diversity occurring in the North Pacific. Here they occupy a variety of ecological habitats and ecological niches.

Many studies (reviewed in Feder & Christensen, 1966) have been made on the diets and feeding methods of sea stars. At least some starfishes are able to locate food from a distance and creep toward it. They eat a wide variety of foods (see, for example, Mauzey, Birkeland & Dayton, 1968). Most species are carnivores, feeding mainly on mollusks (snails, mussels, clams), other echinoderms (sea cucumbers, sea urchins, other sea stars), and crustaceans (especially barnacles). A smaller number of sea stars are omnivores or

Howard M. Feder is Professor of Marine Science at the Institute of Marine Science, University of Alaska, Fairbanks.

scavengers. The methods of obtaining food differ. Some species use cilia and mucus to trap tiny organisms and organic particles suspended in the sea (Anderson, 1959, 1960; MacGinitie & MacGinitie, 1968; Pearse, 1965; Rasmussen, 1965). Some swallow their food whole and digest it internally (see, for example, Christensen, 1970); others partially evert their stomachs and carry out preliminary digestion externally. Species of *Pisaster* (e.g., 8.13–15), *Evasterias* (e.g., 8.12), and some other genera can insert the everted stomach through very small, naturally occurring crevices between the shells of bivalves, or they may create an opening by pulling on the valves with the tube feet (see, for example, Christensen, 1957; Feder, 1955a). At present there is no evidence that toxins are used to paralyze prey prior to feeding. Numerous potential prey organisms, such as sea anemones, other sea stars, sea cucumbers, snails, and scallops, give avoidance responses on contact with, or in the presence of, predatory starfishes (as reviewed in Feder & Christensen, 1966; see also Ansell, 1967; Feder, 1967, 1972; Gore, 1966; Jensen, 1966; Mauzey, Birkeland & Dayton, 1968; Ross, 1974). Saponins, possibly complexed with other molecules, appear to be the causative agents for many of these avoidance responses (see, for example, Mackie, Lasker & Grant, 1968; Rio et al., 1963).

Most sea stars reproduce by shedding gametes into the sea; fertilization is external, and the eggs develop to planktonic larvae, which later metamorphose into bottom-dwelling juveniles. Other species of sea stars—including *Henricia* (e.g., 8.5) and *Leptasterias* (e.g., 8.10, 8.11) on this coast—lay large yolky eggs that are brooded and develop directly to tiny juvenile starfishes. Since the rate of growth to the adult stage depends mainly on temperature and food supply, it is usually not possible to determine an individual's age from its size. In polar waters, for example, growth may slow or even cease during the winter. Sea stars living where food is sparse may be conspicuously smaller than individuals of the same age living where food is plentiful (Feder, 1956; Paine, 1976b), and individuals that have been starved for periods of many months slowly decrease in both weight and size. Growth is usually slow under natural conditions, and life spans may be fairly long (e.g., 5 years for *Astropecten* (e.g., 8.1), at least 20 years for *Pisaster ochraceus* (8.13; see Feder, 1956; Feder & Christensen, 1966; MacGinitie & MacGinitie, 1968).

All sea stars are capable of regenerating damaged or lost arms, but some are more efficient at this than others. *Linckia columbiae* can grow an entire animal from a piece of one arm.

Although predation on the planktonic larval stages is heavy, once sea stars have reached adult size they have few enemies. Some nudibranchs and snails may feed on them (see Arakawa, 1960; Robert, 1902; Semon, 1895), and some starfishes (e.g., *Solaster dawsoni*, 8.7, and *Crossaster papposus*) feed almost exclusively on other sea stars. The king crab, *Paralithodes camtschatica*, eats *Pycnopodia helianthoides* (8.16) in Alaskan waters (G. Powell, pers. comm.). Sea gulls and the sea otter take small numbers of *Pisaster* for food. As Hedgpeth notes (in Ricketts & Calvin, 1968), the main enemies of sea stars along the shore are people: school children, tourists, skin divers, collectors.

The class Asteroidea is divided into several orders and suborders; the classification is based primarily on the skeletal parts, a practice which allows systematists to use the same characters in dealing with fossils (extending back to the Ordovician) and recent forms. Characteristics of soft parts, such as podia, and internal organs may also be used. Taxonomic descriptions usually include size, and in sea stars size is generally given in terms of the arm radius, the distance between the center of the mouth (or of the disk) and the tip of the longest arm. In this chapter, the size listed in the species accounts is usually that of the largest specimen mentioned by Fisher (1911, 1928, 1930), not necessarily a maximum.

The best descriptions and distribution records of west coast sea stars are found in the great monograph of Fisher (1911, 1928, 1930). Useful keys for the identification of species occur in Hopkins & Crozier (1966) and Sutton (1975). Alton (1966) and Hopkins & Crozier (1966) give further data on the subtidal distribution of many species. The general biology of asteroids is well covered in Binyon (1972), Boolootian (1966), Clark (1962), Hyman (1955), Nichols (1969), and Sloan (1977), most of which contain extensive bibliographies.

Phylum Echinodermata / Class Asteroidea / Order Paxillosida
Family Astropectinidae

8.1 Astropecten armatus Gray, 1840
(=A. brasiliensis armatus)

On sand or mud, normally at least half-covered by sediments, very low intertidal zone; common only subtidally, to 60 m; San Pedro (Los Angeles Co.) to Ecuador.

Arm radius to 15 cm; body normally with five arms; aboral surface bearing many paxillae (calcareous elevations each supporting a cluster of small spines) and a prominent row of supramarginal plates along the upper edges of the arms, the plates bearing spines or tubercles; ambulacra with two rows of pointed tube feet lacking suckers.

A second species, **A. verrilli californicus** Fisher (= *A. californicus* Fisher), occurs subtidally (14–488 m) on soft bottoms from Point Reyes (Marin Co.) to Baja California, and is the only *Astropecten* recorded north of San Pedro. Its supramarginal plates do not bear spines or tubercles.

Astropecten armatus can move on or below the surface of sand. It feeds mainly on snails, especially *Olivella biplicata* (13.99), which it detects at a distance. The sea star approaches rapidly and forces its arms into the sand around the snail, shifting if the prey moves. *Olivella* has no clear escape reaction to *Astropecten*, and even large specimens seldom escape. The starfish stomach is not everted, and the snail is quickly swallowed whole. In an aquarium, one hungry *Astropecten* ate one adult *Olivella* a day for 4 or 5 days; other investigators report finding three or four snails (species unspecified) at one time in the stomach. Empty shells are ejected through the mouth. *A. armatus* also feeds on dead fishes, the sea pansy *Renilla köllikeri* (3.26), the sand dollar *Dendraster excentricus* (11.7), and the snail *Nassarius perpinguis*. Other *Astropecten* species eat annelids, clams, scaphopods, small crustaceans, other sea stars, sea cucumbers, sea urchins, brittle stars, and fishes. *A. irregularis* preys selectively on clams whose high metabolic rate precludes their remaining closed in the starfish stomach; they thus gape and are easily digested.

See especially Christensen (1970), Edwards (1969), and Fisher (1911, 1930); see also Anderson (1966), Boolootian (1962), Boolootian & Giese (1958), Burla et al. (1972), Carcelles (1944), Chaet (1966a,b), Doi (1976), Engster & Brown (1972), Fager (1968), Feder & Christensen (1966), Feder, Turner & Limbaugh (1974), Ferlin (1973), Ferlin-Lubini & Ribi (1978), Giese (1966a), Heddle (1967), Hopkins & Crozier (1966), Hulings & Hemlay (1963), Hunt (1925), Johnson & Snook (1927), Kisch (1958), Komatsu (1975), MacGinitie & MacGinitie (1968), Macnae & Kalk (1962), Mori & Matutani (1952), Ribi & Jost (1978), Ribi, Schärer & Ochsner (1977), Rosenthal, Clarke & Dayton (1974), and Wells, Wells & Gray (1961).

Phylum Echinodermata / Class Asteroidea / Order Valvatida
Family Ophidiasteridae

8.2 Linckia columbiae Gray, 1840

On rocks and in tidepools, low intertidal zone; uncommon subtidally, to 73 m; southern California, Baja California, Colombia, and Galápagos Islands.

Arm radius to 9 cm; body rarely symmetrical, normally with five arms, occasionally with one to nine; surface coarsely granular above and below, color variable, generally grayish mottled with dull red or red-brown.

Species of *Linckia* often autotomize one or more arms near the disk; each arm can regenerate a whole new animal, an ability unusual in sea stars.

The diet of this species is not known, but a related species, *L. guildingii*, appears to use mucus moved by flagellated tracts to trap bacteria and fine organic detritus.

See Anderson (1960, 1962a,b), Feder, Turner & Limbaugh (1974), Fisher (1911), Hopkins & Crozier (1966), Hyman (1955), MacGinitie & MacGinitie (1968), Monks (1904), Ricketts & Calvin (1968), Rideout (1978), and Swan (1966).

Phylum Echinodermata / Class Asteroidea / Order Spinulosida
Family Poraniidae

8.3 Dermasterias imbricata (Grube, 1857)
Leather Star

Fairly common on rocky shores, very low intertidal and shallow subtidal waters, occasionally to 91 m; on pilings and sea walls in harbors; Prince William Sound (Alaska) to Point

Loma (San Diego Co.); uncommon intertidally in southern California.

Arm radius to 12 cm; disk relatively convex; body normally with five arms, smooth, leathery, slippery; color blue-gray, mottled with red or orange; animals often with a strong garlic or sulfurous odor.

The leather star feeds largely on sea anemones, especially *Epiactis prolifera* (3.29), but also *Anthopleura xanthogrammica* (3.30), *A. elegantissima* (3.31), *Tealia coriacea* (3.34), and *Metridium senile* (3.37). Where anemones are not abundant, their diet includes sea cucumbers (*Cucumaria lubrica, Psolus*), the purple sea urchin *Strongylocentrotus purpuratus* (11.5), and less often other invertebrates, such as sponges, hydroids, sea pens, bryozoans, and ascidians. The prey is swallowed whole and digested internally. *Dermasterias* also feeds on bacterial films and benthic diatoms (A. J. Paul, University of Alaska, pers. comm.). Contact with the leather star causes some sea anemones to detach from the substratum and swim away. This is apparently a response to substances found in special cells on the aboral surface of the sea star. The purple sea urchin may also exhibit an escape response.

See especially Fisher (1911) and Mauzey, Birkeland & Dayton (1968); see also Alton (1966), Birkeland (1974), Bullock (1965), Chia, Atwood & Crawford (1975), Ferguson (1971), Hopkins & Crozier (1966), Margolin (1964a,b), Nigrelli et al. (1967), Nybakken (1969), Robson (1961a,b), Rosenthal & Chess (1970, 1972), Rosenthal, Clarke & Dayton (1974), Ross (1965, 1974), Ross & Sutton (1964), Ştohler (1930), Sund (1958), Ward (1965a,b), Wobber (1975), and Yentsch & Pierce (1955).

Phylum Echinodermata / Class Asteroidea / Order Spinulosida
Family Asterinidae

8.4 Patiria miniata (Brandt, 1835)
Bat Star, Sea Bat

Among rocks overgrown with surfgrass (*Phyllospadix*), larger algae, sponges, and bryozoans, low intertidal zone; subtidal to 290 m on rock, shell, gravel, hard sand, or algae; Sitka (Alaska) to Baja California and Islas de Revillagigedo (Mexico); common in California north of Point Conception (Santa

Barbara Co.); farther south, individuals are scarce, small, usually under rocks.

Arm radius to 10 cm; disk large; arms short and triangular, usually five in number, sometimes four to nine; primary aboral plates crescentic, with concave margins facing center of disk; pedicellariae lacking; color extremely variable, most commonly red or deep orange, plain or mottled with other colors.

Typically an omnivore and scavenger, this sea star feeds by extending its stomach over a great variety of plants and animals, dead or alive, especially surfgrass, algae, and colonial tunicates. Organic films on rock surfaces also provide food, and several plant carbohydrates, including cellulose, are digested. *Patiria* occasionally eats other sea stars, as well. The presence of certain proteins and their breakdown products in the water causes the stomach to evert, and the gut is sometimes extended to trap organic particles suspended in the sea. Animals that show avoidance responses to other sea stars rarely respond to *Patiria*; an exception is the nudibranch *Dendronotus iris* (14.54), which often commences swimming after contact with *Patiria*.

Spawning occurs mainly from May to July, but there is no clear reproductive cycle and some ripe individuals are found at any season. Ripe adults spawn when exposed to air in the shade for 1 to 4 hours, or when injected with water extracts of the radial nerves or with 1-methyladenine (10^{-6} M). Embryos and larvae are transparent and easily reared, making them an excellent choice for classroom demonstrations and embryological experiments.

The *Patiria* population on the bottom in the much-studied kelp forest off Hopkins Marine Station (Pacific Grove, Monterey Co.) averages 4.5 per m² (J. S. Pearse and C. Harrold, pers. comm.). Individuals encountering one another, here and elsewhere, may engage in gentle combat. Bouts, generally involving arm interaction (pushing, placing an arm over an opponent's arm, etc.) are common, and appear to affect the distribution and feeding activities of the population.

A commensal annelid, *Ophiodromus pugettensis* (18.9), lives on the oral surface of *Patiria*, especially in the ambulacral grooves. As many as 20 worms may occur on a single sea star, the number varying seasonally, with the greatest number present during periods of low temperature.

See especially Araki (1964), Farmanfarmaian et al. (1958), Fisher (1911, as *Asterina miniata*), and Moore (1939, 1941); see also Anderson (1959), Atwood (1973), Atwood & Simon (1973), Boolootian (1962, 1966), Boolootian & Giese (1958), Bullock (1965), Caso (1963), Chaet (1966a,b, 1967), Chia, Atwood & Crawford (1975), Child (1941), Epel (1975), Feder, Turner & Limbaugh (1974), Ferguson (1971), Giese (1966a,b), Harrison (1968), Harrison & Philpott (1966), Hayashi (1935b), Heath (1917), Hopkins & Crozier (1966), Houk (1974), Houk & Epel (1974), Irving (1924, 1926), Kanatani (1969, as *Asterina*), Lande & Reish (1968), Loeb (1905), MacGinitie (1938), MacGinitie & MacGinitie (1968), Margolin (1964a), Moore (1953, 1962), Newman (1921a,b,c, 1922, 1925), Nigrelli et al. (1967), Nimitz (1971), Phillips (1977), Rio et al. (1965), Rosenthal, Clarke & Dayton (1974), Schultz & Lambert (1973), Serences (1978), Stevens (1970), Stewart (1970), Stohler (1930), Strathman (1971), Thompson & Chow (1955), Whittaker (1928, 1931), Witt (1930), and Wobber (1970, 1975).

Phylum Echinodermata / Class Asteroidea / Order Spinulosida
Family Echinasteridae

8.5 **Henricia leviuscula** (Stimpson, 1857)

Common on protected sides of rocks, under rocks, and in caves and pools, most frequently where rock is encrusted with sponges and bryozoans, low intertidal zone; subtidal to over 400 m; Aleutian Islands (Alaska) to Bahía de Tortuga (Baja California).

Arm radius to 8.9 cm, usually less; disk small; arms long, tapering, usually five in number (sometimes four or six); pedicellariae absent; aboral surface tan to orange-red or purple, often banded with darker shades, and bearing many groups of short spinelets arranged in a fine network.

The great variation in form, size, color, and spine shape of this species led Fisher (1911) to establish several subspecies and varieties, some of which are known only subtidally.

The sea star feeds on bacteria and other tiny particles, which are captured in mucus and swept to the mouth by ciliated tracts. It may also feed by applying the stomach to the surfaces of sponges and bryozoa. A related form, *H. sanguinolenta*, absorbs some dissolved organic matter through the skin.

Breeding occurs mainly from December to March in Puget Sound (Washington). The sperm have spherical heads, as in

other starfishes. The eggs are orange-yellow and yolky, and development is direct. Smaller females brood their eggs in a depression of the oral area made by arching the disc; larger females discharge their eggs into the sea. This difference in female reproductive behavior occurs also in the northern European population of *H. sanguinolenta* (F.-S. Chia, pers. comm.).

Electron-microscope studies of the ocelli at the arm tips of *H. leviuscula* indicate that the receptor organelles of retinal cells are modified cilia, as in the retinal cells of vertebrates. Other specialized investigations of this species have been carried out on the histology and regeneration of parts of the gut, on body lipid content, and on righting behavior.

See especially Chia (1966b), Fisher (1911), Polls & Gonor (1975), and Rasmussen (1965); see also Alton (1966), Anderson (1960, 1962b, 1965), Chia (1970), Chia, Atwood & Crawford (1975), Eakin (1963), Eakin & Westfall (1964), Feder, Turner & Limbaugh (1974), Ferguson (1967, 1971), Giese (1966a), Grainger (1966), Hayashi (1935a), Hopkins & Crozier (1966), MacGinitie & MacGinitie (1968), Mauzey, Birkeland & Dayton (1968), Nybakken (1969), Rosenthal, Clarke & Dayton (1974), Thompson & Chow (1955), and Vasserot (1962).

Phylum Echinodermata / Class Asteroidea / Order Spinulosida
Family Solasteridae

8.6 **Solaster stimpsoni** Verrill, 1880
Stimpson's Sun Star

Uncommon, mainly on rocks but also on sand, very low intertidal zone; more commonly subtidal to 60 m; Commander Islands (Bering Sea) to Trinidad Head (Humboldt Co.).

Arm radius to 8.2 cm; arms slim, tapering, usually 10 in number (occasionally 9 or 11); aboral color red, orange, yellow, green, or blue, the disk with a blue-gray spot centrally, continuous with a blue-gray stripe along each arm.

Solaster species somewhat resemble *Pycnopodia helianthoides* (8.16), but have a smaller disk and a rougher appearance owing to the presence of paxillae (ossicles bearing clusters of spines).

In Puget Sound (Washington), *S. stimpsoni* feeds almost exclusively on sea cucumbers (*Cucumaria lubrica*; *C. miniata*, 9.3; *Eupentacta*; and *Psolus chitonoides*, all in rocky areas; *Thyone* on sandy bottoms). In turn, it is the main prey of the sea star *Solaster dawsoni* (8.7) and shows defensive and escape behavior when contacted by this species. (See the remarks under *S. dawsoni*.)

See Birkeland (1974), Ferguson (1971), Fisher (1911), Johnson & Snook (1927), Margolin (1976), and Mauzey, Birkeland & Dayton (1968).

Phylum Echinodermata / Class Asteroidea / Order Spinulosida Family Solasteridae

8.7 Solaster dawsoni Verrill, 1880
Dawson's Sun Star

Uncommon, on mud, sand, gravel, and rocks, low intertidal zone; more commonly subtidal to 420 m; Aleutian Islands and Point Franklin (Alaska) to Monterey Bay.

Arm radius 16 cm or more; arms usually 12–13 in number (occasionally only 8–11); aboral coloration usually gray, cream, yellow, or brown, less often bright red or orange, often with light patches interradially.

In Puget Sound (Washington), this species preys mainly on other sea stars, preferring its relative *Solaster stimpsoni* (8.6), but also taking *Mediaster aequalis*, *Crossaster papposus*, *Evasterias troschelii* (8.12), *Pycnopodia helianthoides* (8.16), *Dermasterias imbricata* (8.3), *Henricia leviuscula* (8.5), and even members of its own species. Most of these species show running responses when touched by *S. dawsoni*, and can be captured only if the predator manages to partially overlap and grasp its prey. *S. dawsoni* moves about, alternately raising and lowering its leading arms. When its tube feet contact the prey from above, the arms are brought down and the attacker lurches forward so rapidly that the prey, although frequently faster when on the run, is generally trapped. When so attacked, *S. stimpsoni* curls all of its arms upward above the disc and is occasionally successful in warding off the attack. *S. dawsoni* also feeds on the sea cucumbers *Cucumaria lubrica* and *Eupentacta quinquesemita* (9.4). In the laboratory it attacks the nudibranch *Tritonia*

festiva (14.51), but the latter swims rapidly when touched and generally escapes (C. Birkeland, pers. comm.).

See Birkeland (1974), Fisher (1911), Hancock (1955, 1958), Margolin (1976), and Mauzey, Birkeland & Dayton (1968).

Phylum Echinodermata / Class Asteroidea / Order Forcipulatida Family Asteriidae

8.8 Astrometis sertulifera (Xantus, 1860)

Common on rock, sand, and kelp holdfasts, very low intertidal zone; under rocks intertidally but on tops of rocks at greater depths; subtidal to 40 m; Santa Barbara to Gulf of California.

Arm radius to 8.1 cm; body flexible, rather slimy, with a small disk and five long arms; arm spines tapering, well separated, surrounded basally by circles of pedicellariae; aboral surface brown to greenish, the spines purple, orange, or blue, with red tips; oral surface orange.

This species, misidentified at the time as *Asterias forreri*, was used in classic behavior studies by Jennings (1907) at La Jolla (San Diego Co.). The animals are capable of fairly rapid movement, and when transferred to lab aquaria they "explore" the new area before settling down. They avoid strong light and generally come to rest in shaded areas; increasing the light stimulates a temporary increase in movement. When turned upside down, the animals usually use the arms closer to the madreporite in righting themselves, but display a remarkable flexibility in details of the behavior. Jennings was able to train some individuals to use certain arms habitually in righting, but found that the sea stars were not apt pupils; habit formation was slow and the effects of training lasted at most a few days. He remarked, somewhat pessimistically, "It would now be possible . . . to develop . . . educational methods for the starfish. By beginning with young specimens, it is probable that striking results would be reached. Such an investigation would require steady application to extremely dull and tedious work, for a long period."

Remarkable among Jennings's findings was the discovery that the forceps-like pedicellariae can be used to capture active prey. Large cushions bearing numerous pairs of these tiny pincers surround the bases of the larger spines. The pincers gape open, and the whole cushion is elevated to the level of the spinetip when the dermal gills are disturbed. The pedicellariae clamp shut their toothed jaws when foreign objects, especially living animals, are encountered. Jennings showed that small crabs, sand crabs (*Emerita analoga*, 24.4), and even fishes as large as the sea star itself are captured and held by the pedicellariae and later transferred to the mouth and eaten. (This type of behavior has now been observed in the field, and studied further, in the related sea star **Stylasterias forreri** (de Loriol) (formerly *Asterias forreri*), which in California generally occurs only below scuba depths). *A. sertulifera* also feeds on chitons, snails, clams, barnacles, the brittle star *Ophiothrix spiculata* (10.6), and occasionally sea urchins.

Astrometis gives an avoidance response to the predatory sea star *Heliaster kubinijii*, whose range it overlaps in Baja California; and in laboratory aquaria, *Heliaster* eats *Astrometis*.

For *Astrometis*, see especially Jennings (1907, as *Asterias forreri*); see also Feder, Turner & Limbaugh (1974), Fisher (1928), Hopkins & Crozier (1966), MacGinitie & MacGinitie (1968), and Rosenthal, Clarke & Dayton (1974). For *Stylasterias forreri*, see Chia & Amerongen (1975), Fisher (1928), and Robilliard (1971).

*Phylum Echinodermata / Class Asteroidea / Order Forcipulatida
Family Asteriidae*

8.9 **Orthasterias koehleri** (de Loriol, 1897)

Uncommon, on mud, sand, rock, and kelp; very low intertidal zone; more common at scuba depths and extending to 250 m; Yakutat Bay (Alaska) to Santa Rosa Island (Channel Islands).

Arm radius to 21 cm; disk small, with five slender arms; aboral surface bearing prominent sharp spines and a conspicuous ruff of pedicellariae; color vivid, varying from rosy pink with a gray mottling, to bright red mottled or banded with yellow, the spines whitish or lilac.

Adults feed on small snails, limpets, clams, scallops, chitons, barnacles, and tunicates. *Orthasterias* can dig clams out of cobbled bottoms, and uses the pull of its tube feet to chip

away the outer layer of a clam shell until a small opening is made between the valves. The stomach is then inserted through the opening and the clam digested.

See Chia, Atwood & Crawford (1975), Fisher (1928), Hopkins & Crozier (1966), and Mauzey, Birkeland & Dayton (1968).

*Phylum Echinodermata / Class Asteroidea / Order Forcipulatida
Family Asteriidae*

8.10 **Leptasterias pusilla** (Fisher, 1930)

Common in tidepools, especially those with a rock-sand bottom exposed to sunlight and harboring small green algae, sea lettuce (*Ulva*), coralline algae, and good populations of small snails (especially *Diala acuta*), middle intertidal zone; known only from Pillar Point (San Mateo Co.) to Monterey Bay.

Arm radius to 2.2 cm; species very similar to *Leptasterias hexactis* (8.11), from which it can usually be distinguished by (1) aboral spines (under hand lens) long and thin (2–3.5 times as long as wide), (2) arms appearing slender and gently tapering when the animal is moving, and (3) an aboral surface usually light gray-brown (occasionally reddish), through which the dark-colored caeca may be seen, providing a regular pattern on the arms (R. H. Smith, pers. comm.; cf. *L. hexactis*). Some specimens may be difficult to identify to species with certainty.

This six-rayed star feeds mainly on small snails (*Diala acuta*), but occasionally eats small limpets. It tolerates water temperatures up to 20°C for long periods, as well as reduced salinity. In the field, animals tend to occur in groups. They attach loosely to rocks and are easily dislodged. They avoid sunlight but come out on rock surfaces at night.

In January, the females lay 40–100 yolky eggs, which are brooded below the central disk around the mouth. Development is direct, and juvenile sea stars are released in about 6 weeks.

See especially R. Smith (1971); see also Anderson (1965), Eakin (1963), Eakin & Westfall (1962), and Fisher (1930), and the references for *L. hexactis* (8.11).

Phylum Echinodermata / Class Asteroidea / Order Forcipulatida
Family Asteriidae

8.11 **Leptasterias hexactis** (Stimpson, 1862)
(=L. aequalis)

Common on rocky shores exposed to surf, middle intertidal zone, often in small beds of the bivalve *Mytilus californianus* (15.9), often sheltered under rocks at very low tides or on sunny days; San Juan Islands and Strait of Juan de Fuca (Washington) to Santa Catalina Island (Channel Islands).

Arm radius to 5.2 cm; species very similar to *Leptasterias pusilla* (8.10), from which it can usually be distinguished, in California, by (1) aboral spines (under hand lens) flattened, mushroom-shaped, and arranged so densely as to largely obscure the aboral body wall and pedicellariae, (2) arms appearing relatively broad and heavy when the animal is moving, and (3) an aboral surface usually deep black or brown, occasionally bright red or greenish (R. H. Smith, pers. comm.; cf. *L. pusilla*). Some smaller individuals may be difficult to identify with certainty.

Fisher (1930) remarked "among the numerous small 6-rayed sea stars of the northwest coast of America specific lines are exceedingly difficult to draw." It now appears (Chia, 1966c) that several of the named forms are variants of *L. hexactis*.

Leptasterias hexactis clings very tenaciously to rocks, its body conforming closely to irregularities in the surface. It is a carnivore, but a food generalist, feeding on sea cucumbers, littorine snails, limpets, chitons, small mussels, barnacles, and many other small animals, including dead ones. In a population of these sea stars studied in the rocky intertidal area in Washington, feeding activity at high tide reached a high point in July and August; there was little feeding in January. In choosing food, *L. hexactis* often selects large, hard-to-capture, calorie-rich prey, and these supply most of the sea star's energy. Experiments indicate that *L. hexactis* is often in direct competition with a larger sea star *Pisaster ochraceus* (8.13), for available food.

Breeding occurs from November to April in Puget Sound (Washington). The eggs are yellow, yolky, about 0.9 mm in diameter, and laid in broods of 52–1,491 eggs, the brood size varying with the size of the female. Before fertilization, the eggs must be handled by the tube feet; after fertilization they form a mutually adhering sticky mass. Because brooding females hold the eggs in the region of the mouth below the central disk, they cannot flatten themselves against the substratum; in their humped-up position they are anchored only by podia on the outer parts of the arms, and are often dislodged by waves, losing their eggs. The egg masses are cleaned by the brooding females and, if deprived of parental care, accumulate debris and soon die. The presence of the brooded eggmass blocks the maternal mouth, and females do not feed while brooding, even when food is immediately available to them. Development is direct, and juvenile sea stars 1.5–2 mm in diameter are released in about 8 weeks in Washington, 6 weeks in California. They reach maturity in about 2 years.

Behavior experiments show that cutting the nerve ring at two opposite points causes the two halves of the animal to walk apart until fission occurs.

See especially Chia (1965, 1966a,c, 1968a,b, 1969), B. Menge (1972a,b, 1974, 1975), and J. Menge & B. Menge (1974); see also Chia (1966b), Chia, Atwood & Crawford (1975), Dayton (1971), Feder, Turner & Limbaugh (1974), Fisher (1930), Gordon (1929), Grainger (1966), Hewatt (1937, 1938), Mauzey, Birkeland & Dayton (1968), Moore (1939, 1941), Nicotri (1974), Nybakken (1969), Phillips (1976), Polls & Gonor (1975), R. Smith (1971), and Thompson & Chow (1955).

Phylum Echinodermata / Class Asteroidea / Order Forcipulatida
Family Asteriidae

8.12 **Evasterias troschelii** (Stimpson, 1862)

On rocks and cobbles, occasionally sand bottoms, sometimes with *Pisaster ochraceus* (8.13) but usually in more protected areas, low intertidal zone; subtidal to 70 m; Alaska—Saint George Island (Pribilof Islands), Unalaska (Aleutian Islands), Cook Inlet, and Prince William Sound—to Monterey Bay; uncommon south of Puget Sound (Washington).

Arm radius to 20.5 cm; body generally with five arms; color ranging from orange or brown to blue-gray; species similar in appearance to *Pisaster ochraceus* (8.13), but distinguished from it by (1) the smaller disk and slimmer, more tapering arms, (2) the clusters of pedicellariae among the spines bordering the ambulacral grooves, and (3) the usual absence of a stellate pattern of spines on the central portion of the aboral surface of the disk. Several forms of the species are described by Fisher (1930).

Evasterias troschelii feeds on a variety of animals, its diet varying according to the relative abundance of the various prey species present in its habitat. Prey organisms include bivalves (primarily *Pododesmus cepio*, 15.18; *Protothaca staminea*, 15.41; *Mytilus edulis*, 15.10), limpets, snails, brachiopods (*Terebratalia transversa*, 7.1), barnacles, and tunicates (especially *Pyura haustor*, 12.37). The sea star can pull open bivalves, but can also extend its stomach into any narrow opening between the shells and digest a closed clam. *Evasterias* elicits avoidance responses from various mollusks; the limpet *Notoacmea scutum* (13.23) moves away, and the keyhole limpet *Diodora aspera* (13.10) extends its mantle to cover the shell, preventing the tube feet from getting a secure grip. Scale worms (Polychaeta: Polynoidae) often live in the sea star's ambulacral grooves or on the body surface (see, for example, Fig. 18.1).

See especially Christensen (1957), Fisher (1930), and Mauzey, Birkeland & Dayton (1968); see also Davenport (1950), Davenport & Hickok (1951), Feder & Christensen (1966), Feder, Turner & Limbaugh (1974), Ferguson (1971), Giese (1966a), Margolin (1964a,b), Mortensen (1921), Nybakken (1969), and Strathmann (1971).

Phylum Echinodermata / Class Asteroidea / Order Forcipulatida Family Asteriidae

8.13 Pisaster ochraceus (Brandt, 1835)
Ochre Starfish

Common, middle and low intertidal zones on wave-swept rocky shores; subtidal on rocks to 88 m; juveniles in crevices and under rocks, seldom seen in central and southern California; Prince William Sound (Alaska) to Point Sal (Santa Barbara Co.); a subspecies, *P. ochraceus segnis* Fisher, extends at least to Ensenada (Baja California).

Average arm radius in Monterey Bay 14 cm, occasionally twice this size; arms stout, tapering, usually five in number but varying from four to seven; aboral surface with many small white spines arranged in detached groups or in a reticulate pattern, generally forming a star-shaped design on central part of disk (see the comparison with *Evasterias troschelii*, 8.12); color yellow or pale orange to dark brown or deep purple.

The ochre starfish has been the subject of many field and laboratory investigations. In central California it feeds mainly on mussels (*Mytilus*, 15.9, 15.10) in mussel beds; where *Mytilus* is absent, it feeds mainly on barnacles (*Balanus glandula*, 20.24, and other species; *Tetraclita rubescens*, 20.18), snails, limpets, and chitons. Smaller mussels and sometimes larger ones are pulled open, but the sea star can insert its stomach into snail shells, or into slits as narrow as 0.1 mm between the shells of bivalves. In central California this sea star feeds all year; in Puget Sound (Washington), fewer animals are seen feeding in winter (mainly on chitons) than in summer (when barnacles and limpets are the main prey). The ochre starfish appears to be in direct competition with the small six-rayed sea star *Leptasterias hexactis* (8.11) for smaller food items.

Among the various mollusks eaten by the ochre sea star, numerous species have evolved avoidance responses or escape responses that reduce predation on them. The animals move away when they are touched by the sea star, but they also detect, by scent, a *Pisaster* present in nearby waters and are stimulated to move away. Avoidance responses to *P. ochraceus*, and often to other sea stars, occur in several abalones (*Haliotis* spp., e.g., 13.1, 13.4), several limpets (*Collisella* spp., e.g., 13.14, 13.16; *Notoacmea scutum*, 13.23), keyhole limpets (e.g., *Diodora aspera*, 13.10), turban snails (*Tegula* spp., e.g., 13.31, 13.32; *Calliostoma* spp., e.g., 13.27), and many other forms. The limpets *Collisella limatula* (13.14) and *Notoacmea scutum* respond to water containing *P. ochraceus* scent by creeping up a submerged vertical surface. When on a horizontal surface, the limpets move downstream when they detect the scent of *Pisaster*.

Adult ochre starfish appear to have few enemies, but some are eaten by sea otters and sea gulls. The otter often eats only part of a sea star before discarding it; thus otter feeding grounds usually contain a fair number of *Pisaster* amputees that are regenerating one or two arms.

Pisaster ochraceus is more tolerant of exposure to air than any other *Pisaster* species. It regularly withstands up to 8 hours' exposure during low spring tides in the field, and is apparently unharmed by up to 50 hours of exposure in the laboratory. The inability to tolerate high water temperatures together with low oxygen levels keeps *P. ochraceus* out of shallow bays and high tidepools (Feder, unpubl.).

Sexual reproductions occurs in the spring or early summer. Mature gonads may account for up to 40 percent of the weight of *Pisaster* in well-fed animals; starved individuals show little or no gonad development. Maturation of the eggs and spawning are under the neurochemical control of substances found in the radial nerves. Spawning occurs in April and May in Monterey Bay, and 1–2 months later in Puget Sound. Ripe animals can be induced to spawn by injection of a solution of 1-methyladenine. The eggs are small, and the sperm, as in most echinoderms, have spherical heads. Fertilization occurs in the sea, and development results in a free-swimming, plankton-feeding larva. Larvae have been reared in the laboratory, and the embryonic development and larval feeding have been studied in detail. Little is known of juvenile life following settlement and metamorphosis. The growth rate in adult animals, studied under both field and laboratory conditions, is dependent on food supply. Laboratory-reared animals provided with abundant food eat more, grow faster, and have a "fatter" appearance than animals living in the field. Animals starved for 2 years in the laboratory gradually declined in both size and weight and moved about very little, but quickly responded when finally fed again. The life span under field conditions is estimated to exceed 20 years.

Several biochemical studies have been carried out on *P. ochraceus*. Among the discoveries is one that the digestive gland tissue of the sea star contains metachromatic granules similar to those found in the cells of the mammalian pancreas, and extracts of the gland yield materials with properties like those of the vertebrate hormone insulin.

See especially Feder (1955a, 1959, 1963, 1970), Fisher (1930), Landenberger (1969), Mauzey (1966), B. Menge (1972a, 1975), J. Menge & B. Menge (1974), and Paine (1969, 1976b); see also Allen & Giese (1966), Anderson (1965), Barnier, Sheehan & Williams (1975), Boolootian (1962), Boolootian & Giese (1958), Bullock (1953), Burnett (1955, 1960), Chaet (1966a,b, 1967), Chia (1977), Chia, Atwood & Crawford (1975), Crawford & Chia (1978), Dayton (1971, 1975), Farmanfarmaian et al. (1958), Feder (1955b, 1956, 1967), Feder & Christensen (1966), Feder & Lasker (1964), Ferguson (1971), Fisher (1926), Giese (1966a), Greenfield (1959), Greenfield et al. (1958), Harger (1967), Irving (1924, 1926), Landenberger (1969), Lavoie (1956), Lavoie & Holz (1955), MacGinitie (1938), Margolin (1964a,b), Mauzey, Birkeland & Dayton (1968), Montgomery (1967), Mortensen (1921), Nicotri (1974), Nigrelli et al. (1967), Nimitz (1971), Paine (1966, 1974, 1976a), Paris (1960), Peng & Williams (1973), Phillips (1975a,b, 1976, 1977), Ricketts & Calvin (1968), Rio et al. (1965), Rodegker & Nevenzel (1964), Schuetz (1969), Schultz & Lambert (1973), Stephenson & Stephenson (1961), Stevens (1970), Stickle & Ahokas (1972), Stohler (1930), Strathmann (1971), Thompson & Chow (1955), Vasu & Giese (1966), Wilson & Falkmer (1965), and Yarnall (1964).

*Phylum Echinodermata / Class Asteroidea / Order Forcipulatida
Family Asteriidae*

8.14 **Pisaster giganteus** (Stimpson, 1857)

Common on rocks and pier pilings, very low intertidal zone to 88 m depth in protected coastal areas; occasionally on sand in subtidal waters; Vancouver Island (British Columbia) to Baja California; Fisher (1930) recognized southern forms as an intergrading subspecies, **P. giganteus capitatus** (Stimpson).

Arm radius to over 30 cm; aboral spines differing from those of *P. ochraceus* (8.13) in being fewer, longer, terminally swollen, rather uniformly spaced (never forming a reticulate pattern or star-shaped design on the disk, and surrounded basally by a zone of blue integument; spines of adults white, of juveniles pink, violet, or blue.

This species is an active predator, feeding on numerous bivalves, snails, chitons, and barnacles. Food-choice experiments show that *P. giganteus* from the area of Santa Barbara, when confronted with seven selected food species together, chose them in the following order: (1) the mussels *Mytilus edulis* (15.10) and *M. californianus* (15.9), (2) the ribbed mussel *Septifer bifurcatus* (15.11), (3) the snail *Nucella emarginata* (13.83), and (4) the snails *Acanthina spirata* (13.81) and *Tegula funebralis* (13.32) and the chiton *Nuttallina californica* (16.17). In the field, diet doubtless varies with the availability of different foods. For example, in the kelp forest off Mussel Point (Pacific Grove, Monterey Co.), where populations of *P. giganteus* average 0.11 per m², the sessile vermetid gastropods *Petaloconchus montereyensis* (13.46) and *Serpulorbis squamigerus* (13.45) are abundant; here these two species average 50 percent of the sea star's diet, which also includes motile snails, barnacles, and the tube-building polychaete *Phragmatopoma californica* (18.30; J. S. Pearse and C. Harrold, pers. comm.).

Subtidally in southern California, this sea star and the carnivorous snail *Kelletia kelletii* (13.87) have been observed feeding together on a common food item. *Kelletia* is occasionally eaten by the starfish but shows no avoidance response to it. However, numerous other mollusks (e.g., abalones, limpets, keyhole limpets, turban snails) give avoidance reactions to *P. giganteus* that probably reduce predation.

Learning experiments indicate that *P. giganteus* can be trained to associate a light stimulus with food.

In the Monterey Bay area, the gonads of this species enlarge during the fall and winter, and spawning occurs in March or April.

See especially Fisher (1930), Landenberger (1966, 1968, 1969), and Rosenthal (1971); see also Boolootian (1962), Boolootian & Giese (1958), Bundy & Gustafson (1973), Chaet (1966a,b), Chia, Atwood & Crawford (1975), Feder (1963), Feder & Christensen (1966), Giese (1966a), Greenfield et al. (1958), Harger (1967), Harrison (1968), Harrison & Philpott (1966), Hopkins & Crozier (1966), Margolin (1964a,b), Pequegnat (1961, 1964), Phillips (1976), Rosenthal, Clarke & Dayton (1974), Stevens (1970), and Thompson & Chow (1955).

*Phylum Echinodermata / Class Asteroidea / Order Forcipulatida
Family Asteriidae*

8.15 **Pisaster brevispinus** (Stimpson, 1857)

Occasional, low intertidal zone, much more commonly subtidal from 0.5 m to 100 m, on sand and mud bottoms, sometimes on rocks and pier pilings in quiet water; Sitka (Alaska) to Mission Bay (San Diego Co.).

Maximum arm radius 32 cm; aboral spines much shorter than those of other *Pisaster* species; pink color diagnostic.

One of the largest sea star species known, *P. brevispinus* is adapted to subtidal life, and desiccates rapidly on exposure to air. It walks easily on sand and mud, and in such environments it feeds on living clams (such as *Saxidomus giganteus*, 15.33, *Protothaca* spp., e.g., 15.40–42, *Humilaria kennerleyi*, *Macoma* spp., e.g., 15.50, 15.51, *Tresus nuttallii*, 15.48, and *Panopea generosa*, 15.70), on snails (such as *Olivella biplicata*, 13.99, and *Nassarius fossatus*, 13.92), and on the sand dollar *Dendraster excentricus* (11.7); it is an opportunistic scavenger on dead fishes and the squid *Loligo opalescens* (17.2). The sea star apparently senses a buried clam while walking on the sand above it. The sea star may dig down to the clam (taking 2 to 3 days) or it may extend the tube feet of the disk down through the sand. Recent experiments (Van Veldhuizen & Phillips, 1978) reveal the extraordinary fact that *P. brevispinus* can extend the tube feet of the disk down into the substratum for a distance roughly equal to the starfish arm radius. One tube foot measured was 20 cm long. Once the tube feet of the disk are attached to the clam they contract, and the clam is lifted out of the substratum, or the stomach may be extended out as much as 8 cm to digest the prey in place. On pilings and rocks, *P. brevispinus* feeds mainly on barnacles (*Balanus crenatus*, 20.25, *B. nubilus*, 20.32), mussels (*Mytilus,* 15.9, 15.10), and tube-dwelling annelids.

Some sand-bottom invertebrates, such as the sand dollar *Dendraster excentricus* and the snail *Olivella biplicata*, sense the presence of the sea star by detecting water-borne substances that diffuse from its body; they avoid contact by burying themselves in the sand. The actual touch of a tube foot of the sea star elicits more active escape responses in some animals: the snail *Nassarius fossatus* sometimes leaps and twists violently; the snail *Olivella*, resting on sand, turns away sharply, thrusts out its foot, and speedily digs. Given no opportunity to burrow, as in an aquarium, *Olivella* may crawl away rapidly, flip over backward, or move 5–10 cm away by a violent "swimming" action of the metapodial lobes of the foot.

Interspecific combat has been observed when *P. brevispinus* is in possession of food and the 20-rayed star *Pycnopodia helianthoides* (8.16) approaches and also attempts to feed. *Pycnopodia* was seen to extend its rays above those of *Pisaster*, then raise and lower them, repeatedly touching *Pisaster*. *Pisaster* responded by grasping the attacker with its pedicellariae and, further, by turning its arms over and applying the pedicellariae-bearing aboral surface to the aboral surface of *Pycnopodia*. The 20-rayed star, appearing injured by this action, in each case withdrew without using its own large and numerous pedicellariae. Further experimental study of this species interaction would be most desirable.

A small snail, *Balcis rutila*, lives partially embedded in the body wall of this sea star in deeper water.

In Monterey Bay, the gonads of *Pisaster brevispinus* enlarge during the winter, and spawning occurs in April; gonads are very small from May through October.

See Alton (1966), Boolootian (1962), Boolootian & Giese (1958), Bullock (1953), Edwards (1969), Farmanfarmaian et al. (1958), Fisher (1930), Giese (1966a), Harrison (1968), Hopkins & Crozier (1966), MacGinitie & MacGinitie (1968), Margolin (1964a,b), Mauzey, Birkeland & Dayton (1968), Nigrelli et al. (1967), Phillips (1977), Rio et al. (1965), Rosenthal, Clarke & Dayton (1974), Schwimer (1973), L. Smith (1961), Stevens (1970), Stohler (1930), Thompson & Chow (1955), Van Veldhuizen & Phillips (1978), and Wobber (1975).

Phylum Echinodermata / Class Asteroidea / Order Forcipulatida
Family Asteriidae

8.16 Pycnopodia helianthoides (Brandt, 1835)
Sunflower Star, Twenty-Rayed Star

Frequent, especially in regions rich in seaweeds, low intertidal zone on rocky shores; subtidal to 435 m on rock, sand, and mud; Alaska—Unalaska (Aleutian Islands), Prince William Sound, and Cook Inlet—to San Diego; uncommon south of Carmel Bay (Monterey Co.).

Arm radius 40 cm or more, occasionally 65 cm; disk broader than in *Solaster* species (e.g., 8.6, 8.7), soft, flexible, bearing up to 24 arms in adults, 5 in small juveniles; aboral surface usually pink, purplish, or brown, less often red, orange, or yellow.

This magnificent animal is the largest, heaviest, and most active of the Pacific coast sea stars. Fisher (1928) said of it, "When excited by food, it moves very rapidly and can execute counter movements more actively than any starfish

which I have observed. When under 'full sail,' with its thousands of tube-feet lashing back and forth, it is an impressive animal, and its numerous cushions of pedicellariae and the wide expanse of its flexible body make it a formidable engine of destruction. The fact that a large *Pycnopodia* can bring over 15,000 sucker feet into action against a struggling fish or crab suggests a reason for its success in competition for place and food." Sea urchins (e.g., *Stronglyocentrotus purpuratus*, 11.5; *S. franciscanus*, 11.6; *S. droebachiensis*), and bivalves (especially *Pecten* spp.) are preferred foods; they are usually swallowed whole and digested internally, though the stomach can be partially everted. Other foods include polychaetes, chitons (*Katharina tunicata*, 16.24; *Mopalia muscosa*, 16.19; and *Tonicella lineata*, 16.11), numerous snails, hermit crabs, grapsoid crabs, the sea cucumber *Psolus*, and the sea stars *Leptasterias* spp., (e.g., 8.10, 8.11), *Solaster dawsoni* (8.7, which occasionally eats *Pycnopodia*), and *Crossaster papposus*. In Monterey Bay, *Pycnopodia* feeds seasonally on dead or dying squid (*Loligo opalescens*, 17.2) after the squid have spawned and lie on the bottom. After the squid is swallowed and digested, the indigestible squid pen, too large to be defecated, is frequently extruded through the soft upper body wall of the sea star. Proximity to or contact with *Pycnopodia* initiates escape responses in many intertidal invertebrates.

When two sunflower stars meet in the field, they frequently display actions (especially various arm movements) that are interpreted as combative (agonistic) behavior; such bouts appear to affect the distribution and feeding activities of the population. Interspecific bouts between *Pycnopodia* and *Pisaster brevispinus* (8.15) or *P. giganteus* (8.14) may take place when *Pisaster* is in possession of food (see the discussion under *Pisaster brevispinus*).

Studies of many types have been carried out on *Pycnopodia*, ranging from ecology and development to biochemistry, but no single aspect of the sea star's biology is well known. Fertilizable eggs have been obtained from December to June. Development results in pelagic, plankton-feeding larvae. Rough handling may cause the adult animals to shed their arms, and in the field many *Pycnopodia* show arms being regenerated.

In Alaskan waters the king crab *Paralithodes camtschatica* attacks and feeds on *Pycnopodia* (G. Powell, pers. comm.).

See especially Fisher (1928), Greer (1961), Kjerskog-Agersborg (1918, 1922), Mauzey, Birkeland & Dayton (1968), Montgomery (1967), Paul & Feder (1975), Phillips (1978), and Ritter & Crocker (1900); see also Alton (1966), Boolootian (1962), Boolootian & Giese (1958), Chia, Atwood & Crawford (1975), Dayton (1975), Feder, Turner & Limbaugh (1974), Ferguson (1971), Giese (1966a), Greer (1962), Hamilton (1921), Hopkins & Crozier (1966), Margolin (1976), Mortensen (1921), Nigrelli et al. (1967), Nybakken (1969), Phillips (1976, 1977), Rio et al. (1963, 1965), Stevens (1970), Stohler (1930), Strathmann (1971), Thompson & Chow (1955), Williams (1952), and Wobber (1970, 1973, 1975).

Literature Cited

Allen, W. V., and A. C. Giese. 1966. An *in vitro* study of lipogenesis in the sea star *Pisaster ochraceus*. Comp. Biochem. Physiol. 17: 23–38.

Alton, M. S. 1966. Bathymetric distribution of the sea stars (Asteroidea) off the northern Oregon coast. J. Fish. Res. Bd. Canada 23: 1673–1714.

Anderson, J. M. 1959. Studies on the cardiac stomach of a starfish, *Patiria miniata* (Brandt). Biol. Bull. 117: 185–201.

――――. 1960. Histological studies on the digestive system of a starfish, *Henricia*, with notes on Tiedemann's pouches in starfishes. Biol. Bull. 119: 371–98.

――――. 1962a. Pyloric caeca and Tiedemann's pouches in *Linckia columbiae*. Amer. Zool. 2: 387 (abstract).

――――. 1962b. Studies on visceral regeneration in sea-stars. I. Regeneration of pyloric caeca in *Henricia leviuscula* (Stimpson). Biol. Bull. 122: 321–42.

――――. 1965. Studies on visceral regeneration in sea-stars. II. Regeneration of pyloric caeca in Asteriidae, with notes on the source of cells in regenerating organs. Biol. Bull. 128: 1–23.

――――. 1966. Aspects of nutritional physiology, pp. 329–57, *in* Boolootian (1966).

Ansell, A. D. 1967. Leaping and other movements in some cardiid bivalves. Anim. Behav. 15: 421–26.

Arakawa, K. Y. 1960. Miscellaneous notes on Mollusca. (1) Feeding habits of some marine molluscs. Venus 21: 66–71.

Araki, G. S. 1964. On the physiology of feeding and digestion in the sea star *Patiria miniata*. Doctoral thesis, Biological Sciences, Stanford University, Stanford, Calif. 182 pp.

Atwood, D. C. 1973. Correlation of gamete shedding with presence of neurosecretory granules in asteroids (Echinodermata). Gen. Comp. Endocrinol. 20: 347–50.

Atwood, D. C., and J. L. Simon. 1973. Histological and histochemical analysis of neurosecretory granules in asteroids (Echinodermata). Trans. Amer. Microscop. Soc. 92: 175–84.

Barnier, G. W., M. V. Sheehan, and D. C. Williams. 1975. The production and secretion of digestive enzymes in the purple seastar *Pisaster ochraceus*. Mar. Biol. 29: 261–66.

Binyon, J. 1972. Physiology of echinoderms. New York: Pergamon. 264 pp.

Birkeland, C. 1974. Interactions between a sea pen and seven of its predators. Ecol. Monogr. 44: 211–32.

Boolootian, R. A. 1962. The perivisceral elements of echinoderm body fluids. Amer. Zool. 2: 275–84.

――――, ed. 1966. Physiology of Echinodermata. New York: Interscience. 822 pp.

Boolootian, R. A., and A. C. Giese. 1958. Coelomic corpuscles of echinoderms. Biol. Bull. 115: 53–63.

Bullock, T. H. 1953. Predator recognition and escape responses of some intertidal gastropods in presence of starfish. Behaviour 5: 130–40.

――――. 1965. Comparative aspects of superficial conduction systems in echinoids and asteroids. Amer. Zool. 5: 545–62.

Bundy, H. F., and J. Gustafson. 1973. Purification and comparative biochemistry of a protease from the starfish *Pisaster giganteus*. Comp. Biochem. Physiol. 44B: 241–51.

Burla, H., V. Ferlin, B. Pabst, and G. Ribi. 1972. Notes on the ecology of *Astropecten aranciacus*. Mar. Biol. 14: 235–41.

Burnett, A. L. 1955. A demonstration of the efficacy of muscular force in the opening of clams by the starfish, *Asterias forbesi*. Biol. Bull. 109: 355.

――――. 1960. The mechanism employed by the starfish, *Asterias forbesi* to gain access to the interior of the bivalve, *Venus mercenaria*. Ecology 41: 583–84.

Carcelles, A. 1944. Nuevos datos sobre el contenído estómacal de "*Astropecten cingulatus*" Sladen. Physis (Revista de la Soc. Argentina de Cien. Nat.), Buenos Aires, 19: 461–72.

Caso, M. E. 1963. Estudios sobre Equinodermos de México: Contribución al conocimiento de los equinodermos de las Islas Revillagigedo. Anales Inst. Biol., Univ. Mexico 33: 293–330.

Chaet, M. E. 1966a. The gamete-shedding substances of starfishes: A physiological-biochemical study. Amer. Zool. 6: 263–71.

――――. 1966b. Neurochemical control of gamete release in starfish. Biol. Bull. 130: 43–58.

――――. 1967. Gamete release and shedding substance of seastars, pp. 13–24, *in* N. Millot, ed., Echinoderm biology. Symp. Zool. Soc. London 20. 240 pp.

Chaet, M. E., and D. E. Philpott. 1960. Secretory structure in the tube feet of starfish. Biol. Bull. 119: 308–9.

Chia, F.-S. 1966a. Brooding behavior of a six-rayed starfish, *Leptasterias hexactis*. Biol. Bull. 130: 304–15.

————. 1966b. The development of two brooding seastars, *Henricia leviuscula* and *Leptasterias hexactis*. Amer. Zool. 6: 331–32 (abstract).

————. 1966c. Systematics of the six-rayed sea star, *Leptasterias*, in the vicinity of San Juan Island, Washington. Syst. Zool. 15: 300–306.

————. 1968a. The embryology of a brooding starfish, *Leptasterias hexactis* (Stimpson). Acta Zool. 49: 321–64.

————. 1968b. Some observations on the development and cyclic changes of the oocytes in a brooding starfish, *Leptasterias hexactis*. J. Zool., London 154: 453–61.

————. 1969. Histology of the pyloric caeca and its changes during brooding and starvation in a starfish, *Leptasterias hexactis*. Biol. Bull. 135: 185–92.

————. 1970. Some observations on the histology of the ovary and RNA synthesis in the ovarian tissues of the starfish, *Henricia sanguinolenta*. J. Zool., London 162: 287–91.

————. 1977. Scanning electron microscopic observations of the mesenchyme cells in the larvae of the starfish *Pisaster ochraceus*. Acta Zool. 58: 45–51.

Chia, F.-S., and H. Amerongen. 1975. On the prey-catching pedicellariae of a starfish, *Stylasterias forreri* (de Loriol). Canad. J. Zool. 53: 748–55.

Chia, F.-S., D. Atwood, and B. Crawford. 1975. Comparative morphology of echinoderm sperm and possible phylogenetic implications. Amer. Zool. 15: 533–65.

Child, C. M. 1941. Differential modifications of coelom development in the starfish, *Patiria miniata*. Physiol. Zool. 14: 449–60.

Christensen, A. M. 1957. The feeding behavior of the sea-star *Evasterias troschelii* Stimpson. Limnol. Oceanogr. 2: 180–97.

————. 1970. Feeding biology of the sea-star *Astropecten irregularis* Pennant. Ophelia 8: 1–134.

Clark, A. M. 1962. Starfishes and their relations. London: British Museum (Natur. Hist.). 119 pp.

Crawford, B. J., and F.-S. Chia. 1978. Coelomic pouch formation in the starfish *Pisaster ochraceus* (Echinodermata: Asteroidea). J. Morphol. 157: 99–119.

Davenport, D. 1950. Studies in the physiology of commensalism. 1. The polynoid genus *Arctonoë*. Biol. Bull. 98: 81–93.

Davenport, D., and J. F. Hickok. 1951. Studies in the physiology of commensalism. 2. The polynoid genera *Arctonoë* and *Halosydna*. Biol. Bull. 100: 71–83.

Dayton, P. K. 1971. Competition, disturbance, and community organization: The provision and subsequent utilization of space in a rocky intertidal community. Ecol. Monogr. 41: 351–89.

————. 1975. Experimental evaluation of ecological dominance in a rocky intertidal algal community. Ecol. Monogr. 45: 137–59.

Defretin, R. 1952a. Étude histochimique des mucocytes des pieds ambulacraires de quelques échinoderms. Rec. Trav. Sta. Mar. Endoume 6: 31–33.

————. 1952b. Sur les mucocytes des podia de quelques échinoderms. Comparison de leur sécrétion avec d'autres mucoprotides. C. R. Acad. Sci. 234: 1806–8.

Doi, T. 1976. Some aspects of feeding ecology of the sea stars, genus *Astropecten*. Publ. Amakusa Mar. Biol. Lab., Kyushu Univ. 4: 1–19.

Eakin, R. M. 1963. Lines of evolution of photoreceptors, pp. 393–425, *in* D. Mazia and A. Tyler, eds., General physiology of cell specialization. New York: McGraw-Hill. 434 pp.

Eakin, R. M., and J. A. Westfall. 1962. Fine structure of photoreceptors in the hydromedusan, *Polyorchis pencillatus*. Proc. Nat. Acad. Sci. 48: 826–33.

————. 1964. Further observations on the fine structure of some invertebrate eyes. Z. Zellforsch. 62: 310–32.

Edwards, D. C. 1969. Predators on *Olivella biplicata*, including a species-specific predator avoidance response. Veliger 11: 326–33.

Engster, M. S., and S. C. Brown. 1972. Histology and ultrastructure of the tube foot epithelium in the phanerozonian starfish, *Astropecten*. Tissue & Cell 4: 503–18.

Epel, D. 1975. The program of and mechanisms of fertilization in the echinoderm egg. Amer. Zool. 15: 507–22.

Fager, E. W. 1968. A sand-bottom epifaunal community of invertebrates in shallow water. Limnol. Oceanogr. 13: 448–64.

Farmanfarmaian, A., A. C. Giese, R. A. Boolootian, and J. Bennett. 1958. Annual reproductive cycles in four species of west coast starfishes. J. Exper. Zool. 138: 355–67.

Feder, H. M. 1955a. On the methods used by the starfish *Pisaster ochraceus* in opening three types of bivalve molluscs. Ecology 36: 764–67.

————. 1955b. The use of vital stains in marking Pacific coast starfish. Calif. Fish & Game 41: 245.

————. 1956. Natural history studies on the starfish *Pisaster ochraceus* (Brandt, 1835) in the Monterey Bay area. Doctoral thesis, Biological Sciences, Stanford University, Stanford, Calif. 294 pp.

————. 1959. The food of the starfish, *Pisaster ochraceus*, along the California coast. Ecology 40: 721–24.

————. 1963. Gastropod defensive responses and their effectiveness in reducing predation by starfishes. Ecology 44: 505–12.

————. 1967. Organisms responsive to predatory sea stars. Sarsia 29: 371–94.

————. 1970. Growth and predation by the ochre sea star, *Pisaster ochraceus* (Brandt), in Monterey Bay, California. Ophelia 8: 161–85.

————. 1972. Escape responses in marine invertebrates. Sci. Amer. 227: 93–100.

Feder, H. M., and A. M. Christensen. 1966. Aspects of asteroid biology, pp. 87–127, *in* Boolootian (1966).

Feder, H. M., and R. Lasker. 1964. Partial purification of a substance from starfish tube feet which elicits escape responses in gastropod molluscs. Life Sci. 3: 1047–51.

Feder, H. M., C. H. Turner, and C. Limbaugh. 1974. Observations on fishes associated with kelp beds in southern California. Calif. Dept. Fish & Game, Fish Bull. 160. 138 pp.

Ferguson, J. C. 1967. Utilization of dissolved exogenous nutrients by the starfishes, *Asterias forbesi* and *Henricia sanguinolenta*. Biol. Bull. 132: 161–73.

———. 1971. Uptake and release of free amino acids by starfishes. Biol. Bull. 141: 122–29.

Ferlin, V. 1973. The mode of dislocation of *Astropecten aranciacus*. Helgoländer Wiss. Meeresunters. 24: 151–56.

Ferlin-Lubini, V., and G. Ribi. 1978. Daily activity pattern of *Astropecten aranciacus* (Echinodermata: Asteroidea) and two related species under natural conditions. Helgoländer Wiss. Meeresunters. 31: 117–27.

Fisher, W. K. 1911. Asteroidea of the north Pacific and adjacent waters. 1. Phanerozonia and Spinulosa. Bull. U.S. Nat. Mus. 76: 1–419.

———. 1926. *Pisaster*, a genus of sea-stars. Ann. Mag. Natur. Hist. (9) 17: 554–66.

———. 1928. Asteroidea of the north Pacific and adjacent waters. 2. Forcipulata (part). Bull. U.S. Nat. Mus. 76: 1–245.

———. 1930. Asteroidea of the north Pacific and adjacent waters. 3. Forcipulata (concluded). Bull. U.S. Nat. Mus. 76: 1–356.

Giese, A. C. 1966a. Lipids in the economy of marine invertebrates. Physiol. Rev. 46: 244–98.

———. 1966b. On the biochemical constitution of some echinoderms, pp. 757–96, *in* Boolootian (1966).

Gordon, I. 1929. Skeletal development in *Arbacia*, *Echinarachnius*, and *Leptasterias*. Phil. Trans. Roy. Soc. London B217: 289–334.

Gore, R. H. 1966. Observations on the escape response in *Nassarius vibex* (Say) (Mollusca: Gastropoda). Bull. Mar. Sci. Gulf Caribb. 16: 423–34.

Grainger, E. H. 1966. Sea stars (Echinodermata: Asteroidea) of arctic North America. Fish. Res. Bd. Canada, Bull. 152. 70 pp.

Greenfield, L. J. 1959. Biochemical and environmental factors involved in the reproductive cycle of the sea star *Pisaster ochraceus* (Brandt). Doctoral thesis, Biological Sciences, Stanford University, Stanford, Calif. 143 pp.

Greenfield, L. G., A. C. Giese, A. Farmanfarmaian, and R. Boolootian. 1958. Cyclic biochemical changes in several echinoderms. J. Exper. Zool. 139: 507–24.

Greer, D. L. 1961. Feeding behavior and morphology of the digestive system of the sea star *Pycnopodia helianthoides* (Brandt) Stimpson. Master's thesis, Zoology, University of Washington, Seattle. 77 pp.

———. 1962. Studies on the embryology of *Pycnopodia helianthoides* (Brandt) Stimpson. Pacific Sci. 16: 280–85.

Hamilton, W. F. 1921. Coordination in the starfish. 1. Behavior of the individual tube feet. J. Compar. Psychol. 1: 473–88.

Hancock, D. A. 1955. The feeding behaviour of starfish on Essex oyster beds. J. Mar. Biol. Assoc. U.K. 34: 313–31.

———. 1958. Notes on starfish on an Essex oyster bed. J. Mar. Biol. Assoc. U.K. 37: 565–89.

Harger, J. R. 1967. Population studies on sea mussels. Doctoral thesis, Biology, University of California, Santa Barbara. 328 pp.

Harrison, G. 1968. Subcellular particles in echinoderm tube feet. J. Ultrastruc. Res. 23: 124–33.

Harrison, G., and D. Philpott. 1966. Subcellular particles in echinoderm tube feet. J. Ultrastruc. Res. 16: 537–47.

Hayashi, R. 1935a. Studies on the morphology of Japanese sea-stars. I. Anatomy of *Henricia sanguinolenta* var. *ohshimai* n. var. J. Fac. Sci., Hokkaido Imp. Univ., (6) Zool. 4: 1–26.

———. 1935b. Studies on the morphology of Japanese sea-stars. II. Internal anatomy of two short-rayed sea-stars, *Patiria pectinifera* (Müller & Troschel) and *Asterina batheri* Goto. J. Fac. Sci., Hokkaido Imp. Univ., (6) Zool. 4: 197–212.

Heath, H. 1917. The early development of a starfish, *Patiria* (*Asterina*) *miniata*. J. Morphol. 29: 461–69.

Heddle, D. 1967. Versatility of movement and the origin of the asteroids, pp. 125–41, *in* N. Millot, ed., Echinoderm biology. Symp. Zool. Soc. London 20. 240 pp.

Hewatt, W. G. 1937. Ecological studies on selected marine intertidal communities of Monterey Bay, California. Amer. Midl. Natur. 18: 161–206.

———. 1938. Notes on the breeding seasons of the rocky beach fauna of Monterey Bay, California. Proc. Calif. Acad. Sci. (4) 19: 283–88.

Hopkins, T. S., and G. F. Crozier. 1966. Observations on the asteroid echinoderm fauna occurring in the shallow water of southern California (intertidal to 60 meters). Bull. So. Calif. Acad. Sci. 65: 129–45.

Houk, M. S. 1974. Respiration of starfish oocytes during meiosis, fertilization, and artificial activation. Exper. Cell Res. 83: 200–206.

Houk, M. S., and D. Epel. 1974. Protein synthesis during hormonally induced meiotic maturation and fertilization in starfish oocytes. Develop. Biol. 40: 298–310.

Hulings, N. C., and D. W. Hemlay. 1963. An investigation of the feeding habits of two species of sea stars. Bull. Mar. Sci. Gulf Caribb. 13: 354–59.

Hunt, O. D. 1925. The food of the bottom fauna of the Plymouth fishing grounds. J. Mar. Biol. Assoc. U.K. 13: 560–99.

Hyman, L. H. 1955. The invertebrates: Echinodermata. Vol. 4. New York: McGraw-Hill. 763 pp.

Irving, L. 1924. Ciliary currents in starfish. J. Exper. Zool. 41: 115–24.

———. 1926. Regulation of the hydrogen ion concentration and its relation to metabolism and respiration in the starfish. J. Gen. Physiol. 10: 345–58.

Jennings, H. S. 1907. Behavior of the starfish, *Asterias forreri* de Loriol. Univ. Calif. Publ. Zool. 4: 53–185.

Jensen, M. 1966. The response of two sea-urchins to the sea-star *Marthasterias glacialis* (L.) and other stimuli. Ophelia 3: 209–19.

Johnson, M. E., and H. J. Snook. 1927. Seashore animals of the Pacific coast. New York: Macmillan. 695 pp.

Kanatani, H. 1969. Mechanism of starfish spawning: Action of neural substance on the isolated ovary. Gen. Comp. Endocrinol. Suppl. 2: 582–89.

Kisch, B. S. 1958. *Astropecten irregularis* précieux auxiliare du malacologiste. Bull. Cent. Étud., Rech. Sci. Biarritz 2: 9–15.

Kjerskog-Agersborg, H. P. 1918. Bilateral tendencies and habits in the twenty-rayed starfish *Pycnopodia helianthoides* (Stimpson). Biol. Bull. 35: 232–54.

———. 1922. The relation of the madreporite to the physiological anterior end in the twenty-rayed starfish, *Pycnopodia helianthoides* (Stimpson). Biol. Bull. 42: 202–16.

Komatsu, M. 1975. On the development of the sea-star, *Astropecten latespinosus* Meissner. Biol. Bull. 148: 49–59.

Lande, R., and D. J. Reish. 1968. Seasonal occurrence of the commensal polychaetous annelid *Ophiodromus pugettensis* on the starfish *Patiria miniata*. Bull. So. Calif. Acad. Sci. 67: 104–11.

Landenberger, D. E. 1966. Learning in the Pacific starfish *Pisaster giganteus*. Anim. Behav. 14: 414–18.

———. 1968. Studies on selective feeding in the Pacific starfish *Pisaster* in southern California. Ecology 49: 1062–75.

———. 1969. The effect of exposure to air on Pacific starfish and its relation to distribution. Physiol. Zool. 42: 220–30.

Lavoie, M. E. 1956. How sea stars open bivalves. Biol. Bull. 111: 114–22.

Lavoie, M. E., and G. G. Holz, Jr. 1955. How sea stars open bivalves. Biol. Bull. 109: 363.

Loeb, J. 1905. Artificial membrane formation and chemical fertilization in a starfish (*Asterina*). Univ. Calif. Publ. Physiol. 2: 147–58.

MacGinitie, G. E. 1938. Notes on the natural history of some marine animals. Amer. Midl. Natur. 19: 207–19.

MacGinitie, G. E., and N. MacGinitie. 1968. Natural history of marine animals. 2nd ed. New York: McGraw-Hill. 523 pp.

Mackie, A. M., R. Lasker, and P. T. Grant. 1968. Avoidance reactions of a mollusc *Buccinum undatum* to saponin-like surface-active substances in extracts of the starfish *Asterias rubens* and *Marthasterias glacialis*. Comp. Biochem. Physiol. 26: 415–28.

Macnae, W., and M. Kalk. 1962. The fauna and flora of sand flats at Inhaca Island, Mozambique. J. Anim. Ecol. 31: 93–128.

Margolin, A. S. 1964a. The mantle response of *Diodora aspera*. Anim. Behav. 12: 187–94.

———. 1964b. A running response of *Acmaea* to seastars. Ecology 45: 191–93.

———. 1976. Swimming of the sea cucumber *Parastichopus californicus* (Stimpson) in response to sea stars. Ophelia 15: 105–14.

Mauzey, K. P. 1966. Feeding behavior and reproductive cycles in *Pisaster ochraceus*. Biol. Bull. 131: 127–44.

Mauzey, K. P., C. Birkeland, and P. K. Dayton. 1968. Feeding behavior of asteroids and escape responses of their prey in the Puget Sound region. Ecology 49: 603–19.

Menge, B. A. 1972a. Competition for food between two intertidal starfish species and its effect on body size and feeding. Ecology 53: 635–44.

———. 1972b. Foraging strategy of a starfish in relation to actual prey availability and environmental predictability. Ecol. Monogr. 42: 25–50.

———. 1974. The effect of wave action and competition on brooding and reproductive effort in the seastar *Leptasterias hexactis*. Ecology 55: 84–93.

———. 1975. Brood or broadcast? The adaptive significance of different reproductive strategies in the two intertidal sea stars *Leptasterias hexactis* and *Pisaster ochraceus*. Mar. Biol. 31: 87–100.

Menge, J. L., and B. A. Menge. 1974. Role of resource allocation, aggression and spatial heterogeneity in coexistence of two competing intertidal starfish. Ecol. Monogr. 44: 189–209.

Monks, S. P. 1904. Variability and autotomy of *Phataria*. Proc. Acad. Natur. Sci. Philadelphia 56: 596–600.

Montgomery, D. H. 1967. Responses of two haliotid gastropods (Mollusca), *Haliotis assimilis* and *Haliotis rufescens* to the forcipulate asteroids (Echinodermata) *Pycnopodia helianthoides* and *Pisaster ochraceus*. Veliger 9: 359–68.

Moore, A. R. 1939. Injury, recovery and function in an aganglionic central nervous system. J. Compar. Psychol. 28: 313–28.

———. 1941. Dysfunction in righting and locomotion in a starfish (*Patiria*) with supernumerary rays. J. Compar. Psychol. 32: 483–87.

———. 1953. Alteration of developmental patterns in *Patiria miniata* by means of trypsin. J. Exper. Zool. 123: 561–69.

———. 1962. Collecting and preserving the developmental stages of the Pacific coast bat starfish, *Patiria miniata*. Ward's Bull. (n.s.) 1: 3–4.

Mori, S., and K. Matutani. 1952. Studies on the daily rhythmic activity of the starfish *Astropecten polyacanthus* Müller et Troschel, and the accompanied physiological rhythms. Publ. Seto Mar. Biol. Lab. 2: 213–25.

Mortensen, T. 1921. Studies of the development and larval forms of echinoderms. Copenhagen: Gad. 261 pp., 33 pls.

Newman, H. H. 1921a. The experimental production of twins and double monsters in the larvae of the starfish *Patiria miniata*, together with a discussion of the causes of twinning in general. J. Exper. Zool. 33: 321–52.

———. 1921b. On the development of the spontaneously parthenogenetic eggs of *Asterina* (*Patiria*) *miniata*. Biol. Bull. 40: 105–17.

———. 1921c. On the occurrence of paired madreporic pores and pore canals in the advanced bipinnaria larvae of *Asterina* (*Patiria*) *miniata*, together with a discussion of the significance of similar structures in other echinoderm larvae. Biol. Bull. 40: 118–25.

———. 1922. Normal versus subnormal development in *Patiria miniata*. A caution to laboratory embryologists. Biol. Bull. 43: 1–9.

———. 1925. An experimental analysis of asymmetry in the starfish, *Patiria miniata*. Biol. Bull. 49: 111–38.

Nichols, D. 1966. Functional morphology of the water-vascular system, pp. 219–44, *in* Boolootian (1966).

———. 1969. Echinoderms. 4th ed. London: Hutchinson University Library. 192 pp.

Nicotri, M. E. 1974. Resource partitioning, grazing activities, and influence on the microflora by intertidal limpets. Doctoral thesis, Zoology, University of Washington, Seattle. 247 pp.

Nigrelli, R. F., M. F. Stempien, Jr., G. D. Ruggieri, V. R. Liguori, and J. T. Cecil. 1967. Substances of potential biomedical importance from marine organisms. Fed. Amer. Socs. Exper. Biol. Fed. Proc. 26: 1197–1205.

Nimitz, Sister M. A. 1971. Histochemical study of gut nutrient reserves in relation to reproduction and nutrition in the sea stars, *Pisaster ochraceus* and *Patiria miniata*. Biol. Bull. 140: 461–81.

Nybakken, J. W. 1969. Pre-earthquake intertidal ecology of Three Saints Bay, Kodiak Island, Alaska. Biol. Pap., Univ. Alaska, 9. 115 pp.

Paine, R. T. 1966. Food web complexity and species diversity. Amer. Natur. 100: 65–75.

———. 1969. The *Pisaster-Tegula* interaction: Prey patches, predator food preference, and intertidal community structure. Ecology 50: 950–61.

———. 1974. Intertidal community structure: Experimental studies on the relationship between a dominant competitor and its principal predator. Oecologia 15: 93–120.

———. 1976a. Biological observations on a subtidal *Mytilus californianus* bed. Veliger 19: 125–30.

———. 1976b. Size-limited predation: An observational and experimental approach with the *Mytilus-Pisaster* interaction. Ecology 57: 858–73.

Paris, O. H. 1960. Some quantitative aspects of predation by muricid snails on mussels in Washington Sound. Veliger 2: 41–47.

Paul, A. J., and H. M. Feder. 1975. The food of the sea star *Pycnopodia helianthoides* (Brandt) in Prince William Sound, Alaska. Ophelia 14: 15–22.

Pearse, J. S. 1965. Reproductive periodicities in several contrasting populations of *Odontaster validus* Koehler, a common antarctic asteroid. Biology of the Antarctic Seas 11. Antarctic Res. Ser. 5: 39–85.

Peng, R. K. Y., and D. C. Williams. 1973. Partial purification and some enzymatic properties of proteolytic enzyme fractions isolated from *Pisaster ochraceus* pyloric caeca. Comp. Biochem. Physiol. 44B: 1207–17.

Pequegnat, W. E. 1961. Life in the scuba zone. II. Natur. Hist. 70: 46–55.

———. 1964. The epifauna of a California siltstone reef. Ecology 45: 272–83.

Phillips, D. W. 1975a. Distance chemoreception-triggered avoidance behavior of the limpets *Acmaea* (*Collisella*) *limatula* and *Acmaea* (*Notoacmea*) *scutum* to the predatory starfish *Pisaster ochraceus*. J. Exper. Zool. 191: 199–210.

———. 1975b. Localization and electrical activity of the distance chemoreceptors that mediate predator avoidance behaviour in *Acmaea limatula* and *Acmaea scutum* (Gastropoda, Prosobranchia). J. Exper. Biol. 63: 403–12.

———. 1976. The effect of species-specific avoidance response to predatory starfish on the intertidal distribution of two gastropods. Oecologia 23: 83–94.

———. 1977. Avoidance and escape responses of the gastropod mollusc *Olivella biplicata* (Sowerby) to predatory asteroids. J. Exper. Mar. Biol. Ecol. 28: 77–86.

———. 1978. Chemical mediation of invertebrate defensive behaviors and ability to distinguish between foraging and inactive predators. Mar. Biol. 49: 237–243.

Polls, I., and J. Gonor. 1975. Behavioral aspects of righting in two asteroids from the Pacific coast of North America. Biol. Bull. 148: 68–84.

Rasmussen, B. 1965. On taxonomy and biology of the north Atlantic species of the asteroid genus *Henricia* Gray. Medd. Danm. Fisk.-Havunders. 4: 157–213.

Ribi, G., and P. Jost. 1978. Feeding rate and duration of daily activity of *Astropecten aranciacus* (Echinodermata: Asteroidea) in relation to prey density. Mar. Biol. 45: 249–254.

Ribi, G., R. Schärer, and P. Ochsner. 1977. Stomach contents and size-frequency distributions of two coexisting sea star species, *Astropecten aranciacus* and *A. bispinosus*, with reference to competition. Mar. Biol. 43: 181–85.

Ricketts, E. F., and J. Calvin. 1968. Between Pacific tides. 4th ed. Revised by J. W. Hedgpeth. Stanford, Calif.: Stanford University Press. 614 pp.

Rideout, R. S. 1978. Asexual reproduction as a means of population maintenance in the coral reef asteroid *Linckia multifora* on Guam. Mar. Biol. 47: 287–295.

Rio, G. J., G. D. Ruggieri, M. F. Stempien, Jr., and R. F. Nigrelli. 1963. Saponin-like toxin from the giant sunburst starfish, *Pycnopodia helianthoides*, from the Pacific Northwest. Amer. Zool. 3: 554–55.

Rio, G. J., M. F. Stempien, Jr., R. F. Nigrelli, and G. D. Ruggieri. 1965. Echinoderm toxins—1. Some biochemical and physiological properties of toxins from several species of Asteroidea. Toxicon 3: 147–55.

Ritter, W. E., and G. R. Crocker. 1900. Multiplication of rays and bilateral symmetry in the 20-rayed star-fish, *Pycnopodia helianthoides*

(Stimpson). Papers from the Harriman Alaska Expedition, III. Proc. Wash. Acad. Sci. 2: 247–74.

Robert, A. 1902. Recherches sur le développement des troques. Arch. Zool. Expér. Gén. (3) 10: 278.

Robilliard, G. A. 1971. Feeding behavior and prey capture in an asteroid *Stylasterias forreri*. Syesis 4: 191–95.

Robson, E. A. 1961a. Some observations on the swimming behavior of the anemone *Stomphia coccinea*. J. Exper. Biol. 38: 343–63.

———. 1961b. The swimming response and its pacemaker system in the anemone *Stomphia coccinea*. J. Exper. Biol. 38: 685–94.

Rodegker. W., and J. C. Nevenzel. 1964. The fatty acid composition of three marine invertebrates. Comp. Biochem. Physiol. 11: 53–60.

Rosenthal, R. J. 1971. Trophic interaction between the sea star *Pisaster giganteus* and the gastropod *Kelletia kelletii*. Fish. Bull. 69: 669–79.

Rosenthal, R. J., and J. R. Chess. 1970. Predation on the purple urchin by the leather star. Calif. Fish & Game 56: 203–4.

———. 1972. A predator-prey relationship between the leather star, *Dermasterias imbricata*, and the purple urchin, *Strongylocentrotus purpuratus*. Fish. Bull. 70: 205–16.

Rosenthal, R. J., W. D. Clarke, and P. K. Dayton. 1974. Ecology and natural history of a stand of giant kelp *Macrocystic pyrifera*, off Del Mar, California. Fish. Bull. 72: 670–84.

Ross, D. M. 1965. The behaviour of sessile coelenterates in relation to some conditioning experiments. Anim. Behav. (Suppl.) 1: 43–53.

———. 1974. Behavior patterns in associations and interactions with other animals, pp. 281–312, *in* L. Muscatine and H. M. Lenhoff, eds., Coelenterate biology, reviews and perspectives. New York: Academic Press. 501 pp.

Ross, D. M., and L. Sutton. 1964. Inhibition of the swimming response by food and of nematocyst discharge during swimming in the sea anemone *Stomphia coccinea*. J. Exper. Biol. 41: 751–57.

Schuetz, A. W. 1969. Induction of oocyte shedding and meiotic maturation in *Pisaster ochraceus*: Kinetic aspects of radial nerve factor and ovarian factor-induced changes. Biol. Bull. 137: 524–34.

Schultz, T. W., and C. C. Lambert. 1973. Changes in adenine nucleotide levels and respiration during 1-methyladenine-induced maturation of starfish oocytes. Exper. Cell. Res. 81: 163–68.

Schwimer, S. R. 1973. Trace metal levels in three subtidal invertebrates. Veliger 16: 95–102.

Semon, R. 1895. Über den Zweck der Ausscheidung von freier Schwefelsaüre bei Meeresschnecken. Biol. Zentralbl. 15: 80–93.

Serences, M. 1978. Aspects of the commensal relationship between the polychaete *Ophiodromus pugettensis* (Johnson) and the batstar *Patiria miniata* (Brandt). Master's thesis, Biological Sciences, Stanford University, Stanford, Calif. 74 pp.

Sloan, N. A. 1977. Coping with stardom: The lives of starfish. Waters, J. Vancouver Aquarium 2: 3–31.

Smith, J. E. 1946. The mechanics and innervation of the starfish tube foot–ampulla system. Phil. Trans. Roy. Soc. London B232: 279–310.

———. 1947. The activities of the tube feet of *Asterias rubens* L. 1. The mechanics of movement and of posture. Quart. J. Microscop. Sci. 88: 1–14.

Smith, L. S. 1961. Clam-digging behavior in the starfish, *Pisaster brevispinus* (Stimpson 1857). Behaviour 18: 148–53.

Smith, R. H. 1971. Reproductive biology of a brooding sea-star from the Monterey Bay Region, *Leptasterias pusilla* (Fisher). Doctoral thesis, Biological Sciences, Stanford University, Stanford, Calif. 229 pp.

Stephenson, T. A., and A. Stephenson. 1961. Life between tidemarks in North America. IVa. Vancouver Island. J. Ecol. 49: 1–29.

Stevens, M. 1970. Procedures for induction of spawning and meiotic maturation of starfish oocytes by treatment with 1-methyladenine. Exper. Cell Res. 59: 482–84.

Stewart, W. C. 1970. A study of the nature of the attractant emitted by the asteroid hosts of the commensal polychaete, *Ophiodromus pugettensis*. Doctoral thesis, Biology, University of California, Santa Barbara. 86 pp.

Stickle, W. B., and R. Ahokas. 1972. The effects of fluctuating salinity upon the blood-osmotic integrity of several molluscs and echinoderms. Amer. Zool. 12: 713 (abstract).

Stohler, R. 1930. Gewichtsverhältnisse bei gewissen marinen Evertebraten. Zool. Anz. 91: 149–55.

Strathmann, R. R. 1971. The feeding behavior of planktotrophic echinoderm larvae: Mechanisms, regulation, and rates of suspension-feeding. J. Exper. Mar. Biol. Ecol. 6: 109–60.

Sund, P. N. 1958. A study of the muscular anatomy and swimming behavior of the sea anemone, *Stomphia coccinea*. Quart. J. Microscop. Sci. 99: 401–20.

Sutton, J. E. 1975. Class Asteroidea, pp. 623–27, *in* R. I. Smith and J. T. Carlton, eds., Light's manual: Intertidal invertebrates of the central California coast. Berkeley and Los Angeles: University of California Press. 716 pp.

Swan, E. F. 1966. Growth, autotomy, and regeneration, pp. 397–434, *in* Boolootian (1966).

Thompson, T. G., and T. J. Chow. 1955. The strontium-calcium atom ratio in carbonate-secreting marine organisms, pp. 20–39, *in* Papers in Marine Biology & Oceanography, Deep-Sea Res. 3 (Suppl.).

Van Veldhuizen, H. D., and D. W. Phillips. 1978. Prey capture by *Pisaster brevispinus* (Asteroidea: Echinodermata) on soft substrate. Mar. Biol. 48: 89–97.

Vasserot, J. 1962. Caractère hautement spécialisé du régime alimentaire chez les asterides *Echinaster sepositus* et *Henricia sanguinolenta*, prédateurs de spongaires. Bull. Soc. Zool. France 86: 796–809.

Vasu, B. S., and A. C. Giese. 1966. Protein and non-protein nitrogen in the body fluid of *Pisaster ochraceus* (Echinodermata) in relation to nutrition and reproduction. Comp. Biochem. Physiol. 19: 351–61.

Ward, J. A. 1965a. An investigation on the swimming reaction of the anemone *Stomphia coccinea*. I. Partial isolation of a reacting substance from the asteroid *Dermasterias imbricata*. J. Exper. Biol. 158: 357–64.

———. 1965b. An investigation on the swimming reaction of the anemone *Stomphia coccinea*. II. Histological location of a reacting substance in the asteroid *Dermasterias imbricata*. J. Exper. Biol. 158: 365–72.

Wells, H. W., M. J. Wells, and I. E. Gray. 1961. Food of the sea-star *Astropecten articulatus*. Biol. Bull. 120: 265–71.

Whittaker, D. M. 1928. Localization in the starfish egg and fusion of blastulae from egg fragments. Physiol. Zool. 1: 55–75.

———. 1931. On the conduction of cortical change at fertilization in the starfish egg. Biol. Bull. 60: 23–29.

Williams, W. 1952. A giant starfish. Natur. Hist. 61: 397–98.

Wilson, S., and S. Falkmer. 1965. Starfish insulin. Canad. J. Biochem. 43: 1615–24.

Witt, S. M. 1930. The early development of the starfish *Patiria miniata*. Master's thesis, Biological Sciences, Stanford University, Stanford, Calif. 53 pp.

Wobber, D. R. 1970. A report on the feeding of *Dendronotus iris* on the anthozoan *Cerianthus* from Monterey Bay, Calif. Veliger 12: 383–87.

———. 1973. Aboral extrusion of squid pens by the sea star, *Pycnopodia helianthoides*. Veliger 16: 203–6.

———. 1975. Agonism in asteroids. Biol. Bull. 148: 483–96.

Yarnall, J. L. 1964. The responses of *Tegula funebralis* to starfishes and predatory snails. Veliger 6 (Suppl.): 56–58.

Yentsch, C. S., and D. C. Pierce. 1955. "Swimming" anemone from Puget Sound. Science 122: 1231–33.

Chapter 9

Holothuroidea: *The Sea Cucumbers*

Joe H. Brumbaugh

At first glance, the holothurians, or sea cucumbers, do not seem to conform to the generalized concept of echinoderms as pentaradiate, spiny-skinned animals. With few exceptions the few hundred species known are soft-bodied, cylindrical creatures with the mouth and anus lying at opposite ends of the elongate body. They are sluggish, and most of them lie with one particular side next to the substratum. This "ventral" side is often flattened and contains three of the five radii. The oral end usually bears ten or more oral tentacles, arranged in a circle around the mouth. These modified podia, small and inconspicuous or large and elaborately branched, are used in feeding. Most holothurians also have locomotory tube feet, which are distributed in one of three basic patterns: they may be restricted to rows along each radius, distributed over the entire body, or concentrated on the ventral side. Often they terminate in adhesive disks. In most holothurians locomotion is accomplished by the tube feet, in conjunction with extensions and contractions of the muscular body. The oral tentacles may aid in locomotion, particularly in those species lacking locomotor tube feet. In some deep-water forms, the podia are enlarged into strong stiltlike papillae, which the animals use in the manner of legs (see, for example, Barham, Ayer & Boyce, 1967). The body wall in the cucumbers lacks the rigidity typical of other echinoderms, for the calcareous plates constituting most of the skeletal system are very small spicules that are not articulated with one another. These spicules vary greatly in size and shape, and it may be necessary to examine them microscopically to make taxonomic identifications. They are easily

extracted by removing some skin or a tube foot from the body and adding a few drops of household chlorine bleach to dissolve the soft tissues.

Holothurians are worldwide in distribution and live at all depths of the ocean. Roughly 700 living species have been described. A few are capable of limited swimming and some are pelagic, but most are bottom dwellers. Those inhabiting soft sand or mud floors live on or in the substratum itself, ingest it, and digest the minute organisms and organic detritus contained in it. Cucumbers that dwell on hard rock extend their tentacles and strain the water for plankton, or brush them over the surface to pick up microscopic organisms. Some species can absorb dissolved organic matter from seawater (see, for example, Fontaine & Chia, 1968). No cucumber species has evolved as either predator or parasite, but at least one commensal species is known, living attached to a deep-water angler fish off California (Martin, 1969a,b).

In most holothurians the sexes are separate. In some species the females produce many small eggs, which are fertilized after release and develop into free-swimming plankton-feeding larvae (indirect development). In other species the females produce only a few rather large yolky eggs, and from these ultimately hatch young cucumbers that resemble the adults (direct development). In the latter cases the eggs are often brooded by a parent.

Economically, holothurians are of little importance in this country, but in the Indo-Pacific certain species are dried and used as food. They are marketed under the names Bêche-de-Mer or Trepang.

Sea cucumbers have few known predators, apart from man and some starfishes (see, for example, Mauzey, Birkeland & Dayton,

Joe H. Brumbaugh is Professor of Biology at Sonoma State University.

1968). Most California shore fishes ignore them, though they are occasionally eaten by the kelp bass and the sand bass (*Paralabrax clathratus* and *P. nebulifer*; see Quast, 1968), and occasionally by the sea otter (J. Vandevere, pers. comm.). Soft-bodied though they are, cucumbers have some defenses against predators. Some species eviscerate easily, leaving the entrails to the predator while the body wall moves off and regenerates a new set. Some species (none occurring locally) eviscerate special Cuvierian organs, which form masses of sticky tubules that entangle predators (see, for example, Endean, 1957; Habermehl & Volkwein, 1971; Mueller & Zahn, 1972; Mueller, Zahn & Schmid, 1970). Many cucumbers secrete toxic substances called "holothurins," which are damaging or lethal to organisms that ingest them (see, for example, Alender & Russell, 1966; Bakus, 1968; Halstead, 1965; Lasley & Nigrelli, 1970, 1971). Some holothurins have been demonstrated to inhibit growth of fungi or of tumors in experimental animals (see, for example, Shimada, 1969).

The class Holothuroidea is divided into several orders, mainly on the basis of the nature of the podia, tentacles, and respiratory tree. The best recent comprehensive account of sea cucumber structure and biology is that of Hyman (1955). Holothuroidean physiology is reviewed in Boolootian (1966), and Binyon (1972); aspects of ecology are covered in Bakus (1968, 1973), Pawson (1966), and Yamanouchi (1956). A key to California shore cucumbers is given by Rutherford (1975). Although there is no single monographic account of the west coast species, the papers of Clark, Deichmann, and Pawson cited in the bibliography will be helpful in confirming the identifications made from the photographs and descriptions presented here.

Phylum Echinodermata / Class Holothuroidea / Order Dendrochirotida Family Cucumariidae

Cucumaria curata Cowles, 1907   Tar Spot
9.1 **Cucumaria pseudocurata** Deichmann, 1938   Tar Spot

These two species, similar externally, are easily confused, and their treatment here differs, in part, from that given in Rutherford (1975). The *C. curata* referred to by Brumbaugh (1965), Filice (1950), Ricketts & Calvin (1968 and earlier editions), and E. Smith (1962) appears to be *C. pseudocurata* (E. Deich-

mann, pers. comm.). California records of the common northern species *C. lubrica* (e.g., Ricketts & Calvin, 1968) probably also represent *C. pseudocurata*.

Cucumaria curata is known with certainty only from tidepools on exposed rocky platforms on the outer coast near Yankee Point and Malpaso Creek (Monterey Co.), where it is locally abundant. *C. pseudocurata* is known from the Queen Charlotte Islands (British Columbia) to Monterey Co., and very likely extends beyond this range. It occurs in large aggregations on both protected and unprotected rocky shores from the upper level of mussel beds (*Mytilus californianus*, 15.9) down to about zero tide level; the lower vertical limit of distribution appears to be controlled by occasional predation by the starfish *Pycnopodia helianthoides* (8.16). *C. pseudocurata* may be found under mussel beds, on rock platforms (clinging to the bases of coralline algae or underneath thick mats of the seaweeds *Rhodomela* and *Odonthalia*), in shallow pools, in rock caves, or on vertical faces of partially protected rocks in surge channels; in the latter situation, densities to 4,000 individuals per /m² may occur.

Maximum length approaches 3.5 cm for both species. The tube feet are soft and retractile, generally occurring in definite rows. The ten tentacles are highly branched. The body is dark brown to black dorsally, lighter ventrally; individuals from deeply shaded areas are much paler than those exposed to sunlight.

The two species are not easily separated on the basis of external features, but the following characters (C. H. Brown, pers. comm.) are helpful in most instances:

C. curata: the three ventral rows of tube feet placed close together, the two dorsal rows widely separated; all ten tentacles approximately equal in size in both juveniles and adults; males usually with several genital papillae, which may be branched dendritically.

C. pseudocurata: the five rows of tube feet about equally spaced; the two ventral tentacles conspicuously smaller than the other eight tentacles in juveniles and some adults; males usually with a single genital papilla, which is unbranched or branched only one or two times.

Critical separation of the two species requires microscopic examination of the spicules (see Cowles, 1907; Deichmann,

1938b). A summary of the discriminating spicular characters is given below:

C. curata: body-wall spicules consisting mostly of small oval buttons usually perforated by four holes but sometimes by six or more; larger plates bearing twenty or more holes occurring less frequently; both buttons and plates with undulating smooth margins; walls of tube feet containing three-armed supporting rods with a large basal hole in the broad third arm and a varying number of holes at the ends of all three arms; simple rod-shaped spicules rare in tube feet, less sturdy than supporting rods in *C. pseudocurata*.

C. pseudocurata: body-wall spicules consisting of comparatively large plates, mostly perforated by two large oblong holes at the middle and several smaller holes at the ends; edges of plates bluntly dentate and often bearing small knobs; walls of tube feet containing spectacle-shaped supporting rods with few to many holes at each end.

Both species feed on phytoplankton and attached diatoms, which are caught by tiny adhesive papillae at the tips of the tentacle branches and transferred to the mouth. The sexes are separate; males bear one or more genital papillae dorsally within or near the tentacular crown, whereas the females lack papillae. Fertilization has not been observed, but in both species the entire population spawns at one time, usually over a period of only a few days. *C. curata* spawns in mid-December; *C. pseudocurata* spawns (in Monterey and Sonoma Cos.) from middle to late January. A unique, bilaterally symmetrical sperm is produced by *C. pseudocurata*; sperm morphology has not been studied in *C. curata*. Females of both species lay eggs about 1 mm in diameter. The eggs of *C. curata* are brownish yellow, those of *C. pseudocurata* bright golden yellow. Both species brood the eggs below the body for about a month. Development is direct, without a pelagic stage; at hatching, yellowish pentactula larvae with five tentacles and two locomotory tube feet are released. Dark pigment is deposited as additional tentacles and tube feet develop. Small living juveniles, mounted in a drop of seawater under a cover glass, make excellent objects for observation under the compound microscope. Size-frequency graphs for populations of *C. pseudocurata* show modes representing animals 1, 2, 3, and 4 or more years old; only animals in the last group are sexually mature.

For *C. pseudocurata* and *C. curata*, see especially Brumbaugh (1965) and Rutherford (1973, 1975); see also Bakus (1974), Chia, Atwood & Crawford (1975), Clark (1901b,c), Cowles (1907), Deichmann (1938b), Filice (1950), Ricketts & Calvin (1968), E. Smith (1962), and R. Turner & Rutherford (1975). For a sampling of work on other *Cucumaria* species, see Atwood (1974a, 1975), Atwood & Chia (1974), Chia & Buchanan (1969), Doyle (1967), Doyle & McNiell (1964), Engstrom (1974), Fish (1967a,b,c), Fontaine & Chia (1968), Harrison (1968), Hinegardner (1974), Manwell (1966), Motohiro (1960), Pawson & Fell (1965), Pople & Ewer (1954, 1955, 1958), Schaller (1973), and Yingst (1972).

Phylum Echinodermata / Class Holothuroidea / Order Dendrochirotida Family Cucumariidae

9.2 **Cucumaria piperata** (Stimpson, 1864)
(= Pentacta piperata)

Uncommon, in holes and crevices on rocky shores, low intertidal zone in northern localities; usually subtidal to 82 m depth in southern areas; British Columbia to Baja California.

Body 5–6 cm long, yellowish white, speckled with brown or black particularly near tentacular crown; tube feet arranged in definite rows; tentacles ten, yellowish, branched, equal in size.

Natural history details on this species are fragmentary. Spawning occurs in the spring in Puget Sound (Washington). The planktonic larvae that appear to belong to *C. piperata* are orange in color, and their tentacles bear small calcareous plates with relatively few perforations. Tests with fishes show that the tissues of the body wall, but not the viscera, contain toxic materials. Other studies have been made on the sperm, the corpuscles of the coelomic fluid, and the nature of the hemoglobin present in some of the cells.

See Bakus (1974), Chia, Atwood & Crawford (1975), Clark (1924), Deichmann (1937), Hetzel (1963, 1965), Johnson & Johnson (1950), Shelford et al. (1935), Stimpson (1864, as *Pentacta piperata*), and Terwilliger & Read (1970).

Phylum Echinodermata / Class Holothuroidea / Order Dendrochirotida
Family Cucumariidae

9.3 **Cucumaria miniata** Brandt, 1835

Fairly common, nestled in crevices and under rocks, low intertidal zone, in Sonoma Co. and northward; subtidal in Monterey Bay; Sitka (Alaska) to Monterey Co.

Body 10–25 cm long, usually brick red, but ranging from pinkish white to purple, the wall thick and tough; tube feet arranged in definite rows, other podia occasionally occurring between rows; tentacles ten, bright orange, branched, equal in size.

Undisturbed animals in crevices often have the body curved in a U shape, so that the mouth and anal openings are exposed to moving water. The tentacles are used to trap small organisms and detritus from the water, withdrawing rapidly when disturbed. *C. miniata* is eaten by the sea stars *Solaster stimpsoni* (8.6) and *S. endeca* in Puget Sound (Washington). Tests show that neither the body wall nor the viscera contain materials toxic to fishes. Eggs, embryos, and larvae are orange in color; the larvae occur in the plankton of Puget Sound in March and April. Laboratory studies have been made on the cells of the coelomic fluid, especially the red cells and the hemoglobin they contain. Other investigations of this species have involved the pharyngeal retractor muscles and physiological responses of the animals to conditions of altered salinity.

See Bakus (1974), Chia, Atwood & Crawford (1975), Clark (1901b,c, 1924), Crescitelli (1945), Fontaine & Lambert (1973), Hall (1927), Hetzel (1963, 1965), Johnson & Johnson (1950), Manwell (1959), Mauzey, Birkeland & Dayton (1968), Selenka (1867), Shelford et al. (1935), Stickle & Ahokas (1972), and Terwilliger & Read (1970).

Phylum Echinodermata / Class Holothuroidea / Order Dendrochirotida
Family Cucumariidae

9.4 **Eupentacta quinquesemita** (Selenka, 1867)
(= Cucumaria chronhjelmi)

Fairly common under rocks and in crevices, low intertidal zone on rocky shores; common on concrete piles and marina floats in Monterey harbor; Vancouver (British Columbia) to Morro Bay (San Luis Obispo Co.).

Body 4–8 cm long, cylindrical, white or cream-colored; tube feet restricted to ambulacra, forming a double row in each radius, strongly supported by skeletal elements and thus relatively rigid and nonretractile; tentacles ten (eight large, two small), branched, yellow.

Adults rarely expose their tentacles during daylight hours. In Puget Sound (Washington), larger *Eupentacta* are eaten by the sea star *Solaster stimpsoni* (8.6), and in some subtidal areas juvenile *Eupentacta* are a staple food of the much smaller starfish *Leptasterias hexactis* (8.11).

Spawning occurs in the spring. Development is indirect; eggs, embryos, and larvae are greenish in color and occur in the Puget Sound plankton from March to May. Laboratory investigations have been carried out on such matters as the presence of tissue substances toxic to fishes (results not clear-cut), the nature of the coelomic corpuscles, the body lipid content, the sperm type, and the DNA content of cells.

See Bakus (1974), Chia, Atwood & Crawford (1975), Clark (1901a,c, 1924), Deichmann (1938b), Giese (1966, as *C. chronhjelmi*), Haderlie (1968), Hetzel (1963, 1965), Hinegardner (1974), Johnson & Johnson (1950, in part as *C. chronhjelmi*), Mauzey, Birkeland & Dayton (1968), Selenka (1867), Shelford et al. (1935), and Theel (1886a, as *C. chronhjelmi*).

Phylum Echinodermata / Class Holothuroidea / Order Dendrochirotida
Family Cucumariidae

9.5 **Pachythyone rubra** (Clark, 1901)
(= Thyone rubra)

Fairly common, especially in the holdfasts of *Laminaria*, *Egregia*, and other large algae, occasionally in sandy deposits around colonial ascidians and the roots of surfgrass (*Phyllospadix*), low intertidal and adjacent subtidal zones on rocky shores; Monterey Bay to southern California and Channel Islands.

Body 2–2.5 cm long, cylindrical, orange to orange-red dorsally, white ventrally, the wall stiffened by many calcareous spicules; tube feet not arranged in obvious rows but scattered over the body, bearing conspicuous end plates; tentacles ten (eight large, two small), branched, rarely extended during daylight hours.

A free-swimming stage is lacking in the life history of this species. Fertilization (method unknown) is internal, and the embryos, which lack pigment, develop in the parental body cavity. In the summer months in Monterey Bay, mothers contain several young, usually ranging from small embryos and pentactula stages (early juveniles with five tentacles) to larger juveniles almost half the length of the parent and bearing ten branched tentacles. The release of the young has not been observed, but presumably occurs by rupture of the body wall or by rupture at the base of the respiratory trees and subsequent birth by way of the cloaca. The tube feet of juveniles are confined to the ambulacra, but additional podia develop later in the interambulacral areas.

See Clark (1901b,c), Deichmann (1939, 1941), MacGinitie & MacGinitie (1968), Pawson & Fell (1965), and Ricketts & Calvin (1968), usually as *Thyone rubra*.

Phylum Echinodermata / Class Holothuroidea / Order Dendrochirotida Family Psolidae

9.6 Lissothuria nutriens (Clark, 1901)
(= Thyonepsolus nutriens)

Fairly common on vertical rock faces and in sandy deposits among algal holdfasts, surfgrass (*Phyllospadix*) roots, sponges, or colonial ascidians, low intertidal zone; subtidal to 20 m; Monterey Bay to southern California and Channel Islands.

Body 1.5-2 cm long, the wall thick but lacking scales, the dorsal surface scarlet or bright orange-red; ventral side pale pink, flattened, bearing three longitudinal rows of tube feet; mouth and anus directed upward; tentacles ten (eight large, two small), branched.

Lissothuria produces yolky eggs about 1 mm in diameter, which pass out the gonopore, are caught by the tentacles, and then passed to tube feet and deposited in shallow pits on the dorsal body surface, where they are brooded. Development has been studied at Monterey Bay, where brooding has been observed from March to November. Pentactula larvae hatch

from the eggs and are eventually released as juveniles with numerous tube feet.

See Clark (1901b,c), Deichmann (1937, 1941), Hinegardner (1974), Ludwig (1904, as *Psolidium nutriens*), Pawson (1967), Pawson & Fell (1965), C. Turner, Ebert & Given (1968), and Wootton (1949).

Phylum Echinodermata / Class Holothuroidea / Order Aspidochirotida Family Stichopodidae

9.7 Parastichopus californicus (Stimpson, 1857)
(= Stichopus californicus)

Often encountered on rocky shores protected from strong wave action and on pilings in open bays, low intertidal and subtidal waters in northern areas; usually subtidal to 90 m in California; British Columbia to Isla Cedros (Baja California).

Body 25–40 cm long; dorsal and lateral surfaces dark red, brown, or yellow, bearing large, stiff, conical papillae or pseudospines often paler in color and tipped with red; tube feet densely arranged on ventral side; mouth directed ventrally at anterior end.

This is the largest sea cucumber on California shores. When disturbed, individuals contract and often squirt a powerful stream of water from the posterior end, a response shared by many smaller cucumbers. The hindgut in this and many other holothurians bears a pair of highly branched diverticula, which project into the coelomic cavity of the body. These so-called respiratory trees serve as "water lungs." Oxygenated water is forcibly pumped into the respiratory trees in several successive inhalations, until they are greatly expanded within the coelom; the deoxygenated water is then expelled in one long, powerful exhalation. Ciliated protozoans are reported to live in the respiratory trees.

Parastichopus californicus feeds on organic detritus and small organisms, which it ingests with bottom sediments. In turn, it is eaten by the sea stars *Pycnopodia helianthoides* (8.16) and *Solaster endeca*, to which it gives a characteristic escape response, "looping" a short distance along the bottom like an oversized inchworm, or flexing the body in a clumsy sort of swimming. The species is occasionally eaten by man and by the sea otter.

Tests with fishes indicate that the body does not store substances toxic to predators, as many tropical sea cucumbers do. A close relative, *Stichopus japonicus*, is commonly eaten in Japan, and its nutritive value has been studied.

Breeding in *P. californicus* occurs in the summer. One hormone has been isolated from the ovary. The sperm have spherical heads (like the sperm of most echinoderms except sea urchins), with a DNA content unusually low for cucumber sperm. Development is indirect; fertilized eggs develop into auricularia larvae, whose structure and feeding behavior have been studied in detail. The auricularia metamorphoses into a doliolaria larva, which swims for a time, then settles. The entire pelagic phase lasted 7–13 weeks in the laboratory.

Populations of adult animals in Puget Sound (Washington) eviscerate in the months of October and November, then proceed to regenerate new sets of internal organs. Viscera may be expelled at other seasons if the animals are kept in warm or stale water. The scale worm *Arctonoe pulchra* (18.2) sometimes occurs as a commensal on *P. californicus*.

For *P. californicus*, see Bakus (1974), Boolootian & Giese (1958), Chia, Atwood & Crawford (1975), Clark (190lb, 1913, 1922), Courtney (1927), Dan (1967), Deichmann (1937, 1938a), Dimock & Davenport (1971), Giese (1966), Harrison (1968), Hetzel (1963, 1965), Hinegardner (1974), Hufty & Schroeder (1974), Johnson & Johnson (1950), Mauzey, Birkeland & Dayton (1968), Margolin (1976), Mortensen (1921), Prosser & Judson (1952), Sheikh & Djerassi (1977), Shelford et al. (1935), Stimpson (1857), Strathman (1971, 1978), Strathman & Sato (1969), Swan (1961), Theel (1886a, as *Stichopus johnsoni*; 1886b, as *S. fuscus*), C. Turner, Ebert & Given (1968), and Webster (1968). For work on related species in the genus *Stichopus*, see Caso (1966), Ciereszko et al. (1962), Crozier (1918), Dawbin (1949), W. Freeman & Simon (1964), Gay & Simon (1964), Hill (1970), Imai (1950), Matsumura (1974), Motohiro (1960), Shimada (1969), Simon, Mueller & Dewhurst (1964), Stevens (1901), Tanaka (1958a,b), Tanikawa (1955a,b), and Tanikawa et al. (1955).

Phylum Echinodermata / Class Holothuroidea / Order Aspidochirotida Family Stichopodidae

9.8 **Parastichopus parvimensis** (Clark, 1913)
(= Stichopus parvimensis)

Common on sandy or sandy-mud surfaces and between rocks, low intertidal zone of bays and well-protected rocky shores; subtidal to at least 27 m, on rocks, pilings, sandy or mud bottoms, and, in tropical regions, in seagrass beds; Monterey Bay to Punta San Bartolomé (Baja California); uncommon and found only subtidally north of Point Conception.

Body to 25 cm long, cylindrical, light chestnut brown dorsally, much paler below; papillae or pseudospines on dorsal side conspicuous, conical, tipped with black; tube feet concentrated on ventral surface.

Somewhat smaller than *P. californicus* (9.7), this species creeps more rapidly than most cucumbers (about 1 m in 15 minutes). It feeds on soft sediments, digesting the organic detritus and small organisms contained therein. Uptake of digested food by the gut wall has been studied. Predators are not known, but tests with fishes show that neither the body wall nor the visera contain toxic materials. The scale worm *Arctonoe pulchra* (18.2) often lives as a commensal on the cucumber's body.

See Bakus (1974), Clark (1913, 1922), Deichmann (1937, 1938a), Dimock (1974), Dimock & Davenport (1971), Harrison (1968), Lawrence et al. (1967), Parker (1921, as *S. panimensis*, a misprint for *S. parvimensis*), and C. Turner, Ebert & Given (1968); see also references under *P. californicus*.

Phylum Echinodermata / Class Holothuroidea / Order Apodida Family Synaptidae

9.9 **Leptosynapta albicans** (Selenka, 1867)
(= L. inhaerens)

Locally common, buried in sand under rocks on outer coast, and in sand and mud flats in areas protected from surf, middle to low intertidal zones and subtidal waters; Puget Sound (Washington) to San Diego.

Body usually 5–15 cm long when fully extended, very

much shorter when contracted, cylindrical, wormlike, pinkish to brownish white, the wall in relaxed individuals thin and translucent, rendering the longitudinal muscle bands and some viscera visible; tentacles 10–12, pinnately branched; other podia lacking.

Leptosynapta, like other members of the order Apodida, lacks tube feet, radial canals, and respiratory trees; the only podia are the oral tentacles. Respiratory gas exchange takes place through the very thin body wall.

Leptosynapta feeds by ingesting fine sediment and digesting the organic detritus and small organisms present in it. The animals are effective burrowers; small anchorlike spicules in the skin aid in traction and may cause specimens to adhere loosely to the fingers when picked up. Both the young and the adults are capable of swimming by flexing and straightening the body. Rough handling causes body contractions that often result in rupture of the body wall, evisceration, and even fragmentation of the body. Regeneration experiments with a related species, *L. crassipatina*, show that any segment of the body can regenerate parts at its own or a more posterior region; only parts containing the anterior oral complex can regenerate a whole animal. Individuals relax completely after a few minutes' immersion in an isotonic solution of magnesium chloride and are excellent for laboratory dissection.

Few studies have been carried out on *L. albicans*, but a variety of investigations have been made of related apodous sea cucumbers.

See Atwood (1973, 1974a,b), Bakus (1974), Berrill (1966), Chia, Atwood & Crawford (1975), Clark (1907, 1924), Costello (1946), Costello & Henley (1971), Freeman (1966), Glynn (1965), Hammen & Osborne (1959), Heding (1928), Hefferman & Wainwright (1974), Hetzel (1963, 1965), Hyman (1955), Reish (1961), Selenka (1867), Shelford et al. (1935), and G. Smith (1971).

Literature Cited

Alender, C. B., and F. E. Russell. 1966. Pharmacology, pp. 529–43, *in* Boolootian (1966).

Atwood, D. G. 1973. Ultrastructure of the gonadal wall of the sea cucumber *Leptosynapta clarki* (Echinodermata: Holothuroidea). Z. Zellforsch. Mikroskop. Anat. 141: 319–30.

————. 1974a. Fine structure of spermatogonia, spermatocytes, and spermatids of the sea cucumbers *Cucumaria lubrica* and *Leptosynapta clarki* (Echinodermata: Holothuroidea). Canad. J. Zool. 51: 323–32.

————. 1974b. Fine structure of the spermatozoon of the sea cucumber *Leptosynapta clarki* (Echinodermata: Holothuroidea). Cell Tissue Res. 149: 223–33.

————. 1975. Fine structure of an elongated dorso-ventrally compressed echinoderm (Holothuroidea) spermatozoon. J. Morphol. 145: 189–208.

Atwood, D. G., and F.-S. Chia. 1974. Fine structure of an unusual spermatozoon of a brooding sea cucumber, *Cucumaria lubrica*. Canad. J. Zool. 52: 519–23.

Bakus, G. J. 1968. Defensive mechanisms and ecology of some tropical holothurians. Mar. Biol. 2: 23–32.

————. 1973. The biology and ecology of tropical holothurians, pp. 325–67, *in* O. A. Jones and R. Endean, eds., Biology and geology of coral reefs, vol. 2, biology 1. New York: Academic Press. 480 pp.

————. 1974. Toxicity in holothurians: A geographical pattern. Biotropica 6: 229–36.

Barham, E. G., N. J. Ayer, Jr., and R. E. Boyce. 1967. Macrobenthos of the San Diego trough: Photographic census and observations from bathyscaphe, *Trieste*. Deep-Sea Res. 14: 773–84.

Berrill, M. 1966. The ethology of the synaptid holothurian, *Opheodesoma spectabilis*. Canad. J. Zool. 44: 457–82.

Binyon, J. 1972. Physiology of echinoderms. New York: Pergamon. 264 pp.

Boolootian, R. A., ed. 1966. Physiology of Echinodermata. New York: Interscience. 822 pp.

Boolootian, R. A., and A. C. Giese. 1958. Coelomic corpuscles of echinoderms. Biol. Bull. 115: 53–63.

Brumbaugh, J. H. 1965. The anatomy, diet, and tentacular feeding mechanism of the dendrochirote holothurian *Cucumaria curata* Cowles, 1907. Doctoral thesis, Biological Sciences, Stanford University, Stanford, Calif. 119 pp.

Caso, M. E. 1966. Contribución al estudio de los Holoturoideos de México: Morfología interna y ecologia de *Stichopus fuscus* Ludwig. An. Inst. Biol. Univ. Nac. Auton. Méx. 37: 175–81.

Chia, F.-S., and J. B. Buchanan. 1969. Larval development of *Cucumaria elongata* (Echinodermata: Holothuroidea). J. Mar. Biol. Assoc. U.K. 49: 151–59.

Chia, F.-S., D. Atwood, and B. Crawford. 1975. Comparative morphology of echinoderm sperm and possible phylogenetic implications. Amer. Zool. 15: 533–65.

Ciereszko, L. S., E. M. Ciereszko, E. R. Harris, and C. A. Lane. 1962. On the occurrence of vanadium in holothurians. Comp. Biochem. Physiol. 7: 127–29.

Clark, H. L. 1901a. Echinoderms from Puget Sound. Proc. Boston Soc. Natur. Hist. 29: 323–37.

———. 1901b. The holothurians of the Pacific coast of North America. Zool. Anz. 24: 162–71.

———. 1901c. Synopses of North American invertebrates. XV. The Holothuroidea. Amer. Natur. 35: 479–96.

———. 1907. The apodous holothurians. A monograph of the Synaptidae and Molpadiidae. Smithsonian Contr. Knowledge 35. 231 pp.

———. 1913. Echinoderms from Lower California with descriptions of new species. Bull. Amer. Mus. Natur. Hist. 32: 185–236.

———. 1922. The holothurians of the genus *Stichopus*. Bull. Mus. Comp. Zool. 65: 39–74.

———. 1924. Some holothurians from British Columbia. Canad. Field Natur. 38: 54–57.

Costello, D. P. 1946. The swimming of *Leptosynapta*. Biol. Bull. 90: 93–96.

Costello, D. P., and C. Henley. 1971. Methods for obtaining and handling marine eggs and embryos. 2nd ed. Marine Biological Laboratory, Woods Hole, Mass. 247 pp.

Courtney, W. D. 1927. Fertilization in *Stichopus californicus*. Publ. Puget Sound Biol. Sta. 5: 257–60.

Cowles, R. P. 1907. *Cucumaria curata*, sp. nov. Johns Hopkins Univ. Circ. 195: 1–2.

Crescitelli, F. 1945. A note on the absorption spectra of the blood of *Eudistylia gigantea* and of the blood pigment in the red corpuscles of *Cucumaria miniata* and *Molpadia intermedia*. Biol. Bull. 88: 30–36.

Crozier, W. J. 1918. The amount of bottom material ingested by holothurians (*Stichopus*). J. Exper. Zool. 26: 379–89.

Dan, J. C. 1967. Acrosome reaction and lysins, pp. 237–93, *in* C. B. Metz and A. Monroy, eds., Fertilization—comparative morphology, biochemistry, and immunology, vol. 1. New York: Academic Press. 489 pp.

Dawbin, W. H. 1949. Auto-evisceration and the regeneration of viscera in the holothurian *Stichopus mollis* (Hutton). Trans. Proc. Roy. Soc. New Zeal. 77: 497–523.

Deichmann, E. 1937. The Templeton Crocker Expedition. IX. Holothurians from the Gulf of California, the west coast of Lower California and Clarion Island. Zoologica 22: 161–76.

———. 1938a. Eastern Pacific expeditions of the New York Zoological Society. XVI. Holothurians from the western coasts of Lower California and Central America, and from the Galápagos Islands. Zoologica 23: 361–87.

———. 1938b. New holothurians from the western coast of North America and some remarks on the genus *Caudina*. Proc. New England Zool. Club 16: 103–15.

———. 1939. A new holothurian of the genus *Thyone* collected on the Presidential cruise of 1938. Smithsonian Misc. Coll. 98: 1–7.

———. 1941. The Holothuroidea collected by the Velero III during the years 1932 to 1938. Part I, Dendrochirota. Allan Hancock Pacific Exped. 8: 61–194.

Dimock, R. V., Jr. 1974. Intraspecific aggression and the distribution of a symbiotic polychaete on its hosts, pp. 29–44, *in* W. B. Vernberg, ed., Symbiosis in the sea. Columbia: University of South Carolina Press. 276 pp.

Dimock, R. V., Jr., and D. Davenport. 1971. Behavioral specificity and the induction of host recognition in a symbiotic polychaete. Biol. Bull. 141: 472–84.

Doyle, W. L. 1967. Vesiculated axons in haemal vessels of an holothurian, *Cucumaria frondosa*. Biol. Bull. 132: 329–36.

Doyle, W. L., and G. F. McNiell. 1964. The fine structure of the respiratory tree in *Cucumaria*. Quart. J. Microscop. Sci. 105: 7–12.

Endean, R. 1957. The cuvierian tubules of *Holothuria leucospilata*. Quart. J. Microscop. Sci. 98: 455–72.

Engstrom, N. A. 1974. Population dynamics and prey-predation relations of a dendrochirote holothurian, *Cucumaria lubrica*, and sea stars in the genus *Solaster*. Doctoral thesis, Zoology, University of Washington, Seattle. 172 pp.

Filice, F. P. 1950. A study of some variations in *Cucumaria curata*. Wasmann J. Biol. 8: 39–48.

Fish, J. D. 1967a. The biology of *Cucumaria elongata* (Echinodermata: Holothuroidea). J. Mar. Biol. Assoc. U.K. 47: 129–43.

———. 1967b. The digestive system of the holothurian, *Cucumaria elongata*. I. Structure of the gut and hemal system. Biol. Bull. 132: 337–53.

———. 1967c. The digestive system of the holothurian *Cucumaria elongata*. II. Distribution of the digestive enzymes. Biol. Bull. 132: 354–61.

Fontaine, A. R., and F.-S. Chia. 1968. Echinoderms: An autoradiographic study of assimilation of dissolved organic molecules. Science 161: 1153–55.

Fontaine, A. R., and P. Lambert. 1973. The fine structure of the haemocyte of the holothurian, *Cucumaria miniata* (Brandt). Canad. J. Zool. 51: 323–32.

Freeman, P. J. 1966. Observations of osmotic relationships in the holothurian *Opheodesoma spectabilis*. Pacific Sci. 20: 60–69.

Freeman, W. P., and S. E. Simon. 1964. The histology of holothuroidean muscle. J. Cell. Compar. Physiol. 63: 25–38.

Gay, W. S., and S. E. Simon. 1964. Metabolic control in holothuroidean muscle. Comp. Biochem. Physiol. 11: 183–92.

Giese, A. C. 1966. Lipids in the economy of marine invertebrates. Physiol. Rev. 46: 244–98.

Glynn, P. W. 1965. Active movements and other aspects of the biology of *Astichopus* and *Leptosynapta* (Holothuroidea). Biol. Bull. 129: 106–27.

Habermehl, G., and G. Volkwein. 1971. Aglycones of the toxins from the cuvierian organs of *Holothuria forskali* and a new nomenclature for the aglycones from Holothuroidea. Toxicon 9: 319–26.

Haderlie, E. C. 1968. Marine fouling organisms in Monterey harbor. Veliger 10: 327–41.

Hall, A. R. 1927. Histology of the retractor muscle of *Cucumaria min-iata*. Publ. Puget Sound Biol. Sta. 5: 205–19.

Halstead, B. W. 1965. Poisonous and venomous marine animals of the world. Vol. 1 (pp. 567–88 refer specifically to holothurians). Washington, D.C.: U.S. Govt. Printing Office. 994 pp.

Hammen, C. S., and P. J. Osborne. 1959. Carbon dioxide fixation in marine invertebrates: A survey of major phyla. Science 130: 1409–10.

Harrison, G. 1968. Subcellular particles in echinoderm tube feet. II. Class Holothuroidea. J. Ultrastruc. Res. 23: 124–33.

Heding, S. G. 1928. Synaptidae. Papers from Dr. Th. Mortensen's Pacific Exped. 1914–1916, No. 46. Vidensk. Medd. Dansk. Naturh. Foren. 85: 105–323.

Hefferman, J. M., and S. A. Wainwright. 1974. Locomotion of the holothurian *Euapta lappa* and redefinition of peristalsis. Biol. Bull. 147: 95–104.

Hetzel, H. R. 1963. Studies on holothurian coelomocytes. I. A survey of coelomocyte types. Biol. Bull. 125: 289–301.

————. 1965. Studies on holothurian coelomocytes. II. The origin of coelomocytes and the formation of brown bodies. Biol. Bull. 128: 102–11.

Hill, R. B. 1970. Effects of some postulated neurohumours on rhythmicity of the isolated cloaca of a holothurian. Physiol. Zool. 43: 109–23.

Hinegardner, R. 1974. Cellular DNA content of the Echinodermata. Comp. Biochem. Physiol. 49: 219–26.

Hufty, H. M., and P. C. Schroeder. 1974. A hormonally active substance produced by the ovary of the holothurian *Parastichopus californicus*. Gen. Comp. Endocrinol. 23: 348–51.

Hyman, L. H. 1955. The invertebrates: Echinodermata. Vol. 4. New York: McGraw-Hill. 763 pp.

Imai, T. 1950. On the artificial breeding of Japanese sea cucumber, *Stichopus japonicus* Selenka. Bull. Inst. Agr. Res. Tôhôku Univ. 2: 269–76.

Johnson, M. W., and L. T. Johnson. 1950. Early history and larval development of some Puget Sound echinoderms with special reference to *Cucumaria* spp. and *Dendraster excentricus*. Scripps Inst. Oceanogr. Contr. 439: 74–84.

Lasley, B. J., and R. F. Nigrelli. 1970. The effect of crude holothurin on leucocyte phagocytosis. Toxicon 8: 301–6.

————. 1971. The effect of holothurin on leucocyte migration. Zoologica 56: 1–12.

Lawrence, D. C., A. L. Lawrence, M. L. Greer, and D. Mailman. 1967. Intestinal absorption in the sea cucumber, *Stichopus parvimensis*. Comp. Biochem. Physiol. 20: 619–27.

Ludwig, H. 1904. Brütpflege bei Echinodermen. Zool. Jahrb., Suppl. 7: 683–99.

MacGinitie, G. E., and N. MacGinitie. 1968. Natural history of marine animals. 2nd ed. New York: McGraw-Hill. 523 pp.

Manwell, C. 1959. Oxygen equilibrium of *Cucumaria miniata* hae-moglobin and the absence of the Bohr effect. J. Cell Compar. Physiol. 53: 75–83.

————. 1966. Sea cucumber sibling species: Polypeptide chain types and oxygen equilibrium of haemoglobin. Science 152: 1393–96.

Margolin, A. S. 1976. Swimming of the sea cucumber *Parastichopus californicus* (Stimpson) in response to sea stars. Ophelia 15: 105–14.

Martin, W. E. 1969a. A commensal sea cucumber. Science 164: 855.

————. 1969b. *Rynkatorpa pawsoni* n. sp. (Echinodermata: Holothuroidea), a commensal sea cucumber. Biol. Bull. 137: 332–37.

Matsumura, T. 1974. Collagen fibrils of the sea-cucumber, *Stichopus japonicus*: Purification and morphological study. Connec. Tissue Res. 2: 117–25.

Mauzey, K. P., C. Birkeland, and P. K. Dayton. 1968. Feeding behavior of asteroids and escape responses of their prey in the Puget Sound region. Ecology 49: 603–19.

Mortensen, T. 1921. Studies on the development and larval forms of echinoderms. Copenhagen: Gad. 261 pp.

Motohiro, T. 1960. Studies on the mucoprotein in marine products. I. Isolation of mucoprotein from the meats of *Stichopus japonicus* and *Cucumaria japonica*. II. Isolation of poly-fucose sulfate from the mucoprotein in *Cucumaria japonica*. III. Crystallization of itin sulfate from the mucoprotein in *Cucumaria japonica*. Bull. Jap. Soc. Sci. Fish. 26: 1171–82.

Mueller, W. E. G., and R. K. Zahn. 1972. The adhesive behavior in cuvierian tubules of *Holothuria forskali*: Biochemical and biophysical investigations. Cytobiologie 5: 335–51.

Mueller, W. E. G., R. K. Zahn, and K. Schmid. 1970. Morphologie und Funktion der Cuvierschen Organe von *Holothuria forskali* Delle Chiaje (Echinodermata: Holothuroidea: Aspidochirota: Holothuriidae). Z. Wiss. Zool. 181: 219–32.

Parker, G. H. 1921. The locomotion of the holothurian *Stichopus panimensis* Clark [misprint for *parvimensis*]. J. Exper. Zool. 33: 205–8.

Pawson, D. L. 1966. Ecology of holothurians, pp. 63–71, in Boolootian (1966).

————. 1967. The psolid holothurian genus *Lissothuria*. Proc. U.S. Nat. Mus. 122: 1–117.

Pawson, D. L., and H. B. Fell. 1965. A revised classification of the dendrochirote holothurians. Breviora 214: 1–7.

Pople, W., and D. W. Ewer. 1954. Studies on the myoneural physiology of Echinodermata. I. The pharyngeal retractor muscle of *Cucumaria*. J. Exper. Biol. 31: 114–26.

————. 1955. Studies on the myoneural physiology of Echinodermata. II. Circumoral conduction in *Cucumaria*. J. Exper. Biol. 32: 59–69.

————. 1958. Studies on the myoneural physiology of Echinodermata. III. Spontaneous activity of the pharyngeal retractor muscle of *Cucumaria*. J. Exper. Biol. 35: 712–30.

Prosser, C. L., and C. L. Judson. 1952. Pharmacology of haemal vessels of *Stichopus californicus*. Biol. Bull. 102: 249–51.

Quast, J. C. 1968. Observations on the food of the kelp-bed fishes, pp. 109–42, *in* W. J. North and C. L. Hubbs, eds., Utilization of kelp-bed resources in southern California. Calif. Dept. Fish & Game, Fish Bull. 139: 1–264.

Reish, D. J. 1961. A study of benthic fauna in a recently constructed boat harbor in southern California. Ecology 42: 84–91.

Ricketts, E. F., and J. Calvin. 1968. Between Pacific tides. 4th ed. Revised by J. W. Hedgpeth. Stanford, Calif.: Stanford University Press. 614 pp.

Rutherford, J. C. 1973. Reproduction, growth, and mortality of the holothurian *Cucumaria pseudocurata*. Mar. Biol. 22: 167–76.

———. 1975. Class Holothuroidea, pp. 634–37, *in* R. I. Smith and J. T. Carlton, eds., Light's manual: Intertidal invertebrates of the central California coast. 3rd ed. Berkeley and Los Angeles: University of California Press. 716 pp.

Schaller, F. 1973. Über die Tentakelbewegungen der dendrochiroten Holothurien *Cucumaria lefevrei* und *Cucumaria saxicola*. Zool. Anz. 191: 162–70.

Selenka, E. 1867. Beiträge zur Anatomie und Systematik der Holothurien. Z. Wiss. Zool. 17: 291–374.

Sheikh, Y. M., and C. Djerassi. 1977. Biosynthesis of sterols in the sea cucumber *Stichopus californicus*. Tetrahedron Letters 36: 3111–14.

Shelford, V. E., A. O. Weese, L. A. Rice, D. I. Rasmussen, A. MacLean, N. M. Wismer, and J. H. Swanson. 1935. Some marine biotic communities of the Pacific coast of North America. Ecol. Monogr. 5: 249–354.

Shimada, S. 1969. Antifungal steroid glycoside from sea cucumber. Science 163: 1462.

Simon, S. E., J. Muller, and D. J. Dewhurst. 1964. Ionic partition in holothuroidean muscle. J. Cell. Compar. Physiol. 63: 77–84.

Smith, E. H. 1962. Studies of *Cucumaria curata* Cowles, 1907. Pacific Natur. 3: 233–46.

Smith, G. N. 1971. Regeneration in the sea cucumber *Leptosynapta*. I. The process of regeneration. II. The regenerative capacity. J. Exper. Zool. 117: 319–42.

Stevens, N. M. 1901. Studies on ciliate Infusoria. Proc. Calif. Acad. Sci. (3) 3: 1–42, 5 pls.

Stickle, W. B., and R. Ahokas. 1972. The effects of fluctuating salinity upon the blood osmotic integrity of several molluscs and echinoderms. Amer. Zool. 12: 713 (abstract).

Stimpson, W. 1857. The Crustacea and Echinodermata of the Pacific shores of North America. J. Boston Soc. Natur. Hist. 6: 84–86.

———. 1864. Description of new species of marine invertebrates from Puget Sound. Proc. Acad. Natur. Sci. Philadelphia 16: 153–61.

Strathmann, R. R. 1971. The feeding behavior of planktotrophic echinoderm larvae: Mechanisms, regulation, and rates of suspension feeding. J. Exper. Mar. Biol. Ecol. 6: 109–60.

———. 1978. Length of pelagic period in echinoderms with feeding larvae from the northeast Pacific. J. Exper. Mar. Biol. Ecol. 34: 23–27.

Strathmann, R. R., and H. Sato. 1969. Increased germinal vesicle breakdown in oocytes of the sea cucumber. *Parastichopus californicus* induced by starfish radial nerve extract. Exper. Cell Res. 54: 127–29.

Swan, E. F. 1961. Seasonal evisceration in the sea cucumber (*Parastichopus californicus* (Stimpson)). Science 133: 1078–79.

Tanaka, Y. 1958a. Feeding and digestive processes of *Stichopus japonicus*. Bull. Fac. Fish. Hokkaido Univ. 9: 14–28.

———. 1958b. Seasonal changes occurring in the gonad of *Stichopus japonicus*. Bull. Fac. Fish. Hokkaido Univ. 9: 29–36.

———. 1955a. Studies on the proteins of the meat of a sea cucumber (*Stichopus japonicus* Selenka). Mem. Fac. Fish. Hokkaido Univ. 3: 1–91.

———. 1955b. Studies on the nutritive value of the meat of a sea cucumber (*Stichopus japonicus* Selenka). I. General introduction and explanation of the plan of investigations. Bull. Fac. Fish. Hokkaido Univ. 5: 338–40.

Tanikawa, E., et al. (various coauthors). 1955. Studies on the nutritive value of the meat of a sea cucumber (*Stichopus japonicus* Selenka). II–VII. Bull. Fac. Fish. Hokkaido Univ. 5: 338–51; 6: 37–51.

Terwilliger, R. C., and K. R. H. Read. 1970. The hemoglobins of the holothurian echinoderms *Cucumaria miniata* Brandt, *Cucumaria piperata* Stimpson, and *Molpadia intermedia* Ludwig. Comp. Biochem. Physiol. 36: 339–51.

Theel, H. 1886a. Report on the Holothuroidea. Report on the scientific results of the voyage of H.M.S. *Challenger* during the years 1873–1876. Pt. 39. Zoology 14: 1–290.

———. 1886b. Report on the results of dredgings by the U.S. Coast Survey steamer *Blake*. XXX. Report on Holothuroidea. Bull. Mus. Comp. Zool. 13: 1–21.

Turner, C. H., E. E. Ebert, and R. R. Given. 1968. The marine environment offshore from Point Loma, San Diego County. Calif. Dept. Fish & Game, Fish Bull. 140: 1–85.

Turner, R. L., and J. C. Rutherford. 1975. Caloric content and organic composition of eggs and pentactulae of the brooding holothuroid *Cucumaria curata*. Amer. Zool. 15: 787 (abstract).

Webster, S. K. 1968. An investigation of the commensals of *Cryptochiton stelleri* (Middendorff, 1847) in the Monterey Peninsula area, California. Veliger 11: 121–25.

Wootton, D. 1949. The development of *Thyonepsolus nutriens* Clark. Doctoral thesis, Biological Sciences, Stanford University, Stanford, Calif. 97 pp., 5 pls.

Yamanouchi, T. 1956. The daily activity rhythms of the holothurians in the coral reef of the Palao Islands. Publ. Seto. Mar. Biol. Lab. 5: 347–62.

Yingst, J. Y. 1972. A new species of rock dwelling dendrochirote holothurian from Catalina Island. Bull. So. Calif. Acad. Sci. 71: 145–50.

Chapter 10

Ophiuroidea: *The Brittle Stars*

William C. Austin and Michael G. Hadfield

Ophiuroids, commonly known as brittle stars or serpent stars, are regarded as either a separate echinoderm class or as a subclass of the Asterozoa (a grouping that also includes the sea stars). Most brittle stars are easily distinguished from sea stars by their arms, which are thin, flexible, and sharply set off from a nearly circular disk. Portions of the arms are easily broken off (hence the name "brittle stars"), but the lost parts are readily regenerated. Whereas in sea stars the tube feet extend from open furrows on the lower surfaces of the body and arms, in most brittle stars the furrows are closed and covered by a series of hard plates.

Ophiuroids are found in all seas, at a wide range of depths, and on varied substrata. They are among the most conspicuous animals on the soft bottoms of the great abyssal plains of the oceans. About 2,000 living ophiuroid species have been described, of which 16 are known from the California intertidal region. Here they are generally inconspicuous, for they are small in size and tend to seclude themselves in protected places during daylight hours or when the tide is out. Brittle stars are more prominent intertidally in tropical or subtropical waters (including the Gulf of California), but even there they are usually hidden during the day. They are often abundant in subtidal situations, and huge populations occur on the continental shelf and in deeper waters off California (see, for example, Barham, Ayer & Boyce, 1966; Barnard & Ziesenhenne, 1961). Both intertidal and subtidal forms play important ecological roles as particle feeders, predators on small animals, scavengers, and detritus feeders. In turn they are fed upon by predaceous fishes, sea stars, and crabs.

The photographs and the species characters used in this chapter should be sufficient to identify adult animals from California intertidal areas. The terminology associated with these characters is complex, and requires some explanation. The following description applies to our local intertidal species, but not necessarily to species elsewhere.

The upper side of the disk is covered by calcareous plates, including a pair of enlarged plates, the radial shields, directly above each arm, which often operate to raise and lower the upper disk wall. The disk plates may be covered by skin, spines, or granules. A series of five, or in some species six, triangular jaws is arranged in a starlike pattern in the center of the lower side of the disk. These jaws can open and close, and along their free margins they bear projecting ossicles or spines that provide useful taxonomic characters. The spines fringing the two flat sides of each jaw are called oral papillae; they are absent in some species, but when present are easily visible when the jaws are closed. The spine rows on the sides of a jaw may meet at the apex as one or two terminal or apical oral papillae. When the jaws are agape and one can look inward toward the true mouth, other spines can be seen projecting medially from the apex of each jaw. The largest ossicles here, usually forming a single row going inward toward the mouth along each jaw apex, are the teeth. In some species the row of large teeth does not extend all the way out from the mouth, and the outer region of the jaw apex (just below a terminal oral papilla if one is present) bears a cluster of small spines, the dental or tooth papillae.

William C. Austin is Senior Research Associate of the Khoyatan Marine Laboratory, Cowichan Station, British Columbia. Michael G. Hadfield is Associate Professor of Cytology and Zoology, Pacific Biomedical Research Center and Department of Zoology, at the University of Hawaii.

One or two bursal or genital slits lie adjacent to, and parallel with, each arm base on the lower side of the disk. These slits open into internal pouches (bursae), which play a role in respiration and in the release of gametes or the brooding of young. The arms are covered by a series of plates: a single row of ventral arm plates, a single row of lateral arm plates on each side, and, typically, a single row of dorsal arm plates. In some species the dorsal plates may be subdivided into a number of smaller plates or may be associated with accessory dorsal arm plates. Each lateral plate carries a vertical row of arm spines, their number, shape, and operation varying among the species. Pairs of tube feet (podia) extend out from pores between the jaws and from pores along the ventral sides of the arms. In brittle stars the podia lack the terminal suckers that occur in many other echinoderms, but they may produce sticky secretions that allow them to cling to various objects. In some species the podia can be retracted into the arms. Their functions vary and can include roles in feeding, defecation, locomotion, burrowing, reception of various stimuli, and probably respiration. One or more small spines called tentacle scales can occur adjacent to each podial pore.

The sexes are usually separate in brittle stars, but males and females typically appear alike externally. During spawning sperm are shed into the genital bursae, from which they emerge by way of the genital slits to the sea. The females likewise shed their eggs to the genital bursae. In species that produce large numbers of tiny eggs, the eggs, like the sperm, pass immediately to the sea. They are fertilized there and develop into free-swimming, plankton-feeding ophiopluteus larvae. After a time they settle to the bottom and undergo metamorphosis. In some species of brittle stars, the females produce a much smaller number of relatively large eggs containing much yolk. These eggs are retained in the genital bursae, are fertilized and brooded there, and develop directly into tiny juvenile brittle stars, bypassing the swimming larval stage. Still other species show an intermediate condition, producing yolky larvae that swim for a time but carry their own food supply with them.

Brittle stars are best studied alive, either in the field or in aquaria provided with refrigerated circulating seawater. However, if they are to be preserved (as in 70 percent alcohol), they should first be anesthetized; otherwise many will readily shed their arms. Anesthesia can be accomplished by exposing the animals for several hours to a mixture of equal parts of seawater and a tap-water solution containing either 7.5 percent by weight magnesium chloride (Glauber's salt) or 20 percent magnesium sulfate (Epsom salts).

Most California species are poorly known. There are large gaps in our knowledge of where the different forms occur, what conditions they can tolerate without damage, what they eat and how they get it, when and how they breed, what their major predators and parasites are, how long they live, how they behave and respond to stimuli, how they function internally, and a host of other things. Where information is lacking on California species, we have included references to studies of related species in other parts of the world. Many of the outstanding papers on structure and development of brittle stars are roughly a century old.

The following sources are helpful for the identification of species and provide an entry to additional taxonomic literature: worldwide, H. L. Clark (1915), Fell (1960); northeast Pacific, Austin & Haylock (1973), H. L. Clark (1911), D'iakonov (1967), Nielsen (1932); Washington and British Columbia, Kozloff (1974), Kyte (1969); central and northern California, May (1924), Sutton (1975); southern California, Boolootian & Leighton (1966), McClendon (1909); Mexico, Brusca (1973), Caso (1951), Steinbeck & Ricketts (1941).

Good reviews containing information on one or more aspects of the biology of ophiuroids include Binyon (1972), Boolootian (1966), A. Clark (1962), Costello & Henley (1971), Cuénot (1948), Dawydoff (1948), Delage & Herouard (1903), Fell (1966), Florkin & Scheer (1969), Hendler (1975), Hyman (1955), Ludwig & Hamann (1901), Millott (1967), Mortensen (1921), Nichols (1964, 1969), Smith (1965), Spencer & Wright (1966), and Strathmann (1971, 1975). Most of these references have extensive bibliographies. See also the excellent bibliography of Holland & Holland (1969).

Phylum Echinodermata / Class Ophiuroidea / Order Ophiurida
Suborder Gnathophiurina / Family Ophiactidae

10.1 **Ophiactis simplex** (Le Conte, 1851)
(= O. arenosa)

Locally abundant on rocky shores, especially in the sponge *Halichondria panicea* (2.15), in masses of the tube snail *Serpulorbis squamigerus* (13.45), and in clumps of coralline and other algae, middle and low intertidal zones to 45 m depth; Santa Cruz Island (Channel Islands) to Panama and Galápagos Islands.

Disk diameter to 6 mm, more commonly 2-3 mm; arms 3.5–5 times as long as disk diameter, typically six, occasionally only five; oral papillae one on each side of each jaw (diagnostic); color as in photograph or with darker brownish green predominating.

One other six-armed species, **O. savignyi** (Müller & Troschel), also reported from southern California, typically has two oral papillae on each side of each jaw.

Ophiactis simplex may occur in large numbers (up to several hundred per 0.1 m²). Like many other species of *Ophiactis*, it can multiply asexually by dividing in two across the disk, each half subsequently regenerating its missing parts (schizogony). Sexual reproduction also occurs. Little is known of feeding, though the animals can absorb some dissolved organic substances from seawater. Several related species feed on small particles that adhere to sticky secretions of the tube feet.

See McClendon (1909, as *O. arenosa*), Nielson (1932), Steinbeck & Ricketts (1941), and Stephens & Virkar (1966). For anatomy and biology of related species, see Boffi (1972), Boolootian (1966), A. Clark (1967), Cuénot (1891), R. Pentreath (1970), Preyer (1886–87), and Simroth (1876, 1877).

*Phylum Echinodermata / Class Ophiuroidea / Order Ophiurida
Suborder Gnathophiurina / Family Ophiactidae*

10.2 Ophiopholis aculeata (Linnaeus, 1767)
var. kennerlyi Lyman, 1860
Daisy Brittle Star

Under stones, in algal holdfasts, and in gravel and shell deposits, low intertidal zone on rocky shores; subtidal to 732 m; the species occurs in northern Atlantic and Pacific waters, including low Arctic; this variety ranges from southeastern Bering Sea (55°N) to south of Santa Barbara, increasingly abundant at higher latitudes.

Disk diameter to 22 mm; arm length 3.5–4 times disk diameter; accessory plates six to ten, surrounding each dorsal armplate; hue and pattern extremely variable.

This variety intergrades with other varieties, two of which occur north of Puget Sound (British Columbia).

This species has been studied extensively in Atlantic waters. Individuals are cryptic and lethargic, remaining motionless for long periods in aquaria. Food particles suspended in the water are captured by adhesive mucus on the podia and transported by the podia along the arms to the mouth. The animals also feed on larger material and detritus. In the North Atlantic they are eaten by fishes, especially cod and flatfishes.

Ophiopholis aculeata has been observed to spawn in the laboratory in January, February, March, July, October, and November at Friday Harbor, Washington (R. Strathmann, pers. comm.), and a lunar cycle in spawning activity is reported for animals in the White Sea (U.S.S.R.). Spawning may be induced by allowing the animals to warm to room temperature for approximately an hour in dry bowls, then immersing them in sea water at temperatures of 8–14°C. Detailed studies have been made of feeding in the eight-armed ophiopluteus larvae. Metamorphosis of the larvae occurs from 83 to 216 days after fertilization.

See especially Gislén (1924), LaBarbera (1978), Olsen (1942), and Strathmann (1971, 1978); see also Blegvad (1914), Boolootian (1966), Buchanan (1962), Costello & Henley (1971), Eichelbaum (1910), Fontaine (1962b, 1963), Kuznetsov & Sokolova (1961), Kyte (1969), May (1924), Mileikovsky (1970), Moment (1962), Roushdy & Hansen (1960), Strathmann (1975), Taylor (1958), Turpaeva (1953), and Wintzell (1918).

*Phylum Echinodermata / Class Ophiuroidea / Order Ophiurida
Suborder Gnathophiurina / Family Amphiuridae*

10.3 Amphipholis squamata (Delle Chiaje, 1829)
(=Axiognathus squamata, Amphiura squamata)

Abundant in algal holdfasts, among branches of coralline algae, under stones, in rock crevices, and in shell and gravel debris in tidepools, middle to low intertidal zone on rocky shores; subtidal to 823 m; nearly worldwide in boreal, temperate, and tropical waters.

Disk diameter to 5 mm; arms whitish, 3–4 times as long as disk diameter; upper surface of disk pale gray, tan, lavender, or pale orange, with white spot at margin just above each arm; ovoviviparous, the young developing within the bursae.

In **A. pugetana** (Lyman), a related California intertidal species, the arms are 7–8 times the disk diameter in larger specimens and gray or banded gray and white; the young are not brooded. Both this species and *A. squamata* are sometimes placed in the genus *Axiognathus* (Thomas, 1966).

Amphipholis squamata is a mobile species; rowing movements of the arms propel the animals on horizontal surfaces, and sticky secretions on the podia enable them to climb glass aquarium walls. Individuals dislodged into the water curl their arms about the disk, making the body a compact mass that sinks rapidly to the bottom (a response shared by some other brittle stars, notably *Amphiodia occidentalis*, 10.4). Strong mechanical or other stimulation causes cells at the bases of the arm spines to emit a yellow-green luminescence, which appears to be intracellular. The photogenic cells have long processes similar to the neurons of the animal's peripheral nervous system. The animal also fluoresces, but only after the onset of luminescence.

Amphipholis squamata feeds on material both in suspension and on the bottom. Its diet includes unicellular algae, protozoans, small animals, and detritus.

It has been known for more than a century that the species broods its young. This occurs throughout the year. Mature eggs are 0.1 mm in diameter, and only one at a time reaches maturity in each of the ten ovaries. Fertilization is internal. The animals are protandric hermaphrodites. The young are brooded until they are large juveniles with well-developed arms; they are "born" by actively creeping out of the parental bursal apertures. They are easily released prematurely by flipping off the top of the parental disk with a needle (this species may shed the top of the disk and adjacent viscera spontaneously). Brooding of the young and the absence of a swimming larva may in part account for the dense populations sometimes seen in restricted areas. The species has been found associated with floating materials, which may help explain why it is probably the most widely distributed of all shallow-water ophiuroids; a single pregnant animal, transported to a new area, could start a new population. The animals are fairly tolerant of environmental change; some populations even occur on shores alternately covered by fresh and salt water.

Amphipholis squamata is host to a curious mesozoan parasite, *Rhopalura ophiocomae*. The rate of infestation is low on the Pacific coast at the two locations where the parasite has been noted, Friday Harbor (Washington) and Pacific Grove (Monterey Co.).

See especially Fell (1946) and R. Pentreath (1970); see also Apostolides (1881), Bernasconi (1926), Boffi (1972), Boolootian (1966), Brehm & Morin (1977), Buchanan (1963), Costello & Henley (1971), Cuénot (1891), Fewkes (1887), Fontaine & Chia (1968), Harvey (1952), Kozloff (1969), Kyte (1969), MacBride (1892), Martin (1968), Nielsen (1932), Russo (1891), and Thomas (1966).

*Phylum Echinodermata / Class Ophiuroidea / Order Ophiurida
Suborder Gnathophiurina / Family Amphiuridae*

10.4 Amphiodia occidentalis (Lyman, 1860)
(=Diamphiodia occidentalis)

Sometimes abundant under rocks resting on sand, in algal holdfasts in protected coastal tidepools with sandy bottoms, in mud flats around roots of the eelgrass *Zostera*, and in holdfasts of large kelps, middle and low intertidal zones; subtidal to 367 m; Kodiak Island (Alaska) to San Diego, possibly also Gulf of California; not reported intertidally south of Monterey.

Disk to 11 mm in diameter, the upper side lacking spines; arms very long, 9–15 times disk diameter, bearing a row of three blunt, flattened arm spines on each side of each segment; no accessory plates between dorsal and lateral arm plates; color as in photograph or grayer.

Amphiodia occidentalis burrows to shallow depths in soft substrata, using the tube feet; burrowing behavior is spectacular and is easily seen if an animal is placed on fine sand under water. The tube feet move up and down vertically between rows of arm spines, digging sand from below the arms and disk and piling it on top, so that the animal appears to sink slowly from sight. In further digging, the disk descends more rapidly than the arms. The podia then transport particles to the arm tips, where the excavated sand forms small mounds at the surface. Each arm now lies in a narrow tunnel leading to the disk. In daylight only the tips of some arms are visible, but at night the arms may extend some distance out of the tunnels. Arms extended horizontally over the bottom are engaged in deposit feeding, whereas those held vertically in the water are involved in suspension feeding. Subsurface browsing also occurs, and the animals are capable of changing position without coming to the surface. Periodically, one or rarely two arms display well-defined up-and-down oscillations; which produce water currents down across the face of the disk and out through the tunnels of non-oscillating arms. These currents, in conjunction with the opening and closing of the bursae and the alternate raising and lowering of the disk, serve a respiratory purpose. Waste material is removed by rapid disk contraction, forcing particles out the open mouth, where they are picked up by podia and transported out of the burrow. Animals that are removed from the sand or otherwise disturbed tend to curl their arms around the disk, forming a tight ball. The arms are readily shed, and many animals exhibit regenerating arm tips.

In a Monterey Bay population of *Amphiodia occidentalis*, reproductive condition was followed through the year. All animals 6 mm or more across the disk were reproductively mature. Most spawning occurred in the late spring and early summer, although isolated individuals were observed spawning in January. During spawning, at least the disk is elevated above the substratum, and the sexual products stream out of the bursae. The eggs develop into swimming larvae.

Along with the typical *A. occidentalis* in the Monterey Bay area, one may find another form, distinguished by its habit of brooding its young, by its maximum disk diameter of only about 4 mm, and by its arms length, which does not exceed about 25 mm, and averages about five times the disk diameter (Cunningham, 1977).

See Barnard & Ziesenhenne (1961), Boolootian (1966), Cunningham (1977), Fager (1968), Fell (1962,) Fontaine (1965), Kyte (1969), May (1924), Ricketts & Calvin (1968), and Sutton (1976).

Phylum Echinodermata / Class Ophiuroidea / Order Ophiurida
Suborder Gnathophiurina / Family Amphiuridae

10.5 **Amphiodia urtica** (Lyman, 1860)
(= Diamphiodia urtica, Amphiodia barbarae)

Uncommon in mud and sandy mud, low intertidal zone; abundant subtidally to 1,624 m; Shumagin Islands (Aleutian Islands, Alaska) to Mexico (16°N).

Disk to 9 mm in diameter, the upper side smooth, with no spines or granules; arms very long, 10–21 times disk diameter, bearing three sharply pointed spines, circular in cross section, on each side of each segment; no accessory plates between dorsal and lateral arm plates; minute spinelets sometimes occurring near genital bursae; disk brownish pink to pink, arms pink to pinkish red.

Amphiodia urtica occurs in huge numbers (up to 1,500 per m²) subtidally along the northeastern Pacific continental shelf, but has been reported in the low intertidal zone only sporadically, in British Columbia, Washington, and southern California. Although not restricted to associations with holdfasts and other hard objects, as in *Amphiodia occidentalis* (10.4), it feeds similarly, exhibiting both deposit feeding and suspension feeding. It also burrows in the same manner, but to a greater depth, and is most difficult to extricate in one piece. Portions of the arms and the top of the disk, along with the gut and gonads, may be cast off and regenerated. The large proportion of regenerated arm segments suggests that some predators may be cropping off portions of the arms rather than eating the whole animal. Predators, at least in British Columbia waters, include the lemon sole (*Parophrys vetulus*), the flathead sole (*Hippoglossoides elassodon*), the sand sole (*Psettichthys melanostictus*), and the sea star *Luidia foliolata*.

A male *A. urtica* was observed spawning in November in southern California, and in British Columbia a population in October contained members with ripe gonads and members with recently emptied gonads.

At least one other long-armed brittle star with a smooth disk may be found in sandy-mud areas, although intertidal occurrence is definitely established only north of Oregon. This species, **A. periercta** (Clark), is yellowish brown with white markings and occasional traces of red. Larger animals cannot be confused with any other long-armed burrowing species, for the disk may reach 20 mm in diameter and the arms 260 mm in length. Still another long-armed amphiurid species, **Ophiocnida hispida** (Le Conte), occurs in the southern California intertidal. It has long spines covering the disk, as in *Ophiothrix* (e.g., 10.6, 10.7,) but it has oral papillae, which *Ophiothrix* lacks.

See Barnard & Ziesenhenne (1961), Kyte (1969), Lie (1968), MacGinitie (1949), MacGinitie & MacGinitie (1968), May (1924), Ricketts & Calvin (1968), and Sutton (1975). For studies on anatomy, behavior, and ecology of other species of *Amphiodia* and related forms, see Apostolides (1881), Boolootian (1966), Buchanan (1962, 1963, 1964), Des Artes (1910), Fedotov (1926), Fell (1962), Fenaux (1970), Fricke (1970), Gislén (1924), Hendler (1973, 1975), Salzwedel (1974), Singletary (1971), Stancyk (1973), Turner (1974), and Woodley (1967, 1975).

Phylum Echinodermata / Class Ophiuroidea / Order Ophiurida
Suborder Gnathophiurina / Family Ophiothricidae

10.6 **Ophiothrix spiculata** Le Conte, 1851

Often seen under rocks, in rock crevices, and in algal holdfasts, low intertidal zone; subtidal on hard substrata to depths of 2,059 m; Moss Beach (San Mateo Co.) to Galápagos Islands and Bahía de Sechura (Peru, 6°S).

Disk diameter to 18 mm; arm length 5–8 times disk diameter; both disk and arms bearing prominent erect spines adorned with rows of thornlike spinelets, the ventral arm spines near the arm tips in the form of toothed hooks; each jaw with a cluster of spines (tooth papillae) at the apex, but lacking oral papillae on sides; coloration conspicuous but quite variable in hue and pattern.

Individuals typically insinuate one or more arms into fissures or crevices, anchoring with the straight and hooked arm spines. The remaining arms project into the water, where particles are caught by sticky secretions of spines and podia. The podia are papillose, pointed, and remarkably prehensile, wrapping around spines to clean off adherent particles, coiling around small organisms to capture them, and acting in a coordinated fashion with other tube feet on the arm to transfer captured and mucus-entrapped food to the jaws. The jaws close several times on each food mass, compacting the material before it is swallowed. The gut lining lacks cilia, and indeed cilia play no important role in the capture and transfer of food. The absence of cilia here is probably related to the utilization of very sticky mucus in feeding. Like *Ophiopteris papillosa* (10.11), *O. spiculata* can feed from the surface film and can capture larger organisms, but the methods employed are different.

Population densities may be very high. A siltstone reef off Newport Bay (Orange Co.) bore up to 80 *O. spiculata* per 0.1 m². On the rocky bottom and kelp holdfasts off La Jolla (San Diego Co.), Limbaugh (1955) noted that this species "occurs in almost unbelievable numbers in certain areas. The bottom in deeper water may be covered to depths of an inch or more by millions of these active animals. Holdfasts have a hairy or mossy appearance owing to the projecting and wavy arms of these animals."

Few details of reproduction are known. However, spawning has often been noted in July at Pacific Grove (Monterey Co.). Both eggs and sperm are shed, and fertilized eggs develop into free-swimming, plankton-feeding larvae. A few juveniles, presumably arising from settled larvae, have been recovered from fouling panels immersed for 1-month intervals in Monterey Bay from April to August.

Predators include the sea star *Astrometis sertulifera* (8.8), which elicits an avoidance or escape response, and fishes, such as the rock wrasse (*Halichoeres semicinctus*), the pile perch (*Rhacochilus vacca*), and the sand bass (*Paralabrax nebulifer*).

See especially Austin (1966); see also Andrews (1945), H. L. Clark (1911), Fox & Scheer (1941), Haderlie (1968, 1969), Hewatt (1937, 1938), Limbaugh (1955), MacGinitie (1949), MacGinitie & MacGinitie (1968), May (1924), McClendon (1909), Mortensen (1921), Pequegnat (1964), Quast (1968), Steinbeck & Ricketts (1941), and Turner, Ebert & Given (1969). For work on other species of *Ophiothrix*, see references under *O. rudis*.

Phylum Echinodermata / Class Ophiuroidea / Order Ophiurida
Suborder Gnathophiurina / Family Ophiothricidae

10.7 **Ophiothrix rudis** Lyman, 1874

Under rocks and in coarse sand, low intertidal zone; subtidal to 9 m; San Luis Obispo Co. to near Acapulco (Mexico), and recently found in Monterey Bay (J. Sutton, pers. comm.).

Disk diameter to 15 mm; arm length 4.5-7 times disk diameter; color variable, the disk often gray-green; tooth papillae and oral papillae as in *O. spiculata* (10.6); adults distinguished from *O. spiculata* by smooth (not spiny) arm spines and by stout rounded cylinders (not thorny spines) covering dorsal surface of disk; juveniles, however, bearing thorny arm spines.

Little information is available on the biology of *O. rudis.* Morphological differences between this species and *O. spiculata* suggest functional differences related to feeding mechanisms and habitat.

For *O. rudis*, see Fox & Scheer (1941), McClendon (1909), and Steinbeck & Ricketts (1941). For structure and biology of the related European species *O. fragilis* and *O. quinquemaculata*, see Allain (1974), Apostolides (1881), Broom (1975), Brun (1969), Buchanan (1962, 1963), Cuénot (1888, 1891), Czihak (1959), Gislén (1924), Guille (1964, 1965), Hamann (1889), Koehler (1887), V. Pentreath & Cottrell (1971), Pequignat (1966), Preyer (1886–87), Rehfeldt (1961), Roubaud (1965), Russo (1895), Smith (1937, 1940), Taylor (1958), Teuscher (1876), Tortonese (1959), Vevers (1956), Von Uexküll (1904), Warner (1971), Warner & Woodley (1975), and Wintzell (1918).

Phylum Echinodermata / Class Ophiuroidea / Order Ophiurida
Suborder Chilophiurina / Family Ophionereidae

10.8 **Ophionereis annulata** (Le Conte, 1851)

Adults commonly under rocks in sand or associated with various algae, low intertidal pools; juveniles often in barnacle beds and sponge masses, middle and low intertidal zones; subtidal to 60 m; San Pedro (Los Angeles Co.) to Esmeraldas (Ecuador, 1°N) and Galápagos Islands.

Disk diameter to 18 mm; arms long, 7–9 times disk diameter, bearing an accessory arm plate between dorsal and lateral arm plates on each side of each arm segment; color pattern variable, the dark arm bands sometimes broader than those in the photograph.

In adults of a related California species, **O. eurybrachyplax** H. L. Clark, the undersides of the arms are solidly colored (rather than banded as in *O. annulata*), and the inner portion of the underside of the disk between the arms bears small rounded granules (which are absent or limited to one or two in *O. annulata*).

Ophionereis annulata moves about by stepping actions of the podia rather than by whole-arm movements. Food may be grasped by an arm tip and brought to the mouth by a coiling of the whole arm; the arms are thinner and more flexible in young animals. Aquarium observations show that the animals can also capture and transport particles to the mouth by elongate tube feet. Little is known of the biology of this species, though some work has been done on the tropical Atlantic species *O. reticulata* and the New Zealand form *O. fasciata.*

See H. L. Clark (1911), Johnson & Snook (1927), MacGinitie & MacGinitie (1968), May (1925), R. Pentreath (1969, 1970), Ricketts & Calvin (1968), and Steinbeck & Ricketts (1941).

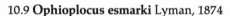

Phylum Echinodermata / Class Ophiuroidea / Order Ophiurida
Suborder Chilophiurina / Family Ophiuridae

10.9 **Ophioplocus esmarki** Lyman, 1874

Uncommon, under rocks, in crevices, and in kelp holdfasts, low intertidal pools; subtidal to 70 m; Tomales Bay (Marin Co.) to San Diego.

Disk diameter to 30 mm; arms 2–3 times as long as disk diameter, the dorsal arm plates not single but consisting of several smaller plates; arm spines very short, capable of being folded down against the arm; color diagnostic, solid brown or red-brown on upper surface of disk and arms (often somewhat darker than in the photograph).

A relatively inactive species with short, stiff arms. In aquaria it moves very slowly over the bottom or up the glass walls. It can capture moribund or slow-moving small animals, which are passed down the arms to the jaws by the adhesive tube feet. During transport the arm may be bent ventrally, bringing the podia along the arm closer together and allowing more of them to contact the food at a time as it moves along. Fertilization is internal, and the eggs, brooded in maternal genital bursae, develop directly into small brittle stars possessing the adult form. In central California, animals with enlarged ovaries have been noted in January, and females brooding up to 100 or more small young have been noted in July.

See Boolootian (1966), MacGinitie & MacGinitie (1968), May (1924), McClendon (1909), and Ricketts & Calvin (1968).

Phylum Echinodermata / Class Ophiuroidea / Order Ophiurida
Suborder Chilophiurina / Family Ophiodermatidae

10.10 **Ophioderma panamense** Lütken, 1859
(= O. panamensis)

Under rocks in pools, in algal holdfasts and coral, low intertidal zone; subtidal to 40 m; continental shelf north of Santa Rosa Island (Channel Islands) to Galápagos Islands and Paita, Peru (5°S); very common in appropriate habitats in southern California and Gulf of California.

Disk to 45 mm in diameter, smooth but adorned with small granules, these also appearing on arm bases in some juveniles; arms 3–6 times as long as disk diameter, banded, with very short spines; bursal slits totaling twenty; adults predominantly olive-green to gray-brown, with white bands on arms; juveniles predominantly ivory.

Ophioncus granulosus Ives is the only other California intertidal species with two bursal slits, rather than one, on each side of each arm under the disk. Reported from Monterey Bay and Laguna Beach (Orange Co.), it is distinguished by the stubby, unbanded arms 1–1.5 times the disk diameter in length, by the color, which varies from solid white to pink with red spots on the disk, and by the disk plates, which bear swollen granules that give the disk a rugose appearance. *Ophioderma teres* (Lyman), occurring in Mexico not far south of San Diego, also has double bursal slits, but its brown arms have no banding.

An active species, *O. panamense* comes out of hiding at night

and moves rapidly over the bottom, using lateral rowing strokes of two to four arms. Fragments of meat or small living prey are grasped in a lateral loop of the arm and brought to the jaws and oral podia by a ventral flexure at the base of the arm. Feeding may continue until the gut is so full that it distends the upper disk surface. Studies on this species include work on feeding, body-fluid corpuscles, differentiation of ova, and biochemical composition. The reproductive season has not been recorded.

See Boolootian (1962, 1966), Giese (1966), Johnson & Snook (1927), Kessel (1968), MacGinitie (1949), MacGinitie & MacGinitie (1968), McClendon (1909), Pequegnat (1961, 1964), Ricketts & Calvin (1968), Schechter & Lucero (1968), and Steinbeck & Ricketts (1941). For studies of anatomy, behavior, and development of the related Atlantic species *Ophioderma brevispina* and *O. longicauda*, see Allee (1927), Allee & Fowler (1932), Brooks & Grave (1899), Costello & Henley (1971), Fenaux (1969, 1972), Frenzel (1892), Preyer (1886–87), and Teuscher (1876).

Phylum Echinodermata / Class Ophiuroidea / Class Ophurida
Suborder Chilophiurina / Family Ophiocomidae

10.11 **Ophiopteris papillosa** (Lyman, 1875)

Fairly common, under rocks and in algal holdfasts, sometimes associated with the purple sea urchin *Strongylocentrotus purpuratus* (11.5), low intertidal pools; subtidal to 140 m; Barkley Sound (Vancouver Island, British Columbia) to Isla Cedros (Baja California, 28°N).

Disk diameter to 45 mm; arm length 3–4.5 times disk diameter; all spines flat and blunt, the upper arm spines very short, the lower arm spines conspicuously longer; color of intertidal specimens usually deep chocolate-brown, with bands on the arms, that of subtidal animals and regenerating arms lighter (see the photograph).

In aquaria, individuals may remain nearly motionless, with their arms flexed up into the water and their tube feet protruding in rows below each arm. Suspended particles adhere to mucus secretions on the spines and tube feet, and are passed down the arms to the mouth by the tube feet. *O. papillosa* also climbs aquarium walls and extends the arms just below the surface; the podia then collect particles from the surface film. This species is an active carnivore, as well, obtaining larger food material in a manner similar to *Ophioderma panamense* (10.10). The arm spines may be used to break off small pieces from larger food items held between the jaws. Some of the above feeding mechanisms are similar to those of *Ophiocomina nigra* (northeastern Atlantic), *Ophiocoma scolopendrina* (Indo-Pacific), and *Ophiopteris fasciata* (southwestern Pacific).

See Fox & Scheer (1941), Johnson & Snook (1927), May (1924), McClendon (1909), and Ricketts & Calvin (1968). For anatomy, development, behavior, and pigments of related species of *Ophiocomina, Ophiocoma,* and *Ophiopteris,* see Apostolides (1881), Boolootian (1966), Buchanan (1962, 1963), Chartock (1972), Cuénot (1888, 1891), Fontaine (1962a, 1964, 1965), Magnus (1962, 1964), Narasimhamurti (1933), Östergren (1904), R. Pentreath (1969, 1970), Smith (1937), and Wintzell (1918).

Phylum Echinodermata / Class Ophiuroidea / Order Ophiurida Suborder Euryalina / Family Gorgonocephalidae

10.12 Gorgonocephalus eucnemis (Müller & Troschel, 1842) (= G. caryi)   Basket Star

Sometimes abundant on rocky bottoms with moderate to strong water currents, or on mud and sand bottoms with projecting boulders, sea fans, and sea pens; subtidal from 10 m to almost 2,000 m, typically 15–150 m; occasionally snagged by fishermen, sometimes washed ashore during storms; Bering Sea to Laguna Beach (Orange Co.); East Siberian Sea, Chukchi Sea, and Japan; North Atlantic south to Cape Cod (Massachusetts) and Faeroe Islands (Denmark).

Disk diameter to 14 cm; arms 4–5 times as long as disk diameter, branching dichotomously, forming thousands of ultimate branchlets; color variable between individuals and often on different parts of the body, ranging from tan, beige, orange-red, and pink to almost white, the disk typically darker than the arms.

Gorgonocephalus eucnemis is one of the largest known ophiuroids, reaching a total diameter over half a meter. The generic name alludes to the gorgons of Greek mythology, whose human heads bore writhing snakes in place of hair. The aptness of the name is readily apparent when one sees the animal in life, for its highly mobile, branching arms give it a distinct and spectacular appearance. When feeding on suspended particles, the animal extends several of these arms out into the water, and the branchlets spread out like a fan, oriented so that they are more or less perpendicular to the water current. Macroscopic zooplankton, such as copepods, chaetognaths, and jellyfish are caught by an array of microscopic hooks on the arms. In laboratory preparations the hooks bend over and trap hairlike objects in a manner analogous to the pedicellariae of some starfish. Branchlets in the vicinity curl and wrap around captured plankters, forming a localized knot. A series of these knots may form at various points on the extended arms over a period of time, and mucus extruded from glands may further immobilize the prey. At some point, knot-laden arms slowly roll in toward the underside of the disk, and the captured plankters are transferred to the mouth in a manner still requiring clarification. *G. eucnemis* sometimes feeds on small benthic animals such as sea pens (*Stylatula elongata*, 3.25).

An association of the young of this species with the alcyonarian soft coral *Gersemia rubiformis* (and related species) has provoked interest since the turn of the century. The ranges of the two species are similar, though *G. rubiformis* enters shallower water and extends less far south on our coast. In shallow British Columbia waters, the presence of a *Gorgonocephalus* population almost ensures the sighting of *Gersemia* colonies nearby. In Puget Sound (Washington), juveniles with unbranched arms are present within the pharynges of *Gersemia* polyps. The juveniles are probably passively ingested at some earlier stage.

Early development is relatively direct. The early bilaterally symmetrical larvae have no locomotory structures, and begin to assume radial symmetry when 5 days old. The young basket stars and polyps appear undamaged by this association, and the basket stars may even obtain food collected by the polyps. After developing for a time within the polyps, the juveniles cling to the outside of the colony for a period and then move on to adult basket stars. Here the young appear to obtain food captured by the adult; they do not leave until the arms are sufficiently long and branched to act as food-capturing nets.

Despite the large numbers of basket stars that may occur in current-swept areas, some locations popular with divers have been "cleaned out." Growth rates of other brittle stars suggest that it would take a number of years to repopulate an area.

See H. L. Clark (1911), Fedotov (1924, 1926a, 1931), Grieg (1900), MacGinitie (1949), May (1924), McDaniel (1975), and Patent (1969, 1970a,b).

Literature Cited

Allain, J.-Y. 1974. Écologie des bancs d'*Ophiothrix fragilis* (Abildgaard) (Echinodermata, Ophiuroidea) dans le Golfe Normanno-Breton. Cah. Biol. Mar. 15: 255–73.

Allee, W. C. 1927. Studies in animal aggregations: Some physiological effects of aggregation on the brittle starfish, *Ophioderma brevispina*. J. Exper. Zool. 48: 475–95.

Allee, W. C., and J. B. Fowler. 1932. Studies in animal aggregations: Further studies on oxygen consumption and autotomy of the brittle star *Ophioderma*. J. Exper. Zool. 64: 33–50.

Andrews, H. L. 1945. The kelp beds of the Monterey region. Ecology 26: 24–37.

Apostolides, N. C. 1881. Anatomie et développement des Ophiures. D. Sci. Thèse. Acad. Paris. Paris: A. Hennuyer. 104 pp.

Austin, W. C. 1966. Feeding mechanisms, digestive tracts and circulatory systems in the ophiuroids *Ophiothrix spiculata* Le Conte, 1851, and *Ophiura luetkeni* (Lyman, 1860). Doctoral thesis, Biological Sciences, Stanford University, Stanford, Calif. 278 pp.

Austin, W. C., and M. P. Haylock. 1973. British Columbia marine faunistic survey report: Ophiuroidea from the northeast Pacific. Fish. Res. Bd. Canada, Tech. Rep. 426. 36 pp.

Barham, E. G., N. J. Ayer, and E. R. Boyce. 1966. Megabenthic fauna of the San Diego Trough: Photographic study from the bathyscaph *Trieste* No. 19, pp. 20–21, *in* 2nd International Oceanographic Congress, Abstracts.

Barnard, J. L., and F. C. Ziesenhenne. 1961. Ophiuroid communities of southern California coastal bottoms. Pacific Natur. 2: 131–52.

Bernasconi, I. 1926. Una ofiura vivipara de Necochea. Ann. Mus. Nac. Hist. Natur. 34: 145–53.

Binyon, J. 1972. Physiology of echinoderms. New York: Pergamon. 264 pp.

Blegvad, H. 1914. Food and conditions of nourishment among the communities of invertebrate animals found on or in the sea bottom in Danish waters. Rep. Danish Biol. Sta. 22: 41–78.

Boffi, E. 1972. Ecological aspects of ophiuroids from the phytal of S. W. Atlantic Ocean warm waters. Mar. Biol. 15: 316–28.

Boolootian, R. A. 1962. The perivisceral elements of echinoderm body fluids. Amer. Zool. 2: 275–84.

———, ed. 1966. Physiology of Echinodermata. New York: Interscience. 822 pp.

Boolootian, R. A., and D. Leighton. 1966. A key to the species of Ophiuroidea (brittle stars) of the Santa Monica Bay and adjacent areas. Los Angeles Co. Mus. Contr. Sci. 93. 20 pp.

Brehm, P., and J. G. Morin. 1977. Localization and characterization of luminescent cells in *Ophiopsila californica* and *Amphipholis squamata* (Echinodermata: Ophiuroidea). Biol. Bull. 152: 12–25.

Brooks, W. K., and C. Grave. 1899. *Ophiura brevispina*. Mem. Nat. Acad. Sci. 8: 79–100.

Broom, D. M. 1975. Aggregation behaviour of the brittle-star *Ophiothrix fragilis*. J. Mar. Biol. Assoc. U.K. 55: 191–97.

Brun, E. 1969. Aggregation of *Ophiothrix fragilis* (Abildgaard) (Echinodermata: Ophiuroidea). Nytt Mag. Zool. 17: 153–60.

Brusca, R. C. 1973. A handbook to the common intertidal invertebrates of the Gulf of California. Tucson: University of Arizona Press. 427 pp.

Buchanan, J. B. 1962. A re-examination of the glandular elements in the tube-feet of some common British ophiuroids. Proc. Zool. Soc. London 138: 645–50.

———. 1963. Mucus secretion within the spines of ophiuroid echinoderms. Proc. Zool. Soc. London 141: 251–59.

———. 1964. A comparative study of some features of the biology of *Amphiura filiformis* and *Amphiura chiajei* (Ophiuroidea) considered in relation to their distribution. J. Mar. Biol. Assoc. U.K. 44: 565–76.

Caso, M. E. 1951. Contribución al conocimiento de los ofiuroideos de México. I. Algunas especies de ofiuroideos litorales. Ann. Inst. Biol. 22: 219–312.

Chartock, M. H. 1972. The role of detritus in a tropical marine ecosystem: Niche separation in congeneric ophiuroids, food partitioning in cryptic invertebrates and herbivore detritus production at Eniwetok, Marshall Islands. Doctoral thesis, Biology, University of Southern California.

Clark, A. M. 1962. Starfishes and their relations. London: British Museum (Natur. Hist.). 119 pp.

———. 1967. Variable symmetry in fissiparous Asterozoa, pp. 143–57, *in* Millott (1967).

Clark, H. L. 1911. North Pacific ophiurans in the collection of the United States National Museum. Bull. U.S. Nat. Mus. 75. 302 pp.

———. 1915. Catalogue of recent ophiurans. Mus. Comp. Zool. Harvard Mem. 25: 165–376.

Costello, D. P., and C. Henley. 1971. Methods for obtaining and handling marine eggs and embryos. 2nd ed. Marine Biological Laboratory, Woods Hole, Mass. 247 pp.

Cuénot, L. 1888. Études anatomiques et morphologiques sur les Ophiures. Arch. Zool. Exper. (3) 6: 33–82.

_____. 1891. Études morphologiques sur les Échinodermes. Arch. Biol. 11: 303–680.

_____. 1948. Anatomie, éthologie et systématique des Échinodermes, pp. 3–272, *in* P.-P. Grassé, ed., Traité de zoologie, vol. 11. Paris: Masson. 1,077 pp.

Cunningham, C. J. B. 1977. Aspects of the reproductive biology of intertidal amphiurids of Monterey Bay. Master's thesis, Biology, San Jose State University, San Jose, Calif.

Czihak, G. 1959. Vorkommen und Lebensweise de *Ophiothrix quinquemaculata* in dem nördlichen Adria bei Rovinj. Thalassia Jugoslavica 1: 19–27.

Dawydoff, C. 1948. Embryologie des échinodermes, pp. 277–363, *in* P.-P. Grassé, ed., Traité de zoologie, vol. 11. Paris: Masson. 1,077 pp.

Delage, Y., and E. Hérouard. 1903. Traité de zoologie concrète. 3. Les Échinodermes. Paris: Librairie H. Le Soudier. 495 pp.

Des Artes, L. 1910. Über die Lebensweise von *Amphiura chiajei*. Bergens Mus. Årb., Naturvid. Rekke 12: 1–10.

D'iakonov, A. M. 1967. Ophiuroids of the U.S.S.R. seas. Zool. Inst. Akad. Sci. U.S.S.R. English transl. Jerusalem: Israel Program of Scientific Translation (I.P.S.T. Press), available through U.S. Dept. Commerce, Washington, D.C. 123 pp.

Eichelbaum, E. 1910. Über Nahrung und Ernährungsorgane von Echinodermen. Wiss. Meeresunters. Kiel 11: 189–275.

Fager, E. W. 1968. A sand-bottom epifaunal community of invertebrates in shallow water. Limnol. Oceanogr. 13: 448–64.

Fedotov, D. M. 924. Einige Beobachtungen über die Biologie und Metamorphose von *Gorgonocephalus*. Zool. Anz. 61: 303–11.

_____. 1926a. Die Morphologie der Euryalae. Z. Wiss. Zool. 127: 403–528.

_____. 1926b. Zur Morphologie einiger typischen, vorzugsweise Lebendiggebärenden, Ophiuren. Trav. Lab. Zool. Sta. Biol. Sebastopol (2) 6: 39–72.

_____. 1931. Über eigenartigen Parasitismus bei Stechelhaütern. Z. Morph. Ökol. Tiere 22: 401–15.

Fell, H. B. 1946. The embryology of the viviparous ophiuroid *Amphipholis squamata* Delle Chiaje. Trans. Proc. Roy. Soc. New Zeal. 75: 419–64.

_____. 1960. Synoptic keys to the genera of Ophiuroidea. Zool. Publ. Victoria University, Wellington, New Zeal. 26. 44 pp.

_____. 1962. A revision of the major genera of amphiurid Ophiuroidea. Trans. Roy. Soc. New Zeal. Zool. 2: 1–26.

_____. 1966. The ecology of ophiuroids, pp. 129–43, *in* Boolootian (1966).

Fenaux, L. 1969. Le développement larvaire chez *Ophioderma longicaudus* (Retzius). Cah. Biol. Mar. 10: 59–62.

_____. 1970. Maturation of the gonads and seasonal cycle of the planktonic larvae of the ophiuroid *Amphiura chiajei* Forbes. Biol. Bull. 138: 262–71.

_____. 1972. Évolution saisonniére des gonades chez l'ophiure *Ophioderma longicauda* (Retzius), Ophiuroidea. Internat. Rev. Ges. Hydrobiol. 57: 257–62.

Fewkes, J. W. 1887. On the development of the calcareous plates of *Amphiura*. Bull. Mus. Harvard Coll. 13: 107–50.

Florkin, M., and B. T. Scheer, eds. 1969. Chemical zoology. 3. Echinodermata, Nematoda, and Acanthocephala. New York: Academic Press. 687 pp.

Fontaine, A. R. 1962a. The colours of *Ophiocomina nigra* (Abildgaard). I. Colour variation and its relation to distribution. II. The occurrence of melanin and fluorescent pigments. III. Carotenoid pigments. J. Mar. Biol. Assoc. U.K. 42: 1–8, 9–31, 33–47.

_____. 1962b. Neurosecretion in the ophiuroid *Ophiopholis aculeata*. Science 138: 908–9.

_____. 1963. A comparative study of the integumentary mucous cells of ophiuroids. Proc. 16th Internat. Congr. Zool., Washington, D.C., 1:87.

_____. 1964. The integumentary mucous secretions of the ophiuroid *Ophiocomina nigra*. J. Mar. Biol. Assoc. U.K. 44: 145–62.

_____. 1965. The feeding mechanisms of the ophiuroid *Ophiocomina nigra*. J. Mar. Biol. Assoc. U.K. 45: 373–85.

Fontaine, A. R., and F.-S. Chia. 1968. Echinoderms: An autoradiographic study of assimilation of dissolved organic molecules. Science 161: 1153–55.

Fox, D. L., and B. T. Scheer. 1941. Comparative studies of the pigments of some Pacific coast echinoderms. Biol. Bull. 80: 441–55.

Frenzel, J. 1892. Beiträge zur vergleichenden Physiologie und Histologie der Verdauung. I. Mitteilung. Der Darmcanal der Echinodermen. Arch. Anat. (Physiol. Abt.) Ann. 1892: 81–114.

Fricke, H. W. 1970. Beobachtungen über Verhalten und Lebensweise des im Sand lebenden Schlangensternes *Amphioplus* sp. Helgoländer Wiss. Meeresunters. 21: 124–33.

Giese, A. C. 1966. On the biochemical constitution of some echinoderms, pp. 757–96, *in* Boolootian (1966).

Gislén, T. 1924. Echinoderm studies. Zool. Bidrag Uppsala 9: 3–316.

Grieg, J. A. 1900. Die Ophiuriden der Arktis. Fauna Arctica Bd. 1 (2): 268–71.

Guille, A. 1964. Contribution a l'étude de la systématique et de l'écologie d'*Ophiothrix quinquemaculata* d. Ch. Vie et Milieu 15: 243–308.

_____. 1965. Observations faites en soucoupe plongeante à la limite inférieure d'un fond à *Ophiothrix quinquemaculata* d. Ch. au large de la côte du Roussillon. Comm. Internat. Explor. Sci. Mer Méditerr., Rap. Procès-Verbaux Réunions 18: 115–18.

Haderlie, E. C. 1968. Marine fouling organisms in Monterey harbor. Veliger 10: 327–41.

_____. 1969. Marine fouling and boring organisms in Monterey harbor. II. Second year of investigation. Veliger 12: 182–92.

Hamann, O. 1889. Anatomie und Histologie der Ophiuren und Crinoiden. Jena Z. Naturwiss. 23: 233–388.

Harvey, E. N. 1952. Bioluminescence. New York: Academic Press. 649 pp.

Hendler, G. L. 1973. Northwest Atlantic amphiurid brittlestars, *Amphioplus abditus* (Verrill), *Amphioplus macilentus* (Verrill), and *Amphioplus sepultus* n. sp. (Ophiuroidea: Echinodermata): Systematics, zoogeography, annual periodicities, and larval adaptations. Doctoral thesis, Zoology, University of Connecticut, Storrs. 328 pp.

———. 1975. Adaptational significance of the patterns of ophiuroid development. Amer. Zool. 15: 691–715.

Hewatt, W. G. 1937. Ecological studies on selected marine intertidal communities of Monterey Bay, California. Amer. Midl. Natur. 18: 161–206.

———. 1938. Notes on the breeding seasons of the rocky beach fauna of Monterey Bay, California. Proc. Calif. Acad. Sci. (4) 23: 283–88.

Holland, N. D., and L. Z. Holland. 1969. A bibliography of echinoderm biology, continuing Hyman's 1955 bibliography through 1965. Pubbl. Staz. Zool. Napoli 37: 441–543.

Hyman, L. H. 1955. The invertebrates: Echinodermata. Vol. 4. New York: McGraw-Hill. 763 pp.

Johnson, M. E., and H. J. Snook. 1927. Seashore animals of the Pacific coast. New York: Macmillan. 659 pp.

Kessel, R. G. 1968. An electron microscope study of differentiation and growth in oocytes of *Ophioderma panamensis*. J. Ultrastruc. Res. 22: 63–89.

Koehler, R. 1887. Recherches sur l'appareil circulatoire des Ophiures. Ann. Sci. Natur. (7) 2: 101–58.

Kozloff, E. N. 1969. Morphology of the orthonectid *Rhopalura ophiocomae*. J. Parasitol. 55: 171–95.

———. 1974. Keys to the marine invertebrates of Puget Sound, the San Juan Archipelago, and adjacent regions. Seattle: University of Washington Press. 226 pp.

Kuznetsov, A. P., and M. N. Sokolova. 1961. On feeding and distribution of *Ophiopholis aculeata* (L.) [In Russian.] Trans. Inst. Okeanol. Akad. Nauk SSSR 46: 98–102.

Kyte, M. A. 1969. A synopsis and key to the recent Ophiuroidea of Washington State and southern British Columbia. Fish. Res. Bd. Canada 26: 1727–41.

LaBarbera, M. 1978. Particle capture by a Pacific brittle star: Experimental test of the aerosol suspension feeding model. Science 201: 1147–49.

Lie, U. 1968. A quantitative study of benthic infauna in Puget Sound, Washington, U.S.A., in 1963–1964. Fiskeridir. Skr. Ser. Havunders. 14: 229–556.

Limbaugh, C. 1955. Fish life in the kelp beds and the effects of kelp harvesting. Univ. Calif. Inst. Mar. Res., IMR Ref. 55–9: 1–158.

Ludwig, C. R., and O. Hamann. 1901. Echinodermen (Stachelhäuter). Die Schlangensterne, pp. 745–966, *in* Dr. H. G. Bronns Klassen und Ordnungen des Tierreichs, Bd. 2, Abt. 3, Buch 3. Leipzig: Winter.

MacBride, E. W. 1892. The development of the genital organs, ovoid gland, axial and aboral sinuses in *Amphiura squamata*, together with some remarks on Ludwig's haemal system in this ophiurid. Quart. J. Microscop. Sci. (2) 34: 129–53.

MacGinitie, G. E. 1949. The feeding of ophiurans. J. Entomol. Zool. Pomona 41: 27–29.

MacGinitie, G. E., and N. MacGinitie. 1968. Natural history of marine animals. 2nd ed. New York: McGraw-Hill. 523 pp.

Magnus, D. B. E. 1962. Über das "Abweiden" der Flutwasseroberflache durch den Schlangenstern *Ophiocoma scolopendrina* (Lamarck). Verhandl. Deut. Zool. Ges. Wien 62: 471–81.

———. 1964. Geseitenströmung und Nahrungsfiltration bei Ophiuren und Crinoiden. Helgoländer Wiss. Meeresunters. 10: 104–17.

Martin, R. B. 1968. Aspects of the ecology and behaviour of *Axiognathus squamata* (Echinodermata, Ophiuroidea). Tane 14: 65–81.

May, R. M. 1924. The ophiurans of Monterey Bay. Proc. Calif. Acad. Sci. (4) 13: 261–303.

———. 1925. Les réactions sensorielles d'une Ophiure (*Ophionereis reticulata* (Say)). Bull. Biol. France Belg. 59: 373–402.

McClendon, J. F. 1909. The ophiurans of the San Diego region. Univ. Calif. Publ. Zool. 6: 33–64.

McDaniel, N. 1975. The basket star. Pacific Diver 1: 36–37.

Mileikovsky, S. A. 1970. Seasonal and daily dynamics in pelagic larvae of marine shelf bottom invertebrates in near shore waters of Kandalaksha Bay (White Sea). Mar. Biol. 5: 180–94.

Millott, N., ed. 1967. Echinoderm biology. Symp. Zool. Soc. London 20. London: Academic Press. 240 pp.

Moment, G. B. 1962. Reflexive selection: A possible answer to an old puzzle. Science 136: 262–63. Comments by Li, Moment, and Spofford. Science 136: 1055–56; 138: 457–58.

Mortensen, T. 1921. Studies of the development and larval forms of echinoderms. Copenhagen: Gad. 261 pp., 33 pls.

Narasimhamurti, N. 1933. The development of *Ophiocoma nigra*. Quart. J. Microscop. Sci. 76: 63–88.

Nichols, D. 1964. Echinoderms: Experimental and ecological. Oceanogr. Mar. Biol. Ann. Rev. 2: 393–423.

———. 1969. Echinoderms. 4th ed. London: Hutchinson University Library. 192 pp.

Nielsen, E. 1932. Ophiurans from the Gulf of Panama, California, and the Strait of Georgia. Vid. Medd. Dansk Naturh. Foren. 91: 241–336.

Olsen, H. 1942. The development of the brittle star *Ophiopholis aculeata* (O. Fr. Müller), with a short report on the outer hyaline layer. Bergens Mus. Årb., Naturvid. Rekke 6: 1–107.

Östergren, Hj. 1904. Funktion der Füsschen bei den Schlangensternen. Biol. Centralbl. 24: 559–65.

Patent, D. H. 1969. The reproductive cycle of *Gorgonocephalus caryi* (Echinodermata, Ophiuroidea). Biol. Bull. 136: 241–52.

———. 1970a. Life history of the basket star, *Gorgonocephalus eucnemis* (Müller & Troschel) (Echinodermata, Ophiuroidea). Ophelia 8: 145–60.

_____. 1970b. The early embryology of the basket star *Gorgonocephalus caryi* (Echinodermata, Ophiuroidea). Mar. Biol. 6: 262–67.

Pentreath, R. J. 1969. Morphology of the gut and a qualitative review of digestive enzymes in some New Zealand ophiuroids. J. Zool., London 159: 413–23.

_____. 1970. Feeding mechanisms and the functional morphology of podia and spines in some New Zealand ophiuroids (Echinodermata). J. Zool., London 161: 395–429.

Pentreath, V. W., and G. A. Cottrell. 1971. "Giant" neurons and neurosecretion in the hyponeural tissue of *Ophiothrix fragilis* Abildgaard. J. Exper. Mar. Biol. Ecol. 6: 249–64.

Pequegnat, W. E. 1961. New world for marine biologists. In shallow offshore waters lie fertile areas for study. Natur. Hist. 70: 8–17.

_____. 1964. The epifauna of a California siltstone reef. Ecology 45: 272–83.

Pequignat, E. 1966. "Skin digestion" and epidermal absorption in irregular and regular urchins and their probable relation to the outflow of spherule-coelomocytes. Nature 210: 397–99.

Preyer, W. 1886–87. Über die Bewegungen der Seesterne. Mittheil. Zool. Sta. Neapel. 7: 27–127, 191–233.

Quast, J. C. 1968. Observations on the food of kelp-bed fishes, pp. 109–42, *in* W. J. North and C. L. Hubbs, eds., Utilization of kelp-bed resources in southern California. Calif. Dept. Fish & Game, Fish Bull. 139: 1–264.

Rehfeldt, N. 1961. Observationer over *Ophiothrix fragilis* til belysning af dens felske yngelpleje. Thesis, University of Copenhagen. 92 pp.

Ricketts, E. F., and J. Calvin. 1968. Between Pacific tides. 4th ed. Revised by J. W. Hedgpeth. Stanford, Calif.: Stanford University Press. 614 pp.

Roubaud, P. 1965. Le tube digestif d'*Ophiothrix quinquemaculata* (Delle Chiaje), étude histologique mise au point technique. Vie et Milieu 16: 757–98.

Roushdy, H. M., and Vagn. Kr. Hansen. 1960. Ophiuroids feeding on phytoplankton. Nature 188: 517–18.

Russo, A. 1891. Embriologia dell *Amphiura squamata*. Atti R. Accad. Napoli (2) 5: 1–24.

_____. 1895. Studii anatomici sulla famiglia Ophiothricidae del Golfo di Napoli. Ricerche Lab. Anat. Roma 4: 157–80.

Salzwedel, H. 1974. Arm-regeneration bei *Amphiura filiformis* (Ophiuroidea). Veroff. Inst. Meeresforsch. Bremerh. 14: 161–67.

Schechter, J., and J. Lucero. 1968. A light and electron microscopic investigation of the digestive system of the ophiuroid *Ophioderma panamensis* (brittle star). J. Morphol. 124: 451–82.

Simroth, H. 1876. Anatomie und Schizogonie der *Ophiactis virens* Sars. Ein Beitrag zur Kenntniss der Echinodermen. Z. Wiss. Zool. 27: 417–85, 555–60.

_____. 1877. Anatomie und Schizogonie der *Ophiactis virens* Sars. Ein Beitrag zur Kenntniss der Echinodermen. 2. Schizogonie. Z. Wiss. Zool. 28: 419–526.

Singletary, R. L. 1971. The biology and ecology of *Amphioplus coniortodes*, *Ophionepthys limicola* and *Micropholis gracillima* (Ophiuroidea: Amphiuridae). Doctoral thesis, University of Miami, Coral Gables. 146 pp.

Smith, J. E. 1937. The structure and function of the tube feet in certain echinoderms. J. Mar. Biol. Assoc. U.K. 22: 345–57.

_____. 1940. The reproductive system and associated organs of the brittle-star *Ophiothrix fragilis*. Quart. J. Microscop. Sci. 82: 267–309.

_____. 1965. Echinodermata, pp. 1519–58, *in* T. H. Bullock and G. A. Horridge, Structure and function in the nervous systems of invertebrates, vol. 2. San Francisco: Freeman. 1,719 pp.

Spencer, W. K., and C. W. Wright. 1966. Asterozoans, pp. U4–107, *in* R. C. Moore, ed., Treatise on invertebrate paleontology, part U, Echinodermata 3. New York: Geol. Soc. Amer.; Lawrence: University of Kansas Press.

Stancyk, S. E. 1973. Development of *Ophiolepis elegans* (Echinodermata: Ophiuroidea) and its implications in the estuarine environment. Mar. Biol. 21: 7–12.

Steinbeck, J., and E. F. Ricketts. 1941. Sea of Cortez. A leisurely journal of travel and research. New York: Viking. 598 pp.

Stephens, G. C., and R. A. Virkar. 1966. Uptake of organic material by aquatic invertebrates. IV. The influence of salinity on the uptake of amino acids by the brittle star, *Ophiactis arenosa*. Biol. Bull. 131: 172–85.

Strathmann, R. 1971. The feeding behavior of planktotrophic echinoderm larvae: Mechanisms, regulation, and rates of suspension-feeding. J. Exper. Mar. Biol. Ecol. 6: 109–60.

_____. 1975. Larval feeding in echinoderms. Amer. Zool. 15: 717–30.

_____. 1978. Length of pelagic period in echinoderms with feeding larvae from the northeast Pacific. J. Exper. Mar. Biol. Ecol. 34: 23–27.

Sutton, J. E. 1975. Class Ophiuroidea, pp. 627–34, *in* R. I. Smith and J. T. Carlton, eds., Light's manual: Intertidal invertebrates of the central California coast. 3rd ed. Berkeley and Los Angeles: University of California Press. 716 pp.

_____. 1976 (publ. 1978). Partial revision of the genus *Amphiodia* (Ophiuroidea: Amphiuridae) from the west coast of North America. Thalassia Jugoslavica 12: 341–48.

Taylor, A. M. 1958. Studies on the biology of the off-shore species of Manx Ophiuroidea. Master's thesis, Zoology, University of Liverpool. 59 pp.

Teuscher, R. 1876. Beiträge zur Anatomie der Echinodermen. 2. Ophiuridea. Jena Z. Naturwiss. 10: 263–80.

Thomas, L. P. 1966. A revision of the tropical American species of *Amphipholis* (Echinodermata: Ophiuroidea). Bull. Mar. Sci. 16: 827–33.

Tortonese, E. 1959. Ecofenotipi e biologia di *Ophiothrix fragilis* (Ab.) nel Golfo di Genova. Doriana 2 (100): 1–9.

Turner, C. H., E. E. Ebert, and R. R. Given. 1969. Man-made reef ecology. Calif. Dept. Fish & Game, Fish Bull. 146: 1–221.

Turner, R. L. 1974. Post-metamorphic growth of the arms in *Ophiophragmus filograneus* (Echinodermata: Ophiuroidea) from Tampa Bay, Florida (U.S.A.). Mar. Biol. 24: 273–77.

Turpaeva, E. P. 1953. Feeding and nutritional groups of the sea benthic invertebrates. [In Russian.] Trans. Inst. Okeanol. Akad. Nauk SSSR 7: 259–99

Vevers, H. G. 1956. Observations on feeding mechanisms in some echinoderms. Proc. Zool. Soc. London 126: 484–85.

Von Uexküll, J. 1904. Studien über den Tonus. II. Die Bewegungen der Schlangensterne. Z. Biol. 46: 1–37.

Warner, G. F. 1971. On the ecology of a dense bed of the brittle star *Ophiothrix fragilis*. J. Mar. Biol. Assoc. U.K. 51: 267–82.

Warner, G. F., and J. D. Woodley. 1975. Suspension-feeding in the brittle-star *Ophiothrix fragilis*. J. Mar. Biol. Assoc. U.K. 55: 199–210.

Wintzell, J. 1918. Bidrag till de Skandinaviska ophiuridernas biologi och fysiologi. Uppsala: Appelbergs Boktryckeri Aktiebolag. 147 pp.

Woodley, J. D. 1967. Problems in the ophiuroid water-vascular system, pp. 75–104, *in* Millott (1967).

———. 1975. The behavior of some amphiurid brittle-stars. J. Exper. Mar. Biol. Ecol. 18: 29–46.

Chapter 11

Echinoidea: *The Sea Urchins*

J. Wyatt Durham, Carol D. Wagner, and Donald P. Abbott

Echinoids are more or less globose or flattened echinoderms with radiating spines. The mouth, surrounded by a soft peristome, lies on the lower (oral) surface of the body. In less specialized, or "regular," sea urchins, the anus lies at the pole opposite the mouth on the upper (aboral) surface. The anal region (periproct) is encircled by an apical ring of plates, which bear the genital pores and the madreporite, or opening of the water-vascular system. In the "irregular" types of urchins, the anus has moved posteriorly out of the apical system of plates, sometimes as far as the oral surface. In "heart urchin" types, the mouth has moved anteriorly as the anus moved posteriorly.

The main part of the skeleton, the shell or "test," is covered externally by a ciliated epidermis, and is built of numerous calcareous plates arranged in 20 columns that extend along meridians from the apical system to the mouth. Five pairs of these columns form the ambulacra, each plate of which is pierced by one or two pores through which the tube feet extend. The intervening pairs of columns are termed the interambulacra. In all types of urchins considered in this work, the plates of the test are rigidly sutured together to form a solid test. The bleached tests of dead urchins are often cast ashore by waves.

The remainder of the skeleton consists of spines borne on the test. These vary from few and elongate to numerous and short, and are variously modified for locomotion, burrowing, and defense. The largest spines are known as primaries, and in the regular types of urchins there is only one primary spine to a plate (or compound plate) in the ambulacra. Smaller spines are termed secondaries and miliaries. In some of the heart urchins, the spines in certain narrow bands (fascioles) on the upper or posterior surface are greatly reduced in size, very numerous, and furnished with abundant cilia. The cilia, beating in unison, create currents over the surface of the test that aid in feeding, respiration, and sanitation.

Within the mouth of the regular urchins, and of such irregular urchins as the sand dollars and the keyhole urchins, there is a complex jaw apparatus known as "Aristotle's lantern." It is built of 40 calcareous ossicles held together by a complex system of muscles. Five of the ossicles are teeth, which work together in nibbling and scraping food, and are used by the common purple sea urchin in excavating burrows. Some echinoids are omnivorous, others feed very largely on plants; the sand dollars gather minute organisms and organic particles; and the heart urchins ingest large quantities of sand or mud and digest the accompanying organic material.

Echinoids occur in all oceans. They live on the sea floor or burrow into the underlying sediments. Some prefer rocky areas; others live on sandy or muddy bottoms, where some types of sand dollars and heart urchins burrow to depths of 15–25 cm. A few echinoids can tolerate salinities as low as 57 percent that of seawater, and others can withstand salinities of over 110 percent. About 800 living species of echinoids are known, and nearly ten times as many fossil species. The fossil record extends back to the Ordovician, but most existing orders arose during the Mesozoic.

J. Wyatt Durham, formerly Curator of Mesozoic and Cenozoic Invertebrates in the Museum of Paleontology, is Professor Emeritus of Paleontology at the University of California, Berkeley. Carol D. Wagner is Curator of the Paleontological Collections at the University of Alaska Museum, Fairbanks. Donald P. Abbott is Professor, Department of Biological Sciences and Hopkins Marine Station, Stanford University.

The sexes are separate in echinoids; hermaphroditism is a rare anomaly. Gametes, usually in immense numbers, are shed into the sea, where fertilization occurs and the pelagic larvae develop. Larval mortality is heavy, and after a period of days or weeks the survivors settle and metamorphose into tiny urchins. The growth rates and life spans of echinoids are not well known, but the sand dollar *Dendraster excentricus* (11.7) occasionally reaches the age of 13 years. Large purple urchins (*Strongylocentrotus purpuratus*, 11.5) are probably 8–10 years old, and growth lines in the plates of large red urchins (*S. franciscanus*, 11.6) suggest that they may be as much as 15 to 20 years old. The relative ages of individuals of the same species may be determined by comparing the numbers of plates or pore pairs per ambulacrum; the number tends to increase with age.

In California inshore waters the main predators on sea urchins are the sea otter, certain starfishes (especially *Pycnopodia helianthoides*, 8.16), and fishes (particularly the California sheephead, but also the sand bass, ocean whitefish, black perch, and others). Along the shore, sea gulls and crows take some, but the main enemy of sea urchins here appears to be man. The gonads or roe of some species, such as the purple urchin and the large red urchin, were used as food by the Indians and are considered delicacies by some today. There has been a spectacular increase in recent years in the number of sea urchins collected in California and shipped primarily to Japan for food. In 1971 only 100 kg were collected, with a value of $36; by 1974 over 3.5 million kg were collected, with a value of $520,215 (McAllister, 1976; Oliphant, 1973). Souvenir hunters and collectors make their inroads as well, and the cleaned tests of urchins are often sold in curio shops. Finally, biologists take a small toll for teaching and research. Echinoid eggs, sperm, and embryos have been used extensively in experimental studies and have contributed greatly to our understanding of basic processes in fertilization and development.

In this chapter we have included all of the species of echinoids that occur in the littoral zone, as well as those apt to be encountered by divers in shallow water. Supplementary references that may prove useful in identifying California echinoids are H. L. Clark (1948), Grant & Hertlein (1938), and Pearse (1975). The biology of most of these species is poorly known, but two forms, the purple urchin and the sand dollar, have been the objects of extraordinary numbers of investigations.

Good short accounts of the echinoids as a group are available in A. Clark (1962) and Nichols (1969); for more comprehensive treatments, see Cuénot (1948), Durham et al. (1966), and Hyman (1955). For information on physiology, biochemistry, ecology, and behavior, see Binyon (1972), Nichols (1964), J. Smith (1965), and reviews by the numerous contributors to Boolootian (1966), Florkin & Scheer (1969), and Millott (1967). A discussion and review of growth in the echinoid skeleton is available in Pearse & Pearse (1975). Descriptive and experimental embryological studies and techniques are covered in Chia & Whiteley (1975), Costello & Henley (1971), Czihak & Peter (1975), Dawydoff (1948), Gustafson (1969), Gustafson & Wolpert (1967), Harvey (1956), Hinegardner (1967), Mortensen (1921), and Tyler & Tyler (1966). For larval biology, see Strathmann (1971, 1974, 1975). For overall classification and phylogeny, see Durham (1966), Durham & Melville (1957), Durham et al. (1966), and Mortensen (1928–51).

Phylum Echinodermata / Class Echinoidea / Order Cidaroida / Family Cidaridae

11.1 **Eucidaris thouarsii** (Valenciennes, 1846)

In holes and crevices and under rocks, low intertidal zone to 140 m depth on rocky shores; Saint Catherine's Bay and Emerald Bay (both Santa Catalina Island, Channel Islands), Isla Guadalupe and other islands off Baja California, and Gulf of California to Ecuador, including offshore Central American islands and Galápagos Islands.

Test commonly 50–70 mm in diameter, heavy and strong, with few large interambulacral plates and numerous small plates in the sinuous ambulacra; primary spines to 16 cm long; interambulacral spines heavy, conspicuous, and often encrusted with algae, sponges, hydroids, bryozoans, worm tubes, etc.; color in living or dried specimens brown to very dark purple, with paler bases on secondary and miliary spines forming light-colored zones along ambulacra and centers of interambulacra.

This is a southern species, only twice collected alive off the California coast (Santa Catalina Island), though fossil specimens are known from central California south. Individuals from the Gulf of California and the Galápagos Islands are generally smaller than those from Isla Guadalupe and other offshore islands, which may be nearly twice as large as Gulf individuals and have a very strong test.

The order Cidaroida, to which *Eucidaris* belongs, is the only order of living urchins with fossil representatives extending back to the mid-Paleozoic. Other existing echinoid orders are traceable as fossils only to the Mesozoic Era, and appear to have originated from cidaroid stocks.

See Boolootian (1966), H. L. Clark (1948), Fitch (1962), Grant & Hertlein (1938), McPherson (1968a,b), Mortensen (1921), Raup (1962, 1966), Steinbeck & Ricketts (1941), and Strachan, Turner & Mitchell (1968).

Phylum Echinodermata / Class Echinoidea / Order Diadematoida
Family Diadematidae

11.2 **Centrostephanus coronatus** (Verrill, 1867)

In shaded and sheltered situations, low intertidal zone on rocky shores; subtidal to 110 m; southern California to Mexico and Galápagos Islands.

Test typically flattened above and below, to 63 mm in diameter, bearing primary spines to 100 mm long; peristome half the diameter of test; interambulacra about 1.5 times as wide as ambulacra at ambitus (widest part of test); both ambulacral and interambulacral plates with prominent primary tubercles covering much of the surface, and with one or more much smaller secondary tubercles in the marginal area and below; the tubercles distinctly perforate and crenulate; distinguished from all other California, shallow-water, "regular" urchins by its hollow, thorny spines; one or two bright red-purple, clavate, spines appearing in each column of spines near the apical system; young individuals commonly with banded spines, but the banding sometimes obscure in adults; adults very dark, sometimes with greenish miliary spines.

This is the only urchin in the order Diadematoida in temperate west coast waters. Specimens should be handled with care, as the spines, though not poisonous, are very sharp and brittle. **Diadema mexicanus** A. Agassiz (= *Centrechinus mexicanus*), a very similar species that occurs in the southern part of the range of *C. coronatus*, is distinguished by its possession of poison-tipped spines, by its lack of red-purple clavate spines near the apical system, and by the presence of spine-

lets on the ten oral plates surrounding the mouth (these plates in *C. coronatus* bear only pedicellariae).

See H. L. Clark (1948), Grant & Hertlein (1938), and Steinbeck & Ricketts (1941).

Phylum Echinodermata / Class Echinoidea / Order Arbacioida
Family Arbaciidae

11.3 **Arbacia stellata** Gmelin, 1872
(= A. incisa)

On hard substrata sheltered from strong wave action, low intertidal zone and subtidal to 40 m, occasionally to 90 m; Newport Bay (Orange Co.) and Gulf of California to Peru and Galápagos Islands.

Test commonly about 25 mm in diameter, bearing primary spines slightly longer than diameter of test; peristome diameter about three-fourths diameter of test; area immediately around anus covered by four large plates; tubercles and spines lacking in middle of ambulacral areas near apical system (as in all *Arbacia* species); tubercles imperforate; young individuals light-colored, becoming reddish, purplish, or even black as adults; test with reddish spots in spineless areas of upper side.

This sharp-spined southern species is very common in the Gulf of California. The only California record is based on six specimens from Newport Bay, Orange Co. (H. L. Clark, 1948). Although little is known of the biology of *A. stellata*, the common east coast species *A. punctulata* has been studied from many points of view, and the eggs have been favorite subjects for three generations of experimental embryologists and cell biologists.

For *A. stellata* (usually as *A. incisa*), see H. L. Clark (1948), Grant & Hertlein (1938), and Steinbeck & Ricketts (1941). For other *Arbacia* species, see especially Harvey (1956), and also Boolootian (1966), Costello & Henley (1971), and Gordon (1929).

Phylum Echinodermata / Class Echinoidea / Order Temnopleuroida
Family Toxopneustidae

11.4 **Lytechinus anamesus** H. L. Clark, 1912
(=L. pictus)

On sandy bottoms at depths of 2–300 m, often in large herds, and on eelgrass in bays of Baja California; Santa Rosa Island (Channel Islands) to south of Bahía Sebastián Vizcaíno (Baja California) and Gulf of California; records from the Panamic region are questionable.

Test moderately thin, distinctly domed and flattened below, to about 40 mm in diameter, bearing primary spines to 25 mm long; peristome about one-third diameter of test; primary interambulacral tubercles small, imperforate; ambulacral pore pairs arranged in oblique rows of three; color of juveniles very pale, of mature individuals yellowish to dull gray-purple with purplish blotches on upper surface.

The variation in color and in the relative length of the primary spines is so great among Pacific coast representatives of *Lytechinus* that some authors have recognized two species: *L. anamesus*, with long slender spines that lack bands in young and adults; and **L. pictus** (Verrill), with shorter spines banded with shades of red to purple in the young, and with more plates per column than in comparable-sized specimens of *L. anamesus*. Within the same limited geographical area (e.g., La Jolla, San Diego Co.), forms with both short and long spines occur, in somewhat different habitats. However, egg diameter is the same (120 μm) in both, and the larvae show no significant differences; reciprocal (hybrid) crosses are 100 percent successful, and the resulting larvae develop normally; moreover, when blastomeres of the two species are dissociated at the morula stage (those from one species stained with neutral red) and mixed together, they reaggregate in mosaics of the two cell types and reconstitute normal blastulae (V. Vacquier, pers. comm.). These facts together suggest that the two types represent merely ecological variants of the same species. Although less important as grazers on large kelps than the larger *Strongylocentrotus* (e.g., 11.5, 11.6), they consume smaller algae so effectively that reestablishment of algal growth on denuded surfaces occurs only when populations are below

about 10 urchins per m². Most spawning occurs from June through September.

Lytechinus anamesus (usually under the name *L. pictus*) has been used in many embryological and physiological investigations, including classic and modern studies of fertilization and research on protein synthesis and immune response. It is one of a small number of sea urchins that has been raised in the laboratory from fertilization to the onset of sexual reproduction. At 24°C, development of the pluteus larva takes about a week to the first changes associated with development of the juvenile urchin. Settlement and the final change to the definitive urchin form occur very rapidly, taking place 1–2 months after fertilization. The newly settled urchin, scarcely 1 mm in diameter, lacks mouth and anus, but these develop shortly, and feeding begins 4–5 days after settling. Fed in the laboratory on sessile diatoms, the juveniles increase in diameter about 1 mm every 18 days. The first ripe gametes are produced in 4–5 months, though the animals are still very far from full grown.

Lytechinus shows immune responses to grafted tissue similar to those found in vertebrates.

For west coast *Lytechinus,* see Boolootian (1966, index), H. L. Clark (1948), Coffaro & Hinegardner (1977), Ebert (1975), Epel (1967, 1977), Hinegardner (1967, 1969, 1974, 1975), Leighton (1966), Leighton, Jones & North (1966), Ricketts & Calvin (1968), Tegner & Epel (1976), Timourian & Watchmaker (1975), and Tyler & Tyler (1966). For Atlantic species of *Lytechinus,* see Boolootian (1966), Lawrence (1975, 1976), Mendes, Abbud & Umiji (1963), H. Moore et al. (1963), H. Moore & McPherson (1965), and Mortensen (1921).

Phylum Echinodermata / Class Echinoidea / Order Echinoida
Family Strongylocentrotidae

11.5 **Strongylocentrotus purpuratus** (Stimpson, 1857)
Purple Sea Urchin

Common in lower intertidal zone on rocky shores and pilings, typically in areas of moderate to strong wave action; subtidal to 160 m; Vancouver Island (British Columbia) to Isla Cedros (Baja California).

Test commonly 50 mm in diameter and occasionally 100

mm, varying from low to high and from circular to pentagonal in outline; body and spines in life typically bright purple, occasionally pale green or greenish tinged with purple (especially in juveniles); ambulacral pore pairs (easily seen on cleaned test) arranged in curving arcs of eight on upper side of test; primary interambulacral tubercles occupying much less than half the area of individual plates, the secondary tubercles numerous and forming quite distinct rows in all columns.

This is one of the best-known sea urchins in the world, and the subject of much research. Individuals commonly inhabit rounded burrows or depressions in the rock, which they may enlarge with their teeth and perhaps spines as they grow; they may even erode depressions in steel pilings by flaking off rust. Broken spines are regenerated, and the teeth grow out at a rate that results in their complete renewal in about 75 days in the laboratory. The tube feet enable the animals to cling firmly to the substratum. They are also used to attach to the aboral surface such materials as shell fragments and algae, which provide both shade and camouflage. A variety of brown and red algae are eaten; however, the kelp *Macrocystis* is first choice, and in places this urchin seriously affects growth of kelp beds. In turn, the urchin is preyed upon by sea stars (mainly *Pycnopodia helianthoides*, 8.16, and *Astrometis sertulifera*, 8.8, occasionally *Pisaster ochraceus*, 8.13, and *Dermasterias imbricata*, 8.3), some fishes (especially the California sheephead, *Pimelometopon pulchrum*), and the sea otter. Pigment from the urchin lodges in the bones of the otter and stains them purple.

The purple urchin tolerates temperatures of 5–23.5°C, and salinities ranging from 80 to 110 percent that of seawater, under laboratory conditions. Temperatures about 26°C in the field cause mass mortality. Gas exchange between the environment and the viscera inside the test takes place largely through the tube feet, rather than through the "gills," and little exchange occurs when the animals are out of water with the tube feet withdrawn. *S. purpuratus* cannot tolerate water low in oxygen, and in nature lives where the sea is well aerated. Ciliated protozoans and the flatworm *Syndisyrinx franciscanus* inhabit the gut. External commensals reported include a purple polychaete, *Flabelligera commensalis*, living among the spines and an isopod crustacean, *Colidotea rostrata* (21.23).

Purple urchins become sexually mature during their second year, when they are 25 mm or more in diameter. From Washington to northern Baja California, most spawning occurs in the first 3 months of the year, but some ripe animals can be found from September to May and occasionally even to July. The sexes are separate, though occasional hermaphrodites occur. Eggs and sperm are easily obtained by injecting ripe animals with a few milliliters of 0.5 M potassium chloride, an extract of the radial nerves, or even 5–10 ml of air. The gametes are widely used in experimental studies, and are convenient for demonstrating fertilization and cleavage in the classroom. Pluteus larvae are easily raised in 2–4 days. The sperm-egg interactions during fertilization have been studied in detail by various methods, including scanning electron microscopy. In the vicinity of sewage outfalls where chlorine is used as a disinfectant, *S. purpuratus* may suffer reproductive failure, for available chlorine in concentrations as low as 0.05 parts per million severely inhibits fertilization by immobilizing the sperm. Eggs of the purple urchin, fertilized with sperm of the sand dollar *Dendraster excentricus* (11.7), give rise to interordinal hybrid larvae that appear to share features of both parental stocks; they do not complete development. Growth of normal (non-hybrid) juveniles after metamorphosis is slow. The larger *S. purpuratus* may be 10 years old, and approximately 5 percent of the animals in some areas may be older than 30 years, but size alone is not a reliable indicator of age.

Taki (1978) has shown (using tetracycline labeling) that growth of the related *S. intermedius* is more rapid, and the growth lines in the plates of the test more visible, when it feeds on the algae *Laminaria japonica* var. *ochotensis* than when it feeds on *Rhodoglossum japonicum* or *Ulva pertusa*.

See Andrews (1945), Bennett & Giese (1955), Berger (1965), Boolootian (1966), Boolootian & Campbell (1964), Boolootian & Giese (1958), Boolootian & Lasker (1964), Boolootian & Moore (1956, 1959), Bullock (1965), Chatlynne (1969), Chia, Atwood & Crawford (1975), Cochran & Engelmann (1972, 1975), Davies, Crenshaw & Heatfield (1972), Dayton (1975), Dixit (1973), Doezema (1969), Doezema & Phillips (1970), Donnay & Pawson (1969), Douglas (1976),

Ebert (1965, 1967a,b, 1968, 1971, 1977), Epel (1975, 1977), Epley
& Lasker (1959), Farmanfarmaian (1966, 1968), Farmanfarmaian
& Giese (1963), Farmanfarmaian & Phillips (1962), Fox (1972),
Giese (1958, 1966, 1967), Giese & Farmanfarmaian (1963), Giese,
Farmanfarmaian et al. (1966), Giese, Greenfield et al. (1959),
Gonor (1968, 1970, 1972), Grant & Hertlein (1938), Haderlie (1968),
Heatfield (1970, 1971a,b), Heatfield & Travis (1975), Hilgard & Phil-
lips (1968), Hinegardner (1967, 1969, 1974), L. Holland, Giese &
Phillips (1967), N. Holland (1965), N. Holland & Lauritis (1968), N.
Holland & Nimitz (1964), Huang & Giese (1958), Irwin (1953), Jen-
sen (1974), Johansen & Vadas (1967), Lasker & Boolootian (1960),
Lasker & Giese (1954), Lawrence (1975), Lawrence, Lawrence &
Giese (1966), Lawrence, Lawrence and Holland (1965), Lehman
(1946), Leighton (1966), Leighton, Jones & North (1966), Lowry &
Pearse (1973), Lynch (1929, 1930), Lyons (1960), Lyons, Bishop &
Bacon (1964), MacGinitie & MacGinitie (1968), Mauzey, Birkeland
& Dayton (1968), Mazia & Dan (1952), McCauley & Carey (1967), A.
Moore (1943, 1957), Muchmore & Epel (1973), North (1964), North
& Pearse (1970), Pearse & Pearse (1975), Pequignat (1966), Phelan
(1977), Raup (1962, 1966), Ricketts & Calvin (1968), Rosenthal &
Chess (1970, 1972), Schatten & Mazia (1976), Speer & Herzberg
(1961), Stohler (1930), Strathmann (1971, 1975), Swan (1952, 1953,
1961), Taki (1978), Tegner & Dayton (1977), Tegner & Epel (1972,
1976), Thompson & Chow (1955), Turner, Ebert & Given (1968),
Tyler & Tyler (1966), Ulbricht (1973), Ulbricht & Pritchard (1972),
Vacquier (1975), Vacquier, Tegner & Epel (1973), Weber et al.
(1969), Webster (1975), Webster & Giese (1975), and Went & Mazia
(1959).

*Phylum Echinodermata / Class Echinoidea / Order Echinoida
Family Strongylocentrotidae*

11.6 **Strongylocentrotus franciscanus** (A. Agassiz, 1863)
Red Sea Urchin

Uncommon, very low intertidal zone on open, coastal rocky
shore; more abundant subtidally, and extending to depths of
90 m; northern Japan and Alaska to Isla Cedros (Baja California).

Test often 100 mm or more in diameter, bearing primary
spines to 50 mm or more in length; color usually red or red-
brown, sometimes bright dark purple; primary spines and tubercles distinctly larger than secondaries, the tubercles
occupying a large part of the surface of aboral plates; ambulacra with eight or nine pore pairs in each arc (easily seen on
the cleaned test), the arcs more nearly vertical than those of
S. purpuratus (11.5); both compound ambulacral plates and interambulacral plates relatively larger than in *S. purpuratus* (see
also Fig. 21.23).

This magnificent urchin reaches a much greater size than
other species in the genus. Large adults may be as much as 20
years old. Like *S. purpuratus*, the red urchin eats a variety of
brown and red algae, but a favorite food is the kelp *Macrocystis*. In a kelp forest where food is abundant, the red urchin
moves about very little, but seaward of a kelp bed off Santa
Cruz, marked red urchins shifted position by an average of 50
cm a day.

Man harvests kelp in southern California, and investigations by Wheeler North, of the California Institute of Technology, and others have shown a tangled network of relationships between sea urchins, man, kelp, abalones, and other
organisms, in which cause and effect are not always clear. It
seems likely that populations of sea urchins and abalones
soared upward in California inshore waters after the sea otter,
a major predator on both, was hunted virtually to extinction
here in the last century. The abalones, which browse on seaweeds, are among the urchins' main competitors for food.
In recent decades, abalones too have been depleted by man,
and at the same time sewage pollution of inshore waters
has increased. Sewage tends to inhibit growth of kelp, but
both red and purple urchins are able to absorb certain dissolved organic substances from water and thus may benefit
from moderate amounts of household sewage; heavy pollution and industrial wastes may, of course, kill kelp and urchins alike. In areas where kelp once flourished and urchins
now abound, even when pollution is abated somewhat, kelp
beds can scarcely get a start, since the urchins consume young
plants as fast as they begin to grow. Field experiments in
areas off San Diego where pollution is not severe have shown
that lowering the urchin population by application of quick-
lime to the bottom (a practice condemned by many marine
biologists) allows kelp to reestablish beds before the urchin
population recovers. Once established, the kelp bed tends to
hold its own. In the Monterey Bay area, a resurgent sea otter
population has virtually eliminated the large beds of red urchins that flourished a decade before at depths of a few me-

ters; in these areas kelp now abounds where formerly it was sparse or absent.

The red urchin breeds in the spring. Gonads enlarge during the winter, and some ripe animals are found from February to July in central California, but the main spawning is in April and May. The season is earlier in southern California, and some winter spawning has been reported. The free-swimming larval period lasts 62–131 days under aquarium conditions. The gonads, eaten raw, are considered a delicacy by man as well as otters.

Juveniles of the red urchin are often found beneath the outspread spines of adults.

Occasional specimens appear to be hybrids between the red and purple urchins. Crosses made in the laboratory, using eggs from the purple urchin and sperm from the red urchin (the reverse cross usually fails), result in pluteus larvae, which in size and skeletal characters are intermediate between the normal larvae of the two species, suggesting inheritance from both parents.

In the Puget Sound area (Washington), a small amphipod crustacean, *Dulichia*, lives on the red urchin. The crustacean builds elongate rods, from detritus and its own fecal pellets cemented together, which extend outward from the tips of the urchin's primary spines. Diatoms flourish on these rods, and the amphipods move up and down, feeding on the diatoms in the summer and keeping the crop tended and weeded of other growth. In winter, when the diatom growth is poor, the amphipods filter detritus from the water.

See Andrews (1945), Baker (1973), Bennett & Giese (1955), Berger (1965), Boolootian (1958, 1966), Boolootian & Giese (1958), Boolootian & Moore (1956), Chia, Atwood & Crawford (1975), H. L. Clark (1948), Ebert (1975, 1977), Fuji (1967), Giese (1958, 1959a,b, 1966), Grant & Hertlein (1938), Greenfield et al. (1958), Jensen (1974), Johansen & Vadas (1967), M. W. Johnson (1930), Lawrence (1975), Lehman (1946), Leighton (1966), Leighton, Jones & North (1966), Lowry & Pearse (1973), Lynch (1929, 1930), Mattison et al. (1977), Mauzey, Birkeland & Dayton (1968), McCauley & Carey (1967), McCloskey (1970), A. Moore (1943), Mortensen (1921), Nissen (1969), North (1964), North & Pearse (1970), Raup (1958), Sandeman (1965), Strathmann (1971, 1978), Swan (1952, 1953, 1961), Tegner & Dayton (1977), Thompson & Chow (1955), Towe (1967), Turner, Ebert & Given (1968), Tyler & Tyler (1966), Ulbricht (1973), Ulbricht & Pritchard (1972), and Weese (1926).

Phylum Echinodermata / Class Echinoidea / Order Clypeasteroida Family Dendrasteridae

11.7 **Dendraster excentricus** (Eschscholtz, 1831) Sand Dollar

Abundant in localized aggregations on sandy or sandy-mud bottoms, low intertidal and subtidal zones in sheltered bays; subtidal to 40 m (rarely to 90 m) in open coastal areas; southeastern Alaska to central west coast of Baja California; Gulf of California records unconfirmed.

Test commonly reaching 75 mm in diameter; color pale gray-lavender, medium brown, red-brown, or dark purplish black, the aboral spine tips sometimes much lighter than rest of body; distinguished from *D. vizcainoensis* (11.8) by its greater eccentricity, more evenly developed spines and tubercles, and more unequally shaped aboral "petals" (best seen on a cleaned test); *D. laevis* (11.9) is usually thinner, lighter in color, with finer and more even spine and tubercle development, and with much smaller and more equal petals.

The clean and empty tests of dead sand dollars cast ashore are a familiar sight on many California beaches. The convex upper (aboral) surface of the test bears a striking flowerlike pattern: at its center lies the madreporite, adjacent to which, on the interambulacra, lie the genital pores; radiating out from this center are the five "petals," which mark the ambulacral radii. Turning the sand dollar over reveals the flattened oral surface, which bears a central hole for the mouth and a smaller hole near the posterior edge for the anus.

This is a highly variable species; populations from quiet water may appear distinctly different from those on adjacent open coastal bottoms, and the tests of individuals living in colder waters and more turbulent environments tend to have heavier walls and internal supports. In quiet waters the animals usually live with the anterior third or more of the body buried and the remainder protruding at an angle into the water; in rougher water the animals lie flat, either at the surface or partly or wholly buried. In the intertidal zone, animals tend to bury themselves as the tide goes out. Subtidal populations sometimes exceed 625 animals per m² of bottom.

The main locomotory organs of adults are the myriad slender spines on the oral surface of the test; operating together in waves, they carry a sand dollar across the substratum or drive it forward edgewise into the sand, while the spines and the tube feet at the leading margin clear the way and pile sand on the disk. Young animals less than 15 mm in diameter may move mainly by means of the podia that bear suckers, and can even climb a glass aquarium wall. Podia are of two sorts: tiny sucker-tipped tube feet which are widely distributed on both body surfaces, and large, highly modified respiratory podia, appearing as broad flanges of soft tissue, which occupy the borders of the petals on the aboral surface.

Three different methods of feeding are used, depending on the nature and size of the food. Organic particles and organisms less than 50 μm in diameter are drawn by water currents between the club-tipped spines on the aboral surface, then swept by the cilia to the margins and around to the oral side, where they enter food grooves leading to the mouth. Fragments of algae more than 50 μm across may be held by the sucker-tipped podia and transported to a food groove. Small active prey (crustacean larvae, small copepods) are trapped on the oral surface of the body by the spines, which converge over the prey, enclosing it in small tepee-like cones. From these cages the prey is grabbed by two-jawed pedicellariae and passed to a food groove by tube feet. In the food grooves small particles move in a mucus stream to the mouth, whereas larger food items are passed along by the tube feet and spines. The jaws are used to chew food for as long as 15 minutes before swallowing occurs, and the ingested food then stays in the gut about 2 days. Considerable sand is taken in with the food. Indeed, juvenile *Dendraster* less than 30 mm in diameter selectively ingest heavy sand grains, which they store in a remarkably long diverticulum of the gut. It has been suggested that the stored sand acts as a "weight belt," helping to stabilize the animal in a shifting sandy environment.

Among the predators on sand dollars are fishes, especially the California sheephead (*Pimelometopon pulchrum*) and the starry flounder (*Platichthys stellatus*), and the large pink sea star *Pisaster brevispinus* (8.15). Proximity to the starfish causes sand dollars to burrow down, and a pink sea star walking through a sand dollar bed is reported to leave a meter-wide swath in its track where sand dollars have gone below. Sea gulls eat some *Dendraster* washed ashore or exposed at low tide. Some sand dollars show healed broken edges, indicating an ability to recover from sizeable breaks or bites; lesions on the test of some specimens lack spines but bear a greenish film of microscopic algae. Occasional acorn barnacles (*Balanus pacificus*, 13.27) are found growing on the aboral test surface of otherwise normal sand dollars.

The major spawning in central and southern California occurs from May through July, but some fertile animals are found throughout the summer. At San Diego, populations living on the open coast and in protected tidal channels showed similar periods of reproductive activity, but individuals in the protected channels grow to a larger size, produce relatively larger gonads, and contain ripe gametes for a longer period of time. Males and females occur in approximately equal numbers and look alike externally. Occasional hermaphrodites are found.

Dendraster are easily kept for long periods in laboratory aquaria, and the gonads ripen in season if the captive animals receive food. Long a favorite source of gametes for experimental embryology, the animals spawn readily if injected with 0.5 M potassium chloride. Ripe spawned eggs have already undergone meiosis. Fertilization is external. Freshly spawned eggs bear a conspicuous jelly coat, which experiments show deters sand dollars from feeding on their own eggs (though later they eat their own larvae without hesitation). Development is rapid in the laboratory, and large four-armed pluteus larvae develop in 2–4 days. The species has been raised through metamorphosis to the juvenile sand dollar stage wholly in the laboratory. In aquaria at fluctuating temperatures of about 7–17°C, the period from fertilization through metamorphosis and settling lasted 68–162 days.

Crosses between male *Dendraster* and female purple urchins (representing different orders of echinoids, which diverged in the Mesozoic) yield hybrid larvae, which in several characters appear intermediate between those of the two parental species.

Studies at Zuma Beach (Los Angeles Co.) show that most *Dendraster* there reach sexual maturity by 4 years of age. The oldest animals in most samples were 6–10 years, but an occasional individual reached 13 years. Average females produced 356,000–379,000 eggs per year, but a good recruitment of juveniles to the benthic stock occurred only every few years. Animals can be aged by counting growth rings on the plates of the test. The relative ages of individuals can be determined by counting the number of pore pairs (or plates) in a petal.

See especially Birkeland & Chia (1971), Chia (1969a,b, 1973), Durham (1955), Merrill & Hobson (1970), A. Moore (1957), Niesen (1977), and Timko (1975, 1976); see also Boolootian (1958, 1964, 1966, index), Boolootian & Giese (1958), Boolootian & Moore (1956), Brookbank (1969), Chia, Atwood & Crawford (1975), H. L. Clark (1935, 1948), Goodbody (1960), Gordon (1929), Grant & Hertlein (1938), Harriss & Pilkey (1966), Hinegardner (1974), Hyman (1958), Ikeda (1939), M. E. Johnson & Snook (1935), MacGinitie & MacGinitie (1968), McCauley & Carey (1967), Mortensen (1921), Orcutt (1950), Parker (1927), Parker & Van Alstyne (1932), Phelan (1977), Pilkey & Hower (1960), Quast (1968), Rappaport (1960), Raup (1958), Reisman (1965), L. Smith (1961), Strathmann (1971, 1978), Thompson & Chow (1955), Tyler & Tyler (1966), Vacquier (1971), and Whiteley et al. (1975).

Phylum Echinodermata / Class Echinoidea / Order Clypeasteroida Family Dendrasteridae

11.8 Dendraster vizcainoensis Grant & Hertlein, 1938
Sand Dollar

On sandy bottoms below low tide to depths of 30 m; sometimes cast ashore by waves; known with certainty only from Baja California (Bahía del Rosario to Bahía Sebastián Vizcaíno), but possible occurring as far north as San Diego and mistaken for *D. excentricus* (11.7).

Test reaching 63 mm in diameter, typically pentamerous in outline, the central aboral area distinctly domed; body brown to purplish with paler spines; enlarged primary spines around much of margin and in rows along food grooves, the grooves extending onto aboral side and into petals; petals relatively shorter and narrower than in *D. excentricus*, longer and wider than those of *D. laevis* (11.9), and typically more open than in either of these species.

See H. L. Clark (1948), Durham (1955), Grant & Hertlein (1938), and the references under *D. excentricus*.

Phylum Echinodermata / Class Echinoidea / Order Clypeasteroida Family Dendrasteridae

11.9 Dendraster laevis H. L. Clark, 1948
Sand Dollar

On the surface or slightly buried in sandy areas, in subtidal water depths of 6–55 m; occasionally washed ashore; reported only along mainland coast from Mugu Slough (Point Mugu, Ventura Co.) to San Diego and offshore from Channel Islands to Los Coronados; possibly as far south as Bahía de Tortuga (Baja California).

Test to 55 mm in diameter, flat and delicate; color in life yellowish and greenish tan on oral surface, tan and sometimes faintly purplish posteriorly on aboral side; spines and tubercles very small, never enlarged marginally or along food grooves as in *D. vizcainoensis* (11.8); apical system (from which petals arise) nearly central; petals much narrower and smaller than in *D. excentricus* (11.7).

Unlike the other species of *Dendraster*, this form is not known to occur in very shallow water. The ranges of *D. laevis* and *D. excentricus* overlap in the area from Point Mugu to La Jolla (San Diego Co.), but *D. laevis* typically lives in deeper water.

See Boolootian (1966), H. L. Clark (1948), and Durham (1955).

Phylum Echinodermata / Class Echinoidea / Order Clypeasteroida Family Mellitidae

11.10 Encope micropora L. Agassiz, 1841
Keyhole Urchin

Commonest in lower intertidal and adjacent subtidal zones on sandy bottoms, but extending to 30 m; southernmost California to Peru.

Test typically 75–100 mm in diameter, easily recognized by the six oval holes (lunules) perforating the test, one at the end of each petal and one between the posterior pair of petals, just back of the anus; color in life varying from yellowish or greenish brown to dark purplish brown, generally darkest around lunules; test very dense, with highly developed internal skeletal supports.

This is the *E. californica* of some authors, but not Verrill, 1870. It is the only species of keyhole urchin on the west coast of Baja California occurring north of 26°N latitude, and only rarely has it been reported as far north as San Diego. It is very common in the Gulf of California, where it occurs in huge beds associated with another keyhole urchin, *Encope grandis*. The function of the lunules in *Encope* and other keyhole urchins has been a topic for more speculation than investigation. However, in *Astriclypeus* the lunules have been shown to play a role in righting and burrowing, and in *Mellita* they provide shorter pathways by which tiny food particles trapped on the aboral surface can be carried to the food grooves and ambulacral grooves leading to the mouth on the oral surface.

Individuals of *E. micropora* lie flat on the bottom with the oral surface down, often partially exposed but covered with a thin layer of sand. Shallowly buried individuals can sometimes be spotted by swimming over submerged sand flats and spotting small pits or depressions corresponding in size and arrangement to the keyholes of an individual.

Encope micropora is commonly host to the commensal crab *Dissodactylus lockingtoni*. The small crustacean occupies the front portion of the central posterior keyhole, just posterior to the anus, and it feeds on the feces of the echinoid.

Encope micropora has been hybridized with the sand dollar *Dendraster excentricus* (11.7) in the laboratory, and the larvae reared through metamorphosis to the juvenile urchin stage. No matter which species was used to provide eggs, both the pluteus larvae and the young urchins showed dominance of the paternal characteristics.

For *E. micropora*, see H. L. Clark (1948), Durham (1950), Glassell (1935), and Hinegardner (1975, as *E. californicus*). For other sand dollars with lunules, see Crozier (1919, 1920), Dexter (1977), Durham (1955), Ebert & Dexter (1975), Goodbody (1960), Grant & Hertlein (1938), Hyman (1958), Ikeda (1939, 1941), Kenk (1944), MacGinitie & MacGinitie (1968), Mortensen (1921), Parker (1927), Parker & Van Alstyne (1932), Phelan (1977), and Salsman & Tolbert (1965).

Phylum Echinodermata / Class Echinoidea / Order Spatangoida Family Brissidae

11.11 **Brissopsis pacifica** (A. Agassiz, 1898)
Heart Urchin

Burrowing in sandy or muddy bottoms, subtidal at 9–75 m depth; occasionally washed ashore; Channel Islands and coast of southern California to Panama and Galápagos Islands.

Test usually 40–50 mm long and 25 mm high; body and spines medium to light fawn-brown in color, contrasting with dark brown peripetalous fasciole (the groove among the spines surrounding petal area on aboral surface, well shown in the photograph); paired petals (as seen on bare test) slightly rounded and distinctly sunken, the anterolateral pair extending slightly over halfway to margin, the posterior pair very close together and extending less than halfway to margin.

See Chesher (1968), H. L. Clark (1948), and Grant & Hertlein (1938); see also the references cited under *Lovenia cordiformis* (11.12).

Phylum Echinodermata / Class Echinoidea / Order Spatangoida Family Loveniidae

11.12 **Lovenia cordiformis** (A. Agassiz, 1872)
Heart Urchin, Sea Porcupine

Locally abundant, usually just below surface in sand or silty sediments, extremely low intertidal zone to 140 m; San Pedro (Los Angeles Co.) and Channel Islands to Panama and Galápagos Islands.

Test commonly to 75 mm long and 13–18 mm high, oval, tapering toward the rear; color variable, off-white, gray-brown, yellowish, rose, or purplish; spines on upper surface uniformly colored or banded; easily distinguished from other west coast heart urchins by the extremely long spines on interambulacra (areas between petals) on aboral surface, each spine attached to a large tubercle so deeply sunken in the test that it appears as a depressed pit on the cleaned skeleton; petals not rounded and only faintly sunken; four genital pores.

Lovenia is the most common heart urchin in shallow water along California shores, and the one most likely to be encountered by the shore observer or swimmer. As in other heart urchins the ovoid test has clear anterior and posterior ends, with the anus more or less posterior; the test is thin and fragile compared with those of the regular urchins of rocky shores. When brought to the surface and disturbed, *Lovenia* may scoot about, using the long lateral spines. The long aboral spines are normally pointed posteriorly, but when aroused, the animal defends itself by erecting these and pointing them toward the site of irritation; the spines can produce painful wounds if not handled with care.

Little is known of the biology of local heart urchins, but studies of other forms (*Echinocardium, Moira*) indicate that, when placed at the surface, they often burrow more or less straight down into the sand or mud for 1 to several centimeters. A small vertical shaft to the surface is kept open by the action of the very long tube feet and by the application of mucus to prevent cave-ins. The animals are detritus feeders, and the elongate tube feet are extended to the surface and used to rake surface deposits (often rich in tiny organisms and organic detritus) down the shaft. Descending particles are caught in mucus and transported forward and down along the anterior ambulacrum to the mouth. Some food particles may be digested on, and absorbed through, the skin. The tube feet near the mouth may also be used to collect food particles from the adjacent burrow walls and floor. A jaw apparatus (Aristotle's lantern) is lacking. The urchins may tunnel forward a short distance and use the long tube feet to dig another shaft to the surface, this act being repeated as often as a new feeding spot is needed.

See Chesher (1963, 1968), H. L. Clark (1948), Gordon (1926), Grant & Hertlein (1938), M. E. Johnson & Snook (1935), MacGinitie & MacGinitie (1968), H. Moore (1936), H. Moore & Lopez (1966), Nichols (1959, 1962), Pearse (1969), Pequignat (1966), Ricketts & Calvin (1968), and Steinbeck & Ricketts (1941).

Phylum Echinodermata / Class Echinoidea / Order Spatangoida Family Spatangidae

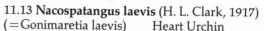

11.13 **Nacospatangus laevis** (H. L. Clark, 1917)
(=Gonimaretia laevis)   Heart Urchin

Burrowing in sandy bottoms at depths of 5–300 m; occasionally washed ashore; Channel Islands to Gulf of California.

Test to 44 mm long and 18 mm high; small individuals rather dark brown, larger animals light fawn-brown; test similar to that of *Lovenia cordiformis* (11.12), but distinguishable from it as follows: (1) anterior petal not sunken, (2) apical system bearing only three genital pores, and (3) aboral surface lacking enlarged tubercles borne in deep depressions in the test, but with a few, slightly enlarged tubercles present along side of anterior ambulacrum aborally.

See H. L. Clark (1948), Grant & Hertlein (1938), and other references listed under *Lovenia cordiformis* (11.12).

Literature Cited

Andrews, H. L. 1945. The kelp beds of the Monterey region. Ecology 26: 24–37.

Baker, S. 1973. Growth of the red sea urchin *Strongylocentrotus franciscanus* (Agassiz) in two natural habitats. Master's thesis, Biology, San Diego State University, San Diego, Calif. 83 pp.

Bennett, J., and A. C. Giese. 1955. The annual reproductive and nutritional cycles in two western sea urchins. Biol. Bull. 109: 226–37.

Berger, J. 1965. The infraciliary morphology of *Euplotes tuffani* n. sp., a commensal in strongylocentrotid echinoids, with comments on echinophilous populations of *Euplotes balteatus* (Dujardin) (Ciliata: Hypotrichida). Protistologica 1: 17–31.

Binyon, J. 1972. Physiology of echinoderms. New York: Pergamon. 264 pp.

Birkeland, C., and F.-S. Chia. 1971. Recruitment risk, growth, age and predation in two populations of sand dollars, *Dendraster excentricus* (Eschscholtz). J. Exper. Mar. Biol. Ecol. 6: 265–78.

Boolootian, R. A. 1958. Notes on an unexpected association between a common barnacle and echinoid. Bull. So. Calif. Acad. Sci. 57: 91–92.

———. 1964. The occurrence of *Balanus concavus* on the test of *Dendraster excentricus*. Bull. So. Calif. Acad. Sci. 63: 185–91.

———, ed. 1966. Physiology of Echinodermata. New York: Interscience. 822 pp.

Boolootian, R. A., and J. L. Campbell. 1964. A primitive heart in the echinoid *Strongylocentrotus purpuratus*. Science 145: 173–75.

Boolootian, R. A., and A. C. Giese. 1958. Coelomic corpuscles in echinoderms. Biol. Bull. 115: 53–63.

Boolootian, R. A., and R. Lasker. 1964. Digestion of brown algae and distribution of nutrients in the purple sea urchin *Strongylocentrotus purpuratus*. Comp. Biochem. Physiol. 11: 273–89.

Boolootian, R. A., and A. R. Moore. 1956. Hermaphroditism in echinoids. Biol. Bull. 111: 328–35.

———. 1959. A case of ovotestes in the sea urchin *Strongylocentrotus purpuratus*. Science 129: 271–72.

Brookbank, J. W. 1969. DNA synthesis during early development of interordinal hybrids. Amer. Zool. 9: 599 (abstract).

Bullock, T. H. 1965. Comparative aspects of superficial conduction systems in echinoids and asteroids. Amer. Zool. 5: 545–62.

Chatlynne, L. G. 1969. A histochemical study of oogenesis in the sea urchin *Strongylocentrotus purpuratus*. Biol. Bull. 136: 167–84.

Chesher, R. H. 1963. The morphology and function of the frontal ambulacrum of *Moira atropos* (Echinoidea: Spatangoida). Bull. Mar. Sci. 13: 549–73.

———. 1968. The sytematics of sympatric species of West Indian spatangoids: A revision of the genera *Brissopsis, Plethotaenia, Paleopneustes,* and *Savinaster*. Stud. Trop. Oceanogr. Miami 7: 1–168.

Chia, F.-S. 1969a. Histology of the pedicellariae of the sand dollar, *Dendraster excentricus* (Echinodermata). J. Zool., London 157: 503–7.

———. 1969b. Some observations on the locomotion and feeding of the sand dollar *Dendraster excentricus* (Eschscholtz). J. Exper. Mar. Biol. Ecol. 3: 162–70.

———. 1973. Sand dollar: A weight belt for the juvenile. Science 181: 73–74.

Chia, F.-S., D. Atwood, and B. Crawford. 1975. Comparative morphology of echinoderm sperm and possible phylogenetic implications. Amer. Zool. 15: 533–65.

Chia, F.-S., and A. H. Whiteley, eds. 1975. Developmental biology of the echinoderms. Amer. Zool. 15: 483–775.

Clark, A. M. 1962. Starfishes and their relations. London: British Museum (Natur. Hist.) 119 pp.

Clark, H. L. 1935. Some new echinoderms from California. Ann. Mag. Natur. Hist. (10) 15: 120–29.

———. 1948. A report on the echini of the warmer eastern Pacific, based on the collections of the Velero III. Allan Hancock Pacific Exped. 8: 225–352.

Cochran, R. C., and F. Engelmann. 1972. Echinoid spawning induced by a radial nerve factor. Science 178: 423–24.

———. 1975. Environmental regulation of the annual reproductive season of *Strongylocentrotus purpuratus* (Stimpson). Biol. Bull. 148: 393–401.

Coffaro, K. A., and R. T. Hinegardner. 1977. Immune response in the sea urchin *Lytechinus pictus*. Science 197: 1389–90.

Costello, D. P., and C. Henley. 1971. Methods for obtaining and handling marine eggs and embryos. 2nd ed. Marine Biological Laboratory, Woods Hole, Mass. 247 pp.

Crozier, W. J. 1919. On the regeneration and the re-formation of lunules in *Mellita*. Amer. Natur. 53: 93–96.

———. 1920. Notes on the bionomics of *Mellita*. Amer. Natur. 54: 435–42.

Cuénot, L. 1948. Anatomie, éthologie et systématique des échinodermes, pp. 3–275, *in* P. Grassé, ed., Traité de zoologie, vol. 11. Paris: Masson. 1,077 pp.

Czihak, G., and R. Peter, eds. 1975. The sea urchin embryo: Biochemistry and morphogenesis. Berlin: Springer-Verlag. 700 pp.

Davies, T. T., M. A. Crenshaw, and B. M. Heatfield. 1972. The effect of temperature on the chemistry and structure of echinoid spine regeneration. J. Paleontol. 46: 874–83.

Dawydoff, C. 1948. Embryologie des échinodermes, pp. 277–363, *in* P. Grassé, ed., Traité de zoologie, vol. 11. Paris: Masson. 1,077 pp.

Dayton, P. K. 1975. Experimental evaluation of ecological dominance in a rocky intertidal algal community. Ecol. Monogr. 45: 137–59.

Dexter, D. M. 1977. A natural history of the sand dollar *Encope stokesi* L. Agassiz in Panama. Bull. Mar. Sci. 27: 544–51.

Dixit, D. 1973. A simple technique for obtaining gametes from the sea urchin, *Strongylocentrotus purpuratus*. BioScience 23: 39.

Doezema, P. 1969. Carbohydrates and carbohydrate metabolism of echinoderms, pp. 101–22, *in* Florkin & Scheer (1969).

Doezema, P., and J. H. Phillips. 1970. Glycogen storage and synthesis in the gut of the purple sea urchin, *Strongylocentrotus purpuratus*. Comp. Biochem. Physiol. 34: 691–97.

Donnay, G., and D. L. Pawson. 1969. X-ray diffraction studies of echinoderm plates. Science 166: 1147–50.

Douglas, C. A. 1976. Availability of drift materials and the covering response of the sea urchin *Strongylocentrotus purpuratus* (Stimpson). Pacific Sci. 30: 83–89.

Durham, J. W. 1950. 1940 E. W. Scripps cruise to the Gulf of California. II. Megascopic paleontology and marine stratigraphy. Geol. Soc. Amer. Mem. 43. 216 pp.

———. 1955. Classification of clypeasteroid echinoids. Univ. Calif. Publ. Geol. Sci. 31: 73–198.

———. 1966. Evolution among the Echinoidea. Biol. Rev. 41: 368–91.

Durham, J. W., and R. V. Melville. 1957. A classification of echinoids. J. Paleontol. 31: 242–72.

Durham, J. W., H. B. Fell, A. G. Fischer, P. M. Kier, R. V. Melville, D. L. Pawson, and C. D. Wagner. 1966. Echinoids, pp. U211–640, *in* R. C. Moore, ed., Treatise on invertebrate paleontology, echinoderms 3, vols. 1 and 2. New York: Geol. Soc. Amer.; Lawrence: University of Kansas Press.

Ebert, T. A. 1965. A technique for the individual marking of sea urchins. Ecology 46: 193–94.

————. 1967a. Growth and repair of spines in the sea urchin *Strongylocentrotus purpuratus* (Stimpson). Biol. Bull. 133: 141–49.

————. 1967b. Negative growth and longevity in the purple sea urchin *Strongylocentrotus purpuratus* (Stimpson). Science 157: 557–58.

————. 1968. Growth rates of the sea urchin *Strongylocentrotus purpuratus* related to food availability and spine abrasion. Ecology 49: 1075–91.

————. 1971. Sea urchin mortality. Bull. Ecol. Soc. Amer. 52: 22.

————. 1975. Growth and mortality of post-larval echinoids. Amer. Zool. 15: 755–75.

————. 1977. An experimental analysis of sea urchin dynamics and community interactions on a rock jetty. J. Exper. Mar. Biol. Ecol. 27: 1–22.

Ebert, T. A., and D. M. Dexter. 1975. A natural history study of *Encope grandis* and *Mellita grantii*, two sand dollars in the northern Gulf of California, Mexico. Mar. Biol. 32: 397–407.

Epel, D. 1967. Protein synthesis in sea urchin eggs: A "late" response to fertilization. Proc. Nat. Acad. Sci. 57: 899–906.

————. 1975. The program of and mechanisms of fertilization in the echinoderm egg. Amer. Zool. 15: 507–22.

————. 1977. The program of fertilization. Sci. Amer. 237: 128–38.

Epley, R. W., and R. Lasker. 1959. Alginase in the sea urchin, *Strongylocentrotus purpuratus*. Science 129: 214–15.

Farmanfarmaian, A. 1966. The respiratory physiology of echinoderms, pp. 245–65, *in* Boolootian (1966).

————. 1968. The controversial echinoid heart and hemal system—function effectiveness in respiratory exchange. Comp. Biochem. Physiol. 24: 855–63.

Farmanfarmaian, A., and A. C. Giese. 1963. Thermal tolerance and acclimation in the western purple sea urchin, *Strongylocentrotus purpuratus*. Physiol. Zool. 36: 327–43.

Farmanfarmaian, A., and J. H. Phillips. 1962. Digestion, storage, and translocation of nutrients in the purple sea urchin (*Strongylocentrotus purpuratus*). Biol. Bull. 123: 105–20.

Fitch, J. E. 1962. A sea urchin, a lobster and a fish, new to the marine fauna of California. Calif. Fish & Game 48: 216–21.

Florkin, M., and B. T. Scheer, eds. 1969. Chemical zoology. 3. Echinodermata, Nematoda, and Acanthocephala. New York: Academic Press. 687 pp.

Fox, D. L. 1972. Chromatology of animal skeletons. Amer. Sci. 60: 436–47.

Fuji, A. 1967. Ecological studies on the growth and food consumption of Japanese common littoral sea urchin, *Strongylocentrotus intermedius* (A. Agassiz). Mem. Fac. Fish. Hokkaido Univ. 15: 83–160.

Giese, A. C. 1958. Incidence of *Syndesmis* in the gut of two species of sea urchins. Anat. Rec. 132: 441–42 (abstract).

————. 1959a. Annual reproductive cycles of marine invertebrates. Ann. Rev. Physiol. 21: 547–76.

————. 1959b. Reproductive cycles of some west coast invertebrates, pp. 625–38, *in* R. B. Withrow, ed., Photoperiodism and related phenomena in plants and animals. Publ. 55. Washington, D.C.: American Association for the Advancement of Science. 903 pp.

————. 1966. Lipids in the economy of marine invertebrates. Physiol. Rev. 46: 244–98.

————. 1967. Changes in body-component indexes and respiration with size in the purple sea urchin *Strongylocentrotus purpuratus*. Physiol. Zool. 40: 194–200.

Giese, A. C., and A. Farmanfarmaian. 1963. Resistance of the purple sea urchin to osmotic stress. Biol. Bull. 124: 182–92.

Giese, A. C., A. Farmanfarmaian, S. Hilden, and P. Doezema. 1966. Respiration during the reproductive cycle in the sea urchin *Strongylocentrotus purpuratus*. Biol. Bull. 130: 192–201.

Giese, A. C., L. Greenfield, H. Huang, A. Farmanfarmaian, R. A. Boolootian, and R. Lasker. 1959. Organic productivity in the reproductive cycle of the purple sea urchin. Biol. Bull. 116: 49–58.

Glassell, S. 1935. New or little known crabs from the Pacific coast of north Mexico. Trans. San Diego Soc. Natur. Hist. 8: 91–106.

Gonor, J. J. 1968. Temperature relations of central Oregon marine intertidal invertebrates. ONR Data Rep. 34, Dept. Oceanogr., Oregon State Univ. Ref. 68-38. 49 pp.

————. 1970. Oregon coastal marine animals. Oregon State Univ. Tech. Rep. 199, Ref. 70-40: 79–102.

————. 1972. Gonad growth in the sea urchin, *Strongylocentrotus purpuratus* (Stimpson) (Echinodermata: Echinoidea) and the assumptions of the gonad index methods. J. Exper. Biol. Ecol. 10: 89–103.

Goodbody, I. 1960. The feeding mechanism in the sand dollar *Mellita sexiesperforata* (Leske). Biol. Bull. 119: 80–86.

Gordon, I. 1926. On the development of the calcareous test of *Echinocardium cordatum*. Phil. Trans. Roy. Soc. London B215: 255–313.

————. 1929. Skeletal development in *Arbacia*, *Echinarachnius*, and *Leptasterias*. Phil. Trans. Roy. Soc. London B217: 289–334.

Grant, U. S., IV, and L. G. Hertlein. 1938. The west American Cenozoic Echinoidea. Univ. Calif. Los Angeles Publ. Math. Phys. Sci. 2: 1–225.

Greenfield, L., A. C. Giese, A. Farmanfarmaian, and R. A. Boolootian. 1958. Cyclic biochemical changes in several echinoderms. J. Exper. Zool. 139: 507–24.

Gustafson, T. 1969. Fertilization and development, pp. 149–206, *in* Florkin & Scheer (1969).

Gustafson, T., and L. Wolpert. 1967. Cellular movement and contact in sea urchin morphogenesis. Biol. Rev. 42: 442–98.

Haderlie, E. C. 1968. Marine fouling organisms in Monterey harbor. Veliger 10: 327–41.

Harriss, R. C., and O. H. Pilkey. 1966. Temperature and salinity control of the concentration of skeletal Na, Mn, and Fe in *Dendraster excentricus*. Pacific Sci. 20: 235–38.

Harvey, E. B. 1956. The American *Arbacia* and other sea urchins. Princeton, N. J.: Princeton University Press. 298 pp.

Heatfield, B. M. 1970. Calcification in echinoderms: Effects of temperature and Diamox on incorporation of calcium-45 *in vitro* by regenerating spines of *Strongylocentrotus purpuratus*. Biol. Bull. 139: 151–63.

———. 1971a. Growth of the calcareous skeleton during regeneration of spines of the sea urchin, *Strongylocentrotus purpuratus* (Stimpson): A light and scanning electron microscopic study. J. Morphol. 134: 57–89.

———. 1971b. Origin of calcified tissue in regenerating spines of the sea urchin, *Strongylocentrotus purpuratus* (Stimpson): A quantitative radioautographic study with tritiated thymidine. J. Exper. Zool. 178: 233–46.

Heatfield, B. M., and D. F. Travis. 1975. Ultrastructural studies of regenerating spines of the sea urchin *Strongylocentrotus purpuratus* I. Cell types without spherules. II. Cell types with spherules. J. Morphol. 145: 13–71.

Hilgard, H. R., and J. H. Phillips. 1968. Sea urchin response to foreign substances. Science 161: 1243–45.

Hinegardner, R. T. 1967. Echinoderms, pp. 139–55, *in* F. H. Wilt and N. K. Wessells, eds., Methods in developmental biology. New York: Crowell. 813 pp.

———. 1969. Growth and development of the laboratory-cultured sea urchin. Biol. Bull. 137: 465–75.

———. 1974. Cellular DNA content of the Echinodermata. Comp. Biochem. Physiol. 49B: 219–26.

———. 1975. Morphology and genetics of sea urchin development. Amer. Zool. 15: 679–89.

Holland, L. Z., A. C. Giese, and J. H. Phillips. 1967. Studies on the perivisceral coelomic fluid protein concentration during seasonal and nutritional changes in the purple sea urchin. Comp. Biochem. Physiol. 21: 361–71.

Holland, N. D. 1965. An autoradiographic investigation of tooth renewal in the purple sea urchin (*Strongylocentrotus purpuratus*). J. Exper. Zool. 158: 275–82.

Holland, N. D., and J. A. Lauritis. 1968. The fine structure of the gastric exocrine cells of the purple sea urchin, *Strongylocentrotus purpuratus*. Trans. Amer. Microscop. Soc. 87: 201–9.

Holland, N. D., and Sister A. Nimitz. 1964. An autoradiographic and histochemical investigation of the gut mucopolysaccharides of the purple sea urchin (*Strongylocentrotus purpuratus*). Biol. Bull. 127: 280–93.

Huang, H., and A. C. Giese. 1958. Tests for digestion of algal polysaccharides by some marine herbivores. Science 127: 475.

Hyman, L. H. 1955. The invertebrates: Echinodermata. Vol. 4. New York: McGraw-Hill. 763 pp.

———. 1958. Notes on the biology of the five-lunuled sand dollar. Biol. Bull. 114: 54–56.

Ikeda, H. 1939. Studies on the pseudofasciole of the scutellids (Echinoidea, Scutellidae). J. Dept. Agr., Kyusyu Imp. Univ., 6: 41–93.

———. 1941. Function of the lunules of *Astriclypeus* as observed in the righting movement (Echinoidea). Annot. Zool. Japon. 20: 78–82.

Irwin, M. C. 1953. Sea urchins damage steel piling. Science 118: 307.

Jensen, M. 1974. The Strongylocentrotidae (Echinoidea), a morphologic and systematic study. Sarsia 57: 113–48.

Johansen, K., and R. L. Vadas. 1967. Oxygen uptake and responses to respiratory stress in sea urchins. Biol. Bull. 132: 16–22.

Johnson, M. E., and H. J. Snook. 1935. Seashore animals of the Pacific coast. New York: Macmillan. 659 pp.

Johnson, M. W. 1930. Notes on the larval development of *Strongylocentrotus franciscanus*. Publ. Puget Sound Biol. Sta. 7: 401–11.

Kenk, R. 1944. Ecological observations on two Puerto-Rican echinoderms, *Mellita lata* and *Astropecten marguinatus*. Biol. Bull. 87: 177–87.

Lasker, R., and R. A. Boolootian. 1960. Digestion of the alga, *Macrocystis pyrifera*, by the sea urchin, *Strongylocentrotus purpuratus*. Nature 188: 1130.

Lasker, R., and A. C. Giese. 1954. Nutrition of the sea urchin, *Strongylocentrotus purpuratus*. Biol. Bull. 106: 328–40.

Lawrence, J. M. 1975. On the relationships between marine plants and sea urchins. Oceanogr. Mar. Biol. Ann. Rev. 13: 213–86.

———. 1976. Covering response in sea urchins. Nature 262: 490–91.

Lawrence, J. M., A. L. Lawrence, and A. C. Giese. 1966. Role of the gut as a nutrient-storage organ in the purple sea urchin (*Strongylocentrotus purpuratus*). Physiol. Zool. 39: 281–90.

Lawrence, J. M., A. L. Lawrence, and N. D. Holland. 1965. Annual cycle in the size of the gut of the purple sea urchin, *Strongylocentrotus purpuratus* (Stimpson). Nature 205: 1238–39.

Lehman, H. E. 1946. A histological study of *Syndisyrinx franciscanus*, gen. et sp. nov., an endoparasitic rhabdocoel of the sea urchin, *Strongylocentrotus franciscanus*. Biol. Bull. 91: 295–311.

Leighton, D. L. 1966. Studies of food preference in algivorous invertebrates of southern California kelp beds. Pacific Sci. 20: 104–13.

Leighton, D. L., L. G. Jones, and W. J. North. 1966. Ecological relationships between giant kelp and sea urchins in southern California, pp. 141–53, *in* E. G. Young and J. L. McLachlan, eds., Proc. 5th Internat. Seaweed Symp. New York: Pergamon. 424 pp.

Lowry, L. F., and J. S. Pearse. 1973. Abalones and sea urchins in an area inhabited by sea otters. Mar. Biol. 23: 213–19.

Lynch, J. E. 1929. Studies on the ciliates from the intestine of *Strongylocentrotus*. I. *Entorhipidium* gen. nov. Univ. Calif. Publ. Zool. 33: 27–56.

_____. 1930. Studies on the ciliates from the intestine of *Strongylocentrotus*. II. *Lechriopyla mystax*, gen. nov., sp. nov. Univ. Calif. Publ. Zool. 33: 307–50.

Lyons, R. B. 1960. Antigenic anatomy of the tissues of the sea urchin *Strongylocentrotus purpuratus*. Master's thesis, Anatomy, University of Oregon, Portland.

Lyons, R. B., W. R. Bishop, and R. L. Bacon. 1964. Fine structure of the scalloped, flagellated cup cells of the sea urchin *Strongylocentrotus purpuratus*. J. Cell Biol. 23: 55A–56A.

MacGinitie, G. E., and N. MacGinitie. 1968. Natural history of marine animals. 2nd ed. New York: McGraw-Hill. 523 pp.

Mattison, J. E., J. D. Trent, A. L. Shanks, T. B. Akin, and J. S. Pearse. 1977. Movement and feeding activity of red sea urchins (*Strongylocentrotus franciscanus*) adjacent to a kelp forest. Mar. Biol. 39: 25–30.

Mauzey, K. P., C. Birkeland, and P. K. Dayton. 1968. Feeding behavior of asteroids and escape responses of their prey in the Puget Sound region. Ecology 49: 603–19.

Mazia, D., and K. Dan. 1952. The isolation and biochemical characterization of the mitotic apparatus of dividing cells. Proc. Nat. Acad. Sci. 38: 826–38.

McAllister, R. 1976. California marine fish landings for 1974. Calif. Dept. Fish & Game, Fish Bull. 166: 1–53.

McCauley, J. E., and A. G. Carey, Jr. 1967. Echinoidea of Oregon. J. Fish. Res. Bd. Canada 24: 1385–1401.

McCloskey, L. R. 1970. A new species of *Dulichia* (Amphipoda, Podoceridae) commensal with a sea urchin. Pacific Sci. 24: 90–98.

McPherson, B. F. 1968a. Contributions to the biology of the sea urchin *Eucidaris tribuloides* (Lamarck). Bull. Mar. Sci. 18: 400–443.

_____. 1968b. Feeding and oxygen uptake of the tropical sea urchin *Eucidaris tribuloides* (Lamarck). Biol. Bull. 135: 308–21.

Mendes, E. G., L. Abud, and S. Umiji. 1963. Cholinergic action of homogenates of sea urchin pedicellariae. Science 139: 408–9.

Merrill, R. J., and E. S. Hobson. 1970. Field observations of *Dendraster excentricus*, a sand dollar of western North America. Amer. Midl. Natur. 83: 595–624.

Millott, N., ed. 1967. Echinoderm biology. Symp. Zool. Soc. London 20. London: Academic Press. 240 pp.

Moore, A. R. 1941. On the mechanics of gastrulation in *Dendraster excentricus*. J. Exper. Zool. 87: 101–11.

_____. 1943. Maternal and paternal inheritance in the plutei of hybrids of the sea urchins *Strongylocentrotus purpuratus* and *Strongylocentrotus franciscanus*. J. Exper. Zool. 94: 211–28.

_____. 1957. Biparental inheritance in an interordinal cross of sea urchin and sand dollar. J. Exper. Zoo. 135: 75–83.

Moore, H. B. 1936. The biology of *Echinocardium cordatum*. J. Mar. Biol. Assoc. U.K. 20: 655–72.

Moore, H. B., and N. N. Lopez. 1966. The ecology and productivity of *Moira atropos* (Lamarck). Bull. Mar. Sci. 16: 648–67.

Moore, H. B., and B. F. McPherson. 1965. A contribution to the study of the productivity of the urchins *Tripneustes esculentus* and *Lytechinus variegatus*. Bull. Mar. Sci. 15: 855–71.

Moore, H. B., T. Jutare, J. C. Bauer, and J. A. Jones. 1963. The biology of *Lytechinus variegatus*. Bull. Mar. Sci. 13: 23–53.

Mortensen, T. 1921. Studies of the development and larval forms of echinoderms. Copenhagen: Gad. 261 pp., 33 pls.

_____. 1928–51. Monograph of the Echinoidea. 5 vols., 17 parts. Copenhagen: Reitzel.

Muchmore, D., and D. Epel. 1973. The effects of chlorination of wastewater on fertilization in some marine invertebrates. Mar. Biol. 19: 93–95.

Nichols, D. 1959. Mode of life and taxonomy in irregular sea-urchins, pp. 61–80, *in* A. J. Cain, ed., Function and taxonomic importance. London: Systematics. 140 pp.

_____. 1962. Differential selection in populations of a heart urchin, pp. 105–18, *in* D. Nichols, ed., Taxonomy and geography. London: Systematics. 158 pp.

_____. 1964. Echinoderms: Experimental and ecological. Oceanogr. Mar. Biol. Ann. Rev. 2: 393–423.

_____. 1969. Echinoderms. 4th ed. London: Hutchinson University Library. 192 pp.

Niesen, T. M. 1977. Reproductive cycles of two populations of the Pacific sand dollar *Dendraster excentricus*. Mar. Biol. 42: 365–73.

Nissen, H. U. 1969. Crystal orientation and plate structure in echinoid skeletal units. Science 166: 1150–52.

North, W. J., ed. 1964. An investigation of the effects of discharged wastes on kelp. Resources Agency of California, Bd., State Water Qual. Control Publ. 26: 1–124.

North, W. J., and J. S. Pearse. 1970. Sea urchin population explosion in southern California coastal waters. Science 167: 209.

Oliphant, M. S. 1973. California marine fish landings for 1971. Calif. Dept. Fish & Game, Fish Bull. 159: 1–49.

Orcutt, H. G. 1950. The life history of the starry flounder *Platichthys stellatus* (Pallas). Calif. Dept. Fish & Game, Fish Bull. 78: 1–64.

Parker, G. H. 1927. Locomotion and righting movements in echinoderms, especially in *Echinarachnius*. Amer. J. Psychol. 39: 167–80.

Parker, G. H., and M. Van Alstyne. 1932. Locomotor organs of *Echinarachnius parma*. Biol. Bull. 62: 195–200.

Pearse, J. S. 1969. Reproductive periodicities of Indo-Pacific invertebrates in the Gulf of Suez. I. The echinoids *Prionocidaris baculosa* (Lamarck) and *Lovenia elongata* (Gray). Bull. Mar. Sci. 19: 323–50.

_____. 1975. Class Echinoidea, pp. 621–22, *in* R. I. Smith and J. T. Carlton, eds., Light's manual: Intertidal invertebrates of the central California coast. 3rd ed. Berkeley and Los Angeles: University of California Press. 716 pp.

Pearse, J. S., and V. B. Pearse. 1975. Growth zones in the echinoid skeleton. Amer. Zool. 15: 731–53.

Pequignat, E. 1966. "Skin digestion" and epidermal absorption in irregular and regular sea urchins and their probable relation to the outflow of spherule-coelomocytes. Nature 210: 397–99.

Phelan, T. F. 1977. Comments on the water vascular system, food grooves, and ancestry of the clypeasteroid echinoids. Bull. Mar. Sci. 27: 400–422.

Pilkey, O. H., and J. Hower. 1960. The effect of environment on the concentration of skeletal magnesium and strontium in *Dendraster*. J. Geol. 68: 203–16.

Quast, J. C. 1968. Observations on the food of the kelp-bed fishes, pp. 109–42, *in* W. J. North and C. L. Hubbs, eds., Utilization of kelp-bed resources in southern California. Calif. Dept. Fish & Game, Fish Bull. 139: 1–264.

Rappaport, R., Jr. 1960. Cleavage of sand dollar eggs under constant tensile stress. J. Exper. Zool. 144: 225–31.

Raup, D. M. 1958. The relation between water temperature and morphology in *Dendraster*. J. Geol. 66: 668–77.

———. 1962. Crystallographic data in echinoderm classification. Syst. Zool. 11: 97–108.

———. 1966. The endoskeleton, pp. 379–95, *in* Boolootian (1966).

Reisman, A. W. 1965. The histology and anatomy of the intestinal tract of *Dendraster excentricus*, a clypeasteroid echinoid. Master's thesis, Zoology, University of California, Los Angeles.

Ricketts, E. F., and J. Calvin. 1968. Between Pacific tides. 4th ed. Revised by J. W. Hedgpeth. Stanford, Calif.: Stanford University Press. 614 pp.

Rosenthal, R. J., and J. R. Chess. 1970. Predation on the purple urchin by the leather star. Calif. Fish & Game 56: 203–4.

———. 1972. A predator-prey relationship between the leather star, *Dermasterias imbricata*, and the purple urchin, *Strongylocentrotus purpuratus*. Fish. Bull. 70: 205–16.

Salsman, G. G., and W. H. Tolbert. 1965. Observations on the sand dollar, *Mellita quinquiesperforata*. Limnol. Oceanogr. 10: 152–55.

Sandeman, D. C. 1965. Electrical activity in the radial nerve cord and ampullae of sea urchins. J. Exper. Biol. 43: 247–56.

Schatten, G. P., and D. Mazia. 1976. The surface events at fertilization: The movements of the spermatozoon through the sea urchin egg surface and the roles of the surface layers. J. Supramolec. Struc. 5: 343–69.

Smith, J. E. 1965. Echinodermata, pp. 1519–58, *in* T. H. Bullock and G. A. Horridge, Structure and function in the nervous systems of invertebrates, vol. 2. San Francisco: Freeman. 1,719 pp.

Smith, L. S. 1961. Clam-digging behavior in the starfish, *Pisaster brevispinus* (Stimpson, 1857). Behaviour 18: 148–53.

Speer, D. P., and F. Herzberg. 1961. Dental methods useful in the study of invertebrate animals with shells. J. So. Calif. State Dental Assoc. 29: 127–33.

Steinbeck, J., and E. F. Ricketts. 1941. Sea of Cortez. New York: Viking. 598 pp.

Stohler, R. 1930. Gewichtsverhältnisse bei gewissen marinen Evertebraten. Zool. Anz. 91: 149–55.

Strachan, A. R., C. H. Turner, and C. T. Mitchell. 1968. Two fishes and a mollusk, new to California's marine fauna, with comments regarding other recent anomalous occurrences. Calif. Fish & Game 54: 49–57.

Strathmann, R. R. 1971. The feeding behavior of planktotrophic echinoderm larvae: Mechanisms, regulation, and rates of suspension feeding. J. Exper. Mar. Biol. Ecol. 6: 109–60.

———. 1974 (publ. 1976). Introduction to function and adaptation in echinoderm larvae. Thalassia Jugoslavica 10: 321–39.

———. 1975. Larval feeding in echinoderms. Amer. Zool. 15: 717–30.

———. 1978. Length of pelagic period in echinoderms with feeding larvae from the northeast Pacific. J. Exper. Mar. Biol. Ecol. 34: 23–27.

Swan, E. F. 1952. Regeneration of spines by sea urchins of the genus *Strongylocentrotus*. Growth 16: 27–35.

———. 1953. The Strongylocentrotidae (Echinoidea) of the northeast Pacific. Evolution 7: 269–73.

———. 1961. Some observations on the growth rate of sea urchins in the genus *Strongylocentrotus*. Biol. Bull. 120: 420–27.

Taki, J. 1978. Formation of growth lines in test plates of the sea urchin, *Strongylocentrotus intermedius*, reared with different algae. Bull. Jap. Soc. Sci. Fish. 44: 955–60.

Tegner, M. J., and P. K. Dayton. 1977. Sea urchin recruitment patterns and implications of commercial fishing. Science 196: 324–26.

Tegner, M. J., and D. Epel. 1972. Sea urchin sperm-egg interaction studied with the scanning electron microscope. Science 179: 685–88.

———. 1976. Scanning electron microscope studies of sea urchin fertilization. I. Eggs with vitelline layers. J. Exper. Zool. 197: 31–58.

Thompson, T. G., and T. J. Chow. 1955. The strontium-calcium atom ratio in carbonate-secreting marine organisms. Pap. Mar. Biol. Oceanogr., Deep-Sea Res. 3 (Suppl.): 20–39.

Timko, P. L. 1975. High density aggregation in *Dendraster excentricus*: Analysis of strategies and benefits concerning growth, age structure, feeding, hydrodynamics, and reproduction. Doctoral thesis, Zoology, University of California, Los Angeles. 323 pp.

———. 1976. Sand dollars as suspension feeders: A new description of feeding in *Dendraster excentricus*. Biol. Bull. 151: 247–59.

Timourian, H., and G. Watchmaker. 1975. The sea urchin blastula: Extent of cellular determination. Amer. Zool. 15: 607–27.

Towe, K. M. 1967. Echinoderm calcite: Single crystal or polycrystalline aggregate. Science 157: 1048–50.

Turner, C. H., E. E. Ebert, and R. R. Given. 1968. The marine environment offshore from Point Loma, San Diego County. Calif. Dept. Fish & Game, Fish Bull. 140: 1–85.

Tyler, A., and B. S. Tyler. 1966. The gametes; some procedures and properties, pp. 639–82, *in* Boolootian (1966).

Ulbricht, R. J. 1973. Effect of temperature acclimation on the metabolic rate of sea urchins. Mar. Biol. 19: 273–77.

Ulbricht, R. J., and A. W. Pritchard. 1972. Effect of temperature on the metabolic rate of sea urchins. Biol. Bull. 142: 178–85.

Vacquier, V. D. 1971. The appearance of beta-1,3-glucanohydrolase activity during the differentiation of the gut of sand dollar plutei. Develop. Biol. 26: 1–10.

————. 1975. The isolation of intact cortical granules from sea urchin eggs: Calcium ions trigger granule discharge. Develop. Biol. 43: 62–74.

Vacquier, V. D., M. J. Tegner, and D. Epel. 1973. Protease released from sea urchin eggs at fertilization alters the vitelline layer and aids in preventing polyspermy. Exper. Cell. Res. 80: 111–19.

Weber, J. N., N. T. Greer, B. Voight, E. White, and R. Roy. 1969. Unusual strength properties of echinoderm calcite related to structure. J. Ultrastruc. Res. 26: 355–66.

Webster, S. K. 1975. Oxygen consumption in echinoderms from several geographical locations, with particular reference to the Echinoidea. Biol. Bull. 148: 157–64.

Webster, S. K., and A. C. Giese. 1975. Oxygen consumption of the purple sea urchin with special reference to the reproductive cycle. Biol. Bull. 148: 165–80.

Weese, A. O. 1926. Food and digestive processes of *Strongylocentrotus dröbachiensis*. Publ. Puget Sound Biol. Sta. 5: 165–79.

Went, H. A., and D. Mazia. 1959. Immunochemical study of the origin of the mitotic apparatus. Exper. Cell. Res. 7 (Suppl.): 200–218.

Whiteley, H. R., S. Mizuno, Y. R. Lee, and A. H. Whiteley. 1975. Transcripts of reiterated DNA sequences in the determination of blastomeres and early differentiation in echinoid larvae. Amer. Zool. 15: 629–48.

Chapter 12
Urochordata: *The Tunicates*

Donald P. Abbott and Andrew Todd Newberry

Of all the animal groups treated in this book, these are our own closest relatives. In their portraits, they show few outward signs of the relationship, but they share with the vertebrates the features diagnostic for the phylum Chordata: they have a dorsal tubular nerve cord, the swollen anterior end of which forms a simple brain; they possess a skeletal supporting rod, the notochord, lying ventral to the nerve cord; the pharynx develops gill slits on each side; and along the pharyngeal floor runs a glandular groove, the endostyle, considered to be the forerunner of the vertebrate thyroid gland. Some of these characters, as in ourselves, are seen clearly only during embryonic stages. Other shared features are even less apparent. Whatever our common heritage with the tunicates, a wide gulf separates us today, for the tunicates, with few exceptions, are animals deeply committed to a life of filter feeding in the sea, of passing a current of water through the body and removing from it planktonic food and other materials essential to life. Roughly 2,000 living specces of tunicates are known. Both benthic and pelagic species occur in all oceans, and a few benthic forms tolerate brackish conditions.

The tunicates covered here are all members of the class Ascidiacea, the ascidians or sea squirts, the group of urochordates most familiar to visitors of the shore. Virtually all are sessile in the adult stages, and most of those pictured here live in low and shaded areas where exposure to sunlight and to heating and desiccation at low tide are minimal. The animals flourish best in regions free from strong wave shock yet bathed in freely flowing, clean, cold sea-

Donald P. Abbott is Professor, Department of Biological Sciences and Hopkins Marine Station, Stanford University. Andrew Todd Newberry is Professor of Biology at Cowell College of the University of California, Santa Cruz.

water. Though ascidian species differ in their tolerance limits, their requirements overlap broadly, and in certain situations, as in rocky tunnels and small sheltered caves, or on rocks swept by currents in the surfgrass (*Phyllospadix*) zone, one may run across "tunicate heavens," where as many as a dozen species grow tightly intermingled.

Ascidians, like ourselves, develop from fertilized eggs. Their embryology, clearly presented by Brien (1948) and Hirai (1968), is grossly similar in many ways to that of fishes and frogs, and generally results in tiny, swimming, tadpole larvae. These vary in size and shape, but most share certain features. They never feed, but live on stored yolk. They swim actively by means of a tail supported by the notochord and flexed by a pair of muscle bands. The muscle cells of the tail contain striated fibrils intimately associated with those of adjacent cells (Berrill & Sheldon, 1964; Cavey & Cloney, 1972, 1974). The tail generally twists slightly as it beats, giving the tadpole a tumbling, barrel-rolling course. The larval brain contains sensory structures, usually a multicellular light receptor (ocellus) and a one-celled statocyst (statocyte or otocyst) by which presumably the animal tells up from down. Anteriorly, the tadpole has adhesive organs, by which it can attach firmly and irrevocably to a submerged substratum.

For the motile larvae of sessile adults, final attachment is a critical act, for if the larva does not settle in a place favorable for the rest of the life span, all is lost. The behavior and responses of the larva usually change markedly as the time for settlement approaches. Experimental studies are few, but they suggest that the larval sensory and action systems are geared to picking a suitable spot; probably light, chemical stimuli, and tactile cues all play roles. Given the opportunity, tadpoles often settle close to adults of the same species, which, by their very presence, indicate that this is a good spot to grow up.

Metamorphosis follows immediately after attachment, and as it proceeds it becomes apparent that the larva consists of two distinct parts. One part is the tadpole body itself, with its larval sensory and motor equipment, the other is the rudiment of the future adult within, which has thus far remained inactive. In some ascidians (e.g., *Distaplia,* 12.21, 12.22), the adult is already well formed in miniature; in others (e.g., *Ciona,* 12.23; *Styela,* 12.33–36; *Boltenia,* 12.39), organogenesis of the adult has scarcely started. Metamorphosis reviewed by Cloney (1978), involves first a withdrawal of all the tail structures into the cavity of the trunk; no longer needed for locomotion, they will undergo autolysis and be used as food. Then the body of the tiny developing adult slowly rotates so that the siphons by which the animal will feed and respire are pointed away from the substratum. The pharynx enlarges steadily until it becomes the largest single organ in the body; its original gill slits divide again and again until the pharyngeal wall is a veritable sieve, pierced by numerous small holes or stigmata. The delicate pharynx lies in a protected space, the atrium, which is enclosed by a double fold of the outer body wall.

The gut further differentiates to form esophagus, stomach, and intestine, and the latter curves on itself to terminate beside the pharynx in the atrium.

A heart develops and links up with the circulatory system, which is formed partly of closed vessels (in pharynx and test) and partly of open passageways in the softer tissues. The heartbeat is extraordinary in tunicates; every few minutes the heart pauses briefly, then reverses the direction of its beat; thus the blood vessels of the whole body serve alternately as arteries and veins. Neither the mechanism of reversal nor its advantages are entirely clear (see Goodbody, 1974; Heron, 1975).

As the animal grows, it continues to secrete outside of its skin a "tunic" or "test," varying in character with the species but always in essence an external connective tissue, formed by the skin and containing living cells that have crept out through the epidermis. The tunic contains both proteins and carbohydrates. Among the latter are polysaccharide fibrils that are very similar, in some ascidians, to plant cellulose.

Soon the juvenile ascidian opens its apertures to the surrounding water and begins to feed. The stigmata perforating the pharyngeal wall are bordered by cilia that beat outward, moving water from pharynx to atrium. As a result, a steady stream of water is drawn into the mouth (where a coarse screen of tentacles prevents the entrance of larger objects or animals), and a corresponding stream exits from the atrium via the atrial siphon.

In the pharynx a midventral groove, the endostyle, secretes two sheets of porous mucus, one on each side. Frontal cilia on the pharyngeal wall move these sheets dorsally, while the pressure of water flow keeps them pressed against the inner pharyngeal surface. Water flows freely through the moving mucous sheets, but particles adhere. At the dorsal midline of the pharynx, ciliated tentacles or a ciliated membrane roll the food-laden mucus into a rope and pass it posteriorly into the esophagus and stomach for digestion. Excellent descriptions of feeding are presented in Jørgensen (1966), MacGinitie (1939a), and Werner & Werner (1954). Waste materials expelled from the intestine to the atrium are whisked away by the exhalant water current. When disturbed, the ascidian contracts its whole body, spurting water from one or both apertures; such body contractions also occur from time to time even without obvious disturbance, discharging any accumulated debris from both pharynx and atrium.

Many species of ascidians remain solitary, each settling tadpole giving rise to a single adult individual. These are the "simple" ascidians. But in many other species the young animal derived from a settled tadpole undergoes an asexual multiplication, or budding, forming a colony or clone of genetically identical individuals (zooids).

Colony formation here is closely related to the ability to regenerate lost parts, e.g., replacement of a lost arm by a starfish. In many species, regenerative ability is so good that even small parts separated from the body can develop into whole new individuals. All that such species need, to progress from regeneration to asexual multiplication, is a process by which the body regularly buds off pieces of itself, or undergoes fission into two or more parts. Asexual reproduction of this sort is widespread in simpler animals, but in the phylum Chordata it is limited very largely to the tunicates; where cloning occurs among the vertebrates, it is usually due to polyembryony (in which the very early embryo subdivides to produce two or more identical individuals, e.g., identical twins in man, identical quadruplets in armadillos) or to development of eggs without fertilization by a sperm (parthenogenesis, gynogenesis), as in some fishes and reptiles.

In colonial tunicates the zooids of a clone, each a complete individual, remain joined together in some manner. In compound (composite) ascidians the zooids of a colony are fully embedded in a common mass of tunic material. In some compound species the zooids are scattered more or less evenly through the common tunic, and the oral and atrial apertures of all zooids protrude at the test surface. In other compound species the zooids are organized into groups (systems) in which each zooid has its mouth opening at the test surface; its atrial siphon empties into an enclosed chamber in the test (common cloacal cavity), which in turn has a single opening to the exterior. In contrast to the compound ascidians (but intergrading with them) are the "social" ascidians, in which the individual zooids are largely separated from each other, but remain joined at their bases by stolons or by a thin, vascularized sheet of common test carpeting the substratum.

Colony formation represents an alternative strategy of growth. In most animals, including simple ascidians, growth involves the enlargement of a single body. However, in colonial ascidians, as in forms such as bryozoans, growth is accompanied by repeated subdivision, so that, instead of one large body, numerous small, genetically identical bodies are produced. Each zooid in the ascidian colony feeds, produces gametes, and reproduces asexually, largely on its own, though more or less in coordination with its clonemates. The zooids may be joined only by the common tunic, or they may be linked by a network of blood vessels running within the tunic.

Colony formation has certain advantages. For one thing, a colony is normally less vulnerable to destruction than a single large individual; consumption of even nine-tenths of the colony by a predator scarcely disturbs the remaining zooids. For another thing, the colonial pattern of growth may help the species secure living space, always at a premium on the low intertidal substrata. The colony may expand laterally, pushing aside or overgrowing and smothering neighboring creatures; alternatively, the colony may extend upward from a small attachment to form a skyscraper full of zooids that towers above the neighbors. Feeding advantages, if any, are hard to estimate, but a thousand small mouths seem to do at least as well as one large mouth. There may be a reproductive advantage, quite aside from the greater chance the colony has of surviv-

ing to sexual maturity; since each zooid in the colony bears gonads and produces gametes in the breeding season, the larger the clone the greater the reproductive potential, and colonies often get larger than most solitary ascidians. Finally, an alternation of sexual and asexual reproductive phases in a life history is of value in an evolutionary sense, in providing a mechanism for more effectively conserving favorable genetic variants. In colonial species, each successful larval settlement is cloned, and the most successful clones grow largest and produce the most offspring. With cloning, the probability of accidental destruction of a favorable genetic strain is decreased. Taken together, the advantages of cloning seem significant. At any rate, not only is cloning widespread in tunicates, it occurs in such varied ways that the animals are suspected of having evolved the colonial way of life more than once, and perhaps several times, from progenitor stocks having great capacity for repair and regeneration. Good accounts and reviews of asexual reproduction in ascidians are those of Berrill (1935, 1951, 1961a) and Brien (1948, 1968, 1972); intriguing experimental studies of budding are found in the papers of Freeman, Mukai, Nakauchi, Sabbadin, Scott, and Watanabe, among others.

Asexual reproduction in ascidians of the suborder Aplousobranchia appears related to another phenomenon—the elongation of the zooid body. In some social ascidians, but more particularly in the compound colonial forms, the zooids often have very long and slender bodies; this allows a large number of them to be packed in side by side, mouths all opening at the surface, without taking up much space on the substratum. In some forms it is the zooid abdomen (containing the loop of the gut posterior to the pharynx) that is greatly elongated (as in *Archidistoma*, 12.15–18; *Pycnoclavella*, 12.14; *Clavelina*, 12.20); in others the abdomen remains relatively short, but the body below the loop of the gut is greatly extended, becoming a postabdomen that contains the gonads and the heart (as, for example, in *Aplidium*, 12.1–4; *Synoicum*, 12.6; *Ritterella*, 12.7–9). A slice through a fresh compound ascidian colony, made parallel to the long axes of the zooids, usually reveals the general arrangement, but anatomical details of the zooids are best seen if the colony is first relaxed in an isotonic solution of magnesium chloride and then preserved in 10 percent formalin.

In colonial species with elongate zooids, asexual reproduction most often takes place by strobilation, that is, by a series of transverse epidermal constrictions that slice the long body (abdomen, postabdomen, or both regions) into a linear series of short sections, each of which develops into a new complete zooid.

All ascidians, whether solitary or colonial, reproduce sexually, usually according to some seasonal pattern. All species are probably hermaphroditic; statements to the contrary should be distrusted unless accompanied by fresh evidence. However, in some species the male gonads develop first and decline markedly as the ovaries arise (e.g., *Distaplia smithi*, 12.22). Self-fertilization is probably uncommon. Some species show a high rate of infertility when the eggs are exposed to sperm from the same individual or colony (e.g., papers of T. H. Morgan; Rosati & De Santis, 1978). Other species show abnormal development following experimental self-fertilization. In a few species the eggs produce materials that attract the sperm (Miller, 1975).

Many simple ascidians discharge both eggs and sperm to the sea, where fertilization and subsequent development occur. However, some simple and most colonial species retain their eggs, and after fertilization brood them—usually in the atrial cavity beside the pharynx, sometimes in a special brood pouch (e.g., *Distaplia*). When the eggs are brooded, the tadpoles are often very large, with the rudiment of the adult ascidian inside fairly well developed; such tadpoles often settle quickly, and have free-swimming periods measured in minutes or a few hours.

From the standpoint of the chemist, tunicates have some remarkable attributes, some of which are reviewed by Barrington (1974) and Goodbody (1974). The production of cellulose and closely allied polysaccharides (see above) is not widespread in metazoans. However, it is not unique to ascidians; cellulose fibers show up sparingly in old beef and even in old human tissues (Hall & Saxl, 1960, 1961). Some tunicates concentrate unusual metals from seawater, such as vanadium, niobium, titanium, and chromium, along with iron and other less esoteric materials (see Carlson, 1975; Ciereszko et al., 1963; Danskin, 1978; Goodbody, 1974; Swinehart et al., 1974). Their function remains uncertain. A remarkable variety of sterols occurs in ascidians (Carlson, 1977; Tan, 1975; Yasuda, 1975), but cholesterol and cholestanol appear to be the most abundant and widespread. Much work has been done surveying the phosphagens of tunicates, vertebrates, and related groups (see especially Watts, 1975). The phenomenon of iodine-binding in the endostyle and tunic, and reports of the presence of thyroxine in ascidians, have aroused great interest; developments here are reviewed by Barrington (1965, 1974, 1975), Goodbody (1974), and Thorpe & Thorndyke (1975). Chitin fibers (Peters, 1966) occur along with iodinated proteins in the peritrophic membranes secreted around food in the gut. Some ascidians produce materials with antibacterial, antifungal, antitumor, and even antiviral activity (see, for example, Carter & Rinehart, 1978; Cheng & Rinehart, 1978).

"What good are tunicates?" some may ask, thinking of direct and immediate human benefits. Their direct economic importance is small. Half a dozen large species are taken as food for man in the Orient, Chile, the Mediterranean, and the Arctic, but only in Japan is there a systematic attempt to culture a species for food. There the large solitary *Halocynthia roretzi*, called the "hoya," is cultured on ropes or twisted vines suspended from floats in bays (see Plate XI in Ooishi & Illg, 1974). In nature the ascidians are eaten by numerous predators besides man, including the sea stars (Mauzey, Birkeland & Dayton, 1968), various gastropods, and even some inshore fishes (see, for example, Quast, 1968; Russo, 1975), but they are not important items in the food chains leading directly to man. Ascidians do often occur in vast populations on floats and pilings in harbors, where, as filter feeders, they consume particles of organic detritus and help keep the water clean. Because of their ability to concentrate certain materials, they may be useful indicators of industrial pollution (see, for example, Papadopoulou & Kanias, 1977). And, as noted earlier, extracts of some ascidians have shown antibacterial or antiviral properties, while substances from other forms have properties that suggest possible use in treating cancer or leukemia. The results so far, however, are strictly preliminary. On the other hand, tunicates can be a nuisance as fouling organisms on ships and docks (see, for example, Haderlie, 1974; Millar, 1971a; Scheer, 1945), and several species are blamed for an allergenic asthma suffered by oyster shuckers in Japan (see *Styela clava*, 12.34, and *S. plicata*, 12.36).

Tunicates have long provided embryologists and cell biologists with materials for investigation, not only of ascidian embryology as such, but of fundamental problems in development and differentiation. The foundation for precise work was established in such classical descriptive accounts as those of Castle (1896) on *Ciona*, and Conklin (1905) on *Styela*. More recent experimental work in development is reviewed by Reverberi (1961, 1971).

The fact that some tunicates produce clones makes it possible to conduct repeated experiments on populations of individuals that are genetically identical; the researcher can rule out genetic differences as a source of undesired variation, or can investigate the separate roles of nature and nurture on development. Further, the fact that in colonial ascidians the development of the egg and the development of the bud follow rather different pathways but lead to very similar end results is a phenomenon that cries out for investigation. The study of tunicate genetics, though still in its infancy, has made a promising start (see, for example, Mukai & Watanabe, 1975a,b; Sabbadin, 1964, 1969, 1973; Sabbadin & Graziani, 1967b), and chromosome counts are available for a number of species (Colombera, 1973, 1974a,b,c).

Our concern so far has been with the sessile filter-feeding ascidians common along our shores, but adaptive radiation in the tunicates has produced some oddly different forms. Among the ascidians on European coasts are a number of tiny species that live in the interstices between grains of sand and gravel. Some have elongate bodies and, being unattached, squirm about in almost worm-like fashion (F. Monniot, 1965). In the deep sea are ascidians whose pharynges are no longer adapted to filtering. Instead, in some the oral siphon is greatly expanded into large lips or petal-like lobes that somehow trap small food; in others the oral siphon has developed into an elongated muscular grasping organ rather like an elephant's trunk, with distal fingers for handling larger animal food (Millar, 1959, 1970a; C. Monniot & F. Monniot, 1975, 1978). A new class of tunicates, the Sorberacea, was recently established for the latter forms (C. Monniot, F. Monniot & Gaill, 1975).

Two other classes of tunicates, the Thaliacea and Larvacea, are planktonic forms, exploiting the open sea as a habitat. The thaliacean *Pyrosoma* is brilliantly luminescent and forms tubular swimming colonies, in California waters seldom larger than an ear of corn, but elsewhere occasionally nearly a meter in diameter and more than 9 m long (Grace, 1971). Smaller thaliaceans of several species are abundant from time to time and place to place in the California Current system, and occasionally come inshore in great numbers. The salps appear to be ecological opportunists in which asexual reproductive processes are geared to produce huge numbers of offspring very rapidly when conditions for salps are favorable (see, for example, Berrill, 1961b; Silver, 1975). Members of the class Larvacea (see Alldredge, 1976) are always present in coastal plankton hauls and may be found by microscopic examination.

The question of the origin of the vertebrates, and of the chordates themselves, has intrigued zoologists for more than a century. Many facts of anatomy, embryology, and biochemistry suggest that echinoderms, hemichordates, chordates, and possibly other groups form together a natural branch of the animal kingdom, but the question of who begat whom is still conjectural. Berrill (1955) has argued that the tunicates were the progenitor stock, at least for the vertebrates. Fossil tunicates are very scarce, but remains reminiscent of the modern ascidian *Botryllus* appear in the upper Cambrian (Müller, 1977). Some other investigators see vertebrates as arising from fossil stocks regarded by most paleontologists as early echinoderms (Eaton, 1970; Gislén, 1930; Jefferies, 1975). There are other views as well; see also Barrington (1965), Bone (1972), W. Garstang (1928), Godeaux (1974), Gutmann (1975), Jollie (1973), Makioka, Hirabayashi & Watanabe (1978), Schmidtke et al. (1977), Tokioka (1971), and contributors in Barrington and Jefferies (1975).

A great deal remains to be done on the systematics and biology of west coast ascidians. Those interested should first consult general works on the group, such as Berrill (1950), Brien (1948), Godeaux (1974), and Huus (1937, old but useful), and two outstanding recent reviews, Millar (1971b) and Goodbody (1974), which together cover much of the natural history and physiology.

Useful general works on taxonomy, providing outlines of the higher classification of ascidians, include Berrill (1950), Kott (1969), Plough (1978), Tokioka (1953), and Van Name (1945). C. Monniot & F. Monniot (1972) provides an excellent tabular key to all existing genera of ascidians generally recognized at that date. And classification in relation to ascidian evolution is considered by Berrill (1936), Kott (1969), Millar (1966a), and Tokioka (1971).

For further help in identifying California ascidians, the most useful references are Abbott (1975), Van Name (1945), and Ritter & Forsyth (1917). Fay & Johnson (1971) discusses distribution of ascidian species in southern California, and Tokioka (1963) compares the ascidian faunas of Japan and the American Pacific coast.

Phylum Chordata / Subphylum Urochordata / Class Ascidiacea
Order Enterogona / Suborder Aplousobranchia
Family Polyclinidae (= Synoicidae)

12.1 **Aplidium californicum** (Ritter & Forsyth, 1917)
(=Amaroucium californicum)

Common on shaded rock surfaces protected from strong surf, lower middle intertidal zone to 30 m depth, occasionally reaching 85 m; Vancouver Island (British Columbia) to La Paz (Baja California); records from Alaska and Galápagos Islands unconfirmed.

Colonies forming encrusting sheets and slabs commonly 15 cm across and 1 cm thick, occasionally to 30 cm across and 3 cm thick; colony surface smooth to irregular, usually not sand-encrusted; tunic gray, yellowish, opalescent white, or transparent; zooids usually yellowish, orange, or reddish brown, arranged in clusters (systems); postabdomen long and slender; stomach with longitudinal ridges; pharynx with 7–15 (usually 8–12) rows of stigmata.

This is the commonest colonial ascidian on semiprotected rocky shores, giving way to *Archidistoma ritteri* (12.15) where wave action is stronger. The population is widespread and highly variable, at one extreme approaching *A. multiplicatum* of Asian waters, and is overdue for more detailed biological study. Zooids are easily revealed in a slice taken through a colony. Their elongate bodies consist from front to rear of three regions: a thorax, bearing the mouth, pharynx, and atrial aperture; an abdomen, in which the esophagus, stomach, and intestine are arranged in a loop; and an elongate postabdomen, appearing like a long tail and bearing proximally the ovaries (usually orange or yellow) and the testes (opaque white), and at the posterior tip the heart.

In central California the gonads enlarge in late winter. Sperm are shed into the sea, but the eggs are retained; these mature one at a time, and 2–6 (occasionally up to 11) developing embryos at different stages are brooded simultaneously in a row in the atrial cavity dorsal to the pharynx. Fully developed tadpole larvae emerge one at a time from the atrial pore, swim until a suitable substratum is contacted, then attach, undergo metamorphosis, commence feeding, and reproduce asexually to form new colonies. In old colonies, the gonads regress in the late summer. Food reserves accumulate in the postabdomen, which becomes opaque, detaches from the body, and cuts itself by epidermal constrictions into a linear series of buds. The abdomen may also form buds. Each bud then regenerates to form a new zooid, and the new zooids organize themselves into systems, perhaps guided by an attractant secreted by the mother zooid that produced the buds (see Nakauchi & Kawamura, 1974a,b, 1978). The maximum life span of a colony is not known, but degenerating colonies are often seen in the winter months.

Chemical investigations on *A. californicum* include studies of its sterols (several present, mostly cholesterol and cholestane), its vanadium content (quite low, less than 10 ppm of the ashed weight), and some major organic constituents of the colony (protein 24.0 percent of dry weight, lipids 6.3 percent).

Other species may be found associated with this ascidian. The amphipod *Polycheria osborni* (22.6) is present in elongate depressions in the test in many colonies. These small crustaceans lie on their backs, gripping the margins of the burrow with the claws of their walking legs and pulling the entrance closed when disturbed. At other times the amphipods beat their three pairs of pleopods (paddle-like abdominal appendages), creating a current past the body and gills. The two pairs of bristly antennae appear to be used to capture food carried in by the water current. When artificially removed from the test, the amphipods crawl about clumsily on their backs or swim for short distances. The same is true of juvenile amphipods newly escaped from the brood pouches of their mothers. The parasitic copepod *Pholeterides furtiva* is reported from the test of *A. californicum* in Puget Sound (Washington), and a somewhat similar copepod is found in the bodies of zooids in central California.

The bat starfish, *Patiria miniata* (8.4), is an occasional predator on this ascidian in the Monterey Bay area. In Puget Sound, *Aplidium* (=*Amaroucium*) spp. are sometimes fed upon by the sea stars *Dermasterias imbricata* (8.3), *Mediaster aequalis*, and *Pteraster tesselatus*.

For *A. californicum*, see Abbott (1975), Coe (1932), Coe & Allen (1937), Fay & Johnson (1971), Giese (1966), Illg (1958), Mauzey, Birkeland & Dayton (1968), Reish (1964b), Ritter & Forsyth (1917), Skogsberg & Vansell (1928), Swinehart et al. (1974), Tan (1975),

Tokioka (1953, 1967), Trason (1959), Turner, Ebert & Given (1968, 1969), and Van Name (1945). For other species, see especially Nakauchi (1966a, 1970) and Nakauchi & Kawamura (1974a,b, 1978); see also references under *A. solidum* (12.2).

Phylum Chordata / Subphylum Urochordata / Class Ascidiacea
Order Enterogona / Suborder Aplousobranchia
Family Polyclinidae (=Synoicidae)

12.2 **Aplidium solidum** (Ritter & Forsyth, 1917)
(=Amaroucium solidum)

Fairly common on rocks and pilings in relatively calm but freely circulating clean water, low intertidal zone to about 40 m depth, most commonly 0–15 m; Vancouver Island (British Columbia) to San Diego.

Small colonies rounded, larger colonies forming massive slabs to 20 cm or more across and 5 cm thick; zooids arranged in clusters (systems); thoracic regions of zooids usually bright red or orange-brown in life; stomach with numerous longitudinal ridges; pharynx with 12–16 (usually 13–15) rows of stigmata.

This species is present the year around in central California, but most abundantly in the spring and summer. Colonies can grow to a diameter of over 7 cm in 3 months. Reproduction has not been followed carefully, but up to ten larvae are brooded at a time in the atrial cavity of a zooid in the spring and summer months in Monterey Bay. Oozooids reared from settled larvae have only four rows of stigmata, but the zooids they produce by budding have seven or eight rows. The oozooid abdomen becomes one large bud, whereas the postabdomen forms two to four smaller buds.

The commensal amphipod *Polycheria osborni* (22.6) rarely occurs in the test. The nudibranch *Hermissenda crassicornis* (14.66) often eats the upper parts of the zooids of *A. solidum* in Monterey harbor.

For *A. solidum*, see Abbott (1975), Fay & Johnson (1971), Haderlie (1968, 1969, 1974), McLean (1962), Millar (1963), Nakauchi (1979), Ritter & Forsyth (1917), Skogsberg & Vansell (1928), Swinehart et al. (1974), Van Name (1945), and Yarnall (1972). For other species, see especially Freeman (1971) and Scott (1945, 1946, 1952, 1954, 1957, 1959, 1962); see also Barnes (1971, 1974), Carter & Rinehart (1978),

Cloney (1966, 1969, 1978), Cloney & Grimm (1970), Colombera (1973), Costello & Henley (1971), Gaill (1972), Gorman, McReynolds & Barnes (1971), C. Grave (1920, 1921), Lash, Cloney & Minor (1973), Lynch (1961), Makioka, Hirabayashi & Watanabe (1978), Nakauchi (1970), Scott & Schuh (1963), Tokioka (1972), Van Name (1945), and references under *A. californicum* (12.1).

Phylum Chordata / Subphylum Urochordata / Class Ascidiacea
Order Enterogona / Suborder Aplousobranchia
Family Polyclinidae (=Synoicidae)

12.3 **Aplidium propinquum** (Van Name, 1945)
(=Amaroucium propinquum)

Low intertidal zone on protected rocky shores and pilings; Puget Sound (Washington) to southern California.

Colonies consisting of clusters of plump, club-shaped, sand-encrusted lobes; individual lobes to about 5 cm long, usually less; stomach of zooid with numerous longitudinal ridges; pharynx with 17–21 rows of stigmata.

This little-known species is common in the Monterey Bay area, but so far has been recorded from few other localities. It needs further study. Gonads are well developed in the summer months. In the Puget Sound area the zooids may be host to the symbiotic copepods *Haplostoma elegans* and *Botryllophilus* sp.

See Abbott (1975), Fay & Johnson (1971), Ooishi & Illg (1977), and Van Name (1945).

Phylum Chordata / Subphylum Urochordata / Class Ascidiacea
Order Enterogona / Suborder Aplousobranchia
Family Polyclinidae (=Synoicidae)

12.4 **Aplidium arenatum** (Van Name, 1945)
(=Amaroucium arenatum)

Low intertidal zone on rocky shores; subtidal to at least 5 m; Puget Sound (Washington) to southern California.

Colonies consisting of clusters of slender, sand-encrusted lobes usually 1–2 cm long, each lobe containing only a few zooids; stomach of zooids with about five longitudinal ridges; pharynx with only five rows of stigmata.

Tadpole larvae are released during the summer in central California.

In the Puget Sound area five species of symbiotic copepods have been found inhabiting zooids of this ascidian: *Haplostoma minutum*, *Haplosaccus elongatus*, and three species of *Haplostomella* (*H. dubia*, *H. distincta*, and *H. oceanica*). The relationship between host and symbionts, and the presence of the copepods in *A. arenatum* in other areas, remain to be determined, as do most other aspects of the biology of this ascidian.

See Abbott (1975), Fay & Johnson (1971), Ooishi & Illg (1977), and Van Name (1945).

Phylum Chordata / Subphylum Urochordata / Class Ascidiacea
Order Enterogona / Suborder Aplousobranchia
Family Polyclinidae (=Synoicidae)

12.5 **Polyclinum planum** (Ritter & Forsyth, 1917)·

Common on rocky shores, usually in regions where water circulates freely but not forcefully, very low intertidal zone to at least 30 m; northern California to Isla San Gerónimo (Baja California).

Larger colonies each consisting of a single, brown, flattened lobe to 20 cm or more across and 4 cm thick but more commonly half that size, attached by a short stalk arising from one margin; zooids arranged in circular systems around common cloacal apertures; stomach smooth-walled; pharynx with 13–17 rows of stigmata.

This is a very distinctive species, excellent for class study. Zooids are easily relaxed by immersing a colony for 2–3 hours in an isotonic solution of magnesium chloride and are then best examined in cross sections of the colony cut with a razor blade or scalpel. The zooids occupy the periphery of the lobe, and the central core is largely tunic material. The body wall is transparent, and details of the large pharynx are especially well seen. The postabdomen is small and much shorter than in most other common members of the family Polyclinidae.

Small colonies of *P. planum* are hemispherical, but take on the distinctive form of the adult by the time they are 4–5 cm

high. Larger colonies are occasionally but not typically pitted with the burrows of the amphipod *Polycheria osborni* (22.6).

The species has received more attention from chemists than from biologists. Its major sterols are cholesterol and cholestane; vanadium is scarce, less than 10 ppm of the ashed weight. Most details of the life history remain to be worked out.

For *P. planum*, see Abbott (1975), Fay & Johnson (1971), McLean (1962), Ritter & Forsyth (1917), Skogsberg & Vansell (1928), Swinehart et al. (1974), Tan (1975), Thompson & Chow (1955), and Van Name (1945). For related species, see Gill (1972) and Krishnan (1975).

Phylum Chordata / Subphylum Urochordata / Class Ascidiacea
Order Enterogona / Suborder Aplousobranchia
Family Polyclinidae (=Synoicidae)

12.6 **Synoicum parfustis** (Ritter & Forsyth, 1917)

Fairly common in some areas, in or near clumps of the surfgrass *Phyllospadix*, very low intertidal zone; subtidal to at least 30 m; northern California to San Diego.

Colonies consisting of few to many lobes, the lobes orange, teardrop-shaped, 4–8 cm long, broad and rounded distally, becoming narrower basally where attached to each other or to the substratum; zooids arranged in systems opening distally on each lobe; stomach smooth, lacking multiple longitudinal ridges; pharynx usually with 14–16 rows of stigmata.

Zooids of this species are excellent for the microscopic study of zooid structure. Lobes from a colony should be immersed for 2–4 hours in an isotonic solution of magnesium chloride ($MgCl_2$); the zooids can then be removed by picking off the surface layer of test at the broad end of a lobe with forceps, then grasping a zooid by its exposed end and withdrawing it slowly and gently from the lobe. The zooids lie parallel to the long axis of the lobe, and a gentle squeeze on the lower portion of the lobe may help eject them. Several zooids can be mounted side by side in isotonic $MgCl_2$ under a single large coverslip and examined using a compound microscope.

Sexual reproduction occurs in the summer; most zooids

then have the postabdomen crowded with white testes, bearing as well a small cluster of red or orange eggs. The eggs mature, and are fertilized, one at a time. Fertilized eggs are retained and passed to the atrial cavity beside the pharynx, where they are brooded until the larvae are ready to swim. As many as ten larvae at various stages of development may be brooded simultaneously by one zooid.

Vanadium content of the colony is low, less than 10 ppm of the ashed weight.

For *S. parfustis*, see Abbott (1975), Fay & Johnson (1971), Ritter & Forsyth (1917, as *Macroclinum parfustis*), Swinehart el at. (1974), and Van Name (1945). For other species, see Gaill (1972).

Phylum Chordata / Subphylum Urochordata / Class Ascidiacea
Order Enterogona / Suborder Aplousobranchia
Family Polyclinidae (=Synoicidae)

12.7 **Ritterella pulchra** (Ritter, 1901)
(=Sigillinaria pulchra)

Under rocks, and on walls of caves and other deeply shaded rock surfaces, lower middle intertidal zone to a few meters depth; Yakutat Bay (Alaska); central and southern California.

Colonies formed of several to many lobes, the lobes bright orange, upright, flat-topped, to about 40 mm high and 15 mm across at the tip, tapering but firmly attached at the base, adjacent ones connected by a continuous strand or sheet of test, all transparent, with little or no encrusting sand, leaving zooids clearly visible from the outside; zooids opening at distal ends of lobes, not arranged in systems; stomach wall with numerous longitudinal ridges; pharynx usually with eight to ten rows of stigmata.

This is a common species in Monterey Bay, and has probably been overlooked in many areas where it is not yet reported. The larger lobes often bear the slitlike burrows of the amphipod *Polycheria osborni* (22.6). Recent studies by Nakauchi indicate that tadpole larvae are liberated at least during March and April. Colonies kept in the dark overnight release swimming larvae during morning hours. These soon attach and undergo metamorphosis. Growth of the juvenile tunicate

(oozooid) to the onset of feeding takes 3–4 days. Oozooids have only four rows of stigmata. Under laboratory conditions the oozooid later produces two to five buds by strobilation of the postabdomen. The ozooid survives the process, and within 6–7 days the buds develop into feeding zooids with seven to eight rows of stigmata.

See Abbott (1975), Fay and Johnson (1971), Nakauchi (1977), Tokioka (1968), and Van Name (1945).

Phylum Chordata / Subphylum Urochordata / Class Ascidiacea
Order Enterogona / Suborder Aplousobranchia
Family Polyclinidae (=Synoicidae)

12.8 **Ritterella rubra** Abbott & Trason, 1968

Under ledges and in shaded caves, low intertidal and adjacent subtidal zones on rocky shores exposed to moderate surf; known only from Monterey Co.

Colonies composed of numerous scarlet to crimson, round-topped lobes to 30 mm high and 15 mm in diameter distally but often smaller; zooids not arranged in systems; external stomach wall usually bearing numerous tubercles; pharynx with 10–13 rows of stigmata.

A very attractive and distinctive species, probably more widespread than present records indicate. Tadpole larvae are formed in the late spring and summer and are brooded beside the pharynx until ready for release. Some colonies bear burrows of the amphipod *Polycheria osborni* (22.6). Comparative life history studies of west coast representatives of the genus would be desirable.

See Abbott (1975), Abbott & Trason (1968), Nakauchi (1977), and Tokioka (1968).

Phylum Chordata / Subphylum Urochordata / Class Ascidiacea
Order Enterogona / Suborder Aplousobranchia
Family Polyclinidae (=Synoicidae)

12.9 **Ritterella aequalisiphonis** (Ritter & Forsyth, 1917)

Most common in locally sheltered situations in areas of strong to moderate surf, low intertidal zone and subtidally to 15 m depth on rocky coasts; Puget Sound (Washington) to San Diego; possibly northern Japan.

Colonies composed of few to many slender capitate lobes, some branched, all nearly equal in length, heavily sand-encrusted, closely packed, the whole cluster forming a compact hemispherical mound commonly 40–80 mm across and 10–15 mm high on substratum; zooids orange-brown; stomach with about six to nine longitudinal ridges; pharynx with seven to ten rows of stigmata.

In colonial form (though not in zooid structure) this species closely resembles the east coast *Aplidium* (= *Amaroucium*) *pellucidum*. In both species, the spaces between lobes provide a protected habitat for small crustaceans.

The gonads of *R. aequalisiphonis* are well developed in the summer in central California, and budding by strobilation of the posterior portions of the body is evident in some colonies collected in winter. The symbiotic copepod *Haplostomides bellus* has been found inhabiting zooids of some colonies in the Puget Sound area. No detailed studies exist on most aspects of the biology of *R. aequalisiphonis*.

See Abbott (1975), Fay and Johnson (1971), Ooishi & Illg (1977), Ritter & Forsyth (1917), Tokioka (1967, 1968), Turner, Ebert & Given (1968), and Van Name (1945), often as *Amaroucium aequalisiphonis* or *Sigillinaria aequalisiphonis*.

Phylum Chordata / Subphylum Urochordata / Class Ascidiacea
Order Enterogona / Suborder Aplousobranchia
Family Polyclinidae (=Synoicidae)

12.10 **Euherdmania claviformis** (Ritter, 1903)

Common, especially in and around beds of the surfgrass *Phyllospadix* in regions of moderate surge, low intertidal zone and subtidal to 12 m; Bodega Bay (Sonoma Co.) to San Diego.

Individual zooids tubular, 2–6 cm in length, 2–4 mm in diameter, connected to one another only basally, sometimes growing in dense clusters, sometimes scattered in rows or small clumps among *Phyllospadix* stolons and colonies of *Distaplia occidentalis* (12.21), *Archidistoma ritteri* (12.15), *Perophora annectens* (12.24), and other ascidians; tunic colorless or gray, often sand-encrusted.

Uncertainty about the evolutionary relationships of *Euherdmania* with other ascidians has stimulated study of its biology. Sexual reproduction occurs in the summer at both Pacific Grove (Monterey Co.) and La Jolla (San Diego Co.). Fertilization is internal, and the zygotes are brooded in an elongate oviduct that extends most of the length of the body. Eggs become mature, and are fertilized, one at a time; thus in summer the oviduct contains a linear series of embryos ranging from early cleavage stages at the proximal end to tadpoles at the distal end that are ready for release into the atrium of the parent. The larvae develop in about 70 days, and are expelled to the sea by parental body contractions reminiscent of those seen during birth in mammals. The tadpole body is elaborate, containing in simple form almost all of the definitive adult organs, along with such purely larval structures as the cerebral sense organs (a statocyst and a lens ocellus), the muscular tail, and the larval attachment organs. The tadpoles usually swim for less than 2 hours, then attach by everting the two sticky tubular attachment organs. The tail is withdrawn into the body within 45 minutes, but metamorphosis and reorientation of the body take 48 hours, and feeding does not commence until about a week after attachment. Budding of the juvenile oozooid has not been described, but in an already established colony, budding involves epidermal constrictions that cut the elongate abdomen transversely into three to five segments. Each segment of the body then regenerates the missing parts, and the growing new zooids dissolve their way out of the old tunic and secrete themselves new tubular sheaths.

The blood of *Euherdmania* contains vanadium (about 475 ppm dry weight) in vanadocyte blood cells. The tunicate itself is often eaten by the sea star *Patiria miniata* (8.4).

See especially Ritter (1903, as *Herdmania claviformis*) and Trason (1957); see also Abbott (1975), Bethune (1955), Coe (1932), Coe & Allen (1937), Fay & Johnson (1971), Goldberg, McBlair & Taylor (1951), Ritter & Forsyth (1917), and Van Name (1945).

Phylum Chordata / Subphylum Urochordata / Class Ascidiacea
Order Enterogona / Suborder Aplousobranchia / Family Didemnidae

12.11 **Didemnum carnulentum** Ritter & Forsyth, 1917

Common on rocks, shells, worm tubes, algae, wharf pilings, and other hard substrata, lower middle intertidal zone to about 30 m depth, along open coastal areas to clean-water bays; Oregon to Panama.

Colonies thin (1–4 mm), flat, relatively smooth, encrusting, to about 12 cm across but usually only 1–3 cm; tunic opaque, generally dense white or gray, often tinged with pink, orange, or flesh color, containing very numerous microscopic calcareous (aragonite) spicules 0.01–0.02 mm in diameter and shaped like spiny globules; pharynx with four rows of stigmata.

Species of the family Didemnidae have very small zooids with short bodies and only a few rows of stigmata. Many produce calcareous spicules. Budding occurs in a manner analogous to that in members of the family Polyclinidae, but the body, being short, constricts into only two parts, an upper thoracic section and a lower abdominal section. Moreover, in this case "regeneration" precedes rather than follows fission; thus, during budding a body is often seen in which a tiny new thorax and a tiny new abdomen are growing from the slender waist of a mature zooid. Division of the body soon occurs, giving rise to two individuals, each with a body half-old and half-new.

Tropical didemnids often culture unicellular algae in their extensive common cloacal cavities. The plant cells of some tropical Pacific didemnids were recently discovered to have the morphology of blue-green algae but the chlorophyll pigments (a and b) of the green algae and higher green plants. These plants are the first known representatives of a newly described division of algae, the Prochlorophyta, a group bound to intrigue those interested in the evolution of chloroplasts and nucleated cells.

Didemnum carnulentum as a species has received scant attention. In southern California, larval release occurs during the late spring and summer months. Colonies on panels at La Jolla (San Diego Co.) grow to "a diameter of several inches" in 1–2 months. A closely similar ascidian is occasionally eaten by the sea stars *Dermasterias imbricata* (8.3) and *Pteraster tesselatus* in Puget Sound (Washington).

For *D. carnulentum*, see Abbott (1975), Coe (1932), Coe & Allen (1937), Eldredge (1966), Fay & Johnson (1971), McLean (1962), Ritter & Forsyth (1917), and Van Name (1945). For other *Didemnum* species, see Colombera (1973, 1974b), Jackson & Buss (1975), Lewin (1975, 1976, 1977), Mauzey, Birkeland & Dayton (1968), Millar (1951), F. Monniot (1970c), Pérès (1948b), Schulz-Baldes & Lewin (1976), and Van Name (1945).

Phylum Chordata / Subphylum Urochordata / Class Ascidiacea
Order Enterogona / Suborder Aplousobranchia / Family Didemnidae

12.12 **Trididemnum opacum** (Ritter, 1907)

On rocks and in caves and crevices sheltered from sunlight and strong wave action, low intertidal zone on open coast and in clean bays; subtidal to 60 m on rocks walls and bases of larger kelps; northern California to San Diego, possibly Cabo San Lucas (Baja California).

Larger colonies forming tough encrusting sheets 10–20 cm across, gray-white to lavender in color, with undulating or wrinkled upper surface and variable thickness (2–15 mm); oral apertures of zooids abundant and close together on test surface, the common cloacal apertures widely spaced; spicules very abundant in test, consisting of spiny globes to about 35 μm in diameter.

A remarkable snail, *Lamellaria stearnsii* (13.64), lives and feeds on *T. opacum*. The mantle folds of the snail, which envelop and conceal the shell, closely match the color of the ascidian colony, and acid glands on the snail's mantle resemble the oral apertures of the zooids in appearance, size, and distribution (see also Fig. 13.64). In laboratory aquaria the snails move about only at night, and during the day they are well camouflaged.

Little is known of the biology of *T. opacum*. However, adult structure, larval morphology, histology of the blood and tunic, spicule formation, spermatogenesis, and chromosome number (2N = 30) have been studied in other *Trididemnum* species.

For *T. opacum*, see Abbott (1975), Eldredge (1966), Fay & Johnson (1971), Ghiselin (1964), McLean (1962), Ritter & Forsyth (1917), and Van Name (1945). For other *Trididemnum* species, see Berrill (1950), Colombera (1973, 1974b), Lafargue (1974), Lafargue & Kniprath (1978), F. Monniot (1970b), Pérès (1948c), and Tuzet, Bogoraze & Lafargue (1974).

Phylum Chordata / Subphylum Urochordata / Class Ascidiacea
Order Enterogona / Suborder Aplousobranchia / Family Didemnidae

12.13 **Diplosoma macdonaldi** Herdman, 1886
(=D. pizoni)

Small colonies (0.5–2 cm across) occurring under sheltered rocks on open coast, larger colonies abundant on pilings and on other fouling organisms in clean quiet waters, low intertidal zone to 50 m depth; common especially in Bodega, San Francisco, and Monterey Bays; Vancouver Island (British Columbia) to San Diego, and possibly South America; widespread in tropical and temperate waters of Atlantic and Pacific Oceans and Indonesian region.

Colonies forming delicate sheets 1–2 mm thick and to 20 cm or more across, composed o a transparent tunic containing very numerous zooids usually less than 2 mm long; spicules absent.

The name *Diplosoma* ("double body") alludes to the fact that the first bud arises precociously, during the embryonic stages; the tadpole larva has a single larval body, brain, and tail, but it has two oral siphons and two conspicuous pharynges. When metamorphosis is complete the juvenile ascidian is already a colony, with the oozooid and the first blastozooid approximately equal in size and appearance and functioning side by side. Subsequent budding takes place as described for *Didemnum* (12.11).

In California, reproduction occurs mainly in the spring and summer; colonies expand rapidly, often overgrowing adjacent organisms but causing little damage. Zooids are not arranged in obvious systems, but their atria open into an extensive common cloacal cavity in the test, and this in turn vents to the outside by occasional, craterlike, common cloacal apertures (three visible in photograph). The transparent tunic permits close observation of living zooids; important details of the

method by which tunicates secure food were worked out first on *D. macdonaldi* and two other California ascidians by Mac-Ginitie (1939b).

Detailed ultrastructural studies have been made of the larva and metamorphosis in this species. The swimming muscles of the larval tail consist of two bilaterally symmetrical muscle bands, each containing roughly 800 uninucleate striated muscle cells reminiscent of vertebrate cardiac muscle cells. Owing to a 90° rotation of the tail on its axis during embryonic development, the muscle bands come to lie dorsally and ventrally in the tail, whereas the dorsal nerve cord lies on the left side. Swimming movements of the tail cease 5–10 seconds after the larva attaches by its adhesive organs, and withdrawal of the tail into the body begins 1.5–2 minutes thereafter and takes 6–9 minutes. Tail withdrawal results from contraction of cytoplasmic microfilaments in the tail epidermis, a process inhibited by the drug cytochalasin B. Metamorphosis can be induced in the tadpole of *D. macdonaldi* simply by pinching its tail.

Studies of other species of *Diplosoma* include work on budding, life history, histology of the blood and tunic, chromosome number (2N=30), larval behavior, and mechanism of larval attachment.

For *D. macdonaldi*, see Abbott (1975), Cavey & Cloney (1976), Cloney (1969, 1978), Eldredge (1966), Fay & Johnson (1971), Lash, Cloney & Minor (1973), MacGinitie (1939a,b, 1968), Reish (1964b), Ritter & Forsyth (1917), and Van Name (1945). For other species, see Berrill (1950b), Burighel, Nunzi & Schiaffino (1977), Colombera (1973, 1974b), Crisp & Ghobashy (1971), Lafargue & Valentinčič (1973), D. Lane (1973), Millar (1951), Pérès (1948c), and Thinh & Griffiths (1977).

Phylum Chordata / Subphylum Urochordata / Class Ascidiacea
Order Enterogona / Suborder Aplousobranchia / Family Polycitoridae

12.14 **Pycnoclavella stanleyi** Berrill & Abbott, 1949

Very low intertidal zone in semiprotected situations on rocky shores where wave action stirs up sand; subtidal to at least 10 m; Vancouver Island (British Columbia) to Isla San Gerónimo (Baja California.)

Submerged colonies easily recognized by the distinctive upright, orange- or gold-striped thoraxes of expanded zooids rising above a background rug of sandy tubes; zooids 8–20 mm long, to 5 mm wide at expanded pharynx, occupying separate, sand-encrusted tubes closely packed together but only sporadically joined, forming small clumps or large mats 50 cm or more across.

The following account is based largely on the excellent study of Trason (1963). The slender zooids have bodies that are divided into a short orange thorax and an elongate worm-like abdomen containing the loop of the gut, the gonads, and the heart. Budding, as in other aplousobranchs, involves transverse fission of the body and development of the fragments into complete new zooids. In *P. stanleyi*, budding begins with formation of a vascular extension at the posterior end of the body and construction of a second beating heart anterior to the original heart. An epidermal constriction extending transversely across the body between the two hearts then begins to cut the abdomen into anterior and posterior segments. Duplication of other structures, both fore and aft of the constriction, is well under way before the fission is completed. The anterior end of the zooid connects the two separate limbs of its gut loop and forms a new stomach on its former esophagus. The posterior bud retreats posteriorly into the newly formed vascular process, completes development of a new thoracic region, and emerges as an independent zooid.

At Pacific Grove (Monterey Co.), some sexual and asexual reproduction occurs throughout the year, but in a large colony the individuals that are sexually reproducing at any particular time tend to occur in clumps, not at random. Male and female gonads develop near the posterior end of the body. In sexual reproduction the eggs are retained, and after fertilization they are brooded, as in *Euherdmania claviformis* (12.10), in a linear series in the long oviduct, the youngest stages near the ovary and the maturing tadpoles nearest the atrium. How fertilization occurs in this species (or for that matter in most other colonial ascidians) is not clear; sperm released by other zooids are thought to enter the atrium and pass down the oviduct to fertilize eggs as they are released by the ovary. A given zooid may have either testes or ovaries (but not both)

fully developed at a given time. Individuals brooding larvae are found in all seasons; tadpoles are known to be released in the summer (especially in July and August), but release has not been studied at other seasons.

The active tadpole larvae have two rows of stigmata on each side of the pharynx and a somewhat simplified light receptor (lens ocellus), but (most unusual) they lack an organ of balance (otocyst or statocyst). A discrete neural gland is lacking in both larva and adult, as in the ascidian family Didemnidae (e.g., 12.11–13) and the genus *Cystodytes* (e.g., 12.19). Of 45 tadpoles artificially liberated from their parents in the laboratory, more than half attached either immediately or within 4 hours of release, but ten tadpoles swam intermittently for 36 hours before settling. Attachment occurs by the sudden eversion of three sticky, tubular attachment organs like the two found in *Euherdmania*. Metamorphosis ensues; tail withdrawal is completed in 45 minutes, and organogenesis and reorientation of the body are accomplished within 48 hours. In laboratory-reared oozooids, feeding most commonly begins 5–6 days after attachment, and budding about 3 weeks after attachment.

Colonies in the field are sometimes preyed upon by the starfish *Patiria miniata* (8.4).

See especially Trason (1963); see also Abbott (1975), Berrill & Abbott (1949), Fay & Johnson (1971), and McLean (1962).

Phylum Chordata / Subphylum Urochordata / Class Ascidiacea
Order Enterogona / Suborder Aplousobranchia / Family Polycitoridae

12.15 Archidistoma ritteri (Van Name, 1945)
(=Eudistoma ritteri)

On rocks near the surfgrass *Phyllospadix*, and on walls of channels and caves in regions of strong surge but not direct wave impact, low intertidal zone and subtidal waters to about 20 m; Vancouver Island (British Columbia) to San Diego.

Colonies quite variable in form: on walls of caves and channels, forming sheets 1–2 cm thick, the sheets bearing smooth lumps or projecting fingerlike lobes to 5 cm long; on horizontal rock surfaces, often forming small mushroomlike, stalked lobes mostly 1–2 cm high; overall color yellow to pale gray or transparent; zooids not in systems; pharynx small, with three rows of stigmata.

In this abundant species the zooids are small and slender, their bodies divided into a short thorax, whose apertures both open at the surface of the colony, and an elongate abdomen containing the gut loop, the hermaphroditic gonads, and the heart.

Sexual and asexual reproduction alternate during the year. Sexual reproduction results in embryos that are retained in the terminal part of the oviduct and the adjacent atrial cavity, as many as eight larvae (usually fewer) in various stages being brooded at one time in a single zooid. Zooids brooding larvae are found from April or May through August or September. When sexual reproduction is completed the gonads and usually the thorax undergo regression, and the abdomen turns opaque as cells containing stored food materials (trophocytes) accumulate there. Budding then occurs by strobilation; a series of one to six (usually three) transverse epidermal constrictions form on the zooid between the base of the old thorax and the anterior end of the stomach, slicing the body into short sections. Reorganization and regeneration begin even before the buds are cut off from one another, and each bud develops into a functioning individual. Budding goes on most actively from September or October through March. By January the male gonads have begun to enlarge; enlargement of the ovaries follows shortly thereafter.

Tadpole larvae are released mainly from June through August. Under laboratory conditions they alternately swam actively and rested on the bottom; most attached and underwent metamorphosis after 6 hours, but the duration of the free-swimming period ranged from 1 to 45 hours. The active tadpole larva has two rows of stigmata on each side (nonfunctional as yet), a well-developed light receptor (lens ocellus), and an organ of balance (statocyst). Attachment is first made by a trio of suckerlike adhesive organs whose structure and operation have been studied in detail. Settling is often gregarious. Metamorphosis ensues, and the oozooid commences feeding 9–10 days after attachment. Budding in the oozooid is like that in blastozooids of mature colonies, but since the oozooid is smaller, it produces fewer buds (usually two or three).

Chance discoveries indicate that young animals can survive severe osmotic stress. Oozooids exposed to tap water for 2 hours at 13°C and then returned to seawater survived, budded within 24 hours, and produced fully functional zooids. In another case, five metamorphosing larvae that were neglected and spent 8 months at 13°C "encrusted by salt crystals in brine" survived this treatment, though the animals became opaque and metamorphosis was arrested. When returned to normal seawater at 14°C, they gradually lost their opacity and in 4–6 weeks completed development into functioning zooids (Levine, 1960, 1962a).

Analyses of *A. ritteri* colonies for chemical constituents reveal that protein forms 24.8 percent of the dry weight, lipids about 4.5 percent, and glycogen 1.2 percent. The body content of certain transition metals is remarkably high: in whole, unpreserved zooids vanadium has been recorded up to 471 ppm organic dry weight, chromium up to 144 ppm, and titanium up to 1,512 ppm; analyses of entire colonies show titanium values only one-tenth this amount. The freshly cut tunic of *A. ritteri* liberates strong sulfuric acid (pH 1–2).

The commensal amphipod *Polycheria osborni* (22.6) is often very abundant in burrows in the test. In colonies from the Puget Sound area, the symbiotic copepods *Haplostoma setiferum*, *Haplostomella oceanica*, and *Botryllophilus* sp. have been found inhabiting some zooids.

For *A. ritteri* (usually as *Eudistoma ritteri*), see especially Levine (1960, 1962b); see also Abbott (1975), Berrill (1947c), Cloney (1978), Fay & Johnson (1971), Giese (1966), Levine (1961), Ooishi & Illg (1977), Skogsberg & Vansell (1928), Swinehart et al. (1974), and Van Name (1945). For other *Archidistoma* (= *Eudistoma*) species, see Berrill (1948b), Gaill (1972), and Nakauchi (1966b).

Phylum Chordata / Subphylum Urochordata / Class Ascidiacea
Order Enterogona / Suborder Aplousobranchia / Family Polycitoridae

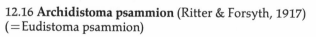

12.16 Archidistoma psammion (Ritter & Forsyth, 1917)
(= Eudistoma psammion)

On rock surfaces exposed to moderate surf, often accompanied by the surfgrass *Phyllospadix*, low intertidal zone on open coast; subtidal to at least 30 m on hard bottom; Olympic Peninsula (Washington) to San Diego.

Colonies forming flat slabs 1–2 cm thick and up to 20 cm or more across; test firm, tough, leathery to touch, with or without encrusting sand, dark brown, purple, maroon, or gray; zooids arranged in circular systems around regularly distributed shallow craters in test surface.

This is a hardy species, strongly constructed and firmly attached, which sustains considerable wave action without damage. Unlike other *Archidistoma* species on the coast, the zooids are arranged in systems, and the atrial siphons of zooids (best seen in vertical slices taken through the systems of the colony) are often much longer than the oral siphons. Asexual and sexual reproduction have not been studied in detail, but budding is probably similar to that in *A. ritteri* (12.15), and zooids brooding tadpoles are reported in the spring.

The commensal amphipod *Polycheria osborni* (22.6) is found in shallow burrows in the tests of some colonies but not others. A second commensal often seen is the clam *Mytilimeria nuttallii* (15.83), which lives with its delicate shell completely embedded in the test of the colony (see Fig. 15.83b). The clam itself has received some study, but its establishment in, and relationship with, the ascidian certainly deserves more attention.

Chemical studies of various sorts have revealed several sterols in the colony (mostly cholesterol and cholestane), strong sulfuric acid in fluid from the freshly cut tunic (one wonders how the *Mytilimeria* shell avoids contact with the acid), and a relatively high content of vanadium (up to 614 ppm dry weight) in zooids removed from the test. Protein forms 24.6 percent of the total dry weight, lipids 4.0 percent, and glycogen 1.2 percent.

Archidistoma psammion is at least occasionally eaten by the sea stars *Dermasterias imbricata* (8.3) and *Patiria miniata* (8.4).

See Abbott (1975), Berrill (1947c, 1948b), Fay & Johnson (1971), Gaill (1972), Giese (1966), Mauzey, Birkeland & Dayton (1968), McLean (1962), Ritter & Forsyth (1917), Skogsberg & Vansell (1928), Swinehart et al. (1974), Tan (1975), Van Name (1945), and Yonge (1952).

Phylum Chordata / Subphylum Urochordata / Class Ascidiacea
Order Enterogona / Suborder Aplousobranchia / Family Polycitoridae

12.17 Archidistoma molle (Ritter, 1900)
(= Eudistoma molle)

Most common in low intertidal beds of the surfgrass *Phyllospadix*, exposed to moderate surf; subtidal to 10 m on rocky shores; abundant at Pacific Grove (Monterey Co.); British Columbia to Carmel Bay (Monterey Co.), but known from relatively few localities.

Colonies very distinctive in life, forming smoothly rounded or oval lobes commonly 5–10 cm across; test an opaque white color, dotted with bright-red anterior ends of zooids; each zooid with two apertures opening close together at the surface.

Colonies of this species live quite well for many months in laboratory aquaria. Both sexual and asexual reproduction occur at all seasons at Pacific Grove (Monterey Co.). Sexual reproduction predominates from May to October, when a large number of zooids in any colony are brooding one to six larvae each. Tadpoles released in the laboratory can be reared to functioning oozooids in vessels of non-circulating seawater. Budding is commonest from November to February. Prior to budding, nutritive reserves accumulate in the very long abdomen (up to 20 mm long), starting at the posterior end and turning the previously transparent zooid white and opaque. Following this the zooid body strobilates, cutting itself into as many as ten segments or buds, each of which regenerates to form a complete zooid. (A juvenile colony is pictured in Fig. 12.14.)

Chemical studies show that freshly cut test surfaces are highly acidic (pH 1–2) and that the vanadium content is high (5,000–6,000 ppm of the ashed weight for zooids alone, 1,000 ppm for the test alone). Protein forms 39.5 percent and lipids 10.5 percent of the organic dry weight of isolated zooids; for the whole colony including test and zooids, protein is only 24.6 percent and lipids 5.7–6.3 percent of dry weight.

For *A. molle*, see Abbott (1975), Giese (1966), Haderlie et al. (1974), Levine (1962a), Swinehart et al. (1974), and Van Name (1945). For other species, see Berrill (1947c, 1948b) and Gaill (1972).

Phylum Chordata / Subphylum Urochordata / Class Ascidiacea
Order Enterogona / Suborder Aplousobranchia / Family Polycitoridae

12.18 **Archidistoma diaphanes** (Ritter & Forsyth, 1917)
(=Eudistoma diaphanes)

Fairly common in crevices, under rocks, and on shaded protected rock walls, low intertidal zone and subtidal waters to at least 20 m on rocky shores; Cape Mendocino to San Diego.

Colonies forming encrusting sheets a few centimeters across or consisting of rounded lobes to 3 cm in diameter; test usually colorless or gray; zooids whitish or yellowish, averaging about 3 mm long, not arranged in systems.

A poorly characterized species about which little is known. The cut surface of the tunic is highly acidic (pH 1–2); the vanadium content of the colony as a whole is low, about 25 ppm of the ashed weight.

See Abbott (1975), Fay & Johnson (1971), Haderlie et al. (1974), Ritter & Forsyth (1917), Skogsberg & Vansell (1928), Swinehart et al. (1974), and Van Name (1945).

Phylum Chordata / Subphylum Urochordata / Class Ascidiacea
Order Enterogona / Suborder Aplousobranchia / Family Polycitoridae

12.19 **Cystodytes lobatus** (Ritter, 1900)

Under boulders and on shaded rocks exposed to moderate surf or surge, very low intertidal zone on open coast; subtidal to 200 m; northern Vancouver Island (British Columbia) to Isla San Gerónimo (Baja California).

Younger colonies forming encrusting sheets 1–15 cm in diameter and 3–10 mm thick, the surface flat or undulating; older colonies often massive, up to 25 cm across and 30 mm thick, the surface bearing elevated knobs, ridges, or convolutions; color gray, whitish, or orange-pink; zooids arranged in systems; spicules (best seen in a slice of the colony) disk-shaped, in younger colonies usually abundant and forming protective capsules around abdominal regions of zooids, in older colonies often sparse and accompanied by irregular calcareous crystals in deeper layers of test.

At present it appears that *C. lobatus* is the only species of

the genus present in shallow California coastal waters. The *Cystodytes* sp. listed in Abbott (1975) represents younger colonies of *C. lobatus*. A cosmopolitan tropical species, **C. dellechiajei** (Della Valle, 1877), common in the Gulf of California and along the Central American Pacific coast, has similar spicules but forms thin (5 mm), flat colonies usually pigmented with black, brown, or purple.

Budding takes place in the same general manner as in *Archidistoma* (e.g., 12.15), by abdominal fission or strobilation accompanied by regeneration, a process in which each existing zooid gives rise to two or more zooids at the same site. Examination of colonies of various sizes shows that the zooids are not becoming steadily more crowded as growth occurs and that the larger colonies do not have several zooids in each capsule of spicules. New spicules and capsules are produced as a colony grows, and the already existing spicules are redistributed at each budding (G. Lambert, pers. comm.).

In the Monterey Bay area the tunic in many colonies contains burrows of the commensal amphipod *Polycheria osborni* (22.6; see also under *Aplidium californicum*, 12.1). Many colonies also conceal specimens of the clam *Mytilimeria nuttallii* (15.83), whose presence is hard to spot unless the clam allows its delicate valves to gape, revealing a slit in the surface of the colony. In the Puget Sound area some zooids are hosts to the symbiotic copepods *Haplostoma dentatum* and *Botryllophilus* sp.

Test fluids, liberated when the tunic of *C. lobatus* is sliced with a knife, contain strong sulfuric acid, at a pH of 1–2. Some acid is liberated when the colonies are preserved, as calcareous spicules tend to dissolve even when colonies are placed in neutral fixatives. Clearly, the acid is normally held in such form or place that it cannot reach the nearby calcareous spicules or the shell of *Mytilimeria*. The role of the tunic acid (also present in numerous ascidians of the family Polycitoridae that lack calcareous spicules) is not known but needs investigation. Perhaps it functions to discourage the activities of some predators or deleterious microorganisms, or has a role in the sequestering of vanadium. A specimen of *C. lobatus* (portion of a whole colony) had a vanadium content of 50 ppm organic dry weight.

The larva of *C. lobatus* has a unique ring-shaped ampulla. *Cystodytes* and *Pycnoclavella* (e.g., 12.14) are the only genera in the family Polycitoridae in which the neural gland is virtually absent as a discrete structure. The diploid chromosome number in *Cystodytes* (based on a count in one species) is 40.

See Abbott (1975), J. Carlisle, Turner & Ebert (1964), Cloney (1978), Fay & Johnson (1971), G. Lambert (1978), McLean (1962), F. Monniot (1970a,b), Skogsberg & Vansell (1928), Swinehart et al. (1974), Turner & Strachan (1969), Van Name (1945), and Yonge (1952).

Phylum Chordata / Subphylum Urochordata / Class Ascidiacea
Order Enterogona / Suborder Aplousobranchia / Family Clavelinidae

12.20 **Clavelina huntsmani** Van Name, 1931
Light-Bulb Tunicate

Most abundant in spring and summer under ledges and on shaded vertical surfaces, low intertidal zone on rocky shores; subtidal to at least 30 m; British Columbia to San Diego.

Zooids borne in separate, transparent (unless overgrown) tubes of tunic 20–50 mm in length and 5–10 mm in diameter, often occurring in dense clusters or beds to 50 cm across; test often showing two orange-pink longitudinal bands near distal end of zooid.

Although *Clavelina* is a colonial ascidian, the zooids are independent of one another in the adult stage. They can be used in biology classes to provide an excellent introduction to tunicate structure and function. The thorax occupies the distal quarter of the body, and the oral and atrial siphons are terminal. Observations of feeding can be made by immersing a zooid completely in a vial of seawater under a dissecting microscope, allowing it to expand, and orienting it so that one can peer down through the mouth and focus on the inner lining of the pharynx. When a suspension of phytoplankton or graphite particles is added to the water, the food-trapping action of mucus sheets on either side of the pharynx is easily seen, as is the formation of the food-laden mucus rope and its passage down to the esophagus.

Below the pharynx, the elongate abdomen occupies most of the rest of the body. The organs displayed there include a long narrow loop of gut (consisting of esophagus, stomach, and intestine), a beating heart in which the direction of blood flow is periodically reversed, and a hermaphroditic gonad with its orange ovary in the center surrounded by a circlet of white testes. Ducts from both parts of the gonad pass anteriorly to open in the atrial cavity, which surrounds the pharynx.

Transparent stolons, simple or branched, may be seen extending outward from the bases of some zooids, especially those at the periphery of the colony. Each stolon consists of a hollow tubular outgrowth of skin and test material, its interior passage divided longitudinally, except at the stolon tip, by a septum of mesenchyme cells. The two channels thus formed in the stolon conduct a continuous flow of blood to and from the zooid body. Buds arise when stolon tips become inflated and cut off as vesicles. They round up and develop into new individuals, the mesenchyme providing the main source of new tissues.

In summer months clusters of developing larvae, bright orange in color, are brooded in the atrial cavity at the base of the pharynx; they are easily visible through the transparent tunic and body wall. They may be removed for examination simply by cutting off the end of the zooid or by opening the atrial cavity with a scalpel and shaking them into a bowl of seawater. Mounted in the seawater on slides, they are worthwhile objects for examination under a compound microscope, though the presence of yolk and orange pigment tends to obscure internal details. Advanced tadpole larvae (each with two clearly visible black spots marking the sense organs and a well-developed tail wrapped about the body), if washed and allowed to remain in a bowl of seawater for a day, often swim actively. The onset of metamorphosis can often be accelerated by osmotic stress, easily provided by placing swimming tadpoles in 50 percent seawater (seawater diluted with an equal amount of fresh water) for 15 minutes, then returning them to seawater.

In late fall the colonies of *C. huntsmani* regress. Although occasional zooids may be found in the winter, the usual overwintering bodies are isolated tips of the stolons. Brought into warmer conditions in the laboratory, *Clavelina* can be raised in the winter.

The amphipod *Polycheria osborni* (22.6) is occasionally found in the test of *C. huntsmani*. Vanadium content of the tissues is relatively low (30 ppm organic dry weight). Several sterols are present, predominantly cholesterol and cholestane.

Much research has been carried out on the European *C. lepadiformis*, a species very similar to ours. The works of Brien provide an excellent introduction here and contain valuable reviews and bibliographies.

For *C. huntsmani*, see Abbott (1975), Alderman (1936), Bradway (1936), Child (1951), Fay & Johnson (1971) Haderlie et al. (1974), Huntsman (1913b,c), McLean (1962), Swinehart et al. (1974), Tan (1975), and Van Name (1945). For other species, see Brien (1948, 1968, 1970a,b, 1971, 1972), Burighel, Nunzi & Schiaffino (1977), Colombera (1973, 1974b), Cowden (1968), Driesch (1902a,b), Fiala-Médioni (1974, 1978a,b), Lyerla & Lyerla (1978), Mukai (1977b), Peters (1966), Ries (1937), Salvatore, Vecchio & Macchia (1961), Weel (1940), and Werner & Werner (1954).

*Phylum Chordata / Subphylum Urochordata / Class Ascidiacea
Order Enterogona / Suborder Aplousobranchia / Family Clavelinidae*

12.21 **Distaplia occidentalis** Bancroft, 1899

Abundant on rocks, often among roots of the surfgrass *Phyllospadix* or in mixed clusters with other colonial ascidians, in regions where surge or currents provide good circulation without strong wave shock, low intertidal zone and subtidally to at least 15 m; Vancouver Island (British Columbia) to San Diego, possibly Kodiak Island (Alaska) to Chile.

Smaller colonies (5–20 mm in diameter) pedunculate and often mushroom-shaped, larger colonies (4–10 cm in diameter) forming flattened lobes with or without short broad stalks; color remarkably variable, white, gray, yellow, pink, red, purple-brown, or combinations of colors (e.g., scarlet and purple, pink and red, gray and white, yellow and brown); zooids embedded in common test and arranged in clusters (systems).

In Puget Sound (Washington), sexual reproduction occurs from early April to late August, peaking from May to July. California populations have not been followed systematically, but tadpole larvae are brooded at least in the summer at Monterey, and new colonies appear on settling panels in the summer at La Jolla (San Diego Co.). Developing eggs are held in a special brood pouch, formed by a diverticulum of the body wall in the posterior thoracic region that encloses a loop of the oviduct. Each hermaphroditic zooid produces two to four eggs. These are several times the diameter of the oviduct, and an egg leaving the ovary becomes squeezed into a slender sausage shape. Sperm have been seen in the oviduct, and fertilization probably occurs there. Once two or three embryos have entered the brood pouch, it detaches from the parent zooid (which soon degenerates) and lies as a free sac in the common tunic. Mature tadpoles later somehow escape from the brood pouch and test, though occasional larvae appear to undergo metamorphosis while still within the test.

The fully developed tadpole larva has three sets of sense organs associated with the hollow cerebral vesicle of the larval brain, two of which are marked by small masses of black pigment barely visible to the naked eye. The larval eye, or photoreceptor, consists of a black-pigmented cup formed of a single cell, some 15–20 retinal cells whose photosensitive processes represent modified cilia, and three transparent lens cells stacked one upon another. A second sense organ situated nearby is the statocyst, consisting of a baglike cell (statocyte) filled with dense black pigment; this projects into the cerebral vesicle but remains attached to the wall by a narrow neck. The black material possibly lends buoyancy rather than added weight to the organ, while the neck of the cell is free to bend as the larva swims its spiral course. The third type of sense organ, also unicellular, bears a highly modified cilium whose free end is inflated like a balloon; the surface of the balloon is invaginated deeply with numerous tubules. On morphological (ultrastructural) grounds it has been suggested that this cilium might function as a pressure receptor.

Under natural conditions 75–85 percent of larval release occurs in the morning; few larvae are released in the dark. Animals kept in the dark for 2 days shed almost no larvae, but released more than 400 larvae in a 3-hour light period that followed and then almost none in a subsequent 15-hour dark period.

The larvae swim actively on release. Swimming movements are accomplished by two muscle bands, one on each side of the tail, together containing 1,500 or more uninucleated muscle cells somewhat reminiscent of vertebrate cardiac muscle.

Their development has been followed recently in exquisite detail. The tail muscles and the contractile cells of the heart are the only striated muscles found in ascidians; the remaining body muscles of larval and adult ascidians are unstriated. The larval tail is stiffened by a single row of some 40 notochord cells, whose central vacuoles have fused to form one long, continous cavity extending the length of the tail.

Attachment and metamorphosis follow after a swimming period ranging from a few minutes to a few hours. Detailed studies have been made of the structure and eversion of the attachment papillae. Metamorphosis of active tadpoles can be initiated within a minute or two by simply pinching the tail with forceps, or by adding the biological dye Janus green B to the seawater to give a final dilution of 5–10 ppm.

In other species of *Distaplia*, and presumably *D. occidentalis* as well, budding starts in the larval stage before the tadpole is released to swim. A diverticulum extends downward from the left epicardium in the tadpole and pushes out the adjacent body wall. The resulting tubular structure constricts at the base, detaches from the main body, and lies in the larval tunic as a "probud." More than one may be formed. A probud consists of two vesicles of simple epithelium, one within the other, the outer one formed of larval skin (ectoderm), the inner of epicardial epithelium (entoderm, since the epicardia arise as diverticula of the pharynx). Sandwiched between the two layers lie a few mesenchyme and blood cells. The probud elongates and constricts to form a few more buds, all with the same basic structure, but further development of the buds awaits metamorphosis of the larva.

The most spectacular event in ascidian metamorphosis is withdrawal of the larval tail. In *Distaplia occidentalis* this takes about 6 minutes. The muscles, notochord, and dorsal nerve cord of the tail are forcibly shoved into the larval body cavity in a coil by strong contraction of the epidermis surrounding the tail. This contraction is brought about by the action of cytoplasmic microfilaments 50–70 Å in diameter. It was studies of tail withdrawal in *D. occidentalis* and in one other ascidian, *Aplidium* (= *Amaroucium*) *constellatum,* that provided the first concrete evidence of the existence of contractile microfilaments. The drug cytochalasin B, which interferes with microfilament action, inhibits tail withdrawal in *D. occidentalis*

and other ascidians. The filaments are composed of, or contain, the muscle protein actin.

Subsequent developments have been followed in some species of the genus but not in *D. occidentalis*. Despite careful study by several investigators, it remains uncertain whether blastozooids in a large colony all arise from the persistent strobilation of the larval probud, or whether, as described by Berrill (1948b), some new buds originate from the epidermis and epicardia of degenerating blastozooids. *D. occidentalis* should provide good material for settling some crucial points.

Chemical studies show that the tunic of *D. occidentalis* is highly acid, with a pH of about 1; vanadium content is in the range of 600–1,200 ppm organic dry weight in whole colonies. Chromosomes have been counted, with conflicting results (2N=24, 2N=30).

Zooids in a colony show some coordination of activity; for example, they contract more or less together when the colony is disturbed. Electrophysiological recordings using whole colonies of *Distaplia* sp. did not reveal any evidence of electrical impulses in either muscle or nerve that might serve to coordinate actions of separate zooids in a colony. Perhaps coordination occurs by the mechanical sensitivity of zooids to body contractions of their neighbors.

Some *D. occidentalis* colonies at Pacific Grove (Monterey Co.) are inhabited by the commensal amphipod *Polycheria osborni* (22.6). In Puget Sound some specimens serve as hosts to the symbiotic copepods *Haplostoma albicatum* and *Haplostomella reducta*. In the same area and in adjacent British Columbia the large polyclad flatworm *Pseudoceros canadensis* often occurs with ascidians, especially *D. occidentalis*.

For *D. occidentalis*, see Abbott (1975), Bancroft (1899b), Cavey & Cloney (1972, 1974), Ching (1978), Cloney (1966, 1969, 1977, 1978), Coe (1932), Coe & Allen (1937), Colombera (1973, 1974b), Eakin & Kuda (1971), Fay & Johnson (1971), Lash, Cloney & Minor (1973), Mackie (1974), McLean (1962), Ritter & Forsyth (1917), Skogsberg & Vansell (1928), Swinehart et al. (1974), Van Name (1945), and Watanabe & Lambert (1973). For other species, see Berrill (1948a), Brien (1939), and Ivanova-Kazas (1967).

Phylum Chordata / Subphylum Urochordata / Class Ascidiacea
Order Enterogona / Suborder Aplousobranchia / Family Clavelinidae

12.22 **Distaplia smithi** Abbott & Trason, 1968

Under ledges, on walls of channels and caves, and in other shaded areas receiving strong current but not direct wave shock, very low intertidal zone and subtidal waters to about 15 m on rocky open coast; Prince William Sound (Alaska) to Monterey Co.

Colonies consisting of clusters of lobes joined by basal stolons; stalk of each lobe slender, cylindrical, 1–5 cm long, expanded distally forming an enlarged and usually somewhat flattened blade bearing one to nine double rows of zooids; color whitish or gray to orange-brown.

This remarkable species, named in honor of the zoology professor Ralph I. Smith of the University of California at Berkeley, is very similar in the shape of the colony to some species of *Sycozoa*. Fully functional zooids occur only in the expanded lobes at the ends of stalks. In these lobes, each double row of zooids forms a system, with the zooids arranged alternately on opposite sides of a tubular common cloacal cavity that empties at the tip of the terminal blade. Along each row, zooids are arranged in order of age, the youngest placed proximally on the blade, the oldest at the distal border. The colony appears to undergo both renewal and attrition continually; new zooids are added at the inner (basal) end of each system, while old and degenerating zooids are lost as the distal margin of the blade is eroded. Budding has not been studied, but buds are thought to arise, as in *Sycozoa*, by the continuous strobilation (somewhere down in the stalk or stolons) of probuds segregated during larval development (see *D. occidentalis*, 12.21). Tiny developing blastozooids can be found in the stalk.

Sexual reproduction occurs largely during the first half of the calendar year. New zooids arising in January and February show developing gonads, the testes making their appearance first, followed by the ovaries. Ripe eggs and embryos are bright orange. Developing larvae are incubated in brood pouches similar to those of *Distaplia occidentalis*. Brood pouches containing one to five embryos at various stages of development are abundant and conspicuous in colonies from

April through June, and most tadpoles have been liberated before July. Tadpole larvae have a light receptor (lens ocellus) and an organ of balance (statocyst).

Parasitic copepods occur in some colonies in central California.

For *D. smithi*, see Abbott & Trason (1968) and McLean (1962). For species of *Sycozoa*, see Brewin (1953), Kott (1969), Millar (1960), and Van Name (1945).

Phylum Chordata / Subphylum Urochordata / Class Ascidiacea
Order Enterogona / Suborder Phlebobranchia / Family Cionidae

12.23 **Ciona intestinalis** (Linnaeus, 1767)

Attached to firm substrata, often fouling pilings, floats, and ships in protected bay waters, low intertidal zone to at least 500 m depth; southern Alaska to San Diego; worldwide in ports and bays.

Body cylindrical, reaching 10–15 cm in length, tunic soft, pale green, translucent to nearly transparent, revealing very long pharynx, gut loop, gonoducts at sexual maturity, and longitudinal muscles; siphons close together at free end of body, their rims somewhat scalloped and carrying pale orange spots.

The wide distribution of this species, its abundance in accessible habitats such as ports, its large size and easy dissection, and its hardiness in aquaria have made it a favorite for physiological research. Most of its organs and most phases of its life history have been scrutinized.

The longevity of *Ciona* varies with temperature; animals in the tropics grow rapidly to sexual maturity and live only several months, whereas subarctic animals live several years. Reproduction may continue in a population all year, the individuals maturing in relay in warm waters, or it may be confined to the summer months in colder habitats. *Ciona* breeds throughout the year in southern California, but apparently only in the warmer months in northern California. Mortality is often heavy in bays when winter rain and runoff reduce the salinity of surface waters.

Mature animals spawn eggs and sperm repeatedly during the reproductive season. Spawning is governed by light and perhaps by other factors as well. Thus, spawning occurs after only 4 minutes' illumination in animals adapted to darkness at Woods Hole (Massachusetts) or after about a half-hour illumination in southern California, but only after several hours of illumination in central Japan during the warmer months. In the laboratory, light-triggered spawning can be induced at any time of day by appropriate prior darkness.

In *Ciona*, sperm in the water swim toward the egg, though they do not swim faster as they approach it. The sperm attractant emitted by the egg is water-soluble and species-specific. Tadpole larvae hatch 12–25 hours after fertilization, depending on the temperature. Much has been learned of the development of *Ciona* since Castle's classic description (1896) of the early embryology. Reverberi's reviews (1961, 1971) of ascidian development introduce some of the experimental investigations of ovogenesis and early embryogenesis, including those pursued in many European laboratories. The diploid chromosome number is 28.

Ciona pumps water through its branchial basket at rates averaging 50 ml per hour per gram of whole wet weight and sometimes reaching 185 ml per hour per gram. It takes up vanadium from water passing through the branchial sac and gut and binds it at levels far above those in the surrounding water. Vanadium concentrations reach 10^{-4} mol per kg of wet weight of the whole animal or 100 ppm by dry weight. Vanadium is concentrated in the blood; hence organs especially richly bathed by blood show high levels of the element. Occasionally the notion is advanced to exploit dense populations of *Ciona* commercially for their vanadium; Elroi and Komarovsky (1961) suggest that "for special grades of vanadium, this source might be competitive" with commercial ores!

The extraordinary reversing heartbeat of ascidians has been most thoroughly examined in *Ciona*; Goodbody (1974) provides an introduction to the extensive literature on this phenomenon. The heartbeat is myogenic. All heart cells are potentially pacemakers, but the rhythm and direction of pumping through the whole structure is controlled by pacemaker regions at either end of the heart, especially at the visceral end. Blood composition, pressure changes, rhythmic interplays, and other factors probably all contribute to the regulation of heart action.

The ground substance of *Ciona*'s tunic appears to be produced by the epidermis, but the tunic contains as well a large population of blood cells that have passed through the epidermis. Zooids divested of their tunics are able to regenerate them to an appreciable depth in about a week. *Ciona* also regenerates excised structures such as the neural complex; indeed, zooids divided transversely can regrow their entire anterior portions if some pharyngeal tissue is present in the stump. Excised siphons regrow in less than a week; muscles and other internal tissues differentiate from hemoblasts derived from the epicardia, and the epidermis that closed the original wound forms both epidermis and tunic for the new siphons. Studies of siphon regeneration in *Ciona* played an important role in the claims of Austrian zoologist Paul Kammerer in the 1920's that he had experimental proof of the inheritance of acquired characters.*

Ciona binds iodine both in its tunic and its endostyle. Whether the iodine is absorbed directly in the tunic or carried there from binding sites elsewhere in the body is uncertain. Endostylar iodine-binding is most intriguing, because it appears to be associated with cellular secretions of iodated organic compounds resembling thyroxine, and reinforces the view that the endostyle is phylogenetically the precursor of the vertebrate thyroid gland. The actual role of bound iodine in the biology of ascidians remains perplexing.

Ciona in Norway is eaten by a variety of benthic fishes and is especially important in the diet of plaice. It is also heavily preyed on by the sea star *Asterias rubens*, which may eat its weight (about 50 g) in *Ciona* in four days. On a dry-weight basis the sea star may devour a third of its biomass in *Ciona* daily. *Asterias* everts its stomach through one of *Ciona*'s siphons, then consumes all the tissues except the tunic.

Ciona intestinalis is one of the best-known ascidians in the world. Although many problems concerning its biology remain unsolved, and even its taxonomic placement is uncertain, the existing literature on the species is so voluminous that it is most sensibly approached through those reviews of

---

*Discredited in his own time, Kammerer is staunchly defended by writer Arthur Koestler (1971; The case of the midwife toad; London: Hutchinson), but attempts to repeat Kammerer's experiments with *Ciona* (e.g., Whittaker, 1975) do not support the original claims.

ascidian biology (cited below) that draw heavily on *Ciona* as a representative ascidian.

For reviews, see Barrington (1965, 1975), Goodbody (1974), Millar (1953, 1971b), and Reverberi (1961, 1971); see also M. Anderson (1968), Barrington & Barron (1960), Barrington & Thorpe (1965a,b), Bellomy (1972), Berg & Baker (1962), Berg & Humphreys (1960), Berrill (1947b), Bouchard-Madrelle & Lender (1972), Burighel, Nunzi & Schiaffino (1977), D. Carlisle (1951), Castle (1896), Cloney (1964), Dodd & Dodd (1966), Dybern (1963, 1965), Eakin & Kuda (1971, 1972), Elroi & Komarovsky (1961), Fiala-Médioni (1974, 1978a,b), Fujita & Nanba (1971), Georges (1969, 1971, 1977), Ghiani & Orsi (1966), Giese (1966), Goldberg, McBlair & Taylor (1951), Gulliksen (1972, 1973), Gulliksen & Skjaeveland (1973), Haderlie (1974), Hoshino (1969), Hoshino & Tokioka (1967), Kessel (1967), Kriebel (1968a,b,c), Krijgsman (1956), Kustin, Ladd & McLeod (1975), Kustin et al. (1974), Ladd (1974), Laird (1971), C. Lambert & Brandt (1967), C. Lambert & Laird (1971), N. Lane (1971, 1972), MacGinitie (1939,a,b), McLeod et al (1975), Millar (1952, 1966b, 1970b), Miller (1975), Morgan (1938–44, 1944, 1945a,b), Nishiyama, Sasaki & Suzuki (1972), Pérès (1948a, 1952, 1954), Peters (1966), Reese (1967), Reish (1963, 1964a,b), Rosati & De Santis (1978), Sabbadin (1957), Scheer (1945), Schmidtke et al. (1977), Shumway (1978), Sutton (1953), Swinehart et al. (1974), Taylor (1967), Thomas (1970a,b), Thorpe (1972), Thorpe & Thorndyke (1975), Whittaker (1975), Whittingham (1967), Woollacott (1974, 1977), Woollacott & Porter (1977), and M. Yamaguchi (1970, 1975).

Phylum Chordata / Subphylum Urochordata / Class Ascidiacea
Order Enterogona / Suborder Phlebobranchia / Family Perophoridae

12.24 **Perophora annectens** Ritter, 1893

Common on rocks, other colonial ascidians (e.g., *Clavelina*, 12.20), red algae such as *Gastroclonium*, and other firm substrata in well-circulated water, lower middle intertidal zone to 30 m depth on open coast; Ucluelet (Vancouver Island, British Columbia) to San Diego.

Colonies to 10 cm across, consisting of clusters of zooids; zooids yellow-green, globular, 2–3 mm in diameter, arising from a network of stolons on substratum; some colonies with zooids laterally fused to form dense plaques, others with zooids well separated, both conditions sometimes present in different parts of the same colony.

This is a superb species for use by invertebrate zoology classes. Individual zooids or small pieces of a colony make excellent living whole mounts, while heart action and feeding are easily observed. Portions of colonies tied to glass slides and kept in a continuous flow of seawater attach readily and continue to grow in the laboratory.

Studies thus far made on this species concern its morphology and some aspects of its asexual reproduction and biochemistry. In the Monterey area colonies appear in the spring and grow rapidly by budding. They reproduce sexually in the late summer and early fall, then regress as winter approaches.

Vanadium is present in this species at truly extraordinary levels: 8,000–9,000 ppm ash-free organic dry weight. Vanadium is concentrated mainly in the green cells and "signet-ring cells" of the blood.

In Ricketts & Calvin (1968, p. 112) *P. annectens* is said to have amazing contractile powers, and when disturbed to withdraw its bright-orange zooids, rapidly changing the color of the colony. These comments, unfortunately repeated in Van Name (1945, p. 168) from an earlier edition of Ricketts & Calvin, apply not to *Perophora* but to *Pycnoclavella stanleyi* (12.14).

Two close relatives, *P. viridis*, common along the Atlantic U.S. coast, and *P. orientalis*, in Japan, have been extensively examined, and much of what we know of them probably applies to *P. annectens* as well. They reproduce asexually by stolonic budding. The stolon, a tubular outgrowth of the ventral body wall, protrudes buds some distance behind its advancing tip. The stolon is divided longitudinally nearly to its tip by a median partition of mesenchyme. Blood flows out on one side of the partition and returns along the other; and blood exchange occurs with all the zooids arranged along the stolon. The stolonic buds extrude as pockets incorporating all tissues of the stolon: tunic, epithelium, mesenchyme, and blood. Blood cells accumulate in these pockets; their roles are disputable, but in irradiated preparations budding proceeds only if non-irradiated dividing blood cells are injected into the area. In particular, the small lymphocytes can form buds by themselves in pockets of nondividing parental cells. In *P. orientalis*, thiourea inhibits growth of stolons but stimulates the eruption of buds, whereas thyroxine and triiodothyronine enhance stolon elongation but suppress budding. Electrical effects on the orientation of stolonic growth are striking; 95 percent of the stolons in one study of *P. viridis* grew toward the anode in an electrically polarized aquarium. The stolons are easily isolated and manipulated, and students of regeneration have compared them with hydroid stolons. The length of the isolated piece of stolon itself influences the cellular rearrangements that lead to bud formation. As in the hydroid *Tubularia* (e.g., 3.2, 3.3), the polarity of pieces of stolon may be reversed, such that buds erupt at the end farthest from what had been the growing tip. Again, as in some hydroids, masses of cells extruded by squeezing the stolon can reorganize themselves in culture into stolons that then form buds.

Although colonies of *P. orientalis* do not fuse in natural conditions, cut pieces of different colonies fuse easily when pressed together, revealing no colony specificity against grafting in this species such as occurs in *Botryllus* (12.29).

The thin tunic of *Perophora viridis* contains cellulose and other polysaccharide fibers as well as proteins, all secreted by the epidermis of the zooid, not by cells resident in the tunic itself. The blood of *P. viridis* contains eight cell types; Overton's (1966) description of them provides a good model for future studies in ascidian hematology.

Individual zooids of *Perophora* are too small for easy physiological study, but one curious phenomenon has been observed: stimulation of the neural ganglion in *P. orientalis* alters the heartbeat in the zooid, and this effect is passed along to other zooids on the same stolon. The blood appears to be the vehicle of transfer; electrostimulation of the stolon itself does not cause this cardiac effect, and the lag time between the stimulus in a zooid and the occurrence of the effect in non-stimulated zooids reflects the time taken by blood to flow out of the electrically affected heart, through the stolonic vessel, and into the heart of the second zooid.

As in other ascidians, the zooids of *Perophora* species are hermaphroditic. In sexual reproduction the sperm are shed to the sea, but the eggs are retained, and developing tadpole larvae are brooded in the atrium until ready to swim. In *P. viridis* the tadpoles are shed a few hours after dawn and swim for only a few hours before attaching to the substratum. Initially, the larvae swim upward and toward the light, but before settling, the responses to light and gravity are re-

versed. Attachment is facilitated by a secretion from the anterior papillae.

For *P. annectens*, see Abbott (1975), Fay & Johnson (1971), Ritter (1893, 1896), Swinehart et al. (1974), and Van Name (1945). For other species, see Deck, Hay & Revel (1966), Deviney (1934), Ebara (1971), Freeman (1964), Fukumoto (1971), George (1930, 1939), Goldin (1948), Goodbody (1974), C. Grave & McCosh (1923), Hirai (1951), Overton (1966), S. Smith (1970), and Van Name (1945).

Phylum Chordata / Subphylum Urochordata / Class Ascidiacea
Order Enterogona / Suborder Phlebobranchia / Family Ascidiidae

12.25 **Ascidia ceratodes** (Huntsman, 1912)

Often present in enormous numbers on pilings, floats, and rocks in bays; more scarce and only in protected situations on open rocky coast; lower middle intertidal zone to 50 m depth; Departure Bay and Northumberland Straits (British Columbia) to Costa Rica and possibly Ecuador and northern Chile.

Body laterally compressed, occasionally to 7 cm long (exclusive of siphons) but usually 3–4 cm; tunic translucent, gray to yellowish green, often fouled or overgrown, usually adhering to substratum by left side; siphons ranging from short to more than half length of body, the rims scalloped, often with orange spots, the oral siphon extending from anterior end, the atrial siphon projecting from mid-body region on dorsal side.

This species is very abundant the year around at Monterey, where its fouling characteristics, reproductive cycle, and growth rate have been studied (Haderlie, 1974; King, 1975). The animals colonize surfaces that have been submerged for about 6 months; on panels submerged for 18 months they "covered practically everything." Adult animals contain some ripe gametes at all seasons. Limited spawning may take place in any month, but most spawning and larval settlement occur in June and July. Animals attain sexual maturity in about 4 months, at a body length of about 2 cm. They reach 4 cm in a year, and growth slows thereafter. Individuals in protected situations have life spans of 18–20 months, but where wave action is more severe, the mortality is high and only 10 percent of those settling survive more than a year.

Ascidia ceratodes flourishes inside the seawater supply pipes and in the flowing runoff from aquaria on sea tables at the Hopkins Marine Station (Pacific Grove, Monterey Co.). Animals exposed to a constant unidirectional flow of water are always oriented with the oral siphon pointing upstream and the atrial siphon curved downstream.

Ascidia ceratodes is an excellent source of gametes for embryological observation and experiment. The adults are best removed from their tunics by slicing through the test with a dull knife. On the naked body (and sometimes even through the tunic) the loop of the intestine can be seen on the left side. Lying in the intestinal loop are the genital ducts where the ripe gametes are stored. The ducts are conveniently color-coded: the oviduct is full of red or pink eggs, and the sperm duct contains a dense white mass of sperm. Simply nicking the wall of either duct with a needle releases the gametes. Although the individuals are hermaphroditic, they usually show a high rate of self-sterility; therefore one should add eggs from one individual and sperm from another to a bowl of seawater. Tadpoles hatch in less than 24 hours at about 16–18°C. European species of *Ascidia* (and the related genus *Phallusia*) have been used extensively for experimental studies of early development (e.g., Reverberi, 1961, 1971).

Vanadium is concentrated by *A. ceratodes* at a very high level: 1,300 ppm ash-free dry weight in whole animals, 6,200 ppm in the blood, and 8,000 ppm in the fluid of the tunic. In another species, *A. nigra*, almost 1.5 percent of the wet weight of blood cells is vanadium! In *A. ceratodes* the vanadium is concentrated especially in blood cells called vanadocytes. In vacuoles of intact living cells the metal occurs as a labile vanadium (III)–aquo complex in a strongly acidic medium. The condition of the element may alter and other artifacts may appear during analytic procedures that cause hemolysis, and Carlson (1975, 1977) suggests that the vanadocyte itself be considered the functional unit of vanadium action. No physiological role is known for the vanadium in *A. ceratodes* (but see below).

In British Columbian waters the hydroid *Endocrypta huntsmani* is sometimes found growing inside the pharynx of *A. ceratodes* (see *Ascidia paratropa*, 12.26).

Other American members of the genus *Ascidia* that have attracted the attention of physiologists (see Goodbody, 1974)

are *A. callosa* of the Pacific Northwest and *A. nigra* (= *A. atra*) in the Caribbean. The blood of *A. nigra* has a nearly neutral plasma, but sulfuric acid makes the cell contents highly acidic. Seven to nine types of blood cells occur, some types probably representing developmental stages of others. A cubic millimeter of blood holds about 53,000 cells, 2 percent of its volume. Experiments with *A. nigra* show that the tunic is toxic to settling larvae, probably through the action of vanadium and acid liberated by degenerating vanadocytes (Stoecker, 1978).

The tropical *A. nigra* reproduces in roughly monthly cycles and settles throughout the year, though most heavily in winter. Animals that survive the first few weeks live for 2–3 years. In *A. callosa* recent studies show that maternal RNA, present in the egg before fertilization, controls developmental events until the embryo hatches as a complex tadpole larva. Thereafter the animal transcribes its own DNA and translates its post-oocyte RNA into materials essential to its metamorphosis and later adulthood. These events may possibly be related to the fact that early ascidian development is highly mosaic, whereas later stages and adults show more developmental lability.

For *A. ceratodes*, see Abbott (1975), Carlson (1975, 1977), Coe (1932), Coe & Allen (1937), Danskin (1978), Fay & Johnson (1971), Goldberg, McBlair & Taylor (1951), Haderlie (1968, 1969, 1974), King (1975), Swinehart et al. (1974), and Van Name (1945). For other species, see Ciereszko et al. (1963), Fiala-Médioni (1978a), Goodbody (1974), Goodbody & Gibson (1974), Hecht (1918a,b), Kustin et al. (1976), C. Lambert (1971), Reverberi (1961, 1971), Stoecker (1978), Tokioka (1972), Vallee (1967–68), and Van Name (1945).

Phylum Chordata / Subphylum Urochordata / Class Ascidiacea
Order Enterogona / Suborder Phlebobranchia / Family Ascidiidae

12.26 **Ascidia paratropa** (Huntsman, 1912)

Attached to assorted hard substrata, rarely very low intertidal, more commonly subtidal, extending to about 80 m depth in well-circulated waters; Unga Strait (Aleutian Islands, Alaska) at least to southern Monterey Co.

Body roughly cylindrical, 5–15 cm long, tunic clean, clear,

transparent, colorless with pink-washed areas, covered with prominent warty tubercles; siphons close together at one end.

Several other invertebrates are sometimes found living inside *A. paratropa*; none of them appears to cause much damage or disadvantage to the ascidian. In northern Washington at least three species of copepod crustaceans live in the pharynx, along with the ciliated protozoan *Euplotaspis cionaecola* and occasionally the tiny crab *Pinnotheres pugettensis*. However, the most extraordinary commensal, found in a small proportion of *A. paratropa* around Friday Harbor (Washington) and southern British Columbia, is the hydroid cnidarian *Endocrypta huntsmani*. The hydroid grows just ahead of the dorsal lamina of the pharynx, forming tuftlike colonies about 1 cm across, visible through the tunic as reddish discolorations. Mature hydroid colonies release free-swimming medusae, which presumably are expelled through the ascidian's oral siphon. The hydroid also occurs in some specimens of *Ascidia callosa* and *Corella willmeriana* in the Puget Sound area of Washington (C. Lambert, pers. comm.) and is reported from *Ascidia ceratodes* (12.25), *Ciona intestinalis* (12.23), and *Halocynthia aurantium* in Departure Bay (Vancouver Island, British Columbia).

Ascidia paratropa probably breeds only in the summer, and it does not brood its eggs. Its diploid chromosome complement is 18. It can be hybridized experimentally with *A. callosa*. The egg envelopes must be removed manually (digesting them off with the enzyme trypsin drastically upsets development). With this precaution about two-thirds of *A. callosa* eggs and one-half of *A. paratropa* eggs fertilized by the other species develop into only slightly distorted tadpoles; a few settle and metamorphose to zooids that appear functionally competent except for aberrant guts. An attempt to raise offspring through sexual maturity should be made.

The tunic of *A. paratropa*, glasslike in its transparency, has received some study. It is very largely seawater, the organic matter constituting only 0.8 percent of tunic wet weight. About half the organic matter is carbohydrate (polysaccharide) resembling plant cellulose, and about half is protein. Investigation of the brain of *A. paratropa* shows the presence of neurosecretory cells of the sort characteristic of the ascidian suborders Aplousobranchia and Phlebobranchia.

See Abbott (1975), Burreson (1973), Cloney (1964), Colombera (1974b), Danskin (1978), Dawson & Hisaw (1964), Faulkner (1970), Huntsman (1911, 1912), Illg (1958), Minganti (1974), Pearce (1966), M. Smith & Dehnel (1971), and Van Name (1945).

Phylum Chordata / Subphylum Urochordata / Class Ascidiacea Order Enterogona / Suborder Phlebobranchia / Family Corellidae

12.27 **Corella** spp.

Attached to firm substrata, including harbor floats and occasionally other organisms, in calm, clean water, low intertidal zone to about 50 m; Loring (Alaska) to Puget Sound (Washington) and subtidally at least as far south as Monterey Bay.

Body oblate, laterally compressed, to 5 cm long; tunic glassy, clear, colorless, slightly roughened by wrinkles or tiny papillae, sometimes sparsely flecked with orange or gold; internal organs easily visible through tunic; stigmata of pharynx arranged in spirals.

Recent study (G. Lambert, pers. comm.) indicates that there are probably two *Corella* species on the American west coast: *C. inflata* Huntsman, a shallow-water species, limited to British Columbia and the San Juan Archipelago (Washington), which broods its developing larvae in an expanded antero-dorsal pocket of the atrium; and **C. willmeriana** Herdman, a deeper-water, non-brooding form, lacking an expanded atrial pocket, occurring from Loring (Alaska) to central or southern California. Both species occur in Puget Sound, where *C. inflata* abounds on floats and extends subtidally to about 20 m; *C. willmeriana* is taken only by dredging or scuba diving and occurs down to about 75 m. Most of the published west coast studies on "*C. willmeriana*" were probably carried out on *C. inflata* (G. Lambert, pers. comm.).

Although many ascidians brood their fertilized eggs in the atrial cavity and release active tadpole larvae, *C. inflata* has evolved an unusual method for doing this. The attached animals are oriented such that the expanded antero-dorsal pocket of the atrium is uppermost, directed toward the surface of the water. The eggs, unlike those of other west coast ascidians, float in seawater (owing to the replacement of some sodium ions by ammonium ions in the vacuoles of the larger outer follicle cells surrounding the eggs) and are thus trapped against the roof of the expanded atrial pocket. They remain there until they hatch. Removal of gastrulae from the atrial chamber of a brooding parent was once reported to result in deformed larvae, but subsequent observations do not confirm this (C. Lambert, pers. comm.).

After leaving the parental atrium, the larvae swim for a few minutes to a few hours, then settle and metamorphose, usually attaching themselves to clean surfaces that have been recently submerged or denuded. Larvae tend to aggregate in settling; thus adults are often found clumped together. Studies of *C. inflata* at Friday Harbor (Washington) show that individuals live for only 5–8 months but attain sexual maturity in 3–4 months (at a body length of about 12 mm). Breeding occurs throughout the year. The short life span and rapid attainment of sexual maturity result in two or more generations per year and a steady supply of young to colonize available surfaces. The diploid chromosome number is 12.

In Puget Sound the polyclad flatworm *Eurylepta leoparda* and the sea stars *Mediaster aequalis*, *Solaster dawsoni* (8.7), and *Pteraster tesselatus* prey on *Corella inflata*. The colonial ascidian *Diplosoma macdonaldi* (12.13) overgrows it in the winter, and diatoms may coat it heavily in the summer.

Studies of the tunic of *Corella* show that it is very largely seawater, with organic matter (two-thirds carbohydrate, one-third protein) constituting only 0.6 percent of the wet weight. The mantle within the tunic often bears white flecks, which in *C. inflata* are uric acid crystals (C. Lambert, pers. comm.). Both mantle and tunic in west coast *Corella* species are so transparent that one can see a good deal of the internal anatomy in living animals, including the extraordinary spiraled stigmata of the pharynx. The genera *Corella* and *Molgula* (e.g., 12.41, 12.42), such different ascidians in many ways, both may have intricately shaped spiral stigmata, whose development has been described but whose functional advantages over the simpler stigmata of most ascidians have yet to be experimentally demonstrated. The ciliary beat that drives water through the stigmata of *Corella* is under nervous control, via fibers of the visceral nerves that extend from the neural ganglion; stimulation of the nerves is associated with an arrest of ciliary action.

Corella concentrates iron in its tissues but does not take up

detectable amounts of vanadium, contradicting the taxonomic generalization that appreciable vanadium uptake is a trait of the suborder Phlebobranchia.

Further detailed developmental studies of the larvae of phlebobranch ascidians are needed, and *Corella* would be a good place to start.

For accounts probably based on *C. inflata*, see Child (1927), Ching (1977), C. Lambert & G. Lambert (1978), G. Lambert (1968), Cloney (1964, 1969), Colombera (1974b), Huntsman (1912, *C. inflata* only), Kriebel (1968a,b), Mackie et al. (1974), Mauzey, Birkeland & Dayton (1968), and Oliphant (1972). Accounts referring to *C. willmeriana* only are Abbott (1975), Haderlie (1971), and Huntsman (1912, *C. willmeriana*). The species of *Corella* in Danskin (1978), Faulkner (1970), and M. Smith & Dehnel (1971) is uncertain. For other species of *Corella*, see Franzén (1976) and Peters (1966).

*Phylum Chordata / Subphylum Urochordata / Class Ascidiacea
Order Enterogona / Suborder Phlebobranchia / Family Corellidae*

12.28 **Chelyosoma productum** Stimpson, 1864

On solid substrata, including pilings and floats, occasionally on other organisms, in calm to moderately rough waters, very low intertidal zone to about 50 m; Prince William Sound (Alaska) to San Diego.

Body to 6 cm high but California specimens usually smaller, the free end terminating in a flat, oval disk bearing siphons; both disk and siphons overlaid by distinctive plates of tunic; tunic thin, hard, tough.

The disk plates of the test of *C. productum* and its relatives show consistent patterns according to species. The plates are thickened regions of the tunic's cuticle, deposited by "bladder cells" that move out into the tunic from the epidermis. The plates articulate with one another where the cuticle is thin. They bear concentric markings, resembling the growth lines of fish scales, which appear to be growth rings. On the assumption that they represent annual rings (as yet unverified), Huntsman (1921) concluded that *C. productum* grows in annual spurts and lives for about 3 years and that its more northern relative *C. macleayanum* lives perhaps for 5 years.

The brain of *C. productum* has neurosecretory cells much

like those of *Ascidia* and other phlebobranch ascidians. However, secretions by the brain and neutral gland do not appear to be related to reproduction, as once supposed; removal of both these organs from *C. productum* after one breeding season does not inhibit gonadal development the following season, even though the excised organs fail to regenerate. Breeding occurs in the spring; both eggs and sperm are shed to the sea.

At least two species of commensal copepod crustaceans live in the branchial sac. The ascidian is preyed on in Puget Sound (Washington) by the sea star *Orthasterias koehleri* (8.9).

Chelyosoma productum has been used in cardiac studies of tunicates, principally in comparison with the much more thoroughly studied *Ciona intestinalis* (12.23). The vanadium content of the body of *C. productum,* with the tunic removed, is 800 ppm dry weight.

See especially Bancroft (1898) and Huntsman (1921); see also Abbott (1975), Cloney (1964), Danskin (1978), Dawson & Hisaw (1964), Dudley (1966), Fay & Johnson (1971), Hisaw, Botticelli & Hisaw (1966), Illg (1958), Kriebel (1968a,b), Mauzey, Birkeland & Dayton (1968), Turner, Ebert & Given (1969), and Van Name (1945).

Phylum Chordata / Subphylum Urochordata / Class Ascidiacea
Order Pleurogona / Suborder Stolidobranchia / Family Styelidae

12.29 **Botryllus** spp.

Attached to pilings, floats, rocks, seaweeds, and other animals, low intertidal and shallow subtidal zones, in calm bays and protected recesses of more exposed habitats; genus distributed throughout California and worldwide.

Colonies encrusting, often several centimeters across; zooids 1–2 mm long as viewed from above, commonly forming circular, oval, or star-shaped clusters around common cloacal apertures in test; colors often striking, most commonly orange but including purple, yellow, brown, and green, the zooids usually contrasting strongly with the tunic.

The several *Botryllus* species on California shores need taxonomic study. One species, **B. tuberatus** Ritter & Forsyth, is easily identified by its four rows of stigmata; it is known from Bodega Bay (Sonoma Co.) to San Diego, and from Japan, the Philippines, the Asian mainland, and several Pacific islands.

Other species in California have more than four rows of stigmata.

Whereas California's species have received scant attention, *B. schlosseri* of Atlantic and Mediterranean waters and *B. primigenus* of the western Pacific have been extensively examined, especially with regard to their asexual and sexual reproduction, genetics, and development; what follows is based largely on investigations of these two species, but it probably applies broadly to the California species, as well.

Although *Botryllus* colonies may live for well over a year, zooids within them persist for only a week to 10 days. One or two "pallial" buds arise on the body wall antero-laterally on each zooid and develop while still attached. By the time the buds' siphons open, the parental zooid has become decrepit, and the buds in turn have already begun to extrude a subsequent asexual generation. Thus a single stellate system in a colony simultaneously contains zooids of three asexual generations. Over entire colonies there are more or less synchronous waves of budding and development. As the number of zooids in a colony increases, there is continual rearrangement of asexual progeny into new systems, so that few systems comprise more than about 20 large, fully functioning zooids (in contrast to the many dozens of zooids found in single systems in *Botrylloides,* 12.30).

Although *Botryllus* zooids are fundamentally bilaterally symmetrical, their right and left sides differ somewhat in anatomy. Sometimes in the process of budding the symmetry gets reversed, producing individuals in mirror image, or situs inversus. Once it arises, this reversed symmetry is transmitted to all asexual descendants of the zooid. Studies by Sabbadin show that the condition can be caused, and reversed, experimentally.

Budding of a different sort, called vascular budding, also occurs in *Botryllus.* Here the buds arise from the very extensive network of blood vessels that ramify through the test and interconnect all feeding zooids. At the margins of the colony, and at scattered points elsewhere, the test-vessel system bears blind tubular enlargements (vascular ampullae), which very slowly pulsate and aid circulation in the tunic. The vascular buds arise from the bases of vascular ampullae and apparently develop zooids there solely from cells of the test-vessel wall (originally derived from ectoderm). Vascular buds themselves have not been seen to reach sexual maturity; they arise naturally but can be provoked experimentally by removing all the zooids from a region of the colony and leaving the test-vessel system functioning by itself. Situs inversus is transmitted by vascular budding as well as by pallial budding.

Botryllus colonies regress in winter but grow vigorously in the spring in temperate waters. Sexual reproduction occurs in the late spring and early summer. Settled larvae give rise to colonies of several hundred zooids in the first few months; the colonies reach sexual maturity at 1–2 months' age. Sperm mature rapidly in each zooid's testes, but the ova maturing in a zooid are derived from oogonia, produced by earlier individuals in the colony, that are carried in the test-vessel system until they implant in their "definitive" ovary. Fertilized eggs are retained and brooded in the atrium until tadpole larvae are ready to swim.

Colonies of *Botryllus* range over an astonishing spectrum of colors. These colors are controlled by only a few genetic factors (alleles), whose action and expression have been examined experimentally, especially by Sabbadin et al. in Italy. Since colonies can be fused, genetically mixed by anatomically united masses can be created experimentally, and the sources, progeny, persistence, compatibility, and lability of various components (e.g., blood cells, germ cells) can be traced. The fusibility of colonies is also controlled by multiple alleles at only a few loci.

Botryllus's small size has discouraged much physiological study, but the ease of its culture, the ready fertilization of its eggs in vitro, the precise regulation of cloning, the fusibility of genetically diverse colonies, the tolerance of surgical disruption (including even wholesale removal of the zooids from the test), and the animal's varied morphogenetic patterns combine to make this genus one of exceptional promise in studies of developmental regulation and genetics.

A fossil, *Palaeobotryllus*, very reminiscent of modern *Botryllus* species in general form, was recently found in upper Cambrian deposits in Nevada and other western localities (Müller, 1977).

See especially Berrill (1941a,b,c), C. Grave (1934), C. Grave & Woodbridge (1924), Millar (1971b, the only recent general review incorporating much material about *Botryllus*), Mukai & Watanabe (1974, 1975a, 1976a,b), Oka & Watanabe (1957), Ritter & Forsyth (1917), Sabbadin (1969, 1971, 1972), Sabbadin, Zaniolo & Majone (1975), and Van Name (1945); see also Abbott (1975), Bancroft (1899a, 1903b), Brunetti (1974), Brunetti & Burighel (1969), Burighel & Brunetti (1971), Burighel & Milanese (1973, 1975), Cloney (1978), Colombera (1969, 1974b), De Santo (1968), De Santo & Dudley (1969), Fay & Johnson (1971), B. Grave (1933), C. Grave & Riley (1935), Herdman (1924), Izzard (1968, 1973, 1974), Katow & Watanabe (1974), Milkman (1967), Millar (1952), Mukai (1974a,b, 1977a), Mukai, Sugimoto & Taneda (1978), Mukai & Watanabe (1977), Müller (1977), Pizon (1892), Sabbadin (1955a,b, 1960, 1964, 1973, 1977), Sabbadin & Graziani (1967a,b), Sabbadin & Tontodonati (1967), Tanaka (1973), Tanaka & Watanabe (1973), Tokioka (1967), Watkins (1958), Watterson (1945), and Zaniolo, Sabbadin & Resola (1976). These papers lead, as well, to a good deal of the other literature on the genus.

Phylum Chordata / Subphylum Urochordata / Class Ascidiacea Order Pleurogona / Suborder Stolidobranchia / Family Styelidae

12.30 **Botrylloides** spp.

On pilings, floats, and other substrata, often with *Botryllus* (12.29), in bays and harbors, low intertidal and shallow subtidal zones; genus distributed throughout California and worldwide.

Colonies encrusting, often several centimeters across; zooids 1–2 mm long as viewed from above; genus most easily distinguished from *Botryllus*, in California forms, by the general shape of the systems, which are elongate and often serpentine (rather than oval or stellate as in *Botryllus*), giving the impression of zooids arranged in long rows or chains; larvae brooded in incubatory pouches (not present in *Botryllus*), one on each side.

Although California species are poorly known, studies of the Atlantic *B. leachi* and the Japanese *B. violaceus* show that *Botrylloides* shares many of the features of budding, sexual reproduction, development, etc., outlined for *Botryllus* (12.29), and is proving to be equally valuable for experimental studies.

See Abbott (1975), Bancroft (1903a), Berrill (1947a, 1950) Brunetti (1976), Burighel, Brunetti & Zaniolo (1976), Burighel, Nunzi & Schiaffino (1977), Coe (1932), Coe & Allen (1937), Colombera (1974b), Fay & Johnson (1971), Garstang & Garstang (1928), Hirai (1951), MacGinitie (1939b), Miyamoto & Freeman (1970), Mukai (1977a), Mukai, Sugimoto & Taneda (1978), Mukai & Watanabe (1974, 1975b), Oka & Watanabe (1959), Pizon (1899), Ritter & Forsyth (1917), Sabbadin & Tontodonati (1967), Van Name (1945), and M. Yamaguchi (1975).

Phylum Chordata / Subphylum Urochordata / Class Ascidiacea
Order Pleurogona / Suborder Stolidobranchia / Family Styelidae

12.31 **Metandrocarpa taylori** Huntsman, 1912

Attached to hard substrata in well-circulated water on rocky shores, low intertidal zone to about 20 m; China Hat (British Columbia) to San Diego.

Colonies up to 20 cm or more across, encrusting; zooids round or oval, usually less than 5 mm high, bright red to pale orange (rarely yellow or green), largely independent of one another and thus appearing like tiny simple ascidians, but joined basally by a thin sheet of tunic encrusting substratum; edges of colony fringed by ampullae (enlarged, blind blood reservoirs) of blood-vessel system of tunic.

Asexual reproduction, bud development, larval traits, and spawning behavior have been fully enough described now to permit substantial experimental work. Because the species is less hardy than *Botryllus* (12.29), studies may be restricted to seaside laboratories. *M. taylori* extrudes buds throughout the year, most vigorously in the winter, and breeds most actively in the spring and summer, but colonies show some asexual and sexual activity at all months. Fertilization is internal, and the developing tadpoles are brooded in the atrium. Light stimulates the release of larvae, which emerge most abundantly a few hours after dawn and swim for a few hours to 2 days before settling and metamorphosing. Development of the oozooid to the onset of feeding takes 11 days at about 8°C, but weeks or even months pass before the protrusion of one to several buds, which then bud in turn at 1 or 2 weeks of age.

Zooids that bud persist, unlike the zooids of botryllids,

which rapidly degenerate after budding. Young colonies have a dendritic appearance, whereas older *M. taylori* colonies are densely packed, as later buds fill in available spaces among older zooids. The tunic has an elaborate system of blood vessels linking all zooids with each other and with the advancing margin of the test at the edge of the colony. This test-vessel system plays a critical role in budding.

Buds arise from the parent's outer body wall; they detach from the parent and come to lie in the sheet of test encrusting the substratum before organogenesis takes place. Detached buds consist of two vesicles of epithelium, one within the other, along with a few mesenchyme and blood cells. Organogenesis of the bud is quite different from that of the embryo and involves profound deformations of the internal vesicle, formerly part of the parental atrial epithelium. Changes taking place progress directly to the adult characters, rather than repeating larval traits en route. The bud pays no attention to the germ-layer origins of its tissues; the atrial epithelium of the parent, which constitutes the principal organ-forming layer of the bud, is ectodermal in origin, but its diverticula form the new gut as well as the nervous system. Development in asexual reproduction often departs from the conservative traditions of the developing egg.

A closely related species, **M. dura** (Ritter), also usually bright red in color, occurs subtidally along the whole range of *M. taylori*. *M. dura* forms more massive encrustations, with the zooids of the colony more thickly embedded in the test; as a result, the colony appears like an *M. taylori* colony in which the tests of all zooids are laterally fused. The difference in appearance relates mostly to *M. dura*'s faster or more abundant proliferation of tunic and to traits that may well derive from this quantitative distinction. *M. dura* also breeds mostly in the spring and summer; the young are brooded in the atrial cavity until they begin to swim.

For *M. taylori*, see Abbott (1953, 1955, 1975), Fay & Johnson (1971), Haven (1971), Newberry (1965a,b), Van Name (1945, including also *M. michaelseni*), Watanabe & Lambert (1973), and Watanabe and Newberry (1976). For *M. dura*, see Abbott (1975), Fay & Johnson (1971), Ritter (1896, as *Goodsiria dura*), and Van Name (1945). For other species, see Makioka, Hirabayashi & Watanabe (1978).

Phylum Chordata / Subphylum Urochordata / Class Ascidiacea
Order Pleurogona / Suborder Stolidobranchia / Family Styelidae

12.32 **Cnemidocarpa finmarkiensis** (Kiaer, 1893)

On hard substrata in well-circulated waters, very low intertidal zone to at least 50 m (540 m in Japan); common from Alaska to Washington, sparingly distributed subtidally to southern Monterey Co.; northwestern Pacific; European and Canadian arctic.

Zooid hemispherical, generally less than 3 cm across base; tunic clean, smooth, opaque, brilliant red to rose pink (white when very small), with pearly luster; siphons far apart and prominent in relaxed animals.

Only bits of the biology of this handsome animal have so far been studied. Adults are sexually mature at least in the summer; they shed both sperm and eggs to the sea. Larvae are easily raised from gametes stripped from the gonads; they settle after a few hours or days of swimming and resorb their tails by active contraction of the cells of the notochord.

The tunic of adults, though rather thin, is firm and tough, with organic matter forming 12.4 percent of its wet weight, a value comparable to that of the much more massive test of *Pyura haustor* (12.37). Slightly more than half the organic matter is carbohydrate (tunicin), the remainder protein. Neurosecretory cells have been identified in the neural ganglion; they are cytologically similar to such cells in other ascidians assigned to the suborder Stolidobranchia. The vanadium content of the body, with test removed, is very low.

Cnemidocarpa finmarkiensis is one of several ascidian species that harbor the copepod *Pygodelphys aquilonaris* in the branchial sac. In Puget Sound (Washington) the ascidian falls prey to the sea star *Orthasterias koehleri*.

See Abbott (1975), Cloney (1964, 1969), Danskin (1978), Dawson & Hisaw (1964), Dudley (1966), Faulkner (1970), Illg (1958), McLean (1962), Millar (1966b), M. Smith & Dehnel (1971), and Van Name (1945).

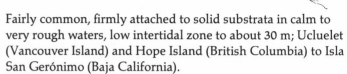

Phylum Chordata / Subphylum Urochordata / Class Ascidiacea
Order Pleurogona / Suborder Stolidobranchia / Family Styelidae

12.33 **Styela montereyensis** (Dall, 1872)

Fairly common, firmly attached to solid substrata in calm to very rough waters, low intertidal zone to about 30 m; Ucluelet (Vancouver Island) and Hope Island (British Columbia) to Isla San Gerónimo (Baja California).

Body elongate, cylindrical, supported on a thinner stalk about equal to body in length, the overall length occasionally exceeding 25 cm in calm habitats but more often 8–15 cm in exposed sites; siphons close together at distal end, the oral siphon recurved, the atrial siphon straight; tunic with prominent longitudinal ridges and grooves running the entire length of animal (representing thick and thin regions of tunic), otherwise relatively smooth, yellow to dark red-brown, often fouled with debris and other organisms in harbors but clean in wave-swept areas; most easily confused with *Styela clava* (12.34).

Ritter and Forsyth (1917) lamented, "That this, one of the earliest known and most familiar ascidian species of Pacific North America, should have remained to this time without a detailed description is one of the vicissitudes in the progress of knowledge of our local marine fauna." Their dismay can be echoed a half-century later; *S. montereyensis* is frequently encountered and often tallied in faunal surveys, but little has been published on its biology.

Breeding and larval settlement occur in the summer. Study of the developing eggs shows that the inner follicle of the oocyte provides the test cells that are enclosed with the ovum inside the chorion. Both eggs and sperm are shed to the sea. R. L. Miller (pers. comm.) finds that the fertilized eggs form at one pole a crescent of green pigment, which behaves just like the yellow crescent described for *Styela partita* by Conklin (1905). In natural situations the larvae settle best on surfaces that have been underwater at least several months. In metamorphosing larvae the tail collapses, as is characteristic in members of the suborder Stolidobranchia, by active contraction of the notochord itself, not (as in Aplousobranchia) by contraction of the caudal epidermis.

In Monterey harbor this is a conspicuous but not very

abundant member of the fouling community. The maximum life span is not known, but one individual, growing submerged at 2 m depth on a polypropylene rope used to suspend experimental fouling panels at Monterey, reached a length of 5 cm in less than 3 months, and at the time of this writing is alive and well at an age of about 3 years and a length of 23 cm (E. C. Haderlie, pers. comm.).

On the open western coasts of Washington and Vancouver Island (British Columbia), *S. montereyensis* harbors the copepod crustacean *Pygodelphys aquilonaris* in its branchial sac. The vanadium content of the body, with test removed, is rather low, only 36–40 ppm dry weight.

See Abbott (1975), Abbott & Johnson (1972), Bancroft (1899a), J. Carlisle, Turner & Ebert (1964), Cloney (1964, 1969), Coe (1932), Coe & Allen (1937), Danskin (1978), Dudley (1966), Fay & Johnson (1971), Haderlie (1968, 1971, 1974), Haderlie et al. (1974), Reish (1964a,b), Ritter & Forsyth (1917), Rosenthal, Clarke & Dayton (1974), Sims (1979), Turner, Ebert & Given (1969), Van Name (1945), and Young (1978).

Phylum Chordata / Subphylum Urochordata / Class Ascidiacea Order Pleurogona / Suborder Stolidobranchia / Family Styelidae

12.34 **Styela clava** Herdman, 1881

Common on rocks, floats, and pilings, low intertidal zone to at least 25 m, usually in calm waters; San Francisco Bay to San Diego Bay; Japan, Asian mainland, Australia; northwestern European harbors, northeastern United States.

Body to 15 cm long, club-shaped, plump, lacking sharp division between broader body and narrower stalk; tunic yellow-gray to brownish, often overgrown with fouling organisms; similar to *S. montereyensis* (12.33) but distinguished as follows: (a) tunic usually bearing conspicuous tubercles, (b) tunic often with irregular longitudinal wrinkles but never with regular ridges and grooves along entire body representing thick and thin regions of test, and (c) both siphons straight, the oral siphon not recurved toward base of stalk.

This species, probably introduced to California waters in the late 1920's, has been mistakenly called "*Styela barnharti*" in numerous biological publications between 1945 and 1972.

S. clava is native to the Orient; its harbor habitat and its pattern of spread to California, the northeastern United States, Europe, and Australia suggest that it has fouled ships' hulls and been carried about in this way.

In a population at Venice (Los Angeles Co.), *S. clava* increased in length at an average rate of 1–1.5 cm per month up to a length of 8.5 cm, after which growth slowed. Individuals lived for 12 to 18 months. Animals that settled in early spring reached 2–3 cm in June, 6 cm by late August, and 9 cm by late October, at which size sexual maturity occurs.

Oogenesis takes several months, and mature ova may be retained in the ovaries for some time; spermatogenesis, from primary spermatocytes to active sperm, takes only about 10 days. As in *S. plicata* (12.36), the ovarian germinal cells divide into oocytes and follicle cells, and these latter proliferate both into more follicle cells and into test cells that are included with the eggs inside the chorion. Spawning in California occurs from early spring through autumn. Both the eggs and sperm are shed to the sea. Once fertilized, the egg rearranges its contents so that, at the four-cell stage, the periphery of the posterior blastomeres contains three times as many mitochondria and eight times as many lipid droplets as the anterior blastomeres; these lipid droplets correspond to the "yellow mitochondria," or yellow crescent, that Conklin (1905) described in his classic study of cell lineage in *S. partita*.

Styela clava individuals are extremely sensitive to vibrations but not especially to direct touch, except that disturbances may provoke partial closure of the oral siphon. Extirpation of the neural ganglion renders the animal much less sensitive to vibration and disrupts control of its atrial siphon in coordinated siphonal behavior. Individuals periodically contract with sufficient coordination of their siphons that water is expelled in reverse through the oral siphon, and the branchial sac is thereby emptied. They also contract periodically with both siphons closed, for reasons as yet unknown but perhaps related to the circulation of blood.

Mature specimens of this species, kept in moving and aerated water at 10°C, filter 78–175 ml of water per minute per gram of dry weight of body, but (like bivalves) they drastically reduce or entirely stop this water flow in standing water.

Recent studies of cell proliferation in the gut of *S. clava* show that mitosis occurs principally around the bases of grooves, as in the folds of the stomach wall, but also in other highly restricted zones. Cells migrate from these areas to renew tissues elsewhere. Cell populations in the gut that are not renewed are rare or absent, and the entire digestive tract is in state of progressive turnover; cell lives are measured in spans of weeks, and aged cells are ejected to the gut lumen. Similar studies of the blood cells show that these, too, are renewed and have lifetimes of several weeks.

A complex array of sterols has been identified from *S. clava*; cholesterol and cholestanol are present in greatest abundance. The diploid chromosome complement is 32.

Styela clava has been blamed for asthma among oyster shuckers in Japan. Several ascidian species grow on the oyster shells, and the shuckers shatter the ascidians into droplets and bits when hammering open the shellfish. Ascidian extracts have proved to be highly allergenic in medical investigations of this industrial problem, which occurs in poorly ventilated and residue-contaminated workhouses around Hiroshima.

Under the local popular name "mideuduck," *S. clava* is a highly prized seafood in southern Korea.

See especially Abbott & Johnson (1972) and Johnson (1971); see also Abbott (1975), Bellomy (1972), Berg & Humphreys (1960, as *S. barnharti*), Bevis & Thorndyke (1978), Colombera (1974b), Conklin (1905), Ermak (1975a,b, 1976a,b, 1977), Hirai (1951), Holmes (1973, 1976), Huwae & Lavaleye (1975), Millar (1970b), Miller (1975), Plough (1978), Scheer (1945, as *S. barnharti*), Sims (1979), Thorndyke (1977, 1978), Van Name (1945, as *S. barnharti*), Z. Yamaguchi (1931), and Yasuda (1975).

Phylum Chordata / Subphylum Urochordata / Class Ascidiacea Order Pleurogona / Suborder Stolidobranchia / Family Styelidae

12.35 **Styela truncata** Ritter, 1901

Common on harbor floats and pilings and on protected coastal rocks, low intertidal zone to about 20 m; Yakutat Bay (Alaska) to San Diego.

Body oval, squat, usually less than twice as high as wide, usually less than 3 cm high; tunic opaque, pale yellow-orange to dull brown; siphons reddish, close together atop body.

Although this ascidian is widespread, both as a fouling organism on docks and as an inhabitant of more natural habitats, it is easily overlooked because of its small size and sometimes cryptic habits. On the open coast, individuals sometimes occur buried under colonies of *Aplidium californicum* (12.1) with only their siphons showing.

Unlike most other members of the genus *Styela*, *S. truncata* broods its young. The tubular ovaries curve downward at their distal ends, rather than pointing upward toward the atrial siphon as in most other *Styela* species. Huntsman (1913a) used this ovarian arrangement, found in both *S. truncata* and *S. yakutatensis*, as the basis for a new genus, *Katatropa*, no longer recognized. In terms of function, the recurved ovaries of these two species merely help ensure that eggs, when emitted by the oviducts, will be deposited in the base of the atrial cavity instead of being swept out the atrial siphon, as happens in oviparous *Styela* species.

The limits of the breeding season are uncertain, although the atrium contains larvae at least in the summer months at Monterey.

See Abbott (1975), Fay & Johnson (1971), Haderlie (1968, 1969, 1971), Huntsman (1912, 1913a), Ritter (1901), and Van Name (1945).

Phylum Chordata / Subphylum Urochordata / Class Ascidiacea Order Pleurogona / Suborder Stolidobranchia / Family Styelidae

12.36 **Styela plicata** (Lesueur, 1823)

On hard substrata in calm waters of bays and harbors, very low intertidal to about 30 m; Santa Barbara to San Diego; worldwide in warm and temperate regions.

Body ovoid, to 9 cm long, commonly 4–7 cm; surface cobbled by rounded lumps and bulges most noticeable around siphons; tunic clean, semi-opaque, whitish to brown, often with darker lines radiating from apertures; siphons when relaxed prominently marked by a square of thick lips, when retracted hidden in lobes of tunic.

This is the *"Styela barnharti"* of Ritter and Forsyth (1917), but not of later authors. Under its proper name, *S. plicata*, it has been the subject of research in several parts of the world, especially with regard to reproduction.

The elongate gonads are attached to the inner surface of the body wall on each side; they are suspended in the atrium and their genital ducts open near the atrial siphon. There are 2–11 (commonly 4–7) gonads on the right side and 1–3 (commonly 2) on the left. Each gonad consists of a sinuous tubular ovary bordered by numerous oval or branched testes. The histology of the gonads and the development of the egg envelopes have been studied. The diploid chromosome number is 32.

Both eggs and sperm are shed to the sea. Sperm-attracting materials have been discovered in the eggs. Spawning is light-triggered, but the delay between signal and release is so long that spawning occurs only toward late afternoon or even sunset in natural habitats. Embryos develop a yellow crescent like that described for *S. partita* by Conklin (1905). The larvae hatch in midwater the following morning, and most then settle during that day.

The reproductive period in southern California is probably broad and may not be very consistent. Thus, new settlement of *S. plicata* was recorded mainly in the summer and early fall in 1960–62 (Reish), but animals spawned most readily from September to March in 1974–75 (West & Lambert). In Japanese waters of comparable conditions, *S. plicata* combines rapid growth to sexual maturity with a short subsequent life span (a few months), so that several generations may relay one another through the year. Sexual activity there occurs in the late spring, producing a generation that reproduces in turn in midsummer, the offspring of which breed in the fall; this last generation overwinters to become sexually active the following spring. In describing this, Yamaguchi (1975) noted that "ascidians as a group might be one of the most eurythermal animals in regard to reproduction." Short life cycles and rapid maturation extend a species' overall breeding season and give its members a substantial advantage as exploiters of

fresh surfaces. In Italy, *S. plicata* lives a full year, breeding several times during this life span.

S. plicata's annual patterns elsewhere suggest the complexity of factors that must be assessed in reckoning seasonal reproductive behavior. The northern U.S. Atlantic limit of the species (Cape Hatteras, North Carolina) appears set by intolerable winter conditions, duplicating there the winter destruction of the species in Italy (Lagoon of Venice). But in North Carolina the species appears in the benthos only in the spring and fall, despite summer conditions that do not inhibit reproduction. Suitable summer food is available, and, at temperatures of 18–28°C, less than half of the nourishment absorbed from the gut is needed for metabolic maintenance; the surplus is ample to support reproduction. The lack of recruitment of new individuals into the population during the summer there seems to reflect not lack of reproduction but rather intense predation upon gametes, larvae, and newly settled young. Where and when settling succeeds, the *S. plicata* population can be impressive; at Pivers Island (North Carolina), this species constituted 70–89 percent of the biomass of the benthic invertebrates present.

As in *S. clava*, water circulation through the pharynx in *S. plicata* is most rapid in an external current and is sharply reduced in still water. Ligature of the oral or atrial siphon, but not both together, leads to rapid regeneration; a new orifice is formed in a few days, and water transport resumes.

The blood of *S. plicata* contains materials that agglutinate some mammalian erythrocytes. *S. plicata*'s hemagglutinin has been characterized as a large, heat-stable molecule, possibly a polysaccharide or mucopolysaccharide. *S. plicata* shares with *S. clava* a complex array of sterols, dominated by cholesterol but containing also significant amounts of cholestanol. Vanadium is concentrated in moderate amounts, as well as niobium and tantalum.

The branchial sac of *S. plicata* harbors several species of copepods, four species in specimens from Florida. Florida specimens also provide shelter in their lobular tunics for the little bivalve *Musculus lateralis*; in one survey, a sample of 12 ascidians held 174 of these mollusks and one ascidian alone carried 90. Some *S. plicata* from Florida have small crabs living in the atrial cavity.

Along with several other ascidian species, *S. plicata* has been implicated in the occurrence of asthma among oyster shuckers in Japan (see the account of *Styela clava*, 12.34).

See Abbott (1975), Abbott & Johnson (1972), Bellomy (1972), Bertrand (1971), Bretz (1972), Burighel, Nunzi & Schiaffino (1977), Colombera (1974b), Fiala-Médioni (1978a,b), Fisher (1976, 1977), Fuke & Sugai (1972), Gaill (1977), George (1930, 1937, 1939), Illg (1958), Kanatani, Hidaka & Onosaka (1964), Kokubu & Hidaka (1965), Miller (1975), Plough (1978), Reish (1964a,b), Ritter & Forsyth (1917, as *S. barnharti*), Sabbadin (1957), Sims (1979), Sutherland (1978), Taylor (1967), Tucker (1942), Van Name (1945), West & Lambert (1976), Williams & Thomas (1967), M. Yamaguchi (1970, 1975), and Yasuda (1975).

Phylum Chordata / Subphylum Urochordata / Class Ascidiacea
Order Pleurogona / Suborder Stolidobranchia / Family Pyuridae

12.37 Pyura haustor (Stimpson, 1864)

On rocks, pilings, floats, mussel beds, kelp holdfasts, and other solid substrata, in calm bays to rough waters of open coast, low intertidal zone to depths exceeding 200 m; Shumagin Islands (Aleutian Islands, Alaska) to San Diego.

Body roughly globular, reaching a diameter of 5–8 cm but usually less; tunic very tough and leathery, sharply ridged and creased even in expanded animals, opaque, pale red-tan to ochre-brown, usually obscured by attached debris; siphons in expanded animals relatively long, stiffly erect, with distinctly red-rimmed orifices.

Ritter (1909) designated the population from Point Conception (Santa Barbara Co.) to San Diego as a separate species (described as *Halocynthia johnsoni* Ritter, 1909), distinguished from more northerly populations by its smoother and less highly vascularized tunic and its larger number of oral tentacles. Later workers have not recognized this separation, but the pattern of variation needs reexamination. The animals still occur widely in southern California, but are apparently much scarcer in harbors and bays than was the case in the early part of the century.

The tough tunic provides an excellent external protection. Roughly 13 percent of its wet weight is organic matter, of which about 40 percent is protein and 60 percent carbohydrate. The pharynx is large, and as in most large ascidians of the suborder Stolidobranchia, the wall bears prominent longitudinal folds that greatly increase the surface area available for respiration and filter feeding. The pharyngeal wall is a multilayered network of blood vessels, easily seen in a simple dissection (see Abbott, 1975, p. 640).

Sexual activity has been noted in all seasons except midwinter. Both eggs and sperm are shed to the sea. The diploid chromosome complement is 22. The larval structure and events of metamorphosis in *P. haustor* resemble those of *Boltenia villosa* (12.39). The withdrawal of the tail at metamorphosis results from active contraction of the notochordal cells themselves, not from contraction of the caudal epidermis.

Neurosecretory cells have been demonstrated in the neural ganglion. Their granules are dispersed in the fashion typical of the order Pleurogona, not compactly associated with the nucleus as is characteristic of species in the order Enterogona. The animals do not concentrate vanadium.

Several species of copepods live in *P. haustor*'s branchial sac, including *Pygodelphys aquilonaris*, which also inhabits several other ascidians. In northern Washington, *P. haustor* harbors protozoans in its pharynx, again species that inhabit other ascidians: the suctorian *Trichophyra salparum* and the ciliates *Euplotaspis cionaecola* and *Parahypocoma rhamphisokarya*. In Puget Sound (Washington), *P. haustor* is preyed on by the sea star *Solaster stimpsoni* (8.6).

See Abbott (1975), Burreson (1973), Cloney (1959, 1964, 1969), Colombera (1974b), Danskin (1978), Dawson & Hisaw (1964), Dudley (1966), Fay & Johnson (1971), Illg (1958), Mauzey, Birkeland & Dayton (1968), C. Monniot (1965), Ritter (1909), M. Smith & Dehnel (1971), Van Name (1945), and Wardrop (1970). For reviews of pyurid biology, see Barrington (1974), Berrill (1975), Goodbody (1974), Millar (1971b), and C. Monniot (1965).

12.38 **Pyura mirabilis** (von Drasche, 1884)

On rocks, or in sand but adhering to shell and gravel, in well-circulated waters, very low intertidal zone to about 60 m; British Columbia to southern California; Japan, Korea.

Body cylindrical, reaching a length of 10 cm or more, with one siphon extending from each end; tunic fairly opaque, gray-white to light brown, clean or slightly fouled by mud.

The living specimen shown in the photograph had two siphons of equal size when captured; one was bitten off near the base by a fish inadvertently placed in the same aquarium.

Our knowledge of this curious species is so far limited to anatomical descriptions of preserved specimens. The functional significance of the extraordinary sausage-like shape of the body, with its oral and atrial ends, is not clear. Pelagic tunicates with oral and atrial apertures at opposite ends use the water currents thereby created for locomotion, but this is not the case here.

The pharyngeal stigmata of *P. mirabilis* are numerous but very small compared to those of other pyurids. As Monniot has remarked, *P. mirabilis* seems to be "a species that has evolved specialized habits adapted to life on shifting substrates." But its adaptations appear to be so different from those of the family Molgulidae, many of whose species live on soft substrata, that the puzzle of its design remains. Even the generic affiliation of this animal is in doubt: Monniot maintains it to be a species of *Pyura*, but Tokioka considers the gonads more characteristic of the genus *Herdmania*.

The species is probably oviparous, and nothing is known of the breeding habits.

See Abbott (1975), C. Monniot (1965), Ritter (1907), Tokioka (1965, 1967), and Van Name (1945).

12.39 **Boltenia villosa** (Stimpson, 1864)

On hard substrata in well-circulated waters, lowest intertidal zone to 100 m; Prince Rupert (British Columbia) to San Diego.

Body globular, 1–3 cm across, usually but not always borne on a thin stalk 1–6 cm long; tunic reddish, orange, pale tan, or brown, armed with many spines, these either smooth or adorned with irregular bristles; siphons short.

Many details of egg development, larval structure and behavior, and the process of metamorphosis are known for this species (and several others) through the excellent studies of Cloney and some of his students at the University of Washington. The developing egg gains its yolk granules not from transformation of mitochondria (as some workers have concluded) but rather from the enlargement of golgi vesicles in the oocyte, though mitochondria evidently contribute to the synthesis of yolk. The nucleus of the developing egg and the nuclei of the embryo bear numerous small structures called "annulate lamellae" that appear to be attached to the inside of the nuclear envelope; their function remains problematic.

Ripe gametes can be obtained at almost any season, but most abundantly in the summer. Both eggs and sperm are shed to the sea, and fertilization occurs there. The diploid chromosome number is 32. The developing egg forms a red-pigmented crescent, which is equivalent to the yellow crescent described for *Styela partita* by Conklin (1905). Larvae develop from fertilized eggs in 31–33 hours at 12°C. The larva is about 0.8 mm long, and its larval brain bears both a light receptor (lens ocellus) and a statocyst. Three anterior attachment organs are present. The notochord cells, about 40 in number, are lined up in single file in the immature larva, but become rearranged to enclose a single extracellular tubular vacuole of fluid running the length of the tail in the active larva. The larval pharynx and siphons are rudimentary. Two types of muscle cells are present in the tail.

Settlement and metamorphosis occur at any time from about 6 hours to 5 days after hatching. Larvae settle more readily on some surfaces (e.g., wax-coated glass) than others (e.g., clean glass). In larvae that have hatched at least 6 hours previously, metamorphosis can usually be initiated within 5–15 minutes by a concentration of 7×10^{-6} Janus green B (a biological dye) in seawater. At metamorphosis the tail is crumpled and pulled into the posterior trunk by contraction of the cytoplasmic microfilaments in the notochord cells. The outer layer of the larval tunic is molted, and the tadpole trunk is gradually developed into a functioning juvenile ascidian.

Boltenia villosa concentrates substantial amounts of vanadium in its tissues (500–750 ppm dry weight of body with tunic removed). This level of concentration was previously thought to occur only in some members of the suborder Phlebobranchia. A few other members of the Stolidobranchia also sequester vanadium (e.g., *Styela montereyensis*, 12.33, and *Halocynthia hilgendorfi igaboja*, 12.40), but at much lower levels.

In Puget Sound (Washington), *B. villosa* harbors three species of copepods in its branchial sac (*Pygodelphys aquilonaris*, *Doropygus fernaldi*, and *Scolecodes huntsmani*), as well as two protozoans (the suctorian *Trichophyra salparum* and the thigmotrich ciliate *Parahypocoma rhamphisokarya*). The ascidian is preyed on in that region by two sea stars, *Dermasterias imbricata* (8.3) and *Orthasterias koehleri* (8.9).

The circumpolar arctic and north Atlantic species *Boltenia ovifera*, the sea potato, is considerably larger than *B. villosa*. It is eaten by Eskimos when it washes ashore. Its heart is a tube 0.3 cm in diameter and 5 cm long, permitting easy dissection and convenient electrophysiological manipulation. The test is spiny but less so than that of *B. villosa*. The test spines of *B. ovifera* are formed by axially arranged bundles of cellulose fibrils lying in the superficial region of the test. How these bundles form is not known; they appear to be well away from cells that might be proliferating or organizing the structure of the tunic.

See Abbott (1975), Burreson (1973), Cloney (1959, 1964, 1969, 1978), Colombera (1974b), Danskin (1978), Dawson & Hisaw (1964), Dudley (1966), Everingham (1968a,b), Faulkner (1970), Fay & Johnson (1971), Hirai (1951), Hsu (1962, 1963, 1967), Hsu & Cloney (1958), Illg (1958), Lash, Cloney & Minor (1973), Mauzey, Birkeland & Dayton (1968), Mishra & Colvin (1969, 1970), C. Monniot (1965, review of family), Plough (1969), Van Name (1945), and Weiss, Goldman & Morad (1976).

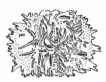

Phylum Chordata / Subphylum Urochordata / Class Ascidiacea Order Pleurogona / Suborder Stolidobranchia / Family Pyuridae

12.40 **Halocynthia hilgendorfi igaboja** Oka, 1906 (= H. igaboja)

On rock or gravel in well-circulated waters, very low intertidal zone to 165 m; Prince Rupert (British Columbia) to Los Angeles and Santa Catalina Island (Channel Islands); Japan and adjacent Asian mainland (but see below).

Body globular, lacking a stalk, to 10 cm in diameter but commonly much less; tunic opaque, dark brown, heavily carpeted with large flexible spines that are typically ringed more or less regularly by circles of recurved thornlike bristles; expanded siphons reddish.

Kott's (1968) revision of the genus *Halocynthia* includes this species in *H. hispida*, reported widely in the Pacific and Indian Oceans. Until more detailed comparisons are made, we retain the name used by Monniot (1965) in his review of the Pyuridae of the world.

This animal is a veritable hedgehog, but its remarkable spines have not been examined with respect to either their development or their role in the life of the animal. In the only close scrutiny so far of the spines of pyurid tunics (Mishra & Colvin), the puzzle remains of how the test can undergo organization far from sites of obvious cell activity. Distribution of cells in the test of *H. hilgendorfi igaboja* has not yet been ascertained.

Reproduction has not been studied in detail, nor natural spawning recorded, but fertile gametes have been obtained by cutting the gonads in May. Eggs are shed to the sea, not brooded. Development of the tadpole larvae, from fertilization to hatching, takes 44–46 hours at 12°C. Active larvae are about 1.5 mm long (2 mm including the caudal fin). The cerebral vesicle of the larval brain contains a light receptor (lens ocellus) and an organ of balance (statocyst or otocyst). Both the pharynx and gut are poorly differentiated in the larva. The notochord consists of a row of about 40 cells; a few hours after hatching these cells arrange themselves around a fluid-filled vacuole extending the length of the tail. At metamorphosis, they contract, withdrawing the tail.

The blood of this species and that of *Styela plicata* (12.36) contain agglutinins for mammalian erythrocytes but not for each other, and the ascidian hemagglutinins are distinguishable from each other by their activity. The body of *H. hilgendorfi igaboja*, with tunic removed, contains a substantial amount of vanadium (175 ppm dry weight, for the one specimen analyzed).

In Puget Sound (Washington) the small crab *Pinnotheres pugettensis* is sometimes found living within the body of this ascidian. At least one species of copepod crustacean, *Doropygopsis longicauda*, inhabits the branchial sac.

See Abbott (1975), Cloney (1959, 1964, 1969), Danskin (1978), Dawson & Hisaw (1964), Fay & Johnson (1971), Fuke & Sugai (1972), Illg (1958), Kott (1968), Mishra & Colvin (1969, 1970), C. Monniot (1965), Pearce (1966), and Van Name (1945).

*Phylum Chordata / Subphylum Urochordata / Class Ascidiacea
Order Pleurogona / Suborder Stolidobranchia / Family Molgulidae*

12.41 **Molgula manhattensis** (DeKay, 1843)

On hard, soft, or mixed substrata in calm or moderately protected waters, often invading estuaries, low intertidal zone to 30 m; Bodega Bay (Sonoma Co.), Tomales Bay (Marin Co.), and San Francisco Bay; Gulf of California; Japan, Australia; Maine to Texas excluding Florida peninsula; Europe from White Sea (U.S.S.R.) to Adriatic Sea; northwestern Africa.

Body globular, somewhat laterally compressed, usually 1–3 cm in diameter, occasionally 5 cm; tunic soft, translucent and yellow-green to gray-green in clean individuals, partly covered with tendrils that trap debris; renal vesicle prominent below gonad on right side; pharynx with both curved and spiral stigmata.

This introduced species was found widespread in San Francisco Bay in the 1950's. It has dispersed wherever calm or estuarine conditions favor it, both here and in ports around the world. It tolerates pollution well: its habitat in Morocco is turbid much of the year; in Japan it was found in a notably polluted euryhaline pond; and it formed 75 percent of the fauna of experimental aquaria containing 10 μg per liter of the polychlorinated biphenyl Aroclor 1254. The beast may not thrive

on stress, but it tolerates it better than most other ascidians, occasionally achieving immense population densities: probably 100 individuals per m² in Chesapeake Bay (Virginia and Maryland) and probably at least that in San Pablo Bay (Marin Co.; see Ganssle's comment under *M. verrucifera*, 12.42).

Studies in other parts of the world suggest that *M. manhattensis* lives about a year, rapidly reaches sexual maturity, and reproduces repeatedly during its lifetime. Sexual activity is intense from spring to fall. Individuals are reported to shed their eggs and sperm at night in nature, but in the laboratory when the light is suddenly increased the animals spawn within half an hour. The diploid chromosome number is 32.

Molgulids form prominent concretions, containing both uric acid and calcareous deposits, in a bladderlike vesicle called a "renal sac" (a loaded term, since the function remains enigmatic after a century's conjecture). The renal sac is pressed closely against the heart. The uric acid precipitated comes from the catabolism of nucleic acids, whereas protein catabolism produces ammonia. The sac is endodermal and wholly closed. In *M. manhattensis* it appears after metamorphosis, and the concretions are visible before the zooid starts feeding and before its heart starts to beat. Fungi of unknown origin and uncertain affinity inhabit the sac in *M. manhattensis*; they appear after the earliest concretions form.

Most work on the ascidian endostyle and its role in iodine-binding has employed *Ciona*, but recently *Molgula* has provided comparisons. Iodine is bound intracellularly in Zone 7 of the endostyle, where this groove joins the barrel of the pharynx; the binding appears mediated by a peroxidase. Some iodine also is bound into the tunic, but its route there is uncertain.

With few exceptions the ascidian propensity to concentrate vanadium is limited to some members of the suborders Aplousobranchia and Phlebobranchia; stolidobranchs, such as *Molgula*, concentrate iron instead. *M. manhattensis* has a vanadium content of less than 20 ppm ash-free dry weight in its tissues, but contains 16,400–26,600 ppm iron. The biological role of neither element is clear.

In California waters *M. manhattensis* is occasionally eaten by the leopard shark (*Triakis semifasciata*) and the brown smoothhound (*Mustelus henlei*).

See Abbott (1975), R. Anderson (1971), Azéma (1937), Berrill (1931), Colombera (1974c), Crawford & Homsher (1975), Das (1948), Dunn (1974, 1975), Ganssle (1966), Godeaux & Firket (1968), Goodbody (1957, 1965, 1974), B. Grave (1933), Hamilton & Yurchyk (1973), Hansen (1974), Illg (1958), Kessel (1967), Kessel & Kemp (1962), Kott (1976), Lash, Cloney & Minor (1973), Millar (1966b, 1970b), Nolfi (1970), Redick (1969), Russo (1975), Sabbadin (1957), Saffo (1976, 1978), Saffo & Lowenstam (1978), Sengel & Kieny (1963), Stephens & Schinske (1961), Swinehart et al. (1974), Tokioka & Kado (1972), Trason (1959), Van Name (1945), and Whittingham (1967).

Phylum Chordata / Subphylum Urochordata / Class Ascidiacea
Order Pleurogona / Suborder Stolidobranchia / Family Molgulidae

12.42 **Molgula verrucifera** Ritter & Forsyth, 1917

On firm substrata in well-circulated waters, often attached to zooids of the ascidian *Euherdmania claviformis* (12.10), low intertidal zone to about 10 m; Santa Barbara to San Diego.

Body globular, about 1 cm across; test translucent gray-green when clean, but often sand-encrusted; individuals often aggregated into clumps; pharynx with most stigmata straight or only slightly curved, not spiral.

Little is yet known of this species, beyond the fact that it broods its young. Ganssle (1966) remarked that enormous populations covered the bottom of San Pablo Bay (Marin Co.) in 1963–64, so abundantly "that it was impossible to haul the trawl aboard by hand." He was almost certainly referring here to *M. manhattensis* and not to *M. verrucifera*.

See Abbott (1975), Berrill (1931), Coe (1932), Coe & Allen (1937), Fay & Johnson (1971), Ganssle (1966, but see above), Ritter & Forsyth (1917), and Van Name (1945).

Literature Cited

Abbott, D. P. 1953. Asexual reproduction in the colonial ascidian *Metandrocarpa taylori* Huntsman. Univ. Calif. Publ. Zool. 61: 1–78.
———. 1955. Larval structure and activity in the ascidian *Metandrocarpa taylori*. J. Morphol. 97: 569–94.
———. 1975. Phylum Chordata: Introduction and Urochordata, pp. 638–55, *in* R. I. Smith and J. T. Carlton, eds., Light's manual: Intertidal invertebrates of the central California coast. Berkeley and Los Angeles: University of California. 716 pp.
Abbott, D. P., and J. V. Johnson. 1972. The ascidians *Styela barnharti, S. plicata, S. clava,* and *S. montereyensis* in Californian waters. Bull. So. Calif. Acad. Sci. 71: 95–105.
Abbott, D. P., and W. B. Trason. 1968. Two new colonial ascidians from the west coast of North America. Bull. So. Calif. Acad. Sci. 67: 143–54.
Alderman, A. L. 1936. Some new and little known amphipods of California. Univ. Calif. Publ. Zool. 41: 53–74.
Alldredge, A. 1976. Appendicularians. Sci. Amer. 235: 95–102.
Anderson, M. 1968. Electrophysiological studies on initiation and reversal of the heart beat in *Ciona intestinalis*. J. Exper. Biol. 49: 363–85.
Anderson, R. S. 1971. Cellular responses to foreign bodies in the tunicate *Molgula manhattensis* (DeKay). Biol. Bull. 141: 91–98.
Azéma, M. 1937. Recherches sur le sang et l'excrétion chez les ascidies. Ann. Inst. Océanogr. 17: 1–150.
Bancroft, F. W. 1898. The anatomy of *Chelyosoma productum* Stimpson. Proc. Calif. Acad. Sci. (3) 1: 309–32, pl. 18.
———. 1899a. A new function of the vascular ampullae in the Botryllidae. Zool. Anz. 22: 450–62.
———. 1899b. Ovogenesis in *Distaplia occidentalis* Ritter (ms.), with remarks on other species. Bull. Mus. Compar. Zool., Harvard, 35: 59–112, pls. 1–6.
———. 1903a. Aestivation of *Botrylloides gascoi* Della Valle. Mark Anniversary Vol.: 147–77, pl. 11.
———. 1903b. Variation and fusion of colonies in compound ascidians. Proc. Calif. Acad. Sci. (3) 3: 137–86, pl. 17.
Barnes, S. N. 1971. Fine structure of the photoreceptor and cerebral ganglion of the tadpole larva of *Amaroucium constellatum* (Verrill) (subphylum Urochordata, class Ascidiacea). Z. Zellforsch. 117: 1–16, 15 pls.
———. 1974. Fine structure of the photoreceptor of the ascidian tadpole during development. Cell Tissue Res. 155: 27–45.
Barrington, E. J. W. 1965. The biology of Hemichordata and Protochordata. San Francisco: Freeman. 176 pp.
———. 1974. Biochemistry of primitive deuterostomians, pp. 61–95, *in* M. Florkin and B. T. Scheer, eds., Chemical zoology, vol. 8, Deuterostomians, cyclostomes, and fishes. New York: Academic Press. 682 pp.
———. 1975. Problems of iodine binding in ascidians, pp. 129–58, *in* Barrington & Jefferies (1975).
Barrington, E. J. W., and N. Barron. 1960. On the organic binding of iodine in the tunic of *Ciona intestinalis*. J. Mar. Biol. Assoc. U.K. 39: 513–23.

Barrington, E. J. W., and R. P. S. Jefferies, eds. 1975. Protochordates. Symp. Zool. Soc. London 36. New York: Academic Press. 361 pp.

Barrington, E. J. W., and A. Thorpe. 1965a. An autoradiographic study of the binding of iodine[125] in the endostyle and pharynx of the ascidian *Ciona intestinalis* (L.). Gen. Comp. Endocrinol. 5: 373–85.

————. 1965b. The identification of monoiodotyrosine, diiodotyrosine, and thyroxine in extracts of the endostyle of the ascidian *Ciona intestinalis* (L.). Proc. Roy. Soc. London B163: 136–49.

Bellomy, M. D. 1972. Sea squirt asthma. Sea Frontiers 18: 36–39.

Berg, W. E., and P. C. Baker. 1962. Antigens in isolated blastomeres of the ascidians *Ciona* and *Styela*. Acta Embryol. Morphol. Exper. 5: 274–79.

Berg, W. E., and W. J. Humphreys. 1960. Electron microscopy of four-cell stages of the ascidians *Ciona* and *Styela*. Develop. Biol. 2: 42–60.

Berrill, N. J. 1931. Studies in tunicate development. II. Abbreviation of development in the Molgulidae. Phil. Trans. Roy. Soc. London B219: 281–346.

————. 1935. Cell division and differentiation in asexual and sexual development. J. Morphol. 57: 353–427.

————. 1936. Studies in tunicate development. V. The evolution and classification of ascidians. Phil. Trans. Roy. Soc. London B226: 43–70.

————. 1941a. The development of the bud in *Botryllus*. Biol. Bull. 80: 169–84.

————. 1941b. Size and morphogenesis of the bud in *Botryllus*. Biol. Bull. 80: 185–93.

————. 1941c. Spatial and temporal growth patterns in colonial organisms. Third Growth Symp. Growth 5: 89–111.

————. 1947a. The developmental cycle of *Botrylloides*. Quart. J. Microscop. Sci. 88: 393–407.

————. 1947b. The development and growth of *Ciona*. J. Mar. Biol. Assoc. U.K. 26: 616–25.

————. 1947c. The structure, development and budding of the ascidian, *Eudistoma*. J. Morphol. 81: 269–81.

————. 1947d. The structure, tadpole and budding of the ascidian *Pycnoclavella aurilucens* Garstang. J. Mar. Biol. Assoc. U.K. 27: 245–51.

————. 1948a. Budding and the reproductive cycle of *Distaplia*. Quart. J. Microscop. Sci. 89: 253–89.

————. 1948b. Structure, tadpole and bud formation in the ascidian *Archidistoma*. J. Mar. Biol. Assoc. U.K. 27: 380–88.

————. 1950. The Tunicata, with an account of the British species. London: Ray Society. 354 pp.

————. 1951. Regeneration and budding in tunicates. Biol. Rev. 26: 456–75.

————. 1955. The origin of vertebrates. Oxford: Clarendon. 257 pp.

————. 1961a. Growth, development, and pattern. San Francisco: Freeman. 555 pp.

————. 1961b. Salpa. Sci. Amer. 201: 150–60.

————. 1975. Chordata: Tunicata, pp. 241–82, *in* A. C. Giese and J. S. Pearse, eds., Reproduction of marine invertebrates, 2, Entoprocts and lesser coelomates. New York: Academic Press. 344 pp.

Berrill, N. J., and D. P. Abbott. 1949. The structure of the ascidian, *Pycnoclavella stanleyi* n. sp. and the nature of its tadpole larva. Canad. J. Res. D27: 43–49.

Berrill, N. J., and H. Sheldon. 1964. The fine structure of the connections between muscle cells in ascidian tadpole larva. J. Cell Biol. 23: 664–69.

Bertrand, G. A. 1971. The ecology of the nest-building bivalve *Musculus lateralis* commensal with the ascidian *Molgula occidentalis*. Veliger 14: 23–29.

Bethune, W. J. 1955. Some phases in the reproduction and development of the colonial ascidian *Euherdmania claviformis*. Master's thesis, Biological Sciences, Stanford University, Stanford, Calif. 98 pp.

Bevis, P. J. R., and M. C. Thorndyke. 1978. Endocrine cells in the esophagus of the ascidian *Styela clava*, a cytochemical and immunofluorescence study. Cell Tissue Res. 187: 153–58.

Bone, Q. 1972. The origin of chordates. Oxford Biol. Readers 18. London: Oxford University Press. 16 pp.

Bouchard-Madrelle, C., and T. Lender. 1972. Mise en évidence de la neurosécrétion et étude de sa variation chez *Ciona intestinalis* (tunicier), au cours du développement et de la régénération des gonades. Ann. Endocrinol. 33: 129–38.

Bradway, W. 1936. The experimental alteration of the rate of metamorphosis in the tunicate, *Clavelina huntsmani* (Van Name). J. Exper. Zool. 72: 213–24.

Bretz, W. L. 1972. Effects of water-current speed on the filtration rate of *Styela plicata*, a tunicate epibenthic suspension feeder. Amer. Zool. 12: 720 (abstract).

Brewin, B. I. 1953. Australian ascidians of the sub-family Holozoinae and a review of the sub-family. Trans. Roy. Soc. New Zeal. 81: 53–64.

Brien, P. 1939. Contribution à l'étude de bourgeonnement et de l'organogénèse du blastozoïde des Distomidae (=Polycitoridae) (*Distaplia magnilarva* = *Holozoa magnilarva*). Ann. Soc. Roy. Zool. Belg. 70: 101–52, pls. 1, 2.

————. 1948. Embranchement des tuniciers: Morphologie et reproduction, pp. 553–894, *in* P.-P. Grassé, ed., Traité de zoologie, vol. 11. Paris: Masson. 1,077 pp.

————. 1968. Blastogenesis and morphogenesis. Advances Morphogen. 7: 151–203.

————. 1970a. Formation de la branchie chez *Clavelina lepadiformis* Müller au cours de la régénération du thorax par un fragment abdominal et dans le bourgeonnement d'un blastozooide. [With English summary.] Arch. Biol., Liège, 81: 633–69.

————. 1970b. Régénération du thorax par un segment abdominal chez *Clavelina lepadiformis* Müller. [With English summary.] Arch. Biol., Liège, 81: 441–90.

————. 1971. Régénération d'un ascidiozoide de *Clavelina lepadiformis* Müller par un fragment de stolon et par un thorax isolé. Mém. Acad. Roy. Belg., Sci. (8) 39: 9–92.

————. 1972. La reproduction asexuée. Un aspect des études faites à la Station Biologique de Roscoff, de 1924 à 1969. Cah. Biol. Mar. 13: 659–79.

Brunetti, R. 1974. Observations on the life cycle of *Botryllus schlosseri* (Pallas) (Ascidiacea) in the Venetian lagoon. Boll. Zool. 41: 225–52.

————. 1976. Biological cycle of *Botrylloides leachi* (Savigny) (Ascidiacea) in the Venetian lagoon. Vie et Milieu 26: 105–22.

Brunetti, R., and P. Burighel. 1969. Sviluppo dell'apparato vascolare coloniale in *Botryllus schlosseri* (Pallas). Pubbl. Staz. Zool. Napoli 37 (Suppl.): 137–48.

Burighel, P., and R. Brunetti. 1971. The circulatory system in the blastozooid of the colonial ascidian *Botryllus schlosseri* (Pallas). Boll. Zool. 38: 273–89.

Burighel, P., R. Brunetti, and G. Zaniolo. 1976. Hibernation of the colonial ascidian *Botrylloides leachi* (Savigny): Histological observations. Boll. Zool. 43: 293–302.

Burighel, P., and C. Milanesi. 1973. The fine structure of the gastric epithelium of the ascidian *Botryllus schlosseri*. Vacuolated and zygomatic cells. Z. Zellforsch. 145: 541–55.

————. 1975. Fine structure of the gastric epithelium of the ascidian *Botryllus schlosseri*. Mucous, endocrine and plicated cells. Cell Tissue Res. 158: 481–96.

Burighel, P., M. G. Nunzi, and S. Schiaffino. 1977. A comparative study of the organization of the sarcotubular system in ascidian muscle. J. Morphol. 153: 205–24.

Burreson, E. M. 1973. Symbiotic ciliates from solitary ascidians in the Pacific northwest, with a description of *Parahypocoma rhamphisokarya*, n. sp. Trans. Amer. Microscop. Soc. 92: 517–22.

Carlisle, D. B. 1951. Corpora lutea in an ascidian, *Ciona intestinalis*. Quart. J. Microscop. Sci. 92: 201–3.

Carlisle, J. G., Jr., C. H. Turner, and E. E. Ebert. 1964. Artificial habitat in the marine environment. Calif. Dept. Fish & Game, Fish Bull. 124: 1–93.

Carlson, R. M. K. 1975. Nuclear magnetic resonance spectrum of living tunicate blood cells and the structure of the native vanadium chromogen. Proc. Nat. Acad. Sci. 72: 2217–21.

————. 1977. I. The structure of the native vanadium chromogen in tunicate blood cells. II. Rapid analysis of marine sterol mixtures through computer-assisted gas chromatography–mass spectrometry. Doctoral thesis, Chemistry, Stanford University, Stanford, Calif. 252 pp.

Carter, G. T., and K. L. Rinehart, Jr. 1978. Aplidiasphingosine, an antimicrobial and antitumor terpenoid from an *Aplidium* sp. (marine tunicate). J. Amer. Chem. Soc. 100: 7441–42.

Castle, W. E. 1896. The early embryology of *Ciona intestinalis* Flemming (L.). Bull. Mus. Comp. Zool. 27: 201–80, pls. 1–13.

Cavey, M. J., and R. A. Cloney. 1972. Fine structure and differentiation of ascidian muscle. I. Differentiated caudal musculature of *Distaplia occidentalis* tadpoles. J. Morphol. 138: 349–74.

————. 1974. Fine structure and differentiation of ascidian muscle. II. Morphometrics and differentiation of the caudal muscle cells of *Distaplia occidentalis* tadpoles. J. Morphol. 144: 23–70.

————. 1976. Ultrastructure and differentiation of ascidian muscle. I. Caudal musculature of the larva of *Diplosoma macdonaldi*. Cell Tissue Res. 174: 289–313.

Cheng, M. T., and K. L. Rinehart, Jr. 1978. Polyandrocarpidines: Antimicrobial and cytotoxic agents from a marine tunicate (*Polyandrocarpa* sp.) from the Gulf of California. J. Amer. Chem. Soc. 100: 7409–11.

Child, C. M. 1927. Developmental modification and elimination of the larval stage in the ascidian, *Corella willmeriana*. J. Morphol. Physiol. 44: 467–514.

————. 1951. Oxidation-reduction indicator patterns in the development of *Clavelina huntsmani*. Physiol. Zool. 24: 353–67.

Ching, H. L. 1977. Redescription of *Eurylepta leoparda* Freeman, 1933 (Turbellaria: Polycladida), a predator of the ascidian *Corella willmeriana* Herdman, 1898. Canad. J. Zool. 55: 338–42.

————. 1978. Redescription of a marine flatworm: *Pseudoceros canadensis* Hyman, 1953 (Polycladida: Cotylea). Canad. J. Zool. 56: 1372–76.

Ciereszko, L. S., E. M. Ciereszko, E. R. Harris, and C. A. Lane. 1963. Vanadium content of some tunicates. Comp. Biochem. Physiol. 8: 137–40.

Cloney, R. A. 1959. Larval morphology and metamorphosis of *Boltenia villosa* and other ascidians. Doctoral thesis, Zoology, University of Washington, Seattle. 106 pp. plus 3 appendixes and 99 figs.

————. 1964. Development of the ascidian notochord. Acta Embryol. Morphol. Exper. 7: 111–30.

————. 1966. Cytoplasmic filaments and cell movements: Epidermal cells during ascidian metamorphosis. J. Ultrastruc. Res. 14: 300–328.

————. 1969. Cytoplasmic filaments and morphogenesis: The role of the notochord in ascidian metamorphosis. Z. Zellforsch. 100: 31–53.

————. 1977. Larval adhesive organs and metamorphosis in ascidians. I. Fine structure of the everting papillae of *Distaplia occidentalis*. Cell Tissue Res. 183: 423–44.

————. 1978. Ascidian metamorphosis: Review and analysis, pp. 255–82, *in* F.-S. Chia and M. E. Rice, eds., Settlement and metamorphosis of marine invertebrate larvae. New York: Elsevier/North Holland. 290 pp.

Cloney, R. A., and L. Grimm. 1970. Transcellular emigration of blood cells during ascidian metamorphosis. Z. Zellforsch. 107: 157–73.

Coe, W. R. 1932. Season of attachment and rate of growth of sedentary marine organisms at the pier of the Scripps Institution of Oceanography, La Jolla, California. Bull. Scripps Inst. Oceanogr., Tech. Ser. 3: 37–88.

Coe, W. R., and W. E. Allen. 1937. Growth of sedentary marine organisms on experimental blocks and plates for nine successive years at the pier of the Scripps Institution of Oceanography. Bull. Scripps Inst. Oceanogr., Tech. Ser. 4: 101–36.

Colombera, D. 1969. The karyology of the colonial ascidian *Botryllus schlosseri* (Pallas). Caryologia 22: 339–49.

———. 1973. The chromosomes of lower chordates, pp. 167–77, *in* A. B. Chiarelli and E. Campanna, eds., Cytotaxonomy and vertebrate evolution. New York: Academic Press. 783 pp.

———. 1974a. Chromosome evolution in the phylum Echinodermata. Z. Zool. Syst. Evolutionsforsch. 12: 299–308.

———. 1974b. Chromosome number within the class Ascidiacea. Mar. Biol. 26: 63–68.

———. 1974c. Chromosomes of *Molgula manhattensis* de Kay (Ascidiacea). Experientia 30: 352–53.

Conklin, E. G. 1905. The organization and cell-lineage of the ascidian egg. J. Acad. Natur. Sci. Philadelphia 13: 1–119, 12 pls.

Costello, D. P., and C. Henley. 1971. Methods for obtaining and handling marine eggs and embryos. 2nd ed. Marine Biological Laboratory, Woods Hole, Mass. 247 pp.

Cowden, R. R. 1968. The embryonic origin of blood cells in the tunicate *Clavelina picta*. Trans. Amer. Microscop. Soc. 87: 521–24.

Crawford, E. A., and P. J. Homsher. 1975. DNA studies of a Chesapeake Bay population of the tunicate *Molgula manhattensis* (DeKay) (Ascidiacea). Chesapeake Sci. 16: 208–11.

Crisp, D. J., and A. F. A. A. Ghobashy. 1971. Responses of the larva of *Diplosoma listerianum* to light and gravity, pp. 443–65, *in* D. J. Crisp, ed., Fourth European Mar. Biol. Symp. Cambridge, Eng.: Cambridge University Press. 599 pp.

Danskin, G. P. 1978. Accumulation of heavy metals by some solitary tunicates. Canad. J. Zool. 56: 547–51.

Das, S. M. 1948. The physiology of excretion in *Molgula* (Tunicata, Ascidiacea). Biol. Bull. 95: 307–19.

Dawson, A. B., and F. L. Hisaw, Jr. 1964. The occurrence of neurosecretory cells in the neural ganglia of tunicates. J. Morphol. 114: 411–24.

Deck, J. D., E. D. Hay, and J.-P. Revel. 1966. Fine structure and origin of the tunic of *Perophora viridis*. J. Morphol. 120: 267–80.

De Santo, R. S. 1968. The histochemistry, the fine structure, and the ecology of the synthesis of the test in *Botryllus schlosseri* (Pallas)

Savigny. Doctoral thesis, Biology, Columbia University, New York. 277 pp.

De Santo, R. S., and P. L. Dudley. 1969. Ultramicroscopic filaments in the ascidian *Botryllus schlosseri* (Pallas) and their possible role in ampullar contractions. J. Ultrastruc. Res. 28: 259–74.

Deviney, E. M. 1934. The behavior of isolated pieces of ascidian (*Perophora viridis*) stolon as compared with ordinary budding. J. Elisha Mitchell Sci. Soc. 49: 185–224, pls. 15–17.

Dodd, J. M., and M. H. I. Dodd. 1966. An experimental investigation of the supposed pituitary affinities of the ascidian neural complex, pp. 233–52, *in* H. Barnes, ed., Some contemporary studies in marine science. London: Allen & Unwin. 716 pp.

Driesch, H. 1902a. Über ein neues harmonisch-äquipotentielles System und über solche Systeme überhaupt. Arch. Entwickelungsmech. Org. 14: 227–46.

———. 1902b. Studien über das Regulationsvermögen der Organismen. 6. Die Restitutionen der *Clavelina lepadiformis*. Arch. Entwickelungsmech. Org. 14: 247–87.

Dudley, P. L. 1966. Development and systematics of some Pacific marine symbiotic copepods. A study of the biology of the Notodelphyidae, associates of ascidians. Univ. Wash. Publ. Biol. 21. 282 pp.

Dunn, A. D. 1974. Ultrastructural autoradiography and cytochemistry of the iodine-binding cells in the ascidian endostyle. J. Exper. Zool. 188: 103–24.

———. 1975. Iodine metabolism in the ascidian *Molgula manhattensis*. Gen. Comp. Endocrinol. 25: 83–95.

Dybern, B. I. 1963. Biotope choice in *Ciona intestinalis* (L.). Influence of light. Zool. Bidrag Uppsala 35: 589–601.

———. 1965. The life cycle of *Ciona intestinalis* (L.) f. *typica* in relation to environmental temperature. Oikos 16: 109–31.

Eakin, R. M., and A. Kuda. 1971. Ultrastructure of sensory receptors in ascidian tadpoles. Z. Zellforsch. 112: 287–312.

———. 1972. Glycogen in lens of tunicate tadpole (Chordata: Ascidiacea). J. Exper. Zool. 180: 267–70.

Eaton, T. E. 1970. The stem-tail problem and the ancestry of the chordates. J. Paleontol. 44: 969–79.

Ebara, A. 1971. Physiological relation of the dorsal ganglion to the heart of a compound ascidian, *Perophora orientalis*. Comp. Biochem. Physiol. 39A: 795–805.

Eldredge, L. G. 1966. A taxonomic review of Indo-Pacific didemnid ascidians and descriptions of twenty-three central Pacific species. Micronesica 2: 161–261.

Elroi, D., and B. Komarovsky. 1961. On the possible use of the fouling organism *Ciona intestinalis* as a source of vanadium, cellulose and other products. Proc. Gen. Fish. Coun. Medit. 6: 261–67.

Ermak, T. H. 1975a. An autoradiographic demonstration of blood cell renewal in *Styela clava* (Urochordata: Ascidiacea). Experientia 31: 837–38.

_____. 1975b. Cell proliferation in the digestive tract of *Styela clava* (Urochordata: Ascidiacea) as revealed by autoradiography with tritiated thymidine. J. Exper. Zool. 194: 449–66.

_____. 1976a. Cell migration kinetics in the stomach of *Styela clava* (Urochordata: Ascidiacea). J. Exper. Zool. 197: 339–45.

_____. 1976b. Renewal of the gonads in *Styela clava* (Urochordata: Ascidiacea) as revealed by autoradiography with tritiated thymidine. Tissue & Cell 8: 471–78.

_____. 1977. Glycogen deposits in the pyloric gland of the ascidian *Styela clava* (Urochordata). Cell Tissue Res. 176: 47–55.

Everingham, J. W. 1968a. Attachment of intranuclear annulate lamellae to the nuclear envelope. J. Cell Biol. 37: 540–50.

_____. 1968b. Intranuclear annulate lamellae in ascidian embryos. J. Cell Biol. 37: 551–54.

Faulkner, G. T. 1970. Formation and fine structure of the larval tunic in simple ascidians. Doctoral thesis, Zoology, University of Wisconsin. 147 pp.

Fay, R. C., and J. V. Johnson. 1971. Observations on the distribution and ecology of the littoral ascidians of the mainland coast of southern California. Bull. So. Calif. Acad. Sci. 70: 114–24.

Fiala-Médioni, A. 1974. Éthologie alimentaire d'invertébrés benthiques filtreurs (ascidies). II. Variations des taux de filtration et de digestion en fonction de l'espèce. Mar. Biol. 28: 199–206.

_____. 1978a. Filter-feeding ethology of benthic invertebrates (ascidians). III. Recording of water current *in situ*—rate and rhythm of pumping. IV. Pumping rate, filtration rate, filtration efficiency. Mar. Biol. 45: 185–90; 48: 243–49.

_____. 1978b. A scanning electron microscope study of the branchial sac of benthic filter-feeding invertebrates (ascidians). Acta Zool. 59: 1–9.

Fisher, T. R. 1976. Oxygen uptake of the solitary tunicate *Styela plicata*. Biol. Bull. 151: 297–305.

_____. 1977. Metabolic maintenance costs of the suspension feeder *Styela plicata*. Mar. Biol. 41: 361–69.

Franzén, Å. 1976. The fine structure of spermatid differentiation in a tunicate, *Corella parallelogramma* (Müller). Zoon 4: 115–20.

Freeman, G. 1964. The role of blood cells in the process of asexual reproduction in the tunicate *Perophora viridis*. J. Exper. Zool. 156: 157–84.

_____. 1971. A study of the intrinsic factors which control the initiation of asexual reproduction in the tunicate *Amaroucium constellatum*. J. Exper. Zool. 178: 433–56.

Fujita, H., and H. Nanba. 1971. Fine structure and functional properties of the endostyle of ascidians *Ciona intestinalis*. A part of phylogenetic studies of the thyroid gland. Z. Zellforsch. 121: 455–69.

Fuke, M. T., and T. Sugai. 1972. Studies on the naturally occurring hemagglutinin in the coelomic fluid of an ascidian. Biol. Bull. 143: 140–49.

Fukumoto, M. 1971. Experimental control of budding and stolon elongation in *Perophora orientalis*, a compound ascidian. Develop., Growth, Differentiation 13: 73–88.

Gaill, F. 1972. Morphologie comparée de la glande pylorique chez quelques aplousobranches (tuniciers). Arch. Zool. Expér. Gén. 113: 295–307.

_____. 1977. Morphologie et histologie de la glande pylorique des Styelidae (ascidies). Bull. Mus. Nat. Hist. Natur. Paris (Zool.) (3) 340: 1041–55.

Ganssle, D. 1966. Fishes and decapods of San Pablo and Suisun Bays, pp. 64–94, *in* D. W. Kelley, comp., Ecological studies of the Sacramento–San Joaquin Estuary. 1. Zooplankton, zoobenthos, and fishes of San Pablo and Suisun Bays, zooplankton and zoobenthos of the Delta. Calif. Dept. Fish & Game, Fish Bull. 133. 133 pp.

Garstang, S. L., and W. Garstang. 1928. On the development of *Botrylloides*. Quart. J. Microscop. Sci. 72: 1–49.

Garstang, W. 1928. The morphology of the Tunicata, and its bearings on the phylogeny of the Chordata. Quart. J. Microscop. Sci. 72: 51–187.

George, W. C. 1930. Further observations on ascidian blood. J. Elisha Mitchell Sci. Soc. 45: 239–44, pl. 29.

_____. 1937. The formation of new siphon openings in the tunicate, *Styela plicata*. J. Elisha Mitchell Sci. Soc. 53: 87–92, pl. 7.

_____. 1939. A comparative study of the blood of the tunicates. Quart. J. Microscop. Sci. 81: 391–428, pls. 22–24.

Georges, D. 1969. Spermatogénèse et spermiogénèse de *Ciona intestinalis* L. observées au microscope électronique. J. Microscop. 8: 391–400.

_____. 1971. Le rythme circadien dans la glande neurale de l'ascidie *Ciona intestinalis*. Étude d'anatomie microscopique. Acta Zool. 52: 257–73.

_____. 1977. Analyse fonctionelle du complex neural chez *Ciona intestinalis* (tunicier, ascidiacé): Le role du ganglion nerveux. Gen. Comp. Endocrinol. 32: 454–73.

Ghiani, P., and L. Orsi. 1966. Le zone cigliate dell'endostilo in *Ciona intestinalis* L.: Aspetti ultrastrutturali e funzionali. Boll. Mus. Ist. Biol. Univ. Genova 34: 227–78.

Ghiselin, M. T. 1964. Morphological and behavioral concealing adaptations of *Lamellaria stearnsii*, a marine prosobranch gastropod. Veliger 6: 123–24, pl. 16.

Giese, A. C. 1966. Lipids in the economy of marine invertebrates. Physiol. Rev. 46: 244–98.

Gislén, T. 1930. Affinities between the Echinodermata, Enteropneusta, and Chordonia. Zool. Bidrag Uppsala 12: 199–304.

Godeaux, J. E. A. 1974. Introduction to the morphology, phylogenesis, and systematics of lower Deuterostomia, pp. 3–60, *in* M. Florkin and B. T. Scheer, eds., Chemical zoology, 8, Deuterostomians, cyclostomes, and fishes. New York: Academic Press. 682 pp.

Godeaux, J., and H. Firket. 1968. Étude au microscope électronique de l'endostyle d'une ascidie stolidobranche *Molgula manhattensis*. Ann. Sci. Nat. Zool., Paris, (12) 10: 163–86.

Goldberg, E. D., W. McBlair, and K. M. Taylor. 1951. The uptake of vanadium by tunicates. Biol. Bull. 101: 84–94.

Goldin, A. 1948. Regeneration in *Perophora viridis*. Biol. Bull. 94: 184–93.

Goodbody, I. 1957. Nitrogen excretion in Ascidiacea. I. Excretion of ammonia and total non-protein nitrogen. J. Exper. Biol. 34: 297–305.

_____. 1965. Nitrogen excretion in Ascidiacea. II. Storage excretion and the uricolytic enzyme system. J. Exper. Biol. 42: 299–305.

_____. 1974. The physiology of ascidians. Adv. Mar. Biol. 12: 1–149.

Goodbody, I., and J. Gibson. 1974. The biology of *Ascidia nigra* (Savigny). V. Survival in populations settled at different times of the year. Biol. Bull. 146: 217–37 (with refs. to prior papers in this series).

Gorman, A. L. F., J. S. McReynolds, and S. N. Barnes. 1971. Photoreceptors in primitive chordates: Fine structure, hyperpolarizing potentials, and evolution. Science 172: 1052–54.

Grace, R. V. 1971. Giant *Pyrosoma* seen in New Zealand seas. Austral. Natur. Hist. Dec. 1971: 118–19.

Grave, B. H. 1933. Rate of growth, age at sexual maturity, and duration of life of certain sessile organisms, at Woods Hole, Massachusetts. Biol. Bull. 65: 375–86.

Grave, C. 1920. *Amaroucium pellucidum* (Leidy) form *constellatum* (Verrill). I. The activities and reactions of the tadpole larva. J. Exper. Zool. 30: 239–57.

_____. 1921. *Amaroucium constellatum* (Verrill). II. The structure and organization of the tadpole larva. J. Morphol. 36: 71–101.

_____. 1934. The *Botryllus* type of ascidian larva. Carnegie Inst. Wash., Pap. Tortugas Lab. 28: 143–58, pls. 1–4.

Grave, C., and G. K. McCosh. 1923. *Perophora viridis* (Verrill). The activities and structure of the free-swimming larva. Wash. Univ. Stud., Sci. Ser. 11: 89–116.

Grave, C., and G. Riley. 1935. Development of the sense organs of the larva of *Botryllus schlosseri*. J. Morphol. 57: 185–211.

Grave, C., and H. Woodbridge. 1924. *Botryllus schlosseri* (Pallas): The behavior and morphology of the free-swimming larva. J. Morphol. Physiol. 39: 207–47.

Gulliksen, B. 1972. Spawning, larval settlement, growth, biomass, and distribution of *Ciona intestinalis* L. (Tunicata) in Borgenfjorden, North-Tröndelag, Norway. Sarsia 51: 83–96.

_____. 1973. The vertical distribution and habitat of the ascidians in Borgenfjorden, North-Tröndelag, Norway. Sarsia 52: 21–28.

Gulliksen, B., and S. H. Skjaeveland. 1973. The sea-star, *Asterias rubens* L., as predator on the ascidian, *Ciona intestinalis* (L.), in Borgenfjorden, North-Tröndelag, Norway. Sarsia 52: 15–20.

Gutmann, W. F. 1975. The tunicate model. Zool. Beitr. 21: 279–303.

Haderlie, E. C. 1968. Marine fouling organisms in Monterey harbor. Veliger 10: 327–41.

_____. 1969. Marine fouling and boring organisms in Monterey harbor. II. Second year of investigation. Veliger 12: 182–92.

_____. 1971. Marine fouling and boring organisms at 100 feet depth in open water of Monterey Bay. Veliger 13: 249–60.

_____. 1974. Growth rates, depth preference and ecological succession of some sessile invertebrates in Monterey harbor. Veliger 17 (Suppl.): 1–35.

Haderlie, E. C., J. C. Mellor, C. S. Minter III, and G. C. Booth. 1974. The sublittoral benthic fauna and flora off Del Monte beach, Monterey, California. Veliger 17: 185–204.

Hamilton, D. H., Jr., and J. Yurchyk. 1973. Biomass characteristics of the ascidian *Molgula manhattensis* (DeKay). Chesapeake Sci. 14: 67–68.

Hansen, D. J. 1974. Aroclor 1254: Effect on composition of developing estuarine animal communities in the laboratory. Contr. Mar. Sci. 18: 19–33.

Hall, D. A., and H. Saxl. 1960. Human and other animal cellulose. Nature 187: 547–50.

_____. 1961. Studies of human and tunicate cellulose and their relation to reticulin. Proc. Roy. Soc. London B155: 202–17.

Haven, N. D. 1971. Temporal patterns of sexual and asexual reproduction in the colonial ascidian *Metandrocarpa taylori* Huntsman. Biol. Bull. 140: 400–15.

Hecht, S. 1918a. The physiology of *Ascidia atra* Lesueur. I. General physiology. II. Sensory physiology. J. Exper. Zool. 25: 229–59, 261–99.

_____. 1918b. The physiology of *Ascidia atra* Lesueur. III. The blood system. Amer. J. Physiol. 45: 157–87.

Herdman, E. C. 1924. *Botryllus*. Liverpool Mar. Biol. Comm. [LMBC] Memoir 26: 1–40, pls. 1–6.

Heron, A. C. 1975. Advantages of heart reversal in pelagic tunicates. J. Mar. Biol. Assoc. U.K. 55: 959–63.

Hirai, E. 1951. A comparative study on the structures of the tadpoles of ascidians. Sci. Rep. Tôhôku Univ. (4) Biol. 19: 79–87.

_____. 1968. Tunicata, pp. 538–77, *in* M. Kume and K. Dan, eds., Invertebrate embryology. Publ. for Nat. Library of Medicine, U.S. Pub. Health Serv., by NOLIT, Belgrade. 605 pp.

Hisaw, F. L., Jr., C. R. Botticelli, and F. L. Hisaw. 1966. A study of the relation of the neural gland–ganglionic complex to gonadal development in an ascidian, *Chelyosoma productum* Stimpson. Gen. Comp. Endocrinol. 7: 1–9.

Holmes, N. 1973. Water transport in the ascidians *Styela clava* Herdman and *Ascidiella aspersa* (Müller). J. Exper. Mar. Biol. Ecol. 11: 1–13.

_____. 1976. Occurrence of the ascidian *Styela clava* Herdman in Hobsons Bay, Victoria: A new record for the southern hemisphere. Proc. Roy. Soc. Victoria (Austral.) 88: 115–16.

Hoshino, Z. 1969. On the development of the circulatory system of the young ascidian, *Ciona robusta*. Publ. Seto Mar. Biol. Lab. 17: 7–17.

Hoshino, Z., and T. Tokioka. 1967. An unusually robust *Ciona* from the northeastern coast of Honsyu Island, Japan. Publ. Seto Mar. Biol. Lab. 15: 275–90.

Hsu, W. S. 1962. An electron microscopic study on the origin of yolk in the oocytes of the ascidian *Boltenia villosa* Stimpson. Cellule 62: 147–55.

———. 1963. The nuclear envelope in the developing oocytes of the tunicate, *Boltenia villosa*. Z. Zellforsch. 58: 660–78.

———. 1967. The origin of annulate lamellae in the oocyte of the ascidian, *Boltenia villosa* Stimpson. Z. Zellforsch. 82: 376–90.

Hsu, W. S., and R. A. Cloney. 1958. Mitochondria and yolk formation in the ascidian, *Boltenia villosa* Stimpson. Cellule 59: 211–24, pls. 1–6.

Huntsman, A. G. 1911. Ascidians from the coasts of Canada. Trans. Canad. Inst. 9: 111–48.

———. 1912. Holosomatous ascidians from the coast of western Canada. Contr. Canad. Biol. 1906–10: 103–85, pls. 10–21.

———. 1913a. The classification of the Styelidae. Zool. Anz. 41: 482–501.

——— 1913b. Protostigmata in ascidians. Proc. Roy. Soc. London B86: 440–53.

———. 1913c. On the origin of the ascidian mouth. Proc. Roy. Soc. London B86: 454–59.

———. 1921. Age-determination, growth and symmetry in the test of the ascidian, *Chelyosoma*. Trans. Roy. Canad. Inst. 13: 27–38, pl. 1.

Huus, J. 1937. Ascidiaceae = Tethyodeae = Seescheiden, pp. 545–692, *in* W. Kükenthal and T. Krumbach, eds., Handbuch der Zoologie, 5(2), Tunicata. Berlin and Leipzig: Gruyter.

Huwae, P. H. M., and M. S. S. Lavaleye. 1975. *Styela clava* Herdman, 1882 (Tunicata, Ascidiacea), new for Netherlands. [In Dutch; English summary.] Zool. Bijdr. 17: 79–81.

Illg, P. L. 1958. North American copepods of the family Notodelphyidae. Proc. U.S. Nat. Mus. 107: 463–649.

Ivanova-Kazas, O. M. 1967. Reproduction asexuée et cycle évolutif de l'ascidie *Distaplia unigermis*. Cah. Biol. Mar. 8: 21–62.

Izzard, C. S. 1968. Migration of germ cells through successive generations of pallial buds in *Botryllus schlosseri*. Biol. Bull. 135: 424 (abstract).

———. 1973. Development of polarity and bilateral asymmetry in the palleal bud of *Botryllus schlosseri*. J. Morphol. 139: 1–26.

———. 1974. Contractile filipodia and *in vivo* cell movement in the tunic of the ascidian, *Botryllus schlosseri*. J. Cell Sci. 15: 513–35.

Jackson, J. B. C., and L. Buss. 1975. Alleopathy and spatial competition among coral reef invertebrates. Proc. Nat. Acad. Sci. 72: 5160–63.

Jefferies, R. P. S. 1975. Fossil evidence concerning the origin of the chordates. Symp. Zool. Soc. London 36: 253–318.

Johnson, J. V. 1971. The annual growth and reproductive cycle of *Styela clava* in the Marina Del Rey, Venice, California. Master's thesis, Zoology, University of Nebraska. 31 pp.

Jollie, M. 1973. The origin of the chordates. Acta Zool. 54: 81–100.

Jorgensen, C. B. 1966. Biology of suspension feeding. Oxford: Pergamon. 357 pp.

Kanatani, H., T. Hidaka, and M. Onosaka. 1964. On the spawning season and growth of the ascidian *Styela plicata* in Aburatsubo Bay. [In Japanese; English summary.] Zool. Mag., Tokyo, 73: 108–11.

Katow, H., and H. Watanabe. 1974. Discrimination and elimination of the degenerated cells of the ampullae in the compound ascidian *Botryllus primigenus* Oka. [In Japanese; English summary.] Zool. Mag., Tokyo, 83: 146–51.

Kessel, R. G. 1967. The origin and fate of secretion in the follicle cells of tunicates. Z. Zellforsch. 76: 21–30.

Kessel, R. G., and N. E. Kemp. 1962. An electron microscope study of the oocyte, test cell and follicular envelope of the tunicate, *Molgula manhattensis*. J. Ultrastruc. Res. 6: 57–76.

King, R. E. 1975. The population biology of *Ascidia ceratodes* in Monterey harbor, California. Master's thesis, Biological Sciences, San Francisco State University. 76 pp.

Kokubu, N., and T. Hidaka. 1965. Tantalum and niobium in ascidians. Nature 205: 1028–29.

Kott, P. 1968. A review of the genus *Halocynthia* Verrill, 1879. Proc. Linn. Soc. New South Wales 93: 76–89.

———. 1969. Antarctic Ascidiacea. Antarc. Res. Ser. 13. Washington, D.C.: Amer. Geophys. Union, NAS-NRC. 239 pp.

———. 1976. Introduction of the North Atlantic ascidian *Molgula manhattensis* (DeKay) to two Australian river estuaries. Mem. Queensland Mus. 17: 449–55.

Kriebel, M. E. 1968a. Studies on cardiovascular physiology of tunicates. Biol. Bull. 134: 434–55.

———. 1968b. Pacemaker properties of tunicate heart cells. Biol. Bull. 135: 166–73.

———. 1968c. Cholinoceptive and adrenoceptive properties of the tunicate heart pacemaker. Biol. Bull. 135: 174–80.

Krijgsman, B. J. 1956. Contractile and pacemaker mechanisms of the heart of tunicates. Biol. Rev. 31: 288–312.

Krishnan, G. 1975. Nature of tunicin and its interaction with other chemical components of the tunic of the ascidian, *Polyclinum madrasensis* Sebastian. Indian J. Exper. Biol. 13: 172–76.

Kustin, K., K. V. Ladd, and G. C. McLeod. 1975. Site and rate of vanadium assimilation in the tunicate *Ciona intestinalis*. J. Gen. Physiol. 65: 315–28.

Kustin, K., K. V. Ladd, G. C. McLeod, and D. L. Toppen. 1974. Water transport rates of the tunicate *Ciona intestinalis*. Biol. Bull. 147: 608–17.

Kustin, K., D. S. Levine, G. C. McLeod, and W. A. Curby. 1976. The blood of *Ascidia nigra*: Blood-cell frequency distribution, morphology, and the distribution and valence of vanadium in living blood cells. Biol. Bull. 150: 426–41.

Ladd, K. V. 1974. The distribution and assimilation of vanadium with respect to the tunicate *Ciona intestinalis*. Doctoral thesis, Brandeis University, Waltham, Mass. 139 pp.

Lafargue, F. 1974. Révision taxonomique des Didemnidae des côtes de France (ascidies composées). Description des espèces de Banyuls-sur-mer. Généralités. Genre *Trididemnum*. Ann. Inst. Océanogr. 50: 173–84.

Lafargue, F., and E. Kniprath. 1978. Formation des spicules de Didemnidae (ascidies composées). 1. L'apparition des spicules chez l'oozoide après la métamorphose. Mar. Biol. 45: 175–84.

Lafargue, F., and T. Valentinčič. 1973. *Diplosoma carnosum* Von Drasche, 1883 (ascidie composée, nord Adriatique) et essai de clé tabulaire des espèces Européennes du genre *Diplosoma*. Biol. Vestn. (Ljubljana) 21: 139–51.

Laird, C. D. 1971. Chromatid structure: Relationship between DNA content and nucleotide sequence diversity. Chromosoma 32: 378–406.

Lambert, C. C. 1971. Genetic transcription during the development and metamorphosis of the tunicate *Ascidia callosa*. Exper. Cell Res. 66: 401–9.

Lambert, C. C., and C. L. Brandt. 1967. The effect of light on the spawning of *Ciona intestinalis*. Biol. Bull. 132: 222–28.

Lambert, C. C., and C. D. Laird. 1971. Molecular properties of tunicate DNA. Biochim. Biophys. Acta 240: 39–45.

Lambert, C. C., and G. Lambert. 1978. Tunicate eggs utilize ammonium ions for flotation. Science 200: 64–65.

Lambert, G. 1968. The general ecology and growth of a solitary ascidian, *Corella willmeriana*. Biol. Bull. 135: 296–307.

———. 1978. Early post-metamorphic growth and spicule formation in the compound ascidian *Cystodytes lobatus*, with observations on predation by a prosobranch. 59th Ann. Meet., West. Soc. Natur., Abs. Symp. Contr. pap., pp. 39–40 (abstract).

Lane, D. J. W. 1973. Attachment of the larva of the ascidian *Diplosoma listerianum*. Mar. Biol. 21: 47–58.

Lane, N. J. 1971. The neural gland in tunicates: Fine structure and intracellular distribution of phosphatases. Z. Zellforsch. 120: 80–93.

———. 1972. Neurosecretory cells in the cerebral ganglion of adult tunicates: Fine structure and distribution of phosphatases. J. Ultrastruc. Res. 40: 480–97.

Lash, J. W., R. A. Cloney, and R. R. Minor. 1973. The effect of cytochalasin B upon tail resorption and metamorphosis in ten species of ascidians. Biol. Bull. 145: 360–372.

Levine, E. P. 1960. Studies on the structure, reproduction, development, and accumulation of metals in the colonial ascidian *Eudi-*

stoma ritteri Van Name, 1945. Doctoral thesis, Biological Sciences, Stanford University, Stanford, Calif. 192 pp.

———. 1961. Occurrence of titanium, vanadium, chromium, and sulfuric acid in the ascidian *Eudistoma ritteri*. Science 133: 1352–53.

———. 1962a. An evaluation of the ordinal division of the ascidian orders Phlebobranchiata and Aplousobranchiata based on a comparison of three species of *Eudistoma* with *Diazona*. Amer. Zool. 2: 424 (abstract).

———. 1962b. Studies on the structure, reproduction, development, and accumulation of metals in the colonial ascidian *Eudistoma ritteri* Van Name, 1945. J. Morphol. 111: 105–37.

Lewin, R. A. 1975. A marine *Synechocystis* (Cyanophyta, Chroococcales) epizoic on ascidians. Phycologia 14: 153–60.

———. 1976. Prochlorophyta as a proposed new division of algae. Nature 261: 697–98.

———. 1977. *Prochloron*, type genus of the Prochlorophyta. Phycologia 16: 217.

Lyerla, T. A., and J. H. Lyerla. 1978. A preliminary study of electrophoretic variation of enzymes in the tunicates *Clavelina picta* and *Clavelina oblonga*. Comp. Biochem. Physiol. 59B: 111–16.

Lynch, W. F. 1961. Extrinsic factors influencing metamorphosis in bryozoan and ascidian larvae. Amer. Zool. 1: 59–66.

MacGinitie, G. E. 1939a. The method of feeding of tunicates. Biol. Bull. 77: 443–47.

———. 1939b. Some effects of fresh water on the fauna of a marine harbor. Amer. Midl. Natur. 21: 681–86.

Mackie, G. O. 1974. Behaviour of a compound ascidian. Canad. J. Zool. 52: 23–27.

Mackie, G. O., D. H. Paul, C. L. Singla, M. A. Sleigh, and D. E. Williams. 1974. Branchial innervation and ciliary control in the ascidian *Corella*. Proc. Roy. Soc. London B187: 1–35.

Makioka, A., T. Hirabayashi, and H. Watanabe. 1978. Distribution of antigenic sites common to vertebrate tropomyosin in tunicate muscle. Mar. Biol. 48: 261–69.

Mauzey, K. P., C. Birkeland, and P. K. Dayton. 1968. Feeding behavior of asteroids and escape responses of their prey in the Puget Sound region. Ecology 49: 603–19.

McLean, J. H. 1962. Sublittoral ecology of kelp beds of the open coast area near Carmel, California. Biol. Bull. 122: 95–114.

McLeod, G. C., K. V. Ladd, K. Kustin, and D. L. Toppen. 1975. Extraction of vanadium (V) from seawater by tunicates: A revision of concepts. Limnol. Oceanogr. 20: 491–93.

Milkman, R. 1967. Genetic and developmental studies on *Botryllus schlosseri*. Biol. Bull. 132: 229–43.

Millar, R. H. 1951. The stolonic vessels of the Didemnidae. Quart. J. Microscop. Sci. 92: 249–54.

———. 1952. The annual growth and reproductive cycle in four ascidians. J. Mar. Biol. Assoc. U.K. 31: 41–61.

————. 1953. *Ciona*. Liverpool Mar. Biol. Comm. [LMBC] Memoir 35: 1–123, pls. 1–19.

————. 1959. Ascidiacea. Galathea Rep. 1: 189–209, pl. 1.

————. 1960. Ascidiacea. Discovery Rep. 30: 1–160, pls. 1–6.

————. 1963. Australian ascidians in the British Museum (Natural History). Proc. Zool. Soc. London 141: 689–746.

————. 1966a. Evolution in ascidians, pp. 519–34, *in* H. Barnes, ed., Some contemporary studies in marine science. London: Allen & Unwin. 716 pp.

————. 1966b. Tunicata. Ascidiacea. Marine invertebrates of Scandinavia No. 1. Oslo: Oslo University Press. 123 pp.

————. 1970a. Ascidians, including specimens from the deep sea, collected by R. V. "Vema" and now in the American Museum of Natural History. Zool. J. Linn. Soc. 49: 99–159.

————. 1970b. British ascidians. Tunicata: Ascidiacea. Keys and notes for the identification of the species. Synopses of the British fauna No. 1. London: Academic Press. 92 pp.

————. 1971a. Ascidians as fouling organisms, pp. 185–95, *in* E. B. G. Jones and S. K. Eltringham, eds., Marine borers, fungi and fouling organisms of wood. Organization for Economic Cooperation and Development (OECD), Paris.

————. 1971b. The biology of ascidians. Adv. Mar. Biol. 9: 1–100.

Miller, R. L. 1975. Chemotaxis of the spermatozoa of *Ciona intestinalis*. Nature 254: 244–45.

Minganti, A. 1974. Hybrid larvae between *Ascidia callosa* Stimpson and *Ascidia paratropa* (Huntsman). Acta Embryol. Exper. 1974: 43–50.

Mishra, A. K., and J. R. Colvin. 1969. The microscopic and submicroscopic structure of the tunic of two ascidians, *Boltenia* and *Molgula*. Canad. J. Zool. 47: 659–63.

————. 1970. Scanning electron microscopy of the spines of the tunic of the ascidian, *Boltenia ovifera*. Canad. J. Zool. 48: 475–77.

Miyamoto, D., and G. Freeman. 1970. The origin of the cells which form zooids during vascular budding in the ascidian *Botrylloides diegense*. Amer. Zool. 10: 533–34 (abstract).

Monniot, C. 1965. Étude systématique et évolutive de la famille des Pyuridae (Ascidiacea). Mém. Mus. Nat. Hist. Natur. (A) 36: 1–203.

Monniot, C., and F. Monniot. 1972. Clé mondiale des genres d'ascidies. Key to ascidian genera of the world. [Parallel texts in French and English.] Arch. Zool. Expér. Gén. 113: 311–67.

————. 1975. Abyssal tunicates: An ecological paradox. Ann. Inst. Océanogr., Paris, 51: 99–129.

————. 1978. Recent work on the deep-sea tunicates. Oceanogr. Mar. Biol. Ann. Rev. 16: 181–228.

Monniot, C., F. Monniot, and F. Gaill. 1975. Les Sorberacea: Une nouvelle classe de tuniciers. Arch. Zool. Expér. Gén. 116: 77–122.

Monniot, F. 1965. Ascidies interstitielles des côtes d'Europe. Mém. Mus. Nat. Hist. Natur. (A) 35: 1–154, pls. 1–10.

————. 1970a. *Cystodytes incrassatus* n. sp., ascidie fossile du Pliocène breton. Nouvelle interprétation des *Neanthozoites* Deflandre-Rigaud. C. R. Acad. Sci., Paris, 271: 2280–82, pl. 1.

————. 1970b. Les spicules chez les tuniciers aplousobranches. Arch. Zool. Expér. Gén. 111: 303–11, pls. 1–6.

Morgan, T. H. 1938–44. The genetics and physiological problems of self-sterility in *Ciona*. I, II (1938): J. Exper. Zool. 78: 271–334; III, IV (1939): J. Exper. Zool. 80: 19–80; V (1942): J. Exper. Zool. 90: 199–228; VI (1944): J. Exper. Zool. 95: 37–59.

————. 1944. Some further data on self fertilization in *Ciona*. J. Exper. Zool. 97: 231–48.

————. 1945a. The conditions that lead to normal or abnormal development of *Ciona*. Biol. Bull. 88: 50–62.

————. 1945b. Normal and abnormal development of eggs of *Ciona*. J. Exper. Zool. 100: 407–16.

Mukai, H. 1974a. A histological study on the degeneration of zooids in a compound ascidian, *Botryllus primigenus*. [In Japanese; English summary.] Zool. Mag., Tokyo, 83: 18–23.

————. 1974b. Photo-induced accumulation of pigment cells in a compound ascidian, *Botryllus primigenus*. Annot. Zool. Japon. 47: 43–47.

————. 1977a. Comparative studies on the structure of reproductive organs of four botryllid ascidians. J. Morphol. 152: 363–79.

————. 1977b. Histological and histochemical studies of two compound ascidians, *Clavelina lepadiformis* and *Diazona violacea*, with special reference to the trophocytes, ovary and pyloric gland. Sci. Rep. Fac. Educ. Gunma Univ. 26: 37–77.

Mukai, H., K. Sugimoto, and Y. Taneda. 1978. Comparative studies on the circulatory system of the compound ascidians *Botryllus*, *Botrylloides* and *Symplegma*. J. Morphol. 157: 49–77.

Mukai, H., and H. Watanabe. 1974. On the occurrence of colony specificity in some compound ascidians. Biol. Bull. 147: 411–21.

————. 1975a. Distribution of fusion incompatibility types in natural populations of the compound ascidian, *Botryllus primigenus*. Proc. Japan Acad. 51: 44–47.

————. 1975b. Fusibility of colonies in natural populations of the compound ascidian, *Botrylloides violaceus*. Proc. Japan Acad. 51: 48–50.

————. 1976a. Relation between sexual and asexual reproduction in the compound ascidian *Botryllus primigenus*. Sci. Rep. Fac. Educ. Gunma Univ. 25: 61–79.

————. 1976b. Studies on the formation of germ cells in a compound ascidian, *Botryllus primigenus* Oka. J. Morphol. 148: 337–61.

————. 1977. Shedding of gametes in the compound ascidian *Botryllus primigenus*. Mar. Biol. 39: 311–17.

Müller, K. J. 1977. *Palaeobotryllus* from the Upper Cambrian of Nevada—a probable ascidian. Lethaia 10: 107–18.

Nakauchi, M. 1966a. Budding and colony formation in the ascidian, *Amaroucium multiplicatum*. Japon. J. Zool. 15: 151–72.

_____. 1966b. Budding and growth in the ascidian, *Archidistoma aggregatum*. Rep. Usa Mar. Biol. Sta. 13: 1–10.

_____. 1970. Asexual reproduction in *Amaroucium yamazii* (a colonial ascidian). Publ. Seto Mar. Biol. Lab. 17: 309–28, pls. 16–17.

_____. 1977. Development and budding in the oozooids of polyclinid ascidians. 2. *Ritterella pulchra*. Annot. Zool. Japon. 50: 151–59.

_____. 1979. Development and budding in the oozooids of polyclinid ascidians. 3. *Aplidium solidum*. Annot. Zool. Japon. 52: 40–49.

Nakauchi, M., and K. Kawamura. 1974a. Behavior of buds during common cloacal system formation in the ascidian, *Aplidium multiplicatum*. Rep. Usa Mar. Biol. Sta. 21: 19–27.

_____. 1974b. Experimental analysis of the behavior of buds in the ascidian, *Aplidium multiplicatum*. I. Rep. Usa Mar. Biol. Sta. 21: 29–38.

_____. 1978. Additional experiments on the behavior of buds in the ascidian, *Aplidium multiplicatum*. Biol. Bull. 154: 453–62.

Newberry, A. T. 1965a. The structure of the circulatory apparatus of the test and its role in budding in the polystyelid ascidian *Metandrocarpa taylori* Huntsman. Mém. Acad. Roy. Belg., Sci. (2) 16: 1–59, figs. 1–42.

_____. 1965b. Vascular structures associated with budding in the polystyelid ascidian *Metandrocarpa taylori*. Ann. Soc. Roy. Zool. Belg. 95: 57–74.

Nishiyama, A., Y. Sasaki, and T. Suzuki. 1972. Tetrodotoxin and *Ciona* heart. Tôhôku J. Exper. Med. 107: 95–96.

Nolfi, J. R. 1970. Biosynthesis of uric acid in the tunicate, *Molgula manhattensis*, with a general scheme for the function of stored purines in animals. Comp. Biochem. Physiol. 35: 827–42.

Oka, H., and H. Watanabe. 1957. Vascular budding, a new type of budding in *Botryllus*. Biol. Bull. 112: 225–40.

_____. 1959. Vascular budding in *Botrylloides*. Biol. Bull. 117: 340–46.

Oliphant, L. W. 1972. Ultrastructure and mechanism of contraction of the ascidian heart. Doctoral thesis, Zoology, University of Washington, Seattle. 193 pp.

Ooishi, S., and P. L. Illg. 1974. *Haplostomella halocynthiae* (Fukui), an ascidicolid copepod associated with a simple ascidian, *Halocynthia roretzi* (Drasche) from Japan. Publ. Seto Mar. Biol. Lab. 21: 365–75, pl. xi.

_____. 1977. Haplostominae (Copepoda, Cyclopoida) associated with compound ascidians from the San Juan Archipelago and vicinity. Spec. Publ. Seto Mar. Biol. Lab., Ser. 5. 154 pp., 1 pl.

Overton, J. 1966. The fine structure of blood cells in the ascidian *Perophora viridis*. J. Morphol. 119: 305–26.

Papadopoulou, C., and G. D. Kanias. 1977. Tunicate species as marine pollution indicators. Mar. Poll. Bull. 8: 229–31.

Pearce, J. B. 1966. The biology of the mussel crab, *Fabia subquadrata*, from the waters of the San Juan Archipelago, Washington. Pacific Sci. 20: 3–35.

Pérès, J.-M. 1948a. Recherche sur la genèse et la régénération de la tunique chez *Ciona intestinalis* L. Bull. Inst. Océanogr. Monaco 936: 1–12.

_____. 1948b. Recherches sur la genèse et la régénération de la tunique chez *Clavelina lepadiformis* Müller. Arch. Anat. Microscop. Morphol. Expér. 37: 230–60.

_____. 1948c. Recherches sur le sang et la tunique commune des ascidies composées. I. Aplousobranchiata (Polyclinidae et Didemnidae). Ann. Inst. Océanogr. 28: 345–473.

_____. 1952. Recherches sur le cycle sexuel de *Ciona intestinalis* (L.). Arch. Anat. Microscop. Morphol. Expér. 41: 153–83.

_____. 1954. Considérations sur le fonctionnement ovarien chez *Ciona intestinalis* (L.). Arch. Anat. Microscop. Morphol. Expér. 43: 58–78.

Peters, W. 1966. Chitin in Tunicata. Experientia 22: 820–21.

Pizon, A. 1892. Histoire de la blastogénèse chez les Botryllidés. Ann. Sci. Natur., Zool. Paleontol. 14: 1–386, pls. 1–9.

_____. 1899. Études biologiques sur les tuniciers coloniaux fixés. Bull. Soc. Sci. Nat. Ouest France 9: 1–56, pls. 1–16.

Plough, H. H. 1969. Genetic polymorphism in a stalked ascidian from the Gulf of Maine. J. Heredity 60: 193–205.

_____. 1978. Sea squirts of the Atlantic continental shelf from Maine to Texas. Baltimore: Johns Hopkins University Press. 118 pp.

Quast, J. C. 1968. Observations on the food of the kelp-bed fishes, pp. 109–42, *in* W. J. North and C. L. Hubbs, eds., Utilization of kelp-bed resources in southern California. Calif. Dept. Fish & Game, Fish Bull. 139: 1–264.

Redick, T. 1969. The effect of temperature on heart rate in *Molgula manhattensis*. Amer. Zool. 9: 589 (abstract).

Reese, J. P. 1967. Photoreceptive regulation of spawning in *Ciona intestinalis* (L.). Master's thesis, Biology, San Diego State College, Calif. 81 pp.

Reish, D. J. 1963. Mass mortality of marine organisms attributed to the "red tide" in southern California. Calif. Fish & Game 49: 265–70.

_____. 1964a. Studies on the *Mytilus edulis* community in Alamitos Bay, California. I. Development and destruction of the community. Veliger 6: 124–31.

_____. 1964b. Studies on the *Mytilus edulis* community in Alamitos Bay, California. II. Population variations and discussion of the associated organisms. Veliger 6: 202–7.

Reverberi, G. 1961. The embryology of ascidians. Advances Morphogen. 1: 55–101.

_____. 1971. Ascidians, pp. 507–50, *in* G. Reverberi, ed., Experimental embryology of marine and fresh-water invertebrates. Amsterdam: North-Holland. 587 pp.

Ries, E. 1937. Die Tropfenzellen und ihre Bedeutung für die Tunica-bildung bei *Clavelina*. Roux Arch. Entwickelungsmech. Org. 137: 363–71.

Ritter, W. E. 1893. Tunicata of the Pacific coast of North America. I. *Perophora annectens* n. sp. Proc. Calif. Acad. Sci. (2) 4: 37–85, pls. 1–3.

———. 1896. Budding in compound ascidians, based on studies on *Goodsiria* and *Perophora*. J. Morphol. 12: 149–238, pls. 12–17. [*Goodsiria* = *Metandrocarpa*.]

———. 1901. Papers from the Harriman Alaska Expedition. 23. The ascidians. Proc. Washington Acad. Sci. 3: 225–66, pls. 27–30.

———. 1903. The structure and affinities of *Herdmania claviformis*, the type of a new genus and family of ascidians from the coast of California. Mark Anniversary Vol.: 237–61, pls. 18–19. [*Herdmania* changed to *Euherdmania* 1904.]

———. 1907. The ascidians collected by the United States Fisheries Bureau steamer Albatross on the coast of California during the summer of 1904. Univ. Calif. Publ. Zool. 4: 1–52, pls. 1–3.

———. 1909. *Halocynthia johnsoni* n. sp., a comprehensive inquiry as to the extent of law and order that prevails in a single animal species. Univ. Calif. Publ. Zool. 6: 65–114, pls. 7–14.

Ritter, W. E., and R. A. Forsyth. 1917. Ascidians of the littoral zone of southern California. Univ. Calif. Publ. Zool. 16: 439–512.

Rosati, F., and R. De Santis. 1978. Studies on fertilization in the ascidians. 1. Self-sterility and specific recognition between gametes of *Ciona intestinalis*. Exper. Cell Res. 112: 111–19.

Rosenthal, R. J., W. D. Clarke, and P. K. Dayton. 1974. Ecology and natural history of a stand of giant kelp, *Macrocystis pyrifera*, off Del Mar, California. U.S. Nat. Mar. Fish. Serv., Fish. Bull., 72: 670–84.

Russo, R. A. 1975. Observations on the food habits of leopard sharks (*Triakis semifasciata*) and brown smoothhounds (*Mustelus henlei*). Calif. Fish & Game 61: 95–103.

Sabbadin, A. 1955a. Il ciclo biologico di *Botryllus schlosseri* (Pallas) [Ascidiacea] nella Laguna di Venezia. Arch. Oceanogr. Limnol. 10: 217–30, pl. 1.

———. 1955b. Studio sulle cellule del sangue di *Botryllus schlosseri* (Pallas) [Ascidiacea]. Arch. Ital. Anat. Embriol. 60: 33–67.

———. 1957. Il ciclo biologico di *Ciona intestinalis* (L.), *Molgula manhattensis* (DeKay) e *Styela plicata* (Lesueur) nella Laguna Veneta. Arch. Oceanogr. Limnol. 11: 1–28, pls. 1–2.

———. 1960. Nuove ricerche sull 'inversione sperimentale del "situs viscerum" in *Botryllus schlosseri*. Arch. Oceanogr. Limnol. 12: 131–43. 1 pl.

———. 1964. The pigments of *Botryllus schlosseri* (Ascidiacea) and their genetic control. Ricerca Sci. 34(IIB): 439–44.

———. 1969. The compound ascidian *Botryllus schlosseri* in the field and in the laboratory. Pubbl. Staz. Zool. Napoli 37 (Suppl.): 62–72.

———. 1971. Self- and cross-fertilization in the compound ascidian *Botryllus schlosseri*. Develop. Biol. 24: 379–91.

———. 1972. Results and perspectives in the study of a colonial ascidian, *Botryllus schlosseri*. Fifth Europ. Mar. Biol. Symp.: 327–34.

———. 1973. Recherches expérimentales sur l'ascidie coloniale *Botryllus schlosseri*. Bull. Soc. Zool. France 98: 417–34.

———. 1977. Linkage between two loci controlling colour polymorphism in the colonial ascidian *Botryllus schlosseri*. Experientia 33: 876–77.

Sabbadin, A., and G. Graziani. 1967a. Microgeographical and ecological distribution of colour morphs of *Botryllus schlosseri* (Ascidiacea). Nature 213: 815–16.

———. 1967b. New data on the inheritance of pigments and pigmentation patterns in the colonial ascidian *Botryllus schlosseri* (Pallas). Rivista Biol. 60: 581–98.

Sabbadin, A., and A. Tontodonati. 1967. Nitrogenous excretion in the compound ascidians *Botryllus schlosseri* (Pallas) and *Botrylloides leachi* (Savigny). Monitore Zool. Ital. (n.s.) 1: 185–90.

Sabbadin, A., G. Zaniolo, and F. Majone. 1975. Determination of polarity and bilateral asymmetry in palleal and vascular buds of the ascidian *Botryllus schlosseri*. Develop. Biol. 46: 79–87.

Saffo, M. B. 1976. Studies on the renal sac of *Molgula manhattensis* DeKay (Ascidiacea, Tunicata, Phylum Chordata). Doctoral thesis, Biological Sciences, Stanford University, Stanford, Calif. 128 pp.

———. 1978. Studies on the renal sac of the ascidian *Molgula manhattensis*. 1. Development of the renal sac. J. Morphol. 155: 287–309.

Saffo, M. B., and H. A. Lowenstam. 1978. Calcareous deposits in the renal sac of a molgulid tunicate. Science 200: 1166–68.

Salvatore, G., G. Vecchio, and V. Macchia. 1961. Sur la présence d'hormones thyroidiennes chez un tunicier, *Clavelina lepadiformis* (M. Edw.) var. *rissoana*. C. R. Soc. Biol. Paris 154: 1380–84.

Scheer, B. T. 1945. The development of marine fouling communities. Biol. Bull. 89: 103–21.

Schmidtke, J., C. Weiler, B. Kunz, and W. Engel. 1977. Isozymes of a tunicate and a cephalochordate as a test of polyploidisation in chordate evolution. Nature 266: 532–33.

Schulz-Baldes, M., and R. A. Lewin. 1976. Fine structure of *Synechocystis didemni* (Cyanophyta: Chroococcales). Phycologia 15: 1–6.

Scott, Sister F. M. 1945. The developmental history of *Amaroecium constellatum*. I. Early embryonic development. Biol. Bull. 88: 126–38.

———. 1946. The developmental history of *Amaroecium constellatum*. II. Organogenesis of the larval action system. Biol. Bull. 91: 66–80.

———. 1952. The developmental history of *Amaroecium constellatum*. III. Metamorphosis. Biol. Bull. 103: 226–41.

———. 1954. Metamorphic differentiation in *Amaroecium constellatum* treated with nitrogen mustard. J. Exper. Zool. 127: 331–66.

———. 1957. Regeneration and differentiation in buds of *Amaroe-*

cium treated with nitrogen mustard. J. Exper. Zool. 135: 557–86.

———. 1959. Tissue affinity in *Amaroecium*. I. Aggregation of dissociated fragments and their integration into one organism. Acta Embryol. Morphol. Exper. 2: 209–26.

———. 1962. Tissue affinity in *Amaroecium*. II. Reaggregation of three partial zooids into functioning Siamese twins. Biol. Bull. 122: 396–416.

Scott, Sister F. M., and J. E. Schuh, S.J. 1963. Intraspecific reaggregation in *Amaroecium constellatum* labeled with tritiated thymidine. Acta Embryol. Morphol. Exper. 6: 39–54.

Sengel, P., and M. Kieny. 1963. Role du complexe formé par la glande neurale, le ganglion nerveux et l'organe vibratile sur la différenciation sexuelle des gonades de *Molgula manhattensis*. Bull. Soc. Zool. France 87: 615–28.

Shumway, S. E. 1978. Respiration, pumping activity and heart rate in *Ciona intestinalis* exposed to fluctuating salinities. Mar. Biol. 48: 235–42.

Silver, M. W. 1975. The habitat of *Salpa fusiformis* in the California Current as defined by indicator assemblages. Limnol. Oceanogr. 20: 230–37.

Sims, L. 1980. Osmoregulatory capabilities of three co-occurring stolidobranch ascidians, *Styela clava*, *S. plicata*, and *S. montereyensis*. Master's thesis, Biology, California State University, Fullerton.

Skogsberg, T., and G. H. Vansell. 1928. Structure and behavior of the amphipod, *Polycheria osborni*. Proc. Calif. Acad. Sci. (4) 17: 267–95.

Smith, M. J., and P. A. Dehnel. 1971. The composition of tunic from four species of ascidians. Comp. Biochem. Physiol. 40B: 615–22.

Smith, S. D. 1970. Effects of electrical fields upon regeneration in the metazoa. Amer. Zool. 10: 133–40.

Stephens, G. C., and R. A. Schinske. 1961. Uptake of amino acids by marine invertebrates. Limnol. Oceanogr. 6: 175–81.

Stoecker, D. 1978. Resistance of a tunicate to fouling. Biol. Bull. 155: 615–26.

Sutherland, J. P. 1978. Functional roles of *Schizoporella* and *Styela* in the fouling community at Beaufort, North Carolina. Ecology 59: 257–64.

Sutton, M. F. 1953. The regeneration of the siphons of *Ciona intestinalis* L. J. Mar. Biol. Assoc. U.K. 33: 249–68.

Swinehart, J. H., W. R. Biggs, D. J. Halko, and N. C. Schroeder. 1974. The vanadium and selected metal contents of some ascidians. Biol. Bull. 146: 302–12.

Tan, W. L. 1975. Natural products from some marine invertebrates. [Part C. Investigation of the sterols from ascidians.] Doctoral thesis, Chemistry, Stanford University, Stanford, Calif. 150 pp.

Tanaka, K. 1973. Allogenic inhibition in a compound ascidian, *Botryllus primigenus* Oka. II. Cellular and humoral responses in "nonfusion" reaction. Cell Immunol. 7: 427–43.

Tanaka, K., and H. Watanabe. 1973. Allogenic inhibition in a compound ascidian, *Botryllus primigenus* Oka. I. Processes and features of "nonfusion" reaction. Cell Immunol. 7: 410–26.

Taylor, K. M. 1967. The chromosomes of some lower chordates. Chromosoma 21: 181–88.

Thinh, L. V., and D. J. Griffiths. 1977. Studies of the relationship between the ascidian *Diplosoma virens* and its associated microscopic algae. I. Photosynthetic characteristics of the algae. Australian J. Mar. Freshwater Res. 28: 673–81.

Thomas, N. W. 1970a. Mucus-secreting cells from the alimentary canal of *Ciona intestinalis*. J. Mar. Biol. Assoc. U.K. 50: 429–38, 4 pls.

———. 1970b. Morphology of cell types from the gastric epithelium of *Ciona intestinalis*. J. Mar. Biol. Assoc. U.K. 50: 737–46, 3 pls., 1 tab.

Thompson, T. G., and T. J. Chow. 1955. The strontium-calcium atom ratio in carbonate-secreting marine organisms. Pap. Mar. Biol. Oceanogr., Deep-Sea Res. 3 (Suppl.): 20–39.

Thorndyke, M. C. 1977. Observations on the gastric epithelium of ascidians with special reference to *Styela clava*. Cell Tissue Res. 184: 539–50.

———. 1978. Evidence for a "mammalian" thyroglobulin in endostyle of the ascidian *Styela clava*. Nature 271: 61–62.

Thorpe, A. 1972. Ultrastructural and histochemical features of the endostyle of the ascidian *Ciona intestinalis* with special reference to the distribution of bound iodine. Gen. Comp. Endocrinol. 19: 559–71.

Thorpe, A., and M. C. Thorndyke. 1975. The endostyle in relation to iodine binding. Symp. Zool. Soc. London 36: 159–77.

Tokioka, T. 1953. Ascidians of Sagami Bay. Tokyo: Iwanami Shoten. 315 pp., 79 pls., 1 map.

———. 1963. Contributions to Japanese ascidian fauna. XX. The outline of Japanese ascidian fauna as compared with that of the Pacific coasts of North America. Publ. Seto Mar. Biol. Lab. 11: 131–56.

———. 1965. Questions concerning the diagnoses of some ascidian genera. Publ. Seto Mar. Biol. Lab. 13: 125–29.

——— 1967. Pacific Tunicata of the United States National Museum. Bull. U.S. Nat. Mus. 251: 1–247.

———. 1968. Contributions to Japanese ascidian fauna. XXIV. On *Sigillinaria clavata* Oka, 1933. Publ. Seto Mar. Biol. Lab. 16: 199–205.

———. 1971. Phylogenetic speculation of the Tunicata. Publ. Seto Mar. Biol. Lab. 19: 43–63.

———. 1972. On a small collection of ascidians from the Pacific coast of Costa Rica. Publ. Seto Mar. Biol. Lab. 19: 383–408.

Tokioka, T., and Y. Kado. 1972. The occurrence of *Molgula manhattensis* (DeKay) in brackish water near Hiroshima, Japan. Publ. Seto Mar. Biol. Lab. 21: 21–29.

Trason, W. B. 1957. Larval structure and development of the oo-zooid in the ascidian *Euherdmania claviformis*. J. Morphol. 100: 509–45.

———. 1959. Brief notes on interesting ascidians. Veliger 2: 19.

———. 1963. The life cycle and affinities of the colonial ascidian *Pycnoclavella stanleyi*. Univ. Calif. Publ. Zool. 65: 283–326.

Tucker, G. H. 1942. The histology of the gonads and development of the egg envelopes of an ascidian (*Styela plicata* Lesueur). J. Morphol. 70: 81–113.

Turner, C. H., E. E. Ebert, and R. R. Given. 1968. The marine environment offshore from Point Loma, San Diego County. Calif. Dept. Fish & Game, Fish Bull. 140: 1–79.

———. 1969. Man-made reef ecology. Calif. Dept. Fish & Game, Fish Bull. 146: 1–221, App. 1.

Turner, C. H., and A. R. Strachan. 1969. The marine environment in the vicinity of the San Gabriel River mouth. Calif. Fish & Game 55: 53–68.

Tuzet, O., D. Bogoraze, and F. Lafargue. 1974. La spermatogénèse de *Polysyncraton lacazei* Giard, 1872 et *Trididemnum cereum* Giard, 1872 (ascidies composées, aplousobranches). Bull. Biol. France Belg. 108: 151–68.

Vallee, J. A., Jr. 1967–68. Studies of the blood of *Ascidia nigra* (Savigny). I. Total blood cell counts, differential blood cell counts, and hematocrit values. II. Vanadocyte agglutination and its effect upon the heart. III. Some aspects of the biochemistry of vanadocyte agglutination. Bull. So. Calif. Acad. Sci. 66: 23–28, 117–24; 67: 89–95.

Van Name, W. G. 1945. The North and South American ascidians. Bull. Amer. Mus. Natur. Hist. 84: 1–476, pls. 1–31.

Wardrop, A. B. 1970. The structure and formation of the test of *Pyura stolonifera* (Tunicata). Protoplasma 70: 73–86.

Watanabe, H., and C. C. Lambert. 1973. Larva release in response to light by the compound ascidians *Distaplia occidentalis* and *Metandrocarpa taylori*. Biol. Bull. 144: 556–66.

Watanabe, H., and A. T. Newberry. 1976. Budding by oozooids in the polystyelid ascidian *Metandrocarpa taylori*. J. Morphol. 148: 161–76.

Watkins, M. 1958. Regeneration of buds in *Botryllus*. Biol. Bull. 115: 147–52.

Watterson, R. L. 1945. Asexual reproduction in the colonial tunicate, *Botryllus schlosseri* (Pallas) Savigny, with special reference to the developmental history of intersiphonal bands of pigment cells. Biol. Bull. 88: 71–103.

Watts, D. C. 1975. Evolution of phosphagen kinases in the chordate line. Symp. Zool. Soc. London 36: 105–27.

Weel, P. B. van. 1940. Beiträge zur Ernährungsbiologie der ascidien. Pubbl. Staz. Zool. Napoli 18: 50–79.

Weiss, J., Y. Goldman, and M. Morad. 1976. Electromechanical properties of the single-layered heart of the tunicate *Boltenia ovifera* (sea potato). J. Gen. Physiol. 68: 503–18.

Werner, E., and B. Werner. 1956. Über den Mechanismus des Nahrungserwerbs der Tunicaten, speziell der Ascidien. Helgoländer Wiss. Meeresunters. 5: 57–92.

West, A. B., and C. C. Lambert. 1976. Control of spawning in the tunicate *Styela plicata* by variations in a natural light regime. J. Exper. Zool. 195: 263–70.

Whittaker, J. R. 1975. Siphon regeneration in *Ciona*. Nature 255: 224–25.

Whittingham, D. G. 1967. Light-induction of shedding of gametes in *Ciona intestinalis* and *Molgula manhattensis*. Biol. Bull. 132: 292–98.

Williams, R. B., and L. K. Thomas. 1967. The standing crop of benthic animals in a North Carolina estuarine area. J. Elisha Mitchell Sci. Soc. 83: 135–39.

Woollacott, R. M. 1974. Microfilaments and the mechanism of light-triggered sperm release in ascidians. Develop. Biol. 40: 186–95.

———. 1977. Spermatozoa of *Ciona intestinalis* and analysis of ascidian fertilization. J. Morphol. 152: 77–88.

Woollacott, R. M., and M. E. Porter. 1977. A synchronized multicellular movement initiated by light and mediated by microfilaments. Develop. Biol. 61: 41–57.

Yamaguchi, M. 1970. Spawning periodicity and settling time in ascidians, *Ciona intestinalis* and *Styela plicata*. Rec. Oceanogr. Works Japan 10: 147–55.

———. 1975. Growth and reproductive cycles of the marine fouling ascidians *Ciona intestinalis*, *Styela plicata*, *Botrylloides violaceus*, and *Leptoclinum mitsukurii* at Aburatsubo-Moroiso Inlet (central Japan). Mar. Biol. 29: 253–59.

Yamaguchi, Z. 1931. Some notes on the physiology of *Styela clava* Herdman. Sci. Rep. Tôhôku Imp. Univ. (4) Biol. 6: 597–607.

Yarnall, J. L. 1972. The feeding behavior and functional anatomy of the gut in the eolid nudibranchs *Hermissenda crassicornis* (Eschscholtz, 1831) and *Aeolidia papillosa* (Linnaeus, 1761). Doctoral thesis, Biological Sciences, Stanford University, Stanford, Calif. 126 pp.

Yasuda, S. 1975. Sterol compositions of sea squirts (Ascidiacea). Comp. Biochem. Physiol. 50B: 399–402.

Yonge, C. M. 1952. Structure and adaptation in *Entodesma saxicola* (Baird) and *Mytilimeria nuttallii* Conrad. Univ. Calif. Publ. Zool. 55: 439–50.

Young, C. M. 1978. Orientation and utilization of ambient water currents by the ascidian *Styela montereyensis*. 59th Ann. Meet., West. Soc. Natur., Abs. Symp. and Contr. Pap., p. 9 (abstract).

Zaniolo, G., A. Sabbadin, and C. Resola. 1976. Dynamics of the colonial cycle in the ascidian *Botryllus schlosseri*. The fate of isolated buds. Acta Embryol. Exper. 1976: 205–13.

Mollusca: *Introduction to the Phylum and to the Class Gastropoda*

Donald P. Abbott and Eugene C. Haderlie

The mollusks most familiar to us are soft-bodied animals with shells, such as oysters, mussels, limpets, snails, and squid. "Molluscus," a Latin term for "soft," has provided the name for the group. Since prehistoric times, man has eaten the larger species, or found in them a source of dye or fiber, while the shells have served as containers, tools, weapons, ornaments, trumpets, windowpanes, and even money.

One must be cautious in generalizing about mollusks, since nearly every broad statement one can make about the group has important exceptions. Thus, on the whole, mollusks possess a body roughly divisible into three regions—a head, a foot, and a visceral mass or hump—yet in bivalves the head, as such, scarcely exists, and in other groups one may look in vain for a clear-cut foot. Again, mollusks characteristically possess a shell covering much or all of the body, yet there are numerous groups and species where the shell in the adult stage is vestigial or lacking. Most mollusks are equipped with hard mouthparts, commonly including a toothed strap (radula) used as a rasp, and often a pair of biting jaws, as well, but such mouthparts are lacking in bivalves and some gastropods. And so it is with numerous other characteristics of mollusks, including some features of the internal anatomy and the pattern of

Donald P. Abbott is Professor, Department of Biological Sciences and Hopkins Marine Station, Stanford University. Eugene C. Haderlie is Professor of Oceanography at the Naval Postgraduate School, Monterey.

embryonic development. Yet such is the ability of the mind and eye to perceive common features amid differences in detail, that even a small child soon learns to recognize most mollusks as such, once a few typical examples have been pointed out.

The mollusks possess a fundamental ground plan of structure and function that has proved highly flexible in an evolutionary sense, permitting an enormous diversification of size, form, and way of life. Modern mollusks range from nearly microscopic species inhabiting the spaces between sand grains to giant cephalopods attaining a length (with tentacles) of 22 m and an estimated weight approaching 2 metric tons. Some mollusks are sessile in the adult stage, but among existing species are those that cling, creep, burrow, walk, leap, float, swim slowly or swiftly, or even occasionally become airborne, jet-propelled for short distances above the water. There are herbivores, carnivores, omnivores, detritus feeders, and even highly modified internal parasites; their methods of feeding include scraping or grazing, browsing, boring, pursuing and capturing animal prey, scavenging on dead remains, consuming bottom deposits, filtering detritus particles and plankton from the water, and even absorbing dissolved organic matter.

The present diversity of molluscan species reflects not only the contemporary success of the group, but a long history extending back for more than half a billion years. The greatest variety of mollusks today exists in the sea, but two molluscan classes have successfully invaded fresh water, and one has done well on land. The number of living species of mollusks is often given as over 100,000, but it is more likely in the range of 45,000–50,000 (46,810 according to the recent critical estimate of Boss, 1971). Even so, in terms of number of existing species the mollusks rank second only to the arthropods among the metazoan phyla. Approximately half of the species of living mollusks are marine; the other half live on land or in fresh water. The marine species are mainly bottom dwellers in the intertidal zones or shallow subtidal waters, but some live on the bottom at great depths, and some are pelagic.

Living mollusks are divided into seven classes: Gastropoda (snails, slugs, and allies), Bivalvia (clams, mussels, oysters), Polyplacophora (chitons), Cephalopoda (octopuses, squids), Scaphopoda (tusk shells), Aplacophora, and Monoplacophora. The last three classes do not occur intertidally in California and are not treated further here. Excellent and readable general accounts of the phylum Mollusca are presented in Morton (1967) and Yonge & Thompson (1976). Valuable shorter treatments are those of Morton & Yonge (1964), Russell-Hunter (1968, pp. 112–69), Stasek (1972), and Yonge (1960). Various aspects of physiology and biochemistry of mollusks are reviewed in Wilbur & Yonge (1964, 1966) and Florkin & Scheer (1972). For systematics, species identification, and illustrations of shells, see Abbott (1974), Keen & Coan (1974), and Smith & Carlton (1975), and the references cited there.

More than 80 percent of all living molluscan species are gastropods. These are the descendants of an ancient molluscan stock, probably already equipped with a symmetrically coiled shell, which long ago underwent an evolutionary change called torsion. In torsion, the visceral hump and shell were rotated 180° counterclockwise (as viewed from above) relative to the head and foot, twisting the relatively narrow "waist" of tissue connecting the visceral hump to the foot, and bringing the mantle cavity containing the anus and gills from the rear to a new position overlying the head. An abbreviated version of this extraordinary evolutionary event occurs during the larval development of most gastropods today. Modern marine snails typically pass through a veliger larval stage, which possesses a recognizable head, foot, and visceral hump, along with the rudiments of most internal organs. The veliger swims by means of a crown of cilia encircling the head and often extending in flaps or lobes reaching outward on either side like gigantic ears. The early veliger larva resembles a tiny snail with a coiled shell, but the mouth and anus are located at opposite ends

of the body, as in other molluscan classes. During the veliger stage torsion is brought about by a combination of asymmetrical muscular contraction and asymmetry in the rate and pattern of growth on the right and left sides, the relative importance of the two processes differing in different groups. At any rate, torsion swings the shell and underlying visceral hump around until the original tail end overlies the head, and it puts a twist in the main lateral nerve cords leading from the head to the viscera. In most gastropods the effects of torsion in the larval stage persist throughout life, but in some (the opisthobranchs) the subsequent developmental changes bring about a degree of untwisting (detorsion) in one or more systems.

Why torsion? What functional advantages accompanied this extraordinary change? The British zoologist Walter Garstang (1929) pointed out the advantage to the veliger larva; before torsion, when disturbed, the larva draws its foot into the shell and blocks the entrance with its head, whereas after torsion, the animal pulls in its head and blocks the entrance with its less vulnerable foot. Others point out advantages accruing to the settled gastropod in postlarval life. For one thing, torsion has brought the gills and associated sense organs around to the front, assuring the former a fresher water supply and the latter a better position from which to sense what lies ahead in the environment (Morton, 1958). For another, the reoriented shell and visceral mass appear better balanced and positioned for carrying about (see, for example, Ghiselin, 1966). Whatever the crucial advantages, torsion brought with it problems as well, particularly a sanitary problem (e.g., Yonge, 1947), for these mollusks no longer deposited their waste products behind them but upon their heads and the feeding area before them. They survived the inconvenience, but early gastropod evolutionary changes relate in part to abating the sanitary problem.

In addition to the unique feature of torsion, certain other characters are widespread among gastropods. Most species have jaws and a radula, and most have eyes borne on tentacles on the head. The ventral foot ("Gastropoda" = "stomach-foot") is commonly broad and muscular, well suited to creeping upon, and clinging to, the substratum. In most snail-like gastropods the visceral hump is elongated dorsally, coiled into a compact asymmetrical spiral and covered by a conforming shell that provides protection and support. The spiral shell is usually "right-handed" or dextral in marine snails; that is, when one holds the shell with spire uppermost and looks into the open aperture, the aperture appears on the right side.

In some gastropods the aperture can be tightly closed by means of a hard operculum or trap door borne on the back of the snail's foot. The shells of many gastropods are beautifully colored and patterned, and some are ornamented with an elaborate surface sculpture of projecting flanges or spines whose adaptive significance is often not clear.

The class Gastropoda consists of three subclasses, the Prosobranchia, the Opisthobranchia, and the Pulmonata. The first of these is considered in the next chapter; the second and third are treated together in the following chapter. Successive chapters deal with the Bivalvia, Polyplacophora, and Cephalopoda.

Literature Cited

Abbott, R. T. 1974. American seashells. 2nd ed. New York: Van Nostrand Reinhold. 663 pp.

Boss, K. J. 1971. Critical estimate of the number of Recent Mollusca. Occas. Pap. Mollusks, Mus. Compar. Zool., Harvard Univ. 3: 81–135.

Florkin, M., and B. T. Scheer, eds. 1972. Chemical zoology. 7. Mollusca. New York: Academic Press. 567 pp.

Garstang, W. 1929. The origin and evolution of larval forms. Rep. Brit. Assoc. Adv. Sci. (Glasgow) 1928: 77–98.

Ghiselin, M. T. 1966. The adaptive significance of gastropod torsion. Evolution 20: 337–48.

Keen, A. M., and E. Coan. 1974. Marine molluscan genera of western North America: An illustrated key. Stanford, Calif.: Stanford University Press. 208 pp.

Morton, J. E. 1958. Torsion and the adult snail; a re-evaluation. Proc. Malacol. Soc. London 33: 2–10.

_____. 1967. Molluscs. 4th ed. London: Hutchinson University Library. 244 pp.

Morton, J. E., and C. M. Yonge. 1964. Classification and structure of the Mollusca, pp. 1–58, *in* Wilbur & Yonge (1964).

Russell-Hunter, W. D. 1968. A biology of lower invertebrates. New York: Macmillan. 181 pp.

Smith, R. I., and J. T. Carlton, eds. 1975. Light's manual: Intertidal invertebrates of the central California coast. 3rd ed. Berkeley and Los Angeles: University of California Press. 716 pp.

Stasek, C. R. 1972. The molluscan framework, pp. 1–44, *in* Florkin & Scheer (1972).

Wilbur, K. M., and C. M. Yonge, eds. 1964. Physiology of Mollusca. Vol. 1. New York: Academic Press. 473 pp.

_____. 1966. Physiology of Mollusca. Vol. 2. New York: Academic Press. 645 pp.

Yonge, C. M. 1947. The pallial organs in the aspidobranch Gastropoda and their evolution throughout the Mollusca. Phil. Trans. Roy. Soc. London B 232: 443–518.

_____. 1960. General characters of Mollusca, pp. I3–36, *in* R. C. Moore and C. W. Pitrat, eds., Treatise on invertebrate paleontology, Mollusca 1. New York: Geol. Soc. Amer.; Lawrence: University of Kansas Press. 351 pp.

Yonge, C. M., and T. E. Thompson. 1976. Living marine molluscs. London: Collins. 288 pp.

Chapter 13

Prosobranchia: *Marine Snails*

Donald P. Abbott and Eugene C. Haderlie

The subclass Prosobranchia is the largest of the three main divisions of the class Gastropoda, and includes most of the common marine snails along our coast. The name of the group ("Prosobranchia" = "forward gills") refers to the fact that after torsion occurs in development, the gills (ctenidia), occupying the spacious mantle cavity above and behind the head, are located anterior to the heart and other viscera. Most prosobranchs have a well-developed shell, and in all but a few species the lateral nerve cords remain twisted following torsion. Typically the sexes are separate, though some prosobranchs are protandric hermaphrodites, undergoing a sex change from male to female during life. Eggs and sperm are shed to the sea in some species; in others fertilization is internal. In some species the eggs hatch into free-swimming veliger larvae; in others the young snails hatch directly from an egg capsule. A few species retain the eggs in the body and bear living young.

The Prosobranchia is an old group, with fossils dating from the lower Cambrian. A careful estimate places the number of living prosobranch species for the world at about 20,000 (Boss, 1971); earlier estimates were often much larger. The great majority are marine, living on the bottom at all depths in all oceans of the world. Some species are pelagic in the sea. Some species live in fresh water and some on land. Many live well in laboratories and at least one species has been kept for 36 years.

The subclass is divided into three orders, the smallest of which is the order Archaeogastropoda. Geologically the oldest group, this also contains the living species that retain, to the greatest degree, those characters considered primitive in the gastropods; in particular, they show the greatest degree of bilateral symmetry in the soft parts, and often possess both members of primitively paired structures. In this respect, among the most old-fashioned gastropods alive today are the abalones and the keyhole limpets, which both have two gills, two mucous glands, two kidneys, two heart auricles, etc. These forms have slits or holes in the shell, which serve for the exit of water that has passed through the gills and carries a burden of waste materials. Also included in the Archaeogastropoda are the more advanced "true" limpets with their cap-shaped shells; these forms have lost the gill on the right side, and retain only the left gill. Some forms have either reduced or lost the right heart auricle, as well, and show other evidences of increasing bilateral asymmetry. Most species are either herbivores or consumers of both plant materials and detritus, but a few of the top shells are carnivores, feeding on sponges, hydroids, and bryozoans. All archaeogastropods found on California shores discharge their sexual products into the sea, where fertilization occurs, and subsequent development normally results in a swimming veliger larva.

The second prosobranch order, Mesogastropoda, is the largest single order of mollusks. Adaptive radiation has resulted in thousands of species that represent an immense diversity of specialization. A large number feed on plants or on organic detritus, but the group includes some common predatory carnivores such as the moon snail *Polinices* (13.62, 13.63), along with some remarkably modified sedentary and sessile filter feeders, and some internal par-

Donald P. Abbott is Professor, Department of Biological Sciences and Hopkins Marine Station, Stanford University. Eugene C. Haderlie is Professor of Oceanography at the Naval Postgraduate School, Monterey.

asites so strangely altered as to be scarcely recognizable as mollusks at all. However, most mesogastropods are snail-like in appearance with a conspicuous shell. Most have a single gill, the left, whose axis is fused to the wall of the mantle cavity and bears only a single row of respiratory filaments or leaflets. Only a single member of most primitively paired organs is retained. In most species the radula is well developed and commonly bears seven plates, or teeth, in each broad row. Some forms have an area of the mantle margin pulled out and rolled up into an elongated siphon that projects forward through a groove or canal in the lip of the aperture; the siphon conducts water into the mantle cavity where it first bathes the gills and then carries off wastes. Most of the common intertidal and subtidal snails of California shores are mesogastropods. The order also includes a number of large tropical forms whose shells are popular with collectors, such as the cowries, conchs, helmets, and tritons.

Members of the third prosobranch order, the Neogastropoda or Stenoglossa, are nearly all carnivores, either predators or scavengers or both. As in the mesogastropods there is only a single gill (the left) and a single member of other primitively paired organs. Usually the animals have a large mobile siphon for water intake. The radula is relatively narrow ("Stenoglossa" = "narrow tongue"), with three or fewer toothed plates per row; in one group it consists of separate teeth that are loaded with poison and used one at a time for harpooning prey. Among the common neogastropods along our shores are *Acanthina, Nucella, Ocenebra, Amphissa, Nassarius,* and *Olivella.* Many of the colorful tropical marine snails such as the cones, augers, murexes, volutes, turrets, and marginellas are also assigned to this order.

While the higher classification of gastropods rests mainly on anatomical features, the lower taxa are defined largely on the features of the shell and radula. Most California prosobranchs can be identified from the shell, but in some of the species included here the color and form of the soft parts are also distinctive. Shell size, too, may be a useful feature in identification. In the measurements given here, for highly flattened forms such as limpets and abalones the term "length" means the greatest diameter of the shell. For all coiled shells with an elevated spire, the term length (=height) refers to the total length along the axis of coiling. Each 360° coil of the shell forms a whorl, and adjacent whorls are separated by a more or less distinct suture. The lowest, largest whorl (called the body whorl) opens in the aperture, whose margin is termed the lip. The outer surface of the shell is sometimes smooth, sometimes decorated by various ridges, grooves, nodules, or even prominent flanges called lamellae or varices. This external sculpturing of the shell in typical snails is either spiral (that is, following the whorls and sutures) or axial (crossing at right angles over the whorls and sutures) or both. On conical shells, such as those of limpets, the sculpturing may be radial (that is, extending from the center or apex of the shell directly to the margin) or concentric (running parallel to the margin). In some snails the shell is partially or completely covered with a tough fibrous outer layer, the periostracum.

The spiraling of most shells occurs around a central pillar, the columella. This is concealed inside for most of its length unless the shell is broken, but is visible where it forms the inner lip of the shell aperture. The exposed columella here may bear ridges or teeth of importance in identification, or it may be thickened with a calcareous deposit called a callus. In meso- and neogastropods the lip of the shell aperture closest to the columella may be extended into a protective siphonal canal. And in some snails the end of the columella closest to the shell aperture may bear an indentation or hole called the umbilicus.

The shell is secreted by the mantle, consisting of the tissues covering the visceral hump (essentially all of the body except the head and foot). At its periphery the mantle is often extended in a fold or flap whose edges lie near the lips of the shell aperture when the animal is extended. The shell grows larger only by calcareous deposition at the lips of the aperture by secretory areas on the mantle margin. In contrast, the older parts of the shell are repaired or thickened from the inside by secretions from the whole outer surface of the mantle. In forms with cap-shaped shells, like the limpets and keyhole limpets, where the shell aperture is essentially as large as the shell itself, the mantle fold usually extends as a wide awning over a pallial groove surrounding the whole body. In such forms the mantle margin often bears conspicuous tentacles, and the mantle fold may provide an important respiratory surface.

The head in living gastropods is often easily seen; it bears the mouth, two cephalic tentacles, and a pair of simple eyes. The head is usually not much help in the identification of snails, but the large and conspicuous foot may show distinctive coloration or form. In some snails (e.g., abalones, keyhole limpets, turban snails) the upper region of the foot bears prominent projections called epipodial tentacles. And many prosobranchs have a tough horny or calcareous operculum, attached to the back of the foot, which is used to close the aperture after the animal has retracted into the shell.

Much of the literature on prosobranchs is sufficiently technical that someone seeking knowledge beyond the level of this book had best start by consulting the references given in the introduction to the mollusks and gastropods (p. 227), or a general zoology text. At a more advanced level, some of the best general references and reviews on prosobranchs are Fretter & Graham (1962), Hyman (1967), Purchon (1968), Webber (1977), Wilbur & Yonge (1964, 1966), and Yonge & Thompson (1976). For central California intertidal snails, a most useful identification aid is the key of Carlton & Roth (1975). Other helpful references are R. Abbott (1974), Dall (1921), Keen & Coan (1974), Keep & Baily (1935), J. McLean (1969), and Oldroyd (1925–27).

Shell collecting as a hobby has attracted so many followers that for some species in some areas the adult snail's worst enemy may be man. We recommend moderation and restraint in collecting living specimens merely to kill them and clean their shells. Learn to admire without being covetous, and to capture with a camera. Further, we suggest that the many who already appreciate the beauty of molluscan shells may find that live-snail watching is an even more exciting and rewarding avocation than shell collecting. The habits of snails, their modes of behavior, their activity patterns under field conditions, and indeed most other features of the natural history of most species have not yet received careful study. This is a field in which a careful amateur can still make a valuable contribution to science.

We wish to acknowledge the considerable help given by Dr. A. Myra Keen in the preparation of this chapter. She made the initial identification of the prosobranchs pictured, settled nomenclatural problems, assembled the basic shell description and geographical distribution for each, and read and commented on the various drafts of the manuscript. She deserves, but has declined, coauthorship.

Phylum Mollusca / Class Gastropoda / Subclass Prosobranchia
Order Archaeogastropoda / Superfamily Pleurotomariacea
Family Haliotidae

13.1 **Haliotis rufescens** Swainson, 1822
Red Abalone

Uncommon, low intertidal zone in rocky areas with heavy surf; more abundant offshore to depths of over 180 m, with maximum concentration in central California between 6 and 17 m depth; Sunset Bay (Oregon) to Bahía de Tortuga (Baja California).

Length to nearly 30 cm; shell exterior usually brick red, the surface irregular, often overgrown by fouling organisms; shell holes oval, externally raised, usually three or four remaining open, occasionally more, rarely none; shell interior iridescent, mostly smooth but with a large, oval, centrally placed muscle scar with a rough surface; tentacles black.

This is California's largest marine snail, and the one most prized as food by man. The name "rufescens" refers to the outer shell color, usually red or pink but quite variable and strongly influenced by the diet. Very small individuals graze on films of microscopic plants, and their shells may be pink, greenish, or white. Larger abalones eat mainly larger algae. When red algae predominate in the diet, the shell laid down is red outside; the red pigment in the shell, rufescine, is a bilin compound, as is phycoerythrin, the red pigment in red algae. In contrast, individuals fed on brown or green algae have shells that range from aquamarine, green, to white. A mixed diet of red and brown algae may yield a somewhat brownish shell or, if red algae are eaten only at considerable intervals, the shell may bear red bands along the growth lines.

Mature abalones are normally sedentary creatures, occupying a permenent position, or scar, on a rock. In regions where sea otters are established, most red abalones live only in deep protected crevices, but north and south of the otter's present range many occupy relatively open rock faces. They do not roam about grazing on attached algae but live almost exclusively on loose plants that lodge near them or are caught with the foot as the plants drift slowly by. Hungry red abalones in calm subtidal waters may lift the front half of the foot off the rock and extend it in the water, ready to grasp plant fragments should they appear. When a drifting plant is contacted, usually by the extended epipodial tentacles, the sides of the foot fold toward the midline, grasping the plant; then the whole foot descends, trapping the plant below.

The holes in all abalone shells lie in a row over the mantle cavity, an elongated chamber that contains the two gills, a pair of large mucous glands, the anus, and the ducts that emit urine and gametes. Strong cilia on the gills create a water current that enters the chamber anteriorly, under the edge of the shell near the head and sometimes also through the largest (most anterior) hole in the shell. The water flows about the gills, then exits through the remaining holes overlying the mantle cavity; feces, urine, and gametes, added at the very back of the cavity, are whisked out in the exhalant water stream. As the abalone grows, new holes are added to the growing margin and old holes become plugged with nacre.

Boring organisms often infest the shells of red abalones. The boring sponge *Cliona celata* (2.20) initially attacks near the apex, then spreads out, riddling the shell with passageways and chambers and reducing it to a frail skeleton. The small clam *Penitella conradi* (15.76) bores into the shell from the outside. As it penetrates, the abalone secretes nacre locally over the inner surface, forming a blister pearl; complete perforation of the shell is uncommon.

The breeding season is not well defined; spawning tends to be greater in the spring and summer, but within a given population some animals are found spawning at any time of year. Spawning may follow sudden changes in water temperature, 1–2 hours of exposure to air at low tide, or any stimulus causing sharp muscular contraction in relaxed ripe animals. Females may spawn in the presence of live sperm in the water. The addition of a little hydrogen peroxide to seawater (approximately a 5 mM solution, in seawater, pH about 9.1) causes synchronous spawning in gravid males and females. The eggs, fertilized externally, sink to the bottom. Swimming trochophore larvae develop in about 12 hours and veliger larvae in 24 hours at 15°C. The swimming period lasts about 5 days at 15°C, and about 14 days at 10°C under laboratory conditions (E. Ebert, pers. comm.). The larvae then settle to the bottom, metamorphose, and begin to graze. They reach 1 mm in length in 40 days, and 10 mm in 3–4 months in the

laboratory at 15°C. During the first 3–4 years, average growth is estimated at about 17 mm per year but varies greatly and may be as much as 48mm in a single year. Growth slows with increasing size and age, and marked specimens 185–224 mm long (about 7–9 inches) living in the field sometimes show no growth for periods of up to 5 years. Size is therefore a poor indication of age, but the largest animals found could be well over 20 years old. Sexual reproduction begins while the animals are still relatively small. Minimum length at the onset of maturity is about 40 mm in both sexes in northern California; spawning has been noted in females 41 mm long and males 47 mm long (E. Ebert, pers. comm.).

Abalones provide one of the few well-documented cases of hybridization in mollusks. In southern California the red abalone frequently hybridizes with the white abalone (**Haliotis sorensoni** Bartsch), occasionally hybridizes with the pink and pinto abalones (*H. corrugata*, 13.3, and **H. kamtschatkana assimilis** Dall), and produces very rare interspecific hybrids with the flat and the green abalones (**H. walallensis** Stearns and *H. fulgens*, 13.2). A hybrid between the red abalone and **H. kamtschatkana kamtschatkana** Jonas is reported from Mendocino Co.

Fecundity is high in the red abalone; a female about 20 cm long was estimated to contain about 12.6 million ripe oocytes. Mortality is also high, probably greatest in the planktonic larval stages. The small percent that survive to a stage where they select a permanent site are still preyed upon by sea stars, crabs, octopuses, fishes, sea otters, and man. The red abalone gives active escape responses to contact with, or proximity of, the sea stars *Pycnopodia helianthoides* (8.16) and *Pisaster ochraceus* (8.13). The responses may include a fast, galloping retreat, often accompanied by rotation of the shell, alternately to right and left, through an arc of 180°. Much mucus is produced, and the epipodium may be extended over the margin of the shell in a manner that dislodges starfish tube feet. Given a vertical surface the abalone may even climb out of water. Sea otters and fishes sometimes catch abalones that have raised the front lip of the shell in reaching out for drifting algae. A sharp bump from the head of a fish like the sheephead (*Pimelometopon*), or a pry with the lower jaw of an otter, may dislodge the snail. Sea otters also use a portable rock to break the top of an abalone shell. Holding the rock with both paws, the otter strikes the shell repeatedly with short strokes. Soon the shell covering the abalone's great shell muscle is smashed, and the abalone is then easily denuded and captured. Octopuses are capable of pulling smaller abalones from the rocks, or of drilling a hole in the shell and injecting a paralyzing venom. Larger abalones often bear one or more shell or tissue scars where a rock crab (*Cancer antennarius*, 25.16) has broken the edge of the shell with a powerful claw, then reached in and tweaked off bits of meat. Storms, too, may cause mortality in shallow water, when waves dislodge abalones from their home scars. Abalones falling on irregular rocks can usually right themselves. Abalones falling right side up on sand can creep to reach a solid substratum; those dropped upside down cannot right themselves, though they can plough slowly along by repeatedly digging the heel of their foot in the sand and giving a push. In this condition, however, they are highly vulnerable to large fishes.

Man is a long-time predator on the red abalone. Studies of shell mounds on the California coast and Channel Islands show that red abalones were eaten and their shells used as tools, vessels, and ornaments by Indians at least 7,000 years ago. The shells were traded to tribes as far east as Colorado and Texas. Red abalones still provide us with food and ornaments, but those wishing to collect the animals for any purpose should check the current "California Sport Fishing Regulations," by the state Department of Fish and Game. As of 1979, one may take red abalones 7 inches (178 mm) or more in length; the bag limit is four abalones of all kinds together.

The commercial catch of red abalones traditionally comes mainly from depths of 7–18 m on the coastal shelf between Cape San Martin (Monterey Co.) and Avila (San Luis Obispo Co.). Yearly catches in this area have declined with the expansion of the sea otter population into the fishing grounds, and other species of abalones now make up two-thirds to three fourths of the annual commercial abalone catch in the state. This averaged 1,670 metric tons (1,840 tons) yearly for the decade 1965–74, but has been declining. The red abalone can be reared in mariculture, but raising it to legal size in culture is not yet profitable. Researchers are presently exploring the feasibility of rearing 1–2-year-old abalones (of several different species) and transplanting them into areas where natural populations have been depleted.

The red abalone has served as a laboratory animal in a wide variety of physiological and biochemical investigations. Among the matters studied have been digestion, absorption from the gut, carbohydrate metabolism, urine formation and excretion, DNA content, hemocyanin blood pigments, antimicrobial activity of tissue fluids, toxic effects of copper, shell pigments, shell proteins, and nacreous-layer deposition.

See especially Cox (1962, including bibliography); see also Bennett & Nakada (1968), Boolootian, Farmanfarmaian & Giese (1962), Carlisle (1945, 1962), Comfort (1950), Cooper, Wieland & Hines (1977), Crofts (1929), Curtis (1966), Giorgi & DeMartini (1977), Hansen (1970), Harrison (1962), Hinegardner (1974), Leighton (1960, 1961, 1966, 1968, 1974), Lowry & Pearse (1973), Martin, Stephenson & Martin (1977), McAllister (1976), Meyer (1967), W. Miller, Nishioka & Bern (1973), Minchin (1975), Montgomery (1967), Morse et al. (1977), Olsen (1968a,b) Prescott & Li (1966), Owen, McLean & Meyer (1971), Talmadge (1977), Wise (1970), and Young & DeMartini (1970).

Phylum Mollusca / Class Gastropoda / Subclass Prosobranchia
Order Archaeogastropoda / Superfamily Pleurotomariacea
Family Haliotidae

13.2 **Haliotis fulgens** Philippi, 1845
Green Abalone

Fairly common in rocky areas, low intertidal zone to 10 m depth, scarce to 18 m, commonest at 2–3 m in deep crevices exposed to strong wave action; Point Conception (Santa Barbara Co.) to Bahía Magdalena (Baja California).

Length to about 25 cm, usually less than 20 cm; shell exterior olive-green to reddish brown, often with fine spiral ribs, often overgrown with other organisms; shell holes circular, slightly elevated, five to seven remaining open; shell interior strongly iridescent, dark green, blue, and lavender; muscle scar pronounced; tentacles olive-green.

Among California abalones, this species is considered to have the most beautiful shell. Adult green abalones are relatively sedentary, though they move about somewhat more than pink and red abalones. They feed almost exclusively on larger drifting algae, which are grasped with the foot and held between the foot and substratum while being rasped by the radula. Red algae, such as *Gelidium, Pterocladia, Plocamium,* and *Gigartina,* are especially selected, but pieces of the larger brown algae, including *Macrocystis* and *Eisenia,* are taken as they drift by.

Spawning occurs from early summer to early fall, and average females shed 2–3.5 (mean 2.7) million eggs per year. The larvae settle after a short pelagic life. Juvenile abalones live in very protected places and move about grazing mainly at night. At 1–2 years of age they settle more or less permanently and develop a home scar. Animals less than 70 mm in shell length grow 14–15 mm a year, larger animals more slowly. Sexual maturity is attained at 5–7 years. The largest animals found may be as much as 20 years old.

Ecological studies suggest that the lower depth-distribution limit for green abalones is set by the availability of drifting red algae, though tolerance for lower temperatures may also be a factor. The population density of adults may depend on the extent of suitable crevice habitats in shallow rough waters. Juvenile green abalones are most abundant where adults abound, suggesting that the larvae tend to settle gregariously in the presence of adults.

Some green abalones are taken by sportsmen and commercial divers. Legal-size animals (1979 state regulations) are those with shells at least 6 inches (about 153 mm) long. The commercial catch for 1972–74 was 193, 71, and 55 metric tons, respectively. The other main predator on both small and large green abalones is the octopus. The abalone apparently derives some protection from the proximity of the moray eel *Gymnothorax mordax,* which often occupies the same crevices. Observations indicate that the eel rarely eats undisturbed green abalones, but readily takes octopods and is a strong deterrent to human abalone hunters. The small shrimp *Betaeus harfordi* (23.3) may occur as a commensal in the mantle cavity of the green abalone, as well as of other abalone species.

Interspecific hybrids between the green and pink abalones, and more rarely between greens and reds, are known from southern California. Studies of the hemocyanin blood pigments of the green and pink abalones show that these are slightly more similar than are the hemocyanins of green and red abalones.

See especially Cox (1962) and Tutschulte (1976); see also Curtis (1966), Hinegardner (1974), McAllister (1976), Leighton (1974), Meyer (1967), Owen, McLean & Meyer (1971), Pilson & Taylor (1961), and Talmadge (1964).

Phylum Mollusca / Class Gastropoda / Subclass Prosobranchia Order Archaeogastropoda / Superfamily Pleurotomariacea Family Haliotidae

13.3 Haliotis corrugata Gray, 1828
Pink Abalone

Uncommon in low intertidal zone, commoner on exposed rock surfaces at 6–60 m depth in protected bays and on open coast, mainly in areas having beds of the kelp *Macrocystis;* Point Conception (Santa Barbara Co.) to Bahía de Tortuga (Baja California); particularly abundant on Santa Barbara and San Clemente Islands (Channel Islands).

Length usually 15–17 cm but reaching 25 cm; shell rounded in outline, flat in juvenile, arched in adult, usually with a scalloped margin, greenish or reddish brown, with irregular diagonal rows of nodes or corrugations, the surface often heavily fouled with marine organisms; shell holes with raised rims, two to four remaining open; shell interior strikingly iridescent, mainly pink and green; muscle scar conspicuous, marked with dark green; tentacles black.

This species occupies quieter, more sheltered waters than the green abalone, less often dwelling in crevices. The lower depth limit is probably set by temperature; the animals seldom settle or thrive in waters cooler than 14°C. Adults are very sedentary, occupying a permanent scar on the rock. They feed wholly on pieces of drifting algae, captured with their foot. The brown alga *Eisenia* and the red *Plocamium* are preferred to some extent, but the pink abalone is not fussy about its food, and the brown algae *Macrocystis, Dictyopteris,* and *Pachydictyon* and many other species of plants are taken as they drift by.

Some evidence indicates that the pink abalone may spawn twice a year, first in late winter, then again in early summer.

Average females produce one to two million eggs a year. The swimming larvae tend to settle where adults are common, and recruitment of young from the plankton is poor in apparently suitable areas where the adult population has been depleted. Juveniles 1–2 years old adopt permanent sites on the substratum and begin the switch from grazing to the capturing of pieces of loose drifting algae. Growth of animals less than 70 mm long averages about 15–17 mm a year; larger animals grow only 6–7 mm a year. Sexual maturity comes at 3–5 years of age, and most animals probably do not survive more than 7 years. The largest animals found are probably about 20 years old.

Sea stars and octopuses take some juveniles and adults as food, but the greatest predator on adult pink abalones at present is probably man equipped with scuba. The pink abalone entered the commercial market in important amounts only after World War II. It underwent peak exploitation in the 1950's; 1,588 metric tons (1,750 tons) were taken in the year 1952 alone. It now ranks third among the abalones in commercial importance in California. The California catch, nearly all taken off the Channel Islands, averaged 186 metric tons (205 tons) a year for the period 1972–74. Commercial and sports divers using scuba have depleted legal-size animals (over 6 in, or 153 mm) at depths down to around 20 m on the more accessible parts of the southern California shelf. However, the pink abalone is one of the species that has been raised to juvenile stages in mariculture. Although it is not now commercially profitable to rear it to eating size, researchers are exploring the feasibility of restoring favorable but presently depleted areas by seeding them with cultured juveniles 1–2 years old.

Interspecific hybrids of the pink abalone occur most often with the red abalone, less commonly with the green and flat abalones, and more rarely still with the white and threaded abalones in southern California. A variety of shell forms and colors results, making recognition of hybrids difficult. Among California abalones the hemocyanin blood pigment of the pink abalone is most similar to that of the green abalone. A subspecies of the pink abalone, *H. corrugata oweni*, has been described from Isla Guadalupe (Mexico).

See especially Cox (1962) and Tutschulte (1976); see also Bartsch (1940), Bonnot (1930, 1940, 1948, 1949), Crofts (1929), Curtis (1966), C. Edwards (1913), Hinegardner (1974), Leighton (1974), G. MacGinitie & N. MacGinitie (1966), Meyer (1967), McAllister (1976), Owen, McLean & Meyer (1971), Palmer (1907), and Talmadge (1966).

Phylum Mollusca / Class Gastropoda / Subclass Prosobranchia
Order Archaeogastropoda / Superfamily Pleurotomariacea
Family Haliotidae

13.4 **Haliotis cracherodii** Leach, 1814
Black Abalone

Common under large rocks and in crevices, high intertidal zone down to 6 m depth, most abundant intertidally; Point Arena (Mendocino Co.) to Cabo San Lucas (Baja California); rare north of San Francisco; Coos Bay (Oregon) record not verified.

Maximum length over 20 cm; shell exterior dark blue, dark green, or nearly black, smooth, usually free of marine growth; outer dark edge of shell extending beyond nacreous area, forming a narrow dark band as seen from below; shell holes with rims not elevated above surface, usually five to seven remaining open, occasionally none; shell interior pearly, with pink and green iridescence; muscle scar faint; tentacles black.

This species occurs higher in the intertidal zone than any other California abalone, rendering it more vulnerable to terrestrial predators but safer from marine hunters. Smaller animals dwell in crevices and may move about in search of food; animals over 90 mm long occupy more-exposed rocks, at least where sea otters are absent and collecting by man is prohibited, and are quite sedentary. Smaller animals graze on diatom films and coralline algae, but larger ones subsist on loose pieces of algae brought in by waves and currents and grasped by the foot. Many plants are eaten, but the bulk of the diet consists of larger brown algae. When animals were fed only a single algal species, growth in the laboratory was greatest on a diet of the brown algae *Macrocystis pyrifera* and *Pelvetia fastigiata* and the red alga *Gigartina canaliculata*. In preference tests, the animals showed a preference for the brown alga *Egregia*. Under crowded conditions in the field, adult black abalones may graze the growth of marine plants from one another's shells.

Mature eggs are most abundant in female animals in the summer (July to September); a smaller peak occurs in January. Mature sperm are present in the male gonads year-round, but peak in the summer, just before spawning, and in midwinter. The summer spawning, in August and September, is conspicuous, and individuals lose some 20 percent of the weight of the soft parts. Spawning is more or less synchronous at a given location, but two populations in the Monterey area, separated by only 11.2 km, spawned 6 weeks apart. Gametogenesis begins immediately after spawning in the fall, slows in the winter, then speeds up in March. A small winter spawning is possible but not confirmed. In the Los Angeles area spawning starts in the late spring and continues through the summer.

Details of the early life history have not been published, but related species settle from the plankton and reach a length of 1 mm in about 3 months and 6 mm in 6 months. Black abalones are probably 20–30 mm long when 1 year old. Growth rates thereafter show great variation. On Santa Cruz Island (Channel Islands) the average growth rate for all animals up to 80 mm long was 14–15 mm per year. Animals 91–120 mm long averaged an increase in length of 2–4 mm a year, whereas those over 121 mm long showed more shell erosion than growth. These growth studies and others show that the growth rate varies with body size, with geographical area, and possibly with season in some places; there may be conspicuous differences in growth rate between similar individuals in the same microhabitat, or in the same individual during successive time intervals. Size is not a reliable indicator of age.

Hybridization of the black abalone with other species is not known. The black abalone does not closely resemble any other local species, and studies of its hemocyanin blood pigment show that it is not much like that of any other local species.

The blue-black coloring of the outer shell of the black abalone results from a bilin pigment, perhaps a cupro-mesobiliviolin. In intertidal specimens the shells are often eroded by the boring sponge *Cliona celata* (2.20). Subtidal individuals sometimes carry a small boring clam (*Penitella conradi*, 15.76)

in their shells, and, like other abalones, they may have one or more small shrimp (*Betaeus harfordi*, 23.3) living in and about the mantle cavity.

The enemies of the black abalone include octopuses, sea stars such as *Pisaster ochraceus* (8.13), fishes such as the cabezon (*Scorpaenichthys marmoratus*), sea otters where present, and man. The black abalone shows the same avoidance responses to contact with the ochre starfish, *P. ochraceus*, that one sees in the red abalone, but the reaction is weaker and less consistent; if contact is prolonged the abalone may cease to respond at all. *Pisaster ochraceus* has been observed to feed on barnacles attached to a black abalone shell without disturbing the abalone. Storm waves may dislodge black abalones living on more exposed rocks, and dislodged animals tend to fall on their backs. They can usually right themselves if they land on rocks but not on sand. However, if they land on sand right side up, they can crawl; if they land upside down, they sometimes propel themselves like clumsy barges by repeatedly digging the tail of their foot into the sand and kicking the shell ahead.

Shells of the black and other abalones occur in the kitchen middens of California coastal Indians over a period extending back more than 7,000 years. The meat was eaten fresh or dried; the shells provided ornaments or, with the holes plugged with tar, served as scoops and bowls. The black abalone was also important in the California commercial abalone fishery before 1900, in a period when the bulk of the catch was taken intertidally by the Chinese and dried for export to the Orient. As shore stocks became depleted, abalone fishing moved offshore, divers took over the collecting, and until very recent times the catch consisted mainly of red and pink abalones. With heavy exploitation of these stocks, the black abalone is again being taken commercially; the average catch of blacks for the years 1972–74 was 616 metric tons (679 tons), and in 1973–74 the harvest of blacks exceeded that of reds.

See especially Cox (1962); see also Behrens (1979), Bergen (1971), Boolootian, Farmanfarmaian & Giese (1962), Campbell (1965), Comfort (1949, 1950), Curtis (1966), Feder (1963), Giese (1969), Leighton & Boolootian (1963), Meyer (1967), Minchin (1975), Owen, McLean & Meyer (1971), Palmer (1907), Tixier & Lederer (1949), Webber & Giese (1969), and Wright (1975).

Phylum Mollusca / Class Gastropoda / Subclass Prosobranchia
Order Archaeogastropoda / Superfamily Fissurellacea
Family Fissurellidae

13.5 Fissurella volcano Reeve, 1849
Keyhole Limpet

Locally abundant on sides and undersurfaces of large boulders, middle intertidal zone; Crescent City (Del Norte Co.) to Bahía Magdalena (Baja California).

Shell 20–35 mm long; apical opening narrow, elongate, placed just anterior of center; sculpture of radiating ribs or striae; shell exterior pink with black or reddish-brown rays, interior greenish; mantle striped with red, foot yellowish. A subspecies, **F. volcano crucifera** Dall (Fig. 13.5b), has four white stripes on shell.

The distinctive "keyhole" at the apex of the shell serves as an exit for wastes and for water that has passed over the gills. The presence of slots or holes in the shell to facilitate the departure of wastes is a primitive feature. Like their relatives the abalones, the keyhole limpets also retain other old-fashioned gastropod features, such as paired gills and paired auricles to the heart. Most modern snails have only one gill (or none) and a single auricle.

The symmetrical, conical shell is a more recent innovation, whose evolution is suggested by the sequence of events occurring during development. In the keyhole limpet the tiny pelagic larva has an asymmetrical, coiled shell without a hole. After the larva settles out of the plankton the shell ceases to coil, and the body whorl grows as a shallow cone. A slit appears anteriorly in the shell but soon closes off to form a hole, and with continued growth the hole gradually "moves" to the shell apex.

Like many other intertidal mollusks, *F. volcano* is preyed upon by *Pisaster ochraceus* (8.13). It detects the sea star at a distance, and shows a characteristic flight response to it. Other aspects of the limpet's biology are poorly known, though related species in the Atlantic have received some study.

See Bullock (1953), Curtis (1966), Eales (1950), Hinegardner (1974), Hughes (1971), Krinsley (1959), Lowenstam (1962), and Ward (1966).

Phylum Mollusca / Class Gastropoda / Subclass Prosobranchia
Order Archaeogastropoda / Superfamily Fissurellacea
Family Fissurellidae

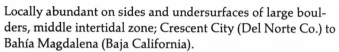

13.6 Lucapinella callomarginata (Dall, 1871)
Southern Keyhole Limpet, Fleshy Keyhole Limpet

Common on or under rocks and on wharf pilings encrusted with mussels, tunicates, and sponges, in lagoons and bays, low intertidal zone and subtidal waters; Morro Bay (San Luis Obispo Co.) to Bahía Magdalena (Baja California); more southerly records dubious.

Shell 21–30 mm long, rather flat and thick, elongate with narrow anterior end, the margin rough and crenulate; apical opening four times as long as wide, wider posteriorly; sculpture of both radial ribs and concentric lamellae; shell exterior cream-colored or buff, with gray radial stripes; mantle fringed; foot orange-brown or orange-red, with dark spots, massive, projecting out posteriorly; excurrent siphon set off by large fleshy papillae.

In bays and channels in southern California this limpet is often found embedded in the tissues of the mud-flat sponge *Tetilla mutabilis*. The limpet attacks the sponge primarily at the base, where the sponge is attached to the substratum, and excavates a cavity large enough to encompass the massive foot of the limpet; it then burrows forward, eating a large groove in the sponge. Some individuals may burrow deeply into the sponge, but they move close to the sponge's osculum prior to spawning. Such limpet activity often conspicuously distorts the sponge's shape. Evidence suggests that several other species of sponges are attacked by *Lucapinella*. The limpets can move over muddy substrata readily and appear to feed on algae films when sponges are not available.

See especially R. Miller (1968); see also J. McLean (1967).

Phylum Mollusca / Class Gastropoda / Subclass Prosobranchia
Order Archaeogastropoda / Superfamily Fissurellacea
Family Fissurellidae

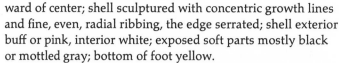

13.7 **Megatebennus bimaculatus** (Dall, 1871)
Two-Spotted Keyhole Limpet

Uncommon, on holdfasts of kelp or under rocks, low intertidal zone to 30 m depth; Forrester Island (Alaska) to southern Baja California.

Shell to 16 mm long, elongate, the lateral margins parallel and elevated, the ends rounded; apical opening up to one-third shell length, elongate, central in position; sculpture of radiating bands and concentric lines; interior shell margin with a shallow encircling groove; shell exterior usually buff, with radiating gray stripes, sometimes whitish with pink or red lines; body soft, much larger than shell, colored red, yellow, or white, often with dark spots or blotches.

This species often occurs on compound ascidians where the color pattern provides good camouflage. In the laboratory *Megatebennus* has been observed feeding on compound ascidians, and sponge spicules have been found in the gut of specimens collected in the field. Yet the digestive tract of this animal possesses a crystalline style, a rod of mucoprotein and other material, which is characteristic of mollusks feeding mainly on phytoplankton. Perhaps in nature the limpet is omnivorous.

See Ghiselin, deMan & Wourms (1975) and J. McLean (1969).

Phylum Mollusca / Class Gastropoda / Subclass Prosobranchia
Order Archaeogastropoda / Superfamily Fissurellacea
Family Fissurellidae

13.8 **Megathura crenulata** (Sowerby, 1825)
Giant Keyhole Limpet

Uncommon, on rocks, low intertidal zone and subtidal waters, occasionally abundant in protected crevices and spaces within breakwaters formed of large broken rocks; Monterey Bay to Isla Asunción (Baja California).

Shell (largely concealed under mantle in life) to 13 cm long, body to about 25 cm long; apical opening large, oval, just forward of center; shell sculptured with concentric growth lines and fine, even, radial ribbing, the edge serrated; shell exterior buff or pink, interior white; exposed soft parts mostly black or mottled gray; bottom of foot yellow.

Although an excellent detailed account of the anatomy of *Megathura* is available (Illingworth, 1902), very little is known about the natural history of the animal except that its diet includes seaweeds and colonial ascidians. It has been used for experimental studies on gamete agglutination. Sexually ripe individuals can be found at any season of the year, and by examining the gonads (after retracting the mantle) the animals can be sexed. Sperm remain viable for over a week when the testes are removed and stored at 4°C. Sperm agglutinated in an irreversible reaction when they came in contact with water previously exposed to eggs of the species. A specific enzyme that dissolves the egg membrane has been extracted from *Megathura* sperm. Numerous details are available on the biochemical constituents of the body of *M. crenulata*.

Coastal Indians used the shells of *Megathura*, along with other mollusk shells, as wampum.

See especially Illingworth (1902); see also Berg (1967), Curtis (1966), Giese (1969), Heller & Raftery (1976), Hinegardner (1974), Stearns (1875), and Tyler (1939, 1940).

Phylum Mollusca / Class Gastropoda / Subclass Prosobranchia
Order Archaeogastropoda / Superfamily Fissurellacea
Family Fissurellidae

13.9 **Diodora arnoldi** McLean, 1966
(= Diodora murina)

On undersides of rocks, rare intertidally but common subtidally to depths of 25 m; Crescent City (Del Norte Co.) to Isla San Martín (Baja California).

Shell 20 mm long, with nearly parallel lateral margins, the posterior slope rounded; apical opening oval, one-third of shell length from anterior end; shell surface sculptured with very fine cancellate lines; shell white, often with gray radiating lines.

See J. McLean (1966a).

Phylum Mollusca / Class Gastropoda / Subclass Prosobranchia
Order Archaeogastropoda / Superfamily Fissurellacea
Family Fissurellidae

13.10 **Diodora aspera** (Rathke, 1833)
Rough Keyhole Limpet

Common under stones and large boulders or under canopy of algae, low intertidal zone; more common subtidally in the south; Afognak Island (Alaska) to Camalú (Baja California).

Shell to 70 mm long, thick, triangular in profile; apical opening circular, slightly anterior to center; sculpture of coarse, cancellate, radial ribs, every fourth rib larger, the concentric sculpture weak; shell color gray with gray-brown radiating bands or with black and white radiating stripes.

Diodora aspera is omnivorous, but encrusting bryozoans are consumed in preference to algae. In turn, the limpet is preyed upon by sea stars, and has developed a remarkable defensive response, which is elicited by contact with *Pycnopodia* (8.16), *Orthasterias* (8.9), *Leptasterias* (8.10, 8.11), and several species of *Pisaster* (8.13–15). Instead of running away, the limpet extends its foot, elevating the shell. The mantle flap, ordinarily underlying the edge of the shell, is divided at its margin into a series of low ridges or folds. Two of these folds become greatly extended; one extends downward, covering the side of the foot, and the other extends up to cover the outside of the shell. The mantle margin lining the keyhole extends outward and upward. The result is that the starfish's tube feet find no ready place to grip the limpet; those tube feet that found attachment sites on the shell early in the attack generally loosen their grip when contacted by the extending mantle fold.

Some limpets have a second line of defense against sea stars, as well. The polychaete *Arctonoe vittata* (18.3) is a common commensal in the mantle cavity and pallial groove of *Diodora aspera*, and the worms may be as long as or longer than the host (both species are shown in Fig. 18.3). Predatory sea stars that approach and grasp a keyhole limpet are often bitten by the worm, causing them to retreat.

The blood of *Diodora aspera* contains a bluish respiratory pigment, hemocyanin. The pH of the blood is unusually low, about 7.1 instead of 7.3–7.8 as in most marine mollusks. Moreover, the respiratory pigment shows no "Bohr effect";

the affinity of the pigment for oxygen shows no change when the pH is altered. In this respect the hemocyanin here resembles that in the gumboot chiton, *Cryptochiton stelleri* (16.25).

See Dimock & Dimock (1969), Feder (1972), Hinegardner (1974), Margolin (1964b), and Redmond (1963).

Phylum Mollusca / Class Gastropoda / Subclass Prosobranchia
Order Archaeogastropoda / Superfamily Patellacea
Family Acmaeidae

13.11 **Acmaea mitra** Rathke, 1833
White-Cap Limpet

Common on rocks bearing coralline algae, low intertidal and shallow subtidal zones in protected sites near areas of heavy surf; Pribilof and Aleutian Islands (Alaska) to Isla San Martín (Baja California).

Shell to 35 mm long and 30 mm high, white, conical, apex almost in center, surface sculptured with fine concentric growth lines and radial striations.

The shell of *Acmaea mitra* is often overgrown with the encrusting thalli and nodular holdfasts of the red coralline algae that form the main food of this species. Fecal pellets consist of perfectly white calcareous strings of material rasped from the plants. The radular teeth of *A. mitra*, like those of some of its relatives that rasp algae from rocks, are capped with goethite, a hard crystalline iron compound unusual among marine invertebrates.

Acmaea mitra breeds in winter along the California coast and spawns when the sea temperature is at or near its minimum (December and January). Growth studies on this limpet have shown that the shell proportions change with growth. Differences in shell shape are not normally related to environmental differences, for example, animals of the same length living at different levels in the intertidal zone do not have shells differing significantly in height.

Laboratory tests show that *A. mitra* usually does not exhibit an escape response to predatory starfishes.

See Fritchman (1961a, 1962), Hinegardner (1974), Lowenstam (1962), Margolin (1964), J. McLean (1966b, 1969), Shotwell

(1950a,b), and Yonge (1962). For a collection of papers on California limpets, see D. Abbott et al. (1968).

Phylum Mollusca / Class Gastropoda / Subclass Prosobranchia
Order Archaeogastropoda / Superfamily Patellacea
Family Acmaeidae

13.12 Collisella digitalis (Rathke, 1833)
Ribbed Limpet

Common on vertical rock faces, upper intertidal and splash zones, often occurring with *C. scabra* (13.13); Aleutian Islands (Alaska) to southern Baja California.

Shell 15–30 mm long, apex near anterior margin, occasionally overhanging; typically concave in profile in front of apex, convex behind; anterior slope and margin of shell smooth, remainder ribbed; exterior highly variable but usually brownish or olive-green with white blotches or dots; interior with brownish patch, often owl-shaped near center; some young individuals closely resembling *C. scabra* but lacking black speckles on side of foot; *C. strigatella* (= *C. paradigitalis*) resembling *C. digitalis* in general shape, but with exterior of shell lacking raised radial ribs and with interior usually lacking brown owl-shaped patch at apex.

At first glance *Collisella digitalis* and *C. scabra* appear to occupy the same niche, for they coexist on rocks in the high intertidal and splash zones over a broad geographical range. However, careful studies of distribution and abundance show that *C. digitalis* is mainly restricted to vertical or overhanging rock faces, whereas *C. scabra* is dominant on gently sloping or horizontal surfaces. The upper limit of distribution, as in many intertidal forms, is determined by the animals' ability to tolerate the effects of prolonged exposure to air. The largest individuals are found at the highest levels.

Both *C. digitalis* and *C. scabra* graze on microscopic films of algae and possess rather thin radulae, though they have the digestive enzymes (laminarinase, K-carrageenase, agarase, and fucoidinase) to utilize larger and coarser algae for food. The grazing activity of these limpets often limits the algal crop on rock faces throughout most of the year, and the competition for available food can limit their growth rates and maximum size.

During periods of low tide, *C. digitalis* individuals remain in place on dry vertical surfaces and tend to orient with the head pointed downward or to the right. When wetted again they begin to move about and forage. During periods of rough surf an entire population may move upward on the rocks. Animals often cluster in shady spots on the rocks during low tide, disperse as the tide rises, then recluster as the tide falls. A cluster of up to 44 individuals was observed over a 32-day period. Daily shifts in cluster position occurred, and some individuals left the cluster to be replaced by outsiders. Homing experiments have given inconsistent results. In some areas the limpet shows no homing response, in other areas up to 54 percent return to a particular home site.

On rocks inhabited by the barnacle *Pollicipes polymerus* (20.11), *C. digitalis* may exhibit two forms. Limpets living on the rock surface appear normal, whereas those living on the shells of the barnacle have shells higher in relation to their length than the others, and their color patterns resemble those of the barnacles.

Spawning has been observed in central California in the winter and spring months and also in June and July. Recruitment of juvenile limpets to the benthic population is high in spring and fall, low in summer, and absent in winter (November through January). Adults may reach an age of 6 years or more.

Collisella digitalis has been the subject of investigations on excretion (ammonia, urea, and uric acid were reported) and on respiration. Both the vascularized mantle flap and the gill are important in gas exchange. When an animal is exposed to air, the mantle fold expands with blood and the gill contracts, but under water the mantle fold is flattened and the gill elongated. The respiration rate of animals exposed to moist air is consistently lower than that of animals under water.

Collisella digitalis may accumulate concentrations of metals to which they are exposed along the California coast; limpets from beneath the Golden Gate Bridge (San Francisco Bay) have exceedingly high levels of lead (up to 900 ppm) in the tissues, probably related to the prevalence of automobile exhaust fumes.

When the shell of this and other limpet species is damaged, the animal can repair the chipped or broken part by laying down new shell material on the inside. The shells are often severely eroded on the top and sides by the shell-inhabiting fungus *Didymella conchae* (an ascomycete).

Large individual *C. digitalis* living well down on vertical rocks often harbor immature gammarid amphipods (*Hyale grandicornis*) in the open spaces below the shell. These small crustaceans are associated with other intertidal limpets as well. Hidden by day, they move out at night and appear to feed on the algae growing on the limpet's shell. The amphipods remain with the limpets until just before they mature, then they leave.

Collisella digitalis is preyed upon by shorebirds, sometimes by the crab *Pachygrapsus crassipes* (25.43), and occasionally by sea stars, to which it usually fails to give any escape response.

See D. Abbott et al. (1968), Baldwin (1968), Baribault (1968), Beppu (1968), Bonar (1936), Breen (1971, 1972), Bulkley (1968), Chapin (1968), Collins (1976), Dayton (1971), Doran & McKenzie (1972), Duerr (1967), Frank (1964, 1965b,c), Fritchman (1961c, 1962), Galbraith (1965), Giesel (1969), Glynn (1965), Graham (1972), Hardin (1968), Haven (1964, 1965, 1971, 1973), Hinegardner (1974), Jessee (1968b), Johnson (1968, 1975), Kingston (1968), Margolin (1964a), Millard (1968), A. Miller (1968), Murphy (1976), Nicotri (1974, 1977), Perkins (1971), Shotwell (1950a,b), Straughan (1971), A. Test (1945, 1946), Villee & Groody (1940), Walker (1968), Wicksten (1978), Willoughby (1973), and Yonge (1962).

Phylum Mollusca / Class Gastropoda / Subclass Prosobranchia
Order Archaeogastropoda / Superfamily Patellacea
Family Acmaeidae

13.13 Collisella scabra (Gould, 1846)
Rough Limpet

Common on horizontal or gently sloping rocks, upper intertidal and splash zones; Cape Arago (Oregon) to southern Baja California.

Shell to about 30 mm long, low in profile; apex well forward of center; surface heavily ribbed and with scalloped margin; exterior mottled greenish or brown, white or gray where eroded; interior white or with irregular brown lines and blotches; side of foot and head pale, speckled with black.

Individuals characteristically occupy a very specific "home site" where the edge of the shell closely matches the contour of the rock. The animals move about at high tide, but each returns to its home site after grazing on the film of algae and diatoms on the rocky surface nearby. How the limpets find their way home is still incompletely understood. In *C. scabra* the homing behavior develops only in animals over 5 mm in length. Experiments indicate that the topography of the rock surface is utilized in homing. The eyes play no role in this behavior, but when the cephalic tentacles are removed the ability to find the home site decreases. Aggressive encounters between two *C. scabra* have been seen, in which the individuals push each other and the "loser" turns and moves away.

Collisella scabra and *C. digitalis* (13.12) compete with each other for food where they occur together, and the two effectively limit the algal film on upper intertidal rocks during most of the year. However, the two species occupy slightly different habitat niches, *C. digitalis* favoring vertical or overhanging rock faces, *C. scabra* predominating on horizontal surfaces and gentle slopes. Of the two, *C. scabra* appears to withstand desiccation and high temperatures better. In both species, the vascularized mantle fold expands and is used as a supplementary respiratory surface when the animals are out of water. Under damp aerial conditions, *C. scabra* shows a much lower rate of respiration than *C. digitalis*.

Studies of reproductive activity of *C. scabra* in central California reveal some differences within and between local populations. Spawning occurs, probably more than once, mainly in the period from January through March. The gonads are depleted before summer, but gonad enlargement begins in August and continues through the fall. Settlement of larvae from the plankton takes place mainly from July to October, though some occurs in almost every month of the year. The *C. scabra* living lower down in the intertidal zone reproduce for a longer period of the year than those higher up on the rocks, and show less marked seasonality in reproductive events.

Both the gonad development and the growth of individual limpets vary with intertidal position and food supply. Animals living higher up on the rocks undergo greater annual growth and reach a larger size than do those lower down; the food supply is greater at lower levels but so is competition for food. The largest *C. scabra* in the high intertidal zone at Bodega Head (Sonoma Co.), with a shell length of 24 mm, are estimated to be about 11 years old, and the largest in the lower intertidal regions (shell length 16 mm) are judged at about 7 years of age. Specimens from the high intertidal zone have a lower metabolic rate and larger glycogen stores than those living lower down. Only lower populations are subject to predation by the sea star *Pisaster ochraceus* (8.13); the limpet shows no escape behavior on contact with the starfish. The lower populations of *C. scabra* also harbor numerous juvenile amphipods (*Hyale grandicornis*) under the shell.

See especially Haven (1964, 1965), and Sutherland (1970, 1972); see also Baldwin (1968), Baribault (1968), Beppu (1968), Bonar (1936), Brandt (1950), Bulkley (1968), Collins (1976), Feder (1963), Fritchman (1961c, 1962), Glynn (1965), Hardin (1968), Hewatt (1938, 1940), Hinegardner (1974), Jessee (1968a), Johnson (1968, 1975), J. McLean (1969), Murphy (1976), Perkins (1971), Stohler (1930), Villee & Groody (1940), Walker (1968), Wells (1917), White (1968), and Yonge (1962).

Phylum Mollusca / Class Gastropoda / Subclass Prosobranchia
Order Archaeogastropoda / Superfamily Patellacea / Family Acmaeidae

13.14 Collisella limatula (Carpenter, 1864)
File Limpet

Abundant on semiprotected rocks, middle to low intertidal zones; Newport (Oregon) to southern Baja California.

Shell 30–45 mm long, low in profile, with strong prickly radial ribs, the margin with sawtooth notching; apex one-third shell length from anterior end; exterior yellow, buff, or greenish brown, sometimes with darker mottling or a mosaic of white angular spots; head and sides of foot gray or black.

In bays, a more inflated form with coarser radial ribbing occurs, which has been referred to as **C. limatula moerchii** Dall in the past.

A variety of features have made this one of the best-studied limpets on our shores. It is common and easily recognized; it lives well under laboratory conditions, yet it is also accessible for field experiments. It is a convenient size for physiological and behavior studies. Its internal organ systems are better

"color coded" than those of many snails, so it is favorable for dissection.

Under field conditions at Pacific Grove (Monterey Co.), active movement and feeding take place only when the animals are submerged or splashed with water, regardless of the time of day, though not all animals become active on submersion. Activity of the population is greatest during rising tide, declines a bit at high tide, increases somewhat during tidal ebb, and virtually halts at low water. Light influences the direction of movement; statistically, the population tends to move downward when submerged by day, and upward when submerged at night. About half the animals show some homing behavior.

These limpets are herbivores, feeding on microscopic algae and on some of the larger algae that form encrusting sheets, such as the softer red algae *Hildenbrandia* and *Peyssonnelia*, and the coralline algae *Lithophyllum* and *Lithothamnium*. The limpet *Collisella pelta* (13.16), which often occurs in the same habitat, eats mainly erect algae, and thus does not compete strongly with the file limpet for food.

Growth and reproduction have been followed at Palos Verdes Peninsula (Los Angeles Co.). Under favorable conditions, up to 2 mm² of shell surface can be added per day. The largest animals are probably 2–3 years old. Somewhat unexpectedly, the growth rate and especially the rate of gonad development increase in winter as the temperature decreases, and decline as temperatures rise in the spring. Spawning of olive-green eggs has been observed in January, February, April, and October, and abrupt increases in shell growth rates often follow the spawning periods. A population of *C. limatula moerchii* in Tomales Bay (Marin Co.) spawned mainly in September, though a small laboratory spawning occurred in March.

Animals living high in the intertidal zone have a slower rate of heartbeat than animals of the same size living lower down, over a wide range of environmental temperatures. Heartbeat is slower in the winter than in the summer, and slower in starved animals than in well-fed ones. Laboratory studies indicate that *C. limatula* significantly alters its heart and respiratory rates in response to altered environmental temperature. There is some seasonal temperature acclimation in the rates of physiological processes. Respiratory gas exchange occurs through both the gill and the mantle flap, the latter becoming more important when the limpets are out of water. Animals exposed to environmental salinities ranging from 25 to 150 percent seawater do not osmoregulate.

On the Monterey Peninsula *Collisella limatula* is commonly preyed upon by the lined shore crab, *Pachygrapsus crassipes* (25.43). The crab can remove a limpet from a rock by placing a pincer (cheliped) under the shell and lifting it off. But if the limpet clamps down on the rock, the crab may open one pincer, and, coming down from above, place the two claw tips on opposite sides of the shell, about halfway to the top. Then, both pinching and exerting pressure downward, the crab pops the top off the shell. The break normally occurs along a circular cleavage zone just above the point of the shell muscle attachment. In many areas up to 14 percent of all limpet shells cast up in beach drift have the tops removed, suggesting that crabs cause significant mortality in limpet populations. Sea stars, especially *Pisaster ochraceus* (8.13), are also important predators. *C. limatula* detects at a distance, and exhibits escape response to, all of the predatory sea stars it normally encounters in the field. In general, when receptors on the mantle margin detect sea star scent in the water, the limpets escape by moving upward on vertical rocks, or downstream on horizontal surfaces.

See Baribault (1968), Beppu (1968), Bonar (1936), Bulkley (1968), Chapin (1968), Eaton (1968), Feder (1963), Feder & Lasker (1964), Fritchman (1961c, 1962), Hinegardner (1974), Johnson (1968), Kingston (1968), Kitting (1979), Lindberg, Kellogg & Hughes (1975), Markel (1974), Phillips (1975a,b, 1976), Ross (1968), Seapy (1966), Segal (1956, 1961, 1962), Segal & Dehnel (1962), and Walker (1968).

Phylum Mollusca / Class Gastropoda / Subclass Prosobranchia
Order Archaeogastropoda / Superfamily Patellacea / Family Acmaeidae

13.15 **Collisella ochracea** (Dall, 1871)

Locally common on undersides of rocks, very low intertidal zone and on subtidal bottoms below kelp beds; Aleutian Islands (Alaska) to Isla Cedros (Baja California).

Shell to 30 mm long, thin, low; apex well forward; sculp-

ture of fine radial ridges; exterior pale brown or greenish with white-checkered pattern; interior with alternating dark and light bands at periphery, dark stain in center.

This fragile species lives farther down on rocks than most other west coast limpets, one reason it has received very little study. Like some European limpets, it uses ciliary currents to clean the pallial grooves and to accumulate sediment therein on the right side. It then directs the exhalant respiratory current along the right pallial groove to flush the sediment out and away from the animal.

See J. McLean (1969), A. Test (1945, 1946), and Yonge (1962).

Phylum Mollusca / Class Gastropoda / Subclass Prosobranchia
Order Archaeogastropoda / Superfamily Patellacea / Family Acmaeidae

13.16 **Collisella pelta** (Rathke, 1833)
Shield Limpet

Common on rocky reefs, middle to low intertidal zones; Aleutian Islands (Alaska) to Bahía del Rosario (Baja California).

Shell to 40 mm long, strong; apex near center; all slopes convex and usually ribbed, sometimes smooth; exterior highly variable, brown or green to nearly black, often checkered with white, or with peripheral rays and bands of white; interior bluish white with brown spot near center.

This species is often associated with brown algae (*Egregia, Postelsia, Pelvetia, Laminaria*) and is common in mussel beds. Small specimens occurring on the stipes and holdfasts of *Egregia* are usually black, with weak ribbing on the shells, and somewhat resemble *Notoacmea insessa* (13.20). They move onto nearby rocks when about 10 mm long, change their pattern of growth, develop shell ribbing, and produce a different color pattern. Animals on flat rocks or under *Pelvetia* have flatter shells than specimens in mussel beds. Few limpets are more widely distributed in the intertidal zone. A small percentage of individuals exhibit homing behavior.

Collisella pelta shows a tidal rhythm of activity, moving about and feeding only when wetted by waves or submerged. Feeding does not occur at every high tide. The limpets consume a great variety of algae, both large and microscopic, but the bulk of the diet usually consists of the common erect algae present, especially the reds *Endocladia*, *Rhodoglossum*, and *Iridaea*, and the browns *Pelvetia*, *Egregia*, and *Postelsia*. Where shield limpets and file limpets occur together they do not compete strongly for food.

Current studies on the population genetics of *C. pelta* and other acmaeid species suggest that different species living together in the same habitat tend to diverge with respect to the exact nature of their digestive enzymes, suggesting that there may be some differences in diet.

The shell color is at least partly a consequence of diet. All feeding is accomplished by the radula. Limpets lack the paired biting jaws present in so many gastropods; instead, there is a single hard palate supporting the roof of the buccal cavity.

Spawning in central California populations occurs throughout the year, though activity is lowest in the summer. The few data available on growth suggest that the animals grow faster than *C. digitalis*, and that typical individuals may reach 30 mm length in roughly 3 years. Shell proportions are not significantly different in animals of the same length from different intertidal elevations.

The limpet gives marked escape (running) responses to the three species of sea stars known to be predators on the *C. pelta* population (*Pisaster ochraceus*, 8.13; *Leptasterias hexactis*, 8.11; and *Evasterias*, 8.12), but it only occasionally exhibits a reaction to other predatory starfishes, such as *Pycnopodia* (8.16) and *Solaster* (8.6, 8.7).

In common with other limpets, *C. pelta* often harbors immature individuals of the commensal amphipod *Hyale grandicornis* in the nuchal cavity and pallial grooves.

See Baribault (1968), Bulkley (1968), Craig (1968), Dayton (1971), Feder (1963), Frank (1965b), Fritchman (1961c, 1962), Glynn (1965), Jobe (1968), Johnson (1968), Kingston (1968), Margolin (1964a), J. McLean (1969), B. Menge (1972), S. L. Miller (1974), Murphy (1976), Nicotri (1974, 1977), Perkins (1971), Ruth (1948), Shotwell (1950a,b), Stohler (1930), A. Test (1945, 1946), Villee & Groody (1940), Walker (1968), Wicksten (1978), and Yonge (1962).

Phylum Mollusca / Class Gastropoda / Subclass Prosobranchia
Order Archaeogastropoda / Superfamily Patellacea / Family Acmaeidae

13.17 Collisella asmi (Middendorff, 1847)
Black Limpet

Nearly always on shells of the snails *Tegula funebralis* (13.32) or *T. gallina* (13.35), common on rocks and in pools, middle intertidal zone on rocky shores; British Columbia to Isla Socorro (Islas de Revillagigedo, Mexico).

Shell to 11 mm long, 8 mm high, apex near center; surface with fine radial striations, usually eroded except at margins; color dark brown to black externally and internally.

Highly restricted in habitat, these small limpets feed exclusively by rasping tiny plants growing on the shells of their snail hosts. The food supply on a single host might soon become exhausted, but *C. asmi* easily transfers from one *Tegula* shell to another when the hosts aggregate in clusters in shaded areas at low tide. In laboratory experiments, most *C. asmi* changed snail hosts at least once a day, and some changed several times. *C. asmi* that were removed from their hosts and placed in dishes of seawater providing a variety of rocks and shells as possible substrata usually selected *Tegula* shells occupied by living animals (either the original snails or hermit crabs—though *C. asmi* is not common on shells occupied by hermit crabs in the field). The claim that the limpet detects host shells at a distance needs verification. Some observations indicate that adult limpets select their host shells only after touching them with their long, mobile tentacles. *Tegula* shells long unoccupied are not selected by limpets, nor are shells (occupied or not) that have been treated with alcohol, and even shells washed with distilled water are rendered less attractive.

Collisella asmi spawns in both the spring and fall. Gametes are shed into the water and develop into swimming larvae with spirally coiled shells. In regular limpets, as in the keyhole limpets, the symmetrical limpet form is acquired only after settling and metamorphosis. How long the larvae swim, whether the hosts are located before or after initial settlement, and how the larvae or young limpets locate their hosts are intriguing matters that remain to be explained.

See Alleman (1968), Eikenberry & Wickizer (1964), Fritchman (1961c, 1962), Murphy (1976), A. Test (1945, 1946), F. Test (1945), and Yonge (1962).

Phylum Mollusca / Class Gastropoda / Subclass Prosobranchia
Order Archaeogastropoda / Superfamily Patellacea / Family Acmaeidae

13.18 **Collisella instabilis** (Gould, 1846)
Unstable Seaweed Limpet

On holdfasts and stipes of the brown alga *Laminaria*, low intertidal zone and subtidal waters; Kodiak Island (Alaska) to San Diego.

Shell to 35 mm long, with lateral margins curved upward near each end (shell rocks back and forth when put on flat surface); apex just forward of center; surface nearly smooth with faint concentric growth lines and radial ribbing near margins; exterior brown, the apex often lighter; interior bluish white with a central brownish blotch.

In central California this limpet occurs only on, and feeds on, the stipes and holdfasts of the brown alga *Laminaria dentigera* (including *L. andersonii*). Little is known of its biology.

See J. McLean (1966b) and A. Test (1945, 1946).

Phylum Mollusca / Class Gastropoda / Subclass Prosobranchia
Order Archaeogastropoda / Superfamily Patellacea / Family Acmaeidae

13.19 **Collisella triangularis** (Carpenter, 1864)
Triangular Limpet

On coralline algae or coralline-encrusted shells of *Tegula brunnea* (13.31), low intertidal zone and subtidal waters; Sitka (Alaska) to Baja California.

Shell to 7 mm long, 4 mm wide, elongate in outline, high and conical; apex forward of center, the anterior and posterior slopes irregularly convex; sculpture of fine radiating lines; color white, with pinkish to brown spots or rays; shell often overgrown with encrusting coralline algae.

Very little is known of the biology of this species. Coralline algae seem to be the only food, and the gut is often full of cal-

cium carbonate particles. Well-developed ciliary currents in the pallial grooves carry sediment and waste materials to the middle of the right side and down the wall of the foot. The mucus-laden mass is then expelled by gentle muscular contractions.

See J. McLean (1966b), A. Test (1945, 1946), and Yonge (1962).

Phylum Mollusca / Class Gastropoda / Subclass Prosobranchia
Order Archaeogastropoda / Superfamily Patellacea / Family Acmaeidae

13.20 **Notoacmea insessa** (Hinds, 1842)
Seaweed Limpet

Common on fronds of the brown alga *Egregia*, low intertidal zone; Wrangell Island (Alaska) to Bahía Magdalena (Baja California).

Shell to 22 mm long, thin, high, the sides parallel, apex slightly forward of center; surface smooth, lustrous, sculptured with fine radial lines; exterior dark brown, sometimes spotted with white near apex; interior brown.

This species lives only on the straplike fronds of the feather boa kelp, *Egregia menziesii*. Young limpets show no particular orientation, but older ones live with the sides of the shell parallel to the long axis of the frond. They occur anywhere along the frond. They eat some epiphytes but feed very largely on the epidermal and cortical tissues of *Egregia*, forming deep depressions into which the limpets fit, and thereby gaining protection from desiccation and mechanical abrasion. The limpets can move from frond to frond on the plant.

Notoacmea insessa is capable of spawning throughout the year, but does so mainly in the spring and summer. Eggs and sperm are released into the sea. Fertilized eggs develop rapidly and hatch as trochophores. Four days after fertilization (at 12°C) the larvae reach the late veliger stage and are ready to settle and metamorphose on a new plant host. The settled young grow rapidly, up to 0.1 mm per day in shell length; when the shell width approaches the width of the *Egregia* frond, the animals grow further only in length and height.

The host plant grows most rapidly during the summer. It declines during the fall and winter through abrasion and destruction by storm waves, and then resumes rapid growth in the spring.

During the summer, along with accelerated plant growth, the number of limpets per plant increases rapidly through the settlement of young. By fall, these new recruits are all mature. Winter storms take a heavy toll, both of *Egregia* fronds (weakened by limpet grazing) and of the limpets themselves, but many animals survive to spawn the following spring. The largest *N. insessa* found are probably not more than 1 year old.

Young *N. insessa* settle preferentially on large, old, post-reproductive *Egregia*, on plants crowded together rather than on isolated plants, and on plants already populated with adult limpets. Young *N. insessa* grow faster and survive better if they settle on scars made by older limpets.

See especially Proctor (1968); see also Andrews (1945), Black (1976), Fritchman (1961b, 1962), J. McLean (1969), A. Test (1945, 1946), Walker (1968), and Yonge (1962).

Phylum Mollusca / Class Gastropoda / Subclass Prosobranchia
Order Archaeogastropoda / Superfamily Patellacea / Family Acmaeidae

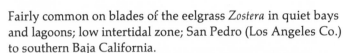

13.21 **Notoacmea paleacea** (Gould, 1853)
Surfgrass Limpet

Locally common on leaves of the surfgrass *Phyllospadix*, low intertidal zone on open coast; Vancouver Island (British Columbia) to Camalú (Baja California).

Shell to 10 mm long, to 3 mm wide, fragile, with parallel sides and with notch in anterior right margin; apex near front margin; sculpture of fine radial riblets; exterior light brown, darker on margins; interior white with bluish cast.

The smallest of the west coast limpets, this species is modified in connection with its life on the thin, straplike, wave-swept leaves of surfgrass. The limpet is usually no wider than the leaf it lives on, and the ends of the shell have saddle-shaped depressions, which allow the animal to fit tightly on a leaf that is a flattened oval in cross section. The slight notch on the right side of the shell allows the escape of water carrying wastes.

In some areas this animal feeds only on the epiphytic algae that grow on the surfgrass leaves; in other areas the radula rasps into the surface of the stout leaves themselves.

Sparse data suggest that breeding occurs throughout the year in central California.

See Fritchman (1961b), J. McLean (1969), and Yonge (1962).

Phylum Mollusca / Class Gastropoda / Subclass Prosobranchia
Order Archaeogastropoda / Superfamily Patellacea / Family Acmaeidae

13.22 **Notoacmea depicta** (Hinds, 1842)
Painted Limpet

Fairly common on blades of the eelgrass *Zostera* in quiet bays and lagoons; low intertidal zone; San Pedro (Los Angeles Co.) to southern Baja California.

Shell to 12 mm long, to 4 mm wide, differing from *N. paleacea* (13.21) in its greater size, its coloration (pale, with radiating brown streaks, the anterior slope brownish), its absence of radiating sculpture, and its habitat.

This is the only California limpet that has left open rocky shores for a more protected bay environment. A relative of *N. paleacea*, it is poorly studied, and with the continuing disruption of bay environment it is in danger of elimination in southern California.

See J. McLean (1969) and A. Test (1945, 1946).

Phylum Mollusca / Class Gastropoda / Subclass Prosobranchia
Order Archaeogastropoda / Superfamily Patellacea / Family Acmaeidae

13.23 **Notoacmea scutum** (Rathke, 1833)
Plate Limpet

Common on rocks protected from strong wave impact, middle intertidal zone; southern Bering Sea to Point Conception (Santa Barbara Co.).

Shell to 63 mm long, low in profile, with rounded apex near center; surface usually smooth, sculptured with coarse, flat-topped ridges; color highly variable, externally often brownish or greenish with white spots or lines, internally light with continuous dark rim and dark blotches near center; shell often bearing tufts of green alga *Enteromorpha* or *Ulva*.

This species shows a tidal rhythm of activity. The population moves up and down the rocks with the rise and fall of the tides, and shows greater activity at night than in daylight. The distance moved by individual limpets between successive low tides at Pacific Grove (Monterey Co.) ranged from 41 to 193 cm (mean 102 cm). Most limpets did not show homing behavior. In laboratory tests submerged limpets moved upward, travelling slowly in light coming from above, more rapidly in light coming from below, and at an intermediate rate in the dark. Unlike most gastropods, this limpet (and perhaps other limpets as well) expands the foot not simply by shifting the position of fluids within the body but by taking seawater directly into sinuses in the foot. The animals cling firmly to rock and cannot creep on sand.

In Puget Sound (Washington) *N. scutum* eats mostly microscopic algae. At Pacific Grove (Monterey Co.), it feeds mainly on the larger crustose red algae of the middle intertidal zone, though some microscopic plants are taken, too. The radula is longer than in any other local limpet examined, measuring about twice the length of the shell, though only a small portion of this at the front end is in use at any one time. Feeding sounds made by the rasping of the radula differ depending on the food being consumed and the depth of the rasps, and have been recorded in connection with studies of diet and feeding strategy (Kitting, 1978).

The gut, as in other true limpets, is long and coiled; as in other species that feed on macroscopic algae, it is relatively thick-walled. The gut harbors the digestive enzymes K-carrageenase, laminarinase, and fucoidinase. Metabolic excretory products include ammonia, urea, and uric acid. Respiratory gas exchange occurs both through the gill and through the mantle fold, the latter playing the major role when the animals are exposed at low tide. In animals exposed to salinities ranging from 50 to 125 percent that of seawater, potassium levels of the blood always remained above those of the medium, whereas blood levels of sodium, calcium, magnesium, and chloride did not.

Predators on *N. scutum* in central California include the crab *Pachygrapsus crassipes* (25.43), which sometimes chips the edge of the shell and pries the limpet up, and—more importantly—predatory sea stars. Running (escape) responses on the part of the limpet are elicited by contact with *Pisaster* (e.g., 8.13), *Leptasterias* (8.10, 8.11), *Evasterias* (8.12), *Orthasterias* (8.9), and *Pycnopodia* (8.16). The limpets also detect sea stars at a distance by means of scent receptors located on the mantle margin, along with photoreceptors, touch receptors, and contact chemoreceptors. The avoidance response has been analyzed in detail. Limpets in flowing water on horizontal surfaces normally move upstream, but when *Pisaster* scent is present they move downstream. Limpets on vertical surfaces generally move upward in the presence of *Pisaster* scent, even when the stimulus was applied in a downward flowing current.

Juveniles of the amphipod *Hyale grandicornis* are often found in the pallial groove under the shell of this limpet.

Central California populations of *N. scutum* contain some sexually ripe animals at all times of the year; natural spawning probably occurs here in all except the summer months. In a closely related east coast species, *N. testudinalis*, the males spawn first and this stimulates the females to release their eggs, which are 0.14 mm in diameter; photopositive trochophore larvae develop 10–13 hours after fertilization and remain trochophores for 19 hours; veligers develop 31–36 hours after fertilization and swim for about 56 hours, becoming increasingly photonegative. This development, from the egg to the metamorphosed juvenile, requires about 6 weeks. In *N. scutum*, eggs removed from ripe ovaries have been fertilized in the laboratory and the resulting larvae subjected to analysis for digestive enzyme activity.

The eggs and larvae of *N. scutum* contain the enzyme laminarinase (β-1,3-glucanhydrolase). This discovery led investigators to look for (and find) this enzyme in sea urchin eggs and larvae.

See Baribault (1968), Beppu (1968), Bulkley (1968), Chapin (1968), Dayton (1971), Feder (1963), Fritchman (1961b, 1962), Johnson (1968), Kessel (1964), Kingston (1968), Kitting (1978, 1979), Lindberg, Kellogg & Hughes (1975), Margolin (1964a), Menge (1972), S. L. Miller (1974), Muchmore (1968), Murphy (1976), Nicotri (1974, 1977), Phillips (1975a,b, 1977a), Rogers (1968), Shotwell (1950a,b), Stohler (1930), Walker (1968), Webber (1970), Webber & Dehnel (1968), and Yonge (1962).

Phylum Mollusca / Class Gastropoda / Subclass Prosobranchia
Order Archaeogastropoda / Superfamily Patellacea / Family Acmaeidae

13.24 Lottia gigantea Sowerby, 1834
Owl Limpet

Common on cliff faces and rocks of surf-beaten shores, high
and middle intertidal zone; Neah Bay (Washington) to Bahía
de Tortuga (Baja California).

Shell to 90 mm long, usually less, oval, low in profile; apex
near anterior margin; surface often rough and eroded; exterior
brown with whitish spots, the brown areas raised relative to
white spots; interior dark with brown margin and prominent
owl-shaped marking within bluish muscle scar; foot gray on
side, orange or yellow on sole.

The anatomy of *Lottia* is described in detail and beautifully
illustrated in a paper published in 1904 by W. K. Fisher, later
Director of the Hopkins Marine Station (Pacific Grove, Monterey Co.). The eggs were the subject of an early study in
chemical embryology by Jacques Loeb (1905). The animals
probably breed in the fall and early winter in California (September through January), and large specimens appear to be
10–15 year old.

Most patellacean limpets use the mantle fold around the
edge of the body as a respiratory surface when they are out of
water. *Lottia* differs anatomically from other California patellacean limpets in having a series of ridges or flaps (pallial
gills) on the lower surface of the mantle fold, which increase
the surface available for aerial respiration. When water
washes over the animal at the ebb and flow of the tide, the
shell margin is elevated a few millimeters above the rock surface and the surging water flushes out the pallial cavity and
grooves. Even in pounding surf this limpet does not clamp the
shell down firmly on the rock as one might expect. When
submerged, *Lottia* uses ciliary currents to pass a stream of
water over the main anterior gill, the respiratory stream entering the nuchal cavity on the left and leaving on the right.
Most animals on vertical or steeply sloping surfaces are found
oriented with the head downward. In this position, after the
tide falls, water running down the rock face and under the
limpet with slightly raised shell supplements ciliary currents
in the nuchal cavity and helps move wastes away from the gill
and to the outside.

At low tide, some individuals of *Lottia* occupy "home scars"
on the rock that exactly fit the margin of the shell; other individuals show no signs of homing to a specific resting site. In
either event, *Lottia* exhibits clear territorial behavior. Studies
on Santa Barbara shores show that each *Lottia* lives within an
area of algal film of approximately 1,000 cm², this area representing the "territory" of the individual. Each *Lottia* grazes
there and keeps its territory free of other animals that settle
there or that move in from adjacent areas. The limpets remove small barnacles by rasping them off with the radula,
and they dislodge smaller limpets, mussels, sea anemones,
and even other *Lottia* by pushing, using the shell like a bull-
dozer. By keeping other grazers outside the boundaries and
by grazing only part of the area at a time, *Lottia* allows the
algal film to grow to a thickness of 1 mm or more. During the
spring and summer the territories of individual *Lottia* can be
spotted from 15 m away. When a *Lottia* is removed from its
territory, other grazing limpets move in; their density soon
equals that in outside areas, and the algal patch, as a distinctive area, entirely disappears. When moved to a new area, a
Lottia drives off other nearby grazers, and within 3 weeks a
new territory of roughly 1,000 cm² is covered with an algal
film. When the algal film of a territory is experimentally reduced or removed, the resident *Lottia* increases the area of its
territory.

Like the limpet *Acmaea mitra* (13.11) and some other forms
that rasp on hard surfaces, *Lottia* precipitates the hard iron
compound goethite (α-Fe$_2$O-H$_2$O) as a cap on the radular
teeth; and in both mature radular teeth and their base plates,
Lottia also precipitates silica in the form of opal (SiO$_2$-nH$_2$O)
in amounts of up to 1.5 mg of opal in each radula.

Lottia gigantea was eaten by the coastal Indians, and the
shells are often found in kitchen middens. It is eaten sporadically today and has sometimes supported a small commercial
fishery; during the years 1919 and 1920 the California commercial shellfish catch included up to 8,200 kg of limpets,
most of them this species.

See especially Fisher (1904) and Stimson (1970, 1973); see also D.
Abbott (1956), Curtis (1966), Galbraith (1965), Hewatt (1934), Hinegardner (1974), Loeb (1905), Lowenstam (1962, 1971), G. MacGinitie
& N. MacGinitie (1968), Richardson (1934), Wells (1917), and Yonge
(1962).

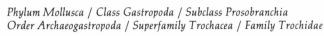

Phylum Mollusca / Class Gastropoda / Subclass Prosobranchia
Order Archaeogastropoda / Superfamily Trochacea / Family Trochidae

13.25 **Calliostoma annulatum** (Lightfoot, 1786)
Purple-Ringed Top Snail

Common on the brown algae *Macrocystis* and *Cystoseira* in offshore kelp forests and on rocks inshore, low intertidal zone; Forrester Island (Alaska) to Isla San Gerónimo (Baja California).

Shell to 30 mm in diameter, nearly evenly conical; surface with beaded spiral ridges; exterior yellow to orange-brown with purple band at suture and next to columella (see also Fig. 24.10); foot bright orange, with brown blotches.

This attractive snail and the related species *C. ligatum* (13.27) and *C. canaliculatum* (13.26) were collected on the third voyage of Captain James Cook and were among the first mollusks named from the west coast of America. All three species live on the stalks and blades of large kelp plants in California, with *C. annulatum* occupying an intermediate position in the forest, below *C. canaliculatum* and above *C. ligatum* and the bottom. *C. annulatum* reportedly moves up the kelp stipes to near the sea surface in "bright weather" and descends under other conditions. The animals can move rapidly; when snails were collected from the kelp, marked, and later released on the bottom, many were 20–30 feet up the kelp stipes within 24 hours.

Calliostoma annulatum is an omnivore. On "clean" kelp in the spring the food is mainly the kelp itself; the snail's digestive gland produces the enzyme laminarinase, which digests the main carbohydrate in kelp (laminarin). However, the snail prefers animal foods when these are available, especially hydroids (e.g., *Obelia*, 3.10; sertularians) and encrusting bryozoans (*Membranipora*, e.g., 6.1; *Hippothoa*). Detritus and some diatoms and copepods are taken, too. *C. annulatum* on the sea floor takes some of the cnidarian *Corynactis californica* (3.41) and scavenges opportunistically on dead fish. In aquaria the snails have been seen to eat hydroids, the anemone *Epiactis prolifera* (3.29), the stalked jellyfish *Haliclystis*, dead nudibranchs (*Polycera atra*, 14.48), dead keyhole limpets (*Fissurella volcano*, 13.5), dead chitons, nudibranch eggs, and other items,

including even canned dog food. Although jaws are often poorly developed in the Trochidae, observations by Perron (1975) suggest that they play an important role here. Hydroid stems in the gut often appeared "neatly cut into short segments." Further, when attacking anemones, *C. annulatum*, after initial contact, "would rear up on its metapodium, expand its lips, and suddenly lunge forward while biting at one of the anemone's tentacles." A dorid nudibranch was also attacked in this way.

A related subtidal species, **C. gloriosum** Dall, feeds mostly on the sponge *Xestospongia diprosopia* under both field and aquarium conditions, though dead animals are also taken. The shell bears a layer of mucus, which makes it slippery and not easily held by potential predators.

See Keen (1975), Keep & Baily (1935), Lowry, McElroy & Pearse (1974), S. L. Miller (1974), Perron (1975), and Sellers (1977).

Phylum Mollusca / Class Gastropoda / Subclass Prosobranchia
Order Archaeogastropoda / Superfamily Trochacea / Family Trochidae

13.26 **Calliostoma canaliculatum** (Lightfoot, 1786)
Channeled Top Snail

Fairly common in California on the uppermost stipes and blades of the kelp *Macrocystis* and large brown alga *Cystoseira*, less common on very low intertidal rocks inshore from kelp beds; Sitka (Alaska) to Camalú (Baja California).

Shell to 35 mm in diameter, conical, sculptured with low, spiral ridges; exterior white to grayish yellow, with brown on interspaces between ribs and a small bluish area near columella; foot tan, with brown spots.

This snail occupies the highest level of the canopy in the kelp forest, and is most abundant in the outer areas of the bed. The diet is very similar to that of its relative *C. annulatum* (13.25), and includes some of the kelp itself along with the attached hydroids, bryozoans, diatoms, and detritus. Dead animal material is eaten under aquarium conditions, but *C. canaliculatum* probably gets relatively little of that in the field, for its is seldom on the bottom. Large populations found in the summer on *Cystoseira* are eating mainly hydroids.

See Keep & Baily (1935), Herzberg (1966), Lowry, McElroy & Pearse (1974), and Sellers (1977).

Phylum Mollusca / Class Gastropoda / Subclass Prosobranchia
Order Archaeogastropoda / Superfamily Trochacea / Family Trochidae

13.27 **Calliostoma ligatum** (Gould, 1849)
(=C. costatum)   Blue Top Snail

On the brown alga *Cystoseira* and the kelp *Macrocystis*, occasionally on low intertidal rocks inshore from kelp beds; Prince William Sound (Alaska) to San Diego; abundant in low intertidal zone in Puget Sound (Washington), uncommon in southern California.

Shell to 25 mm in diameter, conical, chocolate brown with light tan spiral ridges; blue inner nacreous layer showing through to outside in some individuals; foot with orange sole, dark-brown sides.

Of the three species of *Calliostoma* prevalent in central California kelp beds, this species lives closest to the bottom and nearest to shore. On *Macrocystis* it eats both the kelp itself (it digests the main brown algal carbohydrate, laminarin), and adhering bryozoans, hydroids, diatoms, and detritus. On the bottom, surprisingly, it is often found eating the colonial tunicate *Cystodytes lobatus* (12.19), whose tissues and spicules are seen in the snail's gut. The high concentration of sulfuric acid in the ascidian tunic does not appear to deter predation by *C. ligatum*. Around the San Juan Islands (Puget Sound, Washington), the *C. ligatum* in the intertidal zone contained mainly diatoms, whereas the individuals found subtidally contained mainly detritus in the gut. The snails consume hydroids in laboratory aquaria.

This species has a relatively long and narrow foot, with which it moves well on a hard bottom and fairly well on sand. In locomotion the muscular waves travel from the rear to the front of the foot, and the waves on the two sides of the foot are out of phase. When individuals are exposed to a current of water, and a starfish (*Pisaster*, 8.13–8.15) is placed some distance away upstream, the snails move downstream at nearly double their normal speed (normal speed 0.7–1.2 mm per sec; escape speed 1.2–2.9 mm per sec). The increase in speed is due to increases in both velocity and frequency of muscular waves in the foot, but the wavelength (about 0.2 the length of the foot) remains the same. The snails cover the shell with mucus, which deters but by itself does not always prevent capture by sea stars (C. Harrold, pers. comm.).

See Emerson (1965), Lowry, McElroy & Pearse (1974), S. L. Miller (1974), Perron (1975), and Sellers (1977).

Phylum Mollusca / Class Gastropoda / Subclass Prosobranchia
Order Archaeogastropoda / Superfamily Trochacea / Family Trochidae

13.28 **Calliostoma gemmulatum** Carpenter, 1864
Gem Top Snail

Uncommon, rocky areas and wharf pilings, low intertidal zone; Cayucos (San Luis Obispo Co.) to southern Baja California.

Shell to 15 mm in diameter, 20 mm in height, with swollen whorls, deep sutures, two beaded spiral ridges on early whorls, and up to six beaded spiral ridges on body whorl; color greenish with darker spots or longitudinal markings.

Little is known of the biology of this species, but the shell, along with the shells of other related forms, has been photographed using x-rays.

See Herzberg (1966) and J. McLean (1969).

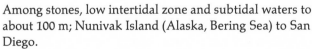

Phylum Mollusca / Class Gastropoda / Subclass Prosobranchia
Order Archaeogastropoda / Superfamily Trochacea / Family Trochidae

13.29 **Margarites pupillus** (Gould, 1849)
Little Margarite

Among stones, low intertidal zone and subtidal waters to about 100 m; Nunivak Island (Alaska, Bering Sea) to San Diego.

Shell to 15 mm in diameter, conical, with spiral ribbing; base of body whorl flattened; color pink or orange, the aperture brilliantly iridescent.

In Puget Sound (Washington), *M. pupillus* is preyed upon

by the snail *Searlesia dira* (13.86) and by the nudibranch *Dirona albolineata* (14.58), which crush the snails in the jaws. *M. pupillus* crawls at a maximum rate of just under 2 mm per sec, using waves of contraction that pass from the rear of the foot to the front (direct waves). Little else is known of the biology of this species, though information is available on British *Margarites*.

See Fretter & Graham (1962), Griffith (1967), Lloyd (1971), S. L. Miller (1974), Robilliard (1971), and Wise (1970).

Phylum Mollusca / Class Gastropoda / Subclass Prosobranchia
Order Archaeogastropoda / Superfamily Trochacea / Family Trochidae

13.30 **Norrisia norrisi** (Sowerby, 1838)
Norris's Top Snail

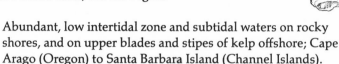

Common on kelp, occasionally washed ashore; Point Conception (Santa Barbara Co.) to Isla Asunción (Baja California).

Shell to 55 mm in diameter, broad, with low spire and swollen smooth whorls; umbilicus conspicuous, open; shell exterior light brown, darker near umbilicus; columella greenish; operculum horny with spiral bands of hairs; soft parts bright red in life.

This species feeds mainly on brown algae. Individuals from the southern California kelp-bed habitat prefer *Egregia* to *Laminaria*, and *Laminaria* to *Macrocystis*, but all three species are eaten. In the *Macrocystis* beds off Santa Catalina Island (Channel Islands) the *Norrisia* population undergoes a diurnal vertical migration. Animals, especially those high in the kelp forest, move down during the day. By 1500 hours, two-thirds of the snails are on the bottom third of the plants. At this time some animals begin to come up, and upward movement increases markedly at dusk. By 2300 hours three-fourths of the snails are above the bottom third of the plants. Laboratory experiments suggest that this movement involves both endogenous factors and external stimuli.

See Hinegardner (1974), Leighton (1966), G. MacGinitie & N. MacGinitie (1968), J. McLean (1969), W. E. Miller (1975), North (1964), and Wise (1970).

Phylum Mollusca / Class Gastropoda / Subclass Prosobranchia
Order Archaeogastropoda / Superfamily Trochacea / Family Trochidae

13.31 **Tegula brunnea** (Philippi, 1848)
Brown Turban Snail, Brown Tegula

Abundant, low intertidal zone and subtidal waters on rocky shores, and on upper blades and stipes of kelp offshore; Cape Arago (Oregon) to Santa Barbara Island (Channel Islands).

Shell to 30 mm in diameter, smooth, rounded-conical; columella usually with one tooth; color orange or bright brown; foot with dark brown or black sides, white or cream color below.

The vertical ranges of the two common turban snails in the intertidal zone in central California scarcely overlap; *Tegula brunnea* lives lower down than *T. funebralis* (13.32), the bulk of the population living subtidally. In the kelp forests at Pacific Grove (Monterey Co.) in August, 83 percent of the population occurred on the kelp *Macrocystis*, 10 percent on the brown alga *Dictyoneuropsis*, and 5 percent on the brown alga *Cystoseira*. On *Macrocystis* the animals were scarcest at the base and most abundant near the top of the kelp-forest canopy. In more exposed regions of the coast the brown turban was commonest on the brown algae *Laminaria* and *Pterygophora*.

Information on *T. brunnea* is fragmentary. Good food studies are lacking, but in laboratory aquaria the animals readily eat the brown algae *Macrocystis* and *Egregia*. The main excretory product is uric acid, curious in an animal that lives submerged nearly all the time. Spawning in laboratory aquaria at 15–16°C was observed in Oregon populations in August. The eggs are bright grass-green in color. Sperm are released as a milky cloud into the water, and the two-celled stage is reached 50 minutes after fertilization.

Both *T. brunnea* and *T. funebralis* cling firmly to the substratum, but are sometimes dislodged. On an irregular rock bottom they readily right themselves, but on a gravel bottom they use pebbles as tools for altering the position of the body. When these snails are inverted on such a bottom, the anterior part of the foot reaches down, grasps pebbles, and transfers them to the posterior end of the foot. When enough pebbles are collected and held aloft the center of gravity of the snail is changed so that it rolls over and rights itself.

Adult brown turban snails on the sea floor are preyed upon by starfishes (*Pisaster,* 8.13, 8.14), to which the snails show an avoidance response. Some *T. brunnea* are taken by sea otters. The crab *Cancer antennarius* (25.16) can also capture *T. brunnea* on the sea floor, but cannot crack the shells as easily as those of *T. funebralis*. The empty shells of all turban snails are used by larger hermit crabs.

Four Japanese species of *Tegula* all have the haploid chromosome number 18; a comparison with the several species on our coast would be of interest.

See Belcik (1965), Bollay (1964), Fritchman (1965), Herzberg (1966), Hinegardner (1974), Leonard (1964), Lowry, McElroy & Pearse (1974), J. McLean (1962b), Merriman (1967), Nishikawa (1962), Nolfi (1964), Patterson (1967), Stohler (1958), Valentine & Meade (1960), and Weldon & Hoffman (1975).

Phylum Mollusca / Class Gastropoda / Subclass Prosobranchia
Order Archaeogastropoda / Superfamily Trochacea / Family Trochidae

13.32 **Tegula funebralis** (A. Adams, 1855)
Black Turban Snail, Black Tegula

Common to very abundant in tidepools and on rocks, middle intertidal zone in protected coastal areas; Vancouver Island (British Columbia) to central Baja California.

Shell usually to 30 mm in diameter, body whorl with weak spiral striations and a spiral band below suture; columella with two teeth; shell dark purple to black, white below near columella; foot black on sides.

This is one of the best-known snails on the coast and has served in many field and laboratory investigations. At low tide *T. funebralis* is sedentary and often aggregates in groups of several hundred individuals in crevices and shaded areas. In some areas the clusters contain individuals all of a characteristic size. Population density seems to decrease as the amount of algal cover increases, thus the animals are characteristically abundant only on open rocky surfaces or in pools not covered with algae. The maximum population density in the Monterey Bay area is 0.6–1.2 m above zero tide level, with the largest individuals being lowest.

During high tide at night (but not during the day) the animals move up to the tops of rocks. Currents and turbulence stimulate them to activity and the animals orient into the flow, but when agitation of the water becomes very strong they head for shelter by moving downward on the rocks, regardless of current direction. The foot is long and relatively narrow. Locomotory waves of activity pass along the foot from front to rear (retrograde waves), and the waves on the two sides of the foot are out of phase. The animals move well on rocks but are clumsy on sand unless they hold the shell forward so that its weight is centered above the foot. Overturned snails can right themselves on rock, and on gravel they pick up pebbles and use them to regain normal stance in the same manner as *T. brunnea* (13.31). The normal speed of locomotion (0.6–0.8 mm per sec) is nearly doubled (to 1.3–1.4 mm per sec) when the animals are contacted by the sea star *Pisaster ochraceus* (8.13). The greater speed is due to increases in the frequency, the velocity, and (unlike *Calliostoma,* 13.25–28) the wavelength of the muscular locomotory waves of the foot.

Black turban snails are negatively phototactic. Parts of the body surface in addition to the eyes are sensitive to light, since even animals that are blind give a withdrawal response when a shadow is cast on them during the day.

Tegula funebralis eats many species of algae, including microscopic films, attached algae, and wrack. It prefers the softer fleshy algae such as *Macrocystis,* *Nereocystis,* and *Gigartina.* The radular teeth are less hard than in the chitons and true limpets, which rasp food from hard rocks. Enzymes in the foregut and midgut can digest not only storage carbohydrates, such as starch and laminarin, but also structural carbohydrates, such as cellulose, alginic acid, agar, fucoidin, and carrageenan, which probably form the bulk of the diet. At least three proteolytic enzymes have also been demonstrated. The ability to make use of a wide variety of algae for food has undoubtedly contributed to the success and prominence of this species.

In the intertidal zone, the animals are often exposed to air for a good part of each day, and their main excretory product, uric acid, suggests adaptation to a life partly out of water. Two kidneys are present, but uric acid has been found only in the right kidney. The main respiratory pigment, a hemocyanin, which colors the blood faintly blue, is very similar to that found in *T. brunnea.*

The sexes are separate. To some extent animals can be sexed by the color of the sole of the foot: those with the palest feet are males, those with darker feet are females. Detailed studies of reproduction are lacking, but females were seen to deposit small gelatinous masses about 3 mm in diameter, each containing several hundred eggs, in April at Pacific Grove (Monterey Co.). The eggs averaged 0.19 mm in diameter. Veliger larvae emerged from the egg mass on the seventh day and settled 12 days after the eggs were laid. Young snails feed on the film of microscopic algae on rocks. Growth studies in the laboratory indicate that young snails grow fairly rapidly; a population of snails 4–5.6 mm in shell diameter, with an average weight of 27 mg in June, showed a size range of 5.6–9.8 mm and an average weight of 177.3 mg the following March. Field studies in Oregon show that larger individuals grow very slowly. Age there can be approximated by counting the more prominent growth lines, which appear to be produced annually. Large individuals occurring in nature may be 20–30 years old; thus the life span may exceed that known for any other snail. Along the Oregon coast at least, shell growth ceases from November through February in animals more than 2 years old. When the animal is growing, the outer lip of the aperture may extend to 6 μm per day. Old snails often have shells badly eroded or pitted, especially at the apex, owing to mechanical abrasion, to the action of an ascomycete fungus, or to the boring or rasping activity of other animals. When the shell is broken, the break can be repaired or healed by secretion of new shell material. The shell often bears colonies of a boring bryozoan, and microscopic algae grow over the black outer surface. The algae are eaten by the commensal limpet *Collisella asmi* (13.17), which moves from one *Tegula* shell to another in its search for food. Five other species of *Collisella* (*C. pelta*, 13.16; *C. digitalis*, 13.12; *C. strigatella*; *C. limatula*, 13.14; and rarely *C. scabra*, 13.13) and *Notoacmea scutum* (13.23) occur here and there on the black turban's shell. The filter-feeding slipper limpets *Crepidula adunca* (13.55) and *C. perforans* (13.59) are also found here, and a single turban may be host to as many as eight of them at a time (see, for example, Fig. 13.59).

In nature, *Tegula funebralis* is preyed upon by many organisms, including sea otters, the red rock crab (*Cancer antennarius*, 25.16), and especially the sea star *Pisaster ochraceus* (8.13). Predation probably accounts for the scarcity of the black turban in subtidal waters. The snails have some defenses against predators. They detect predatory sea stars at a distance by means of sensitive olfactory organs, the bursicles, located on the gill. These organs, now known to be present in many archaeogastropods, were first discovered to be sensory receptors through studies on *T. funebralis*. *Tegula* flees from both scent and touch of predatory sea stars but ignores nonpredaceous ones. It also eludes carnivorous snails by crawling up on top of the predator's shell, where it is out of reach. The black turbans are also collected by man as food, and were used extensively by California Indians. Present (1979) state Fish and Game regulations set a bag limit of 35 on turban snails.

When black turban snails die, their shells are inhabited by hermit crabs. From 75 to 90 percent of the adult hermit crabs *Pagurus granosimanus* (24.13) and *P. samuelis* (24.10) are dependent on *T. funebralis* shells for homes.

Tegula funebralis has been found to contain some unusual nucleotides, as well as low-density compounds, in the DNA. The female gonad is bright green due to droplets of quinone-like pigments in the yolk of the eggs.

See especially D. Abbott et al. (1964); see also D. Abbott, Blinks & Phillips (1964), Belcik (1965), Berrie & Devereaux (1964), Best (1964), Bollay (1964), Brewer (1975), Bullock (1953), Burke (1964), Curtis (1966), Darby (1964), Eikenberry & Wickizer (1964), Emerson (1965), Erickson (1964), Feder (1963), Frank (1965a, 1969, 1975), Fritchman (1965), Galli & Giese (1959), Glynn (1965), Graham (1972), Hewatt (1934), Hinegardner (1974), Kitting (1979), Kosin (1964), Krinsley (1959), Leonard (1964), Lowenstam (1962), MacDonald & Maino (1964), McGee (1964), J. McLean (1962a), Merriman (1967), S. L. Miller (1974), Nolfi (1964), Overholser (1964), Paine (1969, 1971), Peppard (1964), Perkins (1971), Phillips (1978), Pilson & Taylor (1961), Putnam (1964), Schroeder & Cleland (1964), Stohler (1964, 1969b), Szal (1969, 1971), Wara & Wright (1964), Wise (1970), and Yarnall (1964).

Phylum Mollusca / Class Gastropoda / Subclass Prosobranchia
Order Archaeogastropoda / Superfamily Trochacea / Family Trochidae

13.33 **Tegula aureotincta** (Forbes, 1852)
Gilded Turban Snail, Gilded Tegula

Common, low intertidal and adjacent subtidal zones on rocky shores; Ventura Co. to Bahía Magdalena (Baja California).

Shell to 40 mm in diameter, with broad, low, spiral ridges prominent on base, less obvious on body whorl, intersecting fine oblique ridges; shell dark gray or olive above, marked with yellow or orange around umbilicus below.

Studies of the radula and shell suggest that this species is more closely related to *T. brunnea* (13.31) than to other local *Tegula* species, and the two are sometimes placed in a common subgenus. Some specimens of *T. aureotincta* resemble the extreme variants of *T. brunnea*, suggesting that the two species show parallel trends in variation. X-ray photos have been made of the shell, but little else is known about the biology of *T. aureotincta*.

See Fritchman (1965), Herzberg (1966), J. McLean (1969), Merriman (1967), and Stohler (1958).

Phylum Mollusca / Class Gastropoda / Subclass Prosobranchia
Order Archaeogastropoda / Superfamily Trochacea / Family Trochidae

13.34 **Tegula eiseni** Jordan, 1936
Banded Turban Snail, Banded Tegula

Common in rubble and on rocks, middle and low intertidal zones and in subtidal kelp forests; Los Angeles Co. to Bahía Magdalena (Baja California).

Shell to 25 mm in diameter, with inflated whorls bearing several nodular spiral ribs; umbilicus deep; shell brownish.

Little is known regarding the biology of this species. X-ray photos have been made of the shell. The radula closely resembles that of *T. mariana*, a species from Baja California.

See Fritchman (1965), Herzberg (1966), J. McLean (1969), and Merriman (1967).

Phylum Mollusca / Class Gastropoda / Subclass Prosobranchia
Order Archaeogastropoda / Superfamily Trochacea / Family Trochidae

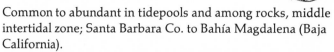

13.35 **Tegula gallina** (Forbes, 1852)
Speckled Turban Snail, Speckled Tegula

Common to abundant in tidepools and among rocks, middle intertidal zone; Santa Barbara Co. to Bahía Magdalena (Baja California).

Shell to 35 mm in diameter, much like that of *T. funebralis* (13.32) but paler gray to greenish with axial stripes of white in a checkered or zigzag pattern, and lacking a distinct, narrow, sculptured, spiral band just below suture.

In the characteristics of both the radula and the shell, this species most closely resembles *T. funebralis*. These two *Tegula* species are sometimes found together where their ranges overlap in southern California and northern Mexico. Comparative biological studies of both species in such areas would be desirable.

See Curtis (1966), Fritchman (1965), J. McLean (1969), Merriman (1967), and Stohler (1950).

Phylum Mollusca / Class Gastropoda / Subclass Prosobranchia
Order Archaeogastropoda / Superfamily Trochacea / Family Trochidae

13.36 **Tegula montereyi** (Kiener, 1850)
Monterey Turban Snail, Monterey Tegula

Fairly common on brown algae, especially *Macrocystis*, in off-shore kelp beds; uncommon in low intertidal zone inshore of kelp beds; Bolinas Bay (Marin Co.) to Santa Barbara Island (Channel Islands).

Shell to 45 mm in diameter, conical, smooth, with flat brown sides; base of shell lighter, with deep open umbilicus, the sloping wall of which bears a spiral rib.

In a kelp forest survey conducted in August at Pacific Grove (Monterey Co.), 51 percent of the *T. montereyi* population occurred on the kelp *Macrocystis*, 21 percent on the brown alga *Cystoseira*, 16 percent on the red alga *Gigartina*, and 12 percent on the brown seaweed *Dictyoneuropsis*. The population showed three size classes, with modes at 13, 18, and 22–33 mm basal shell diameter. Snails of the two larger size classes were mainly on *Macrocystis* (at all levels but mainly on the lower halves of kelp plants), whereas the smallest size class occurred mainly on the other three algal species. On the open coast south of Carmel (Monterey Co.), *T. montereyi* was also found on the large brown algae *Laminaria* and *Pterygophora*. The radula of *T. montereyi* shows considerable variation, but little else has been published on the biology of the species.

See R. Abbott (1974), Andrews (1945), Fritchman (1965), Lowry, McElroy & Pearse (1974), J. McLean (1962a), and Merriman (1967).

Phylum Mollusca / Class Gastropoda / Subclass Prosobranchia
Order Archaeogastropoda / Superfamily Trochacea / Family Trochidae

13.37 **Tegula pulligo** (Gmelin, 1791)
Dusky Turban Snail, Dusky Tegula

Uncommon on low intertidal rocks inshore of kelp beds; more common subtidally on larger algae in kelp forests; Sitka (Alaska) to Baja California.

Shell to 30 mm in diameter, sharply conical, top of inner lip forming a flange on side of open umbilicus next to aperture; shell gray or brownish, sometimes with orange, brown, or white spots; cephalic tentacles and dorsal side of head black; foot black with lavender blotches.

A kelp forest survey made in August at Pacific Grove (Monterey Co.) showed this species to be the commonest large snail among the six trochacean species present. Most specimens occurred on large brown algae (39 percent on *Macrocystis*, 31 percent on *Cystoseira*, 21 percent on *Dictyoneuropsis*), but 9 percent were found on red algae (*Gigartina*). The population showed two clear size classes, with modes at 14 and 20 mm basal shell diameter. The large snails were mostly on the lower parts of the kelp *Macrocystis*; the smaller animals occurred on the other algal species. On the open coast south of Carmel (Monterey Co.), *T. pulligo* juveniles were found on rocks, and adults on the large brown algae *Laminaria* and *Pterygophora*.

In both shell form and habitat this species resembles *T.*

montereyi (13.36), but the radula is rather different. Variations in shell form show some parallels with variation in *T. brunnea* (13.31) and *T. aureotincta* (13.33).

See R. Abbott (1974), Belcik (1965), Fritchman (1965), Griffith (1967), Lowry, McElroy & Pearse (1974), Merriman (1967), and Stohler (1958).

Phylum Mollusca / Class Gastropoda / Subclass Prosobranchia
Order Archaeogastropoda / Superfamily Trochacea / Family Turbinidae

13.38 **Astraea gibberosa** (Dillwyn, 1817)
Red Top Snail, Red Turban Snail

Low intertidal zone and below on rocky shores in British Columbia, usually subtidal in California, to 80 m depth; Queen Charlotte Islands (British Columbia) to Bahía Magdalena (Baja California).

Shell to 50 mm in diameter in California (to 75 mm in British Columbia), conical; body whorl with a regular pattern of bumps or modules lying between intersecting spiral and diagonal grooves; shell base flat with coarse spiral furrows; shell red under brownish periostracum; operculum with shallow groove, calcareous, oval, smooth.

In northern California this species is often washed up into tidepools by storm waves. Little is known of its biology.

See R. Abbott (1974), Griffith (1967), and J. McLean (1969).

Phylum Mollusca / Class Gastropoda / Subclass Prosobranchia
Order Archaeogastropoda / Superfamily Trochacea / Family Turbinidae

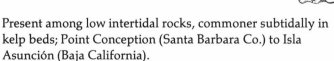

13.39 **Astraea undosa** (Wood, 1828)
Wavy Top Snail, Wavy Turban Snail

Present among low intertidal rocks, commoner subtidally in kelp beds; Point Conception (Santa Barbara Co.) to Isla Asunción (Baja California).

Shell to 110 mm in diameter, heavy, sculptured with undulating ridges and nodes; shell base flat with well-defined spiral cords; color of shell tan; periostracum prominent; operculum ovate, calcareous, with large rough ridges.

Specimens up to 50 mm in diameter occur intertidally, but the larger animals are usually found only subtidally in kelp beds. Little is known of their biology.

Examination of the shell shows that in the undulating ridges and nodes, the outer contours of the shell are formed first, with the nodes as hollow outpocketings that are later filled up by the nacreous layer. In the shell of a related Atlantic species, *A. olfersi*, the outer calcareous layer (just below the fibrous periostracum) is relatively thin, homogeneous, and composed of aragonite with traces of calcite, whereas the bulk of the shell is made up of successive nacreous lamellae of aragonite interleaved with thin layers of organic material. The shell matrix proteins of *Astraea caelata* are similar in amino acid composition to those of the black abalone, *Haliotis cracherodii* (13.4).

See R. Abbott (1974), Curtis (1966), Ghiselin et al. (1967), Hinegardner (1974), Jurberg (1970), J. McLean (1969), North (1964), and Stohler (1959a,b).

*Phylum Mollusca / Class Gastropoda / Subclass Prosobranchia
Order Archaeogastropoda / Superfamily Trochacea / Family Turbinidae*

13.40 Homalopoma luridum (Dall, 1885)
(=H. carpenteri)

Common under rocks, middle and low intertidal zones and subtidal waters; Sitka (Alaska) to Isla San Gerónimo (Baja California).

Shell to 9 mm in diameter, rounded with very low spire; sculpture of rounded spiral ribs; exterior color highly variable, gray, brown, reddish, or solid white, sometimes banded with light colors, interior of aperture nacreous; operculum thickened and calcareous.

Little has been published on the biology of these abundant small snails. Their shells are an important source of homes for juvenile hermit crabs living under intertidal stones on California shores.

See Griffith (1967) and J. McLean (1969).

*Phylum Mollusca / Class Gastropoda / Subclass Prosobranchia
Order Mesogastropoda / Superfamily Littorinacea / Family Lacunidae*

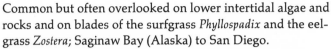

13.41 Lacuna marmorata Dall, 1919
Chink Snail

Common but often overlooked on lower intertidal algae and rocks and on blades of the surfgrass *Phyllospadix* and the eelgrass *Zostera*; Saginaw Bay (Alaska) to San Diego.

Shell to 7 mm high, thin, with surface sculpture of fine spiral lines; columella with a longitudinal groove; umbilicus narrow; shell color brown, often with white markings and white stripe inside base of aperture.

Near Friday Harbor (Washington) four species of *Lacuna* occur, including *L. marmorata*, and some of these hybridize, resulting in a variety of colors and patterns in the shells. The pleustid amphipod *Stenopleustes* sp., found in the same habitat on eelgrass, often mimics the color patterns and locomotor behavior of the snails. Few fishes in eelgrass beds feed on *Lacuna*, and laboratory experiments showed that fishes also ignored *Stenopleustes* as long as the amphipods remained on the eelgrass and did not swim freely in the water. This is one of the few known cases of Batesian mimicry among marine invertebrates (see also *Mitrella carinata*, 13.91).

See Carlton & Roth (1975), Field (1974), and Oldroyd (1927).

*Phylum Mollusca / Class Gastropoda / Subclass Prosobranchia
Order Mesogastropoda / Superfamily Littorinacea
Family Littorinidae*

13.42 Littorina planaxis Philippi, 1847
Eroded Periwinkle

Abundant on rocks, high intertidal and splash zones; Charleston (Oregon; not Puget Sound, Washington, as sometimes recorded) to Bahía Magdalena (Baja California).

Shell to 18 mm high, usually less, strong, oval, often much eroded; columella wide, polished, light in color; exterior gray-brown, sometimes with white spots; interior of aperture brown except for white band curving inward at base.

This periwinkle is of special interest because it occupies a higher vertical position on the shore than any other marine molluscan species in California. It lives out of water most of the time. It can secrete a mucus holdfast around the aperture of the shell, which glues the animal loosely to the rock face. This allows the snail to maintain its position without further expenditure of energy, and to withdraw into its shell and avoid desiccation. The mucus can be secreted in air or under water during unfavorable conditions. It is disposed of by the proboscis and radula before the animal again moves about.

The gill of *L. planaxis* is highly modified for aerial respiration. Adult snails can survive out of water for 2–3 months and perhaps much longer, but can also stand submersion indefinitely in seawater. They do not drown, as sometimes reported. They tolerate temperature changes as well as some California land snails. They can also survive several days' submersion in fresh water. Individuals often cluster together in cracks or slightly shaded areas on the rock surface and may remain motionless for days at a time unless wetted by wave splash. The splash-zone population tends to move downward during neap tides. Juveniles and smaller adults remain lower down, in the zone dominated by the barnacle *Balanus glandula* (20.24) and the red alga *Endocladia*; populations counted in this zone at Pacific Grove (Monterey Co.) showed a range of from zero to 808 *L. planaxis* per 400 cm², the highest population showing a dry-weight biomass of 1,198 mg per 400 cm².

Larger *L. planaxis* live above the zones where most macroscopic algae grow, and thus feed on films of diatoms, blue-green algae, unicellular and filamentous green algae, and occasional small specimens of larger plants. A variety of digestive enzymes has been demonstrated. The digestive efficiency appears to be low, for when fragments of macroscopic algae such as the green *Cladophora* are eaten, they often pass out undigested with the feces. Organic-matter assimilation appears to be only about 7 percent of consumption, and snails 8 mm in height require up to 6 hours to cycle food through the gut. This diet supports growth; under natural conditions in the field in southern California a population studied for more than a year averaged an increase in shell height of about 0.2 mm per month, and an increase of 0.6 mg per snail per month in dry weight of organic matter.

The grazing activities of *L. planaxis* result in considerable erosion of the rock face. It has been estimated that the rasping activity of the radulae of littorines may deepen certain high tidepools by 1 cm every 16 years, an amount equal to that caused by all the other erosive processes in the area. This is the case even though the teeth have a hardness of only 2–3 on the Mohs hardness scale, as compared with a hardness of 5 for *Lottia gigantea* (13.24), *Acmaea mitra* (13.11), and many chitons that graze on rocks. The molluscan radula is worn away in front and must be continually replaced from behind. When the radular ribbon is marked, the replacement rate can be monitored. In some littorines up to seven rows of teeth are replaced daily at 20°C.

Adult *L. planaxis* in the presence of water can go without food for many weeks. A population from San Diego, so starved at 22–24°C and then subjected to analysis for major body constituents, showed no differences from unstarved controls in total nitrogen, total protein, and total polysaccharides, but a large drop in lipids, a drop in free amino acids, and a rise in stored uric acid. Lipids are the main source of energy in starving animals. In normal animals the nitrogenous waste products include ammonia, urea, and uric acid; the enzymes urease, uricase, and xanthine oxidase are found in kidney tissues.

When moving about, individual snails tend to follow mucus trails laid down by other littorines and to move in the direction the secreting snails have taken. The tentacles are necessary for this behavior; animals devoid of tentacles cannot follow mucus trails, and bump into things when they move about, but they can still detect diffusable substances from the predaceous snail *Acanthina spirata* (13.81).

In a related species from the east coast, males produce pheromones during the reproductive season that attract *both* males and females. In *L. planaxis* the males are often found sitting on the shells of females; when sexually mature the females secrete materials that excite males to mating activity. However, it appears that in *L. planaxis* the male, even when roused, cannot distinguish the sex of another snail with which it is in contact until it has attempted to mate with the partner. Fertilization is internal. Individuals copulating or laying eggs can be seen at any time of year in California but especially during the spring and summer. Sometimes two males attempt to mount the same snail at the same time. A battle follows, lasting 0.5–3 minutes, in which the two males push at each other with the front (basal) ends of their shells till one is dislodged. Usually such a battle occurs over a female, but occasionally over a male, when neither battler has bothered to determine the third party's sex before the battle.

In the field, females lay their eggs in high pools, at the air-water interface, or below the surface. Most eggs are laid around dawn or dusk, by animals recently wetted or immersed. Females close to each other tend to deposit eggs at the same time. The eggs are round, pink, 70–84 *μ*m (mean 77 *μ*m) in diameter, and individually packaged in disk-shaped capsules that emerge stacked one on top of another. The egg capsules, deposited in masses of 700–2,000 (usually 1,200–1,300), are clustered in a mucus matrix shaped like a sausage coiled in a spiral of one or two turns. Egg masses laid at the water's edge have adhering bubbles and float; those laid under water do not. Within a day the mucus matrix starts to disappear; when released from the matrix, the individual capsules sink. Cleavage is holoblastic, and swimming veliger larvae emerge in 2–9 days, depending partly on the temperature over a range of 11–25°C. Information available on spawning and development (see Imai, 1964) is still fragmentary, and further study is needed.

Information on predators of *L. planaxis* is also inadequate. The habitat niches of carnivorous snails of the genus *Acanthina* (13.79–81) overlap that of *L. planaxis*, and attacks by the former on the latter have been noted several times. *Acanthina* drills the shell of *L. planaxis* near the columella, and approximately one-third of the *L. planaxis* shells occupied by hermit crabs at Pacific Grove are drilled in this manner. *L. planaxis* under water exhibits an escape response to nearby *Acanthina*. *Littorina planaxis* experimentally transferred to a lower intertidal position and tethered there were eaten by the crabs *Cancer antennarius* (25.16) and *C. productus* (25.22) and by the sea stars *Patiria miniata* (8.4) and *Leptasterias* (8.10, 8.11). Although these forms may consume *L. planaxis* that tumble from the rocks, they live outside the snail's normal vertical range. Experiments suggest that the lower limit of the adult *L. planaxis* population in the intertidal zone is set by wave action and by the inability of the snails to cling firmly enough to keep from being washed away.

Species of *Littorina* occur the world around in intertidal situations. Different species live at different intertidal levels and show corresponding differences in their degrees of adaptation to a semi-terrestrial existence. There is a large and interesting

literature on European, eastern American, Japanese, and Hawaiian forms. A detailed account of the anatomy of **L. littorea** Linnaeus, a common Atlantic species recently reported from California, is available (Fretter & Graham, 1962).

See Ankel (1936, 1937), Behrens (1974), Bigler (1964), Bingham (1972), Bock & Johnson (1967), Bollay (1964), Bonar (1936), Carlton (1969), Costello & Henley (1971), Dahl (1964), Dinter (1974), Dinter & Manos (1972), Foster (1964), Fox (1960), Fretter & Graham (1962), Gibson (1964), Glynn (1965), Hall (1973), Herzberg (1966), Hewatt (1934, 1937), Hopkins Marine Station (1964), Imai (1964), Isarankura & Runham (1968), Lowenstam (1962), J. Menge (1974), Neale (1965), Newell (1958a,b, 1965), North (1954), Perkins (1971), Peters (1964), Urban (1962), and Yamada (1977).

Phylum Mollusca / Class Gastropoda / Subclass Prosobranchia
Order Mesogastropoda / Superfamily Littorinacea / Family Littorinidae

13.43 **Littorina scutulata** Gould, 1849
Checkered Periwinkle

Abundant in high and upper middle intertidal zones on rocky shores; Kodiak Island (Alaska) to Bahía de Tortuga (Baja California).

Shell to 13 mm high, usually less, smooth, conical, usually taller in proportion to diameter than *L. planaxis* (13.42); columella narrow; shell brownish to nearly black, often with lighter bands or spots in a checkered pattern; aperture purplish within, lacking white band curving inward below.

Littorina scutulata lives lower down on the rocks than *L. planaxis*, and migrates up and down with the rising and falling tide. In tidepools the animals tend to emerge by night and submerge by day. They tolerate exposure to air less well than *L. planaxis*, although some animals have been kept out of water for 30 days with no mortality. At low tide they are often found hidden, the smaller animals in holdfasts of the red alga *Endocladia*, the larger ones in deep cracks or crevices or in empty barnacle shells. In general, the largest individuals live in, or move up to, the highest positions on the rocks.

Littorina scutulata creeps about by means of muscular waves, which travel from front to rear in the foot (retrograde waves), and in which the contractions on the two sides of the foot are out of phase (ditaxic waves). The foot clings to rocks with a measured tenacity unusually great for a gastropod; this, together with their small size, helps the animals hold their positions under turbulent conditions. Population densities of this periwinkle in the *Balanus-Endocladia* zone at Pacific Grove (Monterey Co.) ranged from 11 to 1,149 snails per 400 cm² and averaged 456 per 400 cm², representing a dry-weight biomass of 202–10,300 mg (mean 2,909 mg) per 400 cm².

Like *L. planaxis*, *L. scutulata* feeds mainly on films of diatoms, microscopic algae, lichens, etc., that occur on otherwise bare rock surfaces in the high intertidal zone, but it inhabits a zone where macroscopic algae also occur, and it readily eats *Pelvetia*, *Cladophora*, *Ulva*, and other softer-bodied larger plants. Metabolic efficiency appears to be about 7 percent. Experiments on natural populations of *L. scutulata* artificially caged with *L. planaxis* suggest that, in areas where the two species overlap, it is *L. scutulata* that suffers from competition.

Breeding is reported at all seasons except summer in California. Preliminary observations suggest that the females lay their eggs at the waterline or under water in tidepools, usually in the evening or at night. As in *L. planaxis*, the eggs are individually packaged in flattened capsules, which in turn are borne in a sausage-shaped gelatinous mass coiled in a spiral of three to five turns and which may hold upward of 2,000 eggs, generally more than in the *L. planaxis* egg mass (see Imai, 1964). Swimming veliger larvae hatch out in 7–8 days at 13–15°C, but information on development is framentary.

At Pacific Grove, recruitment of the young (up to a shell length of 1.4 mm) to the benthic population was greatest from November through May, but was slight in the summer months. Shell length increased by about 0.3 mm per month, and the life span appeared to be usually less than 2 years.

Predators on *L. scutulata* include the carnivorous gastropods *Acanthina spirata* (13.81) and *A. punctulata* (13.80) and the small starfish *Leptasterias hexactis* (8.11). *L. scutulata* gives an avoidance response to these forms when newly confronted by them, but after a period of cohabitation no longer reacts to the presence of the predator nearby. *Acanthina* consistently drills through the shells of captured littorines at a particular location near the columella on the body whorl; it then inserts its proboscis, and feeds. Behavior studies on *A. punctulata* in southern California show that *L. planaxis* is preferred to *L. scutulata*, and in general, larger littorines are preferred to smaller

ones. At Pacific Grove, in the immediate vicinity of the Hopkins Marine Station, the shells of *L. planaxis* and *L. scutulata* occupied by young hermit crabs were examined in June and September for signs of drilling by *Acanthina*. Of 93 *L. planaxis*, 33.3 percent were drilled; of 221 *L. scutulata*, 12.7 percent were drilled. An earlier study, conducted in May in the same area, reported that 20 percent of 443 *L. planaxis* shells found were drilled, but did not concern itself with *L. scutulata*.

The fine structure of the eye of *L. scutulata* has been investigated by electron microscopy. The anatomy of this organ is remarkably similar to that of the eye of the land snail *Helix aspersa*.

Littorines are often intermediate hosts for a number of parasites, and in Oregon over 10 percent of the *L. scutulata* population is infected with cercariae (fluke larvae).

See Bock & Johnson (1967), Buckland-Nicks & Chia (1976), Castenholz (1961), Chow (1975), Dahl (1964), Duerr (1965), Emerson (1965), Feder (1963), Foster (1964), Glynn (1965), Hayes (1929), Hewatt (1934), Hopkins Marine Station (1964), Humphrey & Macy (1930), Imai (1964), Johnson (1975), Mayes & Hermans (1973), J. Menge (1974), S. L. Miller (1974), Nicotri (1974, 1977), North (1954), Perkins (1971), Spight (1975b), Tittle (1964), Urban (1962), and Valentine & Meade (1960).

Phylum Mollusca / Class Gastropoda / Subclass Prosobranchia
Order Mesogastropoda / Superfamily Rissoacea / Family Vitrinellidae

13.44 **Vitrinella oldroydi** Bartsch, 1907

Present as a commensal in the pallial groove of the chiton *Stenoplax heathiana* (16.8), which occurs on undersides of smooth rocks partly embedded in sand, lower middle and low intertidal zones; Mendocino Co. to Puerto Santo Tomás (Baja California).

Shell about 2 mm in diameter, discoidal; whorls rounded; aperture circular, the umbilicus open; color white or yellowish.

We know very little about the biology of these small snails.

See Carlton & Roth (1975) and J. McLean (1969).

Phylum Mollusca / Class Gastropoda / Subclass Prosobranchia
Order Mesogastropoda / Superfamily Cerithiacea / Family Vermetidae

13.45 Serpulorbis squamigerus (Carpenter, 1857)
(=Aletes squamigerus)   Scaled Worm Snail

Tubular shells common, attached on wharf pilings and on upper surfaces of protected intertidal rocks; offshore to depths of 20 m or more; Monterey Bay to Baja California.

Tube to 12 mm in diameter, 125 mm in length, twisted, wrinkled, with longitudinal cords or ribs; first part of tube irregularly coiled and cemented to substratum, last part usually standing erect; color gray to pinkish gray; body soft, elongate; no operculum.

Serpulorbis squamigerus is a sessile snail. The shell, a long twisted tube firmly anchored in place, resembles the tube of a serpulid worm, but has the three layers characteristic of a molluscan shell. In southern California these gregarious animals live in concentrations of up to 650 individuals per m² of substratum. In central California (e.g., Monterey harbor) they occur singly or as small clusters of tubes.

For most of the animal's life the foot serves no locomotory function, and statocysts are absent. The circular plug that stoppers the aperture when the animal is withdrawn probably represents the metapodium, or posterior part of the foot; what corresponds to the anterior part of the foot of most other snails is represented mostly by a pair of large mucous glands. These glands open through a single aperture lying between a pair of pedal (epipodial?) tentacles, and play a critically important role in feeding.

Unable to roam about, the vermetid gastropods are particle feeders, employing mucus to trap drifting particles and small plankton as food. The mucus secreted by the pedal glands is carried by ciliary currents in the tips of the pedal tentacles, and somehow spun into a delicate, triangular, veil-like net up to 50 cm² in area. Formation of new net takes only 3–4 minutes when the animal is submerged. The sheet may remain suspended in the water for up to half an hour, during which diatoms, other small organisms, and particles of organic detritus adhere to the sticky surface. Then the animal retrieves the net by extending the radula far out of the mouth, grasping the mucus, and pulling it back into the mouth. The jaws hold the engulfed net while it is being swallowed. Turbulent water seems to stimulate the animals to spin nets and start feeding. Some particles are also filtered out by the gills, and possibly provide a supplementary source of food. When disturbed, the animal slowly retracts its foot into the shell.

The males produce sperm of two kinds (see the account of *Petaloconchus montereyensis*, 13.46). Fertilized eggs are laid in the summer in capsules of about 600 eggs each; up to 67 capsules are cemented to the inner side of the tube near the aperture, dorsal to the mantle cavity. The females develop a longitudinal slit in the mantle at that point, permitting them to emerge to feed without detaching the capsules. Nearly all the eggs in each capsule develop normally, and after an undetermined period of brooding they hatch out as swimming larvae and escape to be dispersed by the sea. Later they settle out (usually gregariously) and metamorphose into young snails with typically coiled shells. Soon they cement themselves to rocks or to the tubes of adults, and from then on the shell is secreted as the uncoiled, wormlike tube so characteristic of the adult. At least some individuals undergo a male-to-female sexual change during the winter.

In the kelp forest off Mussel Point (Pacific Grove, Monterey Co.), *Serpulorbis* (along with *Petaloconchus montereyensis*) is preyed upon by the sea star *Pisaster giganteus* (8.14), and vermetids may make up 50 percent of the sea star's diet (C. Harrold, pers. comm.). In southern California the carnivorous snails *Ceratostoma foliatum* (13.70) and *Shaskyus festivus* eat *Serpulorbis*.

Many animals carry fluke cercariae in the abdomen, and these feed on the gonad, eventually causing parasitic castration. The tubes of *Serpulorbis* in the Malibu region in southern California often house the commensal crab *Opisthopus transversus* (25.42), but not those in the Monterey Bay region.

See especially Hadfield (1966, 1970); see also Curtis (1966), Hinegardner (1974), Holmes (1900), Keen (1960, 1961), G. MacGinitie & N. MacGinitie (1968), Morton (1965), Pequegnat (1964), and Yonge & Iles (1939).

Phylum Mollusca / Class Gastropoda / Subclass Prosobranchia
Order Mesogastropoda / Superfamily Cerithiacea / Family Vermetidae

13.46 **Petaloconchus montereyensis** Dall, 1919

Common but patchy in distribution, on undersurfaces of rocks, low intertidal zone; very abundant in places subtidally; Monterey Bay to Isla San Martín (Baja California).

Tube to 2 mm in diameter, 25 mm or more in length; early part of tube with pits or diagonal grooves, projecting part of tube smooth; operculum present.

This species occasionally occurs as tightly coiled, isolated individuals, but more commonly is found as firmly cemented, entangled masses with the distal tubes and apertures projecting outward. The most common habitat in Monterey Bay is on rocks slightly below the zero tide level in areas of heavy but broken wave turbulence, and on the sea floor below kelp forests. In some places rock surfaces of two or more square meters' area may be covered with tubes in concentrations of up to 100,000 snails per m².

Petaloconchus, like *Serpulorbis squamigerus* (13.45), feeds by secreting a large veil of mucus that traps drifting particles and organisms; this is then retrieved by the radula, and eaten. As a subsidiary food source, particles removed from water circulated about the gill are caught in mucus and carried by cilia to a point on the anterior dorsal margin of the foot just in front of the head. Here the material accumulates in a ball, and the snail extends its mouth and picks it up with the radula. Unlike *Serpulorbis*, the foot in this species bears an operculum, which stoppers the tube when the animal is withdrawn. A feature that may be unique among gastropods is the periodic production of a new operculum below the old one, followed by a shedding of the old one.

The *Petaloconchus* population reproduces throughout the year, though not all individuals are necessarily active at one time. The animals are not hermaphroditic, and at any one time about 20–30 percent of the larger snails are brooding females.

The males (as in *Serpulorbis*) produce two types of sperm, normal (eupyrene) sperm used in fertilization, and much larger apyrene sperm, which in *Petaloconchus* lack flagella; the apyrene sperm later disintegrate and may help nourish the eupyrene sperm. The sperm are passed from the males to the females in packages (spermatophores). The eggs are laid in capsules, each containing about 100 eggs; up to 28 capsules are brooded at a time by a female, within the mantle cavity.

Only one offspring emerges from each capsule, for during development one larval snail eats all the other embryos. The latter, called "nurse eggs," may have been fertilized by more than one sperm; in any case they develop slowly and abnormally before being eaten by the single normal embryo.

In the kelp forest off Mussel Point (Pacific Grove, Monterey Co.), *Petaloconchus* is preyed upon by the sea star *Pisaster giganteus* (8.14) and, along with *Serpulorbis squamigerus*, makes up half the sea star's diet (C. Harrold, pers. comm.).

Trematode cercariae are often found in the gonad of this species, and the host may suffer parasitic castration. *Petaloconchus* also hosts a snail (*Odostomia* sp.) that feeds on the fleshy edge of the mantle. Tubes of the worm *Dodecaceria fewkesi* (18.25) are commonly found entwined among tubes of the snail.

See Hadfield (1966, 1969, 1970) and Keen (1961).

Phylum Mollusca / Class Gastropoda / Subclass Prosobranchia
Order Mesogastropoda / Superfamily Cerithiacea / Family Potamididae

13.47 **Cerithidea californica** (Haldeman, 1840)
California Horn Snail

Abundant on mud flats, high intertidal zone in protected bays; Tomales Bay (Marin Co.) to Laguna San Ignacio (Baja California).

Shell to 45 mm high, with tall spire; whorls slightly rounded; axial ribs low but well defined, spiral ribs faint; few rounded varices near base; shell brown with lighter columella; operculum with multiple spirals.

Cerithidea is probably the commonest snail occurring high on lagoon mud flats in southern California, often forming dense aggregations under debris and among plants. In laboratory tests the animals proved tolerant of estuarine conditions. After 12 days' immersion all animals were alive and active at a salinity half that of seawater, and at a salinity 18 percent that of seawater most were alive and some were active. Some animals survived 9 days' immersion in fresh water, but remained withdrawn in their shells the whole time.

Cerithidea californica feeds on fine organic detritus and microorganisms associated with it occurring at the mud surface. This detritus is scooped up with the radula, which is very short (only about one-tenth the shell length) and subject to little wear and tear, for the food is already finely divided. The stomach bears a crystalline style, as in many plankton-feeding mollusks, and the gut is adapted to transporting and digesting a more or less continuous flow of microscopic food. Published accounts of the anatomy are available. The species may have been taken as food by California Indians.

Several studies have been made of the parasites carried by the horn snail. In Newport Bay (Orange Co.), for example, over two-thirds of all individuals are infected with larval trematodes. Eighteen species of cercariae have been isolated from this species, and the snails are a good source of larval trematodes for classroom study. Many individual snails carry multiple infections, and peak infections occur in December and May. The digestive gland and gonadal area of the visceral spiral are often the main loci of infection. Marine birds and fishes are the definitive hosts of these parasites.

See Bequaert (1942), Bright (1958, 1960), Curtis (1966), Driscoll (1972), Hunter (1942), MacDonald (1969a,b), Martin (1955, 1972), J. McLean (1969), Nakadal (1960), Scott & Cass (1977), and Yoshino (1975).

Phylum Mollusca / Class Gastropoda / Subclass Prosobranchia
Order Mesogastropoda / Superfamily Cerithiacea / Family Potamididae

13.48 Batillaria attramentaria (Sowerby, 1855)
(=B. cumingi)

Abundant on soft mud, high to low intertidal zones in quiet bays and estuaries; Boundary Bay (British Columbia) to Elkhorn Slough (Monterey Co.).

Shell usually not over 35 mm high, occasionally to 46 mm, with tall slender spire; sculpture of axial flanges on early whorls but only spiral threads on later whorls; color black to tan with white flecks or bands.

This snail, sometimes misidentified as *B. zonalis*, was introduced to the west coast from Japan and is now very common in bays where Japanese oysters have been planted, especially in Elkhorn Slough and Tomales Bay (Marin Co.). It was first noted on shipments of oyster spat in 1930 and was common by 1950. Populations are highest below the level of mean higher high water, and especially in shallow depressions that retain some water even at low tide; here the snails may occur in densities of 7,000 per m^2. Ordinarily they lie at the surface, but in cold weather they burrow down 1–2 cm into the ooze. The animals tolerate temperatures of at least 1–29°C at Tomales Bay. They survive out of water for at least 16 days, and tolerate a wide range of salinities.

Batillaria feeds mainly on the wet surface film of detritus, microorganisms, and mud, but also takes fresh and decaying green algae such as *Enteromorpha* and *Ulva*. The radula is very short (about one-tenth the shell length), as tends to be the case with snails having soft diets. The digestive tract, studied in detail, has a large crystalline style.

A Tomales Bay population showed breeding in the spring and summer. Mating pairs were seen from March to June, with a peak in May. The smallest animals seen mating had shells 12 mm long, and most were more than 15 mm long. Eggs were laid in strings 1–2 cm long, sheathed with debris and castings. The pale eggs, 0.2 mm in diameter, are in individual capsules, about 13 capsules per mm of egg string. Development (followed in a British Columbian population) is direct, without a planktonic stage; juveniles hatch in about 11 days. In Tomales Bay, juveniles 1–2 mm long are most abundant from June to August, especially in late June. The animals add about 6 mm of shell length every year, and individuals 5 years old are about 30 mm long. In this population the largest animals (about 35 mm) were estimated to be about 10 years old. In other parts of Tomales Bay the animals may reach a length of 46 mm.

Batillaria crawls in a rather clumsy manner, but can shift its position swiftly by attaching its foot to the underside of the water surface film, especially on an incoming tide, and riding the current for a fair distance.

Like other members of the family of horn snails, *Batillaria* is infected with larval trematodes, and when animals carry heavy infections of cercariae in the gonads they produce no viable eggs or sperm.

See especially Driscoll (1972) and Whitlatch (1974); see also Carlton & Roth (1975), Chapman & Banner (1949), Hanna (1966), MacDonald (1969a,b), J. McLean (1960), and Yamada & Sankurathri (1977).

Phylum Mollusca / Class Gastropoda / Subclass Prosobranchia
Order Mesogastropoda / Superfamily Cerithiacea / Family Cerithiidae

13.49 **Bittium eschrichtii** (Middendorff, 1849)
Threaded Bittium

Under rocks and in clean gravel and coarse sand, low intertidal zone and subtidal waters; Sitka (Alaska) to Cabo San Lucas (Baja California).

Shell to 20 mm high, usually less, with tall spire; spiral ribbing distinct; anterior canal short, poorly differentiated; color whitish to gray, often with axial or spiral brown bands.

Little is known of the biology of this animal.

See R. Abbott (1974), Carlton & Roth (1975), Griffith (1967), and S. L. Miller (1974).

Phylum Mollusca / Class Gastropoda / Subclass Prosobranchia
Order Mesogastropoda / Superfamily Epitoniacea (= Ptenoglossa)
Family Epitoniidae

13.50 **Epitonium tinctum** (Carpenter, 1864)
Tinted Wentletrap

Fairly common on or near bases of sea anemones (*Anthopleura elegantissima*, 3.31, and *A. xanthogrammica*, 3.30, in California), low intertidal and shallow subtidal zones; Vancouver Island (British Columbia) to Bahía Magdalena (Baja California).

Shell to 15 mm high in intertidal animals, to 32 mm in subtidal ones; spire with two or three nuclear (larval shell) whorls and seven or eight postnuclear whorls, the whorls rounded, and bearing 9–13 axial ribs or varices continuing across sutures; color white, often with a thin brown or purplish spiral stripe below suture; disturbed animals sometimes secreting a purple dye.

Snails of the family Epitoniidae are called "wentletraps," the name deriving from the Danish term for a winding stair-case. Dating from Mesozoic (Jurassic) times, they are widespread inhabitants of continental shelf waters, and many species occur on or near sea anemones and corals. They are specialized carnivores. The proboscis when extended may be longer than the shell; it bears a radula with numerous similar teeth forming rows of pointed spines. The predaceous snails are so small in relation to their hosts or prey that they can be thought of as temporary parasites.

Epitonium tinctum at Pacific Grove (Monterey Co.) occurs on or near the intertidal anemones *Anthopleura elegantissima* (3.31) and *A. xanthogrammica* (3.30); in the Santa Barbara area *A. elegantissima* is the normal host. The snails do not always remain on the prey. The snails are strongly attracted to both *Anthopleura* species, which are clearly located from a distance by scent. Some snails respond positively to the scent of the anemones *Tealia lofotensis* (3.35) and *Epiactis prolifera* (3.39), although others do not. The snails gave no response to the anthozoan *Corynactis californica* (3.41), and they were repelled by the anemone *Metridium senile* (3.37).

An *E. tinctum* isolated from its *Anthopleura* host for a day, and then replaced in a tank with it, extends its proboscis and waves it about as it approaches the prey. When contact is made the proboscis repeatedly touches a tentacle (arousing little or no response), then the snail suddenly bites the tentacle, either taking off the tip or nipping a chunk from the side. The *Anthopleura* tentacle and the snail's proboscis both retract a bit after the bite. The anemones *Tealia lofotensis*, *T. crassicornis* (3.33), and *Epiactis prolifera* are also eaten by *Epitonium tinctum* but not as readily, and in turn retract more strongly when bitten. *Corynactis* is sometimes explored with the proboscis but not eaten. *Metridium senile* is avoided, and some snails contacting this anemone were killed. *Epitonium tinctum* does not feed continuously; populations on *Anthopleura elegantissima* at Santa Barbara came out to feed twice a day, when the anemones were covered by incoming high tides.

The mild response by *Anthopleura* to *E. tinctum* bites, and observations on other epitoniids, suggests that the snails are delivering to the prey an anesthetic or paralyzing material. Extracts of *Epitonium* experimentally applied to frog nerves inhibited action potentials; applied to isolated anemone tentacles they reduce or inhibited normal contraction responses.

Another wentletrap, **Epitonium indianorum** Carpenter, occurs subtidally at Pacific Grove, usually on *Tealia crassicornis* and *T. lofotensis*; the snail is attracted to these anemones by scent, and feeds on them.

See DuShane (1974), Hochberg (1971), Keep & Baily (1935), J. McLean (1969), Robertson (1963), Salo (1977), and Smith (1977).

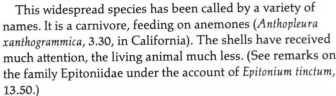

Phylum Mollusca / Class Gastropoda / Order Prosobranchia
Order Mesogastropoda / Superfamily Epitoniacea / Family Epitoniidae

13.51 Opalia chacei Strong, 1937
Chace's Wentletrap

Uncommon under rocks, low intertidal and subtidal zones, usually associated with anemones; Queen Charlotte Strait (British Columbia) to Santa Catalina Island (Channel Islands).

Shell to 30 mm high, with base delimited by a low spiral cord or keel; axial ribs broadly rounded, six to eight per whorl, every third rib larger and running over keel onto base; shell color white to cream.

This species is considered by some to be a variety of **O. wroblewskyi** Mörch, which ranges from Forrester's Island (Alaska) to San Diego at a depth of 0–90 m.

Opalia chacei is probably a specialized predator like *Epitonium tinctum* (13.50) on sea anemones. Little is known of its biology.

See R. Abbott (1974), Carlton & Roth (1975), and Cowan (1964).

Phylum Mollusca / Class Gastropoda / Subclass Prosobranchia
Order Mesogastropoda / Superfamily Epitoniacea / Family Epitoniidae

13.52 Opalia funiculata (Carpenter, 1857)
(= Opalia insculpta)   Sculptured Wentletrap

Fairly common at bases of sea anemones, low intertidal zone; Santa Monica (Los Angeles Co.) to Panama, Ecuador, and Galápagos Islands.

Shell to 20 mm high, with about 14 broad, weak axial ribs; shoulder at top of whorls flat; basal keel strong; color white to golden yellow.

This widespread species has been called by a variety of names. It is a carnivore, feeding on anemones (*Anthopleura xanthogrammica*, 3.30, in California). The shells have received much attention, the living animal much less. (See remarks on the family Epitoniidae under the account of *Epitonium tinctum*, 13.50.)

A variety of features, including the nature of the radula, the diet, and the amino acid composition of the shell-matrix proteins, all link *Epitonium* and its allies with Janthinidae, a family of pelagic, carnivorous, purple-shelled snails that secrete a bubble raft and cruise the high seas, feeding on such animals as the cnidarian *Velella velella* (3.18).

See R. Abbott (1974), DuShane (1974), Ghiselin et al. (1967), J. McLean (1969), Robertson (1963), and Thorson (1957).

Phylum Mollusca / Class Gastropoda / Subclass Prosobranchia
Order Mesogastropoda / Superfamily Hipponicacea / Family Hipponicidae

13.53 Hipponix cranioides Carpenter, 1864
Hoof Snail

Fairly common, often occurring in clusters on undersurfaces of rocks or in deep protected crevices, low intertidal zone in areas of heavy surf and subtidal waters; Vancouver Island (British Columbia) to Baja California, perhaps to Panama and farther south.

Shell to 25 mm in diameter, variable in height and outline, generally cap-shaped with apex near margin; sculptured with flat concentric lamellae and fine radial lines; color whitish under a brown periostracum.

In their cryptic habitat these animals are easily overlooked. For a long time they have been called by the name of the common Atlantic species *H. antiquatus*, but the larval shells of the two species are very distinct.

In post-juvenile stages these limpetlike snails become sessile; the foot secretes a shell base (or "ventral valve") on the rock at the home site and adheres to it. The dorsal and ventral valves fit together perfectly at their margins; thus when the animals clamp down they are tightly closed. The animals feed by actively extending the long, mobile proboscis and grasping fragments of coralline algae and detritus. The variety of material found in the gut shows that the animals take whatever is washed within reach and are not fussy feeders. One might expect such sedentary animals to have evolved the ability to filter feed on plankton, but this has not occurred. The family Hipponicidae is only one of half a dozen different prosobranch gastropod groups that have independently evolved a limpetlike shell.

In clustered populations males and females occur together. Most males are smaller than most females, suggesting that the animals might be functional males when young, then undergo a change to females when they get older. At Pacific Grove (Monterey Co.) in mid-February, in a cluster of 89 *H. cranioides*, only 15 percent were males, and 30 percent of the females were brooding eggs under the shell; no animals were identified as intersexes.

A related form, *H. conicus*, widespread in the Indo-Pacific, has received some study in Australia and Fiji. In the south Australian population, large animals (over 16 mm long) are females. Small animals (2.5–16 mm) are either males, with a developed penis, or animals (up to 13 mm long) showing no clear sexual development. The only males noted were animals collected close to females, suggesting that contact with adult females may cause small limpets to become males. However, some adult females have a rudimentary penis (not yet reported in *H. cranioides* in California), suggesting that at least some females were males earlier in life. The animals probably breed all year in Australia. The eggs are brooded in sacs or capsules; capsules contain 9–24 embryos each, and a female broods 5–10 sacs at a time. The young hatch as creeping juveniles. Fragmentary information on a Fijian population indicates that up to 36 egg capsules, each containing about 300 ova, are carried by a female; the larvae might be liberated here as swimming veligers.

Much remains to be done on California species of *Hipponix*.

See Carlton & Roth (1975), Cernohorsky (1968), Cowan (1974), Laws (1970), and Yonge (1953, 1960).

Phylum Mollusca / Class Gastropoda / Subclass Prosobranchia
Order Mesogastropoda / Superfamily Hipponicacea / Family Hipponicidae

13.54 Hipponix tumens Carpenter, 1864
Ribbed Hoof Snail

Fairly common in protected crevices, low intertidal zone and subtidal waters; Crescent City (Del Norte Co.) to Bahía Magdalena (Baja California).

Shell to 15 mm in diameter; apex curving downward over posterior margin of shell; exterior with coarse radial ridges and concentric growth lines; color white, with a yellowbrown, bearded periostracum.

The biology of this species has not been studied (see the account of *H. cranioides*, 13.53).

See Carlton & Roth (1975), Cowan (1974), Griffith (1967), and J. McLean (1969).

Phylum Mollusca / Class Gastropoda / Subclass Prosobranchia
Order Mesogastropoda / Superfamily Calyptraeacea / Family Calyptraeidae

13.55 Crepidula adunca Sowerby, 1825
Hooked Slipper Snail

On shells of larger intertidal snails, especially *Tegula funebralis* (13.32); Queen Charlotte Islands (British Columbia) to Punta Santo Tomás (Baja California); abundant in northern California, uncommon south of Point Conception (Santa Barbara Co.).

Shell to 25 mm long, oval; apex forming a high, recurved beak overhanging posterior margin; interior of shell with prominent "deck" or shelf curved forward at both ends, giving inverted shell a slipper-like appearance (hence the name "slipper snail"); shell exterior brown, the shelf white.

These commensal limpetlike snails, although commonest on shells of *Tegula funebralis* (13.32), occur also on *T. brunnea* (13.31) and sometimes on *Calliostoma ligatum* (13.27). They are often found in company with the true limpet *Collisella asmi* (13.17), from which they are easily distinguished by the off-center hooked apex. *Tegula* shells (e.g., 13.31, 13.32) occupied by hermit crabs often bear *C. adunca*.

Unlike *Collisella asmi*, *Crepidula adunca* does not rasp algae from the host snail's shell. All of the *Crepidula* species are filter feeders, with a gill highly modified in this connection. Intricate mechanisms involving cilia and mucus strain plankton and organic particles from the water and carry them toward the mouth. The details may be seen beautifully in a small *Crepidula* (of this or any other species) if it is removed from its host, allowed to attach to a glass slide, immersed in seawater, and viewed from below under a microscope. The filtering mechanisms are highly efficient in removing particles from water, so much so that in some species the rate of shell growth decreases in areas of high turbidity, presumably due to clogging of the filtering mechanism. A radula is present, but its use in feeding is restricted largely to picking up pellets of particles entwined in mucus. *Crepidula adunca* individuals are often found sitting on top of one another on a *Tegula*, a smaller animal sitting on a larger one. The animals undergo a sex change as they grow; small individuals are males, middle-sized ones are intersexes, and large individuals are females. The males are attracted to the females; typically they mount on the shells of females, then remain there, feeding and mating but in other respects quite sedentary, until they, in turn, become females.

Some of the classic cell-lineage studies were carried out on the developing embryos of the Atlantic species *C. fornicata*. *C. adunca* breeds the year around in California and has large yolky eggs about 0.3 mm in diameter, which undergo direct development. The developing eggs are brooded in capsules under the mother's shell, and the young are liberated as miniatures of the adult, lacking mainly large size and gonads. Excellent accounts of development of the young and of the anatomy of the adult are available for this species.

When the hatched young are ready for release, the female periodically lifts her shell 1–3 mm above the substratum, and with her head pushes out those young that have come free of the capsules. Not all young are released at once, and those released may be of different ages. When released, the young average 1.9 mm long. They crawl readily but cannot cling well; most fall off the *Tegula* shell and sink to the bottom where they move about actively.

When removed from its host and placed next to it under water, an adult *C. adunca* usually tries to crawl back on, but newly hatched young do not seem to be strongly attracted to either *Tegula* or adult *Crepidula* shells. After the young have grown to a length of about 4 mm, they apparently seek out *Tegula* or adult *Crepidula*, mount them, settle down, and become less active. Some even settle on empty *Tegula* shells. While independent and active, the young use their radulae for rasping microscopic algal films from the rocks, but as they become older and sedentary they become filter feeders.

Intertidal animals of this species have significantly lower respiratory rates than those collected nearby from subtidal areas and measured at the same temperature. This difference may be a compensatory response to the generally higher maximum temperatures in the intertidal zone.

See especially Moritz (1938, 1939) and Putnam (1964); see also Coe (1953), Conklin (1897), Costello & Henley (1971), Harrold (1975), Heath (1916), Johnson (1972), Orton (1912), Werner (1953), and Yonge (1938).

Phylum Mollusca / Class Gastropoda / Subclass Prosobranchia
Order Mesogastropoda / Superfamily Calyptraeacea / Family Calyptraeidae

13.56 **Crepidula aculeata** (Gmelin, 1791)
Spiny Slipper Snail

Uncommon in rock crevices and attached to other shells, low intertidal zone and subtidal waters; Piedras Blancas Point (San Luis Obispo Co.) to Valparaiso (Chile); Atlantic and Indo-Pacific regions.

Shell length to 13 mm in California (to 40 mm farther south); apex of shell posterior, depressed, curved toward one side; spiral ribs rough and prickly; shelf notched at each end; shell and shelf pinkish white to white.

As in other slipper snails the adults are sedentary; the gill and mantle cavity are highly modified in connection with filter feeding on small plankton and detritus.

See Keen (1971) and J. McLean (1969).

Phylum Mollusca / Class Gastropoda / Subclass Prosobranchia
Order Mesogastropoda / Superfamily Calyptraeacea / Family Calyptraeidae

13.57 **Crepidula nummaria** Gould, 1846
White Slipper Snail

Common in localized protected situations, under rocks or in abandoned snail shells or pholad holes, low intertidal zone on outer coast; Bering Sea to southern California.

Shell to over 40 mm long, thick to thin, with apex overhanging posterior margin; shell white, with thick, tattered, yellowish-brown periostracum.

The taxonomy of the white slipper snails on the west coast is still in a confused state. The species present are quite variable, the size and shape of the shell being influenced by environmental conditions. The limits of variation in morphology and in habitat are inadequately known for all species. To further complicate matters, two Atlantic species of *Crepidula* with white or light-colored shells (*C. plana* Say, *C. fornicata* Linnaeus) have been introduced (probably along with oysters) from the east coast; *C. fornicata* usually has brown markings on the shell, but in *C. plana* the shell is white, as in *C. nummaria* and *C. perforans* (13.59). Detailed studies of comparative biology are needed, including experimental examination of the growth characteristics in different habitats and of mating between different populations.

Information in Coe (1953, as *C. nivea*) probably applies to *C. nummaria* and not *C. plana*. According to Coe, *C. nummaria* occurs in two growth forms. One form is found on the outside and inside (often simultaneously), of large snail shells occupied by hermit crabs, and also on rocks. These individuals grow to 40 mm or more in length. Individuals of the other form live inside the apertures of small gastropod shells and seldom exceed 16 mm in length. These small forms, if removed to a larger surface, grow into the larger forms.

Each female releases more than 5,000 eggs at each spawning, which develop into pelagic larvae. When they settle out, if they find themselves near a functional female they develop into males. Of the larvae that settle in isolation from other individuals, about half initially develop into males. The male phase persists only if the individual is continually in association with a functional female. When a male is removed from a female associate, he usually begins to change from a male to a female, but if he is returned to contact with a functional female his male system becomes functional again. More work is needed on this and other *Crepidula* species to get at the mechanisms involved in sex determination and sex change.

Like *C. adunca* (13.55) and other *Crepidula* species, this is a sedentary form that is specialized for filter feeding. It subsists on small plankton and water-borne organic detritus.

See Berry (1955), Carlton & Roth (1975), Coe (1942a,b, 1949, 1953), and references under the account of *C. adunca*.

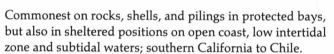

Phylum Mollusca / Class Gastropoda / Subclass Prosobranchia
Order Mesogastropoda / Superfamily Calyptraeacea / Family Calyptraeidae

13.58 **Crepidula onyx** Sowerby, 1824
Onyx Slipper Snail

Commonest on rocks, shells, and pilings in protected bays, but also in sheltered positions on open coast, low intertidal zone and subtidal waters; southern California to Chile.

Shell to 70 mm long, strong, tan to dark brown outside, the interior shelf white; distinguished from *C. adunca* (13.55) by (1) usually larger size, (2) apex low, near margin, curved to one side, not markedly hooked (rather than high, hooked, and centrally placed), (3) interior shelf notched at ends rather than curved forward at both ends, and (4) periostracum more prominent.

These are sedentary feeders on plankton and fine organic detritus, which are collected from the water by methods involving mucus and cilia, as in other species of the genus.

Like *Crepidula nummaria* (13.57), *C. onyx* occurs in two growth forms: broad-shelled animals are seen on rocks, pilings, large shells, and other surfaces allowing room for growth; narrow-shelled animals are formed inside the apertures of small snail shells. As many as ten individuals may be attached to one another. The larger, older individuals at the bottom are females, the smaller ones are males. A change in sex, from male to female, normally occurs in midlife; the anatomical changes involved here are described in a classic paper (Coe, 1942a). When juveniles were raised in isolation, about

90 percent of them developed into males; in *C. nummaria* only about half the isolated young developed into males.

See Coe (1942a,b, 1949, 1953), G. MacGinitie & N. MacGinitie (1968), and J. McLean (1969).

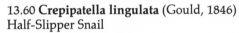

Phylum Mollusca / Class Gastropoda / Subclass Prosobranchia
Order Mesogastropoda / Superfamily Calyptraeacea / Family Calyptraeidae

13.59 **Crepidula perforans** (Valenciennes, 1846)
Western White Slipper Snail

Under rocks, inside apertures of gastropod shells, and in empty pholad burrows in rocks, low intertidal zone and subtidal waters; Vancouver Island (British Columbia) to Baja California.

Shell to 55 mm long, but often very much less, highly variable in form, conforming to contour of substratum, thin and fragile to fairly robust; large shells often conspicuously foliated; shelf notched laterally; shell color white; periostracum thin, smooth, brown, often inconspicuous. (See the remarks under *C. nummaria*, 13.57.)

Although we know of no certain means of distinguishing the white *Crepidula* species, the form inhabiting the apertures of *Tegula* shells (e.g., 13.31, 13.32) that are occupied by hermit crabs of the genus *Pagurus* (e.g., 24.10–13) on the outer coast in central California is the species we have in mind here; specimens looking very much like larger editions of the same occur nearby under rocks and in holes. Coe (1953), who cultured mixed populations of juvenile *C. perforans* (referred to as *C. williamsi*) and *C. nummaria* (13.57; referred to as *C. nivea*) in a jar with an adult female of each species, found that in 3 weeks the juveniles clustered around the female of their own species. Said he, "Conchologists have sometimes been unable to distinguish the two species but that difficulty evidently does not present itself to the young snails."

At Point Pinos (Pacific Grove, Monterey Co.) up to 19 *C. perforans* may occur in a *Tegula funebralis* shell (13.32) carried by a hermit crab. Unpublished studies (C. D. Barbour, pers. comm.) on this population show that females brooding young are present at all times of the year, as are males. Newly

hatched young are about 1 mm long. Males (animals with a penis) are 2–10 mm in length; females, 12–21 mm; animals of undetermined sex cover the entire size range, but are usually less abundant than either males or females except in the intermediate size range, 7–13 mm. Clearly, many of the smaller animals are not males.

This species is abundant, maintained easily (on hermit crab hosts) in the laboratory, and appears well suited to experimental studies.

See Berry (1955), Coe (1953, probably as *C. williamsi*), Haderlie (1976), and J. McLean (1969).

Phylum Mollusca / Class Gastropoda / Subclass Prosobranchia
Order Mesogastropoda / Superfamily Calyptraeacea / Family Calyptraeidae

13.60 **Crepipatella lingulata** (Gould, 1846)
Half-Slipper Snail

On rocks and snail shells, low intertidal zone and below, commoner subtidally; Bering Sea and Alaska to Isla Asunción (Baja California).

Shell to 25 mm long, rounded, fairly flat; apex near margin; interior shelf attached on only one side; shell speckled or radially striped brown and white; periostracum yellowish.

This is a filter feeder, like members of the genus *Crepidula* (e.g., 13.55–59), but detailed studies of its biology are lacking.

See Coe (1942b, 1949), Griffith (1967), and J. McLean (1969).

Phylum Mollusca / Class Gastropoda / Subclass Prosobranchia
Order Mesogastropoda / Superfamily Calyptraeacea / Family Calyptraeidae

13.61 **Crucibulum spinosum** Sowerby, 1824
Cup-and-Saucer Limpet

Common on dead bivalve shells in bays and lagoons, low intertidal zone and subtidal waters; San Pedro (Los Angeles Co.) to Chile.

Shell to 30 mm in diameter, low, conical; apex often sharply pointed and twisted; surface with radial rows of spines; exterior brown to yellowish with darker radial rays; internal cup white, attached along most of one side.

This species is the only California representative of the genus, but numerous species occur in tropical waters farther south on the west coast. Like the species of *Crepidula* (e.g., 13.55–59), *Crucibulum spinosum* is a filter feeder, and individuals undergo a sex change from male to female in midlife. Small males attach themselves to the shells of larger females.

See Coe (1949), G. MacGinitie & N. MacGinitie (1968), and J. McLean (1969).

Phylum Mollusca / Class Gastropoda / Subclass Prosobranchia
Order Mesogastropoda / Superfamily Naticacea / Family Naticidae

13.62 Polinices lewisii (Gould, 1847)
Lewis's Moon Snail

On sand and mud in protected bays, low intertidal zone and below; subtidal to about 150 m on soft substrata off open coast; Vancouver Island (British Columbia) to Isla San Gerónimo (Baja California).

Shell to 130 mm in diameter, globular, with shallow groove at shoulder of whorls; columella with a calcareous callus reaching to edge of deep umbilicus; shell yellowish with thin brown periostracum; operculum horny, dark brown.

These large snails plough forward with the foot just below the surface of the sand or mud. The extended foot is much larger than the shell. It serves as a stable platform for the animal on a shifting bottom, it provides a large surface for locomotion, and it can fold into an effective grasping organ to hold prey. The snail has two means of locomotion, best studied in the east coast *P. duplicatus*. The foot is heavily ciliated and equipped with mucous glands. When the cilia are active they always beat from front to rear, and the animal moves forward at about the same rate that separate particles of sand are carried backward over the foot. There is plenty of reserve power here to help in digging: the ciliary action alone can move weights of up to nearly 0.5 gm per cm² of foot surface, and to creep over a flat surface the animal needs to move only about 0.15 gm per cm². Ciliary creeping is slow; one animal

averaged about 1 cm per 5.7 seconds. The cilia on the foot turn off and on together even in widely separated regions, indicating that nervous control is involved.

The snails also move by pedal muscular waves, which travel from rear to front (direct waves), with the contractions on the two sides of the foot in phase (monotaxic waves). Locomotion is faster this way, the snails moving 1 cm in 1.7–3.5 seconds. Both ciliary and muscular actions may occur together, and in burrowing, the anterior (propodial) part of the foot can be extended hydrostatically like a protruding tongue.

The above description applies well to *P. lewisii*. This moon snail creeps equally well on hard and soft substrata, and can climb the glass walls of an aquarium. The propodium covers the front of the shell, so the burrowing snail tapers like a wedge anteriorly. Snails can bury themselves in 6–8 minutes in loose gravel, more slowly in muddy sand.

When *P. lewisii* is fully expanded the foot is relatively enormous, up to four times the volume of the shell. When disturbed, the animal contracts, ejecting quantities of water from pores at the back of the foot. Eventually it draws the diminished foot inside and closes the operculum. This remarkable performance is made possible by the presence of the aquiferous system, an interconnected series of spongy sinuses in the foot, which open to the outside by 40–80 tiny posterior pores that can be closed by sphincter muscles. A connective tissue lining of the sinuses keeps the seawater from mixing with blood and out of direct contact with muscles and nerve fibers. Seawater can be taken up, expelled, or shifted from place to place in the aquiferous system; when the foot is fully expanded, the seawater in it represents nearly a quarter of the total body weight.

In the region of Nanaimo (Vancouver Island, British Columbia), more animals are at the surface at night than during the day, but nocturnal periodicity is not marked. The snails come up after heavy rains, but stay buried in cold weather. Population densities near shore were 0.85–4 *P. lewisii* per m². The snails generally avoid each other, but a feeding animal attracts other snails.

The moon snail is a predator mainly on clams, which it reaches by digging in the softer substratum with its foot. Prey is probably located mainly by scent. Exposed clam meat is detected up to 2 m away, but uninjured clams can only be detected about 15 cm away. Hunting *P. lewisii* follow a zigzag search pattern. Once a clam is located and grasped with the foot, the snail drills a hole in the shell, usually lateral and ventral to the beak or umbo. The snail has a padlike organ, the accessory boring organ (ABO), on the lower lip of the proboscis. In drilling it alternately rasps with the radula for a period, then places the ABO close to or on the drill site. The radular teeth alone are too weak to drill the shell, so apparently the ABO secretions have a softening effect. Studies on other drilling snails with ABOs suggest that the active agent may be the enzyme carbonic anhydrase. In the Nanaimo area the two commonest clams, *Protothaca staminea* (15.41) and *Saxidomus nuttalli* (15.34), form the bulk of the diet; *Clinocardium nuttallii* (15.29), although available, is rarely taken. The snails, in laboratory tests, each ate a sizable clam about every 4 days. In Bodega Bay harbor (Sonoma Co.), the clam *Macoma nasuta* (15.50) occupies the same range as *Polinices* and makes up 88 percent of the diet of the snail. *Macoma secta* (15.51) also occurs in the Bay but outside the range of *Polinices*. When presented with both species in the laboratory, the snail selects *M. secta*. After a meal, Nanaimo snails rested below the surface for 8–24 hours, then crept away, on a straight course for an average of more than 4 m before engaging in further activities.

The sex ratio for young *P. lewisii* is about 1:1 at Nanaimo, but male mortality is higher, and for animals of 100 mm shell length 90 percent of the population is female. Males are a little smaller than females and have a slightly thicker shell. The snails start mating at a shell length of about 55 mm. The eggs are laid in a broad, curved "sand collar" about 15 cm in outside diameter, shaped around the shell and the foot. This stiff, rubbery collar contains a layer of small eggs (250 μm in diameter) concealed between two layers of sand held together by mucus. The collars are frequently found cast up on the beach. The eggs hatch in midsummer, releasing larvae into the internal space of the collar. Here they may swim for several weeks, until released by disintegration of the collar. Very shortly thereafter they settle out, at Nanaimo, on the green seaweed *Ulva*, at 10–12 m depth. The juveniles at first graze diatoms on the *Ulva*, later eat the seaweed itself, and finally, at an age

of 5–6 months (shell length 5–6 mm), move to the soft substratum and commence to hunt animal prey. At a shell length of about 30 mm they begin the trek to shallower water. Maximum age in this species is probably several years.

In southern California the small crab *Opisthopus transversus* (25.42) sometimes occurs as a commensal of this snail, and empty *P. lewisii* shells provide homes for the hermit crab *Isocheles pilosus* (24.9; both species are shown in Fig. 24.9) and the subtidal hermit crab *Paguristes bakeri*. The empty shells are also found in California Indian middens, and even today the snails are sometimes taken for food. California Fish and Game regulations (1979) set a bag limit of five moon snails, and none may be taken north of the Golden Gate Bridge (San Francisco Bay).

See especially Bernard (1967); see also Copeland (1922), Curtis (1966), Edwards & Heubner (1977), Giese (1969), Giglioli (1955), Griffith (1967), Hopper (1972), G. MacGinitie (1935), G. MacGinitie & N. MacGinitie (1968), J. McLean (1969), S. L. Miller (1974), Schwimer (1973), Talmadge (1972), and Turner (1953).

Phylum Mollusca / Class Gastropoda / Subclass Prosobranchia Order Mesogastropoda / Superfamily Naticacea / Family Naticidae

13.63 Polinices reclusianus (Deshayes, 1839)
Southern Moon Snail

Present on sandy mud bottoms in bays, low intertidal and shallow subtidal zones; Mugu Lagoon (Ventura Co.) to Mazatlán (Mexico).

Shell to 70 mm in diameter, heavy; umbilicus filled by a calcareous pad except for a small opening; shell tan or brown; operculum horny, brown.

Like *P. lewisii* (13.62), this snail is a predator on other mollusks, especially clams, which it opens by drilling a hole through the shell. The shells of *P. reclusianus* appear in some California Indian middens, and some snails were probably eaten. California Fish and Game regulations (1979) prohibit taking moon snails of any sort north of the Golden Gate Bridge (San Francisco Bay); south of there the bag limit is 5, and there is no closed season.

See Curtis (1966), G. MacGinitie & N. MacGinitie (1968), J. McLean (1969), S. L. Miller (1974), and Pilsbry (1929).

Phylum Mollusca / Class Gastropoda / Subclass Prosobranchia
Order Mesogastropoda / Superfamily Lamellariacea / Family Lamellariidae

13.64 **Lamellaria stearnsii** Dall, 1871

Usually on colonial ascidians, under rocks and in sheltered places, low intertidal zone and subtidal waters; Alaska to Islas Marías (Mexico).

Shell to about 15 mm long, thin, white, with widely flaring aperture; right and left mantle lobes fused together dorsally, covering much or all of shell; soft parts pinkish to white with some darker spots.

This is one of several inadequately known species of *Lamellaria* found in California. In *L. stearnsii*, fusion of the mantle folds is incomplete, leaving a dorsal pore that can open to expose the shell when the animal is disturbed. In Monterey Bay, *L. stearnsii* occurs on the colonial ascidian *Trididemnum opacum* (12.12), where it is so beautifully camouflaged that it resembles a bulge on the tunicate colony. The pinkish-white color of the snail's exposed soft parts closely matches the ground color of the ascidian, and the dark acid glands in the snail's mantle resemble, in size and arrangement, the oral apertures of the ascidian zooids. The snail's concealed shell is attached to the body by two shell muscles, one on each side (rather than just one, on the right side, as in most mesogastropods). During the day the snails are rather inactive, but at night they move about and use the proboscis to feed on the ascidian colony. A species of *Lamellaria* occasionally occurs on the tunicate *Ascidia ceratodes* (12.26) in Monterey harbor.

Reproduction and larval life, while unknown in *L. stearnsii*, have been studied in the Atlantic species *L. perspicua*, which also lives on, and feeds on, colonial ascidians. After mating, the female snails puncture the colonial tunic of the host ascidian and deposit within it their globular egg capsules, 2–3 mm in diameter, each containing 1,000–3,000 eggs. The side of the capsule at the surface of the tunic bears a short neck, which remains plugged until the larvae are ready for release. The veliger larvae that hatch out are beautiful, complex creatures that have a relatively long pelagic life. Off Plymouth

(England) they are found in the plankton 12 months of the year, though most abundantly in the summer. The larvae, called "echinospira" larvae (literally "spiny coil"), have a large, lobed velum and a remarkable shell that for nearly a century was regarded as a double shell. The inner shell is lightly calcified and inhabited by the larva. The outer "shell," surrounding the inner shell but separated from it by a large, fluid-filled space, is not calcified. It probably represents a detached periostracum. As the larva grows and continues to coil, the older parts of the outer shell begin to get in the way, and the larva uses the operculum on its small foot to slice off the offending parts. At settling, the larva sheds both periostracum and operculum. These extraordinary happenings, and a fine redescription of the larvae (Lebour, 1935), inspired the British zoologist Walter Garstang (1951) to write his longest biological verse, "Echinospira's double shell," which itself contains several original observations. Echinospira larvae are found in a few related genera that occur in California waters (e.g., *Velutina*, 13.66; *Erato*, 13.67; *Trivia*, 13.68). Little is known of development in any California species, though echinospira larvae occur in the California inshore plankton. All species of *Lamellaria* need further study.

See Fretter & Graham (1962), Garstang (1951), Ghiselin (1964), Lebour (1935), and D. Smith (1977).

Phylum Mollusca / Class Gastropoda / Subclass Prosobranchia
Order Mesogastropoda / Superfamily Lamellariacea / Family Lamellariidae

13.65 **Lamellaria rhombica** Dall, 1871

Uncommon in rocky areas, often associated with compound ascidians, low intertidal zone, commoner subtidally; Neah Bay (Washington) to San Luis Obispo Co. (D. Behrens, pers. comm.).

Shell to 18 mm long, thin, somewhat more quadrate than *Lamellaria stearnsii* (13.64), white in color; soft parts translucent, bluish white or tan, mottled with darker color; part of shell exposed.

In Monterey Bay, *L. rhombica* resembles the ascidian colonies on which it lives and presumably feeds.

See Ghiselin (1964).

Phylum Mollusca / Class Gastropoda / Subclass Prosobranchia
Order Mesogastropoda / Superfamily Lamellariacea / Family Velutinidae

13.66 **Velutina** sp.
(? = V. velutina (Müller, 1776))   Smooth Velutina

Scarce under rocks in protected coves, low intertidal zone; commoner among rocks and gravel subtidally to at least 225 m; Arctic Ocean to Monterey Co.; east coast south to Cape Cod (Massachusetts); northern Europe.

Shell to 22 mm long, globular, delicate; spire low, suture distinct, aperture large and flaring; color pinkish with dark-brown, velvety periostracum.

This species, often called by the invalidated name *V. laevigata* in California, differs from Arctic and European populations of *V. velutina* and may represent a separate species. Little is known of its habitat and biology in California, but the European *V. velutina* has received considerable study. It can withdraw completely into its shell, but in most other respects its structure and biology are reminiscent of *Lamellaria* (e.g., 13.64, 13.65). *V. velutina* in Europe lives and feeds on solitary tunicates of the genera *Ascidia*, *Phallusia*, and *Styela*. It produces a sticky mucus when disturbed (the hypobranchial mucous gland is unusually large). It lays eggs in capsules in ascidian tunics, and the hatched larvae are of the echinospira type (see the discussion under *L. stearnsii*, 13.64). Unlike *Lamellaria* species, *Velutina velutina* is a simultaneous hermaphrodite. The European species *V. plicatilis* feeds on the hydroid *Tubularia*.

See Carlton & Roth (1975), Diehl (1956), Fretter & Graham (1962), Griffith (1967), N. MacGinitie (1959), and Riser (1969).

Phylum Mollusca / Class Gastropoda / Subclass Prosobranchia
Order Mesogastropoda / Superfamily Lamellariacea / Family Triviidae

13.67 **Erato vitellina** Hinds, 1844
Apple-Seed Erato

Fairly common in some areas, on stones, pilings, and ascidians, low intertidal zone and subtidal waters; Bodega Bay (Marin Co.) to Bahía Magdalena (Baja California).

Shell to 16 mm long, pear-shaped, shiny, lacking a periostracum; aperture narrow; columella with five to eight small teeth; outer lip with seven to ten teeth; shell reddish with light margin around aperture; soft parts whitish yellow with yellow, orange, and brown spots.

Little is known about the biology of this species except that it is characteristically associated with ascidians and presumably feeds on them. A relative, *E. voluta* from European waters, is better known. It feeds on zooids of the colonial ascidians *Botryllus* and *Botrylloides.* The snail cruises along, laying a mucus track on the surface of a tunicate colony, its mantle lobes extended to envelop nearly the whole shell, and its long siphon held low and extended ahead, apparently questing for a zooid with its mouth wide open. When one is found, the siphon lifts, and the heretofore retracted proboscis is suddenly everted and rammed "like a piston" into the unfortunate zooid's mouth. The snail's proboscis is long, and carries the jaws and radula at its tip. The jaws take bites of the viscera (not the tunic) and the radula pulls them in. It takes the snail about 20 minutes to eat a zooid, and two zooids may constitute a meal. The sexes are separate in *E. voluta*; the eggs are laid in a capsule, and hatch as echinospira larvae (see the account of *Lamellaria stearnsii*, 13.64).

See especially Fretter (1951) and Fretter & Graham (1962); see also Carlton & Roth (1975) and J. McLean (1969).

Phylum Mollusca / Class Gastropoda / Subclass Prosobranchia
Order Mesogastropoda / Superfamily Lamellariacea / Family Triviidae

13.68 **Trivia solandri** (Sowerby, 1832)
Solander's Trivia

Uncommon under rocks, low intertidal zone and subtidal waters; Palos Verdes (Los Angeles Co.) to Panama.

Shell to 20 mm long with smooth dorsal furrow bordered by ends of spiral ribs, each of which terminates dorsally in a whitish rounded area; shell tan to brown in color, the ribs white.

A second species, **T. californiana** (Gray, 1827), is distinguished by its smaller size (to 11 mm long), darker purplish shell, and spiral ribs running across dorsal furrow.

The *Trivia* shell somewhat resembles that of the cowries (Cypraeacea, e.g., 13.69). In *Trivia* the final whorl of the adult shell expands to cover and hide the earlier whorls, and at this stage the outer lip turns inward; soon the aperture narrows to a midventral slit bordered by thickened lips. In juvenile snails the spire is still visible and the aperture is larger. The big mantle folds extend to cover the shell in undisturbed animals, but can fully retract into the shell along with the head and foot. Two shell muscles are present.

Little is known of the biology of California species of *Trivia*, but the European forms *T. monacha* and *T. arctica* have received some study. These, like their relatives *Lamellaria*, *Velutina*, and *Erato*, eat tunicates, preferring *Botryllus*, *Botrylloides*, *Trididemnum*, *Diplosoma*, and *Polyclinum*. The snails creep along an ascidian colony with the proboscis extended, using the jaws to bite the test and zooids, and the radula to scoop up the parts. Tunic and zooids are both ingested, but only the zooids are digested. The sexes are separate. The eggs are laid in tunicate colonies in globular capsules with a short protruding neck. The embryos hatch as echinospira larvae (see the account of *Lamellaria stearnsii*, 13.64).

See especially Fretter (1951) and Fretter & Graham (1962); see also Keen (1971) and J. McLean (1969).

Phylum Mollusca / Class Gastropoda / Subclass Prosobranchia
Order Mesogastropoda / Superfamily Cypraeacea / Family Cypraeidae

13.69 Cypraea spadicea Swainson, 1823
(=Zonaria spadicea)   Chestnut Cowrie

Along ledges and in crevices shaded by seaweed and eelgrass, low intertidal zone; subtidal to at least 50 m; Monterey to Isla Cedros (Baja California); rare north of Point Conception (Santa Barbara Co.).

Shell of adults 40–120 mm long, but usually not over 65 mm, with smooth polished surface and slitlike aperture bordered by denticles; dorsal surface mostly brown; margins and ventral surface gray or white; mantle lobes orange-brown with darker spots.

The cypraeans, or cowries, whose attractive shells have led to their collection and use as ornaments and money since prehistoric times, are primarily warm-water forms. Numerous species occur along eastern tropical Pacific shores, but only one is common in California. In cypraeans the shell is initially thin-walled and has a wide aperture, but at some point in growth the outer lip commences to grow toward the columella. As this happens the shell thickens, and the aperture narrows and develops greatly thickened margins, usually with teeth. The mantle folds, which can extend over the outer shell surface, deposit additional shell material externally that hides the short spire. With this development, adult growth virtually ceases. Keen (1971, p. 490) notes, "Because of the great variability in the size of fully mature shells of the same species, there has long circulated a story that the animal of *Cypraea* is able to dissolve its shell and resecrete another larger one within a short period of time. This ability would appear to be entirely legendary, although the animal is able to dissolve partially or resorb the early whorls, making more room for its body."

Cypraea spadicea is a carnivore and carnivorous scavenger and is reported to eat such things as anemones, sponges, ascidians, and snail eggs. The eggs are laid in July in southern California, deposited in pointed capsules each holding roughly 800 eggs. The capsules are attached to the substratum in clusters of 100–120. The parent protects the brood until hatching, which takes about 3 weeks. The veliger larvae are not of the echinospira type.

A juvenile 30 mm long taken in southern California in May was maintained in a seawater aquarium for 10 months, along with some adult *C. spadicea*. Both adults and juvenile thrived on a diet of frozen shrimp. The juvenile grew in length to 43 mm in 6 weeks. At this size the outer lip began to turn in. By 8 weeks the lips had formed a narrow, toothed aperture, and the length was 44 mm. Eight months later the shell was only 45 mm long. The healthy adults kept in the same tank showed no growth in 10 months of captivity.

See especially Darling (1965) and Ingram (1938, 1947); see also Donohue (1965a,b), Herzberg (1966), Keen (1971), G. MacGinitie & N. MacGinitie (1968), and Schilder (1965, 1966).

Phylum Mollusca / Class Gastropoda / Order Prosobranchia
Order Neogastropoda / Superfamily Muricacea / Family Muricidae

13.70 Ceratostoma foliatum (Gmelin, 1791)
(=Purpura foliata)　　Leafy Hornmouth

On rock faces, especially near barnacles and bivalves, low in-
tertidal zone; subtidal to at least 65 m; Sitka (Alaska) to San
Diego; uncommon south of Point Conception (Santa Barbara
Co.).

Shell to nearly 10 cm high, usually less than 8 cm; body
whorl with three wide flanges or varices; siphonal canal
closed; outer lip of aperture with strong projecting tooth near
anterior end; exterior gray or white to yellow-brown on body
whorl; interior white; operculum horny.

In members of the family Muricidae the shell is often ornamented with flanges that may bear thorny projections. In this
species the juvenile shell is sculptured with numerous longitudinal ribs crossed at right angles by equally spaced spiral
cords, producing a pattern of squares separated by low ridges.
As the whorls increase in number and size, the longitudinal
ribs become channeled into groups that form projecting
flanges (varices); it is as though the ribs were being piled on
top of one another in narrow piles. On the last and largest
whorl there are only three flanges, two projecting outward to
the right and left of the aperture, and one extending dorsally
on the shell like a sail. The transition from juvenile to adult
shell configuration is gradual. As the number of flanges on
each whorl decreases, the flanges become taller and stronger,
and eventually the distinctive tooth is added on the outer lip
of the aperture. Growth involves not only the addition of new
calcium carbonate, but also a dissolving of a portion of the
older flanges, which would otherwise stand in the way of new
growth.

Why this extraordinary sculpturing on the shell? Various
functions ascribed to the large varices include: strengthening
the shell; protecting the sides of the aperture when the animal
is clinging to a rock; providing a sharp blade to cut the mouth
of a predatory fish; and, as "fins," increasing the chances that
a detached snail falling through the water will land right side
up. Experimental testing of the last hypothesis shows that the
dorsal flange does act as a hydrodynamic destabilizer. A fully

grown snail falling more than 20 shell lengths through the
water is made to tumble such that it lands foot down more
than half the time; with the dorsal flange filed off, the shell
usually lands with the aperture up.

Shell growth usually ceases altogether when the animals
reach a length of about 80 mm (less if food was scarce, more if
the supply was rich) and an age of 4 years. Thereafter, the
shell is more subject to fouling (see, for example, Fig. 6.9),
and to erosion by such organisms as boring sponges (e.g.,
Cliona, 2.21); shell length thus *decreases* year by year, and in
the oldest animals (perhaps 16 years old) may drop to a mere
5 or 6 cm.

Ceratostoma foliatum is a predator, well equipped to bore
through the calcareous shells of its prey. In the intertidal and
shallow subtidal areas where it is best studied (Washington)
it feeds mainly on the barnacles *Balanus glandula* (20.24) and
Semibalanus cariosus (20.23), and on the bivalves *Pododesmus
cepio* (15.18) and *Penitella penita* (15.78). In southern California
it feeds on the sessile gastropod *Serpulorbis squamigerus* (13.45).

In Washington, spawning has been observed in late February and early March. Mature snails, previously rather scattered (1–10 animals per m² in suitable environments) gather
into clusters. After mating, the females deposit their eggs in
yellow, fusiform egg cases averaging about 13 mm in length,
which are attached to rocks or shells subtidally, often at 12–
25 m depth. The egg cases of all females in the cluster are attached side by side in a common mass on the substratum.
Adult females produce an average of 40 egg capsules a year,
each containing 30–80 eggs (mean 51.4 eggs). In the Washington populations studied, egg production per female per
year ranged from 300 to 11,000, and averaged 2,056. The average female spawns yearly for 7.4 years and has an estimated
lifetime egg production of 15,200 eggs.

Development occurs within the egg capsules. Under good
conditions all the eggs develop into larvae and none serve as
nurse eggs, though there is some loss at this stage from nematode infestations and other causes. Veliger larvae are produced but they undergo metamorphosis before release, and
the young emerge from the egg case as juvenile snails. The
time from egg laying to emergence is about 4 months at 10–
12°C. Young juvenile snails kept in the laboratory increased

in shell length 0.7–2.5 mm (usually 1.3–1.6 mm) per month, but the growth rate steadily declined with age; animals 15 mm in initial shell length grew at an average rate of about 14 mm per yr, whereas those 50 mm long added an average of only 8 mm in a year. The mean adult shell length in Washington populations was about 73 mm for males and 79 mm for females.

Studies of the foot of *C. foliatum* show that the animal moves by means of muscular waves traveling from front to rear (retrograde waves), and the waves on the two sides of the foot are out of phase (ditaxic waves). The tenacity with which the foot clings to the substratum when the animal moves or is stationary is much less than that of many snails, suggesting that the species is best adapted to subtidal areas below strong surf and surge.

See especially Spight, Birkeland & Lyons (1974) and Spight & Lyons (1974); see also R. Abbott (1974), Chace (1916), Dayton (1971), Griffith (1967), G. MacGinitie & N. MacGinitie (1968), J. McLean (1962), S. L. Miller (1974), Palmer (1977), Radwin & D'Attilio (1976), and Ricketts & Calvin (1968).

Phylum Mollusca / Class Gastropoda / Subclass Prosobranchia
Order Neogastropoda / Superfamily Muricacea / Family Muricidae

13.71 Ceratostoma nuttalli (Conrad, 1837)
Nuttall's Hornmouth

Among rocky rubble and on wharf pilings, middle and low intertidal zones; Point Conception (Santa Barbara Co.) to Bahía Santa María (Baja California).

Shell to about 65 mm high, like that of *C. foliatum* (13.70) but smaller, with relatively smaller varices (flanges) and with a single large axial rib between flanges, bearing conspicuous nodules; shell white to brown, darker shells often with white bands.

Like *C. foliatum*, this animal is a carnivore, using its radula to bore holes into the mussels *Mytilus edulis* (15.10) and *M. californianus* (15.9) and other mollusks. The shell has been examined with x-rays, and the DNA content of the sperm determined.

See R. Abbott (1974), Harger (1967), Herzberg (1966), Hinegardner (1974), J. McLean (1969), and Radwin & D'Attilio (1976).

Phylum Mollusca / Class Gastropoda / Subclass Prosobranchia
Order Neogastropoda / Superfamily Muricacea / Family Muricidae

13.72 Pteropurpura trialata (Sowerby, 1841)
(=Pterynotus trialatus)   Three-Winged Murex

On rock ledges and on breakwaters at bay entrances, low intertidal zone; Bodega Bay (Marin Co.) to Baja California; scarce north of Palos Verdes (Los Angeles Co.), fairly common in southern California.

Shell to about 85 mm high, with flanges or varices somewhat frilled or fluted; outer lip of aperture smooth, without projecting tooth; shell sometimes solid white, more often whitish with brown between spiral ribs.

These carnivorous animals feed on sessile and sedentary gastropods, including *Serpulorbis squamigerus* (13.45) and *Crepidula* (e.g., 13.55–59). They drill holes through the shells of the prey using the radula. Like other muricids they deposit their eggs in capsules attached to rocks. One individual of this species was observed riding about on a fish.

See R. Abbott (1974), Fitch (1963), G. MacGinitie & N. MacGinitie (1968), J. McLean (1969), and Radwin & D'Attilio (1976).

Phylum Mollusca / Class Gastropoda / Subclass Prosobranchia
Order Neogastropoda / Superfamily Muricacea / Family Muricidae

13.73 Maxwellia gemma (Sowerby, 1879)
Gem Murex

Sometimes locally common in rocky areas, particularly on breakwaters at bay entrances, very low intertidal zone; more commonly subtidal, extending to 55 m; Santa Barbara to Isla Asunción (Baja California).

Shell to about 45 mm high, short-spired, with raised spiral cords and six or seven varices consisting of broad round-topped radial ridges, with little space between successive varices; aperture round; siphonal canal closed; shell white with black or brown spiral cords.

Little is known of the biology of this species except that it is carnivorous. X-ray photographs of the shell have been published.

See R. Abbott (1974), Herzberg (1966), J. McLean (1969), and Radwin & D'Attilio (1976).

Phylum Mollusca / Class Gastropoda / Subclass Prosobranchia
Order Neogastropoda / Superfamily Muricacea / Family Muricidae

13.74 Ocenebra circumtexta Stearns, 1871
Circled Rock Snail

Common on rocks, middle to low intertidal regions along outer coast in areas of heavy surf; Trinidad (Humboldt Co.) to Scammon Lagoon (Baja California).

Shell to about 25 mm high, generally not over 20 mm with several low, rounded axial ribs cut by deeply incised spiral grooves; denticles within outer lip of aperture; shell white or light gray with two bands of squarish brown spots per whorl.

This species is carnivorous. X-rays of the shell have been published.

See Herzberg (1966), J. McLean (1969), and Radwin & D'Attilio (1976).

Phylum Mollusca / Class Gastropoda / Subclass Prosobranchia
Order Neogastropoda / Superfamily Muricacea / Family Muricidae

13.75 Ocenebra interfossa Carpenter, 1864
Sculptured Rock Snail

Locally common under rocks, low intertidal zone and offshore to 100 m depth on rock and shale bottoms; Semidi Islands (Alaska) to Punta Santo Tomás (Baja California).

Shell reaching 40 mm in height but commonly not over 22 mm, sculptured with 8–11 well-developed axial ribs crossed by equally strong but more closely spaced spiral cords; color dull grayish brown or yellow.

Crawling in this species involves waves of muscular contraction that travel from front to rear in the foot, with the waves on the two sides of the foot out of phase. The animals

cling to rock surfaces with fair tenacity, but live in somewhat sheltered microhabitats. They are carnivores, capable of drilling through the calcareous shells of mollusks and barnacles. Growth is most rapid in the late summer (August to October) in Washington; animals 12–21 mm in initial length showed maximum growth rates of 0.2–2 mm per month in this period, with the smallest animals showing the most rapid growth. Animals 18 mm or more in length have been observed to lay eggs, and individuals probably take less than 2 years to reach minimum mature size.

See Bormann (1946), Griffith (1967), S. L. Miller (1974), Radwin & D'Attilio (1976), and Spight, Birkeland & Lyons (1974).

Phylum Mollusca / Class Gastropoda / Subclass Prosobranchia
Order Neogastropoda / Superfamily Muricacea / Family Muricidae

13.76 Ocenebra lurida (Middendorff, 1848)
Lurid Rock Snail

Common on or under rocks, low intertidal zone in northern California, sublittoral zone in southern California; Sitka (Alaska) to Punta Santo Tomás (Baja California).

Shell to 40 mm high, usually less, with six to ten broad, low axial ridges crossed by prominent spiral cords; aperture oval with six or seven small denticles inside outer lip; color white or pale yellowish to dark brown or red, occasionally with darker spiral bands (northern California specimens often dark brown to black; Vancouver Island shells ashy gray to ashy brown).

This species is common north of Point Conception (Santa Barbara Co.), but rare to the south. It clings to rock surfaces more firmly than *O. interfossa* (13.75). It has been observed feeding on the gumboot chiton, *Cryptochiton stelleri* (16.25), where it rasps a pit through the dark outer layers on the dorsal surface into the yellow fleshy mantle covering the valves. Such pits, up to 10 mm in diameter and 3–4 mm deep, are common on the gumboot chitons of several areas in northern California. In Washington, 41 adult snails, confined in an aquarium with mussels (*Mytilus edulis*, 15.10) 1–2 cm long and barnacles (*Balanus glandula*, 20.24) 2–5 mm across the base always ate the barnacles. Growth records are sparse, but indicate mean growth rates of 0.5–0.7 mm per month in the field. The animals are estimated to reach minimum adult size in 1–2 years. Copulation occurs in the spring in Washington, and the eggs are deposited in capsules attached to rocks.

See Griffith (1967), S. L. Miller (1974), Spight, Birkeland & Lyons (1974), and Talmadge (1975).

Phylum Mollusca / Class Gastropoda / Subclass Prosobranchia
Order Neogastropoda / Superfamily Muricacea / Family Muricidae

13.77 **Roperia poulsoni** (Carpenter, 1864)
(=Ocenebra poulsoni)   Poulson's Rock Snail

Generally on rocks in low intertidal pools and on pier pilings, commoner in bays than on open coast; Santa Barbara to Bahía Magdalena (Baja California).

Shell to about 60 mm high, usually less than half that, with heavy axial ridges, crossed by well-spaced white spiral cords, the cords separated by several fine, brown, spiral grooves; siphonal canal open; aperture with denticles on inside of outer lip.

Excellent biological studies of this species have been made near La Jolla (San Diego Co.). The snails are carnivorous, feeding on bivalves (e.g., *Mytilus californianus*, 15.9; *M. edulis*, 15.10; *Penitella penita*, 15.78), barnacles, and snails. They bore through the calcareous shells of the prey and insert the proboscis bearing the radula.

The sexes are separate; females average slightly larger than males and are more numerous among snails more than 25 mm long. Sex reversal is unknown. The snails aggregate for mating and egg laying, which takes place from March through July at La Jolla. Clusters of up to 366 snails have been observed. Two females were seen to deposit a clump of 43 egg cases. The number of eggs per capsule varies but averages nearly 200. Essentially this same number of larvae hatch from each capsule; thus there is little or no loss through failure to fertilize, or through cannibalism within the capsule. Development in the capsule takes 3–4 weeks, and swimming veliger larvae escape to the plankton. Mortality in the plankton is high (above 99 percent). The smallest juvenile seen in the habitat occupied by adults was about 7 mm long.

Snails are estimated to reach a length of 10 mm in 3–6 months. Thereafter, measurements on animals marked and later recaptured in the field show that the growth rate declines with increasing size and age. Animals 10 mm long when marked grew about 7 mm a year; those initially 15 mm long grew about 5 mm a year, and those initially 22 mm or more showed little growth. Snails 30–60 mm in shell length must greatly exceed the growth rates measured. Animals probably attain sexual maturity in their third year. Mean longevity is estimated at 9 years or more, with the oldest animals living up to 15 years. The oldest animals are not the largest ones, since growth ceases after the first few years, and the shells are gradually eroded.

The most important predator on larger *R. poulsoni* in the La Jolla area is the red rock crab, *Cancer antennarius* (25.16), which each year consumes 2–3 percent of the adult population. Adult mortality is highest in the spring, from March through June.

See especially Fotheringham (1971); see also Chace (1916), Harger (1967), Hinegardner (1974), J. McLean (1969), Radwin & D'Attilio (1976), and Spight, Birkeland & Lyons (1974).

Phylum Mollusca / Class Gastropoda / Subclass Prosobranchia
Order Neogastropoda / Superfamily Muricacea / Family Muricidae

13.78 **Urosalpinx cinerea** (Say, 1822)
Atlantic Oyster Drill

Common in low intertidal zone and subtidal waters in quiet bays and brackish estuaries where oysters have been planted; Boundary Bay (British Columbia) to Newport Bay (Orange Co.); east coast from Nova Scotia to northeastern Florida; Great Britain.

Shell usually 20–35 mm high in California with five or six rounded whorls; sculpture of 8–12 axial ridges crossed by spiral cords; suture distinct; aperture oval, with thin outer lip; siphonal canal short, open; shell ash gray, brown, or yellow externally, reddish or purple internally.

This oyster drill was introduced to the west coast from the Atlantic, probably on eastern oysters, sometime in the period 1869–88. It has been common in some years in Tomales Bay

(Marin Co.), San Francisco Bay, and Elkhorn Slough (Monterey Co.). It feeds mainly on barnacles and such bivalves as mussels and oysters, though other snails, even other *U. cinerea*, are sometimes taken. Because of its great economic importance as a predator on commercial oysters in the Atlantic, it has been the object of very numerous studies in the eastern United States and Great Britain, and excellent summaries of its biology in those areas are available.

The snail penetrates the shells of its prey using both the radula and a specialized accessory boring organ (ABO), a suckerlike gland located in the midline on the sole of the foot near the anterior end. The ABO produces a viscous secretion, which at least at times may be acidic (pH as low as 3.8) and which contains the enzyme carbonic anhydrase. The secretion softens both calcareous and protein components of the shell. A feeding snail attaches itself firmly to its prey with its foot. Once the boring site has been selected by the extended proboscis and the periostracum has been rasped away, the snail alternately uses the ABO and the radula. The ABO is pressed firmly against the bore site for a few minutes to nearly an hour; the animal then withdraws the anterior foot and rasps the greatly weakened shell with the radula. Snails with the ABO removed cannot drill until the gland has regenerated.

Like oysters, *U. cinerea* tolerates brackish conditions well, but the snails cease feeding at salinities below about 37 percent that of seawater. New England populations showed maximum rates of feeding at 25°C and a salinity 80 percent that of seawater.

Size at sexual maturity is variable, but spawning females are at least 16 mm long. In New England, gametogenesis occurs throughout the year, but eggs and sperm accumulate in the spring and early summer and are depleted in the late summer and fall as spawning occurs. Mating in *U. cinerea* has been observed in Virginia from February to May and from October to December. The female may exhibit a "precopulatory dance," in which the shell is twisted from side to side several times. Mating usually takes only 4–5 minutes. Females may mate with more than one male, and after copulation they may carry viable sperm in their reproductive tracts for up to 6 months before spawning. Spawning animals move inshore and deposit their eggs in vase-shaped capsules. Small females may deposit capsules at irregular intervals through the year, but larger females probably spawn only once a year. The egg capsule is formed in the oviduct, but is finally molded and attached through the action of the egg-capsule gland, located on the sole of the foot just back of the ABO. Small females (16.5 mm) average 4.7 eggs per capsule, whereas large ones (29 mm) average 11.5. Females lay 7–96 egg capsules per season (generally 20–40); larger females do not necessarily produce more capsules than small ones. The eggs hatch in about 2 months. The young emerge as juvenile snails, and there is no planktonic larval period. Some young are lost during development through infection by a fungus resembling *Plectospira*.

The juvenile snails eat tiny barnacles, small snails, and encrusting bryozoans. They grow at a rapid but constantly diminishing rate (e.g., snails initially 5 mm long grew an average of 12 mm in a year, those initially 15 mm averaged 7 mm growth, those 25 mm initially grew 0–2 mm). In temperate climates, growth is seasonal, and the rate varies greatly with food supply and other conditions. Annual growth lines occur on Connecticut shells. Sexual maturity is reached in 2 years, and most animals probably do not survive to reproduce for more than 2 years. Maximum life span may be 5–8 years.

See especially Carriker (1955) and Carriker & Van Zandt (1972); see also Blake (1960), Carriker (1943, 1957, 1959, 1961, 1969, 1971), Carriker, Schaadt & Peters (1974), Carriker, Scott & Martin (1963), Carriker & Van Zandt (1967), Carriker, Van Zandt & Grant (1978), Chapman & Banner (1949), Cole (1942), Costello & Henley (1971), Federighi (1929, 1931a,b), Franz (1971), Galtsoff, Prytherch & Engle (1937), Ganaros (1957), Hancock (1954, 1956, 1959), Hanks (1957), Hanna (1966), Hargis & MacKenzie (1961), Haskin (1950), Haydock (1964), Hinegardner (1974), Human (1971), Loosanoff (1962), MacKenzie (1962), Manzi (1970), Manzi, Calabrese & Rawlins (1972), Nylen, Provenza & Carriker (1969), Pratt (1974, 1977), Radwin & D'Attilio (1976), Sandeen, Stephens & Brown (1954), Smarsh et al. (1969), Spight, Birkeland & Lyons (1974), Tamarin & Carriker (1968), and Wood (1968).

Phylum Mollusca / Class Gastropoda / Subclass Prosobranchia
Order Neogastropoda / Superfamily Muricacea / Family Thaisidae

13.79 **Acanthina paucilirata** (Stearns, 1871)
Checkered Unicorn

Upper intertidal zone on rocky shores; San Pedro (Los Angeles Co.) to Isla Cedros (Baja California); uncommon north of San Diego.

Shell to about 25 mm high; body whorl with a broad shoulder and about four narrow, white, spiral cords separating spiral bands, the bands checkered alternately with black and white; outer lip of aperture with long spine.

This carnivorous species has received little study to date. An aggregation of snails engaged in reproductive activity was reported from La Jolla (San Diego Co.) in March.

See R. Abbott (1974), Fotheringham (1971), and J. McLean (1969).

Phylum Mollusca / Class Gastropoda / Subclass Prosobranchia
Order Neogastropoda / Superfamily Muricacea / Family Thaisidae

13.80 **Acanthina punctulata** (Sowerby, 1825)
Spotted Unicorn

Common in high intertidal zone on rocky shores exposed to moderate but not strong surf, often in vicinity of the barnacle *Balanus glandula* (20.24), the sea anemone *Anthopleura elegantissima* (3.31), and the snails *Littorina planaxis* (13.42) and *L. scutulata* (13.43); Monterey Bay to Punta Santo Tomás (Baja California).

Shell to about 25 mm high, usually distinguished from *A. spirata* (13.81) by (1) a relatively shorter spire, (2) whorls with shoulders rounded or absent, (3) spiral cords small or absent, and (4) spiral rows of black marks less prominent.

This species and the next, *A. spirata*, formerly united as *A. spirata*, are now considered probably separate species. In fact, the adjacent line drawing and the color photograph 13.80 probably show *A. spirata* (James McLean, pers. comm.). *A. punctulata* occupies a higher intertidal position than *A. spirata*.

At low tide, at Pacific Grove (Monterey Co.), *A. punctulata*

may be exposed high and dry on open rock surfaces bearing barnacles, but larger numbers occur in pools or snuggled down between wet contracted anemones in beds of *Anthopleura elegantissima* (3.31). *Acanthina punctulata* moves up and down with the tide, passing through regions where its food is abundant. At Mussel Point (Pacific Grove), the main food species are barnacles (*Balanus glandula* and *Chthamalus dalli*, see under 20.12) and snails (*Littorina scutulata*, 13.43; *L. planaxis*, 13.42). The last of these lives largely above the region roamed by *Acanthina*, but dislodged individuals that fall into pools occupied by the predator are caught and eaten. This snail also has been observed feeding on small *Nucella emarginata* (13.83) (L. A. West, pers. comm.). *Acanthina* uses the radula in conjunction with an accessory boring organ (see the account of *Urosalpinx cinerea*, 13.78) to drill the shells of its prey. About 20 percent of the shells of *L. planaxis* occupied by hermit crabs at Mussel Point have been drilled by carnivorous snails, probably mainly *Acanthina*. Shells of the small high-intertidal bivalve *Lasaea cistula* (15.21) are also found with drill holes probably made by *Acanthina*. *A. punctulata* readily captures and feeds upon the snail *Tegula funebralis* (13.32) under both field and aquarium conditions.

Quantitative studies of predation at Santa Cruz Island (Channel Islands) show that the biomass of the diet of *A. punctulata* there has the following constituents: 46 percent *Littorina scutulata*, 44 percent *Littorina planaxis*, 6 percent *Balanus glandula*, and 4 percent *Chthamalus fissus* (20.12). Here *Acanthina* moves over wet rocks at low tide, following a more or less random path. Prey are contacted by the outstretched head tentacles and either accepted or passed over. Snails accepted are captured by the uplifted and extended anterior part of the *Acanthina* foot, and are drilled through the columella. It takes *Acanthina* 15–60 hours to drill and eat a littorine, depending on its size. Of the total time spent in both foraging and feeding, *Acanthina* spends about 5 percent in finding and securing prey, and 95 percent in drilling and feeding. Under laboratory conditions, larger littorines are preferred to smaller ones. Littorines pursued are not always caught by *Acanthina*; they may run away or crawl up on top of the predator's shell (they are safe as long as they stay there).

The predators may feed gregariously, several snails feeding on the same littorine.

Predation by *A. punctulata* on barnacles was described by Hewatt (1934) as involving the use of the spine on the outer lip of the aperture: "When attacking a barnacle, the snail assumes a position above the opening of the barnacle shell so that this spine is directly above the line of contact of the closed scutes of the barnacle. The *Acanthina* usually takes this position when the tide is out and the barnacle thus is closed. When the water returns over the area, the natural reaction of the barnacle is to open up and begin the feeding activities. As soon as this occurs, the snail quickly inserts its spine into the opening between the scutes, the proboscis is everted, and the soft parts of the barnacle are consumed. This procedure has been observed both in the natural habitat and in the laboratory." Later observers of this and other *Acanthina* species have not noted this use of the spine, and the matter needs reinvestigation. Toxic material, including a new choline esterase, has been found in the hypobranchial gland, and may play some role in relation to predation.

At Pacific Grove, groups of 30–40 *A. punctulata* have been observed aggregating to copulate and spawn in May and June. Aggregations occurred near the holdfasts of the brown alga *Pelvetia*, and numerous females deposited their egg capsules together, in clusters. Individual capsules, which are flask-shaped, cream-colored, about 5 mm long, and pointed at the tip, contain 400–500 eggs. Only about 10 percent of the eggs complete development into veliger larvae, the remainder serving as nurse eggs. Metamorphosis occurs before hatching, and the juvenile snails that emerge are immediately able to drill prey. Developing from egg deposition to hatching took 16–20 days at ambient temperatures at Pacific Grove. Some new hatchlings are eaten by the starfish *Leptasterias hexactis* (8.11).

For *A. punctulata* (whether as *A. punctulata*, *A. spirata*, or *A. lapilloides*), see Bender et al. (1974), Bigler (1965), Bollay (1964), Glynn (1965), Hewatt (1934, 1937), G. MacGinitie & N. MacGinitie (1968), J. Menge (1974), Perkins (1971), and Yarnall (1964). For accounts probably referring to *A. spirata*, see Haydock (1964) and Spight (1976a,b). For more information on the genus, see also Cooke (1918), Harger (1967), Hemingway (1973, 1976), Murdoch (1969), Paine (1966), and Strong (1924, 1925).

Phylum Mollusca / Class Gastropoda / Subclass Prosobranchia
Order Neogastropoda / Superfamily Muricacea / Family Thaisidae

13.81 Acanthina spirata (Blainville, 1832)
Angular Unicorn

Common in high and middle intertidal zones on protected rocks and pilings; Tomales Bay (Marin Co.) to Camalú (Baja California).

Shell to 40 mm high, usually distinguished from *A. punctulata* (13.80) by a taller spire, by whorls with a prominent shoulder typically bearing a ridge or keel, and by a sculpture of spiral cords.

This species and the preceding one, *A. punctulata*, were long united under the name *A. spirata*. The species either intergrade in characters or come close to it, but at least in central California they show distinctive distributions, and most individuals, unless old and worn, can be assigned to one or the other species with no hesitation. In the Monterey Bay area, where the two species occur on the same rocks, as in Monterey harbor, *A. spirata* occurs lower down.

Most of the studies alleged to be on *A. spirata* probably refer to *A. punctulata*, but studies made north of Monterey Bay are here attributed to *A. spirata* in the sense used above.

Investigation of the egg capsules and development of *A. spirata* in Tomales Bay showed that the egg capsules are about 4–9 mm in length and contained 40–140 eggs. All eggs develop to the gastrula stage, but some stop development there, and these "nurse eggs" are eaten by embryos that go on to reach the veliger larval stage. The number of advanced embryos per capsule ranges from 17 to 46. On the average, there are 1.67 nurse eggs per surviving embryo. The juvenile snails that hatch from the capsule are 0.55–0.75 mm in shell length.

See Haydock (1964) and Spight (1976a,b); see also the account and references for *A. punctulata* (13.80).

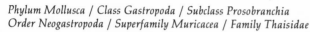

Phylum Mollusca / Class Gastropoda / Subclass Prosobranchia
Order Neogastropoda / Superfamily Muricacea / Family Thaisidae

13.82 Nucella canaliculata (Duclos, 1832)
(=Thais canaliculata)   Channeled Dogwinkle

Locally common on rocks and in mussel beds, middle intertidal zone; Aleutian Islands (Alaska) to Cayucos (San Luis
Obispo Co.).

Shell to about 40 mm high; whorls sculptured with numerous spiral cords separated by narrow grooves bearing tiny
scales; color white or dark orange with darker mottling on
cords.

This species is a characteristic inhabitant of mussel beds,
occupying a lower intertidal position than *N. emarginata*
(13.83) but a higher one than *N. lamellosa* (13.84). In favorable
habitats populations may exceed 120 per m². A carnivore, it
feeds mainly on *Mytilus* and secondarily on barnacles (especially *Semibalanus cariosus*, 20.23). The shells of the prey are
drilled by the radula, used in conjunction with an accessory
boring organ located on the sole of the foot. The snail generally spends 1–2 days drilling and consuming a prey organism.
Individuals at San Juan Island (Washington) were estimated
to consume about 0.13 *Mytilus* per snail per day.

Breeding occurs in the spring and summer. After mating,
the females deposit their eggs in flask-shaped capsules 6–11
mm high, attached in clusters to rocks in shaded situations.
At San Juan Island each capsule contains about 15–55 eggs
averaging 620 μm in diameter; in some localized populations
all eggs develop into veliger larvae and eventually hatch as
small snails, but in other populations nearby 16–92 percent
of the eggs develop normally, while the remainder are abortive and are consumed by the larvae that do develop. In larval
feeding, yolk is flaked off the undeveloped eggs by cilia on
the larval lips. The veligers cannot ingest whole eggs or feed
on each other. Metamorphosis occurs within the capsule, and
the young emerge as juvenile snails about 1.3 mm in shell
length. A study of reproduction reported in *T. canaliculata*
from central California shows significant differences (more
eggs and a higher percentage of nurse eggs); further work is
needed to settle the taxonomic status of this population.

See R. Abbott (1974), Bertness (1977), Dall (1915), Dayton (1971),
Griffith (1967), Houston (1971), Lyons & Spight (1973), S. L. Miller
(1974), Paris (1960), and Spight (1976a,b, 1977).

Phylum Mollusca / Class Gastropoda / Subclass Prosobranchia
Order Neogastropoda / Superfamily Muricacea / Family Thaisidae

13.83 Nucella emarginata (Deshayes, 1839)
(=Thais emarginata)   Emarginate Dogwinkle

High and middle intertidal zones on rocks experiencing
strong to slight wave action, especially in mussel beds and
among barnacles; Bering Sea to northern Baja California; uncommon south of Point Conception (Santa Barbara Co.).

Shell to 40 mm high, usually less than 30 mm, heavy, with
low spire; sculpture very variable, some shells nearly smooth,
others with weak spiral cords (large and small cords alternating in some), the cords with or without nodules; aperture
more than half as long as shell, wide; siphonal canal short;
umbilicus closed; exterior white, yellow, orange, gray, brown,
or black, often with light and dark spiral banding; interior
brownish or purple.

This predatory species may occasionally attain populations
of over 400 snails per m² in favorable spots. It feeds primarily
on mussels (with a preference for *Mytilus edulis*, 15.10, over
M. californianus, 15.9) and barnacles (mainly *Balanus glandula*,
20.24, but also *Pollicipes polymerus*, 20.11, and *Chthamalus dalli*,
20.12). In some areas, such limpets as *Collisella scabra* (13.13)
and *C. limatula* (13.14) are taken in fair numbers (L. A. West,
pers. comm.). Holes are drilled in the shells of the prey by
means of the radula and an accessory boring organ, borne on
the sole of the foot, which softens the shell (see the *Urosalpinx
cinerea* account, 13.78). The proboscis bearing the radula is
then extended a long distance (up to the length of the snail
body) through the hole to permit feeding on the soft parts
within. In some areas more than 86 percent of dead *Mytilus*
collected in the field have been killed by predaceous snails
such as *Nucella*. The predation pressure from snails seems to
be greatest at the lower margin of the mussel bed, and this
(along with predation by the sea star *Pisaster ochraceus*, 8.13,
and the sea otter) may help limit the distribution of *Mytilus* in

lower tidal zones. Predation on *Balanus* species clears so many of these larger barnacles from the rocks that it makes more room for the small barnacle *Chthamalus*. *Nucella emarginata* also preys on herbivorous snails such as *Tegula funebralis* (13.32) and *Littorina planaxis* (13.42), and elicits a "mounting" response from them: when they are contacted by the foot of the predator, they climb on the predator's shell—a comparatively safe spot (the tiger cannot eat you when you are riding on his back).

Detailed studies of marked *N. emarginata* at Pacific Grove (Monterey Co.) show that different individuals in the same small area may differ markedly in their diets, but that individual snails are relatively consistent in what they eat, and each selects a much more limited range of prey species than that consumed, collectively, by the local *N. emarginata* population (L. A. West, pers. comm.).

Studies of the foot of *N. emarginata* show that locomotion is accomplished by waves of muscular activity that move from the front to the rear and are out of phase on the two sides of the foot. The average speed on a hard smooth surface is 2.1 cm per minute. The animals cling very strongly to rocks when stationary, much less strongly when moving. Small individuals live higher in the intertidal zone than larger ones and show greater resistance to thermal stress.

Nucella emarginata spawns sporadically throughout the year in California but most actively from November through March. The females deposit their eggs in elaborate, yellow, vase-shaped egg capsules, commonly in mussel beds (see the photograph). Each capsule has a solid stalk or peduncle and a longitudinal suture that extends from the top of the stalk to the rounded apex. Capsules average about 6 mm high. As many as 300 capsules may be deposited in a cluster. The eggs are 180–210 μm in diameter. There are usually 500–600 eggs per capsule, though capsules may contain from 64 to about 1,000 eggs. Most of the eggs are sterile nurse eggs; they are consumed by the larvae, which develop from the relatively few fertile eggs in each capsule. The veliger larvae swim about in the fluid of the capsule and undergo metamorphosis there, ultimately developing into tiny snails with a creeping foot and a shell 1.1–1.2 mm long. In the Pacific northwest an average of 10–20 (occasionally up to 50) hatchlings leave the

capsule 2.5–4 months after it has been deposited; along the central California coast development may be more rapid.

Nucella emarginata, like other members of the genus on the coast, has a haploid chromosome number of 35. Biochemical analysis of this species has recently shown that the hypobranchial gland produces a biologically active choline ester tentatively identified as N-methylmurexine. Along the California coast, the species carries heavy loads of trace metals: up to 570 ppm copper and 1,700 ppm zinc.

See R. Abbott (1974), Ahmed & Sparks (1970), Bender et al. (1974), Bertness (1977), Bertness & Schneider (1976), Bollay (1964), Carriker (1961), Connell (1970), Costello & Henley (1971), Dayton (1971), Dehnel (1955), Emerson (1965), Emlen (1966), Glynn (1965), Graham (1972), Harger (1967), Hewatt (1934, 1935, 1937), Houston (1971), LeBoeuf (1971), Lyons & Spight (1973), J. McLean (1969), S. L. Miller (1974), Murdoch (1969), Paris (1960), Perkins (1971), Risebrough et al. (1967), Spight (1972, 1975a,b, 1976a,b), Spight, Birkeland & Lyons (1974), Spight & Emlen (1976), West (1978), and Yarnall (1964).

Phylum Mollusca / Class Gastropoda / Subclass Prosobranchia
Order Neogastropoda / Superfamily Muricacea / Family Thaisidae

13.84 Nucella lamellosa (Gmelin, 1791)
(=Thais lamellosa)   Wrinkled Purple, Frilled Dogwinkle

Locally common on low intertidal rocks, especially below mussel beds; Bering Strait to central California.

Shell to 50 mm high, highly variable in shape and sculpture, some specimens nearly smooth, with little or no sculpture, others more or less frilly with well-developed axial ridges and projections or spiral bands; siphonal canal relatively long; color also variable, ranging from white through orange to brown, unicolored or banded.

This snail is not uncommon along the shores of northern California, but in the Pacific northwest it is one of the most abundant intertidal whelks. It lives lower down in the intertidal zone than either *N. emarginata* (13.83) or *N. canaliculata* (13.82). Kincaid (1967) and Spight (1976c) illustrate the remarkable variability in form, size, and color of the shell in this species.

Like other whelks, *N. lamellosa* is a carnivore; it prefers acorn barnacles but also feeds on mussels and other mollusks. The radula is used to penetrate the shells of the prey, with chemical aid from secretions of the accessory boring organ on the sole of the foot. In a related species, *N. lapillus*, the enzyme carbonic anhydrase is involved in the breakdown of calcium carbonate in the shell of the prey.

Breeding occurs in the winter or spring. Breeding animals aggregate in groups of from a few dozen to several hundred individuals. The aggregations normally occur at the low tide level, and the size of the group seems independent of the density of adults on the shore. Long-term studies of marked individuals show that the animals do not become sexually mature until the fourth year. They then often return to their hatching site and join a breeding group. Individuals tend to breed with the same group in successive years; therefore the breeding groups are natural and persistent units.

The eggs are deposited in vase-shaped, yellow egg capsules, each about 1 cm long, which are attached in clusters to the undersides of rocks. In British Columbia these clusters of egg capsules are known as "sea oats." The egg capsules, unlike those of other *Nucella* species on the coast, rarely contain nurse eggs. Spawning is relatively synchronous; 95 percent of the mature females in a breeding group deposit capsules at the same time. The development time varies with the temperature: young snails emerge from the capsules 140 days after deposition at 6–8°C, and 67–91 days after deposition at 9.6–11°C. The capsules are preyed upon by other animals and nearly half of the fertilized eggs fail to reach the hatching stage. Newly hatched snails also suffer high mortality and, although about 1,000 eggs are produced annually by a female, rarely do 10 of these survive and develop to 1 year of age. The growth rate varies greatly with the food supply. As young animals grow, the shape of their shells undergoes progressive change. Shell growth may also be related to diet; snails with abundant barnacles for food produce heavier and stouter shells.

During spawning, the female snails lose from 28–38 percent of the weight of the visceral mass. Deposited egg capsules contain quantities of protein and lipid but little polysaccharide material. Lipid and protein stores provide the main sources of energy during development and also during starvation in adults. *N. lamellosa* has also been studied with respect to seasonal changes in biochemical composition and respiration and to the effects of starvation on metabolism, respiration, and nutrient stores.

See especially Spight (1972, 1973, 1974, 1976c); see also Bertness (1977), Bertness & Schneider (1976), Connell (1970), Dayton (1971), Emerson (1965), Griffith (1967), Kincaid (1957), Lambert & Dehnel (1974), Lyons & Spight (1973), S. L. Miller (1974), Monique & Fournie (1969), Paris (1960), Spight (1975a,b, 1976a), Spight, Birkeland & Lyons (1974), Spight & Emlen (1976), Stickle (1971, 1973, 1975), and Stickle & Duerr (1970).

Phylum Mollusca / Class Gastropoda / Subclass Prosobranchia
Order Neogastropoda / Superfamily Buccinacea / Family Melongenidae

13.85 Busycotypus canaliculatus (Linnaeus, 1758)
(=Busycon canaliculatum)   Channeled Whelk

Common on mud and sand bottoms, low intertidal zone and subtidal waters; San Francisco Bay; Cape Cod (Massachusetts) to St. Augustine (Florida).

Shell to 185 mm high, large, relatively thin, lacking axial sculpture; sutures distinct and channeled; outer lip of aperture smooth; siphonal canal long, open, slightly curved; periostracum yellow-brown, fuzzy or feltlike.

This snail was introduced into San Francisco Bay from the east coast sometime before 1938, and it is now relatively abundant. It is an active carnivore and feeds primarily on young clams. Because of its size and easy availability on Atlantic shores, this species has been used in a variety of anatomical, physiological, biochemical, embryological, and behavioral studies.

On the east coast *Busycotypus* buries itself in mud or sand but emerges to forage and feed, characteristically during the night in the summer and during the day in the winter. When foraging it moves along at rates up to 1 cm per 6–12 seconds and is attracted to living bivalves by the water leaving the clams' exhalant siphons. The osphradium, a sense organ located near the snail's gill, aids in locating prey; snails with the osphradium removed are unable to detect food at a distance in an aquarium. Food items of east coast populations include such bivalves as edible mussels, oysters, quahogs (*Mercenaria mercenaria*, 15.31), and razor clams. Large *Busycotypus* can rapidly dig up a living clam such as a mature quahog. The clam is then held in a hollow in the snail's large foot and rotated until the ventral edges of the valves are brought up against the lip of the snail's shell. The columellar muscles then contract, pressing the lip of the snail's shell tightly against the edge of the clam's valve and chipping the latter. The process is repeated, and eventually it creates a hole large enough for the snail to insert its proboscis and feed on the clam. *Busycotypus* is a serious pest on many clam and oyster beds along the coasts of New Jersey and the Carolinas.

On the east coast *B. canaliculatus* breeds from May or June until November, the season varying somewhat with latitude and temperature regime. The eggs are about 1 mm in diameter and are passed to the outside by way of a temporary groove in the foot. Secretions enclose them in a tough disk-shaped case. Up to nearly 100 egg cases, each containing 20–50 eggs, are strung along a cord, the end of which is firmly attached to some solid object in the water. In the laboratory, spawning animals produce one egg case every 2 hours.

Cleavage in *Busycotypus* closely resembles that in *Crepidula* (e.g., 13.55–59) up to the 60–cell stage. Development is slow, and on the east coast hatching occurs 13 months after spawning. The young leaving the egg cases are fully developed juveniles about 6 mm long.

The main predators on *Busycotypus* are crabs, gulls, and man. The shells are used as ornaments, and the foot of the snail makes a good chowder.

See especially Magalhaes (1948) and Pierce (1950); see also Carlton & Roth (1975), Carriker (1951), Colton (1908), Conklin (1907), Copeland (1918), Costello & Henley (1971), Dakin (1912), Duwe (1954), Fretter & Graham (1962), Hanna (1966), Herrick (1906), Hinegardner (1974), Klotz & Klotz (1955), Kohn (1961), Mendel & Bradley (1905, 1906), Paine (1962), Stohler (1962b), and Strong & Green (1970).

Phylum Mollusca / Class Gastropoda / Subclass Prosobranchia Order Neogastropoda / Superfamily Buccinacea / Family Buccinidae

13.86 Searlesia dira (Reeve, 1846)
Dire Whelk

Middle intertidal zone on rocky shores, from protected outer coast to quiet bays, usually not extending subtidally (but known from Cobb Seamount, 250 miles off Oregon, at 35 m depth); Chirikof Island (Alaska) to Monterey; abundant in Pacific northwest, uncommon in California.

Shell to about 45 mm high, heavy; whorls sculptured with spiral grooves and about nine low, rounded, axial swellings; columella shiny; siphonal canal short; color gray or brownish.

A detailed study of this species has been made in Puget Sound (Washington), where population densities may average 1 or 2 per m² and ordinarily do not exceed 6 per m², but occasionally reach 36 per m² in especially favorable spots. The animals do not migrate up and down with the tide. Large and small animals occur at all intertidal levels, but the larger animals are more common at lower elevations. The animals roam about most actively when submerged in calm water. At low tide or in rough water they are often inactive and sheltered in pools and crevices, or even buried under gravel. Locomotion takes place by means of muscular waves traveling from front to rear on the foot. Their ability to cling to rocks is only moderate, approaching that of *Nucella lamellosa* (13.84).

Searlesia is a carnivore, consuming a very wide variety of the macroscopic animal species in its vicinity, especially snails and limpets (*Littorina scutulata*, 13.43; *L. sitkana*; *Notoacmea scutum*, 13.23; *Collisella pelta*, 13.16; *C. digitalis*, 13.12; *Calliostoma ligatum*, 13.27; *Margarites pupillus*, 13.29; *Lacuna* sp.) chitons (*Ischnochiton*, *Katharina tunicata*, 16.24), and barnacles (especially *Balanus glandula*, 20.24, but also *B. cariosus*). Dead organisms are detected from a distance, and such carrion as crabs and fishes may be fed on by as many as six *Searlesia* at once. The proboscis may extend out fully the length of the shell, which not only facilitates gregarious feeding on large prey but permits the consumption of worms in tubes. It also permits *Searlesia* to feed on prey that are being digested by the everted stomach of the starfish *Pisaster ochraceus* (8.13). The snails cannot drill the shells of prey, but must insert the proboscis past the opercular plates of barnacles and littorines, and under the shell edge of limpets. Dead or injured organisms appear to be given first choice as food, but the possibility that *Searlesia* secretes materials that anesthetize or poison healthy prey needs to be investigated.

The sexes are separate. Males develop a penis when the shell is 15–20 mm long. The females lay eggs in all months except the summer (June through August). The eggs are deposited in low convex capsules usually found attached to rocks in clusters of 5–10 (sometimes as many as 32) in shaded crevices. The capsules measure 6.2–8 mm long and 4.8–6.8 mm wide, and contain 50–175 eggs each. The eggs are about 0.24 mm in diameter, and most of them show abortive development. Only 1–13 eggs in a capsule develop into veliger larvae. These nourish themselves by eating the undeveloped "nurse eggs." The veliger stage, reached in about 1 month, lasts for about 2 months inside the capsule. The larvae then metamorphose into small, actively crawling snails; their shells, about 2 mm long, are just starting to show axial ribs. Details on hatching and early free life are lacking. Sparse growth data indicate that animals 10–25 mm long increase in net shell length (growth minus erosion of spire) by 4–5 mm per year, and that animals more than 30 mm long probably add less than 1 mm per year to net shell length. Longevity is unknown, but animals 40 mm long could be 15 years or more in age.

See especially Lloyd (1971); see also R. Abbott (1974), Dayton (1971), Emerson (1965), Griffith (1967), S. L. Miller (1974), and Nicotri (1974).

Phylum Mollusca / Class Gastropoda / Subclass Prosobranchia
Order Neogastropoda / Superfamily Buccinacea / Family Buccinidae

13.87 **Kelletia kelletii** (Forbes, 1852)
Kellet's Whelk

Rare, under rock ledges, low intertidal zone; common subtidally to 70 m on rocky reefs and gravel bottoms and below offshore kelp beds; Point Conception (Santa Barbara Co.) to Isla Asunción (Baja California).

Shell to about 170 mm high, robust and heavy; sculpture of rounded axial swellings crossed by thin spiral lines; outer lip of aperture sharp; color white or gray.

This is one of the largest gastropods found in southern California. It often invades crab and lobster traps at 20–70 m depth and may kill the trapped crustaceans. Its normal food is dead or injured animals it finds on the sea floor. It often feeds with the predatory sea star *Pisaster giganteus* (8.14) on common food items. The snail displays no avoidance response in the presence of the sea star, despite the fact that when other food is not available *Pisaster* is a major predator on the whelk.

The snails mate in March or April, and spawning usually occurs in April regardless of water temperature. During spawning the snails aggregate into groups of up to 20 individuals. Eggs are deposited in capsules secured to the rocks, and over a 30-day period a single female has been observed to produce as many as 85 egg capsules. Each capsule is flattened and ovoid in shape and may contain up to 1,000 yellow eggs. Fertile eggs develop into veliger larvae in 30–34 days after spawning at 14–17°C, and the larvae swim free when the gelatinous plug in the end of the capsule dissolves.

Although *Kelletia* does not exhibit an escape response to *Pisaster giganteus* in its natural habitat, experiments show that isolated radular muscle from the snail is sensitive to extracts from the sea stars *Pisaster ochraceus* (8.13) and *Pycnopodia helianthoides* (8.16). *Kelletia* is occasionally eaten by man today. The shells have been found in California Indian middens and are thought to have been used for trumpets or bird calls.

See Curtis (1966), Feder & Lasker (1968), Hinegardner (1974), G. MacGinitie (1938), G. MacGinitie & N. MacGinitie (1968), J. McLean (1969), and Rosenthal (1969, 1970, 1971).

Phylum Mollusca / Class Gastropoda / Subclass Prosobranchia
Order Neogastropoda / Superfamily Buccinacea / Family Buccinidae

13.88 **Macron lividus** (A. Adams, 1855)
Livid Macron

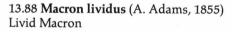

Common under stones, low intertidal zone on rocky shores; Orange Co. to Bahía San Bartolomé (Baja California).

Shell to about 25 mm high, lacking axial sculpture, but with spiral sculpture of fine lines; whorls rounded; base with spiral grooves; periostracum thin, brown, tightly adherent.

Little is known of the biology of this animal.

See J. McLean (1969).

Phylum Mollusca / Class Gastropoda / Subclass Prosobranchia
Order Neogastropoda / Superfamily Buccinacea / Family Columbellidae

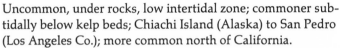

13.89 **Amphissa columbiana** Dall, 1916
Wrinkled Dove Snail

Uncommon, under rocks, low intertidal zone; commoner subtidally below kelp beds; Chiachi Island (Alaska) to San Pedro (Los Angeles Co.); more common north of California.

Shell to 30 mm high, thin, with fine longitudinal ribs running from apex to middle of body whorl and with uniform close-set spiral threads; shell color highly variable, often orange or dull greenish yellow, sometimes pink, mauve, or brown, often with brownish spots.

This species is often abundant under bottom rocks and in kelp holdfasts in *Macrocystis* beds in the Monterey area. Studies of its locomotory behavior show it to be an agile form, propelled by pedal muscular waves that travel from front to rear and extend uniformly across the foot. The animals are excellent climbers. Frequently they rear up, standing only on the back of the foot, then swing around and set off on a new course. The foot is equipped at the rear with a pedal gland, which produces a strong mucus thread capable of suspending the snail like a spider on a strand of silk. In the laboratory the snails show a strong tendency to move upstream toward dead animal matter, but their food habits in nature, and most other aspects of their biology, need further study.

See especially S. L. Miller (1974); see also R. Abbott (1974), Andrews (1945), Griffith (1967), J. McLean (1962), and Pearse & Lowry (1974).

Phylum Mollusca / Class Gastropoda / Subclass Prosobranchia
Order Neogastropoda / Superfamily Buccinacea / Family Columbellidae

13.90 **Amphissa versicolor** Dall, 1871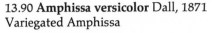
Variegated Amphissa

Common among algae and under rocks, middle and low intertidal zones on protected outer coast and subtidally in kelp holdfasts; Fort Bragg (Mendocino Co.) to Isla San Martín (Baja California).

Shell to about 17 mm high, strong, with axial ribs oriented at a slant on whorls and crossed by fine, close-set spiral cords; inner lip of aperture with spiral striations; color highly variable, including white, gray, yellow, and brown, often mottled or with dark spiral lines.

In its distribution in the Monterey area, this species overlaps *A. columbiana* (13.89), but it is much commoner intertidally and occurs in areas less protected from surge. The eggs, laid in small helmet-shaped capsules about 2 mm in diameter, have been noted in July on the red alga *Iridaea* at Pacific Grove (Monterey Co.). The capsules contained 35–50 veliger larvae.

See R. Abbott (1974), Andrews (1945), Hewatt (1934, 1937), and J. McLean (1962, 1969).

Phylum Mollusca / Class Gastropoda / Subclass Prosobranchia
Order Neogastropoda / Superfamily Buccinacea / Family Columbellidae

13.91 **Mitrella carinata** (Hinds, 1844)
(=Columbella carinata)   Carinated Dove Snail

Common on rocks, algae, and the surfgrass *Phyllospadix*, low intertidal zone; abundant on kelp stipes and holdfasts subtidally; Forrester Island (Alaska) to southern Baja California.

Shell to 11 mm high, shiny; shoulder of main body whorl sometimes round but often with a small carina or keel; outer lip of aperture slightly undulating; shell color variable, dark yellow to brown, occasionally with white or brown splotches; keel of shoulder often lighter in color.

In beds of the kelp *Macrocystis*, *Mitrella* is sometimes the most abundant animal living on the blades and stipes, yet little is known of the biology of the snail. It appears to be a microcarnivore and detritus feeder, eating material adhering to kelp blades rather than the kelp itself. In southern California the amphipod *Pleustes platypa* (22.3) mimics very closely the form and color of *Mitrella carinata* (see Fig. 22.3), particularly when the two are found living together on *Macrocystis*. This is one of the few cases of arthropod-mollusk mimicry known to us (see also *Lacuna marmorata*, 13.41), and its adaptive significance needs further study.

See Andrews (1945, as *Columbella carinata*), Crane (1969), J. McLean (1962, 1969), North (1964), and Pearse & Lowry (1974).

Phylum Mollusca / Class Gastropoda / Subclass Prosobranchia
Order Neogastropoda / Superfamily Buccinacea / Family Nassariidae

13.92 **Nassarius fossatus** (Gould, 1849)
Channeled Nassa

Common on mud and sand in bays and on soft bottoms in protected coastal areas, low intertidal and subtidal zones; Vancouver Island (British Columbia) to Laguna San Ignacio (Baja California).

Shell to about 47 mm high, with about seven whorls; upper whorls and about a third of body whorl bearing axial ribs crossed by narrower spiral ridges; deep groove around base of shell; outer lip of aperture thickened, with interior ridges; siphonal canal short, sharply bent; color gray or brown; callus of columella yellow or orange.

This is the largest nassariid species on the west coast. It is primarily a scavenger, feeding on dead animals. It moves effectively on hard substrata but is especially well adapted to crawling on sand or mud or burrowing just below the surface. Locomotion involves both muscular and ciliary activity of the very large flat foot. The snails exhibit a very rapid righting response when turned over. Confronted by predatory sea stars, such as *Pisaster brevispinus* (8.15), they exhibit escape responses involving turning, rapid rocking of the shell, and even violent somersaulting with the foot. The response has been recorded photographically by Feder (1967) for a related species.

Eggs are deposited in late winter or in the spring. A spawn-

ing animal uses the radula to clean a spot about 1.5 cm² on a leaf of the eelgrass *Zostera* or on a rock surface, then takes about 10 minutes to deposit a single egg capsule. It then moves ahead and repeats the process, depositing a string of up to 45 capsules in about 7 hours (see the photograph). Larvae hatch from the capsules in about 20 days.

Studies of digestion in this animal indicate that the uptake of food occurs in the tubule cells of the digestive gland.

See Demond (1951, 1952), Feder (1967), Fretter & Graham (1962), Griffith (1967), Hinegardner (1974), G. MacGinitie (1931, 1935), G. MacGinitie & N. MacGinitie (1968), N. McLean (1971), and Pilson & Taylor (1961).

Phylum Mollusca / Class Gastropoda / Subclass Prosobranchia
Order Neogastropoda / Superfamily Buccinacea / Family Nassariidae

13.93 **Nassarius mendicus** (Gould, 1850)
Lean Nassa

Common on both hard and soft substrata, low intertidal zone to about 75 m depth along open coast and in bays; Kodiak Island (Alaska) to Isla Asunción (Baja California).

Shell 15–25 mm high, slender, with a distinct furrow around base; sculpture of 7–12 axial ribs intersecting fine spiral lines; exterior color yellow, gray, or brown, variable, with alternately light and dark spiral bands; interior white.

This snail is a scavenger and is often found offshore feeding on bait in shrimp traps or on setlines. On muddy bottoms it ploughs into the substratum and moves forward with the tubular anterior siphon projecting into the water above.

See Demond (1951, 1952), Griffith (1967), and J. McLean (1969).

Phylum Mollusca / Class Gastropoda / Subclass Prosobranchia
Order Neogastropoda / Superfamily Buccinacea / Family Nassariidae

13.94 **Nassarius mendicus cooperi** (Forbes, 1852)
Cooper's Lean Nassa

On mud flats, low intertidal zone and offshore to about 45 m depth; Puget Sound (Washington) to San Diego; commoner in southern part of range.

Shell usually 12–18 mm high, intergrading with that of *N. mendicus mendicus* (13.93), but with larger and more conspicuous axial ribs, these forming prominent nodes on the whorls.

This snail is a scavenger on dead animal material. It occurs together with the more typical *N. mendicus* over a wide part of its range.

See R. Abbott (1974) and Demond (1951, 1952).

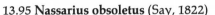
Phylum Mollusca / Class Gastropoda / Subclass Prosobranchia
Order Neogastropoda / Superfamily Buccinacea / Family Nassariidae

13.95 Nassarius obsoletus (Say, 1822)
(= *Ilyannassa obsoleta*)   Eastern Mud Snail

Common on low intertidal mud flats in bays; British Columbia to central California; very abundant in San Francisco Bay; Gulf of St. Lawrence to Florida.

Shell 20–30 mm high; sculpture of faint spiral lines crossed by obscure growth lines and oblique folds; inner lip of aperture with a strong ridge; shell brown to black, sometimes with spiral bands of dark color; aperture black.

This snail was accidentally introduced to California before 1911, along with oysters from the east coast. It is primarily an herbivorous deposit feeder, but dead animals such as worms, mollusks, crustaceans, and fishes are eaten if available. At least one San Francisco Bay population is also carnivorous, feeding on tiny burrowing polychaetes of the family Spionidae; these are rapidly clasped (by bringing together ventrally the anterior right and left sides of the foot) and are held firmly while the proboscis is brought down and the worm ingested by radular action (P. Milburn, pers. comm.). Some proteins in extracts of animal material are detected in the water in very low concentrations, and cause a marked feeding response in the snails. The anatomy of the digestive tract is typical of a carnivorous snail, yet the animals can subsist entirely on a diet of algal detritus. Hydrolytic enzymes for algal breakdown are produced, and absorption occurs in the midgut gland or stomach-intestine. A crystalline style is present.

On the east coast this species has been the subject of many biological studies. The eggs, which are deposited on eelgrass or on empty bivalve shells in shiny, transparent capsules, have long been considered excellent material for developmental studies. During early cleavage successive mitotic divisions require 5–30 minutes at 23°C. The blastomeres and polar lobes are easily isolated for microchemical analysis. During later development the embryos synthesize messenger RNA before and during formation of organ primordia.

After hatching, the planktonic larvae feed selectively on several species of phytoplankton. On the east coast *Cyclotella nana* is the food of choice, followed by species of *Phaeodactylum* and *Dunaliella*. The larvae settle out and metamorphose into tiny snails that feed mainly on bottom debris and detritus. Growth of as much as 1.5 mm in shell length per month occurs in the summer in New England, and animals may live to be 5 years old.

When these snails are found in groups or clusters they often exhibit a sort of "schooling" behavior, with members of a group showing a strong tendency to conform in direction of movement. On damp slopes they all move downhill; in clean seawater they move downstream; addition of tissue fluids from damaged animals causes the group to move away downstream or to bury themselves in sand. Isolated individuals are attracted to water that has passed over other individuals some distance away.

Nassarius obsoletus is commonly infected with larval trematodes, and when the locus of the infection is in the gonads the snail suffers parasitic castration.

See especially Brown (1969); see also Berg & Kato (1959), Cather (1963), Cheng, Sullivan & Harris (1973), Clement (1935, 1956), Collier (1965), Collier & McCann-Collier (1954), Copeland (1918), Costello & Henley (1971), Crisp (1969), Crowell (1944), Demond (1952), Dimon (1905), Griffith (1967), Gurin & Carr (1971), Hanna (1966), Hinegardner (1974), Jacobsen & Boylan (1971), Jenner (1958), Mangum, Kushins & Sassaman (1970), Morgan (1933), Paulson & Scheltema (1967), Raven (1966), Schaefer (1969), Scheltema (1964, 1967), Stephens, Sandeen & Webb (1953), Vassallo (1969), Vernberg, Vernberg & Beckerdite (1969), and Wicksten (1978).

Phylum Mollusca / Class Gastropoda / Subclass Prosobranchia
Order Neogastropoda / Superfamily Buccinacea / Family Nassariidae

13.96 Nassarius tegula (Reeve, 1853)
Covered-Lip Nassa

Exposed at low tide, on mud flats in bays and lagoons; San Francisco Bay to Laguna San Ignacio (Baja California); more common in southern part of range.

Shell to 19 mm high, sculptured with low spiral lines and with nodules on shoulders of whorls; outer lip of aperture thickened, inner lip with broad, white, smooth area or callus; shell gray-brown, often with a thin, purple, spiral band.

In Baja California, groups of *N. tegula* have been observed attacking the large bubble snail *Bullaria gouldiana*, and feeding on the soft parts. The species is also a scavenger, feeding on dead and often putrid animal material.

See Demond (1951, 1952) and J. McLean (1969).

Phylum Mollusca / Class Gastropoda / Subclass Prosobranchia
Order Neogastropoda / Superfamily Buccinacea / Family Fasciolariidae

13.97 Fusinus luteopictus (Dall, 1877)
Painted Spindle

On and under rocks, low intertidal zone; more common offshore to 40 m depth; Monterey Bay to Isla San Gerónimo (Baja California).

Shell to 26 mm high, spindle-shaped; sculpture of spiral cords and axial ribs, the ribs forming strong nodes on early whorls but both ribs and nodes weak or absent on later whorls; siphonal canal broad, well developed; shell dark brown, with a broad lighter band spiraling along middle of each whorl; tip of siphonal canal light in color.

Little is known of the biology of this species except that it deposits its eggs in capsules cemented to rocks.

See Chace (1916) and J. McLean (1969).

Phylum Mollusca / Class Gastropoda / Subclass Prosobranchia
Order Neogastropoda / Superfamily Volutacea / Family Olividae

13.98 Olivella baetica Carpenter, 1864
Little Olive, Beatic Olivella

Common in lagoons, low intertidal zone; abundant offshore on sandy bottoms in shallow water on open coast; Kodiak Island (Alaska) to Cabo San Lucas (Baja California).

Shell to 19 mm high, smaller and more slender than *O. biplicata* (13.99); columella with two folds at base; color generally gray-brown to tan, with purple-brown spots or bands below suture, occasionally with brown longitudinal lines.

This shallow burrower in rubble and soft substrata has been little studied compared to its relative *O. biplicata*, the purple olivella. A comparison of the two species should prove interesting.

Olivella baetica is preyed upon by offshore sea stars.

See Gifford & Gifford (1944), Griffith (1967), Herzberg (1966), and J. McLean (1969).

Phylum Mollusca / Class Gastropoda / Subclass Prosobranchia
Order Neogastropoda / Superfamily Volutacea / Family Olividae

13.99 Olivella biplicata (Sowerby, 1825)
Purple Olive, Purple Olivella

Abundant on sandy bottoms in lagoons, bays, and protected areas along open coast, low intertidal zone; subtidal to about 50 m; Vancouver Island (British Columbia) to Bahia Magdalena (Baja California).

Shell up to about 30 mm high, oval, smooth and shiny, with long aperture; columella with two folds; color varying from nearly all white to lavender or grayish black, occasionally yellow or orange, often purple at base, the base often set off by a dark line.

The purple olive commonly burrows just below the sand surface. It leaves a ploughed track on the surface behind it, and the snail's position may be revealed by a small sandy hummock. The white or cream-colored foot bears specialized regions at the front that extend up over part of the shell anteriorly and laterally when the animal is undisturbed. The large muscular foot, the streamlined and polished shell, and the active cilia and mucous glands on the exposed soft parts of the body all help to make this snail a very effective shallow burrower. A tubular siphon, consisting of a rolled-up fold of the mantle margin that passes through the notch in the aperture of the shell, can be extended above the surface of the sand as a snorkel for water intake. Other processes of the mantle margin include a dark flap arising near the base of the siphon and extending back over the top of the body whorl and a slender tentacle, which when extended winds about the spire of the shell to its tip in an external sutural groove. The function of these processes is uncertain; possibly they help to tell the animal whether the top of its shell is above or below the surface when it is burrowing.

Although some individual snails may be seen at the surface and others are well buried at any hour, the *Olivella* population does show a statistically consistent pattern of activity related to light and darkness. During the day most animals are buried in sand and inactive. They come up around dusk, and at night many are active at the sand surface. They start to bury themselves again well before dawn. Many animals also move up and down the beach with the tide; in general, the larger animals live higher up on the shore than the smaller ones. In many areas the purple olives congregate in clusters, and in Monterey harbor densities of 650 individuals per m² can occur.

This species is probably omnivorous; it has been seen eating blades of kelp (*Macrocystis*) and a variety of fresh and partially decayed animal material, such as the mollusks *Tegula funebralis* (13.32) and *Mytilus californianus* (15.9). Feeding on deposits of particulate detritus may also be important, for a strong ciliary current in a groove on the foot leads to the mouth and the gut often contains some sand.

In California, mating and spawning occur throughout the year. During courtship the males follow the mucus tracks of sexually ripe females. When contact is made, the male places the forward part of his foot on the female's shell and secretes a sticky mass of mucus, which bonds the two animals together. Coupling may last for 3 days and persists even if the animals are rolled in the surf or picked up.

Egg capsules are deposited separately on small stones, empty shells, etc. Each transparent capsule consists of an irregular basal portion and a dome-shaped cap about 0.5 mm in diameter, and contains only one clearly visible, white egg. An individual female has been observed to deposit as many as 4,236 egg capsules in a six-week period. The young develop to the veliger larval stage in the capsules. The development time is highly variable, even at constant temperature. Some young hatch 10 days after egg laying; others are still active inside their capsules after 28 days. Hatching young immediately swim toward the bottom and apparently remain on or near the bottom until metamorphosis into juvenile snails. The young reach a length of about 16 mm the first year. At this size they are sexually mature. They increase in length 1–5 mm per year thereafter. Males grow somewhat faster than females and reach a larger size. The average life span appears to be between 8 and 15 years.

Purple olives are fed upon by octopuses (which drill holes in the shells near the apex); by the snails *Conus californicus* (13.103), *Polinices reclusianus* (13.63), and *Polinices lewisii* (13.62); by the sea stars *Pisaster brevispinus* (8.15) and *Astropecten armatus* (8.1); by gulls and other shorebirds; and probably by crabs and fishes. Man also collects these snails for food and ornamentation; formerly the shells were used in burials and as money by some California coastal Indians.

Olivella exhibits a variety of avoidance and escape responses when confronted with the sea stars *Pisaster brevispinus* (8.15), *P. ochraceus* (8.13), and *Pycnopodia helianthoides* (8.16). The snail senses the presence of a sea star by detecting waterborne substances that diffuse from its body. On a natural sandy bottom the snails avoid contact by turning away or burrowing into the sand. If the tube feet of the sea star make contact with *Olivella*, the snail thrusts out its foot and digs downward very rapidly. In a laboratory aquarium, with no opportunity to burrow, the snail may crawl rapidly away from a sea star, or rear up and fall over backward. This somersaulting is accomplished by a flapping movement of the anterolateral lobes of the foot, which may be so violent that the snail is lifted off the bottom and carried some 5–10 cm away.

In some areas the purple olives are parasitized by trematode larvae in the gonads. Heavy infestation can result in parasitic castration of up to 50 percent of the individuals.

Significant levels of certain metals have been found in the tissues of *Olivella* in the Monterey area: copper and lead are relatively high in animals living near Fisherman's Wharf; silver, cadmium, and zinc in those off the Monterey sewage outfall; and iron in animals from Elkhorn Slough. The animals in all cases appeared normal.

See Burch & Burch (1963), Curtis (1966), Duerr (1965), Edwards (1965, 1968, 1969a,b), Emerson (1965), Farmer (1970), Gifford & Gifford (1941, 1944), Griffith (1967), Hinegardner (1974), Krinsley (1959), Marcus & Marcus (1959), Orr (1941), Phillips (1977b,c), Schwimer (1973), Stohler (1952, 1959a,b, 1960, 1962a, 1969a), Valentine & Meade (1960), and Zell (1955).

Phylum Mollusca / Class Gastropoda / Subclass Prosobranchia
Order Neogastropoda / Superfamily Volutacea / Family Mitridae

13.100 **Mitra idae** Melville, 1893
Ida's Miter

Uncommon among rocky rubble, low intertidal zone; more commonly subtidal under kelp; Crescent City (Del Norte Co.) to Isla Cedros (Baja California).

Shell to nearly 80 mm high, heavy, elongate, with high spire; sculpture of fine axial and spiral lines; columella with three strong folds; color dark brown, under a thick, black periostracum; soft parts white.

This beautiful animal is a specialized carnivore. The snails are relatively sluggish in laboratory aquaria and can persist for several weeks without macroscopic food.

In southern California, breeding occurs from February to July, and in Monterey, spawning has been observed in May. Prior to copulation the female releases a mass of mucus that appears to attract males. While mating, the females remain attached to the substratum, and two males have been observed to copulate simultaneously with a single female. The female may also be spawning during the mating period.

Eggs are deposited on rocks in flattened, elongate, yellow or transparent capsules about 8 mm high. In the laboratory one female was observed to deposit 91 egg capsules in a 4-day period, and another 173 capsules over a period of 17 days. Capsules contain from 100 to nearly 1,000 eggs, each egg being about 0.2 mm in diameter. At 13°C, the first cleavage of the fertilized egg occurs 5 hours after capsule deposition, mobile trochopores are developed after 10 days, and preveligers by 17–19 days. Veligers escape from capsules 24 days after deposition. At temperatures of 17.5 – 18.8°C the early developmental stages are speeded up somewhat, but the hatching of veligers is delayed until 26–27 days after capsule deposition.

See Allison (1959), Cate (1967, 1968), Cernohorsky (1966), Chess & Rosenthal (1971), Coan (1966), Herzberg (1966), Hinegardner (1974), Kellogg & Lindberg (1975), and J. McLean (1969).

Phylum Mollusca / Class Gastropoda / Subclass Prosobranchia
Order Neogastropoda / Superfamily Volutacea / Family Marginellidae

13.101 **Volvarina taeniolata** Mörch, 1860
Banded California Marginella

Common on undersides of rocks, low intertidal pools and subtidal waters; Point Conception (Santa Barbara Co.) to Ecuador.

Shell up to 10 mm high, shiny, with long aperture and four columellar folds; outer lip of aperture smooth and thickened; color ivory or yellow-orange, with lighter, wide, spiral banding.

Little is known of the biology of this common small snail.

See Coan & Roth (1966) and J. McLean (1969).

Phylum Mollusca / Class Gastropoda / Subclass Prosobranchia
Order Neogastropoda / Superfamily Volutacea / Family Turridae

13.102 **Pseudomelatoma torosa** (Carpenter, 1865)
Knobbed Drill

Uncommon under rocks, low intertidal zone and offshore to 80 m depth; possibly Oregon, certainly Monterey Bay to Laguna San Ignacio (Baja California).

Shell to 30 mm high, tall-spired, with broad axial ribs, sinuous growth lines, and a spiral row of knobs; color dark brown to black, the knobs lighter.

Nothing is known of the biology of this species, but x-ray photographs of the shell have been published.

See Herzberg (1966) and Powell (1966).

Phylum Mollusca / Class Gastropoda / Subclass Prosobranchia Order Neogastropoda / Superfamily Volutacea / Family Conidae

13.103 Conus californicus Hinds, 1844
California Cone

Common to scarce on rocky and sandy bottoms, low intertidal and subtidal zones to 30 m; Farallon Islands (San Francisco Co.) to Bahía Magdalena (Baja California).

Shell to 40 mm high, conical, with a low spire and narrow aperture; sculpture of fine spiral lines; color grayish brown, with a chestnut-brown periostracum.

The cone snails form a large group. Many are spectacularly colored and most live in tropical seas. Only this single species is found in California. Rather drab in color, its behavior is anything but dull. In a remarkable series of adaptations, the cones have specialized in the use of poison darts to slay animal prey, including even such active forms as small fishes.

The radula in species of *Conus* is reduced to a series of separate spearlike teeth, which are secreted within the radular sac and passed up one at a time, as needed, to the back of the buccal cavity. Electron microscope studies of the teeth show that each is a thin sheet of horny material rolled into a tube of several whorls, the outer whorl being flattened and pointed at the apex and possessing one or more barbs along the edge. A tooth of *C. californicus* differs from that of other species living farther south in having five barbs, one of which terminates in a short blade.

A hunting *Conus* locates its food from a distance by scent. It moves toward the prey, extending its proboscis as it does. Inside, near its tip, the extended proboscis holds a single hollow tooth whose base is juxtaposed to the opening of the venom gland. The snail touches its prey very gently with the tip of the proboscis, then a sudden internal muscular contraction

thrusts in the harpoon. The poison is a protein. The prey is quickly paralyzed and then enveloped by the proboscis, which expands to take in large food items in a manner reminiscent of a python swallowing its prey.

Conus californicus remains buried in the sand when not feeding, but emerges to capture a great variety of living prey, including gastropods such as *Olivella biplicata* (13.99), *Nassarius fossatus* (13.92), and *Bulla gouldiana* (14.6), bivalves such as *Macoma nasuta* (15.50), *Tagelus californianus* (15.58), and *Solen rosaceus* (15.62), several genera of polychaetes, and many other forms (Kohn, 1966). It has also been observed feeding on dead fishes and octopuses. Reported items of diet include members of six classes in four animal phyla.

In the La Jolla area (San Diego Co.) subtidal populations of this species have a symbiotic relationship with a small brachyuran crab. About 10 percent of the cones carry immature specimens of the crab *Opisthopus transversus* (25.42) lodged in the mantle cavity behind the gill.

The eggs of *C. californicus* are deposited in elaborate capsules, and the larvae are released in about 10 days. The beating of the cilia on the veligers is strongly influenced by such drugs as acetylcholine and serotonin, and the larvae have been used as assay organisms in pharmacological experiments.

See Alexander (1973), Clench & Kondo (1943), Curtis (1966), Hanna (1963), Hanna & Strong (1949), Hinegardner (1958, 1974), Kohn (1956, 1959, 1966), Kohn, Nybakken & VanMol (1972), Kohn, Saunders & Wiener (1960), G. MacGinitie & N. MacGinitie (1968), Nybakken (1967, 1970), Peile (1939), Saunders (1959), Saunders & Wolfson (1961), Vilkas (1974), and Wolfson (1974).

Literature Cited

Abbott, D. P. 1956. Water circulation in the mantle cavity of the owl limpet *Lottia gigantea* Gray. Nautilus 69: 79–87.

Abbott, D. P., L. R. Blinks, and J. H. Phillips. 1964. An experiment in undergraduate teaching and research in the biological sciences. Veliger 6 (Suppl.): 1–6.

Abbott, D. P., L. R. Blinks, J. H. Phillips, and R. Stohler, eds. 1964. The biology of *Tegula funebralis* (A. Adams, 1855). Veliger 6 (Suppl.): 1–82.

Abbott, D. P., D. Epel, J. H. Phillips, I. A. Abbott, and R. Stohler, eds. 1968. The biology of *Acmaea*. Veliger 11 (Suppl.): 1–112.

Abbott, R. T. 1974. American seashells. 2nd ed. New York: Van Nostrand Reinhold. 663 pp.

Ahmed, M., and A. K. Sparks. 1970. A note on the chromosome number and interrelationships in the marine gastropod genus *Thais* of the United States Pacific coast. Veliger 12: 293–94.

Alexander, C. G. 1973. The neuroanatomy of the osphradium in *Conus flavidus* Lamarck. Veliger 16: 68–71.

Alleman, L. L. 1968. Factors affecting the attraction of *Acmaea asmi* to *Tegula funebralis*. Veliger 11 (Suppl.): 61–63.

Allison, E. C. 1959. *Mitra montereyi* Berry from the Monterey Peninsula. Veliger 2: 20.

Anderlini, V. 1974. The distribution of heavy metals in the red abalone, *Haliotis rufescens*, on the California coast. Arch. Environ. Cont. Toxicol. 2: 253–65.

Andrews, H. L. 1945. The kelp beds of the Monterey region. Ecology 26: 24–37.

Ankel, W. E. 1936. Die Frasspuren von *Helcion* und *Littorina* und die Funktion der Radula. Verh. Deutsch. Zool. Ges., Zool. Anz. Suppl. 9: 174–82.

———. 1937. Wie frisst *Littorina*? I. Radula-Bewegung und Fresspuren. Senckenbergiana 19: 317–33.

Baldwin, S. 1968. Manometric measurements of respiratory activity in *Acmaea digitalis* and *Acmaea scabra*. Veliger 11 (Suppl.): 79–82.

Ball, E. G., and B. Meyerhof. 1940. The occurrence of cytochrome and other haemochromogens in certain marine forms. Biol. Bull. 77: 321.

Baribault, W. H. 1968. Nitrogen excretory products in the limpet *Acmaea*. Veliger 11 (Suppl.): 109–12.

Bartsch, P. 1940. The western American *Haliotis*. Proc. U.S. Nat. Mus. 89: 49–59.

Behrens, D. W. 1979. Ambicoloration in the black abalone, *Haliotis cracherodii* Leach. Calif. Fish & Game 65: 54–55.

Behrens, S. 1974. Ecological interactions of three species of *Littorina* (Gastropoda: Prosobranchia) along the west coast of North America. Master's thesis, University of Oregon, Eugene.

Belcik, F. P. 1965. Note on a range extension and observations of spawning in *Tegula*, a gastropod. Veliger 7: 233–34.

Bender, J. A., K. DeRiemer, T. E. Roberts, R. Rushton, P. Boothe, H. S. Moser, and F. A. Fuhrman. 1974. Choline esters in the marine gastropods *Nucella emarginata* and *Acanthina spirata*. Comp. Gen. Pharmacol. 5: 191–98.

Bennett, R. J., and H. I. Nakada. 1968. Comparative carbohydrate metabolism of marine molluscs. I. The intermediary metabolism of *Mytilus californianus* and *Haliotis rufescens*. Comp. Biochem. Physiol. 24: 787–97.

Beppu, W. J. 1968. A comparison of carbohydrate digestion capabilities in four species of *Acmaea*. Veliger 11 (Suppl.): 98–101.

Bequaert, J. 1942. Random notes on American Potamididae. Nautilus 56: 20–30.

Berg, W. E. 1967. Some experimental techniques for eggs and embryos of marine invertebrates, pp. 767–76, in F. H. Wilt and N. K. Wessells, eds., Methods in developmental biology. New York: Crowell. 813 pp.

Berg, W. E., and Y. Kato. 1959. Localization of polynucleotides in the egg of *Illyanassa*. Acta Embryol. Morphol. Exper. 2: 227–33.

Bergen, M. 1971. Growth, feeding and movement in the black abalone (*Haliotis cracherodii* Leach, 1814). Master's thesis, University of California, Santa Barbara.

Bernard, F. R. 1967. Studies on the biology of the naticid clam drill *Polinices lewisi* (Gould) (Gastropoda, Prosobranchiata). Fish. Res. Bd. Canada, Tech. Rep. 42: 1–44.

Berrie, W. R., and M. W. Devereaux. 1964. Identification and location of carbohydrases in the intestinal tract of *Tegula funebralis* (Mollusca: Gastropoda). Veliger 6 (Suppl.): 20–24.

Berry, S. S. 1955. The West Coast's confused and confusing white slipper shells, *Crepidula,* subgenus *Ianacus*. Ann. Rep. Amer. Malacol. Union. Bull. 22: 32.

Bertness, M. D. 1977. Behavioral and ecological aspects of shore-level size gradients in *Thais lamellosa* and *Thais emarginata*. Ecology 58: 86–97.

Bertness, M. D., and D. E. Schneider. 1976. Temperature relations of Puget Sound thaids in reference to their intertidal distribution. Veliger 19: 47–58.

Best, B. A. 1964. Feeding activities of *Tegula funebralis* (Mollusca: Gastropoda). Veliger 6 (Suppl.): 42–45.

Bigler, E. 1964. Attrition on the *Littorina planaxis* population. Research paper, Biol. 175h, library, Hopkins Marine Station of Stanford University, Pacific Grove, Calif. 28 pp.

Bingham, F. O. 1972. The mucus holdfast of *Littorina irrorata* and its relationship to relative humidity and salinity. Veliger 15: 48–50.

Black, R. 1976. The effects of grazing by the limpet, *Acmaea incessa*, on the kelp, *Egregia laevigata*, in the intertidal zone. Ecology 57: 267–77.

Blake, J. W. 1960. Oxygen consumption of bivalve prey and their attractiveness to the gastropod *Urosalpinx cinerea*. Limnol. Oceanogr. 5: 273–80.

Bock, C. E., and R. E. Johnson. 1967. The role of behavior in determining the intertidal zonation of *Littorina planaxis* Philippi, 1847, and *Littorina scutulata* Gould, 1849. Veliger 10: 42–54.

Bollay, M. 1964. Distribution and utilization of gastropod shells by the hermit crabs *Pagurus samuelis, Pagurus granosimanus*, and *Pagurus hirsutiusculus* at Pacific Grove, California. Veliger 6 (Suppl.): 71–76.

Bonar, Lee. 1936. An unusual ascomycete in the shells of marine animals. Univ. Calif. Publ. Botany 19: 187–93.

Bonnot, P. 1930. Abalones in California. Calif. Fish & Game 16: 15–23.

_____. 1940. California abalones. Calif. Fish & Game 26: 200–211.

_____. 1948. The abalone of California. Calif. Fish & Game 34: 141–69.

_____. 1949. Abalone, pp. 160–61, *in* The commercial fish catch of California for the year 1947 with an historical review 1916–1947. Calif. Div. Fish & Game, Fish Bull. 74: 1–267.

Boolootian, R. A., A. Farmanfarmaian, and A. C. Giese. 1962. On the reproductive cycle and breeding habits of two western species of *Haliotis*. Biol. Bull. 122: 183–93.

Bormann, M. 1946. A survey of some west American *Ocenebras,* with description of a new species. Nautilus 60: 37–43, pl. 4.

Boss, K. J. 1971. Critical estimate of the number of recent Mollusca. Occas. Pap. Mollusks, Mus. Comp. Zool. Harvard 40: 81–135.

Brant, D. H. 1950. A quantitative study of the homing behavior of the limpet *Acmaea scabra*. Spec. Prob. Rep. (unpublished). Zoology, University of California, Berkeley.

Breen, Paul A. 1971. Homing behavior and population regulation in the limpet *Acmaea* (*Collisella*) *digitalis*. Veliger 14: 177–83.

_____. 1972. Seasonal migration and population regulation in the limpet *Acmaea* (*Collisella*) *digitalis*. Veliger 15: 133–41.

Brewer, B. A. 1975. Epizoic limpets on the black turban snail *Tegula funebralis* (A. Adams, 1855). Veliger 17: 307–10.

Bright, D. B. 1958. Morphology of the common mudflat snail, *Cerithidea californica*. Bull. So. Calif. Acad. Sci. 57: 127–39.

_____. 1960. Morphology of the common mudflat snail, *Cerithidea californica* II. Bull. So. Calif. Acad. Sci. 59: 9–15.

Brown, S. C. 1969. The structure and function of the digestive system of the mud snail *Nassarius obsoletus* (Say). Malacologia 9: 447–500.

Bukland-Nicks, J. A., and F.-S. Chia. 1976. Fine structural observations of sperm resorption in the seminal vesicle of a marine snail, *Littorina scutulata* (Gould, 1849). Cell Tiss. Res. 172: 503–15.

Bulkley, P. T. 1968. Shell damage and repair in five members of the genus *Acmaea*. Veliger 11 (Suppl.): 64–66.

Bullock, T. H. 1953. Predator recognition and escape responses of some intertidal gastropods in presence of starfish. Behavior 5: 130–40.

Burch, J. Q., and R. L. Burch. 1963. Genus *Olivella* in the Eastern Pacific. Nautilus 77: 1–29.

Burke, W. R. 1964. Chemoreception by *Tegula funebralis* (Mollusca: Gastropoda). Veliger 6 (Suppl.): 17–20.

Campbell, J. L. 1965. The structure and function of the alimentary canal of the black abalone, *Haliotis cracherodii* Leach. Trans. Amer. Microscop. Soc. 84: 376–95.

Carlisle, J. G., Jr. 1945. The technique of inducing spawning in *Haliotis rufescens* Swainson. Science 102: 566–67.

_____. 1962. Spawning and early life history of *Haliotis rufescens* Swainson. Nautilus 76: 44–49.

Carlton, J. T. 1969. *Littorina littorea* in California (San Francisco and Trinidad Bays). Veliger 11: 283–84.

Carlton, J. T., and B. Roth. 1975. Phylum Mollusca: Shelled gastropods, pp. 467–514, *in* R. I. Smith and J. T. Carlton, eds., Light's manual: Intertidal invertebrates of the central California coast. 3rd ed. Berkeley and Los Angeles: University of California Press. 716 pp.

Carriker, M. R. 1943. On the structure and function of the proboscis in the common oyster drill *Urosalpinx cinerea* Say. J. Morphol. 73: 441–506.

_____. 1951. Observations on the penetration of tightly closing bivalves by *Busycon* and other predators. Ecology 32: 73–83.

_____. 1955. Critical review of biology and control of oyster drills *Urosalpinx* and *Eupleura*. U.S. Fish & Wildlife Service, Spec. Sci. Rep. Fish 148: 1–150.

_____. 1957. Preliminary study of the behavior of newly hatched oyster drills, *Urosalpinx cinerea* (Say). J. Elisha Mitchell Sci. Soc. 73: 328–51.

_____. 1959. Comparative functional morphology of the drilling mechanism in *Urosalpinx* and *Eupleura* (muricid gastropods). Proc. 15th Internat. Congr. Zool., London, pp. 373–76.

_____. 1961. Comparative functional morphology of boring mechanisms in gastropods. Amer. Zool. 1: 263–66.

_____. 1969. Excavation of boreholes by the gastropod, *Urosalpinx*: An analysis by light and scanning electron microscopy. Amer. Zool. 9: 917–33.

_____. 1971. Dissolution of the shell of *Mytilus* by the accessory boring organ of *Urosalpinx*. Amer. Zool. 11: 699.

Carriker, M. R., J. G. Schaadt, and V. Peters. 1974. Analysis by slow-motion picture photography and scanning electron microscopy of radular function in *Urosalpinx cinerea follyensis* (Muricidae, Gastropoda) during shell penetration. Mar. Biol. 25: 63–76.

Carriker, M. R., D. B. Scott, and G. N. Martin, Jr. 1963. Demineralization mechanism of boring gastropods, pp. 55–89, *in* R. F. Sognannaes, ed., Mechanisms of hard tissue destruction, publ. 75. Washington, D.C.: American Association for the Advancement of Science.

Carriker, M. R., and D. Van Zandt. 1967. Gastropod *Urosalpinx*: pH of accessory boring organ while boring. Science 158: 920–22.

_____. 1972. Predatory behavior of a shell-boring muricid gastropod, pp. 157–244, *in* H. E. Winn and B. L. Oila, eds., Behavior of marine animals, 1, Invertebrates. New York: Plenum. 244 pp.

Carriker, M. R., D. Van Zandt, and T. J. Grant. 1978. Penetration of molluscan and non-molluscan minerals by the boring gastropod *Urosalpinx cinerea*. Biol. Bull. 155: 511–26.

Castenholz, R. W. 1961. The effect of grazing on marine littoral diatom populations. Ecology 42: 783–94.

Cate, J. M. 1967. The radulae of nine species of Mitridae. Veliger 10: 192–95.

_____. 1968. Mating behavior in *Mitra idae* Melvill, 1893. Veliger 10: 247–52.

Cather, J. N. 1963. A time schedule of the meiotic and early mitotic stages of *Ilyanassa*. Caryologia 16: 663–70.

Cernohorsky, W. O. 1966. A study of mitrid radulae and a tentative generic arrangement of the family Mitridae (Mollusca: Gastropoda). Veliger 9: 101–26.

––––––. 1968. Observations on *Hipponix conicus* (Schumacher, 1817). Veliger 10: 275–80.

Chace, E. P. 1916. Egg-cases of some local gastropoda. Lorquina 1: 9–10.

Chapin, D. 1968. Some observations of predation on *Acmaea* species by the crab *Pachygrapsus crassipes*. Veliger 11 (Suppl.): 67–68.

Chapman, W. M., and A. H. Banner. 1949. Contributions to the life history of the Japanese oyster drill, *Tritonalia japonica*, with notes on other enemies of the Olympia oyster, *Ostrea lurida*. State of Washington Dept. Fish., Biol. Bull. 49-A: 169–200.

Cheng, T. C., J. T. Sullivan, and K. R. Harris. 1973. Parasitic castration of the marine prosobranch gastropod *Nassarius obsoletus* by sporocysts of *Zoogonus rubellus* (Trematoda): Histopathology. J. Invert. Pathol. 21: 183–90.

Chess, J. R., and R. J. Rosenthal. 1971. On the reproductive biology of *Mitra idae* (Gastropoda: Mitridae). Veliger 14: 172–76.

Chow, V. 1975. The importance of size in the intertidal distribution of *Littorina scutulata* (Gastropoda: Prosobranchia). Veliger 18: 69–78.

Clement, A. C. 1935. The formation of giant polar bodies in centrifuged eggs of *Ilyanassa*. Biol. Bull. 69: 403–14.

––––––. 1956. Experimental studies on germinal localization in *Ilyanassa*. II. The development of isolated blastomeres. J. Exper. Zool. 132: 427–45.

Clench, W. J., and Y. Kondo. 1943. The poison cone shell. Amer. J. Trop. Med. 23: 105–20.

Coan, E. 1966. Nomenclatural units in the gastropod family Mitridae. Veliger 9: 127–37.

Coan, E., and B. Roth. 1966. The west American Marginellidae. Veliger 8: 276–99.

Coe, W. R. 1942a. Influence of natural and experimental conditions in determining shape and rate of growth in gastropods of the genus *Crepidula*. J. Morphol. 71: 35–51.

––––––. 1942b. The reproductive organs of the prosobranch mollusk *Crepidula onyx* and their transformation during the change from male to female phase. J. Morphol. 70: 501–12.

––––––. 1949. Divergent methods of development in morphologically similar species of prosobranch gastropods. J. Morphol. 84: 383–400.

––––––. 1953. Influences of association, isolation, and nutrition on the sexuality of snails of the genus *Crepidula*. J. Exper. Zool. 122: 5–20.

Cole, H. A. 1942. The American whelk tingle, *Urosalpinx cinerea* (Say), on British oyster beds. J. Mar. Biol. Assoc. U.K. 25: 477–508.

Collier, J. R. 1965. Ribonucleic acids of the *Ilyanassa* embryo. Science 147: 150–51.

Collier, J. R., and M. McCann-Collier. 1954. Shell gland formation in the *Ilyanassa* embryo. Exper. Cell. Res. 34: 512–14.

Collins, L. S. 1976. Abundance, substrate angle, and desiccation resistance in two sympatric species of limpets. Veliger 19: 199–203.

––––––. 1977. Substrate angle, movement and orientation of two sympatric species of limpets, *Collisella digitalis* and *Collisella scabra*. Veliger 20: 43–48.

Colton, H. S. 1908. How *Fulgur* and *Sycotypus* eat oysters, mussels, and clams. Proc. Acad. Natur. Sci. Philadelphia 60: 3–10.

Comfort, A. 1949. Acid-soluble pigments of molluscan shells. 3. The indigoid character of the blue pigment of *Haliotis cracherodii* Leach. Biochem. J. 45: 204–8.

––––––. 1950. The pigmentation of molluscan shells. Biol. Rev. 26: 285–301.

Conklin, E. G. 1897. The embryology of *Crepidula*. J. Morphol. 13: 1–226.

––––––. 1907. The embryology of *Fulgur*; a study of the influence of yolk on development. Proc. Acad. Natur. Sci. Philadelphia 59: 320–59.

Connell, J. H. 1970. A predator-prey system in the marine intertidal region. I. *Balanus glandula* and several predatory species of *Thais*. Ecol. Monogr. 40: 49–78.

Cooke, A. H. 1918. On the radula of the genus *Acanthina* G. Fischer. Proc. Malacol. Soc. London 13: 6–11.

Cooper, J., M. Wieland, and A. Hines. 1977. Subtidal abalone populations in an area inhabited by sea otters. Veliger 20: 163–67.

Copeland, M. 1918. The olfactory reactions and organs of the marine snails *Alectrion obsoleta* (Say) and *Busycon canaliculatum* (Linn.). J. Exper. Zool. 25: 177–227.

––––––. 1922. Ciliary and muscular locomotion in the gastropod genus *Polinices*. Biol. Bull. 42: 132–42.

Costello, D. P., and C. Henley. 1971. Methods for obtaining and handling marine eggs and embryos. 2nd ed. Marine Biological Laboratory, Woods Hole, Mass. 247 pp.

Cowan, I. M. 1964. New information on the distribution of marine Mollusca on the coast of British Columbia. Veliger 7: 110–13.

––––––. 1974. The west American Hipponicidae and the application of *Malluvium*, *Antisabia*, and *Hipponix* as generic names. Veliger 16: 377–80.

Cox, K. W. 1962. California abalones, Family Haliotidae. Calif. Dept. Fish & Game, Fish Bull. 118: 1–133.

Craig, P. S. 1968. The activity pattern and food habits of the limpet *Acmaea pelta*. Veliger 11 (Suppl.): 13–19.

Crane, J. M., Jr. 1969. Mimicry of the gastropod *Mitrella carinata* by the amphipod *Pleustes platypa*. Veliger 12: 200.

Crisp, M. 1969. Studies on the behavior of *Nassarius obsoletus* (Say) (Mollusca: Gastropoda). Biol. Bull. 136: 355–73.

Crofts, D. R. 1929. *Haliotis*. Liverpool Mar. Biol. Comm. [LMBC] Memoir 29. University Press of Liverpool. 174 pp.

Crowell, J. 1944. The fine structure of the polar lobe of *Ilyanassa obsoleta*. Acta Embryol. Morphol. Exper. 7: 225–34.

Curtis, F. 1966. Molluscan species from early southern California archeological sites. Bull. So. Calif. Acad. Sci. 65: 107–27.

Dahl, A. L. 1964. Macroscopic algal food of *Littorina planaxis* Philippi and *Littorina scutulata* Gould (Gastropoda: Prosobranchiata). Veliger 7: 139–43.

Dakin, W. J. 1912. *Buccinum* (the whelk). Liverpool Mar. Biol. Comm. [LMBC] Memoir 20. University Press of Liverpool. 115 pp.

Dall, W. H. 1915. Notes on the species of the molluscan subgenus *Nucella* inhabiting the northwest coast of America and adjacent regions. Proc. U.S. Nat. Mus. 49: 74–75.

Darby, R. L. 1964. On growth and longevity in *Tegula funebralis* (Mollusca: Gastropoda). Veliger 6 (Suppl.): 6–7.

Darling, S. D. 1965. Observations on the growth of *Cypraea spadicea*. Veliger 8: 14–15.

Dayton, P. K. 1971. Competition, disturbance, and community organization: The provision and subsequent utilization of space in a rocky intertidal community. Ecol. Monogr. 41: 351–89.

Dehnel, P. A. 1955. Rates of growth of gastropods as a function of latitude. Physiol. Zool. 28: 115–44.

Demond, J. 1951. Key to the Nassariidae of the west coast of North America. Nautilus 65: 15–17.

———. 1952. The Nassariidae of the west coast of North America between Cape San Lucas, Lower California, and Cape Flattery, Washington. Pacific Sci. 6: 300–317, 2 pls.

Diehl, M. 1956. Die Raubschnecke *Velutina velutina* das Feind und Bruteinmieter der Ascidie *Styela coriacea*. Kieler Meeresforsch. 12: 180–85.

Dimock, R. V., Jr., and J. G. Dimock. 1969. A possible "defense" response in a commensal polychaete. Veliger 12: 65–68.

Dimon, A. C. 1905. The mud snail: *Nassa obsoleta*. Cold Spring Harbor Monogr. 5: 1–45.

Dinter, I. 1974. Pheromonal behavior in the marine snail *Littorina littorea* Linnaeus. Veliger 17: 37–39.

Dinter, I., and P. J. Manos. 1972. Evidence for a pheromone in the marine periwinkle *Littorina littorea* Linnaeus. Veliger 15: 45–47.

Donohue, J. 1965a. *Cypraea*: A list of the species. Veliger 7: 219–24.

———. 1965b. Size ranges in *Cypraea*. Veliger 8: 22, 1 fig.

Doran, S. R., and D. S. McKenzie. 1972. Aerial and aquatic respiratory responses to temperature variations in *Acmaea digitalis* and *Acmaea fenestrata*. Veliger 15: 38–42.

Driscoll, A. L. 1972. Structure and function of the alimentary tract of *Batillaria zonalis* and *Cerithidea californica*, style-bearing mesogastropods. Veliger 14: 375–86.

Duerr, F. G. 1965. Survey of digenetic trematode parasitism in some prosobranch gastropods of the Cape Arago region, Oregon. Veliger 8: 42.

———. 1967. Patterns of nitrogen excretion in marine prosobranch snails. Amer. Zool. 7: 785 (abstract).

DuShane, H. 1974. The Panamic-Galápagan Epitoniidae. Veliger 16 (Suppl.): 1–84.

Duwe, A. E. 1954. The morphology of the eye of *Busycon canaliculatum*, a large marine gastropod. Turtox News 32: 50–51.

Eales, N. B. 1950. Secondary symmetry in gastropods. Proc. Malacol. Soc. London 28: 185–96.

Eaton, C. M. 1968. The activity and food of the file limpet *Acmaea limatula*. Veliger 11 (Suppl.): 5–12.

Edwards, C. L. 1913. The abalone industry in California. Calif. Div. Fish & Game, Fish Bull. 1: 1–15.

Edwards, D. C. 1965. Distribution patterns within natural populations of *Olivella biplicata*, their underlying behavioral mechanisms, and their ecological significance. Doctoral thesis, University of Chicago.

———. 1968. Reproduction in *Olivella biplicata*. Veliger 10: 297–304.

———. 1969a. Predators on *Olivella biplicata*, including a species-specific predator-avoidance response. Veliger 11: 326–33.

———. 1969b. Zonation by size as an adaption for intertidal life in *Olivella biplicata*. Amer. Zool. 9: 399–417.

Eikenberry, A. B., Jr., and D. E. Wickizer. 1964. Studies on the commensal limpet *Acmaea asmi* in relation to its host, *Tegula funebralis* (Mollusca: Gastropoda). Veliger 6 (Suppl.): 66–70.

Emerson, D. N. 1965. Summer polysaccharide content in seven species of west coast intertidal prosobranch snails. Veliger 8: 62–66.

Emlen, J. M. 1966. Time, energy and risk in two species of carnivorous gastropods. Master's thesis, University of Washington, Seattle.

Erickson, A. D. 1964. Proteolytic enzymes in the gastropod *Tegula funebralis*. Veliger 6 (Suppl.): 14–16.

Farmer, W. M. 1970. Swimming gastropods (Opisthobranchia and Prosobranchia). Veliger 13: 73–89.

Feder, H. M. 1963. Gastropod defensive responses and their effectiveness in reducing predation by starfishes. Ecology 44: 505–12.

———. 1967. Organisms responsive to predatory sea stars. Sarsia 29: 371–94.

———. 1972. Escape responses in marine invertebrates. Sci. Amer. 227: 92–100.

Feder, H. M., and R. Lasker. 1964. Partial purification of a substance from starfish tube feet which elicits escape responses in gastropod molluscs. Life Sci. 3: 1047–51.

———. 1968. A radula muscle preparation from the gastropod *Kelletia kelletia*, for biochemical assays. Veliger 10: 283–85.

Federighi, H. 1929. Rheotropism in *Urosalpinx cinerea* (Say). Biol. Bull. 56: 331–40.

———. 1931a. Salinity death-points of the oyster drill snail *Urosalpinx cinerea* Say. Ecology 12: 346–53.

———. 1931b. Studies on the oyster drill (*Urosalpinx cinerea*, Say). Bull. U.S. Bur. Fish. 47: 85–115.

Field, L. H. 1974. A description and experimental analysis of Batesian mimicry between a marine gastropod and an amphipod. Pacific Sci. 28: 439–47.

Fitch, J. E. 1963. The enigma of a hitch-hiking snail. Leaflets in Malacol. 1: 131–32.

Fisher, W. K. 1904. The anatomy of *Lottia gigantea* Gray. Zool. Jahrb. (Anat.) 20: 1–66.

Foster, M. S. 1964. Microscopic algal food of *Littorina planaxis* Philippi and *Littorina scutulata* Gould (Gastropoda: Prosobranchiata). Veliger 7: 149–52.

Fotheringham, N. 1971. Life history patterns of the littoral gastropods *Shaskyus festivus* (Hinds) and *Ocenebra poulsoni* Carpenter. (Prosobranchia: Muricidae). Ecology 52: 742–57.

Fox, D. L. 1960. Perspectives in marine biochemistry, pp. 617–21, *in* R. F. Nigrelli, ed., Biochemistry and pharmacology of compounds derived from marine organisms. Ann. N.Y. Acad. Sci. 90: 615–950.

Frank, P. W. 1964. On home range of limpets. Amer. Natur. 98: 99–104.

———. 1965a. The biodemography of an intertidal snail population. Ecology 46: 831–44.

———. 1965b. Growth of three species of *Acmaea*. Veliger 7: 201–2.

———. 1965c. Shell growth in a natural population of the turban snail *Tegula funebralis*. Growth 29: 395–403.

———. 1969. Sexual dimorphism in *Tegula funebralis*. Veliger 11: 440.

———. 1975. Latitudinal variation in the life history features of the black turban snail *Tegula funebralis* (Prosobranchia: Trochidae). Mar. Biol. 31: 181–92.

Franz, D. R. 1971. Population age structure, growth and longevity of the marine gastropod *Urosalpinx cinerea* Say. Biol. Bull. 140: 63–72.

Fretter, V. 1951. Some observations on the British cypraeids. Proc. Malacol. Soc. London 29: 14–20.

Fretter, V., and A. Graham. 1962. British prosobranch molluscs. Ray Soc., London, 26: 1–755.

Fritchman, H. K., II. 1961a. A study of the reproductive cycle in the California Acmaeidae (Gastropoda). Part I. Veliger 3: 57–67.

———. 1961b. A study of the reproductive cycle in the California Acmaeidae (Gastropoda). Part II. Veliger 3: 95–101.

———. 1961c. A study of the reproductive cycle in the California Acmaeidae (Gastropoda). Part III. Veliger 4: 41–47.

———. 1962. A study of the reproductive cycle in the California Acmaeidae (Gastropoda). Part IV. Veliger 4: 134–40.

———. 1965. The radulae of *Tegula* species from the west coast of North America and suggested intrageneric relationships. Veliger 8: 11–14.

Galbraith, R. T. 1965. Homing behavior in the limpets *Acmaea digitalis* and *Lottia gigantea*. Amer. Midl. Natur. 74: 245–46.

Galli, D. R., and A. C. Giese. 1959. Carbohydrate digestion in a herbivorous snail, *Tegula funebralis*. J. Exper. Zool. 140: 415–40.

Galtsoff, P. S., H. F. Prytherch, and J. B. Engle. 1937. Natural history and methods of controlling the common oyster drills (*Urosalpinx cinerea* Say and *Eupleura caudata* Say). U.S. Bur. Fish, Fish. Circ. 25: 1–24.

Ganaros, A. E. 1957. Marine fungus infecting eggs and embryos of *Urosalpinx cinerea*. Science 125: 1194.

Garstang, W. 1951. Larval forms. Oxford: Blackwell. 85 pp.

Ghiselin, M. T. 1964. Morphological and behavioral concealing adaptations of *Lamellaria stearnsii*, a marine prosobranch gastropod. Veliger 6: 123–24.

Ghiselin, M. T., E. DeMan, and J. P. Wourms. 1975. An anomalous style in the gut of *Megatebennus bimaculatus*, a carnivorous prosobranch gastropod. Veliger 18: 40–43.

Ghiselin, M.T., E. T. Degens, D. W. Spencer, and R. H. Parker. 1967. A phylogenetic survey of molluscan shell matrix proteins. Breviora Mus. Compar. Zool., Harvard 262: 1–35.

Gibson, D. G. 1964. Mating behavior in *Littorina planaxis* Philippi. Veliger 7: 134–39.

Giese, A. C. 1969. A new approach to the biochemical composition of the mollusc body. Oceanogr. Mar. Biol. Ann. Rev. 7: 175–229.

Giesel, J. T. 1969. Factors influencing the growth and relative growth of *Acmaea digitalis*, a limpet. Ecology 50: 1084–87.

Gifford, D. W., and E. W. Gifford. 1941. Color variation in *Olivella biplicata*. Nautilus 55: 10–12.

———. 1944. Californian *Olivellas*. Nautilus 57: 73–80.

Giglioli, M. E. C. 1955. The egg masses of the Naticidae (Gastropoda). J. Fish. Res. Bd. Canada 12: 287–327.

Giorgi, A. E., and J. D. DeMartini. 1977. A study of the reproductive biology of the red abalone, *Haliotis rufescens* Swainson, near Mendocino, California. Calif. Fish & Game 63: 80–94.

Glynn, P. W. 1965. Community composition, structure, and interrelationships in the marine intertidal *Endocladia muricata–Balanus glandula* association in Monterey Bay, California. Beaufortia 12: 1–198.

Graham, D. L. 1972. Trace metal levels in intertidal molluscs of California. Veliger 14: 365–72.

Griffith, L. M. 1967. The intertidal univalves of British Columbia. British Columbia Prov. Mus., Handbook 26: 1–101.

Gurin, S., and W. E. Carr. 1971. Chemoreception in *Nassarius obsoletus*: The role of specific stimulatory proteins. Science 174: 293–95.

Haderlie, E. C. 1976. Destructive marine wood and stone borers in Monterey Bay, pp. 947–53, *in* J. M. Sharpley and A. M. Kaplan, eds., Proc. 3rd Internat. Symp. on Biodegradation. London: Applied Science. 1138 pp.

Hadfield, M. G. 1966. The reproductive biology of the California vermetid gastropods *Serpulorbis squamigerus* (Carpenter, 1857) and

Petaloconchus montereyensis Dall, 1919. Doctoral thesis, Biological Sciences, Stanford University, Stanford, Calif. 174 pp.

———. 1969. Nurse eggs and giant sperm in the Vermetidae (Gastropoda). Amer. Zool. 9: 1141–42.

———. 1970. Observations on the anatomy and biology of two California vermetid gastropods. Veliger 12: 301–9.

Hall, J. R. 1973. Intraspecific trail-following in the marsh periwinkle *Littorina irrorata* Say. Veliger 16: 72–75.

Hancock, D. A. 1954. The destruction of oyster spat by *Urosalpinx cinerea* (Say) on Essex oyster beds. J. Cons. Internat. Explor. Mer. 20: 186–96.

———. 1956. The structure of the capsule and the hatching process in *Urosalpinx cinerea* (Say). Proc. Zool. Soc. London 127: 565–71.

———. 1959. The biology and control of the American whelk tingle *Urosalpinx cinerea* (Say) on English oyster beds. Fish. Invest. London 22: 1–66.

Hanks, J. E. 1957. The rate of feeding of the common oyster-drill, *Urosalpinx cinerea* (Say), at controlled water temperatures. Biol. Bull. 112: 330–35.

Hanna, G. D. 1963. West American mollusks of the genus *Conus*. II. Occ. Pap. Calif. Acad. Sci. 35: 1–101.

———. 1966. Introduced mollusks of western North America. Occ. Pap. Calif. Acad. Sci. 48: 1–108.

Hanna, G. D., and A. M. Strong. 1949. West American mollusks of the genus *Conus*. Proc. Calif. Acad. Sci. 26: 247–322.

Hansen, J. C. 1970. Commensal activity as a function of age in two species of California abalones (Mollusca: Gastropoda). Veliger 13: 90–94.

Hardin, D. D. 1968. A comparative study of lethal temperatures in the limpets *Acmaea scabra* and *Acmaea digitalis*. Veliger 11 (Suppl.): 83–87.

Harger, J. R. E. 1967. Population studies on *Mytilus* communities. Doctoral dissertation, University of California, Santa Barbara. 328 pp.

Hargis, W. J., Jr., and C. L. MacKenzie, Jr. 1961. Sexual behavior of the oyster drills: *Eupleura caudata* and *Urosalpinx cinerea*. Nautilus 75: 7–16.

Harrison, F. M. 1962. Some excretory processes in the abalone, *Haliotis rufescens*. J. Exper. Biol. 39: 179–92.

Harrold, C. 1975. The effect of habitat and acclimation temperature upon the respiratory rate of the marine gastropod *Crepidula adunca* Sowerby (1825). Master's thesis, Biological Sciences, Stanford University, Stanford, Calif. 26 pp.

Haskin, H. H. 1940. The role of chemotropism in food selection by the oyster drill, *Urosalpinx cinerea* Say. Anat. Rec. 78: 95 (abstract).

———. 1950. The selection of food by the oyster drill *Urosalpinx cinerea* Say. Proc. Nat. Shellfish Assoc. 1950: 62–68.

Haven, S. B. 1964. Habitat differences and competition in two intertidal gastropods in central California. Bull. Ecol. Soc. Amer. 45: 52 (abstract).

———. 1965. Field experiments on competition between two ecologically similar intertidal gastropods. Bull. Ecol. Soc. Amer. 46: 160 (abstract).

———. 1971. Niche differences in the intertidal limpets *Acmaea scabra* and *Acmaea digitalis* (Gastropoda) in central California. Veliger 13: 231–48.

———. 1973. Competition for food between the intertidal gastropods *Acmaea scabra* and *Acmaea digitalis*. Ecology 54: 143–51.

Haydock, C. I. 1964. An experimental study to control oyster drills in Tomales Bay, California. Calif. Fish & Game 50: 11–28.

Hayes, F. R. 1929. Contribution to the study of marine gastropods. III. The development, growth, and behavior of *Littorina*. Contrib. Canad. Biol. and Fish. 6: 415–30.

Heath, H. 1916. The nervous system of *Crepidula adunca* and its development. Proc. Acad. Natur. Sci. Philadelphia 68: 479–85.

Heller, E., and M. A. Raftery. 1976. The vitelline envelope of eggs from the giant keyhole limpet *Megathura crenulata*. II: Products formed by lysis with sperm enzymes and dithiothreitol. Biochemistry 15: 1199–1203.

Hemingway, G. T. 1973. Feeding in *Acanthina spirata*. Master's thesis, Biology, San Diego State University. 88 pp.

———. 1976. A comment on Thaisid "cannibalism." Veliger 18: 341.

Herrick, J. C. 1906. Mechanism of the odontophoral apparatus in *Sycotypus canaliculatus*. Amer. Natur. 40: 707–37.

Herzberg, F. 1966. The use of radiography for the study and identification of spiral marine shells. Veliger 9: 247–49.

Hewatt, W. G. 1934. Ecological studies on selected marine intertidal communities of Monterey Bay. Doctoral thesis, Biological Sciences, Stanford University, Stanford, Calif. 150 pp.

———. 1935. Ecological succession in the *Mytilus californianus* habitat as observed in Monterey Bay, California. Ecology 16: 244–51.

———. 1937. Ecological studies on selected marine intertidal communities of Monterey Bay, California. Amer. Midl. Natur. 18: 161–206.

———. 1938. Notes on the breeding seasons of rocky beach fauna of Monterey Bay, California. Proc. Calif. Acad. Sci. 23: 283–88.

———. 1940. Observations on the homing limpet *Acmaea scabra* Gould. Amer. Midl. Natur. 24: 205–8.

Hinegardner, R. T. 1958. The venom apparatus of the cone shell. Hawaii Med. J. 17: 533–36.

———. 1974. Cellular DNA content of the Mollusca. Comp. Biochem. Physiol. 47A: 447–60.

Hochberg, F. G., Jr. 1971. Functional morphology and ultrastructure of the proboscis complex of *Epitonium tinctum* (Gastropoda: Ptenoglossa). Echo 4, Abs. & Proc. 4th Ann. Meet. West. Soc. Malacol., pp. 22–23.

Holmer, S. J. 1900. The early cleavage and formation of the mesoderm of *Serpulorbis squamigerus* Carpenter. Biol. Bull. 1: 115–21.

Hopkins Marine Station. 1964. The periwinkle *Littorina*. Final papers, Biol. 175h, library, Hopkins Marine Station of Stanford University, Pacific Grove, Calif. 345 pp.

Hopper, C. N. 1972. Aspects of the prey preference and feeding biology of *Polinices lewisii* (Gastropoda). Echo 5, Abs. & Proc. 5th Ann. Meet. West. Soc. Malacol., pp. 30–31.

Houston, R. S. 1971. Reproductive biology of *Thais emarginata* (Deshayes, 1839) and *Thais canaliculata* (Duclos 1832). Veliger 13: 348–57.

Hughes, R. N. 1971. Ecological energetics of the keyhole limpet *Fissurella barbadensis* Gmelin. J. Exper. Mar. Biol. Ecol. 6: 167–78.

Human, V. L. 1971. The occurrence of *Urosalpinx cinerea* (Say) in Newport Bay. Veliger 13: 299.

Humphrey, R. R., and R. W. Macy. 1930. Observations on some of the probable factors controlling the size of certain tide pool snails. Publ. Puget Sound Biol. Sta. 7: 205–8.

Hunter, W. S. 1942. Studies on cercaria of the common mud-flat snail *Cerithidea californica*. Doctoral thesis, University of California, Los Angeles.

Illingworth, J. F. 1902. The anatomy of *Lucapina crenulata* Gray. Zool. Jahrb. (Anat. Ontog.) 16: 449–80.

Imai, K. 1964. Some aspects of spawning behavior and development in *Littorina planaxis* and *Littorina scutulata*. Research paper, Biol. 175h, library, Hopkins Marine Station of Stanford University, Pacific Grove, Calif. 13 pp.

Ingram, W. M. 1938. Notes on the cowry *Cypraea spadicea* Swainson. Nautilus 52: 1–4.

———. 1947. Fossil and Recent Cypraeidae of the western regions of the Americas. Bull. Amer. Paleontol. 31: 47–124.

Isarankura, K., and N. W. Runham. 1968. Studies on the replacement of the gastropod radula. Malacologia 7: 71–91.

Jacobsen, S., and D. Boylan. 1971. Interference with chemotaxis in a marine snail, *Nassarius obsoletus*, by sea water soluble components of kerosine. Amer. Zool. 11: 694 (abstract).

Jenner, C. E. 1958. An attempted analysis of schooling behavior in the marine snail *Nassarius obsoletus*. Biol. Bull. 115: 337–38.

Jessee, W. F. 1968a. New northern limit for the limpet *Acmaea digitalis*. Veliger 11: 144.

———. 1968b. Studies of homing behavior in the limpet *Acmaea scabra*. Veliger 11 (Suppl.): 52–55.

Jobe, A. 1968. A study of morphological variation in the limpet *Acmaea pelta*. Veliger 11 (Suppl.): 69–72.

Johnson, J. K. 1972. Effect of turbidity on the rate of filtration and growth of the slipper limpet *Crepidula fornicata* Lamarck, 1799. Veliger 14: 315–20.

Johnson, S. E. 1968. Occurrence and behavior of *Hyale grandicornis*, a gammarid amphipod commensal in the genus *Acmaea*. Veliger 11 (Suppl.): 56–60.

———. 1975. Microclimate and energy flow in the marine rocky intertidal, pp. 559–87, *in* D. M. Gates and R. B. Schmerl, eds., Per-

spectives in biophysical ecology. New York: Springer-Verlag. 609 pp.

Joselow, M., and C. R. Dawson. 1955. Hemocyanin and radioactive copper. Science 121: 300–302.

Juberg, P. 1970. The shell structure of *Astraea olfersi* (Gastropoda: Turbinidae). Malacologia 10: 415–21.

Keen, A. M. 1960. Vermetid gastropods and marine intertidal zonation. Veliger 3: 1–2.

———. 1961. A proposed reclassification of the gastropod family Vermetidae. Bull. Brit. Mus. (Natur. Hist.), Zool. 7: 181–213.

———. 1971. Sea shells of tropical west America. 2nd ed. Stanford, Calif.: Stanford University Press. 1,064 pp.

———. 1975. On some west American species of *Calliostoma*. Veliger 17: 413–14.

Keep, J., and J. L. Baily, Jr. 1935. West coast shells. Stanford, Calif.: Stanford University Press. 350 pp.

Kellogg, M. G., and D. R. Lindberg. 1975. Notes on the spawning and larval development of *Mitra idae* Melvill (Gastropoda: Mitridae). Veliger 18: 166–67.

Kessel, M. M. 1964. Reproduction and larval development of *Acmaea testudinalis*. Biol. Bull. 127: 294–303.

Kincaid, T. 1957. Local races and clines in the marine gastropod *Thais lamellosa* Gmelin. A population study. Seattle: Calliostoma. 75 pp.

Kingston, R. S. 1968. Anatomical and oxygen electrode studies of respiratory surfaces and respiration in *Acmaea*. Veliger 11 (Suppl.): 73–78.

Kitting, C. L. 1978. Foraging of individuals within the limpet species *Acmaea* (*Notoacmea*) *scutum* at Monterey Bay, and the consequences on the mid-intertidal algae. Doctoral thesis, Biological Sciences, Stanford University, Stanford, Calif. 197 pp.

———. 1979. The use of feeding noises to determine the algal foods being consumed by individual intertidal molluscs. Oecologia 40: 1–18.

Klotz, I. M., and T. A. Klotz. 1955. Oxygen-carrying proteins: A comparison of the oxygenation reaction in hemocyanin and hemerythrin with that in hemoglobin. Science 121: 477–80.

Kohn, A. J. 1956. Piscivorous gastropods of the genus *Conus*. Proc. Natur. Acad. Sci. 42: 168–71.

———. 1959. The ecology of *Conus* in Hawaii. Ecol. Monogr. 29: 47–90.

———. 1961. Chemoreception in gastropod molluscs. Amer. Zool. 1: 291–308.

———. 1966. Food specialization in *Conus* in Hawaii and California. Ecology 47: 1041–43.

Kohn, A. J., J. W. Nybakken, and J. J. VanMol. 1972. Radula tooth structure of the gastropod *Conus imperialis* elucidated by scanning electron microscopy. Science 176: 49–51.

Kohn, A. J., P. R. Saunders, and S. Wiener. 1960. Preliminary studies on the venom of the marine snail *Conus*, pp. 706–25, *in* R. Nigrelli, ed., Biochemistry and pharmacology of compounds derived from marine organisms. Ann. N.Y. Acad. Sci. 90: 615–950.

Kosin, D. F. 1964. Light responses of *Tegula funebralis* (Mollusca: Gastropoda). Veliger 6 (Suppl.): 46–50.

Krinsley, D. 1959. Manganese in modern and fossil gastropod shells. Nature 183: 770–71.

Lambert, P., and P. A. Dehnel. 1974. Seasonal variation in biochemical composition during the reproductive cycle of the intertidal gastropod *Thais lamellosa* Gmelin (Gastropoda: Prosobranchia). Canad. J. Zool. 52: 305–18.

Lash, J. W. 1959. Presence of myoglobin in "cartilage" of the marine snail *Busycon*. Science 130: 334.

LeBoeuf, R. 1971. *Thais emarginata* (Deshayes): Description of the veliger and egg capsule. Veliger 14: 205–11.

Lebour, M. V. 1935. The echinospira larvae (Mollusca) of Plymouth. Proc. Zool. Soc. London 1935: 163–74.

Leighton, D. L. 1960. An abalone lacking respiratory apertures. Veliger 3: 48.

———. 1961. Observations on the effect of diet on shell coloration in the red abalone, *Haliotis rufescens* Swainson. Veliger 4: 29–32.

———. 1966. Studies of food preference in algivorous invertebrates of Southern California kelp beds. Pacific Sci. 20: 104–13.

———. 1968. A comparative study of food selection and nutrition in the abalone, *Haliotis rufescens* (Swainson) and the sea urchin *Strongylocentratus purpuratus* (Stimpson). Contr. Scripps Inst. Oceanogr. 38: 1853–54.

———. 1974. The influence of temperature on larval and juvenile growth in the three species of southern California abalones. Fish Bull. 72: 1137–45.

Leighton, D., and R. A. Boolootian. 1963. Diet and growth in the black abalone, *Haliotis cracherodii*. Ecology 44: 227–38.

Leonard, Y. I. 1964. Excretory products of *Tegula funebralis* and *Tegula brunnea*. Veliger 6 (Suppl.): 28–30.

Lindberg, D. R., M. G. Kellogg, and W. E. Hughes. 1975. Evidence of light reception through the shell of *Notoacmea persona* (Rathke, 1833). Veliger 17: 383–88.

Lloyd, M. C. 1971. The biology of *Searlesia dira* (Mollusca: Gastropoda) with emphasis on feeding. Doctoral thesis, University of Michigan, Ann Arbor. 103 pp.

Loeb, J. 1905. On chemical methods by which the eggs of a mollusk (*Lottia gigantea*) can be caused to become mature. Univ. Calif. Publ. Physiol. 3: 1–8.

Loosanoff, V. L. 1962. Recent advances in the control of shellfish predators. Proc. Gulf & Caribb. Fish. Inst. 15: 113–28.

Lowenstam, H. A. 1962. Goethite in radular teeth of recent marine gastropods. Science 137: 279–80.

———. 1971. Opal precipitation by marine gastropods (Mollusca). Science 171: 487–89.

Lowry, L. F., A. J. McElroy, and J. S. Pearse. 1974. The distribution of six species of gastropod molluscs in a California kelp forest. Biol. Bull. 147: 386–96.

Lowry, L. F., and J. S. Pearse. 1973. Abalones and sea urchins in an area inhabited by sea otters. Mar. Biol. 23: 213–19.

Lyons, A., and T. M. Spight. 1973. Diversity of feeding mechanisms among embryos of Pacific northwest *Thais*. Veliger 16: 189–94.

MacDonald, J. A., and C. B. Maino. 1964. Observations on the epipodium, digestive tract, coelomic derivatives, and nervous system of the trochid gastropod *Tegula funebralis*. Veliger 6 (Suppl.): 50–55.

MacDonald, K. B. 1969a. Molluscan faunas of the Pacific coast salt marshes and tidal creeks. Veliger 11: 399–405.

———. 1969b. Quantitative studies of salt marsh faunas from the North American Pacific coast. Ecol. Monogr. 39: 33–60.

MacGinitie, G. E. 1931. The egg-laying process in *Alectrion fossatus* Gould. Ann. Mag. Natur. Hist. (10) 8: 258–61.

———. 1935. Ecological aspects of a California marine estuary. Amer. Midl. Natur. 16: 629–765.

———. 1938. Notes on the natural history of some marine animals. Amer. Midl. Natur. 19: 207–19.

MacGinitie, G. E., and N. MacGinitie. 1966. Starved abalones. Veliger 8: 313.

———. 1968. Natural history of marine animals. 2nd ed. New York: McGraw-Hill. 523 pp.

MacGinitie, N. 1959. Marine Mollusca of Point Barrow, Alaska. Proc. U.S. Nat. Mus. 109: 59–208.

MacKenzie, C. L., Jr. 1962. Transportation of oyster drills on horseshoe "crabs." Science 137: 36–37.

Magalhaes, H. 1948. An ecological study of snails of the genus *Busycon* at Beaufort, North Carolina. Ecol. Monogr. 18: 377–409.

Mangum, G. P., L. J. Kushins, and C. Sassaman. 1970. Responses of intertidal invertebrates to low oxygen conditions. Amer. Zool. 10: 258–59.

Manzi, J. J. 1970. Combined effects of salinity and temperature on the feeding, reproductive and survival rates of *Eupleura candata* (Say) and *Urosalpinx cinerea* (Say) (Prosobranchia: Muricidae). Biol. Bull. 138: 35–46.

Manzi, J. J., A. Calabrese, and D. M. Rawlins. 1972. A note on gametogenesis in the oyster drills, *Urosalpinx cinerea* (Say) and *Eupleura candata* (Say). Veliger 14: 271–73.

Marcus, Er., and Ev. Marcus. 1959. Studies on Olividae (pp. 99–164, 11 pls.). On the reproduction of *Olivella* (pp. 189–96, 1 pl). Bol. Fac. Filos. Cienc. São Paulo, Brazil, Zool. 22: 99–199.

Margolin, A. S. 1964a. The mantle response of *Diodora aspera*. Anim. Behav. 12: 187–94.

———. 1964b. A running response of *Acmaea* to seastars. Ecology 45: 191–93.

Markel, R. P. 1974. Aspects of the physiology of temperature acclimation in the limpet *Acmaea limatula* Carpenter (1864): An integrated field and laboratory study. Physiol. Zool. 47: 99–109.

Martin, M., M. D. Stephenson, and J. H. Martin. 1977. Copper toxicity experiments in relation to abalone deaths observed in a power plant's cooling waters. Calif. Fish & Game 63: 95–100.

Martin, W. E. 1955. Seasonal infections of the snail *Cerithidea californica* Haldeman, with larval trematodes, pp. 203–10, *in* Essays in natural sciences. Allan Hancock Found., Univ. So. Calif., Los Angeles. 335 pp.

————. 1972. An annotated key to the cercariae that develop in the snail *Cerithidea californica*. Bull. So. Calif. Acad. Sci. 71: 39–43.

Mayes, M., and C. O. Hermans. 1973. Fine structure of the eye of the prosobranch mollusk *Littorina scutulata*. Veliger 16: 166–68.

McAllister, R. 1976. California marine fish landings for 1974. Calif. Dept. Fish & Game, Fish Bull. 166: 1–53 (see also similar reports for earlier years).

McGee, P. 1964. A new pigment from *Tegula funebralis* (Mollusca: Gastropoda). Veliger 6 (Suppl): 25–27.

McLean, J. H. 1960. *Batillaria cumingi*, introduced cerithiid in Elkhorn Slough. Veliger 2: 61–63.

————. 1962a. Manometric measurements of respiratory activity in *Tegula funebralis*. Veliger 4: 191–93.

————. 1962b. Sublittoral ecology of kelp beds of the open coast areas near Carmel, California. Biol. Bull. 122: 95–114.

————. 1966a. A new genus of Fissurellidae and a new name for a misunderstood species of west American *Diodora*. Los Angeles Co. Mus. Contr. Sci. 100: 1–8.

————. 1966b. West American Gastropoda: Superfamilies Patellacea, Pleurotomariacea, and Fissurellacea. Doctoral thesis, Biological Sciences, Stanford University, Stanford, Calif. 255 pp.

————. 1967. West American species of *Lucapinella*. Veliger 9: 349–52.

————. 1969. Marine shells of southern California. Los Angeles Co. Mus. Natur. Hist., Sci. Ser. 24, Zool. 11. 104 pp.

McLean, N. 1971. On the function of the digestive gland in *Nassarius* (Gastropoda: Prosobranchia). Veliger 13: 273–74.

Mendel, L. B., and H. C. Bradley. 1905. Experimental studies in the physiology of the molluscs. Amer. J. Physiol. 13: 17–29; 14: 313–27.

————. 1906. Experimental studies in the physiology of the molluscs. Amer. J. Physiol. 17: 167–76.

Menge, B. A. 1972. Competition for food between two intertidal starfish species and its effect on body size and feeding. Ecology 53: 635–64.

Menge, J. L. 1974. Prey selection and foraging period of the predaceous rocky intertidal snail, *Acanthina punctulata*. Oecologia 17: 293–316.

Merriman, J. A. 1967. Systematic implications of radular structures of west coast species of *Tegula*. Veliger 9: 399–403.

Meyer, R. J. 1967. Hemocyanins and the systematics of California

Haliotis. Doctoral thesis, Biological Sciences, Stanford University, Stanford, Calif. 92 pp.

Millard, C. S. 1968. The clustering behavior of *Acmaea digitalis*. Veliger 11 (Suppl.): 45–51.

Miller, A. C. 1968. Orientation and movement of the limpet *Acmaea digitalis* on vertical rock surfaces. Veliger 11 (Suppl.): 30–44.

Miller, R. L. 1968. Some observations on the ecology and behavior of *Lucapinella callomarginata*. Veliger 11: 130–34.

Miller, S. L. 1974. Adaptive design of locomotion and foot form in prosobranch gastropods. J. Exper. Mar. Biol. Ecol. 14: 99–156.

Miller, W., R. S. Nishioka, and H. A. Bern. 1973. The "juxtaganglionic" tissue and the brain of the abalone *Haliotis rufescens* Swainson. Veliger 16: 125–29.

Miller, W. E. 1975. Environmental cues and the orientation and movement of *Norrisia norrisii*. Veliger 17: 292–95.

Minchin, D. 1975. The righting response of haliotids. Veliger 17: 249–50.

Monique, C., and J. Fournie. 1969. Boring mechanism in *Thais lapillus* (Gastropoda: Muricidae): Intervention of an enzymatic system in active transport of cations during the destruction of $CaCO_3$. Amer. Zool. 9: 1145 (abstract).

Montgomery, D. H. 1967. Responses of two haliotid gastropods (Mollusca), *Haliotis assimilis* and *Haliotis rufescens*, to the forcipulate asteroids (Echinodermata), *Pycnopodia helianthoides* and *Pisaster ochraceus*. Veliger 9: 359–68.

Morgan, T. H. 1933. The formation of the antipolar lobe in *Ilyanassa*. J. Exper. Zool. 64: 433–67.

Moritz, C. E. 1938. The anatomy of the gastropod *Crepidula adunca* Sowerby. Univ. Calif. Publ. Zool. 43: 83–91.

————. 1939. Organogenesis in the gastropod *Crepidula adunca* Sowerby. Univ. Calif. Publ. Zool. 43: 217–48.

Morse, D. E., H. Duncan, N. Hooker, and A. Morse. 1977. Hydrogen peroxide induces spawning in mollusks, with activation of prostaglandin endoperoxide synthetase. Science 196: 298–300.

Morton, J. E. 1965. Form and function in the evolution of the Vermetidae. Bull. Brit. Mus. (Natur. Hist.), Zool. 11: 585–630.

Muchmore, A. V. 1968. Laminarinase and hexokinase activity during embryonic development of *Acmaea scutum*. Veliger 11 (Suppl.): 105–8.

Murdock, W. W. 1969. Switching in general predators: Experiments on predator specificity and stability of prey populations. Ecol. Monogr. 39: 335–54.

Murphy, P. G. 1976. Electrophoretic evidence that selection reduces ecological overlap in marine limpets. Nature 261: 228–30.

Nakadal, A. M. 1960. Carotenoids and chlorophyllic pigments in the marine snail, *Cerithidea californica* Haldeman, intermediate host for several avian trematodes. Biol. Bull. 119: 98–108.

Neale, J. R. 1965. Rheotactic responses in the marine mollusk *Littorina planaxis* Philippi. Veliger 8: 7–10.

Newell, G. E. 1958a. The behaviour of *Littorina littorea* (L.) under natural conditions and its relation to position on the shore. J. Mar. Biol. Assoc. U.K. 37: 229–39.

———. 1958b. An experimental analysis of the behaviour of *Littorina littorea* (L.) under natural conditions and in the laboratory. J. Mar. Biol. Assoc. U.K. 37: 241–66.

———. 1965. The eye of *Littorina littorea*. Proc. Zool. Soc. London 144: 75–85.

Nicotri, M. E. 1974. Resource partitioning, grazing activities, and influence on the microflora by intertidal limpets. Doctoral thesis, Zoology, University of Washington, Seattle.

———. 1977. Grazing effects of four marine intertidal herbivores on the microflora. Ecology 58: 1020–30.

Nishikawa, S. 1962. A comparative study of chromosomes in marine gastropods, with some remarks on cytotaxonomy and phylogeny. J. Shimonoseki College Fisheries 11: 149–86.

Nolfi, J. R. 1964. The hemocyanins of *Tegula funebralis* and *Tegula brunnea* (Mollusca: Gastropoda). Veliger 6 (Suppl.): 11–13.

North, W. J. 1954. Size distribution, erosive activities and gross metabolic efficiency of the marine intertidal snails, *Littorina planaxis* and *L. scutulata*. Biol. Bull. 106: 185–87.

———. 1964. An investigation of the effects of discharged wastes on kelp. State of Calif. Water Quality Control Board Publ. 26: 1–124.

Nybakken, J. 1967. Preliminary observations on the feeding behavior of *Conus purpurascens* Broderip, 1833. Veliger 10: 55–57.

———. 1970. Radular anatomy and systematics of the west American Conidae (Mollusca: Gastropoda). Amer. Mus. Novitates 2414: 1–29.

Nylen, M. U., D. V. Provenza, and M. R. Carriker. 1969. Fine structure of the accessory boring organ of the gastropod *Urosalpinx*. Amer. Zool. 9: 935–65.

Oldroyd, I. S. 1927. The marine shells of the west coast of North America. Stanford Univ. Publ., Univ. Ser., Geol. Sci., Vol. 2, Part 3. 340 pp.

Olsen, David A. 1968a. Banding patterns in *Haliotis*. II. Some behavioral considerations and the effect of diet on shell coloration for *Haliotis rufescens, Haliotis corrugata, Haliotis sorenseni* and *Haliotis assimilis*. Veliger 11: 135–39.

———. 1968b. Banding patterns of *Haliotis rufescens* as indicators of botanical and animal succession. Biol. Bull. 134: 139–47.

Orr, P. C. 1941. Exceptional burial in California. Science 94: 539–40.

Orton, J. H. 1912. The mode of feeding of *Crepidula*, with an account of the current-producing mechanism in the mantle cavity, and some remarks on the mode of feeding in gastropods and lamellibranchs. J. Mar. Biol. Assoc. U.K. 9: 444–78.

Overholser, J. A. 1964. Orientation and response of *Tegula funebralis* to tidal current and turbulence (Mollusca: Gastropoda). Veliger 6 (Suppl.): 38–41.

Owen, B., J. H. McLean, and R. J. Meyer. 1971. Hybridization in the eastern Pacific abalones (*Haliotis*). Bull. Los Angeles Co. Mus. Natur. Hist. Science: No. 9, Jan. 15, 1971.

Paine, R. T. 1962. Ecological diversification in sympatric gastropods of the genus *Busycon*. Evolution 16: 515–23.

———. 1966. Function of labial spines, composition of diet, and size of certain marine gastropods. Veliger 9: 17–24.

———. 1969. The *Pisaster-Tegula* interaction: Prey patches, predator food preference, and intertidal community structure. Ecology 50: 950–61.

———. 1971. Energy flow in a natural population of the herbivorous gastropod *Tegula funebralis*. Limnol. Oceanogr. 16: 86–98.

Palmer, A. R. 1977. Function of shell sculpture in marine gastropods: Hydrodynamic destabilization in *Ceratostoma foliatum*. Science 197: 1293–95.

Palmer, C. F. 1907. The anatomy of California Haliotidae. Proc. Acad. Natur. Sci. Philadelphia 59: 396–407.

Paris, O. H. 1960. Some quantitative aspects of predation by muricid snails on mussels in Washington Sound. Veliger 2: 41–47.

Patterson, C. M. 1967. Chromosome numbers and systematics in streptoneuran snails. Malacologia 5: 111–25.

Paulson, T., and R. S. Scheltema. 1967. Selective feeding on algal cells by the veliger larvae of the gastropod *Nassarius obsoletus* (Say). Amer. Zool. 7: 770–71.

Pearse, J. S., and L. F. Lowry, eds. 1974. An annotated species list of the benthic algae and invertebrates in the kelp forest community at Point Cabrillo, Pacific Grove, California. Coastal Marine Lab., Univ. Calif., Santa Cruz, Tech. Rep. No. 1. 73 pp.

Peile, A. J. 1939. Radula notes VIII. 34. *Conus*. Proc. Malacol. Soc. London 23: 348–55.

Peppard, M. C. 1964. Shell growth and repair in the gastropod *Tegula funebralis*. Veliger 6 (Suppl.): 59–63.

Pequegnat, W. E. 1964. The epifauna of a California siltstone reef. Ecology 45: 272–83.

Perkins, R. J., Jr. 1971. A systems approach to energy flow in a rocky inter-tidal zone. Doctoral thesis, University of Georgia, Athens. 106 pp.

Perron, F. 1975. Carnivorous *Calliostoma* (Prosobranchia: Trochidae) from the northeastern Pacific. Veliger 18: 52–54.

Peters, R. S. 1964. Function of the cephalic tentacles in *Littorina planaxis* Philippi (Gastropoda: Prosobranchiata). Veliger 7: 143–48.

Phillips, D. W. 1975a. Distance chemoreception-triggered avoidance behavior of the limpet *Acmaea (Collisella) limatula* and *Acmaea (Notoacmea) scutum* to the predatory starfish *Pisaster ochraceus*. J. Exper. Zool. 191: 199–210.

———. 1975b. Localization and electrical activity of the distance chemoreceptors that mediate predator avoidance behaviour in *Acmaea limatula* and *Acmaea scutum* (Gastropoda, Prosobranchia). J. Exper. Biol. 63: 403–12.

———. 1976. The effect of a species-specific avoidance response to predatory starfish on the intertidal distribution of two gastropods. Oecologia 23: 83–94.

———. 1977a. A scanning electron microscope study of sensory tentacles on the mantle margin of the gastropod *Acmaea* (*Notoacmea*) *scutum*. Veliger 19: 266–71.

———. 1977b. Activity of the gastropod mollusk *Olivella biplicata* in response to natural light/dark cycle. Veliger 20: 137–43.

———. 1977c. Avoidance and escape response of the gastropod mollusc *Olivella biplicata* (Sowerby) to predatory asteroids. J. Exper. Mar. Biol. Ecol. 28: 77–86.

———. 1978. Chemical mediation of invertebrate defensive behaviors and the ability to distinguish between foraging and inactive predators. Mar. Biol. 49: 237–43.

Pierce, M. E. 1950. *Busycon canaliculatum*, pp. 336–44, *in* F. A. Brown, Jr., ed., Selected invertebrate types. New York: Wiley. 597 pp.

Pilsbry, H. A. 1929. *Neverita reclusiana* (Desh.) and its allies. Nautilus 42: 109–13.

Pilson, M. E. Q., and P. B. Taylor. 1961. Hole drilling by octopus. Science 134: 1366–68.

Powell, A. W. B. 1966. The molluscan families Speightiidae and Turridae. Bull. Auckland Inst. & Mus. 5. 184 pp.

Pratt, D. M. 1974. Attraction to prey and stimulus to attack in the predatory gastropod *Urosalpinx cinerea*. Mar. Biol. 27: 37–45.

———. 1977. Homing in *Urosalpinx cinerea* in response to prey effluent and tidal periodicity. Veliger 20: 30–32.

Prescott, B., and C. P. Li. 1966. Antimicrobial agents from sea food. Malacologia 5: 45–46.

Proctor, S. J. 1968. Studies on the stenotopic marine limpet *Acmaea insessa* (Mollusca: Gastropoda: Prosobranchia) and its algal host *Egregia menziesii* (Phaeophyta). Doctoral thesis, Biological Sciences, Stanford University, Stanford, Calif. 144 pp.

Purchon, R. D. 1968. The biology of the Mollusca. New York: Pergamon. 560 pp.

Putnam, D. A. 1964. The dispersal of young of the commensal gastropod *Crepidula adunca* from its host *Tegula funebralis*. Veliger 6 (Suppl.): 63–66.

Radwin, G. E., and A. D'Attilio. 1976. Murex shells of the world. Stanford, Calif.: Stanford University Press. 284 pp.

Raven, C. P. 1966. Morphogenesis: The analysis of molluscan development. New York: Pergamon. 365 pp.

Richardson, F. 1934. Diurnal movements of the limpets *Lottia gigantea* and *Acmaea persona*. J. Entomol. Zool. (Claremont, Calif.) 26: 53–55.

Risebrough, R. W., D. B. Menzel, D. J. Margin, Jr., and H. S. Olcott. 1967. DDT residues in Pacific sea birds: A persistent insecticide in marine food chains. Nature 216: 589–91.

Riser, N. W. 1969. Feeding behavior of some New England marine gastropods. Nautilus 82: 112–13.

Robertson, R. 1963. Wentletraps (Epitoniidae) feeding on sea anemones and corals. Proc. Malacol. Soc. London 35: 51–63.

Robilliard, G. A. 1971. Predation by the nudibranch *Dirona albolineata* on three species of prosobranchs. Pacific Sci. 25: 429–35.

Rogers, D. A. 1968. The effects of light and tide on movements of the limpet *Acmaea scutum*. Veliger 11 (Suppl.): 20–24.

Rosenthal, R. J. 1969. A method of tagging mollusks underwater. Veliger 11: 288–89.

———. 1970. Observations on the reproductive biology of the Kellet's whelk, *Kelletia kelletii* (Gastropoda: Neptuneidae). Veliger 12: 319–24.

———. 1971. Trophic interaction between the sea star *Pisaster giganteus* and the gastropod *Kelletia kelletii*. Fish. Bull. 69: 669–79.

Ross, T. L. 1968. Light responses in the limpet *Acmaea limatula*. Veliger 11 (Suppl.): 25–29.

Ruth, F. S. 1948. Studies on the natural history of the limpets of the family acmaeidae. Master's thesis, College of the Pacific, Stockton, Calif. 151 pp.

Salo, S. 1977. Observations on feeding, chemoreception, and toxins in two species of *Epitonium*. Veliger 20: 168–72.

Sandeen, M. I., G. C. Stephens, and F. A. Brown, Jr. 1954. Persistent daily and tidal rhythms of oxygen consumption in two species of marine snails. Physiol. Zool. 27: 350–56.

Saunders, P. R. 1959. Some observations on the feeding habits of *Conus californicus* Hinds. Veliger 1: 13–14.

Saunders, P. R., and F. Wolfson. 1961. Food and feeding behavior in *Conus californicus* Hinds, 1844. Veliger 3: 73–75.

Schaefer, C. W. 1969. Feeding and chemoreception in the mud-snail, *Nassarius obsoletus*. Nautilus 82: 108–9.

Scheltema, R. S. 1964. Feeding habits and growth in the mud-snail *Nassarius obsoletus*. Chesapeake Sci. 5: 161–66.

———. 1967. The relationship of temperature to the larval development of *Nassarius obsoletus* (Gastropoda). Biol. Bull. 132: 253–65.

Schilder, F. A. 1965. The geographical distribution of cowries. Veliger 7: 171–83.

———. 1966. The higher taxa of cowries and their allies. Veliger 9: 31–35.

Schroeder, P. C., and C. F. Cleland. 1964. Some properties of DNA from the prosobranch gastropod *Tegula funebralis*. Veliger 6 (Suppl.): 8–10.

Schwimer, S. R. 1973. Trace metal levels in three subtidal invertebrates. Veliger 16: 95–102.

Scott, D. B., and T. L. Cass. 1977. Response of *Cerithidea californica* (Haldeman) to lowered salinities and its paleoecological implications. Bull. So. Calif. Acad. Sci. 76: 60–63.

Seapy, R. R. 1966. Reproduction and growth in the file limpet, *Acmaea limatula* Carpenter, 1864. Veliger 8: 300–310.

Segal, E. 1956. Microgeographic variation as thermal acclimation in an intertidal mollusc. Biol. Bull. 111: 129–52.

———. 1961. Acclimation in molluscs. Amer. Zool. 1: 235–44.

————. 1962. Initial response of the heart rate of a gastropod, *Acmaea limatula*, to abrupt changes in temperature. Nature 195: 674–75.

Segal, E., and P. A. Dehnel. 1962. Osmotic behavior in an intertidal limpet, *Acmaea limatula*. Biol. Bull. 122: 417–30.

Sellers, R. G., Jr. 1977. The diets of four species of *Calliostoma* (Gastropoda, Trochidae) and some aspects of their distribution within a kelp bed. Master's thesis, Biological Sciences, Stanford University, Stanford, Calif. 31 pp.

Shotwell, J. A. 1950a. Distribution of volume and relative linear measurement changes in *Acmaea*, the limpet. Ecology 31: 51–61.

————. 1950b. The vertical zonation of *Acmaea*, the limpet. Ecology 31: 647–49.

Smarsh, A., H. H. Chauncey, M. R. Carriker, and P. Person. 1969. Carbonic anhydrase in the accessory boring organ of the gastropod *Urosalpinx*. Amer. Zool. 9: 967–82.

Smith, C. R. 1977. Chemical recognition of prey by the gastropod *Epitonium tinctum* (Carpenter, 1864). Veliger 19: 331–40.

Smith, D. L. 1977. A guide to marine coastal plankton and marine invertebrate larvae. Dubuque, Iowa: Kendall/Hunt. 161 pp.

Smith, R. I., and J. T. Carlton, eds. 1975. Light's manual: Intertidal invertebrates of the central California coast. 3rd ed. Berkeley and Los Angeles: University of California Press. 716 pp.

Spight, T. M. 1972. Patterns of change in adjacent populations of an intertidal snail *Thais lamellosa*. Doctoral thesis, University of Washington, Seattle. 325 pp.

————. 1973. Ontogeny, environment, and shape of a marine snail, *Thais lamellosa* Gmelin. J. Exper. Mar. Biol. Ecol. 13: 215–28.

————. 1974. Sizes of populations of a marine snail. Ecology 55: 712–29.

————. 1975a. Factors extending gastropod embryonic development and their selective cost. Oecologia 21: 1–16.

————. 1975b. On a snail's chances of becoming a year old. Oikos 26: 9–14.

————. 1976a. Ecology of hatching size for marine snails. Oecologia 24: 283–94.

————. 1976b. Hatching size and the distribution of nurse eggs among prosobranch embryos. Biol. Bull. 150: 491–99.

————. 1976c. Colors and patterns of an intertidal snail, *Thais lamellosa*. Res. Popul. Ecol. 17: 176–90.

————. 1977. Is *Thais canaliculata* (Gastropoda: Muricidae) evolving nurse eggs? Nautilus 91: 74–76.

Spight, T. M., C. Birkeland, and A. Lyons. 1974. Life histories of large and small murexes (Prosobranchia: Muricidae) Mar. Biol. 24: 229–42.

Spight, T. M., and J. Emlen. 1976. Clutch sizes of two marine snails with a changing food supply. Ecology 57: 1162–78.

Spight, T. M., and A. Lyons. 1974. Development and functions of the shell sculpture of the marine snail *Ceratostoma foliatum*. Mar. Biol. 24: 77–83.

Stearns, R. E. C. 1875. Aboriginal shell money. Proc. Calif. Acad. Sci. 5: 113–20, pl. 6.

Stephens, G. C., M. K. Sandeen, and H. M. Webb. 1953. A persistent tidal rhythm of activity in the mud snail *Nassa obsoleta*. Anat. Rec. 117: 635.

Stephens, G. C., and R. A. Schinske. 1961. Uptake of amino acids by marine invertebrates. Limnol. Oceanogr. 6: 175–81.

Stickle, W. B., Jr. 1967. Some physiological effects of starvation upon the foot and visceral mass of *Thais lamellosa* (Gmelin, 1791). Amer. Zool. 74: 785 (abstract).

————. 1970. Seasonal changes in the body components and nutrient deposits (lipid, polysaccharide and protein) of the intertidal prosobranch *Thais lamellosa* (Gmelin, 1791). Amer. Zool. 10: 307 (abstract).

————. 1971. The metabolic effects of starving *Thais lamellosa* immediately after spawning. Comp. Biochem. Physiol. 40A: 627–34.

————. 1973. The reproductive physiology of the intertidal prosobranch *Thais lamellosa* (Gmelin). I. Seasonal changes in the rate of oxygen consumption and body component indexes. Biol. Bull. 144: 511–24.

————. 1975. The reproductive physiology of the intertidal prosobranch *Thais lamellosa* (Gmelin). II. Seasonal changes in biochemical composition. Biol. Bull. 148: 448–60.

Stickle, W. B., and F. G. Duerr. 1970. The effects of starvation on the respiration and major nutrient stores of *Thais lamellosa*. Comp. Biochem. Physiol. 33: 689–95.

Stimson, J. 1970. Territorial behavior of the owl limpet *Lottia gigantea*. Ecology 51: 113–18.

————. 1973. The role of the territory in the ecology of the intertidal limpet *Lottia gigantea* (Gray). Ecology 54: 1020–30.

Stohler, R. 1930. Gewichtsverhältnisse bei gewissen marinen Evertebraten. Zool. Anz. 91: 149–55.

————. 1950. Studies on mollusk populations. I. Nautilus 64: 47–51.

————. 1952. Studies on mollusk populations. II. Nautilus 65: 135–37.

————. 1958. Studies on mollusk populations. IIIa. Nautilus 71: 129–31.

————. 1959a. Studies on mollusk populations. IV. Nautilus 73: 65–72h.

————. 1959b. Two new species of west North American gastropods. Proc. Calif. Acad. Sci. (4) 29: 423–44.

————. 1960. Studies on mollusk populations. VI. Nautilus 73: 95–103.

————. 1962a. Preliminary report on growth studies in *Olivella biplicata*. Veliger 4: 150–51.

————. 1962b. *Busycotypus* (B.) *canaliculatus* in San Francisco Bay. Veliger 4: 211–12.

————. 1964. Studies on mollusk populations. VI. *Tegula funebralis* (A. Adams, 1855) (Mollusca: Gastropoda). Veliger 6 (Suppl.): 77–81.

———. 1969a. Growth studies in *Olivella biplicata* (Sowerby, 1825). Veliger 11: 259–67.

———. 1969b. The type of *Tegula funebralis* (A. Adams, 1855). Veliger 11: 406–7.

Straughan, D. 1971. Breeding and larval settlement of certain intertidal invertebrates in the Santa Barbara Channel following pollution by oil, pp. 223–44, *in* D. Straughan, comp., Biological and oceanographical survey of the Santa Barbara Channel oil spill, 1969–1970, 1, Biology and bacteriology. Allan Hancock Found., Univ. So. Calif., Los Angeles. 426 pp.

Strong, A. M. 1924. Notes on *Acanthina* from California. Nautilus 38: 18–22.

———. 1925. *Acanthina*. Nautilus 38: 104.

Strong, P. L., and J. P. Green. 1970. Digestive enzymes in the marine prosobranch gastropod *Busycon canaliculatum*. Amer. Zool. 10: 508.

Sutherland, J. P. 1970. Dynamics of high and low populations of the limpet *Acmaea scabra* (Gould). Ecol. Monogr. 40: 169–88.

———. 1972. Energetics of high and low populations of the limpet, *Acmaea scabra* (Gould). Ecology 53: 430–37.

Szal, R. A. 1969. Distance chemoreception in a marine snail, *Tegula funebralis*. Doctoral thesis, Biological Sciences, Stanford University, Stanford, Calif. 202 pp.

———.1971. A "new" sense organ of primitive gastropods. Nature 229: 490–92.

Talmadge, R. R. 1964. The races of *Haliotis fulgens* Philippi. Trans. San Diego Soc. Natur. Hist. 13: 369–76.

———. 1966. A new haliotid from Guadalupe Island, Mexico. Los Angeles Co. Mus. Contr. Sci. 109: 1–4.

———. 1972. Notes on some California Mollusca: Geographical, ecological and chronological distribution. Veliger 14: 411–13.

———. 1975. A note on *Ocenebra lurida* (Middendorff). Veliger 17: 414.

———. 1977. Notes on a California hybrid *Haliotis* (Gastropoda: Haliotidae). Veliger 20: 37–38.

Tamarin, A., and M. R. Carriker. 1968. The egg capsule of the muricid gastropod *Urosalpinx cinerea*. J. Ultrastruc. Res. 21: 26–40.

Test, A. R. G. 1945. Ecology of California *Acmaea*. Ecology 26: 395–405.

———. 1946. Speciation in limpets of the genus *Acmaea*. Contrib. Lab. Vert. Biol. Univ. Michigan 31: 1–24.

Test, F. H. 1945. Substrate and movements of the marine gastropod *Acmaea asmi*. Amer. Midl. Natur. 33: 791–93.

Thompson, W. F. 1920. The abalones of northern California. Calif. Fish & Game 6: 45–50.

Thorson, G. 1957. Parasitism in the marine gastropod family Scalidae. Vidensk. Meddel. Dansk Naturhist. Foren. 119: 55–58.

Tittle, K. M. 1964. Chemically stimulated escape responses of *Littorina planaxis* and *Littorina scutulata* to the carnivorous gastropod *Acanthina spirata*. Research paper, Biol. 175h, library, Hopkins Marine Station of Stanford University, Pacific Grove, Calif. 16 pp.

Tixier, R., and E. Lederer. 1949. Sur l'haliotivioline, pigment principal des coquilles d'*Haliotis cracherodii*. C. R. Acad. Sci. Paris 228: 1669–71.

Turner, H. J. 1953. The drilling mechanism of the Naticidae. Ecology 34: 222–23.

Tutschulte, T. C. 1976. The comparative ecology of three sympatric abalones. Doctoral thesis, Marine Biology, University of California, San Diego. 335 pp.

Tyler, A. 1939. Extraction of an egg membrane–lysin from sperm of the giant keyhole limpet (*Megathura crenulata*). Proc. Nat. Acad. Sci. 25: 317–23.

———. 1940. Sperm agglutination in the keyhole limpet, *Megathura crenulata*. Biol. Bull. 78: 159–78.

Urban, E. K. 1962. Remarks on the taxonomy and intertidal distribution of *Littorina* in the San Juan Archipelago, Washington. Ecology 43: 320–23.

Valentine, J. W., and R. F. Meade. 1960. Isotopic and zoogeographic paleotemperatures of Californian Pleistocene Mollusca. Science 132: 810–11.

Vassallo, M. T. 1969. The ecology of *Macoma inconspicua* (Broderip & Sowerby, 1829) in central San Francisco Bay. I. The vertical distribution of the *Macoma* community. Veliger 11: 223–34.

Vernberg, W. B., F. J. Vernberg, and F. W. Beckerdite, Jr. 1969. Larval trematodes: Double infections in the common mud-flat snail. Science 164: 1287–88.

Vilkas, A. G. 1974. Observations on the effect of various drugs on the activity of preoral cilia of the prosobranch veliger, *Conus californicus* Hinds. Veliger 16: 289.

Villee, C. A., and T. C. Groody. 1940. The behavior of limpets with reference to their homing instinct. Amer. Midl. Natur. 24: 190–204.

Walker, C. G. 1968. Studies on the jaw, digestive system, and coelomic derivatives in representatives of the genus *Acmaea*. Veliger 11 (Suppl.): 88–97.

Wara, W. M., and B. B. Wright. 1964. The distribution and movement of *Tegula funebralis* in the intertidal region, Monterey Bay, California (Mollusca: Gastropoda). Veliger 6 (Suppl.): 30–37.

Ward, J. 1966. Feeding, digestion, and histology of the digestive tract in the keyhole limpet *Fissurella barbadensis* Gmelin. Breeding cycle of the keyhole limpet *Fissurella barbadensis* Gmelin. Biol. Mar. Sci. 16: 668–84, 685–95.

Webber, H. H. 1970. Uptake of sea water into the fluid spaces of the prosobranch gastropod *Acmaea scutum*. Veliger 12: 417–20.

———. 1977. Gastropoda: Prosobranchia, pp. 1–97, *in* A. C. Giese and J. S. Pearse, eds., Reproduction of marine invertebrates, vol. 4. New York: Academic Press. 369 pp.

Webber, H. H., and P. A. Dehnel. 1968. Ion balance in the prosobranch gastropod, *Acmaea scutum*. Comp. Biochem. Physiol. 25: 49–64.

Webber, H. H., and A. C. Giese. 1969. Reproductive cycle and gametogenesis in the black abalone *Haliotis cracherodii*. Mar. Biol. 4: 152–59.

Weldon, P. J., and D. F. Hoffman. 1975. Unique form of tool-using in two gastropod molluscs (Trochidae). Nature 256: 720–21.

Wells, M. M. 1917. Behavior of limpets with particular reference to the homing instinct. J. Anim. Behav. 7: 387–95.

Werner, B. 1953. Über den Nahrungserwerb der Calyptraeidae (Gastropoda: Prosobranchia). Morphologie, Histologie und Funktion der am Nahrungserwerb beteiligten Organe. Helgoländer Wiss. Meeresunters. 4: 260–315.

West, L. A. 1978. Prey selection by the carnivorous snail *Nucella* (*Thais*) *emarginata* in a natural rocky intertidal environment. 59th Ann. Meet., West. Soc. Natur., Abs. Symp. & Contr. Pap., p. 7 (abstract).

White, T. J. 1968. Metabolic activity and glycogen stores of two distinct populations of *Acmaea scabra*. Veliger 11 (Suppl.): 102–4.

Whitlatch, R. B. 1974. Studies on population ecology of the salt marsh gastropod *Batillaria zonalis*. Veliger 17: 47–55.

Wicksten, M. K. 1978. Checklist of marine mollusks at Coyote Point Park, San Francisco Bay, California. Veliger 21: 127–30.

Wilbur, K. M., and C. M. Yonge, eds. 1964. Physiology of Mollusca. Vol. 1. New York: Academic Press. 473 pp.

———. 1966. Physiology of Mollusca. Vol. 2. New York: Academic Press. 645 pp.

Willoughby, J. W. 1973. A field study on the clustering and movement behavior of the limpet *Acmaea digitalis*. Veliger 15: 223–30.

Wise, S. W. 1970. Microarchitecture and deposition of gastropod nacre. Science 167: 1486–88.

Wolfson, F. H. 1974. Two symbioses of *Conus* (Mollusca: Gastropoda) with brachyuran crabs. Veliger 16: 427–29.

Wood, L. 1967. Acquisition of prey preference in a marine gastropod. Amer. Zool. 7: 787.

———. 1968. Physiological and ecological aspects of prey selection by the marine gastropod *Urosalpinx cinerea* (Prosobranchia: Muricidae). Malacologia 6: 267–320.

Wright, M. B. 1975. Growth in the black abalone, *Haliotis cracherodii*. Veliger 18: 194–99.

Yamada, S. B. 1977a. Range extension in *Littorina sitkana* Philippi, 1845, and range contraction in *Littorina planaxis* Philippi, 1847. Veliger 19: 368.

———. 1977b. Geographic range limitation of the intertidal gastropods *Littorina sitkana* and *L. planaxis*. Mar. Biol. 39: 61–65.

Yamada, S. B., and C. S. Sankurathri. 1977. Direct development of the intertidal gastropod *Batillaria zonalis* Bruguière, 1792. Veliger 20: 179.

Yarnall, J. L. 1964. The responses of *Tegula funebralis* to starfishes and predatory snails (Mollusca: Gastropoda). Veliger 6 (Suppl.): 56–58.

Yonge, C. M. 1938. Evolution of ciliary feeding in the Prosobranchia, with an account of feeding in *Capulus ungaricus*. J. Mar. Biol. Assoc. U.K. 22: 453–68.

———. 1953. Observations on *Hipponix antiquatus*. Proc. Calif. Acad. Sci. (4) 28: 1–24.

———. 1960. Further observations on *Hipponix antiquatus* with notes on North Pacific pulmonate limpets. Proc. Calif. Acad. Sci. (4) 31: 111–19.

———. 1962. Ciliary currents in the mantle cavity of species of *Acmaea*. Veliger 4: 119–23.

Yonge, C. M., and E. J. Iles. 1939. On the mantle cavity, pedal gland, and evolution of mucous feeding in the Vermetidae. Ann. Mag. Natur. Hist. (11) 3: 536–56.

Yonge, C. M., and T. E. Thompson. 1976. Living marine molluscs. London: Collins. 288 pp.

Yoshino, T. P. 1975. A seasonal and histologic study of larval *Digenea* infecting *Cerithidea californica* (Gastropoda: Prosobranchia) from Goleta Slough, Santa Barbara County, California. Veliger 18: 156–61.

Young, J. S., and J. D. DeMartini. 1970. The reproductive cycle, gonadal histology, and gametogenesis of the red abalone *Haliotis rufescens* (Swainson). Calif. Fish & Game 56: 298–399.

Zell, C. P. B. 1955. The morphology and general histology of the reproductive system of *Olivella biplicata* (Sowerby), with a brief description of mating behavior. Master's thesis, Zoology, University of California, Berkeley.

Chapter 14

Opisthobranchia and Pulmonata:
The Sea Slugs and Allies

Robert D. Beeman and Gary C. Williams

The opisthobranchs, or sea slugs, include some of the most beautiful and delicate organisms in the animal kingdom, and they have long been favorites of tidepool observers. Their external features vary widely. Generally, the head bears two pairs of sensory tentacles, an anterior pair of cephalic or oral tentacles (sometimes fused as a veil), used tactilely, and more posteriorly a conspicuous pair of antennalike rhinophores, of disputed sensory function, which are sometimes lacking. Some opisthobranchs (e.g., *Rictaxis*, 14.1), have a strong, spiral, external shell with an operculum; in *Aplysia* (e.g., 14.8, 14.9) and related forms, the shell is greatly reduced and buried in the tissues; and the nudibranchs (order Nudibranchia) and others completely lack shells as adults, though the dorsal body surface may be stiffened by calcareous spicules.

At one time the opisthobranchs were classified into two orders, the shell-less Nudibranchia and the usually shell-bearing Tectibranchia. The order Nudibranchia is still retained, but "Tectibranchia," clearly an unnatural group, has been replaced by the several orders used in this chapter. Definitions and further information on the orders are available in Beeman (1977), Franc (1968), Ghiselin (1966), Hyman (1967), Odhner (1939), Pruvot-Fol (1954), and Taylor & Sohl (1962).

Opisthobranchs are found in all seas from the polar regions through the tropics. Roughly 3,000 species are known. Species diversity is highest in tropical waters, and the vast Indo-Pacific zoogeographical province (which extends from east Africa to Hawaii) supports the greatest number of species. California shores have a remarkably diverse temperate fauna: approximately 130 opisthobranch species have been recorded for the state.

Opisthobranchs range in size from minute forms living between sand grains (order Acochlidiacea) to the California black sea hare, *Aplysia vaccaria* (14.9), probably the world's largest gastropod, reported to reach 76 cm in length and 15.9 kg in weight. The large size is especially impressive here, for many species are generally "annual" animals, maturing and often dying within a year after hatching from their deposited egg masses as microscopic swimming veliger larvae. The tiny larval shells of opisthobranchs characteristically spiral to the left rather than to the right as in most marine snails. Larval mortality is typically well over 99 percent; a slight reduction results in the population explosions that occur sporadically in various species. Some opisthobranchs, like *Bulla gouldiana* (14.6), are drab and inconspicuous; others, like *Corambe pacifica* (14.35), *Rostanga pulchra* (14.26), and *Phyllaplysia taylori* (14.10) accurately mimic their surroundings. Most nudibranchs, however, are noted for their brilliant and conspicuous markings. In this order, colors and patterns may be warning signals to predators; many opisthobranchs are foul-tasting or toxic, many produce highly acid secretions (e.g., pH 1.0), and some eolid nudibranchs (suborder Aeolidacea) pack their cerata (specialized projections on the dorsal body surface) with unexploded stinging nematocysts

Robert D. Beeman is Professor of Marine Biology at San Francisco State University. Gary C. Williams is a marine biologist at San Francisco State University.

they have isolated from their cnidarian food. Thus, opisthobranchs are preyed upon by very few animals. Recorded opisthobranch predators have usually been other opisthobranchs.

The food habits of most opisthobranchs are very poorly known, but some generalizations can be made. The nudibranchs are ecologically important in that they are primary predators on a wide variety of sessile organisms. The eolid nudibranchs, as noted above, eat mainly coelenterates such as hydroids and sea anemones. Many dorid nudibranchs feed specifically on sponges, whereas other dorids, along with the shell-less pteropods (order Gymnosomata) and some members of the order Cephalaspidea (e.g., *Navanax inermis*, 14.4) feed on mobile animals including other opisthobranchs. With rare exceptions, the sacoglossans (order Sacoglossa) and sea hares (order Anaspidea) are all herbivorous. And many of the tiny shelled species of the order Pyramidellida are external parasites on other invertebrates. Food habits of California nudibranchs are summarized in McDonald & Nybakken (1978). Opisthobranchs are not economically important to man.

Today's diverse opisthobranchs may have arisen from several groups of prosobranch gastropod ancestors, some of which showed a tendency to burrow. Early evolutionary selection of features favorable to burrowing may explain some of the trends that now characterize the opisthobranchs: reduction and loss of the shell, development of a compact body, return of the gill cavity to the rear ("Opisthobranch" = "rear gill") accompanied by untwisting, or detorsion, of the gut and nerve tracts, and reorganization of the gill itself. Gill structure varies greatly in opisthobranchs. Some forms, like *Gastropteron pacificum* (14.5), *Pleurobranchaea californica* (14.14), and their allies, have a gill (ctenidium) on the right side superficially resembling the gill of a prosobranch gastropod. Other opisthobranchs develop secondary gills very different in appearance, such as the ring of plumes about the anus in dorid nudibranchs (e.g., *Archidoris montereyensis*, 14.30), or the lateral rows of fingerlike projections (cerata) of eolid nudibranchs (e.g., *Hermissenda crassicornis*, 14.66). Sacoglossans like *Elysia hedgpethi* (14.12) lack gills altogether, and respire directly through the body surface. Along with these changes, opisthobranchs have developed complex hermaphroditic reproductive systems, reciprocal copulation, and distinctive threadlike sperm with a spiraled nucleus. Today some forms remain burrowers, but opisthobranchs are found in almost every marine habitat.

Many opisthobranch features are shared by another subclass of the gastropods, the Pulmonata. However, the pulmonates, which include the garden snails and slugs, are generally adapted to terrestrial life by having a lung instead of gills. Only a few pulmonates are marine. Six genera occur locally: *Trimusculus, Melampus, Onchidella, Pedipes, Ovatella,* and *Williamia*. Specific examples of the first three are included in this chapter.

Studies of California opisthobranchs to date have been mainly taxonomic and distributional in nature. An unusually large number of species have been called by two or more different names in recent literature; hence we have often cited synonyms, with authors and dates. We know little or nothing of the biology of many species. However, works on functional morphology, physiology, ecology, reproductive biology, and ultrastructure are beginning to appear more often. Especially dramatic contributions to neurophysiology have been made by studies on *Aplysia, Tritonia, Hermissenda, Anisodoris, Pleurobranchaea,* and *Melibe* (see especially Kandel, 1976, 1979). Much of the current work in systematics and natural history is published in the malacozoological journals, including *The Veliger* and *Malacologia,* or is cited in the *Opisthobranch Newsletter*. Recent workers, from whose studies we have drawn much of this chapter's information, include MacFarland (especially his magnificent 1966 posthumous monograph), Er. Marcus (especially 1961), Bonar, Greene, Hurst, Lance, McBeth, Nybakken, Steinberg, and many others. Baba (especially 1949) provides useful information on the related opisthobranchs of Japan. The keys to California opisthobranchs in Smith & Carlton (1975) are comprehensive. Popular accounts of local forms may be found in M. Johnson & Snook (1927), Ricketts & Calvin (1968), and Farmer (1968). The basic references of worldwide application are Franc (1968), Hoffmann (1932–40), Hyman (1967), H. Russell (1971), and T. Thompson (1976). Aspects of the general biology of opisthobranchs are also covered in Beeman (1977), Harris (1973), Kandel (1979), Morton (1967), Purchon (1968), Raven (1958), Wilbur & Yonge (1964, 1966), and Yonge & Thompson (1976). See Gosliner (1979, 1980) and Williams & Gosliner (1979) for recent changes in the scientific names of some California species included below.

Good color photography is the most satisfactory way to record the appearance of opisthobranchs. This chapter illustrates some of the opisthobranchs found intertidally and several of those likely to be found by divers. The reader is cautioned to consider that even specialists, using both external and internal features, have difficulty making definite identifications of many opisthobranchs. Finally, we are pleased to acknowledge our indebtedness to Joan Steinberg, James Carlton, Donald Abbott, and Eugene Haderlie for their generous help in the preparation of this chapter, and to numerous colleagues who provided photographs.

Phylum Mollusca / Class Gastropoda / Subclass Opisthobranchia
Order Cephalaspidea / Family Acteonidae

14.1 **Rictaxis punctocaelatus** (Carpenter, 1864)
(=Acteon punctocaelatus (Carpenter, 1864))
Striped Barrel Snail

Found plowing in sandy flats or in sand around rocks, usually in protected areas, rarely along open coast; uncommon intertidally, common subtidally to about 90 m; southeastern Alaska to Bahía Magdalena (Baja California), including possible subspecies.

Shell averaging about 12 mm long, white with tan to black-brown bands; operculum minute, far smaller than aperture, sometimes lost; radula well developed (contrasting with degenerate radula in *Acteon*).

This little-known species deposits its eggs in a gelatinous egg sac attached to the substratum by a slender thread. Most members of the family to which it belongs are specialized feeders on annelid worms. There are accounts of the anatomy of European relatives.

See Fretter & Graham (1954), Habe (1956), Johansson (1954), Kanafoff & Emerson (1959), Er. Marcus (1958a), Ev. Marcus (1972), and Rudman (1971).

Phylum Mollusca / Class Gastropoda / Subclass Opisthobranchia
Order Cephalaspidea / Family Scaphandridae

14.2 **Cylichnella inculta** (Gould, 1855)
(=Tornatina inculta Gould & Carpenter, 1856; Acteocina inculta (Gould, 1855))   Barrel Bubble Snail

At times extremely abundant about 5–6 mm under surface of syrupy muds, high intertidal mud flats in shallow bays, occasionally offshore; Monterey Bay to Gulf of California.

Shell averaging about 5.5 mm long, 2.5 mm in diameter, barrel-shaped, with external spire (often eroded), colored white, cream, or brown, without transverse stripes.

Like many other cephalaspideans, this species burrows just under the surface of soft substrata, using its fleshy head shield as a plough. It probably feeds on foraminifera; its biology is poorly known.

Two similar forms, **Cylichnella culcitella** (Gould, 1853) and **C. harpa** (Dall, 1871), were both formerly assigned to the genus *Retusa*. *C. culcitella* occurs in sand and mud flats in bays, from Alaska to Bahía de San Quintín (Baja California). Mature animals have shells over 10 mm long, white with brown periostracum, and showing a light-colored, spiraled etching. *C. harpa* is occasionally common in gravel, sand, and mud, low intertidal zone to about 80 m depth, from the Queen Charlotte Islands (British Columbia) to San Diego. The shell is strong, barrel-like, and 3–5 mm long, with an elongate aperture and strongly ridged whorls at one end. The animal feeds on small prosobranch gastropods.

See Gosliner (1979). For *C. inculta*, see also Franz (1967), Keen & Coan (1974), Er. Marcus (1955), McLean (1969), and Steinberg (1963a). For *C. culcitella*, see also McLean (1969) and Shonman & Nybakken (1978), as *Acteocina culcitella*. For *C. harpa*, see also McLean (1969), Roller (1971), and Steinberg (1963a).

Phylum Mollusca / Class Gastropoda / Subclass Opisthobranchia
Order Cephalaspidea / Family Aglajidae

14.3 **Aglaja ocelligera** (Bergh, 1894)
(=Doridium ocelligera Bergh, 1894)   Yellow-Spotted Aglaja

Often buried in surface layers of muddy bottoms, middle and low intertidal zones and subtidal waters; Bering Sea to San Diego.

Length often about 25 mm, but ranging to over 40 mm; color brownish black with very distinctive light yellowish to white spots.

This predaceous form lacks jaws and radula, but feeds on bubble snails (*Haminoea*), discharging the small shells of the prey clean and intact. It spawns pear-shaped egg masses from June to August at Friday Harbor (Washington) and in July and August at Dillon Beach (Marin Co.).

A close relative, **A. diomedea** (Bergh, 1894) (=*A. nana* Steinberg & Jones, 1960), is occasionally found burrowing in the surface layer of intertidal and subtidal mud in bays, from Kodiak Island (Alaska) to Morro Bay (San Luis Obispo Co.). It is smaller than *A. ocelligera* (about 10–20 mm long); the body ranges from uniform brown or black to cream mottled with brown, and generally lacks the light spots characteristic of *A. ocelligera*. However, San Francisco Bay specimens may be translucent gray-white marked with irregular black flecks and small yellow-brown dots. Information on the two species is mixed in MacFarland's monograph (1966; pl. 2, fig. 4 is *A. ocelligera*, not *A. diomedea*). Several other genera of cephalaspideans, mostly poorly known, are reported from California.

For *A. ocelligera*, see Ghiselin (1963), Gonor (1963), MacFarland (1966), Mattox (1958), Steinberg (1963a), Steinberg & Jones (1960), and White (1945). For *A. diomedea*, see Crane (1971), Glynn (1966), Gonor (1963), MacFarland (1966, in part, but not pl. 2, fig. 4), MacGinitie (1935), Er. Marcus (1961), Steinberg (1963a), Steinberg & Jones (1960), and White (1945).

Phylum Mollusca / Class Gastropoda / Subclass Opisthobranchia
Order Cephalaspidea / Family Aglajidae

14.4 **Navanax inermis** (Cooper, 1862)
(=Chelidonura inermis, Navarchus inermis)   Navanax

On sheltered shores, most abundant from low intertidal zone to about 8 m depth; individuals reach maximum size in shallow, protected bays in mud flats and eelgrass beds; small animals (about 50–60 mm, 8 gms) occur in protected areas, low intertidal zone to about 30 m depth; Monterey Bay to Gulf of California.

Body cigar-shaped, typically 100–125 mm long but reaching 200 mm in length and 300 gms in weight; dark velvet brown with blue-violet sheen; markings highly variable but generally including oblong yellow or white streaks or spots; margins trimmed with orange-yellow line spotted with bright blue; head with median notch; tentacles rolled; mouth palps with dense hairlike processes; ventral side rounded without typical appearance of foot; shell internal.

This common form is often found moving up mud-flat

channels on ebbing tides. A predator, it follows the mucus trails of various mollusks, tracking down and swallowing other shelled opisthobranchs (e.g., *Haminoea vesicula*, 14.7; *Bulla gouldiana*, 14.6; small members of its own species), later defecating the empty shells unbroken. When irritated, it exudes a yellowish fluid that causes an alarm response in other *Navanax* individuals. The light-yellow egg strings are laid in skeinlike gelatinous masses that are often seen on eelgrass in California, and on the brown algae *Padina* and *Sargassum* in the Gulf of California (especially during April). The copepod *Pseudomolgus navanaei* is often seen attached to the gills, and the pea crab *Opisthopus transversus* (25.42) sometimes nestles between the lateral flaps. Excellent studies have been made of *N. inermis* with respect to natural history, energetics, and limiting factors.

See Baba & Abe (1959), Barker & Levitan (1971), Bertsch & Smith (1970), Eskin & Harcombe (1977), Gosliner (1980), Gosliner & Williams (1972), Lance (1966), Levitan, Tauc & Segundo (1970), MacFarland (1966), MacGinitie (1935), Er. Marcus (1961), Er. Marcus & Ev. Marcus (1970), Murray (1977), Murray & Lewis (1974), Paine (1963, 1964, 1965), Rudman (1974), Sleeper & Fenical (1977), and Spray et al. (1977).

Phylum Mollusca / Class Gastropoda / Subclass Opisthobranchia
Order Cephalaspidea / Family Gastropteridae

14.5 **Gastropteron pacificum** Bergh, 1893
Pacific Wingfoot Snail

Not common intertidally, on mud, more common subtidally to at least 180 m depth on sandy or rocky bottoms; Aleutian Islands (Alaska) to Point Loma (San Diego Co.).

Body to about 33 mm long, sack-shaped, yellow with red flecks, with two large winglike flaps extending from lateral edges of foot and capable of being folded over body dorsally; gill of 16 – 20 leaflets partially hidden on right side; shell internal, smooth, coiled, nearly transparent.

Although *Gastropteron* moves with a gliding motion and can swim for brief periods, it is not truly pelagic. It is thought to feed upon foraminifera. The clear, gelatinous, almost globular egg mass contains widely separated, rounded capsules of spherical pink eggs. In northern parts of the range the eggs are attacked by the egg-sucking sacoglossan opisthobranch *Olea hansineensis*. MacFarland (1966) gives fine figures of this animal and its delicate internal shell, and provides an outline of the reproductive system. The anatomy and taxonomy of this and some related Japanese forms have received limited study, and the feeding structures have been compared with those of other cephalaspideans.

See Baba (1970), Baba & Tokioka (1965), Bertsch (1969a), Crane (1971), Farmer (1970), Haefelfuiger & Kress (1967), Hurst (1965, 1967), MacFarland (1966), Ev. Marcus (1971a), and Tokioka & Baba (1964).

Phylum Mollusca / Class Gastropoda / Subclass Opisthobranchia
Order Cephalaspidea / Family Bullidae

14.6 Bulla gouldiana Pilsbry, 1893
Cloudy Bubble Snail

Often very abundant, plowing in surface layers of mud and sandy mud just below lowest tide level in bays; Morro Bay (San Luis Obispo Co.) and Gulf of California to Ecuador.

Shell to about 55 mm in length, 37 mm in diameter, thin-walled, reddish brown or mottled grayish-brown spotted with black dots and light V-shaped markings; shell spire sunken; body yellow-brown to orange.

This is the largest of the California bubble snails. The mantle almost covers the shell in active animals. Note the commensal *Crepidula* snail on the shell of the specimen pictured. The commensal pea crab *Opisthopus transversus* (25.42) sometimes occurs in the mantle cavity. The diet of *B. gouldiana* is not well known, but members of the family Bullidae are generally considered to be herbivores, despite reports that some species swallow whole mollusks and crush them internally. *B. gouldiana* is one of the species eaten by the predaceous opisthobranch *Navanax inermis* (14.4).

The functional anatomy of the reproductive system of *B. gouldiana* has been recently investigated. The eggs are deposited in long, tangled, yellow egg strings on mud or in eelgrass during the summer, at least in southern California. The an-

imals are reported to live about a year. A short account of the natural history is available for a British form, *B. hydatis*.

See Berrill (1931), Guiart (1901), Keen (1971), Er. Marcus (1961), Oldroyd (1927), Robles (1975), Steinberg (1963a), and Zilch (1959–60).

Phylum Mollusca / Class Gastropoda / Subclass Opisthobranchia
Order Cephalaspidea / Family Atyidae

14.7 Haminoea vesicula (Gould, 1855)
White Bubble Snail

Seasonally abundant intertidally and subtidally on mud flats and sandy-mud areas of bays, and on boat floats; Alaska to Gulf of California.

Shell usually less than 15 mm long, fragile, translucent, whitish or greenish yellow, with sunken spire; shell aperture half the diameter of shell, with strongly curving internal profile and widely flaring anterior lip.

This form, whose shell is too small to contain the body, burrows just below the surface of sand or mud. It is preyed upon by the predaceous opisthobranch *Navanax inermis* (14.4). The eggs are deposited in a thick, coiled, deep-yellow egg string about 10 mm wide and up to 20 cm long.

A related species, **H. virescens** (Sowerby, 1833) (= *H. cymbiformis* Carpenter, 1853; *H. strongi* Naker and Hanna, 1927), is seasonally common in upper intertidal pools on rocky shores from Alaska to Panama. Its shell is larger (usually over 15 mm long), thicker, and whitish and more opaque; its aperture is more than half the diameter of the shell.

See Costello & Henley (1971), Keen (1971), Er. Marcus & Ev. Marcus (1967), McLean (1969), Morton (1972), Perrier & Fischer (1914), Smallwood (1904), Spicer (1933), Steinberg (1963a), and Vincente (1962, 1966, 1969b).

Phylum Mollusca / Class Gastropoda / Subclass Opisthobranchia
Order Anaspidea / Family Aplysiidae

14.8 Aplysia californica Cooper, 1863
(=Tethys californica (Cooper, 1863); T. ritteri Cockerell, 1901)
California Brown Sea Hare

Among seaweed, sometimes in kelp canopy, low intertidal
zone to depths of over 18 m on sheltered shores with light to
moderate wave action, sometimes displaced to sandy areas
and points of strong surf; also on mud flats and bottoms of
shallow bays, estuaries, and harbors; Humboldt Bay to Gulf
of California; sporadic in northern part of range.

Large specimens exceeding 400 mm in length and weighing
several kilograms; mature forms reddish, brownish, and/or
greenish, overlaid with a network of dark lines arising from
scattered round spots (yellow line on Fig. 14.8a is a "spaghetti" marking tag); young specimens usually reddish; parapodia (dorsal flaps) not joined to each other at rear, marked
with banding on upper, inner surfaces; shell internal, with an
accessory plate.

In terms of what can be learned in the laboratory, this is
one of the best-known marine animals. Knowledge is expanding more rapidly now than ever before, and the present
account scarcely scratches the surface of what is known.

Individuals move about more by day than by night. Locomotion involves a series of actions reminiscent of walking in
the inchworm. Starting with its foot on the ground, the sea
hare raises its front end clear of the substratum, extends it
well forward, brings it down again and reattaches it, then
releases the foot posteriorly and draws the rear part forward.
This is repeated time and again, the direction of movement
changing if the anterior body is inclined to the right or left as
it is extended. Hungry animals generally move more than
those that are well fed. The rate of locomotion is generally
slow, but speeds of up to 1.3 m per minute have been noted
for short periods. The sea hares give an avoidance response to
contact with some sea stars (e.g., *Astrometis sertulifera*, 8.8) and
predaceous opisthobranchs (*Navanax inermis*, 14.4; *Pleurobranchaea californica*, 14.14), rapidly withdrawing the head, turning
sharply, and literally galloping away. They are unable to
swim, though some related species can do so by flapping the
parapodia like wings or using them to expel a water jet.

Very few animals are known to prey on post-larval juvenile
and adult *A. californica*, but in some bay situations (e.g., Lun-
ada Bay, Palos Verdes, Los Angeles Co.) the giant green
anemone *Anthopleura xanthogrammica* (3.30) is reported to take
substantial numbers. Contact with the anemone elicits escape
behavior enabling most but not all *Aplysia* to elude their
captors. The sea hares are ingested whole, but the remains are
expelled when digestion is only 67–85 percent complete. The
digestive gland of *Aplysia*, comprising roughly 10 percent of
the body weight, contains toxic materials, and when digestion
by the anemone reaches a point where the gland is exposed,
the anemone regurgitates its meal.

Aplysia californica is herbivorous. Individuals feed mainly by
day, and graze on a variety of algae and the eelgrass *Zostera*.
Food is scented at a distance underwater by receptor organs
on the posterior and probably anterior tentacles. When food
is held a few centimeters away, the animals may extend the
neck region and wave the head and tentacles about to localize
the source of the scent. Taste receptors (contact chemoreceptors) occur on the anterior tentacles and oral veil. In feeding,
an individual uses its toothed radula to grasp and tear off
sizable pieces of plant material. These are swallowed and held
temporarily in a distensible region of the esophagus (the
crop), then passed to a muscular gizzard lined with strong
pyramidal teeth where they are ground up and mixed with
enzymes from various sources. The anus is located dorsally,
near the rear of the parapodial lobes, and fecal pellets emerge
by way of the siphon. The siphon also expels water that has
already passed through the gill. The gill itself is hidden between the parapodial folds and covered by a flap of tissue
containing the small, flattened shell.

The head bears a pair of simple eyes that detect changes in
intensity of white light; sensitivity is poor at the red end of
the visible spectrum, better in the near ultraviolet. In general,
the animals respond less strongly to light than to chemical
stimuli. The eye shows a circadian rhythm in nervous (electrical) activity; the biological clock is contained in the eye itself,
and in the eyes removed from the body and maintained in
organ culture it can be reset and entrained. The osphradium,
a receptor organ lying within the mantle cavity near the gill,

detects osmotic changes in the sea hare's aquatic environment.

When irritated, the animals exude a dark-purple fluid, whose source is the purple gland. The ink gets its purple color from a derivative of phycoerythrin, the red pigment in the red algae composing part of the sea hare's diet. Animals that are milked of their purple secretions and then fed only on brown and green algae produce no purple dye. To some degree, body color, too, varies with the diet.

All adult individuals are simultaneous hermaphrodites, but they cannot fertilize their own eggs and must mate with other individuals. In mating, any individual may act only as a male, or only as a female, or as both sexes simultaneously; sometimes whole chains and circles of mating animals are seen in the field. The sperm in aplysiids resembles an attenuated corkscrew; it consists of a long flagellum, around the anterior portion of which are twisted a long helical nucleus and a pair of even longer enveloping mitochondria. After mating, the animals lay long, tangled, yellowish-green, spaghetti-like egg strings containing up to a million eggs. These are deposited among rocks and seaweeds, intertidally and subtidally, and hatch in about 12 days. In Monterey harbor the eggs have been found in August and December.

Several species of *Aplysia* (including *A. californica*) and allied genera have now been raised through their whole life histories in the laboratory. In *A. californica* the hatched veliger larvae swim for 34 days or more. They settle and undergo metamorphosis consistently only on red algae. Following metamorphosis, feeding juveniles double their weight every 10 days for 3 months before growth begins to slow down. Some individuals are sexually mature 120 days after hatching. though they are not fully grown. The life span, often a year or less, is short for so large an animal. Populations in the field fluctuate greatly from year to year, as often happens in species producing large numbers of eggs.

The anatomy and embryology of *Aplysia* species are reported in large, older works, but both structure and development are being reinvestigated with modern tools and methods. And many recent publications reflect the extensive use of species of *Aplysia* in neurobiological studies.

Aplysia californica has turned out to be nearly ideal for research aimed at relating the overt behavior of an animal to the structure and function of the cells and cell networks that make up its nervous system. The central nervous system of *A. californica* is composed of nine major ganglia linked together by nerve fiber tracts. The nerve cell bodies (as in most invertebrates) occupy the peripheries of the ganglia, and some of them are extrordinarily large—nearly a millimeter in diameter. Their large size makes these neurons easier to study by such techniques as the electrical recording of nervous activity, the injection of dyes and drugs directly into the cell, and chemical analysis of the contents and secretions of individual cells. Not only are some neurons conveniently large, but they are so consistently placed and so distinctly color-coded that the same corresponding cell can be recognized in different individuals. Thus one can carry out experiments on a particular nerve cell in different animals, just as one can experiment with the same organ in different individuals. Moreover, the number of neurons in an average individual ganglion of *Aplysia* (roughly 2,000), and indeed in the whole nervous system, although large, is still far smaller than that in the simplest vertebrates, and the connections between cells are generally easier to determine and map out. The repertoire of behavior of *Aplysia* is less elaborate than that in many higher organisms, and is sufficiently stereotyped to make behavior studies simpler.

The various ways in which these attributes and other features of *Aplysia* and its relatives have been used to elucidate key questions and problems in the neurobiological basis of behavior are presented with simplicity, clarity, and elegance in two recent books (Kandel, 1976, 1979); much good natural history is presented in a matrix of ideas and experiments relating to a very active and exciting frontier of modern biology.

See especially Kandel (1976, 1979, both containing excellent bibliographies); for anesthesia and dissection techniques, see Beeman (1968a) and Purchon (1968); see also Arch (1976), Batham (1961), Bebbington & Thompson (1969), Beeman, (1963a, 1968b, 1972, 1977), Bern (1967), Brandriff & Beeman (1973), Brenori et al. (1968), Bullock & Horridge (1965), Carefoot (1967, 1970), Castellucci et al. (1970), Chapman & Fox (1969), Coggeshall (1967, 1969, 1970), Darling & Cosgrove (1966), Eales (1921, 1960), Frazier et al. (1967),

Gardner (1971), Howells (1942), Hughes & Tauc (1962), Jacklet (1969a,b), Jahan-Parwar (1972), P. Johnson & Chapman (1970), Kandel (1970), Kandel, Frazier & Coggeshall (1967), Kandel et al. (1967), Kay (1964), Kennedy (1971), Krakauer (1974), Kriegstein (1977a,b), Kriegstein, Castellucci & Kandel (1974), Kupfermann (1972), Kupfermann & Kandel (1969), Kupfermann et al. (1970), Lasek & Dower (1971), Lewis, Everhart & Zeevi (1969), Lickey & Berry (1966), Linton (1966), MacFarland (1966), MacGinitie (1934), Er. Marcus (1958a, 1961), Mazzarelli (1893), McCaman & Dewhurst (1970), Morton (1972), Peretz (1969), Peterson (1970), Pinsker et al. (1970), Raven (1958, 1964), Strenth & Blankenship (1978), Switzer-Dunlap (1978), Switzer-Dunlap & Hadfield (1977), T. Thompson & Bebbington (1969, 1970), Tobach, Gold & Ziegler (1965), Toevs & Brackenbury (1969), Usuki (1970), Vicente (1966, 1969a,b), Wachtel & Kandel (1967), Waser (1968), Wilson (1971), Winkler (1955, 1957), Winkler & Dawson (1963), Winkler & Tilton (1962), and the references under *A. vaccaria* (14.9).

Phylum Mollusca / Class Gastropoda / Subclass Opisthobranchia
Order Anaspidea / Family Aplysiidae

14.9 **Aplysia vaccaria** Winkler 1955
California Black Sea Hare

Sporadically common, low intertidal zone in rocky and sandy areas, and subtidal waters especially around kelp beds; Morro Bay (San Luis Obispo Co.) to Bahía de los Ángeles (Baja California).

Probably the world's largest gastropod, commonly under 300 mm but attaining a length of over 760 mm (ca 30 inches), and a weight of 15.9 kg (ca 35 lbs); body dark reddish brown to black, with white-speckled patches more or less visible; parapodia broadly joined to each other posteriorly, overlapping dorsally; shell internal, saucerlike, without an accessory plate; membrane covering shell with a distinct dorsal opening; purple secretion absent (compare *A. californica*, 14.8).

This species appears to feed almost entirely on the large brown kelp *Egregia*. Its large, tangled, pinkish-white egg strings are deposited both intertidally and subtidally. The pinnotherid crab, *Opisthopus transversus* (25.42) is sometimes found as a commensal living under the parapodial folds. Experiments with the closely related *Aplysia juliana* of Hawaii show that the cephalic tentacles are involved in tasting food and suggest that the rhinophores detect currents.

See Bartsch & Rehder (1939), Beeman (1968a), Beondé (1968), Bern (1967), Carefoot (1967, 1970), Frings & Frings (1965), Konigsor & Hunsaker (1971), Lance (1967), Lickey (1968), Van Weel (1957), Winkler (1955, 1957), and Winkler & Dawson (1963), and the references under *A. californica* (14.8).

Phylum Mollusca / Class Gastropoda / Subclass Opisthobranchia
Order Anaspidea / Family Dolabriferidae

14.10 **Phyllaplysia taylori** Dall, 1900
(=Phyllaplysia zostericola McCauley, 1960; Petalifera taylori (Dall, 1900))   Taylor's Sea Hare

Sometimes common on blades of the eelgrass *Zostera*, low intertidal zone in bays and estuaries; Nanaimo (British Columbia) to San Diego.

Body usually about 25–45 mm long but ranging to over 75 mm, dorsoventrally flattened; coloration mimicking that of *Zostera*, a bright-green base color with brown-black and white spots and stripes; sometimes with small, clear, flat shell on mantle shelf above gill cavity.

Phyllaplysia taylori is the smallest of the California sea hares and is an ideal animal for many experimental studies. It lives almost entirely upon *Zostera*, and at low tide can most often be found wedged between the lower sections of the eelgrass blades. The animals graze on the film of diatoms and other small organisms that cover the eelgrass. As in other sea hares, the stomach contains a set of hardened teeth, which help grind up the food ingested. The siliceous shells of diatoms are abrasive, requiring a complete renewal of the stomach teeth in as little as 25 days. Eggs are deposited in rectangular, adhesive packets, which are attached to the *Zostera*, mainly in the summer. Growth is very rapid, and the life span may be as short as 3–8 months. Specimens from Elkhorn Slough (Monterey Co.) are usually largest from April to October. Tolerance to changes in salinity and temperature is evidently great.

Sea hares, like other opisthobranchs, are hermaphroditic, producing eggs and sperm at the same time. Fertilization is internal, and all incoming and outgoing reproductive products must pass through a single duct. Because copulation is often reciprocal and the animals may be laying eggs at the same time, the single duct may simultaneously contain incoming sperm, outgoing sperm, and an exiting string of fertilized eggs—a plumber's nightmare. Obviously the system works, but just how it does was a matter of controversy for many years. Recent studies, using radioactive labels to distinguish an animal's own sperm from that of its mate, have settled some problems and provided a detailed picture of the functional anatomy of the reproductive system of *P. taylori*. Autoradiographic study has also elaborated the stages of spermatogenesis and indicated that the entire process takes about 20 days.

See Beeman (1963b, 1968a,b , 1969, 1970a,b,c,d, 1972, 1977), Brandriff & Beeman (1973), Bridges (1975), Krakauer (1971), MacFarland (1966), Er. Marcus (1961), McCauley (1960), T. Thompson (1970), and G. Williams & Gosliner (1973b).

Phylum Mollusca / Class Gastropoda / Subclass Opisthobranchia
Order Sacoglossa / Suborder Polybranchiacea / Family Hermaeidae

14.11 Stiliger fuscovittatus Lance, 1962
Streaked Stiliger

Very abundant seasonally (April through June), exclusively on filamentous red algae, low intertidal zone in bays; San Juan Island (Washington) to Bahía de los Ángeles (Baja California).

Sexually active animals averaging about 8 mm in length, ranging to over 10 mm; body translucent with reddish-brown spots and lines; cerata large, with white spots; rhinophores tapered, smooth, not grooved, with a brownish line running down from near tip, sometimes continuing on body between eyes.

This small form lives and feeds on the red alga *Microcladia* (not *Polysiphonia pacifica*, as originally reported) in Puget Sound (Washington). In Monterey Bay, it is always found on the red alga *Callithamnion*. As in other sacoglossans, the radula has relatively few teeth, arranged singlefile. Only one tooth at a time is operational, and this is used to slit open plant cells so that their contents can be sucked up. Old worn or broken teeth are retained and stored in a sack below the mouth ("Sacoglossa" = "tongue sack"). The spiral egg strings, which

contain sausage-shaped sections, are variable in length; they are laid on algae mainly from May to July. Most details of anatomy and natural history are still lacking.

The defense responses of a related Atlantic species, *S. vanellus*, include shedding the cerata and secreting from them fluids that might deter predators.

Several other sacoglossans, all diagnosed in McDonald (1975), may be found on California marine plants. **Alderia modesta** (Lovén, 1844) occurs on the alga *Vaucheria* in *Salicornia* marshes; **Aplysiopsis smithi** (Marcus, 1961) is found on the green alga *Enteromorpha*; **Aplysiopsis oliviae** (MacFarland, 1966) is found on brown algae; and **Placida dendritica** (Adler & Hancock, 1843) occurs on the green algae *Bryopsis* and (more rarely) *Codium*.

For *S. fuscovittatus*, see Clark (1971), Edmunds (1966a), Fretter (1943), Kay (1968), Lance (1962b), and H. Russell (1946). For other species, see Gonor (1961; for *Aplysiopsis*, see *Hermaeina*), Greene (1968), Hand & Steinberg (1955), Keen (1973, 1974), Long (1969a), MacFarland (1966; for *Aplysiopsis*, see *Hermaeina*), Er. Marcus (1961), McDonald (1975), Pruvot-Fol (1954), Roller (1970b), and G. Williams & Gosliner (1973a).

Phylum Mollusca / Class Gastropoda / Subclass Opisthobranchia
Order Sacoglossa / Suborder Elysiacea / Family Elysiidae

14.12 Elysia hedgpethi Marcus, 1961
(= E. bedeckta MacFarland, 1966)   Hedgpeth's Elysia

Seasonally common on the green algae *Codium fragile* and *Bryopsis corticulans*, low intertidal and subtidal mud flats; San Juan Island (Washington) to Puertecitos (Baja California).

Body to at least 35 mm long, olive-green with large lateral flaps bearing conspicuous light-blue flecks and a few small white spots.

Elysia hedgpethi feeds on green algae by piercing the cell walls with the radula and sucking out the cell contents. Although the animals closely resemble the algae on which they feed, they are easily collected by placing *Bryopsis* or *Codium* in a bowl of seawater and allowing the sacoglossans time to crawl off. The egg strings are white and cylindrical, and arranged in a counterclockwise spiral 4–6 mm in diameter.

They have been noted on *Codium fragile*. Reproduction, larval features, and ecology have been studied in several other species of *Elysia*.

It has been known for a century that the green color in the European *E. viridis* was chlorophyll. Electron microscopy and physiological studies have recently shown that *E. viridis*, *E. hedgpethi*, and several other sacoglossans examined store functional chloroplasts within the cells of the branched digestive glands. The chloroplasts, derived from the algal food, carry on photosynthesis for as long as 10 days after ingestion in *E. hedgpethi*, but in *E. viridis* the chloroplasts live and carry out photosynthesis for at least 3 months after ingestion, provided the animals are kept in the light.

See Baba (1957b), Bailey & Bleakney (1967), Clark (1971), Farmer (1967), Fretter (1940), Greene (1968, 1970a,b,c), Kawaguti & Yanasu (1965), Lance (1966), MacFarland (1966), Er. Marcus (1955, 1961), Muscatine & Greene (1973), Reid (1964), H. Russell (1946), D.L. Taylor (1967, 1968, 1970), T. Thompson (1961a), and G. Williams & Gosliner (1973a).

Phylum Mollusca / Class Gastropoda / Subclass Opisthobranchia
Order Notaspidea / Family Tylodinidae

14.13 **Tylodina fungina** Gabb, 1856
Yellow Sponge Tylodina

Sometimes abundant on the yellow sponge *Aplysina fistularis* (2.7), occasionally on rocks or kelp, low intertidal zone to about 9 m depth along rocky shores and in bays; Cayucos (San Luis Opispo Co.) to Guaymas (Mexico); Galápagos Islands.

Body typically about 35 mm long; shell external, limpetlike, "thatched" in appearance, about 23 mm long; body and shell yellow; mantle cavity lacking; gill well developed, halfway back under right edge of shell; anus just above gill; genital groove anterior to right rhinophore; rhinophores hollow, slit near tips.

This species generally matches the color of its sponge host. The mollusk excavates depressions in the sponge that just fit its foot.

See Bertsch (1970), Burn (1960), Dushane (1966), Keen (1971), MacFarland (1966), Ricketts & Calvin (1968), Sphon & Mulliner (1972), T. Thompson (1970), and Vayssiére (1898).

Phylum Mollusca / Class Gastropoda / Subclass Opisthobranchia
Order Notaspidea / Family Pleurobranchidae

14.14 **Pleurobranchaea californica** MacFarland, 1966
Pleurobranchaea

Locally abundant on fine sand and green mud, shallow subtidal bottoms to depths of about 340 m; often taken on set lines and in crab pots; mouth of Klamath River (Humboldt Co.) to San Diego and Santa Cruz Island (Channel Islands).

Body to about 360 mm long, mottled brown with large, irregular, translucent white patches; shell absent; large gill under right edge of fleshy dorsal shield; oral veil above mouth broad, with many short, branched processes extending anteriorly.

Despite its size, this bulky notaspidean is almost neutrally buoyant when submerged. It creeps actively (up to 1 m per minute), apparently by ciliary action alone, and can move upside down on the underside of a water surface film. When irritated, it swims in a clumsy fashion by strongly flexing the body alternately to the right and left. Other defenses include biting a source of prolonged irritation and secreting strongly acid materials from the dorsal surface. Under laboratory conditions, *Pleurobranchaea* is a very active predator and scavenger, feeding on almost any meat available, including the cephalaspidean *Navanax inermis* (14.4) the sea anemone *Anthopleura elegantissima* (3.31), dead squid, and even other members of its own species. The huge eversible proboscis is equipped with powerful lateral jaws and a very large radula; thus large food items can be engulfed and swallowed whole.

Behavior studies show numerous, rather stereotyped patterns of response. Animals turned upside down usually right themselves within a half a minute. Mating responses, though normally occurring only when two animals are placed together (and not always then), are often displayed by solitary individuals placed in water that recently contained mating pairs. The feeding response is a dominant one, preempting most other activities except egg laying: animals that have been turned upside down and fed usually remain on their backs until the food is gone; copulating animals separate and commence to eat when food is presented. Hormones found in the nervous system and blood of animals laying eggs raise the threshold of the feeding response greatly, an effect that presumably reduces the chances that the animals will eat their own eggs. The consistency of the response patterns plus the simplicity of the nervous system, the ease with which it can be exposed or even removed from the body, and the large size of the nerve cells make *Pleurobranchaea* exceptionally suitable for studies of the physiological basis of behavior and learning.

See Chivers (1967), Coan (1964), Davis & Gillette (1978), Davis & Mpitsos (1971), Davis, Mpitsos & Pinneo (1974), Davis, Siegler & Mpitsos (1973), Davis et al. (1974), Gillette, Kovac & Davis (1978), Kandel (1979), MacFarland (1966), Mpitsos & Collins (1975), Mpitsos, Collins & McClellan (1978), Mpitsos & Davis (1973), Odhner (1926), Ram & Davis (1977), Siegler (1977), and T. Thompson (1969, 1970).

Phylum Mollusca / Class Gastropoda / Subclass Opisthobranchia
Order Notaspidea / Family Pleurobranchidae

14.15 **Berthella californica** (Dall, 1900)
(=Pleurobranchus californicus Dall, 1900)   White Berthella

Occasional under low intertidal rocks, uncommon subtidally to 33 m; Crescent City (Del Norte Co.) to San Diego.

Body usually about 20 mm in length, but ranging to about 50 mm, white; dorsum lacking turbercles; large gill with smooth stem under right edge of dorsal shield; internal shell thin, white, extending at least one-half length of body (determined by feeling through fleshy dorsal shield).

The anatomy of a relative, *B. granulata*, has been reported in some detail. The animals probably feed on sponges. Another relative, **Pleurobranchus strongi** MacFarland, 1966, is occasionally found in rocky tidepools and subtidal waters from Moss Beach (San Mateo Co.) to San Pedro (Los Angeles Co.). It reaches a length of about 20 mm; the dorsum is pale yellow and bears numerous fine tubercles delicately tipped and cored with yellow.

Some members of the family Pleurobranchidae repel enemies with extremely strong acid secretions from glands in the dorsum.

For *B. californica*, see Baba (1969), Bertsch (1970), Hill (1962), Lance (1966), MacFarland (1966), Er. Marcus & Ev. Marcus (1955), Ev. Marcus (1971b), and Mattox (1953). For other species of *Berthella*, see T. Thompson (1960a) and T. Thompson & Slinn (1959). For *Pleurobranchus strongi*, see MacFarland (1966), Morton (1972), and Roller (1970b).

Phylum Mollusca / Class Gastropoda / Subclass Opisthobranchia
Order Nudibranchia / Suborder Doridacea / Superfamily Eudoridoidea
Family Conualeviidae

14.16 **Conualevia alba** Collier & Farmer, 1964
Smooth-Horned Dorid

Uncommon under stones, low intertidal zone; subtidal to 9 m; Pacific Grove (Monterey Co.) to Bahía de Tortuga (Baja California).

Body usually 12–35 mm long, completely white with opaque white glands at edge of dorsum; rhinophores long and tapering, very smooth throughout.

See Collier & Farmer (1964) and Lance (1966).

Phylum Mollusca / Class Gastropoda / Subclass Opisthobranchia
Order Nudibranchia / Suborder Doridacea / Superfamily Eudoridoidea
Family Cadlinidae

14.17 **Cadlina limbaughi** Lance, 1962
Limbaugh's Cadlina

Uncommon under stones, low intertidal zone; subtidal to 9 often seen in subtidal zone, 10–45 m depth; Santa Barbara to Isla Coronados (Baja California).

Body typically 15–31 mm long; dorsum translucent white with numerous scattered opaque spots, these white to lemon-yellow in color; gills and rhinophores tipped with black to dark reddish brown; gills numbering 6–8, bipinnate or tripinnate.

Nothing has been reported on the biology of this species.

See Lance (1962a) and Sphon & Lance (1968).

Phylum Mollusca / Class Gastropoda / Subclass Opisthobranchia
Order Nudibranchia / Suborder Doridacea / Superfamily Eudoridoidea
Family Cadlinidae

14.18 **Cadlina flavomaculata** MacFarland, 1905
Yellow-Spotted Cadlina

Uncommon in most areas, but typically found in groups under rocks and in rocky tidepools, low intertidal zone; subtidal to 20 m; Vancouver Island (British Columbia) to Punta Eugenia (Baja California); rare in Gulf of California.

Body 15–20 mm long, white to yellowish with 6–11 yellow spots in a row on each side; dorsum densely spiculate; rhinophores very dark brown-black; gills 10–12, white to yellow.

These extremely sluggish animals have been seen feeding on the sponge *Aplysilla glacialis* (2.5), but little is known of their biology.

See Lance (1962a), MacFarland (1906, 1966), Er. Marcus (1961), McDonald & Nybakken (1978), and Nybakken (1978).

Phylum Mollusca / Class Gastropoda / Subclass Opisthobranchia
Order Nudibranchia / Suborder Doridacea / Superfamily Eudoridoidea
Family Cadlinidae

14.19 **Cadlina luteomarginata** MacFarland, 1966
(=C. marginata MacFarland, 1905)   Yellow-Edged Cadlina

Common, especially under rocky ledges with sponges and tunicates, low intertidal zone and subtidal waters to 45 m depth; Vancouver Island (British Columbia) to Punta Eugenia (Baja California); not known intertidally in southern part of range.

Body usually 25–40 mm long, but ranging to 80 mm, whitish to light yellow; narrow band around margin and tips of low tubercles bright yellow; bipinnate gills numbering six; as in other *Cadlina*, dorsum feels warty and gritty to touch, owing to dense, freely projecting spicules.

This species is reported to feed on sponges, including *Halichondria panicea* (2.15), *Myxilla incrustans*, and *Higginsia* sp.

See Belcik (1975), Bertsch et al. (1972), Lance (1962a), MacFarland (1966), McDonald & Nybakken (1978), and Nybakken (1978).

Phylum Mollusca / Class Gastropoda / Subclass Opisthobranchia
Order Nudibranchia / Suborder Doridacea / Superfamily Eudoridoidea
Family Cadlinidae

14.20 **Cadlina modesta** MacFarland, 1966   Modest Cadlina

Sporadically common among sponges in rocky areas, low intertidal zone to 50 m depth; Vancouver Island (British Columbia) to La Jolla (San Diego Co.).

Body usually 10 to 45 mm long, pale yellow, with continuous, irregular series of yellow dots along sides; dorsum tuberculate, densely spiculate; rhinophores pale to light brown; gills 10–12, pale yellow to whitish.

Observers have seen this nudibranch feeding on the sponge *Aplysilla glacialis* (2.5).

See Bertsch (1969a), MacFarland (1966), McDonald & Nybakken (1978), Nybakken (1978), and Robilliard (1971c).

Phylum Mollusca / Class Gastropoda / Subclass Opisthobranchia
Order Nudibranchia / Suborder Doridacea / Superfamily Eudoridoidea
Family Cadlinidae

14.21 **Cadlina sparsa** (Odhner, 1921)
Dark-Spotted Cadlina

Rare, low intertidal zone on rocky shores and subtidal waters to about 40 m depth; Monterey Bay to Islas Juan Fernández (Chile).

Body usually 15–25 mm long; dorsum light yellow to yellowish pink, with small, irregularly spaced, black or brown spots with or without yellow centers (the spots are dark glands with yellow outlets); rhinophores ridged with 12 leaflets on each side; gills 12 in number, unipinnate.

A related form, *C. laevis*, has been shown to develop directly into a miniature of the adult, without intervention of the free-swimming veliger larval stage typical of most opisthobranchs.

See Er. Marcus (1959, 1961), Nybakken (1978), and T. Thompson (1967).

Phylem Mollusca / Class Gastropoda / Subclass Opisthobranchia
Order Nudibranchia / Suborder Doridacea / Superfamily Eudoridoidea
Family Chromodorididae

14.22 Chromodoris macfarlandi Cockerell, 1902
(=Glossodoris macfarlandi (Cockerell, 1902))
MacFarland's Chromodoris

Typically subtidal on rocky bottoms at north end of range, more common in low intertidal zone in south; Monterey Bay to around Isla Cedros (Baja California).

Body usually less than 35 mm long, but reaching 60 mm, brilliant purple with yellow margin, and with three longitudinal, golden-yellow stripes and a yellow stripe on middle of foot tip (virtually colorless after preservation).

Like other *Chromodoris*, this species feeds on sponges. It can repel predators such as *Navanax inermis* (14.4) with noxious non-acid secretions.

See Aboul-Ela (1959), Bertsch (1977), Cockerell (1901), Engel & Nijssen-Meyer (1964), Farmer (1963), MacFarland (1966), McDonald & Nybakken (1978), and Young (1967, 1969).

Phylum Mollusca / Class Gastropoda / Subclass Opisthobranchia
Order Nudibranchia / Suborder Doridacea / Superfamily Eudoridoidea
Family Chromodorididae

14.23 Chromodoris porterae Cockerell, 1902
(=Hypselodoris porterae, Mexichromis porterae, Glossodoris porterae)   Porter's Nudibranch

Uncommon, rocky low intertidal zone in summer; subtidal to 18 m; Monterey to Isla Cedros (Baja California); rare at northern end of range.

Body usually 15–30 mm long, smooth, ultramarine in color, with a pair of longitudinal golden-yellow lines broken at rhinophores, lacking yellow spots.

The specific name, *porterae*, comes from Mrs. Cockerell's maiden name.

The animals have been seen feeding on the sponge *Dysidea amblia*.

See Bertsch (1977), Cockerell (1901), MacFarland (1966), McDonald (1975), McDonald & Nybakken (1978), and the references under *C. macfarlandi* (14.22) and *Hypselodoris californiensis* (14.24).

Phylum Mollusca / Class Gastropoda / Subclass Opisthobranchia
Order Nudibranchia / Suborder Doridacea / Superfamily Eudoridoidea
Family Chromodorididae

14.24 Hypselodoris californiensis (Bergh, 1879)
(=Glossodoris californiensis (Bergh, 1879); Chromodoris californiensis Bergh, 1879)   Blue-and-Gold Nudibranch

Sometimes common, often under rocks in low intertidal zone and in subtidal waters to 30 m depth on rocky coasts; Monterey Bay to Gulf of California; rare at northern end of range.

Body to about 67 mm long, deep ultramarine-blue with many golden-yellow or orange spots; color in alcohol uniformly greenish blue.

This species feeds on sponges (*Steletta estrella*, *Dysidea amblia*, and *Haliclona* sp.) and can repel predators by means of noxious secretions. The taxonomic status of the Gulf of California form is controversial.

See D. Anderson & Lane (1963), Bertsch (1973, 1977), Burn (1961, 1962, 1966), Crozier & Arey (1919), MacFarland (1966), McBeth (1970, 1971), McDonald & Nybakken (1978), H. Russell (1968), Sphon (1971), and Young (1967, 1969).

Phylum Mollusca / Class Gastropoda / Subclass Opisthobranchia
Order Nudibranchia / Suborder Doridacea / Superfamily Eudoridoidea
Family Actinocyclidae

14.25 Hallaxa chani Gosliner & Williams, 1975
Chan's Nudibranch

Generally uncommon, low intertidal zone on rocky shores, sometimes locally common in tidepools in summer; Tomales Point (Marin Co.) to Shell Beach (San Luis Obispo Co.).

Body usually 10–20 mm long; ground color pale lemon yellow; central hepatic region brownish, with many small brownish flecks and larger paired dark spots in front of branchial plume; rhinophores pale yellow, tipped with brownish maroon; gills numbering 12–14, unipinnate.

Feeding has been reported on the colonial tunicate *Didemnum carnulentum* (12.11), but virtually nothing else is known of the biology of this species. It is named for biologist Gordon L. Chan.

See Gosliner & Williams (1970, as *Doris* sp.; 1975), McDonald (1975, as *Hallaxa* sp.), McDonald & Nybakken (1978), and Nybakken (1978).

Phylum Mollusca / Class Gastropoda / Subclass Opisthobranchia
Order Nudibranchia / Suborder Doridacea / Superfamily Eudoridoidea
Family Rostangidae

14.26 **Rostanga pulchra** MacFarland, 1905
Red Sponge Nudibranch

Abundant low intertidal zone, uncommon subtidally to 18 m, usually found on almost identically colored red siliceous sponges that encrust undersides of rocky ledges; Vancouver Island (British Columbia) to Puertecitos (Baja California) and Chile; not Japan, as sometimes reported.

Body usually 10–30 mm long, usually bright red but varying from light yellow-red to deep scarlet, with minute brown and black spots sprinkled everywhere between dorsal papillae; oral tentacles finger-shaped; leaflets on rhinophores vertical; gills numbering six to nine, same color as body.

Rostanga feeds largely on the red siliceous sponges *Ophlitaspongia pennata* (2.14), *Esperiopsis originalis*, and *Plocamia karykina*, and reportedly also on *Acarnus erithacus* (2.8) and *Isociona lithophoenix*. The nudibranch can locate sponges from a distance by scent. It moves about sluggishly on the sponge and removes the top layer of sponge tissue, leaving a shallow groove about 0.75 mm wide or a series of pits as a trail. Spicules characteristic of the food sponge have been found in the gut and the fecal material of the nudibranch. When individual *Rostanga* were marked, by cutting small notches in the edge of the mantle, some animals were found to move about considerably, while others remained on the same sponge for several days. In one case a marked individual was found on the same sponge 37 days after being marked.

The similarity in color between *Rostanga*, its egg masses, and the sponges on which it feeds is remarkable. The nudibranch apparently incorporates the pigments from its food sources non-selectively, and all the major carotenoids contributing to its color are found in the sponges on which it feeds.

In the Monterey Bay area *Rostanga* breeds the year around. Mating and egg laying occur on or near one of the preferred food sponges. The egg strands are cylindrical and laid in a spiral pattern. Although the jelly matrix of the strands is transparent, the eggs are brightly pigmented and the egg masses closely resemble the sponge substratum in color. The average length of the egg strands is about 5.8 cm, and a strand contains from 2,000 to over 13,000 oval egg capsules. Each capsule normally holds one egg measuring about 73 μm in diameter, but occasionally two somewhat smaller eggs occupy a single capsule.

In laboratory aquaria with water at 15°C, adult *Rostanga* produce one egg mass per day during the spawning period. Embryological studies show that the first cleavage spindle occurs within 1 hour of oviposition, the first cell division takes place at about 6 hours, the blastula is complete after 2 days, gastrulation occurs on day 4, the early veliger appears on day 7, and hatching occurs from days 10 to 13. In water at 8–10°C the process is slowed and hatching occurs 30 days after egg laying. As development proceeds, the distinct reddish pigment found in the eggs becomes paler, and by the time the veliger is formed the pale pigmentation remains only in the digestive gland. In the laboratory, up to 95 percent of the embryos completed development to the hatching stage. After hatching, the actively swimming veliger larvae feed on planktonic algae, but as they mature they settle to the bottom and crawl about.

Rostanga pulchra larvae have been reared through metamorphosis in the laboratory at Friday Harbor (Washington), and the morphology and behavior of the veliger have been studied in some detail. During the first week of pelagic life the larvae are attracted to light and swim very actively. Thereafter, swimming activity lessens, and in the laboratory the animals spend more time moving about the bottoms of their containers, as if searching for suitable substratum. On a diet of the phytoplankters *Monochrysis lutheri* and *Isochrysis galbana* the animals increased in shell length from 150 to 300 μm in 35–40 days at 10–15°C. At this stage, most larvae are ready to undergo metamorphosis, and do so if a suitable sponge substratum (e.g., *Ophlitaspongia pennata*) is placed in the container. Metamorphosis takes about 24 hours. Juveniles, reared on sponges in the laboratory, reached a length of 4.5 mm in 70 days.

Rostanga appears to have few predators in nature, and in the laboratory a variety of starfishes, crabs, and fishes normally found in the intertidal zone with *Rostanga* did not feed on the nudibranchs when these were made available to them. The predaceous cephalaspidean *Navanax inermis* (14.4) is repelled by *Rostanga*, apparently by non-acid secretions.

See especially E. Anderson (1971) and Chia & Koss (1978); see also Ayling (1968), Bertsch et al. (1972), Cook (1962), De Laubenfels (1929), Hurst (1967), Lance (1966), MacFarland (1905, 1966), Er. Marcus (1959, 1961), McDonald & Nybakken (1978), Nybakken (1978), and Sphon (1972).

Phylum Mollusca / Class Gastropoda / Subclass Opisthobranchia Order Nudibranchia / Suborder Doridacea / Superfamily Eudoridoidea Family Aldisidae

14.27 Aldisa sanguinea (Cooper, 1862)
Blood Spot

Occasionally common on rocks , low intertidal zone at Point Pinos (Pacific Grove, Monterey Co.), otherwise rare intertidally; Anguilar Point (Barkley Sound, British Columbia) to Isla San Diego (Baja California); Japan.

Body usually about 17–25 mm long, bright red, usually with one to four blackish markings (sometimes none or five); oral tentacles ear-shaped (auriform); ridges on rhinophores placed at an angle; larger dorsal processes lacking spine-studded basal constrictions; specimens from San Luis Obispo Co. reported to have a T-shaped pattern of yellow on posterior half.

Aldisa feeds on red and orange-red sponges similar to itself in color, including *Ophlitaspongia pennata* (2.14) and *Hymendesmia brepha*. On this background it lays a matching egg mass consisting of a spiral ribbon with a single egg in each capsule.

Similar species include **Platydoris macfarlandi** Hanna, 1951 (rare, subtidal to about 150 m, known from Pismo Beach (San Luis Obispo Co.); adults over 30 mm long, body very flat, dorsum dark red and velvety), and **Thordisa bimaculata** Lance, 1966 (locally common, rocky intertidal and subtidal waters to 10 m, Carmel (Monterey Co.) to Isla Natividad (Baja California); adults about 25–28 mm long, orange to dull yellow or whitish with two brownish spots on dorsal midline; dorsum thickly set with "hairy" papillae with constricted bases, inflated middles, and tapering tips).

For *Aldisa sanguinea*, see Baba, Hamatani & Hisai (1956), Cooper (1863), Ferreira & Bertsch (1975), MacFarland (1966), Er. Marcus (1961), McDonald & Nybakken (1978), Nybakken (1978), Robilliard & Baba (1972), and Roller (1969a). For other species, see Hanna (1951) and Lance (1966).

Phylum Mollusca / Class Gastropoda / Subclass Opisthobranchia Order Nudibranchia / Suborder Doridacea / Superfamily Eudoridoidea Family Archidorididae

14.28 Atagema quadrimaculata Collier, 1963
(=Petelodoris spongicola MacFarland, 1966)
Hunchback Nudibranch

On encrusting sponges under large rocks, very low intertidal zone; Monterey to San Diego.

Body about 12–65 mm long, narrow, firm, pale brownish with many minute black to brown spots, closely mimicking appearance of sponge host; surface hard and "crunchy" in texture, owing to contained spicules; dorsum with an irregular, median longitudinal ridge rising over a distinctive hump just ahead of gills; contracted gill covered by three distinctive valvelike flaps.

This little-known species is probably uncommon; however, it is easily overlooked because of its close resemblance to the sponges upon which it lives.

See Collier (1963).

Phylum Mollusca / Class Gastropoda / Subclass Opisthobranchia
Order Nudibranchia / Suborder Doridacea / Superfamily Eudoridoidea
Family Archidorididae

14.29 Archidoris odhneri (MacFarland, 1966)
(= Austrodoris odhneri MacFarland, 1966)
White-Knight Nudibranch

Rare, low intertidal zone, occasionally abundant subtidally in rocky areas; Vancouver Island (British Columbia) to San Diego.

Body typically 60–90 mm long, but ranging to over 150 mm, pure white (occasionally cadmium yellow), tuberculate, the tubercles low, thickly set, rounded to conical, spiculate to the touch; rhinophores large, conical, deeply retractile; gills numbering seven, forming a white plume.

This handsome species, present subtidally in some areas throughout the year, feeds on several sponges, among them *Halichondria panicea* (2.15) and representatives of the genera *Myxilla*, *Mycale*, *Stylissa*, *Tedania*, *Craniella*, and *Syringella*.

The nudibranch deposits its eggs in areas of strong currents. The egg mass forms a thin, wide ribbon, arranged in an oval spiral and attached by one edge to the substratum. Each capsule in the ribbon usually contains 8–12 eggs.

See Belcik (1975), Burn (1968), Hurst (1967), MacFarland (1966), McDonald & Nybakken (1978), Odhner (1934), Potts (1966), and Robilliard (1971c).

Phylum Mollusca / Class Gastropoda / Subclass Opisthobranchia
Order Nudibranchia / Suborder Doridacea / Superfamily Eudoridoidea
Family Archidorididae

14.30 Archidoris montereyensis (Cooper, 1862)
Monterey Dorid

Common, low intertidal zone on hard substrata in north end of range; rare intertidally and uncommon subtidally to 50 m at southern end of range; Alaska to San Diego.

Body usually 25–50 mm long, but ranging to over 125 mm, bright yellow to yellow-orange with dark markings both on and between conical dorsal tubercles; oral tentacles ear-shaped; gills numbering seven, yellowish.

This species feeds on the yellow crumb-of-bread sponge, *Halichondria panicea* (2.15), which it evidently cannot locate by scent alone. Other sponges are also eaten.

Reproduction (or at least egg laying) occurs throughout the year. Individual animals are simultaneous hermaphrodites, but, as in other opisthobranchs studied, they do not fertilize their own eggs with their own sperm. When mating occurs, an individual receives non-motile sperm from its partner, and stores it in a seminal receptacle (spermatocyst), where the sperm lie with their heads embedded in the epithelial cells of the lining. After an interval (duration unknown) the sperm become motile and capable of fertilizing eggs. The ripe eggs, about 80 μm in diameter, are fertilized just before they are spawned, and before meiosis has occurred.

Fertilization occurs deep within the body, after which the eggs are encased in capsules containing an average of three eggs (range 1–18 eggs). The capsules are formed into a narrow cord, which in turn is folded back and forth on itself to produce a gelatinous ribbon. The egg ribbon is bright yellow to pale cream in color; it occurs in a fairly close coil and is attached to the substratum along one margin, which is slightly shorter than the free edge. The whole ribbon may contain about two million eggs.

Meiosis begins about 4 hours after egg laying. The first polar body departs from the egg and divides again; the second polar body remains at the egg surface. The first cleavage occurs 6–7 hours after spawning, the second at 10 hours, the third at 12 hours, at a temperature of 17°C. At first the divisions occur synchronously within each localized region of the egg ribbon, but as time goes on the embryos near the center of the cluster develop more slowly than those at the margins. By 6 days after spawning the embryos have become early trochophore larvae (gastrula stage). Over the next few days cilia develop, and the larvae begin to swim within their capsules in the egg ribbon. In 14 days the early veliger larval stage is reached, and by 20 days the larvae are late veligers. Hatching of the larvae from their capsules, accompanied by a gradual disintegration of the gelatinous egg ribbon, occurs 20–25 days after egg laying. The larvae swim for a period, but most settle to the bottom within 2 hours after hatching. Later development has not yet been followed.

See Bertsch et al. (1972), Cook (1962), Costello (1938), Crane (1971), Forrest (1953), Hinegardner (1974), Hurst (1967), Kay & Young (1969), MacFarland (1966), Er. Marcus (1961), McDonald & Nybakken (1978), McGowan & Pratt (1954), Miller (1961), Nicaise (1969), Nybakken (1978), Potts (1966), Sphon (1972), T. Thompson (1959), and Young (1969).

Phylum Mollusca / Class Gastropoda / Subclass Opisthobranchia
Order Nudibranchia / Suborder Doridacea / Superfamily Eudoridoidea
Family Discodorididae

14.31 **Anisodoris nobilis** (MacFarland, 1905)
Sea Lemon

Common, low intertidal zone on rocky shores and harbor pilings; subtidal to 35 m; Vancouver Island (British Columbia) to Isla Coronados (Baja California); rare intertidally in southern part of range.

Body commonly 25–75 mm in length, but ranging to over 260 mm, bright orange to light yellow or sometimes white, with dark markings only between (not on) the club-shaped dorsal tubercles; oral tentacles finger-shaped; gills numbering six, tipped with white.

This is the largest and one of the most conspicuous of California nudibranchs. It has a penetrating, persistent, fruity odor, which seems to discourage predators. It feeds by rasping on a variety of sponges, including *Axocielita originalis* (2.9), *Astylinifer arndti*, *Hymenamphiastra cyanocrypta* (2.10), *Lissodendoryx firma*, *Halichondria panicea* (2.15), *Haliclona permollis* (2.16), and species in the genera *Mycale*, *Zygherpa*, *Parasperella*, and *Prianos*. Quantities of organic detritus are also ingested. Studies of marked animals at Pacific Grove (Monterey Co.) show that individual nudibranchs are quite conservative in their food habits; they tend to keep on eating the same food species, even if they are transferred to other sites.

Mating is reciprocal, sometimes occurring between partners of vastly different size. The spawning period varies with the locality, but extends through several months (e.g., November to March in Monterey Bay). The abundance of this species, the large size and accessibility of the nerve cells, and the relatively simple organization of the nervous system have made this species useful for neurophysiological research.

See Bertsch et al. (1972), Gorman & Marmor (1970), Gorman & Mirolli (1969), Kitting (1978), MacFarland (1966), Er. Marcus (1961), McBeth (1970, 1971), and Nybakken (1978).

Phylum Mollusca / Class Gastropoda / Subclass Opisthobranchia
Order Nudibranchia / Suborder Doridacea / Superfamily Eudoridoidea
Family Discodorididae

14.32 **Diaulula sandiegensis** (Cooper, 1862)
Ring-Spotted Dorid

Seldom abundant but found regularly on large rocks and in pools, low intertidal zone and subtidal to 35 m along rocky shores; Japan and Alaska to Cabo San Lucas (Baja California) and Puerto Peñasco (Mexico); main population intertidal at north end of range, subtidal at south end.

Body usually 25–90 mm long, but occasionally reaching 125 mm, white or gray to light brown, with few to many prominent, dark-brown or black, ring-shaped markings on dorsum (northern specimens seeming to have more spots); texture hard and "gritty," owing to dense arrangement of spicules in the epidermis; gills numbering six or seven, on posterior part of body.

This species feeds mainly on sponges; it has been found on several species, among them *Halichondria panicea* (2.15), *Haliclona permollis*, and representatives of the genera *Myxilla* and *Petrosia*. Spawning occurs almost throughout the year in Monterey Bay. The egg ribbon is white, relatively narrow, and attached by one margin under rock ledges. Completed egg ribbons typically form oval spirals with three to eight whorls. They contain one or two eggs per capsule, and up to 16 million eggs per ribbon. Certain aspects of embryology were studied at the end of the last century.

See Bertsch et al. (1972), Costello (1938), Hurst (1967), MacFarland (1897, 1966), Er. Marcus (1961), Nybakken (1978), and Sphon (1972).

*Phylum Mollusca / Class Gastropoda / Subclass Opisthobranchia
Order Nudibranchia / Suborder Doridacea / Superfamily Eudoridoidea
Family Discodorididae*

14.33 Discodoris heathi MacFarland, 1905
Gritty Dorid

Middle and low intertidal zones on rocky shores, common
during summer but uncommon at other seasons; Vancouver
Island (British Columbia) to Bahía de San Quintín (Baja California); rare intertidally in southern part of range.

Body usually 20–35 mm long, soft but gritty to touch,
white or yellowish to pale sandy brown, with very small dark
markings often concentrated as a dark blotch just in front of
the eight to ten gills; dorsum bearing indistinct tubercles; foot
papillose on upper surface, not smooth.

Observers have found this species clinging to the sponges
Halichondria panicea (2.15) and *Myxilla incrustans*, and feeding
on the sponge *Adocia gellindra*. Most details of the biology are
lacking.

See Farmer (1967), MacFarland (1906, 1966), Er. Marcus (1961),
McDonald & Nybakken (1978), and Nybakken (1978).

*Phylum Mollusca / Class Gastropoda / Subclass Opisthobranchia
Order Nudibranchia / Suborder Doridacea / Superfamily Porodoridoidea
Family Dendrodorididae*

14.34 Doriopsilla albopunctata (Cooper, 1863)
(=Dendrodoris fulva (MacFarland, 1905); Dendrodoris
albopunctata (Cooper, 1863); Doriopsilla reticulata Eliot, 1906)
Salted Dorid

Abundant on rocky shores, low intertidal zone and subtidal
waters to 45 m; Van Damme Beach State Park (Mendocino
Co.) to Puertecitos (Baja California).

Body to about 70 mm long, varying from bright yellow to
warm red-brown, with darker hues commoner at southern
end of range (northern form may be taxonomically distinct);
dorsum glandular, the glands fine, white, dotlike, more conspicuous on darker specimens; oral region usually plain, with
concealed, porelike mouth; no radula.

This species feeds successfully on sponges, despite the ab-

sence of a radula. The sponges reportedly taken include
Acarnus erithacus (2.8), *Cliona celata* (2.20), *Ficulina suberea*, and
Suberites sp. The food is first macerated externally with secretions from a tubular structure everted from the mouth, then
the resulting soft material is sucked in. Spawning occurs
throughout the year, but mainly in the summer. The eggs are
laid in closely coiled yellow ribbons fastened to rocks or
algae.

See Ghiselin (1964), MacFarland (1966), McBeth (1971), McDonald & Nybakken (1978), Nybakken (1978), and Steinberg (1961).

*Phylum Mollusca / Class Gastropoda / Subclass Opisthobranchia
Order Nudibranchia / Suborder Doridacea / Superfamily Anadoridoidea
Tribe Suctoria / Family Corambidae*

14.35 Corambe pacifica MacFarland & O'Donoghue, 1929
Frost Spot

Seasonally common, low intertidal zone, on the bryozoan
Membranipora encrusting the offshore kelp *Macrocystis* and on
the eelgrass *Zostera*; Vancouver Island (British Columbia) to
Punta Eugenia (Baja California).

Body to about 10 mm long, almost transparent; dorsal cuticle bearing lines resembling pattern of *Membranipora*; dorsum
deeply notched in middle of rear margin; rhinophores low,
blunt, grooved longitudinally; up to 14 gills on each side between dorsum and foot, the gills with opposite ridging.

Doridella steinbergae (Lance, 1962) (=*Corambella bolini*
MacFarland, 1966; *C. steinbergae* Lance, 1962) is similar, but
has long smooth rhinophores and an unnotched dorsum.

Corambe pacifica feeds on bryozoans by rasping the colonies
and sucking out the soft parts. Most bryozoan colonies show
damage from these small but voracious predators. The narrow, white, coiled egg bands of *C. pacifica* are common on
Membranipora colonies and adjacent kelp surfaces. *Doridella
steinbergae* also occurs on bryozoan-encrusted kelp often in
enormous numbers, and feeds on *Membranipora*.

See G. Anderson (1971), Franz (1967), Lance (1962b), MacFarland
(1966), MacFarland & O'Donoghue (1929), McBeth (1968), and
McDonald & Nybakken (1978).

Phylum Mollusca / Class Gastropoda / Subclass Opisthobranchia
Order Nudibranchia / Suborder Doridacea / Superfamily Anadoridoidea
Tribe Suctoria / Family Okeniidae

14.36 Ancula pacifica MacFarland, 1905
Pacific Ancula

Present sporadically during summer on hydroids and bryozoans, low intertidal rocky pools; sometimes common in summer among filamentous diatoms or mussels in San Francisco Bay; San Juan Islands (Washington) to Point Loma (San Diego Co.).

Body to about 16 mm long, cream-colored (photograph too blue), with three longitudinal yellow-orange lines; dorsum bearing two delicate extra appendages beside each rhinophore and six to ten extra appendages along gills; appendages usually yellow near ends, tipped with orange.

This graceful species is active in aquaria and sometimes swims at the surface. It has been observed feeding on the entoproct *Barentsia ramosa*. Its internal anatomy is incompletely known, and most natural history details are lacking.

From time to time related species may be encountered. A. **lentiginosa** Farmer & Sloan, 1964, occurs intertidally and subtidally on rocky shores and boat docks, from Monterey Bay to Bahía de los Ángeles (Baja California). **Trapania velox** (Cockerell, 1901) (= *Drepania, Drepanida,* and *Thecacera velox*) is found from San Luis Obispo to San Diego Bay. Both these species reach lengths of over 20 mm, and both have two pairs of extra appendages, one pair next to the rhinophores and one pair flanking the gills.

Two smaller relatives, neither usually over 10 mm long, are **Okenia plana** Baba, 1960, which is found in San Francisco Bay, on mud-flat rocks at China Camp (Marin Co.), and in Japan, and **O. angelensis** Lance, 1966, which is found on boat floats and rocky shores from Duxbury Reef (Marin Co.) to Bahía de los Ángeles (Baja California), commonly on the brown alga *Sargassum*. Both have extra appendages along the sides and one or more on the dorsum. All are diagnosed in the key in McDonald (1975).

For *A. pacifica,* see Behrens (1971a), Farmer & Sloan (1964), MacFarland (1906, 1966), MacFarland & O'Donoghue (1929), Er. Marcus (1961), McDonald & Nybakken (1978), Nybakken (1978), and Robilliard (1971c). For *A. lentiginosa,* see Farmer & Sloan (1964), Lance (1966), and Nybakken (1978). For *Okenia plana,* see Baba (1960a), Morse (1972), Steinberg (1963b), and Vogel & Schultz (1970). For *O. angelensis,* see Lance (1966) and Morse (1972). For *Trapania velox,* see Kress (1970), Lance (1966), and MacFarland (1929, 1966).

Phylum Mollusca / Class Gastropoda / Subclass Opisthobranchia
Order Nudibranchia / Suborder Doridacea / Superfamily Anadoridoidea
Tribe Suctoria / Family Okeniidae

14.37 Hopkinsia rosacea MacFarland, 1905
Hopkins's Rose

Seasonally common on low intertidal rocks, and in subtidal waters toward northern end of range; common intertidally and rare subtidally to 6 m in southern part of range; Coos Bay (Oregon) to Puerto Santo Tomás (Baja California).

Body averaging nearly 30 mm long; entire animal a very distinctive rose-pink color; rhinophores and ring of gills almost hidden by many long fingerlike processes covering dorsal surface; body internally stiffened by numerous calcareous spicules.

This beautiful species feeds mainly, perhaps exclusively, on the pink encrusting bryozoan *Eurystomella bilabiata* (6.9). An entrance hole in the bryozoan's outer casing is made by rasping, and the soft parts are then sucked out. The rosy color is due to the carotenoid pigment, hopkinsiaxanthin. Eggs are laid in a narrow rose-colored ribbon spiraling counterclockwise. Predators such as *Navanax inermis* (14.4) are repelled, apparently by a non-acid secretion. The genus *Hopkinsia* was dedicated to Timothy Hopkins, railroad executive and friend of Leland Stanford, whose gifts established, and helped endow, the Hopkins Marine Station at Pacific Grove (Monterey Co.).

See Johannes (1963), MacFarland (1905, 1966), Er. Marcus (1961), McBeth (1970, 1971), Nybakken (1978), and Strain (1949).

Phylum Mollusca / Class Gastropoda / Subclass Opisthobranchia
Order Nudibranchia / Suborder Doridacea / Superfamily Anadoridoidea
Tribe Suctoria / Family Onchidorididae

14.38 Acanthodoris lutea MacFarland, 1925
Orange-Peel Nudibranch

Most common in rocky pools, low intertidal zone and adjacent subtidal areas during summer months; Dillon Beach (Marin Co.) to Cabo Colnett (Baja California).

Body typically about 22 mm long, bright orange, usually spotted with lemon yellow on dorsum; gills gray; tips of rhinophores port-wine color; odor very distinctive, fruity.

This species secretes a material toxic to other animals confined in the same container. Members of the genus *Acanthodoris* are reported to feed mainly on tunicates and bryozoans, sucking out the body contents with an oral pump mechanism.

See Bertsch et al. (1972), Forrest (1953), MacFarland (1926, 1966), and G. Williams & Gosliner (1979).

Phylum Mollusca / Class Gastropoda / Subclass Opisthobranchia
Order Nudibranchia / Suborder Doridacea / Superfamily Anadoridoidea
Tribe Suctoria / Family Onchidorididae

14.39 Acanthodoris nanaimoensis O'Donoghue, 1921
(=Acanthodoris columbina MacFarland, 1926)
Wine-Plume Dorid

Common during summer months, on low intertidal rocks and in subtidal waters; Vancouver Island (British Columbia) to Shell Beach (San Luis Obispo Co.).

Body about 15–30 mm long, whitish, usually with yellow-tipped tubercles covering dorsum; adult specimens usually with brownish speckling; port-wine coloration both on tips of rhinophores and on margin of gills.

Adults of this species have been observed eating colonial ascidians. The egg capsules are extruded in a yellow to creamy-white ribbon attached by one margin to the substratum and arranged in two to three very closely and neatly coiled whorls; the free margin of the ribbon has a narrow transparent edge.

See Hurst (1967), MacFarland (1926), McDonald & Nybakken (1978), Sphon (1972), Steinberg (1963b), and G. Williams & Gosliner (1979).

Phylum Mollusca / Class Gastropoda / Subclass Opisthobranchia
Order Nudibranchia / Suborder Doridacea / Superfamily Anadoridoidea
Tribe Suctoria / Family Onchidorididae

14.40 Acanthodoris rhodoceras Cockerell & Eliot, 1905
Black-and-White Dorid

Common in summer on bay boat landings, in rocky pools, and on mud flats, low intertidal zone; rare subtidally to 18 m; Dillon Beach (Marin Co.) to Los Coronados (Baja California).

Body usually about 12–25 mm long, white to gray-white on dorsum; markings quite variable, typically black on tips of tubercles and along body margin, often edged with yellow outside the black line; rhinophores and gills rust-brown to black.

Two other species of *Acanthodoris*, both scarce, may occasionally be encountered. **A. brunnea** MacFarland, 1905, ranges from Vancouver Island (British Columbia) south to Santa Monica Bay (Los Angeles Co.): body 15–23 mm long; dorsum brown with dark brown or black blotches and scattered yellow spots; tips of gills and margin of body lemon yellow; rhinophores deep blue-black tipped with yellow. It is reported to feed on bryozoans. **A. hudsoni** MacFarland, 1905, occurs intertidally from Vancouver Island (British Columbia) to Santa Barbara: length 10–20 mm; dorsum yellowish white to pinkish white; tips of gills and tubercles and along body margin lemon yellow.

Almost nothing is known of the natural history of these species. Other species of *Acanthodoris* are reported to feed mainly on tunicates, bryozoans, and occasionally acorn barnacles, sucking out the body contents with an oral pump mechanism.

For *A. rhodoceras*, see Forrest (1953), MacFarland (1926), and Er. Marcus (1961). For *A. brunnea*, see MacFarland (1905, 1926, 1966), McDonald & Nybakken (1978), Robilliard (1971b), and Sphon & Lance (1968). For *A. hudsoni*, see Bertsch et al. (1972), Gosliner & Williams (1973b), MacFarland (1905, 1926), McDonald (1970), Sphon (1972), and G. Williams & Gosliner (1979).

Phylum Mollusca / Class Gastropoda / Subclass Opisthobranchia
Order Nudibranchia / Suborder Doridacea / Superfamily Anadoridoidea
Tribe Suctoria / Family Onchidorididae

14.41 Onchidoris bilamellata (Linnaeus, 1767)
(=Onchidoris fusca (Müller, 1776))   Many-Gilled Onchidoris

Seasonally common, on pilings and rocks in mud of estuaries, low intertidal zone; sometimes very abundant subtidally and on bay boat landings during summer months; Alaska to Morro Bay (San Luis Obispo Co.); circumboreal.

Body usually 10–30 mm long, yellow-white to light rust-brown or deep chocolate with lighter papillae; dorsum bearing short, club-shaped, spiculate papillae, and 16–32 or more simple pinnate gills arranged in two half-circles; dark dorsal markings quite variable, generally two or three irregular longitudinal stripes; gills not retractile into cavity beneath surface of body.

Sometimes seen in enormous numbers by divers, this nudibranch feeds on *Balanus crenatus* (20.25) and other acorn barnacles, sucking out the body contents by means of an oral pumping mechanism. Eggs have been seen on the shores of Vancouver Island (British Columbia) in February, May, and June, but are common only in winter. The egg ribbons, attached along one margin, usually form irregular curves, flaring out along the longer free margin. Capsules contain one to three eggs each, and entire ribbons bear an average of 60,000 eggs.

See Barnes & Powell (1954), Belcik (1975), Forrest (1953), Hurst (1967), Er. Marcus (1958b, 1961), McCance & Masters (1937), McDonald (1970), McDonald & Nybakken (1978), and Potts (1966, 1970).

Phylum Mollusca / Class Gastropoda / Subclass Opisthobranchia
Order Nudibranchia / Suborder Doridacea / Superfamily Anadoridoidea
Tribe Suctoria / Family Onchidorididae

14.42 Onchidoris hystricina (Bergh, 1878)
White Onchidoris

Occasionally very common on rocky shores, low intertidal zone and subtidal waters to 18 m; Kyska (Aleutian Islands, Alaska) to Cabo Colnett (Baja California).

Adults less than 12 mm long; dorsum white with many opaque white flecks, sometimes with a bluish sheen; dorsal papillae very numerous, short, club-shaped, with many spicules; gills 8–12, simple, pinnate, arranged in a circle.

Little is known of this species, and only the external features and radula have ever been illustrated.

See Er. Marcus (1961) and Nybakken (1978).

Phylum Mollusca / Class Gastropoda / Subclass Opisthobranchia
Order Nudibranchia / Suborder Doridacea / Superfamily Anadoridoidea
Tribe Non-Suctoria / Family Triophidae

14.43 Triopha catalinae (Cooper, 1863)
(=Triopha carpenteri (Stearns, 1873))   Sea-Clown Nudibranch

Very common, middle intertidal zone to 35 m depth on rocky shores; often on kelp subtidally; Aleutian Islands (Alaska) to Baja California; Japan.

Body typically about 25 mm long, ranging to about 150 mm, robust, translucent white to pale yellow with deep-orange spots; orange spots also on gills and all appendages; gills numbering five, not retractile below body surface; oral veil with short branched appendages.

This is one of the largest nudibranchs able to crawl on the underside of the surface film of tidepools. It feeds on several genera of arborescent bryozoans by digesting the soft parts from branches that are ingested intact. Tidepool fishes avoid it, suggesting that some sort of chemical repellant is released.

Eggs have been observed in April and June in Washington. The white or cream-colored egg ribbon forms a loose coil, which is attached to the substratum along its shorter edge. The free edge is wavy, and the ribbon itself appears striated, owing to the internal arrangement of the egg strings in rows.

See Baba (1957a), Bertsch et al. (1972), Costello (1938), Ferreira (1977), Hurst (1967), MacFarland (1966), Er. Marcus (1961), McBeth (1970, 1971), McDonald & Nybakken (1978), Nybakken (1978), Nybakken & Eastman (1977), and Sphon (1972).

Phylum Mollusca / Class Gastropoda / Subclass Opisthobranchia
Order Nudibranchia / Suborder Doridacea / Superfamily Anadoridoidea
Tribe Non-Suctoria / Family Triophidae

14.44 Triopha maculata MacFarland, 1905
(=Triopha grandis MacFarland, 1905) Spotted Triopha

In rocky tidepools, low intertidal zone and subtidal waters to depth of 33 m; Bodega Bay (Sonoma Co.) to Cabo San Quintín (Baja California).

This is a variable species, and three main forms are usually encountered. (1) Young red individuals (cited as *Triopha* sp. in many works), generally 12 mm long or less, are most often seen in rocky tidepools; they occur throughout the geographic range all year, but most abundantly in the summer. (2) Individuals of the most common form of the species (Fig. 14.44a) are usually 15–40 mm long, but occasionally reach 50 mm; the body is dark brown or reddish brown with many small, white to bluish-white spots on polygonally shaped raised areas on dorsum and sides, and with five small red spots forming a line from the gills to the rhinophore area; the gills, rhinophores, and appendages are red to orange-red; the oral veil bears short, branched appendages. This form is particularly abundant in summer months and reportedly feeds on several bryozoans. (3) Individuals of the largest form of the species (formerly known as *T. grandis*; Fig. 14.44b) are usually 40–75 mm long, but reach about 150 mm; they are light yellowish brown with bluish-white spots or irregular raised areas on the dorsum and sides; the rhinophores and gills are light yellow-gray. This form is most commonly seen during summer months on offshore kelp; it is not an inhabitant of rocky tidepools, but occurs on mud flats and in estuaries, often in *Zostera* beds.

In life the colors and patterns of the animal often resemble those of older blades of the kelp *Macrocystis*, which are often lighter in color than young blades and are covered with white encrusting colonies of bryozoans and hydroids. The coiled egg ribbons are commonly seen on the subtidal kelps *Macrocystis* and *Nereocystis*.

See Farmer (1967), Ferreira (1977), MacFarland (1906, 1966), McDonald & Nybakken (1978), Nybakken (1978), and Steinberg (1961).

Phylum Mollusca / Class Gastropoda / Subclass Opisthobranchia
Order Nudibranchia / Suborder Doridacea / Superfamily Anadoridoidea
Tribe Non-Suctoria / Family Triophidae

14.45 Crimora coneja Marcus, 1961
Rabbit Nudibranch

Reported only from low intertidal rocks at Point Loma (San Diego Co.).

Body 9–19 mm long, white or yellowish-white; dorsum with numerous orange processes often tipped with black, especially those near midline; rhinophores orange, retractile into cavities; lateral processes on body often forked; spicules buried in back, few in number but enormous in size.

Little is known of this species beyond the brief notes and sketches of Er. Marcus (1961).

Phylum Mollusca / Class Gastropoda / Subclass Opisthobranchia
Order Nudibranchia / Suborder Doridacea / Superfamily Anadoridoidea
Tribe Non-Suctoria / Family Aegiretidae

14.46 Aegires albopunctatus MacFarland, 1905
Salt-and-Pepper Nudibranch

Seasonally common under rocky ledges, in rocky pools, and on bay boat landings, usually on white sponges, low intertidal zone; subtidal to 30 m; Vancouver Island (British Columbia) to Bahía de los Ángeles (Baja California).

Body usually about 10 mm long, but reaching 20 mm, very firm, white to yellowish with opaque salt-and-pepper spots; dorsum thickly set with blunt tubercles arranged in irregular rows; gills small numbering three, each one protected by a tuberculate lobe.

This species is abundant at Newport Bay (Orange Co.) and La Jolla (San Diego Co.) in early summer. It is very sluggish, shuns the light, and soon dies in captivity. It feeds on white sponges, which it closely matches in color. The eggs are laid in short, white, slightly spiraled bands. Predators such as *Navanax inermis* (14.4) and *Pleurobranchaea californica* (14.14) are repelled, apparently with a non-acid secretion.

See MacFarland (1905, 1966), Er. Marcus (1961), and Nybakken (1978).

Phylum Mollusca / Class Gastropoda / Subclass Opisthobranchia
Order Nudibranchia / Suborder Doridacea / Superfamily Anadoridoidea
Tribe Non-Suctoria / Family Polyceridae

14.47 **Polycera atra** MacFarland, 1905
Sorcerer's Nudibranch

Common from April to October in bays on boat landings, boats, rocks, kelp, and pilings, especially on the bryozoan *Bugula*; subtidal to 50 m; Limantour Estero (Marin Co.) to Los Coronados (Baja California) and Puerto Peñasco (Mexico).

Body usually about 15 mm long, occasionally reaching about 50 mm, smooth, humped in front, bearing 8–11 non-retractile gills; color gray, marked by distinctive pattern of longitudinal black stripes with oblong orange spots (occasionally fusing as lines) between them (colors lighter in south); rhinophores, gills, and frontal veil also showing orange spots; frontal veil bearing six to eight long pointed processes.

Polycera atra feeds on the branching bryozoan *Bugula*; the jaws and radula break open the bryozoan's shell, and an oral pumping mechanism sucks out the soft bodies of the zooids. *P. atra* is also reported to feed on the bryozoan *Membranipora* and on the superficial tissues of the subtidal cnidarian *Lophogorgia chilensis*.

See Lewbel & Lance (1975), MacFarland (1905, 1966), McDonald & Nybakken (1978), Nybakken (1978), Odhner (1941), Pohl (1905), and Robilliard (1972).

Phylum Mollusca / Class Gastropoda / Subclass Opisthobranchia
Order Nudibranchia / Suborder Doridacea / Superfamily Anadoridoidea
Tribe Non-Suctoria / Family Polyceridae

14.48 **Polycera hedgpethi** Marcus, 1964
Hedgpeth's Polycera

In shaded areas on and under boat floats in bays, especially on the bryozoan *Bugula*; also subtidal; Tomales Bay (Marin Co.) to Bahía de los Ángeles (Baja California).

Body often 12–25 mm in length, sometimes reaching about 50 mm; ground color light gray but body appearing dark brown to black, owing to numerous dark flecks; two lateral yellow to orange-yellow lines on body; dorsum and sides

with numerous tubercles usually tipped with yellow-orange; gills tripinnate, seven to nine in number; three fingerlike processes on each side of branchial plume joined at bases; papillae on frontal veil numbering four to six.

Less common than *P. atra* (14.47), this species is found only occasionally, usually in the summer. It tends to avoid light, even in aquaria. It has been observed feeding on *Bugula pacifica*, and it probably eats other bryozoans as well. The egg strings, similar to those of *P. atra*, are found on or near *Bugula* colonies.

Two other species of *Polycera* may also be encountered. **P. zosterae** O'Donoghue, 1924, is uncommon, occurring from the Vancouver Island region (British Columbia) to Bodega Bay (Marin Co.), usually on the bryozoan *Bowerbankia gracilis* (6.18); it feeds on *Bowerbankia* and species of *Membranipora*. Adults are about 10 mm long; body with scattered tubercles, three to five non-retractile gills, three to eight short blunt projections beside the gills; veil papillae numbering 10–14, bump-like; color dark brown with bright-yellow spots in groups on the body and on the tips of the tubercles and gills. **P. tricolor** Robilliard, 1971, which is occasionally common on brown laminarian algae from shallow subtidal waters to at least 60 m depth, ranges from Bamfield (Vancouver Island, British Columbia) to Scripps Submarine Canyon (off La Jolla, San Diego Co.). Adults are 30–36 mm long; body smooth with a few small papillae, five or six stout gills, and 8–12 long fingerlike projections from the dorsum in the gill area; veil papillae elongate, numbering 8–11; color translucent grayish white, the appendages generally black with chrome-yellow tips.

For *P. hedgpethi*, see Farmer (1968), Lance (1966), Er. Marcus (1964), and McDonald & Nybakken (1978). For *P. tricolor*, see Robilliard (1971a). For *P. zosterae*, see Gosliner & Williams (1973a) and O'Donoghue (1924).

Phylum Mollusca / Class Gastropoda / Subclass Opisthobranchia
Order Nudibranchia / Suborder Doridacea / Superfamily Anadoridoidea
Tribe Non-Suctoria / Family Polyceridae

14.49 Laila cockerelli MacFarland, 1905
Cockerell's Nudibranch

In tidepools, low intertidal zone, and adjacent subtidal waters on rocky shores; commonest in California in winter and spring; Vancouver Island (British Columbia) to Bahía de los Ángeles (Baja California); populations at south end of range uncommon in winter and generally subtidal to 35 m.

Body usually 12–30 mm long, white to yellowish, with distinctive red-orange color on marginal, club-shaped processes and on rhinophores; spicules below translucent body surface producing an irregular network of transparent lines; gills obscure and contractile, but not retractile below body surface.

This species is named for the English and American entomologist and malacologist T. D. A. Cockerell (1866–1948). Its anatomy was described in some detail early in the century. The animals have been observed feeding on the bryozoan *Hincksina velata*.

In the Vancouver area, spawning occurs from May to June. The pale-pink egg string spirals tightly and contains only about 6,500 eggs.

See Guernsey (1913), MacFarland (1966), Er. Marcus (1961), McDonald & Nybakken (1978), Nybakken (1978), and Sphon (1972).

Phylum Mollusca / Class Gastropoda / Subclass Opisthobranchia
Order Nudibranchia / Suborder Dendronotacea / Family Tritoniidae

14.50 Tritonia festiva (Stearns, 1873)
(=Duvaucelia festiva; Lateribranchaea festiva; Sphaerostoma undulata O'Donoghue, 1924)   Diamondback Nudibranch

Occasional, with hydroids and sponges, low intertidal zone under rocky ledges; subtidal to 50 m; Vancouver Island (British Columbia) to Los Coronados (Baja California); Japan.

Body usually about 20 mm long, but attaining 70 mm, white to pale orange-pink or even light brown; fine white lines usually forming a diamond pattern on dorsum but occa-

sionally almost invisible; up to 12 processes on head veil; up to 16 gill plumes per side.

This species has been observed in the subtidal zone feeding on the pink gorgonian *Lophogorgia chilensis* and the orange sea pen *Ptilosarcus gurneyi,* and in the intertidal zone feeding on *Clavularia* (3.23). All these prey are cnidarians in the subclass Alcyonaria.

Species of *Tritonia* have recently shown their value for neurophysiological studies. As in many opisthobranchs, the nervous system is simple in plan, and the major ganglia contain some nerve cells of immense size. These are so consistent in their size, color, and position that homologous cells can be recognized in different individuals. Attempts are under way to link specific brain cells with specific actions of the body. Several studies have been made of the neural elements triggering the complex actions of bending, curling, turning, and flexing that the animals normally use in swimming away from the predaceous sea star *Pycnopodia helianthoides* (8.16).

The functional anatomy of the reproductive system and the embryological development of the related species *T. hombergi* have been reported in two fine British papers (Thompson 1961b, 1962).

Three related species, all occurring in both Japanese and American Pacific waters, may be encountered occasionally in California. All are diagnosed in the key in McDonald (1975), and all are illustrated in MacFarland (1966) and Thompson (1971). **Tochuina tetraquetra** (Pallas, 1788) (also placed in *Tritonia* and several other genera) gets as far south as Monterey Bay. A giant yellow form, it is the world's largest nudibranch, exceeding 300 mm in length and 1.4 kg wet weight. It feeds on the large subtidal sea pen *Ptilosarcus gurneyi* and the soft coral (alcyonacean) *Gersemia rubiformis.* **Tritonia diomedea** Bergh, 1894, also extending south to Monterey (though rare), is a pinkish monster up to 220 mm long that reportedly feeds on another sea pen, *Virgularia.* **Tritonia exsulans** Bergh, 1894 (not the *T. exsulans* of Marcus, 1961, which is *T. diomedea*), sometimes found as far south as Punta Santo Domingo (Baja California), is another large pink species, differing from *T. diomedea* in details of the reproductive system. It, too, is said to feed on sea pens.

See Birkeland (1974), Field & MacMillan (1973), Getting (1976, 1977), Goddard (1973), Gomez (1973), Konishi (1971), MacFarland (1966), Er. Marcus (1961), McDonald & Nybakken (1978), Nybakken (1978), Sphon (1972), Steinberg (1961), T. Thompson (1961b, 1962, 1964, 1971), Willows (1967, 1971), and Willows & Hoyle (1967).

Phylum Mollusca / Class Gastropoda / Subclass Opisthobranchia
Order Nudibranchia / Suborder Dendronotacea / Family Dendronotidae

14.51 **Dendronotus subramosus** MacFarland, 1966
Stubby Dendronotus

Sporadically common on colonies of the hydroid *Aglaophenia struthionides* (3.14) growing on rocks or algae, low intertidal pools or offshore to 120 m depth; found only subtidally at northern end of range, on hydroids in areas swept by strong currents; San Juan Islands (Washington) to Los Coronados (Baja California).

Body typically about 15–25 mm long, ranging to 65 mm, stout, variably colored but with distinct brown longitudinal lines running length of body near bases of dorsal processes; anterior margin of oral veil above mouth with two or three pairs of stout, blunt, simply branched tentacles or papillae; rhinophores retractile into elongate tubular sheaths, the sheaths bearing no lateral tentacles halfway up their sides; digestive gland invading only first pair of dorsal body processes (cerata).

This species closely resembles the hydroid host to which it firmly clings and on which it feeds. The eggs, reported in June and July and again in October and November in Washington, are often conspicuous on the upper parts of *Aglaophenia*. The long cylindrical egg strand, colored pale pink to white, is attached to the hydroid as an overlapping series of serpentine loops.

Two related forms might be confused with this species. **D. frondosus** (Ascanius, 1774) (= *D. venustus* MacFarland, 1966), cosmopolitan in the northern hemisphere, is common both intertidally and subtidally, especially in the summer. It lives on and feeds on, a wide variety of hydroids, including species of *Tubularia*, *Hydractinia*, *Sarsia*, *Obelia*, *Sertularia*, *Abietinaria*, *Aglaophenia*, and others. It also eats the colonial ascidian *Bo-*

tryllus (12.29). *D. frondosus* differs from *D. subramosus* in its larger size (up to 115 mm long), the absence of longitudinal brown lines on the body, the presence of lateral papillae halfway up the rhinophore sheaths, and the possession of four pairs of tentacles, some highly branched, on the oral veil. Also, the digestive diverticula in this species invade the first four or five pairs of dorsal processes.

The generic name, *Dendronotus*, means "tree-back," in reference to the branched processes on the frontal veil and dorsal surface.

The second species superficially resembling *D. subramosus* is **Hancockia californica** MacFarland, 1923. It is known from Bodega Bay (Sonoma Co.) south to Punta Abreojos (Baja California), but it is rare in the northern part of its range. Instead of an oral veil, the animal has a large palmate lobe on either side anteriorly, each with several fingerlike branches. The digestive diverticula invade all of the dorsal processes and bear chambers for the storage of nematocysts derived from the diet (probably hydroids).

For *D. subramosus*, see MacFarland (1966) and Robilliard (1970). For *D. frondosus*, see McDonald & Nybakken (1978), Robilliard (1970), T. Thompson (1960b), and L. Williams (1971); descriptions of *D. frondosus* in MacFarland (1966) and Er. Marcus (1961) confuse three or four distinct species. For *Hancockia californica*, see MacFarland (1923, 1966).

Phylum Mollusca / Class Gastropoda / Subclass Opisthobranchia
Order Nudibranchia / Suborder Dendronotacea / Family Dendronotidae

14.52 **Dendronotus albus** MacFarland, 1966
White Dendronotus

Rare during summer in southern part of range, on kelp or in rocky pools, low intertidal zone; common during fall and winter at northern end of range, subtidally on hydroids to 30 m depth in rocky, current-swept areas; Victoria (Vancouver Island, British Columbia) to Los Coronados (Baja California).

Body about 15–40 mm long, elongate, translucent white with white line down back and top of tail; dorsal processes (cerata) in four to eight pairs, delicately branched, normally tipped with orange or yellow, the anterior three to six pairs containing branches of the digestive gland.

Dendronotus albus is best known from studies in Puget Sound (Washington), where it feeds on the upper portions of hydroids, especially *Tubularia*. In California it has been noted on *Plumularia*. Copulation has been observed in March and egg laying in October and November in Washington. The eggs are laid in a narrow white band, looped back and forth usually in a simple spiral on hydroid tips. Individuals of *D. albus* swim well.

A very similar (and possibly not distinct) species is **Dendronotus diversicolor** Robilliard, 1970, known from Vancouver Island (British Columbia) to Pismo Beach (San Luis Obispo Co.). It differs from *D. albus* in having only four (occasionally five) pairs of cerata, with only the first two pairs containing branches of the digestive gland, and in having a median dorsal white line on the tail only. It may be mainly white or have bright-orange tips on its body processes. In California it has been reported on hydroids, including *Sertularella tricuspidata* and *Abietinaria* spp.

See MacFarland (1966), McDonald & Nybakken (1978), Nybakken (1978), and Robilliard (1970, 1974).

Phylum Mollusca / Class Gastropoda / Subclass Opisthobranchia
Order Nudibranchia / Suborder Dendronotacea / Family Dendronotidae

14.53 Dendronotus iris Cooper, 1863
(=Dendrodoris giganteus O'Donoghue, 1921)
Giant Dendronotus

Uncommon on low intertidal mud flats; sporadically more abundant subtidally throughout the year to 200 m depth on muddy or sandy bottoms; Unalaska (Aleutian Islands, Alaska) to Los Coronados (Baja California).

One of the largest nudibranchs, reaching a length of over 29 cm, but usually 6–10 cm; body massive; dorsal processes in three to eight pairs, stout, highly branched, only the anterior pair containing branches of the digestive gland; white line along foot margin; two color phases in the Monterey Bay population, one with salmon-red bodies bearing orange-

ended, white-tipped processes (most animals, see the photograph), the other with light milky-purple bodies bearing light-orange processes tipped with white (Washington specimens reportedly gray, orange, or deep maroon-purple, usually with dorsal processes not tipped with white).

Dendronotus iris has jaws relatively much larger than those of other species of *Dendronotus*. It feeds on the tentacles of tube-dwelling ceriantharian sea anemones, and perhaps also on nemerteans. It forages from one *Pachycerianthus fimbriatus* (3.43) to another, leaving behind anemones that appear to have had uneven "haircuts" but are never actually killed. The nudibranchs are sometimes pulled into the tubes when the anemones contract, but seem unharmed by this. Adults are able to swim by side-to-side gyrations of the body, the motion becoming more violent in response to the touch of the predaceous sea star *Pycnopodia helianthoides* (8.16).

The long, loosely looped, white egg strings laid by *D. iris* are common high on ceriantharian tubes in Monterey Bay throughout the year. They contain 50 or more eggs per capsule, and veliger larvae hatch out 10–20 days after oviposition. The two color phases of *D. iris* freely mate with each other.

Related to the dendronotids is the cosmopolitan pelagic planktonic nudibranch *Cephalopyge* (= *Phyllirhoë*) *trematoides* (Chun, 1889), whose flattened body and glasslike transparency reveal its internal organs without dissection.

For *D. iris*, see Ägersborg (1922), Farmer (1970), Gosliner & Williams (1973b), Harris (1973), Hurst (1967), MacFarland (1966), McDonald & Nybakken (1978), Odhner (1936), Robilliard (1970), Sakharov (1962), T. Thompson (1960b), L. Williams (1971), and Wobber (1970). For *Cephalopyge*, see Dales (1953).

Phylum Mollusca / Class Gastropoda / Subclass Opisthobranchia
Order Nudibranchia / Suborder Dendronotacea / Family Tethyidae

14.54 Melibe leonina (Gould, 1852)
(=Chioraera leonina Gould, 1852; C. dalli Heath, 1917)   Melibe

Common in warmer months in California in localized areas on kelp (*Macrocystis*), both in offshore beds and in harbors; in Puget Sound commonly inhabiting intertidal beds of eelgrass

(*Zostera*), especially March through September; Alaska to Gulf of California.

Body to over 100 mm long; species readily recognized by its huge oral hood, which may make up half the body length; overall color yellowish brown to olive-green; dorsal processes (cerata) in five or six pairs, leaflike; rhinophores borne on hood, on stalks with sail-like extensions.

This remarkable nudibranch lacks a radula but has the oral veil and adjacent wall of the head expanded into a unique oral hood, used in trapping small animal plankton and rapidly moving benthos. Crustaceans provide the main staples, especially copepods, ostracods, caprellids, and gammarid amphipods, but molluscan post-larval stages and other small animals are taken, too. In feeding, the animal remains firmly anchored to the substratum by its narrow foot. The oral hood is then elevated and expanded by hydrostatic pressure in the hood's extensive blood sinuses. The hood may remain motionless for a time, then makes muscular sweeps downward or to the right or left. When anything contacts the lower surface of the hood, the two sides are swept together and the fringing tentacles interdigitate, locking in the prey. The whole hood then contracts, forcing the prey into the mouth. Smaller animals (less than 25 mm long) eat more benthic organisms than larger adults. Large *Melibe* feed mainly in the evening; juveniles feed also during the day.

Melibe generally remain firmly attached, but if dislodged swim actively (and usually upside down) by flexions of the body alternately to the right and left. The hood is unimportant in swimming. In Monterey Bay, the kelp crab *Pugettia producta* (25.4) sometimes eats *Melibe* (and stimulates a swimming response), but the sea star *Pycnopodia helianthoides* (8.16) is repelled, on contact, apparently by non-acid secretions of the nudibranch. The scale worm *Halosydna brevisetosa* (18.4) sometimes occurs as a commensal on *Melibe*, feeding on the sea slug's fecal pellets. Some specimens of *Melibe* are reported to harbor symbiotic algae.

The eggs of *Melibe* are laid subtidally in broad, usually cream-colored ribbons 2.5–12.5 cm long, which are attached to kelp and eelgrass during the summer. The ribbon, attached along its shorter margin, forms a tight coil or a series of wavy folds. The large, oval, frequently flattened egg capsules con-

tain 15–25 eggs each. Excellent studies have been made of the morphology of *M. leonina* and *M. rosea*, and the animals have recently been used for neurophysiological research.

For *M. leonina*, see Ägersborg (1923), Ajeska & Nybakken (1976), Guberlet (1928), Heath (1917), Hurst (1967, 1968), MacFarland (1966), McDonald & Nybakken (1978), S. Thompson (1976), and Waidhofer (1969). For *M. rosea*, see De Vries (1963).

*Phylum Mollusca / Class Gastropoda / Subclass Opisthobranchia
Order Nudibranchia / Suborder Dendronotacea / Family Dotonidae*

14.55 Doto amyra Marcus, 1961
(=Doto varians MacFarland, 1966, in part; D. wara Marcus, 1961)
Hammerhead Doto

Uncommon, low intertidal zone on rocky shores and on boat landings in bays; Dillon Beach (Marin Co.) to Mission Bay (San Diego Co.); Puerto Peñasco (Mexico).

Body about 6–12 mm long, uniformly white (lacking black markings); dorsal processes (cerata) in five to seven pairs, very large, pinkish, resembling clusters of grapes; rhinophores smooth with funnel-shaped sheaths; oral veil extending forward from head forming a square lobe on each side.

Doto amyra feeds on hydroids, including species of *Obelia* and *Abietinaria*. Several species of *Doto* are reported from this coast; little is known about their biology, and species identifications are often difficult to make. Species similar to *D. amyra* are **D. columbiana** O'Donoghue, 1921, which occurs from Vancouver Island (British Columbia) to Duxbury Reef (Marin Co.), and **D. kya** Marcus, 1961 (=*D. varians* MacFarland, 1966, in part), which is found from Duxbury Reef to Shell Beach (San Luis Obispo Co.). The differences between the species are described in McDonald (1975).

See Kress (1968), MacFarland (1966), Er. Marcus (1961), McDonald (1975), McDonald & Nybakken (1978), Odhner (1936), and Roller (1970b).

Phylum Mollusca / Class Gastropoda / Subclass Opisthobranchia
Order Nudibranchia / Suborder Arminacea / Superfamily Euarminoidea
Family Arminidae

14.56 Armina californica (Cooper, 1862)
(=Pleurophyllidia californica Cooper, 1862; Armina
vancouverensis (Bergh, 1876); A. columbiana O'Donoghue, 1924;
A. digueti Pruvot-Fol, 1955)   Striped Armina

Rare intertidally, more commonly found burrowing just
below surface of sandy and muddy bottoms, shallow subtidal
waters to about 230 m; Vancouver Island (British Columbia)
to Panama; rare in Gulf of California.

Body 25–50 mm long, flattened; dorsum with longitudinal
light stripes on grayish to brown background; no dorsal appendages except rhinophores; lamellate gills under body margin along length of body on both sides; frontal veil distinct.

When this burrower is at rest, only the tips of the rhinophores are visible at the sand surface. The animals feed using
suction to ingest pieces of the sea pansy *Renilla köllikeri* (3.26),
which responds with a bioluminescent display. Sea pens (e.g.,
Ptilosarcus gurneyi) may also be eaten. The eggs are laid in a
pale, pink-brown, spirally wound, convoluted chain of capsules, each capsule containing 3–22 eggs. Fertilization of the
eggs of this species was reported in a fine paper by MacFarland at the turn of the century.

See Bertsch (1968) Birkeland (1974), Hurst (1967), Lance (1962c),
MacFarland (1897, as *Pleurophyllidia*; 1966), Er. Marcus (1961),
McDonald & Nybakken (1978), and Pruvot-Fol (1955).

Phylum Mollusca / Class Gastropoda / Subclass Opisthobranchia
Order Nudibranchia / Suborder Arminacea / Superfamily Metarminoidea
Family Dironidae

14.57 Dirona albolineata Cockerell & Eliot, 1905
Chalk-Lined Dirona

Sporadically common on mudflats, in rocky tidepools, and on
bay boat landings, low intertidal zone; rarely subtidal to 35 m;
Vancouver Island (British Columbia) to La Jolla (San Diego
Co.).

Body typically 20–42 mm long, ranging to about 120 mm,

pale translucent white, occasionally yellowish to orange,
marked with conspicuous opaque white lines; processes
(cerata) on dorsum large, inflated, smooth, breaking off easily;
dorsal ridge on tail white; anterior end with conspicuous, undulating oral veil.

These very beautiful and delicate animals feed on prosobranch gastropods, such as *Margarites pupillus* (13.29) and *Lacuna carinatus*, and a variety of other invertebrates including
sponges, hydroids, ectoprocts, tunicates, and crustaceans.
They produce an egg mass containing about 350,000 eggs. Of
internal organs, only the radula and reproductive system have
been illustrated. The description of a new species, *D. aurantia*
Hurst, 1966, dredged from Washington subtidal waters, includes a fine drawing of the nervous system.

See Hurst (1966, 1967), MacFarland (1912, 1966), Er. Marcus
(1961), McDonald & Nybakken (1978), Nybakken (1978), and Robilliard (1971b).

Phylum Mollusca / Class Gastropoda / Subclass Opisthobranchia
Order Nudibranchia / Suborder Arminacea / Superfamily Metarminoidea
Family Dironidae

14.58 Dirona picta MacFarland in Cockerell & Eliot, 1905
Spotted Dirona

Common in summer, rare in other seasons, rocky low intertidal areas; subtidal to 9 m; Charleston (Oregon) to Puerto
Peñasco (Mexico).

Body commonly 20–40 mm long, ranging to 100 mm, typically translucent light yellow-green to rust brown, but highly
variable due to numerous brown, green, yellow, cream, or
pink spots; dorsal processes (cerata) large, inflated, easily detached, each with a pale-red spot on lower-outer surface and
tubercles on inner surface; dark green digestive gland visible
through body wall.

Dirona picta feeds on bryozoans, such as *Celleporella hyalina*
and probably *Thaumatoporella*, and on the hydroid *Aglaophenia*.

See Belcik (1975), Cockerell & Eliot (1905), Farmer & Collier
(1963), MacFarland (1912, 1966), Er. Marcus (1961), McDonald &
Nybakken (1978), and Nybakken (1978).

Phylum Mollusca / Class Gastropoda / Subclass Opisthobranchia
Order Nudibranchia / Suborder Arminacea / Superfamily Metarminoidea
Family Zephyrinidae

14.59 **Antiopella barbarensis** (Cooper, 1863)
(=Janolus barbarensis (Cooper, 1863); J. fuscus O'Donoghue, 1924; A. aureocincta MacFarland, 1966)   Cockscomb Nudibranch

Common from May to August on floats and pilings, and on rocks and kelp, low intertidal zone in bays and lagoons; rarely subtidal to about 30 m; Vancouver Island (British Columbia) to Isla San Diego (Baja California).

Body typically 13–50 mm long, distinguished from *Hermissenda crassicornis* (14.66) by its bright red "cockscomb" arising between sheathless rhinophores, and by the absence of blue lines on body; dorsal processes (cerata) very numerous, arising both in front of and behind rhinophores, easily detached, the tips orange and blue or white (generally bluish in southern areas); anterior oral veil small.

This is a most fragile animal: the cerata may begin to fall off as soon as the animal is picked up. It has been observed feeding on the hydroid *Corymorpha palma* (3.4) and the bryozoan *Bugula californica* (6.4).

See Ferreira & Bertsch (1975), MacFarland (1966), McDonald & Nybakken (1978), Nybakken (1978) and Steinberg (1963b).

Phylum Mollusca / Class Gastropoda / Subclass Opisthobranchia
Order Nudibranchia / Suborder Aeolidacea / Superfamily Eueolidoidea
Tribe Pleuroprocta / Family Corphellidae

14.60 **Coryphella trilineata** O'Donoghue, 1921
(=C. fisheri MacFarland, 1966; C. piunca Marcus, 1961)
Three-Lined Nudibranch

Common on bay boat landings, and on rocky areas and mud flats, low intertidal zone; subtidal at northern end of range; Vancouver Island (British Columbia) to Los Coronados (Baja California); rare in Gulf of California.

Body to about 27 mm long, variable in color, depending on color of food in digestive diverticula, generally translucent white, with three longitudinal opaque white lines; interior of cerata orange to light red; oral tentacles and rhinophores usually yellow to orange.

Coryphella trilineata feeds on hydroids, especially those growing on algae. It has also been observed on the larger hydroids *Tubularia crocea* (3.3) and *Eudendrium.* Spiral egg bands with up-and-down secondary loops are at first laid free in the water, held by the curved body of the animal, and then deposited on seaweeds.

Related forms, all rare, may be encountered occasionally and are diagnosed in McDonald (1975). **C. pricei** MacFarland, 1966, occurs in very low rocky tidepools, from Duxbury Reef (Marin Co.) to Monterey Bay. **C. cooperi** Cockerell, 1901, is known from Monterey Bay to San Pedro (Los Angeles Co.), usually on the hydroid *Tubularia crocea* (3.3). And **Cumanotus beaumonti** (Eliot, 1906) is reported on mud flats from San Juan Islands (Washington) to San Diego, and also from the northeastern Atlantic.

All of the species dealt with above feed on cnidarians for at least a part of their diets, and store in the tips of their cerata some of the still unexploded nematocysts from their prey. Experiments on an east coast form, *Coryphella verrucosa*, show that there is a turnover in the stored nematocysts every 3–5 days, even when the nudibranchs are not using their cerata defensively. This short residence time suggests that, at least under the conditions prevailing in the cerata, the nematocysts may have a very limited "shelf life".

For *Coryphella trilineata*, see Bridges & Blake (1972), MacFarland (1966), McDonald & Nybakken (1978), Miller (1971), Morse (1969), Nybakken (1974), Roller (1972), and Steinberg (1963b). For *C. pricei*, see MacFarland (1966). For *C. cooperi*, see MacFarland (1966) and Er. Marcus (1961). For *Cumanotus beaumonti*, see especially Gosliner & Williams (1970) and Hurst (1967); see also Day & Harris (1978).

Phylum Mollusca / Class Gastropoda / Subclass Opisthobranchia
Order Nudibranchia / Suborder Aeolidacea / Superfamily Eueolidoidea
Tribe Pleuroprocta / Family Flabellinidae

14.61 **Flabellinopsis iodinea** (Cooper, 1862)
(=Flabellina iodinea (Cooper, 1862))   Elegant Eolid

Sometimes on pilings, low intertidal zone, but more commonly on kelp to 15 m depth at northern end of range; uncommon intertidally but common subtidally to 35 m depth toward southern end of range; Vancouver Island (British Columbia) to Puerto Peñasco (Mexico).

Body typically 25–40 mm long, ranging to about 90 mm, translucent and vivid purple, with red rhinophores and orange cerata.

This gaudy species feeds on the polyps of the hydroid *Eudendrium ramosum* and sometimes on colonies of the ascidian *Diplosoma macdonaldi* (12.13). When disturbed, it can swim by rapidly flexing the body laterally into a U-shape, first to one side, then the other. Potential predators such as *Navanax inermis* (14.4) are repelled, apparently with a non-acid secretion. *F. iodinea* lays salmon-pink, gelatinous egg masses; in Monterey Bay, eggs are found on kelp in December.

See Goodwin & Fox (1955), Grigg & Kiwala (1970), Hinegardner (1974), Kitting (1975), MacFarland (1966), MacGinitie & MacGinitie (1968), McBeth (1970, 1971), Miller (1971), and L. Russell (1929).

Phylum Mollusca / Class Gastropoda / Subclass Opisthobranchia Order Nudibranchia / Suborder Aeolidacea / Superfamily Eueolidoidea Tribe Acleioprocta / Family Cuthonidae

14.62 **Cuthona lagunae** (O'Donoghue, 1926)
(=Trinchesia lagunae (O'Donoghue, 1926); Catriona lagunae (O'Donoghue, 1926); Cratena rutila MacFarland, 1966; Catriona ronga Marcus, 1961)   Orange-Faced Nudibranch

Common during summer, often on fine algae, low intertidal zone on rocky shores; Palomarin (Marin Co.) to Rosarito Beach (Baja California).

Body typically 5–14 mm in length; dorsal processes (cerata) long, cylindrical, arranged in rows of distinct clusters; rhinophores, top of head between rhinophores, and tips of cerata orange-red; oral tentacles white; digestive diverticula inside cerata black.

This is one of a whole series of slim, beautiful, and very similar eolid nudibranchs, which are easily confused. Several species in each of the genera *Cuthona* and *Eubranchus*, and one species in each of the genera *Catriona* and *Tenellia*, are involved. All of them have a pair of long slender rhinophores, all but *Tenellia* have a pair of long slender oral tentacles, and all have numerous, more or less tentacular dorsal processes

(cerata). All are diagnosed in McDonald (1975), and most are illustrated in color in MacFarland (1966). Several of these species have been noted on hydroids, which very likely are important in the diet. In all species, each of the cerata contains a branch of the digestive gland that terminates in a specialized compartment, the cnidosac, containing stored nematocysts derived from the cnidarian food.

For *C. lagunae*, see Baba & Abe (1964), Baba & Hamatani (1963), Long (1969a), MacFarland (1966, as *Cratena rutila*), Roller (1969b), Schmekel & Wechsler (1967), and Schonenberger (1969). For other species of *Cuthona* (formerly *Trinchesia*), see Hurst (1967), Long (1969a,b), MacFarland (1966), McDonald (1975), Roller (1969a,b), and Sphon (1972). For *Eubranchus*, see Baba (1960b, 1971), Behrens (1971b), Edmunds & Kress (1969), Lance (1961), MacFarland (1966), McDonald (1975), Er. Marcus (1961), Robilliard (1971c), Roller (1970a,b), and Roller & Long (1969). For *Catriona*, see Baba & Hamatani (1963), Lance (1966), MacFarland (1966), McDonald (1975), Robilliard (1971c), and Roller (1969b). For *Tenellia*, see McDonald (1975), Pruvot-Fol (1954), and Steinberg (1963b). For a revision of the family Cuthonidae, see G. Williams & Gosliner (1979).

Phylum Mollusca / Class Gastropoda / Subclass Opisthobranchia Order Nudibranchia / Suborder Aeolidacea / Superfamily Eueolidoidea Tribe Acleioprocta / Family Cuthonidae

14.63 **Cuthona divae** (Marcus, 1961)
(=Precuthona divae Marcus, 1961; Cuthona rosea MacFarland, 1966)   Rose-Pink Cuthona

Sometimes common in summer on colonies of pink hydroid *Hydractinia* on rocks, low intertidal zone; subtidal on rocks and mud to 20 m depth; San Juan Islands (Washington) to Los Angeles Co.

Body typically about 15–25 mm long, ranging to about 34 mm, cream-colored to pink; dorsal processes (cerata) long, cylindrical, very numerous, translucent to slightly blue externally, often pink or brown within, arranged in distinct, slightly oblique rows; rhinophores long, smooth, blunt; oral veil small.

This species feeds on *Hydractinia*, which occurs under rocks in low tidepools and on holdfasts of the brown alga *Cystoseira* subtidally. The white, tightly spiraled egg ribbons are reported in August off the San Juan Islands (Washington). Though little is known of this animal, there are short reports of the anatomy, development, and life history of related species of *Cuthona*.

See Baba (1963), MacFarland (1966), Er. Marcus (1961), Rao (1961), Robilliard (1971c), and G. Williams & Gosliner (1979).

Phylum Mollusca / Class Gastropoda / Subclass Opisthobranchia
Order Nudibranchia / Suborder Aeolidacea / Superfamily Eueolidoidea
Tribe Acleioprocta / Family Fionidae

14.64 Fiona pinnata (Eschscholtz, 1831)
Fiona

An open ocean form, occasionally cast ashore on floating debris with gooseneck (lepadid) barnacles, or on the pelagic cnidarian *Velella velella* (3.18), especially in summer; found cast ashore at Limantour Estero (Marin Co.) and many other localities on the Pacific Coast; probably cosmopolitan in warmer seas.

Body typically 15–35 mm long, translucent gray to cream and pale umber in thick regions, differing from *Aeolidia* (e.g., 14.67) and *Eubranchus* by possessing membranous flaps running along length of dorsal processes (cerata); digestive diverticula in dorsal processes brown to raw umber, with trace of green.

There is only one species of *Fiona*. It lives a pelagic, gregarious existence on flotsam, driftwood, floating seaweeds and animals, and occasionally the hulls of ships. A carnivore, it feeds on hydroids, the pelagic hydrozoans *Velella* and *Porpita*, the pelagic barnacles *Lepas* and *Alepas*, and the pelagic snail *Ianthina*. In feeding on the gooseneck barnacle (*Lepas* (*Lepas*) *anatifera*, 20.4), the predator uses its jaws to clasp the prey near the junction of stalk and shell plates, then rasps actively with its radula. Bending and shaking by the barnacle may dislodge the predator, but if this does not occur the barnacle soon ceases all movement and gapes open. The nudibranch then enters between the gaping plates and consumes the prey.

Growth under laboratory conditions is extremely rapid; a group of 12 animals captured with *Lepas* (*Lepas*) *anatifera* on

floating driftwood increased in average length from 3.4 mm to 30 mm in 24 days, an average gain in length of 1.11 mm per day. Mutual copulation was noted on the seventeenth day (at an average body length just below 20 mm), and egg laying started 12.5 hours later. In the field the white egg strings are laid on cleaned *Velella* floats or other floating material. Eggs laid in the laboratory were 495 μm in diameter, and hatched as veliger larvae in 5 days. A classic cell-lineage study was made on this species early in the century by Casteel.

See especially Holleman (1972); see also Bieri (1966, 1970), Burn (1967), Casteel (1904), Kropp (1931), MacFarland (1966), Er. Marcus (1961), McDonald & Nybakken (1978), and L. Russell (1929).

Phylum Mollusca / Class Gastropoda / Subclass Opisthobranchia
Order Nudibranchia / Suborder Aeolidacea / Superfamily Eueolidoidea
Tribe Cleioprocta / Family Phidianidae

14.65 Phidiana pugnax Lance, 1962
(=*P. nigra* MacFarland, 1966)   Fighting Phidiana

Occasionally on kelp, low intertidal and subtidal rocky areas to depths of 35 m; seasonally common in January and July at northern end of range; Monterey Bay to Puerto Rompiente (Baja California).

Body typically 30–37 mm in length, ranging to about 63 mm; dorsal processes (cerata) densely arranged, containing black digestive diverticula within, and tipped with red and white bands; rhinophores with orange-red ring around base; head tentacles joined by a line of bright orange-red; dorsum lacking blue-bordered orange areas.

As the species name implies, this form is extremely pugnacious, attacking and dismembering other eolid nudibranchs. Most specimens show signs of combat. The cerata bend forward and wave violently when the animal is disturbed. *P. pugnax* has been observed eating the hydroid *Hydractinia*, and it probably feeds mainly on cnidarians, but the diet seems to include almost any animal tissue.

See Lance (1962a), MacFarland (1966), McDonald & Nybakken (1978), and Nybakken (1978).

Phylum Mollusca / Class Gastropoda / Subclass Opisthobranchia
Order Nudibranchia / Suborder Aeolidacea / Superfamily Eueolidoidea
Tribe Cleioprocta / Family Facelinidae

14.66 Hermissenda crassicornis (Eschscholtz, 1831)
Hermissenda

Very common, especially in spring and summer, in varied habitats (rocky pools, marina floats, pilings, mud flats), low intertidal zone; subtidal to 35 m depth; Sitka (Alaska) to Puertecitos (Baja California); commonest in center of range.

Body to about 80 mm long, always recognizable, even in juvenile stages, by presence of orange areas on back bordered by bright light-blue lines; color and pattern otherwise extremely variable, especially in younger animals; cerata present only posterior to rhinophores, characteristically with white tips ringed below with orange.

Hermissenda is one of the most abundant and conspicuous nudibranchs in California. In central California it feeds mainly on hydroids, but the varied diet also includes small sea anemones, bryozoans, colonial ascidians (*Aplidium solidum*, 12.2; botryllids, 12.29, 12.30), annelids, small crustacea, tiny clams, dead animals of all sorts, and even other *Hermissenda*. In Puget Sound (Washington) it is a major predator on the sea pen *Ptilosarcus gurneyi* in subtidal waters.

Mating animals and egg masses are most often seen in southern California in winter but are found year around in Puget Sound. The pink egg string, laid in narrow coils, is regularly constricted and resembles linked pink sausages. It is commonly attached to algae and to blades of the eelgrass *Zostera*. Individual egg capsules in the string usually contain one egg, but sometimes up to four.

Recent investigations of this species include studies of the eggs and larvae, the gut, food habits, storage of nematocysts in the cerata, the fine structure and physiology of the eye, associative learning processes, and intraspecific fighting behavior. The eye is of special interest to neurophysiologists because it contains only five cells, each about 75 μm in diameter, large enough to receive a recording electrode. The eyes are of the rhabdomeric type and contain bodies suspected of being symbiotic fungi.

Hermissenda is an aggressive creature. Encounters between two individuals frequently result in fights involving lunging and biting. Mutual head-on contact is the type of encounter most likely to provoke a battle. An animal whose head encounters the side or tail of a second animal usually gets in the first bite, and usually emerges the winner. The copepod *Hemicyclops thysanotus* is often found adhering to the dorsal surface of *Hermissenda*.

A related nudibranch, **Emarcusia morroensis** Roller, 1972, is occasionally found among hydroids on harbor floats from Elkhorn Slough (Monterey Co.) to San Diego. It is 4–16 mm long; the body is a translucent gray with two median, oval, cadmium-orange patches, one in front and one in back of the rhinophores. The terminal thirds of the fingerlike rhinophores and oral tentacles are covered with opaque white dots, and the grayish cerata have three dark brownish-black bands.

See Ägersborg (1925), Barth (1964), Behrens & Tuel (1977), Bertsch et al. (1972), Birkeland (1974), Bürgin (1963), Costello (1938), Crane (1971), Crow & Alkon (1978), Dennis (1967), Eakin, Westfall & Dennis (1967), Graham (1938), Hurst (1967), Lance (1966), MacFarland (1966), Er. Marcus (1961), Nybakken (1978), Roller (1972), Stensaas et al. (1969), Yarnall (1972), and Zack (1974, 1975a,b).

Phylum Mollusca / Class Gastropoda / Subclass Opisthobranchia
Order Nudibranchia / Suborder Aeolidacea / Superfamily Eueolidoidea
Tribe Cleioprocta / Family Aeolidiidae

14.67 Aeolidia papillosa (Linnaeus, 1761)
Shag-Rug Nudibranch

Sporadically common on rocky shores, harbor pilings, and mud flats, low intertidal zone; subtidal to about 760 m; found all along California; also north Pacific, north Atlantic, cosmopolitan.

Larger specimens usually 38–64 mm in length, sometimes reaching 100 mm; basic color white to brown; dorsum covered with cerata, appearing like shag rug; cerata flattened, lacking conspicuous longitudinal membranes, taking on color of food; triangular area between cephalic tentacles and base of rhinophores distinctive, usually opaque white.

Aeolidia papillosa feeds mainly on sea anemones. Different species are taken in different areas; for a comprehensive list, see McDonald & Nybakken (1978). In the Monterey Bay area, *Metridium senile* (3.37) is preferred, but other species such as *Anthopleura elegantissima* (3.31), *Epiactis prolifera* (3.29), *Tealia coriacea* (3.34), and *T. crassicornis* (3.33) are also eaten, along with occasional sea pens and hydroids. Food is located from a distance, and *Aeolidia* shows a preference for anemones already damaged. The nudibranch may be injured and even killed by the stinging nematocysts of the prey, and attacks on large anemones are carried out cautiously. Mucus is first spread over the base of the column of the prey, then *Aeolidia* moves closer, its cerata held forward over the head, and bites large chunks from the base. The predator's buccal cavity and esophagus are lined with a protective cuticle, which limits damage from nematocysts there. The digestive glands in the cerata soon take on the color of the food; animals feeding on *Metridium* closely mimic the appearance of the prey, in both color and form. As in numerous other eolids, unexploded nematocysts from the food are stored in special compartments (cnidosacs) at the tips of the cerata, and are extruded if *Aeolidia* itself is attacked. Studies of an east coast population feeding on *Metridium senile* showed that, of the several different types of nematocysts found in the anemone, only two types were stored in quantity in the cnidosacs.

The eggs are laid in thin-walled egg capsules crowded into a large, untidy, pink or white egg string. *Aeolidia* living on mud flats often lay their eggs on the eelgrass *Zostera*. Adult animals have been used in a variety of experimental studies.

A similar form **Aeolidiella takanosimensis** Baba, 1930, is found seasonally on docks and intertidal rocks in harbors and bays from Palos Verdes Peninsula (Los Angeles Co.) to San Diego and in Japan, from where it may have been introduced. Typically, the body is slender and 17–30 mm long; the front corners of the foot are angular; the oral tentacles, rhinophores, head, and heart area are tinged with vermilion; the cerata are veined with chocolate.

For *Aeolidia papillosa*, see Baba (1930), Bertsch (1974), Braanis & Geelen (1953), Day & Harris (1978), Edmunds (1966b), Graham (1938), Harris (1973), Hinegardner (1974), Hurst (1967), MacFarland (1966), Er. Marcus (1961), McDonald & Nybakken (1978), Nybak-

ken (1978), H. Russell (1942), Stehouwer (1952), Streble (1968), Waters (1973), Wolter (1967), and Yarnall (1972). For *Aeolidiella*, see Baba (1930).

Phylum Mollusca / Class Gastropoda / Subclass Opisthobranchia
Order Nudibranchia / Suborder Aeolidacea / Superfamily Eueolidoidea
Tribe Cleioprocta / Family Spurillidae

14.68 **Spurilla chromosoma** Cockerell & Eliot, 1905
Frosted Spurilla

Uncommon, on boat floats and in low intertidal zone on rocky shores throughout the year in southern California; very abundant in rocky regions, Gulf of California; San Pedro (Los Angeles Co.) to San Felipe (Baja California) and Jalisco (Mexico).

Body typically 12–28 mm long, pale orange with medial patches of opaque white; dorsal processes (cerata) pinkish, tipped with white.

This species has received little attention so far; only the radula and external features have been illustrated. The animals are reported to feed on the sea anemone *Metridium senile* (3.37), using the attack behavior described under *S. oliviae* (14.69). Er. Marcus (1961) provides a key to the world's genera in this family (under Aeolidiidae).

See Harris (1973), Keen (1971), and Er. Marcus (1961).

Phylum Mollusca / Class Gastropoda / Subclass Opisthobranchia
Order Nudibranchia / Suborder Aeolidacea / Superfamily Eueolidoidea
Tribe Cleioprocta / Family Spurillidae

14.69 **Spurilla oliviae** (MacFarland, 1966)
(=Aeolidiella oliviae MacFarland, 1966)   Olive's Spurilla

Of sporadic occurrence in shaded pools, low intertidal and subtidal zones on rocky shores; Duxbury Reef (Marin Co.) to Point Loma (San Diego Co.); rare at north end of range.

Body typically 20–30 mm in length, light cream color; cerata thickly set, vivid orange to light rose color, with white-frosted tips and sometimes with orange-vermillion streaks; rhinophores distinctive, "bushy" in appearance, brilliant orange-vermillion to red with white tips.

This species, named for Mrs. Olive MacFarland, moves away from bright light and is generally nocturnal in its activity. It eats sea anemones, especially *Metridium senile* (3.37), *Epiactis prolifera* (3.29), and *Corynactis californica* (3.41). In attacking *Metridium*, the nudibranch first bites the anemone's column, then retreats as the anemone extrudes from its body numerous long, filamentous acontia (see under 3.37), heavily laden with nematocysts. The sea slug then advances and eats most of the acontia before further attacking the main body of the anemone.

See Harris (1973), MacFarland (1966), McDonald & Nybakken (1978), and G. Williams (1971).

Phylum Mollusca / Class Gastropoda / Subclass Pulmonata
Order Basommatophora / Family Trimusculidae

14.70 Trimusculus reticulatus (Sowerby, 1835)
(=Gadinia reticulata Sowerby, 1835)   Reticulate Button Snail

Locally common, but hidden and not often encountered; attached upside down on roofs of dark sea caves and deeply shaded horizontal crevices, and in abandoned burrows of rock-boring animals, middle intertidal zone on rocky shores; Coos Bay (Oregon) to Acapulco (Mexico).

Shell about 10–20 mm in diameter, circular, tent-shaped, marked with radiating lines and some concentric striations, white in color, sometimes tinted orange, pink, or green; muscle scar on inside surface of shell horseshoe-shaped; eyes at bases of tentacles; two soft, rounded oral lobes sometimes projecting from under anterior edge of shell.

Although capable of creeping, these animals are virtually sessile and may not change position for months on end. They spend more time exposed to air than submerged. Gills are lacking, and the mantle cavity and mantle fold apparently act as respiratory surfaces. Recent studies show that the animals do not graze on encrusting algal films, as once believed, but feed exclusively on particles (chiefly diatoms) suspended in the water. Glands on the head, anterior foot, and adjacent mantle fold produce a sustained flow of mucus. Food particles are brought in by the water, which covers the habitat at high tide and enters at other times by surge and splash. The particles adhere to the mucus and are ingested by sustained rhythmic motions of the odontophore, that lobe of muscle and cartilage that bears the radula. The radula itself is very tiny, with microscopic teeth, and is not protruded from the mouth; instead, ridges on the odontophore facilitate the intake of mucus. Jaws are lacking and salivary glands vestigial. Animals have been maintained in the laboratory for more than 2 years on a phytoplankton diet.

The egg masses are gelatinous and petal-like, and are attached in rosettes on the rock around the animal that laid them. Egg deposition occurs in April at Tomales Head (Marin Co.). The young hatch as free-swimming, feeding veliger larvae.

See Dall (1870), Hubendick (1947), Imamura (1969), K. Johnson (1968), Schumann (1911), Simroth & Hoffman (1928), Walsby (1975), and Yonge (1958).

Phylum Mollusca / Class Gastropoda / Subclass Pulmonata
Order Basommatophora / Family Melampidae

14.71 Melampus olivaceus Carpenter, 1857
Olive Ear Snail

On mud flats in bays and brackish estuaries, usually out of water at high tide among mats of salt grass (*Salicornia*); Mugu Lagoon (Ventura Co.) to Scammon Lagoon (Baja California), and Gulf of California to Mazatlán (Mexico); common in southern part of range.

Shell pear-shaped, about 10–20 mm long, smooth, brownish or olive-drab in color, with lighter spiral stripes; aperture toothed or bearing parallel ridges and grooves within.

This snail is an air breather, living under semi-terrestrial conditions. It moves by a "looping" motion of its divided foot, the forward part serving as a plow, pulling the body along the surface on the broad, sledlike rear part of the foot. Very little is known of this particular species, but ecological notes have been published on *M. coffeus*, and the life history and reproductive biology of *M. bidentatus* have recently been studied in detail. Evolution in this family of snails (formerly Ellobiidae) has been treated at some length.

Three other marine pulmonate (air-breathing) snails may be encountered between the tidemarks: **Pedipes unisulcatus** Cooper, 1866, on splash-zone rocks, **Ovatella myosotis** (Draparnaud, 1801) in salt marshes and crevices of docks and pilings, and **Williamia peltoides** (Carpenter, 1864) on low intertidal and subtidal rocks. All are illustrated in McLean (1969). Smith & Carlton (1975) diagnoses the distinguishing characteristics of many of the marine pulmonates.

See Apley (1970), Apley, Russell-Hunter & Avolizi (1967), Avolizi (1967), Broek (1950), Donohue (1965), Golley (1960), Hausman (1932), Holle & Dineen (1957, 1959), Keen (1971), Knipper & Meyer (1956), Kuschel (1963), MacDonald (1969), Er. Marcus & Ev. Marcus (1965), McLean (1969), Morton (1955a,b), and Smith & Carlton (1975).

Phylum Mollusca / Class Gastropoda / Subclass Pulmonata
Order Stylommatophora / Suborder Onchidiacea / Family Onchidellidae

14.72 Onchidella borealis Dall, 1871
Leather Limpet

Fairly common but inconspicuous on the red alga *Odonthalia floccosa*, on the brown alga *Laminaria*, in mussel beds, and in rocky crevices and depressions, high to low intertidal zones; Alaska to San Luis Obispo Co.

Body to about 12 mm long, limpetlike in appearance but actually shell-less, with a leathery texture; color deep reddish-brown or brown mottled with black and white; eyes at tips of tentacles; breathing pore (pneumostome), anus, and female genital pore located posteriorly, male pore below right sensory lobe and tentacle.

These mollusks feed by scraping the diatom film from rocks and algae when the tide is out. They are said to respire well both when exposed to air and when submerged, but animals placed in a dish of water usually crawl out with little delay. On incoming tides they cluster in algal holdfasts and rock crevices (perhaps around trapped bubbles), possibly using taste for guidance and tidal rhythm for timing. Active movement often stops when the tide is in.

See Abbott (1974), Duncan (1961), Fretter (1943), Gardiner (1928), Ghiselin (1966), Hoffman (1928), Er. Marcus (1961), Er. Marcus & Ev. Marcus (1956), Solem (1959), and Stringer (1963).

Literature Cited

Abbott, R. T. 1974. American seashells. 2nd ed. New York: Van Nostrand. 663 pp.

Aboul-Ela, I. A. 1959. On the food of nudibranchs. Biol. Bull. 117: 439–42.

Adjeska, R. A., and J. Nybakken. 1976. Contributions to the biology of *Melibe leonina* (Gould, 1852) (Mollusca: Opisthobranchia). Veliger 19: 19–26.

Ägersborg, H. P. K. 1922. Notes on the locomotion of the nudibranchiate mollusk *Dendronotus giganteus* O'Donoghue. Biol. Bull. 42: 257–66.

———. 1923. The morphology of the nudibranchiate mollusc *Melibe* (syn. *Chioraera leonina* (Gould)). Quart. J. Microscop. Sci. 67: 508–92.

———. 1925. The sensory receptors and the structure of the oral tentacles of the nudibranchiate mollusk *Hermissenda crassicornis* (Eschscholtz, 1831) syn. *Hermissenda opalescens* Cooper 1862, 1863. Acta Zool. 6: 167–82.

Anderson, D. G., and C. E. Lane. 1963. A water soluble pigment of the nudibranch, *Hypselodoris edenticulata*. Bull. Mar. Sci. Gulf Caribb. 13: 262–66.

Anderson, E. S. 1971. The association of the nudibranch *Rostanga pulchra* MacFarland, 1905, with the sponges *Ophlitaspongia pennata*, *Esperiopsis originalis*, and *Plocamia karykina*. Doctoral thesis, Biology, University of California, Santa Cruz. 151 pp.

Anderson, G. B. 1971. A contribution to the biology of *Doridella steinbergae* and *Corambe pacifica*. Master's thesis, California State University, Hayward. 48 pp.

Apley, M. L. 1970. Field studies on life history, gonadal cycle and reproductive periodicity in *Melampus bidentatus* (Pulmonata: Ellobiidae). Malacologia 10: 381–97.

Apley, M. L., W. D. Russell-Hunter, and R. J. Avolizi. 1967. Annual reproductive turnover in the salt-marsh pulmonate snail, *Melampus bidentatus* (Say). Biol. Bull. 133: 455–56.

Arch, S. 1976. Neuroendocrine regulation of egg laying in *Aplysia californica*. Amer. Zool. 16: 167–75.

Avolizi, R. J. 1967. Annual reproductive turnover in the salt-marsh pulmonate snail, *Melampus bidentatus* (Say). Biol. Bull. 133: 455–56.

Ayling, A. M. 1968. The feeding behavior of *Rostanga rubicunda* (Mollusca, Nudibranchia). Tane 14: 25–42.

Baba, K. 1930. Studies on Japanese nudibranchs. 3. Phyllidae, Aeolididae. Venus 2: 117–25.

———. 1949. Opisthobranchia of Sagami Bay collected by His Majesty the Emperor of Japan. Tokyo: Iwanami Shoten. 194 pp.

———. 1957a. A revised list of the species of Opisthobranchia from the northern part of Japan, with some additional descriptions. J. Fac. Sci., Hokkaido Univ., (4) Zool. 13: 8–14.

———. 1957b. The species of the genus *Elysia* from Japan. Publ. Seto Mar. Biol. Lab. 6: 69–74.

———. 1960a. The general *Okenia*, *Goniodoridella* and *Goniodoris* from Japan (Nudibranchia, Goniodorididae). Publ. Seto Mar. Biol. Lab. 8: 79–83.

———. 1960b. Two new species of the genus *Eubranchus* from Japan (Nudibranchia, Eolidacea). Publ. Seto Mar. Biol. Lab. 8: 299–302.

———. 1963. The anatomy of *Cuthona futario* n. sp. (=*C. bicolor* of Baba, 1933) (Nudibranchia—Eolidoidea). Publ. Seto Mar. Biol. Lab. 6: 109–20.

———. 1969. List of the Pleurobranchidae and the Pleurobranchaeidae from Japan. [In Japanese.] Collecting and Breeding 31: 190–91.

———. 1970. List of the Gastropteridae and the Runcinidae from Japan. [In Japanese.] Collecting and Breeding 32: 46–48.

———. 1971. Review of the anatomical aspects of *Eubranchus misakiensis* Baba, 1960, from Mukaishima, Japan (Nudibranchia: Eolidoidea: Eubranchidae). Venus 30: 63–66.

Baba, K., and T. Abe. 1959. The genus *Chelidonura* and a new species *C. tsurugensis*, from Japan. Publ. Seto Mar. Biol. Lab. 7: 279–80.

———. 1964. A catrionid, *Catriona beta* n. sp., with a radula of *Cuthona* type (Nudibranchia, Eolidoidea). Ann. Rep. Noto Mar. Lab., Fac. Sci., Univ. Kanazawa 4: 9–14.

Baba, K., and I. Hamatani. 1963. A cuthonid, *Cuthona alpha* n. sp., with a radula of *Catriona* type (Nudibranchia, Eolidoidea). Publ. Seto Mar. Biol. Lab. 11: 339–43.

Baba, K., I. Hamatani, and K. Hisai. 1956. Observations on the spawning habits of some of the Japanese Opisthobranchia (II). Publ. Seto Mar. Biol. Lab. 5: 209–20.

Baba, K., and T. Tokioka. 1965. Two more new species of *Gastropteron* from Japan, with further notes on *G. flavum* T. & B. (Gastropoda: Opisthobranchia). Publ. Seto Mar. Biol. Lab. 12: 363–79.

Bailey, K. H., and J. S. Bleakney. 1967. First Canadian report of the sacoglossan *Elysia chlorotica* Gould. Veliger 9: 353–54.

Barker, J. L., and H. Levitan. 1971. Salicylate: Effect on membrane permeability of molluscan neurons. Science 172: 1245–47.

Barnes, H., and H. T. Powell. 1954. *Onchidoris fusca* (Muller), a predator of barnacles. J. Anim. Ecol. 23: 361–63.

Barth, J. 1964. Intracellular recording from photoreceptor neurons in the eyes of a nudibranch mollusc (*Hermissenda crassicornis*). Comp. Biochem. Physiol. 11: 311–15.

Bartsch, P., and H. A. Rehder. 1939. Mollusks collected on the presidential cruise of 1938. Smithsonian Misc. Collec. 98: 1–18.

Batham, E. J. 1961. Infoldings of nerve fibre membranes in the opisthobranch mollusc *Aplysia californica*. J. Biophys. Biochem. Cytol. 9: 490–92.

Bebbington, A., and T. E. Thompson. 1969. Reproduction in *Aplysia* (Gastropoda, Opisthobranchia). Malacologia 9: 253 (abstract).

Beeman, R. D. 1963a. Notes on the California species *Aplysia* (Gastropoda: Opisthobranchia). Veliger 5: 145–47.

———. 1963b. Variation and synonymy of *Phyllaplysia* in the northeastern Pacific (Mollusca: Opisthobranchia). Veliger 6: 43–47.

———. 1968a. The order Anaspidea. Veliger 3 (Suppl): 87–102.

———. 1968b. The use of succinylcholine and other drugs for anesthetizing or narcotizing gastropod mollusks. Pubbl. Staz. Zool. Napoli 36: 267–70.

———. 1969. An autoradiographic demonstration of stomach tooth renewal in *Phyllaplysia taylori* Dall, 1900 (Gastropoda: Opisthobranchia). Biol. Bull. 136: 141–46.

———. 1970a. The anatomy and functional morphology of the reproductive system in the opisthobranch mollusk *Phyllaplysia taylori* Dall, 1900. Veliger 13: 1–31.

———. 1970b. An autoradiographic and phase contrast study of spermatogenesis in the anaspidean opisthobranch *Phyllaplysia taylori* Dall, 1900 (Gastropoda: Opisthobranchia). Arch. Zool. Expér. Gén. 111: 5–22.

———. 1970c. An autoradiographic study of sperm exchange and storage in a sea hare, *Phyllaplysia taylori*, a hermaphroditic gastropod (Opisthobranchia: Anaspidea). J. Exper. Zool. 175: 125–32.

———. 1970d. An ecological study of *Phyllaplysia taylori* Dall, 1900 (Gastropoda: Opisthobranchia), with an emphasis on its reproduction. Vie et Milieu (A) Biol. Mar. 21: 189–211.

———. 1972. Sperm biology in anaspidean mollusks. Echo 5: 19–21.

———. 1977. Gastropoda: Opisthobranchia, pp. 115–79, *in* A. C. Giese and J. S. Pearse, eds., Reproduction of marine invertebrates, vol. 4. New York: Academic Press. 369 pp.

Behrens, D. W. 1971a. The occurrence of *Ancula pacifica* MacFarland in San Francisco Bay. Veliger 13: 297–98.

———. 1971b. *Eubranchus misakiensis* Baba, 1960 (Nudibranchia: Eolidacea), in San Francisco Bay. Veliger 14: 214–15.

Behrens, D. W., and M. Tuel. 1977. Notes on the opisthobranch fauna of south San Francisco Bay. Veliger 20: 33–36.

Belcik, F. P. 1975. Additional opisthobranch mollusks from Oregon. Veliger 17: 276–77.

Beondé, A. C. 1968. *Aplysia vaccaria*, a new host for the pinnotherid crab *Opisthopus transversus*. Veliger 10: 375–78.

Bern, H. 1967. On eyes that may not see and glands that may not secrete. Amer. Zool. 7: 815–21.

Berrill, N. J. 1931. The natural history of *Bulla hydatis* Linn. J. Mar. Biol. Assoc. U.K. 17: 567–71.

Berry, A. J. 1977. Gastropoda: Pulmonata, pp. 181–226, *in* A. C. Giese and J. S. Pearse, eds., Reproduction of marine invertebrates, vol. 4. New York: Academic Press. 369 pp.

Bertsch, H. 1968. Effect of feeding by *Armina californica* on the bioluminescence of *Renilla koellikeri*. Veliger 10: 440–41.

———. 1969a. A note on the range of *Gastropteron pacificum* (Opisthobranchia: Cephalaspidea). Veliger 11: 431–33.

————. 1969b. *Cadlina modesta*: A range extension, with notes on habitat and a color variation. Veliger 12: 231–32.

————. 1970. Opisthobranchs from Isla San Francisco, Gulf of California, with description of a new species. Contr. Sci. Santa Barbara Mus. Natur. Hist. 2: 1–16.

————. 1973. Distribution and natural history of opisthobranch gastropods from Las Cruces, Baja California del Sur, Mexico. Veliger 16: 105–11.

————. 1974. Descriptive study of *Aeolidia papillosa* with scanning electron micrographs of the radula. Tabulata 7: 3–6.

————. 1977. The Chromodoridinae nudibranchs from the Pacific coast of America. 1. Investigative methods and supra-specific taxonomy. Veliger 20: 107–18.

Bertsch, H., and A. A. Smith. 1970. Observations on opisthobranchs of the Gulf of California. Veliger 13: 171–74.

Bertsch, H., T. Gosliner, R. Wharton, and G. Williams. 1972. Natural history and occurrence of opisthobranch gastropods from the open coast of San Mateo County, California. Veliger 14: 302–14.

Bieri, R. 1966. Feeding preferences and rates of the snail, *Ianthina prolongata*, the barnacle, *Lepas anserifera*, the nudibranchs, *Glaucus atlanticus* and *Fiona pinnata*, and the food web in the marine neuston. Publ. Seto Mar. Biol. Lab. 14: 161–70.

————. 1970. The food of *Porpita* and niche separation in three neuston coelenterates. Publ. Seto Mar. Biol. Lab. 17: 305–7.

Birkeland, C. B. 1974. Interactions between a sea pen and seven of its predators. Ecol. Monogr. 44: 211–32.

Bonar, D. B. 1978. Morphogenesis at metamorphosis in opisthobranch molluscs, pp. 177–96, *in* F.-S. Chia and M. E. Rice, eds., Settlement and metamorphosis of marine invertebrate larvae. New York: Elsevier. 290 pp.

Braanis, W. G., and F. M. Geelen. 1953. The preference of some nudibranchs for certain coelenterates. Arch. Néerl. Zool. 10: 241–64.

Brandriff, B., and R. D. Beeman. 1973. Observations on the gametolytic gland in the anaspidean opisthobranchs *Phyllaplysia taylori* and *Aplysia californica*. J. Morphol. 141: 395–410.

Brenori, M., E. Anlonini, D. Fasella, J. Wyman, and A. R. Fanelli. 1968. Reversible thermal denaturation of *Aplysia* myoglobin. J. Molec. Biol. 34: 497–504.

Bridges, C. S. 1975. Larval development of *Phyllaplysia taylori* Dall, with a discussion of development in the Anaspidea (Opisthobranchiata: Anaspidea). Ophelia 14: 161–84.

Bridges, C. S., and J. A. Blake. 1972. Embryology and larval development of *Coryphella trilineata* O'Donoghue, 1921 (Gastropoda: Nudibranchia). Veliger 14: 293–97.

Broek, A. N. 1950. On some brackish water mollusca from the Lake of Maracaibo. Zool. Meded. 31: 79–87.

Bullock, T. H., and G. A. Horridge. 1965. Structure and function in the nervous system of invertebrates. 2 vols. San Francisco: Freeman. 1,719 pp.

Bürgin, U. F. 1963. The color pattern of *Hermissenda crassicornis* (Eschscholtz, 1831) (Gastropoda: Opisthobranchia: Nudibranchia). Veliger 7: 205–15.

Burn, R. 1960. On *Tylodina corticalis* (Tate), a rare opisthobranch from south-eastern Australia. J. Malacol. Soc. Australia 4: 64–69.

————. 1961. A new doridid nudibranch from Torquay, Victoria. Veliger 4: 55–56.

————. 1962. Descriptions of Australian Eolidaceae (Mollusca: Opisthobranchia). J. Malacol. Soc. Australia 9: 25–34.

————. 1966. On three new Chromodorinae from Australia (Opisthobranchia: Nudibranchia). Veliger 8: 191–96.

————. 1967. First record of a pelagic eolid from Victoria. Victoria Natur. 84: 116–17.

————. 1968. *Archidoris odhneri* (MacFarland, 1966) comb. nov. with some comments on the species of the genus on the Pacific coast of North America. Veliger 11: 90–92.

Carefoot, T. H. 1967. Growth and nutrition of *Aplysia punctata* feeding on a variety of marine algae. J. Mar. Biol. Assoc. U.K. 47: 565–89.

————. 1970. A comparison of absorption and utilization of food energy in two species of *Aplysia*. J. Exper. Mar. Biol. Ecol. 5: 47–62.

Casteel, D. B. 1904. The cell-lineage and early larval development of *Fiona marina*, a nudibranchiate mollusk. Proc. Acad. Natur. Sci. Philadelphia 56: 325–405.

Castellucci, V., H. Pinsker, I. Kupfermann, and E. R. Kandel. 1970. Neuronal mechanisms of habituation and dishabituation of the gill-withdrawal reflex in *Aplysia*. Science 167: 1745–48.

Chapman, D. J., and D. L. Fox. 1969. Bile pigment metabolism in the seahare *Aplysia*. J. Exper. Mar. Biol. Ecol. 4: 71–78.

Chia, F.-S. 1978. Development and metamorphosis of the planktotrophic larvae of *Rostanga pulchra* (Mollusca: Nudibranchia). Mar. Biol. 46: 109–19.

Chia, F.-S., and R. Koss. 1978. Development and metamorphosis of the planktotrophic larvae of *Rostanga pulchra* (Mollusca: Nudibranchia). Mar. Biol. 46: 109–19.

Chivers, D. D. 1967. Observations on *Pleurobranchaea californica* MacFarland, 1966 (Opisthobranchia, Notaspidea). Proc. Calif. Acad. Sci. (4) 32: 515–21.

Clark, K. B. 1971. Life cycles of southern New England nudibranch molluscs. Doctoral thesis, University of Connecticut, Storrs.

Coan, E. 1964. A note on the natural history of *Pleurobranchaea* sp. (Gastropoda: Opisthobranchia). Veliger 6: 173.

Cockerell, T. D. A. 1901. Pigments of nudibranchiate Mollusca. Nature 65: 79–80.

Cockerell, T. D. A., and C. Eliot. 1905. Notes on a collection of Californian nudibranchs. J. Malacol. 12: 31–53.

Coggeshall, R. E. 1967. A light and electron microscope study of the abdominal ganglion of *Aplysia californica*. J. Neurophysiol. 30: 1263–87.

———. 1969. A fine structural analysis of the statocyst in *Aplysia californica*. J. Morphol. 127: 113–31.

———. 1970. A cytologic analysis of the bag cell control of egg laying in *Aplysia*. J. Morphol. 132: 461–85.

Collier, C. L. 1963. A new member of the genus *Atagema* (Gastropoda: Nudibranchia), a genus new to the Pacific northeast. Veliger 6: 73–75.

Collier, C. L., and W. M. Farmer. 1964. Additions of the nudibranch fauna of the east Pacific and the Gulf of California. San Diego Soc. Natur. Hist. Trans. 13: 377–96.

Cook, E. F. 1962. A study of food choices of two opisthobranchs, *Rostanga pulchra* MacFarland and *Archidoris montereyensis* Cooper. Veliger 4: 194–96.

Cooper, J. G. 1863. Some new genera and species of California mollusca. Proc. Calif. Acad. Natur. Sci. 2: 202–7.

Costello, D. P. 1938. Notes on the breeding habits of the nudibranchs of Monterey Bay and vicinity. J. Morphol. 63: 319–43.

Costello, D. P., and C. Henley. 1971. Methods for obtaining and handling marine eggs and embryos. 2nd ed. Marine Biological Laboratory, Woods Hole, Mass. 247 pp.

Crane, S. 1971. The feeding and reproductive behaviour of the sacoglossan gastropod *Olea hansineensis* Ägersborg, 1923. Veliger 14: 57–59.

Crow, T. J., and D. L. Alkon. 1978. Retention of an associative behavioral change in *Hermissenda*. Science 201: 1239–41.

Crozier, W. J., and L. B. Arey. 1919. Sensory reactions of *Chromodoris zebra*. J. Exper. Zool. 29: 261–310.

Dales, R. P. 1953. Northeast Pacific Phylliroidae. Ann. Mag. Natur. Hist. (12) 6: 193–94.

Dall, W. H. 1870. Materials towards a monograph of the Gadiniidae. Amer. J. Conchol. 6: 8–22.

Darling, S. D., and R. E. Cosgrove. 1966. Marine natural products. I. The search for *Aplysia* terpenoids in red algae. Veliger 8: 178–80.

Davis, W. J., and R. Gillette. 1978. Neural correlate of behavioral plasticity in command neurons of *Pleurobranchaea*. Science 199: 801–4.

Davis, W. J., and G. J. Mpitsos. 1971. Behavioral choice and habituation in the marine mollusk *Pleurobranchaea californica* MacFarland (Gastropoda, Opisthobranchia). Z. Vergl. Physiol. 75: 207–32.

Davis, W. J., G. J. Mpitsos, and J. M. Pinneo. 1974. The behavioral hierarchy of the mollusk *Pleurobranchaea*. I. The dominant position of the feeding behavior. II. Hormonal suppression of feeding associated with egg-laying. J. Compar. Physiol. 90: 207–43.

Davis, W. J., M. V. S. Siegler, and G. J. Mpitsos. 1973. Distributed neuronal oscillators and efference copy in the feeding system of *Pleurobranchaea*. J. Neurophysiol. 36: 258–74.

Davis, W. J., G. J. Mpitsos, M. V. S. Siegler, J. M. Pinneo, and K. B. Davis. 1974. Neuronal substrates of behavioral hierarchies and associative learning in *Pleurobranchaea*. Amer. Zool. 14: 1037–50.

Day, R. M., and L. G. Harris. 1978. Selection and turnover of coelenterate nematocysts in some aeolid nudibranchs. Veliger 21: 104–9.

Dennis, M. J. 1967. Electrophysiology of the visual system in a nudibranch mollusc. J. Neurophysiol. 30: 1439–65.

De Laubenfels, M. W. 1929. The red sponges of Monterey Peninsula. Ann. Mag. Natur. Hist. (9) 19: 258–66.

De Vries, J. B. 1963. Contributions to the morphology and histology of the nudibranch *Melibe rosea* Rang. Ann. Univ. Stellenbosch A38: 105–53.

Donohue, J. 1965. Concerning *Williamia peltoides* (Carpenter). Veliger 8: 19–21.

Duncan, C. J. 1961. The evolution of the pulmonate genital system. Proc. Zool. Soc. London 134: 601–9.

Dushane, H. 1966. Range extension for *Tylodina fungina* Gubb, 1865 (Gastropoda). Veliger 9: 86.

Eakin, R. M., J. A. Westfall, and M. J. Dennis. 1967. Fine structure of the eye of a nudibranch mollusc, *Hermissenda crassicornis*. J. Cell. Sci. 2: 349–58.

Eales, N. B. 1921. *Aplysia*. Liverpool Mar. Biol. Comm. [LMBC] Memoir 24: 1–84.

———. 1960. Revision of the world species of *Aplysia* (Gastropoda, Opisthobranchia). Bull. Brit. Mus. (Natur. Hist.), Zool. 5: 267–404.

Edmunds, M. 1966a. Defensive adaptations of *Stiliger vanellus* Marcus, with a discussion on the evolution of "nudibranch" molluscs. Proc. Malacol. Soc. London 37: 73–81.

———. 1966b. Protective mechanisms in the Eolidacea (Mollusca, Nudibranchia). J. Linn. Soc. London (Zool.) 47: 27–71.

Edmunds, M., and A. Kress. 1969. On the European species of *Eubranchus* (Mollusca, Opisthobranchia). J. Mar. Biol. Assoc. U.K. 49: 879–912.

Engel, H., and J. Nijssen-Meyer. 1964. On *Glossodoris quadricolor* (Ruppell and Leuckart, 1828) (Mollusca, Nudibranchia). Beaufortia 11: 27–32.

Eskin, A., and E. Harcombe. 1977. Eye of *Navanax*: Optic activity, circadian rhythm and morphology. Comp. Biochem. Physiol. 57A: 443–49.

Farmer, W. M. 1963. Two new opisthobranch mollusks from Baja California. San Diego Soc. Natur. Hist., Trans. 13: 81–84.

———. 1967. Notes on the opisthobranchia of Baja California, Mexico, with range extensions—II. Veliger 9: 340–42.

———. 1968. Tidepool animals from the Gulf of California. San Diego, Calif.: Wesword. 68 pp.

———. 1970. Swimming gastropods (Opisthobranchia and Prosobranchia). Veliger 13: 73–89.

Farmer, W. M., and A. J. Sloan. 1964. A new opisthobranch mollusk from La Jolla, California. Veliger 6: 148–50.

Ferreira, A. J. 1977. A review of the genus *Triopha* (Mollusca: Nudibranchia). Veliger 19: 387–402.

Ferreira, A. J., and H. Bertsch. 1975. Anatomical and distributional observations of some opisthobranchs from the Panamic faunal province. Veliger 17: 323–30.

Field, L. H., and D. L. MacMillan. 1973. An electrophysiological and behavioural study of sensory responses in *Tritonia* (Gastropoda, Nudibranchia). Mar. Behav. Physiol. 2: 171–85.

Forrest, J. E. 1953. On the feeding habits and the morphology and mode of functioning of the alimentary canal in some littoral dorid nudibranchiate mollusca. Proc. Linn. Soc. London 164: 225–35.

Franc, A. 1968. Sous-classe des Opisthobranches, pp. 608–893, 1079–82, *in* P.-P. Grassé, ed., Traité de zoologie, vol. 11. Paris: Masson. 1,083 pp.

Franz, D. R. 1967. On the taxonomy and biology of the dorid nudibranch *Doridella obscura*. Nautilus 80: 73–79.

Frazier, W. T., E. R. Kandel, I. Kupfermann, R. Waziri, and R. E. Coggeshall. 1967. Morphological and functional properties of identified neurons in the abdominal ganglion of *Aplysia californica*. J. Neurophysiol. 30: 1288–1351.

Fretter, V. 1940. On the structure of the gut of the sacoglossan nudibranchs. Proc. Zool. Soc. London B110: 185–98.

———. 1943. Studies in the functional morphology and embryology of *Onchidella celtica* (Forbes and Hanley). J. Mar. Biol. Assoc. U.K. 25: 685–720.

Fretter, V., and A. Graham. 1954. Observations on the opisthobranch mollusc *Actaeon tornatilis* (L.). J. Mar. Biol. Assoc. U.K. 33: 565–85.

Frings, H., and C. Frings. 1965. Chemosensory bases of food-finding and feeding in *Aplysia juliana* (Mollusca: Opisthobranchia). Biol. Bull. 128: 211–17.

Gardiner, A. 1928. Notes on British Mollusca. J. Conchol. 18: 249–50.

Gardner, D. 1971. Bilateral symmetry and interneuronal organization in the buccal ganglia of *Aplysia*. Science 173: 550–53.

Getting, P. A. 1976. Afferent neurons mediating escape swimming of the marine mollusc, *Tritonia*. J. Compar. Physiol. 110A: 271–86.

———. 1977. Neuronal organization of escape swimming in *Tritonia*. J. Compar. Physiol. 121A: 325–42.

Ghiselin, M. T. 1963. On the functional and comparative anatomy of *Runcina setoensis* Baba, an opisthobranch gastropod. Publ. Seto Mar. Biol. Lab. 11: 219–28.

———. 1964. Feeding of *Dendrodoris* (*Doriopsilla*) *albopunctata*, an opisthobranch gastropod. Ann. Rep. Amer. Malacol. Union 31: 45–46 (abstract).

———. 1966. Reproductive function and the phylogeny of opisthobranch gastropods. Malacologia 3: 327–78.

Gillette, R., M. P. Kovac, and W. J. Davis. 1978. Command neurons in *Pleurobranchaea* receive synaptic feedback from the motor network they excite. Science 199: 798–801.

Glynn, P. W. 1966. Community composition, structure, and interrelationships in the marine intertidal *Endocladia muricata–Balanus glandula* association in Monterey Bay, California. Beaufortia 12: 1–198.

Goddard, J. 1973. Opisthobranchs of San Francisco Bay. Tabulata 6: 8–10.

Golley, F. B. 1960. Ecologic notes on Puerto Rican Mollusca. Nautilus 73: 152–55.

Gomez, E. D. 1973. Observations on feeding and prey specificity of *Tritonia festiva* (Stearns) with comments on other tritoniids (Mollusca: Opisthobranchia). Veliger 16: 163–65.

Gonor, J. J. 1961. Observations on the biology of *Hermaeina smithi*, a sacoglossan opisthobranch from the west coast of North America. Veliger 4: 85–98.

———. 1963. Structure and function in the digestive gland of a marine snail *Aglaja diomedea* Bergh (Opisthobranchia: Cephalaspidea). Internat. Congr. Zool. Proc. 16: 43.

Goodwin, T. W., and D. L. Fox. 1955. Some unusual carotenoids from two nudibranch slugs and a lamprid fish. Nature 175: 1086–87.

Gorman, A. L. F., and M. F. Marmor. 1970. Contributions of sodium pump and ionic gradients to the membrane potential of a molluscan neuron. J. Physiol. 210: 897–917.

Gorman, A. F., and M. Mirolli. 1969. The input-output organization of a pair of giant neurones in the mollusc, *Anisodoris nobilis* (MacFarland). J. Exper. Biol. 51: 615–34.

Gosliner, T. M. 1979. A review of the systematics of *Cylichnella* Gabb (Opisthobranchia: Scaphandridae). Nautilus 93: 85–92.

———. 1980. Systematics and phylogeny of the Aglajidae (Opisthobranchia: Mollusca). Zool. J. Linn. Soc. (in press).

Gosliner, T. M., and G. C. Williams. 1970. The opisthobranch mollusks of Marin County, California. Veliger 13: 175–80.

———. 1972. A new species of *Chelidonura* from Bahía San Carlos, Gulf of California, with a synonymy of the family Aglajidae. Veliger 14: 424–36.

———. 1973a. The occurrence of *Polycera zosterae* O'Donoghue, 1924, in the Bodega Bay region, California, with notes on its natural history. Veliger 15: 252–53.

———. 1973b. Additions to the opisthobranch mollusk fauna of Marin County, California, with notes on their natural history. Veliger 15: 352–54.

———. 1975. A genus of dorid nudibranch previously unrecorded from the Pacific coast of the Americas, with the description of a new species. Veliger 17: 396–405.

Graham, A. 1938. The structure and function of the alimentary canal of aeolid mollusca. Trans. Roy. Soc. Edinburgh 59: 267–307.

Greene, R. W. 1968. The egg masses and veligers of southern California sacoglossan opisthobranchs. Veliger 11: 100–104.

————. 1970a. Symbiosis in sacoglossan opisthobranchs: Functional capacity of symbiotic chloroplasts. Mar. Biol. 7: 138–42.

————. 1970b. Symbiosis in sacoglossan opisthobranchs: Symbiosis with algal chloroplasts. Malacologia 10: 357–68.

————. 1970c. Symbiosis in sacoglossan opisthobranchs: Translocation of photosynthetic products from chloroplast to host tissue. Malacologia 10: 369–80.

Grigg, R. W., and R. S. Kiwala. 1970. Some ecological effects of discharged wastes on marine life. Calif. Fish & Game 56: 145–55.

Guberlet, J. E. 1928. Observations on the spawning habits of Melibe. Publ. Puget Sound Biol. Sta. 6: 262–70.

Guernsey, M. 1913. The anatomy of Laila cockerelli. J. Entomol. Zool. Pomona 5: 137–57.

Guiart, J. 1901. Contributions à l'étude des gastéropodes opisthobranches et en particulier des céphalaspides. Mém. Soc. Zool. France 14: 5–219.

Habe, T. 1956. Notes on the systematic position of three American seashells. Venus 19: 95–100.

Haefelfuiger, H. R., and A. Kress. 1967. Der Schwimmvorgang bei Gasteropteron rubrum (Rafinesque, 1814) (Gastropoda, Opisthobranchiata). Rev. Suisse Zool. 74: 547–54.

Hamilton, P. V., and H. W. Ambrose III. 1975. Swimming and orientation in Aplysia brasiliana (Mollusca; Gastropoda). Mar. Behav. Physiol. 3: 131–44.

Hand, C., and J. Steinberg. 1955. On the occurrence of the nudibranch Alderia modesta (Lovén, 1844) on the central California coast. Nautilus 69: 22–28. (See also C. Hand, p. 72.)

Hanna, G. D. 1951. A new west American nudibranch mollusk. Nautilus 65: 1–3.

Harris, L. G. 1973. Nudibranch associations, pp. 213–315, in T. C. Cheng, ed., Current topics in comparative pathobiology, vol. 2. New York: Academic Press. 334 pp.

Hausman, S. A. 1932. A contribution to the ecology of the saltmarsh snail Melampus bidentatus Say. Amer. Natur. 66: 541–45.

Heath, H. 1917. The anatomy of an eolid, Chioraera dalli. Proc. Acad. Natur. Sci. Philadelphia 69: 137–48.

Hill, B. J. 1962. Contributions to the morphology and histology of the tectibranch Berthella granulata (Krause). Ann. Univ. Stellenbosch A38: 155–85.

Hinegardner, R. 1974. Cellular DNA content of the Mollusca. Comp. Biochem. Physiol. 47A: 447–60.

Hoffmann, H. 1928. Zur Kenntniss der Oncidiiden. Zool. Jahrb. (Syst.) 55: 29–118.

————. 1932–40. Opisthobranchia, pp. 641–864. in Dr. H. G. Bronns Klassen und Ordnungen des Tierreichs, Bd. 3, Abt. 2, Buch 3. Leipzig: Winter. 1,377 pp., pls.

Holle, P. A., and C. F. Dineen. 1957. Life history of the saltmarsh snail Melampus bidentatus (Say). Nautilus 70: 91–95.

————. 1959. Studies on the genus Melampus (Pulmonata). Nautilus 73: 28–35, 46–51.

Holleman, J. J. 1972. Observations on growth, feeding, reproduction, and development in the opisthobranch, Fiona pinnata (Eschscholtz). Veliger 15: 142–46.

Howells, H. H. 1942. The structure and function of the alimentary canal of Aplysia punctata. Quart. J. Microscop. Sci. 83: 357–97.

Hubendick, B. 1947. Phylogenie and Tiergeographie der Siphonariidae. Zur Kenntnis der Phylogenie in der Ordnung Basommatophora und des Ursprungs der Pulmonatengruppe. Zool. Bidrag Uppsala 24: 1–216.

Hughes, G. M., and L. Tauc. 1962. Aspects of the organization of central nervous pathways in Aplysia depilans. J. Exper. Biol. 39: 45–69.

Hurst, A. 1965. Studies on the structure and function of the feeding apparatus of Philine aperta with a comparative consideration of some other opisthobranchs. Malacologia 2: 281–347.

————. 1966. A description of a new species of Dirona from the northeast Pacific. Veliger 9: 9–15.

————. 1967. The egg masses and veligers of thirty northeast Pacific opisthobranchs. Veliger 9: 255–88.

————. 1968. The feeding mechanism and behavior of the opisthobranch Melibe leonina. Symp. Zool. Soc. London 22: 151–66.

Hyman, L. H. 1967. The invertebrates: Mollusca I. Vol. 6. New York: McGraw-Hill. 792 pp.

Inamura, E. 1969. Trimusculus reticulata (Sowerby): A dissection guide. Unpubl. paper, Zool. 113, San Francisco State University.

Jacklet, J. W. 1969a. Circadian rhythm of optic nerve impulses recorded in darkness from isolated eye of Aplysia. Science 164: 562–63.

————. 1969b. Electrophysiological organization of the eye of Aplysia. J. Gen. Physiol. 53: 21–42.

Jahan-Parwar, B. 1972. Behavioral and electrophysiological studies on chemoreception in Aplysia. Amer. Zool. 12: 525–37.

Johannes, R. E. 1963. A poison-secreting nudibranch (Mollusca: Opisthobranchia). Veliger 5: 104–5.

Johansson, J. 1954. On the pallial gonoduct of Actaeon tornatilis (L.) and its significance for the phylogeny of the Euthyneura. Zool. Bidrag Uppsala 30: 223–32.

Johnson, K. J. 1968. Studies on the feeding, movement, and respiration of the pulmonate limpet Trimusculus (Gadinia) reticulatus (Sowerby) with notes on general morphology, egg masses, and veliger larva. Unpubl. student paper (abstract), Marine Science Center, Newport, Oregon.

Johnson, M. E., and H. J. Snook. 1927. Seashore animals of the Pacific Coast. New York: Macmillan. 659 pp.

Johnson, P. T., and F. A. Chapman. 1970. Comparative studies on the in vitro response of bacteria to invertebrate body fluids. II.

Aplysia californica (sea hare) and *Ciona intestinalis* (tunicate). J. Invert. Pathol. 16: 259–67.

Kanafoff, G. P., and W. K. Emerson. 1959. Late Pleistocene invertebrates of the Newport Bay area, California. Los Angeles Co. Mus. Contr. Sci. 31: 1–47.

Kandel, E. R. 1970. Nerve cells and behavior. Sci. Amer. 223: 57–70.

———. 1976. Cellular basis of behavior: An introduction to behavioral neurobiology. San Francisco: Freeman. 727 pp.

———. 1979. Behavioral biology of *Aplysia*. San Francisco: Freeman. 463 pp.

Kandel, E. R., W. T. Frazier, and R. E. Coggeshall. 1967. Opposite synaptic actions mediated by different branches of an identifiable interneuron in *Aplysia*. Science 155: 346–49.

Kandel, E. R., W. T. Frazier, R. Waziri, and R. E. Coggeshall. 1967. Direct and common connections among identified neurons in *Aplysia*. J. Neurophysiol. 30: 1352–76.

Kawaguti, S., and T. Yanasu. 1965. Electron microscopy on symbiosis between an elysioid gastropod and chloroplasts of a green alga. Biol. J. Okayama Univ. 11: 57–65.

Kay, E. A. 1964. The Aplysiidae of the Hawaiian Islands. Proc. Malacol. Soc. London 36: 173–90.

———. 1968. A review of the bivalved gastropods and a discussion of evolution within the Sacoglossa, pp. 109–34, *in* V. Fretter, ed., Studies in the structure, physiology, and ecology of mollusks. Symp. Zool. Soc. London 22. New York: Academic Press. 377 pp.

Kay, E. A., and D. K. Young. 1969. The Doridaceae (Opisthobranchia: Mollusca) of the Hawaiian Islands. Pacific Sci. 23: 172–231.

Keen, A. M. 1971. Sea shells of tropical west America. 2nd ed. Stanford, Calif.: Stanford University Press. 1,064 pp.

———. 1973. Some nomenclatural problems in Sacoglossa. Veliger 16: 238.

———. 1974. Re *Laura* Trinchese, 1872 (Gastropoda: Opisthobranchia). Veliger 16: 426.

Keen, A. M., and E. Coan. 1974. Marine molluscan genera of western North America: An illustrated key. 2nd ed. Stanford, Calif.: Stanford University Press. 208 pp.

Kennedy, D. 1971. Nerve cells and behavior. Amer. Sci. 59: 36–42.

Kitting, C. L. 1975. A subtidal predator-prey system: *Flabellinopsis iodinea* (Mollusca: Nudibranchia) and its hydroid prey. J. Undergrad. Res. Biol. Sci., Univ. Calif. Irvine 4: 407–20.

———. 1978. Intraspecific resource partitioning: Food selection of adjacent, individual *Anisodoris nobilis* (Mollusca: Nudibranchia) eating subtidal sponges. 59th Ann. Meet., West. Soc. Natur., Abs. Symp. Contr. Pap., pp. 42–43 (abstract).

Knipper, H., and K. O. Meyer. 1956. Biologisch und anatomische Betrachtungen an ostafrikanischen Ellobiiden. [With English summary.] Zool. Jahrb. (Syst.) 84: 99–112.

Konigsor, R. L., Jr., and D. Hunsaker II. 1971. Cellulase from the crop of *Aplysia vaccaria* Winkler, 1955. Veliger 13: 285–89.

Konishi, M. 1971. Ethology and neurobiology. Amer. Sci. 59: 56–63.

Kovac, M. P., and W. J. Davis. 1977. Behavioral choice: Neural mechanisms in *Pleurobranchaea*. Science 198: 632–34.

Krakauer, J. M. 1971. The feeding habits of aplysiid opisthobranchs in Florida. Nautilus 85: 37–38.

———. 1974. A method for estimating live weight and body length from the shell of *Aplysia willcoxi* Heilprin, 1886. Veliger 16: 396–98.

Kress, A. 1968. Untersuchungen zur Histologie, Autotomie und Regeneration drier Doto-Arten *Doto coronata*, *D. pinnatifida*, *D. fragilis* (Gastropoda, Opisthobranchiata). Rev. Suisse Zool. 75: 235–303.

———. 1970. A new record of *Trapania pallida* (Opisthobranchia, Gastropoda) with a description of its reproductive system and a comparison with *T. fusca*. Proc. Malacol. Soc. London 39: 111–16.

Kriegstein, A. R. 1977a. Development of the nervous system of *Aplysia californica*. Proc. Nat. Acad. Sci. 74: 375–78.

———. 1977b. Stages in the post-hatching development of *Aplysia californica*. J. Exper. Zool. 199: 275–88.

Kriegstein, A. R., V. Castellucci, and E. R. Kandel. 1974. Metamorphosis of *Aplysia californica* in laboratory culture. Proc. Nat. Acad. Sci. 71: 3654–58.

Kropp, B. 1931. The pigment of *Velella spirans* and *Fiona marina*. Biol. Bull. 60: 120–23.

Kupfermann, I. 1972. Studies on the neurosecretory control of egg laying in *Aplysia*. Amer. Zool. 12: 513–19.

Kupfermann, I., and E. R. Kandel. 1969. Neuronal controls of a behavioral response mediated by the abdominal ganglion of *Aplysia*. Science 164: 847–50.

Kupfermann, I., V. Castellucci, H. Pinsker, and E. Kandel. 1970. Neuronal correlates of habituation and dishabituation of the gill-withdrawal reflex in *Aplysia*. Science 156: 1743–45.

Kuschel, G. 1963. Composition and relationships of the terrestrial faunas of Easter, Juan Fernandez, Desventuradus and Galápagos Islands. Calif. Acad. Sci. Occas. Pap. 44: 79–95.

Lance, J. R. 1961. A distributional list of southern California opisthobranchs. Veliger 4: 64–68.

———. 1962a. Two new opisthobranch mollusks from southern California. Veliger 4: 155–59.

———. 1962b. A new *Stiliger* and a new *Corambella* (Mollusca: Opisthobranchia) from the northwestern Pacific. Veliger 5: 33–38.

———. 1962c. A new species of *Armina* (Gastropoda, Nudibranchia) from the Gulf of California. Veliger 5: 51–54.

———. 1966. New distributional records of some northeastern Pacific Opisthobranchiata (Mollusca, Gastropoda) with descriptions of two new species. Veliger 9: 69–81.

———. 1967. Northern and southern range extensions of *Aplysia vaccaria* (Gastropoda: Opisthobranchia). Veliger 9: 412.

Lasek, R., and W. Dower. 1971. *Aplysia californica*: Analysis of nuclear DNA in individual nuclei of giant neurons. Science 172: 279–80.

Levitan, H., L. Tauc, and J. P. Segundo. 1970. Electrical transmission among neurons in the buccal ganglion of a mollusc, *Navanax inermis*. J. Gen. Physiol. 55: 484–96.

Lewbel, G. S., and J. R. Lance. 1975. Detached epidermal sheaths of *Lophogorgia chilensis* as a food source for *Polycera atra* (Mollusca: Opisthobranchia). Veliger 17: 346.

Lewis, E. R., T. E. Everhart, and Y. Y. Zeevi. 1969. Studying neural organization in *Aplysia* with the scanning electron microscope. Science 165: 1140–43.

Lickey, M. E. 1968. A learned behavior in *Aplysia vaccaria*. J. Compar. Physiol. Psychol. 66: 712–18.

Lickey, M. E., and R. W. Berry. 1966. Learned behavioral discrimination of food objects by *Aplysia californica*. Physiologist 9: 230 (abstract).

Linton, D. 1966. Grazing mollusks in the weeds. Natur. Hist. 75: 59–61.

Long, S. J. 1969a. Records of *Trinchesia virens*, *Trinchesia fulgens*, and *Placida dendritica* from San Luis Obispo County, California. Tabulata 2: 9–12.

———. 1969b. Two new records of *Cratena abronia*. Veliger 11: 281.

MacDonald, K. B. 1969a. Molluscan faunas of Pacific coast salt marshes and tidal creeks. Veliger 11: 399–405.

———. 1969b. Quantitative studies of salt marsh mollusc faunas from the North American Pacific coast. Ecol. Monogr. 39: 33–60.

MacFarland, F. M. 1897. Cellulare Studien an Mollusken-Eiern. I. Zur Befruchtung des Eies von *Pleurophyllidia californica* (Cooper) Bergh. II. Die Centrosomen bei der Richtung der Korperbildung im Ei von *Diaulula sandiegensis* (Cooper) Bergh. Zool. Jahrb. 10: 227–64.

———. 1905. A preliminary account of the Dorididae of Monterey Bay, California. Proc. Biol. Soc. Washington 18: 35–54.

———. 1906. Opisthobranchiate Mollusca from Monterey Bay, California and vicinity. Bull. U.S. Bur. Fish. 25: 109–51.

———. 1912. The nudibranch family Dironidae. Zool. Jahrb. Suppl. 15: 515–36.

———. 1923. The morphology of the nudibranch genus *Hancockia*. J. Morphol. 38: 65–104.

———. 1926. The Acanthodorididae of the California coast. Nautilus 39: 49–65.

———. 1929. *Drepania*, a genus of nudibranchiate mollusks new to California. Proc. Calif. Acad. Sci. (4) 18: 485–96.

———. 1966. Studies of opisthobranchiate mollusks of the Pacific coast of North America. Mem. Calif. Acad. Sci. 6: 1–546.

MacFarland, F. M., and C. H. O'Donoghue. 1929. A new species of *Corambe* from the Pacific coast of North America. Proc. Calif. Acad. Sci. 28: 1–27.

MacGinitie, G. E. 1934. The egg-laying activities of the sea hare, *Tethys californicus* (Cooper). Biol. Bull. 67: 300–303.

———. 1935. Ecological aspects of a California marine estuary. Amer. Midl. Natur. 16: 629–765.

Marcus, Er. 1955. Opisthobranchia from Brazil. Bol. Fac. Filos. Cienc. Univ. S. Paulo, Zoologia 20: 89–262.

———. 1958a. Notes on Opisthobranchia. Bol. Inst. Oceanogr. Univ. Sao Paulo 7: 31–79.

———. 1958b. On western Atlantic opisthobranchiate gastropods. Amer. Mus. Novitates 1906: 2–82.

———. 1959. Opisthobranchia aus dem roten Meer und von den Malediven. Akad. Wiss. Lit. Mainz, Abh. Math.-Naturwiss. Klasse 12: 873–933.

———. 1961. Opisthobranch mollusks from California. Veliger 3 (Suppl.): 1–85.

———. 1964. A new species of *Polycera* (Nudibranchia) from California. Nautilus 77: 129–31.

Marcus, Er., and Ev. Marcus. 1955. Sea hares and side-gilled slugs from Brazil. Bol. Inst. Oceanogr. Univ. Sao Paulo 6: 3–48.

———. 1956. Zwei atlantische Onchidellen (Ergebnisse der Reiser A. Remane's nach Brasilien und den Kanaren). Kieler Meeresforsch. 12: 76–84.

———. 1965. On two Ellobiidae from southern Brazil. Bol. Fac. Filos. Cienc. Univ. S. Paulo, Zoologia 25: 425–53.

———. 1967. American opisthobranch mollusks. Stud. Trop. Oceanogr. Miami 6: 1–256.

———. 1970. Some gastropods from Madagascar and West Mexico. Malacologia 10: 181–224.

Marcus, Ev. 1971a. Range of *Gastropteron pacificum* Bergh, 1893. Veliger 13: 297.

———. 1971b. Translation of "Key of genera and species of the Pleurobranchidae—from Odhner, 1926." Molluscan Digest 1: A5–6.

———. 1972. On some Acteonidae (Gastropoda, Opisthobranchia). Papéis Avulsos Zool. 25: 167–88.

Mattox, N. T. 1953. A new species of *Pleurobranchus* from the Caribbean (Tectibranchiata). Nautilus 66: 109–14.

———. 1958. Studies on the Opisthobranchiata. II, A new tectibranch of the genus *Philine*. Bull. So. Calif. Acad. Sci. 57: 98–104.

Mazzarelli, G. F. 1893. Monografia delle Aplysiidae del Golfo di Napoli. Mem. Soc. Ital. Sci. Natur. 3: 1–222.

McBeth, J. W. 1968. Feeding behavior of *Corambella steinbergae*. Veliger 11: 145–46.

———. 1970. The deposition and biochemistry of carotenoid pigments in nudibranchiate molluscs. Doctoral thesis, University of California, San Diego. 173 pp.

———. 1971. Studies on the food of nudibranchs. Veliger 14: 158–61.

McCaman, R. E., and S. A. Dewhurst. 1970. Choline acetyltransferase in individual neurons of *Aplysia californica*. J. Neurochem. 17: 1421–26.

McCance, R. A., and M. Masters. 1937. The chemical composition and the acid-base balance of *Archidoris britannica*. J. Mar. Biol. Assoc. U.K. 22: 273–79.

McCauley, J. E. 1960. The morphology of *Phyllaplysia zostericola*, new species. Proc. Calif. Acad. Sci. (4) 29: 549–76.

McDonald, G. R. 1970. Range extensions for *Acanthodoris hudsoni* MacFarland, 1905, and *Onchidoris bilamellata* (Linnaeus, 1767). Veliger 12: 375.

————. 1975. Sacoglossa and Nudibranchia, pp. 522–41, *in* Smith & Carlton (1975).

McDonald, G. R., and J. W. Nybakken. 1978. Additional notes on the food of some California nudibranchs with a summary of known food habits of California species. Veliger 21: 110–19.

McGowan, J. A., and I. Pratt. 1954. The reproductive system and early embryology of the nudibranch *Archidoris montereyensis* (Cooper). Bull. Mus. Compar. Zool., Harvard, 111: 261–76.

McLean, J. H. 1969. Marine shells of southern California. Los Angeles Co. Mus. Natur. Hist., Sci. (24) Zool. 2: 1–104.

Miller, M. C. 1961. Distribution and food of the nudibranchiate Mollusca of the south of the Isle of Man. J. Anim. Ecol. 30: 95–116.

————. 1971. Aeolid nudibranchs (Gastropoda: Opisthobranchia) of the families Flabellinidae and Eubranchidae from New Zealand waters. Zool. J. Linn. Soc. 50: 311–37.

Morse, M. P. 1969. On the feeding of the nudibranch *Coryphella verrocosa rufibranchialis*, with a discussion of its taxonomy. Nautilus 83: 37–40.

————. 1972. Biology of *Okenia ascidicola* spec. nov. (Gastropoda: Nudibranchia). Veliger 15: 97–101.

Morton, J. E. 1955a. The evolution of the Ellobiidae with a discussion on the origin of the Pulmonata. Proc. Zool. Soc. London 125: 127–68.

————. 1955b. The functional morphology of the British Ellobiidae (Gastropoda, Pulmonata) with special reference to the digestive and reproductive systems. Phil. Trans. Roy. Soc. London B239: 89–160.

————. 1967. Molluscs. London: Hutchinson University Library. 244 pp.

————. 1972. The form and functioning of the pallial organs in the opisthobranch *Akera bullata* with a discussion on the nature of the gill in Notaspidea and other tectibranchs. Veliger 14: 337–49.

Mpitsos, G. J., and S. D. Collins. 1975. Learning: Rapid aversive conditioning in the gastropod mollusk *Pleurobranchaea*. Science 188: 954–57.

Mpitsos, G. J., S. D. Collins, and A. D. McClellan. 1978. Learning: A model system for physiological studies. Science 199: 497–506.

Mpitsos, G. J., and W. J. Davis. 1973. Learning: Classical and avoidance conditioning in the mollusk *Pleurobranchaea*. Science 180: 317–20.

Murray, M. J. 1977. Predatory behavior in *Navanax inermis*. Veliger 20: 55.

Murray, M. J., and E. R. Lewis. 1974. Sensory control of prey capture in *Navanax inermis*. Veliger 17: 156–58.

Muscatine, L., and R. W. Green. 1973. Chloroplasts and algae as symbionts in molluscs. Internat. Rev. Cytol. 36: 137–69.

Nicaise, G. 1969. Détection histochimique de cholinestérases dans les cellules gliales et interstitielles des Doridiens. Compt. Rend. Soc. Biol. Paris 163: 2600–2604.

Nybakken, J. 1974. A phenology of the smaller dendronotacean, armiṇacean and aeolidacean nudibranchs at Asilomar State Beach over a twenty-seven month period. Veliger 16: 370–73.

————. 1978. Abundance, diversity and temporal variability in a California intertidal nudibranch assemblage. Mar. Biol. 45: 129–46.

Nybakken, J., and J. Eastman. 1977. Food preferences, food availability and resource partitioning in *Triopha maculata* and *Triopha carpenteri* (Opisthobranchia: Nudibranchia). Veliger 19: 279–89.

Odhner, N. H. 1926. Die Opisthobranchien, pp. 1–100, *in* Further Zool. Res. Swedish Antarct. Exped. 1901–03, vol. 2.

————. 1934. The Nudibranchiata. Brit. Antarct. ("Terra Nova") Exped. 1910, Zool. 7: 229–309.

————. 1936. Nudibranchia Dendronotacea, a revision of the system. Mém. Mus. Roy. Hist. Natur. Belg. (2) 3: 1057–1128.

————. 1939. Opisthobranchiate Mollusca from the western and northern coasts of Norway. Kongelige Norske Vidensk. Selsk. Skr. 1: 1–93.

————. 1941. New polycerid nudibranchiate Mollusca and remarks on this family. Meddel. Gøteb. Mus. Zool. Avdel. 91: 1–20.

O'Donoghue, C. H. 1924. Notes on the nudibranchiate Mollusca from the Vancouver Island region. Trans. Roy. Canad. Inst. 15: 1–33.

Oldroyd, I. S. 1927. The marine shells of the west coast of North America. Stanford Univ. Publ., Geol. Sci. 2: 23–52.

Paine, R. T. 1963. Food recognition and predation on opisthobranchs by *Navanax inermis* (Gastropoda: Opisthobranchia). Veliger 6: 1–9.

————. 1964. Ash and calorie determinations of sponge and opisthobranch tissues. Ecology 45: 384–87.

————. 1965. Natural history, limiting factors and energetics of the opisthobranch *Navanax inermis*. Ecology 46: 603–19.

Peretz, B. 1969. Central neuron initiation of periodic gill movements. Science 166: 1167–72.

Perrier, R., and H. Fischer. 1914. Sur l'existence des spermatophores chez quelques opisthobranches. Compt. Rend. Acad. Sci. Paris 158: 1366–69.

Peterson, R. P. 1970. RNA in single identified neurons of *Aplysia*. J. Neurochem. 17: 325–38.

Pinsker, H., I. Kupfermann, V. Castellucci, and E. Kandel. 1970. Habituation and dishabituation of the gill-withdrawal reflex in *Aplysia*. Science 167: 1740–42.

Pohl, H. 1905. Über den feineren Bau des Genitalsystems von *Polycera quadrilineata*. Zool. Jahrb. Anat. 21: 427–52.

Potts, G. W. 1966. The respiratory anatomy and physiology of two dorid nudibranchs, with information on their ecology. Doctoral thesis, University of London.

———. 1970. The ecology of *Onchidoris fusca* (Nudibranchia). J. Mar. Biol. Assoc. U.K. 50: 269–92.

Pruvot-Fol, A. 1954. Mollusques opisthobranches. Faune de France 58: 1–460.

———. 1955. Les Arminiadae. Bull. Mus. Hist. Natur. (2) 27: 462–68.

Purchon, R. D. 1968. The biology of the Mollusca. New York: Pergamon. 560 pp.

Ram, J. L., and W. J. Davis. 1977. Mechanisms underlying "singleness of action" in the feeding behavior of *Pleurobranchaea californica* (MacFarland, 1966). Veliger 20: 55–56.

Rao, K. V. 1961. Development and life history of a nudibranchiate gastropod *Cuthona adyarensis*. J. Mar. Biol. Assoc. India 3: 186–97.

Raven, C. P. 1958. Morphogenesis: The analysis of molluscan development. New York: Pergamon. 311 pp.

———. 1964. Development, pp. 165–96, *in* Wilbur & Yonge (1964).

Reid, J. D. 1964. The reproduction of the sacoglossan opisthobranch *Elysia maoria*. Proc. Zool. Soc. London 143: 365–93.

Ricketts, E. F., and J. Calvin. 1968. Between Pacific tides. 4th ed. Revised by J. W. Hedgpeth. Stanford, Calif.: Stanford University Press. 614 pp.

Robilliard, G. A., and K. Baba. 1972. *Aldisa sanguinea cooperi* subspec. ecology of the genus *Dendronotus* (Gastropoda: Nudibranchia). Veliger 12: 433–79.

———. 1971a. A new species of *Polycera* (Opisthobranchia: Mollusca) from the northeastern Pacific, with notes on the other species. Syesis 4: 235–43.

———. 1971b. Predation by the nudibranch *Dirona albolineata* on three species of prosobranchs. Pacific Sci. 25: 429–35.

———. 1971c. Range extensions of some northeast Pacific nudibranchs (Mollusca: Gastropoda: Opisthobranchia) to Washington and British Columbia, with notes on their biology. Veliger 14: 162–65.

———. 1974. Range extensions for *Dendronotus diversicolor* (Mollusca: Opisthobranchia). Veliger 16: 335–36.

Robilliard, G. A., and K. Baba. 1972. *Aldisa sanguinea cooperi* subspec. nov. from the coast of the state of Washington, with notes on its feeding and spawning habits (Nudibranchia: Dorididae: Aldisinae). Publ. Seto Mar. Biol. Lab. 19: 409–14.

Robles, L. J. 1975. The anatomy and functional morphology of the reproductive system of *Bulla gouldiana* (Gastropoda: Opisthobranchia). Veliger 17: 278–91.

Roller, R. A. 1969a. A color variation of *Aldisa sanguinea*. Veliger 11: 280–81.

———. 1969b. Nomenclatural changes for the new species assigned to *Cratena* by MacFarland, 1966. Veliger 11: 421–23.

———. 1970a. A list of recommended nomenclatural changes for MacFarland's "Studies of opisthobranchiate mollusks of the Pacific coast of North America." Veliger 12: 371–74.

———. 1970b. A supplement to the annotated list of opisthobranchs from San Luis Obispo County, California. Veliger 12: 482–83.

———. 1971. Notes on the anatomy of two species of *Acteocina*. Echo 3: 31–32.

———. 1972. Three new species of eolid nudibranchs from the west coast of North America. Veliger 14: 416–23.

Roller, R. A., and S. J. Long. 1969. An annotated list of opisthobranchs from San Luis Obispo County, California. Veliger 11: 424–30.

Rudman, W. B. 1971. The family Acteonidae in New Zealand. J. Malacol. Soc. Australia 2: 205–14.

———. 1974. A comparison of *Chelidonura*, *Navanax*, and *Aglaja* with the other genera of the Aglajiidae (Opisthobranchia: Gastropoda). Zool. J. Linn. Soc. 54: 185–212.

Russell, H. D. 1942. Observations on the feeding of *Aeolidia papillosa* L. with notes on the hatching of the veligers of *Cuthona amoena* A. & H. Nautilus 55: 80–82.

———. 1946. Ecological notes concerning *Elysia chlorotica* Gould and *Stiliger fuscata* Gould. Nautilus 59: 95–97.

———. 1968. *Chromodoris californiensis* and *C. calensis*. Nautilus 81: 140–41.

———. 1971. Index Nudibranchia: A catalog of the literature, 1554–1965. Delaware Mus. Natur. Hist., Greenville. 141 pp.

Russell, L. 1929. The comparative morphology of the elysioid and aeolidioid types of the molluscan nervous system. Proc. Zool. Soc. London 1929: 197–233.

Sakharov, D. A. 1962. Giant nerve cells in nudibranchiate mollusks *Aeolidea papillosa* and *Dendronotus frondosus*. [In Russian, with English summary.] Zh. Obshehai Biol. 23: 308–11.

Schmekel, L., and W. Wechsler. 1967. Elektronenmikroskopische Untersuchungen uber Struktur und Entwicklung der Epidermis von *Trinchesia granosa* (Gastrop. Opisthobr.). Z. Zellforsch. 77: 95–114.

Schonenberger, N. 1969. Beitrage zur Entwicklung und Morphologie von *Trinchesia granosa* Schmekel (Gastropoda, Opisthobranchia). Pubbl. Staz. Zool. Napoli 37: 236–92.

Schumann, W. 1911. Über die Anatomie und die systematische Stellung von *Gadinia peruviana* Sowerby und *Gadinia garnoti* Payrandeau. Zool. Jahrb. Suppl. 13: 1–88.

Shonman, D., and J. W. Nybakken. 1978. Food preferences, food availability and food resource partitioning in two sympatric species of cephalaspidean opisthobranchs. Veliger 21: 120–26.

Siegler, M. V. S. 1977. Neuronal basis of *Pleurobranchaea* feeding. Veliger 20: 59–60.

Simroth, H., and H. Hoffmann. 1908–28. Pulmonata, *in* Dr. H. G. Bronns Klassen und Ordnungen des Tierreichs, Bd. 3, Abt. 2, Buch 2. Leipzig: Winter. 1,354 pp., pls.

Sleeper, H. L., and W. Fenical. 1977. Navenones A-C: Trail-breaking alarm pheromones from the marine opisthobranch *Navanax inermis*. J. Amer. Chem. Soc. 99: 2367–68.

Smallwood, W. M. 1904. The maturation, fertilization, and early cleavage of *Haminea solitaria* (Say). Bull. Mus. Compar. Zool., Harvard, 45: 259–318.

Smith, R. I., and J. T. Carlton, eds. 1975. Light's manual: Intertidal invertebrates of the central California coast. 3rd ed. Berkeley and Los Angeles: University of California Press. 716 pp.

Solem, A. 1959. Systematics and zoogeography of the land and freshwater Mollusca of the New Hebrides. Fieldiana: Zool. 43: 1–359.

Sphon, G. G. 1971. The reinstatement of *Hypselodoris agassizi* (Bergh, 1894) (Mollusca: Opisthobranchia). Veliger 14: 214.

———. 1972. Some opisthobranchs (Mollusca: Gastropoda) from Oregon. Veliger 15: 153–57.

Sphon, G. G., and J. R. Lance. 1968. An annotated list of nudibranchs and their allies from Santa Barbara County, California. Proc. Calif. Acad. Sci. (4) 36: 73–83.

Sphon, G. G., and D. K. Mulliner. 1972. A preliminary list of known opisthobranchs from the Galápagos Islands, collected by the Ameripagos Expedition. Veliger 15: 147–52.

Spicer, V. D. P. 1933. Report on a colony of *Haminoea* at Ballast Point, San Diego, California. Nautilus 47: 52–54.

Spray, D. C., M. E. Spira, and M. V. L. Bennett. 1977. Feeding in *Navanax inermis*, neurophysiological aspects. Veliger 20: 59.

Stehouwer, H. 1952. The preference of the sea slug *Aeolidia papillosa* (L.) for the sea anemone *Metridium senile* (L.). Arch. Néerl. Zool. 10: 161–70.

Steinberg, J. E. 1961. Notes on the opisthobranchs of the west coast of North America. I. Nomenclatural changes in the order Nudibranchia (southern California). Veliger 4: 57–63.

———. 1963a. Notes on the opisthobranchs of the west coast of North America. II. The order Cephalaspidea from San Diego to Vancouver Island. Veliger 5: 114–17.

———. 1963b. Notes on the opisthobranchs of the west coast of North America. III. Further nomenclatural changes in the order Nudibranchia. Veliger 6: 63–67.

Steinberg, J. E., and M. L. Jones. 1960. A new opisthobranch of the genus *Aglaja* in San Francisco Bay. Veliger 2: 73–74.

Stensaas, L. J., S. S. Stensaas, and O. Trujillo-Cenoz. 1969. Some morphological aspects of the visual system of *Hermissenda crassicornis* (Mollusca: Nudibranchia). J. Ultrastruc. Res. 27: 510–32.

Strain, H. H. 1949. Hopkinsiaxanthin, a xanthophyll of the sea slug *Hopkinsia rosacea*. Biol. Bull. 97: 207–9.

Streble, H. 1968. Bau und Bedeutung der Nesselsäcke *Aeolidia papillosa* L., der Breitwarzigen Fadenschnecke (Gastropoda, Opisthobranchia). Zool. Anz. 180: 356–472.

Strenth, N. E., and J. E. Blankenship. 1978. Laboratory culture, metamorphosis and development of *Aplysia brasiliana* Rang, 1828 (Gastropoda: Opisthobranchia). Veliger 21: 99–103.

Stringer, B. L. 1963. Embryology of the New Zealand Onchidiidae and its bearing on the classification of the group. Nature 197: 621–22.

Switzer-Dunlap, M. 1978. Larval biology and metamorphosis of aplysiid gastropods, pp. 197–206, *in* F.-S. Chia and M. E. Rice, eds., Settlement and metamorphosis of marine invertebrate larvae. New York: Elsevier. 290 pp.

Switzer-Dunlap, M., and M. G. Hadfield. 1977. Observations on development, larval growth, and metamorphosis of four species of Aplysiidae (Gastropoda, Opisthobranchia) in laboratory culture. J. Exper. Mar. Biol. Ecol. 29: 245–61.

Taylor, D. L. 1967. The occurrence and significance of endosymbiotic chloroplasts in the digestive glands of herbivorous opisthobranchs. J. Phycol. 3: 234–35.

———. 1968. Chloroplasts as symbiotic organelles in the digestive gland of *Elysia viridis* (Gastropoda: Opisthobranchia). J. Mar. Biol. Assoc. U.K. 48: 1–15.

———. 1970. Chloroplasts as symbiotic organelles. Internat. Rev. Cytol. 27: 29–64.

Taylor, D. W., and N. F. Sohl. 1962. An outline of gastropod classification. Malacologia 1: 7–32.

Thompson, S. H. 1976. Membrane currents underlying bursting in molluscan pacemaker neurons. Doctoral thesis, University of Washington, Seattle. 134 pp.

Thompson, T. E. 1959. Feeding in nudibranch larvae. J. Mar. Biol. Assoc. U.K. 38: 239–48.

———. 1960a. Defensive adaptations in opisthobranchs. J. Mar. Biol. Assoc. U.K. 39: 123–34.

———. 1960b. On a disputed feature of the anatomy of the nudibranch *Dendronotus frondosus* Ascanius. Proc. Malacol. Soc. London 34: 24–25.

———. 1961a. The importance of the larval shell in the classification of the Sacoglossa and the Acoela (Gastropoda, Opisthobranchia). Proc. Malacol. Soc. London 34: 233–39.

———. 1961b. The structure and mode of functioning of the reproductive organs of *Tritonia hombergi* (Gastropoda: Opisthobranchia). Quart. J. Microscop. Sci. 102: 1–14.

———. 1962. Studies on the ontogeny of *Tritonia hombergi* Cuvier (Gastropoda, Opisthobranchia). Phil. Trans. Roy. Soc. London B245: 171–218.

———. 1964. Grazing and the life cycles of British nudibranchs. Symp. Brit. Ecol. Soc. 4: 275–97.

———. 1967. Direct development in a nudibranch *Cadlina laevis*, with a discussion of developmental processes in Opisthobranchia. J. Mar. Biol. Assoc. U.K. 47: 1–22.

―――. 1969. Acid secretion in Pacific Ocean gastropods. Australian J. Zool. 17: 755–64.

―――. 1970. Eastern Australian Pleurobranchomorpha (Gastropoda, Opisthobranchia). J. Zool., London 160: 173–98.

―――. 1971. Tritoniidae from the North American Pacific coast (Mollusca: Opisthobranchia). Veliger 13: 333–38.

―――. 1976. Biology of the opisthobranch molluscs. Vol. 1. London: Ray Society no. 151. 206 pp.

Thompson, T. E., and A. Bebbington. 1969. Structure and function of the reproductive organs of three species of Aplysia (Gastropoda: Opisthobranchia). Malacologia 7: 347–80.

―――. 1970. A new interpretation of the structure of the aplysiid spermatozoan (Gastropoda, Opisthobranchia). Arch. Zool. Expér. Gén. 111: 213–16.

Thompson, T. E., and D. J. Slinn. 1959. On the biology of the opisthobranch Pleurobranchus membranaceus. J. Mar. Biol. Assoc. U.K. 38: 507–24.

Tobach, E., P. Gold, and A. Ziegler. 1965. Preliminary observations of the inking behavior of Aplysia (Varria). Veliger 8: 16–18.

Toevs, L. A., and R. W. Brackenbury. 1969. Bag cell–specified proteins and the humoral control of egg-laying in Aplysia californica. Comp. Biochem. Physiol. 29: 207.

Tokioka, T., and K. Baba. 1964. Four new species and a new genus of the family Gastropteridae from Japan (Gastropoda: Opisthobranchia). Publ. Seto Mar. Biol. Lab. 12: 201–99.

Usuki, I. 1970. Studies on the life history of Aplysidae and their allies in the Sado District of the Japan Sea. Sci. Rep. Niigata Univ. (D) Biol. 7: 91–105.

Van Weel, P. B. 1957. Observations on the osmoregulation in Aplysia juliana Pease (Aplysiidae, Mollusca). Z. Vergl. Physiol. 39: 492–506.

Vayssière, A. 1898. Monographie de la famille des Pleurobranchidés. Ital. Ann. Sci. Natur. Zool. (8) 8: 208–402.

Vicente, N. 1962. Particularitiés histologiques des cellules nerveuses et notamment des cellules neurosécrétrices chez Haminoea navicula (Da Costa) (Mollusque, Opisthobranche). Bull. Rec. Trav. Sta. Mar. d'Endoume 25: 293–304.

―――. 1966. Sur les phénomènes neurosécrétoires chez les gastéropodes opisthobranches. Compt. Rend. Acad. Sci. Paris 263: 382–85.

―――. 1969b. Étude histologique et histochimique du système nerveux central, des rhinophores et de la gonade chez les gastéropodes opisthobranches. Tethys 1: 833–74.

―――. 1969a. Corrélations neuroendocrines chez Aplysia rosea ayant subi l'ablation de divers ganglions nerveux. Tethys 1: 875–900.

Vogel, R. M., and L. P. Schultz. 1970. Cargoa cupella, new genus and new species of nudibranch from Chesapeake Bay and the generic status of Okenia Menke, Idalia Leuckart, and Idalla Ørsted. Veliger 12: 388–93.

Wachtel, H., and E. R. Kandel. 1967. A direct synaptic connection mediating both excitation and inhibition. Science 158: 1206–8.

Waidhofer, C. 1969. Anatomische Untersuchungen des Zentralnervensystems von Fimbria fimbria und Melibe leonina. Malacologia 9: 295–96 (abstract).

Walsby, J. R. 1975. Feeding and the radula in the marine pulmonate limpet Trimusculus reticulatus. Veliger 18: 139–45.

Waser, P. M. 1968. The spectral sensitivity of the eye of Aplysia californica. Comp. Biochem. Physiol. 27: 339–47.

Waters, V. L. 1973. Food-preference of the nudibranch Aeolidia papillosa and the effect of the defenses of the prey on predation. Veliger 15: 174–92.

White, K. M. 1945. On two species of Aglaja from the Andaman Islands. Proc. Malacol. Soc. London 26: 91–102.

Wilbur, K. M., and C. M. Yonge, eds. 1964. Physiology of Mollusca. Vol. 1. New York: Academic Press. 473 pp.

―――, eds. 1966. Physiology of Mollusca. Vol. 2. New York: Academic Press. 645 pp.

Williams, G. C. 1971. New record of a color variation in Spurilla oliviae. Veliger 14: 215–16.

Williams, G. C., and T. M. Gosliner. 1973a. Range extensions for four sacoglossan opisthobranchs from the coast of California and the Gulf of California (Mollusca: Gastropoda). Veliger 16: 112–16.

―――. 1973b. A new species of anaspidean opisthobranch from the Gulf of California. Veliger 16: 216–32.

―――. 1979. Two new species of nudibranchiate mollusks from the west coast of North America, with a revision of the family Cuthonidae. Zool. J. Linn. Soc. (in press).

Williams, L. G. 1971. Veliger development in Dendronotus frondosus (Ascanius, 1774) (Gastropoda: Nudibranchia). Veliger 14: 166–71.

Willows, A. O. D. 1967. Behavioral acts elicited by stimulation of single, identifiable brain cells. Science 157: 570–74.

―――. 1971. Giant brain cells in mollusks. Sci. Amer. 224: 68–75.

Willows, A. O. D., and G. Hoyle. 1967. Correlation of behavior with the activity of single identifiable neurons in the brain of Tritonia. Symp. Neurobiol. Invert. Hungarian Acad. Sci. (1967): 443–61.

Wilson, D. L. 1971. Molecular weight distribution of proteins synthesized in single, identified neurons of Aplysia. J. Gen. Physiol. 57: 26–40.

Winkler, L. R. 1955. A new species of Aplysia on the southern California coast. Bull. So. Calif. Acad. Sci. 54: 5–7.

―――. 1957. The biology of the California sea hares of the genus Aplysia. Doctoral thesis, University of Southern California, Los Angeles. 201 pp.

Winkler, L. R., and E. Y. Dawson. 1963. Observations and experiments on the food habits of California sea hares of the genus Aplysia. Pacific Sci. 17: 102–5.

Winkler, L. R., and B. E. Tilton. 1962. Predation on the California sea hare, *Aplysia californica* Cooper, by the solitary great green sea anemone, *Anthopleura xanthogrammica* (Brandt), and the effect of sea hare toxin and acetylcholine on anemone muscle. Pacific Sci. 16: 286–90.

Wobber, D. R. 1970. A report on the feeding of *Dendronotus iris* on the anthozoan *Cerianthus* sp. from Monterey Bay, California. Veliger 12: 383–87.

Wolter, H. 1967. Beiträge zur Biologie, Histologie, and Sinnesphysiologie (insbesondere der Chemorezeption) einiger Nudibranchier (Mollusca, Opisthobranchia) der Nordsee. Z. Morphol. Ökol. Tiere 60: 275–337.

Yarnall, J. L. 1972. The feeding behavior and functional anatomy of the gut in the eolid nudibranchs *Hermissenda crassicornis* (Eschscholtz, 1831) and *Aeolidia papillosa* (Linnaeus, 1761). Doctoral thesis, Biological Sciences, Stanford University, Stanford, Calif. 126 pp.

Yonge, C. M. 1958. Observations in life on the pulmonate limpet *Trimusculus* (*Gadinia*) *reticulata* (Sowerby). Proc. Malacol. Soc. London 33: 31–37.

Yonge, C. M., and T. E. Thompson. 1976. Living marine molluscs. London: Collins. 288 pp.

Young, D. K. 1967. New records of Nudibranchia (Gastropoda: Opisthobranchia) from the central and west-central Pacific with a description of new species. Veliger 10: 159–73.

———. 1969. The functional morphology of the feeding apparatus of some Indo-West-Pacific dorid nudibranchs. Malacologia 9: 421–66.

Zack, S. 1974. The effects of food deprivation on agonistic behavior in an opisthobranch mollusc, *Hermissenda crassicornis*. Behav. Biol. 12: 223–32.

———. 1975a. A description and analysis of agonistic behavior patterns in an opisthobranch mollusc, *Hermissenda crassicornis*. Behaviour 53: 238–67.

———. 1975b. A preliminary study of the effects of nematocyst removal on agonistic behavior in *Hermissenda*. Veliger 17: 271–75.

Zilch, A. 1959–60. Gastropoda Euthyneura, pp. 1–834, *in* W. Wenz, ed., Handbuch der Paläozoologie, vol. 6, part 2.

Chapter 15

Bivalvia: *The Clams and Allies*

Eugene C. Haderlie and Donald P. Abbott

A smaller and less diverse group than the gastropods, the Bivalvia (=Pelecypoda) nevertheless show considerable variety of form and a remarkable series of specializations. In addition to the clams, scallops, oysters, and mussels, the group includes others like the piddocks and shipworms that bore into hard substrata. Bivalves are bilaterally symmetrical mollusks, in which the mantle fold has extended downward on both sides, enclosing not only the visceral mass but also the head and foot in a cavernlike mantle cavity. The shell consists of right and left calcified valves, and an uncalcified hinge and elastic ligament uniting the two valves dorsally. In most animals the two valves are approximately mirror images of one another, but in some bivalves one is quite different from the other. Nearly totally enclosed within the protective mantle and shell, the bivalved mollusks have developed a method of feeding that in all except the most primitive forms involves passing water by ciliary action through an enlarged pair of gills (ctenidia) that filter out particulate food. The definitive bivalve way of life has evolved around this method of feeding, accompanied by a loss of the jaws and radula and most of the other structures of the molluscan head. The ctenidia continue to function also as respiratory organs.

Although most bivalves are relatively sedentary, some can use their muscular foot to burrow actively into mud or sand, and in others the foot anchors the body during boring into wood or stone. Still other clams use the foot to glide about on the substratum more or less in the manner of snails. A few bivalves can swim by re-

peated rapid clapping together of the valves. Many bivalves attach themselves to solid substrata by organic (byssal) threads secreted by glands at the base of the foot; others, such as oysters, cement themselves by one valve to rocks or pilings. Strong adductor muscles (one or two) running from valve to valve internally can close the shell tightly. When bivalves are feeding, the valves usually gape slightly along the margins, but the free edges of the mantle are normally kept together, forming a mantle cavity more or less closed except where water enters and leaves. The mantle folds on the two sides may be fused to some degree, especially posteriorly, to form a ventral and a dorsal siphon, serving respectively for the entrance and exit of the water current essential for respiration and feeding. In some clams the siphons are very short or nonexistent, in others they are long but separate, and in still others they may be fused into a long double tube that cannot be completely withdrawn into the shell. Elongated siphons make it possible for some clams to live deeply buried in sediment and still retain contact with the water above. Ciliated tissues in the siphons, as well as on the mantle and ctenidia, aid in circulating water through the mantle cavity and gills. Food particles filtered out by the ctenidia are trapped in a mucus sheet overlying the gill, and special ciliary tracts move this food-laden sheet toward the mouth. A pair of elaborate folds of ciliated tissue (the labial palps), present on each side of the mouth, actively collect food and mucus from the anterior edges of the gills, sort it (discarding inedible debris), and move it to the mouth.

Most bivalves have separate sexes, but in many oysters and scallops some sort of hermaphroditism is the rule. Sexual products are normally released into the sea, where fertilization takes place, but in a few species the sperm enter the siphon of a female and fertilize

Eugene C. Haderlie is Professor of Oceanography at the Naval Postgraduate School, Monterey. Donald P. Abbott is Professor, Department of Biological Sciences and Hopkins Marine Station, Stanford University.

the eggs internally. During development, most marine bivalves pass through trochophore and veliger larval stages. The bivalve veliger lacks the elaborate ciliated head lobes so characteristic of gastropods, and has a bivalved rather than a coiled univalved shell.

Clams, oysters, and mussels often live in dense concentrations along the rocky seashore or on soft bottoms in the shallow water of bays and estuaries, and man has used them for food since prehistoric times. Near Emeryville (Alameda Co.), on the east shore of San Francisco Bay, for example, there is an Indian shell mound consisting of thousands of cubic meters of discarded bivalve shells, which accumulated over a period of 3,500 years. Thousands of kilograms of bivalves are now collected yearly along our coast by commercial and sport fishermen for use as food or bait. Some bivalves are poisonous during certain seasons; west coast mussels, for example, are dangerous to eat from late spring to early fall. Other bivalves, such as the shipworms, cause considerable damage to wharves, pilings, and wooden ships and boats.

The taxonomy of bivalves is based mainly on the characteristics of the shell, including its general form and structure, and of the form of the dorsal ligament and dentition at the hinge. Other features, such as the nature of the ctenidia and siphons, the shape of the foot, and the arrangement of adductor muscles may be used, but most identification manuals devote attention almost exclusively to the shell or other hard parts.

After metamorphosing from larval states, young clams increase the shell size by concentric growth at the anterior, ventral, and posterior margins of the two minute valves. The oldest parts of the adult clam valves are therefore the basal projections called beaks. The highest and most prominent point on each valve, usually near or coinciding with the beak, is called the umbo (plural, umbones). The outside of the shell may be smooth or variously sculptured, showing growth lines (concentric lines, ridges, lamellae, or grooves) and/or fine or coarse radial striations or ridges extending from the beaks to or toward the margins. Externally the valves assume a variety of colors and patterns, and may be covered with a horny layer (the periostracum). The interior of the valves is lined by a lamellated calcareous layer, laid down by the mantle to increase the shell in thickness; this may be nacreous (pearly), porcellaneous, or chalky in texture and color. This layer also shows adductor muscle scars and a pallial line, paralleling the shell margin and marking the line along which the pallial muscles attach the mantle to the shell.

Dorsally, the elastic hinge ligament opens the valves when the adductor muscles relax, and interlocking teeth prevent the two valves from slipping against one another. In some of the boring clams the shell may include a number of accessory plates dorsally, and may possess on its inner surface near the hinge a calcareous shelf (chondrophore). In many rock-boring piddocks a wide gape between the valves at the anterior end is eventually closed, when the animals cease boring, by a calcareous dome (callum). In shipworms, accessory calcareous plates (pallets) are located far from the main valves and serve to plug the burrow entrance when the siphons are withdrawn.

General references on the Bivalvia include Keen & Coan (1974), Keep (1935), R. Moore et al. (1969, 1971), Morton (1967), Purchon (1968), Wilbur & Yonge (1964, 1966), and Yonge & Thompson (1976). Fitch (1953) is a good reference for California marine bivalves, and a key to those found along the central California coast is found in Coan & Carlton (1975). Reproduction is covered in Giese & Pearse (1979).

We wish to acknowledge the considerable help given by Dr. A. Myra Keen in the preparation of this chapter. She made the initial identifications of the bivalves pictured, settled nomenclatural problems, assembled the basic shell descriptions and geographical distributions for each, and assisted in many other ways, including reading and commenting on various drafts of the manuscript. She deserves, but has declined, coauthorship.

Phylum Mollusca / Class Bivalvia / Subclass Pteriomorpha / Order Arcoida Family Arcidae

15.1 **Barbatia bailyi** (Bartsch, 1931) Baily's Ark

Common on undersurfaces of rocks, low intertidal zone; Los Angeles Co. to Panama.

Shell to 9 mm long, quadrate in form, sculptured with coarse radial ribs and concentric lines; prominent muscle scars internally; row of small hinge teeth; shell exterior brownish or tan.

These small clams are attached to rocks by byssal threads.

See Bartsch (1931), Bretsky (1967), Heath (1941), and Rost (1955).

Phylum Mollusca / Class Bivalvia / Subclass Pteriomorpha / Order Mytiloida Family Mytilidae

15.2 Adula falcata (Gould, 1851)
(=Botula falcata)   Rough Pea-Pod Borer

Uncommon in intertidal zone, more commonly subtidal, boring in clay and shale; Coos Bay (Oregon) to Baja California.

Shell to 80 mm long, elongate, cylindrical, fragile, with beaks near anterior end, a prominent oblique ridge, and a sculpture of vertical or oblique wrinkles; periostracum dark brown, glossy, often worn away at beaks; interior of shell grayish with green iridescence.

This species is commonly found along with other boring bivalves (*Lithophaga plumula*, 15.4, and members of the family Pholadidae, 15.71, 15.73–79) in subtidal shale. Two short siphons extend from the mouth of the burrow, bringing in fresh sea water and planktonic food, and expelling wastes. The animal enlarges its burrow by attaching byssal threads in two groups, such that alternate contractions of the anterior and posterior byssal retractor muscles pull the animal backward and forward. When the adductor muscles of the shell relax, the powerful elastic ligament on the dorsal side opens the valves slightly, so that their filelike outer surfaces rub against the walls of the burrow, abrading the shale. The animal does not rotate as it bores, so the shape of the burrow closely matches that of the shell. The thick periostracum somewhat protects the shell from wear except on the dorsal surface near the beaks, where both periostracum and shell may be deeply eroded. Boring movements extend the burrow ever deeper into the rock and enlarge its diameter. The body grows mainly by addition at the posterior end, so this always lies near the outer rock surface.

As the shale is slowly worn away by boring activity, the loosened rock particles are removed. First, ciliary tracts, mainly on the very mobile foot, carry particles from all parts of the burrow into the mantle cavity. Here the coarse particles, trapped in mucus on the mantle, are carried by cilia to the incompletely fused lower siphon, where water expelled by periodic closure of the valves sweeps them away. The fine, suspended particles are swept along the ciliary tracts on the gills and palps and are ingested along with food material, ultimately to appear in the fecal pellets.

In many areas *A. falcata* bores only into clay or relatively soft shale, but subtidally in Monterey Bay it is found also in hard siliceous chert.

See Burnett (1972), Dall (1916), Fankboner (1971), Fitch (1953), Haderlie (1976), Haderlie et al. (1974), Hinegardner (1974), Oberling (1964), Soot-Ryen (1955), and Yonge (1955).

Phylum Mollusca / Class Bivalvia / Subclass Pteriomorpha / Order Mytiloida Family Mytilidae

15.3 Adula californiensis (Philippi, 1847)
California Pea-Pod Borer

Relatively uncommon, boring in soft shale, intertidal zone to depths of 20 m; Vancouver Island (British Columbia) to San Diego.

Shell to about 45 mm long, elongate, smooth, tapering posteriorly, with beaks just back of anterior end; periostracum well developed, hairy, with encrusted mud and debris on posterodorsal slope; shell white under brown periostracum, often eroded near beaks.

This species is smaller than *Adula falcata* (15.2), but presumably uses the same method of mechanical boring. In Oregon, populations produce ripe eggs from June to October. Developing embryos are easily reared in the laboratory. The first and second cleavages occur 1.5 and 2.5 hours after fertilization at 15°C and a salinity 95 percent that of seawater. The trochophore stage is reached by 15 hours, and the shell gland becomes apparent at 31 hours. The shell completely encloses the soft parts by 72 hours. The rate of development varies linearly with temperature and salinity. Trochophore larvae are active and swim in all directions; in contrast, the veliger larvae swim up and down but not horizontally. In the laboratory, larvae settle out of the plankton 3 days after fertilization. They have not been reared through metamorphosis and the early boring stages.

See Burnett (1972), Dall (1916), Dinamani (1967), Haderlie (1976), Lough & Gonor (1971), Soot-Ryen (1955), and Yonge (1955).

Phylum Mollusca / Class Bivalvia / Subclass Pteriomorpha / Order Mytiloida
Family Mytilidae

15.4 **Lithophaga plumula** (Hanley, 1843)
Date Mussel

Common, boring into shale and mollusk shells, low intertidal zone to 30 m depth; Mendocino Co. to Peru.

Shell to about 55 mm long, rounded at anterior end, tapered at posterior end, with two grooves extending from beaks to posterior end; posterodorsal slope of shell with ridged, feathery, or chalky incrustation and transverse wrinkling; periostracum glossy brown; interior of shell metallic gray.

The date mussel occupies a snugly fitting burrow in shale and attaches to the burrow wall with two small groups of byssal threads. By alternately contracting the anterior and posterior byssal retractor muscles the animal can move backward and forward to a limited extent; it can also rotate within the burrow, forming new byssal threads and breaking old ones.

The method by which these forms bore into rock is still problematical. Older studies reported that *Lithophaga* bored only into calcareous rock, and postulated that the pallial glands on the edge of the mantle secreted an acid which dissolved the limestone. More recent studies on the pallial glands show that they secrete not an acid but a neutral mucoprotein with a calcium-binding ability. It appears that removal of calcium by this chelating agent may explain boring into calcareous rocks. Yet, over the years, other investigators have reported seeing *Lithophaga* boring into non-calcareous rock, and in the shallow subtidal waters of Monterey Bay *L. plumula* is very common in exceedingly hard and flintlike non-calcareous, siliceous chert. Animals removed from the chert are of normal size and shape in every way and show little or no shell abrasion. The chert is much harder than the animal's shell, so mechanical boring would appear to be ruled out.

Mature *Lithophaga* often secrete smooth carbonate linings 4–5 mm thick on the burrow walls, especially near the burrow opening. This lining is apparently harder than some types of shale penetrated by the borers, and as the shale erodes away the tubular calcareous linings project above the substratum surface.

The shell of *Lithophaga* is very fragile and has within it especially conspicuous sublayers of the protein conchiolin. Dead shells are not commonly found and seem to disintegrate rapidly, possibly because of the additional protein layers.

See Amemiya & Amemiya (1923), Berry (1907), Burnett (1972), Fitch (1953), Haas (1942a,b, 1943), Haderlie (1976), Haderlie et al. (1974), Hanna (1928), Hinegardner (1974), Hodgkin (1962), Jaccarini, Bannister & Micallef (1968), MacGinitie (1935), Oberling (1964), Soot-Ryen (1955), Stasek (1963a,b), R. Turner & Boss (1962), and Yonge (1955).

Phylum Mollusca / Class Bivalvia / Subclass Pteriomorpha / Order Mytiloida
Family Mytilidae

15.5 **Modiolus capax** (Conrad, 1837)
(=Volsella capax)   Fat Horse Mussel

Common on pilings in bays, also found as solitary individuals on rocks on semiprotected shores, low intertidal zone and offshore to 50 m depth; Santa Cruz to Peru.

Shell length to about 100 mm; valves in some areas tufted with dense, coarse, dark-brown periostracal hairs; shell bright orange-brown or reddish under periostracum, white internally; soft parts bright orange.

Modiolus capax attaches to the substratum by byssal threads. On rocks on a muddy bottom, only the posterior tips of the shell and the orange mantle protrude above the surface.

See Fitch (1953) and Soot-Ryen (1955).

Phylum Mollusca / Class Bivalvia / Subclass Pteriomorpha / Order Mytiloida
Family Mytilidae

15.6 **Modiolus rectus** (Conrad, 1837)
(=Volsella recta)   Straight Horse Mussel

Uncommon, on mud flats, low intertidal zone and subtidal to 50 m; Vancouver Island (British Columbia) to Gulf of California.

Shell normally to 120 mm long but reaching 230 mm, elongate, thin, narrow, with beaks well back from anterior margin; periostracum thick, shiny, brown, heavily beaded on posterior end; shell bluish under periostracum, white internally; soft parts yellow-orange.

Modiolus rectus normally lives embedded in mud, sand, or gravel with only the posterior tip of the shell and the yellow mantle protruding. Below the surface it is attached to buried rock by strong byssal threads.

See Fitch (1953) and Quayle (1973).

Phylum Mollusca / Class Bivalvia / Subclass Pteriomorpha / Order Mytiloida Family Mytilidae

15.7 Ischadium demissum (Dillwyn, 1817)
Ribbed Horse Mussel

Common, forming clusters attached to rocks and marsh plants on mud flats in bays, high to middle intertidal zones; San Francisco Bay, Los Angeles harbor, and upper Newport Bay (Orange Co.); Atlantic coast.

Shell to about 100 mm long, with many radiating ribs especially on posterior end, scalloped around posterior margin; periostracum thick, dark, often eroded near beaks; interior of shell iridescent; purplish; soft parts yellowish.

Ribbed mussels were introduced accidentally into San Francisco Bay along with east coast oysters before 1890. They attach by byssal threads and often occur in mats of several hundred individuals. Although edible, they are rarely eaten by man, but on the east coast they are used for poultry feed (rich in vitamin D). In San Francisco Bay they are a major food of the clapper rail, which is occasionally trapped when the shells close on its toes.

The taxonomic affinities of the ribbed mussel have been troublesome for a long time. The species has at various times been assigned to the genera *Volsella*, *Modiolus*, *Arcuatula*, *Brachydontes*, and *Geukensia*, and anatomical and physiological studies have been made in an attempt to clarify its systematic position.

The animal may use atmospheric air during periods of exposure, as do many marine forms, but it can also respire an-

aerobically for prolonged periods. It tolerates temperatures of up to 56°C, and salinities up to 200 percent that of pure seawater, and it has a high tolerance for dehydration.

The microstructure of the shell, investigated in east coast populations of *I. demissum*, shows growth increments correlated with tidal exposure. The growth lines appear to correspond to alternating periods of shell deposition during aerobic respiration and of shell dissolution during anaerobic respiration.

See Baginski & Pierce (1974, 1977), Booth & Mangum (1978), Fitch (1953), Hanna (1966), Kuenzler (1961a,b), Lent (1968, 1969), Loosanoff & Davis (1963), Lutz & Rhoads (1977), Murphy (1974), and Pierce (1969, 1973).

Phylum Mollusca / Class Bivalvia / Subclass Pteriomorpha / Order Mytiloida Family Mytilidae

15.8 Musculus pygmaeus Glynn, 1964

Common, attached to blades and holdfasts of the red alga *Endocladia muricata*, on high intertidal rocks; at least Monterey Bay to Cayucos (San Luis Obispo Co.).

Shell to 4.5 mm long, resembling that of tiny *Mytilus edulis* (15.10), but with diagonal ridge on older animals and with prominent anteroventral protuberance; shell surface smooth except for concentric growth lines, chocolate-brown, reddish where periostracum eroded.

This tiny mussel, like larger species, attaches by means of byssal fibers. The animal apparently breeds throughout the year. The males release sperm, but the females retain their eggs and brood their young in spaces between the gills. An adult 2 mm long may brood up to 132 young at a time. Development is completed in the mantle cavity, and the released young resemble miniature adults. The absence of a pelagic larval stage permitting dispersal results in an erratic distribution; the little clams are abundant in some algal patches (up to 100 individuals per 100 cm²) and totally absent in others in very similar localities nearby.

See Glynn (1964, 1965).

Phylum Mollusca / Class Bivalvia / Subclass Pteriomorpha / Order Mytiloida Family Mytilidae

15.9 **Mytilus californianus** Conrad, 1837
California Mussel, California Sea Mussel

Abundant, attached in massive beds on surf-exposed rocks and wharf piles, mainly in uppermiddle intertidal zone on outer coast; subtidal and offshore to 24 m depth; Aleutian Islands (Alaska) to southern Baja California.

Shell to about 130 mm long in most California intertidal animals (to 238 mm along Baja California shores, to 251 mm in subtidal specimens from Johnson Seamount, off Baja California), thick, pointed at anterior end, broadening posteriorly, sculptured with strong radial ribs and irregular growth lines, often with surface eroded or worn; periostracum heavy, blue-black in color; interior of shell blue-gray, somewhat iridescent, darker at margins.

The California sea mussel is one of the commonest invertebrates on surf-swept rocks along California shores. The animals are firmly rooted to the rocks and to each other by byssal fibers secreted by a gland at the base of the foot. Byssal attachment is easily seen in smaller mussels placed in a dish of seawater. If undisturbed, the shell opens slightly and the slender orange foot extends and grips the substratum. A liquid secretion passes along a groove on the back of the foot down to the substratum; when the groove opens to admit seawater, it exposes a strong solidified thread. The foot is then withdrawn to repeat the process. Medical and dental researchers, seeking an effective underwater cement, have looked at byssal fiber formation. Collagenlike proteins and phenols are involved and the process is catalyzed by enzymes. Large mussels are very firmly and probably permanently attached, but small animals seem to be able to break the byssal threads and move about a bit, both up and down and horizontally in the beds.

When submerged, the animals open the valves slightly and use ciliary currents to move water through the gills. Here suspended food is filtered out, caught in mucus, and transported by other cilia to the palps (for preliminary sorting) and then to the mouth. Mussels of average size filter 2–3 liters of water per hour when feeding. Animals living at the lowest part of the bed in the intertidal zone are submerged more of the time, get more food, have higher respiratory and metabolic rates, grow more rapidly (up to 7 mm per month), reach larger size, and have relatively heavier shells than those high in the bed. In southern California, mussels may reach a length of 86 mm within a year of settlement, 120 mm after 2 years, 150 mm at the end of 3 years. Along the Pacific coast there is an inverse relationship between rate of growth and water temperature. In the laboratory, mussels have been observed to feed at temperatures as low as 7°C and as high as 28°C, and to survive salinities of 50–125 percent that of seawater.

Young growing California mussels have about the same nutritional requirement for amino acids as does man. The major food of the mussels seems to be fine organic detritus suspended in the seawater, but living plankton, particularly dinoflagellates, is also important, and there is often a positive correlation between growth rate and abundance of dinoflagellates. In some areas, mussels also remove quantities of algal spores from the water and may deter recruitment of intertidal algae such as *Postelsia*. *M. californianus* can accumulate both dissolved and particulate carbon compounds exuded from the kelp *Macrocystis integrifolia*. At La Jolla (San Diego Co.), it has been estimated that a moderate-sized colony of one million mussels during their second year of life filters 24 million metric tons of sea water and strains out about 36 metric tons of food material. From this, the colony gains about 4.4 metric tons in organic dry weight, indicating a conversion efficiency of about 12 percent.

Some spawning occurs all year, but breeding peaks are in July and December in California. Temperature does not seem to provide a major stimulus for spawning, but mechanical stimuli (such as pulling on the byssal threads) may initiate gamete release, and the spawning of one animal may stimulate others to spawn.

Studies of shell microstructure in this animal show that three calcified layers are present: an outer layer of calcite, a middle nacreous aragonite layer, and an inner prismatic calcite layer. Some animals form small (commercially worthless) pearls in the mantle tissue, and blister pearls are common as bulges in the nacreous layer. The shells have a high content of lead in some areas along the coast.

Beds of *M. californianus* on the open coast are often badly damaged by storm waves, which tear loose and cast ashore large aggregations of mussels. The beds are more susceptible to wave damage if the animals have been heavily preyed upon by sea stars. The bay mussel, *M. edulis* (15.10), is sometimes found mixed with *M. californianus*; such mixed populations are less resistant to wave action because *M. edulis* has weaker byssal threads. A rock face that has been completely cleared of mussels takes about 2.5 years to reestablish the complex and diverse climax community.

California mussels are subject to diseases and infestations by other organisms. A microorganism infecting the digestive gland and kidneys of animals in southern California causes swelling but no neoplasms. In central California, parasitic isopod larvae (epicarids), normally associated with crustacean hosts, are often found in the mantle cavity. One species of pycnogonid (*Achelia chelata*) may parasitize up to 50 percent of the mussels in a group and cause obvious destruction of the hosts' gonads and gill tissue. The crystalline style, a secreted structure lying in the anterior intestine, is often inhabited by *Cristispira*, a very large spirochaete bacterium (also found in the styles of some other clams). Single female individuals of the pea crab *Fabia subquadrata* (25.31) may be found living within the mantle cavity (see Fig. 25.31). The crab orients herself on one of the gills, clasps several gill filaments, and feeds by drawing the mucus food strands produced by the mussel into her mouth. The crab robs the mussel of some food and damages one gill, but otherwise causes little harm to her host.

Small California mussels are preyed upon by many species of intertidal crabs and shore birds, and by predatory snails (*Nucella emarginata*, 13.83; *Ceratostoma nuttalli*, 13.71; *Roperia poulsoni*, 13.77), which drill holes into the shell and feed on the soft parts. Larger mussels are fed upon by sea stars, particularly *Pisaster ochraceus* (8.13), a single individual of which may consume up to 80 mussels per year. The southern sea otter has devastated many formerly extensive mussel beds on the Monterey Peninsula (as at the Hopkins Marine Station, Pacific Grove). Coastal Indians made extensive use of mussels for food, and man still collects them to eat. Because mussels may become poisonous during summer months, they are quarantined in California between May 1 and October 31.

The toxin, which may cause paralysis and death, is found in the dinoflagellate *Gonyaulax catanella*, a planktonic microorganism that may be ingested by the mussel in enormous numbers. The exact chemical nature of the *Gonyaulax* toxin is still unknown.

See especially Bayne (1976), Coe & Fox (1942, 1944), Fox (1936), Fox & Coe (1943); see also Ahmed & Sparks (1970), Anderson (1975), Bartlett (1972), Bayne et al. (1976a,b), Bennett & Nokada (1968), Benson & Chivers (1960), Berry (1954), Burke et al. (1960), Carlson (1905, 1906), Chan (1973), Chew, Sparks & Katkansky (1964), Coe (1948, 1956), Curtis (1966), Dayton (1971, 1973), Dehnel (1956), Dodd (1964, 1966), Fankboner, Blaylock & de Burgh (1978), Feder (1955, 1956, 1959, 1970), Fitch (1953), Fox (1941, 1955, 1960), Fox, Sverdrup & Cunningham (1937), Glynn (1965), D. Graham (1972), Greenberg & Windsor (1962), Haas (1942a), Hare (1963), Harger (1967, 1968, 1970a,b,c, 1972a,b,c), Harger & Landenberger (1971), Hargreaves (1975), Harrison (1975), Hewatt (1935), Hinegardner (1974), Hubbs, Bien & Suess (1960), Jegla & Greenberg (1968), W. Kennedy, Taylor & Hall (1969), Kuhn, Lasnik & Rubenstein (1968), Landenberger (1968), MacGinitie (1939, 1941), Marks & Fox (1934), Moon & Pritchard (1970), Oberling (1964), Paine (1976a,b), Paris (1960), Pequegnat (1964b), Petraitis (1978), Pickens (1965), Pilot & Ryter (1965), Pilson & Taylor (1961), Quayle (1973), Rao (1953a,b, 1954), Reish (1964c), Richards (1928, 1946), Risebrough et al. (1967), Rodegker & Nevenzel (1964), Schantz (1960), Scheer (1940), Sommer & Meyer (1935), Soot-Ryen (1955), Stohler (1930), R. Taylor (1966), Temnikow (1974), Whedon (1936, 1938), Worley (1944), Wourms (1968), Young (1941, 1942, 1945, 1946, 1951), and ZoBell & London (1937).

Phylum Mollusca / Class Bivalvia / Subclass Pteriomorpha / Order Mytiloida Family Mytilidae

15.10 **Mytilus edulis** Linnaeus, 1758
Bay Mussel

Common, often in clusters, attached by byssus to rocks and especially to wharf pilings, low intertidal zone and subtidal to 40 m in bays and sheltered areas; Arctic Ocean to Isla Cedros (Baja California); west coast of South America; Japan, Australia; North Atlantic.

Shell to about 100 mm long, thin, nearly smooth and lacking radiating ridges, with three small teeth below beaks, pointed at anterior end, very broadly rounded at posterior end; periostracum bluish black; interior of shell dull blue.

Mytilus edulis has a much broader geographic distribution and is adapted to a wider variety of habitats than *M. californianus* (15.9). In general *M. edulis* prefers quieter water and lives lower in the intertidal area. It is a common fouling organism on buoys and floats, and in the seawater piping systems on ships and in seaside laboratories. On the California coast it is sometimes found on outer coastal rocks and wharf pilings, but here only in mixed populations with *M. californianus*. Small individuals can withstand wave impact about as well as the California mussel, but larger ones cannot, owing to relatively weak attachment. The shell shape varies greatly with age, density of the bed, rate of growth, and height in the intertidal zone. Animals exposed to wave action grow more slowly and have thicker shells. Animals continuously submerged are generally larger and grow faster than those exposed at low tide, and are "meatier" in relation to the weight of the shell. Those living in shaded areas grow faster than those in full sunlight or darkness. *M. edulis* needs to be submerged at least half the time to show any significant growth; optimum temperature range for growth is 10–20°C.

In mixed populations of the two mussels, *M. edulis* shows considerably more mobility than *M. californianus*. Small individuals tend to crawl to the edge of the colony, a behavior that ensures that *M. edulis* becomes arranged on the outer edges of a mixed clump. This mobility also helps keep the mussels above the surface of the mud in harbor areas.

Byssal threads are formed in the same manner as in *M. californianus*. More byssal threads are formed at night than during the day, more are produced when the mussels are crowded, and there is a direct correlation between the number of threads produced and the temperature and salinity of the water. Post-larval *M. edulis* secretes long single, unattached, byssal threads, which increase drag and allow the young mussels to be carried along by relatively weak currents.

Bay mussels feed nearly continuously when submerged. Even at temperatures near freezing they still filter water through the gills. Detritus particles and planktonic organisms down to 4–5 μm in diameter are trapped by the mucus sheets moving over the gills, but the animals have the ability to sort particles and reject non-food items. As in *M. californianus*, organic detritus seems to be the major food. Some direct absorption of dissolved organic matter may occur. Experiments show that the animals can absorb fish hemoglobin directly from seawater, also dissolved and particulate organic compounds released by the kelp *Macrocystis integrifolia*, and up to 46 percent of the amino acid glycine from experimental solutions. Digestion is both intra- and extracellular; carbohydrate splitting enzymes from the crystalline style digest carbohydrates in the gut, but digestion of proteins and fats is intracellular.

Most of the calcium used in shell growth seems to be absorbed directly from the seawater rather than taken in with the food. Calcium salts are precipitated on the inner side of the periostracum at the shell edge. Calcification occurs over a wide temperature range, but the rate of calcification varies with temperature and salinity. It has been suggested that fossil shells of *M. edulis* might be useful in paleotemperature and paleosalinity determinations.

Mytilus edulis has been used extensively in laboratory experiments involving embryology, muscle physiology, and respiratory and metabolic activity, and has recently been widely studied in regard to heavy metal and hydrocarbon uptake. Heavy metals in solution, particularly mercury and copper, inhibit byssal-thread formation. Lead is taken up at a linear rate dependent on the lead concentration in seawater. This makes *M. edulis* an ideal indicator organism for lead pollution in the marine environment. Cytological studies indicate that *M. edulis* has a diploid chromosome number of 28, the same as *M. californianus*.

Spawning occurs in the late fall and winter along the central California coast. In southern California some ripe males occur all year, but mature ova are present in females only from November through May. Larvae swim freely for about 4 weeks and settle mainly in the summer in southern California and in winter and late spring farther north. On the east coast, the larvae settle from mid-June to mid-August.

In southern California, *M. edulis* settles in massive numbers in some years, sparsely in others. After settlement they grow to an average length of 76 mm after 1 year, and 96 mm after 2 years.

When removed to the laboratory, *M. edulis* has been reported to exhibit a tidal rhythm in the rate of water propulsion, pumping most frequently when the tide is high. This rhythm persists in the laboratory in phase with the external tide cycle for over 4 weeks in continuous light, continuous darkness, or natural light conditions. The rhythm is independent of temperature. East coast mussels brought to California by air were out of phase with the local tidal cycle by 6.5 hours, but when put in the sea off Corona del Mar promptly shifted to the local tidal cycle.

The bay mussel is preyed upon by sea stars, predaceous snails, crabs, and birds. In a mixed population of *M. californianus* and *M. edulis* the sea star *Pisaster ochraceus* (8.13) and the crabs *Pachygrapsus crassipes* (25.43) and *Cancer antennarius* (25.16) selected *M. edulis* over the California mussel, as did several other predators. Predation by *Pisaster* and snails (e.g., *Nucella canaliculata*, 13.82, and *N. emarginata*, 13.83) may limit the downward distribution of *M. edulis* in the intertidal zone, as it does with *M. californianus*. In areas where the Pacific and Japanese oysters are grown, *M. edulis* may be parasitized by the copepod *Mytilicola orientalis*, which lives in the digestive tract of the mussel.

The bay mussel is cultivated extensively for food in Europe, but along California shores it is rarely used by man. Along with the California mussel, it may become toxic during the summer months due to ingestion of dinoflagellates (*Gonyaulax catanella*).

For a recent review of the ecology and physiology of mussels, see Bayne (1976); old but full accounts of the anatomy, biology, and economic value of *M. edulis* will be found in Field (1922) and K. White (1937); see also Abd-el-Wehab & Pantelouris (1957), Ahmed & Sparks (1970), Aiello (1960), Aiello & Guideri (1964), Baird & Drinnan (1957), Bayne (1963, 1964, 1965, 1973), Bayne & Thompson (1970), Bayne et al. (1975), W. Berg (1950, 1954), W. Berg & Kutsky (1951), W. Berg & Prescott (1958), Boëtius (1962), Bouffard (1969), Bradley & Siebert (1978), Brown & Newell (1972), Carr & Reish (1978), Chipperfield (1953), Clay (1965), Coe (1945, 1946, 1948, 1956), Cole & Hepper (1954), Costello & Henley (1971), Coulthard (1929), Courtright, Breese & Krueger (1971), Craig & Hallam (1963), Digby (1969), Dodd (1964, 1966), Dral (1966), Ebling et al. (1964), Engle & Loosanoff (1944), Fankboner, Blaylock & de Burgh (1978), Felton & Aiello (1971), Fitch (1953), Glaus (1968), D. Graham (1972),

H. Graham & Gay (1945), Harger (1967, 1968, 1970b,c, 1972a,b,c), Hinegardner (1974), Hobden (1969), Humphreys (1964), W. Johnson, Kahn & Szent-Györgyi (1959), Jørgensen (1949, 1950), Landenberger (1968), Lee, Sauerheber & Benson (1972), Levinton (1973), Longo & Dornfeld (1967), Loosanoff (1942), Loosanoff & Davis (1963), Loosanoff, Davis & Chaney (1966), Maheo (1970), Malanga (1971), Malone & Dodd (1967), Martella (1974), Martin, Piltz & Reish (1975), D. Moore & Reish (1969), Newcombe (1935), Paris (1960), Pentreath (1973), Pequegnat (1964b), Petraitis (1978), Pickens (1965), Pilson & Taylor (1961), Potts (1954), Quayle (1973), Rao (1953b, 1954), Rattenbury & Berg (1954), Read (1962), Reish (1963, 1964a,b), Reish & Ayers (1968), Richards (1946), Rosen et al. (1978), Ross & Goodman (1974), Rossi & Reish (1976), Schulz-Baldes (1974), Scott & Major (1972), Seed (1968, 1969), Sergy & Evans (1965), Sigurdsson, Titman & Davies (1976), S. Smith & Haderlie (1969), Soot-Ryen (1955), Stubbings (1954), Thompson et al. (1974), Van Winkle (1970), Waldron, Packie & Roberts (1976), Widdows & Bayne (1971), and Yonge (1976).

Phylum Mollusca / Class Bivalvia / Subclass Pteriomorpha / Order Mytiloida Family Mytilidae

15.11 **Septifer bifurcatus** (Conrad, 1837)
Branch-Ribbed Mussel

Common, attached by byssus to rocks and pilings, and at bases of *Mytilus californianus* (15.9) clusters, intertidal zone on exposed shores; Crescent City (Del Norte Co.) to Cabo San Lucas (Baja California).

Shell to about 45 mm long, with prominent radiating ribs, fine and closely spaced on anterior ventral surface, coarser on posterodorsal surface; anterior tip of each valve bridged within by a small flange of shell material; shell black externally, purple internally.

In northern California *Septifer* usually occurs in the low intertidal zone under rocks, but in southern California it is most abundant in the middle intertidal area, where it nestles in nooks and crevices; in mixed beds of *Mytilus californianus* (15.9) and *M. edulis* (15.10), it often lives right next to the rock face. When removed to a more exposed position *Septifer* reacts by drawing itself down into any pit or depression available.

This animal is preyed upon by the sea star *Pisaster ochraceus* (8.13), but in laboratory experiments *Mytilus edulis* and *M. californianus* are selected for food in preference to *Septifer*.

See Coe (1956), Haas (1942a,b), Harger (1967, 1968), Landenberger (1968), Soot-Ryen (1955), and Yonge & Campbell (1968).

Phylum Mollusca / Class Bivalvia / Subclass Pteriomorpha / Order Pterioida Family Ostreidae

15.12 Ostrea lurida Carpenter, 1864
Native Oyster, Olympia Oyster

Rare to common, attached (cemented by one valve) to rocks, oyster shells, and concrete pilings, low intertidal zone in quiet bays and estuaries; Sitka (Alaska) to Cabo San Lucas (Baja California).

Shell to 60 mm long, relatively thin, generally circular, sometimes serrated on edges; valves nearly equal in size, flat; shell exterior gray to bluish black, the interior white or greenish yellow.

Oysters of several species have been eaten by man since prehistoric times, and cultivated at least since the Roman period. They are discussed in literally thousands of technical papers, and from a scientific point of view no marine invertebrates are better known. Yet today we can still agree with Korringa (1952) that "we have but a poor understanding of many important factors in the oyster's biology."

The native oyster, called the Olympia oyster where it is cultivated in the Puget Sound region (Washington), has a fine flavor but is smaller than its main California competitor, the introduced Pacific or Japanese oyster *Crassostrea gigas* (15.13). *Ostrea lurida* was once widely distributed in California bays and at times was an important food of California Indians. The shells in kitchen middens have aided anthropologists in interpreting cultural sequences in the San Francisco Bay area. Now pollution, filling and draining of tidal flats, dredging, and embanking shores for recreational purposes have restricted the area where the native oyster can flourish. Since 1930, the California Department of Fish and Game has fostered its culture by private companies in Humboldt Bay, with some success. Attempts to introduce the species to the east coast and to Japan have failed; the oysters cannot withstand the winter temperatures.

The native oyster reverses sex rhythmically and functions alternately as male and female. Succeeding sexual phases overlap, so a population contains many intersexual individuals and even some functional hermaphrodites. Sperm are shed, in the form of clusters of up to 2,000 mature sperm cells, but ovulated eggs are retained in the mantle cavity where they are fertilized, brooded 10–14 days, then shed as tiny bivalved larvae, 250,000–300,000 to a brood. The larvae swim for another 30–40 days, then settle and grow. They reach marketable size (shells 3–5 cm long) in 3–5 years in California. They become sexually mature 5 months after settling and may produce two broods of larvae a year. In southern California, spawning occurs from April to November, and the water temperature must be at least 16°C for gametes to be released. In Puget Sound, spawning normally occurs during a 6-week period in the spring, when the water reaches 13°C.

Ostrea lurida feeds on coarse plankton and organic detritus and is unable to trap the finest planktonic organisms (nannoplankton) from suspension.

Native oysters are preyed upon by the bat stingray *Myliobatis californicus* and predatory snails (*Urosalpinx cinerea*, 13.78, and *Acanthina spirata*, 13.81). In San Francisco Bay they are parasitized by the copepod *Mytilicola orientalis*.

See especially Barrett (1963), Korringa (1952), and Yonge (1960); see also Bonnot (1935, 1937, 1940, 1949b), Bradley & Siebert (1978), Chapman & Banner (1949), Coe (1930, 1931a,b, 1932a,b, 1934), Davis (1955), Elsey (1935), Fitch (1953), Hertlein (1959), Hinegardner (1974), Hopkins (1934a,b, 1935, 1936, 1937), Hori (1933), Hubbs, Bien & Suess (1960), Kellogg (1915), Loosanoff & Davis (1963), MacGinitie (1941), Quayle (1973), Ricketts & Calvin (1968), Stafford (1914), and Wicksten (1978).

Phylum Mollusca / Class Bivalvia / Subclass Pteriomorpha / Order Pterioida Family Ostreidae

15.13 Crassostrea gigas (Thunberg, 1795)
(=Ostrea gigas)   Pacific Oyster, Japanese Oyster

Common, cemented to rocks and shells in bays and on mud flats, low intertidal zone; British Columbia to southern California; introduced from Japan.

Shell to 300 mm long; shape varying from long and thin

when crowded on soft substratum to round and deep when attached singly on hard substratum; lower (left) valve cemented down, cupped; upper valve usually flat; outer surface very rough, irregular, fluted; shell exterior chalky tan or white, often with radiating purple streaks, the interior shiny white.

Cultured in Japan for 300 years, this oyster was introduced to the west coast in 1902 or even earlier. Populations established in Puget Sound (Washington) and in British Columbia reproduce regularly, spawning in July or August and settling 15–30 days later. In California, mature animals spawn in late July or early August when the water temperature reaches 18°C, but the larvae usually do not survive. The causes of larval mortality are not understood but possibly relate to water turbidity. A few naturally "set" oysters have been found in Humboldt Bay and Tomales Bay (Marin Co.), but young or "seed" oysters must be imported annually to maintain the commercial crop. Seed oysters, growing on old oyster shells, are crated in Japan; traveling as deck cargo on ships, they are sprayed frequently with seawater during the 2–3 weeks' transport, and then are planted in California bays around March each year. Rigorous inspection in Japan and California helps prevent introduction of eggs and young of the oyster drill (a snail, *Ceratostoma inornatum*), a destructive pest which in early years was introduced with oysters into Tomales Bay.

The Pacific oyster can filter the finest nannoplankton from the water. It feeds more efficiently and grows more rapidly than the native oyster, and reaches market size (10–12 cm) in 2–3 years in California. These oysters are sold mainly to restaurants, but some are packaged and sold frozen in retail stores. In the mid-1970's over 680,000 kg of oyster meat was produced annually in Humboldt Bay, Tomales Bay, Drakes Estero (Marin Co.), and Morro Bay (San Luis Obispo Co.), over 86 percent coming from Humboldt Bay.

The sexes of *C. gigas* are separate but may change over the winter, and a few individuals in a population are hermaphroditic. In Japan the oysters spawn in the summer, in repeated bursts, the females producing many millions of eggs. The larvae are free-swimming in the water for up to 4 weeks.

In California, Pacific oysters that have not been harvested may live for 20 years or more. They are preyed upon by the bat stingray *Myliobatis californicus*. Under rare conditions *C. gigas* has been known to ingest quantities of the dinoflagellate *Gonyaulax catanella* and become toxic to man.

See Amemiya (1929), Barrett (1963), C. Berg (1969, 1971), Bernard (1973, 1974a,b), Bonnot (1949b), Boyden, Watling & Thornton (1975), Calabrese & Davis (1969), California (1962), Davis & Hidu (1969), Elsey (1935), Farley (1969), Fitch (1953), Frey (1971), Galtsoff (1964), Gunter (1950), Hancock (1960), Hanna (1966), Haydock (1964), Hopkins (1934a,b), Kobayashi (1959), Korringa (1952, 1953), Loosanoff (1949), Loosanoff & Davis (1963), L. Miller (1962), Mix (1971), Quayle (1952b, 1956a, 1963, 1969, 1973), Rawson (1968), Reinke (1960), Ricketts & Calvin (1968), Risebrough et al. (1967), Span (1978), Sparks et al. (1962), Steele (1964), Stenzel (1964), and Walne (1974).

Phylum Mollusca / Class Bivalvia / Subclass Pteriomorpha / Order Pterioida Family Pectinidae

15.14 Hinnites giganteus (Gray, 1825)
(=Hinnites multirugosus)   Rock Scallop

Common, in rock crevices along exposed outer coast, underneath floats and on pilings in bays, low intertidal zone to subtidal depths of 50 m; adults cemented to substratum, juveniles free; Queen Charlotte Islands (British Columbia) to Punta Abreojos (Baja California).

Juvenile shells to 45 mm in diameter, ribbed, yellowish or orange; adult shells reaching 45–150 mm intertidally (250 mm subtidally); valves thick and heavy, the lower (right) valve taking shape of substratum to which it is cemented but retaining juvenile shape and pattern near umbo, the upper (left) valve with many strong ribs each with many short fluted spines, often badly eroded or penetrated with burrows of boring organisms; shell exterior brownish, the interior pearly white with distinct purple blotch on central part of hinge; mantle orange, the margin bearing numerous blue eyes and slender sensory tentacles.

The juvenile stages of *Hinnites* can swim, like other scallops, by clapping the valves together repeatedly and spurting jets of water outward on either side of the hinge. Swimming is easily started by touching the clam with the arm of a predaceous sea star such as *Pisaster ochraceus* (8.13). At rest a juvenile may attach temporarily by secreting a few byssal threads; these are accommodated by the deep byssal notch on the anterior ear of the right valve (see the photograph, right side). After attaining a diameter of about 25 mm the animal cements the right valve to the rocks and thereafter grows and lives as a sessile organism. Attachment is accomplished by extending the right mantle edge to touch the rock; the outer mantle lobe then secretes new shell substance, which flows onto and fuses with the substratum. Once attached, the scallop grows slowly and may take 25 years or more to reach full size.

In central California, *Hinnites* has been observed to spawn in April. The sexes are separate and can be distinguished by gonad color, the testis being white, the ovary bright red.

Collectors in the low intertidal area are sometimes startled by the noise made when the animal claps down its free upper valve, closing the shell. The rock scallop is highly prized as food, the single large adductor muscle (sometimes 5 cm in diameter) being eaten either raw or fried. Heavy collecting by clammers and divers has depleted the population in many areas.

See Fitch (1953), Grau (1959), Hertlein & Grant (1972), Hinegardner (1974), Hubbs, Bien & Suess (1960), Quayle (1973), Ricketts & Calvin (1968), Weymouth (1920), and Yonge (1936, 1951c).

Phylum Mollusca / Class Bivalvia / Subclass Pteriomorpha / Order Pterioida Family Pectinidae

15.15 **Argopecten aequisulcatus** (Carpenter, 1864)
Speckled Scallop

On sandy or muddy bottoms of bays, in shallow water to 50 m depth; Santa Barbara (California) to Cabo San Lucas (Baja California).

Shell to about 80 mm in diameter; valves convex, sculptured with 19–22 flattened radial ribs; raised concentric growth lines between ribs; anterior and posterior ears about equal in size; shell exterior gray, orange, or reddish with dark spots or blotches; soft parts yellow; edge of mantle with short tentacles and numerous tiny eyes.

This form is considered by some to be a northern subspe-

cies of the Panamic *Argopecten circularis*. Its generic placement has been uncertain, and at various times it has been assigned to the genera *Plagioctenium*, *Aequipecten*, and *Pecten*.

The animals are normally unattached, but they can anchor themselves temporarily to solid substrata by secreting a few byssal threads. When free, they can swim distances of up to a meter at a time by rapidly and repeatedly clapping the valves together. They are collected by man for food, either the single large adductor muscle or the entire animal being eaten.

See Bonnot (1940), Cox (1957), Fitch (1953), Grau (1959), Gutsell (1931), Keen (1971), MacGinitie (1941), McLean (1969), Nelson (1949), Williamson (1902), and Yonge (1936).

Phylum Mollusca / Class Bivalvia / Subclass Pteriomorpha / Order Pterioida Family Pectinidae

15.16 **Leptopecten latiauratus** (Conrad, 1837)
Broad-Eared Pecten

Uncommon, attached to rocks, pilings, and eelgrass in bays, and to shells, worm tubes, and pebbles on offshore bottoms, low intertidal zone to 50 m depth; Point Reyes (Marin Co.) to Cabo San Lucas (Baja California) and Gulf of California.

Shell to about 20 mm in diameter; valves thin, only slightly convex, each bearing 12–16 flat-topped ribs with strong concentric lamellae; hinge line long; shell color bright, ranging from yellow to terra-cotta and brown with lighter or darker V-shaped markings or blotches.

Little is known about the biology of this scallop. The larvae sometimes enter the seawater system of the Kerckhoff Marine Laboratory (Corona del Mar, Orange Co.), and tiny scallops appear by the hundreds in aquaria. Seawater temperature differences influence the ultimate morphology of the shell.

In a related southern California subtidal species, *Pecten diegensis*, approximately daily growth lines can be seen on the shell. Twelve juvenile scallops kept in the Kerckhoff laboratory for 51 days showed numbers of rings formed in that period ranging from 34 to 51.

See Chipman & Hopkins (1954), G. Clark (1968, 1971), Cronly-Dillon (1966), Dakin (1909), Grau (1959), MacGinitie (1938), MacGinitie & MacGinitie (1968), McLean (1969), and W. Miller (1958).

Phylum Mollusca / Class Bivalvia / Subclass Pteriomorpha / Order Pterioida
Family Limidae

15.17 Lima hemphilli Hertlein & Strong, 1946
(=Lima dehiscens of some authors)   File Shell

Fairly common, under rocks on rocky shores and nestling among fouling organisms on wharf pilings, low intertidal zone to 50 m depth; Monterey Bay to Acapulco (Mexico).

Shell to about 30 mm long, with fine radial sculpture and very short ears; valves gaping (unable to close tightly) at anterior and posterior ends; shell exterior white; soft parts orange to pink; mantle fringed with long mobile marginal tentacles.

Lima lives in a "nest" which it constructs of rubble and byssal fibers. The animal can leave the nest and swim rapidly by repeated clapping of the two valves, a movement that ejects water in jets on each side of the hinge. The marginal tentacles are sticky and may be shed when the animal is attacked.

A related species (*Lima scabra*) has been the subject of neurophysiological studies on the eyes, which are located on the mantle edge.

See Baily (1950), Gilmour (1967), Hinegardner (1974), MacGinitie & MacGinitie (1968), Merrill & Turner (1963), Mpitsos (1973), Seilacher (1954), and Stasek (1967).

Phylum Mollusca / Class Bivalvia / Subclass Pteriomorpha / Order Pterioida
Family Anomiidae

15.18 Pododesmus cepio (Gray, 1850)
(=P. macrochisma, Monia macrochisma of some authors)
Abalone Jingle

Common, attached in protected nooks on breakwaters and wave-swept outer coastal rocks, also on wharf pilings and undersides of harbor floats (as in Monterey harbor), and on red abalone shells, near low tide level and subtidally; southern Alaska to Cabo San Lucas (Baja California) and Gulf of California.

Shell to about 80 mm in diameter, rounded in outline, translucent, sculptured with irregular branching radial ribs, attached on one side to substratum by tissues projecting

through opening near hinge of lower (right) valve; valves unequal, the lower valve thin and conforming to substratum, the upper valve convex above, its inner surface highly polished, iridescent green, and bearing two well-marked muscle scars; soft parts bright orange.

This remarkable and specialized clam is distantly related to the scallops and oysters and more closely allied to the tropical windowpane shell, *Placenta (Placuna) placenta*. It is edible but rarely used as food by man.

The conspicuous hole in the lower (right) valve is not a true perforation of the valve but a deep embayment originally evolved from a byssal notch akin to that in the right valve of *Hinnites* (see Fig. 15.14, right side), which becomes centrally located through asymmetrical growth of the shell. The byssal attachment to the substratum is large and calcified. Anchored to it are strong foot and byssal retractor muscles; at their other ends these muscles attach to the upper (left) valve, thus their action is similar to that of adductor muscles, for which they are sometimes mistaken. The true posterior adductor muscle, smaller and separate, lies nearby. It connects the upper and lower valves but is not visible from outside the shell.

Within the shell are exceptionally large and active mucous glands, and Kellogg, who studied feeding and waste removal in many species of clams, remarked (1915), "In no other case has a more furious ciliary action been observed than on the gills and mantle of this form."

The green color inside the upper (and sometimes lower) valve is due, at least in part, to minute algae living within the shell. On the outside, the upper valve is often badly eroded by boring sponges, polychaete worms, and pholads. Studies of reproduction in a population at Tomales Bay (Marin Co.) showed that formation of gametes began in late November, gonads enlarged in the spring, and spawning occurred in July and August.

See Bonnot (1940), Fitch (1953), Frizzell (1930a), Kellogg (1915), Leonard (1969), MacGinitie (1935), MacGinitie & MacGinitie (1968), Quayle (1973), Ricketts & Calvin (1968), and Yonge (1953, 1977).

Phylum Mollusca / Class Bivalvia / Subclass Heterodonta / Order Veneroida
Family Chamidae

15.19 **Chama arcana** Bernard, 1976
(=Chama pellucida of some authors)

Common, cemented to undersides of protected rocks and harbor floats and on pilings, middle intertidal zone to subtidal depths along open coast and in bays; Oregon to Bahía San Juanico (Baja California).

Shell to about 60 mm in diameter, heavy, attached by left valve; surface roughened by concentric, translucent frills, often badly eroded; beaks coiled to the right (clockwise) as viewed from above; lower valve deeply concave internally, upper valve smaller, nearly flat; shell exterior white, sometimes stained with red, pink, or orange; mantle and siphons white, the double row of tentacles yellow.

After the larva of *Chama* settles and the left valve becomes attached to the substratum, growth is accompanied by a slow counterclockwise rotation, giving the shell its characteristic spiral form. When feeding, the animal opens its valves only slightly, and the small, separate siphons extend well above the substratum. Occasionally the small foot protrudes through a pedal gape in the mantle folds, which are otherwise fused along their edges except for the siphonal apertures. The adductor muscles are large, and when they contract, quantities of mucus and foreign material cleaned off the ctenidia and mantle are expelled. The ctenidia themselves are large and elaborately folded, giving an extensive filtering surface. Particulate matter reaching the ctenidia is sorted, and only the finest particles are carried along the marginal food grooves to the palps and mouth.

The sexes of *Chama* are separate, but nothing is known about spawning or early development.

On siltstone reefs at depths of 10 m off southern California, individuals of this species pile on top of one another, forming thick crusts, with population densities sometimes exceeding 400 per 0.1 m^2.

Chama is occasionally taken by clammers, who esteem its flavor despite its small size. It is also taken by the southern sea otter from the floats in the Monterey marina.

See especially Yonge (1967); see also Bernard (1976), Fitch (1953), Grieser (1913), Oberling (1964), Odhner (1919), Pequegnat (1964a), Ricketts & Calvin (1968), and J. Taylor & Kennedy (1969).

Phylum Mollusca / Class Bivalvia / Subclass Heterodonta / Order Veneroida
Family Chamidae

15.20 **Pseudochama exogyra** (Conrad, 1837)
Reversed Chama

Common on rocky reefs, middle intertidal zone along open coast; Oregon to southern Baja California.

Shell to about 50 mm in diameter, cemented to stones by right valve; beaks coiling to left (counterclockwise) as viewed from above; shell otherwise similar to *Chama arcana* (15.19) but frills on valves often badly worn or completely gone; shell color dull white, occasionally greenish; exposed mantle and siphons orange, the tentacles white.

This species is often found in large clusters, along with *Chama arcana* on rocks along the open coast. The valves are unusually thick and dense, apparently lacking calcite. The chamas are related to a now-extinct group of bivalves called rudists (Hippuritacea), which had somewhat similar but less dense shells and formed massive reefs in the intertidal zone in Mesozoic time.

Except for the different symmetry brought about through attachment to the substratum by the right valve, the functional anatomy of *P. exogyra* is much like that described for *Chama arcana*.

See especially Yonge (1967); see also Bernard (1976) and Fitch (1953).

Phylum Mollusca / Class Bivalvia / Subclass Heterodonta / Order Veneroida
Family Erycinidae

15.21 **Lasaea cistula** Keen, 1938

Very abundant, attached to the red alga *Endocladia muricata*, in dead shells of the barnacle *Balanus glandula* (20.24), among mussels, and under sea palm (*Postelsia*) holdfasts, high to middle intertidal zones on surf-swept rocky shores; northern California to Cabo San Lucas (Baja California).

Shell averaging 2 mm long but up to 3.5 mm, oval-oblong to quadrate in outline, plump, with wavy concentric sculpture; shell exterior usually medium to dark red, sometimes white; soft parts white.

In the *Endocladia-Balanus* community in the high intertidal zone at Pacific Grove (Monterey Co.), these tiny clams sometimes occur in numbers over 11,000 per 400 cm². Present in all seasons, they are generally attached to each other and to adjacent structures by byssal fibers, but they can detach themselves and crawl about actively on an extended foot somewhat like snails.

The adults are hermaphroditic, and the large yolky eggs are retained in the mantle cavity of the parent where fertilization and development to the juvenile stage occur. There is no swimming larval stage. In all months of the year some individual clams may be found with developing young, but more brood young in the summer; a single adult produces about 40 young at a time. Juveniles grow to 0.5–0.6 mm in length before being released as miniatures of the adult.

The food of *L. cistula* seems to be mainly organic detritus. In a related species, *L. rubra*, it was determined that, although certain golden-brown algae were easily digested, other diatoms and dinoflagellates were not.

See especially Glynn (1964, 1965); see also Ballantine & Morton (1956), Haas (1942a), Keen (1938), and Oberling (1964).

Phylum Mollusca / Class Bivalvia / Subclass Heterodonta / Order Veneroida Family Erycinidae

15.22 Lasaea subviridis Dall, 1899

Common, nestling among byssal threads in beds of *Mytilus californianus* (15.9), middle intertidal rocks on open coast; Humboldt Co. to northern Baja California.

Shell to 3.5 mm long, slightly flatter than in *L. cistula* (15.21), the anterior end relatively elongate; valves sculptured with wavy, concentric lines; shell exterior green or yellowish with pink on upper parts.

This tiny nestling clam is often abundant in mussel beds. It broods its young, and the mantle spaces of an adult may be almost completely filled with developing juveniles. The species is difficult to distinguish from *Lasaea cistula* (15.21), and the two may well prove to represent a single variable species.

See Keen (1938, 1971) and McLean (1969).

Phylum Mollusca / Class Bivalvia / Subclass Heterodonta / Order Veneroida Family Carditidae

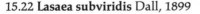

15.23 Glans carpenteri (Lamy, 1922)
(=Glans subquadrata)   Little Heart Clam

Common, attached by byssus to undersurfaces of rocks, low intertidal zone and subtidal; Queen Charlotte Islands (British Columbia) to Camalú (Baja California).

Shell to about 15 mm long, stout, quadrate or rectangular with beaks well forward, sculptured with 14–17 strong, beaded, radiating ribs; shell exterior tan or whitish with brown, pink, or yellowish markings, the interior white with brown in posterodorsal area.

This small clam has a relatively large foot. If detached it crawls about and reattaches by byssal threads, usually in a small depression in the rock surface.

Yolky eggs 120 µm in diameter are released over an extended period into the exhalant portion of the mantle cavity, where they are fertilized and brooded during development. Before liberation the young are 0.8 mm long and have relatively heavy shells. Much too large to be discharged through the exhalant siphon, they must rupture through the gill and leave the parent by way of the inhalant mantle cavity. Released young are often found attached to the outside of the adult animal's shell.

Fossil relatives of this species (*Cardita planicosta* and *C. crenata*), along with other fossil bivalves, provide, through their growth increments, some evidence to support the idea that the earth's rotation is slowing down at a rate of about 2 milliseconds per century, owing to tidal torque.

See Coan (1977a,b), Pannella, MacClintock & Thompson (1968), and Yonge (1969).

Phylum Mollusca / Class Bivalvia / Subclass Heterodonta / Order Veneroida
Family Kelliidae

15.24 **Kellia laperousii** (Deshayes, 1839)
Smooth Kelly Clam

Common, nestling in dark protected places, such as in fouling growth on wharf pilings, in mussel beds, in vacant mussel and barnacle shells, and in unoccupied pholad holes in rocks, middle to low intertidal zones and subtidal waters; Bering Sea to Cabo San Lucas (Baja California) and Gulf of California.

Shell to about 25 mm long, thin, smooth, sculptured with uneven concentric lines; periostracum yellowish green; shell interior dull white.

Kellia is normally attached by up to four fine byssal threads. When free, it moves about very actively, creeping on the sole of its long thin foot. The fleshy mantle is often extended out over the shell; anteriorly it forms a long and large inhalant siphon, an adaption for living in deep crevices or holes. The exhalant siphon is short and lies posteriorly. All of the soft tissues can be withdrawn within the shell.

Kellia is viviparous, and the young are brooded in the mantle cavity. When released they often settle near the parent, and up to a dozen or more animals of varying sizes may be found packed in a single pholad hole. *Kellia* is often found as a fouling organism in seawater piping systems, and a sizable population lives and multiplies in the seawater pipes of the Bodega Marine Laboratory (Sonoma Co.) and the Hopkins Marine Station (Monterey Co.).

See Howard (1953), MacGinitie (1935), Quayle (1973), Stasek (1963a,b), and Yonge (1952c).

Phylum Mollusca / Class Bivalvia / Subclass Heterodonta / Order Veneroida
Family Lucinidae

15.25 **Epilucina californica** (Conrad, 1837)
California Lucine

Common in rocky rubble and sand, low intertidal zone to offshore depths of 80 m; Crescent City (Del Norte Co.) to Laguna San Ignacio (Baja California).

Shell to about 35 mm in diameter, rounded in outline, with beaks centrally placed on dorsal margin; external ligament in a deep pit; outer surfaces of valves sculptured with fine concentric lines and a low ridge extending posteriorly from beaks; shell exterior whitish to yellowish.

Almost nothing is known of the biology of this species.

See Coan & Carlton (1975), McLean (1969), and Wicksten (1978).

Phylum Mollusca / Class Bivalvia / Subclass Heterodonta / Order Veneroida
Family Ungulinidae

15.26 **Diplodonta orbella** (Gould, 1851)
Round Diplodonta

Fairly common in rock rubble and under rocks, low intertidal zone and subtidal waters; Pribilof Islands (Alaska) to southern Baja California.

Shell to about 30 mm in diameter, nearly spherical, with concentric growth lines; shell exterior whitish under a dark-yellow periostracum.

This animal builds a "nest" or covering of sand, shell fragments, or seaweed strands cemented together. Little is known of its biology.

See Haas (1942a,b, 1943), MacGinitie & MacGinitie (1968), Quayle (1973), and Seapy & Kitting (1978).

Phylum Mollusca / Class Bivalvia / Subclass Heterodonta / Order Veneroida
Family Cardiidae

15.27 **Trachycardium quadragenarium** (Conrad, 1837)
Spiny Cockle

Usually uncommon to rare, in firm sand or mud of bays, low intertidal zone to subtidal depths of 20 m; Monterey Bay to southern Baja California.

Shell to about 80 mm in diameter, large, oval, with 41–44 strong radial ribs each with a row of sharp triangular spines; margins of shell deeply crenulated; shell exterior dull grayish brown, buff, or yellow.

This large cockle lives shallowly buried, with just the tips of the valves projecting above the substratum. Sometimes it reacts to nearby footsteps on the sand by popping to the surface. It is rarely used as food because of its scarcity in most intertidal areas.

See Fitch (1953) and Hinegardner (1974).

Phylum Mollusca / Class Bivalvia / Subclass Heterodonta / Order Veneroida Family Cardiidae

15.28 **Laevicardium substriatum** (Conrad, 1837)
Egg Cockle

Common, buried shallowly in mud or sand, low intertidal zone in bays and offshore to depths of 50 m; Mugu Bay (Ventura Co.) to southern Baja California.

Shell to 37 mm in diameter, thin, smooth or finely ribbed, with fine crenulations on margin except posteriorly; shell whitish, gray, or pale yellow flecked with brown.

Nothing is known of the biology of this species. Offshore in southern California its relative, **L. elatum** (Sowerby), grows to over 150 mm in length and is the largest of the cockle family.

See Fitch (1953), McLean (1969), and Seapy & Kitting (1978).

Phylum Mollusca / Class Bivalvia / Subclass Heterodonta / Order Veneroida Family Cardiidae

15.29 **Clinocardium nuttallii** (Conrad, 1837)
(=Cardium corbis)   Basket Cockle, Heart Cockle

Rare to relatively common, in mud or muddy sand of bays, sloughs, and estuaries, low intertidal zone to depths of 200 m offshore in suitable substrata; Bering Sea to San Diego.

Shell to 80 mm in diameter, globular, with 34 radial ribs crossed by concentric growth lines, the ribs and grooves at the edges of the two valves interlocking; periostracum brown; shell buff mottled with brown or red; soft parts yellowish white.

This attractive large cockle lives on or just beneath the surface of fine sediments. It is collected and marketed in the Puget Sound area, where it is much more abundant than in California.

Clinocardium has a powerful foot (protruding in the photograph), and displays a remarkable, leaping escape reaction to predatory sea stars such as *Pisaster brevispinus* (8.15) and *Pycnopodia helianthoides* (8.16). If a cockle, with siphons extended, is contacted on the soft parts by the arm and tube feet of a sea star, it opens the valves widely, extends its long pointed foot, and vigorously pushes itself away from the sea star. *Clinocardium* lives well in laboratory aquaria and is one of the clams best suited for student dissections.

The shells of this cockle on the Oregon coast often show growth bands correlated with tidal cycles; widely spaced bands are deposited during the 24.8-hour cycle of high spring tides, whereas narrowly spaced bands are formed during the 12.4-hour neap tides, when the animals are exposed to air twice as often. In northern waters growth is interrupted in winter, resulting in strong growth rings not seen in most California specimens. Growth rates may be correlated with air temperature rather than seawater temperature. The animals are hermaphroditic, and breeding occurs in the summer in individuals two years old or older. Most animals live no longer than 7 years, but very large specimens may be as old as 16 years.

Clinocardium has numerous tiny eyes, located on optic tentacles on the mantle margin. Each eye has a reflector cup, inside of which are cells whose photoreceptive surfaces are modified cilia. Studies have also been made on the neural pathways involved in regulating cardiac activity.

See Ansell (1967b), Barber & Wright (1969), Cooke (1975), Edmondson (1919, 1920a), Evans (1975), Feder (1967), Fitch (1953), Fraser (1931), Jegla & Greenberg (1968), Johnstone (1899), MacGinitie (1935), Mitchell (1935), Quayle (1973), Ray (1959), Silvey (1968), Stasek (1962, 1963a,b), Talmadge (1972), C. Taylor (1960), Weymouth & Thompson (1931), and Wourms (1968).

Phylum Mollusca / Class Bivalvia / Subclass Heterodonta / Order Veneroida
Family Veneridae

15.30 **Gemma gemma** (Totten, 1834)
Gem Clam

Abundant to rare on mud bottoms, low intertidal zone and subtidal waters; Puget Sound (Washington) to San Diego; Nova Scotia to Gulf of Mexico.

Shell to 5 mm long, tiny, thin, laterally convex; exterior shiny with fine concentric riblets; shell white to buff, with pink or purple on beaks and posterior.

This small clam was introduced to the west coast from the Atlantic in the late 1800's, probably with eastern oysters. It is now common in Tomales Bay (Marin Co.), but is less abundant in San Francisco Bay and the bays of southern California. It occurs subtidally at depths of 30–100 m in Monterey Bay and San Diego Bay. The animal usually lives in sandy or muddy areas in estuarine situations. On the east coast, population densities can exceed 200,000 clams per m².

The females brood the young on the gills. When released, the young juvenile clams do not swim but are dispersed by waves and currents. Young 0.4 mm long are liberated (about 200 per adult) in the summer; by fall they are 2 mm long, and adult size is reached the following summer. Maximum life span seems to be about 2 years.

Experiments indicate that *G. gemma* can survive catastrophic burial to a depth of 230 mm in sand or 57 mm in silt. After burial the clams burrow vertically upward at rates of up to 35 mm per hour and may survive burial for up to 6 days. The clams are preyed upon by sea anemones, crabs, bottom-feeding fishes, ducks, and many shore birds. They are parasitized by trematode larvae.

See R. Green & Hobson (1970), Jackson (1968), Narchi (1971), Nichols (1977), Oberling (1964), Sellmer (1967), Shulenberger (1970), Vassallo (1969), Welch (1969), and Wicksten (1978).

Phylum Mollusca / Class Bivalvia / Subclass Heterodonta / Order Veneroida
Family Veneridae

15.31 **Mercenaria mercenaria** (Linnaeus, 1758)
(=Venus mercenaria)   Quahog

Scarce, in sand or sandy mud, low intertidal zone; Humboldt Bay, San Francisco Bay, and Colorado Lagoon (Long Beach, Los Angeles Co.); Gulf of St. Lawrence to Florida.

Shell to 130 mm long, oval, heavy, compact, sculptured with concentric ridges; external hinge ligament prominent; inner margin of shell crenulate; shell exterior whitish, covered with yellow-brown periostracum, the interior white with violet stains near muscle scars and purple border on ventral margin.

The quahog was introduced in the 1800's from the Atlantic into Humboldt Bay, and into Colorado Lagoon (Los Angeles Co.) in 1940. Small breeding colonies now appear to be established in both localities. It is one of the most important commercial bivalves on the east coast. It was eaten there even in pre-Columbian times, and the shells were used by east coast Indians for wampum and decoration. The common name is a contraction of the American Indian term "poquauhock."

On the New England coast and in southern California this species spawns from June through August, when the water temperature is 24–25°C.

This clam has been investigated from many points of view. Recent studies have been carried out on the physiology of "catch" muscles which maintain closure of the shell, the biochemistry of the blood and digestive processes, and the cytology and ultrastructure of various parts. The quahog is one of the few bivalves from which biologically active antitumor agents have been isolated.

The quahog, and some other clams, produce daily and even subdaily growth lines in the shell. Both 14-day tidal cycles and seasons can be detected by the relative thickness of the daily increments. Studies of daily growth lines in extinct fossil bivalves (and in some other forms producing hard calcareous skeletons, such as corals) indicate that the earth is gradually slowing its rate of rotation; 500 million years ago there were 2 more days per month than there are now. The rate of slowing appears not to have been entirely uniform.

While the daily growth rings reflect external environmental conditions, they are also under the influence of an internal biological clock. When clams are kept in constant light, or constant darkness, the experimental animals continue for a time to maintain a 24-hour rhythm of shell deposition.

See especially Belding (1911); see also Calabrese & Davis (1966), Cheng & Foley (1971), Crane, Allen & Eisemann (1975), Crenshaw & Watanabe (1969), Davis & Hidu (1969), Fitch (1953), Gordon & Carriker (1978), Greenberg & Jegla (1962), Hamwi & Haskins (1969), Hanna (1966), Hegyeli (1964), Hidu & Hanks (1968), Hillman (1961), Hinegardner (1974), Jegla & Greenberg (1968), W. Johnson, Kahn & Szent-Györgyi (1959), Kellogg (1915), J. Loesch & Haven (1973), Loosanoff (1937), Loosanoff & Davis (1950, 1963), Loosanoff, Davis & Chanley (1966), Menzel & Menzel (1965), Pannella, MacClintock & Thompson (1968), Pratt (1953), Salchak & Haas (1971), Savage (1976), Schmeer (1964), Thompson (1975), Walne (1974), Wells (1957), Wicksten (1978), and Yonge & Crenshaw (1971).

Phylum Mollusca / Class Bivalvia / Subclass Heterodonta / Order Veneroida Family Veneridae

15.32 **Tivela stultorum** (Mawe, 1823)
Pismo Clam

Rare to common, low intertidal zone and offshore to 25 m depth on broad sandy beaches exposed to strong surf; Half Moon Bay (San Mateo Co.) to Bahía Magdalena (Baja California).

Shell to 150 mm or more in length, strong, heavy, smooth but sculptured with fine concentric growth lines; beaks nearly central; ligament obvious, elongate, set in deep groove; periostracum shiny, greenish to brownish; shell pale buff to dark chocolate, occasionally marked with brown or purple-brown radiating bands.

This clam has a fine flavor. The best intertidal collecting localities in the past were the beaches of Monterey Bay from Santa Cruz south to Elkhorn Slough, and those at Pismo Beach and Morro Bay (San Luis Obispo Co.). From 1916 to 1947 some 50,000 clams per year were collected by commercial diggers, and in one record year (1918) 350,000 clams were taken, mainly from Pismo Beach and Morro Bay. In 1949, when a previously closed beach in the Pismo Beach area was opened to sport fishermen for the first time, 50,000 clams were taken daily for $2\frac{1}{2}$ months. In recent years there has been a disastrous decline in the intertidal population of these clams; in 1973 only 21 legal-sized clams were taken at Pismo Beach and fewer still at Morro Bay. The Pismo clam population in Monterey Bay dropped rapidly in the 1970's under attacks by both man and the sea otter. Other predators, such as crabs, moon snails, sharks, rays, and shorebirds, also take a toll. Subtidal populations of Pismo clams have not been extensively exploited; off Zuma Beach (Los Angeles Co.) and other areas in southern California dense populations are still found in water 10 m deep.

Tivela orients vertically in the sand, with the hinge and exhalant siphon facing the ocean. Sometimes uncovered by ocean waves, it rapidly reburies itself. Water is jetted from the anterior end of the shell while the foot digs in, and the weight of the heavy shell helps carry the animal down. The clam burrow is shallow, and the posterior tip of the shell remains virtually at the surface. It sometimes bears a tuft of the hydroid *Clytia bakeri* (3.9; see Fig. 3.9).

The clam's food is mainly very fine detritus particles and minute planktonic organisms. Digestion seems to be mainly intracellular, accomplished by phagocytic cells in the digestive diverticula. Some extracellular digestion occurs in the gut, and the crystalline style provides amylases and glycogenases. A 75 mm clam is capable of filtering up to 60 liters of water daily. Growth is continuous, with some slowing in the winter; increase in shell length averages 21 mm per year for the first 3 years, then continues at a somewhat slower rate. It takes 4–6 years for the clam to reach the legal length of 115 mm, and up to 20 years to reach 150 mm. Some very large clams have been estimated to be 43–50 years old.

The sexes are separate, and females first spawn in the second or third summer after hatching. The eggs are small (70 μm in diameter), and the number produced is related to the size of the parent; up to 75 million eggs have been found in the ovary of a single clam. Most spawning in California occurs from June to December.

The shell of the Pismo clam contains variable amounts of magnesium and fairly constant amounts of strontium. The large adductor muscles are dark pink in color, owing to the presence of myoglobin.

In Monterey Bay clams, larval tapeworms have been found in tissues near the gut. They occur in yellowish-white cysts 3.2–3.8 mm in diameter, and have been identified as the larvae of tapeworms whose adult stages inhabit stingrays and skates.

See especially Coe & Fitch (1950), Fitch (1950, 1961), Weymouth (1923b); see also Albrecht (1921), Armstrong (1965), Baxter (1962), Carlisle (1973), Coe (1947, 1948, 1953, 1956), Dushane (1966), Fitch (1953, 1957, 1965), Giese (1958), Gillilan (1964), Grabyan (1973), Hawbecker (1939), Herrington (1929, 1930), Hinegardner (1974), Hubbs, Bien & Suess (1960), MacGinitie (1935), Ricketts & Calvin (1968), Stephenson (1977), Tomlinson (1968), Warner & Katkansky (1969b), and Weymouth (1923a,b).

Phylum Mollusca / Class Bivalvia / Subclass Heterodonta / Order Veneroida Family Veneridae

15.33 **Saxidomus giganteus** (Deshayes, 1839)
Butter Clam, Smooth Washington Clam

Fairly common, buried to depths of 25–35 cm in mud, muddy sand, or muddy gravel, low intertidal zone to subtidal depths of 30 m in bays or on well-protected beaches; Aleutian Islands (Alaska) to San Francisco Bay; rare south of Humboldt Bay.

Shell to 130 mm long, heavy, solid, square to oval in outline, sculptured with fine concentric growth lines (not strong ridges, as in *S. nuttalli*, 15.34); shell exterior yellow in young animals, gray-white in older ones, the interior white.

The butter clam is the most important commercial clam in British Columbia, and in some years 3,300 metric tons of the clam meat are canned. It is also common in many areas of Washington and Oregon, and a small fishery is supported in Humboldt Bay. The clams breed in the summer, but spawning success varies from year to year. They live for 20 years or more, and in certain areas of British Columbia they occur in dense populations. Experiments show that harvesting these clams once every 7 years is less productive than once every 1, 2, or 3 years.

The crystalline style of this clam contains many large spirochaete bacteria (*Cristispira* sp.), which swim freely in all except the functional end of the style. The style releases amylase and oxidase enzymes. Tissue extracts from the butter clam agglutinate human red blood cells of types A_1 and A_1B. The mantle cavity of this species occasionally harbors commensal pinnotherid crabs (adult *Fabia subquadrata*, 25.31, and young *Pinnixa littoralis*, 25.36).

See Berkeley (1962), Fitch (1953), Fraser (1929), Fraser & Smith (1928a), H. Johnson (1964), Marriage (1954), Neave (1942), Quayle (1952a, 1973), and Ricketts & Calvin (1968).

Phylum Mollusca / Class Bivalvia / Subclass Heterodonta / Order Veneroida Family Veneridae

15.34 **Saxidomus nuttalli** Conrad, 1837
Washington Clam

Common, buried to depths of 30 cm or more in mud or sand, low intertidal zone in bays and lagoons, and in sandy areas near rocks on outer coast; Humboldt Bay to Isla San Gerónimo (Baja California).

Shell to 150 mm long, thick, gaping slightly at siphonal end, with growth lines in the form of strong, raised, concentric ridges (rather than fine concentric lines, as in *S. giganteus*, 15.33); shell exterior tan, the interior white with a purple stain at posterior end; species often confused with *Protothaca tenerrima* (15.42), whose shell lacks a posterior gape.

The Washington clam makes a very tasty dish, and is one of the more important clams in the sheltered waters north of Morro Bay (San Luis Obispo Co.).

The antigenic qualities of extracts of muscle tissues of the Washington and related clams have been investigated from the standpoint of bivalve systematics, and the results tend to confirm conventional taxonomy. A larval tapeworm (*Echeneibothrium* sp.) is found in the clam's intestine. The adult tapeworms probably occur in bottom-feeding elasmobranch fishes. Commensal or semi-parasitic copepods have also been found associated with the hard plates at the tips of the siphons of this clam.

See Bonnot (1949a), Fisher (1969), Fitch (1953), Frey (1971), Hinegardner (1974), Illg (1949), Katkansky, Warner & Poole (1969), and Seapy & Kitting (1978).

Phylum Mollusca / Class Bivalvia / Subclass Heterodonta / Order Veneroida
Family Veneridae

15.35 Irus lamellifer (Conrad, 1837)
Rock Venus

Fairly common, nestling in crevices and empty pholad burrows in outer coastal rocks, and among fouling organisms on pilings in bays, low intertidal zone and subtidal waters to 40 m; Monterey Bay to San Diego.

Shell to 50 mm long, white, nearly quadrate in outline, with spaced concentric ribs or flanges often recurved or shelflike; shell white.

Little is known of the biology of this nestling clam. The ability to secrete strong flanges on the shell enables the animal to brace itself firmly in the center of a secluded hole considerably larger than itself.

See Haderlie (1976), McLean (1969), and Oberline (1964).

Phylum Mollusca / Class Bivalvia / Suborder Heterodonta / Order Veneroida
Family Veneridae

15.36 Tapes japonica Deshayes, 1853
(=T. semidecussata)   Japanese Littleneck Clam

Common, in coarse sandy mud of bays, sloughs, and estuaries, buried 2–4 cm below surface, occasionally attached to stone by byssus, middle to low intertidal zones; British Columbia to southern California; introduced from Japan.

Shell to 50 mm long, elongate, oval, sculptured with radiating ribs and less-pronounced concentric ridges; shell color highly variable, usually yellow or buff, with patterns or lines of brown or black.

At various times this species has been assigned to the genera *Paphia*, *Venerupis*, *Protothaca*, and *Ruditapes*. It was accidentally introduced to the west coast in the 1930's, probably with Japanese oysters. It is now extremely common in many areas from British Columbia to San Francisco Bay and is highly prized for food. Techniques have been developed for mass rearing of larvae of the Japanese clam, and there is a possibility that it may be introduced successfully into bays and estuaries where it does not live at present. It tolerates salinities as low as 30–50 percent that of pure seawater, and lives in highly polluted water in San Francisco Bay, where it must be carefully cleaned before it is used for food. Breeding occurs in the summer, and the larvae are very similar in form to those of *Protothaca staminea* (15.41). *T. japonica* grows slowly, particularly where it is overcrowded. It often harbors the commensal pinnotherid crabs *Pinnixa faba* (25.35) and *P. littoralis* (25.36).

See Cheney (1971), Chiba & Ohshima (1957), Fitch (1953), Frey (1971), Hanna (1966), Loosanoff & Davis (1963), Painter (1966), Quayle (1941, 1973), Ricketts & Calvin (1968), Seilacher (1972), Shaw (1956), and Wicksten (1978).

Phylum Mollusca / Class Bivalvia / Subclass Heterodonta / Order Veneroida
Family Veneridae

15.37 Chione californiensis (Broderip, 1835)
(=C. succincta, Venus succincta)   California Chione

Fairly common, buried just below surface on sandy mud flats in bays and sloughs, low intertidal zone and offshore to 50 m depth; Carpinteria (Santa Barbara Co.) to Panama.

Shell to 60 mm long, with strong, widely spaced, concentric lamellae overriding radial riblets; periostracum yellowish brown; shell whitish, stained purplish brown near hinge area.

Little is known about this clam, which is sometimes taken by clam diggers along with other, more abundant chiones. Mass mortality of the species has been observed in bays in Baja California, but this seems to be independent of the red tides that occur there periodically.

See Fitch (1953), Hubbs, Bien & Suess (1960), Parker (1949), and Stohler (1960).

Phylum Mollusca / Class Bivalvia / Subclass Heterodonta / Order Veneroida
Family Veneridae

15.38 Chione fluctifraga (Sowerby, 1853)
Smooth Chione

Fairly common, just below surface on mud and sand flats of bays, low intertidal zone; Mugu Lagoon (Ventura Co.), to Gulf of California.

Shell to 60 mm long, thick, with smooth concentric ribs irregularly broken by radial ribs especially on posterior slope; periostracum heavy, gray or brownish; shell creamy white externally, white internally with purple-brown or blue stain on hinge and posterior area, often eroded and chalky.

This clam is highly regarded as food and supports an extensive sport and commercial fishery in southern California. As it is limited mainly to muddy, back-bay environments, it is in danger of extinction in many areas where harbors are being developed. Little is known of its biology.

See Fitch (1953), Frey (1971), Parker (1949), and Pilson & Taylor (1961).

Phylum Mollusca/ Class Bivalvia / Subclass Heterodonta / Order Veneroida Family Veneridae

15.39 **Chione undatella** (Sowerby, 1835)
Wavy Chione

Abundant just below surface on sand and mud flats, nearer entrance of bays than *C. fluctifraga* (15.38), low intertidal zone and offshore to 50 m depth; Goleta (Santa Barbara Co.) to Peru.

Shell to 60 mm long, similar to that of *C. californiensis* (15.37) but more elongate and with concentric ribs thinner and sharper; radiating ribs conspicuous over entire shell.

This clam is taken extensively for food in southern California. As with other chiones, little is known of its biology. A comparative natural history study of the various species in the genus would be a highly desirable project.

See Fitch (1953), Parker (1949), and Pilson & Taylor (1961).

Phylum Mollusca / Class Bivalvia / Subclass Heterodonta / Order Veneroida Family Veneridae

15.40 **Prototheca laciniata** (Carpenter, 1864)
Rough-Sided Littleneck Clam

In sand or muddy sand in bays, low intertidal zone and adjacent shallow subtidal waters; Monterey Bay to Estero Todos

Santos (Baja California); uncommon in California except in Alamitos Bay (Los Angeles Co.).

Shell to 80 mm long; valves thick, strongly convex, round in outline, sculptured with equally developed radiating ribs and concentric ridges, the intersections of ribs and ridges with small, rough spines; shell buff to orange-brown, often with dark stains.

In various earlier publications this clam has been assigned to the genera *Tapes, Venerupis,* and *Paphia.*

The species is edible but scarce. Small individuals are shallowly buried, but large animals may be 25 cm below the surface. In Morro Bay (San Luis Obispo Co.), it has found infected with larval tapeworms (*Echeneibothrium* sp.), whose definitive hosts are probably bottom-feeding sharks, skates, or rays.

See Fitch (1953), Katkansky & Warner (1969), and Seapy & Kitting (1978).

Phylum Mollusca / Class Bivalvia / Subclass Heterodonta / Order Veneroida Family Veneridae

15.41 **Prototheca staminea** (Conrad, 1837)
(=Paphia staminea, Venerupis staminea)
Common Littleneck Clam

Abundant, in shallow burrows 3–8 cm below surface in coarse sand or sandy mud in bays or coves, and in gravel under larger rocks on open coast, middle to low intertidal zones; Aleutian Islands (Alaska) to Cabo San Lucas (Baja California).

Shell to 70 mm long, oval in outline, with fine radial ribs crossed by numerous weak concentric ridges; shell whitish or tan with angular pattern of chocolate-brown.

This is one of the most abundant west coast clams and is prized as food everywhere.

The sexes are separate. Studies in British Columbia show that the gonads enlarge in the winter, peak in March, and release gametes from April to September. The swimming larvae settle, and young clams 1 mm long develop rudimentary gonads, but sexual differentiation occurs later at a shell length of 15–30 mm (in the second or third year). Animals reach sexual maturity when 22–35 mm long.

Animals in Prince William Sound (Alaska) show well-marked annual growth lines, with measured increments of up to 6 mm per year in shell length. Alaskan animals require an average of 8 years to attain a length of 30 mm (harvestable size), whereas British Columbian animals achieve this size in only 3 years.

Studies in Mugu Lagoon (Ventura Co.) show that mortality rates are high in young and old *P. staminea*, with the standing crop consisting mainly of young adults. This species also shows distinct annual growth lines on the shell.

As in related clams, the common littleneck often contains large numbers of larval tapeworms. Killed in cooking, these are harmless to the gourmet and incapable of infecting man even when alive. They utilize the bat stingray (*Myliobatus californicus*) and possibly other elasmobranchs as definitive hosts.

Shells intermediate in appearance between those of this species and *Protothaca tenerrima* (15.42) have been found as far south as Half Moon Bay (San Mateo Co.), suggesting that hybridization may occur. Similar evidence suggests that the species may also hybridize with the introduced *Tapes japonica* (15.36) in Washington.

In empty pholad holes in offshore shale one sometimes finds small *P. staminea* that have raised, concentric lamellae on the shell, and in that respect resemble *Irus lamellifer* (15.35).

See Bonnot (1949a), Feder & Paul (1973a) Fitch (1953), Fraser (1929), Fraser & Smith (1928b), Frey (1971), Hanna (1966), Hinegardner (1974), Katkansky & Warner (1969), Kincaid (1947), Marriage (1954), Paul & Feder (1973), Paul, Paul, & Feder (1976a,b), Quayle (1943, 1973), Ricketts & Calvin (1968), Schmidt & Warme (1969), Seapy & Kitting (1978), Sparks & Chew (1966), and Warner & Katkansky (1969a).

Phylum Mollusca / Class Bivalvia / Subclass Heterodonta / Order Veneroida Family Veneridae

15.42 **Protothaca tenerrima** (Carpenter, 1857)
Thin-Shelled Littleneck Clam

Uncommon, in burrows to 40 cm below surface in firm sandy mud of bays, low intertidal zone and offshore to 50 m depth;

Vancouver Island (British Columbia) to Bahía Magdalena (Baja California).

Shell to 110 mm long, large, flattened, thin, strongly sculptured on exterior with prominent concentric ridges and fine radiating lines; periostracum gold-colored; shell dull gray; species distinguished from *Saxidomus nuttalli* (15.34), with which it is often confused, by its smooth shell margin and lack of posterior gape.

In the past, various authors have assigned this species to the genera *Tapes, Paphia,* and *Venerupis*.

The clams are occasionally taken along with the Washington clam. They are often preyed upon by moon snails (*Polinices*, e.g., 13.62, 13.63), which drill a hole in the shell, insert the proboscis, and eat the contents. Such bored shells often wash ashore. The species appears to hybridize with *Protothaca staminea* (15.41) north of Half Moon Bay (San Mateo Co.).

See Coan & Carlton (1975), Fitch (1953), Frizzell (1930b), and Quayle (1973).

Phylum Mollusca / Class Bivalvia / Subclass Heterodonta / Order Veneroida Family Petricolidae

15.43 **Petricola pholadiformis** Lamarck, 1818
False Angel Wing

Uncommon, burrowing into mud, clay, and soft rock in bays, low intertidal zone; on west coast found only in Willapa Bay harbor (Washington) and San Francisco Bay, introduced from the North Atlantic; Gulf of St. Lawrence to Gulf of Mexico; Europe.

Shell to 50 mm long, elongate, with beaks well forward, sculptured with coarsely noded radial ribs, exterior grayish white.

This introduced species was first noted on the west coast at Lake Merritt (Oakland, Alameda Co.) in 1934, and at Willapa Bay (Washington) in 1947.

Studies in England show that the animals burrow by first anchoring the foot, then contracting the posterior pedal retractor muscle. This forces the anterior margins of the shell downward and across the base of the burrow. Sharp chisel-like projections on the shell wear away the rock. In soft rock the animals burrow actively at intervals of about 30 seconds, but in harder substrata the intervals are longer. Hydraulic pressures of up to 20 cm of water are developed in the mantle cavity during effective digging strokes.

East coast animals have been conditioned to spawn in the laboratory from December until summer. Clams less than 1 year old are sexually mature. The eggs are 52 µm in diameter, and over 1 million may be discharged at one spawning. Techniques for the mass rearing of larvae have been developed.

See Ansell (1970), Hanna (1966), Hinegardner (1974), Loosanoff & Davis (1963), Owen (1958), and Purchon (1955).

Phylum Mollusca / Class Bivalvia / Subclass Heterodonta / Order Veneroida Family Petricolidae

15.44 **Petricola carditoides** (Conrad, 1837)

Fairly common, nestling in empty pholad holes and rock crevices, low intertidal and shallow subtidal waters; Vancouver Island (British Columbia) to Bahía Magdalena (Baja California).

Shell to 50 mm long, generally quadrangular but often taking shape of cavity in which animal grows, thick, with irregular growth lines; surface white and chalky; juvenile shell showing fine radial lines, sometimes in a zigzag pattern.

Unlike *P. pholadiformis* (15.43), this species is a nestler, rather than an active borer, and merely occupies or at best enlarges the holes it finds for itself. The young stages are attached by a byssus, but the adults are free in the burrows. All stages secrete great quantities of mucus.

See Burnett (1972), Haderlie (1976), and Yonge (1958).

Phylum Mollusca / Class Bivalvia / Subclass Heterodonta / Order Veneroida Family Petricolidae

15.45 **Petricola californiensis** Pilsbry & Lowe, 1932

Fairly common, nestling among mussels on pilings and floats and in empty pholad holes, low intertidal and shallow subti-

dal waters; Playa del Rey (Los Angeles Co.) to Laguna San Ignacio (Baja California).

Shell to 50 mm long, thin, elongate, rounded at both ends, sculptured with fine radial lines most prominent anteriorly and with irregular concentric growth lines; shell white, with purple-brown stains near hinge area and sometimes across dorsum. Little is known of the biology of this clam.

See McLean (1969) and Willett (1931).

Phylum Mollusca / Class Bivalvia / Subclass Heterodonta / Order Veneroida Family Mactridae

15.46 **Mactra californica** Conrad, 1837
California Mactra

Uncommon, buried to depth of 15 cm in intertidal sand flats or sandy mud in bays, also on sandy bottoms offshore; British Columbia to Costa Rica.

Shell to 50 mm long, small, elongate, marked with concentric rings forming a ridge along posterodorsal edge; periostracum fibrous, golden brown; shell white.

This clam is edible but is relatively scarce. Little is known of its biology, but a related east coast species (*M. solidissima*) has been reared through the larval stages in mass laboratory culture.

See Fitch (1953), Loosanoff & Davis (1963), McLean (1969), Strong (1925), and Trueman (1968b).

Phylum Mollusca / Class Bivalvia / Subclass Heterodonta / Order Veneroida Family Mactridae

15.47 **Tresus capax** (Gould, 1850)
(=Schizothaerus capax)   Fat Gaper, Horse Clam

Buried to depths of up to 1 m in fine sandy mud of bays, middle and low intertidal zones; Kodiak Island (Alaska) to central California; common in northwest, uncommon in California.

Shell to 200 mm long, large, oblong, varying from thin and brittle to thick and heavy; posterior end with broad siphonal gape; periostracum dark brown or black in living specimens, shell white or yellow.

The long siphons or "necks" of these large clams extend up to the surface of the substratum in which they are deeply buried. Separate inhalant and exhalant tubes within the neck provide the animal below with a continuous flow of seawater, which brings oxygen and food (suspended diatoms, flagellates, dinoflagellates, and fine detritus) and removes wastes. Both the siphon of the clam and the body within the shell provide good eating, and the clams are commonly dug by man. They are also sought as food by the moon snail *Polinices lewisii* (13.62), the Dungeness crab, *Cancer magister* (25.20), and the sea star *Pisaster brevispinus* (8.15); thus adult mortality is very high in some clam beds.

Two species of commensal pinnotherid crabs (never both present in the same host at the same time) inhabit the mantle cavity of the fat gaper: *Pinnixa faba* (25.35) and *P. littoralis* (25.36). They cause little harm to the clam. The small male crabs move about, but the larger females remain sheltered by a fringe of tissue (the visceral skirt) attached to the clam's visceral mass. It is from this skirt, rather than from the clam's more delicate gills or palps, that the crabs scrape a share of the plankton, entrapped in mucus, that the clam captured for itself.

In Humboldt Bay the fat gaper spawns in the winter at the time of lowest water temperature, and young, newly settled clams are found in the spring. Growth occurs mainly during late spring and summer when planktonic food is most plentiful, and reserves are stored as glycogen and fat. The glycogen stored in the gonads is depleted during the winter months. Fats, stored in the digestive diverticula, are called upon for energy only after several months of food scarcity.

See Bourne & Smith (1972a,b), Fitch (1953), Keen (1962), Machell & DeMartini (1971), Pearce (1965), Quayle (1973), Reid (1968, 1969), Stout (1967), Swan & Finucane (1951), and Wendell et al. (1976).

Phylum Mollusca / Class Bivalvia / Subclass Heterodonta / Order Veneroida Family Mactridae

15.48 Tresus nuttallii (Conrad, 1837)
(=Schizothaerus nuttallii)   Gaper

Common, burrowing to depths of 1 m or more in firm to loose sandy mud, low intertidal zone in bays and on sheltered bottoms offshore to 30 m depth; Strait of Georgia (British Columbia) to Scammon Lagoon (Baja California).

Shell to 200 mm long, large, oblong, with broad siphonal gape at posterior end, yellowish, similar to *T. capax* (15.47) but proportionately longer and with the ventral margin more smoothly curved; periostracum brown, flaking off readily; inhalant and exhalant siphons fused, covered by periostracum, not retractable into shell.

These animals are easily located on mud flats by the jets of water emitted by the siphons when nearby footsteps disturb the clams. Young animals inhabit the upper layers of the substratum and they can burrow actively. Older individuals live deeper, and those over 60 mm long are very slow burrowers. However, the larger animals tend more and more to orient their shells at an angle to the siphons and thus become more firmly anchored basally. In Puget Sound (Washington), where *T. nuttallii* and *T. capax* (15.47) occur together, *T. nuttallii* burrows more deeply and thus better avoids the temporary freezing that sometimes occurs.

The tip of the elongated siphon is protected by two leatherlike siphonal plates, which can be drawn together tightly to seal the openings. These plates provide a hard substratum in an environment (the surface of mud flats) where hard substrata are scarce. Some 50 species, representing 10 animal phyla and several different plant groups, have been found growing on the plates, and in fact the clams can often be located in shallow water by the tufts of plants growing on the siphonal plates.

In central California, spawning takes place mainly when water temperatures are lowest, from February to April, but some spawning occurs throughout the year, and newly settled young can be found in any month. Clams 4 mm long grow at a rate of 0.25 mm per day. One year after settling young clams average 50 mm in length, and after 2 years the females average 70 mm and have mature ova in the gonads.

The gaper clam is edible, both the siphons and other soft parts being used as food, and it is heavily fished during low spring tides on many California mud flats. It is eaten, too, by the same invertebrate predators that eat the fat gaper (15.47). The sea star *Pisaster brevispinus* (8.15) captures buried gapers by extending the central tube feet down into the sediment for distances about equal to the radius of the sea star, then lifting the clam up to its mouth. Small clams are preyed on by the starry flounder.

The gaper is often heavily infected with larval tapeworms (*Echeneibothrium* sp.), which occur in cysts embedded in the flesh. As many as 140 of these cysts have been found in a single clam from Elkhorn Slough (Monterey Co.). They are harmless to man. The host of the adult tapeworm is the bat stingray *Myliobatus californicus*.

See Addicott (1963), Cheng (1967), P. Clark (1972), P. Clark, Nybakken & Laurent (1975), Edmondson (1920a), Fitch (1953), Keen (1962), Laurent (1971), MacGinitie (1935, 1941), MacGinitie & MacGinitie (1968), Marriage (1954), Pearce (1965), Pohlo (1964), Quayle (1973), Seapy & Kitting (1978), L. Smith & Davis (1965), Stout (1970), Swan & Finucane (1951), Van Veldhuizen & Phillips (1978), and Wourms (1968).

Phylum Mollusca / Class Bivalvia / Subclass Heterodonta / Order Veneroida
Family Tellinidae

15.49 **Tellina bodegensis** Hinds, 1845
Bodega Tellen

Scarce to common, on sand flats and beaches, open coast and especially near entrances to bays, low intertidal zone and offshore to 96 m depth; Queen Charlotte Islands (British Columbia) to Bahía Magdalena (Baja California).

Shell to 60 mm long, flat, rounded at anterior end, pointed or blunt at posterior end, thick for its size, glossy, with fine concentric ribs; periostracum thin, retained at shell margin; shell cream-colored.

This clam normally lies on its left side in its burrow a few centimeters down in the substratum, while the two separate siphons are extended upward to the overlying water. The inhalant siphon is much longer than the exhalant and is highly mobile; it moves about and sucks up fine detrital particles from the sand surface. If the siphons are disturbed, portions of them may be autotomized. Animals in sand flats exposed at low tides for prolonged periods may come to the surface.

In a related species from the north Atlantic (*T. tenuis*) the mechanism of burrowing has been investigated. The foot is fully extended into the substratum, then dilates at the tip to form an anchor. The pedal retractor muscles then pull the shell downward, the anterior retractor contracting slightly ahead of the posterior retractor, so that the shell descends with a rocking motion that slices into the sand effectively. Oblique ridges on the shell grip the sand during these shell-rocking movements. The Atlantic species exhibits no growth during winter months in areas off Scotland, and in the summer grows less at the lowest tidal levels than farther up the beach.

The tellinids, and most other bivalves, have a layer of calcium carbonate just under the periostracum that has, for each species, a distinct surface pattern under the electron microscope. These species-specific patterns can be used to identify shell fragments and are useful in determining taxonomic affinities.

Although the Bodega tellen is edible and has an excellent flavor, it is not often collected because of its scarcity.

See Coan (1971), Dinamani (1967), Fitch (1953), Hamilton (1969), MacGinitie (1935), Maurer (1967a,b), Oberling (1964), Quayle (1973), Stanley (1969), Stephens (1929), Trueman (1966a), and Trueman, Brand & Davis (1966).

Phylum Mollusca / Class Bivalvia / Subclass Heterodonta / Order Veneroida
Family Tellinidae

15.50 **Macoma nasuta** (Conrad, 1837)
Bent-Nosed Clam

Common in gravel, sand, mud, or muddy clay, 10–20 cm below surface of substratum, intertidal zone to 50 m depth offshore, mainly in sheltered bays; Kodiak Island (Alaska) to Cabo San Lucas (Baja California).

Shell usually to 60 mm long, occasionally to 110 mm, thin, flattened, bent toward right side at posterior end; external hinge ligament relatively long; shell white; periostracum grayish or brown; siphons separate, white when extended, orange when contracted.

Burrowing with the foot is aided by an anterior-posterior rocking motion of the shell, which helps the animal cut into the substratum. When buried, the clam lies on its left side with the shell somewhat tilted. At high tide the inhalant siphon extends up to several centimeters out of the burrow, then curves downward and sweeps the bottom surface like a vacuum cleaner, sucking up fine detritus and also a great deal of mud and sand. Occasionally the flow is reversed to expel sand. The exhalant siphon, kept just below the surface of the substratum, creates a small fountain of water and sand. Undigested particles are compacted into platter-shaped fecal pellets too large to be easily picked up and reingested; an average-sized clam may produce roughly a million of these each year, amounting to some 200 cm³ of compacted fine particles, evidence that the surface detritus is heavily exploited as a source of food where these clams are thickly distributed.

The bent-nosed clam is reported to spawn in the spring or early summer in Oregon. In central California it apparently breeds the year around; the clam population at Elkhorn Slough (Monterey Co.) exhibits a continuous variation in body size, rather than a series of more or less discrete size groups suggestive of age classes (e.g., 1-year-olds, 2-year-olds, etc.).

The mantle cavity in some animals may house commensal pinnotherid crabs (*Pinnixa faba*, 25.35, or *P. littoralis*, 25.36), and the commensal nemertean worm *Malacobdella grossa* (5.11). The clams themselves are preyed upon by the moon snail *Polinices lewisii* (13.62). In the past they were eaten by California Indians. Relatively few are taken for food today, for the guts are usually filled with mud or sand and are difficult to clean.

See Addicott (1952, 1968), Coan (1971), Edmondson (1920a), Filice (1958), Fitch (1953), Gallucci & Hylleberg (1976), Hinegardner (1974), Hylleberg & Gallucci (1975), Jegla & Greenberg (1968), MacGinitie (1935), Oberling (1964), Pohlo (1969), Quayle (1973), Rae (1979), Reish (1961), Seapy & Kitting (1978), Wicksten (1978), and Yonge (1949).

Phylum Mollusca / Class Bivalvia / Subclass Heterodonta / Order Veneroida Family Tellinidae

15.51 **Macoma secta** (Conrad, 1837)
White Sand Clam

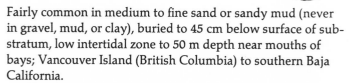

Fairly common in medium to fine sand or sandy mud (never in gravel, mud, or clay), buried to 45 cm below surface of substratum, low intertidal zone to 50 m depth near mouths of bays; Vancouver Island (British Columbia) to southern Baja California.

Shell to 90 mm long, thin, flattened, broadly truncated but not bent toward right side at posterior end, with a diagonal ridge extending from umbo to posteroventral margin; periostracum thin, gray, brownish, or olive; shell white; siphons separate, white when contracted.

Like *M. nasuta* (15.50), this species is mainly a deposit feeder. When buried, it normally lies on its left side. The inhalant and exhalant siphons are usually separated by 3–5 cm where they protrude from the substratum. The oral siphon sucks up organic detritus deposited on the sand surface. Along with the food, quantities of sandy mud are ingested, and the gut is often so full of soil particles that one investigator compared the clam with an earthworm. The absence of *M. secta* in areas of very fine sediments may be related to the fact that mud passed through the gut is not compacted into fecal pellets and can be picked up again by the inhalant siphon.

A pinnotherid crab (see Fig. 25.36b) or the commensal nemertean *Malacobdella grossa* (5.11) sometimes occurs in the mantle cavity.

Macoma secta lives so deeply buried that it appears to have few predators. Its sand-filled gut makes it unattractive for human consumption. In laboratory experiments it is selectively taken as food by the moon snail *Polinices lewisii* (13.62), but in nature the two species are rarely found together. Marine biology classes will find this an excellent clam for dissection.

See Addicott (1952, 1968), Coan (1971), Fitch (1953), Hinegardner (1974), Hopper (1972), Kellogg (1915), Oberling (1964), Pohlo (1969), Quayle (1973), Rae (1979), Reid (1971), Reid & Dunnill (1969), Reid & Reid (1969), Seapy & Kitting (1978), Seilacher (1972), and Yonge (1949).

Phylum Mollusca / Class Bivalvia / Subclass Heterodonta / Order Veneroida
Family Tellinidae

15.52 Macoma balthica (Linnaeus, 1758)
(=M. inconspicua)   Baltic Macoma

Common in mud or silt, middle and low intertidal zone of bays and estuaries, and offshore to 37 m depth; Bering Sea to San Francisco Bay, sporadic records south to San Diego; circumpolar in the Arctic.

Shell to 30 mm long, rarely 45 mm, oval, compressed, smooth, with inconspicuous concentric lines, dull white with pink tinge.

This species may have been introduced to California from farther north along the coast. It is now abundant on intertidal mud flats in San Francisco Bay, where it is found between the surface of the substratum and a depth of 20 cm. It feeds by sweeping the extended siphons over the surface of the mud and sucking up detritus. When the oxygen supply of the surrounding water drops below a critical level, the clam may move up to the surface and expose its mantle margins to the water. On the east coast this species often suffers extensive mortality from neoplasms (abnormal growths) caused by disease organisms.

See Brafield (1963), Brafield & Newell (1961), Christensen, Farley, & Kern (1974), Hinegardner (1974), Newell (1965), Nichols (1977) Quayle (1973), Vassallo (1969, 1971), Wicksten (1978) and Yonge (1949).

Phylum Mollusca / Class Bivalvia / Subclass Heterodonta / Order Veneroida
Family Tellinidae

15.53 Leporimetis obesa (Deshayes, 1855)
Yellow Metis

Small specimens fairly common in low intertidal zone in sand or gravel in bays, larger animals more common offshore to 46 m depth; Point Conception (Santa Barbara Co.) to Bahía Magdalena (Baja California).

Shell to 70 mm long, occasionally to 100 mm, moderately convex, the posterior slope set off by a flexure that is stronger

in right valve; periostracum brown, usually eroded except at margins; shell yellowish, with well-marked growth lines, often of different colors.

This clam has also been called *Florimetis obesa*, *Metis alba*, and *Apolymetis biangulata*.

Little is known of its biology. It is edible but not often taken for food.

See Coan (1971), Fitch (1953), McLean (1969), and Seapy & Kitting (1978).

Phylum Mollusca / Class Bivalvia / Subclass Heterodonta / Order Veneroida
Family Donacidae

15.54 Donax gouldii Dall, 1921
Bean Clam

Abundant in some years, shallowly buried in exposed sandy beaches, middle intertidal zone to 30 m depth on open coast; Santa Cruz Co. to southern Baja California; not common north of Point Conception (Santa Barbara Co.).

Shell to 25 mm long, triangular in shape, with beaks closest to posterior end, sculptured with low radial ribs forming crenulations at margin and with clearly marked concentric growth lines; periostracum thin, transparent; shell color variable, usually buff or yellowish, sometimes with darker rays, highly polished.

The clams tend to remain at a relatively fixed intertidal level on a beach (+75 cm tidal level in Baja California). When waves strike the beach they react by digging deeper; if unearthed by waves they extend the siphons and foot, which provide enough drag to prevent the animals from being carried the full advance of the swash. Related species on the east coast (*D. variabilis*, *D. fossor*) emerge from the sand and let the waves carry them up or down the beach so that they remain in the swash zone regardless of the tidal level. Clams of the species *D. variabilis* respond to acoustical shock, such as wave impact or even the striking of the sand with a shovel, by popping up out of the sand. They rapidly dig into very wet sand between waves.

The population of *D. gouldii* on any one beach may vary

enormously from year to year. At La Jolla (San Diego Co.), resurgent populations appear at irregular intervals of from 2 to 14 years. After a year or more of scarcity, suddenly billions of tiny clams settle from swarms of pelagic larvae originating elsewhere. One year after arrival, the clams, now averaging 12 mm long, may number up to 20,000 per m², and form a band 2–5 m wide for up to 8 km along the shore. Clams are so crowded that they cannot dig properly or get enough suspended plankton for nourishment, so mortality is high and usually within 3 years most are dead. Even on an uncrowded beach the life span is rarely more than 3 years. Females mature after 1 year. They may produce 50,000 or more yellowish eggs at each spawning, and may spawn several times a year.

The posterior end of the shell often bears tufts of the hydroid *Clytia bakeri* (3.9), which rise above the sand and betray the presence of the buried clam. Some clams harbor the larvae of flukes that in the adult stage may have fishes as definitive hosts. Moon snails, gulls, and rays are the main predators on *Donax*; man takes some, too, for these clams make an excellent broth.

See Coan (1973c), Coe (1953, 1955, 1956), Fitch (1953, 1960), Irwin (1973), Jegla & Greenberg (1968), P. Johnson (1966), H. Loesch (1957), Oberling (1964), Pohlo (1967), Strong (1924), Tiffany (1971), Trueman (1968a), Wade (1967), and Yonge (1949).

Phylum Mollusca / Class Bivalvia / Subclass Heterodonta / Order Veneroida
Family Donacidae

15.55 Donax californicus (Conrad, 1837)
Wedge Clam

Common in sandy beaches, low intertidal zone of bays and protected outer coast; Santa Barbara to Bahía Magdalena (Baja California).

Shell to 25 mm long, with beaks only slightly posterior of center, relatively smooth, faintly sculptured, finely crenulate at interior of margins; periostracum heavy, yellowish; shell exterior buff in color, the interior white and stained purple near hinge; siphons short, separate.

This small clam is not a very active burrower and lives near the surface in localities not exposed to pounding surf. Like

Donax gouldii (15.54), it may be extremely common on a particular beach one year, then rare for a number of years. It is sometimes collected by fishermen to make clam broth.

See Coe (1956) and Fitch (1953).

Phylum Mollusca / Class Bivalvia / Subclass Heterodonta / Order Veneroida
Family Psammobiidae

15.56 Gari californica (Conrad, 1837)
Sunset Clam

In sand or gravel near rocky areas, low intertidal zone and offshore to 50 m depth on open coast and at entrances to bays; living animals not often seen, empty valves common; Shelikof Strait (Alaska) to Bahía Magdalena (Baja California).

Shell to 100 mm long, oval, with narrow but definite posterior gape, smooth except for fine concentric growth lines; periostracum strong, yellowish; shell yellow-white, sometimes with radiating pink or red bands.

The common name "sunset clam" relates to the color pattern of the shell. This species normally burrows to depths of 15–20 cm and has separate siphons that reach the surface of the substratum. Little is known of its biology.

See Coan (1973b), Fitch (1953), McLean (1969), and Quayle (1973).

Phylum Mollusca / Class Bivalvia / Subclass Heterodonta / Order Veneroida
Family Psammobiidae

15.57 Nuttallia nuttallii (Conrad, 1837)
(=Sanguinolaria nuttallii)   Purple Clam

Fairly common, buried 30–40 cm deep in sand or gravel, low intertidal zone along outer coast and in bays with strong tidal currents; Bodega Bay harbor (Sonoma Co.) to Bahía Magdalena (Baja California).

Shell to 90 mm long, smooth; right valve flatter than left; ligaments large, projecting; periostracum closely adherent, shiny, dark; shell exterior grayish white with weak purplish rays; interior of valves purple.

This clam lies deeply buried, resting on its right side, with its two long and separate siphons extending to the surface of the substratum. It is a non-selective suspension feeder. Little else is known of its biology.

See Coan (1964, 1973b), Dinamani (1967), Fitch (1953), MacGinitie (1935), Oberling (1964), Pohlo (1972), and Seapy & Kitting (1978).

Phylum Mollusca / Class Bivalvia / Subclass Heterodonta / Order Veneroida Family Psammobiidae

15.58 **Tagelus californianus** (Conrad, 1837)
California Jackknife Clam

Locally abundant in burrows 10–50 cm deep on sandy mud flats of bays, low intertidal zone; Humboldt Bay to Panama; uncommon north of Monterey Bay.

Shell to 100 mm long, the dorsal and ventral margins nearly parallel; beaks nearly central; valves thin, flat; periostracum brownish gray, usually worn away except at the valve margins; shell grayish white, sometimes with yellow bands.

The jackknife clam moves up and down in a permanent burrow. When feeding, it occupies the upper end, about 10 cm below the substratum surface, and extends the two siphons up into the water through separate openings. It moves quickly down the burrow when disturbed.

Although the results of feeding studies are somewhat contradictory, the jackknife clam seems to be a suspension feeder, drawing water into the mantle cavity and filtering out suspended organic particles and planktonic organisms.

The species is edible, but in California it is more often used for fish bait.

See Coan (1973b), Fitch (1953), Hinegardner (1974) Kellogg (1915), Pohlo (1966, 1969), Seapy & Kitting (1978), and Yonge (1952a).

Phylum Mollusca / Class Bivalvia / Subclass Heterodonta / Order Veneroida Family Psammobiidae

15.59 **Tagelus subteres** (Conrad, 1837)

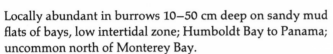

Locally common, burrowed in sand or mud, low intertidal zone in bays; Morro Bay (San Luis Obispo Co.) to Laguna San Ignacio (Baja California).

Shell to 50 mm long, similar to *T. californianus* (15.58) but smaller; periostracum olive to brown, glossy except in posterior area where it is thick and in layers; shell exterior white with violet rays, the interior stained with violet.

Like *T. californianus* (15.58), this clam moves up and down in a smooth-lined permanent burrow. Little is known of its biology.

See Bloomer (1907), Ghosh (1920), Jegla & Greenberg (1968), and McLean (1969).

Phylum Mollusca / Class Bivalvia / Subclass Heterodonta / Order Veneroida Family Semelidae

15.60 **Semele rupicola** Dall, 1915
Semele-of-the-Rocks

Fairly common, nestling in crevices, debris, and mussel beds, low intertidal and shallow subtidal zones; Santa Cruz to Cabo San Lucas (Baja California).

Shell to 45 mm long, rounded or oval, sometimes distorted owing to nestling habit, sculptured with irregular concentric lines crossed by radial riblets; shell exterior yellowish white, the interior with red or orange area on hinge and margin.

Very little is known of the biology of this species.

See Coan (1973a), Dall (1915), and Oberling (1964).

Phylum Mollusca / Class Bivalvia / Subclass Heterodonta / Order Veneroida
Family Semelidae

15.61 **Semele decisa** (Conrad, 1837)
Clipped Semele

Uncommon, in coarse sand or gravel of outer coast, low intertidal zone to 30 m depth; Santa Barbara Co. to Cabo San
Lucas (Baja California).

Shell to 90 mm long, thick and stout sculptured with deep,
unequal and somewhat irregular concentric ridges and
grooves crossed by fine irregular radial grooves; shell exterior
grayish white with purple in grooves, the interior white with
pink or purple stain near hinge and around margins.

The biology of this species has not been investigated.

See Dall (1915), Fitch (1953), and Oberling (1964).

Phylum Mollusca / Class Bivalvia / Subclass Heterodonta / Order Veneroida
Family Solenidae

15.62 **Solen rosaceus** Carpenter, 1864
Rosy Razor Clam

Uncommon, in burrows to 30 cm deep, low intertidal zone in
sandy mud in protected bays; Humboldt Bay to Mazatlan
(Mexico).

Shell to 75 mm long, thin, fragile, elongate, tubular, with
dorsal margin straight and beaks very close to anterior end;
anterior margin of shell bluntly rounded, often with foot protruding; posterior end tapered and rounded, often with fused
siphons projecting; periostracum thin, glasslike; shell rosy or
gray-white, pale pink at beaks.

This clam occupies a permanent burrow. When undisturbed and covered by the tide, it feeds by projecting the annulated siphon up through the substratum to the water above
and drawing down water bearing plankton and suspended organic detritus. If the siphon is injured, it may be shed and
later regenerated. The animal moves rapidly down the burrow when disturbed. When unearthed, it digs in by pushing
the pointed muscular foot into the substratum, dilating the tip
of the foot to serve as an anchor, and finally pulling the shell

downward. This cycle of activity is repeated until the animal
is reburied.

See Fitch (1953), Pohlo (1963), Stasek (1963a), and Trueman
(1966b).

Phylum Mollusca / Class Bivalvia / Subclass Heterodonta / Order Veneroida
Family Solenidae

15.63 **Solen sicarius** Gould, 1850
Sickle Razor Clam

Uncommon, burrowing in firm sediments in sheltered bays,
especially in beds of the broad-leafed eelgrass *Zostera*, low intertidal and shallow subtidal waters; Vancouver Island (British Columbia) to Bahía de San Quintín (Baja California).

Shell to 125 mm long, elongate more or less cylindrical,
gaping at both ends, with dorsal margin slightly concave and
beaks near anterior end; anterior margin of shell somewhat
blunt, often with dark foot protruding; posterior margin
rounded, often with siphons extended; periostracum greenish
yellow or brownish; shiny, shell white.

This clam forms a permanent vertical burrow 30–35 cm
deep, in which it can move up and down rapidly. It can dig
deeper when disturbed, and if unearthed and exposed on the
surface can bury itself again in 30 seconds. The siphons are
joined and can be autotomized in sections if injured. The
clam can leap to a height of several centimeters, or "swim" by
jetting water from the siphons or through the anterior gape
around the foot. The anterior jetting of water is doubtless
used primarily to soften the soil ahead of the animal for more
rapid digging.

See Drew (1907), Fitch (1953), MacGinitie (1935), Pohlo (1963),
and Trueman (1966b).

*Phylum Mollusca / Class Bivalvia / Subclass Heterodonta / Order Veneroida
Family Solenidae*

15.64 Siliqua lucida (Conrad, 1837)
Bright Razor Clam

Uncommon, in burrows on exposed sandy beaches and in loose muddy sand in bays, low intertidal zone and subtidal to 50 m; Tomales Bay (Marin Co.) to Bahía de Todos Santos (Baja California).

Shell to 45 mm long, elongate, thin, rounded at ends, each valve with a prominent vertical internal rib descending from beak; shell whitish, with wide, concentric, brown or yellowish bands and with wide whitish rays extending from beaks to margin; siphons fused.

This small clam is an active digger, but little is known about its biology. It is preyed upon by the starry flounder.

See Hertlein (1961), MacGinitie (1935), and Orcutt (1950).

*Phylum Mollusca / Class Bivalvia / Subclass Heterodonta / Order Veneroida
Family Solenidae*

15.65 Siliqua patula (Dixon, 1789)
Northern Razor Clam

Common in sand on open, flat beaches receiving strong wave action, low intertidal zone to shallow subtidal depths; Alaska to Pismo Beach (San Luis Obispo Co.).

Shell to 170 mm long, thin, rounded at ends, smooth, with beaks at anterior third, each valve with a prominent vertical internal rib descending from beak to margin; valves gaping all around except at hinge; periostracum strong, heavy, shiny, yellow; shell exterior white, interior pinkish.

In contrast to *Solen* (e.g., 15.62, 15.63), this razor clam forms no permanent burrow, but lives in shifting sand on surf-swept beaches. It is an active digger, with a muscular flap on the foot that acts as an anchor. A clam exposed on the surface extends its foot outward the length of the shell, buries it in one quick thrust, and proceeds downward very rapidly. In the sand the animal orients vertically with the two short siphons (fused except at their tips) extending into the water or air above. When disturbed, even by heavy footsteps on the sand, the animal withdraws the siphons and digs deeper into the sand.

Spawning has been observed in Washington when the water warms to 13°C, from mid-May to mid-June. The larvae swim sporadically for about 8 weeks, but spend most of their time resting on soft bottom sediments. After metamorphosis, the clams in Washington grow to an average length of 20 mm after 1 year, 130 mm after 5 years, and 160 mm after 13 years. Growth studies are possible because the shell shows clear annual rings. In Alaska, *S. patula* up to 19 years old have been found, but they grow very little after about 15 years. On beaches in California, the largest animals found are about 9 years old.

Some clams harbor the commensal nemertean *Malacobdella grossa* (5.11) in the mantle cavity.

The northern razor clam is highly regarded as food. It is taken by sport fishermen in California, and by commercial diggers from Alaska to Oregon. The starry flounder also includes razor clams in its diet.

See Fitch (1953), Fraser (1930), Hertlein (1961), Hirschhorn (1962), McMillin (1924), McMillin & Weymouth (1935), Orcutt (1950), Pohlo (1963), Quayle (1973), Ricketts & Calvin (1968), C. Taylor (1959), Weymouth, McMillin & Holmes (1925, 1931), Weymouth, McMillin & Rich (1931), and Yonge (1952a).

*Phylum Mollusca / Class Bivalvia / Subclass Heterodonta / Order Myoida
Family Myidae*

15.66 Cryptomya californica (Conrad, 1837)
False Mya

Fairly common, buried in sand or mud to depths of up to 50 cm, low intertidal zone in bays and lagoons; subtidal along open coast; Chicagof Island (Alaska) to northern Peru.

Shell to 30 mm long, smooth except for fine concentric sculpturing, resembling that of *Mya arenaria* (15.67) but smaller; valves gaping slightly at siphonal end; periostracum thin, brown; shell white.

The short siphons of this clam project, not to the surface of the mud, but into the tunnels of the burrowing shrimps *Upo-*

gebia pugettensis (24.1) and *Callianassa californiensis* (24.2) and the echiurid worm *Urechis caupo* (19.7). In *Upogebia* and *Urechis* tunnels the clam is competing with the burrow owners for suspended food particles, and perhaps to compensate for this the clam has extra-large gills, providing an unusually great water-filtering surface. The clam is relatively mobile, and laboratory observations show that it often changes its feeding sites.

See MacGinitie (1935), Quayle (1973), Wicksten (1978), and Yonge (1951a).

Phylum Mollusca / Class Bivalvia / Subclass Heterodonta / Order Myoida
Family Myidae

15.67 Mya arenaria Linnaeus, 1758
Soft-Shelled Clam

Fairly common, 25 cm or more below surface in thick, dark mud, middle to low intertidal zone in the upper reaches of bays, including brackish areas; southern Alaska to Elkhorn Slough (Monterey Co.); both coasts of Atlantic.

Shell to 150 mm long, soft, easily broken; valves oval, rounded at anterior end, somewhat pointed at posterior end, with gape at each end and with uneven concentric sculpturing; left valve internally with a prominent, triangular, spoon-shaped projection in hinge area; periostracum yellow or brown; shell white or gray; siphons dark.

This common east coast clam may have been introduced to California with the eastern oyster. It was first noted in San Francisco Bay in 1874 and has since spread northward; it reached Alaska in the late 1950's. It was formerly cultivated for food in San Francisco Bay, where it grows optimally at 30 cm above the zero tide level on mud flats that often have salinities as low as 23 percent that of pure seawater.

In New England, *Mya* spawns from June through August, and in the laboratory can be induced to spawn at other times. The eggs are 60–80 μm in diameter. Methods for mass culture of the larvae have been developed.

On the east coast and in Europe, *M. arenaria* has been studied extensively with regard to ecology, growth, burrowing

mechanics, digestive physiology, and pathology. The animals can survive anaerobic conditions for several days, and under these conditions the shell serves as an alkaline reserve to neutralize the lactic acid produced by anaerobic glycolysis. Studies of the ultrastructure of the crystalline style show that it consists in good part of small-bore microtubular elements; animals can survive extraction of the style and regenerate a new one in about 74 days.

See Bonnot (1932, 1949a), Breum (1970), Coe & Turner (1938), Collip (1920), Dow & Wallace (1961), Edmondson (1920a,b), Feder & Paul (1973b), Fitch (1953), J. Green (1964), Gross (1967), Hanna (1966), Hinegardner (1974), Kelley (1966), Kellogg (1899), Kofoid & Bush (1936), Laursen (1966), Loosanoff & Davis (1963), Loosanoff, Davis & Chanley (1966), MacGinitie (1935), MacNeil (1965), Matthiesson (1961), Newcombe (1936), Newcombe & Kessler (1936), Painter (1966), Pfitzenmeyer & Shuster (1960, 1965), Quayle (1973), Reid (1966, 1968), Ricketts & Calvin (1968), Ropes & Stickney (1965), Swan (1952a,b), Trueman (1954, 1966b), H. Turner, Ayres & Wheeler (1948), Vassallo (1969, 1971), Wicksten (1978), Wourms (1968), and Yonge (1923, 1962).

Phylum Mollusca / Class Bivalvia / Subclass Heterodonta / Order Myoida
Family Myidae

15.68 Platyodon cancellatus (Conrad, 1837)
Checked Borer

Fairly common, boring into hard-packed clay, mudstone, or soft sandstone, middle intertidal zone near bay entrances and in protected areas of outer coast; Queen Charlotte Islands (British Columbia) to San Diego.

Shell to 70 mm long, with regular, fine, concentric sculpturing, rounded at anterior end, truncate and widely gaping at posterior end; periostracum thick at margins and near posterior end of shell; shell whitish; siphons long, united, darkly pigmented, armed with two pointed pads near tips.

This clam lives in burrows to 13 cm deep, which it excavates, at least in part, by alternate contractions of the front and rear adductor muscles, rocking the valves on the hinge fulcrum somewhat after the manner of the shipworms *Teredo* and *Bankia* (e.g., 15.80). The small foot seems to be of little aid in boring. The clam digs directly into the substratum without rotating the body; thus the burrow assumes the shape of the shell, and is unlike the circular bore holes of most boring mollusks. Tentacles around the inhalant siphon tip are numerous and complex and undoubtedly detect and prevent entry of larger objects in the water.

The intertidal mudstone reefs and cliffs near Santa Cruz are often honeycombed with the bore holes of *Platyodon*. This weakens the substratum so that waves can break loose massive chunks, contributing to erosion of the coastline. In Los Angeles harbor, this species has been reported boring into low-grade concrete piles.

See Fitch (1953), Hinegardner (1974), MacGinitie & MacGinitie (1968), Quayle (1973), Ricketts & Calvin (1968), and Yonge (1951b).

Phylum Mollusca / Class Bivalvia / Subclass Heterodonta / Order Myoida Family Hiatellidae

15.69 **Hiatella arctica** (Linnaeus, 1767)
(=Saxicava arctica, ?H. pholadis)   Little Gaper

Common, nestling in cracks and crevices in rocks, in empty pholad holes, and among fouling growth on piles, low intertidal zone and subtidal waters; Arctic Ocean to Panama; both coasts of Atlantic.

Shell to 50 mm long, highly variable in shape, usually conforming to surroundings, sculptured with irregular concentric lines or ridges; valves thick, with beaks near anterior end and with a ridge running diagonally downward from beaks; periostracum yellowish, with a row of prickles; shell chalky white; siphons red-tipped.

Post-larval individuals secrete single, long byssal threads, which allow the baby clams to be dispersed by relatively weak currents. The phenomenon (which occurs also in *Mytilus edulis*, 15.10) is reminiscent of the manner in which some baby spiderlings are dispersed by the wind on land, borne aloft on a gossamer strand of silk.

The young of *H. arctica* settling on hard, creviced rock or among fouling growth attach by the byssus, and remain attached as nestlers throughout life. If settlement occurs on soft homogeneous rock, the animals lose the byssus and bore into

the surface by mechanically rocking the valves and pressing the anterior end of the shell against the rock by expansion of the siphons at the rear. In England these animals severely undercut intertidal limestone outcrops. On the Pacific coast they are common inhabitants of the seawater pipes at marine laboratories.

See Ali (1970), Burnett (1972), Haderlie (1976), Hunter (1949), Lewis (1964), Quayle (1973), Sigurdsson, Titman & Davies (1976), and Yonge (1971).

Phylum Mollusca / Class Bivalvia / Subclass Heterodonta / Order Myoida Family Hiatellidae

15.70 **Panopea generosa** (Gould, 1850)
(=Panope generosa)   Geoduck

In burrows, low intertidal zone and subtidal waters in sandy mud of protected bays; Forrester Island (Alaska) to Scammon Lagoon (Baja California); uncommon in California (occasionally found in Morro Bay, San Luis Obispo Co., and at San Pedro, Los Angeles Co.); common farther north.

Shell to 180 mm long, rounded at anterior end, truncate at posterior end; valves gaping on all sides, the hinge exposing reddish-brown soft parts; periostracum thin, light brown; shell white, with irregular growth lines; siphons fused, forming a dark muscular tube to 1 m or more long, lacking cutaneous flaps at tip.

This magnificent beast is the largest and deepest-dwelling of the California burrowing clams, sometimes found in permanent burrows 1.3 m deep, with the long siphons reaching to the mud surface. It is prized for food though difficult to unearth. Although the animal cannot dig rapidly, it appears to, because the siphons keep retracting as the clam digger works down. The clam cannot withdraw the siphon fully into its shell.

In British Columbia these clams spawn in April and early May. Growth is slow, and the animals may live for 15–16 years and weigh 6 kg or more. The name "geoduck" or "gueduc" appears to be derived from the Nisqually Indians' phrase for "dig deep."

See Fitch (1953), Goodwin (1977), MacGinitie (1935), Marriage (1954), Milne & Milne (1948), Quayle (1973), and Ricketts & Calvin (1968).

Phylum Mollusca / Class Bivalvia / Subclass Heterodonta / Order Myoida
Family Pholadidae

15.71 Barnea subtruncata (Sowerby, 1834)
(=B. pacifica)   Mud Piddock

Uncommon, boring into compact mud or shale, low intertidal and shallow subtidal waters in bays and lagoons; Newport (Oregon) to Chile.

Shell to 70 mm long, the siphons to 300 mm long; shell elongate, cylindrical, gaping at both ends, sculptured with raised concentric and radial ridges on anterior end, nearly smooth on posterior end; single dorsal calcareous accessory plate covering area anterior to beaks; periostrocum thin, brownish; shell white; siphons united, mottled dark red-brown grading to white at tip, with a crown of ten, reddish, unbranched papillae surrounding inhalant siphon.

Barnea is most often observed by divers, who see the long, white-tipped siphons with the characteristic, reddish papillae projecting out from deep burrows in soft shale or mudstone. Although the siphons cannot be retracted into the shell, the animal is sensitive to high-frequency pressure changes and will withdraw the siphons into the burrow in response to a diver's exhaled bubbles of air. These animals can extend their siphons up through 15–20 cm of sand when their burrows get buried by sediments as a result of storms. Little is known of the biology of the species.

See Booth (1972), Fitch (1953), Haderlie et al. (1974), and Kellogg (1915)

Phylum Mollusca / Class Bivalvia / Subclass Heterodonta / Order Myoida
Family Pholadidae

15.72 Zirfaea pilsbryi Lowe, 1931
Rough Piddock

Locally common, burrowing to depths of 50 cm or more in heavy mud, sticky clay, and soft shale, low intertidal zone and subtidal waters in bays, lagoons, and estuaries; Nunivak Island (Alaska) to Bahía Magdalena (Baja California).

Shell to 130 mm long, gaping widely both anteroventrally and posteriorly; valves sturdy, with a groove running from beak to ventral margin; posterior surface smooth except for concentric growth lines; anterior surface with coarse concentric ridges bearing short spines in radial rows; small triangular calcareous accessory plate posterior to beaks; periostracum dark brown; shell grayish white.

These are slow-moving animals that burrow in a manner rather like that of shipworms, to which they are related. The suckerlike foot, extending out from the anterior gape, takes a firm grip on the base of the burrow and retracts, pressing the valves firmly against the substratum. The valves are then rocked by alternate contractions of the posterior and anterior adductor muscles, the dorsal and ventral margins of the shell in the hinge area providing the fulcrum. After each rasping stroke by the rough anterior edges of the valves, the animal rotates slightly in the burrow before the next stroke; this results in a cylindrical burrow. At times the animal moves up in its burrow, extending the siphons 15 cm or more above the mud. Loosened bits of mud or clay are periodically blown out of the inhalant siphon by sudden body contractions. There are a few records on the Pacific coast of boring by *Zirfaea* into wood that was buried in mud.

Subtidally, this clam bores vertically downward into soft shale to average depths of 48 cm. The rocks where it lives may periodically become buried by as much as 30 cm of sand, especially during storms; when this occurs, the animal can elongate its cream-colored siphons, extending them up to the surface of the sediment to contact the water above.

This species feeds on suspended particles and plankton brought in by the inhalant siphon and filtered by very large gills that extend far back into the exposed siphon. *Zirfaea* apparently have life spans of 7–8 years.

See Booth (1972), Emerson (1951), Fitch (1953), Haderlie et al. (1974), Kennedy (1974), Lowe (1931), MacGinitie (1935), Quayle (1973), Ricketts & Calvin (1968), Röder (1977), and R. Turner (1954).

Phylum Mollusca / Class Bivalvia / Subclass Heterodonta / Order Myoida
Family Pholadidae

15.73 Netastoma rostrata (Valenciennes, 1846)
(=Nettastomella rostrata)   Beaked Piddock

Common, boring into shale, low intertidal and adjacent sublittoral waters; British Columbia to Isla Cedros (Baja California).

Shell to 20 mm long, small, delicate, white, characterized by formation of a long, tapering, calcareous extension (siphonoplax) on posterior end, often curved or bent to fit burrow; anterior end widely gaping, bordered by a thin, shell-like extension of valves (callum); posterior slope sculptured with thin, raised, concentric ridges.

Netastoma are often found burrowing along with other pholads, and in Monterey Bay they may be the most abundant pholads in soft shale from shallow subtidal depths. They can also bore into the much harder shale called chert, but not as effectively as *Penitella* (15.76–79) or *Chaceia ovoidea* (15.74). *Netastoma* apparently contribute significantly to the erosion of the exposed vertical walls of submarine canyons, such as the Scripps Submarine Canyon, off La Jolla (San Diego Co.).

See Booth (1972), Burnett (1972), Haderlie (1976), Haderlie et al. (1974), Kennedy (1974), Röder (1977), R. Turner (1954), and Warme, Scanland & Marshall (1971).

Phylum Mollusca / Class Bivalvia / Subclass Heterodonta / Order Myoida
Family Pholadidae

15.74 Chaceia ovoidea (Gould, 1851)
(=Pholadidea ovoidea)   Wart-Necked Piddock

Boring into hard clay and shale on open coast and in exposed bays; uncommon in intertidal zone, commoner subtidally; Santa Cruz to Bahía San Bartolomé (Baja California).

Shell to 115 mm long, white, similar in shape to that of *Zirfaea pilsbryi* (15.72) but more rounded; siphons long and joined except for posterior 5 cm, whitish in color, with flecks of brown or orange chitinous material scattered over surface; exhalant siphon bell-shaped when open; inhalant siphon tube-shaped; both siphon tips a uniform deep mahogany-red.

This species is one of the largest of the pholad family of boring clams, and with the siphon fully extended it can reach a length of almost 1 m. It apparently bores in the same manner as *Zirfaea pilsbryi*, using its suckerlike foot as an anchor and eroding the walls of its burrow with the rough shell valves. In southern Monterey Bay it is most abundant in soft shale below kelp beds. Divers report dense concentrations in areas where the animals can bore horizontally into shale ledges, and holes 60 cm deep can be found that often severely undercut the ledges. After reaching maturity the clam ceases boring and secretes a shell-like plate (callum) that partially closes the anterior gape. Subsequent erosion of the rock on the outside often leaves the animals in burrows only 15–25 cm deep, with their long wrinkled siphons exposed and dangling from the bore holes. The siphons are light-sensitive, and in the beam of an underwater lamp they close and contract into the burrow. The clams also bore downward into shale not periodically covered with sand, but the vertical bore holes are usually shallower than the horizontal ones. Soft shale is not the only substratum attacked; specimens with a shell diameter of 8 cm and an overall shell and extended siphon length of 45 cm have been found, in otherwise typical burrows, in hard, flinty chert. In experimental panels of relatively soft Monterey shale, *Chaceia* bored to a depth of 15 cm in less than 10 months.

See Adegoke (1966), Booth (1972), Burnett (1972), Evans (1967b), Fitch (1953), Haderlie (1976), Haderlie et al. (1974), G. Kennedy (1974), Röder (1977), and R. Turner (1955).

Phylum Mollusca / Class Bivalvia / Subclass Heterodonta / Order Myoida
Family Pholadidae

15.75 Parapholas californica (Conrad, 1837)
Scale-Sided Piddock

Boring into stiff clay, soft sandstone, and shale, on open coast and exposed bays; fairly common in low intertidal zone, more abundant at shallow subtidal depths. Bodega Bay harbor (Sonoma Co.) to Bahía San Bartolomé (Baja California).

Shell length to 150 mm; valves divided into three differently sculptured areas: anterior with close-set concentric ridges and radial ribs, middle with concentric growth lines, posterior slope covered with a series of angled, overlapping, periostracal plates; shell dirty white; siphons united, flat-tipped, white to dark reddish brown, the inhalant siphon of larger diameter.

In southern Monterey Bay, this species is the most obvious and widely distributed pholad in shale under kelp beds. In many areas the animals average 50 per m² and, although they do not bore more deeply than about 30 cm, obviously destroy rocky reefs. Divers often see the calcareous "chimneys" of *Parapholas* extending 2–5 cm above the surface of eroding soft shale. These tubular structures are composed of fine rocky material ejected by the siphons and cemented together; they provide protective housings into which the siphons can withdraw. The siphons can be extended some distance out of the burrow, as in the case of clams observed living in reefs covered with 15 cm of sand. *Parapholas* is most abundant in softer shale, but it can penetrate hard siliceous chert. In southern California it is a major cause of erosion on the rim of the Scripps Submarine Canyon, off La Jolla (San Diego Co.). Mature animals that have ceased to bore and have sealed off the anterior gape with a calcareous callum may nevertheless persist for extended periods. Such a specimen, with a shell length of 11 cm, has lived in an aquarium at the Hopkins Marine Station (Pacific Grove, Monterey Co.) for over 7 years, and is still alive and healthy (1979).

See Booth (1972), Burnett (1972), Fitch (1953), Haderlie (1976), Haderlie et al. (1974), G. Kennedy (1974), Minter (1971), Röder (1977), and R. Turner (1955).

Phylum Mollusca / Class Bivalvia / Subclass Heterodonta / Order Myoida
Family Pholadidae

15.76 Penitella conradi Valenciennes, 1846
(=Pholadidea parva, Pholadidea sagitta, Navea subglobosa)
Little Piddock, Abalone Piddock

Fairly common, boring into abalone and mussel shells and shale, low intertidal and shallow subtidal waters along open coast and exposed bays; Esperanza Inlet (Vancouver Island, British Columbia) to Bahía San Bartolomé (Baja California).

Shell to 33 mm long in rock, to 10 mm in abalone shells, oval, gaping widely anteriorly in immature boring stages, sealed in adult by calcareous plate (callum); flaring extension (siphonoplax) on posterior end of shell, composed of dark-brown periostracum, lined internally with white granular calcareous material; dorsal accessory plate flat, rounded posteriorly, pointed anteriorly; shell white; siphons united, small, white, capable of being retracted within shell.

Penitella conradi bores into the shells of at least four species of Pacific coast abalones, most often the red abalone (*Haliotis rufescens*, 13.1). Individuals are also found in up to 20 percent of the California sea mussels at the lower edge of mussel beds in Oregon. Boring rates of up to 19 μm/day have been recorded. As the pholad bores into an abalone shell from the outside, the abalone secretes more nacreous material inside the boresite, forming a blister pearl. Occasionally the borer completely penetrates an abalone or mussel shell. When this occurs the host generally seals the hole internally with a leathery layer of conchiolin, and the borer closes the anterior gape of its own shell with a callum. There is some evidence that *P. conradi* use a chemical method of eroding calcareous shells and rocks, yet large *P. conradi* are also very common borers into both non-calcareous shale and siliceous chert in the subtidal kelp beds of southern Monterey Bay.

See Booth (1972), Burnett (1972), Cowan (1964), Haderlie (1976), Haderlie et al. (1974), Hansen (1970), G. Kennedy (1974), Meredith (1968), Röder (1977), E. Smith (1969), and R. Turner (1955).

Phylum Mollusca / Class Bivalvia / Subclass Heterodonta / Order Myoida
Family Pholadidae

15.77 **Penitella gabbii** (Tryon, 1863)

Fairly common, boring into stiff clay or shale, low intertidal zone on open coast and in exposed bays; more commonly subtidal; Drier Bay (Alaska) to San Pedro (Los Angeles Co.).

Shell to 55 mm long, globular, gaping widely at anterior end in young animals, closed by a curved calcareous plate (callum) in adult; dorsal calcareous accessory plate with lateral wings; shell resembling that of *P. conradi* (15.76) but lacking posterior flaring extension (siphonoplax); siphons up to twice length of shell, white, with distinct papillae on surface.

This species is found in subtidal shale, including hard siliceous chert, together with *P. conradi*, and is equally abundant. In some shale samples dredged from southern Monterey Bay, the burrows of *P. gabbii* often contained black, sulfurous mud. The animals apparently tolerate anaerobic conditions within the burrow outside the shell; however, they receive oxygenated water within the mantle cavity, for the siphons extend into the well-aerated water outside the burrow.

See Booth (1972), Burnett (1972), Haderlie (1976), Haderlie et al. (1974), G. Kennedy (1974), Röder (1977), and R. Turner (1955).

Phylum Mollusca / Class Bivalvia / Subclass Heterodonta / Order Myoida
Family Pholadidae

15.78 **Penitella penita** (Conrad, 1837)
(=Pholadidea penita)   Flap-Tipped Piddock

Common, boring into stiff clay, sandstone, shale, and concrete, middle and low intertidal zones and subtidal waters to 90 m depth on open coast and in exposed bays; Chirikof Island (Alaska) to Bahía San Bartolomé (Baja California).

Shell to 70 mm long, thin, globular; posterior end with concentric ridges; anterior region roughly sculptured, gaping with foot protruding in young burrowing animals, the gape sealed with a calcareous plate (callum) in mature animals; dorsal hinge covered with a triangular calcareous accessory plate; periostracum brown, continuing as a leathery flap at

posterior end of each valve; shell white; siphons white, smooth.

Immature specimens bore mechanically, like *Zirfaea pilsbryi* (15.72), *Parapholas californica* (15.75), *Chaceia ovoidea* (15.74), and *Bankia setacea* (15.80), using the toothed valves as a rasp. The suckerlike foot grasps the bottom of the burrow and pulls the serrated edges of the valves against the burrow walls, the anterior and posterior adductor muscles then alternately contract, rocking the shell valves around a fulcrum formed dorsally by the hinge and ventrally by the condyle on the central margin of each valve. Abrasion occurs as the posterior adductor contracts and the filelike valves scrape upward. Periodically the animal shifts the position of its foot and rotates the shell a bit so that a circular bore hole is formed. Water currents flush away the eroded material. Animals boring in harder rocks develop larger muscles and heavier shells, and grow more slowly, than those in soft stone. An Oregon population in soft stone in the intertidal zone reached maturity in 3 years; one in rock four times as hard took an estimated 21 years. Boring rates recorded were 4–50 mm per year, with the slowest rates in the hardest rock. At maturity boring ceases, a calcareous dome, or callum, is formed, closing the shell anteriorly, the foot atrophies, and sexual reproduction begins. Mature, non-boring animals may live for many years and survive as long as 5 months in burrows buried in sand. Erosion of the rock surface eventually shortens the burrows, and as the mature animals cannot dig deeper, they are finally exposed and die.

This clam produces planktonic larvae, which swim freely for approximately 2 weeks. After settlement, metamorphosis, and the initiation of burrow excavation, the young pholads form approximately 22 growth bands on the shell per year. Each band seems to represent a period of active boring followed by a period of shell deposition. In small, rapidly growing animals even daily growth increments are often detectable in the shell.

Penitella penita is probably the single most important species contributing to the biological erosion of coastal shale along the Pacific coast of North America north of California. Along the Oregon coast, up to 90 percent of all boring clams in the low intertidal zone are of this species. In California the species is fairly common in intertidal sedimentary rock reefs, but in subtidal shale it is less abundant than *P. conradi* (15.76) and *P. gabbii* (15.77). *P. penita* also occasionally bores into concrete and has been reported in poor-quality concrete encasing wooden piles in the Los Angeles harbor.

See Booth (1972), Burnett (1972), G. Clark (1978), Evans (1965, 1966, 1967a, 1968a,b,c, 1970), Evans & Fisher (1966), Evans & Le-Messurier (1972), Fitch (1953), Haderlie (1976), Haderlie et al. (1974), G. Kennedy (1974), Kofoid et al. (1927), MacGinitie & Mac-Ginitie (1968), Quayle (1973), Ricketts & Calvin (1968), Röder (1977), Snoke & Richards (1956), Springer & Beeman (1960), and R. Turner (1955).

Phylum Mollusca / Class Bivalvia / Subclass Heterodonta / Order Myoida
Family Pholadidae

15.79 **Penitella fitchi** Turner, 1955

Uncommon, boring into soft shale and hard chert, low inter-tidal zone to 25 m depth in exposed bays; Monterey Bay to Bahía San Bartolomé (Baja California).

Shell to about 50 mm long, 35 mm high, gaping widely at anterior end in young animals, the gape partially closed in adults by heavy, solid calcareous plate (callum); anterior area, not enclosed by callum, broad, oval, covered by heavy gray-brown periostracum, with only minute pore remaining in adult; dorsal accessory plate heavy, more or less triangular in outline, rounded posteriorly, pointed anteriorly; flexible ex-tension (siphonoplax) on posterior end of shell composed of numerous leaflike layers of gray-brown periostracum, the outer leaves largest; shell white; siphons united, white, smooth, capable of complete retraction into shell.

Nothing is known about the biology of this species. For years it was reported only from Bahía San Bartolomé, but re-cently it has been found in Monterey Bay boring into hard subtidal chert along with *Penitella conradi* (15.76), *P. gabbii* (15.77), and *Chaceia ovoidea* (15.74).

See Haderlie (1979), Kennedy (1974), and R. Turner (1955).

Phylum Mollusca / Class Bivalvia / Subclass Heterodonta / Order Myoida
Family Teredinidae

15.80 **Bankia setacea** (Tryon, 1863)
Shipworm

Common, boring into wharf pilings, ship hulls, and other floating and submerged (but not buried) wood, low intertidal zone and subtidal depths to 70 m or more along open coast and in bays where salinity is above 50 percent that of pure seawater; Bering Sea to southern Baja California.

Shell to 20 mm in diameter, widely gaping at each end; valves with articulating condyles on ventral margins; valve exterior exhibiting three distinctive regions, the anterior one bearing sharp cutting teeth; shell white; body wormlike, to 1 m long, elongate, projecting posteriorly from shell, naked, whitish, the posterior end with separate retractile siphons and carrying two featherlike calcareous processes (pallets).

This animal and its relatives in the family Teredinidae (the shipworms or pileworms) are the most highly modified of the boring clams. The pelagic larvae must settle on wood or they will perish. Settling larvae are only 0.25 mm long and at that stage appear as typical, symmetrical little clams. The shell shape changes markedly in post-larval growth; the shell be-comes an effective drilling tool, the animal begins to excavate a tunnel, and the body elongates behind the shell. This elon-gation is critical, for the siphons must remain at the burrow entrance to ensure a stream of seawater through the body at all times.

While boring, the animal clings to the wood at the inner extremity of its burrow with its rounded, suckerlike foot, pressing the shell against the wood. The anterior toothed parts of the valves are then rasped against the wood by alter-nate contractions of the anterior and posterior adductor mus-cles, which rock the shell valves back and forth over the fulcrum created by the dorsal hinge and ventral condyles of the valves. The effective rasping stroke results from contrac-tion of the enlarged posterior adductor muscle, while contrac-tion of the smaller anterior adductor powers the recovery stroke. As boring proceeds, the animal rotates slowly in the burrow; thus the tunnel excavated is perfectly cylindrical. The clam can only get rid of the wood scrapings by eating them, and as the wood fragments pass through the intestine, some of the wood is digested. Wood, however, does not supply all the clam's nutritional requirements (especially for amino acids), and the animal must supplement its diet by feeding on plankton. The siphons extend out into the water, and an inhalant current brings in food and oxygen. As in most other clams the gills are important in both feeding and respiration, and in the shipworms they extend nearly the full length of the long body. In some species of wood borers, bacteria in the gut fix additional nitrogen, and fungi in the wood may enrich the diet a bit. These are probably of value for animals feeding primarily on wood, which is low in combined nitrogen compounds.

Young boring *Bankia*, after penetrating the wood surface, normally turn and follow the grain of the wood. The burrows of adjacent shipworms parallel one another but do not intersect; each individual somehow detects, and avoids entering, the burrows of others, leaving paper-thin lamellae of wood between them. The ultimate size an individual attains is determined by the amount of wood available and by crowding. When individuals reach mature size, or cannot grow and lengthen the burrow because of crowding, the mantle deposits a thin calcareous lining to the burrow. Infested wood where the shipworms have died often exhibits these shell-lined tunnels. When crowded, *Bankia* grow and lengthen their burrows at an average rate of 43 mm per month, but when not crowded they can achieve a maximum growth rate of over 74 mm per month. Untreated soft wood is often completely destroyed in less than a year when severely infected with *Bankia*, and harder woods do not last much longer.

All young *B. setacea* develop first into functional males. Later about half of these change into females and produce eggs. The eggs and sperm are released into the sea, where fertilization normally occurs. However, several observers have noted that functional males occasionally insert the exhalant siphon into the inhalant siphon of an adjacent female and discharge sperm, and when this occurs fertilization may occur prior to the release of the eggs. In southern California, *B. setacea* spawns during most months of the year, but not in late summer or early fall. Farther north spawning occurs only in the colder months of the year. The eggs are 40–60 μm in diameter. Following fertilization a free-swimming blastula develops in 4–5 hours and the trochophore stage is reached in 12–14 hours.

See Carpenter & Culliney (1975), Coe (1941), Durham & Zullo (1961), Grave (1928), Greenfield & Lane (1953), Haderlie (1976), Haderlie & Mellor (1973), Hill & Kofoid (1927), Isham, Moore & Smith (1951), Lane (1961), Liu & Townsley (1968), Liu & Walden (1970), Menzies, Mohr & Wakeman (1963), Miller (1924, 1926), Miller & Boynton (1926), Morton (1978), Quayle (1953, 1955, 1956a,b, 1959, 1973), Röder (1977), Townsley & Lee (1967), Townsley & Richy (1965), Townsley, Richy & Trussell (1966), C. Turner, Ebert & Given (1969), R. Turner (1966), R. Turner & Johnson (1971), Walden, Allen & Trussell (1966), and F. White (1929).

Phylum Mollusca / Class Bivalvia / Subclass Anomalodesmata Order Pholadamyoida / Family Lyonsiidae

15.81 Entodesma saxicola (Baird, 1863)
(=Lyonsia saxicola)   Rock Entodesma

Fairly common, attached by a byssus in crevices, low intertidal zone on rocky shores, or embedded among fouling growth on wharf piles in bays; Aleutian Islands (Alaska) to Cabo Colnett (Baja California).

Shell to 100 mm long, irregular in shape, thin, calcareous; interior iridescent; exterior completely covered with a thick, olive-brown, periostracum continuing across dorsal side and extending beyond gaping posterior end.

The calcareous shell of this species, consisting very largely of the nacreous layer, is so thin and fragile that it shatters easily if the strong periostracum is allowed to dry out. The animal's foot is small, and functions only in byssus formation in the adult.

Entodesma are hermaphroditic; the paired testes lie ventral to the paired ovaries, and each gonad has its own duct. In spawning, the animals alternately release sperm and eggs. The eggs are small and produced in large numbers.

See Kellogg (1915), Quayle (1973), and Yonge (1952b).

Phylum Mollusca / Class Bivalvia / Subclass Anomalodesmata
Order Pholadamyoida / Family Lyonsiidae

15.82 Lyonsia californica Conrad, 1837

Fairly common, shallowly buried in mud or attached by byssus under edges of rocks resting on sand, low intertidal zone and subtidal waters in bays and offshore; Sitka (Alaska) to Laguna Manuela (Baja California).

Shell to 35 mm long, elongate, broadest in beak area (near anterior end); valve exterior white, opalescent, the interior nacreous; periostracum yellowish, thin; siphons separate; mantle often with minute red spots dorsally and posteriorly, visible through shell.

This species is limited in its distribution to areas of soft substrata. The stomach has well-developed areas for sorting ingested material, and large particles or excessive quantities of sand can be rejected. The mantle lobes are largely fused except at the siphons, a small ventral hole, and a modest pedal gape anteroventrally; thus sediments are effectively kept from entering the mantle cavity. The slender foot enables the animal to burrow, but the byssal gland remains active, permitting attachment to stones buried in the mud. The siphons are short, so the clam cannot burrow deeply.

This bivalve, like most of its near relatives, is a simultaneous hermaphrodite. The ovaries are located dorsolaterally, anterior to the heart, and the testes lie below them in the foot. Eggs and sperm are released alternately and discharged through the exhalant siphon. Ripe animals were observed in August at Pacific Grove (Monterey Co.).

See Ansell (1967a), Maurer (1967c, 1969), Narchi (1968), Wicksten (1978) and Yonge (1952b).

Phylum Mollusca / Class Bivalvia / Subclass Anomalodesmata
Order Pholadamyoida / Family Lyonsiidae

15.83 Mytilimeria nuttallii Conrad, 1837
Sea-Bottle Clam

Common, usually embedded in colonies of compound ascidians, especially *Archidistoma psammion* (12.16) and *Cystodytes lobatus* (12.19; both species are shown in Fig. 12.19) in central

California, low intertidal and shallow subtidal waters; Forrester Island (Alaska) to Isla Rondo (Baja California).

Shell to 40 mm long, thin, fragile, globular; periostracum light brown; shell exterior white, the interior lustrous white.

Life history studies are lacking, but apparently the larval clams of this species settle on established ascidian colonies and attach by a byssus. Eventually the clam comes to be surrounded by the ascidian colony, with only a narrow slit left through which the siphons may emerge. Live clams are difficult to detect unless the siphons are extended. However, the ascidian host usually outlives the clam, and in a dead clam the valves gape, making a conspicuous slit (see the photograph).

M. nuttallii is hermaphroditic; eggs and sperm are spawned alternately.

See Kellogg (1915), Oberling (1964), Quayle (1973), and Yonge (1952b).

Literature Cited

Abd-el-Wahab, A., and E. M. Pantelouris. 1957. Synthetic processes in nucleated and non-nucleated parts of *Mytilus* eggs. Exper. Cell Res. 13: 78–82.

Addicott, W. O. 1952. Ecological and natural history studies of the pelecypod genus *Macoma* in Elkhorn Slough, California. Master's thesis, Biological Sciences, Stanford University, Stanford, Calif. 89 pp.

_____. 1963. An unusual occurrence of *Tresus nuttallii* (Conrad, 1837) (Mollusca: Pelecypoda). Veliger 5: 143–45.

_____. 1968. Additional Pacific coast *Malacobdella grossa*. Nautilus 81: 144.

Adegoke, O. S. 1966. Silicified sand-pipes belonging to *Chaceia* (?) (Pholadidae: Martesiinae) from the late Miocene of California. Veliger 9: 233–35.

Ahmed, M., and A. K. Sparks. 1970. Chromosome number and autosomal polymorphism in the marine mussels *Mytilus edulis* and *Mytilus californianus*. Biol. Bull. 138: 1–13.

Aiello, E. L. 1960. Factors affecting ciliary activity on the gill of the mussel *Mytilus edulis*. Physiol. Zool. 33: 120–35.

Aiello, E. L., and G. Guideri. 1964. Nervous control of ciliary activity. Science 146: 1692–93.

Albrecht, P. G. 1921. Chemical studies of several marine molluscs of the Pacific coast. J. Biol. Chem. 45: 395–405.

Ali, R. M. 1970. The influence of suspension density and temperature on the filtration rate of *Hiatella arctica*. Mar. Biol. 6: 291–302.

Amemiya, I. 1929. On the sex change in the Japanese common oyster (*Ostrea gigas*). Proc. Imp. Acad. (Tokyo) 5: 284–86.

Amemiya, I., and O. Amemiya. 1923. Note on the habit of rock-boring molluscs on the coast of central Japan. Proc. Imp. Acad. (Tokyo) 9: 120–23.

Anderson, G. L. 1975. The effects of intertidal height and the parasitic crustacean *Fabia subquadrata* Dana on the nutrition and reproductive capacity of the California sea mussel *Mytilus californianus* Conrad. Veliger 17: 299–306.

Ansell, A. D. 1967a. Burrowing in *Lyonsia norvegica* (Gmelin) (Bivalvia: Lyonsiidae). Proc. Malacol. Soc. London 37: 387–93.

———. 1967b. Leaping and other movements of some cardiid bivalves. Anim. Behav. 15: 421–26.

———. 1970. Boring and burrowing mechanisms in *Petricola pholadiformis* Lamarck. J. Exper. Mar. Biol. Ecol. 4: 211–20.

Armstrong, L. R. 1965. Burrowing limitations in Pelecypoda. Veliger 7: 195–200.

Baginski, R. M., and S. K. Pierce. 1974. Anaerobiosis: The key to high salinity acclimation in molluscs. Amer. Zool. 14: 1260 (abstract).

———. 1977. The time course of intracellular free amino acid accumulation in tissues of *Modiolus demissus* during high salinity adaptation. Comp. Biochem. Physiol. 57A: 407–12.

Baily, J. L. 1950. Locomotion in *Lima*. Nautilus 63: 112–13.

Baird, R. H., and R. E. Drinnan. 1957. The ratio of shell to meat in *Mytilus* as a function of tidal exposure to air. J. Conseil. Internat. Explor. Mer. 22: 329–36.

Ballantine, D., and J. E. Morton. 1956. Filtering, feeding, and digestion in the lamellibranch *Lasaea rubra*. J. Mar. Biol. Assoc. U.K. 35: 241–74.

Barber, V. C., and D. E. Wright. 1969. The fine structure of the eye and optic tentacle of the mollusc *Cardium edule*. J. Ultrastruc. Res. 26: 515–28.

Barrett, E. M. 1963. The California oyster industry. Calif. Dept. Fish & Game, Fish Bull. 123: 1–103.

Bartlett, B. R. 1972. Reproductive ecology of the California sea mussel, *Mytilus californianus* Conrad. Master's thesis, University of Pacific, Stockton, Calif.

Bartsch, Paul. 1931. The west American mollusks of the genus *Acar*. Proc. U.S. Nat. Mus. 80; Art. 9: 1–4.

Baxter, J. L. 1962. The Pismo clam in 1960. Calif. Fish & Game 48: 35–37.

Bayne, B. L. 1963. Responses of *Mytilus edulis* larvae to increases in hydrostatic pressure. Nature 198: 406–7.

———. 1964. Responses of the larvae of *Mytilus edulis* to light and gravity. Oikos 15: 162–74.

———. 1965. Growth and the delay of metamorphosis of the larvae of *Mytilus edulis* (L.). Ophelia 2: 1–47.

———. 1973. Physiological changes in *Mytilus edulis* L. induced by temperature and nutritive stress. J. Mar. Biol. Assoc. U.K. 53: 39–58.

———, ed. 1976. Marine mussels, their ecology and physiology. London: Cambridge University Press. 506 pp.

Bayne, B. L., P. A. Gabbott, and J. Widdows. 1975. Some effects of stress in the adult on the eggs and larvae of *Mytilus edulis* L. J. Mar. Biol. Assoc. U.K. 55: 675–89.

Bayne, B. L., and R. J. Thompson. 1970. Some physiological consequences of keeping *Mytilus edulis* in the laboratory. Helgoländer Wiss. Meeresunters. 20: 526–52.

Bayne, B. L., C. J. Bayne, T. C. Carefoot, and R. J. Thompson. 1976a. The physiological ecology of *Mytilus californianus* Conrad. I. Metabolism and energy balance. Oecologia 22: 211–28.

———. 1976b. The physiological ecology of *Mytilus californianus* Conrad. 2. Adaptations to low oxygen tension and air exposure. Oecologia 22: 229–50.

Belding, D. L. 1911. The life history and growth of the quahaug (*Venus mercenaria*). Rep. Mass. Comm. Fish & Game 1910: 18–128.

Bennett, R., Jr., and H. I. Nokada. 1968. Comparative carbohydrate metabolism of marine molluscs. I. The intermediary metabolism of *Mytilus californianus* and *Haliotis rufescens*. Comp. Biochem. Physiol. 24: 787–97.

Benson, P. H., and D. C. Chivers. 1960. A pycnogonid infestation of *Mytilus californianus*. Veliger 3: 16–18.

Berg, C. J., Jr. 1969. Seasonal gonadal changes of adult oviparous oysters in Tomales Bay, California. Veliger 12: 27–36.

———. 1971. A review of possible causes of mortality of oyster larvae of the genus *Crassostrea* in Tomales Bay, California. Calif. Fish & Game 57: 67–75.

Berg, W. E. 1950. Lytic effects of sperm extracts on the eggs of *Mytilus edulis*. Biol. Bull. 98: 128–38.

———. 1954. Large-scale constriction and segregation of polar lobes from the eggs of *Mytilus edulis*. Exper. Cell Res. 6: 162–71.

Berg, W. E., and P. B. Kutsky. 1951. Physiological studies of differentiation in *Mytilus edulis*. I. The oxygen uptake of isolated blastomeres and polar lobes. Biol. Bull. 101: 47–61.

Berg, W. E., and D. M. Prescott. 1958. Physiological studies of differentiation in *Mytilus edulis*. II. Accumulation of phosphate in isolated blastomeres and polar lobes. Exper. Cell Res. 14: 402–7.

Berkeley, C. 1962. Toxicity of plankton to *Cristispira* inhabiting the crystalline style of a mollusk. Science 135: 664–65.

Bernard, F. R. 1973. Crystalline style formation and function in the oyster *Crassostrea gigas* (Thunberg, 1795). Ophelia 12: 159–70.

———. 1974a. Annual biodeposition and gross energy budget of mature Pacific oysters, *Crassostrea gigas*. J. Fish. Res. Bd. Canada 31: 185–90.

———. 1974b. Particle sorting and labial palp function in the Pacific oyster *Crassostrea gigas* (Thunberg, 1795). Biol. Bull. 146: 1–10.

_____. 1976. Living Chamidae of the Eastern Pacific (Bivalvia: Heterodonta). Los Angeles Co. Mus. Contr. Sci. 278: 1–43.

Berry, S. S. 1907. Molluscan fauna of Monterey Bay, California. Nautilus 21: 17–21.

_____. 1954. On the supposed stenobathic habitat of the California sea mussel, *Mytilus californianus*. Calif. Fish & Game 40: 69–73.

Bloomer, H. W. 1907. On the anatomy of *Tagelus gibbus* and *T. divisus*. Proc. Malacol. Soc. London 7: 215–25.

Boëtius, I. 1962. Temperature and growth in a population of *Mytilus edulis* (L.) from the northern harbour of Copenhagen (the Sound). Medd. Dansk. Fisk, Havund. (n.s.) 3: 339–45.

Bonnot, P. 1932. Soft-shell clam beds in the vicinity of San Francisco Bay. Calif. Fish & Game 18: 64–66.

_____. 1935. The California oyster industry. Calif. Fish & Game 21: 65–80.

_____. 1937. Settling and survival of spat of the Olympia oyster, *Ostrea lurida*, on upper and lower horizontal surfaces. Calif. Fish & Game 23: 224–28.

_____. 1940. The edible bivalves of California. Calif. Fish & Game 26: 212–39.

_____. 1949a. Clams, pp. 161–64, *in* The commercial fish catch of California for the year 1947 with an historical review, 1916–1947. Calif. Div. Fish & Game, Fish Bull. 74.

_____. 1949b. Oysters, pp. 167–70, *in* The commercial fish catch of California for the year 1947 with an historical review, 1916–1947. Calif. Div. Fish & Game, Fish Bull. 74.

Booth, C. E., and C. P. Mangum. 1978. Oxygen uptake and transport in the lamellibranch mollusc *Modiolus demissus*. Physiol. Zool. 51: 17–35.

Booth, G. S. 1972. The ecology and distribution of rock-boring pelecypods off Del Monte Beach, Monterey, California. Master's thesis, Naval Postgraduate School, Monterey, California. 106 pp.

Bouffard, T. 1969. Cholinesterase localization in two bivalve species. Amer. Zool. 9: 1108 (abstract).

Bourne, N., and D. W. Smith. 1972a. The effect of temperature on the larval development of the horse clam, *Tresus capax* (Gould). Nat. Shellfish Assoc. Proc. 62: 35–37.

_____. 1972b. Breeding and growth of the horse clam, *Tresus capax* (Gould), in southern British Columbia, Nat. Shellfish Assoc. Proc. 62: 38–46.

Boyden, C. R., H. Watling, and I. Thornton. 1975. Effect of zinc on the settlement of the oyster *Crassostrea gigas*. Mar. Biol. 31: 227–34.

Bradley, W., and A. E. Siebert, Jr. 1978. Infection of *Ostrea lurida* and *Mytilus edulis* by the parasitic copepod *Mytilicola orientalis* in San Francisco Bay, California. Veliger 21: 131–34.

Brafield, A. E. 1963. The effects of oxygen deficiency on the behaviour of *Macoma balthica* (L). Anim. Behav. 11: 345–46.

Brafield, A. E., and G. E. Newell. 1961. The behavior of *Macoma balthica* (L.). J. Mar. Biol. Assoc. U.K. 41: 81–87.

Bretsky, S. S. 1967. Environmental factors influencing the distribution of *Barbatia domingensis* (Mollusca, Bivalvia) on the Bermuda platform. Postilla, Yale Univ., 108: 1–14.

Breum, O. 1970. Stimulation of burrowing activity by wave action in some marine bivalves. Ophelia 8: 197–207.

Brown, B. E., and R. C. Newell. 1972. The effect of copper and zinc on the metabolism of the mussel *Mytilus edulis*. Mar. Biol. 16: 108–18.

Burke, J. M., J. Marchisotto, J. J. A. McLaughlin, and L. Provasoli. 1960. Analysis of the toxin produced by *Gonyaulax catanella* in axenic culture, pp. 837–42, *in* R. Nigrelli, ed., Biochemistry and pharmacology of the compounds derived from marine organisms. Ann. N.Y. Acad. Sci. 90: 615–950.

Burnett, N. A. 1972. The ecology of the benthic community of bivalve molluscs in the shale at the Monterey sewer outfall. Master's thesis, California State University, San Francisco. 56 pp.

Calabrese, A., and H. C. Davis. 1966. The pH tolerance of embryos and larvae of *Mercenaria mercenaria* and *Crassostrea virginica*. Biol. Bull. 131: 427–36.

_____. 1969. Spawning of the American oyster *Crassostrea virginica* at extreme pH levels. Veliger 11: 235–36.

California. 1962. Oysters cause "mussel poisoning" for first time in California. Calif. Health, State Dept. Pub. Health 20: 35.

Carlisle, J. G., Jr. 1973. Results of 1971 Pismo clam census. Calif. Fish & Game 59: 138–39.

Carlson, A. J. 1905. Comparative physiology of the invertebrate heart. I. Biol Bull. 8: 123–68.

_____. 1905. Comparative physiology of the invertebrate heart. II. Amer. J. Physiol. 13: 396–426.

_____. 1906. Comparative physiology of the invertebrate heart. V. Amer. J. Physiol. 16: 47–66.

Carpenter, E. J., and J. L. Culliney. 1975. Nitrogen fixation in marine shipworms. Science 187: 551–52.

Carr, R. S., and D. J. Reish. 1978. Studies on the *Mytilus edulis* community in Alamitos Bay, California. VII. The influence of water soluble petroleum hydrocarbons on byssal thread formation. Veliger 21: 283–87.

Chan, G. L. 1973. Subtidal mussel beds in Baja California with a new record size for *Mytilus californianus*. Veliger 16: 239–40.

Chapman, W. M., and A. H. Banner. 1949. Contributions to the life history of the Japanese oyster drill (*Tritonalia japonica*) with notes on other enemies of the Olympian oyster (*Ostrea lurida*). Washington Dept. Fish. Biol. Rep. 49A: 167–200.

Cheney, D. P. 1971. A summary of invertebrate leucocyte morphology with emphasis on blood elements of the Manila clam *Tapes semidecussata*. Biol. Bull. 140: 353–68.

Cheng, T. C. 1967. Marine molluscs as hosts for symbiosis with a review of known parasites of commercially important species. Adv. Mar. Biol. 5: 1–424.

Cheng, T. C., and D. A. Foley. 1971. Blood cells of pelecypods. Amer. Zool. 11: 699–700.

Chew, K. K., A. K. Sparks, and S. C. Katkansky. 1964. First record of *Mytilicola orientalis* Mori in the California mussel *Mytilus californianus* Conrad. J. Fish. Res. Bd. Canada 21: 205–7.

Chiba, K., and Y. Ohshima. 1957. Effect of suspended particles on the pumping and feeding of marine bivalves, especially of the Japanese littleneck clam. Bull. Jap. Soc. Sci. Fish. 23: 340–53.

Chipman, W. A., and J. G. Hopkins. 1954. Water filtration by the bay scallop, *Pecten irradians*, as observed with the use of radioactive plankton. Biol. Bull. 107: 80–91.

Chipperfield, P. N. J. 1953. Observations on the breeding and settlement of *Mytilus edulis* (L.) in British waters. J. Mar. Biol. Assoc. U.K. 32: 449–76.

Christensen, D. J., C. A. Farley, and F. G. Kern. 1974. Epizootic neoplasms in the clam *Macoma balthica* (L.) from Chesapeake Bay. Nat. Cancer Inst. 52: 1739–49.

Clark, G. R., II. 1968. Mollusk shell: Daily growth lines. Science 161: 800–802.

———. 1971. The influence of water temperature on the morphology of *Leptopecten latiauratus* (Conrad, 1837). Veliger 13: 269–72.

Clark, G. W. 1978. Rock boring bivalves and associated fauna and flora of the intertidal terrace at Santa Cruz, California. Master's thesis, Naval Postgraduate School, Monterey, California. 78 pp.

Clark, P. 1972. The growth rate of the juvenile gaper clam, *Tresus nuttallii*, of Elkhorn Slough. Echo 5, Abs. & Proc. 5th Ann. Meet., West. Soc. Malacol., p. 24.

Clark, P., J. Nybakken, and L. Laurent. 1975. Aspects of the life history of *Tresus nuttallii* in Elkhorn Slough. Calif. Fish & Game 61: 215–27.

Clay, E. 1965. Literature survey of the common fauna of estuaries, 14. *Mytilus edulis*. Imperial Chemical Industries, U.K., Paint Div. 78 pp.

Coan, E. 1964. The Mollusca of the Santa Barbara County area. I. Pelecypoda and Scaphopoda. Veliger 7: 29–33.

———. 1971. The northwest American Tellinidae. Veliger 14 (Suppl.): 1–63.

———. 1973a. The northwest American Semelidae. Veliger 15: 314–29.

———. 1973b. The northwest American Psammobiidae. Veliger 16: 40–57.

———. 1973c. The northwest American Donacidae. Veliger 16: 130–39.

———. 1977a. Preliminary review of the northwest American Carditidae. Veliger 19: 375–86.

———. 1977b. *Glans carpenteri* vs. *Glans subquadrata*: The rules concerning renamed transient secondary homonyms. Veliger 20: 63.

Coan, E., and J. T. Carlton. 1975. Phylum Mollusca: Bivalvia, pp. 543–78, *in* R. Smith & Carlton (1975).

Coe, W. R. 1930. Life cycle of the California oyster (*Ostrea lurida*). Anat. Rec. 47: 359.

———. 1931a. Sexual rhythm in the California oyster (*Ostrea lurida*). Science 74: 247–49.

———. 1931b. Spermatogenesis in the California oyster (*Ostrea lurida*). Biol. Bull. 61: 309–15.

———. 1932a. Season of attachment and rate of growth of sedentary marine organisms at the pier of the Scripps Institution of Oceanography, La Jolla, California. Bull. Scripps Inst. Oceanogr., Tech. Ser. 3: 37–86.

———. 1932b. Development of the gonads and the sequence of the sexual phases in the California oyster (*Ostrea lurida*). Bull. Scripps Inst. Oceanogr., Tech. Ser. 3: 119–44.

———. 1934. Alternation of sexuality in oysters. Amer. Natur. 48: 236–51.

———. 1941. Sexual phases in wood-boring mollusks. Biol. Bull. 81: 168–76.

———. 1945. Nutrition and growth of the California bay-mussel (*Mytilus edulis diegensis*). J. Exper. Zool. 99: 1–14.

———. 1946. A resurgent population of the California bay-mussel (*Mytilus edulis diegensis*). J. Morphol. 78: 85–104.

———. 1947. Nutrition, growth and sexuality of the Pismo clam, *Tivela stultorum*. J. Exper. Zool. 104: 1–24.

———. 1948. Nutrition, environmental conditions, and growth of marine bivalve mollusks. J. Mar. Res. 7: 586–601.

———. 1953. Resurgent populations of littoral marine invertebrates and their dependence on ocean currents and tidal currents. Ecology 34: 225–29.

———. 1955. Ecology of the bean clam, *Donax gouldi*, on the coast of California. Ecology 36: 512–14.

———. 1956. Fluctuations in populations of littoral marine invertebrates. J. Mar. Res. 15: 212–32.

Coe, W. R., and J. E. Fitch. 1950. Population studies, local growth rates and reproduction of the Pismo clam (*Tivela stultorum*). J. Mar. Res. 9: 188–210.

Coe, W. R., and D. L. Fox. 1942. Biology of the California sea mussel (*Mytilus californianus*). I. Influence of temperature, food supply, sex and age on the rate of growth. J. Exper. Zool. 90: 1–30.

———. 1944. Biology of the California sea mussel (*Mytilus californianus*). III. Environmental conditions and rate of growth. Biol. Bull. 87: 59–72.

Coe, W. R., and H. J. Turner. 1938. Development of the gonads and gametes in the soft-shell clam (*Mya arenaria*). J. Morphol. 62: 91–111.

Cole, H. A., and B. T. Hepper. 1954. The use of neutral red solution for the comparative study of filtration rates of lamellibranchs. J. Conseil. Internat. Explor. Mer. 20: 197–203.

Collip, J. B. 1920. Studies on molluscan coelomic fluid: Anaerobic respiration in *Mya arenaria*. J. Biol. Chem. 45: 23–49.

Cooke, W. J. 1975. The occurrence of an endozoic green alga in the marine mollusc *Clinocardium nuttallii* (Conrad, 1837). Phycologia 14: 35–39.

Costello, D. P., and C. Henley. 1971. Methods for obtaining and handling marine eggs and embryos. 2nd ed. Marine Biological Laboratory, Woods Hole, Mass. 247 pp.

Coulthard, H. S. 1929. Growth of the sea mussel. Contr. Canad. Biol. (n.s.) 4: 123–36.

Courtright, R. C., W. P. Breese, and H. Krueger. 1971. Formulation of a synthetic seawater for bioassays with *Mytilus edulis* embryos. Water Res. 5: 877–88.

Cowan, I. M. 1964. New information on the distribution of marine Mollusca on the coast of British Columbia. Veliger 7: 110–13.

Cox, Ian, ed. 1957. The scallop: Studies of a shell and its influences on humankind. London: "Shell" Transport & Trading. 135 pp.

Craig, G. Y., and A. Hallam. 1963. Size-frequency and growth-ring analyses of *Mytilus edulis* and *Cardium edule*, and their paleontological significance. Paleontology 6: 731–50.

Crane, J. M., L. G. Allen, and C. Eisemann. 1975. Growth rate, distribution, and population density of the northern quahog, *Mercenaria mercenaria*, in Long Beach, California. Calif. Fish & Game 61: 68–81.

Crenshaw, M. A., and N. Watabe. 1969. The muscle attachment to the shell of *Mercenaria mercenaria*. Amer. Zool. 9: 1139 (abstract).

Cronly-Dillon, J. R. 1966. Spectral sensitivity of the scallop *Pecten maximus*. Science 151: 345–46.

Curtis, F. 1966. Molluscan species from early southern California archaeological sites. Bull. So. Calif. Acad. Sci. 65: 107–27.

Dakin, W. S. 1909. Pecten. Liverpool Mar. Biol. Comm. [LMBC] Memoir 17. 136 pp.

Dall, W. H. 1915. Notes on the Semelidae of the west coast of America, including some new species. Proc. Acad. Natur. Sci. Philadelphia 67: 25–28.

————. 1916. Notes on the Californian species of *Adula*. Nautilus 30: 1–3.

Davis, H. C. 1955. Mortality of Olympia oysters at low temperatures. Biol. Bull. 109: 404–6.

Davis, H. C., and H. Hidu. 1969. Effect of turbidity-producing substances in sea water on eggs and larvae of three genera of bivalve molluscs. Veliger 11: 316–23.

Dayton, P. K. 1971. Competition, disturbance, and community organization: The provision and subsequent utilization of space in a rocky intertidal community. Ecol. Monogr. 41: 351–89.

————. 1973. Dispersion, dispersal, and persistence of the annual intertidal alga *Postelsia palmaeformis* Ruprecht. Ecology 54: 433–38.

Dehnel, P. A. 1956. Growth rates in latitudinally and vertically separated populations of *Mytilus californianus*. Biol. Bull. 110: 43–53.

Digby, P. S. B. 1969. Structure and mode of formation of lamellibranch shell. Amer. Zool. 9: 592 (abstract).

Dinamani, P. 1967. Variation in the stomach structure of the Bivalvia. Malacologia 5: 225–68.

Dodd, J. R. 1964. Environmentally controlled variation in the shell structure of a pelecypod species. J. Paleontol. 38: 1065–71.

————. 1966. Diagenetic stability of temperature-sensitive skeletal properties in *Mytilus* from the Pleistocene of California. Bull. Geol. Soc. Amer. 77: 1213–24.

Dow, R. L., and D. E. Wallace. 1961. The soft-shell clam industry of Maine. U.S. Fish & Wildlife Serv., Circ. 110: 1–36.

Dral, A. D. G. 1966. Movement and co-ordination of the latero-frontal cilia of the gill filaments of *Mytilus*. Nature 210: 1170–71.

Drew, G. A. 1907. The habits and movements of the razor-shell clam, *Ensis directus* Con. Biol. Bull. 12: 127–40.

Durham, J. W., and V. A. Zullo. 1961. The genus *Bankia* Gray (Pelecypoda) in the Oligocene of Washington. Veliger 4: 1–4.

Dushane, H. 1966. Erroneous range extension for *Tivela stultorum* (Mawe, 1823). Veliger 9: 86–87.

Ebling, F. J., J. A. Kitching, W. Muntz, and C. M. Taylor. 1964. The ecology of Lough Ine. 13. Experimental observations of the destruction of *Mytilus edulis* and *Nucella lapillus* by crabs. J. Anim. Ecol. 33: 73-83.

Edmondson, C. H. 1919. *Cardium corbis*, a monoecious bivalve. Science 49: 402.

————. 1920a. Edible Mollusca of the Oregon coast. Occas. Pap. Bernice P. Bishop Mus. 7: 179–201.

————. 1920b. The reformation of the crystalline style in *Mya arenaria* after extraction. J. Exper. Zool. 30: 259–91.

Elsey, C. R. 1935. On the structure and function of the mantle and gills of *Ostrea gigas* (Thunberg) and *Ostrea lurida* Carpenter. Trans. Roy. Soc. Canada 79: 131–60.

Emerson, W. K. 1951. An unusual habitat for *Zirfaea pilsbryi*. Bull. So. Calif. Acad. Sci. 50: 89–91.

Engle, J. B., and V. L. Loosanoff. 1944. On season of attachment of larvae of *Mytilus edulis* Linn. Ecology 25: 433–40.

Evans, J. W. 1965. The growth rate of the rock-boring clam *Penitella penita*. Amer. Zool. 5: 645–46.

————. 1966. The ecology of the rock-boring clam *Penitella penita*, Doctoral thesis, University of Oregon, Eugene.

————. 1967a. Relationship between *Penitella penita* (Conrad, 1837) and other organisms of the rocky shore. Veliger 10: 148–51.

————. 1967b. A re-interpretation of the sand-pipes described by Adegoke. Veliger 10: 174–75.

————. 1968a. Factors modifying the morphology of the rock-boring clam *Penitella penita* (Conrad, 1837). Proc. Malacol. Soc. London 38: 111–19.

————. 1968b. The role of *Penitella penita* (Conrad, 1837) (Family Pholadidae) as eroder along the Pacific coast of North America. Ecology 49: 156–59.

————. 1968c. Growth rate of the rock-boring clam *Penitella penita*

(Conrad, 1837) in relation to hardness of rock and other factors. Ecology 49: 619–28.

———. 1970. Sexuality in the rock-boring clam *Penitella penita*. Canad. J. Zool. 48: 625–27.

———. 1975. Growth and micromorphology of two bivalves exhibiting non-daily growth lines, pp. 119–34, *in* G. D. Rosenberg and S. K. Runcorn, eds., Growth rhythms and the history of the earth's rotation. New York: Wiley-Interscience. 560 pp.

Evans, J. W., and D. Fisher. 1966. A new species of *Penitella* (Family Pholadidae) from Coos Bay, Oregon. Veliger 8: 222–24.

Evans, J. W., and M. H. LeMessurier. 1972. Functional morphology and circadian growth of the rock-boring clam *Penitella penita*. Canad. J. Zool. 50: 1251–58.

Fankboner, P. V. 1971. The ciliary currents associated with feeding, digestion, and sediment removal in *Adula* (*Botula*) *falcata* Gould, 1851. Biol. Bull. 140: 28–45.

Fankboner, P. V., W. M. Blaylock, and M. E. de Burgh. 1978. Accumulation of ^{14}C-labelled algal exudate by *Mytilus californianus* Conrad and *Mytilus edulis* Linnaeus, an aspect of interspecific competition. Veliger 21: 276–82.

Farley, C. A. 1969. Probable neoplastic disease of the hematopoietic system in oysters, *Crassostrea virginica* and *Crassostrea gigas*. Nat. Cancer Inst. Monogr. 31: 541–55.

Feder, H. M. 1955. On the methods used by the starfish *Pisaster ochraceus* in opening three types of bivalve molluscs. Ecology 36: 764–66.

———. 1956. Natural history studies on the starfish *Pisaster ochraceus* (Brandt, 1835) in the Monterey Bay area. Doctoral thesis, Biological Sciences, Stanford University, Stanford, Calif. 294 pp.

———. 1959. The food of the starfish *Pisaster ochraceus* along the California coast. Ecology 40: 721–24.

———. 1967. Organisms responsive to predatory sea stars. Sarsia 29: 371–94.

———. 1970. Growth and predation by the ochre sea star, *Pisaster ochraceus* (Brandt), in Monterey Bay, California. Ophelia 8: 161–85.

Feder, H. M., and A. J. Paul. 1973a. Abundance estimations and growth-rate comparisons for the clam *Protothaca staminea* from three beaches in Prince William Sound, Alaska, with additional comments on size-weight relationships, harvesting and marketing. Inst. Mar. Sci., Univ. Alaska, IMS Tech. Rep. R73-3: 1–34.

———. 1973b. Age, growth and size-weight relationships of the soft-shell clam *Mya arenaria* in Prince William Sound, Alaska. Nat. Shellfish Assoc. Proc. 64: 45–52.

Felton, B. H., and E. Aiello. 1971. The relationship between ions and ciliary activity in the gill of *Mytilus edulis*. Amer. Zool. 11: 665 (abstract).

Field, I. A. 1922. Biology and economic value of the sea mussel *Mytilus edulis*. Bull. U.S. Bur. Fish. 38: 125–257.

Filice, F. 1958. Invertebrates from the estuarine portion of San Francisco Bay and some factors influencing their distributions. Wasmann J. Biol. 16: 159–211.

Fisher, L. 1969. An immunological study of pelecypod taxonomy. Veliger 11: 434–38.

Fitch, J. E. 1950. The Pismo clam. Calif. Fish & Game 36: 285–312.

———. 1953. Common marine bivalves of California. Calif. Dept. Fish & Game, Fish Bull. 90: 1–102.

———. 1957. The plight of the Pismo clam. Outdoor Calif. 18: 3–4.

———. 1960. Exploiting the "unexploited." Outdoor Calif. 21: 6–7.

———. 1961. The Pismo clam. Calif. Dept. Fish & Game, Mar. Res. Leaflet 1. 23 pp.

———. 1965. A relatively unexploited population of Pismo clams, *Tivela stultorum* (Mawe, 1823) (Veneridae). Proc. Malacol. Soc. London 36: 309–12.

Fox, D. L., ed. 1936. The habitat and food of the California sea mussel. Bull. Scripps Inst. Oceanogr., Tech. Ser. 4: 1–64.

———. 1941. Changes in the tissue chloride of the California seamussel to response to heterosmotic environments. Biol. Bull. 80: 111–20.

———. 1955. Organic detritus in the metabolism of the sea. Sci. Monthly 80: 256–59.

———. 1960. Perspectives in marine biochemistry, pp. 617–21, *in* R. F. Nigrelli, ed., Biochemistry and pharmacology of compounds derived from marine organisms. Ann. N.Y. Acad. Sci. 90: 615–950.

Fox, D. L., and W. R. Coe. 1943. Biology of the California sea mussel (*Mytilus californianus*). II. Nutrition, metabolism, growth and calcium deposition. J. Exper. Zool. 93: 205–49.

Fox, D. L., H. N. Sverdrup, and J. P. Cunningham. 1937. The rate of water propulsion by the California mussel. Biol. Bull. 72: 417–38.

Fraser, C. M. 1929. The spawning and free-swimming larval periods of *Saxidomus* and *Paphia*. Trans. Roy. Soc. Canada 23: 195–98.

———. 1930. The razor clam, *Siliqua patula* Dixon, of Graham Island, Queen Charlotte Group. Trans. Roy. Soc. Canada 24: 141–54.

———. 1931. Notes on the ecology of the cockle *Cardium corbis* Martyn. Trans. Roy. Soc. Canada 25: 59–72.

Fraser, C. M., and G. M. Smith. 1928a. Notes on the ecology of the butter clam, *Saxidomus giganteus*. Trans. Roy. Soc. Canada 22: 271–86.

———. 1928b. Notes on the ecology of the little neck clam (*Paphia staminea*) Conrad. Trans. Roy. Soc. Canada 22: 249–69.

Frey, H. W., ed. 1971. California living marine resources and their utilization. State of California, Resources Agency, Dept. Fish & Game, Sacramento. 140 pp.

Frizzell, D. L. 1930a. *Pododesmus macroschisma* Deshayes. Nautilus 43: 104.

———. 1930b. The status of *Protothaca tenerrima alta* Waterfall. Nautilus 44: 48–50.

Gallucci, V. F., and J. Hylleberg. 1976. A quantification of some aspects of growth in the deposit-feeding bivalve *Macoma nasuta*. Veliger 19: 59–67.

Galtsoff, P. S. 1964. The American oyster, *Crassostrea virginica* Gmelin. U.S. Fish & Wildlife Serv., Fish Bull. 64: 1–480.

Ghosh, E. 1920. Taxonomic studies on the soft parts of the Solenidae. Rec. Indian Mus. 19: 47–78.

Giese, A. C. 1958. Myoglobin in the adductor of *Tivela*. Anat. Rec. 132: 442 (abstract).

Giese, A. C., and J. S. Pearse, eds. 1979. Reproduction of marine invertebrates. 5. Molluscs: Pelecypods and lesser classes. New York: Academic Press. 369 pp.

Gillilan, W. 1964. Pismo clam survey. Outdoor Calif. 25: 10–11.

Gilmour, T. H. J. 1967. The defensive adaptations of *Lima hians* (Mollusca: Bivalvia). J. Mar. Biol. Assoc. U.K. 47: 209–21.

Glaus, K. J. 1968. Factors influencing the production of byssal threads in *Mytilus edulis*. Biol. Bull. 135: 420.

Glynn, P. W. 1964. *Musculus pygmaeus* spec. nov., a minute mytilid of the high intertidal zone at Monterey Bay, California (Mollusca: Pelecypoda). Veliger 7: 121–28.

———. 1965. Community composition, structure, and interrelationships in the marine intertidal *Endocladia muricata–Balanus glandula* association in Monterey Bay, California. Beaufortia 12: 1–198.

Goodwin, L. 1977. The effects of season on visual and photographic assessment of subtidal geoduck clam (*Panope generosa* Gould) populations. Veliger 20: 155–58.

Gordon, J., and M. R. Carriker. 1978. Growth lines in a bivalve mollusk: Subdaily patterns and dissolution of shell. Science 202: 519–21.

Grabyan, R. J. 1973. Variations in concentrations of magnesium and strontium in recent shells of *Tivela stultorum*. Bull. So. Calif. Acad. Sci. 72: 42–48.

Graham, D. L. 1972. Trace metal levels in intertidal mollusks in California. Veliger 14: 365–72.

Graham, H. W., and H. Gay. 1945. Season of attachment and growth of sedentary marine organisms at Oakland, California. Ecology 26: 375–86.

Grau, G. 1959. Pectinidae of the eastern Pacific. Allan Hancock Pacific Exped. 23: 1–308.

Grave, B. H. 1928. Natural history of the shipworm *Teredo navalis* at Woods Hole, Mass. Biol. Bull. 55: 260–82.

Green, J. P. 1964. A bacterial disease of *Mya arenaria*. Amer. Zool. 4: 95 (abstract).

Green, R. H., and K. D. Hobson. 1970. Spatial and temporal structure in a temperate intertidal community with special emphasis on *Gemma gemma* (Pelecypoda: Mollusca). Ecology 51: 999–1011.

Greenberg, M. J., and T. C. Jegla. 1962. The pharmacology of the rectum of *Mercenaria mercenaria*. Amer. Zool. 2: 412 (abstract).

Greenberg, M. J., and D. A. Windsor. 1962. Action of acetylcholine in bivalve hearts. Science 137: 534–35.

Greenfield, L. J., and C. E. Lane. 1953. Cellulose digestion in *Teredo*. J. Biol. Chem. 204: 669–72.

Grieser, E. 1913. Über die Anatomie von *Chama pellucida* Broderip. Zool. Jahrb. Suppl. 13: 207–80.

Gross, J. B. 1967. Note on the northward spreading of *Mya arenaria* Linnaeus in Alaska. Veliger 10: 203.

Gunter, G. 1950. The generic status of living oysters and the scientific name of the common American species. Amer. Midl. Natur. 43: 438–39.

Gutsell, J. S. 1931. Natural history of the bay scallop. Bull. U.S. Bur. Fish. 46: 569–632.

Haas, F. 1942a. The habits of life of some west coast bivalves. Nautilus 55: 109–13.

———. 1942b. The habits of life of some west coast bivalves. Nautilus 56: 30–33.

———. 1943. Malacological notes. III. The boring of *Lithophaga*. Field Mus. Natur. Hist., Zool. Ser. 29: 1–23.

Haderlie, E. C. 1976. Destructive marine wood and stone borers in Monterey Bay, pp. 947–53, *in* J. M. Sharpley and A. M. Kaplan, eds., Proc. 3rd Internat. Symp. on Biodegradation. London: Applied Science Publ. 1,138 pp.

———. 1979. Range extension for *Penitella fitchi* Turner, 1955 (Bivalvia: Pholadidae). Veliger 22: 85.

Haderlie, E. C., and J. C. Mellor. 1973. Settlement, growth rates and depth preference of the shipworm *Bankia setacea* (Tryon) in Monterey Bay. Veliger 15: 265–86.

Haderlie, E. C., J. C. Mellor, C. S. Minter III, and G. C. Booth. 1974. The sublittoral benthic fauna and flora off Del Monte Beach, Monterey, California. Veliger 17: 185–204.

Hamilton, G. H. 1969. The taxonomic significance and theoretical origin of surface patterns on a newly discovered bivalve shell layer, the mosaicostracum. Veliger 11: 185–94.

Hamwi, A., and H. H. Haskin. 1969. Oxygen consumption and pumping rates in the hard clam *Mercenaria mercenaria*: A direct method. Science 146: 823–24.

Hancock, D. A. 1960. Symposium: Edible molluscs. 2. The ecology of the molluscan enemies of the edible mollusc. Proc. Malacol. Soc. London 34: 123–43.

Hanna, G. D. 1928. The Monterey shale of California at its type locality with a summary of its fauna and flora. Bull. Amer. Assoc. Petrol. Geol. 12: 969–83.

———. 1966. Introduced mollusks of western North America. Occas. Pap. Calif. Acad. Sci. 48: 1–108.

Hansen, J. C. 1970. Commensal activity as a function of age in two species of California abalones (Mollusca: Gastropoda). Veliger 13: 90–94.

Hare, P. E. 1963. Amino acids in the proteins from aragonite and calcite in the shells of *Mytilus californianus*. Science 139: 216–17.

Harger, J. R. E. 1967. Population studies on *Mytilus* communities. Doctoral thesis, University of California, Santa Barbara. 318 pp.

———. 1968. The role of behavioral traits in influencing the distribution of two species of sea mussel, *Mytilus edulis* and *Mytilus californianus*. Veliger 11: 45–49.

———. 1970a. The effect of wave impact on some aspects of the biology of sea mussels. Veliger 12: 401–14.

———. 1970b. Comparisons among growth characteristics of two species of sea mussel, *Mytilus edulis* and *Mytilis californianus*. Veliger 13: 44–56.

———. 1970c. The effects of species composition on the survival of mixed populations of the sea mussels *Mytilus californianus* and *Mytilus edulis*. Veliger 13: 147–52.

———. 1972a. Variation and relative "niche" size in the sea mussel *Mytilus edulis* in association with *Mytilus californianus*. Veliger 14: 275–82.

———. 1972b. Competitive coexistence: Maintenance of interacting associations of the sea mussels *Mytilus edulis* and *Mytilus californianus*. Veliger 14: 387–410.

———. 1972c. Competitive coexistence among intertidal invertebrates. Amer. Sci. 60: 600–607.

Harger, J. R. E., and D. L. Landenberger. 1971. The effect of storms as a density dependent mortality factor on populations of sea mussels. Veliger 14: 195–201.

Hargreaves, B. 1975. Laboratory feeding rates of *Mytilus californianus* with automatic dosing and recording apparatus. Amer. Zool. 15: 816 (abstract).

Harrison, C. 1975. The essential amino acids of *Mytilus californianus*. Veliger 18: 189–93.

Hawbecker, A. C. 1939. Feeding of gulls on Pismo clams. Condor 41: 120.

Haydock, C. I. 1964. An experimental study to control oyster drills in Tomales Bay, California. Calif. Fish & Game 50: 11–28.

Heath, H. 1941. The anatomy of the pelecypod family Arcidae. Trans. Amer. Philos. Soc. (n.s.) 31: 287–319.

Hegyeli, A. 1964. Temperature dependence of the activity of the antitumor factor in the common clam. Science 146: 77–78.

Herrington, W. C. 1929. The Pismo clam. Calif. Div. Fish & Game, Fish Bull. 18: 1–69.

———. 1930. The Pismo clam, further studies of its life history and depletion. Calif. Div. Fish & Game, Fish Bull. 18: 1–69.

Hertlein, L. G. 1959. Notes on California oysters. Veliger 2: 5–9.

———. 1961. A new species of *Siliqua* (Pelecypoda) from western North America. Bull. So. Calif. Acad. Sci. 60: 12–18.

Hertlein, L. G., and U. S. Grant IV. 1972. The geology and paleontology of the marine Pliocene of San Diego, California (Paleontology: Pelecypoda). San Diego Soc. Natur. Hist., Mem. 2, pt. 2B. 409 pp.

Hewatt, W. G. 1935. Ecological succession in the *Mytilus californianus* habitat as observed in Monterey Bay, California. Ecology 16: 244–51.

Hidu, H., and J. E. Hanks. 1968. Vital staining of bivalve mollusk shells with alizarin sodium monosulfate. Proc. Nat. Shellfish Assoc. 58: 37–41.

Hill, C. L., and C. A. Kofoid, eds. 1927. Marine borers and their relation to marine construction on the Pacific coast. Final Rep., San Francisco Bay Marine Piling Committee, San Francisco. 357 pp.

Hillman, R. E. 1961. Formation of the periostracum in *Mercenaria mercenaria*. Science 134: 1754–55.

Hinegardner, R. 1974. Cellular DNA content of the Mollusca. Comp. Biochem. Physiol. 47A: 447–60.

Hirschhorn, G. 1962. Growth and mortality rates of the razor clam (*Siliqua patula*) on Clatsop beaches, Oregon. Oregon Fish Comm., Contr. 27. 55 pp.

Hobden, D. J. 1969. Iron metabolism in *Mytilus edulis*. II. Uptake and distribution of radioactive iron. J. Mar. Biol. Assoc. U.K. 49: 661–68.

Hodgkin, N. D. 1962. Limestone boring by the mytilid *Lithophaga*. Veliger 4: 123–29.

Hopkins, A. E. 1934a. Accessory hearts in the oyster. Science 80: 411–12.

———. 1934b. Accessory hearts in the oyster *Ostrea gigas*. Biol. Bull. 67: 346–55.

———. 1935. Attachment of larvae of the Olympia oyster, *Ostrea lurida*, to plane surfaces. Ecology 16: 82–87.

———. 1936. Ecological observations on spawning and early development in the Olympia oyster (*Ostrea lurida*). Ecology 17: 551–66.

———. 1937. Experimental observations on spawning, larval development and settling in the Olympia oyster, *Ostrea lurida*. Bull. U.S. Bur. Fish. 48: 439–503.

Hopper, C. N. 1972. Aspects of the prey preference and feeding biology of *Polinices lewisii* (Gastropoda). Echo 5, Abs. & Proc. 5th Ann. Meet., West. Soc. Malacol., pp. 30–31.

Hori, J. 1933. On the development of the Olympia oyster, *Ostrea lurida* Carpenter, transplanted from United States to Japan. Bull. Jap. Soc. Sci. Fish. 1: 269–76.

Howard, A. D. 1953. Some viviparous pelecypod mollusks. Wasmann J. Biol. 11: 233–40.

Hubbs, C. L., G. S. Bien, and H. E. Suess. 1960. La Jolla natural radiocarbon measurements. Amer. J. Sci., Radiocarbon Suppl., 2: 197–223.

Humphreys, W. J. 1964. Electron microscope studies of the fertilized egg and the two-cell stage of *Mytilus edulis*. J. Ultrastruc. Res. 10: 244–62.

Hunter, W. R. 1949. The structure and behavior of *Hiatella gallicana* (Lamarck) and *H. arctica* (L.) with special reference to the boring habit. Proc. Roy. Soc. Edinburgh B72: 271–89.

Hylleberg, J., and V. F. Gallucci. 1975. Selectivity in feeding by the deposit feeding bivalve *Macoma nasuta*. Mar. Biol. 32: 167–78.

Illg, P. L. 1949. A review of the copepod genus *Paranthessius* Claus. Proc. U.S. Nat. Mus. 99: 391–428.

Irwin, T. H. 1973. The intertidal behavior of the bean clam, *Donax gouldii* Dall, 1921. Veliger 15: 206–12.

Isham, L. B., H. B. Moore, and F. G. W. Smith. 1951. Growth-rate measurements of shipworms. Bull. Mar. Sci. 1: 136–47.

Jaccarini, V., W. H. Bannister, and H. Micallef. 1968. The pallial glands and rock boring in *Lithophaga lithophaga* (Lamellibranchia, Mytilidae). J. Zool., London 154: 397–401.

Jackson, J. B. C. 1968. Bivalves: Spatial and size-frequency distributions of two intertidal species. Science 161: 479–80.

Jegla, T. C., and M. J. Greenberg. 1968. Structure of the bivalve rectum. I. Morphology. Veliger 10: 253–63.

Johnson, H. M. 1964. Human blood group A, specific agglutinin of the butter clam *Saxidomus giganteus*. Science 146: 548–49.

Johnson, P. T. 1966. On *Donax* and other sandy-beach inhabitants. Veliger 9: 29–30.

Johnson, W. H., J. S. Kahn, and A. G. Szent-Györgyi. 1959. Paramyosin and contraction of "catch muscles." Science 130: 160–61.

Johnstone, J. 1899. *Cardium*. Liverpool Mar. Biol. Comm. [LMBC] Memoir 2. Liverpool. 84 pp.

Jørgensen, C. B. 1949. The rate of feeding by *Mytilus* in different kinds of suspension. J. Mar. Biol. Assoc. U.K. 28: 333–43.

————. 1950. Efficiency of growth in *Mytilus edulis* and two gastropod veligers. Nature 170: 714.

Katkansky, S. C., and R. W. Warner. 1969. Infestation of the rough-sided little-neck clam, *Protothaca laciniata*, in Morro Bay, California, with larval cestodes (*Echeneibothrium* sp.). J. Invert. Pathol. 13: 125–28.

Katkansky, S. C., R. W. Warner, and R. L. Poole. 1969. On the occurrence of larval cestodes in the Washington clam, *Saxidomus nuttalli*, and the gaper clam, *Tresus nuttallii*, from Drakes Estero, California. Calif. Fish & Game 55: 317–22.

Keen, A. M. 1938. New pelecypod species of the genera *Lasaea* and *Crassinella*. Proc. Malacol. Soc. London 23: 18–32.

————. 1962. Nomenclatural notes on some west American mollusks, with proposal of a new species name. Veliger 4: 178–80.

————. 1971. Seashells of tropical west America: Marine mollusks from Baja California to Peru. 2nd ed. Stanford, Calif.: Stanford University Press. 1,064 pp.

Keen, A. M., and E. Coan. 1974. Marine molluscan genera of western North America. 2nd ed. Stanford, Calif.: Stanford University Press. 208 pp.

Keep, J. 1935. West coast shells. Stanford, Calif.: Stanford University Press. 350 pp.

Kelley, D. W. 1966. Ecological studies of the Sacramento–San Joaquin estuary. I. Zooplankton, zoobenthos, and fishes of San Pablo and Suisun Bays, zooplankton and zoobenthos of the Delta. Calif. Dept. Fish & Game, Fish Bull. 133: 1–133.

Kellogg, J. L. 1899. The life history of the common clam *Mya arenaria*. Comm. Inland Fish., Rhode I. Ann. Rep. 29: 78–96.

————. 1915. Ciliary mechanisms in lamellibranchs with descriptions of anatomy. J. Morphol. 26: 625–701.

Kennedy, G. L. 1974. West American Cenozoic Pholadidae (Mollusca: Bivalvia). San Diego Natur. Hist. Mus., Mem. 8: 1–127.

Kennedy, W. J., J. D. Taylor, and A. Hall. 1969. Environmental and biochemical controls on bivalve shell mineralogy. Biol. Rev. 44: 499–530.

Kincaid, T. 1947. The acclimatization of marine animals in Pacific northwest waters. Min. Conchol. Club So. Calif. 72: 1–3.

Kobayashi, H. 1959. Bipolarity of egg of oyster (*Gryphaea gigas* [Thunberg]). Cytologia 24: 237–43.

Kofoid, C. A., and M. Bush. 1936. The life cycle of *Parachaenia myae* gen. nov., sp. nov., a ciliate parasitic in *Mya arenaria* Linn. from San Francisco Bay, California. Bull. Mus. Roy. Hist. Natur. Belg. 12: 1–15.

Kofoid, C. A., R. C. Miller, E. L. Lazier, W. H. Dore, H. F. Blum, and E. VanSlyke. 1927. Biology section, pp. 188–333, *in* San Francisco Bay Marine Piling Committee, Marine borers and their relation to marine construction on the Pacific coast, San Francisco.

Korringa, P. 1952. Recent advances in oyster biology. Quart. Rev. Biol. 27: 266–308; 339–65.

————. 1953. Oysters. Sci. Amer. 189: 86–91.

Kuenzler, E. J. 1961a. Structure and energy flow on a mussel population in a Georgia salt marsh. Limnol. Oceanogr. 6: 191–204.

————. 1961b. Phosphorus budget of a mussel population. Limnol. Oceanogr. 6: 400–415.

Kuhn, D. A., G. L. Lasnik, and W. D. Rubenstein. 1968. *Cristispira* in intertidal mollusks of southern California. Bacteriol. Proc. 1968: 33.

Landenberger, D. E. 1968. Studies on selective feeding in the Pacific starfish *Pisaster* in southern California. Ecology 49: 1062–75.

Lane, C. E. 1961. The Teredo. Sci. Amer. 204: 132–42.

Laurent, L. L. 1971. The spawning cycle and juvenile growth rates of the gaper clam, *Tresus nuttallii*, of Elkhorn Slough, California. Echo 4, Abs. & Proc. 4th Ann. Meet., West. Soc. Malacol., pp. 24–25.

Laursen, D. 1966. The genus *Mya* in the Arctic region. Malacologia 3: 399–418.

Lee, R. F., R. Sauerheber, and A. A. Benson. 1972. Petroleum hydrocarbons: Uptake and discharge by the marine mussel *Mytilus edulis*. Science 177: 344–46.

Lent, C. M. 1968. Air-gaping in the ribbed mussel *Modiolus demissus* (Dillwyn): Effects and adaptive significance. Biol. Bull. 134: 60–73.

————. 1969. Adaptations of the ribbed mussel, *Modiolus demissus* (Dillwyn), to the intertidal habitat. Amer. Zool. 9: 283–92.

Leonard, V. K. 1969. Seasonal gonadal changes in two bivalve mollusks in Tomales Bay, California. Veliger 11: 382–90.

Levinton, J. 1973. Genetic variation in a gradient of environmental variability: Marine Bivalvia (Mollusca). Science 180: 75–76.

Lewis, J. R. 1964. The ecology of rocky shores. London: English University Press. 326 pp.

Liu, D. L., and T. M. Townsley. 1968. Glucose metabolism in the caecum of the marine borer *Bankia setacea*. J. Fish. Res. Bd. Canada 25: 853–62.

Liu, D. L., and C. C. Walden. 1970. Enzymes of glucose metabolism in the caecum of the marine borer *Bankia setacea*. J. Fish. Res. Bd. Canada 27: 1141–46.

Loesch, H. C. 1957. Studies of the biology of the species of *Donax* on Mustang Island, Texas. Univ. Texas, Publ. Inst. Mar. Sci. 4: 201–27.

Loesch, J. G., and D. S. Haven. 1973. Estimated growth functions and size-age relationships of the hard clam *Mercenaria mercenaria* in the York River, Virginia. Veliger 16: 76–81.

Longo, T. J., and E. J. Dornfeld. 1967. The fine structure of spermatid differentiation in the mussel *Mytilus edulis*. J. Ultrastruc. Res. 24: 462–80.

Loosanoff, V. L. 1937. Spawning of *Venus mercenaria*. Ecology 18: 506–15.

———. 1942. Shell movements of the edible mussel *Mytilus edulis* in relation to temperature. Ecology 23: 231–34.

———. 1949. On the food selectivity of oysters. Science 110: 122.

Loosanoff, V. L., and H. C. Davis. 1950. Conditioning *Venus mercenaria* for spawning in winter and breeding its larvae in the laboratory. Biol. Bull. 98: 60–65.

———. 1963. Rearing of bivalve molluscs. Adv. Mar. Biol. 1: 1–136.

Loosanoff, V. L., H. C. Davis, and P. E. Chanley. 1966. Dimensions and shapes of larvae of some marine bivalve molluscs. Malacologia 4: 351–435.

Lough, R. G., and J. J. Gonor. 1971. Early embryonic stages of *Adula californiensis* (Pelecypoda: Mytilidae) and the effect of temperature and salinity on development rate. Mar. Biol. 8: 118–25.

Lowe, H. N. 1931. Notes on the west coast *Zirfaea*. Nautilus 45: 52–53.

Lutz, R. A., and D. C. Rhoads. 1977. Anaerobiosis and a theory of growth line formation. Science 198: 1222–26.

MacGinitie, G. E. 1935. Ecological aspects of a California marine estuary. Amer. Midl. Natur. 16: 629–765.

———. 1938. Notes on the natural history of some marine animals. Amer. Midl. Natur. 19: 207–19.

———. 1939. Littoral marine communities. Amer. Midl. Natur. 21: 28–55.

———. 1941. On the method of feeding of four pelecypods. Biol. Bull. 80: 18–25.

MacGinitie, G. E., and N. MacGinitie. 1968. Natural history of marine animals. 2nd ed. New York: McGraw-Hill. 523 pp.

Machell, J. R., and J. D. DeMartini. 1971. An annual reproductive cycle of the gaper clam, *Tresus capax* (Gould), in south Humboldt Bay, California. Calif. Fish & Game 57: 274–82.

MacNeil, F. S. 1965. Evolution and distribution of the genus *Mya*, and Tertiary migrations of Mollusca. U.S. Geol. Surv. Prof. Paper 483-G.

Mahéo, R. 1970. Étude de la pose et de l'activité de sécrétion du bissus de *Mytilus edulis* L. Cah. Biol. Mar. 11: 475–83.

Malanga, C. J. 1971. Effects of dopamine and L-dopa on ciliary activity and anaerobic glycolysis in bivalve gills. Amer. Zool. 11: 661 (abstract).

Malone, P. G., and J. R. Dodd. 1967. Temperature and salinity effects on calcification rate in *Mytilus edulis* and its paleoecological implications. Limnol. Oceanogr. 12: 432–36.

Marks, G. W., and D. L. Fox. 1934. Studies on catalase from the California mussel. Bull. Scripps Inst. Oceanogr., Tech. Ser. 3: 297–310.

Marriage, L. D. 1954. The bay clams of Oregon. Their economic importance, relative abundance, and general distribution. Fish Comm. Oregon, Contr. 20. 44 pp.

Martella, T. 1974. Some factors influencing byssal thread production in *Mytilus edulis* (Mollusca: Bivalvia) Linnaeus, 1758. Water, Air & Soil Pollution 3: 171–77.

Martin, J. M., F. M. Piltz, and D. J. Reish. 1975. Studies on the *Mytilus edulis* community in Alamitos Bay, California. V. The effects of heavy metals on byssal thread production. Veliger 18: 183–88.

Matthiesson, G. C. 1961. Intertidal zonation in populations of *Mya arenaria*. Limnol. Oceanogr. 5: 381–88.

Maurer, D. 1967a. Filtering experiments on marine pelecypods from Tomales Bay, California. Veliger 9: 305–9.

———. 1967b. Burial experiments on marine pelecypods from Tomales Bay, California. Veliger 9: 376–81.

———. 1967c. Mode of feeding and diet, and synthesis of studies on marine pelecypods from Tomales Bay, California. Veliger 10: 72–76.

———. 1969. Pelecypod-sediment association in Tomales Bay, California. Veliger 11: 243–49.

McLean, J. M. 1969. Marine shells of southern California. Los Angeles Co. Mus. Natur. Hist., Sci. Ser. 24, Zool. 11. 104 pp.

McMillin, H. C. 1924. The life history and growth of the razor clam. State of Washington Dept. Fish. 52 pp.

McMillin, H. C., and F. W. Weymouth. 1935. Growth rate and variance in the razor clam. Proc. Soc. Exper. Biol. Med. 32: 935–37.

Menzel, R. W., and M. Y. Menzel. 1965. Chromosomes of two species of quahog clams and their hybrids. Biol. Bull. 129: 181–88.

Menzies, R. J., J. Mohr, and C. M. Wakeman. 1963. The seasonal settlement of wood-borers in Los Angeles–Long Beach harbors. Wasmann J. Biol. 21: 97–120.

Meredith, S. E. 1968. Notes on the range extension of the boring clam *Penitella conradi* (Valenciennes) and its occurrence in the shell of the California mussel. Veliger 10: 281.

Merrill, A. S., and R. D. Turner. 1963. Nest building in the bivalve mollusk genera *Musculus* and *Lima*. Veliger 6: 55–59.

Miller, L. R. 1962. Shellfish sanitation in California. Calif. Health, State Dept. Pub. Health 20: 33–34.

Miller, R. C. 1924. The boring mechanism of *Teredo*. Univ. Calif. Publ. Zool. 26: 41–80.

———. 1926. Ecological relations of marine wood-boring organisms in San Francisco Bay. Ecology 7: 247–54.

Miller, R. C., and L. C. Boynton. 1926. Digestion of wood by the shipworm. Science 63: 524–25.

Miller, W. H. 1958. Derivatives of cilia in the distal sense cells of the retina of *Pecten*. J. Biophys. Biochem. Cytol. 4: 227–28.

Milne, L. J., and M. J. Milne. 1948. We go gooeyeducking. Natur. Hist. 57: 162–67.

Minter, C. S., III. 1971. Sublittoral ecology of the kelp beds off Del Monte Beach, Monterey, California. Master's thesis, Naval Postgraduate School, Monterey, California. 177 pp.

Mitchell, H. D. 1935. The microscopic structure of the shell and ligament of *Cardium* (*Cerastoderma*) *corbis* Martyn. J. Morphol. 58: 211–20.

Mix, M. C. 1971. Cell renewal systems in the gut of the oyster *Crassostrea gigas* (Mollusca: Bivalvia). Veliger 14: 202–3.

Moon, T. W., and A. W. Pritchard. 1970. Metabolic adaptations in vertically-separated populations of *Mytilus californianus* Conrad. J. Exper. Mar. Biol. Ecol. 5: 35–46.

Moore, D. R., and D. J. Reish. 1969. Studies on the *Mytilus edulis* community in Alamitos Bay, California. IV. Seasonal variation in gametes from different regions in the Bay. Veliger 11: 250–55.

Moore, R. C., et al., eds. 1969, 1971. Treatise on invertebrate paleontology. Part N. Mollusca, Bivalvia. New York: Geol. Soc. Amer.; Lawrence: University of Kansas Press.

Morton, B. 1978. Feeding and digestion in shipworms. Oceanogr. Mar. Biol. Ann. Rev. 16: 107–44.

Morton, J. E. 1967. Molluscs. 4th ed. London: Hutchinson University Library. 244 pp.

Mpitsos, G. J. 1973. Physiology of vision in the mollusk *Lima scabra*. Neurophysiol. 36: 371–83.

Murphy, D. J. 1974. Freezing tolerance in intertidal molluscs: Dependence of tolerance to cell dehydration. Amer. Zool. 14: 1250 (abstract).

Narchi, W. 1968. The functional morphology of *Lyonsia californica* Conrad, 1837 (Bivalvia). Veliger 10: 305–13.

———. 1971. Structure and adaptation in *Transennella tantilla* (Gould) and *Gemma gemma* (Totten) (Bivalvia: Veneridae). Bull. Mar. Sci. 21: 866–85.

Neave, F. 1942. The butter-clam (*Saxidomus giganteus* Deshayes): Studies in productivity. Rep. British Columbia Dept. Fish. for 1941: J 79–81.

Nelson, H. L. 1949. Scallops, p. 179, *in* The commercial fish catch of California for the year 1947 with an historical review, 1916–1947. Calif. Div. Fish & Game, Fish Bull. 74.

Newcombe, C. L. 1935. A study of the community relationships of the sea mussel *Mytilus edulis* L. Ecology 16: 234–43.

———. 1936. Validity of concentric rings of *Mya arenaria* L. for determining age. Nature 137: 191–92.

Newcombe, C. L., and H. Kessler. 1936. Variations in growth indices of *Mya arenaria* L. on the Atlantic coast of North America. Ecology 17: 429–43.

Newell, R. 1965. The role of detritus in the nutrition of two marine deposit feeders, the prosobranch *Hydrobia ulvae* and the bivalve *Macoma balthica*. Proc. Zool. Soc. London 144: 24–45.

Nichols, F. H. 1977. Infaunal biomass and production on a mudflat, San Francisco Bay, California, pp. 339–57, *in* B. C. Coull, ed., Ecology of marine benthos. Belle W. Baruch Library in Marine Science, no. 6. Columbia: University of South Carolina Press. 467 pp.

Oberling, J. J. 1964. Observations on some structural features of the pelecypod shell. Mitt. Naturf. Ges. Bern. 20: 1–63.

Odhner, N. H. 1919. Studies on the morphology, taxonomy and relations of recent Chamidae. K. Svenska Vetensk-Akad. Handl. 59: 1–102.

Orcutt, H. G. 1950. The life history of the starry flounder *Platichthys stellatus* (Pallas). Calif. Div. Fish & Game, Fish Bull. 78: 1–64.

Owen, G. 1958. Shell form, pallial attachment and the ligament in the Bivalvia. Proc. Zool. Soc. London 131: 637–48.

Paine, R. T. 1976a. Biological observations on a subtidal *Mytilus californianus* bed. Veliger 19: 125–30.

———. 1976b. Size-limited predation: An observational and experimental approach with the *Mytilus-Pisaster* interaction. Ecology 57: 858–73.

Painter, R. E. 1966. Zoobenthos of the San Pablo and Suisun Bays, pp. 40–56, *in* D. W. Kelley, ed., Ecological studies of the Sacramento–San Joaquin Estuary. 1. Zooplankton, zoobenthos, and fishes of San Pablo and Suisun Bays; zooplankton and zoobenthos of the Delta. Calif. Dept. Fish & Game, Fish Bull. 133: 1–133.

Pannella, G., C. MacClintock, and M. N. Thompson. 1968. Paleontological evidence of variations in length of synodic month since late Cambrian. Science 162: 792–96.

Paris, O. H. 1960. Some quantitative aspects of predation by muricid snails on mussels in Washington Sound. Veliger 2: 41–47.

Parker, P. 1949. Fossil and Recent species of the pelecypod genera *Chione* and *Securella* from the Pacific coast. J. Paleontol. 23: 577–93.

Paul, A. J., and H. M. Feder. 1973. Growth, recruitment, and distribution of the little-neck clam *Protothaca staminea* in Galena Bay, Prince William Sound, Alaska. Fish. Bull. 71: 665–77.

Paul, A. J., J. M. Paul, and H. M. Feder. 1976a. Recruitment and growth in the bivalve *Protothaca staminea*, at Olsen Bay, Prince William Sound, ten years after the 1964 earthquake. Veliger 18: 385–92.

————. 1976b. Growth of the littleneck clam *Protothaca staminea* on Porpoise Island, Southeast Alaska. Veliger 19: 163–66.

Pearce, J. B. 1965. On the distribution of *Tresus capax* and *Tresus nuttallii* in the waters of Puget Sound and the San Juan Archipelago (Pelecypoda: Mactridae). Veliger 7: 166–70.

Pentreath, R. J. 1973. The accumulation from water of ^{65}Zn, ^{54}Co, and ^{59}Fe by the mussel *Mytilus edulis*. J. Mar. Biol. Assoc. U.K. 53: 127–44.

Pequegnat, L. H. 1964a. The epifauna of a California siltstone reef. Ecology 45: 272–83.

————. 1964b. The facts about mussel poisoning. Pacific Discov. 17: 20–27.

Petraitis, P. S. 1978. Distribution patterns of juvenile *Mytilus edulis* and *Mytilus californianus*. Veliger 21: 288–92.

Pfitzenmeyer, H. T., and C. N. Shuster. 1960. A partial bibliography of the soft-shell clam *Mya arenaria* L. Chesapeake Biol. Lab., Contr. 123: 1–29.

————. 1965. Annual cycle of gametogenesis of the soft-shelled clam *Mya arenaria* at Solomons, Maryland. Chesapeake Sci. 6: 52–59.

Pickens, P. E. 1965. Heart rate of mussels as a function of latitude, intertidal height, and acclimation temperature. Physiol. Zool. 38: 390–405.

Pierce, S. K., Jr. 1969. Volume control in the ribbed mussel, *Modiolus demissus* (Bivalvia: Mytilidae). Amer. Zool. 9: 1091 (abstract).

————. 1973. The rectum of "*Modiolus*" *demissus* (Dillwyn) (Bivalvia: Mytilidae): A clue to solving a troubled taxonomy. Malacologia 12: 283–93.

Pilot, J., and A. Ryter. 1965. Structure des Spirochaetes: Étude des genre *Cristispira* au microscope optique et au microscope electronique. Ann. Inst. Pasteur 109: 551–62.

Pilson, M. E. Q., and P. B. Taylor. 1961. Hole drilling by octopus. Science 134: 1366–68.

Pohlo, R. H. 1963. Morphology and mode of burrowing in *Siliqua patula* and *Solen rosaceus* (Mollusca: Bivalvia). Veliger 6: 98–104.

————. 1964. Ontogenetic changes of form and mode of life in *Tresus nuttallii* (Bivalvia: Mactridae). Malacologia 1: 321–30.

————. 1966. A note on the feeding behavior in *Tagelus californianus* (Bivalvia: Tellinacea). Veliger 8: 225.

————. 1967. Aspects of the biology of *Donax gouldi* and a note on evolution in Tellinacea (Bivalvia). Veliger 9: 330–37.

————. 1969. Confusion concerning deposit feeding in the Tellinacea. Proc. Malacol. Soc. London 38: 361–64.

————. 1972. Feeding and associated morphology in *Sanguinolaria nuttallii* (Bivalvia: Tellinacea). Veliger 14: 298–301.

Potts, W. T. W. 1954. The inorganic composition of the blood of *Mytilus edulis* and *Anodonta cygnaea*. J. Exper. Biol. 31: 376–85.

Purchon, R. D. 1955. The functional morphology of the rockboring lamellibranch *Petricola pholadiformis* Lamarck. J. Mar. Biol. Assoc. U.K. 34: 257–78.

————. 1968. The biology of the Mollusca. London: Pergamon Press. 560 pp.

Pratt, D. M. 1953. Abundance and growth of *Venus mercenaria* and *Callocardia morrhuana* in relation to the character of the bottom sediments. J. Mar. Res. 12: 60–74.

Quayle, D. B. 1941. The Japanese "littleneck" clam accidentally introduced into British Columbia waters. Pacific Biol. Sta. Progr. Rep. 48: 17–18.

————. 1943. Sex, gonad development and seasonal gonad changes in *Paphia staminea* Conrad. J. Fish. Res. Bd. Canada 6: 140–51.

————. 1952a. The relation of digging frequency to productivity of the butter clam, *Saxidomus giganteus* Desh. J. Fish. Res. Bd. Canada 8: 369–73.

————. 1952b. The seasonal growth of the Pacific oyster (*Ostrea gigas*) in Ladysmith Harbour. Rep. British Columbia Dept. Fish., for 1950: L85–90.

————. 1953. The larvae of *Bankia setacea* Tryon. Rep. British Columbia Dept. Fish. for 1951: 88–91.

————. 1955. The British Columbia shipworm. Rep. British Columbia Dept. Fish. for 1955: K92–104.

————. 1956a. Growth of the British Columbia shipworm. Prog. Rep. Pacific Coast Stas., Fish. Res. Bd. Canada 105: 3–5.

————. 1956b. Pacific oyster culture in British Columbia. Prov. Dept. Fish., Victoria, B.C. 33 pp.

————. 1959. The growth rate of *Bankia setacea* Tryon, pp. 175–83, *in* D. L. Ray, ed., Marine boring and fouling organisms. Seattle: University of Washington Press.

————. 1963. Mortality in Pacific oyster seed. Proc. Nat. Shellfish Assoc. 52: 53–63.

————. 1969. Pacific oyster culture in British Columbia. Bull. Fish. Res. Bd. Canada 169: 1–192.

————. 1973. The intertidal bivalves of British Columbia. British Columbia Prov. Mus., Victoria, B.C., Handbook 17: 1–104.

Rae, J. G. 1979. The population dynamics of two sympatric species of *Macoma* (Mollusca: Bivalvia). Veliger 21: 384–99.

Rao, K. P. 1953a. Rate of water propulsion in *Mytilus californianus* as a function of latitude. Biol. Bull. 104: 171–81.

————. 1953b. Shell weight as a function of intertidal height in a population of littoral pelecypods. Experientia 9: 465–66.

————. 1954. Tidal rhythmicity of rate of water propulsion in *Mytilus*, and its modifiability by transplantation. Biol. Bull. 106: 353–59.

Rattenbury, J. C., and W. E. Berg. 1954. Embryonic segregation during early development of *Mytilus edulis*. J. Morphol. 95: 393–414.

Rawson, G. 1968. Les Huîtres. Biologie—Culture. Bibliographie. Bull. Inst. Océanogr. Monaco 67: 1–51.

Ray, D. L. 1959. Trends in marine biology, pp. 1–8, *in* I. Pratt and

J. E. McCauley, eds., Marine biology, 20th Ann. Biol. Colloq., Oregon State College, Corvallis. 87 pp.

Read, K. R. H. 1962. Respiration of the bivalved molluscs *Mytilus edulis* and *Brachiodontes demissus plicatulus* as a function of size and temperature. Comp. Biochem. Physiol. 7: 89–101.

Reid, R. G. B. 1966. Digestive tract enzymes in the bivalves *Lima hians* (Gmelin) and *Mya arenaria* L. Comp. Biochem. Physiol. 17: 417–33.

———. 1968. The distribution of digestive tract enzymes in lamellibranchiate bivalves. Comp. Biochem. Physiol. 24: 727–44.

———. 1969. Seasonal observations on diet, and stored glycogen and lipids in the horse clam, *Tresus capax* (Gould, 1850). Veliger 11: 378–81.

———. 1971. Criteria for categorizing feeding types in bivalves. Veliger 13: 358–59.

Reid, R. G. B., and R. M. Dunnill. 1969. Specific and individual differences in the esterases of members of the genus *Macoma* (Mollusca: Bivalvia). Comp. Biochem. Physiol. 29: 601–10.

Reid, R. G. B., and A. Reid. 1969. Feeding processes of members of the genus *Macoma* (Mollusca: Bivalvia). Canad. J. Zool. 47: 649–57.

Reinke, E. A. 1960. Shellfish control in the State of California. Calif. Health, State Dept. Pub. Health 17: 148–49.

Reish, D. J. 1961. A study of benthic fauna in a recently constructed boat harbor in southern California. Ecology 42: 84–91.

———. 1963. Mass mortality of marine organisms attributed to the "red tide" in southern California. Calif. Fish & Game 49: 265–70.

———. 1964a. Studies on the *Mytilus edulis* community in Alamitos Bay, California. I. Development and destruction of the community. Veliger 6: 124–31.

———. 1964b. Studies on the *Mytilus edulis* community in Alamitos Bay, California. II. Population variations and discussion of the associated organisms. Veliger 6: 202–7.

———. 1964c. Discussion of the *Mytilus californianus* community on newly constructed rock jetties in southern California. Veliger 7: 95–101.

Reish, D. J., and J. L. Ayers, Jr. 1968. Studies on the *Mytilus edulis* community in Alamitos Bay, California. III. The effects of reduced dissolved oxygen and chlorinity concentration on survival and byssal thread production. Veliger 10: 384–88.

Richards, O. W. 1928. The growth of the mussel *Mytilus californianus*. Nautilus 41: 99–101.

———. 1946. Comparative growth of *Mytilus californianus* at La Jolla, California, and *Mytilus edulis* at Woods Hole, Mass. Ecology 27: 370–73.

Ricketts, E. F., and J. Calvin. 1968. Between Pacific tides. 4th ed. Revised by J. W. Hedgpeth. Stanford, Calif.: Stanford University Press. 614 pp.

Risebrough, R. W., D. B. Menzel, D. J. Martin, Jr., and H. S. Olcott.

1967. DDT residues in Pacific sea birds: A persistent insecticide in marine food chains. Nature 216: 589–91.

Rodegker, W., and J. C. Nevenzel. 1964. The fatty acid composition of three marine invertebrates. Comp. Biochem. Physiol. 11: 53–60.

Röder, H. 1977. Zur Beziehung zwischen Konstruktion und Substrat bei mechanisch bohrenden Bohrmuscheln (Pholadidae, Teredinidae). Senckenbergiana Maritima 9: 105–214.

Ropes, J. W., and A. P. Stickney. 1965. Reproductive cycle of *Mya arenaria* in New England. Biol. Bull. 128: 315–27.

Rosen, M. C., C. R. Stasek, and C. O. Hermans. 1978. The ultrastructure and evolutionary significance of the cerebral ocelli of *Mytilus edulis*, the bay mussel. Veliger 21: 10–18.

Ross, J. R. P., and D. Goodman. 1974. Vertical intertidal distribution of *Mytilus edulis*. Veliger 16: 388–95.

Rossi, S. S., and R. J. Reish. 1976. Studies on the *Mytilus edulis* community of Alamitos Bay, California. VI. Regulation of anaerobiosis by dissolved oxygen concentration. Veliger 18: 357–60.

Rost, H. 1955. A report on the family Arcidae. Allan Hancock Pacific Exped. 20: 177–236.

Salchak, A., and J. Haas. 1971. Occurrence of the northern quahog, *Mercenaria mercenaria*, in Colorado Lagoon, Long Beach, California. Calif. Fish & Game 57: 126–28.

Savage, N. B. 1976. Burrowing activity in *Mercenaria mercenaria* (L.) and *Spisula solidissima* (Dillwyn) as a function of temperature and dissolved oxygen. Mar. Behav. Physiol. 3: 221–34.

Schantz, E. J. 1960. Biochemical studies on paralytic shellfish poisons, pp. 843–55, *in* R. Nigrelli, ed., Biochemistry and pharmacology of compounds derived from marine organisms. Ann. N.Y. Acad. Sci. 90: 615–950.

Scheer, B. T. 1940. Some features of the metabolism of the carotinoid pigment of the California sea-mussel *Mytilus californianus*. J. Biol. Chem. 136: 275–99.

Schmeer, M. R. 1964. Growth-inhibiting agents from *Mercenaria* extracts: Chemical and biological properties. Science 135: 413–14.

Schmidt, R. R., and J. E. Warme. 1969. Population characteristics of *Protothaca staminea* (Conrad) from Mugu Lagoon, California. Veliger 12: 193–99.

Schulz-Baldes, M. 1974. Lead uptake from sea water and food, and lead loss in the common mussel *Mytilus edulis*. Mar. Biol. 25: 177–93.

Scott, D. M., and C. W. Major. 1972. The effect of copper (II) on survival, respiration, and heart rate of the common blue mussel, *Mytilus edulis*. Biol. Bull. 143: 679–88.

Seapy, R. R., and C. L. Kitting. 1978. Spatial structure of an intertidal molluscan assemblage on a sheltered sandy beach. Mar. Biol. 46: 137–45.

Seed, R. 1968. Factors influencing shell shape in the mussel *Mytilus edulis*. J. Mar. Biol. Assoc. U.K. 48: 561–84.

————. 1969. The ecology of *Mytilus edulis* (L.) (Lamellibranchiata) on exposed rocky shores. I. Breeding and settlement. II. Growth and mortality. Oecologia 3: 277–350.

Seilacher, A. 1954. Okologie der triassischen Muschel *Lima lineata* (Schloth.) und ihrer Epöken. Neu. Jahrb. Geol. Paläontol., Stuttgart 4: 163–83.

————. 1972. Divaricate patterns in pelecypod shells. Lethaia 5: 325–43.

Sellmer, G. P. 1967. Functional morphology and ecological life history of the gem clam, *Gemma gemma* (Eulamellibranchia: Veneridae). Malacologia 5: 137–223.

Sergy, G. A., and J. W. Evans. 1965. The settlement and distribution of marine organisms fouling a seawater pipe system. Veliger 18: 87–92.

Shaw, R. F. 1956. The polymorphism of the Japanese littleneck clam. Nautilus 70: 53–59.

Shulenberger, E. 1970. Responses of *Gemma gemma* to a catastrophic burial (Mollusca: Pelecypoda). Veliger 13: 163–70.

Sigurdsson, J. B., C. W. Titman, and P. A. Davies. 1976. The dispersal of young post-larval bivalve molluscs by byssal threads. Nature 262: 386–87.

Silvey, G. E. 1968. Interganglionic regulation of heartbeat in the cockle *Clinocardium nuttallii*. Comp. Biochem. Physiol. 25: 257–69.

Smith, E. H. 1969. Functional morphology of *Penitella conradi* relative to shell-penetration. Amer. Zool. 9: 869–80.

Smith, L. S., and J. C. Davis. 1965. Haemodynamics in *Tresus nuttallii* and certain other bivalves. J. Exper. Biol. 43: 171–80.

Smith, R. I., and J. T. Carlton, eds. 1975. Light's manual: Intertidal invertebrates of the central California coast. 3rd ed. Berkeley and Los Angeles: University of California Press. 716 pp.

Smith, S. V., and E. C. Haderlie. 1969. Growth and longevity of some calcareous fouling organisms, Monterey Bay, California. Pacific Sci. 23: 447–51.

Snoke, L. R., and A. P. Richards. 1956. Marine borer attack on lead cable sheath. Science 124: 443.

Sommer, H., and K. F. Meyer. 1935. Mussel poisoning. Calif. & West. Med. 42: 423–26.

Soot-Ryen, T. 1955. A report on the family Mytilidae. Allan Hancock Pacific Exped. 20: 1–174.

Span, J. A. 1978. Successful reproduction of the giant Pacific oysters in Humboldt Bay and Tomales Bay, California. Calif. Fish & Game 64: 123–24.

Sparks, A. K., and K. K. Chew. 1966. Gross infestation of the littleneck clam *Venerupis staminea* with a larval cestode (*Echeneibothrium* sp.). J. Invert. Pathol. 8: 413–16.

Sparks, A. K., D. Pereyra, I. E. Ellis, and A. R. Sullard, Jr. 1962. Studies in oyster pathology. Univ. Washington Coll. Fish. Contr. 139: 37–38.

Springer, V. G., and E. R. Beeman. 1960. Penetration of lead by the wood piddock *Martesia striata*. Science 131: 1378–79.

Stafford, J. 1914. The native oyster of British Columbia. Rep. British Columbia Comm. Fish., for 1914: R79–102.

Stanley, S. M. 1969. Bivalve mollusk boring aided by discordant shell ornamentation. Science 166: 634–35.

Stasek, C. R. 1962. The form, growth and evolution of the Tridacnidae (giant clams). Arch. Zool. Exper. Gén. 101: 1–40.

————. 1963a. Orientation and form in the bivalved Mollusca. J. Morphol. 112: 195–214.

————. 1963b. Synopsis and discussion of the association of ctenidia and labial palps in the bivalved Mollusca. Veliger 6: 91–97.

————. 1967. Of life and limb in mollusks. Pacific Discov. 20: 11–14.

Steele, E. N. 1964. The immigrant oyster (*Ostrea gigas*) now known as the Pacific oyster. Olympia, Wash.: Warren's Quick Print. 179 pp.

Stenzel, H. B. 1964. Oysters: Composition of the larval shell. Science 145: 155–56.

Stephens, A. C. 1929. Notes on the growth of *Tellina tenuis* da Costa in the Firth of Clyde. J. Mar. Biol. Assoc. U.K. 16: 117–29.

Stephenson, M. D. 1977. Sea otter predation on Pismo clams in Monterey Bay. Calif. Fish & Game 63: 117–20.

Stohler, R. 1930. Beitrag zur kenntnis des Geschlechtszyklus von *Mytilus californianus* Conrad. Zool. Anz. 90: 263–68.

————. 1960. Fluctuations in mollusk populations after a red tide in the Estero de Punta Banda, Lower California, Mexico. Veliger 3: 22–28.

Stout, W. E. 1967. A study of the autecology of the horse-neck clams, *Tresus capax* and *Tresus nuttallii* in South Humboldt Bay, California. Master's thesis, Humboldt State College, Arcata, Calif.

————. 1970. Some associates of *Tresus nuttallii* (Conrad, 1837) (Pelecypoda: Mactridae). Veliger 13: 67–70.

Straughan, D. 1971. Breeding and larval settlement of certain intertidal invertebrates in the Santa Barbara Channel following pollution by oil, pp. 223–44, *in* D. Straughan, comp., Biological and oceanographical survey of the Santa Barbara Channel oil spill, 1969–1970, vol. 1, Biology and bacteriology. Allan Hancock Found., Univ. So. Calif., Los Angeles.

Strong, A. M. 1924. Notes on the *Donax* of California. Nautilus 37: 81–84.

————. 1925. Notes on the mactras of the west coast. Nautilus 38: 98–102.

Stubbings, H. G. 1954. Biology of the common mussel in relation to fouling problems. Research, London 7: 222–29.

Swan, E. F. 1952a. Growth indices of the clam *Mya arenaria*. Ecology 33: 365–74.

————. 1952b. The growth of the clam *Mya arenaria* as affected by the substratum. Ecology 33: 530–34.

Swan, E. F., and J. H. Finucane. 1951. Observations on the genus *Schizothaerus*. Nautilus 66: 19–26.

Talmadge, R. R. 1972. Notes on some California Mollusca: Geographical, ecological and chronological distribution. Veliger 14: 411–13.

Taylor, C. C. 1959. Temperature, growth and mortality—the Pacific razor clam. J. Conseil. Internat. Explor. Mer. 25: 93–100.

———. 1960. Temperature, growth and mortality—the Pacific cockle. J. Conseil. Internat. Explor. Mer. 26: 117–24.

Taylor, J. D., and W. J. Kennedy. 1969. The shell structure and mineralogy of *Chama pellucida* Broderip. Veliger 11: 391–98.

Taylor, R. L. 1966. *Haplosporidium tumefacientis* sp. n., the etiologic agent of a disease of the California sea mussel *Mytilus californianus* Conrad. J. Invert. Pathol. 8: 109–21.

Temnikow, N. K. 1974. Epicardium larvae in *Mytilus californianus* (Mollusca: Bivalvia). Veliger 16: 413–14.

Thompson, I. 1975. Biological clocks and shell growth in bivalves, pp. 149–61, *in* G. D. Rosenberg and S. K. Runcorn, eds., Growth rhythms and the history of the earth's rotation. New York: Wiley-Interscience. 560 pp.

Thompson, R. J., N. A. Ratcliffe, and B. L. Bayne. 1974. Effects of starvation on structure and function in the digestive gland of the mussel (*Mytilus edulis* L.). J. Mar. Biol. Assoc. U.K. 54: 699–712.

Tiffany, W. J., III. 1971. The tidal migration of *Donax variabilis* Say (Mollusca: Bivalvia). Veliger 14: 82–85.

Tomlinson, P. K. 1968. Mortality, growth and yield per recruit for Pismo clams. Calif. Fish & Game 54: 100–107.

Townsley, P. M., and E. G. H. Lee. 1967. Response of fertilized eggs of the mollusk *Bankia setacea* to aflatoxin. J. Assoc. Offic. Analyt. Chem. 50: 361–63.

Townsley, P. M., and R. A. Richy. 1965. Marine borer aldehyde oxidase. Canad. J. Zool. 43: 1011–19.

Townsley, P. M., R. A. Richy, and P. C. Trussell. 1966. The laboratory rearing of the shipworm, *Bankia setacea* (Tryon). Proc. Nat. Shellfish. Assoc. 56: 49–52.

Trueman, E. R. 1954. Observations on the opening of the valves of a burrowing lamellibranch *Mya arenaria*. J. Exper. Biol. 31: 291–305.

———. 1966a. Bivalve mollusks: Fluid dynamics of burrowing. Science 152: 523–25.

———. 1966b. The dynamics of burrowing in *Ensis* (Bivalvia). Proc. Roy. Soc. London 166: 459–76.

———. 1968a. The burrowing activities of bivalves. Symp. Zool. Soc. London 22: 167–86.

———. 1968b. A comparative account of the burrowing process of species of *Mactra* and other bivalves. Proc. Malacol. Soc. London 38: 139–51.

Trueman, E. R., A. R. Brand, and P. Davis. 1966. The dynamics of burrowing of some common littoral bivalves. J. Exper. Biol. 44: 469–92.

Turner, C. H., E. E. Ebert, and R. R. Given. 1969. Man-made reef ecology. Calif. Dept. Fish & Game, Fish Bull. 146: 1–221.

Turner, H. J., J. C. Ayres, and C. L. Wheeler. 1948. Report on investigations of the propagation of the soft-shelled clam *Mya arenaria*. Woods Hole Oceanogr. Inst. Contr. 462: 1–61.

Turner, R. D. 1954–55. The family Pholadidae in the western Atlantic and eastern Pacific. I. Pholadinae. II. Martesiinae, Jouannetiinae and Xylophaginae. Johnsonia 3: 1–160.

———. 1966. A survey and illustrated catalogue of the Teredinidae (Mollusca: Bivalvia). Mus. Comp. Zool. Harvard Univ., Cambridge, Mass. 265 pp.

Turner, R. D., and K. Boss. 1962. The genus *Lithophaga* in the western Atlantic. Johnsonia 4: 81–116.

Turner, R. D., and A. C. Johnson. 1971. Biology of wood-boring molluscs, pp. 259–301, *in* E. B. G. Jones and S. K. Eltringham, eds., Marine borers, fungi and fouling organisms of wood. Organization for Economic Co-operation and Development, Paris.

Van Veldhuizen, H. D., and D. W. Phillips. 1978. Prey capture by *Pisaster brevispinus* (Asteroidea: Echinodermata) on soft substrate. Mar. Biol. 48: 89–97.

Van Winkle, W., Jr. 1970. Effect of environmental factors on byssal thread production. Mar. Biol. 7: 143–48.

Vassallo, M. T. 1969. The ecology of *Macoma inconspicua* (Broderip & Sowerby, 1829) in central San Francisco Bay. I. The vertical distribution of the *Macoma* community. Veliger 11: 223–34.

———. 1971. The ecology of *Macoma inconspicua* (Broderip and Sowerby, 1829) in central San Francisco Bay. II. Stratification of the *Macoma* community within the substrate. Veliger 13: 279–84.

Wade, B. A. 1967. On the taxonomy, morphology, and ecology of the beach clam, *Donax striatus* Linné. Bull. Mar. Sci. 17: 723–40.

Walden, C. C., I. V. F. Allen, and P. C. Trussell. 1966. Estimation of marine-borer attack on wooden surfaces. J. Fish. Res. Bd. Canada 24: 261–72.

Waldron, M., R. M. Packie, and F. L. Roberts. 1976. Pigment polymorphism in the blue mussel, *Mytilus edulis*. Veliger 19: 82–83.

Walne, P. R. 1974. Culture of bivalve molluscs: 50 years' experience at Conway. West Byfleet, Eng.: Fishing News (Books). 173 pp.

Warme, J. E., T. B. Scanland, and N. F. Marshall. 1971. Submarine canyon erosion: Contribution of marine rock burrowers. Science 173: 1127–29.

Warner, R. W., and S. C. Katkansky. 1969a. Infestation of the clam *Protothaca staminea* by two species of tetraphyllidian cestodes (*Echeneibothrium* spp.). J. Invert. Pathol. 13: 129–33.

———. 1969b. A larval cestode from the Pismo clam, *Tivela stultorum*. Calif. Fish & Game 55: 248–51.

Welch, L. D. 1969. Distribution of *Gemma gemma* in Walker Creek delta of Tomales Bay, California. Student Res. Rep., Pacific Marine Station, Dillon Beach, Calif.

Wells, H. W. 1957. Abundance of the hard clam *Mercenaria mercenaria* in relation to environmental factors. Ecology 38: 123–28.

Wendell, F., J. D. deMartini, P. Dinnel, and J. Siecke. 1976. The ecology of the gaper or horse clam, *Tresus capax* (Gould, 1850) (Bivalvia: Mactridae), in Humboldt Bay, California. Calif. Fish & Game 62: 41–64.

Weymouth, F. W. 1920. The edible clams, mussels and scallops of California. Calif. Fish & Game Comm., Fish Bull. 4: 1–74.

——. 1923a. Certain features of the physiology of growth as illustrated by the lamellibranch *Tivela*. Amer. J. Physiol. 63: 412.

——. 1923b. The life-history and growth of the Pismo clam (*Tivela stultorum* Mawe). Calif. Fish & Game Comm., Fish Bull. 7: 1–120.

Weymouth, F. W., H. C. McMillin, and H. B. Holmes. 1925. Growth and age at maturity of the Pacific razor clam *Siliqua patula* (Dixon). Bull. U.S. Bur. Fish. 41: 201–36.

——. 1931. Relative growth and mortality of the Pacific razor clam (*Siliqua patula* Dixon), and their bearing on the commercial fishery. Bull. U.S. Bur. Fish. 46: 543–67.

Weymouth, F. W., H. C. McMillin, and W. H. Rich. 1931. Latitude and relative growth of the razor clam *Siliqua patula*. J. Exper. Biol. 8: 228–49.

Weymouth F. W., and S. H. Thompson. 1931. The age and growth of the Pacific cockle (*Cardium corbis* Martyn). Bull. U.S. Bur. Fish. 46: 633–41.

Whedon, W. F. 1936. Spawning habits of the mussel *Mytilus californianus* Conrad with notes on possible relation to mussel poisoning. Univ. Calif. Publ. Zool. 41: 35–44.

——. 1938. The digestive system of *Mytilus californianus* Conrad. Z. Vergl. Physiol. 25: 509–22.

White, F. D. 1929. Studies on marine wood borers. III. A note on the breeding season of *Bankia* (*Xylotrya*) *setacea* in Departure Bay, B.C. Contr. Canad. Biol. & Fish. (n.s.) 4: 19–25.

White, K. M. 1937. *Mytilus*. Liverpool Mar. Biol. Comm. [LMBC] Memoir 31. Liverpool. 117 pp.

Wicksten, M. K. 1978. Checklist of marine mollusks at Coyote Point Park, San Francisco Bay, California. Veliger 21: 127–30.

Widdows, J., and B. L. Bayne. 1971. Temperature acclimation of *Mytilus edulis* with reference to its energy budget. J. Mar. Biol. Assoc. U.K. 51: 827–43.

Wilbur, K. M., and C. M. Yonge, eds. 1964. Physiology of Mollusca. Vol. 1. New York: Academic Press. 473 pp.

——, eds. 1966. Physiology of Mollusca. Vol. 2. New York: Academic Press. 645 pp.

Willett, G. 1931. *Psephis* (*Petricola*) *tellimyalis* (Cpr.) not the young of *Petricola denticulata* Sby. Bull. So. Calif. Acad. Sci. 30: 39.

Williamson, M. B. 1902. A monograph on *Pecten aequisulcatus* Cpr. Bull. So. Calif. Acad. Sci. 1: 50–51.

Worley, L. G. 1944. Studies of the vitally stained Golgi apparatus. II. Yolk information and pigment concentration in the mussel *Mytilus californianus*. J. Morphol. 75: 77–95.

Wourms, J. P. 1968. Tubular components in the ultrastructural organization of the molluscan crystalline style. Amer. Zool. 8: 803 (abstract).

Yonge, C. M. 1923. Studies on the comparative physiology of digestion. I. The mechanism of feeding, digestion, and assimilation in the lamellibranch *Mya*. J. Soc. Exper. Biol. 1: 15–63.

——. 1936. The evolution of the swimming habit in the Lamellibranchia. Mém. Mus. Roy. Hist. Natur. Belg. 3: 77–100.

——. 1949. On the structure and adaptations of the Tellinacea, deposit-feeding Eulamellibranchia. Phil. Trans. Roy. Soc. London B234: 29–76.

——. 1951a. Studies on Pacific coast mollusks. I. On the structure and adaptations of *Cryptomya californica* (Conrad). Univ. Calif. Publ. Zool. 55: 395–400.

——. 1951b. Studies on Pacific coast mollusks. II. Structure and adaptations for rock boring in *Platyodon cancellatus* (Conrad). Univ. Calif. Publ. Zool. 55: 401–7.

——. 1951c. Studies on Pacific coast mollusks. III. Observations on *Hinnites multirugosus* (Gale). Univ. Calif. Publ. Zool. 55: 409–20.

——. 1952a. Studies on Pacific coast mollusks. IV. Observations on *Siliqua patula* Dixon and on evolution within the Solenidae. Univ. Calif. Publ. Zool. 55: 421–38.

——. 1952b. Studies on Pacific coast mollusks. V. Structure and adaptation in *Entodesma saxicola* (Baird) and *Mytilimeria nuttallii* Conrad, with a discussion on evolution within the family Lyonsiidae (Eulamellibranchia). Univ. Calif. Publ. Zool. 55: 439–50.

——. 1952c. Studies on Pacific coast mollusks. VI. A note on *Kellia laperousii* (Deshayes). Univ. Calif. Publ. Zool. 55: 451–54.

——. 1953. The monomyarian condition in the Lamellibranchia. Trans. Roy. Soc. Edinburgh 62: 443–78.

——. 1955. Adaptation to rock boring in *Botula* and *Lithophaga* (Lamellibranchia, Mytilidae) with a discussion on the evolution of this habit. Quart. J. Microscop. Sci. 96: 383–410.

——. 1958. Observations on *Petricola carditoides* (Conrad). Proc. Malacol. Soc. London 33: 25–31.

——. 1960. Oysters. London: Collins. 209 pp.

——. 1962. On the primitive significance of the byssus in Bivalvia and its effects in evolution. J. Mar. Biol. Assoc. U.K. 42: 113–25.

——. 1967. Form, habit and evolution in the Chamidae (Bivalvia) with reference to conditions in the rudists (Hippuritacea). Phil. Trans. Roy. Soc. London B252: 49–105.

——. 1969. Functional morphology and evolution within the Carditacea (Bivalvia). Proc. Malacol. Soc. London 38: 493–527.

——. 1971. On functional morphology and adaptive radiation in the bivalve superfamily Saxicavacea (*Hiatella* (=*Saxicava*), *Saxicavella, Panomya, Panope, Cyrtodaria*). Malacologia 11: 1–44.

——. 1976. The "mussel" form and habit, pp. 1–12, *in* B. L. Bayne, ed., Marine mussels. London: Cambridge University Press.

——. 1977. Form and evolution in the Anomiacea (Mollusca: Bivalvia)—*Pododesmus, Anomia, Patro, Enigmonia* (Anomiidae): *Placun-*

anomia, Placuna (Placunidae fam. nov.). Phil. Trans. Roy. Soc. London B276: 453–527.

Yonge, C. M., and J. I. Campbell. 1968. On the heteromyarian condition in Bivalvia with special reference to *Dreissena polymorpha* and certain Mytilacea. Trans. Roy. Soc. Edinburgh 68: 21–43.

Young, R. T. 1941. The distribution of the mussel (*Mytilus californianus*) in relation to the salinity of its environment. Ecology 22: 379–86.

———. 1942. Spawning season of the California mussel *Mytilus californianus*. Ecology 23: 490–92.

———. 1945. Stimulation of spawning in the mussel (*Mytilus californianus*). Ecology 26: 58–69.

———. 1946. Spawning and settling season of the mussel *Mytilus californianus*. Ecology 27: 354–63.

———. 1951. Another *Mytilus* hermaphrodite. Nautilus 64: 105.

Young, S. D., and M. A. Crenshaw. 1971. Synthesis of extrapallial proteins in the clam *Mercenaria mercenaria*. Amer. Zool. 11: 655 (abstract).

ZoBell, C. E., and W. A. London. 1937. Bacterial nutrition of the California mussel. Proc. Soc. Exper. Biol. Med. 36: 607–9.

Chapter 16

Polyplacophora: *The Chitons*

Eugene C. Haderlie and Donald P. Abbott

Chitons occur from high in the intertidal zone to depths of over 4,000 m, and from the coldest polar seas to the tropics. They are most abundant on temperate rocky shores, where they find a hard substratum for attachment and usually food in abundance. Some species live on the outer coast and are able to withstand the buffeting of heavy surf; others occupy habitats where the water is quieter, behind offshore reefs or rocks, in embayments, and in tidepools.

Chitons occur as fossils at least as far back as the Upper Cambrian; thus, as a group, they span nearly half a billion years. Long a relatively homogeneous group, the chitons have diversified more rapidly in recent (Cenozoic) times, and today there are approximately 600 living species worldwide. Fully one-fifth of these can be found along the west coast from Alaska to southern California, more than on any coast of comparable length in the world. Fifty species and subspecies are known from Monterey Bay alone, and as many as 15 species have been brought up in a single dredge haul. The seas around Australia and New Zealand also support a large and diverse chiton population.

The chitons differ from all other mollusks in several anatomical features. Most strikingly, the shell consists of eight somewhat overlapping calcareous plates or valves, instead of a single shell or pair of shells as in the gastropods and bivalves. These plates are embedded in a tough, muscular girdle, which holds them in place while allowing considerable freedom of movement. The shell plates are covered by the girdle more extensively in some chitons

Eugene C. Haderlie is Professor of Oceanography at the Naval Postgraduate School, Monterey. Donald P. Abbott is Professor, Department of Biological Sciences and Hopkins Marine Station, Stanford University.

than in others; in *Cryptochiton stelleri* (16.25), the gumboot chiton, the shell plates are completely covered and internal. The small head is concealed when the animals are attached, as is the large and powerful foot, which is used in both clinging and creeping. Numerous gills are arranged in rows in the pallial grooves on either side of the foot. The ability to roll up in a ball, like a pill bug, when disturbed or removed from the substratum, has given rise to the name "sea cradle" for them.

Chitons are generally sluggish animals and move slowly over the substratum. Many are nocturnal and remain concealed under rocks in the daytime. The food of most species consists of algal films scraped off the rocks by the radula, or of small algal clumps, but animal matter makes up much of the diet in others (see Barnawell, 1960; Barnes, 1972; Boolootian, 1964; Demopulos, 1975; Fulton, 1975). The diets of individuals of any one species may differ in different habitats in the intertidal zone (Nishi, 1975; Robb, 1975). One species, *Placiphorella velata* (16.23), is a carnivore and captures crustaceans and other small moving animals. The teeth on the radula in chitons are capped with hard materials containing so much iron that the radula can be picked up with a magnet (Tomlinson, 1959).

The sexes are typically separate, and gametes are usually released into the sea; free-spawned eggs hatch as swimming larvae, which ultimately settle to the bottom and metamorphose into miniature adults. A few species brood their eggs in the grooves beside the foot; here a pelagic stage is usually lacking. The reproduction, early embryology, and metamorphosis of several species have been investigated (Heath, 1899, 1905a; Okuda, 1947; Pearse, 1979; A. Smith, 1966; Thorpe, 1962; Tucker, 1960; Watanabe & Cox, 1975).

The best recent account of chitons as a group is that of Hyman (1967). The natural history of several species of California chitons is treated in Burnett et al. (1975), but little is known of most aspects of the biology of many of the species along our shores.

Chitons have been the subject of several physiological investigations on muscular responses, temperature tolerance, osmotic stress, and blood and tissue chemistry (see several papers in Burnett et al., 1975; see also Giese & Araki, 1962; Giese & Hart, 1967; Greer & Lawrence, 1967).

Identification of species often requires a close look at the girdle and the shell plates. Exposed portions of the shell may be sculptured with ribs or nodules. The portions of the valves covered by the girdle (the anterior edge of the head valve, the posterior margin of the tail valve, and the lateral extensions of the intermediate valves) are called insertion plates and often bear notches or slits, which have been used as taxonomic characters. The girdle may be smooth or bear spines, bristles, or scales. For a key to the intertidal chitons of the central California coast, see A. Smith (1975); for additional illustrations beyond those provided here, see Abbott (1974) and Burghardt & Burghardt (1969). The higher classification used here follows that in A. Smith (1960). An extensive bibliography on chitons is given in A. Smith (1973), and a rectification of west coast chiton nomenclature is available in A. Smith (1977).

The late Allyn G. Smith was generous with his help in the preparation of this chapter. He confirmed the identifications of the species shown in the photographs, helped in sorting out synonyms, and supplied drafts for the species descriptions. He shares no blame for any errors or shortcomings in the chapter. We acknowledge his contribution with pleasure, and remember him with affection.

Phylum Mollusca / Class Polyplacophora / Order Neoloricata
Suborder Lepidopleurina / Family Lepidopleuridae

16.1 Leptochiton rugatus (Pilsbry, 1892)
(=Lepidopleurus rugatus)

Uncommon, generally under well-buried rocks, low intertidal zone; Mendocino Co. to Scammon Lagoon (Baja California) and Gulf of California.

Body to 1.25 cm long; valves rounded dorsally, the lateral

areas slightly raised, the whole surface with fine, regularly arranged granulations; girdle narrow, appearing smooth or sandy, set with delicate scales visible only under magnification; overall color yellowish to orange-yellow; foot liver-colored.

See K. Palmer (1958), Pilsbry (1892), and A. Smith (1947a, 1975).

Phylum Mollusca / Class Polyplacophora / Order Neoloricata
Suborder Ischnochitonina / Family Ischnochitonidae

16.2 Ischnochiton interstinctus (Gould, 1846)
(=I. radians)

Fairly common under rocks, low intertidal zone; Aleutian Islands and Sitka (Alaska) to central California; subtidal off southern California.

Body to 2.5 cm long, 1.5 cm wide, oval, depressed; valves with dorsal longitudinal ridge, lusterless, extremely variable in color ranging from pure white to almost black, often spotted, occasionally with some valves bright blue and the rest another color (in Pacific northwest, especially Puget Sound, Washington, animals generally with brown-red valves); exposed surfaces of valves sculptured with indistinct radiating riblets; interior of valves dark blue; insertion plates of intermediate valves with single slits; girdle often speckled, covered with shining, weakly striated, overlapping, convex scales.

In California, this species spawns in February.

See Leloup (1940a), Pilsbry (1892), and Thorpe (1962).

Phylum Mollusca / Class Polyplacophora / Order Neoloricata
Suborder Ischnochitonina / Family Ischnochitonidae

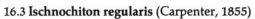

16.3 Ischnochiton regularis (Carpenter, 1855)

Fairly common, often in clusters, on bottoms of smooth rocks resting on hard substrata, low intertidal zone in areas of gentle surf; Union Landing (Mendocino Co.) to Punta Gorda (Monterey Co.)

Body 3–5 cm long, oblong; valves with middorsal longitudinal ridge, relatively smooth; end valves and lateral areas of intermediate valves with many delicate radiating threads; central areas of valves with numerous, longitudinal, somewhat beaded threads, separated by flat intervals; insertion plates of intermediate valves with two or three slits on each side; girdle wide, flat, with closely overlapping striated scales, often slightly lighter in color than valves; overall color uniformly slate-gray or slate-blue, rarely light ultramarine-blue; interior of valves light blue.

These animals avoid direct sunlight. Spawning occurs in February in California.

See Andrus & Legard (1975), Heath (1905a), K. Palmer (1958), and Thorpe (1962).

Phylum Mollusca / Class Polyplacophora / Order Neoloricata
Suborder Ischnochitonina / Family Ischnochitonidae

16.4 **Lepidozona mertensii** (Middendorff, 1846)

Locally abundant, low intertidal zone and subtidal waters; Auke Bay (Alaska) to Isla Guadalupe (Mexico).

Body 2.5–5 cm long; valves elevated, with angular dorsal ridge; end valves and elevated lateral areas of intermediate valves bearing diagonal lines of prominent, rather widely spaced pustules; central areas of valves with latticed sculpturing; overall color varying from orange-red through brick-red to crimson, sometimes with white blotches; girdle scaled, reddish with narrower light bands, the scales regular, overlapping, oval, shining, smooth to lightly striated.

This species is easily confused with *L. cooperi* (16.5) and *L. sinudentata* (16.6).

A small bryozoan (Ctenostomata) is sometimes found growing on the ventral surface of the girdle. *L. mertensii* spawns in February in California.

See Heath (1905a), Helfman (1968), Pilsbry (1892), and Thorpe (1962).

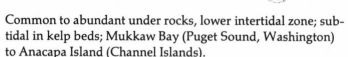

Phylum Mollusca / Class Polyplacophora / Order Neoloricata
Suborder Ischnochitonina / Family Ischnochitonidae

16.5 **Lepidozona cooperi** (Pilsbry, 1892)
(=Ischnochiton cooperi)

Common to abundant under rocks, lower intertidal zone; subtidal in kelp beds; Mukkaw Bay (Puget Sound, Washington) to Anacapa Island (Channel Islands).

Body to 5 cm long; similar in size and shape to *L. mertensii* (16.4) but differing in color (dull olivaceous or dull blackish brown, never reddish), in less-prominent pustular sculpturing on central areas of intermediate valves, and in smaller size of girdle scales.

See Andrews (1945), Ferreira (1978), K. Palmer (1958), Pilsbry (1892), and Stohler (1930).

Phylum Mollusca / Class Polyplacophora / Order Neoloricata
Suborder Ischnochitonina / Family Ischnochitonidae

16.6 **Lepidozona sinudentata** (Carpenter in Pilsbry, 1892)
(=Ischnochiton sinudentatus, ?I. berryi, ?I. decipiens, ?I. (Lepidozona) gallina, ?I. listrum)

Uncommon under rocks, low intertidal zone; commoner subtidally, especially at 50–70 m; Little River (Mendocino Co.) to southern California and Channel Islands; only in deeper water at southern end of range.

Body to 2 cm long; oval, fairly highly arched and ridged along back, differing from more common *L. mertensii* (16.4) principally in the following characters: central areas of intermediate valves with about 12 diagonal granulated ribs replacing lines of prominent pustules; color light or whitish with greenish of brownish-green markings, never reddish; girdle light-colored with alternating bands of darker overlapping scales.

See Ferreira (1978), K. Palmer (1958), and Pilsbry (1892).

Phylum Mollusca / Class Polyplacophora / Order Neoloricata
Suborder Ischnochitonina / Family Ischnochitonidae

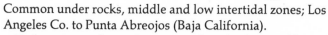

16.7 **Lepidozona californiensis** (Berry, 1931)

Common under rocks, middle and low intertidal zones; Los Angeles Co. to Punta Abreojos (Baja California).

Body to 3.8 cm long, elongate-oval; end valves and lateral areas of intermediate valves sculptured with tubercular diagonal ribs; central areas latticed with closely spaced longitudinal and transverse tubercular ribs of lesser strength; insertion plates of intermediate valves with single slit on each side; girdle covered with overlapping, faintly rib-striated scales becoming smaller toward margin; valves and girdle greenish or brownish yellow, olive-green, or brown, often with scattered dark or light markings.

In California, this species spawns in December.

See Berry (1931), Ferreira (1978), McLean (1969), and Thorpe (1962).

Phylum Mollusca / Class Polyplacophora / Order Neoloricata
Suborder Ischnochitonina / Family Ischnochitonidae

16.8 **Stenoplax heathiana** Berry, 1946
(=Ischnochiton heathiana)

Fairly common especially under rocks embedded in sand or fine gravel, middle and low intertidal zones; Union Landing (Mendocino Co.) to Puerto Santo Tomás (Baja California).

Body to 11 cm long, twice as long as wide; insertion plates of intermediate valves with two or more slits; valves pale grayish or brownish cream, often spotted with light to dark greenish gray; girdle moderately wide, clothed with scales, these fine, crowded, erect, incurved, loosely overlapping, blunt, finely striated; girdle color buff to brown with darker cloudy areas.

A primarily nocturnal species, *S. heathiana* remains buried in sand on the undersides of rocks during the day, but comes out at night and may be found exposed at dawn. It is sensitive to changes in illumination and responds to a shadow passing over it or to a sudden increase in light by clamping its foot down firmly. The animals feed at night on drift algae that lodge at the bases of rocks.

Details of reproduction and early development are described and beautifully illustrated in a classic cell-lineage study made at Pacific Grove (Monterey Co.) near the turn of the century by Stanford Professor Harold Heath (for whom *S. heathiana* is named; see Heath, 1899). Spawning occurs in conjunction with early morning low tides in May and June. The females extrude eggs in a pair of elongate, helically coiled strings of jelly (an unusual feature; in most chitons the eggs are not attached together) averaging nearly 80 cm long, each containing an average of nearly 116,000 eggs. Each egg is about 0.4 mm in diameter and is surrounded by an external membrane or chorion. Fertilization takes place after the eggs are laid. The first cleavage occurs more than 2 hours after fertilization, the second a half hour later, and the larvae begin to swim inside the chorion 24 hours after egg laying. On the seventh day after fertilization the jelly cords break up, and the larvae hatch from the chorion. They swim for 15 minutes to 2 hours, then settle on rocks or seaweeds. Over the next 2 weeks the shell plates form (the tail valve forming last), and the girdle extends outward and downward on all sides to enclose the head and the foot. The larval eyes, now covered by the girdle, commence degeneration. The young chiton begins to feed, using its radula, and is on its way to becoming an adult.

Large numbers of the small commensal snail *Vitrinella oldroydi* (13.44) occasionally occur in the mantle folds of mature animals.

See especially Heath (1899); see also Andrus & Legard (1975), Bauer (1975), Berry (1946), Linsenmeyer (1975), and A. Smith (1947b).

Phylum Mollusca / Class Polyplacophora / Order Neoloricata
Suborder Ischnochitonina / Family Ischnochitonidae

16.9 Stenoplax fallax (Pilsbry, 1892)
(=Ischnochiton fallax)

Occurring sparsely along with *Stenoplax heathiana* (16.8) under smooth rocks embedded in sand, low intertidal zone; more commonly subtidal; Union Landing (Mendocino Co.) to Punta Arbolitos (Baja California).

Similar to *S. heathiana* (16.8) in shape and size but differing as follows: central areas of intermediate valves pitted; insertion plates with single slits; valve color more brilliant and variable, with pink to crimson markings on a background of cream, green, light brown, or beige; young animals generally pink to reddish overall; inside of valves more brilliantly colored; girdle generally darker-colored, with closely packed, exceedingly small scales.

See K. Palmer (1958), Pilsbry (1892), and A. Smith (1947b).

Phylum Mollusca / Class Polyplacophora / Order Neoloricata
Suborder Ischnochitonina / Family Ischnochitonidae

16.10 Stenoplax conspicua (Pilsbry, 1892)
(=Ischnochiton conspicuus)

Common under rocks, sometimes in tidepools nestling in rounded depressions vacated by the purple sea urchin *Strongylocentrotus purpuratus* (11.5), low intertidal zone; Carpinteria (Santa Barbara Co.) to Isla Cedros (Baja California); a southerly form, *S. conspicua sonorana* Berry, occurs in Gulf of California.

Body to 10 cm long, half as wide, moderately elevated; central valve areas prominently raised, sculptured with radiating riblets and concentric ridges; front slope of head valve concave; valves green, brownish, or mottled, sometimes lighter with pink apexes in old eroded animals; valves internally white or tinged with pink and greenish blue; girdle wide and thick, densely covered with short, recurved bristles giving it a velvety appearance.

Stenoplax conspicua is occasionally preyed upon by octo-

puses, which drill small oval holes through the shell plates and presumably kill the chitons with a toxin. The small gastropods *Teinostoma invallata* and *Vitrinella oldroydi* (13.44) and the bivalve *Pristes oblongus* are occasionally found living as commensals under the girdle of this species.

See Albrecht (1921), K. Palmer (1958), Pilsbry (1892), Pilson & Taylor (1961), and A. Smith (1947b).

Phylum Mollusca / Class Polyplacophora / Order Neoloricata
Suborder Ischnochitonina / Family Ischnochitonidae

16.11 Tonicella lineata (Wood, 1815)
Lined Chiton

Scarce to common on rock faces covered with coralline algae, low intertidal zone and subtidal waters; Aleutian Islands (Alaska) to San Miguel Island (Channel Islands); Sea of Okhotsk (U.S.S.R.), northern Japan.

Intertidal-zone forms to 5 cm long; body elongate-oval; valves low, rounded, smooth, shining, with lateral areas not distinctly demarcated, usually light reddish, marked with sinuous or zigzag lines of alternating colors in combinations of dark and light red, dark or light blue and red, or whitish and red, occasionally blotched with some valves uncolored; valve interiors tinged with rose; girdle nude and leathery, yellowish or greenish, often alternately banded.

This handsomely marked animal resembles no other Pacific coast species. It generally occurs on rocks having a growth of erect or crustose coralline algae, which constitute the major food. On a background of pink corallines the colors blend well with the substratum, and the sinuous markings disrupt the lines between adjacent plates. The camouflage is effective, and probably represents a protective device against visual predators. On the Monterey Peninsula the intertidal and subtidal animals show consistent differences, the latter being smaller (1–2 cm long) with purple lines on the girdle.

Along the Oregon coast the species is often found living under purple sea urchins (*Strongylocentrotus purpuratus*, 11.5) or in urchin burrows, sometimes on bare patches of rock to which the individuals seem to home. There the sea stars *Pisas-*

ter ochraceus (8.13) and *Leptasterias hexactis* (8.11) are important predators, but in Monterey Bay sea stars rarely feed on *Tonicella* unless the chitons are removed from the rocks.

Activity patterns of the animals vary with habitat. Near Monterey, intertidal animals remain stationary when exposed at low tide, and move about only when covered by the rising tide; in contrast, the subtidal animals follow a diurnal rhythm with twice as much movement at night as during the day.

Laboratory studies of respiration in *Tonicella* show that oxygen consumption of animals exposed to air is only 73 percent that of submerged individuals; the animals out of water at low tide probably incur an oxygen debt, which is "paid back" on resubmersion. There is no apparent endogenous rhythm of fluctuation in respiratory rate, either tidal or diurnal. As in many other chitons, there may be an asymmetry in the numbers of gills on the two sides of the body.

The species shows latitudinal differences in the timing of its reproductive cycle; in California and central Oregon, eggs are released in April, but on San Juan Island (Washington), release occurs in May and June.

In a detailed study of *Tonicella* in Oregon, Barnes (1972) observed spawning and early development. Green-colored eggs stream from the posterior girdle cleft in groups of two or three, covered by mucus, and intermittent spawning continues for 1.5 hours. In laboratory experiments, naturally spawned eggs and sperm remain viable and capable of normal fertilization for up to 14 hours after release. Following fertilization the first polar body is produced in 0.5 hour, the second after about 1.5 hours. The first cleavage occurs 2 hours after fertilization, and the embryo reaches the 64–72-cell stage in 12 hours. This is followed by gastrulation. The trochophore larvae begin development between 16 and 24 hours (depending on temperature), and hatching occurs 43–44 hours after fertilization. The development of the trochophore stops 150–160 hours after fertilization, and further development occurs only after the larva has settled on a substratum containing crustose coralline algae (or a substratum experimentally treated with an extract of coralline algae). The larvae undergo metamorphosis within 12 hours after settlement and take on the shape of miniature chitons. The developing shell valves are at first lightly calcified and very flexible. By about

5.5 days after settlement the posterior valve has developed, and by the sixth day the small chiton possesses all definitive external morphological features except the gills. Thirty days after settlement the young have fully developed radulae and actively feed on corallines.

See especially Barnes (1972); see also Andrus & Legard (1975), Barnes & Gonor (1973), Demopulos (1975), Johnson (1969), Kincannon (1975), Leloup (1945), Pilsbry (1892), B. Robbins (1975), Seiff (1975), A. Smith (1947a), and Thorpe (1962).

*Phylum Mollusca / Class Polyplacophora / Order Neoloricata
Suborder Ischnochitonina / Family Ischnochitonidae*

16.12 **Cyanoplax hartwegii** (Carpenter, 1855)

Common on rocks protected from strong surf, upper middle intertidal zone, especially under cover of the brown alga *Pelvetia fastigiata* and in high tidepools; Monterey Bay to Punta Abreojos (Baja California).

Body to 5 cm long, ovate, depressed; valves rounded, sculptured with closely spaced granulations with some interspersed wartlike granules, more numerous on front and rear valves; head valve insertion plate with 9–12 slits, tail valve insertion plate with 8–11 slits; valves usually dull olive-green or olive-brown, with a series of white flecks on trailing edges, often with white stripe on valve ridges, flanked by pairs of black blotches; valve interiors blue-green; girdle darker than valves, covered with very fine, dense granular scales; distinguished from *C. dentiens* (16.13) by its much larger adult size, by its more oval and less elongate shape, and by the presence of scattered, warty protuberances on fine granular background of valves.

The diet of *C. hartwegii* is varied, but in beds of the seaweed *Pelvetia*, where populations of up to 30 chitons per m² occur, this brown alga constitutes 80 percent of the diet; in crevice habitats the red algae *Hildenbrandia*, *Petrocelis*, and *Endocladia* and the green alga *Cladophora* make up 90 percent of the food eaten. The animals are highly nomadic, though individuals may return to the same home site every day for several days in a row. They are active mainly at night, when the rocks are dry or awash; they cling to rocks less firmly than other chiton species, and even in their relatively protected habitats they move about mainly when least affected by surge. In general they avoid exposure to strong light.

In high tidepools *C. hartwegii* can survive wide fluctuations in salinity, and laboratory experiments have shown that the body fluids of the animal conform to those of the aquatic environment at salinities ranging from 75 to 125 percent that of normal seawater.

On rocks densely covered with *Pelvetia*, the alga appears to protect this chiton from predation by sea stars and birds.

See Andrus & Legard (1975), Connor (1975), DeBevoise (1975), Lyman (1975), McGill (1975), K. Palmer (1958), Pilsbry (1892, as *Trachydermon hartwegii*), Robb (1975), and A. Smith (1947a).

Phylum Mollusca / Class Polyplacophora / Order Neoloricata
Suborder Ischnochitonina / Family Ischnochitonidae

16.13 Cyanoplax dentiens (Gould, 1846)
(=C. raymondi)

Common in rock crevices, on exposed rock surfaces, on shells or stones, or hidden under algae or under rocks, middle intertidal zone; the only chiton occurring in the *Endocladia-Balanus* association in the Monterey area; Auke Bay (Alaska) to La Jolla (San Diego Co.).

Body to 2 cm long, small, elongate, low-arched, round-backed; valve surfaces granular, lateral areas indistinct; valve color extremely variable (see photographs), sometimes brilliant, usually dark olive-green or brown, often marked with lighter or darker streaks or spots or with crimson blotches; valve interiors bluish green; girdle narrow, covered with very fine sandy scales, dark in color with white spots or blotches; distinguished from *C. hartwegii* (16.12) by being smaller, smoother, and more elongate.

When spawning, several animals may be closely grouped. Fertilized eggs are retained and the young are brooded at least through the trochophore stage: up to 200 young may be held in the pallial grooves adjacent to the foot.

In some localities in the Monterey area, *C. dentiens* often uses another chiton, *Nuttallina californica* (16.17), as a micro-

habitat, moving under the larger chiton to occupy space in the pallial grooves. This may give protection from the dangers of desiccation, insolation, and predation.

See Berry (1917, as *Cyanoplax raymondi*; 1948), Glynn (1965), Gomez (1975, misprinted in title as *C. hartwegii*), Heath (1904, 1905a, 1907, all as *Trachydermon raymondi*), Higly & Heath (1912, as *T. raymondi*), Kues (1974), Pilsbry (1892, as *T. raymondi*), and A. Smith (1947a).

Phylum Mollusca / Class Polyplacophora / Order Neoloricata
Suborder Ischnochitonina / Family Ischnochitonidae

16.14 Basiliochiton heathii (Pilsbry, 1898)
(=Trachydermon heathii)

Rare, very low intertidal zone on rocky shores; more commonly subtidal; New Masset (Graham Island, British Columbia) to Monterey Bay.

Body to 3 cm long; valves elongate-oval with indistinct longitudinal dorsal ridge, lateral areas not well defined; valve surfaces smooth, with very little sculpturing except for very fine, overall granulation; valve color variable, olive-green with some light brownish spots, olive-green with end valves purplish red or crimson, or purplish red or crimson on all valves; valve interiors bright red shading to lighter near margins; girdle smooth to naked eye, minutely granular under magnification, with long fleshy hairs at sutural areas, most conspicuous at posterior end; girdle vivid orange with cream blotches; foot gray; gills darker.

See Berry (1911, as *Mopalia heathii*; 1917, 1925), Leloup (1940a), and Pilsbry (1898).

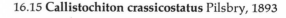

Phylum Mollusca / Class Polyplacophora / Order Neoloricata
Suborder Ischnochitonina / Family Callistoplacidae

16.15 Callistochiton crassicostatus Pilsbry, 1893

Under rocks in protected areas, middle and low intertidal zones and subtidal waters; Monterey Bay to Isla Cedros (Baja California).

Body to 5 cm long, oblong, highly arched; lateral areas of valves each with one prominent rib and sculptured with concentric granular riblets cut by one or several radiating, impressed lines; central areas of valves with strong longitudinal ribs, the intervals between them finely latticed; head valve with seven strong ribs, tail valve with five; valve color mottled grayish green to light brown; girdle finely scaled, narrow, cream-colored with white bands; distinguished from *C. palmulatus* (16.16) by having seven ribs on head valve instead of nine, a more highly arched body, and a single strong rib (rather than double ribs) on each lateral area.

See Leloup (1953) and Pilsbry (1893).

Phylum Mollusca / Class Polyplacophora / Order Neoloricata Suborder Ischnochitonina / Family Callistoplacidae

16.16 Callistochiton palmulatus Carpenter in Pilsbry, 1893

Fairly common under rocks, middle and low intertidal zones and subtidal waters; Moss Beach (San Mateo Co.) to San Diego.

Similar to *C. crassicostatus* (16.15) but smaller; lateral areas of intermediate valves with two nodulose ribs separated by a deep radiating channel; head valve normally with nine strong, tuberculate ribs; tail valve typically somewhat swollen, high, thick and enormously enlarged, a condition perhaps accentuated with advancing age; valves tan with darker markings; girdle narrow with striated scales, dark tan with orange bands.

See Leloup (1953) and Pilsbry (1893).

Phylum Mollusca / Class Polyplacophora / Order Neoloricata Suborder Ischnochitonina / Family Callistoplacidae

16.17 Nuttallina californica (Reeve, 1847)

Very common, clinging tightly to rocks exposed to strong surf, often in crevices or depressions, hidden under coralline algae, or wedged between mussels and barnacles, high to middle intertidal zones; Puget Sound (Washington) to San Diego.

Body to 5 cm long, elongate, moderately elevated; intermediate valves beaked in young or unworn specimens; valve surfaces granular or corrugated, lateral areas not raised but set off by low, curved, diagonal rib; intermediate valves with two slits on each side; valve color dark brown, olive-brown, or blackish, uniformly colored or with whitish stripes or blotches; valve interiors bluish, with black stains at their centers; girdle wide, covered with short, rigid, brownish spinelets with a few intermingled white spines, dull-colored, often with alternating dark and light bands; foot orange-yellow ventrally.

The valves and girdle of *Nuttallina* are often covered with a profuse growth of algae, including *Endocladia*, *Corallina*, *Polysiphonia*, *Cladophora*, and others. The valves of *Nuttallina* individuals living in depressions in beds of the worm *Dodecaceria fewkesi* (18.25) are often badly worn and pitted by a blue-green alga, *Entophysalis deusta*, which gives the chiton a coloration matching the background.

Nuttallina living on steep and somewhat shaded rock faces often harbor the small chiton *Cyanoplax dentiens* (16.13) below the girdle along the pallial grooves. In this microhabitat the smaller chiton apparently receives protection from desiccation, excessive sunlight, and predators.

Nuttallina feed by grazing on larger algae, preferring erect corallines but also taking the red algae *Endocladia muricata* and *Gelidium* and the green alga *Cladophora columbiana*. Pieces of algae as large as 20 percent of the animal's body length are sometimes found in the gut.

Nuttallina have a flexible girdle equipped with a curtainlike inner mantle fold, which permits an animal out of water at low tide to control exposure of the gills and perhaps to retain some water in the pallial grooves. (The inner fold also prevents spinelets on the inner surface of the girdle from scraping against the gills during locomotion.) Laboratory studies have shown that in *Nuttallina* out of water the respiratory rate drops to 73 percent of the rate shown when the animals are submerged. These chitons apparently do not incur an oxygen debt when out of water, but it is possible that metabolism becomes lowered then.

Laboratory studies on the ionic composition of the blood of *Nuttallina* exposed to salinities 50–150 percent that of normal seawater indicate that the blood levels of potassium and calcium are regulated actively, while those of sodium and magnesium are not and vary instead with the environment. *Nuttallina* living in the high intertidal zone are able to cope with osmotic stress better than those living further down.

Western gulls are important predators on *N. californica*. In some localities as much as 15 percent of the chiton population may be taken during the birds' breeding season.

See Andrus & Legard (1975), Berry (1935), Gomez (1975), Hewatt (1935), Linsenmeyer (1975), Moore (1975), Nishi (1975), Pilsbry (1892), Piper (1975), B. Robbins (1975), and Simonsen (1975).

Phylum Mollusca / Class Polyplacopohora / Order Neoloricata
Suborder Ischnochitonina / Family Chaetopleuridae

16.18 Chaetopleura gemma Dall, 1879

Under rocks, middle and low intertidal zones; Pachena Bay (Graham Island, British Columbia) to Bahía Magdalena (Baja California); fairly common in Monterey Bay area.

Body to 2 cm long; central areas of valves with longitudinal lines of granules; lateral areas sculptured with prominent, raised pustules; valves generally greenish or brownish, in some local populations brick red or orange-red with black tail valves; girdle leathery, with sparse, short, translucent spicules; girdle light orange to ashy green.

Along the California coast, *C. gemma* spawns in June.

See Pilsbry (1892) and Thorpe (1962).

Phylum Mollusca / Class Polyplacophora / Order Neoloricata
Suborder Ischnochitonina / Family Mopaliidae

16.19 Mopalia muscosa (Gould, 1846)
Mossy Chiton

Common on rocks and in tidepools, middle and low intertidal zones in regions of low to moderate surf; Queen Charlotte Islands (British Columbia) to Isla Cedros (Baja California).

Body to 9 cm long, oval, depressed, with a moderately developed dorsal ridge; valve surfaces lusterless, sometimes sculptured with wavy, crenulated riblets, but often eroded or overgrown by marine organisms; valves generally dull brown, blackish olive, or grayish, rarely brilliantly tinted with red, orange, or green; valve interiors bluish green tinted with lilac at their centers; girdle tan or cream-colored, densely covered with stiff, round, brownish-red bristles.

This large species tolerates a wide variety of environmental conditions and is one of the few west coast chitons that does well in estuaries. Movement occurs only at night when the animals are wetted or submerged. They remain in place during the day or whenever exposed by low tides. Many animals show homing behavior and move about within a radius of about 50 cm from home base, following set pathways on both outbound and return journeys. The home ranges of individuals generally do not overlap. Individuals living permanently submerged in tidepools were not observed to home or to follow regular pathways.

The animals feed primarily on the red algae *Gigartina papillata* and *Endocladia muricata* and the green alga *Cladophora* when available, but in some areas they appear to be unselective herbivores, and the gut may contain up to 15 percent animal matter.

Spawning has been noted in April and May in Monterey Bay, and more generally from July to September in central and northern California; in Santa Monica Bay (Los Angeles Co.) a winter and early spring spawning has been reported. The eggs and sperm are often shed in large tidepools. Some animals spawn in laboratory aquaria, but a consistent method of inducing spawning in the laboratory is lacking. The eggs, green or golden brown in color, are 0.29 mm in diameter and borne inside a bristly chorion. After fertilization, at 14–16°C the first cleavage occurs in 1 hour, the second in 1.5 hours, and the third in 2.5 hours; hatching occurs in 20 hours. The larvae swim freely for several days, during which time the mantle, foot, and larval eyes develop. Spicules form on the head and form a ring defining the edges of the future girdle by 6.5 days after fertilization. Settlement occurs about 11.5 days after fertilization, provided an appropriate substratum is present. The larvae settle readily in the laboratory if provided with rocks or shells covered with a greenish algal film. The first seven shell plates are visible at 13.5 days. They enlarge as the girdle extends outward, but the eighth (tail) plate forms only about 6 weeks after fertilization.

In San Francisco Bay, animals increase in length 15–34 mm per year; in southern California, growth slows or stops during the winter months. The animals attain sexual maturity in 2 years.

See Anderson (1969), Andrus & Legard (1975), Barnawell (1960), Boolootian (1964), Fitzgerald (1975), Heath (1905a), Hewatt (1938), Hinegardner (1974), Linsenmeyer (1975), Monroe & Boolootian (1965), Pilsbry (1892), S. Smith (1975), Terwilliger & Reed (1970), Thorpe (1962), Watanabe & Cox (1975), and Westersund (1975).

Phylum Mollusca / Class Polyplacophora / Order Neoloricata Suborder Ischnochitonina / Family Mopaliidae

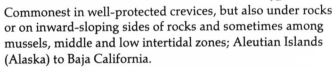

16.20 Mopalia ciliata (Sowerby, 1840)

Commonest in well-protected crevices, but also under rocks or on inward-sloping sides of rocks and sometimes among mussels, middle and low intertidal zones; Aleutian Islands (Alaska) to Baja California.

Body to 7.5 cm long, oval to elongate-oval, depressed; valves sculptured with low, rounded, coalescing beads, the lateral areas set off by a prominent beaded rib; valves variously colored but often with brilliant color patterns consisting of patches of green, red, brown, or occasionally yellow in combination, sometimes with blotches or streaks of white or dark color; valve interiors whitish tinged with pink and blue; girdle rather wide, thick, encroaching between valves at sutures, with a distinct caudal notch, decorated with many curled, brown hairs with groups of short, white spines at their bases; girdle color highly variable, dark green or pinkish to nearly white.

Mopalia ciliata is an omnivorous feeder; in San Francisco Bay, animal material (hydroids, sponges, and bryozoans) makes up 45 percent of the diet, the remainder being diatoms and red and brown algae. Body length increases 11–40 mm per year.

Animals collected on the California coast from Marin Co.

southward spawned in the laboratory at different times: February, March, July, and September through November. The eggs are gray-green in color and 0.2 mm in diameter. The first cleavage occurs 1–1.5 hours after fertilization, the second after 2 hours, and the third after 3–6 hours. Gastrulation begins at 10–12 hours and is complete at 24 hours. Larvae developing at 12–16°C emerge from egg cases 36–42 hours after fertilization. They swim freely up to the eighth day after fertilization, and by that time all eight valves are present. After settlement on the eight day, they undergo metamorphosis and are miniature adults by the sixteenth day.

See Barnawell (1954, 1960), Fitzgerald (1975), Pilsbry (1892), and Thorpe (1962).

Phylum Mollusca / Class Polyplacophora / Order Neoloricata Suborder Ischnochitonina / Family Mopaliidae

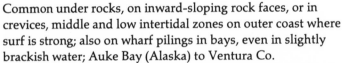

16.21 Mopalia hindsii (Reeve, 1847)

Common under rocks, on inward-sloping rock faces, or in crevices, middle and low intertidal zones on outer coast where surf is strong; also on wharf pilings in bays, even in slightly brackish water; Auke Bay (Alaska) to Ventura Co.

Body to 10 cm long, oval, depressed; valves sculptured with low, granular corrugations more or less zigzagged or crisscross on central areas and oblique on lateral areas; valves light in color to dark olive-gray or brownish, sometimes with white markings along tops of valves; valve interiors white; girdle wide, thick, encroaching at sutures between valve margins, notched at tail end, bearing scattered short hairs.

This is one of the largest and commonest chitons in northern California and the Pacific northwest. Individuals on shaded pilings often have the valves overgrown by bryozoans, hydroids, barnacles, and small tube-forming annelids. Worms of the polychaete family Spionidae may bore into the valves. In California the pallial groove may harbor the marine triclad worm *Nexilis epichitonius* (see Holleman & Hand, 1962).

As much as 59 percent of the diet of *M. hindsii* in San Francisco Bay consists of animal material such as bryozoans and young barnacles, including cyprid larvae. At Monterey the chitons are absent on outer wharf pilings, which are subjected to strong light and support an abundant growth of algae; they are found instead on inner pilings, where the light is subdued and where bryozoans, hydroids, and anemones grow. In this habitat, they feed mainly on small encrusting animal material.

The gonads enlarge, and eggs and sperm develop from October to March, the exact timing of the reproductive cycle varying somewhat from year to year. Spawning has been observed in October in California. In San Francisco Bay, growth rates of the adults range from 20 to 50 mm per year.

Laboratory experiments have shown that, during the reproductive cycle of *M. hindsii*, changes occur in the lipid, protein, and glycogen content of the gonads and sometimes other organs. Lipids are abundant in all tissues except the mantle.

See Andrus & Legard (1975), Barnawell (1954, 1960), Carefoot (1965), Giese & Araki (1962), Giese, Tucker & Boolootian (1959), Holleman & Hand (1962), Pilsbry (1892), and Thorpe (1962).

Phylum Mollusca / Class Polyplacophora / Order Neoloricata Suborder Ischnochitonina / Family Mopaliidae

16.22 Mopalia lignosa (Gould, 1846)

Fairly common, generally on sides of or under large boulders supported by stones above a coarse sand bottom, middle and low intertidal zones on open coast; in bays in Pacific northwest; Prince William Sound (Alaska) to Point Conception (Santa Barbara Co.).

Body to 7 cm long, oval, somewhat elevated, with a longitudinal middorsal ridge; valve surfaces apparently smooth but magnification reveals close, fine pitting in central areas; valves grayish, bluish, or greenish, with streaks and lines of brown, purple-brown, mahogany, or occasionally white; valve interiors white or light blue; girdle rather narrow, cream, brown, or purple-brown above, orange below, with sparsely scattered, short, thick, recurved bristles.

Stomach analyses show that *M. lignosa* feeds on more than two dozen types of algae, but diatoms and the green alga *Ulva* (sea lettuce) appear to be preferred foods. Animal material (small crustaceans, bryozoans, foraminifera) averages 2.7 per-

cent of the gut contents of animals from the Monterey Peninsula area.

As in other chitons, small light-sensitive organs (aesthetes) occur in the shell plates; experimental illumination from the side or from either end caused the animals to retreat.

Spawning has been observed at various times from February to June in California specimens removed to the laboratory and placed in non-circulating seawater with slightly elevated temperature. Males spawn by intermittent spurts of sperm, which gradually diffuse into the surrounding water. Eggs flow out of the females in single file and pile up behind the animal. These are light green, about 0.2 mm in diameter, very yolky, and surrounded by a frilly chorion. Following fertilization, at 14–16°C the first cleavage occurs at 1–1.5 hours, the second at 1.5–2 hours, the third at 2.5–3 hours. Gastrulation occurs at 10–12 hours after fertilization, and hatching at 19–24 hours. Hatching takes about 5 minutes, and the newly hatched free-swimming trochophores are egg-shaped. On the third day after fertilization the larval eyes appear, and after 5 days ridges and grooves appear that correspond to the future shell plates. After about 5.5 days the larvae settle, if shells covered with algal films are provided as a substratum, and a day later the rudiments of the first seven shell plates appear; the eighth (tail) plate does not develop until approximately 6 weeks after fertilization. By 8 weeks after fertilization, ctenidia have started to form and the juvenile chitons have the essential external characteristics of adults.

In laboratory experiments the respiratory rate of *M. lignosa* increased with rising temperature; for animals immersed in normal seawater, oxygen consumption was 14.3 µl per gm wet weight per hour at 8°C, and 36.2 µl per gm wet weight per hour at 19°C. Raising or lowering the salinity was accompanied by a decrease in respiratory rate.

See Andrus & Legard (1975), Boyle (1972), Fulton (1975), Heath (1905a), Hinegardner (1974), Lebsack (1975), Linsenmeyer (1975), Omelich (1967), Pilsbry (1892), Thorpe (1962), and Watanabe & Cox (1975).

Phylum Mollusca / Class Polyplacophora / Order Neoloricata
Suborder Ischnochitonina / Family Mopaliidae

16.23 Placiphorella velata Dall, 1879
Veiled Chiton

Uncommon, in shaded depressions or crevices on or under rocks, usually associated with crustose coralline algae, very low intertidal zone; more commonly subtidal on rocks; Forrester Island (Alaska) to Isla Cedros (Baja California) and upper Gulf of California.

Body to 5 cm long; valves wide and short, brownish or reddish, variously mottled and streaked with white, beige, green, and occasionally black (juveniles often with brightly colored spots); valve interiors bluish white; girdle broad and remarkably expanded into a "head flap" in front, rather narrow behind, the dorsal surface bearing sparse, elongate scaly hairs or setae, the head flap ventrally papillate with several long, fleshy precephalic tentacles arising anterior and lateral to the head; girdle generally cream or beige in color dorsally, head flap ventrally pigmented with red and green.

This extraordinary species can act either as a grazer, scraping food from the rocks with its radula like most other chitons, or as a predatory carnivore, trapping active animals for food. The adult chiton is rather sedentary, moving little on its small foot. Most of the time it remains in place and elevates the head flap of the girdle above the substratum into the water. The flap, pigmented on both sides and fringed with bristles, resembles a blade of red algae, and the rest of the body, often overgrown with algae and bryozoans, is well camouflaged. When small crustaceans and worms wander below the upraised flap, it is quickly lowered, trapping them in less than a second. Small crustaceans (2 mm and below) are swallowed whole; larger prey, including crabs up to 1 cm across, are torn up by the radula and ingested piecemeal.

Placiphorella belongs to the same chiton family (Mopaliidae) as species of *Mopalia*, which ingest some animal food along with plants. Some freshly collected *Placiphorella*, particularly intertidal animals, have the gut full of algae; others contain both plant and animal matter. However, *Placiphorella* represents the only chiton genus adapted to the capture of active animal prey.

In California, spawning occurs in September.

See McLean (1962), Pilsbry (1892), and Thorpe (1962).

Phylum Mollusca / Class Polyplacophora / Order Neoloricata
Suborder Ischnochitonina / Family Mopaliidae

16.24 Katharina tunicata (Wood, 1815)

Common, middle and low intertidal zones, clinging to sides and upper surfaces of rocks exposed to strong wave action and often direct sunlight, on unprotected outer coast; also in inland marine waters associated with swift currents in Pacific northwest; Aleutian Islands (Alaska) to Point Conception (Santa Barbara Co.); Kamchatka.

Body to 12 cm long, elongate-oval; all valves deeply embedded in girdle, exposed only in middorsal area; interior of valves white; girdle thick, shiny, black, leathery.

Katharina feed on brown and red algae and on benthic diatoms; young animals isolated in pools grow to a length of 25 mm in 1 year and 55 mm in 3 years. They reach sexual maturity at a weight of 4 gms (33–36 mm). Spawning takes place from March to July depending on latitude, with the later spawning occurring in colder regions. The eggs are green. The life span seems to be about 3 years.

This species is one of the few chitons on which a chromosome count has been made; *K. tunicata* has a diploid number of 12. Digestive gland tissue is the best material for obtaining chromosomes.

A variety of plants and animals attach to the exposed parts of the valves of *Katharina*, including coralline algae, hydroids, bryozoans, and barnacles.

Katharina have been investigated with regard to organic composition and changes in major biochemical components during the life cycle. Lipids are very abundant in the tissues, but decrease in amount during spawning and during prolonged starvation. Blood proteins are present in large amounts during most of the year but decrease to minimum levels of 180 mg percent just before spawning. The only organs that show major changes during the seasonal cycle are the gonads and the digestive gland. As with other chitons, shell plates occasionally become broken, and the breaks are repaired with a conchiolin membrane after a few weeks. The radular myoglobins have been isolated and purified, and the amino acid composition determined. The denticle caps of the radula teeth contain the hard iron compound magnetite.

See Caplan (1970), Comfort (1957), Dolph & Humphrey (1970), Giese (1969), Giese & Araki (1962), Giese & Hart (1967), Giese, Tucker & Boolootian (1959), Heath (1905a), A. Lawrence, J. Lawrence & Giese (1965), J. Lawrence & Giese (1969), Linsenmeyer (1975), Lowenstam (1962), Nimitz (1963), Nimitz & Giese (1964), Pilsbry (1892), Terwilliger & Read (1970), and Tucker & Giese (1959).

Phylum Mollusca / Class Polyplacophora / Order Neoloricata Suborder Acanthochitonina / Family Acanthochitonidae

16.25 **Cryptochiton stelleri** (Middendorff, 1846) Gumboot Chiton

Rare to common, usually next to deep pools or channels, often exposed to direct sunlight, low intertidal zone in areas protected from strong surf; more often subtidal in kelp beds; Aleutian Islands (Alaska) to San Miguel Island and San Nicolas Island (Channel Islands); northern Hokkaido Island (Japan), Kurile Islands, Kamchatka.

Largest chiton species in the world, to 33 cm long; mantle leathery and thick, completely covering all valves; overall color brick-red or reddish brown, rarely with blotches of lighter color, covered with closely spaced fascicles of very short, spreading spines or spicules; valves white, butterfly-shaped, seen only after dissection (see photograph).

These animals do not attach firmly to rocks, as other chitons do when disturbed, and are sometimes dislodged and washed ashore during storms. The shell plates are often broken, but the animal can repair such breaks. Juvenile forms are often found subtidally but rarely in the intertidal zone.

Large gumboots have up to 80 ctenidia in each pallial groove. When exposed to air during low tide the animals can respire by absorbing oxygen from the atmosphere, but at rates a fourth or a fifth that when submerged. When in air, but protected from direct sunlight, *Cryptochiton* often raise the margin of the mantle so that the ctenidia are exposed. In direct sunlight the animals press the mantle edges tightly against the substratum.

Spawning occurs between March and May in California; it may last for over a week, during which time individuals lose up to 5 percent of their body weight. The eggs are laid in gelatinous spiral strings up to 1 m long and cinnamon-red in color. Since the egg strings do not stick to the substratum, they are broken up by waves. The release of eggs by females triggers the release of sperm by males. Larvae 0.5 mm long are liberated from the eggs 5 days after fertilization; after a free-swimming stage lasting 20 hours, the larvae settle and begin to metamorphose.

The adults are not highly nomadic, though they appear to migrate inshore to breed. Off the Oregon coast, marked individuals remained within 20 m of the point of release even after 2 years. The animals seem to grow slowly and live for 20 years or more.

The gumboot feeds mainly on red algae, especially *Gigartina, Iridaea, Plocamium,* and various corallines, but will also eat *Ulva, Macrocystis,* and small *Laminaria.*

In Monterey Bay the gumboot often carries commensal animals in its pallial grooves. The polynoid worm *Arctonoe vittata* (18.3) occurred in 22–60 percent of the gumboots examined. The highest incidence was found between October and December and the lowest between May and August. The pea crab *Opisthopus transversus* (25.42) occurred as a commensal in 20–40 percent of the chitons examined throughout the year. As many as three crabs were found in the pallial grooves of each *Cryptochiton,* but never more than one worm was present. The commensals appear to feed on plankton and detritus brought in by the respiratory currents of the host, and there is no evidence that the relationship harms the chiton. Laboratory experiments show that both commensals detect *Cryptochiton* from a distance and move toward it. In addition to these animal commensals, 24 species of algae have been identified as epibionts on large subtidal specimens of this species (I. A. Abbott, pers. comm.).

This chiton apparently has few enemies, but in northern

California the predaceous snail *Ocenebra lurida* (13.76) rasps pits 1 cm in diameter and 3–4 mm deep into the dorsal surface of the body, exposing the yellow flesh covering the valves. Surprisingly, the gumboot is largely ignored by the sea otter as a source of food. However, it was once used as food by west coast Indians, and valves are frequently found in prehistoric kitchen middens.

Blood circulation has been studied and found to be very sluggish. The blood volume is large, averaging 44 percent of the wet body weight in adults. The blood glucose concentration in *C. stelleri* ranges from 4 to 12 mg percent, and experimental injection of insulin, norepinephrine, and serotonin has no influence on blood-sugar levels.

Laboratory studies on isolated portions of the intestine of *Cryptochiton* show that the amino acids alanine, glycine, proline, and lysine are actively transported across membranes but that glutamic acid is not. The cells lining the intestine can also transport carbohydrates against a concentration gradient. Physiological studies have also been made on muscle activity and excretion.

The radular muscle myoglobins have been studied, and the amino acid composition determined. An iron-protein complex (ferritin) has been isolated from columnar epithelial cells of the radula. The radular teeth are capped with magnetite.

See Greer & Lawrence (1967), Harrison (1975), Heath (1897, 1903, 1905a,b), A. Lawrence & J. Lawrence (1967), A. Lawrence, J. Lawrence & Giese (1965), A. Lawrence & Mailman (1967), A. Lawrence, Mailman & Puddy (1972), Leloup (1940b, 1955), Lowenstam (1962), MacGinitie & MacGinitie (1968), Michael (1975), Okuda (1947), J. Palmer & Frank (1974), Peterson & Johansen (1973), Pilsbry (1892), Puddy (1970), K. Robbins (1975), Talmadge (1975), Terwilliger & Read (1970), Towe & Lowenstam (1967a,b), Towe, Lowenstam & Nesson (1963), Tucker (1960), Tucker & Giese (1962), Vasu (1965), and Webster (1968).

Literature Cited

Abbott, R. T. 1974. American seashells. 2nd ed. New York: Van Nostrand Reinhold. 663 pp.

Albrecht, P. G. 1921. Chemical study of several marine mollusks of the Pacific coast. J. Biol. Chem. 45: 395–405.

Anderson, E. 1969. Oocyte–follicle cell differentiation in two species of amphineurans (Mollusca), *Mopalia muscosa* and *Chaetopleura apiculata*. J. Morphol. 129: 89–126.

Andrews, H. L. 1945. The kelp beds of the Monterey region. Ecology 26: 24–37.

Andrus, J. K., and W. B. Legard. 1975. Description of the habitats of several intertidal chitons (Mollusca: Polyplacophora) found along the Monterey Peninsula of central California. Veliger 18 (Suppl.): 3–8.

Barnawell, E. B. 1954. The biology of the genus *Mopalia* in San Francisco Bay. Master's thesis, Zoology, University of California, Berkeley.

———. 1960. The carnivorous habit among the Polyplacophora. Veliger 2: 85–88.

Barnes, J. R. 1972. Ecology and reproductive biology of *Tonicella lineata* (Wood, 1815) (Mollusca: Polyplacophora). Doctoral thesis, Oregon State University, Corvallis. 148 pp.

Barnes, J. R., and J. J. Gonor. 1973. The larval settling response of the lined chiton *Tonicella lineata*. Mar. Biol. 20: 259–64.

Bauer, K. 1975. A behavioral light response in the chiton *Stenoplax heathiana*. Veliger 18 (Suppl.): 74–78.

Berry, S. S. 1911. A new California chiton. Proc. Acad. Natur. Sci. Philadelphia 1911: 487–92.

———. 1917. Notes on west American chitons I. Proc. Calif. Acad. Sci. 7: 229–48.

———. 1925. The species of *Basiliochiton*. Proc. Acad. Natur. Sci. Philadelphia 77: 23–29.

———. 1931. A redescription, under a new name, of a well-known California chiton. Proc. Malacol. Soc. London 19: 255–58.

———. 1935. A further record of a chiton (*Nuttallina*) with nine valves. Nautilus 48: 89–90.

———. 1946. A re-examination of the chiton, *Stenoplax magdalenensis* (Hinds), with description of a new species. Proc. Malacol. Soc. London 26: 161–66.

———. 1948. Two misunderstood west American chitons. Leaflets in Malacol. 1: 13–15.

Boolootian, R. A. 1964. On growth, feeding and reproduction in the chiton *Mopalia muscosa* of Santa Monica Bay. Helgoländer Wiss. Meeresunters. 11: 186–99.

Boyle, P. R. 1972. The aesthetes of chitons. Mar. Behav. Physiol. 1: 171–84.

Burghardt, G. E., and L. E. Burghardt. 1969. A collector's guide to west coast chitons. Spec. Publ. 4, San Francisco Aquarium Soc., Calif. 45 pp., 4 pls.

Burnett, R., D. P. Abbott, I. Abbott, C. Baxter, F. A. Fuhrman, M. Gilmartin, C. Harrold, G. Mpitsos, J. Phillips, B. Lyman, and R. Stohler, eds. 1975. The biology of chitons. Veliger 18 (Suppl.): 1–128.

Caplan, Ronald I. 1970. Bioenergetics, growth and population dynamics of an intertidal herbivore, *Katharina tunicata* Wood, 1815

(Mollusca, Amphineura). Doctoral thesis, Oceanography, Oregon State University, Corvallis.

Carefoot, T. H. 1965. Magnetite in the radula of the Polyplacophora. Proc. Malacol. Soc. London 36: 203–12.

Comfort, A. 1957. The duration of life in molluscs. Proc. Malacol. Soc. London 32: 219–41.

Connor, M. S. 1975. Niche apportionment among the chitons *Cyanoplax hartwegii* and *Mopalia muscosa* and the limpets *Collisella limatula* and *Collisella pelta* under the brown alga *Pelvetia fastigiata*. Veliger 18 (Suppl.): 9–17.

DeBevoise, A. E. 1975. Predation on the chiton *Cyanoplax hartwegii* (Mollusca: Polyplacophora). Veliger 18 (Suppl.): 47–50.

Demopulos, P. A. 1975. Diet, activity and feeding in *Tonicella lineata* (Wood, 1815). Veliger 18 (Suppl.): 42–46.

Dolph, C. I., and D. G. Humphrey. 1970. Chromosomes of the chiton *Katharina tunicata*. Trans. Amer. Microscop. Soc. 89: 229–32.

Ferreira, A. J. 1978. The genus *Lepidozona* (Mollusca: Polyplacophora) in the temperate eastern Pacific, Baja California to Alaska, with the description of a new species. Veliger 21: 19–44.

Fitzgerald, W. J. 1975. Movement patterns and phototactic response of *Mopalia ciliata* and *Mopalia muscosa* in Marin County, California. Veliger 18: 37–39.

Fulton, F. T. 1975. The diet of the chiton *Mopalia lignosa* (Gould, 1846) (Mollusca: Polyplacophora). Veliger 18 (Suppl.): 38–41.

Giese, A. C. 1969. A new approach to the biochemical composition of the mollusc body. Oceanogr. Mar. Biol. Ann. Rev. 7: 175–229.

Giese, A. C., and G. Araki. 1962. Chemical changes with reproductive activity of the chitons, *Katharina tunicata* and *Mopalia hindsii*. J. Exper. Zool. 151: 259–67.

Giese, A. C., and M. A. Hart. 1967. Seasonal changes in component indices and chemical composition in *Katharina tunicata*. J. Exper. Mar. Biol. Ecol. 1: 34–46.

Giese, A. C., J. S. Tucker, and R. A. Boolootian. 1959. Annual reproductive cycles of the chitons, *Katharina tunicata* and *Mopalia hindsii*. Biol. Bull. 117: 81–88.

Glynn, P. W. 1965. Community composition, structure and interrelationships in the marine intertidal *Endocladia muricata–Balanus glandula* association in Monterey Bay, California. Beaufortia 12: 1–198.

Gomez, R. L. 1975. An association between *Nuttallina californica* and *Cyanoplax hartwegii*, two west coast polyplacophorans (chitons). [*Cyanoplax hartwegii* is an error; read *C. dentiens*.] Veliger 18 (Suppl.): 28–29.

Greer, M. L., and A. L. Lawrence. 1967. The active transport of selected amino acids across the gut of the chiton (*Cryptochiton stelleri*). 1. Mapping determinations and effects of anaerobic conditions. Comp. Biochem. Physiol. 22: 665–74.

Harrison, J. T. 1975. Isometric responses of somatic musculature of *Cryptochiton stelleri* (Mollusca: Polyplacophora). Veliger 18 (Suppl.): 79–82.

Heath, H. 1897. External features of young *Cryptochiton*. Proc. Acad. Natur. Sci. Philadelphia 1897: 299–302.

———. 1899. The development of *Ischnochiton*. Zool. Jahrb. 12: 567–656.

———. 1903. The function of the chiton subradular organ. Anat. Anz. 23: 92–95.

———. 1904. The larval eye of chitons. Proc. Acad. Natur. Sci. Philadelphia 1904: 257–59.

———. 1905a. The breeding habits of chitons of the California coast. Zool. Anz. 29: 390–93.

———. 1905b. The excretory and circulatory systems of *Cryptochiton stelleri* Midd. Biol. Bull. 9: 213–25.

———. 1907. The gonad in certain species of chitons. Zool. Anz. 32: 10–12.

Helfman, E. S. 1968. A ctenostomatous ectoproct epizoic on the chiton *Ischnochiton mertensii*. Veliger 10: 290–91.

Hewatt, W. G. 1935. Ecological succession in *Mytilus californianus* habitats as observed in Monterey Bay. Ecology 16: 244–51.

———. 1938. Notes on the breeding seasons of the rocky beach fauna of Monterey Bay, California. Proc. Calif. Acad. Sci. 23: 283–88.

Higley, R. M., and H. Heath. 1912. The development of the gonad and gonaducts in two species of chitons. Biol. Bull. 22: 95–97.

Hinegardner, R. 1974. Cellular DNA content of the Mollusca. Comp. Biochem. Physiol. 47A: 447–60.

Holleman, J. T., and C. Hand. 1962. A new species, genus and family of marine flatworms (Turbellaria: Tricladida, Maricola) commensal with mollusks. Veliger 5: 20–22.

Hyman, L. H. 1967. Polyplacophora, pp. 70–142, *in* The invertebrates: Mollusca. Vol. 6. New York: McGraw-Hill. 792 pp.

Johnson, K. M. 1969. Quantitative relationships between gill number, respiratory surface, and cavity shape in chitons. Veliger 11: 272–76.

Kincannon, E. A. 1975. The relations between body weight and habitat temperature and the respiratory rate of *Tonicella lineata* (Wood, 1815) (Mollusca: Polyplacophora). Veliger 18 (Suppl.): 87–93.

Kues, B. S. 1974. A new subspecies of *Cyanoplax dentiens* (Polyplacophora) from San Diego, California. Veliger 16: 297–300.

Lawrence, A. L., and D. C. Lawrence. 1967. Sugar absorption in the intestine of the chiton, *Cryptochiton stelleri*. Comp. Biochem. Physiol. 22: 341–57.

Lawrence, A. L., J. M. Lawrence, and A. C. Giese. 1965. Cyclic variations in the digestive gland and glandular oviduct of chitons (Mollusca). Science 147: 508–10.

Lawrence, A. L., and D. S. Mailman. 1967. Electrical potentials and ion concentrations across the gut of *Cryptochiton stelleri*. J. Physiol. 193: 535–45.

Lawrence, A. L., D. S. Mailman, and R. E. Puddy. 1972. The effect of carbohydrates on the intestinal potentials of *Cryptochiton stelleri*. J. Physiol. 225: 515–27.

Lawrence, J. M., and A. C. Giese. 1969. Changes in lipid composition of the chiton *Katharina tunicata* with reproductive and nutritive state. Physiol. Zool. 42: 353–60.

Lebsack, C. S. 1975. Effect of temperature and salinity on the oxygen consumption of the chiton *Mopalia lignosa*. Veliger 18 (Suppl.): 94–97.

Leloup, E. 1940a. Charactères anatomiques de certains chitons de la côte Californienne. Mém. Mus. Roy. Hist. Natur. Belg. 17: 1–41.

———. 1940b. Sur la présence d'aesthetes chez *Cryptochiton stelleri* (Middendorff, 1846). Bull. Mus. Roy. Hist. Natur. Belg. 16: 1–7.

———. 1945. A propos de certains chitons du genre *Tonicella* Carpenter, 1873. Bull. Mus. Roy. Hist. Natur. Belg. 21: 1–15.

———. 1953. Charactères anatomique de certains Callistochitons. Bull. Mus. Roy. Hist. Natur. Belg. 29: 1–19.

———. 1955. A propos du *Cryptochiton stelleri* (Middendorff, 1846). Bull. Mus. Roy. Hist. Natur. Belg. 31: 1–3.

Linsenmeyer, T. A. 1975. The resistance of five species of polyplacophorans to removal from natural and artificial surfaces. Veliger 18 (Suppl.): 83–86.

Lowenstam, H. A. 1962. Magnetite in denticle capping in Recent chitons (Polyplacophora). Bull. Geol. Soc. Amer. 73: 435–38.

Lyman, B. W. 1975. Activity patterns of the chiton *Cyanoplax hartwegii* (Mollusca: Polyplacophora). Veliger 18 (Suppl.): 63–69.

MacGinitie, G. E., and N. MacGinitie. 1968. Notes on *Cryptochiton stelleri* (Middendorff, 1846). Veliger 11: 59–61.

McGill, V. L. 1975. Response to osmotic stress in the chiton *Cyanoplax hartwegii* (Mollusca: Polyplacophora). Veliger 18 (Suppl.): 109–12.

McLean, J. H. 1962. Feeding behavior of the chiton *Placiphorella*. Proc. Malacol. Soc. London 35: 23–26.

———. 1969. Marine shells of southern California. Los Angeles Co. Mus. Natur. Hist., Sci. Ser. 24, Zool. 11. 104 pp.

Michael, P. F. 1975. Blood glucose concentration and regulation in *Cryptochiton stelleri* (Mollusca: Polyplacophora). Veliger 18 (Suppl.): 117–21.

Monroe, H. C., and R. H. Boolootian. 1965. Reproductive biology of the chiton *Mopalia muscosa*. Bull. So. Calif. Acad. Sci. 64: 223–28.

Moore, M. M. 1975. Foraging of the western gull *Larus occidentalis* and its impact on the chiton *Nuttallina californica*. Veliger 18 (Suppl.): 51–53.

Nimitz, M. A. 1963. The histology and histochemistry of the chiton *Katharina* in relation to the reproductive cycle. Doctoral thesis, Biological Sciences, Stanford University, Stanford, Calif. 197 pp.

Nimitz, M. A., and A. C. Giese. 1964. Histochemical changes correlated with reproductive activity and nutrition in the chiton *Katharina tunicata*. Quart. J. Microscop. Sci. 105: 481–95.

Nishi, R. 1975. The diet and feeding habits of *Nuttallina californica* (Reeve, 1847) from two contrasting habitats in central California. Veliger 18 (Suppl.): 30–33.

Okuda, S. 1947. Notes on the post-larval development of the giant chiton, *Cryptochiton stelleri* (Midd.). J. Fac. Sci., Hokkaido Univ., Zool. 9: 267–75.

Omelich, P. 1967, The behavioral role and the structure of the aesthetes of chitons. Veliger 10: 77–82.

Palmer, J. B., and P. W. Frank. 1974. Estimates of growth of *Cryptochiton stelleri* (Middendorff, 1846). Veliger 16: 301–4.

Palmer, K. V. W. 1958. Type specimens of marine Mollusca described by P. P. Carpenter from the west coast (San Diego to British Columbia). Geol. Soc. Amer. Mem. 76: 1–376.

Pearse, J. S. 1979. Polyplacophora, pp. 27–85, *in* A. C. Giese and J. S. Pearse, eds., Reproduction of marine invertebrates, vol. 5, Molluscs: Pelecypods and lesser classes. New York: Academic Press. 369 pp.

Peterson, J. A., and K. Johansen. 1973. Gas exchange in the giant sea cradle *Cryptochiton stelleri* (Middendorff). J. Exper. Mar. Biol. Ecol. 12: 27–43.

Pilsbry, H. A. 1892–94. The manual of conchology. Vols. 14 and 15. Philadelphia Acad. Natur. Sci.

———. 1898. Chitons collected by Dr. Harold Heath at Pacific Grove, near Monterey, California. Proc. Acad. Natur. Sci. Philadelphia 1898: 287–90.

Pilson, M. E. Q., and P. B. Taylor. 1961. Hole drilling by octopus. Science 134: 1366–68.

Piper, S. C. 1975. The effects of air exposure and external salinity change on the blood ionic composition of *Nuttallina californica*. Veliger 18 (Suppl.): 103–8.

Puddy, R. E. 1970. Ion and carbohydrate absorption in the posterior intestine of *Cryptochiton stelleri*. Doctoral thesis, University of Houston, Texas. 106 pp.

Ricketts, E. F., and J. Calvin. 1968. Between Pacific tides. 4th ed. Revised by J. W. Hedgpeth. Stanford, Calif.: Stanford University Press. 614 pp.

Robb, M. F. 1975. The diet of the chiton *Cyanoplax hartwegii* in three intertidal habitats. Veliger 18 (Suppl.): 34–37.

Robbins, B. A. 1975. Aerial and aquatic respiration in the chitons *Nuttallina californica* and *Tonicella lineata*. Veliger 18 (Suppl.): 98–102.

Robbins, K. B. 1975. Active absorption of D-glucose and D-galactose by intestinal tissue of the chiton *Cryptochiton stelleri* (Middendorff, 1846). Veliger 18 (Suppl.): 122–27.

Seiff, S. R. 1975. Predation upon subtidal *Tonicella lineata* of Mussel Point, California. (Mollusca: Polyplacophora). Veliger 18 (Suppl.): 54–56.

Simonsen, M. 1975. Response to osmotic stress in vertically separated populations of an intertidal chiton, *Nuttallina californica* (Mollusca; Polyplacophora). Veliger 18 (Suppl.): 113–16.

Smith, A. G. 1947a. Class Amphineura, Order Polyplacophora, Family Lepidopleuridae. Conchol. Club. So. Calif., Minutes 66: 3–16.

———. 1947b. Genus *Ischnochiton* Gray, 1847, subgenus *Stenoplax* Dall, 1879. Conchol. Club So. Calif., Minutes 68: 3–11.

———. 1960. Amphineura, pp. I 41–76, *in* P. C. Moore, ed., Treatise on invertebrate paleontology, part I, Mollusca 1. New York: Geol. Soc. Amer.; Lawrence: University of Kansas Press.

———. 1966. The larval development of chitons (Amphineura). Proc. Calif. Acad. Sci. 32: 433–66.

———. 1973. Polyplacophora—a selected bibliography. Of Sea and Shore 4: 201–6, 208.

———. 1975. Class Polyplacophora (Chitons), pp. 457–66, *in* R. I. Smith and J. T. Carlton, eds., Light's manual: Intertidal invertebrates of the central California coast. 3rd ed. Berkeley and Los Angeles: University of California Press. 716 pp.

———. 1977. Rectification of west coast chiton nomenclature (Mollusca: Polyplacophora). Veliger 19: 215–58.

Smith, S. Y. 1975. Temporal and spatial activity patterns of the intertidal chiton *Mopalia muscosa*. Veliger 18 (Suppl.): 57–62.

Stohler, R. 1930. Gewichtsverhältnisse bei gewisser marinen evertebraten. Zool. Anz. 91: 149–55.

Talmadge, R. T. 1975. A note on *Ocenebra lurida* (Middendorff). Veliger 17: 414.

Terwilliger, R. C., and K. R. H. Read. 1970. The radular muscle myoglobins of the amphineuran mollusks *Katharina tunicata* Wood, *Cryptochiton stelleri* Middendorff, and *Mopalia muscosa* Gould. Internat. J. Biochem. 1: 281–91.

Thorpe, S. R., Jr. 1962. A preliminary report on the spawning and related phenomena in California chitons. Veliger 4: 202–10.

Tomlinson, J. T. 1959. Magnetic properties of chiton radulae. Veliger 2: 36.

Towe, K. M., and H. A. Lowenstam. 1967a. Magnetite in denticle capping in Recent chitons (Polyplacophora). Bull. Geol. Soc. Amer. 73: 435–38.

———. 1967b. Ultrastructure and development of iron mineralization in the radular teeth of *Cryptochiton stelleri* (Mollusca). J. Ultrastruc. Res. 17: 1–13.

Towe, K. M., H. A. Lowenstam, and M. H. Nesson. 1963. Invertebrate ferritin: Occurrence in Mollusca. Science 142: 63–64.

Tucker, J. S. 1960. Studies on the reproduction of the amphineuran mollusk *Cryptochiton stelleri* (Middendorff). Doctoral thesis, Biological Sciences, Stanford University, Stanford, Calif. 115 pp.

Tucker, J. S., and A. C. Giese. 1959. Shell repair in chitons. Biol. Bull. 116: 318–22.

———. 1962. Reproductive cycle of *Cryptochiton stelleri* (Middendorff). J. Exper. Zool. 150: 33–43.

Vasu, B. S. 1965. Variations in the body fluid nitrogenous constituents of *Pisaster ochraceus* (Echinodermata) and *Cryptochiton stelleri* (Mollusca) in relation to nutrition and reproduction. Doctoral thesis, Biological Sciences, Stanford University, Stanford, Calif. 142 pp.

Watanabe, J. M., and L. R. Cox. 1975. Spawning behavior and larval development in *Mopalia lignosa* and *Mopalia muscosa* (Mollusca: Polyplacophora) in central California. Veliger 18 (Suppl.): 18–27.

Webster, S. K. 1968. An investigation of the commensals of *Cryptochiton stelleri* (Middendorff, 1847) in the Monterey Peninsula area, California. Veliger 11: 121–25.

Westersund, K. R. 1975. Exogenous and endogenous control of movement in the chiton *Mopalia muscosa* (Mollusca: Polyplacophora). Veliger 18 (Suppl.): 70–73.

Yonge, C. M. 1960. Class Polyplacophora, pp. 9–12, *in* R. C. Moore, ed., Treatise on invertebrate paleontology, part I, Mollusca 1. New York: Geol. Soc. Amer.; Lawrence: University of Kansas Press.

Chapter 17

Cephalopoda: *The Squids and Octopuses*

F. G. Hochberg, Jr., and W. Gordon Fields

Anyone who has marveled at the beauty of living cephalopods will realize that they have evolved a long way from their sedentary molluscan relatives. Colors change in rippling waves over the surface of the soft body, providing instant camouflage or bold courtship patterns. The octopus moves with a fluid grace along the sea floor, while the more active squid darts through open waters offshore. The ability of these animals to perceive the world around them, to discriminate and learn, and to respond quickly in non-stereotyped ways marks them as the most highly organized and intelligent of invertebrates. The large head bears a complex brain encased in cartilage, two large eyes somewhat resembling our own, and other special sense organs. Portions of the foot are modified to form eight sucker-bearing arms, which radiate from the head region, hence the name Cephalopoda ("head-footed ones"). In squids two additional tentacles are present, which shoot out suddenly to grasp prey and then rapidly retract.

All cephalopods swim by a jet propulsion system. Water forcefully expelled from the mantle cavity emerges through the siphon or pallial funnel. This lies on the ventral side and has evolved, like the arms, from the ancestral molluscan foot. The siphon is flexible; hence the stream of water can be pointed in any direction. In squids, swimming is aided by a fin on each side of the body; in many species the body is elongated and streamlined for speed, and the shell is reduced to an internal horny pen that extends the length of the mantle. In the more sedentary octopuses the pen is reduced to a pair of pins, the saclike body lacks fins, and walking with the arms is a second mode of locomotion.

Most cephalopods are active predators, well equipped to sight prey and to catch and hold it with their arms and tentacles (e.g., Fields, 1965; Messenger, 1968). Injection of a salivary toxin helps to subdue the captured animal (Ballering et al., 1972; J. Young, 1965). Beaklike jaws are used for preliminary rendering of a carcass, whereas the radula is used for more delicate cleaning (Altman & Nixon, 1970). In octopuses, the radula may be used to drill holes in mollusk shells (Arnold & Arnold, 1969; Pilson & Taylor, 1961; Wodinsky, 1969). Live cephalopods should be handled with care, for even small individuals can bite with painful results (Berry & Halstead, 1954; Halstead, 1949; High, 1976b; Oglesby, 1972; Snow, 1970). Tetrodotoxin, a potent neurotoxin, has been isolated from the posterior salivary glands of an Australian octopus (Sheumack et al., 1978).

Cephalopods produce a magnificent repertoire of colors, patterns, and textures through a complex interplay of various elements in the skin, including muscles, chromatophores, iridophores, and leucophores (e.g., Arnold, 1967; Brocco, 1975, 1976; Brocco, O'Clair & Cloney, 1974; Denton & Land, 1971; Florey, 1969; Froesch & Messenger, 1978; Kawaguti & Mabuchi, 1970; Kawaguti & Ohgishi, 1962; Mirow, 1972). Color patterns that contrast with the background are used for social signaling (i.e., courtship, schooling, etc.) or for defense. In addition, cephalopods have the remarkable ability to camouflage themselves by matching and thus blending with the hue, the intensity, and even the texture of the background (Boycott, 1953; Hochberg & Packard, 1974; Holmes, 1955; Messen-

F. G. Hochberg, Jr., is Curator of Invertebrate Zoology at the Santa Barbara Museum of Natural History. W. Gordon Fields is Professor Emeritus of Biology at the University of Victoria.

ger, 1974; Messenger, Wilson & Hedge, 1973; Packard, 1974; Packard & Hochberg, 1977; Packard & Sanders, 1971; Warren, Scheier & Riley, 1974). Dark ink stored in a special sac can be released when the animals are threatened or pursued, providing a smoke screen or decoy to distract a predator. Some mid-water squids also possess an elaborate array of light-producing photophores, which may be useful for sight, feeding, defense, mutual recognition, and/or countershading; some squids even emit luminous secretions (Arnold & Young, 1974; Arnold, Young & King, 1974; Berry, 1920; Cousteau, 1954; Dilly, 1973; Haneda, 1956; Herring, 1977; R. Young, 1973, 1975, 1977; R. Young & Roper, 1976, 1977).

The majority of the numerous west coast cephalopod species are pelagic and remain in offshore waters (Anderson, 1978; M. Clarke, 1966; Laevastu & Fiscus, 1978; McGowan, 1967; Okutani & McGowan, 1969; Pearcy, 1965; Pearcy et al., 1977; Roper & Young, 1975; R. Young, 1972). Only one species of squid, *Loligo opalescens* (17.2), is abundant inshore. At certain times of the year immense schools migrate into shallow waters to mate, spawn, and die (Fields, 1950, 1965; Hobson, 1965; McGowan, 1954; Recksiek & Frey, 1978). In California this squid is commercially exploited for bait, fertilizer, and food for man and his pets. The largest squid frequenting the west coast, *Moroteuthis robusta* (17.4), often taken in deep water by trawling gear, is occasionally found washed up on beaches or swimming near the surf line. *Rossia pacifica* (17.1), a small squid, lives on or near the bottom in coastal waters. It is often found in commercial catches of shrimps or fishes, and may be referred to the intertidal biologist for identification.

Many benthic, or bottom-dwelling, cephalopods live at considerable depths. The most conspicuous forms encountered in the intertidal area belong to the genus *Octopus*. Normally inhabiting subtidal regions, they often move into rocky tidepool areas to feed or spawn. The old practice of squirting noxious or irritating chemicals into holes in rocky reefs to drive out the octopuses is highly destructive to other shore life and is now forbidden by law.

Some *Octopus* species lay large eggs (10–17 mm in length); their young hatch as fully formed juveniles resembling miniature adults and remain on the bottom in the vicinity of the parent. Other octopuses lay small eggs (2–6 mm); these give rise to swimming larvae, which feed and grow in the plankton for several weeks or months before settling.

The identification of cephalopods, and especially octopuses, is complicated by the lack of hard parts to count or measure. Additional problems arise from complexes of sibling species, the members of which are outwardly quite similar. These taxonomic difficulties can only be overcome by examination of many features. A composite picture must be formed before specific determination of any one animal can be certain.

If studying a live animal, one should examine the texture and color pattern of the skin. Observers of *Octopus* species should note the size, shape, and location of the frontal and mantle white spots and the presence or absence of dark eyelike spots (ocelli) on the mantle below the real eyes. Routine measurements of size include the dorsal mantle length (body length measured from the apex of the mantle to the edge of the mantle in squid, or from the apex to a point midway between the eyes in octopuses), and the length of the arms in relation to the body length (for reference purposes, the arms are numbered 1 to 4 on each side, starting with the dorsal arms).

To determine the sex of an octopus, examine the third right arm. In males this bears a fold along its entire length and a grooved, spade-shaped tip without suckers. This modified, or hectocotylized, arm is used to transfer packets of sperm (spermatophores) to the mantle cavity or the oviduct of the female during copulation. One should note the length of this grooved tip, the hectocotylus, relative to the length of the arm. In male squids, too, one or two arms may exhibit specialization for spermatophore transfer. Arms are not hectocotylized in females. If brooding females are found, the eggs should be measured, since their size is a key point in separating members of sibling species.

If desired, octopuses can be narcotized utilizing ethyl alcohol. During the ensuing few minutes the sex can be verified by examination of the internal organs, and the gill lamellae can be counted. Organs of the mantle cavity can be exposed by cutting the muscular septum, which connects the mantle to the viscera, and then turning the mantle back. This operation does not appear to harm the animal, which recovers quickly when replaced in fresh seawater. Counts of the number of gill lamellae, or leaflets, on both sides of each gill appear to be reliable indicators of species.

Dicyemid mesozoans are host-specific parasites of the kidneys of cephalopods, and serve as additional aids in identification (Bogolepova-Dobrokhotova, 1960, 1962; Hoffman, 1965; MacConnaughey, 1941, 1949a, b, 1957, 1959, 1960; Nouvel, 1947). If the animal is to be killed, these and other parasites should be recovered and identified (Overstreet & Hochberg, 1977). The beak and radula should also be removed, as these may help in species determination (Adam, 1941; Aldrich, Barber & Emerson, 1971; M. Clarke, 1962, 1977; Iverson & Pinkas, 1971; Mangold & Fioroni, 1966; Solem & Roper, 1975).

Cephalopods have proved to be invaluable laboratory animals for physiological, biochemical, and behavioral research. Readily available in many areas, they can often be collected in large numbers. Octopuses are especially hardy and can be kept for long periods in tanks in the laboratory. In addition, they are easily anesthetized for operation. *Octopus dofleini* (17.7), because of its large size, has been used extensively for studies of excretion, respiration, circulation, reproduction, and fine structure. *Octopus bimaculatus* has also been used for experimental studies but less extensively. The giant nerve fibers found in squids (including *Loligo opalescens* and its east coast relative *L. pealei*) have contributed greatly to our understanding of the generation and conduction of nerve impulses; see the reviews by Baker (1966), Bullock (1948, 1965), Keynes (1958), and Wells (1978) and the laboratory guide of Arnold et al. (1974). Similarly, studies of normal and experimentally altered octopuses and cuttlefish have helped us begin to grasp the complex relations between brain structure and behavioral capabilities (Wells, 1962, 1978; J. Young, 1961, 1971).

More ecological and behavioral work on west coast forms is needed; many studies have been conducted in Europe, the Caribbean, Hawaii, and Japan (Altman, 1966; Boletzky, 1972, Hamabe & Shimizu, 1966; Hochberg & Couch, 1971; Tinbergen & Verway, 1945; Wells, 1962; Yarnall, 1969).

Good general accounts relating to the cephalopod group as a whole can be found in Akimushkin (1965), Bartsch (1916), Boycott (1965), Cousteau & Diole (1973), Lane (1960), Lee (1875), Morton (1958), Nixon & Messenger (1977), Packard (1972), Packard & Sanders (1969, 1971), Robson (1929, 1932), Voss (1973), Voss & Sisson (1967, 1971), Wells (1961, 1962, 1978), Wilbur & Yonge (1964, 1966), and Yonge & Thompson (1976). For identification of west coast forms, see especially Berry (1921a,b, 1953), Packard & Hochberg (1977), Iverson & Pinkas (1971), Phillips (1933b), Pickford (1964), Pickford & MacConnaughey (1949), Roper, Young & Voss (1969), Sasaki (1929), and R. Young (1972). For reproduction, development, rearing, and maintenance, see Arnold, Singley &

Williams-Arnold (1972), Arnold & Williams-Arnold (1977), Boletzky (1974a,b, 1977), Boletzky et al. (1971), Choe (1966), Hanlon (1977), Hanlon et al. (1979), Hurley (1976), Itami et al. (1963), Joll (1976, 1978), LaRoe (1971), Mangold & Froesch (1977), Overath & Boletzky (1974), Summers & McMahon (1974), Summers, McMahon & Ruppert (1974), Taki (1941), Van Heukelem (1973, 1977), Walker, Longo & Bitterman (1970), and Wells & Wells (1959, 1972a,b, 1977).

Phylum Mollusca / Class Cephalopoda / Order Decapoda / Suborder Sepioidea Family Sepiolidae

17.1 **Rossia pacifica** Berry, 1911
Stubby Squid, Short Squid

Benthic in coastal waters, subtidal (16–370 m) but on rare occasions at night found swimming at shore in the intertidal zone; fairly common on bottoms of sand or muddy sand; north Pacific rim from Japan to southern California.

Maximum dorsal mantle length about 5 cm in females, about 3–4 cm in males; total length (including arms but not tentacles) of females about 11 cm, of males about 9 cm; mantle 1.5–2 times as long as wide, flattened above and below, rounded behind, not fused to head in front; fins round, with broad free lobes, almost as long as mantle; head large; arm lengths variable, usually dorsal arms shortest and third arms longest; arm suckers arranged in two rows in proximal and distal portions of each arm, and in two, three, or four rows in middle portion, the suckers on all arms alike in size, except that dorsal arms of male are hectocotylized with much smaller suckers; tentacles may be retracted, or extended longer than the body, with slender terminal clubs bearing centrally up to eight partial rows of small suckers; color in life reddish brown above and pale below, or wholly opalescent greenish gray if disturbed.

These small sepiolids crawl on their arms or swim, and dig shallow depressions in the sea floor, in which they rest with their arms rolled under their heads. They inhabit shrimp beds; over 80 percent of their diet consists of shrimp, although crabs, mysids, small fishes, and cephalopods are also eaten. Spawning occurs in the summer and fall in deep water;

each egg (4–5 mm in diameter) is contained in a large (8 mm by 15 mm) capsule. The capsules are attached singly or in small groups to seaweeds or other objects on the bottom.

Nine species of dicyemid mesozoans have been recorded from the kidneys and branchial hearts of *R. pacifica*. Of these, only two are known to occur off the west coast: *Dicyemennea brevicephaloides* and *D. parva*.

See especially Brocco (1970); see also Akimushkin (1965), Aldrich, Barber & Emerson (1971), Anderson (1978), Berry (1912b), Bogolepova-Dobrokhotova (1960, 1962), Fields & Thompson (1976), Hoffman (1965), Martin & Aldrich (1970), and Sasaki (1929).

Phylum Mollusca / Class Cephalopoda / Order Decapoda / Suborder Teuthoidea Family Loliginidae

17.2 Loligo opalescens Berry, 1911
Market Squid, Common Squid, Opalescent Squid, Sea Arrow, Calamary, Calamari

Pelagic in open coastal waters, returning to school and spawn on muddy sand in shallow inshore areas; common from southern British Columbia to Isla Guadalupe (Mexico) and Bahía Asuncíon (Baja California).

Dorsal mantle length of males 13–19 cm, of females 12–18 cm; total length (including arms but not tentacles) of males 27–28 cm, of females about 20 cm; males with larger heads and arms than females; mantle 4–5 times as long as wide, cylindrical, tapering to a blunt or rounded point behind, not fused to head in front; fins about half as long as mantle, slightly lobed in front; head small, eyelids not perforated; arms with swimming webs and with suckers alternating in two rows; terminal third of left ventral arm hectocotylized in males; clubs at tips of tentacles with two marginal rows of small and two medial rows of larger suckers; color translucent bluish white, changing to mottled gold and brown in the light or to dark brown or dark red when animal is feeding, frightened, or excited.

Adults feed mainly on shrimplike crustaceans (euphausiids, mysids, etc.) fishes, benthic polychaete worms, and their own young; they in turn are eaten by many fishes, birds, and marine mammals. Southern California squid populations spawn mainly in the winter (December to March), but they may also spawn in July. In Monterey Bay, major spawning occurs in May and June, with a minor spawning in November; Washington populations spawn in the summer (July to September). Mating behavior is elaborate, and the females produce numerous, large, cylindrical capsules each containing 180–300 eggs. The first capsules to be laid are anchored to the bottom; later capsules are attached to those already present, forming large clusters. The eggs are small, about 2.3 mm long by 1.5 mm in greatest diameter. Development is direct, and small juveniles emerge in 3–5 weeks. Adults die after spawning. The habits of immature squids, and the movements of populations between the time of hatching and the return to the spawning grounds, are virtually unknown. Most spawning males and females appear to be three years old. Squids are seined commercially on the spawning grounds; about 6,000 metric tons are taken yearly for human food and bait.

An occasional squid is infected with larval tapeworms and nematodes. A single species of dicyemid mesozoan, *Dicyemennea nouveli*, has been recorded from this squid; this report remains unconfirmed, and may in fact be based on material from *Octopus* and not *Loligo*.

Several species of *Loligo* have been used extensively in neurophysiological research.

For comprehensive accounts and bibliographies, see Fields (1965) and Recksiek & Frey (1978); see also Ally, Evans & Thompson (1975), Anderson (1978), Arnold (1965, 1971), Arnold et al. (1974), Berry (1911b, 1912b), Bullock (1965), Classic (1929), Cloney & Florey (1968), Cohen (1976), Dailey (1969), Decourt & Decourt (1968), Fields (1950), Fields & Thompson (1976), Fiscus & Baines (1966), Florey (1966, 1969), Florey & Kriebel (1969), Giese (1966, 1969), Hanlon et al. (1979), Hartline & Lange (1971, 1974) Hobson (1965), Hurley (1974, 1976), Kato & Hardwick (1975), Lange (1974), Lange & Hartline (1974), Longhurst (1969), Loukashkin (1976), MacConnaughey (1959), Macfarlane & Yamamoto (1974), MacGinitie & MacGinitie (1968), McGowan (1954, 1967), Mirow (1972), Okutani & McGowan (1969), Radhakrishnamurthy et al. (1970), Recksiek (1978), Scofield (1924), Srinivasan et al. (1969), Talmadge (1967), Turner & Sexsmith (1964), and Williams (1909).

Phylum Mollusca / Class Cephalopoda / Order Decapoda / Suborder Teuthoidea
Family Ommastrephidae

17.3 Dosidicus gigas (d'Orbigny, 1835)
(= Ommastrephes gigas)   Jumbo Squid, Humboldt Squid

Epipelagic to depths of several hundred meters, rising to surface by day or at night; wide-ranging and extremely abundant off Baja California, west coast of Central and South America, Juan Fernández Islands (Chile) and Galápagos Islands; occasionally found off Australia; in some years abundant along the California coast, when thousands of squid often come ashore in pursuit of spawning grunion.

Animal robust, massive; California specimens averaging 60 cm in total length (excluding tentacles) and about 3 kg in weight, but never more than 150 cm or 15 kg; mantle 5 times as long as wide, smooth, cylindrical, tapering to a blunt point, reddish brown in life; fins large, broadly sagittate, about half as long as mantle; head small; eyelid margin free all around; arms thick at base, the tips in large animals attenuated, the third pair with broad swimming web; arms about equal in length (less that one-half mantle length) except that the dorsal arms are shortest, all armed with numerous large, strongly toothed suckers; either left or right ventral arm of male hectocotylized; club at tentacle tip large, with four rows of suckers, and equipped proximally with a "fixing apparatus" consisting of a single row of about seven alternating fleshy pads and small suckers.

With the exception of *Architeuthis japonica* and *Moroteuthis robusta* (17.4), the jumbo squid is the largest squid known along the west coast. Although native to waters off Peru and Chile, where individuals reach overall lengths of 360 cm and weights up to 90 kg, this species ranges widely (to at least 120°W at the equator) and periodically invades the eastern North Pacific. When present in large numbers, the animals are regarded as pests of both commercial and sport fisheries. *Dosidicus* feeds at the surface both day and night. A good swimmer, it sometimes leaps above the sea surface, apparently continuing to accelerate by ejecting water from the siphon while in the air. It feeds on numerous species of fishes (anchovies, grunions, skippers, myctophids), crustaceans (such as the red galatheid crab *Pleuroncodes planipes*), and mol-

lusks (pteropods, heteropods, and cephalopods—including other *Dosidicus*). In turn, it is an important element in the diet of sperm whales, billfishes, and tunas and is also fished to some extent for human consumption. A Japanese squid jigging research vessel off Baja California in 1971 caught 23,000 kg of this squid in 27 days.

The life cycle is thought to be 3–4 years. Males mature at 1 year and are smaller than females, which mature the following year. Ova measure 0.9 mm by 1.1 mm. In fully mature females the eggs are concentrated in the oviducts and are apparently discharged simultaneously. Spawning occurs in the spring and summer on the continental slope along the coast of Peru. Larvae are planktonic and have a short proboscis or snout. Following spawning, the larvae are dispersed by the Humboldt Current to the north and west over a wide area. A southward migration takes place in the summer and fall. During the fall and winter the squid stay close to shore off Chile. Finally, in the spring, they move back northward to the spawning area. A majority of the adults die after spawning.

See Anderson (1978), Bartsch (1935), Berry (1911a, 1912b), Clark & Phillips (1936), M. Clarke (1965, 1966), M. Clarke, MacLeod & Paliza (1976), Cole & Gilbert (1970), Croker (1937), Duncan (1941), Holder (1899), Nesis (1970, 1973), Okutani (1977), Phillips (1933a), Roper & Young (1975), Sato (1976), Voss & Sisson (1967), and Wormuth (1976).

Phylum Mollusca / Class Cephalopoda / Order Decapoda / Suborder Teuthoidea
Family Onychoteuthidae

17.4 Moroteuthis robusta (Verrill, 1876)
North Pacific Giant Squid

Occasionally seen swimming at the surface near surf line or stranded ashore; pelagic in coastal waters at depths of 100–600 m, frequently caught by trawlers; considered to be abundant along North Pacific rim, Japan to southern California, and 650 km off British Columbia at Weather Station PAPA.

Dorsal mantle length commonly to 120 cm, the total length (including tentacles) to 335 cm; mantle 4–5 times as long as wide, cylindrical, tapering to a sharp point behind, with a reddish surface bearing many fine longitudinal ridges; fins large, rhombic, more than half mantle length; head relatively small, the eyelid margin free all around; arms pointed and unequal in length, the dorsal pair shortest (half mantle length), the ventral pair longest (two-thirds mantle length), the lateral pairs intermediate and equal in length; arm suckers small, sparsely alternating in two rows; club at tentacle tip narrow, bearing two rows of hooks, and equipped proximally with an oval shaped "fixing apparatus" with diagonal rows of alternating suckers and pads complementing those on other club, the clubs thereby locking together at their bases when thrust out for seizing prey; chitinous pen continuous posteriorly with a distinctive internal cartilaginous cone.

Moroteuthis robusta has usually been considered to be the largest invertebrate on the west coast. The dorsal mantle length of the largest reported specimen was 232.4 cm, with a body diameter of 45.7 cm. This was one of the three animals discovered by Dall in 1872, from which the original description of the species was prepared by Verrill. The tentacles were damaged, but normal proportions suggest a total length (including tentacles) of nearly 7 m. Recently, a specimen of *Architeuthis japonica* was found washed ashore on a beach in Oregon. Known to be as large or larger than *Moroteuthis*, the Japanese giant squid was previously thought to occur only off Japan.

Buoyancy in *Moroteuthis* is provided by substituting ammonium ions for some of the much heavier sodium ions in the tissue fluids; this accounts for the animal's low density, and for the bitter taste of its flesh. Throughout its range *Moroteuthis* is a favored item in the diet of sperm whales. Its suckers are believed responsible for many of the circular scars found on whales' heads. Food habits are uncertain. Mature ova in an ovary measure 1.0 mm by 0.76 mm; the nidamental glands, which play a major role in secreting the outer layers of egg masses, are very large, suggesting that an immense egg mass is laid. Several *M. robusta* have been taken off Monterey and Carmel (Monterey Co.), Santa Barbara, and Laguna Beach (Orange Co.).

See Akimushkin (1965), Anderson (1978), Berry (1912b), Betesheva & Akimushkin (1955), M. Clarke (1966), Croker (1934), Dall (1873), Denton, Gilpin-Brown & Shaw (1969), Hochberg (1974), Kawakami (1976), Kodolov (1970), Okutani & Nemoto (1964), Oku-

tani et al. (1976), Pattie (1968), Pearcy (1965), Phillips (1933a, 1961), Pike (1950), Rice (1963), Robbins, Oldham & Geiling (1937), Sasaki (1929), A. Smith (1963), Talmadge (1967), Van Hyning & Magill (1964), Verrill (1876, 1882), and Zuev & Nesis (1971).

Phylum Mollusca / Class Cephalopoda / Order Octopoda / Family Octopodidae

17.5 **Octopus bimaculoides** Pickford & MacConnaughey, 1949
Two-Spotted Octopus, Mud-Flat Octopus

In protected holes and crevices in pools, middle and low intertidal zones, also on mud flats; older animals often subtidal to depths of 20 m on rocks or in kelp beds; San Simeon (San Luis Obispo Co.) and Channel Islands to Ensenada (Baja California).

Dorsal mantle length at sexual maturity 5–20 cm; body pyriform, variable in color, the upper surface generally dark gray, brown, red, or olive, mottled with black, the lower surface lighter; ocelli (eyelike spots) conspicuous, blue-black, one below each eye near bases of arms 2 and 3; skin sculptured, with abundant cirri; arms 2.5–3.5 times body length; adult males with one or two specialized, enlarged suckers on arms 2 and 3; hectocotylus minute, 1/50 arm length; gill lamellae numbering 14–20; ink black; eggs large (10–17 mm by 4–5 mm), borne on short stalks, laid in small clusters (compare with *O. bimaculatus*, below).

Females lay their eggs under rocks from late winter to early summer, and brood them continuously for 2–4 months. The young remain on the bottom after hatching, often moving into the intertidal. Some are taken by man for food. The adults feed on mollusks, crustaceans, and occasionally fishes. Toxic secretions of the posterior salivary glands are used to paralyze the prey. In the case of shelled mollusks the poison is injected through a small hole rasped in the shell by the radula. In the rocky intertidal zone *O. bimaculoides* drills and feeds principally on various limpets (*Collisella* and *Notoacmea* spp.), the black abalone, *Haliotis cracherodii* (13.4), the snails *Olivella biplicata* (13.99), *Tegula funebralis* (13.32), and *T. gallina* (13.35), the clam *Protothaca staminea* (15.41), and several species of hermit crabs (*Pagurus*, 24.10–13) inhabiting empty gastropod shells. In mud-flat regions it is known to feed on the bivalves *Chione undatella* (15.39), *Mytilus edulis* (15.10), *Leptopecten monotimerus*, and *Argopecten aequisulcatus* (15.15); occasionally it captures small fishes such as the blennies *Hypsoblennius gilberti* and *H. gentilis*. In subtidal waters it has been observed feeding on the abalones *Haliotis rufescens* (13.1), *H. fulgens* (13.2), and *H. corrugata* (13.3), the snails *Kelletia kelletii* (13.87), *Astraea undosa* (13.39), and *Norrisia norrisi* (13.30), and the hermit crab *Paguristes ulreyi*. Under laboratory conditions the octopus will eat almost any shelled mollusk offered.

Of the seven dicyemid mesozoan parasites known to inhabit the kidneys, two are specific for this host species: *Dicyemennea californica* and *Dicyema sullivani*.

A sibling species, **O. bimaculatus** Verrill, 1883, also called the two-spotted octopus, occurs slightly farther off shore, from the lowest intertidal zone where the brown alga *Laminaria* occurs to depths of 50 m. It is found from Santa Barbara and the Channel Islands to the southern tip of Baja California, and in the Gulf of California from San Felipe to La Paz. Intertidal specimens are commonest in the southern part of the range. The two species are so similar in most respects that for more than 60 years the two-spotted octopus was thought to represent a single species, easily distinguished from all other west coast octopods by the presence of two blue-black spots, one below each eye near the bases of arms 2 and 3.

The initial clues to the real situation were uncovered by MacConnaughey at the Scripps Institution of Oceanography, during a study of the mesozoan parasites of cephalopods. He discovered that in the two-spotted octopus population there were two types of females, that differed not only in the size of the eggs they produced but also in the constellations of mesozoan parasite species infesting their kidneys. Further study showed other subtle differences, in structure, behavior, and habitat (Pickford & MacConnaughey, 1949). The similarities are so great that not all specimens, especially males, can be identified to species with certainty, but *O. bimaculatus* can often be distinguished by its somewhat larger overall size, its longer arms (4–5 times mantle length), and its much smaller eggs (2–4 mm by 1–1.5 mm). The difficulties experienced by man in distinguishing the sibling species do not extend to the animals themselves; mating experiments involving similar and mixed couples show that male and female two-spotted octopods show an interest only in members of their own species.

Females of *O. bimaculatus* lay eggs in the spring and summer. Thousands of egg capsules are deposited, their long stalks twisted into cords. Females guard their egg strings and carefully clean them during the 1–2.5 months required for development. Upon hatching, the young pass through a short planktonic stage before settling to the bottom.

Adult *O. bimaculatus* feed largely on mollusks and crustaceans, though they can also capture juveniles of the venomous-spined scorpionfish (*Scorpaena guttata*). In turn, the octopus is eaten by adult scorpionfish, by moray eels, and to some extent by man.

Several different dicyemid mesozoans are found in the kidneys of *O. bimaculatus*, and one species, *Dicyemennea abelis*, occurs only in this octopus.

For both octopus species, see Edwards (1969), Hochberg (1976), MacConnaughey (1949a,b), and Pickford & MacConnaughey (1949); for *O. bimaculoides*, see also Cushing, Calaprice & Trump (1963), Hartline & Lange (1971), Lange & Hartline (1974), MacConnaughey (1941, 1951, 1960), MacGinitie (1938), Packard & Hochberg (1977), Peterson (1959), and Pilson & Taylor (1961); for *O. bimaculatus*, see also Burrage (1964), Fotheringham (1974), Fox (1938), Fox & Crane (1942), Fox & Updegraff (1943), Longo & Anderson (1970), MacGinitie & MacGinitie (1968), Muller (1971), Parker (1921), Rosenthal (1971), Rosenthal, Clarke & Dayton (1974), Taylor & Chen (1969), Turner, Ebert & Given (1969), and Turner & Sexsmith (1964).

Phylum Mollusca / Class Cephalopoda / Order Octopoda / Family Octopodidae

17.6 **Octopus rubescens** Berry, 1953
Red Octopus

Common offshore in kelp beds or on bottoms of sandy mud to depths of 200 m; occasionally under stones in low intertidal zone; juveniles often washed ashore in kelp holdfasts; the commonest small octopus in shallow subtidal waters from Alaska to Scammon Lagoon (Baja California) and Gulf of California.

Dorsal mantle length 5–10 cm; body round to ovoid, dull red or reddish brown often mottled with white; ocelli absent;

skin papillate, often with cirri; arms about 4 times body length, the sixth pair of suckers enlarged on all but ventral arms of males; hectocotylus conspicuous, 1/10 length of arm; gill lamellae 22–26; ink reddish or red-brown; eggs small (3–4 mm by 1.5–2 mm), the capsules with long stalks twisted into cords, laid in festoons.

Older accounts that refer to *Octopus punctatus*, *O. apollyon*, and *O. hongkongensis* probably refer in part to *O. rubescens* and in part to *O. dofleini* (17.7)—it is often difficult to tell which.

Female *O. rubescens* protecting and grooming their egg clusters are found from late spring through early winter in rocky areas in the intertidal and shallow subtidal zones. Peaks in breeding occur in August and September. The young hatch in 6–8 weeks, spend a brief period in the plankton, undergo metamorphosis in surface waters, and eventually settle as juveniles in the kelp beds. They feed for a time in and around the kelp holdfasts before migrating farther offshore to bottoms of sandy mud. Mating occurs in deep water in late winter and early spring, after which the population (males first) moves inshore to the spawning grounds.

Adults feed mainly on crustaceans, mollusks, and fishes. Aquarium specimens have been known to drill and eat a variety of gastropods. In the field, small crabs and hermit crabs seem to be preferred. When caught, these prey are killed with secretions from the salivary glands and then opened at the junction between carapace and abdomen. After the viscera are eaten, the legs are pulled off and cleaned out one by one. *Octopus rubescens* is one of the principal foods of various bass (*Paralabrax* spp.) and rockfishes (*Sebastodes* spp.).

From Point Conception south, three mesozoans are typically found in the kidneys: *Dicyemennea adscita*, *Dicyema balamuthi*, and *Conocyema adminicula*. In the northern part of the range these are replaced by two others: *Dicyemennea brevicephala* and *Dicyema apollyoni*.

See Ballering et al. (1972), Berry (1953), Berry & Halstead (1954), T. Clarke, Flechsig & Grigg (1967), Dorsey (1976), Fisher (1923, 1925), Hochberg (1976), MacConnaughey (1941, 1949b), MacGinitie (1938), McGowan (1967), Nouvel (1947), Packard & Hochberg (1977), Turner, Ebert & Given (1969), Warren, Sheier & Riley (1974), and R. Young (1972).

Phylum Mollusca / Class Cephalopoda / Order Octopoda / Family Octopodidae

17.7 **Octopus dofleini** (Wülker, 1910)
North Pacific Giant Octopus, Giant Pacific Octopus

Smaller animals occasionally in low intertidal pools on rocky shores, larger individuals generally subtidal to depths of 100 m; along North Pacific rim from northern Asia to California; subspecies **O. dofleini martini** Pickford only from British Columbia to Monterey; a distinct, but undescribed, subspecies has been taken off the Channel Islands to depths of 500 m.

One of the largest octopus species known, the largest specimen on record with a total arm spread of 9.6 m and a weight of 272 kg; dorsal mantle length usually over 20 cm; body ovoid, with extensive skin folds, red to reddish brown above (often with black reticulation), pale below; ocelli absent; arms 3–5 times body length, lacking specialized enlarged suckers; hectocotylus large, one-fifth length of arm; gill lamellae 25–29; ink dark brown; eggs measuring 6–8 mm by 2–3 mm, borne in capsules on long stalks, these entangled and cemented together to form long festoons.

This species was often called *Paroctopus* or *Octopus apollyon*, *O. punctatus*, or *O. hongkongensis* by earlier authors, though these names were also applied to *O. rubescens* (17.6).

Although *O. dofleini* has been used extensively in laboratory studies, its natural history is still poorly known. The life cycle is thought to be 4–5 years. Eggs are laid throughout the year, though mainly in the winter, and development takes 5–7 months; hatching peaks in the early spring. The young are pelagic for a short period, probably about 1 month.

The adults feed on crustaceans (shrimps and crabs), mollusks (scallops, clams, abalones, moon snails, and small octopuses), and fishes (rockfishes, flat fishes, and sculpins). Large crabs are stalked and then caught with a sudden flick of one or more arms; empty crab carapaces, shiny shells, and bones often litter the entrance to a lair. The octopus takes smaller shrimps and fishes by slowly arching its body over a seaweed bed, then suddenly pouncing, and enclosing the area in a canopy formed by the web membrane that joins the basal parts of adjacent arms. The sensitive arm tips are then inserted into the impounded area to search for food.

Two mesozoans characteristically occur in the kidneys of *O. dofleini*: *Dicyemennea abreida* and *Conocyema deca*.

This octopus is fed upon by seals, sea otters, dogfish sharks, lingcod, and man. It supports small commercial fisheries in Alaska, Canada, Washington, Oregon, and northern California. In Washington the annual catch for the decade 1965–75 ranged between 11,000 and 18,000 kg; most of the individuals taken weighed less than 30 kg.

See especially Gabe (1975), Johnson (1942), MacConnaughey (1957), Pickford (1964), and Winkler & Ashley (1954); see also Anderson (1978), Bayne (1973), Brocco (1975), Brooks et al. (1971), Cox (1949), Emanuel (1957), Emanuel & Martin (1956), Giese (1966), Goddard (1968, 1969), Green (1973), Hanson, Mann & Martin (1973), Harrison & Martin (1965), Hartwick (1977), Hartwick & Thorarinsson (1978), Hartwick, Thorarinsson & Tulloch (1978a,b), Hatano (1958, 1960), High (1976a,b), Hochberg (1976), Johansen (1965a,b), Johansen & Huston (1962), Johansen & Lenfant (1966), Johansen & Martin (1962), Kyte & Courtney (1977), Lenfant & Johansen (1965), Loe & Florey (1966), Mann, Karagiannidis & Martin (1973), Mann, Martin & Thiersch (1966, 1970), Martin & Aldrich (1970), Martin, Harrison et al. (1958), Martin, Lutwak-Mann et al. (1973), Martin, Thiersch et al. (1970), Mottet (1974), Newman (1963), Nigrelli (1960), Packard & Hochberg (1977), Pennington (1979), Potts (1965), Potts & Todd (1965), Pritchard, Huston & Martin (1963), Ruggieri & Rosenberg (1974), L. Smith (1962, 1963), Snow (1970), Wells (1962), Witmer (1975), Witmer & Martin (1973), Wood (1971), and Yamamoto et al. (1965).

Phylum Mollusca / Class Cephalopoda / Order Octopoda / Family Octopodidae

17.8 Octopus micropyrsus Berry, 1953

Offshore in holdfasts of giant kelp (*Macrocystis*) or in large empty gastropod shells, occasionally encountered in low intertidal zone, often washed ashore in kelp holdfasts; Santa Barbara, San Diego, and Channel Islands.

One of the smallest octopus species known, dorsal mantle length 2–2.5 cm; body round, bright red dorsally, flesh-colored ventrally; skin smooth, without cirri; ocelli absent; arms 2.5–3 times body length, with one of the suckers enlarged on each arm except the ventral arms in both sexes; hectocotylus well developed, 1/15 arm length; gill lamellae 10–12; ink black; eggs large (10–12 mm by 5–6 mm), borne on short stalks, laid singly or in small clusters.

Eggs are laid in the spring and early summer in empty shells, under rocks, or in kelp holdfasts. Development time is not known, but the young resemble miniature adults and remain on the bottom after hatching. Adults feed on small crustaceans and mollusks found in kelp holdfasts.

A new species of the mesozoan genus *Dicyemennea* occurs in the kidneys (Hochberg, unpublished).

See Berry (1953) and Hochberg (1976).

Literature Cited

Adam, W. 1941. Notes sur les Céphalopodes. XV. Sur la valeur diagnostique de la radule chez les Céphalopodes Octopodes. Bull. Mus. Roy. Hist. Natur., Belgique 17: 1–19.

Akimushkin, I. I. 1965. Cephalopods of the U.S.S.R. Translation of 1963 Russian edition by A. Mercado. Jerusalem. 223 pp.

Aldrich, M. M., V. C. Barber, and C. J. Emerson. 1971. Scanning electron microscope studies of some cephalopod radulae. Canad. J. Zool. 49: 1589–94.

Ally, J. R. R., R. G. Evans, and T. W. Thompson. 1975. The results of an exploratory fishing cruise for *Loligo opalescens* in southern and central California, June 5–25, 1974. Moss Landing Marine Laboratories Tech. Publ. 75-2: 1–22.

Altman, J. S. 1966–67. The behaviour of *Octopus vulgaris* Lam. in its natural habitat: A pilot study. Underwater Assoc. Rep. 1966–67: 77–83.

Altman, J. S., and M. Nixon. 1970. Use of the beaks and radula by *Octopus vulgaris* in feeding. J. Zool., London 161: 25–38.

Anderson, M. E. 1978. Notes on the cephalopods of Monterey Bay, California, with new records for the area. Veliger 21: 255–62.

Arnold, J. M. 1965. Normal embryonic stages of the squid, *Loligo pealii* (Lesueur). Biol. Bull. 128: 24–32.

———. 1967. Organellogenesis of the cephalopod iridophore: Cytomembranes in development. J. Ultrastruc. Res. 20: 410–20.

———. 1971. Cephalopods, pp. 265–311, in G. Reverberi, ed., Experimental embryology of marine and fresh-water invertebrates. Amsterdam: North-Holland. 587 pp.

Arnold, J. M., and K. O. Arnold. 1969. Some aspects of hole boring predation by *Octopus vulgaris*. Amer. Zool. 9: 991–96.

Arnold, J. M., C. T. Singley, and L. D. Williams-Arnold. 1972. Embryonic development and post-hatching survival of the sepiolid squid *Euprymna scolopes* under laboratory conditions. Veliger 14: 361–64.

Arnold, J. M., and L. D. Williams-Arnold. 1977. Cephalopoda: Decapoda, pp. 243–90, *in* A. C. Giese and J. S. Pearse, eds., Reproduction of marine invertebrates, vol. 4. New York: Academic Press. 369 pp.

Arnold, J. M., and R. E. Young. 1974. Ultrastructure of a cephalopod photophore. I. Structure of the photogenic tissue. Biol. Bull. 147: 507–21.

Arnold, J. M., R. E. Young, and M. V. King. 1974. Ultrastructure of a cephalopod photophore. II. Iridophores as reflectors and transmitters. Biol. Bull. 147: 522–34.

Arnold, J. M., W. C. Summers, D. L. Gilbert, R. S. Manalis, N. W. Daw, and D. L. Gilbert. 1974. A guide to laboratory use of the squid *Loligo pealei*. Marine Biological Laboratory, Woods Hole, Mass. 74 pp.

Baker, P. F. 1966. The nerve axon. Sci. Amer. 214: 74–82.

Ballering, R. G., M. A. Jalving, D. A. VenTresca, L. E. Hallacher, J. T. Tomlinson, and D. R. Wobber. 1972. Octopus evenomation through a plastic bag via a salivary proboscis. Toxicon 10: 245–48.

Bartsch, P. 1916. Pirates of the deep—stories of the squid and octopus. Ann. Rep., Smithsonian Inst. 1916: 347–75.

———. 1935. An invasion of Monterey Bay by squids. Nautilus 48: 107–8.

Bayne, C. J. 1973. Internal defense mechanisms of *Octopus dofleini*. Malacol. Rev. 6: 13–17.

Berry, S. S. 1911a. Notes on some cephalopods in the collection of the University of California. Univ. Calif. Publ. Zool. 8: 301–10.

———. 1911b. Preliminary notices of some new Pacific cephalopods. Proc. U.S. Nat. Mus. 40: 589–92.

———. 1912a. On a cephalopod new to California with a note on another species. 1st Ann. Rep., Laguna Marine Lab., pp. 83–87.

———. 1912b. A review of the cephalopods of western North America. Bull. U.S. Bur. Fish. 30: 269–336.

———. 1920. Light production in cephalopods. Biol. Bull. 38: 141–95.

———. 1953. Preliminary diagnoses of six west American species of *Octopus*. Leaflets in Malacol. 1: 51–58.

Berry, S. S., and B. W. Halstead. 1954. Octopus bites—a second report. Leaflets in Malacol. 1: 59–66.

Betesheva, E. I., and I. I. Akimushkin. 1955. Food of the sperm whale (*Physeter catodon*) in the Kurile Islands region. [In Russian.] Trudy Inst. Okeanol. 18: 86–94.

Bogolepova-Dobrokhotova, I. I. 1960. Dicyemidae of far-eastern seas. I. New species of the genus *Dicyema*. [In Russian; English summary.] Zool. Zhur. S.S.S.R. 39: 1293–1302.

———. 1962. Dicyemidae of far-eastern seas. II. New species of the genus *Dicyemenna*. [In Russian; English summary.] Zool. Zhur. S.S.S.R. 41: 503–18.

Boletzky, S. von. 1972. A note on aerial prey-capture by *Sepia officinalis* (Mollusca, Cephalopoda). Vie et Milieu 23A: 133–40.

———. 1974a. Élevage de Céphalopodes en aquarium. Vie et Milieu 24A: 309–40.

———. 1974b. The "larvae" of Cephalopoda: A review. Thalassia Jugoslavica 10: 45–76.

———. 1977. Post-hatching behaviour and mode of life in cephalopods. Symp. Zool. Soc. London 38: 557–67.

Boletzky, S. von, M. V. von Boletzky, D. Frösch, and V. Gatzi. 1971. Laboratory rearing of Sepiolinae (Mollusca: Cephalopoda). Mar. Biol. 8: 82–87.

Boycott, B. B. 1953. The chromatophore system of cephalopods. Proc. Linn. Soc. London 164: 235–40.

———. 1965. Learning in the octopus. Sci. Amer. 212: 42–50.

Brocco, S. L. 1970. Aspects of the biology of the sepiolid squid *Rossia pacifica* Berry. Master's thesis, Biology, University of Victoria, British Columbia, Canada. 151 pp.

———. 1974a. A comparative study of the morphology of cephalopod reflecting cells. 55th Ann. Meet., West. Soc. Natur., p. 26 (abstract).

———. 1974b. The structure and function of the cephalopod integument. 55th Ann. Meet., West. Soc. Natur., pp. 26–27 (abstract).

———. 1975. The fine structure of the frontal and mantle white spots of *Octopus dofleini*. Amer. Zool. 15: 782 (abstract).

———. 1976. The ultrastructure of the epidermis, dermis, iridophores, leucophores and chromatophores of *Octopus dofleini martini* (Cephalopoda: Octopoda). Doctoral dissertation, Zoology, University of Washington, Seattle.

———. 1978. Chemical and physical properties of the iridosomal platelets of *Octopus dofleini*. 59th Ann. Meet., West. Soc. Natur., p. 10 (abstract).

Brocco, S. L., R. M. O'Clair, and R. A. Cloney. 1974. Cephalopod integument: The ultrastructure of Kölliker's organs and their relationship to setae. Cell Tissue Res. 151: 293–308.

Brooks, D. E., C. Lutwak-Mann, T. Mann, and A. W. Martin. 1971. Motility and energy rich phosphorus compounds in spermatozoa of *Octopus dofleini martini*. Proc. Roy. Soc. London B178: 151–60.

Bullock, T. H. 1948. Synaptic transmission, squid mantle ganglion. J. Neurophysiol. 11: 343–64.

———. 1965. Mollusca: Cephalopods, pp. 1433–1515, *in* T. H. Bullock and G. A. Horridge, Structure and function in the nervous systems of invertebrates, vol. 2. San Francisco: Freeman.

Burrage, B. R. 1964. The possibility of paralytic effects of selected sea stars and brittle-stars on *Octopus bimaculatus*. Trans. Kan. Acad. Sci. 67: 496–98.

Choe, S. 1966. On the eggs, rearing, habits of the fry, and growth of some Cephalopoda. Bull. Mar. Sci. 16: 330–48.

Clark, F. N., and J. B. Phillips. 1936. Commercial use of the jumbo squid, *Dosidicus gigas*. Calif. Fish & Game 22: 143–44.

Clarke, M. R. 1962. The identification of cephalopod "beaks" and the relationship between beak size and total body weight. Bull. Brit. Mus. (Natur. Hist.), Zool. 8: 421–80.

———. 1965. Large light organs on the dorsal surface of the squids *Ommastrephes pteropus, Symplectoteuthis oualaniensis,* and *Dosidicus gigas.* Proc. Malacol. Soc. London 36: 319–21.

———. 1966. A review of the systematics and ecology of oceanic squids. Adv. Mar. Biol. 4: 91–300.

———. 1977. Beaks, nets and numbers. Symp. Zool. Soc. London 38: 89–126.

Clarke, M. R., N. MacLeod, and O. Paliza. 1976. Cephalopod remains from the stomach of sperm whales caught off Peru and Chile. J. Zool., London 180: 477–93.

Clarke, T. A., A. O. Flechsig, and R. W. Grigg. 1967. Ecological studies during project Sealab II. Science 157: 1381–89.

Classic, R. F. 1929. Monterey squid fishery. Calif. Fish & Game 15: 317–20.

Cloney, R. A., and E. Florey. 1968. Ultrastructure of cephalopod chromatophore organs. Z. Zellforsch. Mikroskop. Anat. 89: 250–80.

Cohen, A. C. 1976. The systematics and distribution of *Loligo* (Cephalopoda, Myopsida) in the western North Atlantic, with descriptions of two new species. Malacologia 15: 299–367.

Cole, K. S., and D. L. Gilbert. 1970. Jet propulsion of squid. Biol. Bull. 138: 245–46.

Cousteau, J.-Y. 1954. To the depths of the sea by bathyscaphe. Nat. Geogr. 106: 67–79.

Cousteau, J.-Y., and P. Diole. 1973. Octopus and squid: The soft intelligence. Garden City, N.Y.: Doubleday. 304 pp.

Cox, K. W. 1949. Octopus, pp. 171–72, *in* The commercial fish catch of California for the year 1947 with an historical review 1916–1947. Calif. Dept. Fish & Game, Fish Bull. 74: 1–267.

Croker, R. S. 1934. Giant squid taken at Laguna Beach. Calif. Fish & Game 20: 297.

———. 1937. Further notes on the jumbo squid, *Dosidicus gigas.* Calif. Fish & Game 23: 246–47.

Cushing, J. E., N. L. Calaprice, and G. Trump. 1963. Blood group reactive substances in some marine invertebrates. Biol. Bull. 125: 69–80.

Dailey, M. D. 1969. A survey of helminth parasites in the squid, *Loligo opalescens;* smelt, *Osmerus mordox;* Jack mackerel, *Trachurus symmetricus;* and Pacific mackerel, *Scomber japonicus.* Calif. Fish & Game 55: 221–26.

Dall, W. H. 1873. Aleutian cephalopods. Amer. Natur. 7: 484–85.

Decourt, S., and B. Decourt. 1968. Thousands of these spawning cephalopods provide San Diego divers with unique photos. Skin Diver 17: 44–45.

Denton, E. J., J. B. Gilpin-Brown, and T. I. Shaw. 1969. A buoyancy mechanism found in cranchid squid. Proc. Roy. Soc. London B174: 271–79.

Denton, E. J., and M. F. Land. 1971. Mechanism of reflexion in silvery layers of fish and cephalopods. Proc. Roy. Soc. London B178: 43–61.

Dilly, P. N. 1973. The enigma of colouration and light emission in deep sea animals. Endeavour 32: 25–29.

Dorsey, E. M. 1976. Natural history and social behavior of *Octopus rubescens* Berry. Master's thesis, Zoology, University of Washington, Seattle. 44 pp.

Duncan, D. 1941. Fishing giants of the Humboldt. Nat. Geogr. 79: 373–400.

Edwards, D. C. 1969. Predators on *Olivella biplicata,* including species-specific predator avoidance response. Veliger 11: 326–33.

Emanuel, C. F. 1957. The composition of octopus renal fluid. II. A chromatographic examination of the organic constituents. III. The isolation and chemical properties of dicyemin. IV. Isolation and identification of the methanol soluble substance. Z. Vergl. Physiol. 39: 477–91; 40: 1–7.

Emanuel, C. F., and A. W. Martin. 1956. The composition of octopus renal fluid. I. Inorganic constituents. Z. Vergl. Physiol. 39: 226–34.

Fields, W. G. 1950. A preliminary report on the fishery and on the biology of the squid, *Loligo opalescens.* Calif. Fish & Game 36: 366–77.

———. 1965. The structure, development, food relations, reproduction, and life history of the squid *Loligo opalescens* Berry. Calif. Dept. Fish & Game, Fish Bull. 131: 1–108.

Fields, W. G., and K. A. Thompson. 1976. Ultrastructure and functional morphology of spermatozoa of *Rossia pacifica* (Cephalopoda, Decapoda). Canad. J. Zool. 54: 908–32.

Fiscus, C. H., and G. A. Baines. 1966. Food and feeding behavior of Steller and California sea lions. J. Mammal. 47: 195–200.

Fisher, W. K. 1923. Brooding habits of a cephalopod. Ann. Mag. Natur. Hist. (9) 12: 147–49.

———. 1925. On the habits of an octopus. Ann. Mag. Natur. Hist. (9) 15: 411–14.

Florey, E. 1966. Nervous control and spontaneous activity of the chromatophores of a cephalopod, *Loligo opalescens.* Comp. Biochem. Physiol. 18: 305–24.

———. 1969. Ultrastructure and function of cephalopod chromatophores. Amer. Zool. 9: 429–42.

Florey, E., and M. E. Kriebel. 1969. Electrical and mechanical responses of chromatophore muscle fibers of the squid, *Loligo opalescens,* to nerve stimulation and drugs. Z. Vergl. Physiol. 65: 98–130.

Fotheringham, N. 1974. Trophic complexity in a littoral boulderfield. Limnol. Oceanogr. 19: 84–91.

Fox, D. L. 1938. An illustrated note on the mating and egg-brooding habits of the two-spotted octopus. Trans. San Diego Soc. Natur. Hist. 9: 31–34.

Fox, D. L., and S. C. Crane. 1942. Concerning the pigments of the two-spotted octopus and the opalescent squid. Biol. Bull. 82: 284–91.

Fox, D. L., and D. M. Updegraff. 1943. Adenochrome, a glandular pigment in the branchial hearts of the octopus. Arch. Biochem. Biophys. 1: 339–56.

Froesch, D., and J. B. Messenger. 1978. On leucophores and the chromatic unit of *Octopus vulgaris*. J. Zool., London 186: 163–73.

Gabe, S. H. 1975. Reproduction in the giant octopus of the North Pacific, *Octopus dofleini martini*. Veliger 18: 146–50.

Giese, A. C. 1966. Lipids in the economy of marine invertebrates. Physiol. Rev. 46: 244–98.

——. 1969. A new approach to the biochemical composition of the mollusc body. Oceanogr. Mar. Biol. Ann. Rev. 7: 175–229.

Goddard, C. K. 1968. Studies on the blood sugar of *Octopus dofleini*. Comp. Biochem. Physiol. 27: 275–85.

——. 1969. Effect of tissue extract on the blood sugar of *Octopus dofleini*. Comp. Biochem. Physiol. 28: 271–91.

Green, M. 1973. Taxonomy and distribution of planktonic octopods in the northeastern Pacific. Master's thesis, Fisheries, University of Washington, Seattle. 98 pp.

Halstead, B. W. 1949. Octopus bites in human beings. Leaflets in Malacol. 1: 17–22.

Hamabe, M., and T. Shimizu. 1966. Ecological studies of the common squid *Todarodes pacificus* Steenstrup, mainly in the southwestern waters of the Japan Sea. [In Japanese.] Bull. Jap. Sea Reg. Fish. Lab. 16: 13–55.

Haneda, Y. 1956. Squid producing an abundant luminous secretion found in Suruga Bay, Japan. Sci. Rep. Yokosuka City Mus. 1: 27–32.

Hanlon, R. T. 1977. Laboratory rearing of the Atlantic reef octopus, *Octopus briareus* Robson, and its potential for mariculture. Proc. 8th Ann. Meet., World Maricult. Soc. 8: 471–82.

Hanlon, R. T., R. F. Hixon, W. H. Hulet, and W. T. Yang. 1979. Rearing experiments on the California market squid *Loligo opalescens* Berry, 1911. Veliger 21: 428–31.

Hanson, D., T. Mann, and A. W. Martin. 1973. Mechanism of the spermatophoric reaction in the giant octopus of the North Pacific, *Octopus dofleini martini*. J. Exper. Biol. 58: 711–23.

Harrison, F. M., and A. W. Martin. 1965. Excretion in the cephalopod *Octopus dofleini*. J. Exper. Biol. 42: 71–98.

Hartline, P. H., and G. D. Lange. 1971. A comparative study in the neurophysiology of vision in squid and octopus. Proc. 1st Ann. Meet., Soc. Neurosci., p. 47 (abstract).

——. 1974. Optic nerve responses to visual stimuli in squid. J. Compar. Physiol. 93: 37–54.

Hartwick, E. B. 1977. Population and behavior studies of *Octopus dofleini*. 58th Ann. Meet., West. Soc. Natur., p. 7 (abstract).

Hartwick, E. B., and G. Thorarinsson. 1978. Den associates of the giant Pacific octopus, *Octopus dofleini* (Wülker). Ophelia 17: 163–66.

Hartwick, E. B., G. Thorarinsson, and L. Tulloch. 1978a. Antipredator behavior in *Octopus dofleini* (Wülker). Veliger 21: 263–64.

——. 1978b. Methods of attack by *Octopus dofleini* (Wülker) on captured bivalve and gastropod prey. Mar. Behav. Physiol. 5: 193–200.

Hatano, M. 1958. Lipids from the liver of *Octopus dofleini*. I. Composition of fatty acids of acetone-soluble lipid. II. On the lower fatty acids of acetone-soluble lipid. Bull. Fac. Fish. Hokkaido Univ. 9: 207–17.

——. 1960. Lipids from the liver of *Octopus dofleini*. III. On the unsaponifiable matter. Bull. Fac. Fish. Hokkaido Univ. 11: 218–21.

Herring, P. J. 1977. Luminescence in cephalopods and fish. Symp. Zool. Soc. London 38: 127–59.

High, W. L. 1976a. Escape of Dungeness crabs from pots. Mar. Fish. Rev. 38: 19–23.

——. 1976b. The giant Pacific octopus. Mar. Fish. Rev. 38: 17–22.

Hobson, E. S. 1965. Spawning in the Pacific Coast squid, *Loligo opalescens*. Underwater Natur. 3: 20–21.

Hochberg, F. G. 1974. Southern California records of the giant squid, *Moroteuthis robusta*. Tabulata 7: 83–85.

——. 1976. Benthic cephalopods of the eastern Pacific. Proc. Tax. Stand. Prog. 4: 3–8, 14–25.

Hochberg, F. G., and J. A. Couch. 1971. Biology of cephalopods, pp. 221–28, *in* J. W. Miller, J. G. Vanderwalker, and R. A. Waller, eds., Scientists-in-the-sea. Washington, D.C.: U.S. Dept. Interior.

Hochberg, F. G., and A. Packard. 1974. Color and patterning in cephalopods. 55th Ann. Meet., West. Soc. Natur., p. 21 (abstract).

Hoffman, E. G. 1965. Mesozoa of the sepiolid, *Rossia pacifica* (Berry). J. Parasitol. 51: 313–20.

Holder, C. F. 1899. Some Pacific cephalopods. Sci. Amer. 80: 253.

Holmes, W. 1955. The colour changes of cephalopods. Endeavour 14: 78–82.

Hurley, A. C. 1974. Squid social behavior. 55th Ann. Meet., West. Soc. Natur., p. 22 (abstract).

——. 1976. Feeding behavior, food consumption, growth, and respiration of squid *Loligo opalescens* raised in the laboratory. Calif. Dept. Fish & Game, Fish Bull. 74: 176–82.

Itami, K., Y. Izawa, S. Maeda, and K. Nakai. 1963. Notes on the laboratory culture of the octopus larvae. [In Japanese; English summary.] Bull. Jap. Soc. Sci. Fish. 29: 514–20.

Iverson, I. L. K., and L. Pinkas. 1971. A pictorial guide to beaks of certain eastern Pacific cephalopods. Calif. Dept. Fish & Game, Fish Bull. 152: 83–105.

Johansen, K. 1965a. An apparatus for continuous recording of respiratory exchange in a cephalopod. Comp. Biochem. Physiol. 14: 377–81.

——. 1965b. Cardiac output in the large cephalopod *Octopus dofleini*. J. Exper. Biol. 42: 475–80.

Johansen, K., and M. J. Huston. 1962. Effects of some drugs on the circulatory system of the intact non-anaesthetized cephalopod, *Octopus dofleini*. Comp. Biochem. Physiol. 5: 177–84.

Johansen, K., and C. Lenfant. 1966. Gas exchange in the cephalopod, *Octopus dofleini*. Amer. J. Physiol. 210: 910–18.

Johansen, K., and A. W. Martin. 1962. Circulation in the cephalopod, *Octopus dofleini*. Comp. Biochem. Physiol. 5: 161–76.

Johnson, M. W. 1942. Some observations on the feeding habits of the octopus. Science 95: 478–79.

Joll, L. M. 1976. Mating, egg laying, and hatching of *Octopus tetricus* (Mollusca: Cephalopoda) in the laboratory. Mar. Biol. 36: 327–33.

———. 1978. Observations on the embryonic development of *Octopus tetricus* (Mollusca: Cephalopoda). Australian J. Mar. Freshwater Res. 20: 19–30.

Kato, S., and J. E. Hardwick. 1975. The California squid fishery. F.A.O. Fish. Rep. 170, Suppl. 1: 107–27.

Kawaguti, S., and K. Mabuchi. 1970. Electron microscopy on the chromatophore muscle in cephalopods. Biol. J. Okayama Univ. 16: 1–10.

Kawaguti, S., and S. Ohgishi. 1962. Electron microscopic study on iridophores of a cuttlefish, *Sepia esculenta*. Biol. J. Okayama Univ. 8: 115–29.

Kawakami, T. 1976. Squids found in the stomach of sperm whales in the northwestern Pacific. Sci. Rep. Whales Res. Inst. 28: 145–51.

Keynes, R. D. 1958. The nerve impulse and the squid. Sci. Amer. 199: 83–90.

Kodolov, L. S. 1970. Squids of the Bering Sea. [In Russian.] Inst. Morsk. Rybn. Khoz. Okeanogr. 70 (Izv. Tikhookean. Nauchoissled. Rybn. Khoz. Okeanogr. 72): 162–65. English transl., 1972, pp. 157–60, *in* P. A. Moiseev, ed., Soviet fisheries investigations in the northeastern Pacific, part 5. Jerusalem: Israel Program of Scientific Translation (I.P.S.T. Press), available through U.S. Dept. Commerce, Washington, D.C.

Kyte, M. A., and G. W. Courtney. 1977. A field observation on aggressive behavior between two North Pacific octopus, *Octopus dofleini martini*. Veliger 19: 427–28.

Laevastu, T., and C. Fiscus. 1978. Review of cephalopod resources in the eastern North Pacific. Northwest & Alaska Fisheries Center Processed Report. 15 pp.

Lane, F. W. 1960. Kingdom of the octopus: The life-history of the Cephalopoda. New York: Sheridan. 300 pp. (Paperback ed., New York: Pyramid.)

Lange, G. D. 1974. Retinal responses in cephalopods. 55th Ann. Meet., West. Soc. Natur., p. 11 (abstract).

Lange, G. D., and P. H. Hartline. 1974. Retinal responses in squid and octopus. J. Compar. Physiol. 93: 19–36.

LaRoe, E. T. 1971. The culture and maintenance of the loliginid squids *Sepioteuthis sepioidea* and *Doryteuthis plei*. Mar. Biol. 9: 9–25.

Lee, H. 1875. The octopus; or the "Devil-fish" of fiction and of fact. London: Chapman & Hall. 114 pp.

Lenfant, C., and K. Johansen. 1965. Gas transport by hemocyanin-containing blood of the cephalopod *Octopus dofleini*. Amer. J. Physiol. 209: 991–98.

Loe, R. R., and E. Florey. 1966. The distribution of acetylcholine and cholinesterase in the nervous system and innervated organs of *Octopus dofleini*. Comp. Biochem. Physiol. 17: 509–22.

Longhurst, A. R. 1969. Pelagic invertebrate resources of the California Current. Calif. Coop. Oceanic Fish. Invest. Rep. 13: 60–62.

Longo, F. J., and E. Anderson. 1970. Structural and cytochemical features of the sperm of the cephalopod *Octopus bimaculatus*. J. Ultrastruc. Res. 32: 94–106.

Loukashkin, A. S. 1976. On biology of the market squid, *Loligo opalescens*; a contribution toward the knowledge of its food habits and feeding behavior. Calif. Coop. Oceanic Fish. Invest. Rep. 18: 109–11.

MacConnaughey, B. H. 1941. Two new Mesozoa from California, *Dicyemennea californica* and *Dicyemennea brevicephala* (Dicyemidae). J. Parasitol. 27: 63–69.

———. 1949a. *Dicyema sullivani*, a new mesozoan from lower California. J. Parasitol. 35: 122–24.

———. 1949b. Mesozoa of the family Dicyemidae from California. Univ. Calif. Publ. Zool. 55: 1–34.

———. 1951. The life cycle of the dicyemid Mesozoa. Univ. Calif. Publ. Zool. 55: 295–336.

———. 1957. Two new Mesozoa from the Pacific northwest. J. Parasitol. 43: 358–61.

———. 1959. *Dicyemennea nouveli*, a new mesozoan from central California. J. Parasitol. 45: 533–37.

———. 1960. The rhombogen phase of *Dicyema sullivani* MacConnaughey. J. Parasitol. 46: 608–10.

Macfarlane, S. A., and M. Yamamoto. 1974. The squid of British Columbia as a potential fishery resource—a preliminary report. Fish. Res. Bd. Canada, Tech. Rep. 447: 1–35.

MacGinitie, G. E. 1938. Notes on the natural history of some marine animals. Amer. Midl. Natur. 19: 207–19.

MacGinitie, G. E., and N. MacGinitie. 1968. Natural history of marine animals. 2nd ed. New York: McGraw-Hill. 523 pp.

Mangold, K., and P. Fioroni. 1966. Morphologie et biométrie des mandibules de quelques Céphalopodes méditerranéens. Vie et Milieu 17: 1139–96.

Mangold, K., and D. Froesch. 1977. A reconsideration of factors associated with sexual maturation. Symp. Zool. Soc. London 38: 541–55.

Mann, T., A. Karagiannidis, and A. W. Martin. 1973. Glycosidases in the spermatophores of the giant octopus, *Octopus dofleini martini*. Comp. Biochem. Physiol. 44A: 1377–86.

Mann, T., A. W. Martin, and J. B. Thiersch. 1966. Spermatophores and the spermatophoric reaction in the giant octopus of the North Pacific, *Octopus dofleini martini*. Nature 211: 1279–82.

———. 1970. Male reproductive tract, spermatophores and spermatophoric reaction in the giant octopus of the North Pacific, *Octopus dofleini martini*. Proc. Roy. Soc. London B175: 31–61.

Martin, A. W., and F. A. Aldrich. 1970. Comparison of hearts and branchial heart appendages in some cephalopods. Canad. J. Zool. 48: 751–56.

Martin, A. W., F. M. Harrison, M. J. Huston, and D. M. Stewart. 1958. The blood volumes of some representative molluscs. J. Exper. Biol. 35: 260–79.

Martin, A. W., C. Lutwak-Mann, J. E. A. McIntosh, and T. Mann. 1973. Zinc in the spermatozoa of the giant octopus, *Octopus dofleini martini*. Comp. Biochem. Physiol. 45A: 227–33.

Martin, A. W., J. B. Thiersch, H. M. Dott, R. A. P. Harrison, and T. Mann. 1970. Spermatozoa of the giant octopus of the North Pacific, *Octopus dofleini martini*. Proc. Roy. Soc. London B175: 63–68.

McGowan, J. A. 1954. Observations on the sexual behavior and spawning of the squid *Loligo opalescens*, at La Jolla, California. Calif. Fish & Game 40: 47–54.

———. 1967. Distributional atlas of pelagic molluscs in the California Current region. Calif. Coop. Oceanic Fish. Invest. Atlas 6: 1–218.

Messenger, J. B. 1968. The visual attack of the cuttlefish, *Sepia officinalis*. Anim. Behav. 16: 342–57.

———. 1974. Reflecting elements in cephalopod skin and their importance for camouflage. J. Zool., London 174: 387–95.

Messenger, J. B., A. P. Wilson, and A. Hedge. 1973. Some evidence for colour blindness in *Octopus*. J. Exper. Biol. 59: 77–94.

Mirow, S. 1972. Skin colors in the squids *Loligo pealii* and *Loligo opalescens*. I. Chromatophores. II. Iridophores. Z. Zellforsch. Mikroskop. Anat. 125: 143–75, 176–90.

Morton, J. E. 1958. Molluscs. London: Hutchinson. 232 pp. (American paperback edition, New York: Harper, 1960.)

Mottet, M. G. 1975. A technical report on the fishery biology of *Octopus dofleini*. Washington Dept. Fish. Tech. Rep. 16: 1–39.

Muller, A. 1971. Characteristics of coloration in *Octopus bimaculatus*, the Sea of Cortez blue-dot octopus. Sea of Cortez, Inst. Biol. Res. Newsl. 5: 1–5.

Nesis, K. N. 1970. The biology of the giant squid of Peru and Chile, *Dosidicus gigas*. Okeanologiia 10: 108–18.

Newman, M. A. 1963. "Marijean" octopus expedition. Aquar. Newsl., Vancouver Pub. Aquar. 7: 1–8.

Nigrelli, R. F., ed. 1960. Biochemistry and pharmacology of compounds derived from marine organisms. [Papers by several authors.] Ann. N.Y. Acad. Sci. 90: 615–950.

Nixon, M., and J. B. Messenger, eds. 1977. The biology of cephalopods. Symp. Zool. Soc. London 38. New York: Academic Press. 614 pp.

Nouvel, H. 1947. Les dicyémides. Première partie: Systématique, générations vermiformes, infusorigène et sexualité. Arch. Biol., Paris 58: 59–219.

Oglesby, L. C. 1972. Octopus bites in California. 53rd Ann. Meet., West. Soc. Natur., p. 3 (abstract).

Okutani, T. 1977. Stock assessment of cephalopod resources fished by Japan. F.A.O. Fish. Tech. Pap. 173: 1–62.

Okutani, T., and J. A. McGowan. 1969. Systematics, distribution, and abundance of the epiplanktonic squid (Cephalopoda, Decapoda) larvae of the California Current, April 1954–March 1957. Bull. Scripps Inst. Oceanogr. 14: 1–90.

Okutani, T., and T. Nemoto. 1964. Squids as the food of sperm whales in the Bering Sea and Alaskan Gulf. Sci. Rep. Whales Res. Inst. 18: 111–22.

Okutani, T., Y. Satake, S. Ohsumi, and T. Kawakami. 1976. Squids eaten by sperm whales caught off Joban District, Japan, during January-February, 1976. Bull. Tokai Reg. Fish. Res. Lab. 87: 67–113.

Overath, H., and S. von Boletzky. 1974. Laboratory observations on spawning and embryonic development of a blue-ringed octopus. Mar. Biol. 27: 333–37.

Overstreet, R. M., and F. G. Hochberg. 1975. Digenetic trematodes in cephalopods. J. Mar. Biol. Assoc. U.K. 55: 893–910.

Packard, A. 1972. Cephalopods and fish: The limits of convergence. Biol. Rev. 47: 241–307.

———. 1974. Chromatophore fields in the skin of the octopus. Proc. Physiol. Soc. 238: 38–40.

Packard, A., and F. G. Hochberg. 1977. Skin patterning in *Octopus* and other genera. Symp. Zool. Soc. London 38: 191–231.

Packard, A., and G. Sanders. 1969. What the octopus shows to the world. Endeavour 28: 92–99.

———. 1971. Body patterns of *Octopus vulgaris* and maturation of the response to disturbance. Anim. Behav. 19: 780–90.

Parker, G. H. 1921. The power of adhesion of the suckers of *Octopus bimaculatus* Verrill. J. Exper. Zool. 33: 391–94.

Pattie, B. H. 1968. Notes on giant squid *Moroteuthis robusta* (Dall) Verrill trawled off the southwest coast of Vancouver Island, Canada. Washington Dept. Fish. Res. Pap. 3: 47–50.

Pearcy, W. G. 1965. Species composition and distribution of pelagic cephalopods from the Pacific Ocean off Oregon. Pacific Sci. 19: 261–66.

Pearcy, W. G., E. E. Krygier, R. Mesecar, and F. Ramsey. 1977. Vertical distribution and migration of oceanic micronekton off Oregon. Deep-Sea Res. 24: 223–45.

Pennington, H. 1979. New fishery for Alaskans: The giant Pacific octopus. Alaska Seas and Coasts 7: 1–3, 12.

Peterson, R. P. 1959. The anatomy and histology of the reproductive systems of *Octopus bimaculoides*. J. Morphol. 104: 61–81.

Phillips, J. B. 1933a. Description of a giant squid taken at Monterey with notes on other squid off the California coast. Calif. Fish & Game 19: 128–36.

———. 1933b. Octopi of California. Calif. Fish & Game 21: 20–29.

————. 1961. Two unusual cephalopods taken near Monterey. Calif. Fish & Game 47: 416–17.

Pickford, G. E. 1964. *Octopus dofleini* (Wülker), the giant octopus of the North Pacific. Bull. Bingham Oceanogr. Coll. 19: 1–70.

Pickford, G. E., and B. H. MacConnaughey. 1949. The *Octopus bimaculatus* problem: A study in sibling species. Bull. Bingham Oceanogr. Coll. 12: 1–66.

Pike, G. C. 1950. Stomach contents of whales off the coast of British Columbia. Fish. Res. Bd. Canad., Prog. Rep. Pacif. Coast Sta. 83: 27–28.

Pilson, M. E. Q., and P. B. Taylor. 1961. Hole drilling by *Octopus*. Science 134: 1366–68 (and cover photo).

Potts, W. T. W. 1965. Ammonia excretion in *Octopus dofleini*. Comp. Biochem. Physiol. 14: 339–55.

Potts, W. T. W., and M. Todd. 1965. Kidney function in the octopus. Comp. Biochem. Physiol. 16: 479–89.

Pritchard, A. W., M. J. Huston, and A. W. Martin. 1963. Effects of *in vitro* anoxia on metabolism of octopus heart tissue. Proc. Soc. Exper. Biol. Med. 112: 27–29.

Radhakrishnamurthy, B., S. R. Srinivasan, E. R. Dalferes, and G. S. Berenson. 1970. Composition of glycopeptides from chondroitin sulfate-protein complex from squid skin. Comp. Biochem. Physiol. 36: 107–17.

Recksiek, C. W. 1978. California's market squid. Pacific Discov. 31: 19–27.

Recksiek, C. W., and H. W. Frey, eds. 1978. Biological, oceanographic, and acoustic aspects of the market squid, *Loligo opalescens* Berry. Calif. Dept. Fish & Game, Fish Bull. 169: 1–185.

Rice, D. W. 1963. Progress report on biological studies of the larger Cetacea in the waters off California. Norsk. Hvalfangst-Tidende 52: 181–87.

Robbins, L. L., F. K. Oldham, and E. M. K. Geiling. 1937. The stomach contents of sperm whales caught off the west coast of British Columbia. Rep. Prov. Mus. Natur. Hist. British Columbia, pp. 19–20.

Robson, G. C. 1929. A monograph of the recent Cephalopoda. Part 1. London: British Museum (Natur. Hist.). 236 pp.

————. 1932. A monograph of the recent Cephalopoda. Part 2. London: British Museum (Natur. Hist.). 359 pp.

Roper, C. F. E., and R. E. Young. 1975. Vertical distribution of pelagic cephalopods. Smithsonian Contr. Zool. 209: 1–51.

Roper, C. F. E., R. E. Young, and G. L. Voss. 1969. An illustrated key to the families of the order Teuthoidea (Cephalopoda). Smithsonian Contr. Zool. 13: 1–32.

Rosenthal, R. J. 1971. Trophic interaction between the sea star *Pisaster giganteus* and the gastropod *Kelletia kelletii*. Calif. Dept. Fish & Game, Fish Bull. 69: 669–79.

Rosenthal, R. J., W. D. Clarke, and P. K. Dayton. 1974. Ecology and natural history of a stand of giant kelp, *Macrocystis pyrifera*, off Del Mar, California. Calif. Dept. Fish & Game, Fish Bull. 72: 670–84.

Ruggieri, G. D., and N. D. Rosenberg. 1974. The octopus, "cowardly lion" of the sea. Oceans 4: 50–55.

Sasaki, M. 1929. A monograph of the dibranchiate cephalopods of the Japanese and adjacent waters. J. Fac. Agr., Hokkaido Imp. Univ., 20 (Suppl.): 1–357.

Sato, T. 1976. Results of exploratory fishing for *Dosidicus gigas* (d'Orbigny) off California and Mexico. F.A.O. Fish. Rep. 170, Suppl. 1: 61–7.

Scofield, W. L. 1924. Squid at Monterey. Calif. Fish & Game 10: 176–82.

Sheumack, D. D., M. E. H. Howden, I. Spence, and R. J. Quinn. 1978. Maculotoxin: A neurotoxin from the venom glands of the octopus *Hapalochlaena maculosa* identified as tetrodotoxin. Science 199: 188–89.

Smith, A. G. 1963. More giant squids from California. Calif. Fish & Game 49: 209–11.

Smith, L. S. 1962. The role of venous peristalsis in the arm circulation of *Octopus dofleini*. Comp. Biochem. Physiol. 7: 269–76.

————. 1963. Circulatory anatomy of the octopus arm. J. Morphol. 113: 261–66.

Snow, C. D. 1970. Two accounts of the northern octopus *Octopus dofleini*, biting scuba divers. Res. Rep. Fish. Comm. Oregon 2: 103–4.

Solem, A., and C. F. E. Roper. 1975. Structures of recent cephalopod radulae. Veliger 18: 127–33.

Srinivasan, S. R., B. Radhakrishnamurthy, E. R. Dalferes, and G. S. Berenson. 1969. Glycosaminoglycans from squid skin. Comp. Biochem. Physiol. 28: 169–76.

Summers, W. C., and J. J. McMahon. 1974. Studies on the maintenance of adult squid (*Loligo pealei*). I. Factorial survey. Biol. Bull. 146: 279–90.

Summers, W. G., J. J. McMahon, and G. N. P. A. Ruppert. 1974. Studies on the maintenance of adult squid (*Loligo pealei*). II. Empirical extensions. Biol. Bull. 146: 291–301.

Taki, I. 1941. On keeping octopods in an aquarium for physiological experiments, with remarks on some operative techniques. Venus 10: 140–56.

Talmadge, R. R. 1967. Notes on cephalopods from northern California. Veliger 10: 200–202.

Taylor, P. B., and L.-C. Chen. 1969. The predator-prey relationship between the octopus (*Octopus bimaculatus*) and the California scorpionfish (*Scorpaena guttata*). Pacific Sci. 23: 311–16.

Tinbergen, L., and J. Verway. 1945. Zur Biologie von *Loligo vulgaris* Lam. Arch. Neerl. Zool. 7: 213–86. (The biology of *Loligo vulgaris* Lam. Trans. 2733, Dept. Environment Fisheries and Marine, Ottawa, Canada.)

Turner, C. H., E. E. Ebert, and R. R. Given. 1969. Man-made reef ecology. Calif. Dept. Fish & Game, Fish Bull. 146: 1–221.

Turner, C. H., and J. C. Sexsmith. 1964. Marine baits of California. Calif. Dept. Fish & Game. 71 pp.

Van Heukelem, W. F. 1973. Growth and life span of *Octopus cyanea* (Mollusca: Cephalopoda). J. Zool., London 169: 299–315.

———. 1977. Laboratory maintenance, breeding, rearing and biomedical research potential of the Yucatan octopus *Octopus maya*. Lab. Anim. Sci. 27: 852–59.

Van Hyning, J. M., and A. R. Magill. 1964. Occurrence of the giant squid (*Moroteuthis robusta*) off Oregon. Oregon Fish Comm. Res. Briefs 10: 67–68.

Verrill, A. E. 1876. Note on gigantic cephalopods, a correction. Amer. J. Sci. & Arts (3) 12: 236–37.

———. 1882. Report on the cephalopods of the northeastern coast of America. Rep. U.S. Comm. Fish. Ann. 1879: 211–455.

Voss, G. L. 1973. Cephalopod resources of the world. F.A.O. Fish Circ. 149: 1–75.

Voss, G. L., and R. F. Sisson. 1967. Squids: Jet-powered torpedoes of the deep. Nat. Geogr. 131: 386–411.

———. 1971. Shy monster, the octopus. Nat. Geogr. 140: 776–99.

Walker, J. J., N. Longo, and M. E. Bitterman. 1970. The octopus in the laboratory: Handling, maintenance, training. Behav. Res. Meth. Instrum. 2: 15–18.

Warren, L. R., M. F. Sheier, and D. A. Riley. 1974. Color changes of *Octopus rubescens* during attacks on unconditioned and conditioned stimuli. Anim. Behav. 22: 211–19.

Wells, M. J. 1961. What the octopus makes of it: Our world from another point of view. Amer. Sci. 49: 215–27.

———. 1962. Brain and behaviour in cephalopods. Stanford, Calif.: Stanford University Press. 171 pp.

———. 1978. Octopus: Physiology and behaviour of an advanced invertebrate. London: Chapman & Hall. 417 pp.

Wells, M. J., and J. Wells. 1959. Hormonal control of sexual maturity in *Octopus*. J. Exper. Biol. 36: 1–33.

———. 1972a. Optic glands and the state of the testes in *Octopus*. Mar. Behav. Physiol. 1: 71–83.

———. 1972b. Sexual displays and mating of *Octopus vulgaris* Cuvier and *O. cyanea* Gray and attempts to alter performance by manipulating the glandular condition of the animals. Anim. Behav. 20: 293–308.

———. 1977. Cephalopoda: Octopoda, pp. 291–336, *in* A. C. Giese and J. S. Pearse, eds., Reproduction of marine invertebrates, vol. 4. New York: Academic Press. 369 pp.

Wilbur, K. M., and C. M. Yonge, eds. 1964. Physiology of Mollusca. Vol. 1. New York: Academic Press. 473 pp.

———, eds. 1966. Physiology of Mollusca. Vol. 2. New York: Academic Press. 645 pp.

Williams, L. W. 1909. The anatomy of the common squid *Loligo pealii* Lesueur. Leiden: Brill. 92 pp.

Winkler, L. R., and L. M. Ashley. 1954. The anatomy of the common octopus of northern Washington. Publ. Dept. Biol. Sci., Walla Walla College 10: 1–29.

Witmer, A. 1975. The fine structure of the renopericardial cavity of the cephalopod *Octopus dofleini martini*. J. Ultrastruc. Res. 53: 29–36.

Witmer, A., and A. W. Martin. 1973. The fine structure of the branchial heart appendage of the cephalopod *Octopus dofleini martini*. Z. Zellforsch. 136: 545–68.

Wodinsky, J. 1969. Penetration of the shell and feeding on gastropods by octopus. Amer. Zool. 9: 991–96.

Wood, F. G. 1971. An octopus trilogy. Natur. Hist. 80: 14–24, 84–87.

Wormuth, J. H. 1976. The biogeography and numerical taxonomy of the oegopsid squid family Ommastrephidae in the Pacific Ocean. Bull. Scripps Inst. Oceanogr. 23: 1–90.

Yamamoto, T. K., K. Tasaki, T. Sugawara, and A. Tonosaki. 1965. Fine structure of the octopus retina. J. Cell. Biol. 25: 345–59.

Yarnall, J. L. 1969. Aspects of the behaviour of *Octopus cyanea* Gray. Anim. Behav. 17: 747–54.

Yonge, C. M., and T. E. Thompson. 1976. Living marine molluscs. London: Collins. 288 pp.

Young, J. Z. 1961. Learning and discrimination in the *Octopus*. Biol. Rev. 36: 32–96.

———. 1965. The nervous pathways for poisoning, eating and learning in *Octopus*. J. Exper. Biol. 43: 581–93.

———. 1971. The anatomy of the nervous system of *Octopus vulgaris*. Oxford: Clarendon. 690 pp.

Young, R. E. 1972. The systematics and areal distribution of pelagic cephalopods from the seas off southern California. Smithsonian Contr. Zool. 97: 1–159.

———. 1973. Information feedback from photophores and ventral countershading in mid-water squid. Pacific Sci. 27: 1–7.

———. 1975. *Leachia pacifica* (Cephalopoda, Teuthoidea): Spawning habitat and function of the brachial photophores. Pacific Sci. 29: 19–25.

———. 1977. Ventral bioluminescent countershading in midwater cephalopods. Symp. Zool. Soc. London 38: 161–90.

Young, R. E., and C. F. E. Roper. 1976. Bioluminescent countershading in midwater animals: Evidence from living squid. Science 191: 1046–48.

———. 1977. Intensity regulation of bioluminescence during countershading in living midwater animals. Fish. Bull. 75: 239–52.

Zuev, G. V., and K. N. Nesis. 1971. The role of squids in the food chains of the ocean [In Russian], pp. 78–83, *in* Squids—biology and fishery. Moscow: Izdatel'stvo Pishchevaya Promyshlennost. 360 pp.

The Annelid-Arthropod Complex: *Introduction*

Donald P. Abbott and Eugene C. Haderlie

Annelids generally have soft, slender, wormlike bodies that lack jointed appendages. Arthropods, in contrast, usually have shorter, chunkier bodies, a stiff outer cuticle or exoskeleton, and conspicuous jointed legs. There are other, less immediately obvious differences between annelids and arthropods, but the striking fact about the two groups is not their superficial difference but their fundamental underlying similarity. They share many features of structure, function, and development. Their close affinities have been recognized since the days of Lamarck (1744–1829) and Cuvier (1769–1832), when the phyla were placed together in the group Annulata or Articulata, and today most zoologists would agree that the arthropods arose from early annelidan ancestors.

In animals of both phyla the body consists of segments arranged in linear sequence. Hypotheses concerning the origin of segmentation are critically reviewed by Clark (1964). Typically in annelids the segments are much alike, and most segments bear projecting bristles or setae that help grip the substratum in locomotion; appendages, if present, are simple or elaborate lobes representing outpocketings of the body wall called parapodia. The segmented worms early evolved a wonderful diversity in the sea (Chapter 18), and some forms spread to brackish and fresh water. With the evolution of vegetation on land, the annelids invaded the damper terrestrial substrata and helped form modern soils.

In some ancient Pre-Cambrian annelids the soft cuticle became stiffer and ultimately developed into an external armor. Other metazoan groups were beginning to develop fossilizable hard parts at about the same time. Perhaps the advent of hard parts was related to the evolution of predators (e.g., Hutchinson, 1959), or to the fact that the animals were becoming larger in size and needed better skeletel support (e.g., Nicol, 1966). Or the change may have accompanied a shift in habitat requiring increasingly better protection from physical or chemical environmental factors. In any event, from one or more stocks of annelids the early arthropods arose. In terms of numbers of species they soon became the dominant group of multicellular animals on the earth, and have remained so from the Cambrian to the present (Nicol, 1962).

The differences between annelids and arthropods as we know them today are mostly consequences or correlatives of the hardening of the cuticle into a supporting skeleton in the arthropods. Most annelids have a thin, soft cuticle, but the body in life is rather turgid. Fluids in the interior body cavity or coelom, under pressure from the surrounding tissues, provide a sort of hydrostatic skeleton, which supports the body both at rest and in motion. In polychaetes with sizable parapodia the appendages may be the main locomotory organs in slow walking. In rapid locomotion, however, the movements of the appendages are less important than the coordinated actions of the great circular and longitudinal muscles of the body wall; these actions result in serpentine wriggling or peristaltic movements that are in no way inhibited by the flexible annelid cuticle. In contrast, most adult arthropods have exoskeletons so stiff that locomotion by wriggling or body peristalsis is impossible. Instead, the animals move about by means of articulated appendages in which flexor and extensor muscles are arranged antagonistically at each joint. Arthropod somatic musculature is typically striated,

Donald P. Abbott is Professor, Department of Biological Sciences and Hopkins Marine Station, Stanford University. Eugene C. Haderlie is Professor of Oceanography at the Naval Postgraduate School, Monterey.

and thus better adapted to rapid activity than the smooth muscles of the annelids. It consists of a large number of discrete muscle bundles that are usually anchored to the exoskeleton.

The annelids and some of the more primitive arthropods have elongate bodies in which successive segments and their appendages are much alike, a condition described as homonomous metamerism. The limbs tend to be multipurpose appendages, often serving such functions as locomotion, sense reception, and respiration more or less simultaneously. Body regions, except at head and tail ends, are ill defined. In contrast, most arthropods have shorter, more compact bodies, in which whole groups of adjacent segments are fused together and modified to form distinctive body regions, or tagmata, such as the head, thorax, and abdomen. In many arthropods (for example, crabs, mites, beetles, or ants) the original segmentation is obscured in the adult stages and often in the larvae as well. Moreover, the arthropods have tended to diversify their appendages, specializing particular limbs for specific functions such as sensory reception, biting, clinging, walking, swimming, cleaning the body, respiration, egg carrying, and the like.

Numerous other differences separate most annelids from most arthropods. Annelids typically have closed circulatory systems, with smaller arteries and veins connected by fine capillary vessels; the arthropods, in contrast, have open circulatory systems, in which major arteries, when present, terminate in large blood sinuses that bathe the viscera and conduct blood back to the heart. The coelomic cavity, the main internal space other than the gut in annelids, is much reduced in arthropods, its functions taken over in part by the big blood sinuses; portions of the coelom remain, associated with reproduction and excretion. The arthropods have lost nephridia (important excretory organs in many annelids), and have virtually lost cilia, as well. And in arthropods the stiff outer cuticle must be molted at intervals to allow for growth. Finally, the sense receptors of arthropods, especially the eyes, are usually more elaborate than those of annelids, and are associated with better perception and coordination, more rapid responses, and often more complex behavior.

Annelids have their early ancestral roots in the sea, and most of them are still marine today. Of the 9,000 or more annelid species described, more than 60 percent occur only in the sea. The dominant marine forms are members of the class Polychaeta, though a few representatives of the classes Oligochaeta (earthworms and allies) and Hirudinea (leeches) are present as well. Conversely, the oligochaetes and leeches are the dominant annelids in fresh water and in moist terrestrial habitats, where the polychaetes are scarce or absent.

Sharing features of anatomy, development, and biochemistry with the annelids or arthropods or both are several small phyla of metazoans, including two groups of marine worms, the Echiura and the Sipuncula. The echiurans usually possess setae, and the larvae or adults of several species exhibit serially repeated anatomical parts that are plausibly interpreted as vestiges of metamerism. Biochemically, echiurans appear closest to the annelids (Van Thoai, 1976). They probably represent an offshoot of the annelids in which segmentation has been suppressed, a view supported by Hatschek (1880) a century ago, but they seem best treated as a separate phylum. The sipunculans, in their development, in the structure of the brain and associated structures (Åkesson, 1958), and in various biochemical characteristics (Florkin, 1976; Van Thoai, 1976), show their closest affinities to the annelids, though the differences here are greater than the differences between echiurans and annelids. The absence of clear traces of metamerism makes it more difficult to argue for an annelidan origin of sipunculans (e.g., Clark, 1969). However, the modifications exhibited in members of two modern polychaete families, the Sternaspidae (e.g., Dahl, 1955) and the Caobangidae (Jones, 1974), are very suggestive of the changes that might have occurred in the early evolution of sipunculans from annelids, and at least show the capacity of the annelid body to become modified in very sipunculid-like ways. In any event, the sipunculans are best treated as a separate phylum, allied to the large annelid-arthropod complex.

Arthropods, like annelids, have their remote ancestral roots in the sea. They abound in the sea today, but they are also plentiful in fresh water and on land, and many are temporary inhabitants of the air. Starting in the Silurian period of the Paleozoic, there were many different invasions of the land by various arthropod stocks, some arriving by way of fresh water and damp soil, others departing from the sea through marine swamps or over seashore rocks and beaches. The hard exoskeleton not only provided support for the body out of water, but, properly waterproofed, reduced water loss through evaporation, and in some groups provided a tracheal system of tiny internal air pipes that conduct oxygen to the innermost body cells. On land, the dominant arthropod groups are the insects and arachnids. Many hundreds of new species are described by zoologists each year, and it will not be long before the number of terrestrial arthropod species known and named passes the million mark.

The diversity of arthropods is so vast, and the differences between the largest subdivisions of the group seem so great, that the question is often asked: Do the arthropods form a natural group, a single large branch on an evolutionary tree? Are all modern and fossil arthropods descended from some single original arthropod stock, or was "arthropodization" an idea whose time had come, an evolutionary trend exploited independently (but with numerous parallel changes) in two or more different ancient annelid groups? Our discussion of this is deferred to the introduction to the Crustacea and other marine arthropods (p. 499).

Literature Cited

Åkesson, B. 1958. A study of the nervous system of the Sipunculoideae with some remarks on the development of the two species *Phascolion strombi* Montagu and *Golfingia minuta* Keferstein. Undersökningar över Öresund 38: 1–249.

Clark, R. B. 1964. Dynamics in metazoan evolution. The origin of the coelom and segments. Oxford, Eng.: Clarendon. 313 pp.

———. 1969. Systematics and phylogeny: Annelida, Echiura, Sipuncula, pp. 1–68, *in* M. Florkin and B. T. Scheer, eds., Chemical zoology, vol. 4. New York: Academic Press. 548 pp.

Dahl, E. 1955. On the morphology and affinities of the annelid genus *Sternaspis*. Report of the Lund University Chile Expedition 1948–49. Vol. 21. Lunds Univ. Årsskrift. N.F. Avd. 2. 51: 1–22.

Florkin, M. 1976. Biochemical evidence for the phylogenetic relationships of the Sipuncula, pp. 95–108, *in* M. E. Rice and M. Todorović, eds., Proc. Internat. Symp. Biol. Sipuncula and Echiura. II. Belgrade: Naućno Delo. 204 pp.

Hatschek, B. 1880. Über die Entwicklungsgeschichte von *Echiurus* und die systematische Stellung der Echiuridae (Gephyrei Chaetiferi). Arb. Zool. Inst. Univ. Wien 3: 45–78.

Hutchinson, G. E. 1959. Homage to Santa Rosalia or why are there so many kinds of animals? Amer. Natur. 93: 145–59.

Jones, M. L. 1974. On the Caobangiidae, a new family of the Polychaeta, with a redescription of *Caobangia billeti* Giard. Smithsonian Contr. Zool. 175: 1–55.

Nicol, D. 1962. The Arthropoda, the dominant phylum since the Cambrian period. Syst. Zool. 11: 176–77.

———. 1966. Cope's rule and Precambrian and Cambrian invertebrates. J. Paleontol. 40: 1397–99.

Van Thoai, N. 1976. The biochemical relationship of echiurans and sipunculans to annelids and mollusks, pp. 67–76, *in* M. E. Rice and M. Todorović, eds., Proc. Internat. Symp. Sipuncula and Echiura. II. Belgrade: Naućno Delo. 204 pp.

Chapter 18

Polychaeta: *The Marine Annelid Worms*

Donald P. Abbott and Donald J. Reish

Polychaetes, along with earthworms, leeches, and their relatives, are assigned to the phylum Annelida, the segmented worms. These are forms in which the elongate body is composed of a chain of similar units called somites or metameres. The somites are separated from one another externally by grooves encircling the body ("Annelida" = "little rings"); internally they are usually separated by more or less well-developed transverse septa, or tissue partitions. The head region consists of a prostomium (a lobe lying above and anterior to the mouth) and a peristomium formed by one or more modified segments lying just posterior to the mouth. A specialized region at the posterior end (the pygidium) usually bears the anus. Between head and tail ends lie the trunk segments, varying in number with different species from a few to several hundred. The brain lies dorsal to the gut, usually in the prostomium where it is associated with sense receptors (visible eyes and tentacles in some species). Connectives on either side of the pharynx join the brain with the paired ventral nerve cords (usually fused together), which most commonly bear a pair of ganglia in each segment, at least in regions of the body where the septa are well developed. Each body segment also contains a section of the elongate gut, which is separated from the muscular body wall by the body cavity or coelom. Many segments bear setae or chaetae (bristles projecting laterally), and many also bear tubules connecting the coelom to the outside, which may represent nephridia (excretory tubules), genital ducts, or both combined (Goodrich, 1945). Blood vessels are usually present (Hanson, 1949a).

Some 9,000 species of annelids have been described thus far (Reish, 1974a). They fall into three classes: the Polychaeta (marine worms), the Oligochaeta (earthworms and their allies), and the Hirudinea (leeches). Other annelid groups formerly recognized as separate classes are generally placed in the Polychaeta today (e.g., Fauchald, 1974; Hermans, 1969). Although oligochaetes and leeches do occur in the sea, they are far commoner in fresh water and on land, and are not dealt with further here. Polychaetes, on the other hand, are nearly all marine, though some very interesting forms are found in brackish and fresh water in California and elsewhere (see, for example, J. A. Hanson, 1972; M. Jones, 1967; R. Smith, 1950, 1953, 1958, 1959, 1963, 1964; Oglesby, 1965, 1968, 1969, 1978), and a few species occur in damp soil on land (see, for example, Storch & Welsch, 1972). About 6,000 species are known; thus, in terms of numbers of species, the polychaetes outnumber the other classes combined.

In the Polychaeta (meaning "many setae") the bristlelike setae are commonly both numerous and conspicuous, projecting from the body wall on either side of most trunk segments. Somites bearing setae are called setigerous segments or chaetigers. The setae are generally borne on paired lateral extensions of the body (parapodia). Most commonly each parapodium bears setae in two bundles or sheafs, a dorsal or notopodial bundle ("notum" = "back"), and a ventral or neuropodial bundle (i.e., the bundle closest to the ventral nerve cord). The parapodial lobes are reinforced by one or more especially heavy setae (the aciculae), which serve as skeletal rods.

Donald P. Abbott is Professor, Department of Biological Sciences and Hopkins Marine Station, Stanford University. Donald J. Reish is Professor of Biology at California State University, Long Beach.

The setae are secreted by epidermal cells located around the aciculae; a single cell, the chaetoblast, plays the major role in forming each individual seta (see, for example, Bobin, 1944; O'Clair & Cloney, 1974). Setae are extremely variable in form in the polychaetes (e.g., Blake, 1975c). It may be necessary to examine them under a compound microscope to confirm identification of some species. If so, parapodia bearing setae are easily removed with forceps from worms that have been anesthetized for a few minutes by immersion in a solution of 75 gms magnesium chloride dissolved in a liter of tap water.

Polychaetes occur in all environments in the ocean. Some are planktonic throughout life and inhabit the open sea (e.g., Dales 1955b, 1957a), but most species, as adults, are benthic, dwelling on or in the bottom at various depths. Along the shore, polychaetes are particularly abundant in situations where they receive some protection from the surf, as in mussel beds, in fronds and holdfasts of marine plants, among fouling growth encrusting pilings and floats in harbors, under rocks, or burrowed in sand or mud. Well over 700 species of polychaetes have been described from California inshore and offshore waters (Hartman, 1968, 1969). Even those forms frequently encountered in the intertidal zone are far too numerous to cover here, and most of them are small and not easily recognized from photographs. This chapter presents a sample of the more conspicuous forms that may be encountered along the shores, representing some (but by no means all) of the more important families. Species and genera of polychaetes often present problems in identification, but with a little effort almost anyone can learn to place many of the polychaetes in their proper families.

The anterior end varies considerably from one polychaete family to another. In some it lacks appendages and appears much like the head of an earthworm, as in the families Arabellidae (e.g., 18.18) and Capitellidae (e.g., 18.26). From two to several antennae, and frequently dark eyespots, may be present on the prostomium, as in the families Phyllodocidae (e.g., 18.8), Nereidae (e.g., 18.10–12), Glyceridae (e.g., 18.13), Onuphidae (e.g., 18.17), and Dorvilleidae (e.g., 18.19). In some families the anterior end of the body is virtually concealed by numerous tentacles and/or gills (e.g., 18.33–44). The tentacles or gills in the families Sabellidae and Serpulidae are plumelike and sometimes brilliantly colored (e.g., 18.37–44).

The differences exhibited by various polychaete families reflect differences in ecological roles or ways of life, particularly differ-

ences in food and type of habitat. Forms with elaborate tentacles tend to be sedentary or sessile, many of them living in tubes. The tubes may be soft and mucoid or parchmentlike, or constructed of sand grains cemented together, or even formed largely of secreted calcium carbonate. The elongate appendages may be used for respiration and/or food gathering. Some worms feed on small organisms and bits of organic detritus gathered from nearby bottom deposits or sediments; others filter suspended microorganisms and detritus particles from the surrounding water and may circulate the fluid to ensure a regularly renewed supply. Worms without elongate feeding tentacles may have an eversible proboscis or fleshy lips with which they scoop up and ingest nutritious deposits. Dense populations of worms may cycle sizable fractions of the substratum in a beach or mud flat through their guts every few weeks (e.g., D. Fox, Crane & MacConnaughey, 1948; MacConnaughey & Fox, 1949). Numerous polychaetes have formidable jaws and teeth. These may be seen through the body wall in very small worms when the animals are mounted on slides and viewed with a compound microscope. In larger polychaetes they are easily seen after the worms are anesthetized in a magnesium chloride solution; forceps can then be inserted into the mouth and the jaws gripped and gently pulled out for closer study. These toothed jaws may be used in feeding on plant or animal material or on detritus; the worms often also use them in fighting members of their own and other species, especially in defending small territories against intruders.

Some polychaetes are carnivorous predators, some are herbivores, and still others may be omnivores, scavengers, filter feeders, deposit feeders, and so on (see, for example, Jumars & Fauchald, 1977). In turn, polychaetes are eaten by a great variety of invertebrates (e.g., Fields, 1965; Mauzey, Birkeland & Dayton, 1968), fishes (e.g., Quast, 1968), and shorebirds (e.g., Recher, 1966; Reeder, 1951), and in some parts of the world selected species are prized as food by man (e.g., Smetzer, 1969). Some species of polychaetes secrete toxic materials, which perhaps discourage predators (e.g., Hashimoto & Okaichi, 1960); one such substance (a neurotoxin) provides the basis for a Japanese insecticide (brief review in Ruggieri, 1976).

Many polychaetes have marked regenerative capacities and may replace a lost head or tail end (e.g., Berrill, 1951, 1961; Boilly, 1969; Herlant-Meewis, 1964). The remarkable ability not only to regenerate, but also to reorganize remaining parts of the body in the process, has led to classic, and still continuing, experimental studies (e.g., Berrill & Mees, 1936a,b; Fitzharris, 1973, 1976) on polychaetes in which the trunk is divided into two or more quite distinctive regions. Polychaetes have also provided excellent research material for biologists interested in the study of simpler nervous systems (e.g., Bullock, 1953, 1965; various chapters in Mill, 1978), and in the experimental analysis of behavior and brain function (e.g., R. Clark, 1959a,c, 1960, 1966; Dyal, 1973; Evans, 1963, 1971, 1973).

Typically, in polychaetes the sexes are separate. Gametes mature in the coelomic cavities of some of the body segments and are generally spawned either through ducts to the outside or by rupture of the body wall. Fertilization is usually external. The eggs usually undergo a classical spiral cleavage (e.g., Costello & Henley, 1976; E. Wilson, 1892), resulting in a trochophore larva with an apical tuft of cilia and a girdle of powerful swimming cilia (the prototroch) encircling the body just anterior to the mouth. A bit later the first segments and setae begin to appear; later on, in benthic species, the tiny juvenile worms settle from the plankton and take up an existence on or in the substratum. Growth continues, both by the enlargement of existing somites and by the addition of new somites. New body segments are always added in linear sequence from a special growth zone just anterior to the tail section (pygidium). This pattern of growth by the addition of new segments from a posterior growth zone, called teloblastic growth, is a pervasive growth feature in both annelids and arthropods. Reviews of various aspects of reproduction and development may be found in D. Anderson (1973), R. Clark (1965), Golding (1972), Needham (1969), Olive & Clark (1978), and especially Schroeder & Hermans (1975). Specific directions for obtaining and fertilizing eggs and for maintaining developing embryos for class use or research are given in Costello & Henley (1971). The establishment of laboratory colonies of several species in U.S. and European laboratories permits many types of study heretofore impossible.

Those interested in more detailed studies should consult an invertebrate text for fundamentals. Dales (1967) presents a good semi-popular account of annelids that deals with systematics, morphology, physiology, development, and phylogeny in the phylum. Excellent reviews of many aspects of polychaete physiology are available in Florkin & Scheer (1969) and Mill (1978). Most of the papers cited in this introduction are themselves reviews of certain aspects of annelid biology. A collection of 33 papers in honor of Dr.

Olga Hartman, dean of west coast polychaete biologists, provides almost a textbook of current thought on polychaetes (see Reish & Fauchald, 1977). Those wishing an overview of polychaete taxonomy and diagnoses of higher taxa should consult the excellent works of Day (1967) and Fauchald (1977). The polychaete orders used in this chapter are those defined in Fauchald (1977). Polychaete phylogeny is considered by R. Clark (1969) and Fauchald (1974). The older literature on polychaetes is conveniently referenced in Hartman (1951). For identification of California polychaetes, one should consult Blake (1975c), Hartman (1968, 1969), and Light (1978). Regional references, some of which are valuable far beyond the borders of the areas for which they were designed, include, among others, Berkeley & Berkeley (1952, 1954) for British Columbia, Day (1967) for South Africa, Fauvel (1923, 1927) for France, Fauvel (1953) for India, and Hartman-Schröder (1971) for Germany. Fauchald's (1977) key to the orders, families, and genera of the polychaetes of the world should benefit workers everywhere, especially those in areas not covered by existing monographs.

Phylum Annelida / Class Polychaeta / Order Phyllodocida
Suborder Aphroditiformia / Family Polynoidae

18.1 **Arctonoe fragilis** (Baird, 1863)
Scale Worm

Living in ambulacral grooves on undersides of several species of starfishes, low intertidal zone on rocky shores; subtidal to 275 m; Alaska Peninsula at least to San Francisco Bay, probably to Baja California; uncommon in California.

Length to 85 mm, usually not over 55 mm; dorsal scales 29–34, oval, with ruffled margins, arranged in two rows.

In the Pacific northwest, *A. fragilis* occurs most commonly on the sea star *Evasterias troschelii* (8.12, which it often matches closely in color), less commonly on *Leptasterias aequalis*, *L. hexactis* (8.11), *Orthasterias koehleri* (8.9), *Solaster dawsoni* (8.7), *Stylasterias forreri*, and *Luidia foliolata*. In California it occurs occasionally on *Pisaster ochraceus* (8.13). This worm was the subject of classic studies to determine how the commensal worms locate their starfish hosts. Worms removed from *Evasterias* and placed in plain seawater were found to be attracted by seawater in which *Evasterias* had been sitting when this was introduced into the tank, but did not respond to the addition of untreated water or of water exposed to *Pisaster* or to the sea cucumber *Parastichopus*. Water exposed to injured *Evasterias* did not attract the worms. In general, worms removed from a particular species of sea star and then exposed in a tank to both that species and to an alternative sea star species (a species used as a host by some individuals of *A. fragilis*) chose as host the species from which they had originally been removed.

See Davenport (1950), Davenport & Hickok (1951), Hartman (1968), Hickok & Davenport (1957), MacGinitie & MacGinitie (1968), and Pettibone (1953).

Phylum Annelida / Class Polychaeta / Order Phyllodocida
Suborder Aphroditiformia / Family Polynoidae

18.2 **Arctonoe pulchra** (Johnson, 1897)
Scale Worm

Fairly common, free-living or associated as a commensal with several other invertebrates, low intertidal rocks to 295 m depth; Gulf of Alaska to Isla Cedros (Baja California).

Length to 70 mm, usually less; dorsal scales in 20–33 pairs, smooth, each bearing a dark spot.

In California, *A. pulchra* occurs most commonly as a commensal of the giant chiton *Cryptochiton stelleri* (16.25), the giant keyhole limpet *Megathura crenulata* (13.8), the sea cucumbers *Parastichopus californicus* (9.7) and *P. parvimensis* (9.8), and the polychaete *Loimia montagui* (family Terebellidae). In addition, the worm has been reported on seven other gastropod species, another polychaete, and seven species of sea stars (among them *Petalaster* (=*Luidia*) *foliolata*; *Solaster stimpsoni*, 8.6; *Pteraster tesselatus*; and *Dermasterias imbricata*, 8.3).

Experiments show that worms removed from a host are attracted by water-soluble compounds diffusing from the body of the host; these substances help the worms locate a host in the first place, and aid displaced worms in finding new homes. Water extracts of injured hosts do not attract the worms. A commensal worm often has a body color fairly closely matching that of the host. Worms removed from their hosts generally return to the host species when given a choice

between it and an alternative species (a species occupied as a host by some members of the *A. pulchra* population). Worms transferred to alternative species of hosts tend to remain on the new hosts; at first, if given a chance, they generally return to their original host species, but after 1–3 weeks they come to prefer the new host to the original host species.

Marked territorial behavior and intraspecific aggression on the part of *A. pulchra* usually limit the population density of the worm on a host. A large *Megathura* host rarely bears more than one worm, and the even larger *Parastichopus parvimensis* infrequently supports more than two. If more worms are added, combat between worms quickly ensues. Individuals of *A. pulchra* 20 mm or more in length fight each other upon confrontation, inflicting major damage with their powerful toothed jaws.

Biochemical studies show the major phosphagen present is probably phosphoglycocyamine.

See Davenport (1950), Davenport & Hickok (1951), Dimock (1971, 1974), Dimock & Davenport (1971), Hartman (1968), MacGinitie & MacGinitie (1968), Pettibone (1953), Robin (1964), and Skogsberg, (1942).

Phylum Annelida / Class Polychaeta / Order Phyllodocida
Suborder Aphroditiformia / Family Polynoidae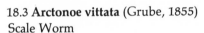

18.3 **Arctonoe vittata** (Grube, 1855)
Scale Worm

Free-living or dwelling commensally with snails, chitons, sea stars, and tube-forming annelids, middle and low intertidal zones on rocky shores; subtidal to 275 m; Bering Strait to Ecuador; Japan.

Length to 100 mm, usually less; dorsum bearing at least 30 pairs of smooth scales, the surface across setigerous segments 7 and 8 with a dark band.

In California waters *A. vittata* occurs most commonly in the pallial grooves of the keyhole limpet *Diodora aspera* (13.10) and the giant chiton *Cryptochiton stelleri* (16.25), and on the starfish *Dermasterias imbricata* (8.3). Over its whole range in the Pacific northwest and Japan it is reported from nine sea stars (including *Henricia leviuscula*, 8.5; *Petalaster* (=*Luidia*) *fo-*

liolata; Pteraster tesselatus; and three species of *Solaster*), one chiton, eight gastropods (mostly limpets, keyhole limpets, and the abalone *Haliotis kamtschatkana*), and two polychaetes (family Terebellidae: *Thelepus crispus*, 18.35; *Neoamphitrite robusta*).

Worms removed from *Diodora* and allowed a choice of hosts usually selected *Diodora* rather than another host species on which some individuals of *A. vittata* are known to live commensally. The scent of a host is detected at a distance underwater by receptor organs on the three prostomial antennae. Hosts are recognized on contact by sense organs located on the two palps. Worms often closely match their hosts in body color.

Studies of *A. vittata* on *Cryptochiton* hosts in Monterey Bay show that the worms are present on chitons living at 4–40 m depth. The incidence of worms on hosts varies seasonally: 60 percent of the host population bears a worm from October to December, but only 22–25 percent do from May to August.

Reproduction of the worms has not been followed, but they probably spawn in spring or early summer; the larvae are planktonic. Juvenile worms are commonest on hosts in August and September, when 75 percent of the chitons bear worms less than 3 mm long. A given chiton host may have an *A. vittata* on one side and a commensal crab (*Opisthopus transversus*, 25.42) on the other, but it never bears more than one adult worm. It seems likely that this results from fighting between adult worms (see *A. pulchra*, 18.2). In the pallial groove of the host the worms normally face into the water current, that is, their heads point toward the head of the chiton. They feed on particles of detritus. Adult worms use oxygen at the rate of 27–167 µl of oxygen per gm of wet weight per hour at 13°C.

Commensal worms on *Diodora* may act to protect their hosts. When *Diodora* is attacked by the sea star *Pisaster ochraceus* (8.13), the worm moves around the pallial groove, seeks out the tube feet of the starfish, then bites these or the adjacent ambulacral area, usually causing the sea star to withdraw its attack.

See Davenport & Hickok (1951), Dimock & Dimock (1969), Gerber & Stout (1967), Hartman (1968), Pettibone (1953), Skogsberg (1942), and Webster (1968, 1975).

Phylum Annelida / Class Polychaeta / Order Phyllodocida Suborder Aphroditiformia / Family Polynoidae

18.4 **Halosydna brevisetosa** Kinberg, 1855
Scale Worm

Common, either free-living in shelter of mussel beds, plant holdfasts, and encrusting growth on pilings, or commensal in tubes of large annelids, low intertidal zone on rocky coast and in bays; subtidal to 545 m; Kodiak Island (Alaska) to Baja California.

Commensals to 110 mm long, free-living worms to 60 mm; dorsal scales in 18 pairs; ventral setae entire (not bifid) at tip; in southern California intergrading with the form described as **H. johnsoni** (Darboux, 1899), which has the ventral setae bifid at tips.

Although the worms are found in bays, they prefer cleaner water. In the Los Angeles–Long Beach harbor area they seldom occur where dissolved oxygen levels drop below 2.5 mg per liter.

Commensal specimens of *H. brevisetosa* have been found inhabiting the occupied tubes of the polychaetes *Pista pacifica* (18.34), *P. elongata* (18.33), *Thelepus crispus* (18.35), *Neoamphitrite robusta,* and *Loimia montagui* (all family Terebellidae), and *Platynereis bicanaliculata* (18.12). Host-commensal relationships have not been studied. In general the commensal worms are larger than those found free-living, suggesting that the commensal habit is adopted later in life, but specimens as small as 6.5 mm long have been taken commensally in Elkhorn Slough (Monterey Co.). *H. brevisetosa* also occurs as a commensal in the umbilicus of moon snail shells occupied by the subtidal hermit crab *Paguristes bakeri.*

Studies of the function of the dorsal scales show that in this species, and in other scale worms, the dorsal scales overlie and shield a ciliated dorsal body surface along which water may be circulated freely even when the animals are tightly wedged into crannies or jammed into clusters of other organisms. Ciliary tracts move a current of seawater dorsally between adjacent parapodia on each side, and the scales direct the flow posteriorly and confine it in a narrow tunnel between the scales and the dorsal body surface. The internal body fluid in the coelom, stirred by cilia, flows in a countercurrent to the external water flow, facilitating gas exchange through thin areas in the body wall.

The sexes are separate. The gonads occur in segments 12–34; sperm are white and the eggs a pale green. In the Monterey area worms are mature in the spring and summer. Eggs ripening in the body cavity are clearly visible through the body wall. Gametes are released through the nephridia (myxonephridia), and fertilization is external. Some species of scale worms brood their embryos for a time below the dorsal scales, but not *H. brevisetosa*. At Tomales Bay (Marin Co.), planktonic larvae are commonly encountered in September and October, less frequently in May and August. The larvae, 0.4 mm long, have a characteristically polynoid, flattened anterior end, and two pairs of black eyes. They lack setae, but the rudiments of eight segments and five pairs of scales (elytra) can be seen. Larvae 0.55 mm long have developed setae on the eight segments. Newly settled juveniles often measure 0.9 mm long, are divided into 11 setigerous segments, and exhibit elytra on segments 2, 4, 5, 7, and 9 and dorsal cirri on segments 3, 6, 8, 10, and 12; adult setae develop following settlement. Juveniles of *H. brevisetosa* have been recorded on settling panels in the Los Angeles area in most months of the year, but most abundantly from June to October.

Relaxed for a short time in a 7.3 percent solution of magnesium chloride, or preserved in formalin, *H. brevisetosa* is an excellent worm for study by dissection. The eversible pharynx is equipped with powerful jaws, with a dorsal and ventral curved tooth on each side. The species has been described as a scavenger and detritus feeder, but food studies are lacking. As in other scale worms, the intestine in most body segments bears remarkable paired lateral caeca; in a related species studies showed that these are involved in both secretion and absorption, exhibiting no clear-cut difference in function from the main intestine. The anus is curiously placed, opening dorsally *anterior* to the pygidium, or posterior terminal unit of the body. The central nervous system, usually colored red with muscle hemoglobin, is often visible through the skin. A neuroendocrine complex is associated with the brain (supraesophageal ganglion). The major phosphagen in the body is phosphoglycocyamine.

For *H. brevisetosa*, see Andrews (1945), Baskin (1971), Blake (1975a), Davenport & Hickok (1951), Gaffney (1973), Hartman (1968), Hewatt (1937), Lwebuga-Mukasa (1970), MacGinitie & MacGinitie (1968), Pettibone (1953), Reish (1971b), Robin (1964), Rossi (1976), and Skogsberg (1942). For information on related forms, see Åkesson (1963), Cazaux (1968), Dales (1962b), and Welsch & Storch (1970).

Phylum Annelida / Class Polychaeta / Order Phyllodocida
Suborder Aphroditiformia / Family Polynoidae

18.5 **Harmothoe imbricata** (Linnaeus, 1767)
Scale Worm

Uncommon in California, but perhaps the most commonly encountered scale worm elsewhere in northern hemisphere; either commensal with tube-dwelling polychaetes or free-living in protected crevices or algal holdfasts, low intertidal zone in brackish or open coastal waters to ocean depths of 3,710 m; Point Barrow (Alaska) to southern California; Pacific, Atlantic, Indian, and Arctic Oceans.

Length to 65 mm, but usually only half that; dorsal scales in 15 pairs; ventral setae bifid at tips.

On the west coast, *H. imbricata* occurs as a commensal in tubes of the polychaetes *Thelepus crispus* (18.35) and *Neoamphitrite robusta* (both family Terebellidae) and *Diopatra ornata* (family Onuphidae).

The development of *H. imbricata* has been studied in several areas well separated geographically, yielding somewhat different results. At Tomales Bay (Marin Co.), female worms have been observed brooding developing eggs on the dorsal body surface below the scales, from September through January; these eggs measure 0.12 mm in diameter and are released into the sea at the trochophore stage. At Arcachon (France), at 8–9°C, the spawned eggs, 80 μm in diameter, are never brooded, and development is entirely pelagic. The developmental pattern in populations in England and northern Europe is similar to that in Tomales Bay; in Greenland the developing stages are probably brooded below the scales until the juvenile worms are ready to settle, and thus a pelagic stage is lacking.

Details of maturation and spawning are best known in

England, where events may approximate those in the Pacific. The sexes are separate, males forming sperm in segments 12–22 and females developing eggs from segment 8 to near the posterior end. Gamete development begins in September and continues through the winter; eggs develop in two separate batches that yield two successive spawnings in one season. Mature gametes of both sexes accumulate in the excretory tubules and at spawning pass out through the nephridiopores ventrally. During most of the year the adult worms are solitary and avoid one another, but they form pairs during the spawning season. Males cling to the backs of females that have not yet spawned (never females brooding their young) and may remain there for up to a month. In males the nephridiopores are directed ventrally; in females, however, the papillae bearing the nephridiopores are elongate and curve dorsally, so that when the eggs are shed they are passed upward and lodge under the dorsal scales. Fertilization occurs there. The number of eggs spawned ranges from less than 5,000 to over 40,000, depending on the size of the female. Larvae are brooded for about 16 days at 3–4°C, then released; a second batch is spawned about two weeks later. Larvae develop in the plankton until the seven- or eight-segment stage (when they have four pairs of scales), then settle to a benthic life. New segments are added rapidly; the adult number of 37 or 38 segments is reached when the worms are 15–20 mm long—thereafter they simply increase in overall size.

Other studies with *H. imbricata* include work on external ciliary currents, bioluminescence, the phosphagens (probably mainly phosphocreatine), general neurophysiology and behavior, the neuroendocrine system, and the fine structure and sensory role of the dorsal cirri.

See especially Daly (1972, 1973a,b), Daly, Evans & Morley (1972), Korn (1958a,b), Lawry (1967), Pettibone (1953), and Rasmussen (1956, 1973); see also Åkesson (1963), Baskin (1971), Blake (1975a,c), Cazaux (1968), Cormier (1969), Costello & Henley (1971), Evans (1971), Hartman)1968), Holborow & Laverack (1972), Horridge (1959), MacGinitie & MacGinitie (1968), J. Nicol (1958), Robin (1964), Seagrove (1938), and Thorson (1946).

Phylum Annelida / Class Polychaeta / Order Phyllodocida
Suborder Aphroditiformia / Family Polynoidae

18.6 **Hesperonoe adventor** (Skogsberg, 1928)
Scale Worm

A fairly common commensal in the burrows of the echiuroid worm *Urechis caupo* (19.7), low intertidal mud flats in protected bays and subtidal waters; Alaska to southern California.

Length to 50 mm or more; dorsal scales in 15 pairs, with smooth margins; ventral setae bearing many fine marginal hairs.

This very active and beautiful scale worm shares the burrow of *Urechis* along with several other boarders, including a small fish, a crab, a clam, and often other organisms; the species name *adventor* means "the guest at an inn." *Hesperonoe* is apparently always a commensal, not known to occur in a free-living state. It feeds on small animals or organic fragments that are carried into the host's burrow by water currents, and sometimes takes food from the mucus net that the host secretes in order to filter feed. The worm usually clings to the burrow wall beside or above the host, a position in which it is safe from the commensal crab *Scleroplax granulata* (25.41). The worm always faces in the same direction as the host; when the host changes position or direction, so does the worm.

Hesperonoe adventor actively fights other members of its own species. No more than one polychaete at a time is found in a *Urechis* burrow. Other *Hesperonoe* experimentally introduced are driven out or killed, and other invading polychaete species are usually killed and eaten.

Hesperonoe reportedly release eggs and sperm in March in central California (Elkhorn Slough, Monterey Co.). Other studies on the species relate to its phosphagen (probably phosphoglycocyamine), the neuroendocrine system associated with its brain, and habituation.

A related species, **H. complanata** (Johnson), lives in the tunnels of the burrowing shrimps *Callianassa californiensis* (24.2), and *Upogebia pugettensis* (24.1), also on protected mud flats.

See especially MacGinitie (1935); see also Baskin (1971), Dyal & Hetherington (1968), Filice (1958), Fisher (1946), Fisher & MacGinitie (1928), Hartman (1968), MacGinitie & MacGinitie (1968), Robin (1964), and Skogsberg (1928).

Phylum Annelida / Class Polychaeta / Order Phyllodocida
Suborder Aphroditiformia / Family Polynoidae

18.7 **Lepidasthenia longicirrata** Berkeley, 1923
Scale Worm

In soft sediments, very low intertidal zone and subtidal waters to 333 m; British Columbia to southern California.

Length to 120 mm; dorsal scales in 35–43 pairs; small papillae in transverse row on ventral side of neuropodium just medial to ventral cirrus.

This species is found free-living, but it also occurs as a commensal in the tubes of the polychaetes *Praxillella affinis pacifica* and *Asychis* sp. (both family Maldanidae). Studies are lacking on Pacific forms, but the commensal behavior of a related species, *L. argus*, has been studied in England by Davenport (1953).

See Hartman (1968) and Pettibone (1953).

Phylum Annelida / Class Polychaeta / Order Phyllodocida
Suborder Phyllodociformia / Family Phyllodocidae

18.8 **Anaitides medipapillata** (Moore, 1909)

On and under rocks, low intertidal zone, central California to Mexico.

Length to 170 mm; color in life iridescent purple-blue; everted proboscis bearing rows of papillae, 10–12 papillae per row.

The animals are usually seen crawling over rocks at low tide. They secrete quantities of mucus. Almost nothing is known of their biology.

See Hartman (1968).

Phylum Annelida / Class Polychaeta / Order Phyllodocida
Suborder Nereidiformia / Family Hesionidae

18.9 **Ophiodromus pugettensis** (Johnson, 1901)
(=Podarke pugettensis)

Commensal in ambulacral grooves of sea stars, mainly *Patiria miniata* (8.4) in California; free-living among fouling organisms on floats and pilings, also on mud bottoms in quiet waters, lower intertidal zone and continental shelf; British Columbia to Gulf of California; Peru; Japan.

Body to 40 mm long, usually dark brown to black, occasionally pale; peristomium with six pairs of tentacles (cirri); dorsal setae represented only by small tufts emerging from bases of dorsal cirri.

Ophiodromus pugettensis has an eversible proboscis that lacks toothed jaws. Free-living animals appear to range about as omnivores; fecal pellets from such animals in the Los Angeles harbor area contained diatoms, crustacean exoskeletons, and miscellaneous debris. Worms living on *Patiria* hosts in the Monterey Bay area fed on plant material but also scavenged on animal matter, showing the same broad omnivorous habits as their sea star hosts.

Although *Patiria miniata* is the main host in California, *O. pugettensis* has been found living on the sea stars *Petalaster* (=*Luidia*) *foliolata* and *Pteraster tesselatus* in the Pacific northwest, and *Oreaster occidentalis* in the Gulf of California; it also occurs on moon snail shells occupied by hermit crabs in British Columbia.

The worms are facultative commensals. Given the opportunity, worms that have been experimentally removed from hosts usually remount the host, which they locate by scent. If water previously exposed to *Patiria* is added to an aquarium containing worms recently removed from *Patiria* hosts, the worms appear attracted by it and move more rapidly; their behavior suggests they are able to detect very small differences in concentration of "host factor" on the two sides of the head. An attempt to isolate and characterize the host factor from *Patiria miniata* shows that it contains two active components with molecular weights of less than 300; one factor serves as an attractant, and the other apparently stimulates non-directional locomotory activity. Individuals of *O. puget-*

tensis captured as free-living animals show no tendency to mount hosts, and remain unattracted by extracts of various species used as hosts by other *O. pugettensis*.

A given *Patiria miniata* host in California may bear up to 20 commensal worms of various sizes; commensals on the same host do not fight each other. Year-round studies of commensal populations on *Patiria* at Dana Point (Orange Co.) showed that the worms were largest and most numerous in midwinter (November to February), and both scarcer and smaller in size in midsummer (June and July).

Worms occupying a given host come and go, with considerable turnover. In the Monterey Bay area it appears that nearly half the worms leave each host each day, usually switching to another *Patiria* in an environment (kelp bed) where encounters between different *Patiria* individuals occur frequently. In both field and laboratory observations, the half-life of a worm on a host was less than a day.

Studies of worms settling on suspended wooden test blocks in the Los Angeles–Long Beach harbor area showed seasonal peaks in recruitment in February and May. Bottles placed in the water to collect suspended sediments in the same area also collected settling larvae in all months of the year. The heaviest settlement occurred from March to June. All worms observed in the harbor area were free-living; none were commensal. Worms were taken in small numbers in heavily polluted sections, even where dissolved oxygen was virtually absent, but populations were much larger in cleaner waters.

Some breeding probably occurs all year, at least in southern California. In Puget Sound (Washington), ripe animals have been observed swarming to a night light in July. Newly shed eggs are 86 μm in diameter. A swimming trochophore larva develops in 24 hours after fertilization at 13°C; the 76-hour trochophore has two red eyes and a ciliated mouth; the 4-day larva has active protonephridia. Worms have been raised in culture from fertilization to the young adult stage at the Pacific Marine Station (Dillon Beach, Marin Co.). Larvae were taken from the spring through late summer with a peak in September. Early growth was rapid, and at 15°C metamorphosis occurred 10 days following the early trochophore stage. A 14-segmented juvenile resembled the adult in color and morphological details.

See Berkeley & Berkeley (1948), Blake (1972, 1975a), Costello & Henley (1971), Dales (1962b), Davenport, Camougis & Hickok (1960), Davenport & Hickok (1957), Hickok & Davenport (1957), Lande & Reish (1968), MacGinitie & MacGinitie (1968), Reish (1954b, 1959, 1961, 1971b) Serences (1978), and Stewart (1970).

Phylum Annelida / Class Polychaeta / Order Phyllodocida
Suborder Nereidiformia / Family Nereidae

18.10 Nereis grubei (Kinberg, 1866)

Common in secreted mucus tubes among seaweed fronds and holdfasts and in mussel beds, upper middle intertidal zone (*Endocladia* region) to shallow subtidal waters on rocky shores; British Columbia to Mexico; Peru.

Length to 100 mm; dorsal lobes of posterior appendages greatly enlarged and with convex upper margin; dorsolateral areas on each side near base of everted proboscis (polychaete specialists call this region area 6) with four paragnaths (tooth-like projections) in a diamond arrangement.

Plants and sessile animals provide both shelter and food for *N. grubei*, which is not finicky in its requirements for either. In nature a wide variety of plants may serve as food, along with small crustaceans, bryozoans, sponges, and detritus; yet laboratory populations thrive on a diet of dried *Cladophora columbiana* (= *C. trichotoma*). Two populations in California have received extensive study: those inhabiting the algae *Cladophora columbiana* and *Egregia* at Point Fermin (Los Angeles Co.), and those living about holdfasts of *Gastroclonium* and *Egregia* in Carmel Bay (Monterey Co.).

The worms normally seen in tubes are sexually immature. They are solitary and very aggressive toward one another; in confrontations between individuals the palps flare out laterally, the jaws extend, and each snaps at and bites the other.

The immature state is maintained, as it is in the larvae of insects, by a "juvenile hormone," secreted in both worms and insects by the brain. When secretion stops, *N. grubei* undergoes a metamorphosis analogous to that of a caterpillar becoming a butterfly. Changes are of two sorts: first, the gametes, already present though immature, enlarge and mature; second, the body of the worm posterior to segment 14 in males and segment 16 in females undergoes drastic changes that adapt it to swimming rather than to crawling and tube-dwelling. The gut and many muscles are virtually dissolved, and some new muscles form. The parapodia enlarge and become leaflike, while the old pointed setae are shed and replaced by sheafs of setae shaped like canoe paddles. The eyes enlarge, and the greenish color of the body in the metamorphosed "natatory region" changes to pink in females and red in males as more blood vessels develop. In this metamorphosed state the sexually mature individuals are called "epitokes." With the gut atrophied the worms are unable to feed, but the remaining climactic phase of life is a short one. Their activities timed by subtle environmental signals and perhaps by built-in timing mechanisms, the worms leave their tubes in swarms and swim at the surface of the sea in a nuptial dance. The rapidly wriggling females release their eggs through paired ruptures in the body wall, completing spawning in a few seconds. Males swim in small circles near the spawning females, each emitting a cloud of sperm from a specialized rosette of tissue encircling the anus. Life for the spawned epitoke is short; those that do not die shortly from the trauma of spawning are probably quickly eaten by fishes. But as the old lives end, new ones begin. The fertilized eggs sink and early development occurs inside a protective membrane. The first three body segments develop simultaneously and are visible at 31 hours. Hatching occurs 32–42 hours after fertilization, and the released larvae swim actively by means of cilia. By 48 hours they are capable of strong muscular movements and have setae, three pairs of eyes, prostomial tentacles, and anal cirri. At 7 days of age the larvae have jaws, palps, well-developed parapodia, and a fourth body segment. At 9 days the proboscis everts and feeding begins. In southern California occasional epitokes matured in laboratory culture as early as 96–98 days after fertilization, though most did not complete metamorphosis until 28–29 weeks. And in southern California, some epitokes can be collected at a night light in all months of the year. In Carmel Bay epitokes swarm monthly, from the middle of February to the middle of June, but the earliest swarms are the heaviest. Development to maturity is probably slower, and spawning worms at Carmel are at least 1 year old, and more probably 2 years.

See especially Reish (1954a) and Schroeder (1968); see also Glynn (1965), Golding (1972), Gould & Schroeder (1969), Hartman (1968, as *N. mediator*), Reish (1954b, 1957, 1967, 1970), Reish & Alosi (1968), and Schroeder (1966, 1967, 1968). For general discussions of the phenomenon of epitoky, see R. Clark (1961) and Durchon (1965).

Phylum Annelida / Class Polychaeta / Order Phyllodocida
Suborder Nereidiformia / Family Nereidae

18.11 Neanthes brandti (Malmgren, 1866)

Burrowing in intertidal and subtidal mud flats in bays, and in very low intertidal and subtidal sand bottoms on protected outer shores; British Columbia to southern California; Siberian Pacific.

Body to 1 m long (possibly twice that when fully relaxed), 20–30 mm in diameter, distinguished by gigantic size, leaflike dorsal lobes on posterior appendages, and *very* numerous paragnaths (toothlike projections) on everted proboscis.

This giant polychaete is seldom seen until mature worms have undergone metamorphosis and become epitokous (see the account of *Nereis grubei*, 18.10). Both sexes then leave their burrows and swarm at the sea surface, spawn, and die. Ripe worms found cast ashore have very soft bodies, which usually rupture when handled, releasing either quantities of green eggs or clouds of whitish sperm. Spawning animals are seen most often in the spring and early summer in central California, but systematic observations are lacking. Epitokous animals are attracted to a night light.

Immature worms feed mainly on the green algae *Ulva* and *Enteromorpha* in Puget Sound (Washington), and they may be omnivorous. Immature worms as well as epitokes can swim, and speeds of 50–80 mm per second have been recorded. The major phosphagen is phosphoglycocyamine, but phosphocreatine is also present in smaller amounts.

Neanthes brandti is closely similar in most respects to the well-known clam worm *Neanthes virens* of Atlantic shores. The two populations may intergrade, and much of what has been published on the behavior, development, structure, and physiology of *N. virens* may apply without great modification to *N. brandti*.

For *N. brandti*, see Blake (1975c), R. Clark & Tritton (1970), Hartman (1968), MacGinitie & MacGinitie (1968), and Robin (1964). Behavior of *N. virens* is reviewed in Dyal (1973) and Evans (1971).

Phylum Annelida / Class Polychaeta / Order Phyllodocida
Suborder Nereidiformia / Family Nereidae

18.12 Platynereis bicanaliculata (Baird, 1863)

Inhabiting mucoid tubes among algal blades and holdfasts or among strands of surfgrass on protected shores; also in upper layers of fine sandy mud or sand in bays, low intertidal zone to 35 m depth; British Columbia to Mexico; Hawaii; Japan, China, Australia.

Length to 70 mm; dorsal lobes of appendages in midbody region each bearing a stout simple seta tipped with a hook or sickle; peristomial tentacles long; everted proboscis bearing transverse rows of small teeth.

The worms construct tubes using secretions from glands in the dorsal lobes of their appendages. They have powerful jaws, which they use for both fighting and feeding. Worms in confrontation posture aggressively, extending and snapping their jaws toward one another; such behavior helps limit the number of worms occupying a given area. The jaws are also used to capture and feed on algae. Worms may extend as much as half a body length from their tubes to catch drifting algae. In some situations the larger pieces are drawn back and fastened to the inside of the free end of the burrow, where they permanently attach and grow. The worms graze on the distal ends of these fronds, receiving from the plants both food and a protective cover for the tubes.

The sexes are separate in *P. bicanaliculata*, and maturation is accompanied by metamorphosis to an epitokous, or swimming, stage (see the account of *Nereis grubei*, 18.10). Worms ready to spawn leave their burrows and swim to the sea surface; they are easily attracted by a light at night. Males in spawning condition (shown in the photograph) release sperm through special papillae; spawning is stimulated by proximity to mature females or by the presence of sperm in the water. Eggs are shed by swimming female epitokes, and fertilization takes place in the water. Swimming larvae are nourished entirely by yolk stored in the egg; they settle from the plankton as juvenile worms 7–10 days after fertilization. Monthly collections of larvae made in the Los Angeles–Long Beach harbor area show that some breeding goes on all year, but settlement of larvae is greatest in March and April and again in June and July. Studies in Puget Sound (Washington) show that spawning which occurs in early August, is nearly synchronous in the population. Within 3 weeks after settling the young worms build tubes, and defensive behavior is well developed by the time juveniles are 20 segments long. At Tomales Bay (Marin Co.) epitokes were collected in May, June, and August. Unfertilized eggs measured 156–180 μm in diameter. At 16°C trochophores appeared 22 hours following fertilization, larvae with three setigerous segments were present after 2 days, and settlement occurred at the six-segment larval stage at an age of at least 21 days. Tube construction followed shortly thereafter. *P. bicanaliculata* has been raised in the laboratory to sexual maturity, a stage attained in 9 months or less.

Many aspects of other species of *Platynereis*, especially *P. dumerilii* in European waters, have been investigated, including structure, behavior, development, physiology, and biochemistry. Among other findings, it is noted that the body color darkens during the day and becomes pale at night; the change persists even in conditions of constant light and constant darkness, and even after experimental removal of eyes and cerebral ganglion.

For *P. bicanaliculata*, see Blake (1972, 1975a,c), Blake & Lapp (1974), Guberlet (1934), Hartman (1968), Lie (1968), MacGinitie & MacGinitie (1968), Reish (1957, 1961, 1964, 1971b), Reish & Alosi (1968), Roe (1975), and Woodin (1974, 1977). For related species, see Baskin (1976), Cazaux (1969), Costello & Henley (1971), Dyal (1973), Evans (1971, 1973), Fischer (1965), Fischer & Brökelmann (1965, 1966), Gwilliam (1969), Pettibone (1963), Rasmussen (1973), Reish (1954a, 1957), and Robin (1964).

Phylum Annelida / Class Polychaeta / Order Phyllodocida
Suborder Glyceriformia / Family Glyceridae

18.13 **Glycera americana** Leidy, 1855

Burrowing in muddy sand and mud, low intertidal zone on protected shores; subtidal to 315 m; British Columbia to Baja California; Peru, Strait of Magellan; Japan, Australia, New Zealand; Atlantic coast, Canada to Brazil.

Length to 350 mm; posteriodorsal margins of most parapodia bearing conspicuous eversible branched gills; fully everted proboscis with four hooks on tip.

Glycera americana, like other species of the family Glyceridae, burrows very quickly with a series of jerky motions. The enormous proboscis (roughly one-third the length of the body when extended) everts rapidly, inserting itself easily even into well-packed sand; the proboscis tip then swells to form an anchor, and proboscis retractor muscles draw the body down after it. A few extensions and retractions of the proboscis suffice to place the body well underground. Burrowing is easily observed by placing the worm in a glass bowl half-filled with wet sand, then watching the bottom of the bowl as the proboscis extends and retracts. The proboscis is also used in hunting prey. The four hooklike teeth at the proboscis tip are equipped with poison glands; the toxin, studied in European species, contains serotonin and proteolytic enzymes, harmful to small prey and sometimes moderately irritating to man.

Reproduction occurs in the summer. Mature worms undergo metamorphosis to a swimming epitokous stage (see the account of *Nereis grubei*, 18.10), then leave their burrows, and enter the water to spawn and die.

Glycerids are often called "bloodworms," for if the body is punctured (or is cut with a shovel when one is digging in sand), quantities of bright red fluid are released. In a technical sense this is not blood at all, for the glycerids lack a separate circulatory system, but the coelomic fluid contains much hemoglobin in red corpuscles.

The biology of *G. americana* is not well known, but numerous studies are available on other glycerids. A related species, *G. dibranchiata*, is a valuable marine bait worm taken especially in Maine and Nova Scotia. The minimum market-able length is 15 cm. Worms are packed in the brown alga *Fucus* and shipped by air freight to all coasts of the United States.

For *G. americana*, see Hartman (1968), MacGinitie & MacGinitie (1968), Pettibone (1963), and Robin (1964). For other species of *Glycera*, see R. Clark (1962b), Dales (1962b), Hoffman & Mangum (1970), Klawe & Dickie (1957), Manaranche (1966), Mangum (1970), Mangum & Miyamoto (1970), Mettam (1967), Michel (1966, 1970), Ockelmann & Vahl (1970), Simpson (1962), Stolte (1932), Thorson (1946), Turner & Sexsmith (1964), Vahl (1976), and Wells (1937).

Phylum Annelida / Class Polychaeta / Order Phyllodocida / Family Nephtyidae

18.14 **Nephtys californiensis** Hartman, 1938

Forming shallow subsurface burrows in clean sandy beaches protected from the full force of open coastal waves, commonest where median sand-grain size is 0.3–0.6 mm, low intertidal zone; British Columbia to Baja California.

Body to 30 cm long, usually less; dorsal side of prostomium bearing pigmented design resembling an eagle with spread wings.

This handsome species is widespread but never very abundant, the greatest densities found (at Moss Landing, Monterey Co.) being four to eight worms per m². Like other members of the family Nephtyidae, it is probably mainly carnivorous, but may ingest some detritus as well. The worms burrow by everting the powerful proboscis, then withdrawing it and walking ahead into the pathway thus opened up in the sand. Extension of the proboscis results from hydrostatic pressure on the coelomic fluid, caused by contraction of the longitudinal and perhaps dorsoventral body muscles (since *Nephtys* lacks a circular layer of body muscles). Worms displaced from the sand into the water can swim rapidly, using sinusoidal wriggling (the waves passing forward along the body), accompanied by powerful paddling actions of the parapodia. Swimming is stimulated by the absence of sand cover, by exposure to light, and by increasing hydrostatic pressure. Displaced worms swim down and bury themselves using wriggling movements. Small spiral gills project into the open space between the dorsal and ventral parapodial lobes, in a position unique among polychaetes. In animals below the sand the parapodia are folded back along the body, and the setae, projecting posteriorly in overlapping fan-shaped bundles, prevent the sand from entering the space between the dorsal and ventral parapodial lobes where the gills lie. In this space strong cilia create a current of water that moves swiftly from front to rear along the body, providing for respiration. *Nephtys* species differ in their ability to keep sediments of different particle-size grades from entering the lateral respiratory groove: for example, **N. caecoides** Hartman can live buried in the finer sediments; it occurs in sandy muds in bays and lagoons and can tolerate brackish conditions one-half the salinity of the sea.

Worms of the genus *Nephtys* are among the best-known polychaetes, thanks to the numerous studies of R. Clark and others. Four small ocelli are present, invisible from the outside, and the brain has enormous neurosecretory lobes. Study of these lobes in several species led R. Clark to challenge the current concept of the evolution of neurosecretory tissue. He points out that ectoderm is primitively secretory; neurosecretory cells retain the primitive secretory function, and neurosecretion is greatest in the most primitive parts of the system. Specialization proceeds from more generalized cells secreting materials like glycoproteins, to the formation of neurons, which secrete only substances active at synapses, and finally to cells secondarily modified and enlarged in connection with secretion of materials like adrenalin.

The circulatory system of *N. californiensis*, known in some detail, is closed, and the blood pigment hemoglobin is present in the plasma. The dominant phosphagen is phosphoglycocyamine.

For *N. californiensis*, see R. Clark (1955, 1956a,c, 1957, 1958a,b, 1959b), R. Clark & M. Clark (1960), R. Clark & Haderlie (1962), Hartman (1938), and Robin (1964). For accounts of closely related forms, see M. Clark & Clark (1962), R. Clark (1956c, 1958c), R. Clark & Tritton (1970), Evans (1971), Filice (1958), J. Jones (1954), Kirkegaard (1970), Metlam (1967), E. Morgan (1969), Rasmussen (1973), Thorson (1946), Welsch & Storch (1970), and D. Wilson (1936ba).

Phylum Annelida / Class Polychaeta / Order Amphinomida
Family Amphinomidae

18.15 **Pareurythoe californica** (Johnson, 1897)

In sand or mud under rocks, low intertidal and shallow subtidal waters on rocky shores; central and southern California.

Length to about 110 mm; head with a dorsal ridge or caruncle extending from base of median prostomial antenna to second body segment.

Pareurythoe californica individuals are usually found in pairs. They should be handled with forceps, not fingers. The term "fire worm" sometimes applies to this and other members of the family Amphinomidae refers to the burning sensations of those who handle them. The sharp, hollow, chitinous setae are provided with an irritating toxin; they penetrate skin easily and break off, causing itching that may last a day or two.

Amphinomid polychaetes lack biting jaws, but prey on sedentary invertebrates by sucking and sometimes by rasping with the roughened ventral buccal cushion inside the mouth.

The major phosphagen of *P. californica* is probably phosphocreatine, but almost nothing else is known of the biology of this species.

See Dales (1962b), Hartman (1968), MacGinitie & MacGinitie (1968), and Robin (1964).

Phylum Annelida / Class Polychaeta / Order Amphinomida
Family Euphrosinidae

18.16 **Euphrosine hortensis** Moore, 1905

In areas where rocks and soft sediments occur together, low intertidal zone and subtidal waters to 203 m; Alaska to San Diego.

Length to 60 mm; caruncle present, extending as middorsal longitudinal fold from head back to segment 6; some ventral setae deeply incised at tips.

Nothing is known of the biology of this species.

See Hartman (1968).

Phylum Annelida / Class Polychaeta / Order Eunicida / Family Onuphidae

18.17 **Diopatra splendidissima** Kinberg, 1865

Inhabiting erect parchment tubes projecting 1–2 cm above the surfaces of sand and mud flats and extending deeply into the substratum, intertidal zone to depths of 30 m in protected bays; southern California to Baja California; Gulf of California, Panama, Ecuador.

Length to 200 mm; tube distinctive; appendages bearing pectinate (comblike) setae with seven or eight teeth.

Little is known of the biology of *D. splendidissima*. Like other *Diopatra* species it has powerful jaws consisting of a pair of curved tongs and smaller toothed plates. It appears that the exposed tube tips are often nipped off by predaceous fishes and crustaceans, for numerous worms are found with regenerating head ends. A related species, **D. ornata** Moore, is abundant in subtidal waters of the open California coast where kelp and bottom rubble occur. *D. ornata*, like several other *Diopatra* species, attaches bits of shell and algae to the exposed part of the tube, a practice that may help reduce predation a bit. A gregarine protozoan, *Cochleomeritus emersoni*, inhabits the gut of *D. ornata*.

An east coast species, *D. cuprea*, has received more study than west coast species of the genus. Its tube approaches a meter in length and is decorated with debris at its exposed end. The worm can turn about in its tube, and individuals showing both anterior and posterior regeneration are found. Gentle up-and-down movements of the worm create a current entering the tube, which irrigates the gills and brings to the scent receptors information about food available near the tube mouth. The worms eat small invertebrates, assorted microorganisms, and plant fragments, ingesting a great variety of small food items, either alive or recently dead.

The production of eggs in maturing females of *D. cuprea* is a remarkable process. Loops of cells resembling bead necklaces arise from the caudal surfaces of septa in segments toward the rear of the body. A single large oocyte, or future egg, occupies the center of the loop, like the pendant of a necklace, with a chain of nurse cells on each side that pass materials on to the enlarging egg. The loops detach and float in the coelom, and

the mature eggs come free of the nurse cells before spawning. Eggs are laid in gelatinous strings that persist for several days. Fertilization is external, and development is rapid; trochophore larvae are formed in 24 hours, small worms with three setigerous segments in 2 days, and worms with seven chaetigers in 13–17 days.

For *D. splendidissima*, see Fauchald (1968), Hartman (1944, 1968), MacGinitie & MacGinitie (1968), and Robin (1964). For accounts of *D. cuprea* and other forms, see Allen (1959), E. Anderson & Huebner (1969), Brenchley (1976), Costello & Henley (1971), Huebner & Anderson (1976), G. Jones (1969), Levine (1973), Mangum & Cox (1971), Mangum, Kushins & Sassaman (1970), Mangum, Santos & Rhodes (1968), Pettibone (1963), and Webster (1975).

Phylum Annelida / Class Polychaeta / Order Eunicida / Family Arabellidae

18.18 **Arabella iricolor** (Montagu, 1804)

Locally common in various habitats, in central California often found burrowing in sand on protected beaches or nestled in rock crevices, also crawling among tubes of *Phyllochaetopterus prolifica* (18.22) on harbor floats; low intertidal zone to 92 m depth; distributed around the world in tropical and temperate seas; Pacific coast from Canada to Mexico; Strait of Magellan.

Length usually 15–20 cm, to 60 cm; prostomium bearing four dark eyes in transverse row; parapodia uniramous (unbranched); species superficially resembling those in the family Lumbrineridae, which lack eyes and have conspicuously different jaws.

The adults of *A. iricolor* roam about, burrowing or invading protected crevices, forming no permanent tubes although sometimes inhabiting the same mucus-lined burrows for up to a few days. Adults behave aggressively when they encounter one another. They are carnivores, probably mainly scavenging dead animal matter, but ingesting considerable detritus as well.

Breeding occurs during the summer on the east coast, and very likely in California as well. Juvenile worms have been discovered living as parasites inside the coelomic cavity of the tube-dwelling polychaete *Diopatra ornata*. In turn, the gut of

A. iricolor is inhabited by the gregarine protozoan *Lecudina arabellae.*

The phosphagen in the tissues is phosphocreatine.

See especially Pettibone (1963); see also Blake (1975c), Dales (1962b), Hartman (1968), Levine (1976), Mark & Whitaker (1976), Pettibone (1957), and Thoai (1968).

Phylum Annelida / Class Polychaeta / Order Eunicida / Family Dorvilleidae

18.19 **Dorvillea moniloceras** (Moore, 1909)
Candy-Striped Worm

Common among worm tubes and tunicates encrusting floats and pilings in Monterey Harbor; less common subtidally in kelp (*Macrocystis*) holdfasts; central and southern California; on Hawaiian reefs associated with coralline algae and dead coral.

Length to about 75 mm; coloration in life distinctive, consisting of alternating transverse pink and white bands on dorsal surface; palps longer than antennae.

The mouth cavity of *Dorvillea* contains a ventral buccal bulb provided with two rows of hooked maxillary teeth, each row mounted on a muscular lobe. The whole structure somewhat resembles certain molluscan radulae, as was noted long ago by the malacologist Pelseneer (1899). *D. moniloceras* in Monterey harbor browses on the short (1–3 mm high) fuzz that flourishes on hard surfaces of floats and other organisms. Gut contents of the worms include important amounts of detritus, attached diatoms, small hydroids, the bryozoan *Bowerbankia* (e.g., 6.18), small crustaceans (especially copepods and ostracods), and the red alga *Polysiphonia*. The hydroids and bryozoan are rasped up in sizable clumps. Some worms contain extra sets of teeth, ready to replace worn-out rows. Other parts are regenerated as well: when worms are cut in two, both the head and tail sections survive and replace the missing parts, and lost antennae and palps are rapidly regenerated.

A related species, **Schistomeringos longicornis** (Ehlers), also called *Dorvillea articulata* and *Stauronereis rudolphi* in California literature, is abundant in bays and harbors of southern California, and occurs offshore, low intertidal zone to 575 m depth. It tolerates semi-polluted conditions and exhibits some uptake of dissolved organic matter from seawater. The complete life cycle took 1 year to complete under laboratory conditions.

See Blake (1975a,c), Cowan (1976), Fauchald (1970), Hartman (1968), Heider (1922), Jumars (1974), Marshall (1976), Pelseneer (1899), Reish (1967, 1970), Richards (1967), and Stephens (1968).

Phylum Annelida / Class Polychaeta / Order Spionida
Suborder Chaetopteriformia / Family Chaetopteridae

18.20 **Chaetopterus variopedatus** (Renier, 1804)
Parchment-Tube Worm

Occupying parchmentlike tubes that are open at both ends and are usually buried to form U-shaped burrows in sandy mud or mud, low intertidal or subtidal bottoms on protected shores; also among fouling organisms on floats and pilings; cosmopolitan in warm and temperate seas; common in central and southern California.

Body to 250 mm long, with three distinct regions, the center one bearing dorsally some large projecting transverse flaps.

These remarkable and specialized worms have inspired numerous investigations of such matters as feeding, tube secretion, irrigation of the burrow, sexual reproduction, development, regeneration, and neuromuscular phenomena. Feeding involves both muscular and ciliary movements and the secretion of mucus. The dorsal flaps, representing enlarged parapodia of segments 14–16, beat back and forth, driving a current of water from front to rear through the burrow. A pair of enlarged dorsal parapodial lobes (the aliform processes), on segment 12, secrete a mucus net that trails backward and filters the water stream, capturing plankton and detritus particles like a plankton net. The net is secreted continuously at its anterior end, while the posterior end of the net, clogged with food, is continuously rolled up by a ciliated cup projecting from the dorsal side of segment 13. When the rolled-up net reaches the size of a BB shot, net secretion stops, water pumping pauses, and the ball of food and mucus is carried forward to the mouth by a ciliated groove running along the center of the back. This happens about every 18 minutes in southern California worms.

The worms can move freely back and forth in their tubes, and even turn around; alternatively, they can hold themselves firmly in one spot by small suction cups on the ventral sides of segments 12–14. A worm often extends its head end out of the burrow, and it may be grasped by a predator. When this happens, the worm cuts off its own anterior end by constricting between segments 12 and 13. As a biologist early in this century (Enders, 1909) remarked, "In this way the genital and current-producing segments would be saved at the expense of the less valuable head region"! The head end is regenerated easily by the remaining posterior body if the body is divided at any point anterior to segment 14, but never if the break occurs at segment 14 or any more-posterior point. Conversely, the front end of the worm regenerates a new posterior region if the body is divided at any point back of segment 12. Segment 14, isolated by itself, can regenerate *both* ends, producing a whole new worm. Segment 14 in the old worm remains segment 14 in the new animal; in a manner not yet clear, the segment "knows its number" and contains the blueprint of a whole worm and the capability to construct it.

The animals are brilliantly luminescent; luminous mucus is produced by epidermal cells, and the secretion is under nervous control. The luminous material is a protein, unrelated to the classical luciferin-luciferase system responsible for luminescence in so many other marine animals.

In sexual reproduction the gametes arise in segments of the third body region, behind the segments bearing fans (14–16). Gametes mature in the coelom and are expelled through the nephridia. The eggs, slightly over 100 μm in diameter when ripe, enter the metaphase of the first maturation division on spawning, and remain there until fertilized. Trochophore larvae develop in 8–24 hours, depending on the temperature. The larvae lack a prototroch (a belt of extra-large cilia anterior to the mouth), but are ciliated all over and soon develop rings of larger cilia posterior to the mouth. Larvae begin to show signs of adopting the adult form at 1.5 mm length, and by the time they reach 2 mm they have nine chaetigers and spend most of their time on the bottom.

Adult worms are used by anglers as bait for perch, rockfish, and cabezon.

See especially Berrill (1928), Costello & Henley (1971), Dales (1967), Enders (1909), and MacGinitie (1939); see also Barnes (1965), Bhaud (1966), Blake (1975c), Brown (1975, 1977), Cormier (1969), Faulkner (1932), Hartman (1969), Henley & Costello (1965), MacGinitie & MacGinitie (1968), T. Morgan (1937), J. Nicol (1952), Scheltema (1974), Shimomura & Johnson (1968), Stephens & Schinske (1961), Thorson (1946), Turner & Sexsmith (1964), and Wells & Dales (1951).

Phylum Annelida / Class Polychaeta / Order Spionida
Suborder Chaetopteriformia / Family Chaetopteridae

18.21 **Mesochaetopterus taylori** Potts, 1914

Inhabiting distinctive, tough parchment tubes lacking annulations, coated with, and buried in, sand, often among roots of the broad eelgrass *Zostera*, low intertidal zone; British Columbia to Dillon Beach (Marin Co.).

Body to at least 300 mm long; head bearing palps but no other appendages; notopodia in midbody region unilobed.

The MacGinities (1968) report following *M. taylori* tubes for a distance of 1.8–2.4 m and to a depth of 1.2 m in mud without coming to the lower end. Studies of feeding, although fragmentary, indicate that at least part of the time two mucus bags are formed, by the second and third segments of the midbody region, which collect plankton in the same way as does the single mucus bag in *Chaetopterus variopedatus* (18.20); a posteriorly directed water current passing through the bags is created by a peristaltic pumping action of the most posterior segments of the midbody region. Alternatively, at other times it appears the worms may use a forward-flowing current along the dorsum to trap food particles in a mucus string secreted in a middorsal groove. The palps contain ciliated grooves associated with both collection and rejection of materials. Clearly the species needs more study. For those to whom deep excavation on mud flats lacks appeal, the MacGinities note that, though whole worms are hard to get, partial worms are not, and they regenerate easily in the laboratory to form whole animals.

See Barnes (1965), Berkeley & Berkeley (1952), Blake (1975c), Hartman (1969), and MacGinitie & MacGinitie (1968).

Phylum Annelida / Class Polychaeta / Order Spionida
Suborder Chaetopteriformia / Family Chaetopteridae

18.22 **Phyllochaetopterus prolifica** Potts, 1914

Inhabiting stiff but flexible tubes generally borne in large clusters, very abundant on protected pilings and floats in Monterey harbor from mean lower low water to 3 m depth, elsewhere low intertidal zone to 450 m depth; British Columbia to southern California.

Tube generally 10–15 cm long, 1–1.5 mm in diameter; prostomium bearing one pair of short tentacles and one pair of slender, elongate palps.

The species name *prolifica* is apt, for the immature worms reproduce asexually, by repeated fission and regeneration. As many as six juvenile worms may occupy the same tube, and new tubes may be started as branches arising from the walls of older tubes. To a large but unknown degree, clusters of tubes represent clones of worms. The clusters provide a protected environment for a host of benthic associates, including hydroids, bryozoans, and other polychaetes.

Several methods of feeding have been described in this or related species: the palps, extended from the tube tip and waved about, catch microscopic plankton and detritus, which they pass to the mouth by a ciliated groove along each palp; or water may be circulated through the tube by cilia on the flaplike parapodia, and food particles are trapped from the stream either by mucus bags or by a mucus rope, and then passed forward to the mouth by a ciliated groove along the the middorsal line.

Sexually reproducing worms are larger in size than those that reproduce asexually; they also secrete larger tubes (1.5 mm in diameter).

See especially Bloom, Bowers & Cullenward (1976); see also Barnes (1965), Berkeley & Berkeley (1952), Bhaud (1966), Donat (1975), Haderlie (1974), Hartman (1969), and MacGinitie & MacGinitie (1968).

Phylum Annelida / Class Polychaeta / Order Spionida / Suborder Cirratuliformia
Family Cirratulidae

18.23 **Cirriformia luxuriosa** (Moore, 1904)

Common in holes and crevices in rocky tidepools, in mussel beds, and among roots of the surfgrass *Phyllospadix*, middle to low intertidal zones and subtidal waters to about 20 m depth on rocky or soft bottoms; central California to Baja California.

Body to 150 mm long, bearing long slender tentacles of two types, reddish branchiae (usually one pair per segment in anterior body) and orange-red tentacular filaments (arising in a dense cluster of 20–40 pairs on dorsum of segments 5–7, each filament bearing a whitish, longitudinal, ciliated groove); spines heavy, black, few in number, arising from sides of body from middle to posterior end.

These sedentary worms remain concealed except for their extended branchiae and tentacular filaments, which are used, respectively, for respiration and for gathering minute organisms and particles of organic detritus as food. In the Los Angeles–Long Beach harbors, post-larval worms are found in all months of the year except May and June; large adults occur on mud bottoms in unpolluted and slightly polluted regions, as well as within clumps of the bay mussel *Mytilus edulis* (15.10) attached to pilings and floats.

The complete life cycle has been followed in southern California. Spawning occurs in Alamitos Bay (Los Angeles Co.) in June and July. Eggs measure 130 μm in diameter. Development is rapid at 20°C, the first three cleavages taking place within 7 minutes and the trochophores occurring by 16 hours. Settlement of larvae to the bottom, and feeding, commenced at 8 days. Two setigers were present at 11 days, and by 27 days the larvae had 11 setigers as well as bifid setae, which were not previously known in the genus. Immature eggs and sperm were seen in adults at 9 months.

The major body phosphagen is probably phosphocreatine.

See Blake (1975b,c), Crippen & Reish (1969), Hartman (1969), MacGinitie & MacGinitie (1968), D. Morgan (1975), Reish (1959, 1961, 1977), and Robin (1964).

Phylum Annelida / Class Polychaeta / Order Spionida / Suborder Cirratuliformia Family Cirratulidae

18.24 **Cirriformia spirabrancha** (Moore, 1904)

Burrowing in sand and mud bottoms with only green branchiae exposed, often very abundant in protected bays and harbors; sometimes on outer coast among roots of eelgrass *Zostera*; low intertidal zone and offshore in subtidal waters to 23 m; central California to Bahía de San Quintín (Baja California).

Similar to *C. luxuriosa* (18.23) but differing in habitat and in possession of numerous yellow setae, rather than few black setae, in posterior parapodia; tentacular filaments in life dark green (rather than orange-red as in *C. luxuriosa*).

This species, typical of bays rather than wave-swept open coast, invades estuaries, often residing in black, sulfide-rich mud. It tolerates both brackish and hypersaline conditions; half the worms experimentally exposed to 23 percent seawater and 220 percent seawater for 6 hours survived. Population densities in Tomales Bay mud flats (Marin Co.) reach 1,800 worms per m² in the lowest intertidal zone (0.5 m below zero tide level); there they form an important food for certain shorebirds, especially the marbled godwit and the short-billed dowitcher, and sometimes the western willet. The worms themselves feed on tiny organisms and organic particles, which are brought to the anterior end of the body by the ciliated grooves of the tentacular filaments. They are ingested by movements of the large ciliated lips flanking the mouth or by a shoveling action of the ventral lip. Nitrogenous wastes are eliminated as ammonia. The main phosphagen is phosphocreatine.

Life history studies have been carried out in two areas. Mature gametes were present from March to May in Alamitos Bay (Los Angeles Co.), and from June to July in colder Tomales Bay. The complete life cycle was followed in southern California and the first 61 days in northern California. Eggs at both localities measure about 130 μm in diameter. Early developmental rates are similar at 17.5°C and 20°C: the 32-cell stage is reached in 4 hours, and by 24 hours there is an actively swimming trochophore larva with mouth, eyes, and apical tuft of cilia. The larvae are nourished by stored yolk.

In the southern population larvae settle on the sixth day and setae appear on the eleventh day. In the northern population, larvae (with five segments but no setae) settle on the seventh day, and setae appear on the thirteenth day. The first larval setae are bifid, but no bifid setae occur in the adult worm. Branchiae appeared at 21 days and no tentacles were present even at 61 days at 17.5°C, whereas branchiae appeared at 13 days and tentacles appeared at 34 days at 20°C.

See especially Blake (1975b,c), Dice (1969), Kennedy (1980), and D. Morgan (1975); see also R. Clark & Haderlie (1962), Courtney (1958), George (1964, 1967), Hartman (1969), Hult (1969), Reeder (1951), Reish (1963), Robin (1964), Thoai (1968) and D. Wilson (1936a).

Phylum Annelida / Class Polychaeta / Order Spionida / Suborder Cirratuliformia Family Cirratulidae

18.25 **Dodecaceria fewkesi** Berkeley & Berkeley, 1954 (=D. pacifica and D. fistulicola of some authors)

Locally abundant, forming fused clusters of calcareous tubes, middle intertidal zone on protected rocky shores and dock pilings; subtidal to about 20 m; British Columbia to southern California; fossil tubes in several west coast deposits: Pliocene at Coos Bay (Oregon), Miocene in San Luis Obispo Co., and Pleistocene in San Pedro (Los Angeles Co.).

Clusters of tubes ranging from a few centimeters across to massive sheets or rounded heads more than 1 m across; mature worm to 40 mm long, dark brown to black; tentacles long, dark, consisting of one pair of grooved prostomial palps and up to 11 pairs of branchial filaments.

Colonies of *Dodecaceria* start with a single tube, formed at settlement by a larval worm developed from a fertilized egg. The founding father (or mother), when grown, breaks in half, and both halves regenerate the missing parts. Fission and regeneration occur repeatedly, and as the worms multiply new tubes are laid down. In larger clusters the tubes lie parallel to one another and are solidly fused together, forming stony masses that, when exposed at low tide, resemble great boulders whose surfaces bear countless holes. Covered with water by incoming tides, the colonies take on the appearance of heads of dark hair as the worms extend their branchiae and palps for respiration and feeding. The paired palps capture tiny organisms and organic particles and move these via a ciliated groove to the burrow entrance, where they are actively picked up by the mouth. From one to several worms, fully developed or in various stages of regeneration, may occupy the same tube at the same time. Each worm lies "bent double," with both head and tail ends protruding from the tube. Sessile ciliated protozoans are commonly borne as commensals on the tentacles.

Adult worms spawn within their tubes and gametes are washed to sea. The worms of each colony, being formed by fission, are all the same sex, so successful reproduction must require spawning of adjacent colonies at nearly the same time. Mature eggs, about 100 μm in diameter, develop into swimming trochophore larvae in 24 hours. Some details of larval development are known. The larval worms do not feed until they settle from the plankton.

The cosmopolitan *Dodecaceria concharum*, which commonly bores into calcareous substrates, has been reported from our coast, but specimens boring into the wall plates of *Balanus nubilus* (20.32) in Monterey harbor are *D. fewkesi* (pers. comm., J. A. Blake).

See especially Berkeley & Berkeley (1954); see also Blake (1975b,c), Courtney (1958), Donat (1975), Hartman (1969), Martin (1933), Reish (1952), and Thompson & Chow (1955).

Phylum Annelida / Class Polychaeta / Order Capitellida / Family Capitellidae

18.26 Capitella capitata (Fabricius, 1780)

In various habitats, often abundant at or near surface of black, sulfide-rich sediments in areas polluted by domestic sewage or cannery wastes, often present in large numbers in estuarine waters free of pollution, occasionally associated with commensal pea crabs (*Pinnixa faba*, 25.35; *P. littoralis*, 25.36); cosmopolitan from cold waters to tropics, but more commonly encountered in temperate waters of both hemispheres.

Body usually to 40 mm long, rarely reaching 100 mm, red, earthwormlike; males with specialized genital hooks on dor-

sum of segments 8 and 9; females with a middorsal pore between segments 8 and 9.

As in members of the family Glyceridae, a circulatory system is lacking, but the coelomic fluid, bright red from the presence of hemoglobin in corpuscles, is moved rapidly along the body by peristalsis. The worms burrow and ingest the substratum, rich in organic detritus, by eversion and retraction of the pharynx. They thrive under conditions of pollution that few other animals can tolerate and, in an inactive state, can even survive several days under anaerobic conditions. In laboratory tests, 50 percent of the worms survived for 28 days in an environment containing only 1.5 mg of dissolved oxygen per liter. Reproduction in specimens exposed to selected field conditions occurred at 3.5 mg of dissolved oxygen per liter, but not at 2.9 mg per liter. The worms have been used in many pollution studies.

Reproduction in California appears to go on all year, with some tendency toward shallow peaks in the summer and winter. Males transfer sperm to females in spermatophores (packets of sperm), which can be stored by the females until the eggs are ripe. The eggs are laid in the tube occupied by the female, and early development occurs there. Two patterns of development exist, one in which the larvae emerge in about 5 days as metatrochophores possessing 8 or 9 setigerous segments, the other in which the larvae hatch at about 7–14 days as juveniles possessing 13 setigerous segments. Settlement may occur within minutes after hatching or be delayed for up to a few days. Growth is rapid and sexual maturity is attained within 1 month at 20°C.

In San Francisco Bay, *C. capitata* is uncommon, but occurs near domestic sewage outfalls and has been reported as far inland as Antioch (Contra Costa Co.) in the Sacramento River delta area. In Humboldt Bay the worm is sometimes found in clusters in mucoid tubes attached to the bodies of female pea crabs (*Pinnixa littoralis*, 25.36) living commensally in the mantle cavities of the clam *Tresus capax* (15.47).

Speciation in the genus *Capitella* has been the subject of many recent studies and is still controversial. If we examine preserved specimens of *Capitella* using a classical morphological approach, we find at least eight known species segregated on the basis of the presence or absence of genital hooks in all specimens or only in males, and the position anteriorly at which hooded hooks first appear. On the other hand, we find that individuals from the same inbred laboratory population may include several of these "morphological species," including both males and females with genital hooks on setigerous segments 8 and 9. Reproductive experiments carried out in the laboratory show that several "sibling species" can be separated on the basis of (1) reproductive isolation, (2) the distribution of hooded hook setae, and (3) the distribution of particular isoenzymes. Clearly the problem of species in *Capitella* is not yet settled; a somewhat similar case occurs in the genus *Ophryotrocha* (family Dorvilleidae), in which the various species can be separated on the basis of chromosome number and egg-laying habits.

One of the species belonging to the *Capitella* complex is **C. ovincola** Hartman (= *C. capitata ovincola*), which is known only from the gelatinous egg capsules of the squid *Loligo opalescens* (17.2) along the coast from Monterey to San Diego. (Another species is known from squid egg capsules in the Mediterranean Sea.) The worms do not attack the eggs or larvae themselves, but they multiply rapidly and when very numerous may disrupt the capsule before the developing squid are ready to hatch.

For *C. capitata*, see Bellan, Reish & Foret (1971, 1972), Blake (1975c), R. Clark (1956b,c), Filice (1958, 1959), Grassle & Grassle (1976, 1977), Haffner (1930), Hartman (1969), Rasmussen (1956), Reish (1959, 1961, 1967, 1970, 1971a,b, 1974b, 1977, 1979), Reish & Barnard (1960), Reish & Richards (1966), Stone & Reish (1965), Thorson (1946), and Warren (1976a,b). For related species, see R. Clark (1956b,c), Fields (1965), Hartman (1969), MacGinitie & MacGinitie (1968), and McGowan (1954).

Phylum Annelida / Class Polychaeta / Order Capitellida / Family Arenicolidae

18.27 Arenicola brasiliensis Nonato, 1958
(= A. caroledna; called A. cristata by some authors)   Lugworm

Burrowing in sand and mud, low intertidal zone in bays; both coasts of North and South America; Hawaii; Japan, Australia; cosmopolitan in warmer seas.

Body to 175 mm long, with 17 segments bearing setae; gills in 11 pairs, on segments 7–17.

Like other members of the family Arenicolidae (the lugworms), *A. brasiliensis* occupies a J-shaped or L-shaped burrow in a soft substratum. The head end, kept well below the surface at the toe of the "J," ingests quantities of sand, resulting in a closed, funnel-shaped depression in the sand surface above the head. Organic matter ingested with the sand is rapidly digested, and the cleaned sand is periodically defecated at the surface near the tail end of the burrow in a distinctive, long, coiled casting. A related species, **A. cristata** Stimpson, deposits fecal wastes in sandy sheets. The branched gills in arenicolids receive circulating blood, which contains hemoglobin in large molecular aggregates. Gas exchange in the burrow is facilitated by pistonlike movements of the body, which stir the water and irrigate the burrow.

The sexes are separate. After the gametes mature in the coelom they are discharged through six pairs of tubular ducts (myxonephridia). In *A. brasiliensis*, mature eggs are about 150 μm in diameter. On discharge they enter the metaphase of the first maturation division and remain at this stage until fertilized. In Japan, where detailed studies were made, spawning occurs in the summer (July to September), usually during neap tides, and egg release may be stimulated by the presence of sperm in the water. Eggs are shed in the burrow, where fertilization occurs. A fertilization membrane is formed and two polar bodies are given off. A few hours after fertilization the eggs are embedded in a firm jelly mass, 2–5 cm in diameter but up to 15 cm long, and extruded to the sand surface, remaining firmly rooted in the burrow by a tenacious stalk. Cleavage is regular and spiral, a trochophore larva forms, and segments are formed one by one, all inside the fertilization membrane within the jelly mass. Hatching usually occurs on the fourth day after fertilization (at 23–25°C), when three to five pairs of setae are present, but the larvae remain for a time in the jelly mass (as it slowly breaks down), and some are not liberated until they have 12–16 somites. They swim for a short time, then settle and begin to burrow.

In addition to the two species of *Arenicola*, two species of the related genus *Abarenicola* (*A. pacifica* and *A. vagabunda*) occur in California. Little is known even of their distribution in this state, but recent studies have been made of their biology at San Juan Island, Washington. All four worms are large and plump and well suited to laboratory studies. Various species in the family have been investigated extensively, especially from the standpoints of behavior and physiology. Their phosphagens (phosphotaurocyamine and phosphohypotaurocyamine) are uncommon in polychaetes and more typical of sipunculid worms.

Lugworms are used as bait by sportfishermen in Europe, Asia, South Africa, and to a limited extent in the Gulf of Mexico. A commercially feasible lugworm aquaculture has been described by d'Asaro and Chen (1976), but thus far it has not been tried. To be successful such a venture will probably have to await greater acceptance of the lugworm as bait by American fishermen.

For *A. brasiliensis*, see Blake (1975c), Hartman (1969, as *A. cristata brasiliensis*), Okada (1941, as *A. cristata*), and Wells (1962, as *A. caroledna*; 1963, as *A. brasiliensis*). For other species of *Arenicola* and *Abarenicola*, see Ashworth (1904), Blake (1975c), Child (1900), Costello & Henley (1971), Dales (1957b, 1967), Evans (1971), Healy & Wells (1959), Hobson (1967), Hylleberg (1975), Krüger (1971), MacGinitie & MacGinitie (1968), May (1972), Mill (1978), Newell (1948, 1949), Oglesby (1973), Thoai (1968), Trueman (1966), and Wells (1945, 1949a,b, 1950, 1951, 1952a, 1954, 1959, 1962).

Phylum Annelida / Class Polychaeta / Order Flabelligerida
Family Flabelligeridae

18.28 **Pherusa capulata** (Moore, 1909)

Burrowed in soft substrata with only the anterior end exposed, low intertidal zone; subtidal on continental shelf and below; southern California to Bahía de San Quintín (Baja California).

Length to about 110 mm; setae of segments 1–4 long, directed anteriorly, together forming a "cephalic cage"; setae of all other segments directed laterally; ventral setae on most parapodia bifid at tips.

The gregarine protozoans *Lecudina pherusae* and *L. zimmeri* inhabit the gut of this worm (Levine 1974b, 1976), but little else is known of the biology of the species.

See the account and references under *Pherusa papillata* (18.29).

Phylum Annelida / Class Polychaeta / Order Flabelligerida
Family Flabelligeridae

18.29 **Pherusa papillata** (Johnson, 1901)

Typically burrowed into sand with only anterior setae exposed; sometimes present in very low intertidal zone, but more commonly subtidal to depths beyond continental shelf; in Monterey Bay common among *Diopatra* tubes in sea bottom offshore of *Macrocystis* (kelp) beds; Alaska to southern California.

Length to 90 mm; setae of segments 1–3 projecting forward to form a "cephalic cage", those of other segments projecting laterally; papillae conspicuous on body surface.

In this and allied members of the family Flabelligeridae the setae of the first one or more chaetigers are elongated and project anteriorly to form a protective cage within which the retractile head with its tentaclelike branchiae can be extended more safely. Two elongate palps reach out over the surface of the substratum well beyond the limits of the cage. Ciliated grooves running along the ventral surfaces of the palps bring a stream of finer sediments to the ciliated lips of the mouth. *P. papillata* ingests much fine sand along with organic detritus. After cleaning the sand surface of detritus in one area, the worm may back down into the sand and emerge later a few centimeters away in a fresh feeding area. Ciliary tracts on the four pairs of branchiae create an effective respiratory current through the cephalic cage. Flabelligerid polychaetes possess a green respiratory blood pigment, chlorocruorin, also found in the polychaete families Serpulidae and Sabellidae.

See Berzins (1976), Blake (1975c), Günther (1912), Hartman (1969), Rasmussen (1973), Schlieper (1927), and Spies (1971, 1975).

Phylum Annelida / Class Polychaeta / Order Terebellida / Family Sabellariidae

18.30 **Phragmatopoma californica** (Fewkes, 1889)

Worms locally abundant, forming tubes of cemented sand grains, middle intertidal zone and below on rocky shores; rarely subtidal to 75 m; central California to Ensenada (Baja California).

Tubes rarely solitary, more often in masses to 2 m or more across, the tubes regularly placed in a honeycomb arrangement, each tube with a flared rim; individual worms to 50 mm long, anterior end bearing a crown of lavender tentacles and a circular operculum, formed of heavy dark setae, serving to block tube when worm is withdrawn.

These are beautiful animals, whose bodies are curiously modified in connection with a tube-dwelling life. Studies of development have clarified the nature of the modifications. The operculum, borne on a dorsal stalklike extension of body segments 1 and 2, is formed from the dorsal setae of these segments. The crown of ciliated tentacles extending from the tube when the animal is submerged at high tide derives from the first body segment. These tentacles trap plankton, organic detritus, and sand grains from surging water. Ciliated grooves at the base of the tentacular crown carry food to the mouth. The sand is transported to a pocketlike building organ posterior to the mouth, which in turn applies a liquid cement and builds selected grains into the wall of the tube. Farther back, the body shows distinct thoracic and abdominal regions, and posteriorly narrows to form a tubular caudal region that curves sharply forward below the abdomen. Excellent preparations permitting observation of the worm in its tube can be made by removing all but a short piece of tube from about the body, placing the worm and short tube on a glass plate or petri dish submerged in seawater, and sprinkling the plate with a light coating of sand. Under these conditions the worm forms a new tube using the glass plate or dish as one side of the tube, thus permitting clear vision of its activities inside the tube. This technique can be used for observing many of the tube-building polychaetes.

Worms removed from their tubes usually spawn spontaneously in the summer. The lavender eggs are about 75 μm in diameter; fertilized, they form swimming trochophore larvae within 24 hours, and by 36 hours the larval worms bear on each side a sheaf of long barbed setae that can be held close to the body or spread fanwise to discourage predators and possibly to slow the rate of sinking.

Metamorphosis associated with settlement involves conspicuous changes in form. Settlement occurs 34–39 days after fertilization at 17–18°C and 18–25 days after fertilization at

21–23°C. The larval development of a very similar east coast species (*P. lapidosa*) has been studied with the scanning electron microscope.

The gut of *P. californica* is inhabited by the gregarine protozoan *Selenidium fauchaldi*.

See especially Dales (1952), Eckelbarger (1977, 1978), and Roy (1974); see also Blake (1975c), Eckelbarger (1975), Eckelbarger & Chia (1976), Hartman (1969), Kirtley (1968), Levine (1974a), and Scholl (1958).

Phylum Annelida / Class Polychaeta / Order Terebellida / Family Sabellariidae

18.31 **Sabellaria cementarium** Moore, 1906
Honeycomb Worm

Worms forming clusters of tubes of cemented sand grains, on and under low intertidal and subtidal rocks to depth of at least 80 m along open coast; Alaska to southern California; Japan.

Tubes sometimes solitary, more often cemented together in a regular "honeycomb" arrangement to form large reefs; individual worms to 80 mm long; operculum, used to block entrance to tube, formed of heavy yellow setae (unusually dark in the photograph).

This is one of several species of *Sabellaria*, distinguished mainly by opercular characters, reported from California waters. None has been studied extensively, but *S. cementarium* has been recommended for use in embryological and biochemical studies.

In Puget Sound (Washington), sexually ripe individuals are numerous in all months of the year. Cracking the tubes and releasing the worms stimulates spawning. Spawned eggs rapidly pass to metaphase of the first meiotic division and remain in this stage until fertilized. Extremely dilute suspensions of sperm are needed to prevent several sperm from entering the egg. Development is slow, the first cleavage occurring more than 2 hours after fertilization (at 10°C); blastulae swim at 15 hours, trochophore larvae are well developed at 70 hours, and larvae with two setigers occur by the sixth day. Planktonic life probably lasts for months. Settlement of larvae on panels in Monterey Bay was noted in late summer and fall.

Other species of *Sabellaria*, especially in Europe and on the east coast of North America, have been studied extensively on such matters as pattern and rate of development, selection by larvae of a place to settle (usually on existing colonies of their own species), the formation and erosion of *Sabellaria* reefs, tube-building activities, the resistance of larvae to pollutants, and the phosphagen compounds present in the body (mainly phosphotaurocyamine).

For *S. cementarium*, see Blake (1975c), Haderlie (1969), Hartman (1969), and Winesdorfer (1967). For other species, see Blake (1975c), Costello & Henley (1971), Dales (1967), Eckelbarger (1974, 1975), Hartman (1969), Kalmus (1931), Kirtley (1968), Kittredge et al. (1962), Seagrove (1938), Thoai (1968), Vovelle (1965), and D. Wilson (1929, 1968a,b,c,d, 1969, 1970a,b).

Phylum Annelida / Class Polychaeta / Order Terebellida / Family Pectinariidae

18.32 **Pectinaria californiensis** Hartman, 1941

Inhabiting a straight, conical tube of cemented sand grains, small end projecting just above surface of substratum, very low intertidal zone in bays to outer continental shelf on open coast; Puget Sound (Washington), and possibly Alaska, to Baja California.

Tubes solitary, to 80 mm long, one sand grain in thickness; worms to 60 mm long, bearing numerous feeding tentacles; cephalic spines (setae) in 13–14 pairs, long, copper-colored.

The distinctive tubes formed by *P. californiensis* and related species are masterpieces of masonry, with each sand grain carefully fitted to its neighbors so that the use of cement is minimal, and with the grains at any particular level along the length of the tube carefully selected for uniform size. The strong anterior setae are used as digging forks, throwing sand to left and right, but they can also fold inward to form an operculum over the head. The ciliated tentacles are extended outward for several centimeters; they carry sand and detritus to the mouth region, where much of it is swallowed. Some sand grains are passed back to the ventral cement glands and building organ for incorporation into the tube. Pistonlike movements of the body pump both water and sand up through the tube and out the upper end. The removal of sand creates about the head a small cavern subject to repeated cave-ins. The water circulating past the body and gills facilitates respiration and carries sand-laden fecal material out of the tube. Gametes, too, are emitted from the upper end of the tube. Within the body the worms lack septa in most regions, and the gut is long and coiled. Relaxed or preserved specimens are excellent for dissection. Small crabs (*Pinnixa*) have been reported as commensals in tubes containing living *Pectinaria californiensis*.

The splendid account of *Pectinaria* natural history by Watson (1928) shows that worm watching can be just as esthetically pleasing and biologically rewarding as bird watching.

See Blake (1975c), Costello & Henley (1971), Dales (1967), Filice (1958), Hartman (1941, 1969), MacGinitie (1935), MacGinitie & MacGinitie (1968), Malawista, Sato & Bensch (1968), Reish (1959), Robin (1964), Thorson (1946), Watson (1928), and D. Wilson (1936b).

Phylum Annelida / Class Polychaeta / Order Terebellida / Family Terebellidae

18.33 **Pista elongata** Moore, 1909

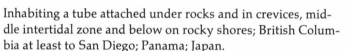

Inhabiting a tube attached under rocks and in crevices, middle intertidal zone and below on rocky shores; British Columbia at least to San Diego; Panama; Japan.

Worm to 200 mm long, distinguished from other California *Pista* species by the distinctive tube, the tube terminating at exposed end in a reticulate spongelike network of fibers.

Like other members of the family Terebellidae, these worms are particle feeders. Elongated extensile tentacles, covered on one side with ciliated tracts that beat toward the mouth, creep out of the mouth of the tube, and extend over the rock surface, picking up deposits and carrying them back to the tube.

Commensal animals sometimes found living in the tubes with *P. elongata* include the scale worms *Halosydna brevisetosa* (18.4) and *Lepidasthenia* sp. and pea crabs (*Pinnixa* spp.).

The major phosphagen found in the worms is probably phosphotaurocyamine.

See Blake (1975c), Hartman (1969), MacGinitie (1935), MacGinitie & MacGinitie (1968), and Robin (1964).

Phylum Annelida / Class Polychaeta / Order Terebellida / Family Terebellidae

18.34 **Pista pacifica** Berkeley & Berkeley, 1942

Inhabiting a vertical tube projecting above the substratum and extending deeply down into it, low intertidal sand and mud flats in protected bays; British Columbia to southern California.

Tubes sandy, distinctive, the upper end terminating in a triangular overhanging hood bearing many marginal tendrils; worm to 370 mm long.

This worm is a particle feeder like *P. elongata* (18.33), picking up fine surface deposits by means of ciliated tentacles extended from the mouth of the tube over the mud surface. The scale worm *Halosydna brevisetosa* (18.4) may occur as a commensal in some tubes.

See Berkeley & Berkeley (1952), Blake (1975c), Hartman (1969), and MacGinitie & MacGinitie (1968).

Phylum Annelida / Class Polychaeta / Order Terebellida / Family Terebellidae

18.35 **Thelepus crispus** Johnson, 1901

Common in some areas, inhabiting a thin, membranous, sand-encrusted tube under rocks, middle to low intertidal zone on rocky shores; Alaska to southern California; India.

Length to 280 mm; anterior body with three pairs of gills and numerous elongate feeding tentacles; posterior thoracic uncini (small, hooklike setae) arranged in a narrow oval on each parapodium.

Like other members of the family Terebellidae, the worms inhabit permanent burrows and use the ciliated feeding tentacles to transport fine, particulate surface deposits to the mouth. They circulate a current of water through their burrows and remove 50–60 percent of the dissolved oxygen in respiration. The major phosphagen found in the worms is phosphoargenine; phosphotaurocyamine is also present. The scale worms *Halosydna brevisetosa* (18.4) and *Hololepidella tuta* sometimes occur as commensals in the burrows of *T. crispus*.

See Blake (1975c), Dales (1955a, 1957b, 1967), Hartman (1969), MacGinitie & MacGinitie (1968), and Robin (1964).

Phylum Annelida / Class Polychaeta / Order Terebellida / Family Terebellidae

18.36 **Thelepus setosus** (Quatrefages, 1865)

In a thin, debris-encrusted tube under rocks, low intertidal zone and adjacent subtidal waters on rocky shores; British Columbia to southern California; Japan; widely distributed in Pacific, Atlantic, and Indian Oceans, Mediterranean Sea, Red Sea.

Length to 200 mm; anterior body with three pairs of gills and numerous elongate feeding tentacles; posterior thoracic uncini (small, hooklike setae) arranged in single row on each parapodium.

The natural history account of *T. crispus* (18.35) applies to this species also. The scale worm *Lepidasthenia* may occur in the tube as an additional commensal.

See the references under *T. crispus*.

Phylum Annelida / Class Polychaeta / Order Sabellida / Family Sabellidae

18.37 **Eudistylia polymorpha** (Johnson, 1901)
Plume Worm, Feather-Duster Worm

Locally common, inhabiting a tough, parchmentlike tube extending deeply into protected crevices and under boulders on rocky shores; large clumps of tubes often attached to pilings in bays; low intertidal zone to 450 m depth; Alaska to southern California.

Body to 250 mm long; anterior end (only portion usually exposed from tube) with a plume of branched gills (prostomial tentacles); three distinct color varieties known: maroon, orange, and brown; each variety sometimes showing alternating bands of lighter color.

The exposed gill plumes of this and other feather-duster worms are familiar and colorful sights in tidepools. The stiff but flexible gills are supported by an internal skeleton formed of cartilage and a material resembling unmineralized bone.

When disturbed, the animals retract instantly into their tubes. The gill plumes bear small eyespots sensitive to light; when a sudden shadow (such as might be caused by an approaching predator or observer) falls on the plume, the worm pops back into its tube. Extremely rapid withdrawal is made possible by the presence of giant nerve fibers, 0.2–0.3 mm in diameter, in the ventral nerve cord, which conduct impulses very rapidly (up to 7 m per second) and enable the whole body to contract as a unit. The gills possess cilia that circulate seawater upward through the plume. The blood to be aerated ebbs and flows in a single longitudinal vessel inside each branchial gill; the blood is green, its color due to the respiratory pigment chlorocruorin. Other cilia, in conjunction with mucus, capture small organisms and particles from the respiratory water flow and pass them downward along ciliated tracts that end at the mouth.

Worms collected in the field are easily maintained in running-seawater aquaria. If a worm is removed from its own tube and placed in a suitably sized glass tube, its activity can be easily observed. The worm secretes a more or less transparent parchment tube within the glass tube and extends it beyond the free end of the glass tube. The worms can live for long periods in aquaria without special feeding. This species is a favorite one for displays in commercial aquaria along the Pacific coast.

The major phosphagen of *E. polymorpha* is believed to be phosphocreatine.

In a related sabellid, *Branchiomma* (= *Megalomma*) *vesiculosum*, the fine structure of the eye suggests that it might function as an analyzer of polarized light (Krasne & Lawrence, 1966; Lawrence & Krasne, 1965); Eakin (1968) comments, "A very interesting eye indeed!"

For *E. polymorpha*, see Berkeley & Berkeley (1952), Blake (1975c), Bullock (1953), Dales (1962a), Hartman (1969), MacGinitie & MacGinitie (1968), Person & Matthews (1967), and Robin (1964). For related species, see also Crescitelli (1945), Eakin (1968), M. Jones (1969), Krasne & Lawrence (1966), Lawrence & Krasne (1965), Terwilliger & Terwilliger (1975), and Zoond (1931).

Phylum Annelida / Class Polychaeta / Order Sabellida / Family Sabellidae

18.38 Sabella crassicornis Sars, 1851
Plume Worm, Feather-Duster Worm

Inhabiting a membranous, sand-encrusted tube extending from secluded crevices, low intertidal zone on rocky shores and subtidal rocks or mixed deposits on continental shelf; Alaska to southern California; northern Europe.

Length to 50 mm; distal end bearing plume of feathery gills, maroon in color, dotted with paired eyespots; gill filaments rectangular in cross section.

The natural history of this species is similar to that of *Eudistylia polymorpha* (18.37). The feathery gills are used in both respiration and filter feeding. A related species, *Sabella pavonina*, has been the subject of many physiologically oriented studies, especially by British workers. Sabellids were found to be inefficient filter feeders, as measured by their ability to capture a known quantity of suspended graphite particles 1–2 μm in diameter from the water of an aquarium, when compared to such other filter feeders as oysters, mussels, and tunicates. In feeding, the largest particles captured by the gills are rejected, intermediate-sized particles are incorporated into the parchment tube, and only the smallest particles are eaten. Sorting of particles by size is accomplished at the base of each gill plume, which bears a longitudinal groove that is V-shaped in cross section. The largest particles will not fit into the V and drop off; medium-sized particles, fitting halfway into the V-shaped groove, are mixed with mucus and transferred to the tube margin by the collarlike membrane at the front edge of the thorax; only the smallest particles, which can fit down into the apex of the V, are carried by cilia to the mouth. The food consists of diatoms, algal spores, protozoans and fine detritus, accompanied by very fine sand and mud. It takes nearly a day for food to pass through the gut at 16°C. Fecal pellets are egested at the posterior end of the worm at the bottom of the tube. They are then carried forward (up) along a midventral, longitudinal, ciliated groove on the abdominal region to the posterior margin of the thorax, at which point the ciliated groove curves around one side and becomes a middorsal groove extending forward to the base of the gill plume. Here the pellets are dumped outside the tube.

The need of sabellid polychaetes to extend the gills out of the tube in order to feed and respire effectively, plus the soft nature of the tube, renders the worms vulnerable to predation by such fast-moving carnivores as fishes. On the east coast, nearly a third of the subtidal *Sabella pavonina* population may show evidence of damage and repair. The sabellids, however, have several defenses. (1) When disturbed, as by a touch, a vibration in the water nearby, or even a shadow cast upon the extended gills (which are usually equipped with small eyes), the worm withdraws very quickly into its tube. The response, mediated by giant nerve fibers that transmit impulses very rapidly, involves an almost simultaneous contraction and shortening of the whole body. (2) The body of the worm is twisted through 180° at the junction of thorax and abdomen (although the nerve cord is in a ventral position all along the body). The twisting appears to be related to a division of labor between appendages of the anterior and posterior body regions. The parapodia and setae in the abdomen are used for anchoring the body in the tube and for backward locomotion; those in the thorax are used for forward locomotion, which brings the anterior end of the worm to the mouth of the tube. The thorax is always short (seldom more than eight segments), whereas the abdomen may be much longer. The functional emphasis is thus on anchorage; even if the whole front of the body is nipped off, most of the abdomen survives, deep within the tube. (3) The isolated abdominal region shows excellent regenerative capabilities. A new head is regenerated, and the first several abdominal segments behind the new head then transform themselves into thoracic segments through a remarkable process of reorganization (see especially papers by Berrill and Fitzharris, below).

For *S. crassicornis*, see Berkeley & Berkeley (1952), Blake (1975c), and Hartman (1969). For other species of *Sabella*, see Berrill (1931, 1951, 1961, 1977), Berrill & Mees (1936a,b), R. Clark (1962b), Dales (1957b, 1967), Fitzharris (1973, 1976), H. Fox (1938), Gotto (1959), Hill (1970), Lee, Gilchrist & Dales (1967), E. Nicol (1930), Thomas (1940), and Wells (1951, 1952b).

Phylum Annelida / Class Polychaeta / Order Sabellida / Family Sabellidae

18.39 **Potamilla occelata** (Moore, 1905)
(= Pseudopotamilla occelata)   Feather-Duster Worm

Occupying a tough, membranous tube, which is usually translucent proximally and sand-encrusted distally, the tubes sometimes occurring in dense clusters, low intertidal zone on rocky shores locally protected from wave shock; subtidal to continental slope depths; Alaska to San Diego.

Length to 150 mm, but usually much less; color of branchial crown variable, purple, orange, alternating purple and orange bands, or colorless; branchial plumes (radioles) 14–17 on each side, not dichotomously branched, each radiole with 7–12 conspicuous brown eyespots in a row dorsally and distally, fewer ventrally.

The general natural history of this species is similar to that of *Eudistylia polymorpha* (18.37), though detailed studies are lacking. One specimen settled on a subtidal test panel in Monterey harbor in May.

A remarkable hydroid, *Proboscidactyla circumsabella*, lives in symbiotic relationship with this worm. The stolons of the hydroid colony form a loose network around the distal region of the tube, and zooids arise from the stolons. The feeding zooids of the hydroid arise in a ring at the distal margin of the worm tube. Each zooid has only two tentacles, arising well back from the mouth. In its appearance and behavior the ring of zooids is reminiscent of a ring of little men; they bow inward, lean back, wave their arms about or clasp them around an extended tentacle of the worm. They feed on particles, some caught in the incoming current, others taken from the stream of mucus passing down the worm's tentacle toward its mouth or from the worm's rejection current.

The hydroid colony buds off free-swimming medusae, which in turn reproduce sexually to produce swimming larvae. Thus in each generation the hydroid larvae must find anew the proper worm tubes to settle on. The manner in which they do this was observed in a related hydroid, *Proboscidactyla flavicirrata*, which occurs on a variety of sabellid worm hosts. The swimming planula larvae of the hydroid establish themselves only on tubes occupied by living worms, and they do not settle on the tubes directly. Instead, the larva is drawn passively by the feeding current created by the tentacular crown of the worm, and soon it strikes a worm tentacle. When this occurs the larva discharges nematocysts, anchoring itself to the plume. It secretes a sticky mucus, and later, when the worm withdraws into its tube, the larva comes off and sticks to the edge of the tube. Stolons and zooids form, and the colony grows as the worm tube elongates, so the feeding zooids always occupy the rim of the tube. If the worm is experimentally removed from its tube, the hydroid colony de-differentiates to a network of stolons, then differentiates zooids again if a worm is reintroduced to the tube. Clearly, the presence of the worm somehow induces differentiation of the hydroid colony.

See Blake (1975c), Dales (1962a), Haderlie (1969), Hand (1954), Hand & Hendrickson (1950), and Hartman (1969). For hydroids living on other sabellids, see Campbell (1968a,b) and Strickland (1971).

Phylum Annelida / Class Polychaeta / Order Sabellida / Family Sabellidae

18.40 **Myxicola infundibulum** (Renier, 1804)

Inhabiting a thick tubular mass of soft, transparent jelly, which is usually either wedged between fouling organisms (mussels, tube worms, tunicates) on floats in protected bays and harbors, or embedded vertically in sediment on soft bottoms, ranging from shallow bays out to edge of continental shelf, often forming large local populations; Bering Sea to California; abundant in Monterey harbor.

Length to 90 mm; gill plumes yellow, green, or brown, in a distal whorl, differing from those of other feather-duster worms in being connected by a continuous circular membrane for most of their length, thereby forming a sort of branchial funnel.

In other sabellids, the branchial plumes are united only basally, and, in feeding, the water is drawn upward between adjacent plumes. The circular membrane, which prohibits this in *Myxicola*, appears to be an adaptation to life in soft substrata; it obstructs ciliary currents that, in a bottom dweller, might otherwise sweep a cloud of fine sediments up into the

branchial crown. The animal seems to have paid a price for this protection; on a per weight basis, it is a less-efficient filter feeder than other sabellids and serpulids. The name *Myxicola* means "slime-dweller," referring to the gelatinous secreted tube.

Myxicola has been the subject of numerous physiological studies, especially in England. Populations there are readily accessible throughout the year. A feature especially attractive to the neurobiologist is the relatively gigantic size of the giant axon in the ventral nerve cord. Most of the dorsal region of the ventral cord consists of one huge axon, up to 1.7 mm in diameter in places, which extends from the brain to the posterior end of the body. This single axon, formed by fusion of many cells, constitutes some 27 percent of the volume of the whole central nervous system; the only other metazoan nerves of comparable size are the giant axons of the stellate ganglion in the squid *Loligo* (e.g., 17.2). Nerve impulses move along the axon at speeds of 6–20 m per second, many times faster than in the ventral nerve cords of polychaetes lacking giant axons. The axon mediates a quick, total withdrawal response of the worm into its tube, facilitating the ejection of waste materials from the branchial crown and providing essential quick retreat from predators. The crown plays an essential role in respiration; it is not autotomized (voluntarily shed) when the animal is disturbed, as occurs in some other sabellids, and amputation of the crown by predator or experimentor is usually lethal.

In other respects the structure and biology of *Myxicola* is rather like that of *Eudistylia polymorpha* (18.37). The main phosphagen is probably phosphotaurocyamine.

See especially Dales (1957c, 1962a), J. Nicol (1948), J. Nicol & Whitteridge (1955), Peter (1970), Roberts (1962), and Wells (1952b); see also Binstock & Goldman (1967, 1969), Blake (1975c), Hartman (1969), Lee, Gilchrist & Dales (1967), MacGinitie & MacGinitie (1968), Okada (1932), Person & Matthews (1967), Thoai (1968), and Ward (1977).

Phylum Annelida / Class Polychaeta / Order Sabellida / Family Serpulidae

18.41 **Mercierella enigmatica** Fauvel, 1923

Worm occupying calcareous tubes, often occurring in masses, attached to pilings and floats in estuaries and harbors where brackish conditions prevail at least periodically, low intertidal zone and shallow subtidal waters; in California widespread only within San Francisco Bay (e.g., Aquatic Park (Berkeley); Lake Merritt (Oakland), duck pond at Palo Alto Yacht Harbor); worldwide in shallow, brackish situations, north temperate and warmer waters, Pacific, Atlantic, and Indian Oceans.

Tube to 100 mm long, often shorter, white, brittle, cylindrical, bearing calcareous flanges (peristomes), these well separated, expanded, reminiscent of the nodes on bamboo but more conspicuous; branchial crown green to brown, bearing six to ten radioles (plumes) on each side; operculum (formed from a modified radiole on one side of crown) prominent, funnel-shaped, bearing spines forming an expanded disk.

This species was first reported from the Normandy coast of Europe in 1922. Two years later it was reported from the London docks, and it spread rapidly along the coast of Europe. It is believed to have originated in the Indian Ocean and to have been spread by ships. It survives oceanic conditions on ship hulls during transport, but rapid growth and reproduction occur only at temperatures above 18°C and at lowered salinities (30–90 percent the salinity of the open sea). *Mercierella* does not occur in southern California, probably because a permanent brackish environment is lacking.

The first studies of calcareous tube formation in serpulid worms, using radioactive tracers, were conducted on this species at the University of California, Berkeley, using radioactive strontium rather than calcium. The tube material was found to originate in a pair of glands located in the collar.

As in many tube-dwelling annelids, these worms jerk back very rapidly into their tubes when disturbed, a withdrawal response mediated by giant nerve fibers. In many annelids the withdrawal response has been treated as an "all or nothing" reflex. However, in *Mercierella* and some other species, there is a gradual loss of the response (habituation) upon repeated stimulation; the initial quick total response is replaced by slower, lesser contractions, possibly mediated not by the giant axons but by other nerves in the ventral nerve cord.

The worms are useful for demonstrating fertilization in the laboratory. The sexes are separate. Gametes arise from the peritoneum and mature in the coelomic cavity. Worms can be made to shed eggs or sperm simply by cracking and partially removing their tubes. The ripe eggs are orange and about 60 μm in diameter. Fertilization occurs in the water. The first and second polar bodies are given off 10 and 55 minutes after fertilization, respectively; the first cleavage occurs in 80–90 minutes, the 16-cell stage at 3–4 hours, and young swimming trochophores at 20 hours. Trochophore larvae 12 days old feed on plant plankton. At 18–20 days metatrochophore larvae are 0.16 mm long and bear three pairs of minute juvenile setae. At 3 weeks of age the larvae have three well-developed setigerous segments and the body is elongating. Settlement occurs 20–25 days after fertilization. Feeding in the adults is much as described under *Eudistylia polymorpha* (18.37).

The animals have a diploid chromosome number of 26. The major phosphagen is phosphotaurocyamine, but phosphocreatine and phosphoargenine are also present.

See Dasgupta & Austin (1960), Day (1967), Dyal (1973), Evans (1971), Fischer-Piette (1937), Gee (1967), Hall (1954), Hartman (1969), Hartman-Schröder (1967), Robin (1964), Rullier (1955), Straughan (1970), and Swan (1950).

Phylum Annelida / Class Polychaeta / Order Sabellida / Family Serpulidae

18.42 **Serpula vermicularis** Linnaeus, 1767

Occupying white calcareous tubes on undersurfaces of rocks, low intertidal zone; on floats and pilings in harbors; on exposed subtidal rocks to over 100 m depth on open coast; Alaska to San Diego; Pacific, Atlantic, and Indian Oceans; common in central and northern California, generally subtidal in southern California.

Tube to 100 mm long, chalky white, often coiled, cylindrical, smooth or with longitudinal ridges, branchial crown red, pink, orange, or banded with white, consisting of 40 pairs of plumes (radioles); operculum a symmetrical funnel with up to 160 crenulations along margin.

This species, first described by Linnaeus from western Europe, is the "type species" of the genus *Serpula* and the family Serpulidae. Serpulid polychaetes share many features with the sabellid feather-duster worms, including the presence on the head of a tuft of plumelike gills that also serve to collect microscopic food from the water and pass it down to the mouth by the action of cilia. Serpulids differ from sabellids in producing tubes of calcium carbonate instead of organic materials, and many of them have modified one of the branchial plumes into an operculum that is used to plug the tube when the worm is withdrawn.

Some aspects of tube formation have been studied in this species. When living worms are removed from their tubes, two conspicuous white sacs on either side of the midventral line in the posterior region of the peristomium can be seen giving off a thick white material for several seconds. These ventral calcium sacs store material from a second pair of structures, the calcium-secreting glands; the latter open to the outside of the body on the ventral shields, the broad glandular pads on the lower surface of the anterior thoracic segments. Studies on this and other serpulids indicate that the organic components of the tube are secreted by the ventral shields. By contrast, the calcium of the tube is probably laid down by both the ventral calcium sacs and the calcium-secreting glands, the whole being troweled and shaped by the ventral shields and the collar just back of the head. Serpulid tubes contain calcium carbonate in two forms: calcite, which is probably secreted by the ventral shield along with mucus, and aragonite, which is probably secreted by the calcium-secreting glands. Small amounts of strontium occur in the tube, as well.

In the serpulids, blood flows posteriorly in a ventral blood vessel; a dorsal vessel with an anterior blood flow, such as occurs in most other polychaete families, is lacking. Instead, forward flow, from the tail to the thorax, occurs in a blood sinus (lined by an endothelial tissue layer) that completely surrounds the entodermal lining of the gut and separates it from the connective tissue and muscular layers. Most of the peripheral body circulation, in gill plumes and elsewhere, follows an ebb-and-flow pattern.

Development has been more thoroughly studied in such serpulids as *Hydroides*, *Pomatoceros*, and *Spirorbis* and its allies than in *Serpula vermicularis*, but some details are available. Spawning animals have been observed at least in the summer in England and California. On the Mediterranean coast of France, at Sète, ripe individuals were commonest in spring and fewest in the winter, but fertilization was obtained artificially throughout the year. Eggs developed best at temperatures of 12–15°C; cleavage is total, gastrulation occurs within 10 hours, and trochophores swim at 15 hours. The pelagic stage is believed to be relatively short. The trochophore larva has two protonephridia, each composed of two cells: a flame cell with 24–28 cilia, the latter surrounded by a collar of microvilli reminiscent of those on the collar cells of sponges; and a tube cell, at one ending to the exterior, at the other end terminating in a collar of microvilli that overlap those of the flame cell. It has been suggested that ultrafiltration may occur in the area where the two collars of microvilli overlap.

The main phosphagen is probably phosphotaurocyamine.

See Blake (1975c), Dales (1962a), Day (1967), Fauvel (1927), Hanson (1948, 1949b, 1950), Hartman (1969), Hedley (1956), Nelson-Smith (1967), Pemerl (1965), Swan (1950), R. Terwilliger (1977), Thoai (1968), Thompson & Chow (1955), and Thorson (1946). For related serpulids, see Costello & Henley (1971, *Hydroides*), Dyal (1973), Neff (1966, 1969a,b), and Thomas (1940, *Pomatoceros*).

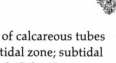

Phylum Annelida / Class Polychaeta / Order Sabellida / Family Serpulidae

18.43 **Salmacina tribranchiata** (Moore, 1923)

Locally abundant, forming tangled masses of calcareous tubes on undersides of rocks, lower middle intertidal zone; subtidal in protected habitats on rocky shores; British Columbia to southern California.

Tube whitish, less than 1 mm in diameter, but forming masses 20 cm or more across; worm red or pinkish, to 5 mm long; thorax consisting of 8–11 segments; operculum absent.

When *Salmacina* reproduces sexually the sexes are normally separate, although hermaphroditism has been reported. In southern California two rows of developing coral-colored eggs were observed near the center of the body within living worms, five to seven eggs per row. Fertilized eggs were retained within the tube, and the young escaped as trochophore larvae.

Asexual reproduction also occurs by a combined process of budding and transverse fission, followed in detail in European species. In *S. tribranchiata*, different stages in the process can be seen in living worms, which can be made to leave their tubes by allowing the surrounding water to become stagnant. The first indications of division are a constriction in one of the abdominal segments (usually abdominal segment 7, but any one from 6 to 13) and development of a band of red pigment in the constricted segment. Then buds of primordial tentacles appear on the segment posterior to the constriction, and before long the constriction effectively separates the worm into front and rear parts. This is followed by the shedding of the long, slender, simple setae and the stout uncini (setae shaped like hooks) from the anterior segment of the detached abdomen; this segment becomes the future prostomium of the "new" worm. The setae and uncini are then shed from the next posterior segment (future collar), and collar setae emerge from just below the budding branchiae. One by one, in order from front to rear, the other abdominal segments lose their old setae and develop into thoracic segments with new setae, until the new worm has transformed about nine of the old abdominal segments into new thoracic segments; the animal then adds new abdominal segments at the pygidium. Meanwhile the "parent" worm (the anterior half) forms a new pygidium at the site of constriction and adds new posterior segments to the abdomen. At this stage the parent and young are connected at the constriction only by a thin tubular strand through which runs the common gut. Violent contractions finally break the two apart, after which the parent develops a new anus on the terminal segment. In the calcerous tubes, the posteriorly placed, newly formed young worm creates a new opening to the outside through the wall of the tube, and from this point secretes its own branch tube. Tubes being actively secreted may increase in length from 5 to 25 mm in 8 days, and during asexual reproduction the size of the cluster of tubes may increase rapidly.

See especially Peabody (1976); see also Blake (1975c), Cresp (1964), Day (1967), Hartman (1969), MacGinitie & MacGinitie (1968), and Nelson-Smith (1967).

Phylum Annelida / Class Polychaeta / Order Sabellida / Family Serpulidae

18.44 Spirorbids

Especially under rocks, also on algae, middle intertidal zone to subtidal waters along rocky shores; in bays and harbors attached to pilings and floats, algae, or mussels; cosmopolitan.

Tube small, calcareous, about 0.5–1 mm in diameter at opening, white, coiled; tentacles orange to bright red; operculum (a modified branchial plume) small.

"Spirorbids" is the common name applied to the numerous species of small, tightly coiled worms in the family Serpulidae (some workers consider the spirorbids to constitute a separate family, the Spirorbidae). The taxonomy of the several species occurring in California has long been confused. Earlier genera were based on the direction of coiling of the tube (i.e., clockwise or counterclockwise); more recent studies show that both types of coiling occur in the offspring of a single worm, and that other characters of the tubes once considered useful in taxonomy are highly variable. Spirorbids of the Pacific coast are under study by E. W. Knight-Jones, P. Knight-Jones, and others (eight new species were recently described from California and Baja California; P. Knight-Jones, 1978). Separation of species is now based on the characteristics of the entire worm as these vary in the species population, rather than on features of the tube alone.

Spirorbids have been investigated especially with respect to reproduction, brooding, larval settlement, and tube secretion. Hermaphroditism is the rule in the spirorbids, the sperm forming in the posterior region of the abdomen and the eggs forming anteriorly. Self-fertilization has occurred in isolated specimens under laboratory conditions, but the process of fertilization has not been observed; it may be either internal or external within the tube. In all forms some type of parental care occurs. Young may be brooded along the abdominal groove, in a brood chamber, or within the operculum. The operculum is hollow and covered with a thin calcareous covering. Exactly how the embryos enter the opercular chamber is unknown, but entrance occurs prior to the first cleavage in at least one species, and the opercular chamber connects to the outside by a small pore in at least two species. The ciliated larvae escape the chamber through a break that somehow forms in the opercular plate. A new opercular chamber is formed beneath the former one, and as many as three empty opercular chambers, one on top of another, have been observed in laboratory-reared specimens.

The complete life history has been worked out for **Janua (Dexiospira) brasiliensis** (Grube) in southern California. This species breeds throughout the year with no apparent seasonal peak; over 60 percent of the individuals collected at any one time are brooding young. The brooding period lasts for 4–5 days with a new brood commencing 8 days later. At 20°C successive generations complete the life cycle approximately every 33 days.

In Monterey Bay, spirorbids often settle on the blades of the brown alga *Macrocystis*, which persist for only a few months. Indirect evidence indicates that these spirorbids live about 2 months.

Among European spirorbids some species occur only on certain specific substrata (e.g., certain species of coralline algae), and studies show that the settling larvae are attracted to extracts of these substrata. Other species are less choosy.

See Bergen (1953), Evans (1971), Gee (1964, 1965, 1967), Gee & Williams (1965), Haderlie (1969, 1974), Hartman (1969), Hedley (1956), King, Bailey & Babbage (1969), E. Knight-Jones (1951), P. Knight-Jones (1978), Nelson-Smith (1967), Potswald (1967, 1968, 1978), Shisko (1975), Silva (1962), S. Smith & Haderlie (1969), Thorson (1946), and Williams (1964).

Literature Cited

Åkesson, B. 1963. The comparative morphology and embryology of the head in scale worms (Aphroditidae, Polychaeta). Ark. Zool. 16: 125–63.

Allen, M. J. 1959. Embryological development of the polychaetous annelid, *Diopatra cuprea* (Bosc). Biol. Bull. 116: 339–61.

Anderson, D. T. 1973. Embryology and phylogeny of annelids and arthropods. New York: Pergamon. 495 pp.

Anderson, E., and E. Huebner. 1969. Development of the oocyte and its accessory cells of the polychaete *Diopatra cuprea* (Bosc). J. Morphol. 126: 163–98.

Andrews, H. L. 1945. The kelp beds of the Monterey region. Ecology 26: 24–37.

Ashworth, J. H. 1904. *Arenicola* (the lug-worm). Liverpool Mar. Biol. Comm. [LMBC] Memoir 11. 118 pp.

Banse, K. 1954. Über Morphologie und Larvalentwicklung von *Nereis (Neanthes) succinea* (Leuckart) 1847. (Polychaeta errantia). Zool. Jahrb. Anat. 74: 160–71.

Barnes, R. D. 1965. Tube-building and feeding in chaetopterid polychaetes. Biol. Bull. 129: 217–33.

Baskin, D. G. 1970. Studies on the infracerebral gland of the polychaete annelid, *Nereis limnicola*, in relation to reproduction, salinity and regeneration. Gen. Comp. Endocrinol. 15: 352–60.

_____. 1971. A possible neuroendocrine system in polynoid polychaetes. J. Morphol. 133: 93–103.

_____. 1976. Neurosecretion and the endocrinology of nereid polychaetes. Amer. Zool. 16: 107–24.

Baskin, D. G., and D. W. Golding. 1970. Experimental studies on the endocrinology and reproductive biology of the viviparous polychaete annelid, *Nereis limnicola* Johnson. Biol. Bull. 139: 461–75.

Bellan, G., D. J. Reish, and J.-P. Foret. 1971. Action toxique d'un détergent sur le cycle de développement de la polychète *Capitella capitata* (Fab.). C. R. Acad. Sci., Paris, 272: 2476–79.

_____. 1972. The sublethal effects of a detergent on the reproduction, development, and settlement in the polychaetous annelid *Capitella capitata*. Mar. Biol. 14: 183–88.

Bergan, P. 1953. On the anatomy and reproductive biology of *Spirorbis* Daudin. Nytt. Mag. Zool. 1: 1–26.

Berkeley, E., and C. Berkeley. 1948. Annelida: Polychaeta Errantia. Canad. Pacific Fauna 9b(1): 1–100.

_____. 1952. Annelida: Polychaeta Sedentaria. Canad. Pacific Fauna 9b(2): 1–139.

_____. 1954. Notes on the life-history of the polychaete *Dodecaceria fewkesi* (nom. n.). J. Fish. Res. Bd. Canada 11: 326–34.

Berrill, N. J. 1928. Regeneration in the polychaet *Chaetopterus variopedatus*. J. Mar. Biol. Assoc. U.K. 15: 151–58.

_____. 1931. Regeneration in *Sabella pavonina* (Sav.) and other sabellid worms. J. Exper. Zool. 58: 495–523.

_____. 1951. Regeneration and budding in worms. Biol. Rev. 27: 401–38.

_____. 1961. Growth, development, and pattern. San Francisco: Freeman. 555 pp.

_____. 1977. Functional morphology and development of segmental inversion in sabellid polychaetes. Biol. Bull. 153: 453–67.

Berrill, N. J., and D. Mees. 1936a. Reorganization and regeneration in *Sabella*. I. Nature of gradient, summation, and posterior reorganization. J. Exper. Zool. 73: 67–83.

_____. 1936b. Reorganization and regeneration in *Sabella*. II. The influence of temperature. III. The influence of light. J. Exper. Zool. 74: 61–89.

Berzins, I. K. 1976. Functional morphology and behavior of the structures associated with the head of the polychaetous annelid *Pherusa papillata* (Flabelligeridae). Research paper, Biol. 175h, library, Hopkins Marine Station of Stanford University, Pacific Grove, Calif. 47 pp.

Bhaud, M. 1966. Étude du développement et de l'écologie de quelques larves de Chaetopteridae (Annélides polychètes). Vie et Milieu 17: 1087–1120.

Binstock, L., and L. Goldman. 1967. Giant axon of *Myxicola*: Some membrane properties as observed under voltage clamp. Science: 158: 1467–69.

————. 1969. Current- and voltage-clamped studies on *Myxicola* giant axons. Effect of tetrodotoxin. J. Gen. Physiol. 54: 730–40.

Bishop, J. 1974. Observations on the swarming of a nereid polychaete, "*Neanthes succinea*," from the northern Gulf of Mexico. Tech. Rep. NOAA-75041803, Dept. Mar. Sci., Louisiana State University. 5 pp.

Blake, J. A. 1972. Polychaete larvae from the northern California coast. Amer. Zool. 12: 724 (abstract).

————. 1975a. The larval development of Polychaeta from the California coast. III. Eighteen species of Errantia. Ophelia 14: 23–84.

————. 1975b. The larval development of Polychaeta from the northern California coast. 1. *Cirriformia spirabrancha* (family Cirratulidae). Trans. Amer. Microscop. Soc. 94: 179–88.

————. 1975c. Phylum Annelida: Class Polychaeta, pp. 151–243, *in* R. I. Smith and J. T. Carlton, eds., Light's manual: Intertidal invertebrates of the central California coast. 3rd ed. Berkeley and Los Angeles: University of California Press. 716 pp.

Blake, J. A., and D. L. Lapp. 1974. Reproductive morphology, swarming behavior and larval development of *Platynereis bicaniculata* (Polychaeta) in an artificial salt water pond. Amer. Zool. 14: 1265 (abstract).

Bloom, D. E., B. A. Bowers, and M. J. Cullenward. 1976. Feeding, regeneration, and colony formation in the polychaete *Phyllochaetopterus prolifica* (Chaetopteridae). Research paper, Biol. 175h, library, Hopkins Marine Station of Stanford University, Pacific Grove, Calif. 45 pp.

Bobin, G. 1944. Morphogénèse des sois chez les annélides polychètes. Ann. Inst. Océanogr., Monaco, 22: 1–106, 6 pls.

Boilly, B. 1969. Sur l'origine des cellules régénératrices chez les annélides polychètes. Arch. Zool. Expér. Gén. 110: 127–43.

Brenchley, G. A. 1976. Predator detection and avoidance: Ornamentation of tube-caps of *Diopatra* spp. (Polychaeta: Onuphidae). Mar. Biol. 38: 179–88.

Brown, S. C. 1975. Biomechanics of water-pumping by *Chaetopterus variopedatus* Renier: Skeletomusculature and kinematics. Biol. Bull. 149: 136–50.

————. 1977. Biomechanics of water-pumping by *Chaetopterus variopedatus* Renier: Kinetics and hydrodynamics. Biol. Bull. 153: 121–32.

Bullock. T. H. 1953. Properties of some natural and quasi-artificial synapses in polychaetes. J. Compar. Neurol. 98: 37–68.

————. 1965. Annelida, pp. 661–789, *in* T. H. Bullock and G. A. Horridge, Structure and function in the nervous systems of invertebrates, vol. 1. San Francisco: Freeman. 798 pp.

Campbell, R. D. 1968a. Colony growth and pattern in the two-tentacled hydroid, *Proboscidactyla flavicirrata*. Biol. Bull. 135: 96–104.

————. 1968b. Host specificity, settling, and metamorphosis of the two-tentacled hydroid *Proboscidactyla flavicirrata*. Pacific Sci. 22: 336–39.

Cazaux, C. 1968. Étude morphologique du développement larvaire d'annélides polychètes (Bassin d'Arcachon). I. Aphroditidae, Chrysopetalidae. Arch. Zool. Expér. Gén. 109: 477–543.

————. 1969. Étude morphologique du développement larvaire d'annélides polychètes (Bassin d'Arcachon). II. Phyllodocidae, Syllidae, Nereidae. Arch. Zool. Expér. Gén. 110: 145–202.

Child, C. M. 1900. The early development of *Arenicola* and *Sternaspis*. Arch. Entwickelungsmech. Org. 9: 587–723, 5 pls.

Clark, M. E., and R. B. Clark. 1962. Growth and regeneration in *Nephtys*. Zool. Jahrb. Physiol. 70: 24–90.

Clark, R. B. 1955. The posterior lobes of the brain of *Nephtys* and the mucus-glands of the prostomium. Quart. J. Microscop. Sci. 96: 545–65.

————. 1956a. The blood vascular system of *Nephtys* (Annelida, Polychaeta). Quart. J. Microscop. Sci. 97: 235–49.

————. 1956b. *Capitella capitata* as a commensal, with a bibliography of parasitism and commensalism in the polychaetes. Ann. Mag. Natur. Hist. (12) 9: 433–48.

————. 1956c. The eyes and photonegative behaviour of *Nephtys* (Annelida, Polychaeta). J. Exper. Biol. 33: 461–77.

————. 1956d. On the origin of neurosecretory cells. Ann. Sci. Natur., Zool. 11: 199–207.

————. 1957. The influence of size on the structure of the brain of *Nephtys*. Zool. Jahrb. Physiol. 67: 261–82.

————. 1958a. The gross morphology of the anterior nervous system of *Nephtys*. Quart. J. Microscop. Sci. 99: 205–20.

————. 1958b. The micromorphology of the supra-oesophageal ganglion of *Nephtys*. Zool. Jahrb. Physiol. 68: 261–96.

————. 1958c. The "posterior lobes" of *Nephtys*: Observations on three New England species. Quart. J. Microscop. Sci. 99: 505–10.

————. 1959a. How much can a ragworm learn? New Scientist 9: 603–5.

————. 1959b. The neurosecretory system of the supra-oesophageal ganglion of *Nephtys* (Annelida; Polychaeta). Zool. Jahrb. Physiol. 68: 395–424.

————. 1959c. The tubicolous habit and the fighting reactions of the polychaete *Nereis pelagica*. Anim. Behav. 7: 85–90.

————. 1960. Habituation of the polychaete *Nereis* to sudden stimuli. I. General properties of the habituation process. II. Biological significance of habituation. Anim. Behav. 8: 82–91, 92–103.

————. 1961. The origin and formation of the heteronereis. Biol. Rev. 36: 199–236.

————. 1962a. Observations on the food of *Nephtys*. Limnol. Oceanogr. 7: 380–85.

————. 1962b. On the structure and functions of polychaete septa. Proc. Zool. Soc. London 138: 543–78.

————. 1965. Endocrinology and the reproductive biology of polychaetes. Oceanogr. Mar. Biol. Ann. Rev. 3: 211–55.

————. 1966. The integrative action of a worm's brain, pp. 345–79, *in* G. M. Hughes, ed., Nervous and hormonal mechanisms of integration. Symp. Soc. Exper. Biol. 20: 1–565.

————. 1969. Systematics and phylogeny: Annelida, Echiura, and Sipuncula, pp. 1–68, *in* Florkin & Scheer (1969).

————. 1978. Composition and relationships, pp. 1–32, *in* Mill (1978).

Clark, R. B., and M. E. Clark, 1960. The ligamentary system and the segmental musculature of *Nephtys*. Quart. J. Microscop. Sci. 101: 149–76.

Clark, R. B., and E. C. Haderlie. 1962. The distribution of *Nephtys californiensis* and *N. caecoides* on the Californian coast. J. Anim. Ecol. 31: 339–57.

Clark, R. B., and D. J. Tritton. 1970. Swimming mechanisms on nereidiform polychaetes. J. Zool., London 161: 257–71.

Cormier, M. J. 1969. Luminescence in annelids, pp. 467–79, *in* Florkin & Scheer (1969).

Costello, D. P., and C. Henley. 1971. Methods for obtaining and handling marine eggs and embryos. 2nd ed. Marine Biological Laboratory, Woods Hole, Mass. 247 pp.

————. 1976. Spiralian development: A perspective. Amer. Zool. 16: 277–91.

Courtney, W. A. M. 1958. Certain aspects of the biology of the cirratulid polychaetes. Doctoral thesis, Zoology, University of London. 177 pp.

Cowan, M. E. 1976. Regeneration in the polychaete annelid *Dorvillea moniloceras* (Eunicea: Dorvilleidae). Research paper, Biol. 175h, library, Hopkins Marine Station of Stanford University, Pacific Grove, Calif. 34 pp.

Crescitelli, F. 1945. A note on the absorption spectra of the blood of *Eudistylia gigantea* and the blood pigment in the red corpuscles of *Cucumaria miniata* and *Molpadia intermedia*. Biol. Bull. 88: 30–36.

Cresp, J. 1964. Études expérimentales et histologiques sur la régénération et la bourgeonnement chez les serpulids *Hydroides norvegica* (Gunn.) et *Salmacina incrustans* (Clap.). Bull. Biol. France Belg. 98: 3–152.

Crippen, R. W., and D. J. Reish. 1969. An ecological study of the polychaetous annelids associated with fouling material in Los Angeles harbor with special reference to pollution. Bull. So. Calif. Acad. Sci. 68: 170–87.

Dales, R. P. 1950. The reproduction and larval development of *Nereis diversicolor* O. F. Müller. J. Mar. Biol. Assoc. U.K. 29: 321–60.

————. 1952. The development and structure of the anterior region of the Sabellariidae, with special reference to *Phragmatopoma californica*. Quart. J. Microscop. Sci. 93: 435–52.

————. 1955a. Feeding and digestion in terebellid polychaetes. J. Mar. Biol. Assoc. U.K. 34: 55–79.

————. 1955b. The pelagic polychaetes of Monterey Bay, California. Ann. Mag. Natur. Hist. (12) 8: 434–44.

————. 1957a. Pelagic polychaetes of the Pacific Ocean. Bull. Scripps Inst. Oceanogr. Univ. Calif. 7: 99–168.

————. 1957b. Preliminary observations on the role of the coelomic cells in food storage and transport in certain polychaetes. J. Mar. Biol. Assoc. U.K. 36: 91–110.

————. 1957c. Some quantitative aspects of feeding in sabellid and serpulid fan worms. J. Mar. Biol. Assoc. U.K. 36: 309–16.

————. 1962a. The nature of the pigments in the crowns of sabellid and serpulid polychaetes. J. Mar. Biol. Assoc. U.K. 42: 259–74.

————. 1962b. The polychaete stomodeum and the inter-relationships of the families of Polychaeta. Proc. Zool. Soc. London 139: 389–428.

————. 1967. Annelids. 2nd ed. London: Hutchinson University Library. 200 pp.

Daly, J. M. 1972. The maturation and breeding biology of *Harmothoë imbricata* (Polychaeta: Polynoidae). Mar. Biol. 12: 53–66.

————. 1973a. Some relationships between the process of pair formation and gamete maturation in *Harmothoe imbricata* (L.) (Annelida: Polychaeta). Mar. Behav. Physiol. 1: 277–84.

————. 1973b. The ability to locate a source of vibrations as a prey-capture mechanism in *Harmothoe imbricata* (Annelida: Polychaeta). Mar. Behav. Physiol. 1: 305–22.

Daly, J. M., S. M. Evans, and J. Morley. 1972. Changes in behavior associated with pair-formation in the polychaete *Harmothoe imbricata* (L.). Mar. Behav. Physiol. 1: 49–69.

d'Asaro, C. N., and H. C. K. Chen. 1976. Lugworm aquaculture. Florida Sea Grant Prog., Rep. 16. 114 pp.

Dasgupta, S., and A. P. Austin. 1960. The chromosome numbers and nuclear cytology of some common British serpulids. Quart. J. Microscop. Sci. 101: 395–400.

Davenport, D. 1950. Studies in the physiology of commensalism. I. The polynoid genus *Arctonoë*. Biol. Bull. 98: 81–93.

————. 1953. Studies in the physiology of commensalism. III. The polynoid genera *Achloë, Gattyana* and *Lepidasthenia*. IV. The polynoid genera *Polynoë, Lepidasthenia* and *Harmothoë*. J. Mar. Biol. Assoc. U.K. 32: 161–73, 273–88.

Davenport, D., G. Camougis, and J. F. Hickok. 1960. Analyses of the behaviour of commensals in host-factor. 1. A hesionid polychaete and a pinnotherid crab. Anim. Behav. 8: 209–18.

Davenport, D., and J. F. Hickok. 1951. Studies in the physiology of commensalism. 2. The polynoid genera *Arctonoë* and *Halosydna.* Biol. Bull. 100: 71–83.

————. 1957. Notes on the early states of the facultative commensal *Podarke pugettensis* Johnson (Polychaeta: Hesionidae). Ann. Mag. Natur. Hist. (12) 10: 625–31.

Day, J. H. 1967. A monograph on the Polychaeta of southern Africa. Brit. Mus. (Natur. Hist.) Publ. 656. Part I, Errantia; Part II, Sedentaria. 2 vols. 878 pp.

Dice, J. F. 1969. Osmoregulation and salinity tolerance in the polychaete annelid *Cirriformia spirabrancha* (Moore, 1904). Comp. Biochem. Physiol. 28: 1331–43.

Dimock, R. V. 1971. Intraspecific aggression and density regulation by a symbiotic polychaete. Amer. Zool. 11: 693–94 (abstract).

————. 1974. Intraspecific aggression and the distribution of a symbiotic polychaete on its hosts, pp. 29–44, *in* W. B. Vernberg, ed., Symbiosis in the sea. Columbia: University of South Carolina Press. 276 pp.

Dimock, R. V., and D. Davenport. 1971. Behavioral specificity and the induction of host recognition in a symbiotic polychaete. Biol. Bull. 141: 472–84.

Dimock, R. V., and J. G. Dimock. 1969. A possible "defense" response in a commensal polychaete. Veliger 12: 65–68, pl. 4.

Donat, W. 1975. Subtidal concrete piling fauna in Monterey harbor, Monterey, California. Master's thesis, Oceanography, Naval Postgraduate School, Monterey. 85 pp.

Durchon, M. 1965. Sur l'évolution phylogénétique et ontogénétique de l'épitoquie chez les néréidiens. Zool. Jahrb., Syst. 92: 1–12.

Dyal, J. A. 1973. Behavior modification in annelids, pp. 225–90, *in* W. C. Corning, J. A. Dyal, and A. O. D. Willows, eds., Invertebrate learning, 1, Protozoans through annelids. New York: Plenum. 296 pp.

Dyal, J. A., and K. Hetherington. 1968. Habituation in the polychaete, *Hesperonoë adventor.* Psychon. Sci. 13: 263–64.

Eakin, R. M. 1968. Evolution of photoreceptors, pp. 194–242, *in* T. Dobzhansky, M. K. Hecht, and W. C. Steere, eds., Evolutionary biology, vol. 2. New York: Appleton-Century-Crofts. 452 pp.

Eakin, R. M., G. G. Martin, and C. T. Reed. 1977. Evolutionary significance of fine structure of archiannelid eyes. Zoomorphologie 88: 1–18.

Eakin, R. M., and J. W. Westfall. 1964. Further observations on the fine structure of some invertebrate eyes. Z. Zellforsch. 62: 310–32.

Eckelbarger, K. J. 1975. Developmental studies of the post-settling stages of *Sabellaria vulgaris* (Polychaeta: Sabellariidae). Mar. Biol. 30: 137–49.

————. 1977. Larval development of *Sabellaria floridensis* from Florida and *Phragmatopoma californica* from southern California (Polychaeta: Sabellariidae), with a key to the sabellariid larvae of Florida and a review of development in the family. Bull. Mar. Sci. 27: 241–55.

————. 1978. Metamorphosis and settlement in the Sabellariidae, pp. 145–64, *in* F.-S. Chia and M. E. Rice, eds., Settlement and metamorphosis of marine invertebrate larvae. New York: Elsevier/North Holland. 290 pp.

Eckelbarger, K. J., and F.-S. Chia. 1976. Scanning electron microscope observations of the larval development of the reef-building polychaete *Phragmatopoma lapidosa.* Canad. J. Zool. 54: 2082–88.

————. 1978. Morphogenesis of larval cuticle in the polychaete *Phragmatopoma lapidosa.* Cell Tissue Res. 186: 187–201.

Enders, H. E. 1909. A study of the life-history and habits of *Chaetopterus variopedatus* Renier et Claparède. J. Morphol. 20: 479–531, 3 pls.

Evans, S. M. 1963. The effect of brain extirpation on learning and retention in nereid polychaetes. Anim. Behav. 11: 172–78.

————. 1971. Behavior in polychaetes. Quart. Rev. Biol. 46: 379–405.

————. 1973. A study of fighting reactions in some nereid polychaetes. Anim. Behav. 21: 138–46.

Fauchald, K. 1968. Onuphidae (Polychaeta) from western Mexico. Allan Hancock Monogr. Mar. Biol. 3: 1–82.

————. 1970. Polychaetous annelids of the families Eunicidae, Lumbrineridae, Iphitimidae, Arabellidae, Lysaretidae and Dorvilleidae from western Mexico. Allan Hancock Monogr. Mar. Biol. 5: 1–335.

————. 1974. Polychaete phylogeny: A problem in protostome evolution. Syst. Zool. 23: 493–506.

————. 1977. The polychaete worms: Definitions and keys to the orders, families, and genera. Los Angeles Co. Mus. Natur. Hist., Sci. Ser. 28. 190 pp.

Faulkner, G. H. 1930. The anatomy and the histology of bud-formation in the serpulid *Filograna implexa,* together with some cytological observations on the nuclei of the neoblasts. J. Linn. Soc. (Zool.) 37: 109–90, 2 pls.

————. 1932. The histology of posterior regeneration in the polychaete *Chaetopterus variopedatus.* J. Morphol. 53: 23–58.

Fauvel, P. 1923. Faune de France. 5. Polychètes errantes. Paris: Lechevalier. 488 pp.

————. 1927. Faune de France. 16. Polychètes sédentaires. Addenda aux errantes, archiannélides, myzostomaires. Paris: Lechevalier. 494 pp.

————. 1953. The fauna of India: Annelida Polychaeta. Allahabad: Indian Press. 507 pp.

Fields, W. G. 1965. The structure, development, food relations, reproduction, and life history of the squid *Loligo opalescens* Berry. Calif. Dept. Fish & Game, Fish Bull. 131: 1–108.

Filice, F. P. 1958. Invertebrates from the estuarine portion of San Francisco Bay and some factors influencing their distribution. Wasmann J. Biol. 16: 159–211.

————. 1959. The effect of wastes on the distribution of bottom invertebrates in the San Francisco Bay estuary. Wasmann J. Biol. 17: 1–17.

Fischer, A. 1965. Über die Chromatophoren und den Farbwechsel bei dem Polychäten *Platynereis dumerilii*. Z. Zellforsch. 65: 290–312.

Fischer, A., and J. Brökelmann. 1965. Morphology and structural changes of the eye of *Platynereis dumerilii* (Polychaeta), pp. 171–74, *in* J. W. Rohen, ed., The structure of the eye, II, Symposium. Stuttgart: Schattauer-Verlag.

———. 1966. Das Auge von *Platynereis dumerilii* (Polychaeta). Sein Feinbau in ontogenetischen und adaptiven Wandel. [With English summary.] Z. Zellforsch. 71: 217–44.

Fischer-Piette, E. 1937. Sur la biologie du serpulien d'eau saumâtre (*Mercierella enigmatica* Fauvel). Bull. Soc. Zool. France 62: 197–208.

Fisher, W. K. 1946. Echiuroid worms of the North Pacific Ocean. Proc. U.S. Nat. Mus. 96: 215–92, pls. 20–37.

Fisher, W. K., and G. E. MacGinitie. 1928. The natural history of an echiuroid worm. Ann. Mag. Natur. Hist. (10) 1: 204–13, pl. X.

Fitzharris, T. P. 1973. Control mechanisms in regeneration and expression of polarity. Amer. Sci. 61: 456–62.

———. 1976. Regeneration in sabellid annelids. Amer. Zool. 16: 593–616.

Florkin, M., and B. T. Scheer, eds. 1969. Chemical zoology. 4. Annelida, Echiura, and Sipuncula. New York: Academic Press. 548 pp.

Fox, D. L., S. C. Crane, and B. H. MacConnaughey. 1948. A biochemical study of the marine annelid worm, *Thoracophelia mucronata*: Its food, biochromes, and carotenoid metabolism. J. Mar. Res. 7: 567–85.

Fox, H. H. 1938. On the blood circulation and metabolism of sabellids. Proc. Roy. Soc. London B125: 554–69.

Gaffney, P. M. Setal variation in *Halosydna brevisetosa*, a polynoid polychaete. Syst. Zool. 22: 171–75.

Gee, J. M. 1963. On the taxonomy and distribution in South Wales of *Filograna*, *Hydroides* and *Mercierella* (Polychaeta: Serpulidae). Ann. Mag. Natur. Hist. (13) 6: 705–15.

———. 1964. The British Spirorbinae (Polychaeta: Serpulidae) with a description of *Spirorbis cuneatus* sp. n. and a review of the genus *Spirorbis*. Proc. Zool. Soc. London 143: 405–41.

———. 1965. Chemical stimulation of settlement in larvae of *Spirorbis rupestris* (Serpulidae). Anim. Behav. 13: 181–86.

———. 1967. Growth and breeding of *Spirorbis rupestris* (Polychaeta: Serpulidae). J. Zool., London 152: 235–44.

Gee, J. M., and G. B. Williams. 1965. Self and cross-fertilization in *Spirorbis borealis* and *S. pagenstecheri*. J. Mar. Biol. Assoc. U.K. 45: 275–85.

George, J. D. 1964. The life history of the cirratulid worm, *Cirriformia tentaculata*, on an intertidal mudflat. J. Mar. Biol. Assoc. U.K. 44: 47–65.

———. 1967. Cryptic polymorphism in the cirratulid polychaete, *Cirriformia tentaculata*. J. Mar. Biol. Assoc. U.K. 47: 75–79.

Gerber, H. S., and J. F. Stout. 1967. Sensory basis of the symbiotic relationship of *Arctonoë vittata* (Grube) (Polychaeta, Polynoidae) to the keyhole limpet *Diadora aspera*. Physiol. Zool. 40: 169–79.

Glynn, P. W. 1965. Community composition, structure and interrelationships in the marine intertidal *Endocladia muricata–Balanus glandula* association in Monterey Bay, California. Beaufortia 12: 1–198.

Golding, D. W. 1972. Studies in the comparative neuroendocrinology of polychaete reproduction. Gen. Comp. Endocrinol. Suppl. 3: 580–90.

Golding, D. W., D. G. Baskin, and H. A. Bern. 1968. The infracerebral gland—a possible neuroendocrine complex in *Nereis*. J. Morphol. 124: 187–216.

Goodrich, E. S. 1945. The study of nephridia and genital ducts since 1895. Quart. J. Microscop. Sci. 86: 113–392.

Gotto, R. V. 1959. Observations on the orientation and feeding of the copepod *Sabelliphilus elongatus* M. Sars on its fan-worm host. Proc. Zool. Soc. London 133: 619–28, pl. 1.

Gould, M. C., and P. C. Schroeder. 1969. Studies on oögenesis in the polychaete annelid *Nereis grubei* Kinberg. I. Some aspects of RNA synthesis. Biol. Bull. 136: 216–25.

Grassle, J. F., and J. P. Grassle. 1977. Temporal adaptations in sibling species of *Capitella*, pp. 177–89, *in* B. C. Coull, ed., Ecology of marine benthos. Belle W. Baruch Library in Marine Science 6. Columbia: University of South Carolina Press. 467 pp.

Grassle, J. P., and J. F. Grassle. 1976. Sibling species in the marine pollution indicator *Capitella* (Polychaeta). Science 192: 567–69.

Guberlet, J. E. 1934. Observations on the spawning and development of some Pacific annelids. Proc. 5th Pacific Sci. Congr., Canada, 1933, Univ. Toronto 5: 4213–20.

Günther, K. 1912. Beitrage zur Systematik der Gattung *Flabelligera* und Studien über den Bau von *Flabelligera* (*Siphonostoma*) *diplochaitus*. Z. Naturwiss., Jena, n.s. 48: 93–186.

Gwilliam, G. F. 1969. Electrical responses to photic stimulation in the eyes and nervous system of nereid polychaetes. Biol. Bull. 136: 385–97.

Haderlie, E. C. 1969. Marine fouling and boring organisms in Monterey harbor. II. Second year of investigation. Veliger 12: 182–92.

———. 1974. Growth rates, depth preference and ecological succession of some sessile marine invertebrates in Monterey harbor. Veliger 17 (Suppl.): 1–35.

Haffner, K. von. 1930. Die Blutbewegung der gefässlosen Capitelliden. Z. Wiss. Zool. 136: 108–39.

Hall, J. H. 1954. The feeding mechanism in *Mercierella enigmatica* Fauvel (Polychaeta: Serpulidae). Wasmann J. Biol. 12: 203–22.

Hand, C. 1954. Three Pacific species of "Lar" (including a new species), their hosts, medusae, and relationships (Coelenterata, Hydrozoa). Pacific Sci. 8: 51–67.

Hand, C., and J. R. Hendrickson. 1950. A two-tentacled, commensal hydroid from California (Limnomedusae, Proboscidactyla). Biol. Bull. 99: 74–87.

Hanson, J. 1948. Transport of food through the alimentary canals of aquatic annelids. Quart. J. Microscop. Sci. 89: 47–51.

———. 1949a. The histology of the blood system in Oligochaeta and Polychaeta. Biol. Rev. 24: 127–73.

———. 1949b. Observations on the branchial crown of the Serpulidae (Annelida, Polychaeta). Quart. J. Microscop. Sci. 90: 221–33.

———. 1950. The blood-system in the Serpulimorpha (Annelida, Polychaeta). I. The anatomy of the blood-system in the Serpulidae. Quart. J. Microscop. Sci. 91: 111–29.

Hanson, J. A. 1972. Tolerance of high salinity by the pileworm, *Neanthes succinea*, from the Salton Sea, California. Calif. Fish & Game 58: 152–54.

Hartman, O. 1938. Review of the annelid worms of the family Nephtyidae from the northeast Pacific, with descriptions of five new species. Proc. U.S. Nat. Mus. 85: 143–58.

———. 1941. Polychaetous annelids. IV. Pectinariidae, with a review of all species from the western hemisphere. Allan Hancock Pacific Exped. 7: 325–45.

———. 1944. Polychaetous annelids. V. Eunicea. Allan Hancock Pacific Exped. 10: 1–238, 8 pls.

———. 1951. Literature of the polychaetous annelids. Los Angeles: privately printed. 290 pp.

———. 1968. Atlas of the errantiate polychaetous annelids from California. Allan Hancock Foundation, University of Southern California. 828 pp.

———. 1969. Atlas of the sedentariate polychaetous annelids from California. Allan Hancock Foundation, University of Southern California. 812 pp.

Hartmann-Schröder, G. 1967. Zur Morphologie, Ökologie und Biologie von *Mercierella enigmatica* (Serpulidae, Polychaeta) und ihre Röhre. Zool. Anz. 179: 421–56.

———. 1971. Annelida, Bostenwürmer, Polychaeta. Die Tierwelt Deutschlands, Part 58. Jena: Gustav Fischer Verlag. 594 pp.

Hashimoto, Y., and T. Okaichi. 1960. Some properties of Nereistoxin. Ann. N.Y. Acad. Sci. 90: 667–73.

Healy, E. A., and G. P. Wells. 1959. Three new lugworms (Arenicolidae, Polychaeta) from the North Pacific area. Proc. Zool. Soc. London 133: 315–35, pls. 1–4.

Hedley, R. H. 1956. Studies of serpulid tube formation. I. The secretion of the calcareous and organic components of the tube by *Pomatoceros triqueter*. II. The calcium-secreting glands in the peristomium of *Spirorbis*, *Hydroides*, and *Serpula*. Quart. J. Microscop. Sci. 97: 411–27.

Heider, K. 1922. Über Zahnwechsel bei polychäten Anneliden. Akad. Wiss. Berlin, Phys.-Math. Kl. Sitzber, Ann. 1922: 488–91.

Henley, C., and D. P. Costello. 1965. The cytological effects of podophyllin and podophyllotoxin on the fertilized eggs of *Chaetopterus*. Biol. Bull. 128: 369–91.

Herlant-Meewis, H. 1964. Regeneration in annelids, pp. 155–215, *in* M. Abercrombie and J. Brachet, eds., Advances in morphogenesis, vol. 4. New York: Academic Press.

Hermans, C. C. 1969. The systematic position of the Archiannelida. Syst. Zool. 18: 85–102.

Hewatt, W. G. 1937. Ecological studies on selected marine intertidal communities of Monterey Bay, California. Amer. Midl. Natur. 18: 161–206.

Hickok, J. F., and D. Davenport. 1957. Further studies in the behavior of commensal polychaetes. Biol. Bull. 113: 397–406.

Hill, S. D. 1970. Origin of the regeneration blastema in polychaete annelids. Amer. Zool. 10: 101–12.

Hobson, K. D. 1967. The feeding and ecology of two North Pacific *Abarenicola* species (Arenicolidae, Polychaeta). Biol. Bull. 133: 343–54.

Hoffman, R. J., and C. P. Mangum. 1970. The function of coelomic cell hemoglobin in the polychaete *Glycera dibranchiata*. Comp. Biochem. Physiol. 36: 211–28.

Holborow, P. L., and M. S. Laverack. 1972. Presumptive photoreceptor structures of the trochophore of *Harmothoe imbricata* (Polychaeta). Mar. Behav. Physiol. 1: 139–56.

Horridge, G. A. 1959. Analysis of the rapid responses of *Nereis* and *Harmothoe* (Annelida). Proc. Roy. Soc. London B150: 245–62.

Huebner, E., and E. Anderson. 1976. Comparative spiralian oogenesis—structural aspects: An overview. Amer. Zool. 16: 315–43.

Hult, J. E. 1969. Nitrogenous waste products and excretory enzymes in the marine polychaete *Cirriformia spirabrancha* (Moore, 1904). Comp. Biochem. Physiol. 31: 15–24.

Hylleberg, J. 1975. Selective feeding by *Abarenicola pacifica* with notes on *Abarenicola vagabunda* and a concept of gardening in lugworms. Ophelia 14: 113–37.

Jones, G. F. 1969. The benthic macrofauna of the mainland shelf of southern California. Allan Hancock Monogr. Mar. Biol. 4: 219.

Jones, J. D. 1954. Observations on the respiratory physiology and on the hemoglobin of the polychaete genus *Nephthys*, with special reference to *N. hombergii* (Aud. et M.-Edw.). J. Exper. Biol. 32: 110–25.

Jones, M. L. 1967. On the morphology of the nephridia of *Nereis limnicola* Johnson. Biol. Bull. 132: 362–80.

———. 1969. Boring of shell by *Caobangia* in freshwater snails of southeast Asia. Amer. Zool. 9: 829–35.

Jumars, P. A. 1974. A generic revision of the Dorvilleidae (Polychaeta), with six new species from deep North Pacific. Zool. J. Linn. Soc. 54: 101–35.

Jumars, P. A., and K. Fauchald. 1977. Between-community contrasts in successful polychaete feeding strategies, pp. 1–20, *in* B. C. Coull, ed., Ecology of marine benthos. Belle W. Baruch Library in Marine Science 6. Columbia: University of South Carolina Press. 467 pp.

Kalmus, H. 1931. Bewegungsstudien an den Larven von *Sabellaria spinulosa* Lueck. Z. Vergl. Physiol. 15: 164–92.

Kennedy, B. 1980. The functional anatomy of *Cirriformia spirabrancha*, and the gills and tentacles of cirratulid polychaetes. Doctoral thesis, Biology, University of California, Santa Cruz.

King, P. E., J. H. Bailey, and P. C. Babbage. 1969. Vitellogenesis and formation of the egg chain in *Spirorbis borealis* (Serpulidae). J. Mar. Biol. Assoc. U.K. 49: 141–50.

Kirkegaard, J. B. 1970. Age determination of *Nephtys* (Polychaeta: Nephtyidae). Ophelia 7: 277–81.

Kirtley, D. W. 1968. The reef-builders. Natur. Hist. 77: 40–45.

Kittredge, J. S., D. G. Simonsen, E. Roberts, and B. Jelinek. 1962. Free amino acids of marine invertebrates, pp. 176–86, *in* J. T. Holden, ed., Amino acid pools: Distribution, formation, and function of free amino acids. Amsterdam: Elsevier. 815 pp.

Klawe, W. L., and L. M. Dickie. 1957. Biology of the bloodworm, *Glycera dibranchiata* Ehlers, and its relation to the bloodworm fishery of the maritime provinces. Bull. Fish. Res. Bd. Canada 115: 1–37.

Knight-Jones, E. W. 1951. Gregariousness and some other aspects of the settling behaviour of *Spirorbis*. J. Mar. Biol. Assoc. U.K. 30: 201–22.

Knight-Jones, P. 1978. New Spirorbidae (Polychaeta: Sedentaria) from the east Pacific, Atlantic, Indian and Southern Oceans. Zool. J. Linn. Soc. 64: 201–40.

Korn, H. 1958a. Vergleichend-embryologische Untersuchungen an *Harmothoe* Kinberg, 1857 (Polychaeta, Annelida). Organogenesis und Neurosekretion. Z. Wiss. Zool. 161: 347–443.

———. 1958b. Zur Unterscheidung der Larven von *Harmothoe* Kinberg, 1857. Kieler Meeresforsch. 14: 177–86.

Krasne, F. B., and P. A. Lawrence. 1966. Structure of the photoreceptors in the compound eyespots of *Branchiomma vesiculosum*. J. Cell. Sci. 1: 239–48.

Krüger, F. 1971. Bau und Leben des Wattwurmes *Arenicola marina*. Helgoländer Wiss. Meeresunters. 22: 149–200.

Lande, R., and D. J. Reish. 1968. Seasonal occurrence of the commensal polychaetous annelid *Ophiodromus pugettensis* on the starfish *Patiria miniata*. Bull. So. Calif. Acad. Sci. 67: 104–11.

Lawrence, P. A., and F. B. Krasne. 1965. Annelid ciliary photoreceptors. Science 148: 965–66.

Lawry, J. V., Jr. 1967. Structure and function of the parapodial cirri of the polynoid polychaete, *Harmothoë*. Z. Zellforsch. 82: 345–61.

Lee, W. L., B. M. Gilchrist, and R. P. Dales. 1967. Carotenoid pigments in *Sabella penicillus*. J. Mar. Biol. Assoc. U.K. 47: 33–37.

Levine, N. D. 1973. *Cochleomeritus emersoni* sp. n. (Protozoa, Apicomplexa), lecudinid gregarine from the Pacific Ocean polychaete *Diopatra ornata* Moore, 1911. J. Protozool. 20: 546–48.

———. 1974a. *Selenidium fauchaldi* sp. n. and *S. telepsavi* (Stuart) comb. nov. (Protozoa, Apicomplexa), gregarines from polychaetes. J. Protozool. 21: 8–9.

———. 1974b. Gregarines of the genus *Lecudina* (Protozoa, Apicomplexa) from Pacific Ocean polychaetes. J. Protozool. 21: 10–12.

———. 1976. Revision and checklist of the species of the aseptate gregarine genus *Lecudina*. Trans. Amer. Microscop. Soc. 95: 695–702.

Lie, U. 1968. A quantitative study of benthic infauna in Puget Sound, Washington, U.S.A., in 1963–1964. With a section on polychaetes by Karl Banse, Katharine D. Hobson, and Frederick H. Nichols. Fiskeridir. Skr. Ser. Havunders. 14: 229–556.

Light, W. J. 1978. Spionidae: Polychaeta: Annelida. 1. Invertebrates of the San Francisco Bay estuary system. Pacific Grove, Calif.: Boxwood. 211 pp.

Lwebuga-Mukasa, J. 1970. The role of elytra in the movement of water over the surface of *Halosydna brevisetosa* (Polychaeta: Polynoidae). Bull. So. Calif. Acad. Sci. 69: 154–60.

MacConnaughey, B. H., and D. L. Fox. 1949. The anatomy and biology of the marine polychaete *Thoracophelia mucronata* (Treadwell) Opheliidae. Univ. Calif. Publ. Zool. 47: 319–39.

MacGinitie, G. E. 1935. Ecological aspects of a California marine estuary. Amer. Midl. Natur. 16: 629–765.

———. 1939. The method of feeding in *Chaetopterus*. Biol. Bull. 77: 115–18.

MacGinitie, G. E., and N. MacGinitie. 1968. Natural history of marine animals. 2nd ed. New York: McGraw-Hill. 523 pp.

Malawista, S. E., H. Sato, and K. G. Bensch. 1968. Vinblastine and griseofulvin disrupt the living mitotic spindle. Science 160: 770–72.

Manaranche, R. 1966. Anatomie du ganglion cérébroïde de *Glycera convoluta* Keferstein (Annélide Polychète), avec quelques remarques sur certains organes prostomiaux. Cah. Biol. Mar. 7: 259–80.

Mangum, C. P. 1970. Respiratory physiology in annelids. Amer. Sci. 58: 641–47.

Mangum, C. P., and C. D. Cox. 1971. Analysis of the feeding response in the onuphid polychaete *Diopatra cuprea* (Bosc). Biol. Bull. 140: 215–29.

Mangum, C. P., L. J. Kushins, and C. Sassaman. 1970. Responses of intertidal invertebrates to low oxygen conditions. Amer. Zool. 10: 516–17 (abstract).

Mangum, C. P., and D. M. Miyamoto. 1970. The relation between spontaneous activity cycles and diurnal rhythms of metabolism in the polychaetous annelid *Glycera dibranchiata*. Mar. Biol. 7: 7–10.

Mangum, C. P., S. L. Santos, and W. R. Rhodes, Jr. 1968. Distribution and feeding in the onuphid polychaete, *Diopatra cuprea* (Bosc). Mar. Biol. 2: 33–40.

Mark, P., and E. M. Whitaker. 1976. Distribution, feeding, and behavior of the polychaetous annelids *Arabella iricolor*, *A. semimaculata* (family Arabellidae), *Lumbrinereis zonata*, and *L. erecta* (family Lumbrineridae) in Monterey Bay. Research paper, Biol. 175h, library, Hopkins Marine Station of Stanford University, Pacific Grove, Calif. 36 pp.

Marshall, W. F., Jr. 1976. Feeding habits of *Dorvillea moniloceras*. Research paper, Biol. 175h, library, Hopkins Marine Station of Stanford University, Pacific Grove, Calif. 19 pp.

Martin, E. A. 1933. Polymorphism and methods of asexual reproduction in the annelid, *Dodecaceria*, of Vineyard Sound. Biol. Bull. 65: 99–105.

Mauzey, K. P., C. Birkeland, and P. K. Dayton. 1968. Feeding behavior of asteroids and escape responses of their prey in the Puget Sound region. Ecology 49: 603–19.

May, D. R. 1972. The effects of oxygen concentration and anoxia on respiration of *Abarenicola pacifica* and *Lumbrinereis zonata* (Polychaeta). Biol. Bull. 142: 71–83.

McGowan, J. A. 1954. Observations on the sexual behavior and spawning of the squid, *Loligo opalescens*, at La Jolla, California. Calif. Fish & Game 40: 47–54.

Mettam, C. 1967. Segmental musculature and parapodial movement of *Nereis diversicolor* and *Nephthys hombergi* (Annelida: Polychaeta). J. Zool., London 153: 245–75.

Michel, C. 1966. Mâchoires et glandes annexes de *Glycera convoluta* (Keferstein), annélide polychète Glyceridae. Cah. Biol. Mar. 7: 367–73.

———. 1970. Rôle physiologique de la trompe chez quatre annélides polychètes appartenant aux genres: *Eulalia, Phyllodoce, Glycera* et *Notomastus*. Cah. Biol. Mar. 11: 209–28.

Mill, P. J., ed. 1978. Physiology of annelids. London: Academic Press. 683 pp.

Morgan, D. E. 1975. Life histories of the cirratulid worms *Cirriformia luxuriosa* and *Cirriformia spirabrancha* (Annelida: Polychaeta). Master's thesis, California State University, Long Beach. 172 pp.

Morgan, E. 1969. The responses of *Nephtys* (Polychaeta: Annelida) to changes in hydrostatic pressure. J. Exper. Biol. 50: 501–13.

Morgan, T. H. 1937. The factors locating the first cleavage plane in the egg of *Chaetopterus*. Cytologia (Fujii Jubilee Vol.): 711–32.

Needham, A. E. 1969. Growth and development, pp. 377–441, *in* Florkin & Scheer (1969).

Neff, J. M. 1966. Ultrastructure of the calcium-secreting glands of *Pomatoceros caeruleus* (Schmarda). (Annelida: Polychaeta: Serpulidae). Amer. Zool. 6: 555 (abstract).

———. 1969a. Mineral regeneration by serpulid polychaete worms. Biol. Bull. 136: 76–90.

———. 1969b. Ultrastructure of the ventral shield epithelium of the serpulid *Pomatoceros caeruleus* (Schmarda). Amer. Zool. 9: 1145–46 (abstract).

Nelson-Smith, A. 1967. Catalogue of main marine fouling organisms. 3. Serpulids. Paris: Organization for Economic Cooperation and Development. 79 pp.

Newell, G. E. 1948. A contribution to our knowledge of the life his-

tory of *Arenicola marina* L. J. Mar. Biol. Assoc. U.K. 27: 554–80.

———. 1949. The later larval life of *Arenicola marina* L. J. Mar. Biol. Assoc. U.K. 28: 635–39.

Nicol, E. A. T. 1930. The feeding mechanism, formation of the tube and physiology of digestion in *Sabella pavonina*. Proc. Roy. Soc. Edinburgh 56: 537–98.

Nicol, J. A. C. 1948. The giant nerve-fibres in the central nervous system of *Myxicola* (Polychaeta, Sabellidae). Quart J. Microscop. Sci. 89: 1–45, 3 pls.

———. 1952. Studies on *Chaetopterus variopedatus* (Renier). I. The light-producing glands. II. Nervous control of light production. J. Mar. Biol. Assoc. U.K. 30: 417–52.

———. 1958. Luminescence in polynoids. IV. Measurements of light intensity. J. Mar. Biol. Assoc. U.K. 37: 33–41.

Nicol, J. A. C., and D. Whitteridge. 1955. Conduction in the giant axon of *Myxicola infundibulum*. Physiol. Comp. Oecol. 4: 101–12.

Ockelmann, K. W., and O. Vahl. 1970. On the biology of the polychaete *Glycera alba*, especially its burrowing and feeding. Ophelia 8: 275–94.

O'Clair, R. M., and R. A. Cloney. 1974. Patterns of morphogenesis mediated by dynamic microvilli: Chaetogenesis in *Nereis vexillosa*. Cell Tissue Res. 151: 141–58.

Oglesby, L. C. 1965. Water and chloride fluxes in estuarine nereid polychaetes. Comp. Biochem. Physiol. 16: 437–55.

———. 1968. Responses of an estuarine population of the polychaete *Nereis limnicola* to osmotic stress. Biol. Bull. 134: 118–38.

———. 1969. Inorganic components and metabolism; ionic and osmotic regulation: Annelida, Sipuncula, Echiura, pp. 211–310, *in* Florkin & Scheer (1969).

———. 1973. Salt and water balance in lugworms (Polychaeta: Arenicolidae) with particular reference to *Abarenicola pacifica* in Coos Bay, Oregon. Biol. Bull. 145: 180–99.

———. 1978. Salt and water balance, pp. 555–658, *in* Mill (1978).

Okada, Y. K. 1932. Les possibilités de la régénération de la tête chez le polychète, *Myxicola aesthetica* (Clap.). Annot. Zool. Japon. 13: 535–50.

———. 1941. The gametogenesis, the breeding habits, and the early development of *Arenicola cristata* Stimpson, tubicolous polychaete. Sci. Rep. Tôhôku Imp. Univ. (4) Biol. 16: 99–145, pls. IV–VI (*A. brasiliensis*).

Olive, P. J. W., and R. B. Clark. Physiology and reproduction, pp. 271–368, *in* Mill (1978).

Painter, R. E. 1966. Zoobenthos of San Pablo and Suisun Bays, pp. 40–56, *in* D. W. Kelley, ed., Ecological studies of the Sacramento–San Joaquin estuary. Calif. Dept. Fish & Game, Fish Bull. 133: 1–133.

Paris, J. 1955. Commensalisme et parasitisme chez les annélides polychètes. Vie et Milieu 6: 525–36.

Peabody, J. W. 1976. Asexual reproduction and bud development in the Californian marine annelid *Salmacina tribranchiata* (Polychaeta: Serpulidae). Research paper, Biol. 175h, library, Hopkins Marine Station of Stanford University, Pacific Grove, Calif.

Pelseneer, P. 1899. Recherches morphologiques et phylogénétiques. Mém. couronnés et Mém. des savants étrangers, Acad. Roy. Belgique 57: 1–113, pls. 1–24.

Pemerl, Sister J. 1965. Ultrastructure of the protonephridium of the trochophore larva of *Serpula vermicularis* (Annelida, Polychaeta). Amer. Zool. 5: 666–67 (abstract).

Person, P., and M. B. Matthews. 1967. Endoskeletal cartilage in a marine polychaete, *Eudistylia polymorpha*. Biol. Bull. 132: 244–52.

Peter, W. G. 1970. Giant *Myxicola* axon used to study nerve membrane. BioScience 20: 235–36.

Pettibone, M. H. 1953. Some scale-bearing polychaetes of Puget Sound and adjacent waters. Seattle: University of Washington Press. 89 pp., 40 pls.

———. 1957. Endoparasitic polychaetous annelids of the family Arabellidae with descriptions of new species. Biol. Bull. 113: 170–87.

———. 1963. Marine polychaete worms of the New England region. 1. Aphroditidae through Trochochaetidae. Bull. U.S. Nat. Mus. 227: 1–356.

Potswald, H. E. 1967. Observations on the genital segments of *Spirorbis* (Polychaeta). Biol. Bull. 132: 91–107.

———. 1968. The biology of fertilization and brood protection in *Spirorbis* (*Laeospira*) *morchi*. Biol. Bull. 135: 208–22.

———. 1978. Metamorphosis in *Spirorbis* (Polychaeta), pp. 127–43, *in* F.-S. Chia and M. E. Rice, eds., Settlement and metamorphosis in marine invertebrate larvae. New York: Elsevier/North Holland. 290 pp.

Quast, J. C. 1968. Observations on the food of the kelp-bed fishes, pp. 109–42, *in* W. J. North and C. L. Hubbs, eds., Utilization of kelp-bed resources in southern California. Calif. Dept. Fish & Game, Fish Bull. 139: 1–264.

Rasmussen, E. 1956. Faunistic and biological notes on marine invertebrates. III. The reproduction and larval development of some polychaetes from the Isefjord, with some faunistic notes. Biol. Medd. Danske Vid. Selsk. 23: 1–84.

———. 1973. Systematics and ecology of the Isefjord marine fauna (Denmark). Ophelia 11: 1–495.

Recher, H. F. 1966. Some aspects of the ecology of migrant shorebirds. Ecology 47: 393–407.

Reeder, W. G. 1951. Stomach analysis of a group of shorebirds. Condor 53: 43–45.

Reish, D. J. 1952. Discussion of the colonial tube-building polychaetous annelid *Dodecaceria fistulicola* Ehlers. Bull. So. Calif. Acad. Sci. 51: 103–7.

———. 1954a. The life history and ecology of the polychaetous annelid *Nereis grubei* (Kinberg). Allan Hancock Found. Occas. Pap. 14: 1–75.

———. 1954b. Polychaetous annelids as associates and predators of the crustacean wood borer, *Limnoria*. Wasmann J. Biol. 12: 223–26.

———. 1957. The life history of the polychaetous annelid *Neanthes caudata* (Delle Chiaje), including a summary of development in the family Nereidae. Pacific Sci. 11: 216–28.

———. 1959. An ecological study of pollution in Los Angeles–Long Beach harbors, California. Allan Hancock Found. Occas. Pap. 22: 1–119.

———. 1961. The use of the sediment bottle collector for monitoring polluted marine waters. Calif. Fish & Game 47: 261–72.

———. 1963. A quantitative study of the benthic polychaetous annelids of Bahía de San Quintín, Baja California. Pacific Natur. 3: 99–436.

———. 1964. Studies on the *Mytilus edulis* community in Alamitos Bay, California. II. Population variations and discussion of associated organisms. Veliger 6: 202–7.

———. 1967. Relationship of polychaetes to varying dissolved oxygen concentrations, pp. 199–214, *in* Advances in water pollution research, vol. 3. Washington, D.C.: Water Pollution Control Federation.

———. 1970. The effects of varying concentrations of nutrients, chlorinity, and dissolved oxygen on polychaetous annelids. Water Res. 4: 721–35.

———. 1971a. Effect of pollution abatement in Los Angeles harbours. Mar. Poll. Bull. 2: 71–74.

———. 1971b. Seasonal settlement of polychaetous annelids on test panels in Los Angeles–Long Beach harbors, 1950–1951. J. Fish. Res. Bd. Canada 28: 1459–67.

———. 1974a. Annelida, pp. 927–37, *in* Encyclopaedia Britannica, 15th ed., vol. 1.

———. 1974b (publ. 1976). The establishment of laboratory colonies of polychaetous annelids. Thalassia Jugoslavica 10: 181–95.

———. 1977. The role of life history studies in polychaete systematics, *in* Reish & Fauchald (1977).

———. 1979. Bristle worms (Annelida: Polychaeta), pp. 77–125, *in* C. J. Hart, Jr., and S. L. H. Fuller, eds., Pollution ecology of estuarine invertebrates. New York: Academic Press.

Reish, D. J., and M. C. Alosi. 1968. Aggressive behavior in the polychaetous annelid family Nereidae. Bull. So. Calif. Acad. Sci. 67: 21–28.

Reish, D. J., and J. L. Barnard. 1960. Field toxicity tests in marine waters utilizing the polychaetous annelid *Capitella capitata* (Fabricius). Pacific Natur. 1: 1–8.

Reish, D. J., and K. Fauchald, eds. 1977. Essays on polychaetous annelids in memory of Dr. Olga Hartman. Allan Hancock Foundation, University of Southern California, Los Angeles. 604 pp.

Reish, D. J., and T. L. Richards. 1966. A culture method for maintaining large populations of polychaetous annelids in the laboratory. Turtox News 44: 16–17.

Richards, T. L. 1967. Reproduction and development of the polychaete *Stauronereis rudolphi*, including a summary of development in the superfamily Eunicea. Mar. Biol. 1: 124–33.

Roberts, M. B. V. 1962. The rapid response of *Myxicola infundibulum* (Grübe). J. Mar. Biol. Assoc. U.K. 42: 527–39.

Robin, Y. 1964. Biological distribution of guanidines and phosphagens in marine Annelida and related phyla from California, with a note on pluriphosphagens. Comp. Biochem. Physiol. 12: 347–67.

Rossi, M. M. 1976. Observations on the life history of *Halosydna johnsoni* (Polychaeta: Polynoidae). Master's thesis, California State University, Long Beach. 132 pp.

Roy, P. A. 1974. Tube dwelling behavior in the marine annelid *Phragmatopoma californica* (Fewkes) (Polychaeta: Sabellariidae). Bull. So. Calif. Acad. Sci. 73: 117–25.

Ruggieri, G. D. 1976. Drugs from the sea. Science 194: 491–97.

Rullier, F. 1948. La vision el l'habitude chez *Mercierella enigmatica* Fauvel. Bull. Lab. Mar. Dinard 30: 21–27.

———. 1955. Développement du serpulien *Mercierella enigmatica* Fauvel. Vie et Milieu 6: 225–40.

Scheltema, R. S. 1974 (publ. 1976). Relationship of dispersal to geographical distribution and morphological variation in the polychaete family Chaetopteridae. Thalassia Jugoslavica 10: 297–312.

Schlieper, C. 1927. *Stylaroides plumosus* eine monographische Darstellung. Z. Morphol. Ökol. Tiere 7: 320–83.

Scholl, D. W. 1958. Effects of an arenaceous tube-building polychaete upon the sorting of a beach sand at Abalone Cove, California. Compass [Sigma Gamma Epsilon] 35: 276–83.

Schroeder, P. C. 1966. A histological and autoradiographic study of normal and induced metamorphosis in the nereid polychaete *Nereis grubei* (Kinberg). Doctoral thesis, Biological Sciences, Stanford University, Stanford, Calif. 170 pp.

———. 1967. Morphogenesis of epitokous setae during normal and induced metamorphosis in the polychaete annelid *Nereis grubei* (Kinberg). Biol. Bull. 133: 426–37.

———. 1968. On the life history of *Nereis grubei* (Kinberg), a polychaete annelid from California. Pacific Sci. 22: 476–81.

Schroeder, P. C., and C. O. Hermans. 1975. Annelida: Polychaeta, pp. 1–213, *in* A. C. Giese and J. S. Pearse, eds., Reproduction of marine invertebrates, vol. 3, Annelids and echiurans. New York: Academic Press. 343 pp.

Segrove, F. 1938. An account of surface ciliation in some polychaete worms. Proc. Zool. Soc. London 108: 85–107.

Serences, M. 1978. Aspects of the commensal relationship between the polychaete *Ophiodromus pugettensis* (Johnson) and the batstar *Patiria miniata* (Brandt). Master's thesis, Biological Sciences, Stanford University, Stanford, Calif. 74 pp.

Shimomura, O., and F. H. Johnson. 1968. *Chaetopterus* photoprotein: Crystallization and cofactor requirements for bioluminescence. Science 159: 1239–40.

Shisko, J. F. 1975. The life history of the annelid (Polychaeta: Serpulidae) *Janua* (*Dexiospira*) *brasiliensis* (Grube). Master's thesis, California State University, Long Beach. 85 pp.

Silva, P. H. D. H. de. 1962. Experiments on choice of substrate by *Spirorbis* larvae (Serpulidae). J. Exper. Biol. 39: 483–90.

———. 1967. Studies on the biology of Spirorbinae (Polychaeta). J. Zool., London 152: 269–79.

Simpson, M. 1962. Reproduction of the polychaete *Glycera dibranchiata* at Solomons, Maryland. Biol. Bull. 123: 396–411.

Skogsberg, T. 1928. A commensal polynoid worm from California. Proc. Calif. Acad. Sci. (4) 17: 253–65.

———. 1942. Redescription of three species of the polychaetous family Polynoidae from California. Proc. Calif. Acad. Sci. (4) 23: 481–502, pl. 43.

Smetzer, B. 1969. Night of the Palolo. Natur. Hist. 78: 64–71.

Smith, R. I. 1950. Embryonic development in the viviparous nereid polychaete, *Neanthes lighti* Hartman. J. Morphol. 87: 417–65.

———. 1953. The distribution of the polychaete *Neanthes lighti* in the Salinas River estuary, California, in relation to salinity, 1948–1952. Biol. Bull. 105: 335–47.

———. 1958. On reproductive pattern as a specific characteristic among nereid polychaetes. Syst. Zool. 7: 60–73.

———. 1959. The synonymy of the viviparous polychaete *Neanthes lighti* Hartman (1938) with *Nereis limnicola* Johnson (1903). Pacific Sci. 13: 349–50.

———. 1963. A comparison of salt loss rate in three species of brackish-water nereid polychaetes. Biol. Bull. 125: 332–43.

———. 1964. D_2O uptake rate in two brackish-water nereid polychaetes. Biol. Bull. 126: 142–49.

Smith, S. V., and E. C. Haderlie. 1969. Growth and longevity of some calcareous fouling organisms, Monterey Bay, California. Pacific Sci. 23: 447–51.

Spies, R. B. 1971. Contributions to the functional anatomy and larval development of *Flabellidermata commensalis* (Moore, 1904). Doctoral thesis, Biology, University of Southern California, Los Angeles. 145 pp.

———. 1975. Structure and function of the head in flabelligerid polychaetes. J. Morphol. 147: 187–208.

Stephens, G. C. 1968. Dissolved organic matter as a potential source of nutrition for marine organisms. Amer. Zool. 8: 95–106.

Stephens, G. C., and R. A. Schinske. 1961. Uptake of amino acids by marine invertebrates. Limnol. Oceanogr. 6: 175–81.

Stewart, W. C. 1970. A study of the nature of the attractant emitted by the asteroid hosts of the commensal polychaete, *Ophiodromus pugettensis*. Doctoral thesis, Biology, University of California, Santa Barbara. 86 pp.

Stolte, H.-A. 1932. Untersuchungen über Bau und Funktion der Sinnesorgane der Polychätengattung *Glycera* Sav. Z. Wiss. Zool. 140: 421–538.

Stone, A. N., and D. J. Reish. 1965. The effect of fresh-water run-off on a population of estuarine polychaetous annelids. Bull. So. Calif. Acad. Sci. 64: 111–19.

Storch, V., and U. Welsch. 1972. Ultrastructure and histochemistry of the integument of air-breathing polychaetes from mangrove swamps of Sumatra. Mar. Biol. 17: 137–44.

Straughan, D. 1970. Establishment of non-breeding population of *Mercierella enigmatica* (Annelida: Polychaeta) upstream from a breeding population. Bull. So. Calif. Acad. Sci. 69: 169–75.

Strickland, D. L. 1971. Differentiation and commensalism in the hydroid *Proboscidactyla flavicirrata*. Pacific Sci. 25: 88–90.

Swan, E. F. 1950. The calcareous tube secreting glands of the serpulid polychaetes. J. Morphol. 86: 285–314.

Terwilliger, R. C. 1977. The unusual respiratory pigment of *Serpula vermicularis*. 59th Ann. Meet., West. Soc. Natur., Abs. Symp. Contr. Pap., p. 30 (abstract).

Terwilliger, R. C., and N. B. Terwilliger. 1975. Quaternary structures of annelid chlorocruorin and polymeric hemoglobins. Amer. Zool. 15: 807 (abstract).

Thoai, N. v. 1968. Homologous phosphagen phosphokinases, pp. 199–229, *in* N. v. Thoai and J. Roche, eds., Homologous enzymes and biochemical evolution. New York: Gordon & Breach Sci. Publ. 436 pp.

Thomas, J. G. 1940. *Pomatoceros, Sabella* and *Amphitrite*. Liverpool Mar. Biol. Comm. [LMBC] Memoir 33. 88 pp.

Thompson, T. G., and T. J. Chow. 1955. The strontium-calcium atom ratio in carbonate-secreting marine organisms. Pap. Mar. Biol. Oceanogr., Deep-Sea Res. 3 (Suppl.): 20–39.

Thorson, G. 1946. Reproduction and larval development of Danish marine bottom invertebrates, with special reference to the planktonic larvae in the sound (Øresund). Medd. Komm. Danmarks Fiskeri- og Havundersøg. (Plankton) 4: 1–523.

Trueman, E. R. 1966. Observations on the burrowing of *Arenicola marina* (L.). J. Exper. Biol. 44: 93–118.

————. 1978. Locomotion, pp. 243–69, *in* Mill (1978).

Turner, C. H., E. E. Ebert, and R. R. Given. 1969. Man-made reef ecology. Calif. Dept. Fish & Game, Fish Bull. 146: 1–221, App. 1.

Turner, C. H., and J. C. Sexsmith. 1964. Marine baits of California. Calif. Dept. Fish & Game. 71 pp.

Vahl, O. 1976. On the digestion of *Glycera alba* (Polychaeta). Ophelia 15: 49–56.

Vovelle, J. 1958. Remarques sur la structure du tube de *Sabellaria alveolata* (L.) et les formations glandulaires impliquées dans son édification. Arch. Zool. Expér. Gén. 95: 52–68.

————. 1965. Le tube de *Sabellaria alveolata* (L.) annélide polychète Hermellidae et son ciment, étude écologique, expérimentale, histologique et histochimique. Arch. Zool. Expér. Gén. 106: 101–80.

Vuillemin, S. 1965. Contribution à l'étude écologique du Lac de Tunis. Biologie de *Mercierella enigmatica* Fauvel. Thesis, Fac. Sci., Paris (A) no. 4622.

Ward, J. A. 1977. Chemoreception of heavy metals by the polychaetous annelid *Myxicola infundibulum* (Sabellidae). Comp. Biochem. Physiol. 58C: 103–6.

Warren, L. M. 1976a. A population study of the polychaete *Capitella capitata* at Plymouth. Mar. Biol. 38: 209–16.

————. 1976b. A review of the genus *Capitella* (Polychaeta: Capitellidae). J. Zool., London 180: 195–209.

Watson, A. T. 1928. Observations on the habits and life-history of *Pectinaria* (*Lagis*) *koreni* Mgr. Proc. Liverpool Biol. Soc. 42: 25–60.

Webster, S. K. 1968. An investigation of the commensals of *Cryptochiton stelleri* (Middendorff, 1847) in the Monterey Peninsula area, California. Veliger 11: 121–25.

————. 1975. Oxygen consumption in echinoderms from several geographical locations, with particular reference to the Echinoidea. Biol. Bull. 148: 157–64.

Wells, G. P. 1937. The movements of the proboscis in *Glycera dibranchiata* Ehlers. J. Exper. Biol. 14: 290–301.

————. 1945. The mode of life of *Arenicola marina* L. J. Mar. Biol. Assoc. U.K. 26: 170–207.

————. 1949a. Respiratory movements of *Arenicola marina* L.: Intermittent irrigation of the tube, and intermittent aerial respiration. J. Mar. Biol. Assoc. U.K. 28: 447–64.

————. 1949b. The behaviour of *Arenicola marina* L. in sand, and the role of spontaneous activity cycles. J. Mar. Biol. Assoc. U.K. 28: 465–78.

————. 1950. Spontaneous activity cycles in polychaete worms, pp. 127–42, *in* J. F. Danielli and R. Brown, eds., Symp. Soc. Exper. Biol. 4. 483 pp.

————. 1951. On the behaviour of *Sabella*. Proc. Roy. Soc. London B138: 278–99.

————. 1952a. The proboscis apparatus of *Arenicola*. J. Mar. Biol. Assoc. U.K. 31: 1–28, 4 pls.

————. 1952b. The respiratory significance of the crown in the polychaete worms *Sabella* and *Myxicola*. Proc. Roy. Soc. London B140: 70–82.

————. 1954. The mechanism of proboscis movement in *Arenicola*. Quart. J. Microscop. Sci. 95: 251–70.

————. 1959. The genera of Arenicolidae (Polychaeta). Proc. Zool. Soc. London 133: 301–14, pls. 1–2.

————. 1962. The warm-water lugworms of the world (Arenicolidae, Polychaeta). Proc. Zool. Soc. London 138: 331–53.

————. 1963. Barriers and speciation in lugworms, pp. 79–98, *in* J. P. Harding and N. Tebble, eds., Speciation in the sea. Syst. Assoc. Publ. 5: 1–199.

Wells, G. P., and R. P. Dales. 1951. Spontaneous activity patterns in animal behaviour: The irrigation of the burrow in the polychaetes *Chaetopterus variopedatus* Renier and *Nereis diversicolor* O. F. Müller. J. Mar. Biol. Assoc. U.K. 29: 661–80.

Welsch, U., and V. Storch. 1970. Histochemical and fine structural observations on the alimentary tract of Aphroditidae and Neph-tyidaé (Polychaeta Errantia). Mar. Biol. 6: 142–47.

Williams, G. B. 1964. The effect of extracts of *Fucus serratus* in promoting the settlement of larvae of *Spirorbis borealis* (Polychaeta). J. Mar. Biol. Assoc. U.K. 44: 397–414.

Wilson, D. P. 1929. The larvae of the British sabellarians. J. Mar. Biol. Assoc. U.K. 15: 221–69.

———. 1936a. The development of *Audouinia tentaculata* (Montagu). J. Mar. Biol. Assoc. U.K. 20: 567–79.

———. 1936b. Notes on the early stages of two polychaetes, *Nephthys hombergi* Lamarck and *Pectinaria koreni* Malmgren. J. Mar. Biol. Assoc. U.K. 21: 305–10.

———. 1968a. Long-term effects of low concentrations of an oil-spill remover ("detergent"): Studies with the larvae of *Sabellaria spinulosa*. J. Mar. Biol. Assoc. U.K. 48: 177–82.

———. 1968b. Temporary adsorption on a substrate of an oil-spill remover ("detergent"): Tests with larvae of *Sabellaria spinulosa*. J. Mar. Biol. Assoc. U.K. 48: 183–86.

———. 1968c. Some aspects of the development of eggs and larvae of *Sabellaria alveolata* (L.). J. Mar. Biol. Assoc. U.K. 48: 367–86.

———. 1968d. The settlement behaviour of the larvae of *Sabellaria alveolata* (L.). J. Mar. Biol. Assoc. U.K. 48: 387–435.

———. 1969. The honeycomb worm. Sea Frontiers 15: 322–29.

———. 1970a. Additional observations on larval growth and settlement of *Sabellaria alveolata*. J. Mar. Biol. Assoc. U.K. 50: 1–31.

———. 1970b. The larvae of *Sabellaria spinulosa* and their settlement behaviour. J. Mar. Biol. Assoc. U.K. 50: 33–52.

Wilson, E. B. 1892. The cell-lineage of *Nereis*. J. Morphol. 6: 361–480.

Winesdorfer, J. E. 1967. Marine annelids: *Sabellaria*, pp. 157–62, *in* F. H. Wilt and N. K. Wessells, eds., Methods in developmental biology. New York: Crowell. 813 pp.

Woodin, S. A. 1974. Polychaete abundance patterns in a marine soft-sediment environment: The importance of biological interactions. Ecol. Monogr. 44: 171–87.

———. 1977. Algal "gardening" behavior by nereid polychaetes: Effects on soft-bottom community structure. Mar. Biol. 44: 39–42.

Zoond, A. 1931. Studies in the localization of respiratory exchange in invertebrates. II. The branchial filaments of the sabellid, *Bispira voluticornis*. J. Exper. Biol. 8: 258–62.

Chapter 19

Sipuncula and Echiura

Mary E. Rice

The Sipuncula and Echiura are unsegmented coelomate marine worms, usually recognized as separate phyla. In the past they have often been grouped together as a class of Annelida, the Gephyrea; however, this classification has been discarded by most modern zoologists, who consider the similarities between the two groups to be mostly superficial (Hyman, 1959; Stephen & Edmonds, 1972). Although acknowledged as distinct phyla, the Sipuncula and Echiura are conveniently considered together because of their historic association and combined treatment in the literature.

The sipunculans are a small but unique group of marine worms, characterized by a total lack of segmentation and a division of the body into a thick, often bulbous, posterior trunk and a thinner anterior introvert—so called because it can be retracted, or introverted, within the trunk. When the introvert is retracted, the body, in many species, assumes the shape of a peanut seed, hence sipunculans are commonly known as the "peanut worms." Sipunculans range in length from 1 cm to more than 60 cm, but the majority are between 2.5 and 15 cm long. Typically, the cuticle is tough, leathery, and covered with minute glandular papillae, often more numerous on the anal and posterior regions of the body. One or more rows of tentacles encircle the anterior tip of the introvert; usually within this circlet, but sometimes ventral to it, is the mouth. Immediately behind the tentacles, the cuticle of the introvert is smooth, but more posteriorly the cuticle may be armed with rows of small chitinous hooks or scattered spines. The thick body wall, composed

Mary E. Rice is Curator of Invertebrate Zoology at the National Museum of Natural History, Smithsonian Institution.

of both circular and longitudinal muscle layers, encloses a spacious, fluid-filled coelom, within which freely floating coelomocytes and developing gametes are suspended. Traversing the coelom, one or two pairs of long retractor muscles extend from an attachment in the head region to a more posterior attachment on the body wall. When these muscles are contracted, the introvert is pulled into the trunk. When the main musculature of the body wall contracts, the pressure within the coelom is increased, and the introvert is everted. The digestive tract is distinguished by a long, narrow esophagus, which extends the length of the introvert and continues into a spiraled, recurved intestine with a descending loop and an ascending loop, coiled around one another. The ascending loop leads into the rectum, which opens to the exterior through a dorsal anus, located anteriorly at the base of the introvert. A pair of nephridia, reduced in some species to a single nephridium, opens to the outside by way of ventrolateral nephridiopores near the level of the anus. The nervous system is similar to that of annelids, except that the nerve cord is unpaired and unsegmented.

Although numbering only 320 known species, the sipunculans are nevertheless widely distributed throughout the oceans of the world, from the tropics to the poles and from the intertidal shores to the abyssal depths. Sedentary for the greater part of their life history, sipunculans commonly burrow into the substratum. Along the shores of California, they may be found wedged in the crevices of rocks, or in abandoned holes of boring bivalves, among the roots of surfgrass or byssal threads of mussels, under rocks, or burrowed in sand, gravel, or mud. Sipunculans may feed by engulfing large quantities of the substratum in which they live and digesting from it the organic matter, or they may ingest small particulate matter directed into the mouth by the ciliary activity of the tentacles. Sipunculans have many known predators, including numerous species of fish and a few gastropod mollusks. In the tropical Indo-West Pacific region, sipunculans are used by man as food.

With few known exceptions, the sexes are separate in sipunculans. Gametes are spawned via the nephridiopores into the surrounding seawater, where fertilization takes place. Eggs vary considerably in shape, size, and yolk content, ranging from spherical cells to flattened ellipsoids and measuring at their greatest dimensions from 100 to 280 μm. Those eggs with the greatest concentrations of yolk develop directly into juvenile worms with no swimming stages, whereas those with less yolk may give rise to two larval stages, a trochophore, similar to that of polychaetes, and a pelagosphera larva, which succeeds the trochophore and has a well-developed locomotory band of metatrochal cilia. The latter stage may live for several months in the plankton before undergoing settlement and metamorphosis. Although the life span of a sipunculan is not known with certainty, it has been estimated to be 25 years.

The echiurans are unsegmented marine worms with a sausage-shaped, saccular body and a remarkably extensile, ventrally grooved proboscis. Because the shape of the proboscis when contracted resembles that of a spoon, the echiurans are commonly called "spoon worms." Echiurans feed on tiny organisms and particles of organic detritus suspended in the water or resting on the bottom. The mouth is located ventrally at the base of the proboscis, and posterior to the mouth are two retractable setae, reminiscent of the setae of annelids. In some species one or two rings of additional setae may encircle the posterior body in the vicinity above the terminal anus. Characteristically the body is covered by mucus-secreting papillae either irregularly scattered over the cuticle or concentrated in bands around the body.

The muscular body wall, covered by a thin cuticle, encloses a spacious, fluid-filled coelom within which the internal organs are loosely suspended. The highly differentiated digestive tract loops back and forth, exceeding the body length several times. The number of nephridia most commonly varies from one to four pairs, but a few species have as many as 100 pairs. A pair of anal sacs occurs as a diverticulum on either side of the rectum. These organs presumably function in excretion and are peculiar to echiurans. A brain is lacking and there is no indication of ganglia in the nerve cord.

Development in echiurans follows a course parallel to that in annelids. In some species there is a segmentation of the coelom and an incipient segmentation of the nerve cord, but in later development all traces of segmentation disappear.

Approximately 130 species of echiurans are known. The animals are exclusively marine and are found at all latitudes, either burrowed in sand or mud or in crevices of rocks or in corals. Most frequently collected in littoral and sublittoral habitats, they have been dredged in the North Pacific by the Russian "Vitiaz" Expedition at depths as great as 9,000 m (Zenkevitch, 1958).

The echiurans and the sipunculans, both unsegmented worms of similar shape, have distinctive characters that provide the basis for their classification as separate phyla. The echiurans are distinguished from the sipunculans by the terminal (rather than dorsoanterior) position of the anus, the presence of a proboscis with a basal mouth (rather than an introvert with a terminal mouth), the presence of setae and anal sacs, and the absence of a supraesophageal brain. The presence of setae in echiurans and the transient segmentation in their development suggest a close alliance of echiurans to the annelid line and their derivation from a segmented ancestor. The sipunculans, with no indication of segmentation in the embryo, are assumed to be a very primitive group that probably evolved from a common ancestor of the annelids and mollusks before the advent of segmentation.

For more detailed accounts of the groups covered in this chapter, see Dawydoff (1959), Hyman (1959), and Tétry (1959). Systematics of west coast forms, including full descriptions of species, are included in Fisher (1946, 1952); Rice (1975c) presents a key to California shore forms. More general matters of systematics and phylogeny are dealt with in Clark (1969), Stephen (1964), and Stephen & Edmonds (1972). Physiology and biochemistry of the groups are reviewed in Florkin & Scheer (1969); reproduction, development (including asexual reproduction), and larval biology are covered in Gould-Somero (1975), Needham (1969), Rajulu (1975), Rice (1970, 1975a,b, 1976, 1978), and Scheltema & Hall (1975). Reports of recent research on sipunculans and echiurans appear in Rice & Todorović (1975, 1976).

Phylum Sipuncula / Family Phascolosomatidae

19.1 **Phascolosoma agassizii** Keferstein, 1866
(=Physcosoma agassizii)

Abundant, buried in sand, or under rocks, in crevices of rocks, in roots of surfgrass and holdfasts of kelp, in mussel beds, and among fouling growth on pilings, middle to low intertidal zone on open coast and protected shores; Kodiak Island (Alaska) to Bahía de San Quintín (Baja California).

Length to 12 cm; introvert nearly half of total body length, with irregular dark brown bands on lighter brown background, distal end bearing 15–25 rings of small hooks; trunk cylindrical, pale sepia to dark brown, frequently with brown to purplish spots; skin rough in texture, owing to conical papillae that are largest at posterior extremity and in pre-anal region; tentacles filiform, forming a circle of up to 24 dorsal to mouth; interior with four retractor muscles, origin of dorsal retractors anterior to origin of ventrals; longitudinal muscle layer divided into anastomosing muscle bands, 20–30 in number at level of retractor origin.

Phascolosoma agassizii is the most common sipunculan of California shores. In Monterey Bay this species breeds from March through May, whereas in the San Juan Islands (Washington) it breeds from June to September. When sexually mature, the gametes may make up as much as 37 percent of the dry mass of the animal. The developmental history is marked by a long-lived planktotrophic larva. Related species on the east coast give rise to long-lived larval stages that are believed to survive transport by currents across the Atlantic to the Azores before metamorphosing.

Study of the fine structure of the cerebral eyes shows that the photoreceptor cells are basically of the type characteristic of flatworms, annelids, mollusks, and their allies.

For *P. agassizii*, see Fisher (1952), Hermans & Eakin (1969, 1975), Manwell (1958), Rice (1967, 1973, 1974, 1975b), Robin (1964), Towle & Giese (1966, 1967), and Wourms (1969). For other *Phascolosoma* species, see Rice & Todorović (1975).

Phylum Sipuncula / Family Sipunculidae

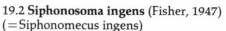

19.2 **Siphonosoma ingens** (Fisher, 1947)
(=Siphonomecus ingens)

Uncommon, in sandy mud and among eelgrass (*Zostera*) roots, low intertidal zone in bays and estuaries; Bodega Bay (Sonoma Co.) to Newport Bay (Orange Co.).

Length to 50 cm; body long and cylindrical, tan or flesh-colored; introvert one-fifth to one-third of total body length, lacking hooks and not easily distinguished from trunk externally; tentacles filiform, surrounding mouth in 12 double rows, 10–12 tentacles per row in large specimens, fewer in smaller animals; interior with four retractor muscles, origin of dorsal retractors anterior to origin of ventrals; longitudinal muscle layer divided into 20–25 bands.

These burrowing animals obtain their nourishment from the organic matter contained in the sediments that they ingest in large amounts as they move along.

The tentacles are rather thick-walled and apparently do not play a very important role in respiration. Exchange of respiratory gases appears to take place through the skin by way of integumental coelomic sacs, which are fluid-filled diverticula of the coelomic cavity that extend into the body wall close to the epidermis.

See Fisher (1947, as *Siphonomecus ingens*; 1952), Manwell (1960), Robin (1964), and Wourms (1969).

Phylum Sipuncula / Family Golfingiidae

19.3 **Themiste pyroides** (Chamberlin, 1919)
(=Dendrostomum pyroides, D. petraeum)

Fairly common under rocks, in crevices, and in abandoned holes made by rock-boring clams, low intertidal zone and adjacent subtidal waters on open coast; Vancouver Island (British Columbia) to Bahía de San Quintín (Baja California).

Length averaging 10–12 cm but reaching 20 cm; body thickset, pear-shaped; introvert one-third to one-half of total body length, lacking prominent papillae, the anterior third reddish to purple-brown bounded posteriorly by a narrow purple zone, the middle third with prominent dark-brown spines; tentacles arising from four roots, highly branched; trunk region smooth, peppered with minute brown spots; interior with two retractor muscles, attaching posteriorly to body wall one-sixth of total body length from posterior end.

This species breeds in late February and early March in Monterey Bay, and from mid-March to early August in northern Washington and at Vancouver Island (British Columbia). Development is direct; the egg adheres to the substratum by a sticky jelly coat, and the embryo hatches from the jelly in 9 days (at 12°C) as a small crawling worm.

The gut of this species often harbors the small rhabdocoel flatworm *Collastoma pacifica*.

See Fisher (1952, as *D. pyroides*) and Rice (1967, 1974, 1975b).

Phylum Sipuncula / Family Golfingiidae

19.4 Themiste zostericola (Chamberlin, 1919)
(=Dendrostomum zostericolum)

Fairly common in low intertidal sand, especially under rocks and among roots of the eelgrass *Zostera* in bays and lagoons; Bodega Bay (Sonoma Co.) to Ensenada (Baja California).

Length averaging about 15 cm but reaching 25 cm; body long, cylindrical, buff with dark-gray lines; introvert one-fifth of total body length, lacking spines but bearing small club-shaped papillae; tentacles pale yellow, brown, red, or dark purple, highly branched, arising from six roots, the dorsal roots often smaller than the others and sometimes asymmetrical; interior with two retractor muscles, attaching posteriorly to body wall one-fourth to one-third of total body length from posterior end.

The body of *T. zostericola* is normally buried in the sand or mud, but the highly branched tentacles are extended out over the surface of the substratum. Oxygen from the seawater is taken in through the thin walls of the tentacles, and small particles of food are moved along the ciliated surfaces of the tentacles into the mouth. The tentacles may also respond to food stimuli by infolding and transferring the food into the mouth, in the manner of a sea anemone. Still another method

by which the worm may secure food is the ingestion of large quantities of sand, from which it digests the organic matter. Experimental studies have shown that if the sand contains no organic matter, it will not be swallowed by the worm.

Other experiments have demonstrated that *T. zostericola* is highly resistant to oxygen deprivation. Worms have been maintained in the laboratory under anaerobic conditions as long as 1 week, suggesting that this species is able to inhabit an environment in which the oxygen content is very low.

See Fisher (1952), Gross (1954), Manwell (1960), Oglesby (1969), and Peebles & Fox (1933), all as *D. zostericolum*.

Phylum Sipuncula / Family Golfingiidae

19.5 Themiste dyscrita (Fisher, 1952)
(=Dendrostomum dyscritum)

Fairly common in rocky crevices and in vacated burrows of rock-boring clams and sea urchins, low intertidal zone and subtidal waters on open coast and protected shores; Boiler Bay (Oregon) to Point Conception (Santa Barbara Co.).

Mature specimens 5–18 cm long; body thickset, pear-shaped but sometimes pointed at posterior tip; introvert one-fourth of total body length, lacking spines, bearing circular glands slightly larger and more sharply defined than in *T. zostericola* (19.4), the anterior fifth or fourth of introvert reddish purple, the posterior part yellowish or gray; tentacles highly branched, arising from six roots, the two dorsal roots shorter than the others; trunk brownish or dark olive-green to black (specimens preserved in formalin may be reddish gray or brown, those in alcohol golden brown or beige to gray); interior with two retractor muscles attaching posteriorly to body wall one-sixth of total body length from posterior end.

The natural history of this species is not well known. Spawning has been observed in specimens collected in late February and early March. Laboratory studies of animals exposed to ambient salinities of 50–125 percent that of normal seawater show that, although the animals tolerate these conditions, they do not regulate osmotically, but conform to the osmotic fluctuations of the environment. However, in worms acclimated to 50–115 percent seawater, the level of chloride ions in the body fluid remained about 6 percent below the level in the environment.

See Fisher (1952, as *D. dyscritum*), Oglesby (1968, 1969), and Robin (1964).

Phylum Sipuncula / Family Golfingiidae

19.6 Golfingia margaritacea californiensis Fisher, 1952

Uncommon in crevices of rocks, low intertidal zone; known only from Pacific Grove, Monterey Bay, and Carmel Bay (both Monterey Co.).

Length less than 25 mm; body slender, pale yellowish brown in life, the skin smooth, without papillae; introvert nearly one-half of total body length, lacking hooks; tentacles filiform, in a single ring of 16 or fewer surrounding the mouth, whitish; interior with four retractor muscles.

Mature specimens collected in November and December contained eggs 0.35 mm in diameter. Little else is known of the biology of this small species, but numerous studies have been made of the structure, development, and physiology of various other species of *Golfingia*.

The curious generic name was created by the British zoologist Sir E. Ray Lankester for the species *Golfingia macintoshii* in 1885, "commemorating," notes Professor W. K. Fisher (1950), "a holiday with Professor MacIntosh at St. Andrews. If the elements of humour are inherent in the incongruous, something may be said for the association of golf and a sipunculid worm. Scarcely, however, were good taste and euphony advanced thereby."

For *G. margaritacea californiensis,* see Fisher (1952). For other species of *Golfingia*, see, for example, Åkesson (1958, 1961), Cole (1952), Ernst (1968), Gerould (1907), Gonse (1956, 1957), Matsumoto & Abbott (1968), Rice (1967, 1974, 1975b), Sawada, Noda & Ochi (1968), and Virkar (1966).

Phylum Echiura / Order Xenopneusta / Family Urechidae

19.7 Urechis caupo Fisher & MacGinitie, 1928
Innkeeper Worm, Fat Innkeeper

Common in burrows in sand and sandy mud, low intertidal and adjacent subtidal zones in bays and estuaries; only subtidal on continental shelf along open coast; documented distribution from Humboldt Bay to Tijuana Slough (San Diego Co.), but range very likely much greater.

Length averaging 15–18 cm but reaching 50 cm; body sausage-like, pinkish; proboscis short; two anterior digging setae on ventral side near mouth; posterior setae numbering 10 or 11, forming a ring around anus, the ring interrupted by a midventral gap, the dorsal setae longer than the ventral; interior with large cloaca adapted for respiration and three pairs of nephridia used to store ripe gametes.

These are the most common of the echiurans of the California coast, and the only ones occurring intertidally. They make relatively permanent U-shaped tunnels in soft substrata, digging and enlarging them with the aid of the anterior setae. The two entrances of a tunnel are separated by a distance of 40–100 cm, depending on the size of the animal, and the burrow may extend from the surface of the mud to a depth of 10–45 cm. Water is circulated through the tunnel by the rhythmic peristaltic-like contractions of the worm's body.

The proboscis of *Urechis* is short and, in contrast to that of other echiurans, is not well adapted for the collection of plankton and detritus. However, *Urechis* has evolved a highly effective and specialized manner of feeding. A slime net, which serves to trap small food particles, is secreted by a circle of mucous glands near the anterior end of the worm. The net, produced in the shape of a funnel and 5–20 cm in length, is attached at its wider end to the sides of the burrow near the entrance and at its narrower end to the circle of glands near the anterior end of the animal. All of the water passing through the tunnel (measured at 29 liters in 24 hours for an average *Urechis*) is strained through the net, and food particles as small as 40 Å are retained. When the mucus net becomes loaded with food, it is loosened from its attachments and the entire net with its burden of food is swallowed by the worm. The animals periodically reverse themselves in their burrows (folding the body double to do so); thus each end of the tunnel serves alternately as entrance and exit.

At least four species of animals are known to dwell within the tube of *Urechis* as commensals. Because of these frequent guests, *Urechis caupo* has been given the name "innkeeper worm" ("caupo" = "innkeeper"). The small goby *Clevelandia ios* (shown in the photograph) uses the burrow of *Urechis* as a refuge, where it retreats from enemies and from desiccation at low tide. As many as five gobies have been found at one time within a single burrow of *Urechis*. Other commensals inhabiting the tube of *Urechis* on a more permanent basis are the polynoid annelid *Hesperonoe adventor* (18.6) and two species of pinnotherid crabs, *Scleroplax granulata* (25.41) and *Pinnixa franciscana* (25.38). The commensals enjoy not only the protective advantages of the tube, but also the food and oxygen provided by the water circulating through the tube and the large particles of food rejected by *Urechis* as it swallows the slime net. Another animal commonly associated with *Urechis* burrows is the small clam *Cryptomya californica* (15.66), which is found buried deeply in mud with its siphons projecting into the burrows.

The sexes are separate in *Urechis*, but males and females look much alike externally. Developing gametes are found in the coelomic fluid at all seasons of the year, in company with blood cells. Long ciliated tendrils extending from the nephridia into the coelom select only the ripe gametes (details of the selection process are fascinating) and transport them to the nephridia, where they are stored until used. Ripe eggs, about 0.13 mm in diameter, are indented on one side and show a clear nucleus. Spawning has been reported in the late spring and early summer, but mature gametes may be found in the nephridia during most of the year. Gametes are easily removed for experimental studies by inserting a smooth-tipped probe or pipette about 0.75 mm in diameter into one of the external nephridial pores, rotating it gently, and withdrawing it. A gush of sperm (whitish) or eggs (pale pinkish, yellowish, or pale olive) follows. Eggs that have been fertilized with a very dilute suspension of sperm complete meiosis, and then cleave. They develop into swimming trochophore larvae, which begin to feed at about 40 hours (at 17°C). Metamorphosis into small burrowing worms occurs about 60 days after fertilization. Development has been studied in detail, and the gametes of *Urechis* have been used in a large number of experimental studies. The adult animals, as well, have been the subjects of a variety of physiological and biochemical investigations.

Intertidal populations of *Urechis* are exploited by man for bait, and in some areas have been severely depleted. However, the bulk of the *Urechis* population probably occurs subtidally, where, safe from man, it is exploited by several bottom-feeding fishes. Starry flounders (*Platichthys stellatus*) taken near Bodega Bay harbor (Sonoma Co.) and Monterey harbor frequently contain *Urechis* within the gut, likewise the diamond turbot (*Hypsopsetta guttulata*) caught in Monterey Bay (C. Hand and E. C. Haderlie, pers. comms). A study of the stomachs of 367 leopard sharks (*Triakis semifasciata*) caught at Elkhorn Slough (Monterey Co.) showed that although the diet was varied, *Urechis* was the single most important food species for sharks over 90 cm long. The *Urechis* in the gut rarely showed tooth marks, suggesting that they may have been sucked from their burrows, a feeding technique used by some sharks. In addition to fishes, the sea otter *Enhydra lutris* has been observed feeding on *Urechis*.

The subtidal distribution of *Urechis* may extend well beyond the present confirmed geographic limits. In 1952, *Urechis caupo* (identified by W. K. Fisher) were found in the stomach of a black sea bass (*Stereolepis gigas*) taken at 35–40 m depth off the entrance to Scammon Lagoon, Baja California (C. Hand, pers. comm.). A gregarine protozoan, *Echiurocystis* (formerly *Enterocystis*) *bullis*, inhabits the intestine of *Urechis caupo*.

See especially Fisher (1946), Fisher & MacGinitie (1928a,b), and Newby (1940); see also Chapman (1968), Das (1968, 1976), Davis & Wilt (1972), Engstrom (1971), Gould (1967, 1969), Gould-Somero (1975), Hall (1931), Lawry (1966), MacGinitie (1935, 1937, 1945), MacGinitie & MacGinitie (1968), Miller (1973), Newby (1941), Noble (1938a,b), Paul (1975), Paul & Gould-Somero (1976), Redfield & Florkin (1931), Robin (1964), Russo (1975), Sawada & Noda (1963), Shimek (1977), Talent (1976), and Tyler (1931).

Literature Cited

Åkesson, B. 1958. A study of the nervous system of the Sipunculoideae with some remarks on the development of the two species *Phascolion strombi* Montagu and *Golfingia minuta* Keferstein. Undersökningar över Öresund 38: 1–249.

———. 1961. The development of *Golfingia elongata* Keferstein (Sipunculidea) with some remarks on the development of neurosecretory cells in sipunculids. Ark. Zool. (2) 13: 511-31.

Chapman, G. 1968. The hydraulic system of *Urechis caupo* Fisher & MacGinitie. J. Exper. Biol. 49: 757–67.

Clark, R. B. 1969. Systematics and phylogeny: Annelida, Echiura, Sipuncula, pp. 1–68, *in* Florkin & Scheer (1969).

Cole, J. B. 1952. The morphology of *Golfingia pugettensis*: A sipunculid worm. Master's thesis, Zoology, University of Washington, Seattle. 78 pp.

Das, N. K. 1968. Developmental features and synthetic patterns of male germ cells of *Urechis caupo*. Arch. Entwickelungsmech. Org. 161: 325–35.

———. 1976. Cytochemical and biochemical analysis of development of *Urechis* oocytes. Amer. Zool. 16: 345–62.

Davis, F. C., Jr., and F. H. Wilt. 1972. RNA synthesis during oogenesis in the echiuroid worm *Urechis caupo*. Develop. Biol. 27: 1–12.

Dawydoff, C. 1959. Classe des echiuriens, pp. 855–907, *in* P.-P. Grassé, ed., Traité de zoologie, vol. 5, part 1. Paris: Masson. 1,053 pp.

Engstrom, W. S. 1971. Removal of the fertilization membrane of fertilized eggs of *Urechis caupo* and development of "membraneless" embryos. Biol. Bull. 140: 369–75.

Ernst, V. V. 1970. The structure and function of the proboscis retractor muscle of the sipunculid, *Golfingia gouldii*. Doctoral thesis, Biology, University of Louisville, Louisville, Kentucky. 184 pp.

Fisher, W. K. 1946. Echiuroid worms of the North Pacific Ocean. Proc. U.S. Nat. Mus. 96: 215–92.

———. 1947. New genera and species of echiuroid and sipunculoid worms. Proc. U.S. Nat. Mus. 97: 351–72.

———. 1950. The sipunculid genus *Phascolosoma*. Ann. Mag. Natur. Hist. (12) 3: 547–52.

———. 1952. The sipunculid worms of California and Baja California. Proc. U.S. Nat. Mus. 102: 371–450.

Fisher, W. K., and G. E. MacGinitie. 1928a. A new echiuroid worm from California. Ann. Mag. Natur. Hist. (10) 1: 199–204.

———. 1928b. The natural history of an echiuroid worm. Ann. Mag. Natur. Hist. (10) 1: 204–13.

Florkin, M., and B. T. Scheer, eds. 1969. Chemical Zoology. 4. Annelida, Echiura, and Sipuncula. New York: Academic Press. 548 pp.

Gerould, J. H. 1907. The development of *Phascolosoma*. (Studies on the embryology of the Sipunculidae II.) Zool. Jahrb. Anat. 23: 77–162.

Gonse, P. 1956. L'ovogénèse chez *Phascolosoma vulgare*. I. Définition cytologique des stades de croissance des ovocytes. II. Recherches biométriques sur les ovocytes. Acta. Zool. 37: 193–224, 225–33.

———. 1957. L'ovogénèse chez *Phascolosoma vulgare*. III. Respiration exogène et endogène de l'ovocyte. Effet de l'eau de mer. IV. Étude chromatique des sucres du plasma, action de différents substrats et du malonate sur la respiration de l'ovocyte. Biochim. Biophys. Acta 24: 267–78, 520–31.

Gould, M. C. 1967. Echiuroid worms: *Urechis*, pp. 163–71, *in* F. H. Wilt and N. K. Wessells, ed., Methods in developmental biology. New York: Crowell. 813 pp.

———. 1969. A comparison of RNA and protein synthesis in fertilized and unfertilized eggs of *Urechis caupo*. Develop. Biol. 19: 482–97.

Gould-Somero, M. 1975. Echiura, pp. 277–311, *in* A. C. Giese and J. S. Pearse, eds., Reproduction of marine invertebrates, vol. 3. New York: Academic Press. 343 pp.

Gross, W. J. 1954. Osmotic responses in the sipunculid *Dendrostomum zostericolum*. J. Exper. Biol. 31: 402–23.

Hall, V. E. 1931. The muscular activity and oxygen consumption of *Urechis caupo*. Biol. Bull. 61: 400–416.

Hermans, C. O., and R. M. Eakin. 1969. Fine structure of the cerebral ocelli of a sipunculid, *Phascolosoma agassizii*. Z. Zellforsch. 100: 325–39.

———. 1975. Sipunculan ocelli: Fine structure in *Phascolosoma agassizii*, pp. 229–37, *in* Rice & Todorović (1975).

Hyman, L. H. 1959. The invertebrates: Smaller coelomate groups. Vol. 5. New York: McGraw-Hill. 783 pp.

Lawry, J. V. 1966. Neuromuscular mechanisms of burrow irrigation in the echiuroid worm *Urechis caupo* Fisher & MacGinitie. I. Anatomy of the neuromuscular system and activity of intact animals. II. Neuromuscular activity of dissected preparations. J. Exper. Biol. 45: 343–56, 357–68.

MacGinitie, G. E. 1935. Normal functioning and experimental behavior of the egg and sperm collectors of the echiuroid, *Urechis caupo*. J. Exper. Zool. 70: 341–55.

———. 1937. The use of mucus by marine plankton feeders. Science 86: 398–99.

———. 1945. The size of the mesh openings in mucous feeding nets of marine animals. Biol. Bull. 88: 107–11.

MacGinitie, G. E., and N. MacGinitie. 1968. Natural history of marine animals. New York: McGraw-Hill. 523 pp.

Manwell, C. 1958. Oxygen equilibrium of *Phascolosoma agassizii* hemerythrin. Science 127: 592–93.

———. 1960. Histological specificity of respiratory pigments. II. Oxygen transfer systems involving hemerythrins in sipunculid worms of different ecologies. Comp. Biochem. Physiol. 1: 277–85.

Matsumoto, Y., and B. C. Abbott. 1968. Folding muscle fibers of the *Golfingia gouldii*. Comp. Biochem. Physiol. 26: 927–36.

Miller, J. H. 1973. An investigation of the microtubule protein in mature oocytes of *Urechis caupo*. Exper. Cell Res. 81: 342–50.

Needham, A. E. 1969. Growth and development, pp. 377–441, *in* Florkin & Scheer (1969).

Newby, W. W. 1940. The embryology of the echiuroid worm *Urechis caupo*. Mem. Amer. Philos. Soc. 16. 219 pp.

———. 1941. The development and structure of the slime-net glands of *Urechis*. J. Morphol. 69: 303–16.

Noble, E. R. 1938a. The life cycle of *Zygosoma globosum* sp. nov., a gregarine parasite of *Urechis caupo*. Univ. Calif. Publ. Zool. 43: 41–65.

———. 1938b. A new gregarine from *Urechis caupo*. Trans. Amer. Microscop. Soc. 57: 142–46.

Oglesby, L. C. 1968. Some osmotic responses of the sipunculid worm *Themiste dyscritum*. Comp. Biochem. Physiol. 26: 155–77.

———. 1969. Inorganic components and metabolism; ionic and osmotic regulation: Annelida, Sipuncula, and Echiura, pp. 211–310, *in* Florkin & Scheer (1969).

Paul, M. 1975. Release of acid and changes in light scattering properties following fertilization of *Urechis caupo* eggs. Develop. Biol. 43: 299–312.

Paul, M., and M. Gould-Somero. 1976. Evidence for a polyspermy block at the level of sperm-egg plasma membrane fusion in *Urechis caupo*. J. Exper. Zool. 196: 105–12.

Peebles, F., and D. L. Fox. 1933. The structure, functions, and general reactions of the marine sipunculid worm *Dendrostoma zostericola*. Bull. Scripps. Inst. Oceanogr., Tech. Ser. 3: 201–24.

Rajulu, G. S. 1975. Asexual reproduction by budding in the Sipuncula, pp. 177–82, *in* Rice & Todorović (1975).

Redfield, A. C., and M. Florkin. 1931. The respiratory function of the blood of *Urechis caupo*. Biol. Bull. 61: 185–210.

Rice, M. E. 1967. A comparative study of the development of *Phascolosoma agassizii*, *Golfingia pugettensis*, and *Themiste pyroides* with a discussion of developmental patterns in the Sipuncula. Ophelia 4: 143–71.

———. 1970. Asexual reproduction in a sipunculan worm. Science 167: 1618–20.

———. 1973. Morphology, behavior, and histogenesis of the pelagosphera larva of *Phascolosoma agassizii* (Sipuncula). Smithsonian Contr. Zool. 132: 1–51.

———. 1974. Gametogenesis in three species of Sipuncula: *Phascolosoma agassizii*, *Golfingia pugettensis*, and *Themiste pyroides*. La Cellule 70: 295–313.

———. 1975a. Observations on the development of six species of Caribbean Sipuncula with a review of development in the phylum, pp. 141–60, *in* Rice & Todorović (1975).

———. 1975b. Sipuncula, pp. 67–127, *in* A. C. Giese and J. S.

Pearse, eds., Reproduction of marine invertebrates, vol. 2. New York: Academic Press. 344 pp.

———. 1975c. Unsegmented coelomate worms, pp. 128–34, *in* R. I. Smith and J. T. Carlton, eds., Light's manual: Intertidal invertebrates of the central California coast. 3rd ed. Berkeley and Los Angeles: University of California Press. 716 pp.

———. 1976. Larval development and metamorphosis in Sipuncula. Amer. Zool. 16: 563–71.

———. 1978. Morphological and behavioral changes at metamorphosis in the Sipuncula, pp. 83–102, *in* F.-S. Chia and M. E. Rice, eds., Settlement and metamorphosis in marine invertebrate larvae. New York: Elsevier/North Holland. 290 pp.

Rice, M. E., and M. Todorović, eds. 1975. Proc. Internat. Symp. on the Biology of Sipuncula and Echiura. Vol. 1. Belgrade: Naučno Delo Press. 355 pp.

———. 1976. Proc. Internat. Symp. on the Biology of Sipuncula and Echiura. Vol. 2. Belgrade: Naučno Delo Press. 204 pp.

Robin, Y. 1964. Biological distribution of guanidines and phosphagens in marine Annelida and related phyla from California, with a note on pluriphosphagens. Comp. Biochem. Physiol. 12: 347–67.

Russo, R. A. 1975. Observations on the food habits of leopard sharks (*Triakis semifasciata*) and brown smoothhounds (*Mustelus henlei*). Calif. Fish & Game 61: 95–103.

Sawada, N., and Y. Noda. 1963. An electron microscope study on the *Urechis* egg. Mem. Ehime Univ. (Japan), Biol. B4: 539–49.

Sawada, N., Y. Noda, and O. Ochi. 1968. An electron microscope study on the oogenesis of *Golfingia ikedai*. Mem. Ehime Univ. (Japan), Biol. B6: 25–39.

Scheltema, R. S., and J. R. Hall. 1975. The dispersal of pelagosphera larvae by ocean currents and the geographical distribution of sipunculans, pp. 103–15, *in* Rice & Todorović (1975).

Shimek, S. J. 1977. The underwater foraging habits of the sea otter, *Enhydra lutris*. Calif. Fish & Game 63: 120–22.

Stephen, A. C. 1964. A revision of the classification of the phylum Sipuncula. Ann. Mag. Natur. Hist. (13) 7: 457–62.

Stephen, A. C., and S. J. Edmonds. 1972. The phyla Sipuncula and Echiura. London: British Museum (Natur. Hist.). 528 pp.

Talent, L. G. 1976. Food habits of the leopard shark, *Triakis semifasciata*, in Elkhorn Slough, Monterey Bay, California. Calif. Fish & Game 62: 286–98.

Tétry, A. 1959. Classe des sipunculiens, pp. 785–854, *in* P.-P. Grassé, ed., Traité de zoologie, vol. 5, part 1. Paris: Masson. 1,053 pp.

Towle, A., and A. C. Giese. 1966. Biochemical changes during reproduction and starvation in the sipunculid worm *Phascolosoma agassizii*. Comp. Biochem. Physiol. 19: 667–80.

———. 1967. The annual reproductive cycle of the sipunculid *Phascolosoma agassizii*. Physiol. Zool. 40: 229–37.

Tyler, A. 1931. The production of normal embryos by artificial parthenogenesis in the echiuroid, *Urechis*. Biol. Bull. 60: 187–211.

Virkar, R. A. 1966. The role of free amino acids in the adaptation to reduced salinity in the sipunculid *Golfingia gouldii*. Comp. Biochem. Physiol. 18: 617–25.

Wourms, J. P. 1969. Ultrastructural analysis of oocyte differentiation and primary envelope formation in sipunculids. Amer. Zool. 9: 1120 (abstract).

Zenkevitch, L. A. 1958. The deep-sea echiurids of the north-western part of the Pacific Ocean. [In Russian.] Trudy Inst. Okeanol. 27: 192–203.

Crustacea and Other Arthropods: *Introduction*

Donald P. Abbott and Eugene C. Haderlie

The arthropods were introduced earlier (The Annelid-Arthropod Complex: Introduction, p. 445) as descendants of annelids that have stiffened the external cuticle into a supporting exoskeleton, evolved jointed limbs, and developed distinctive body regions (tagmata) composed of groups of functionally related segments. With these and related changes, the arthropods have been an enormously successful group, invading all the inhabitable realms of the earth and producing a number of species greater than that in all the rest of the metazoan phyla combined.

Do the arthropods represent a single branch on the evolutionary tree of animals, in the sense that some single annelid stock evolved the basic arthropod traits and then diversified into the main arthropod stocks of ancient and modern times? The question is not an easy one to answer, and indeed it is only in the last few years that a real comparative overview of the arthropods has been possible. The group is so vast that arthropod scholars have tended to specialize in single subgroups, with little concern for other branches. Snodgrass (1951) noted wryly that "arthropodists are divided into the same classes as the arthropods themselves," and later (1952) remarked, "The arthropods are a group of related invertebrates; arthropodists, for the most part, are a group of unrelated vertebrates."

The situation has improved in the past 30 years. Knowledge of most groups has greatly increased, and a growing number of scholars, refusing to limit themselves to one arthropod group, have tried to put the facts of comparative functional anatomy, comparative embryology, and paleontology together in an effort to trace the broad historical outline of arthropod evolution. Major contributions have been made by such workers as Snodgrass (e.g., 1938, 1951, 1952, 1958), Anderson (1973), Størmer (1944), and above all, the late Sidnie Manton (1953, 1963, 1964, 1969, 1970, 1972, 1973, 1977; Tiegs & Manton, 1958). The evolutionary hypotheses that are emerging are sufficiently detailed to better define gaps in knowledge and to direct attention to important areas for further work.

It appears that arthropodization may have begun independently in at least two or three different stocks of segmented worms that, if present today, would probably all be classed as annelids. In this view, arthropods represent a polyphyletic group rather than a natural branch on an evolutionary tree. Consensus is lacking (e.g., Gupta, 1979), but there is a growing tendency to treat the major arthropod subgroups as separate phyla (e.g., Manton, 1977).

One of these subgroups, the phylum Uniramia, probably shares a remote common ancestry with modern earthworms and leeches. The Uniramia arose as a group of walking worms with relatively simple, lobelike appendages; the present Onychophora may represent an early offshoot of this lineage. The stock diversified to form the centipedes and the millipedes (both groups still retaining a somewhat wormlike appearance, though highly evolved in numerous other respects), and ultimately produced the insects and their allies. Most Uniramia beyond the Onychophora today possess a head tagma equipped with one pair of preoral antennae and three pairs of postoral mouthparts, the anterior mouthparts consisting of mandibles. Evolution in the Uniramia accompanied conquest of the land, and today most species are terrestrial, though some insects have secondarily invaded fresh water and a very few have entered

Donald P. Abbott is Professor, Department of Biological Sciences and Hopkins Marine Station, Stanford University. Eugene C. Haderlie is Professor of Oceanography at the Naval Postgraduate School, Monterey.

the fringes of the sea or skitter over the air/sea interface. Representative seashore species of Uniramia are covered in Chapter 28.

Another great lineage, the Trilobita, contains the first arthropods represented in the fossil record. Trilobites had the armor of the dorsal surface divided into a central and two lateral areas, giving the body a trilobed appearance. They possessed a head with a pair of preoral antennae and three pairs of postoral cephalic appendages none of which were specially modified as mouthparts (Cisne, 1974). No trilobites survived the Paleozoic period, but another great lineage that still flourishes today, the Chelicerata, probably originated from early trilobite stock. Trilobites and chelicerates together are sometimes referred to as the Arachnomorpha.

The tagmata of the Chelicerata differ from those of other arthropods. The anterior region, or prosoma, does not correspond to the head tagma found in members of the Uniramia and Crustacea. It lacks antennae and bears in their place a pair of preoral pincers, the chelicerae. Mandibles are lacking, and the prosoma bears five pairs of postoral appendages variously involved in locomotion, feeding, and sometimes other activities. The second body region, or opisthosoma, retains limbs used for swimming and respiration in some contemporary aquatic forms (the Xiphosura, or horseshoe crabs), but in terrestrial chelicerates these limbs are highly modified or lacking. The aquatic chelicerates had their heyday in the first half of the Paleozoic. There are few surviving descendants today: the horseshoe crabs (not really crabs at all; common in New England inshore waters but absent from the west coast) and the pycnogonids (Chapter 27). However, the terrestrial branch of the chelicerate complex, the Arachnida, has done extraordinarily well. Consisting of the spiders, scorpions, mites, solpugids, whip scorpions, and related forms, the arachnids are typically predaceous, and they are the dominant invertebrate carnivores in the terrestrial realm (Manton, 1977). A few assorted arachnids have learned to live in shallow fresh water and along the shores of the sea (Chapter 28), but only the mites have reinvaded the sea itself.

The remaining great arthropod lineage is the Crustacea, considered by some (e.g., Manton, 1977) as deserving phylum status. Crustaceans are arthropods with a head tagma bearing two pairs of antennae, the first pair preoral, the second pair postoral, and three pairs of postoral mouthparts. The anterior mouthparts are mandibles, used for biting, grinding, or compressing food. These are followed by two pairs of maxillae, variously modified for use in the capture or ingestion of food. The remainder of the body forms a trunk region terminating in the telson (comparable to the annelid pygidium), which bears the anus. Some primitive crustaceans have as many as 40 segments in the body and a series of trunk appendages that are much alike and serve multiple functions. Most modern crustaceans have shortened the body to 20 segments or less and show a marked tendency to specialize particular appendages in connection with specific functions. Often the trunk segments are grouped into two definite body regions, usually called thorax and abdomen, though these two tagmata are not homologous in all crustacean groups. In some crustaceans, one or more thoracic segments may fuse with the head without leaving a clear external trace of this fusion. We have used the term "head" in conventional fashion to include only the segment bearing the second maxillae and all parts anterior to this. Some crustaceans possess paired appendages on all trunk segments, others lack limbs on some or all of the abdominal segments, and in some the limbs are reduced or missing on some cephalic or thoracic segments. Appendages of the anterior thoracic segments that are modified as additional mouthparts are termed maxillipeds. Many crustaceans are equipped with a carapace, defined as a fold of the body wall extending backward from the posterior border of the head, which may cover part or all of the trunk region and may fuse dorsally to one or more thoracic segments.

The affinities of the Crustacea with other large arthropod lineages have been a matter of dispute among authorities. Snodgrass, impressed with the similarities between insects, myriapods, and crustaceans, supported the view that they should be placed together in a large subphylum Mandibulata (1938). Hessler and Newman (1975) present the case for a Trilobite-Chelicerate affinity for the Crustacea. Manton (1977 and earlier papers) expressed the opinion that the Crustacea should be treated as a third phylum, equal in status with, and not closely allied to, the Uniramia and the Chelicerata. Whatever the final outcome here, it is apparent that there has been a great deal of convergence in the evolution of different arthropod structures, such as limbs, mouthparts, compound eyes, and tracheal systems (Manton, 1977; Tiegs & Manton, 1958), and homologies are often difficult to establish or disprove without extensive comparative anatomical and developmental study.

In any event, in contrast to the Chelicerata and the Uniramia, which have had their great successful evolutionary radiations on land, the Crustacea have remained primarily aquatic in habitat and adaptation. Some crustaceans have come ashore (see Bliss & Mantel, 1968), but the great majority of species inhabit the sea or fresh water.

Crustaceans are overwhelmingly the dominant arthropods in the sea today, though other forms such as mites, pycnogonids, horseshoe crabs, and even a few insects occur along with them. The number of arthropod species in the sea is far smaller than that on land: known marine crustacean species number roughly 30,000–32,000 (Gordon, 1964), whereas there are more than 300,000 species of beetles alone on land, but the number of individual arthropods present in the marine realm is astronomical. One biologist estimated that there are more individuals of the single crustacean subclass Copepoda in the sea than there are individuals of all other metazoans on earth put together (Hardy, 1956).

In ecological and economic terms, crustaceans are enormously important in the sea. The smaller species are the most abundant of the small metazoan herbivores that graze directly on the unicellular plant plankton of the sea, and in turn they form the most important food for a host of smaller fishes and invertebrate carnivores. Among the larger crustaceans we find herbivores, carnivores, scavengers, filter feeders, and deposit feeders, and varied forms occur in the sea from the intertidal zone to abyssal depths. The larger crustaceans supply important food for larger fishes and whales, and provide such gourmet dishes as shrimp, lobster, and crab for man.

In this book, for practical reasons the crustaceans are treated as a class of the phylum Arthropoda. Living crustaceans are divided, in conventional fashion, into eight subclasses (e.g., Kaestner, 1970; Moore & McCormick, 1969; Waterman & Chace, 1960): Cephalocarida, Branchiopoda, Ostracoda, Mystacocarida, Copepoda, Branchiura, Cirripedia, and Malacostraca. Of these subclasses, only four are of major importance in the sea: the Ostracoda (seed shrimps), the Copepoda, the Cirripedia (barnacles), and the Malacostraca.

Ostracods are not covered in this book. They are small forms, mostly less than 2 mm long, with the body completely enclosed in a bivalved carapace. They are usually overlooked by seashore visitors, though they are present in fair numbers in the interstices of coarse sediments, and in clusters of seaweeds, sponges, and tunicates in the low intertidal zone.

Copepods are treated briefly in Chapter 26. Their small size, their immense populations in the upper levels of the sea, and their key position in the food web (many are grazers on phytoplankton, and in turn serve as food for small carnivores) make them among the most important animals in the economy of the sea. They are abundant but rather inconspicuous inhabitants of the intertidal zone.

The Cirripedia, or barnacles, are both numerous and prominent in the intertidal zone on many rocky shores. They are treated in Chapter 20. We have included there not only the barnacles that are regular inhabitants of the intertidal zone but also a variety of species that attach to floating logs or live on various pelagic animals (from jellyfishes to whales) that sometimes wash into the intertidal zone.

Finally, most of the larger and more conspicuous crustaceans, all over the world, are members of the crustacean subclass Malacostraca. In these crustaceans, the main divisions, or tagmata, of the body are a head, a thorax consisting of eight segments, and an abdomen comprised of six or seven segments plus the terminal telson. Reproductive ducts of the female usually open on the sixth thoracic segment, those of the male on the eighth. The subclass is divided into six superorders, of which three are treated in this book. A single representative of the superorder Hoplocarida, containing the fast and predaceous mantis shrimps, is treated in Chapter 26. The Peracarida includes those forms that brood their young in a special pouch that is situated below the thorax and is formed of leaflike plates (oostegites) projecting medially from the bases of one to several pairs of thoracic limbs. Intertidal peracaridans are represented in Chapter 21 (Isopoda and Tanaidacea) and Chapter 22 (Amphipoda).

The last superorder of Malacostraca, the Eucarida, includes those malacostracans having a carapace that covers the thorax dorsally and laterally and is fused to all of the thoracic segments. This results in an anterior tagma, the cephalothorax, in which the segments are tightly joined. In the largest order, the Decapoda, the first three pairs of thoracic appendages are all modified as mouthparts (maxillipeds), leaving only ten thoracic limbs that are more or less leg-like ("deca-poda"). Female decapods usually carry their fertilized eggs attached to their abdominal appendages until the young hatch as swimming larvae. Decapod crustaceans are treated in Chapters 23–25.

Among the most useful references on arthropods are those by Manton (1969, 1970, 1973, 1977) dealing with structure, functional anatomy, and phylogeny. Also excellent are Anderson (1973), which relates pattern of development to phylogeny, Clark (1973), which reviews some aspects of arthropod biology, and Florkin & Scheer (1970, 1971), which covers numerous aspects of arthropod physiology and biochemistry.

There are several excellent works on crustaceans that will be especially useful to users of this book who wish further details. Readable general accounts of crustacean biology are found in Calman (1911), Gordon (1964), Green (1961), and Schmitt (1965). For more technical surveys of all or part of the crustacea, see Calman (1909), Glaessner (1969), Kaestner (1970), contributions in Whittington & Rolfe (1963), and appropriate sections in Moore (1969). Physiological and biochemical aspects of crustacean biology are treated in Huggins and Munday (1968), Florkin & Scheer (1970, 1971), and Waterman (1960, 1961). Adaptations of crustaceans to life on land are covered in Bliss & Mantel (1968). For excellent keys and illustrations useful for identifying California crustaceans (including a great many of the smaller forms not considered here), see Smith & Carlton (1975).

Literature Cited

Anderson, D. T. 1973. Embryology and phylogeny in annelids and arthropods. New York: Pergamon. 495 pp.

Bliss, D. E., and L. H. Mantel, eds. 1968. Terrestrial adaptations in Crustacea. Amer. Zool. 8: 307–685.

Calman, W. T. 1909. Crustacea, *in* R. Lankester, ed., A treatise on zoology, 7(3). London: Adam & Charles Black. 346 pp.

———. 1911. The life of Crustacea. London: Methuen. 289 pp.

Cisne, J. L. 1974. Trilobites and the origin of arthropods. Science 186: 13–18.

Clarke, K. U. 1973. The biology of the Arthropoda. London: Edw. Arnold. 270 pp.

Florkin, M., and B. T. Scheer, eds. 1970. Chemical zoology. 5. Arthropoda part A. New York: Academic Press. 460 pp.

———, eds. 1971. Chemical zoology. 5. Arthropoda part B. New York: Academic Press. 484 pp.

Glaessner, M. F. 1969. Decapoda, pp. R399–533 (and bibliography, pp. R552–66), *in* R. C. Moore, ed., Treatise on invertebrate paleontology, part R, Arthropoda 4, vol. 2. New York: Geol. Soc. Amer.; Lawrence: University of Kansas Press. 253 pp.

Gordon, I. 1964. Crustacea—general considerations, pp. 31–86, *in* C.

S. A. specialist meeting on crustaceans. Conseil Scientifique pour l'Afrique, Organization de l'Unité Africaine, No. 96. London: Publications Bureau, Watergate House, York Bldgs. 648 pp.

Green, J. 1961. A biology of Crustacea. London: Witherby. 180 pp.

Gupta, A. P., ed. 1979. Arthropod phylogeny. New York: Van Nostrand Reinhold. 762 pp.

Hardy, A. C. 1956. The open sea. Its natural history: The world of plankton. Boston: Houghton Mifflin. 335 pp.

Hessler, R. R., and W. A. Newman. 1975. A trilobitomorph origin for the Crustacea. Fossils and Strata, Oslo 4: 437–59.

Huggins, A. K., and K. A. Munday. 1968. Crustacean metabolism, pp. 271–378, *in* O. Lowenstein, ed., Advances in comparative physiology and biochemistry, vol. 3. New York: Academic Press. 416 pp.

Kaestner, A. 1968. Invertebrate zoology. 2. Arthropod relatives, Chelicerata, Myriapoda. (Translated and adapted by H. W. Levi & L. R. Levi.) New York: Wiley-Interscience. 472 pp.

———. 1970. Invertebrate zoology. 3. Crustacea. (Translated and adapted by H. W. Levi & L. R. Levi.) New York: Wiley-Interscience. 523 pp.

Manton, S. M. 1953. Locomotory habits and the evolution of the larger arthropodan groups, pp. 339–76, *in* Symp. Soc. Exper. Biol. 7, Evolution. Cambridge, Eng.: University Press.

———. 1963. Jaw mechanisms of Arthropoda with particular reference to the evolution of Crustacea, pp. 111–44, *in* Whittington & Rolfe (1963).

———. 1964. Mandibular mechanisms and the evolution of the arthropods. Phil. Trans. Roy. Soc. London B247: 1–183.

———. 1969. Introduction to classification of Arthropoda. Evolution and affinities of Onychophora, Myriapoda, Hexapoda, and Crustacea, pp. R3–56, *in* R. C. Moore, ed., Treatise on invertebrate paleontology, part R, Arthropoda 4, vol. 1. New York: Geol. Soc. Amer.; Lawrence: University of Kansas Press. 398 pp.

———. 1970. Arthropods: Introduction, pp. 1–34, *in* Florkin & Scheer (1970).

———. 1972. The evolution of arthropodal locomotory mechanisms. 10. J. Linn. Soc. Zool. 51: 203–400.

———. 1973. Arthropod phylogeny—a modern synthesis. J. Zool., London 171: 111–30.

———. 1977. The Arthropoda: Habits, functional morphology, and evolution. Oxford: Oxford University Press (Clarendon Press). 527 pp.

Moore, R. C., and L. McCormick. 1969. General features of Crustacea, pp. R57–120, *in* R. C. Moore, ed., Treatise on invertebrate paleontology, part R, Arthropoda 4, vol. 1. New York: Geol. Soc. Amer.; Lawrence: University of Kansas Press. 398 pp.

Schmitt, W. L. 1965. Crustaceans. Ann Arbor, Mich.: University of Michigan Press. 204 pp.

Smith, R. I., and J. T. Carlton, eds. 1975. Light's manual: Intertidal invertebrates of the central California coast. Berkeley and Los Angeles: University of California Press. 716 pp.

Snodgrass, R. E. 1938. Evolution of the Annelida, Onychophora, and Arthropoda. Smithsonian Misc. Coll. 97(6): 1–159.

———. 1951. Comparative studies on the head of mandibulate arthropods. Cornell Studies in Entomology 1. Ithaca, N.Y.: Comstock. 118 pp.

———. 1952. A textbook of arthropod anatomy. Ithaca, N.Y.: Comstock (Cornell University Press). 363 pp.

———. 1956. Crustacean metamorphoses. Smithsonian Misc. Coll. 131(10): 1–78.

———. 1958. Evolution of arthropod mechanisms. Smithsonian Misc. Coll. 138(2): 1–77.

Størmer, L. 1944. On the relationships and phylogeny of fossil and recent Arachnomorpha. Skrift. Norske Videnskaps-Akad. Oslo 5: 1–158.

Tiegs, O. W., and S. M. Manton. 1958. The evolution of the Arthropoda. Biol. Rev. 33: 255–337.

Waterman, T. H., ed. 1960. The physiology of Crustacea. 1. Metabolism and growth. New York: Academic Press. 670 pp.

———, ed. 1961. The physiology of Crustacea. 2. Sense organs, integration, and behavior. New York: Academic Press. 681 pp.

Waterman, T. H., and F. A. Chace, Jr. 1960. General crustacean biology, pp. 1–33, *in* Waterman (1960).

Whittington, H. B., and W. D. I. Rolfe, eds. 1963. Phylogeny and evolution of Crustacea. Cambridge, Mass.: Spec. Publ. Mus. Compar. Zool., Harvard Univ. 192 pp.

Chapter 20

Cirripedia: *The Barnacles*

William A. Newman and Donald P. Abbott

The Cirripedia are crustaceans that as adults are usually sessile, attached to hard substrata or to other organisms. The carapace (mantle) completely envelops the body, and in most forms it secretes a calcareous shell—a trait that led some earlier zoologists to place barnacles in the phylum Mollusca. In adapting to life in the protected environment of the mantle cavity, the barnacle body has undergone an evolutionary reduction or loss of such features as the compound eyes and the abdomen and its appendages, but most adult cirripeds are still easily recognizable as crustaceans upon dissection. This is not the case with the more highly modified parasitic species, whose true cirriped nature is clear only from a study of their larvae.

The barnacles are an ancient group. There are published accounts of Silurian fossil remains, and barnacles have been discovered very recently in the Burgess Shale deposits in British Columbia (D. H. Collins, pers. comm.), beds that date from the middle Cambrian period of the Paleozoic. Barnacles remain a very successful group today, both in number of species—about 1,445 living species are known—and abundance. Charles Darwin, whose superb monographs on Cirripedia (1851, 1854) are still valuable references, once remarked that the present epoch may go down in the fossil record as the "Age of Barnacles," so abundant and widespread are their remains. Barnacle shells make up 50 percent of the carbonate sediments on the Florida shelf (Milliman, 1974), and, in many parts of the world, rock surfaces in the middle intertidal zone bear exten-

sive carpets of barnacles (Stephenson & Stephenson, 1972). Those seeking recreation at the shore may regard barnacles as an unnecessarily abrasive nuisance, but to the biologist interested in ecology and evolution they are among the most intriguing of all invertebrates.

The subclass Cirripedia encompasses four orders: the Ascothoracica, Acrothoracica, Thoracica, and Rhizocephala. Representatives of each order occur in waters off California. The Thoracica, or true barnacles, are ecologically more important by virtue of their abundance and conspicuousness. Therefore only a few comments on the other orders will be made here, a fuller understanding being obtainable from the sources given in the bibliography.

Species of Thoracica occur in virtually all marine environments, from the highest reaches of the tides to the depths of the oceans. Some occur in estuaries, but none completes its life cycle in fresh water. Diversity is greatest in the tropical Indo-Pacific, less in the northeast Pacific, and far less in the North Atlantic (Cornwall, 1951, 1955; Henry, 1940, 1942; Ross, 1962; Zullo, 1966). Size is generally moderate, a few centimeters in greatest dimension, but some barnacles exceed 10 cm and others are but a few millimeters. All are permanently attached as adults. Most retain their calcareous shells throughout life and strain food (small plankton and edible detritus particles) from the water with biramous thoracic limbs called cirri. In some, the cirri are modified for grasping and rasping. Many of the Thoracica form intricate symbiotic associations with larger organisms such as whales, sea snakes, lobsters, medusae, corals, and sponges. A few have become nutritionally dependent on sharks, worms, or corals, which in earlier stages of their evolution they exploited simply for support or protection (Krüger, 1940; Newman,

William A. Newman is Professor of Biological Oceanography at the Scripps Institution of Oceanography of the University of California, San Diego. Donald P. Abbott is Professor, Department of Biological Sciences and Hopkins Marine Station, Stanford University.

Zullo & Withers, 1969). Although most barnacles living symbiotically with other organisms do not occur in the intertidal zone, several species from California's offshore and subtidal waters have been included in this chapter because of their very great biological interest. Some of these forms are periodically washed ashore, and others may be brought in by divers.

The order Thoracica is divided into three suborders: the Lepadomorpha (stalked barnacles), the Verrucomorpha (asymmetrical sessile barnacles), and the Balanomorpha (symmetrical sessile barnacles). In the stalked barnacles the body is divided into a capitulum containing feeding appendages (cirri and mouthparts) and most other organs, and a peduncle or stalk by which the animal is attached to the substratum. The capitulum and sometimes the peduncle are armored with characteristic calcareous plates. The sessile barnacles differ from stalked barnacles mainly in having lost the peduncle, bringing what was the capitulum into direct contact with the substratum. In the process, certain of the capitular plates became articulated to form a rigid wall; others (anteriorly the scuta, posteriorly the terga) formed the movable lid or operculum. This arrangement is generally considered better adapted to shore environments, but radiation of sessile barnacles into a wide array of habitats indicates other advantages. Most Verrucomorpha are found in deep water, and one yet undescribed species is known from the west coast of North America. The Balanomorpha, commonly called acorn barnacles, constitute the majority of species found along the shore.

In general, the true barnacles are classified by the number, arrangement, and specialization of the plates and the details of the appendages (Darwin, 1851, 1854; Krüger, 1940; Newman & Ross, 1976; Newman, Zullo & Withers, 1969; Pilsbry, 1907, 1916). Most common forms are cross-fertilizing hermaphrodites, each barnacle reaching out and depositing sperm in the mantle cavity of another barnacle nearby and in turn, but not simultaneously, receiving sperm. However, there are species in which the hermaphrodites are accompanied by dwarf, complemental males and still others in which the sexes are separate, the female being accompanied by one or more dwarf males (Darwin, 1851, 1854).

Individual barnacles brood their fertilized eggs in the mantle cavity within the shell. Nauplius larvae (nauplii), bearing three pairs of appendages, hatch out, swim, feed, grow, undergo a series of molts, and then metamorphose into cyprid larvae. The cyprid has a bi-valved shell and a body organization somewhat like that of the adult. It cannot feed, but swims using its six pairs of thoracic legs. Periodically it settles to the bottom and crawls about on its first antennae, testing the substratum for suitable places to attach. Settlement in a spot favorable for the rest of the life history is important for sessile animals, and the cyprid's "choice" of settlement site is influenced by a variety of physical and chemical stimuli (reviewed by Lewis, 1978). The cyprid larvae of most species are especially attracted by the presence of attached adults of their own kind. When ready to settle, the cyprids attach their heads to the substratum by their first antennae (which contain cement glands), and then undergo a second metamorphosis to become juvenile barnacles, essentially miniatures of the adults. Adults of some species live only a matter of weeks or a few months; in other species they may persist for several years to a decade or more.

The remaining three orders may be treated more briefly. The Ascothoracica contains forms ranging from semi-predaceous carnivores to obligate parasites on various corals and echinoderms (Newman, 1974; Wagin, 1946). It is an obscure group, but important in understanding the characteristics that distinguish cirripeds from other crustaceans as well as the generalized form from which the extraordinary radiation and diversity of the subclass has evolved. Ascothoracicans differ from all other adult cirripeds in having a bivalved shell, in using their thoracic limbs for swimming rather than feeding, and in being unable to attach permanently by their first antennae. A few rather specialized, wholly parasitic species are known from deeper waters off California (Fisher, 1911). None is included in this chapter.

The Acrothoracica are generally very small. They most commonly burrow in calcareous substrata and show their greatest diversity in coral seas (Tomlinson, 1969). These burrowing barnacles differ from the true barnacles (Thoracica) in having limbs situated at the end of the thorax rather than evenly distributed along it. Otherwise their biologies are similar. All species have separate sexes, the female being accompanied by a dwarf non-feeding male. One highly specialized species, *Trypetesa lateralis* (20.1), is known from California.

The Rhizocephala are highly modified parasites, primarily on decapod crustaceans. The adults are unrecognizable as crustaceans, much less cirripeds, but life histories include the planktonic nauplius and/or cyprid larval stages characteristic of the subclass

(Boschma, 1953; Smith, 1906; Yanagimachi, 1961). In general, a female cyprid attaches and injects its cellular content into the host, where it develops into a complex of tubular rhizoids (the interna). After the interna has become established, a reproductive body develops and appears as the externa, usually on or under the abdomen of the host after a molt. A male cyprid then attaches and injects its undifferentiated content into the virgin externa. This eventually differentiates into the male gametes that fertilize the next generation of eggs. The cycle starts over with the liberation of larvae. The presence of the parasite generally castrates the host and in males feminizes the form of certain appendages and the abdomen. Several rhizocephalans are known from California; accounts of three species (20.34–36) are included in this chapter.

Barnacles have been the subjects of a broad range of scientific research. The true barnacles have long caused problems as fouling organisms on ships and docks, where they add weight and resistance and can cause deterioration of submerged surfaces. Much study has therefore been devoted to understanding their habits and seeking methods to control their settling and growth. On the other hand, barnacles have their direct uses to man, as well. In Japan certain species are cultivated for fertilizer, and in countries like Chile, Spain, and Portugal, some species are prized as food. The ability of barnacle larvae to attach to wet surfaces has recently led to investigations of "barnacle cement" in dental research (anon., 1968; Manly, 1970; Walker, 1972).

Many barnacle species are used in basic biological research. Their reproductive relationships are of interest in studies of sexual mechanisms and determination (Gomez, 1975; Henry & McLaughlin, 1967; Landau, 1976; McLaughlin & Henry, 1972; Tomlinson, 1966; Yanagimachi, 1961). Organic acids have been found to stimulate mating behavior (Collier, Ray & Wilson, 1956), and proteinaceous substances in the shells of attached individuals encourage cyprid larvae to settle (Crisp & Meadows, 1963). Insect juvenile hormone and juvenile hormone mimics induce abnormal metamorphosis (Gomez et al., 1973; Ramenofsky, Faulkner & Ireland, 1974). Many species have very simple photoreceptors (Gwilliam, 1965); others have the largest single muscle fibers known (Hoyle & Smyth, 1963), and these are useful in physiological research (Hagiwara & Takahashi, 1967; Krebs & Schaten, 1976). Barnacles often figure prominently in ecological, biogeographical, and evolutionary studies, and in his work on cirripeds, before professing the theory of natural selection, Darwin (1851, 1854) formulated such principles as neoteny and specialization through simplification.

In terms of taxonomy and geographical distribution the barnacles are among the best-known marine invertebrates of the west coast. They thus contribute readily to the delineation of biogeographical provinces along the shore, providing a model probably applicable to most other benthic invertebrate groups as well. The figure on the facing page summarizes the distribution of those shore

---

Latitudinal ranges of benthic cirripeds of the Oregonian and Californian Provinces (from Newman, 1979). The 23 species shown are divided into four groups; see text.

Group I
1. *Semibalanus cariosus* (20.23)
2. *Chthamalus dalli* (see under 20.12)
3. *Balanus glandula* (20.24)
4. *Pollicipes polymerus* (20.11)
5. *Chthamalus fissus* (20.12)
6. *Tetraclita rubescens* (20.18)

Group II
1. *Solidobalanus hesperius* (20.20)
2. *Balanus crenatus* (20.25)
3. *Balanus nubilus* (20.32)
4. *Balanus aquila* (20.31)
5. *Arcoscalpellum californicum* (20.10)
6. *Megabalanus californicus* (20.33)
7. *Balanus pacificus* (20.27)
8. *Balanus trigonus* (20.26)
9. *Balanus regalis* (20.30)

Group III
1. *Balanus improvisus* (20.29)
2. *Balanus amphitrite amphitrite* (20.28)

Group IV
1. *Trypetesa lateralis* (20.1)
2. *Armatobalanus nefrens* (20.19)
3. *Conopea galeata* (20.22)
4. *Octolasmis californiana* (20.3)
5. *Membranobalanus orcutti* (20.21)
6. *Oxynaspis rossi* (20.2)

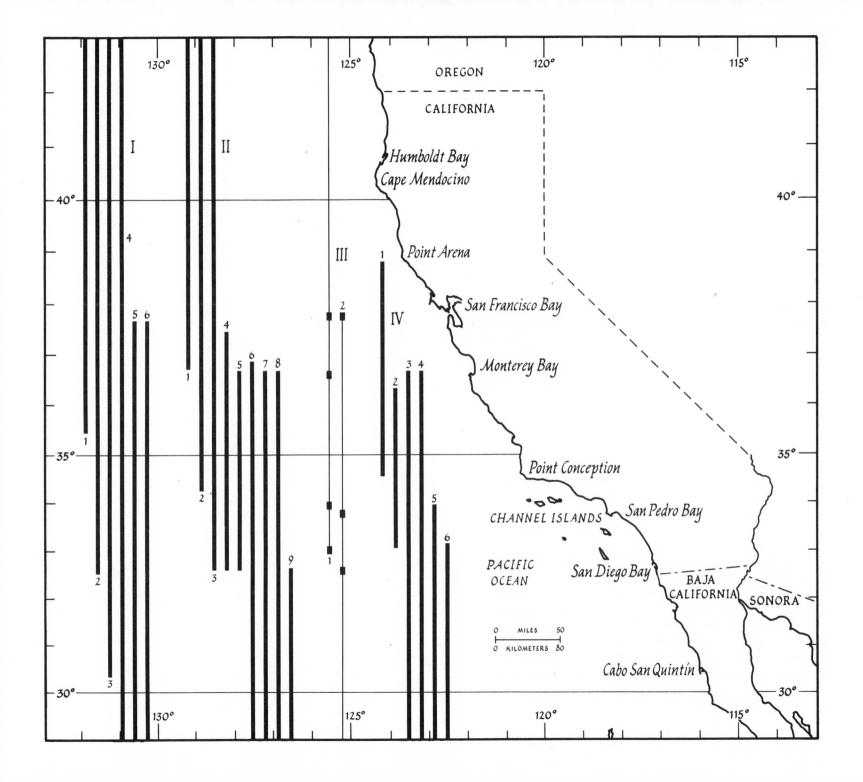

barnacles treated in this chapter that are attached to fixed substrata such as rocks and pilings, or to benthic organisms that move about little if at all (forms attached to such moving hosts as whales, turtles, or ships are omitted).

It is clear from the figure that most of the species encountered near Cape Mendocino in northern California are different from those found in the vicinity of San Diego, and they are usually designated as members of the Oregonian and Californian Provinces, respectively (see, for example, Newell, 1948; Valentine, 1966). The species of each province extend well to the north or south of the political boundaries of California. The region of overlap, the Californian Transition Zone, centering on Point Conception, contains a mixture of species that totals more than the number of species found immediately to the north or south. In addition, the Transition Zone contains a number of short-range endemic species.

There is a marked climatic difference between Oregonian and Californian Provinces, and the change in the Transition Zone is rather abrupt. Differences in ocean temperature are believed to play an important role in sustaining the position of biogeographical boundaries. This conclusion is borne out by the finding that high-latitude shallow-water species occur in deeper (therefore colder) water at the southern ends of their ranges, and by the discovery of cold-water species well to the south of their normal distributional ranges in isolated areas where upwelling of cold deep-ocean water occurs. The importance of temperature in sustaining the boundary is also shown by laboratory studies of thermal tolerances, the temperature optima for enzyme action, and the role of temperature regimes in regulating reproductive cycles. Moreover, studies of paleoecology (e.g., Addicott, 1966) demonstrate that the boundary shifted its latitudinal position with rapid climatic changes during the Pleistocene.

The benthic cirripeds shown in the figure have been divided into four ecological groups: (I) intertidal species (those species native to the coast that are largely restricted to the intertidal zone), (II) shallow-water species (species that, though sometimes found in the low intertidal zone, occur mainly at subtidal depths on the continental shelf), (III) introduced bay and estuarine species (species occurring in both intertidal and shallow subtidal waters), and (IV) obligate commensal species (forms living on or in other living organisms that are themselves sessile or relatively sedentary).

Group I contains the only highly eurytopic form, *Pollicipes poly-*

merus, which ranges through both provinces. All other species in Groups I–IV terminate one end of their latitudinal range in, or reside wholly in, the Transition Zone.

All Group I species are represented by aggregations of individuals that are community dominants. *Chthamalus* generally ranges higher into the intertidal zone than *Balanus* and *Tetraclita*. The northern forms, *Chthamalus dalli* and *Semibalanus cariosus*, are largely replaced in the south by *C. fissus* and *Tetraclita rubescens*, respectively. *Balanus glandula*, occurring for the most part between the northern and southern species in vertical distribution, ranges farther south than *S. cariosus* and is eventually replaced in part by *B. regalis*. As adults, in various species combinations, these species compete for space and food along both horizontal and vertical gradients.

Group II, from continental-shelf waters, follows a similar pattern, the northern *Solidobalanus hesperius*, *B. crenatus*, and *B. nubilus* being replaced by *Megabalanus californicus*, *B. pacificus*, and *B. trigonus* in the south. All these species, with the possible exception of *B. nubilus* and *B. crenatus* in the northern parts of their ranges, are relatively opportunistic. That is, they neither dominate, nor become longstanding members of, the subtidal community.

Two of the four short-range endemic species of the Transition Zone, *B. aquila* and *Arcoscalpellum californicum*, are encountered in this group. The former occurs with *B. nubilus* in the northern part of its range, but both undergo submergence south of Point Conception, where they are replaced by a close relative of *B. aquila*, *B. regalis*. Being at the ends of their respective ranges, apparently neither *B. nubilus* nor *B. regalis* is capable of fully exploiting the available resources, and this has left space for *B. aquila*. The situation for *Arcoscalpellum* appears quite different. The genus occurs primarily in deep water, but *A. californicum* ranges up into shallower water (less than 30 m) in the Transition Zone.

Group III contains two bay and estuarine species. Both have been introduced to California, probably by ships or along with transplanted oysters. *Balanus improvisus*, from the North Atlantic, ranges into the Transition Zone from the north. *B. amphitrite amphitrite*, probably from the Indo-West Pacific, does likewise but from the south.

Members of Group IV, being obligate commensals, do not interact with other barnacle species, at least in the adult stage. They receive varying degrees of protection by their hosts, and are usually limited by host distribution. All are warm-temperate/subtropical derivatives that have cognates in other regions. All are the sole representatives of their groups in the eastern Pacific. Two of the species, *Armatobalanus nefrens* and *Trypetesa lateralis*, are short-range endemics. *A. nefrens* occurs on hydrocorals, sometimes *Errinopora pourtalesii* but more commonly *Allopora californica*, which is itself a short-range endemic. *Trypetesa*, an acrothoracican, burrows in the interior of gastropod shells inhabited by hermit crabs. Although hermit crabs range widely in temperate and tropical regions, and into deep water, *T. lateralis* is found only in the northern part of the Transition Zone.

Thus it can be seen that the Californian Transition Zone is rich in barnacle species, owing to the overlap of the Oregonian and Californian Provinces and to the presence of the short-range endemics. This is the case with the mollusks and undoubtedly many other invertebrates. The disharmony between overlapping provincial communities caused by the steep latitudinal temperature gradient in the vicinity of Point Conception apparently results in incomplete utilization of the resources upon which the short-range endemics depend (Newman, 1979). Identifying these resources and how they are shared by so many species offers a whole area of interesting research.

Phylum Arthropoda / Class Crustacea / Subclass Cirripedia
Order Acrothoracica / Suborder Apygophora / Family Trypetesidae

20.1 **Trypetesa lateralis** Tomlinson, 1953

Common in some localities, burrowed into shells of gastropods (particularly *Tegula* species) occupied by hermit crabs, middle and low intertidal zones; Point Arena (Mendocino Co.) to Point Conception (Santa Barbara Co.); the only member of the order known from north of Mexico on the Pacific coast.

Females to 5 mm long, distinguished from all other California cirripeds by habitat, and in having but three pairs of uniramous cirri arranged on posterior end of thorax; males microscopic.

The females of this remarkable species form slitlike burrows parallel to the shell surface in shells of gastropods *Tegula brunnea* (13.31) and *T. funebralis* (13.32), and occasionally *Cal-*

liostoma costatum and *Acanthina spirata* (13.81). They are best seen by breaking the tops (apices) off shells occupied by hermit crabs (*Pagurus samuelis*, *P. hirsutiusculus*, *P. hemphilli*, and *P. granosimanus*, 24.10–24.13) and removing the crabs. If the shells are then illuminated by a bright spot of light from below (aperture side), and the columella and adjacent areas on the "floor" of the whorls are viewed from above (apical side), the burrows appear as translucent areas in the wall of the shell, often colored yellow, orange, or red.

The body of the female barnacle is partly encased in a muscular mantle provided externally with chitinous teeth that are renewed at each molt. Shell boring involves the rasping action of the teeth and probably chemical action as well; in a related species, *T. nassaroides*, the enzyme carbonic anhydrase reaches high levels in the body during periods of boring activity. The cirri of the female are greatly reduced in *T. lateralis*, but appear to be involved in feeding as in the more familiar acorn barnacles. They are not constantly active, but when in motion they beat regularly, about once every 2 seconds at 17°C, accompanied by regular pulsations of the mantle. These actions create a flow of water in and out of the mantle cavity and past the mouth, a circulation probably involved in both feeding and respiration. The food is uncertain, but probably consists of minute organic particles dropped by the hermit crab in its feeding, or brought in with currents created by movements of the crab within its shell. The gut of the female barnacle is remarkable in lacking both intestine and anus; the muscular esophagus opens into a stomach that is a large, lobulated, blind sac. Experiments show that elimination of undigested particles is probably by regurgitation.

On reaching adult size, females form an external mantle flap, a tongue of tissue provided with rasping teeth, that bores a hole through to the outside of the gastropod shell, leaving the burrow open to the sea on both inner and outer surfaces of the host shell. Once the new hole is completed, water is pumped in a jerky flow from outside to inside the shell, the flow reversing only if the outer hole becomes clogged. The flow provides males and embryos with fresh seawater.

The males are tiny, non-feeding dwarfs, averaging about 0.4 mm in greatest length. They are smaller in size than even the cyprid larvae from which they develop, and lack both gut and feeding appendages. Their most prominent organs are a single testis and seminal vesicle, a nauplius eye, and paired first antennae. In some species of *Trypetesa* the mature males develop a penis much longer than the body, but this is lacking in *T. lateralis*. The sperm are threadlike, and become motile after release into the seawater filling the mantle cavity. They need move only a millimeter or two to reach the eggs. One to several males may be present with a given female, either partly embedded in the mantle of the female or attached by their first antennae to the wall of the cavity near the female's external mantle flap.

The females discharge their eggs into the mantle cavity, where they are brooded. The eggs are oval, averaging about 0.25 mm in greatest length; at least 180 have been observed in a single brood. They develop past the nauplius stage within the egg membranes, hatching as non-feeding, weak-swimming cyprid larvae about 0.5 mm long. The lack of free nauplius stages, together with a tendency of hermit crabs to remain within a localized area, may account for the rather patchy distribution of *T. lateralis* along the shore in areas where it is known to occur. The greatest settlement of the larvae is in the winter (November through January), with a smaller peak in June. The life span, which is not known, may depend on the durability of the shell; the animals have been kept in seawater in a cold box for more than a year without noticeable growth.

The genus *Trypetesa* is a small one, including two known species from the North Atlantic, one from California, one from Japan, and one from Madagascar. The species are generally endemic to a limited geographical area that marks the transition between major biogeographical provinces. It is curious that apparently suitable gastropod shells, inhabited by hermit crabs and ranging from north to south and into deep water, are not inhabited by these barnacles.

For *T. lateralis*, see Tomlinson (1953, 1955, 1960, 1969a,b). For related species, see Darwin (1854), Tomlinson (1969a,b), Turquier (1971), Utinomi (1964), and White (1970).

Phylum Arthropoda / Class Crustacea / Subclass Cirripedia / Order Thoracica Suborder Lepadomorpha / Family Oxynaspididae

20.2 **Oxynaspis rossi** Newman, 1972

On black coral in deep water; recorded from 183 m depth off Santa Catalina Island (Channel Islands) and south and from 55 m depth off Baja California (22°55′N).

Small, to 2 cm in length; distantly related to *Lepas* (e.g., 20.4–6) and *Octolasmis* (e.g., 20.3), but differing from both in having the growth center (umbo) of the carina subcentral rather than basal.

With the exception of one species that lives on a sponge, all members of this genus live on black or horny corals (Cnidaria: Anthozoa: Antipatharia). The group to which *Oxynaspis* belongs is an ancient one, considered by some workers to be transitional between lepadid and scalpellid barnacles. Because of their deep-water habitat, nothing is known of their biology.

See Newman (1972).

Phylum Arthropoda / Class Crustacea / Subclass Cirripedia / Order Thoracica Suborder Lepadomorpha / Family Poecilasmatidae

20.3 **Octolasmis californiana** Newman, 1960

In gill chambers of large decapod crustaceans such as *Panulirus*, *Loxorhynchus grandis* (25.10), *Pugettia producta* (25.4), and *Cancer* spp. (e.g., *C. antennarius*, 25.16, and *C. productus*, 25.22), low intertidal zone and continental shelf; Monterey Co. to Panama.

Small, to 1.5 cm in length; similar to lepadids in having five capitular plates, these reduced in extent, the scutum with two slender arms; differing from lepadids in having uniarticulate caudal appendages.

No other representatives of the diverse tropical family to which this species belongs are known from the northeastern Pacific. *Octolasmis* feeds on plankton carried in the respiratory current of the host. The barnacles are shed along with the exoskeleton of the host when the host molts. Nothing else is known about the biology of the animal.

A closely related east coast species occurs on the gills of blue crabs. The particular portion of the gills that becomes infected is related to the distribution and effectiveness of the host's gill-cleaning mechanisms. The reason *Octolasmis* does not settle on caridean and anomuran crustaceans is probably that these animals have more efficient methods of gill cleaning.

See Newman (1960). For related species, see Bowers (1968) and Walker (1974, on east coast species).

Phylum Arthropoda / Class Crustacea / Subclass Cirripedia / Order Thoracica Suborder Lepadomorpha / Family Lepadidae

20.4 Lepas (Lepas) anatifera Linnaeus, 1758
Goose Barnacle

Common, oceanic, gregarious; numerous individuals attaching in clusters to such floating objects as logs, bottles, glass fishing floats, and shells of the pelagic snail *Janthina*; not usually seen washed ashore south of Point Conception (Santa Barbara Co.); cosmopolitan.

Length to about 15 cm; species distinguished from the smaller *L. (L.) pacifica* (20.5) in having a thicker, opaque scutum and two filament-like appendages at the base of each first cirrus; species determination generally requires dissection.

All lepadids, or goose barnacles, differ from scalpellids in having no more than five capitular plates and a naked peduncle. *Lepas (Lepas)* is distinguished from the subgenus *Dosima* by its heavier valves and in not producing a float of its own.

The many fine setae on the cirri of species in the subgenus *Lepas* permit the capture of tiny organisms and particles. Relatively large planktonic organisms can also be captured and eaten, and material too bulky to be completely devoured by one individual may be relinquished to another. The diet includes young fishes, and large pelagic hydrozoan cnidarians such as *Physalia*, *Porpita*, and *Velella* (3.18). *Lepas* attached to the pelagic gastropod *Janthina* and dependent on it for flotation reach further and hold prey more tightly than *Lepas* attached to other substrata. Such *Lepas* benefit their hosts by

passing on to them surplus or partially consumed prey; thus the association with *Janthina* appears to be mutualistic.

Lepas (Lepas) anatifera individuals have been observed to settle gregariously on the less brightly illuminated portions of floating survey beacons, and they prefer the rougher surfaces. When the barnacles are closely crowded, growth of the capitulum appears to be retarded, whereas that of the peduncle is accelerated. Ovigerous lamellae commonly appear in individuals about 23 mm in capitular length, several weeks after settling. Nauplius larvae appear a week or so thereafter, from individuals with 27 mm capitula.

Adults survive and breed in the laboratory if fed an appropriate diet and maintained in running seawater. After copulation, the individual acting as the female sheds ova into the mantle cavity by contraction of the peduncle. Ovigerous lamellae are attached to ovigerous frenae (membranous extensions of the interior of the mantle) whereas in *Pollicipes* (e.g., 20.11) they are free within the mantle cavity. Embryos develop normally at 19–25°C, and gravid females refrain from molting until during or shortly after the release of nauplii. Molting rate increases with temperature between 10 and 25°C, but falls with further rise in temperature.

See Barnes & Klepal (1971), Bieri (1966), Bigelow (1902), Boëtius (1952), Burnett (1975), Crisp & Southward (1961), Gwilliam (1963), Howard & Scott (1959), E. Jones (1968), Lacombe & Liguori (1969), Mahmoud (1959), Patel (1959), Pilsbry (1907), and Skerman (1958a,b).

Phylum Arthropoda / Class Crustacea / Subclass Cirripedia / Order Thoracica Suborder Lepadomorpha / Family Lepadidae

20.5 Lepas (Lepas) pacifica Henry, 1940

Oceanic, in clusters on floating objects such as *Lepas (Dosima) fascicularis* (20.6), *Velella velella* (3.18), seaweeds, wood, feathers, cardboard; Alaska to San Diego and south.

Length to about 2.5 cm; similar to *L. (L.) anatifera* (20.4) but distinguished from other species of the subgenus *Lepas* in California by small size, thin shell with underlying bluish tissues visible through scutum, and only a single short filamentary appendage at base of each first cirrus.

This species, along with *L. (D.) fascicularis*, is commonly stranded by seasonal onshore winds in southern California. This has provided an opportunity to observe large numbers of the two species and to analyze their settling preferences. In general *L. pacifica* is more often found attached to tar and less often to floating feathers, eelgrass blades, and bits of intertidal brown algae. Occasionally it is found growing on the back of the northern elephant seal, *Mirounga angustirostris*. Curiously, the two barnacle species are found attached in about equal numbers to debris of terrestrial origin.

The circulatory system of *L. (L.) pacifica*, like that of the other lepadids (*L. (D.) fascicularis*, *L. (L.) anatifera* (20.4), and *Conchoderma virgatum*, 20.7), is relatively simple and primitive compared to that of more advanced barnacles, such as *Pollicipes* (e.g., 20.11) and *Megabalanus* (e.g., 20.33). The rostral vessel has been interpreted as a vestige of the heart, and the rostral sinus equivalent to the pericardial sinus of ordinary crustaceans. Curiously, the smaller lepadid species, *L. (L.) pacifica* and *L. (D.) fascicularis*, have larger blood vessels in proportion to body size than do the larger species *L. (L.) anatifera* and *Conchoderma virgatum*. The ladderlike nervous system of *Lepas* is also relatively generalized. In higher forms ganglia coalesce into a large neural mass.

Experiments have shown that *Lepas* can live, grow, and reproduce in certain situations along the shore, but only if protected from predation. Apparently this ancient stock has found its last refuge in the open sea.

Little is known of the natural history of this species.

See Baldridge (1977), Burnett (1975), Cheng & Lewin (1976), Cornwall (1953), and Henry (1940).

Phylum Arthropoda / Class Crustacea / Subclass Cirripedia / Order Thoracica Suborder Lepadomorpha / Family Lepadidae

20.6 Lepas (Dosima) fascicularis Ellis & Solander, 1786

Common, oceanic; cyprid larvae settling on small objects, commonly feathers; developing juveniles and adults forming gas-filled floats of their own; cosmopolitan.

To 5–6 cm in length; distinguished from species of *Lepas*

(*Lepas*) in having thin, papery calcareous plates, in not being as markedly laterally flattened, and in secreting a float from the base of the peduncle rather than depending on floating objects as adults.

Although adults of the subgenera *Lepas* and *Dosima* use different methods for staying afloat, their sizes are generally comparable, and one might expect them to share many of the same resources. However, even though members of both subgenera can handle larger prey, in *Lepas* (*Lepas*) species the cirri are equipped for straining out tiny particles, whereas those in *L. (Dosima)* species are shorter and stouter and are equipped with spinelike setae unsuited for fine filtering. The stomachs of *L. (Dosima)* species are commonly packed with copepods, whose pigment is the source of the carotenoprotein providing the barnacle's blue color.

The abundance of *Lepas* (*Lepas*) is in good part limited by the availability of floating objects (feathers, wood, seaweed, pumice, coconuts, tar, etc.). *Lepas* (*Dosima*) has partially circumvented this limitation. Initially it attaches to a small floating object and then, as the barnacle grows further, it forms a gas-filled float of its own. If the cyprid formed a float during metamorphosis, the species could become completely independent of floating objects. Although this evolutionary step has not been made, it would appear to be an easy one, because the cement by which the cyprid attaches itself to an object, and the substance forming the float, are believed to come from the same glands.

See Ankel (1962), Bainbridge & Roskell (1966), Ball (1944), Barnes & Klepal (1971), Boëtius (1952), Cheng & Lewin (1976), Darwin (1851), Fox & Crozier (1967), Fox, Smith & Wolfson (1967), Knudsen (1963), and Thorner & Ankel (1966).

Phylum Arthropoda / Class Crustacea / Subclass Cirripedia / Order Thoracica Suborder Lepadomorpha / Family Lepadidae

20.7 Conchoderma virgatum (Spengler, 1790)

Oceanic, on a wide variety of objects including ships, buoys, telegraph cables, seaweeds, decapod crustaceans, fishes, marine reptiles, and whales; cosmopolitan.

To about 5 cm in length, with five capitular plates much reduced in extent; capitulum and peduncle leathery, usually with several purplish-brown, longitudinal stripes; species differing from *C. auritum* (20.8) in having five rather than two plates, and in lacking fleshy perforate lobes at top of capitulum.

This is an opportunistic species that settles on any hard substratum away from the shore. It attaches to the parasitic copepod *Pennella* that infects flying fish, and to the hard surfaces of swimming crabs and turtles. The occurrence of *C. virgatum* on fishes is unusual, but it has been reported on species of *Mola, Gymnothorax, Tylosurus, Diodon,* and *Remora,* the *Remora* having been removed from the blue shark *Prionace glauca.* A close relative of *Conchoderma virgatum, C. auritum,* occurs on whales, and a distant relative, *Anelasma,* occurs on certain sharks, but these barnacles are obligate commensals; the occurence of *C. virgatum* on fishes appears to be largely fortuitous.

Conchoderma virgatum grows rapidly. Specimens became mature in less than 17 days after settling on a buoy at approximately 30°S off Australia; the largest animals had capitular lengths approaching 20 mm and contained ovigerous lamellae. During Darwin's visit to the Galápagos Islands, *C. virgatum* seriously fouled the *Beagle* in 33 days. In another case, recorded here with some skepticism, *Conchoderma* individuals were reported to reach a total length of 13–19 mm on instruments immersed for 2.5 days at 50–150 m depth in cold waters off Greenland. On the same occasion, *Balanus amphitrite* was reported to have reached a diameter of 13 mm.

See Clarke (1966), Darwin (1851), E. Jones, Rothschild & Shomura (1968), MacIntyre (1966), and Roskell (1969).

Phylum Arthropoda / Class Crustacea / Subclass Cirripedia / Order Thoracica Suborder Lepadomorpha / Family Lepadidae

20.8 Conchoderma auritum (Linnaeus, 1758)
Rabbit-Eared Barnacle

Attached to the sessile barnacle *Coronula* (e.g., 20.15) on baleen whales, or to teeth of the sperm whale or other toothed cetaceans; cosmopolitan.

Length to 12 cm; species differing from *C. virgatum* (20.7) in

having two fleshy tubular structures projecting from top of capitulum, connecting mantle cavity with exterior, and in having two rather than five calcareous plates.

This species is wholly adapted to living on cetaceans, although it occasionally settles on other organisms and sometimes even on the hulls of submarines. Apparently unable to attach directly to whale skin, it requires a hard surface such as a barnacle or the whale's teeth, where these are continuously exposed due to lip or jaw damage. Both *Conchoderma auritum* and the sessile whale barnacle *Coronula diadema,* to which it most commonly attaches, orient themselves so the cirral net faces in the direction the whale swims, taking full advantage of the water flow. Water bearing food enters the anterior slit and strained water exits by way of the paired tubes, or "ears." As much as 98 percent of the humpback whale population carries some *Conchoderma.* The incidence is higher in female whales, presumably owing to their longer sojourns from high latitudes into warm waters.

Nothing is known of the biology of *C. auritum,* although individuals are apparently reproductively inactive when the hosts migrate into higher latitudes.

See Barnes & Klepal (1971), Clarke (1966), Cornwall (1924), Darwin (1851), MacIntyre (1966), and Petriconi (1969).

Phylum Arthropoda / Class Crustacea / Subclass Cirripedia / Order Thoracica Suborder Lepadomorpha / Family Lepadidae

20.9 Alepas pacifica Pilsbry, 1907

Oceanic, attached to the bells of large jellyfishes; occasionally washed inshore; Indian, Pacific, and South Atlantic Oceans.

To about 6 cm in length; species distinguished from other lepadids in having the number of plates reduced, from a normal complement of five, to a pair of Y-shaped scuta; capitulum globular; entire animal translucent.

Aside from its habit of attaching to jelly fishes, and its occurrence in the few widely distributed localities where it has been reported, nothing is known about this species. As in *Lepas (Dosima),* the cirri are short and spiny rather than long and setose, but how this characteristic relates to its food habits is not understood.

See Pilsbry (1907) and Utinomi (1958).

Phylum Arthropoda / Class Crustacea / Subclass Cirripedia / Order Thoracica
Suborder Lepadomorpha / Family Scalpellidae

20.10 Arcoscalpellum californicum (Pilsbry, 1907)
(=Scalpellum)

On rocks and other organisms at subtidal depths of 18–400 m; in shallow water ranging from Monterey (but usually south of Point Conception, Santa Barbara Co.) to San Diego; in deep water ranging from approximately Point Arena (Mendocino Co.) to Bahía Magdalena (Baja California).

Length to 4 cm; like *Pollicipes polymerus* (20.11) in having capitulum covered with more than five plates, but with a single whorl of basal plates and a peduncle armored by conspicuous scales.

This species appears to be synonymous with *A. osseum* (Pilsbry), which differs mainly in being more heavily calcified. Nothing is known of *A. californicum* other than its geographical distribution. It appears to be hermaphroditic. In related deep-water forms the hermaphrodite may be accompanied by complemental males; or the sexes may be separate, with large females accompanied by dwarf males.

See Cornwall (1951) and Pilsbry (1907).

Phylum Arthropoda / Class Crustacea / Subclass Cirripedia / Order Thoracica
Suborder Lepadomorpha / Family Scalpellidae

20.11 Pollicipes polymerus Sowerby, 1833
(=Mitella polymerus)   Leaf Barnacle

Common, usually in clusters but also mixed with *Mytilus californianus* (15.9), middle intertidal zone on wave-swept rocky shores; British Columbia south at least to Punta Abreojos (Baja California); replaced along western Mexico by *P. elegans*.

To 8 cm in length; capitulum covered with more than five plates and surrounded basally by several whorls of imbricate scales; peduncle tough, roughened by inconspicuous calcareous spicules.

This species feeds by directing its cirral net into currents, usually the backwash of waves. It captures mainly large particles of detritus, but crustaceans up to 1 cm long are also commonly ingested.

The barnacle's body temperature is frequently lower in air than would be expected from the heat load, and evaporation from the peduncle is responsible, a loss of 35–40 percent of the body water being tolerated during periods of exposure of less than 9 hours. The species is an osmoconformer: the salinity of the body fluids increases in sunshine and decreases in rain, and the urine is isosmotic with the blood. Blood pressure averages 250 cm of water, remarkably high for a crustacean. *Pollicipes* has a four-part circulatory system and a blood pump formed by three pairs of skeletal muscles, which are apparently unstriated.

Reproduction, at least in northern California and Washington, occurs in the summer. At Monterey, the seminal vesicles begin to enlarge with spring warming (above 12°C) and decrease in size in the fall, bracketing the occurrence of embryos. The fertilized eggs, expelled on either side of the thorax, adhere to one another and form a pair of flattened disks (ovigerous lamellae). The lamellae are brooded in the mantle cavity within the shell, one on each side. The time from fertilization to the release of swimming nauplius larvae is about 30 days. An individual may produce three to seven broods a year, with 100,000–240,000 larvae per brood, depending on age and size. Only early development is especially sensitive to temperature changes. Growth is slow in attached juvenile barnacles over 10 mm in length. It has been inferred that maturity is reached in 5 years and that fully grown individuals may be 20 years old.

This is a hermaphroditic species, but cross fertilization is thought to be the rule, since isolated individuals have not been found to contain embryos.

Although it is not known to foul ships, the leaf barnacle has been reported on a humpback whale, associated with *Conchoderma* (e.g., 20.8) attached to *Coronula* (e.g., 20.15), and it occasionally becomes established in large numbers in the laboratory seawater system at Scripps Institution of Oceanography at La Jolla (San Diego Co.). The closely related European species, *Pollicipes pollicipes*, is cooked and served as a delicacy in

Portugal and Spain, but it has long been in short supply. In recent years *P. polymerus* has been exported from British Columbia to these countries. Gibbons (1964) pronounces it moderately good eating, and provides recipes.

See Barnes (1959, 1960), Barnes & Barnes (1959a), Barnes & Gonor (1958a,b), Barnes & Klepal (1971), Barnes & Reese (1959, 1960), Burnett (1972), Cornwall (1936, 1951, 1953), Darwin (1851), Dayton (1971), Dudley (1973), H. Fyhn, Petersen & Johansen (1972, 1973), Gwilliam (1963), Hilgard (1960), Holter (1969), Howard & Scott (1959), Lewis (1975a,b, 1977), Lewis, Chia & Schroeder (1973), Paine (1974, 1979), Petersen, Fyhn & Johansen (1974), Rice (1930), and Seapy & Littler (1978).

Phylum Arthropoda / Class Crustacea / Subclass Cirripedia / Order Thoracica Suborder Balanomorpha / Superfamily Chthamaloidea / Family Chthamalidae

Chthamalus dalli Pilsbry, 1916

20.12 Chthamalus fissus Darwin, 1854
(=C. microtretus Cornwall)

Chthamalus dalli: common on rocks, pier pilings, and hard-shelled organisms, high and upper middle intertidal zones, Alaska to San Diego; northern Japan. *C. fissus*: common in similar habitats; San Francisco to Baja California.

Both species small, to 8 mm in diameter, externally similar; wall of shell with both end plates overlapped by adjacent plates (thus differing from balanids, in which one end plate, the rostrum, overlaps adjacent plates).

Reliable identification of these species usually requires removal of the scutal plates and examination of their inner surfaces under a good hand lens. The two scuta are triangular, with their apices meeting at the peak of the operculum. The lateral scutal depressor muscles attach near the basal angles of the scuta laterally. In *C. dalli* the muscle attachment scar bears strong ridges or crests, whereas in *C. fissus* it is a smooth-walled depression. Another character separating the species is the nature of the setae on the ends of the second cirri; under microscopic examination the setae of *C. dalli* appear finely bipectinate, but those of *C. fissus* are coarsely bipectinate and have a pair of basal pectinations enlarged as guards. Speci-

mens of *C. fissus* with a narrow, slitlike orifice were formerly called *C. microtretus*.

Chthamalus individuals can occupy higher intertidal situations than any other acorn barnacles, spending considerably more than half their lives out of water. Some species occur where they are only wetted by splash. On California shores, *Balanus glandula* (20.24) nearly equals *Chthamalus* in resistance to desiccation. Where both *Chthamalus* and *B. glandula* occur, the latter dominates, since a growing shell of *Balanus* pushes adjacent *Chthamalus* from the rocks. However, *Balanus* is preferred over *Chthamalus* by such predators as the gastropod *Nucella emarginata* (13.83) and the starfish *Pisaster ochraceus* (8.13), and this selective predation on *Balanus* apparently makes more space available for *Chthamalus*. Under optimum conditions, densities of 70,000 per m² are found.

Chthamalus is frequently parasitized by the epicaridean isopod *Cryptothir* (=*Hemioniscus*) *balani*; infected barnacles are infertile.

From field sampling, it has been inferred that both *C. dalli* and *C. fissus* that are separated by more than 5 cm from their nearest neighbor fertilize themselves, but cross-fertilization normally occurs. *C. fissus* in central California produces about 16 small broods from spring through fall, with reduced brooding activity in the winter. Brood production is limited by food availability, and depending on the size of the parent, 200–3,000 nauplius larvae are released per brood. Settlement occurs erratically year-round at all tide levels. Mortality in newly settled *C. fissus* on rock surfaces is apparently little affected by the grazing of limpets, owing to the small size of the cyprids and their ability to settle in tiny cracks and minute depressions. *C. fissus* individuals in central California reach reproductive maturity when 2–3 mm in diameter and about 2 months old. They may live to about 3 years, with better survival rate in the high than the low intertidal zone.

Adult *Chthamalus* feed by extending their cirri while splash or wave wash flows over them and retracting the cirri and closing the operculum when the water has flowed away. Although the activity of cirri has not been studied in local species, studies on other forms show it to be temperature-dependent, and its rate apparently differs among local races.

See Augenfeld (1967), Barnes & Barnes (1958), Barnes & Gonor (1958a), Barnes & Klepal (1971), Connell (1970), Cornwall (1953, 1955a), Dayton (1971), Henry (1960), Hines (1976, 1978, 1979), Klepal & Barnes (1975a,b), Newman & Ross (1976), Paine (1974), Pilsbry (1916), Rice (1930), Ross (1962), Seapy & Littler (1978), and Stallcup (1953). For related species, see Monterosso (1933) and Southward (1962, 1975).

Phylum Arthropoda / Class Crustacea / Subclass Cirripedia / Order Thoracica Suborder Balanomorpha / Superfamily Coronuloidea / Family Coronulidae

20.13 Chelonibia testudinaria (Linnaeus, 1758)
Turtle Barnacle

On shells of sea turtles; uncommon in southern California, common in all warm seas including eastern Pacific.

To 80 mm in diameter; wall smooth, low, dome-shaped, broadly attaching to turtle shell, supported internally by numerous septal buttresses.

There is a great diversity of barnacles on turtles. *Chelonibia* is the largest and least specialized genus. Superficially, the wall appears to be made up of six plates, but close inspection reveals that one, the rostrum, is formed of three fused together. The "tripartite" rostrum of *Chelonibia* is a vestige of the eight-plated ancestry of all balanomorphs.

See Darwin (1854), Newman & Ross (1976), Ross & Newman (1967), and Zullo (1966).

Phylum Arthropoda / Class Crustacea / Subclass Cirripedia / Order Thoracica Suborder Balanomorpha / Superfamily Coronuloidea / Family Coronulidae

20.14 Platylepas hexastylos (Fabricius, 1798)

Uncommon on turtles in southern California; commoner elsewhere attached to skin of turtles, of mammals such as the American manatee and the dugong, and of gar fish (*Lepidosteus*); cosmopolitan in warm seas.

To 18 mm in diameter; species superficially resembling *Chelonibia testudinaria* (20.13) but differing markedly in structure of wall, in which six internal buttresses extend down to the membranous basis.

Even though this species is common on turtles, nothing is known about its biology.

See Darwin (1854), Newman & Ross (1976), Pilsbry (1916), and Utinomi (1970).

Phylum Arthropoda / Class Crustacea / Subclass Cirripedia / Order Thoracica Suborder Balanomorpha / Superfamily Coronuloidea / Family Coronulidae

20.15 Coronula diadema (Linnaeus, 1767)

Found locally on the humpback whale (*Megaptera novaeangliae*), especially on the fins, the lips, the long grooves of the throat region, and the lips of the genital aperture; also reported from the fin, blue, and sperm whales; cosmopolitan.

To about 85 mm in diameter, white, dome-shaped, firmly anchored in skin of whale but not (like *Cryptolepas*, e.g., 20.16) completely embedded in it; bearing a single pair of opercular valves suspended in opercular membrane.

This species has undergone striking modification in connection with its life on whales. The outer wall of the barnacle shell bears large radiating ridges or buttresses, each of which develops, on its distal margin, flanges that meet and fuse with similar flanges on the buttresses on either side of it. The result is a roofing over of the buttresses, except on the side of the barnacle adjacent to the whale. The skin of the whale is pulled up between the concealed buttresses as the barnacle grows, firmly anchoring the shell to the whale.

The barnacles on the whale are generally oriented such that the rostrum, and the short extended cirri, face into the current as the whale swims along. This is surely an advantage in feeding. *C. diadema* are thought to live for only 1 year. Their shells frequently form an attachment surface for *Conchoderma auritum* (20.8) and, less often, *Conchoderma virgatum* (20.7). These pedunculate barnacles are borne on the posterior (carinal) edge of the *Coronula* shell, where they do not interfere with *Coronula*'s feeding.

See Cornwall (1955a,b), Crisp & Stubbings (1957), Darwin (1854), and Newman & Ross (1976).

Phylum Arthropoda / Class Crustacea / Subclass Cirripedia / Order Thoracica
Suborder Balanomorpha / Superfamily Coronuloidea / Family Coronulidae

20.16 **Cryptolepas rachianecti** Dall, 1872

On the fins and head of the California gray whale (*Eschrichtius robustus*); Bering Sea to Baja California; Korea; not Hawaii, as once reported.

Shell to 60 mm in diameter, white, depressed, deeply embedded in skin of whale, often with only the bright-yellow opercular membrane and a few ridges of shell visible.

As in *Coronula diadema* (20.15), the wall of the shell in this species develops great radiating ridges or buttresses, but they do not form flanges, so the spaces between buttresses remain open (not roofed over). The shell grows down into the whale skin, as in *Coronula*, but the outer shell surface erodes away until it is virtually flush with the surface of the whale. This is a valuable adaptation, since gray whales have been noted to rub against objects on the sea floor and near shore.

More than a century ago the naturalist and geologist W. H. Dall observed the barnacles while still alive and wrote, "This species is found sessile on the California Gray whale. . . . I have observed them on specimens of that species hauled up on the beach at Monterey for cutting off the blubber . . . and the animal removed from its native element—protruding its bright yellow hood in every direction, to a surprising distance, as if gasping for breath—presented a truly singular appearance."

The surface of the shell of *C. rachianecti* is friable and easily eroded, which may help explain why the pedunculate barnacle *Conchoderma* is not found attached to it. On the other hand, the caprellid amphipod *Cyamus scammoni* (often called the "whale louse") is commonly found clinging to *Cryptolepas* (see the photograph).

See Cornwall (1955a), Dall (1872), Kasuya & Rice (1970), Newman & Ross (1976), and Pilsbry (1916).

Phylum Arthropoda / Class Crustacea / Subclass Cirripedia / Order Thoracica
Suborder Balanomorpha / Superfamily Coronuloidea / Family Coronulidae

20.17 **Xenobalanus globicipitis** Steenstrup, 1851

On various whalebone whales including the sei whale and common finback, also on the blackfish or Pacific pilot whale, usually attached to the fin tips; cosmopolitan.

Length to about 75 mm; species superficially resembling a stalked barnacle, but the six plates of the balanoid wall visible, forming a star-shaped anchor, embedded in skin of host.

This remarkable species has, through convergent evolution, come to resemble a pedunculate barnacle. The elongate tube is not actually a peduncle, but represents the greatly drawn-out opercular membrane. Opercular plates are lacking, but distally the membrane forms a hood with reflexed lips from which the short cirri protrude. This elongate form apparently represents an adaptation to feeding in the turbulent waters behind the trailing edge of whale fins. Little else is known of the biology of this unique species.

See Cornwall (1955a), Darwin (1854), and Newman & Ross (1976).

Phylum Arthropoda / Class Crustacea / Subclass Cirripedia / Order Thoracica
Suborder Balanomorpha / Superfamily Coronuloidea / Family Tetraclitidae

20.18 **Tetraclita rubescens** Darwin, 1854
(=T. squamosa rubescens)

Common, middle and low intertidal zones on rocks exposed to strong surf; occasionally subtidal on hard-shelled organisms such as abalones; San Francisco Bay to Cabo San Lucas (Baja California).

Diameter usually to 30 mm, rarely to 50 mm; wall consisting of only four plates (only balanomorph on this coast in which this is the case); shell of adult reddish, appearing "thatched" externally; shell of young (uneroded) individuals white, superficially resembling shell of *Semibalanus cariosus* (20.23) but with four wall plates instead of six.

Tetraclita is remarkable in that it relies mostly on erosion at the top of the wall rather than diametric growth to enlarge the orifice of the shell. Most balanomorphs grow laterally between wall plates, as well as upward from the base. Most *Tetraclita* do not, and if there is little erosion, the orifice remains remarkably small.

In central California this species produces about three broods in the summer, releasing 1,000–50,000 nauplius larvae per brood, depending on the size of the parent. Individuals do not begin to reproduce until they are about 18 mm in diameter and 2 years old. Like *Semibalanus cariosus*, its northern ecological equivalent, *T. rubescens* is an effective competitor for space in the lower intertidal zone, and individuals grow to a size large enough to exempt them from predation by many gastropods and sea stars. Some may live as long as 10–15 years.

Tetraclita rubescens var. **elegans** Darwin, 1854 (Fig. 20.18b), has the same general distribution as the common form, although it is more prevalent in the lower intertidal and subtidal waters and in the southern half of the geographical range. Up to 20 mm in diameter, it is distinguished from the type variety by its white uneroded wall and peltate form. It is often found on shelled mollusks and crabs that spend much of their time subtidally, and also on wharf piles in Monterey harbor. It is an ecotype, rather than a genetically distinct population.

See Barnes (1959), Barnes & Klepal (1971), Cornwall (1951), Darwin (1854), Emerson (1956), Henry (1960), Hewatt (1946), Hines (1976, 1978, 1979), Newman & Ross (1976), Pilsbry (1916), Rasmussen (1935), Ross (1962), Seapy & Littler (1978), and Willett (1937).

Phylum Arthropoda / Class Crustacea / Subclass Cirripedia / Order Thoracica Suborder Balanomorpha / Superfamily Balanoidea / Family Archaeobalanidae

20.19 **Armatobalanus nefrens** (Zullo, 1963)
(=Balanus nefrens)

On and embedded in the hydrocorals *Allopora californica* and *Errinopora pourtalesii*, subtidal to 64 m; Monterey to Channel Islands.

Shell to 10 mm in diameter, white; only barnacle species in California known to infect hydrocorals.

Balanids have adapted in a variety of ways to living embedded in corals, and some forms found in the tropics are highly modified. However, apart from its specificity for hydrocorals, *A. nefrens* is hardly specialized at all, and one could envisage its living equally well on a rock. The methods it uses to get through the coral's defenses and establish itself, as well as other aspects of its biology, are unknown.

See Ross & Newman (1973) and Zullo (1963).

Phylum Arthropoda / Class Crustacea / Subclass Cirripedia / Order Thoracica Suborder Balanomorpha / Superfamily Balanoidea / Family Archaeobalanidae

20.20 **Solidobalanus hesperius** (Pilsbry, 1916)
(=Balanus hesperius)

On hard-shelled organisms usually from soft bottoms, subtidal from 18 to 64 m depth; Bering Sea to Monterey Bay.

To 20 mm in diameter; walls dirty white, usually ribbed; species similar in appearance to *Balanus crenatus* (20.25), *B. glandula* (20.24), and when smooth (southern end of range), to *B. improvisus* (20.29), but distinguishable from all three in having strongly elevated callus above articular ridge on internal surface of scutum; distinguished from all except mature *B. glandula* in having the wall solid rather than permeated by longitudinal canals.

This species is rarely seen alive unless one resorts to dredging or diving. However, it occurs on mollusks and crabs and is commonly found on the Dungeness crab, *Cancer magister* (25.20), in markets or when washed ashore.

See Barnes & Barnes (1959c), Cornwall (1955a), Newman (1975), and Newman & Ross (1976).

Phylum Arthropoda / Class Crustacea / Subclass Cirripedia / Order Thoracica Suborder Balanomorpha / Superfamily Balanoidea / Family Archaeobalanidae

20.21 **Membranobalanus orcutti** (Pilsbry, 1907)
(=Balanus orcutti)   Sponge Barnacle

In sponges, very low intertidal zone and subtidal waters to 40 m; Point Conception (Santa Barbara Co.) to Cabo San Lucas (Baja California).

To 15 mm in length, white, thin-walled, with membranous basis; rostral plate much elongated and recurved basally.

Balanids frequently become overgrown and smothered by sponges, but they are usually attached to the substratum before the sponge becomes established. *M. orcutti*, on the other hand, settles on established sponges. After metamorphosis the barnacle becomes embedded in sponge tissue with only the opercular aperture exposed. The anterior cirri are armed with special hooks that apparently aid in removing sponge tissue that would otherwise tend to clog the aperture.

Sponge barnacles of the genus *Membranobalanus* are found in all tropical seas of the world but virtually nothing has been published on their biology. However, an intensive study of the sponge-barnacle relationship has recently been made, and the following are some of the results (L. L. Jones, pers. comm.). The apical portions of the shell plates in *M. orcutti* have rows of upturned spines that apparently prevent the young barnacles from pushing themselves out of the sponge as growth at the basal margins of the shell extends those margins deeper into the sponge. The barnacle is found in only two sponge species in California: the loggerhead sponge *Spheciospongia confoederata* (2.21), in shallow water (low intertidal zone to 18 m), and the boring sponge *Cliona celata* var. *californiana* (2.20), in deeper waters (18–40 m). Cyprid larvae show preference for, and actively select, the host species of their parents. Laboratory experiments indicate that the attractiveness of the alternate host species can be increased by exposure of the developing barnacle eggs to the alternate host.

See L. L. Jones (1978), Newman & Ross (1976), and Pilsbry (1907).

Phylum Arthropoda / Class Crustacea / Subclass Cirripedia / Order Thoracica Suborder Balanomorpha / Superfamily Balanoidea / Family Archaeobalanidae

20.22 **Conopea galeata** (Linnaeus, 1771)
(=Balanus galeatus)

Embedded in sea fans (gorgonians), subtidal to 90 m; Monterey to Central America; Caribbean Sea.

To 15 mm in length; individuals aligning themselves with long axis of gorgonian branch, forming a boat-shaped basis attached to axial skeleton, and becoming overgrown by gorgonian tissue.

This species is remarkable not only in being an obligate commensal of gorgonians but in being one of the few balanomorphs in which the normal hermaphrodite is accompanied by small complemental males. Usually several of these males can be found attached within the hermaphrodite's orifice, mainly on the scuta. Males do not feed and apparently serve to ensure cross-fertilization, since the hermaphrodites are commonly widely separated on the gorgonian host. A larval cyprid sex ratio of one male to three hermaphrodites is evidently genetically determined. This suggests that the probability of male propagules finding a place to settle is much greater than that of hermaphroditic propagules. Hermaphrodite cyprids settle directly on the gorgonian axial skeleton, where it has been denuded of tissue by grazing gastropods or other predators. Male cyprids settle and metamorphose only on the hard parts of established hermaphrodites. However, mature cyprids can be induced to metamorphose in vitro without attaching to anything when treated with insect juvenile hormone and certain compounds that mimic them. Barnacle and insect hormones show many similarities.

See Bebbington & Morgan (1976), Cornwall (1951), Darwin (1854), Gomez (1973, 1975), Gomez et al. (1973), McLaughlin & Henry (1962), Molenock & Gomez (1972), Newman & Ross (1976), Pilsbry (1916), and Ross (1962).

Phylum Arthropoda / Class Crustacea / Subclass Cirripedia / Order Thoracica Suborder Balanomorpha / Superfamily Balanoidea / Family Archaeobalanidae

20.23 **Semibalanus cariosus** (Pallas, 1788)
(=Balanus cariosus)

Common in low intertidal zone on rocks along exposed shores; Bering Sea to Morro Bay (San Luis Obispo Co.); Japan.

To 60 mm in diameter; superficially resembling *Tetraclita rubescens* (20.18), especially when young, because of "thatched" texture of wall, but with six plates rather than four making up the wall; species distinguished from various California *Balanus* species by the white or gray, thatched wall permeated by numerous rows of longitudinal tubes, and by the membranous basis.

The thatched appearance of the walls derives from a series of basally directed, riblike buttresses that detach periodically as they are carried upward with growth. In central California individuals grow more or less separately, but in the Pacific northwest they form crowded colonies in which the thatched appearance is not as well developed.

The species is known to brood in winter, and larvae settle in the spring. Individuals grow large enough to be exempt from attack by many predators, and may live up to 10–15 years.

See Barnes (1959), Barnes & Klepal (1971), Batzli (1969), Connell (1970), Cornwall (1955a), Darwin (1854), Dayton (1971), Fahrenbach (1965), Gwilliam (1963, 1965), Gwilliam & Bradbury (1971), Millecchia & Gwilliam (1972), Newman & Ross (1976), Paine (1974), Pilsbry (1916), Rice (1930), Seapy & Littler (1978), Southward & Crisp (1965), Stickle (1973), Worley (1939), and Zullo (1969a).

Phylum Arthropoda / Class Crustacea / Subclass Cirripedia / Order Thoracica Suborder Balanomorpha / Superfamily Balanoidea / Family Balanidae

20.24 **Balanus glandula** Darwin, 1854

Abundant on rocks, pier pilings, and hard-shelled animals, high and middle intertidal zones in bays and along outer coast; Aleutian Islands (Alaska) to Bahía de San Quintín (Baja California); recently reported to be introduced at Puerto de Mar del Plata (Argentina).

Shell to 22 mm in diameter, white to gray, variable in shape and texture but comparable in size and general form to *B. crenatus* (20.25) and to some extent to *Solidobalanus hesperius* (20.20); usually distinguishable from both by the dark area seen externally on each scutum, where underlying pigmented tissues show through thin central portion of shell plate, and by the more heavily ribbed wall.

This is probably the most common intertidal balanid along our shores. It is the ecological counterpart of *Semibalanus balanoides*, which occurs in both the North Pacific and Atlantic. Although *Balanus glandula* is found in bays and the polyhaline portions of estuaries, the main population occurs on the open coast where, throughout its latitudinal range, it forms a band a third of a meter or more in width, in association with the red alga *Endocladia muricata*. At Monterey this band centers at about 1.4 m above mean lower low water, and individuals are submerged 27 percent of the time. Their longest periods of continuous exposure to air and of submergence are about 20.25 and 6.5 hours, respectively. The maximum period of exposure above water at the upper limit of the band is in the order of 24 hours, if the sea is calm and without swell. *B. glandula* is moderately resistant to desiccation compared to other species of *Balanus* so far studied, losing less than 50 percent of its total water in 24 hours. On the other hand, it apparently osmoregulates more strongly than *B. amphitrite amphitrite* (20.28) and *B. improvisus* (20.29), although the last is more tolerant of dilute salinities. Nearly a hundred species of multicellular organisms live in the *Balanus-Endocladia* association. The average number of individuals of the 31 commonest species in the band has been estimated at 210,000 per m², with a dry weight biomass of 2,640 gms (157 gms of protein), of which *Balanus glandula* and *Endocladia muricata* form the greatest proportion. It is not known whether it can deposit salts of heavy metals, such as zinc, as granules in tissues surrounding the midgut; such deposition occurs in *S. balanoides*, where the amount deposited is proportional to environmental levels.

Balanus glandula produces two to six broods during the winter and spring in British Columbia and in central and southern California. It stores yolk in the summer and remains ripe in the fall until cold temperatures induce brooding. Depending on the size of the parent, 1,000–30,000 nauplius larvae are produced per brood. Settlement occurs in all but the highest part of the intertidal zone in the spring and summer. Peak settlement usually occurs in the spring and may result in dense "sets" of *B. glandula* smothering and crowding other organisms, especially *Chthamalus* species (e.g., 20.12). However, the presence of the crustose red alga *Petrocelis middendorffii* apparently inhibits barnacle settlement. Growth rates are variable but tend to be more rapid in the lower intertidal zone, if crowding is not too severe. Basal diameters reach 7–12 mm in 1 year, 10–16 mm in 2 years, and 14–17 mm in 3, with a maximum diameter of about 22 mm. *B. glandula* that settled in the spring are reproductive their first winter and may live up to 8–10 years.

Balanus glandula survives best in the narrow band on the high shore. The upper intertidal limit of the band is set by the species' ability to withstand desiccation; the lower limit appears to be set by both competition and predation. *B. glandula* tends to be crowded out at the bottom of the band by the mussel *Mytilus californianus* (15.9), in more northerly areas by the barnacle *Semibalanus cariosus* (20.23), and in more southerly areas by *Tetraclita rubescens* (20.18). Also, in this zone at the bottom of the band, barnacles can be preyed on effectively by the gastropod *Nucella emarginata* (13.83), and the sea stars *Pisaster ochraceus* (8.13) and *Leptasterias hexactis* (8.11). All three predators prefer *Balanus* to *Chthamalus* as prey; thus their depredation helps to counterbalance the physical dominance of *Balanus* over *Chthamalus*. The effect of limpets feeding on, or bulldozing off, cyprids and juvenile barnacles is more pronounced on *Balanus* than on *Chthamalus*.

See Augenfeld (1967), Barnes & Barnes (1956), Barnes & Gonor (1958a), Barnes & Healy (1969), Barnes & Klepal (1971), Batzli (1969), Bergen (1968), Connell (1970), Cornwall (1955a), Darwin (1854), Dayton (1971), Glynn (1965), Hines (1976, 1978, 1979), Johnson & Miller (1935), Newman (1967), Newman & Ross (1976), Paine (1974), Paine, Slocum & Duggins (1979), Pilsbry (1916), Rice (1930), Seapy & Littler (1978), Spight (1973), Stephenson & Stephenson (1972), Stickle (1973), Walker et al. (1975), Worley (1939), and Zullo (1969a).

Phylum Arthropoda / Class Crustacea / Subclass Cirripedia / Order Thoracica Suborder Balanomorpha / Superfamily Balanoidea / Family Balanidae

20.25 **Balanus crenatus** Bruguière, 1789

Uncommon on rocks and, from Monterey Bay north, on deeply shaded pier pilings, low intertidal zone; common at subtidal depths to 182 m on various objects including seaweeds; northern Japan and Alaska south to Santa Barbara; North Atlantic.

Shell to 20 mm in diameter, white; species separable from *B. glandula* (20.24) and *Solidobalanus hesperius* (20.20), which it resembles, by (1) the smoother wall plates in California speci-

mens of *B. crenatus*, and (2) the presence of large, regular, longitudinal tubes permeating the wall plates of *B. crenatus* (best seen from below, or in broken plates).

Balanus crenatus is in some respects the subtidal equivalent of *B. glandula* in the North Pacific and of *Semibalanus balanoides* of the North Pacific and Atlantic. Size for size *B. crenatus* is more susceptible to water loss through desiccation than some intertidal rock barnacles, especially *Chthamalus* (e.g., 20.12). This and its greater fragility correlates with its intertidal occurrence in relatively quiet waters. The species is found in polyhaline reaches of estuaries on both coasts of North America and on the west coast of Europe. In New Brunswick (Canada), for example, adults are abundant in Miramichi Bay but rare in the adjoining gulf. Plankton sampling revealed that early nauplius larval stages remain high in the water column and therefore move seaward, but later larvae descend deeper and undergo a net movement landward; the cyprids, being the deepest, are carried far into the bay with the saline wedge. Larvae likely spend 2–3 weeks in the plankton, and before they settle they are transported to areas where conditions approach the barnacle's lower salinity tolerance limits.

In years of exceptionally great larval "set," or settlement, extreme crowding occurs, with the result that barnacles with tall columnar walls develop. The individuals and clumps formed have small areas of attachment proportional to their size and are easily broken loose; consequently, relatively few members of heavy sets survive a season. Experiments involving containers suspended from a barge have shown that light has no significant effect on growth rate, fertilization, or development of embryos.

Balanus crenatus is sparsely distributed in the subtidal polyhaline reaches of San Francisco Bay, but it becomes an important subtidal species, in terms of numerical abundance, in the northwest. In Monterey Bay, *B. crenatus* settles erratically during most months of the year. After settlement, the animals can grow to a basal diameter of 20 mm in 1 year.

See Addicott (1966), Austin, Crisp & Patil (1958), Barnes (1953a,b,c, 1959), Barnes & Healy (1969), Barnes & Klepal (1971), Barnes, Klepal & Munn (1971), Barnes & Powell (1950), Barnes, Read & Topinka (1970), Bousefield (1955), Cornwall (1955a), Crisp (1955), Crisp & Barnes (1954), Crisp & Southward (1961), Crisp &

Stubbings (1957), Darwin (1854), Foster (1969, 1971), Haderlie (1974), Herz (1933), Kauri (1962, 1966), Newman (1967), Newman & Ross (1976), Southward & Crisp (1965), and Zullo (1969a).

Phylum Arthropoda / Class Crustacea / Subclass Cirripedia / Order Thoracica Suborder Balanomorpha / Superfamily Balanoidea / Family Balanidae

20.26 **Balanus trigonus** Darwin, 1854

On hard substrata, including hard-shelled invertebrates, or embedded in sponges or corals, low intertidal zone to 90 m depth; Monterey to Peru; cosmopolitan in warm seas.

Rarely more than 15 mm in diameter, usually less than 10 mm; wall ribbed, white usually mottled with some shade of red; orifice triangular in outline; exterior of scuta with one to several rows of pits, distinguishing it from all other species of *Balanus* occurring locally.

This species may form facultative associations with other organisms, and sometimes lives buried in corals or sponges. It has been found on the whale barnacle *Coronula* (e.g., 20.15), and it is therefore likely that whales have been in part responsible for its wide distribution. Its fossil record goes back to the Miocene. It arrived in the Hawaiian archipelago no later than the Pleistocene, at least a million years before the Polynesians. It is one of the few cosmopolitan shore barnacles not suspected of having become so via transport by ships.

This species has generally escaped the attention of west coast naturalists other than in Panama, where it contributes significantly to the sediments formed by coral reefs on offshore islands. Laboratory studies in Florida indicate that it and *Tetraclita stalactifera* are relatively intolerant of high and low temperature extremes (mid-40's and 0°C), compared with *Balanus amphitrite amphitrite* (20.28) and *Chthamalus stellatus*. Settling experiments and field observations indicate that this is primarily a subtidal species; there is no evidence that the larvae settle intertidally only to die later.

See Barnes & Klepal (1971), Darwin (1854), Henry (1960), Newman & Ross (1976), Pilsbry (1916), Ross, Cerame-Vivas & McCloskey (1964), and Werner (1967).

Phylum Arthropoda / Class Crustacea / Subclass Cirripedia / Order Thoracica Suborder Balanomorpha / Superfamily Balanoidea / Family Balanidae

20.27 **Balanus pacificus** Pilsbry, 1916

Occasionally on rocks, usually on other organisms, low intertidal zone to 73 m depth; Monterey Bay to Baja California; replaced in Gulf of California by *B. pacificus mexicanus*, a form ranging south to Panama and perhaps Peru.

To 35 mm in diameter; wall smooth, white with reddish stripes; species distinguished from *B. amphitrite amphitrite* (20.28) by longitudinal striations across growth lines of scuta.

Balanus pacificus produces many broods throughout the year in southern California. The larvae settle year-round on ephemeral substrata in subtidal sandy areas, rather than on rock bottom bearing well-established communities. In many sand dollar populations, up to 15–25 percent of the adults are fouled by *B. pacificus*, and the barnacles themselves become substrata for bryozoans, algae, and especially the hydroid *Clytia bakeri* (3.9). Many living sand dollars washed ashore during storms are fouled, and field observations indicate that the proportion of fouled sand dollars in established populations increases shorewards after storms. Thus, it appears that barnacle fouling increases the tendency of shoreward transport, adding to both sand dollar and barnacle mortality.

By settling on other organisms, *B. pacificus* tends to avoid many predators such as asteroids, gastropods, crustaceans, and fishes. Settling on organisms living in sand may be especially helpful in this regard. An unusual and apparently major cause of mortality is predation by two polyclad flatworms, *Stylochus tripartitus* (4.2) and *Notoplana inquieta*, against which the barnacle has no defenses. The flatworms can reach the barnacles as planktonic larvae. They enter the shell through the opercular aperture and eat the barnacle at a rate (in the laboratory) of 0.05 gms barnacle tissue per gm flatworm tissue per day. *Balanus pacificus* is opportunistic, taking advantage of surfaces freshly placed in the water or freshly cleaned by disturbances, but it does not form a permanent part of the bottom community.

See Cornwall (1962), Darwin (1854), Giltay (1934), Henry (1960), Houk & Duffy (1972), Hurley (1975, 1976), Merrill & Hobson (1970), Newman & Ross (1976), Pilsbry (1916), Ross (1962), and Zullo (1969b).

Phylum Arthropoda / Class Crustacea / Subclass Cirripedia / Order Thoracica
Suborder Balanomorpha / Superfamily Balanoidea / Family Balanidae

20.28 Balanus amphitrite amphitrite Darwin, 1854

Common on hard substrata such as rocks, shells, and pier pilings, low intertidal zone to 18 m depth in bays and estuaries; San Francisco to Panama; introduced into Salton Sea (Imperial and Riverside Cos.); cosmopolitan in warm seas.

To 20 mm in diameter, white with reddish-brown or purplish stripes; resembling young *B. pacificus* (20.27), but lacking longitudinal striations on scuta; similar to *B. improvisus* (20.29), but the latter never has dark-purple stripes on wall.

Although *B. amphitrite* is easy to identify in California, care must be taken to distinguish it from *B. inexpectatus* farther south, and from *B. reticulatus* in all tropical regions of the world, for the species have been much confused in the past.

Balanus amphitrite is a bay form, but does not survive long in very dilute seawater. A fouling barnacle, it is widely distributed in harbors of the world by ships. It becomes established only in bays in which the summer water temperatures reach 20°C, permitting breeding to take place.

Certain Atlantic invertebrates were introduced to the west coast through the commercial importation of the east coast oyster *Crassostrea virginica*. *Balanus improvisus* (20.29) was apparently introduced in this way before 1853. *B. amphitrite* also grows on *C. virginica*, but the first records of this barnacle on our coast are those at San Diego (1921) and San Francisco (1939). It did not arrive with the oysters, apparently because the oysters were imported from north of Cape Hatteras (North Carolina), the barnacle's northern limit of distribution on the east coast. It seems likely that *B. amphitrite* came later, transported by ships. It was first collected in Hawaii and the Philippines between 1914 and 1916, at the Suez Canal in 1924, and in northern Europe in 1929. However, only in the past decade has this species been clearly defined. Therefore, only on the west coast and in northern Europe, where there are no closely related species, do the earlier records tell us clearly the approximate times of introduction.

The Salton Sea population (Imperial and Riverside Cos.) apparently became established early in World War II, when buoys from San Diego Bay used to mark seaplane lanes were hastily transported there. It differs somewhat in form from the coastal populations and has been formally considered a distinct subspecies. The differences could be genetic, owing to rapid selection in an artificial sea with an exceptionally short faunal list. Alternatively, the differences could be environmentally induced, for although the salinity of the Salton Sea approaches that of the ocean, the ionic composition is different, sulfates being markedly higher and chlorides proportionately lower than in normal seawater. The case for the Salton Sea barnacle being an ecotype rather than a genetically distinct form presently seems stronger, since a similar population has been identified statistically in Wilmington harbor (Los Angeles Co.) on the outer coast.

In general, larvae select suitable settling sites in relation to depth, illumination, surface contour, previous settlements of individuals of the same or related species, and the velocity of the currents. Cyprids of *Balanus amphitrite* settle with their tails to the current, in the absence of other stimuli, which orients the cirral net in the right direction (facing upstream) following metamorphosis. Although the cyprids are not responsive to the direction of light when searching for a suitable spot, they head into light while attaching, regardless of the normal current direction. Above all, the larva seeks a small pit in which to settle. Although some reorientation of the body to the current can be accommodated by differential growth after metamorphosis, it is limited after initial attachment has been made.

Shallow-water barnacles generally withdraw and close up when suddenly placed in shadow. Since the compound eyes of the cyprid are shed with the exoskeleton during metamorphosis, it has been presumed that the photoreceptors, or ocelli, are persisting elements of the nauplius larval eye. These ocelli can be seen, one on each side under the basal margins of the scuta, in freshly metamorphosed juveniles of *Balanus*, but their position becomes obscured with calcification, and dissection is necessary to locate them in adults. Each ocellus is composed of three large photoreceptor cells without ommatidial organization. The cell bodies have projections bearing microvilli and a large axon (15–20 μm in diameter). It has been shown electrophysiologically in a close relative, *B. eburneus*, that the primary event at the photoreceptor is an "on" response similar in form to that described for the retinular cells in a number of arthropods.

See Barnes, Klepal & Munn (1971), Barnes, Read & Topinka (1970), Costlow & Bookhout (1956, 1958a,b), Crisp & Costlow (1963), Crisp & Southward (1961), Davis, Fyhn & Fyhn (1973), Fahrenbach (1965), U. Fyhn & Costlow (1975, 1977), Graham & Gay (1945), Henry & McLaughlin (1975), Hillman et al. (1973), Lacombe (1970), Newman (1967), Newman & Ross (1976), Pilsbry (1916), and Zullo, Beach & Carlton (1972).

Phylum Arthropoda / Class Crustacea / Subclass Cirripedia / Order Thoracica Suborder Balanomorpha / Superfamily Balanoidea / Family Balanidae

20.29 Balanus improvisus Darwin, 1854

On rocks, pilings, and hard-shelled organisms, low intertidal zone to subtidal depths in estuaries; particularly tolerant of brackish waters; Columbia River (Oregon) to Salinas River (Monterey Co.), occasionally in harbors south of Point Conception (Santa Barbara Co.); Ecuador; Japan, Australia; North and South Atlantic; introduced into North Pacific.

Small, to 10 mm in diameter; closely related to *B. amphitrite amphitrite* (20.28), but shell wholly white, lacking purple stripes.

Another white, brackish-water form, *B. eburneus*, has been introduced from the east coast to Central America, Hawaii, and Japan, but it is not yet known from the west coast of North America. It is distinguished from both *B. improvisus* and *B. amphitrite* by the longitudinal striations of the exterior of the scuta.

Balanus improvisus is a truly estuarine barnacle, apparently introduced from the east coast before 1853, along with the eastern oyster, *Crassostrea virginica*. In the northern reaches of San Francisco Bay the barnacle spends as much as 10 months of the year in fresh water. Here, contrary to previous reports, it has recently been shown to osmoregulate if given sufficient time to acclimate. Specimens have been found attached to other freshwater organisms such as crayfish and water beetles in various parts of the world. In the North Sea this species is suffering from competition with a recently introduced Australian form, *Elminius modestus*. Comparisons of Baltic Sea *B.*

improvisus living at salinities 75–90 percent that of ocean water with those living at salinities 15–18 percent of seawater show no significant differences between the two populations in a number of respects, including shell characters, ripe embryos, first nauplius larvae, ratio of shell volume to basal diameter, and ratio of shell weight to dry body weight. *Balanus improvisus* is evidently a well-adjusted estuarine species, in contrast to *Semibalanus balanoides*, which is not known to osmoregulate and which terminates its range into estuaries at salinities ranging from 30 to 75 percent that of normal seawater, accompanied by a marked reduction in size.

Studies of relative growth show that in *B. improvisus* the shell grows faster than the body, which provides the adult with a mantle cavity with plenty of room to accommodate developing embryos. In the Miramichi Bay (New Brunswick), larval planktonic life is estimated to last 18 days, the cyprid stage being reached by the fourteenth day. The stages show some vertical stratification with depth according to tidal stage. In general, cyprids are found nearer the surface at the head of the bay during flood tides, which ensures their transport into the intertidal reaches of the estuary where the adults are most commonly found.

In laboratory tests, *B. improvisus* grew more rapidly when fed the alga *Chlamydomonas* alone than a mixture of *Chlamydomonas* and *Nitzschia*, but growth was less than under field conditions. On the average, molting occurred every 2–3 days. Metamorphosed individuals passed through a stage possessing four wall plates in 2 days; this was followed by a stage showing the typical six-plated wall. Carbonic anhydrase inhibitors prevented the transition from four to six plates. At later stages the application of these inhibitors reduced the rate of growth, but individuals removed from the presence of inhibitors resumed normal growth. Treatment of the earliest juvenile, however, precluded further development, suggesting that mechanisms regulating growth must be initiated shortly after metamorphosis, and once blocked are not triggered again.

Three closely related barnacles, including *B. improvisus*, have been used in tissue and organ culture studies. Bovine embryo extract and yeastolate added to *Balanus* saline were favorable for cell outgrowth and organ maintenance; other additions had neutral or negative effects. Organs cultured include mantle parenchyma, ovarioles, and cement glands. Germinative tissue retained high mitotic activity for at least 7 days. Epithelial and fibroblast-like cells underwent extensive growth and migration. Success in this study may in part reflect the wise choice of materials; all three of the species selected are eurytopic, and *B. improvisus* is especially tolerant of estuarine conditions.

See Barnes & Barnes (1961), Barnes & Healy (1969), Barnes & Klepal (1971), Barnes, Klepal & Munn (1971), Barnes, Read & Topinka (1970), Bartha & Henriksson (1971), Bousfield (1955), Carlton & Zullo (1969), Costlow (1956, 1959), Costlow & Bookhout (1953, 1957), Crisp (1953), Crisp & Southward (1961), Darwin (1854), H. Fyhn (1976), U. Fyhn & Costlow (1975), Graham & Gay (1945), L. W. Jones & Crisp (1954), Kauri (1962, 1966), Meith-Avčin (1974), Newman (1967), Newman & Ross (1976), Törnävä (1948), Visscher (1928), and Zullo, Beach & Carlton (1972).

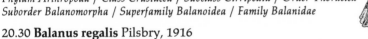

Phylum Arthropoda / Class Crustacea / Subclass Cirripedia / Order Thoracica Suborder Balanomorpha / Superfamily Balanoidea / Family Balanidae

20.30 **Balanus regalis** Pilsbry, 1916

Low intertidal and subtidal zones on surf-washed rocks; Point Conception (Santa Barbara Co.) to western Mexico.

To 60 mm in diameter, closely related and similar in form to the larger northern species *B. aquila* (20.31), but distinguished by having fine, uniform, whitish ribs separated by brown to reddish-brown grooves on the walls, rather than a pure-white wall.

This little-known species, one of the more ornamental balanids, was only recently reported inhabiting wave-swept rocky outcrops in southern California.

See Henry (1960), Newman & Ross (1976), Pilsbry (1916), and Ross (1962).

Phylum Arthropoda / Class Crustacea / Subclass Cirripedia / Order Thoracica Suborder Balanomorpha / Superfamily Balanoidea / Family Balanidae

20.31 **Balanus aquila** Pilsbry, 1907

On rocks, pier pilings, and abalone shells, low intertidal zone to 18 m depth; San Francisco to San Diego.

To 130 mm in diameter; species often confused with another large white species, *B. nubilus* (20.32), but distinguished from it by (1) the small rather than flaring orifice, (2) the beaked terga, and (3) the longitudinal striations on the scuta; distinguished from *B. regalis* (20.30) by the coarsely ribbed white wall.

This species, endemic to a limited region of California, is one of the last surviving members of the *concavus* group, known as fossils from the Oligocene-Pleistocene period in Europe and in North and South America and from the Pliocene period in northern Japan.

Because suppliers of biological materials have often misidentified *B. aquila* and sent it to investigators ordering *B. nubilus*, much of the membrane work reported for *B. nubilus* was actually done on *B. aquila*. Both species occur on pier pilings at Monterey. If there is an "endangered" barnacle species in California, it is *B. aquila*.

See Baskin et al. (1969), Cornwall (1960), Newman (1979), Newman & Ross (1976), Pilsbry (1916), and Zullo (1966).

Phylum Arthropoda / Class Crustacea / Subclass Cirripedia / Order Thoracica Suborder Balanomorpha / Superfamily Balanoidea / Family Balanidae

20.32 **Balanus nubilus** Darwin, 1854

Common on rocks, pier pilings, and hard-shelled animals, low intertidal zone to 90 m depth; southern Alaska to La Jolla (San Diego Co.).

To 110 mm in diameter; not easily confused with any other species except *B. aquila* (20.31), from which it is distinguished by a large, flaring (rather than small) aperture, and by the lack of longitudinal striations on the scuta.

Barnacles of this species are reputed to be eaten by the natives of the northwest, who cook them whole on open fires and then remove the operculum and eat them out of the shell. The animal contains the largest individual muscle fibers known to science and ostensibly has enjoyed considerable popularity in physiological research. However, much of this work was actually performed on *B. aquila*, through misidentification.

See Addicott (1966), Arvy & Lacombe (1968), Arvy, Lacombe & Shimony (1968), Arvy & Liguori (1968), Ashley, Ellory & Hainaut (1974), Barnes (1959), Barnes & Barnes (1959b), Barnes & Gonor (1958a), Carderelli (1968), Cornwall (1936, 1951), Darwin (1854), Devillez (1975), Emerson & Hertlein (1960), Hagiwara & Takahashi (1967), Hagiwara et al. (1968), Harnden (1968), Houk & Duffy (1972), Hoyle, McNeill & Selverston (1973), Hoyle & Smythe (1963), Lacombe (1970), Newman & Ross (1976), Pilsbry (1916), Ross (1962), Shelford et al. (1935), Tait & Emmons (1925), Whitney (1970), and Zullo (1969a,b).

Phylum Arthropoda / Class Crustacea / Subclass Cirripedia / Order Thoracica
Suborder Balanomorpha / Superfamily Balanoidea / Family Balanidae

20.33 **Megabalanus californicus** (Pilsbry, 1916)
(= Balanus tintinnabulum californicus)

Uncommon in low intertidal zone; commonly subtidal to 9 m depth on rocks, pilings, buoys, kelp, mussels, and other hard-shelled organisms; Humboldt Bay to Guaymas (Mexico); very uncommon north of Monterey.

To 60 mm in diameter, with longitudinal red and white stripes; the only *Megabalanus* species known from California; genus distinguished from all other balanid genera in having radii permeated by transverse tubes rather than being solid.

This is one of the most colorful balanids and tends to be very gregarious. Crowding causes the basis to elongate into a deep cup so that the plates remain undistorted and individuals form spectacular clusters. A related species is considered a delicacy in Chile and is sold fresh or canned.

See Aleem (1957), Barnes & Klepal (1971), Coe (1932), Coe & Allen (1937), Graham & Gay (1945), Henry (1960), Hewatt (1946), Kanakoff & Emerson (1959), Newman & Ross (1976), Pilsbry (1916), Rasmussen (1935), Willett (1937), and Zullo (1968).

Phylum Arthropoda / Class Crustacea / Subclass Cirripedia
Order Rhizocephala / Suborder Kentrogonida / Family Sacculinidae

20.34 **Heterosaccus californicus** Boschma, 1933

Parasitic on the spider crabs *Loxorhynchus grandis* (25.10) and *L. crispatus* (25.11), *Pugettia producta* (25.4), and *Taliepus nuttallii* (25.3); known only from Monterey Bay to San Diego.

Externally visible portion of parasite (the "externa") attached to ventral surface of crab's abdomen, in same position that fertilized eggs are carried by ripe female crab; externa to 40 mm wide, globular, saclike, with a central orifice; similar to externa of another rhizocephalan barnacle reported from California, **Loxothylacus panopaei** (Gissler), which differs in parasitizing only crabs of the family Xanthidae.

Little has been published on the biology of *Heterosaccus*. However, its general life history probably resembles that of *Sacculina carcini*, a similar form carefully studied by the French zoologist Yves Delage and others nearly a century ago. In *Sacculina* the larvae hatch as nauplii that differ from those of free-living barnacles mainly in lacking a mouth and gut. The cyprid larval stage also lacks a gut and has only rudimentary mouthparts. Female cyprid larvae locate host crabs (*Carcinus*, *Portunus*, etc.) and attach themselves by one antennule at a place where the host's exoskeleton is very thin and soft. The cyprid then sheds much of its larval body, including thoracic limbs, muscles, and bivalved carapace. The small mass of tissue that survives this metamorphosis becomes surrounded with a new cuticle, one feature of which is a hollow needle or dart. This larval stage, the "kentrogon," extends its dart through the host cuticle and by muscular action injects the larval tissues into the crab's body. The kentrogon tissue then develops the "interna," a central lobe on the host's gut just back of the stomach, and, extending out from this lobe, a branching network of threadlike nutritive rootlets that range throughout the blood spaces of the crab's body. Some 9 months after initial penetration, the interna extends a lobe that emerges on the ventral abdomen of the crab, forming the sacklike externa. The externa contains the large ovary and a sack containing sperm probably derived from a male cyprid (see the account of *Peltogasterella gracilis*, 20.35, and Ichikawa & Yanagimachi, 1960). Eggs and sperm are both shed into an enclosed space, the mantle cavity, where fertilization occurs, and the embryos are brooded until they hatch as nauplius larvae and escape by the orifice. The presence of the parasite inhibits molting and normal sexual development of the host, but seems to cause little damage in other respects. After the parasite reproduces, the externa is sometimes shed or lost, but may persist on the host for at least a year and possibly a lifetime.

All of the rhizocephalans included in this chapter are "kentrogonids," that is, they pass through a kentrogon stage and undergo a period of internal development between initial infection of the crab and the emergence of the externa. This is in contrast to the "akentrogonid" type of rhizocephalan, recently described by Bocquet-Védrine (1972), in which the parasite always remains in communication with the exterior.

See Bocquet-Védrine (1972), Boschma (1953), Day (1935), Delage (1884), Foxon (1940), Fratello (1968), Ichikawa & Yanagimachi (1960), and O'Brien (1977).

Phylum Arthropoda / Class Crustacea / Subclass Cirripedia
Order Rhizocephala / Suborder Kentrogonida / Family Peltogasteridae

20.35 **Peltogasterella gracilis** (Boschma, 1927)

On abdomens of hermit crabs (*Pagurus* spp.); Bering Sea to Chile; Japan; only member of family Peltogasteridae known from California.

External portion of parasite ("externa") to 10 mm long, with mantle aperture terminal (rather than medial as in other California rhizocephalans); one host usually bearing several externae.

Classical studies of rhizocephalan barnacles suggested that they were self-fertilizing hermaphrodites. Some were known to be accompanied by males, presumed to be non-functional. Recent studies of this species in Japan show that the stage parasitizing crabs is not hermaphroditic, but female. The so-called "testes" are actually pockets in the female (analogous to seminal receptacles) in which male tissue has come to reside following the attachment of a male cyprid.

In *Peltogasterella gracilis* two types of females have been re-ported. Their externae are similar, but one type produces only small eggs (140–150 μm diameter). When fertilized, they give rise to small nauplius larvae, and later small cyprids. The cyprid locates a crab host, attaches by an antennule, sheds its thoracic limbs and carapace, and metamorphoses into a kentrogon larva (see the account of *Heterosaccus californicus*, 20.34). The kentrogon penetrates the host with a hollow dart, and injects tissues that proliferate to form the interna, or internal portion, of the parasite. The central cell mass of the interna grows in the abdomen and sends nutritive rootlets ramifying into the rest of the host body. Eventually the rudiment of the externa arises as a bud on the central cell mass, grows, and pushes out through the abdominal wall of the host, while remaining firmly attached to the root system. The female parasite develops ovaries in the externa, and also a pair of small sacs of "male-cell receptacles" that lack gametes.

A second type of female parasite, from another crab host, produces only large eggs (160–170 μm diameter). When these are fertilized, they yield large nauplius larvae and later large cyprids, which become males. The large male cyprids swim and settle, not on crab hosts like the small cyprids, but on the apertures of the mantle cavities of *Peltogasterella gracilis* externae, which are all females. A large cyprid does not metamorphose into a kentrogon stage, but through one of its antennules it injects a cell mass into the mantle cavity of the female parasite. Both types of females receive such male-cell masses, which migrate through the mantle cavity and take up residence in the male-cell receptacles. Here the male cells differentiate to form sperm.

Female parasites may produce several successive batches of eggs, after which the externa dies and is shed, leaving the interna intact. New externae may develop from the same interna. When several externae are present on a host, they are probably (but not certainly) all budded from the same interna. Occasional exceptional female parasites were found that produce both large and small eggs, yielding, respectively, large and small larvae; it is possible that the female actually alternates between large and small cyprid production, as is the case in *Lernaeodiscus porcellanae* (20.36).

See Bocquet-Védrine (1972), Reinhard (1942), and Yanagimachi (1961a,b).

Phylum Arthropoda / Class Crustacea / Subclass Cirripedia
Order Rhizocephala / Suborder Kentrogonida / Family Lernaeodiscidae

20.36 **Lernaeodiscus porcellanae** F. Müller, 1862

On the porcelain crab *Petrolisthes cabrilloi* (24.17), middle and low intertidal zones on rocky shores; La Jolla (San Diego Co.), only member of family Lernaeodiscidae known from west coast of North America; if identification is correct, previously known only from southern Brazil.

To 10 mm wide, distinguished from other solitary California rhizocephalans by fluted edges of external sac.

Although the species was only recently discovered in California, an intensive study of the host-parasite relationship has been made by the late L. Ritchie. The following are some of his results (pers. comm.).

Sexes are separate in *Lernaeodiscus*; the smaller female-producing cyprid enters the gill chamber of the crab, where it attaches to a gill lamella by both first antennae. In the next few days the kentrogon stage forms (see the account of *Heterosaccus californicus*, 20.34). Unless the crab has successfully removed the parasite in the process of cleaning its gills, the latter invades the host through a trough formed by the cyprid labrum (upper lip), rather than by a tube inserted through one of its first antennae. Gill cleaning is accomplished by the posterior pair of thoracic legs, especially modified in connection with this cleaning function. The mere presence of female cyprids initiates vigorous gill cleaning in *P. cabrilloi*, but not in other porcelain crabs of the region. Between 5 and 60 percent of the host population can be infected, the incidence of infection being significantly increased in the laboratory by damaging or removing the last thoracic legs. Infection castrates both sexes and feminizes male crabs. Such non-reproductive individuals have normal life expectancies and continue to draw on the resources of the crab population. Upon completion of maturation the tissues of the parasite inside the crab's body develop an incipient externa, between the proximal abdominal sterna. This forms the externa on the next molt of the host. One to several male cyprids of *L. porcellanae* locate the virgin externa, attach, and extrude their cellular contents into it. These male cells come to reside in receptacles and differentiate into spermatozoa in time to fertilize the first and successive broods of eggs produced in the externa. The maturing externa takes on the general form, and occupies the same position on the crab's abdomen, as the eggs being brooded by a normal female crab; and it is cared for and cleaned by the last pair of thoracic legs as if it were indeed the crab's egg mass. Mature externae wither and die if experimentally deprived of this care, in good part for lack of assistance in molting. Lost externae are regenerated, but normally they are long-lived, producing successive broods of nauplius larvae that develop into female, mixed female and male, or male cyprids, in a succession that is apparently seasonal.

See Boschma (1969).

Literature Cited

Addicott, W. O. 1966. Late Pleistocene marine paleoecology and zoogeography in central California. Geol. Surv. Prof. Pap. 523-C: 1–21.

Aleem, A. A. 1957. Succession of marine fouling organisms on test panels immersed in deep water at La Jolla, California. Hydrobiologia 11: 40–58.

Ankel, W. E. 1962. Die blaue Flottè. Natur und Museum 92: 351–66.

Anon. 1968. Science and the citizen: Barnacle cement. Sci. Amer. 219: 46.

Arvy, L., and D. Lacombe. 1968. Activités enzymatiques traceuses dans "l'appareil cémentaire" des Balanidae (Crustacea, Cirripedia). C. R. Acad. Sci., Paris, 267: 1326–28.

Arvy, L., D. Lacombe, and T. Shimony. 1968. Studies on the biology of barnacles: Alkaline phosphatase activity histochemically detectable in the cement apparatus of the Balanidae (Crustacea, Cirripedia). Amer. Zool. 8: 817 (abstract).

Arvy, L., and V. R. Liguori. 1968. Studies on the biology of barnacles: Differences in muscular cytochrome oxidase activity histochemically detectable in some Balanidae (Crustacea, Cirripedia). Amer. Zool. 8: 817 (abstract).

Ashley, C. C., J. C. Ellory, and K. Hainaut. 1974. Calcium movement in single crustacean muscle fibres. J. Physiol., London, 242: 255–72.

Augenfeld, J. M. 1967. Respiratory metabolism and glycogen storage in barnacles occupying different levels of the intertidal zone. Physiol. Zool. 40: 92–96.

Austin, A. P., D. J. Crisp, and A. M. Patil. 1958. The chromosome numbers of certain barnacles in British waters. Quart. J. Microscop. Sci. 99: 497–504.

Bainbridge, V., and J. Roskell. 1966. A re-description of the larvae of *Lepas fascicularis* Ellis and Solander with observations on the distribution of *Lepas* nauplii in the northeastern Atlantic, pp. 67–82, *in* H. Barnes, ed., Some contemporary studies in marine science. London: Allen & Unwin.

Baldridge, A. 1977. The barnacle *Lepas pacifica* and the alga *Navicula grevillei* on northern elephant seals, *Mirounga angustirostris*. J. Mammal. 58: 428–29.

Ball, E. G. 1944. A blue chromoprotein found in the eggs of the goose-barnacle. J. Biol. Chem. 152: 627–34.

Barnes, H. 1953a. The effect of lowered salinity on some barnacle nauplii. J. Anim. Ecol. 22: 328–30.

———1953b. Size variation in the cyprids of some common barnacles. J. Mar. Biol. Assoc. U.K. 32: 297–304.

———1953c. The effect of light on the growth rate of two barnacles *Balanus balanoides* (L.) and *B. crenatus* Brug. under conditions of total submergence. Oikos 4: 104–11.

———1959. Stomach contents and microfeeding of some common cirripedes. Canad. J. Zool. 37: 231–36.

———1960. The behaviour and ecology of *Pollicipes polymerus*. Rep. Challenger Soc. 3: 30.

Barnes, H., and M. Barnes. 1956. The general biology of *Balanus glandula* Darwin. Pacific Sci. 10: 415–22.

———. 1958. Further observations on self fertilization in *Chthamalus* sp. Ecology 39: 550.

———. 1959a. The effect of temperature on the oxygen uptake and rate of development of the egg masses of two common cirripedes, *Balanus balanoides* (L.) and *Pollicipes polymerus* J. B. Sowerby. Kieler Meeresforsch. 15: 242–51.

———. 1959b. The naupliar stages of *Balanus nubilus* Darwin. Canad. J. Zool. 37: 15–23.

———. 1959c. The naupliar stages of *Balanus hesperius* Pilsbry. Canad. J. Zool. 37: 237–44.

———. 1961. Salinity and the biometry of *Balanus improvisus*. Soc. Sci. Fennica Comm. Biol. 24: 4–7.

Barnes, H., and J. J. Gonor. 1958a. Neurosecretory cells in some cirripedes. Nature 181: 194.

———. 1958b. Neurosecretory cells in the cirripede *Pollicipes polymerus* J. B. Sowerby. J. Mar. Res. 17: 81–102.

Barnes, H., and M. J. R. Healy. 1969. Biometrical studies on some cirripedes. II. Discriminant analysis of measurements on the scuta and terga of *Balanus balanus* (L.), *B. crenatus* Brug., *B. improvisus* Darwin, *B. glandula* Darwin, and *B. amphitrite stutsburi* Darwin (*B. pallidus stutsburi*). J. Exper. Mar. Biol. Ecol. 4: 51–70.

Barnes, H., and W. Klepal. 1971. The structure of the pedicel of the penis in cirripedes and its relation to other taxonomic characters. J. Exper. Mar. Biol. Ecol. 7: 71–94.

Barnes, H., W. Klepal, and E. A. Munn. 1971. Observations on the form and changes in the accessory droplet and motility of the spermatozoa of some cirripedes. J. Exper. Mar. Biol. Ecol. 7: 173–96.

Barnes, H., and H. T. Powell. 1950. The development, general morphology and subsequent elimination of barnacle populations, *Balanus crenatus* and *B. balanoides*, after a heavy initial settlement. J. Anim. Ecol. 19: 175–79.

Barnes, H., R. Read, and J. A. Topinka. 1970. The behaviour on impaction by solids of some common cirripedes and relation to their normal habitat. J. Exper. Mar. Biol. Ecol. 5: 70–87.

Barnes, H., and E. S. Reese. 1959. Feeding in the pedunculate cirripede *Pollicipes polymerus* J. B. Sowerby. Proc. Zool. Soc. London 132: 569–85.

———. 1960. The behaviour of the stalked intertidal barnacle *Pollicipes polymerus* J. B. Sowerby, with special reference to its ecology and distribution. J. Anim. Ecol. 29: 169–85.

Bartha, S. D., and S. Henriksson. 1971. The growth of sea organisms and the effect on the corrosion resistance of stainless steel and titanium. Cent. Rech. Études Oceanogr. Paris, Travaux 11: 7–20.

Baskin, R. J., W. C. Stanford, P. D. Morse, and M. L. Biggs. 1969. The formation of filaments from barnacle myosin. Comp. Biochem. Physiol. 29: 471–74.

Batzli, George O. 1969. Distribution of biomass in rocky intertidal communities on the Pacific coast of the United States. J. Anim. Ecol. 38: 531–46.

Bebbington, P. M., and E. D. Morgan. 1977. Detection and identification of moulting hormone (ecdysones) in the barnacle *Balanus balanoides*. Comp. Biochem. Physiol. 56B: 77–79.

Bergen, M. 1968. The salinity tolerance limits of the adults and early-stage embryos of *Balanus glandula* Darwin, 1854 (Cirripedia, Thoracica). Crustaceana 15: 229–34.

Bieri, R. 1966. Feeding preferences and rates of the snail, *Ianthina prolongata*, the barnacle, *Lepas anserifera*, the nudibranchs, *Glaucus atlanticus* and *Fiona pinnata*, and the food web in the marine neuston. Publ. Seto Mar. Biol. Lab. 14: 161–70, 2 pls.

Bigelow, M. A. 1902. The early development of *Lepas*. A study of cell lineage and germ layers. Bull. Mus. Compar. Zool., Harvard, 40: 61–144.

Bocquet-Védrine, J. 1972. Les Rhizocephales. Cah. Biol. Mar. 13: 615–26.

Boëtius, J. 1952–53. Some notes on the relation to the substratum of *Lepas anatifera* L. and *Lepas fascicularis* E. et S. Oikos 4: 112–17.

Boschma, H. 1953. The Rhizocephala of the Pacific. Zool. Meded. 32: 185–201.

———. 1969. Notes on rhizocephalan parasites of the genus *Lernaeodiscus*. K. Nederl. Akad. Weten. Amsterdam Proc. (C) 72: 413–19.

Bousfield, E. L. 1955. Ecological control of the occurrence of barnacles in the Miramichi estuary. Bull. Nat. Mus. Canada 137: 1–69.

Bowers, R. L. 1968. Observations on the orientation and feeding behavior of barnacles associated with lobsters. J. Exper. Mar. Biol. Ecol. 2: 105–12.

Burnett, B. R. 1972. Aspects of the circulatory system of *Pollicipes polymerus* J. B. Sowerby (Cirripedia: Thoracica). J. Morphol. 136: 79–108.

————. 1975. Blood circulation in four species of barnacles (*Lepas*, *Conchoderma*: Lepadidae). Trans. San Diego Soc. Natur. Hist. 17: 293–304.

Carderelli, N. F., preparer. 1968. Barnacle cement as dental restorative adhesive. [With selective bibliography.] Nat. Inst. Dental Res. Publ. 151: 1–49.

Carlton, J. T., and V. A. Zullo. 1969. Early records of the barnacle *Balanus improvisus* Darwin from the Pacific coast of North America. Occas. Pap. Calif. Acad. Sci. 75: 1–6.

Cheng, L., and R. A. Lewin. 1976. Goose barnacles (Cirripedia: Thoracica) on flotsam beached at La Jolla, California. Fish. Bull. 74: 212–17.

Clarke, R. 1966. The stalked barnacle *Conchoderma*, ectoparasitic on whales. Norsk. Hvalfangst-Tidende 55: 153–68.

Coe, W. R. 1932. Season of attachment and rate of growth of sedentary marine organisms at the pier of the Scripps Institution of Oceanography, La Jolla, California. Bull. Scripps Inst. Oceanogr., Tech. Ser. 3: 37–74.

Coe, W. R., and W. E. Allen. 1937. Growth of sedentary marine organisms on experimental blocks and plates for nine successive years at the pier of the Scripps Institution of Oceanography. Bull. Scripps Inst. Oceanogr., Tech. Ser. 4: 101–36.

Collier, A., S. Ray, and W. B. Wilson. 1956. Some effects of specific organic compounds on marine organisms. Science 124: 220.

Connell, J. H. 1970. A predator-prey system in the marine intertidal region. I. *Balanus glandula* and several predatory species of *Thais*. Ecol. Monogr. 40: 49–78.

Cornwall, I. E. 1924. Notes on west American whale barnacles. Proc. Calif. Acad. Sci. 13: 421–31.

————. 1936. On the nervous system of four British Columbian barnacles (one new species). J. Biol. Bd. Canada 1: 469–75.

————. 1951. The barnacles of California. Wasmann J. Biol. 9: 311–46.

————. 1953. The central nervous system of barnacles (Cirripedia). J. Fish. Res. Bd. Canada 10: 76–84.

————. 1955a. Canadian Pacific fauna. 10, Arthropoda; 10e, Cirripedia. Ottawa: Fish Res. Bd. Canada. 49 pp.

————. 1955b. The barnacles of British Columbia. British Columbia Prov. Mus., Handbook 7: 1–69.

————. 1960. Barnacle shell figures and repairs. Canad. J. Zool. 38: 827–32.

————. 1962. The identification of barnacles, with further figures and notes. Canad. J. Zool. 40: 621–29.

Costlow, J. D. 1956. Shell development in *Balanus improvisus* Darwin. J. Morphol. 99: 359–415.

————. 1959. Effects of carbonic anhydrase inhibitors on shell development and growth of *Balanus improvisus* Darwin. Physiol. Zool. 32: 177–84.

Costlow, J. D., and C. G. Bookhout. 1953. Moulting and growth in *Balanus improvisus*. Biol. Bull. 105: 420–33.

————. 1956. Molting and shell growth in *Balanus amphitrite niveus*. Biol. Bull. 110: 107–16.

————. 1957. Body growth versus shell growth in *Balanus improvisus*. Biol. Bull. 113: 224–32.

————. 1958a. Larval development of *Balanus amphitrite* var. *denticulata* Broch reared in the laboratory. Biol. Bull. 114: 284–95.

————. 1958b. Molting and respiration in *Balanus amphitrite* var. *denticulata* Broch. Physiol. Zool. 31: 271–80.

Crisp, D. J. 1953. Changes in the orientation of barnacles of certain species in relation to water currents. J. Anim. Ecol. 22: 331–43.

————. 1955. The behaviour of barnacle cyprids in relation to water movement over a surface. J. Exper. Biol. 32: 569–90.

Crisp, D. J., and H. Barnes. 1954. The orientation and distribution of barnacles at settlement with particular reference to surface contour. J. Anim. Ecol. 23: 142–62.

Crisp, D. J., and J. D. Costlow. 1963. The tolerance of developing cirripede embryos to salinity and temperature. Oikos 14: 22–34.

Crisp, D. J., and P. S. Meadows. 1963. Adsorbed layers: The stimulus to settlement in barnacles. Proc. Roy. Soc. London B158: 364–87.

Crisp, D. J., and A. J. Southward. 1961. Different types of cirral activity of barnacles. Phil. Trans. Roy. Soc. London B243: 271–308.

Crisp, D. J., and H. G. Stubbings. 1957. The orientation of barnacles to water currents. J. Anim. Ecol. 26: 179–96.

Dall, W. H. 1872. On the parasites of the cetaceans of the N.W. coast of America, with descriptions of new forms. Proc. Calif. Acad. Sci. 4: 299–301.

Darwin, C. 1851. A monograph on the sub-class Cirripedia, with figures of all the species. The Lepadidae; or, pedunculated cirripedes. London: Ray Society. 400 pp.

————. 1854. A monograph on the sub-class Cirripedia with figures of all the species. The Balanidae, the Verrucidae, etc. London: Ray Society. 684 pp.

Davis, C. W., U. E. H. Fyhn, and H. J. Fyhn. 1973. The intermolt cycle of cirripeds: Criteria for its stages and its duration in *Balanus amphitrite*. Biol. Bull. 145: 310–22.

Day, J. H. 1935. The life history of *Sacculina*. Quart. J. Microscop. Sci. 77: 549–83.

Dayton, P. K. 1971. Competition, disturbance, and community organization: The provision and subsequent utilization of space in a rocky intertidal community. Ecol. Monogr. 41: 351–89.

Delage, Y. 1884. Évolution de la Sacculine. Arch. Zool. Expér. Gén. 2: 417–738.

Devillez, E. J. 1975. Observations on the proteolytic enzymes in the digestive fluid of the barnacle *Balanus nubilus*. Comp. Biochem. Physiol. 51A: 471–74.

Dudley, P. L. 1973. Synaptonemal polycomplexes in spermatocytes of the gooseneck barnacle, *Pollicipes polymerus* Sowerby (Crustacea: Cirripedia). Chromosoma 40: 221–42.

Emerson, W. K. 1956. Pleistocene invertebrates from Punta China, Baja California, Mexico. With remarks on the composition of the Pacific coast Quaternary faunas. Bull. Amer. Mus. Natur. Hist. 111: 313–42.

Emerson, W. K., and L. G. Hertlein. 1960. Pliocene and Pleistocene invertebrates from Punta Rosalia, Baja California, Mexico. Amer. Mus. Novitates 2004: 1–8.

Fahrenbach, W. H. 1965. The micromorphology of some simple photoreceptors. Z. Zellforsch. 66: 233–54.

Fisher, W. K. 1911. Asteroidea of the North Pacific and adjacent waters. Bull. U.S. Nat. Mus. 76: 1–419, 122 pls. (cirripeds pp. 237, 264, 404; pl. 111).

Foster, B. A. 1969. Tolerance of high temperatures by some intertidal barnacles. Mar. Biol. 4: 326–32.

———. 1971. Desiccation as a factor in the intertidal zonation of barnacles. Mar. Biol. 8: 12–29.

Fox, D. L., and G. F. Crozier. 1967. Astaxanthin in the blue oceanic barnacle *Lepas fascicularis*. Experientia 23: 12–14.

Fox, D. L., V. E. Smith, and A. A. Wolfson. 1967. Disposition of carotenoids in the blue goose barnacle *Lepas fascicularis*. Experientia 23: 965–70.

Foxon, G. E. H. 1940. Notes on the life history of *Sacculina carcini* Thompson. J. Mar. Biol. Assoc. U.K. 24: 253–64.

Fratello, B. 1968. Cariologia e tassonomia dei Sacculinidi (Cirripedi, Rizocefali). Caryologia 21: 359–67.

Fyhn, H. J. 1976. Holeuryhalinity and its mechanisms in a cirriped crustacean, *Balanus improvisus*. Comp. Biochem. Physiol. 53A: 19–30.

Fyhn, H. J., J. A. Peterson, and K. Johansen. 1972. Eco-physiological studies of an intertidal crustacean, *Pollicipes polymerus* (Cirripedia, Lepadomorpha). I. Tolerance to body temperature change, desiccation and osmotic stress. J. Exper. Biol. 57: 83–102.

———. 1973. Heart activity and high-pressure circulation in Cirripedia. Science 180: 513–15.

Fyhn, U. E. H., and J. D. Costlow. 1975. Tissue cultures of cirripeds. Biol. Bull. 149: 316–30.

———. 1977. Histology and histochemistry of the ovary and oogenesis in *Balanus amphitrite* and *B. eburneus* Gould (Cirripedia, Crustacea). Biol. Bull. 152: 351–59.

Gibbons, E. 1964. Stalking the blue-eyed scallop. New York: McKay. 332 pp. (Hunting the wild goose barnacle (*Lepas* and *Mitella* species), pp. 211–13.)

Giltay, L. 1934. Note sur l'association de *Balanus concavus pacificus*

Pilsbry (Cirripède) et *Dendraster excentricus* (Eschscholtz) (Echinoderme). Bull. Mus. Roy. Hist. Natur. Belg. 10: 1–7.

Glynn, P. W. 1965. Community composition, structure, and interrelationships in the marine intertidal *Endocladia muricata–Balanus glandula* association in Monterey Bay, California. Beaufortia 12: 1–198.

Gomez, E. D. 1973. Observations on feeding and prey specificity of *Tritonia festiva* (Stearns) with comments on other tritoniids (Mollusca: Opisthobranchia). Veliger 16: 163–65.

———. 1975. Sex determination in *Balanus* (*Conopea*) *galeatus* (L.) (Cirripedia Thoracica). Crustaceana 28: 105–7.

Gomez, E. D., D. J. Faulkner, W. A. Newman, and C. Ireland. 1973. Juvenile hormone mimics: Effect on cirriped crustacean metamorphosis. Science 179: 813–14.

Graham, H. W., and H. Gay. 1945. Season of attachment and growth of sedentary marine organisms at Oakland, California. Ecology 26: 375–86.

Gwilliam, G. F. 1963. The mechanism of the shadow reflex in Cirripedia. I. Electrical activity in the supraesophageal ganglion and ocellar nerve. Biol. Bull. 125: 470–85.

———. 1965. The mechanism of the shadow reflex in Cirripedia. II. Photoreceptor cell response, second-order responses, and motor cell output. Biol. Bull. 129: 244–56.

Gwilliam, G. F., and J. C. Bradbury. 1971. Activity patterns in the isolated central nervous system of the barnacle and their relation to behavior. Biol. Bull. 141: 502–13.

Haderlie, E. C. 1974. Growth rates, depth preference and ecological succession of some sessile marine invertebrates in Monterey harbor. Veliger 17 (Suppl.): 1–35.

Hagiwara, S., and K. Takahashi. 1967. Surface density of calcium ions and calcium spikes in the barnacle muscle fiber membrane. J. Gen. Physiol. 50: 583–601.

Hagiwara, S., R. Gruener, H. Hayashi, H. Sakata, and A. D. Grinnell. 1968. Effect of external and internal pH changes on K and Cl conductances in the muscle fiber membrane of a giant barnacle. J. Gen. Physiol. 52: 773–92.

Harnden, D. G. 1968. Digestive carbohydrases of *Balanus nubilus* (Darwin, 1854). Comp. Biochem. Physiol. 25: 303–9.

Henry, D. P. 1940. Notes on some pedunculate barnacles from the North Pacific. Proc. U.S. Nat. Mus. 88: 225–36.

———. 1942. Studies on the sessile Cirripedia of the Pacific coast of North America. Univ. Washington Publ. Oceanogr. 4: 95–134.

———. 1960. Thoracic Cirripedia of the Gulf of California. Univ. Washington Publ. Oceanogr. 4: 135–58.

Henry, D. P., and P. A. McLaughlin. 1967. A revision of the subgenus *Solidobalanus* Hoek (Cirripedia Thoracica) including a description of a new species with complemental males. Crustaceana 12: 43–58.

———. 1975. The barnacles of the *Balanus amphitrite* complex (Cirripedia, Thoracica). Zool. Verhandel. 141: 1–254.

Herz, L. E. 1933. The morphology of the later stages of *Balanus crenatus* Bruguière. Biol. Bull. 64: 432–42.

Hewatt, W. G. 1946. Marine ecological studies on Santa Cruz Island, California. Ecol. Monogr. 16: 185–208.

Hilgard, G. H. 1960. A study of reproduction in the intertidal barnacle, *Mitella polymerus*, in Monterey Bay, California. Biol. Bull. 119: 169–88.

Hillman, P., F. A. Dosge, S. Hochstein, B. W. Knight, and B. Minke. 1973. Rapid dark recovery of the invertebrate early receptor potential. J. Gen. Physiol. 62: 77–86.

Hines, A. H. 1976. The comparative reproductive ecology of three species of intertidal barnacles. Doctoral thesis, Zoology, University of California, Berkeley. 259 pp.

————. 1978. Reproduction in three species of intertidal barnacles from central California. Biol. Bull. 154: 262–81.

————. 1979. The comparative reproductive ecology of three species of intertidal barnacles, pp. 213–34, *in* S. E. Stancyk, ed., Reproductive ecology of marine invertebrates, vol. 9. Belle W. Baruch Library in Marine Science. Columbia: University of South Carolina Press.

Holter, A. R. 1969. Carotenoid pigments in the stalked barnacle *Pollicipes polymerus*. Comp. Biochem. Physiol. 28: 675–84.

Houk, J. L., and J. M. Duffy. 1972. Two new sea urchin–acorn barnacle associations. Calif. Fish & Game 58: 321–23.

Howard, G. K., and H. C. Scott. 1959. Predaceous feeding in two common gooseneck barnacles. Science 129: 717–18.

Hoyle, G., P. A. McNeill, and A. I. Selverston. 1973. Ultrastructure of barnacle giant muscle fibers. J. Cell. Biol. 56: 74–91.

Hoyle, G., and T. Smyth. 1963. Giant muscle fibers in a barnacle, *Balanus nubilus* Darwin. Science 139: 49–50.

Hurley, A. C. 1975. The establishment of populations of *Balanus pacificus* Pilsbry (Cirripedia) and their elimination by predatory Turbellaria. J. Anim. Ecol. 44: 521–32.

————. 1976. The polyclad flatworm *Stylochus tripartitus* Hyman as a barnacle predator. Crustaceana 31: 110–11.

Ichikawa, A., and R. Yanagimachi. 1960. Studies on the sexual organization of the Rhizocephala. II. The reproductive function of the larval (cypris) males of *Peltogaster* and *Sacculina*. Annot. Zool. Japon. 33: 42–56.

Johnson, M. W., and R. C. Miller. 1935. The seasonal settlement of shipworms, barnacles, and other wharf-pile organisms at Friday Harbor, Washington. Univ. Washington Publ. Oceanogr. 2: 1–18.

Jones, E. C. 1968. *Lepas anserifera* Linné (Cirripedia, Lepadomorpha) feeding on fish and *Physalia*. Crustaceana 14: 312–13.

Jones, E. C., B. J. Rothschild, and R. S. Shomura. 1968. Additional records of the pedunculate barnacle, *Conchoderma virgatum* (Spengler), on fishes. Crustaceana 14: 194–96.

Jones, L. L. 1978. The life history patterns and host selection behavior of a sponge symbiont, *Membranobalanus orcutti* (Pilsbry)

(Cirripedia). Doctoral thesis, Oceanography, University of California, San Diego. 112 pp.

Jones, L. W. G., and D. J. Crisp. 1954. The larval stages of the barnacle *Balanus improvisus* Darwin. Proc. Zool. Soc. London 123: 765–80.

Kanakoff, G. P., and W. K. Emerson. 1959. Late Pleistocene invertebrates of the Newport Bay area, California. Los Angeles Co. Mus. Contr. Sci. 31: 1–47.

Kasuya, T., and D. W. Rice. 1970. Notes on baleen plates and on arrangement of parasitic barnacles of gray whale. Sci. Rep. Whales Res. Inst. 22: 39–43.

Kauri, T. 1962. On the frontal filaments and nauplius eye in *Balanus*. Crustaceana 4: 131–42.

————. 1966. On the sensory papilla X-organ in cirriped larvae. Crustaceana 11: 115–22.

Klepal, W., and H. Barnes. 1975a. A histological and scanning electron microscope study of the formation of the wall plates in *Chthamalus depressus* (Poli). J. Exper. Mar. Biol. Ecol. 20: 183–98.

————. 1975b. The structure of the wall plate in *Chthamalus depressus* (Poli). J. Exper. Mar. Biol. Ecol. 20: 265–85.

Knudsen, J. W. 1963. Notes on the barnacle *Lepas fascicularis* found attached to the jellyfish *Velella*. Bull. So. Calif. Acad. Sci. 62: 130–31.

Krebs, W., and B. Schaten. 1976. The lateral photoreceptor of the barnacle, *Balanus eburneus*: Quantitative morphology and fine structure. Cell Tissue Res. 168: 193–207.

Krüger, P. 1940. Cirripedia, *in* Dr. H. G. Bronns Klassen und Ordnungen des Tierreichs, Bd. 5, Abt. 1, Buch 3. Leipzig: Winter. 560 pp.

Lacombe, D. 1970. A comparative study of the cement glands in some balanid barnacles (Cirripedia, Balanidae). Biol. Bull. 139: 164–79.

Lacombe, D., and V. R. Liguori. 1969. Comparative histological studies of the cement apparatus of *Lepas anatifera* and *Balanus tintinnabulum*. Biol. Bull. 137: 170–80.

Landau, M. 1976. A comment on self-fertilization in the barnacle *Balanus eburneus* Gould (Cirripedia Thoracica). Crustaceana 30: 105–6.

Lewis, C. A. 1975a. Some observations on factors affecting embryonic and larval growth of *Pollicipes polymerus* (Cirripedia: Lepadomorpha) in vitro. Mar. Biol. 32: 127–39.

————. 1975b. Development of the gooseneck barnacle *Pollicipes polymerus* (Cirripedia: Lepadomorpha): Fertilization through settlement. Mar. Biol. 32: 141–53.

————. 1977. Ultrastructure of a fertilized barnacle egg (*Pollicipes polymerus*) with peristaltic constrictions. Wilhelm Roux Archivs 181: 333–55.

————. 1978. A review of substratum selection in free-living and symbiotic cirripeds, pp. 207–18, *in* F.-S. Chia and M. E. Rice, eds.,

Settlement and metamorphosis of marine invertebrate larvae. New York: Elsevier/North Holland. 290 pp.

Lewis, C. A., F.-S. Chia, and T. E. Schroeder. 1973. Peristaltic constrictions in fertilized barnacle eggs (*Pollicipes polymerus*). Experientia 29: 1533–35.

MacIntyre, R. J. 1966. Rapid growth in stalked barnacles. Nature 212: 637–38.

Mahmoud, M. F. 1959. The structure of the outer integument of some pedunculate cirripedes (barnacles). Proc. Egypt. Acad. Sci. 14: 61–69.

Manly, R. S., ed. 1970. Adhesion in biological systems. New York: Academic Press. 302 pp.

McLaughlin, P. A., and D. P. Henry. 1972. Comparative morphology of complemental males in four species of *Balanus* (Cirripedia Thoracica). Crustaceana 22: 13–30.

Meith-Avčin, N. 1974. DDT and the rugophilic response of settling barnacles *Balanus improvisus*. J. Fish. Res. Bd. Canad. 31: 1960–63.

Merrill, R. J., and E. S. Hobson. 1970. Field observations of *Dendraster excentricus*, a sand dollar of western North America. Amer. Midl. Natur. 83: 595–624.

Millecchia, R., and G. F. Gwilliam. 1972. Photoreception in a barnacle: Electrophysiology of the shadow reflex pathway in *Balanus cariosus*. Science 177: 438–41.

Milliman, J. D. 1974. Marine carbonates. [Part 1, Recent sedimentary carbonates.] New York: Springer-Verlag. 375 pp.

Molenock, J., and E. D. Gomez. 1972. Larval stages and settlement of the barnacle *Balanus* (*Conopea*) *galeatus* (L.) (Cirripedia Thoracica). Crustaceana 23: 100–108.

Monterosso, B. 1933. L'anabiosi nei cirripedi e il problema della vita latente (ipobiosi). Richerche morfologiche, biologiche e sperimentali in *Chthamalus stellatus* (Poli) var. *depressa* Darwin. Arch. Zool. Ital. 19: 17–379.

Newell, I. M. 1948. Marine molluscan provinces of western North America: A critique and a new analysis. Proc. Amer. Philos. Soc. 92: 155–66.

Newman, W. A. 1960. *Octolasmis californiana* spec. nov., a pedunculate barnacle from the gills of the California spiny lobster. Veliger 3: 9–11.

———. 1967. On physiology and behaviour of estuarine barnacles. Symp. Crustacea, Proc. Mar. Biol. Assoc. India, Part 3: 1038–66.

———. 1972. An oxynaspid (Cirripedia, Thoracica) from the eastern Pacific. Crustaceana 23: 202–8.

———. 1974. Two new deep-sea Cirripedia (Ascothoracica and Acrothoracica) from the Atlantic. J. Mar. Biol. Assoc. U.K. 54: 437–56.

———. 1975. Phylum Arthropoda: Crustacea, Cirripedia, pp. 259–69, *in* R. I. Smith and J. T. Carlton, eds., Light's manual: Intertidal invertebrates of the central California coast. 3rd ed. Berkeley and Los Angeles: University of California Press. 716 pp.

———. 1979. The Californian Transition Zone: Significance of short-range endemics, pp. 399–416, *in* J. Gray and A. Boucot, eds., Historical biogeography, plate tectonics, and the changing environment. 37th Ann. Biol. Colloq. Corvallis: Oregon State University Press.

Newman, W. A., and A. Ross. 1976. Revision of the balanomorph barnacles; including a catalog of the species. San Diego Natur. Hist. Mus., Mem. 9: 1–108.

Newman, W. A., V. A. Zullo, and T. H. Withers. 1969. Cirripedia, pp. R206–95, *in* R. C. Moore, ed., Treatise on invertebrate paleontology, Arthropods 4, vol. 1. New York: Geol. Soc. America; Lawrence: University of Kansas Press.

O'Brien, J. 1977. Observations on *Heterosaccus californicus* Boschma (Cirripedia: Sacculinidae) in individuals and populations of *Pugettia producta* (Randall) (Decapoda: Majidae). 58th Ann. Meet., West. Soc. Natur., Abs. Symp. & Contr. Pap., pp. 51–52 (abstract).

Paine, R. T. 1974. Intertidal community structure. Experimental studies on the relationship between a dominant competitor and its principal predator. Oecologia 15: 93–120.

———. 1979. Disaster, catastrophe, and local persistence of the sea palm *Postelsia palmaeformis*. Science 205: 685–87.

Paine, R. T., C. J. Slocum, and D. O. Duggins. 1979. Growth and longevity in the crustose red alga *Petrocelis middendorffii*. Mar. Biol. 51: 185–92.

Patel, B. 1959. The influence of temperature on the reproduction and moulting of *Lepas anatifera* L. under laboratory conditions. J. Mar. Biol. Assoc. U.K. 38: 589–97.

Petersen, J. A., H. J. Fyhn, and K. Johansen. 1974. Eco-physiological studies of an intertidal crustacean, *Pollicipes polymerus* (Cirripedia, Lepadomorpha): Aquatic and aerial respiration. J. Exper. Biol. 61: 309–20.

Petriconi, V. 1969. Vergleichend anatomische Untersuchungen an Rankenfüssern (Crustacea, Cirripedia). Zur Funktionsmorphologie der Mundwerkzeuge der Cirripedia. Verhandl. Deut. Zool. Ges. Wien 33: 539–47.

Pilsbry, H. A. 1907. The barnacles (Cirripedia) contained in the collections of the U.S. National Museum. Bull. U.S. Nat. Mus. 60: 1–122.

———. 1916. The sessile barnacles (Cirripedia) contained in the collections of the U.S. National Museum; including a monograph of the American species. Bull. U.S. Nat. Mus. 93: 1–366.

Ramenofsky, M., D. J. Faulkner, and C. Ireland. 1974. Effect of juvenile hormone on cirriped metamorphosis. Biochem. Biophys. Res. Commun. 60: 172–78.

Rasmussen, D. I. 1935. Southern California *Balanus-Littorina* communities; Effects of wave action and friable material, pp. 304–8, *in* Shelford et al. (1935).

Reinhard, E. G. 1942. The reproductive role of the complemental males of *Peltogaster*. J. Morphol. 70: 389–402.

Rice, L. 1930. Peculiarities in the distribution of barnacles in communities and their probable causes. Publ. Puget Sound Biol. Sta. 7: 249–57.

Roskell, J. 1969. A note on the ecology of *Conchoderma virgatum* (Spengler, 1790) (Cirripedia, Lepadomorpha). Crustaceana 16: 103–4.

Ross, A. 1962. Results of the Puritan-American Museum of Natural History Expedition to western Mexico. 15. The littoral balanomorph Cirripedia. Amer. Mus. Novitates 2084: 1–44.

Ross, A., M. J. Cerame-Vivas, and L. R. McCloskey. 1964. New barnacle records for the coast of North Carolina. Crustaceana 7: 312–13.

Ross, A., and W. A. Newman. 1967. Eocene Balanidae of Florida, including a new genus and species with a unique plan of "turtle barnacle" organization. Amer. Mus. Novitates 2288: 1–21.

————. 1973. Revision of the coral-inhabiting barnacles (Cirripedia: Balanidae). Trans. San Diego Soc. Natur. Hist. 17: 137–74.

Seapy, R. R., and M. A. Littler. 1978. The distribution, abundance, community structure, and primary productivity of macroorganisms from two central California rocky intertidal habitats. Pacific Sci. 32: 293–314.

Shelford, V. E., A. O. Weese, L. A. Rice, D. I. Rasmussen, A. MacLean, N. M. Wismer, and J. H. Swanson. 1935. Some marine biotic communities of the Pacific coast of North America. Ecol. Monogr. 5: 249–354.

Skerman, T. M. 1958a. Notes on settlement in *Lepas* barnacles. New Zeal. J. Sci. 1: 383–90.

————. 1958b. Rates of growth in two species of *Lepas* (Cirripedia). New Zeal. J. Sci. 1: 402–11.

Smith, G. 1906. Rhizocephala. Fauna und Flora des Golfes von Neapel. 29: 1–123.

Southward, A. J. 1962. On the behaviour of barnacles. IV. The influence of temperature on cirral activity and survival of some warm-water species. J. Mar. Biol. Assoc. U.K. 42: 163–77.

————. 1975. Intertidal and shallow water Cirripedia of the Caribbean. Studies on the fauna of Curaçao and other Caribbean Islands 46, no. 150 (Uitg. Natuurwet. Stud. Suriname Ned. Antillen 82): 1–53.

Southward, A. J., and D. J. Crisp. 1965. Activity rhythms of barnacles in relation to respiration and feeding. J. Mar. Biol. Assoc. U.K. 45: 161–85.

Spight, T. M. 1973. Ontogeny, environment and shape of a marine snail *Thais lamellosa* Gmelin. J. Exper. Mar. Biol. Ecol. 13: 215–28.

Stallcup, W. B. 1953. Distribution of the barnacle *Chthamalus dalli* Pilsbry at Cabrillo Point, Monterey Bay, California. Field & Lab. (Contr. Dept. Sci. So. Methodist Univ.) 21: 143–46.

Stephenson, T. A., and A. Stephenson. 1972. Life between tide-marks on rocky shores. San Francisco: Freeman. 425 pp.

Stickle, W. B. 1973. The reproductive physiology of the intertidal

prosobranch *Thais lamellosa* (Gmelin). I. Seasonal changes in the rate of oxygen consumption and body component indexes. Biol. Bull. 144: 511–24.

Tait, J., and W. F. Emmons. 1925. Experiments and observations on Crustacea. VI. The mechanism of massive movement of the operculum of *Balanus nubilis*. Proc. Roy. Soc. Edinburgh 45: 42–47.

Thorner, E., and W. E. Ankel. 1966. Die Entenmuschel *Lepas fascicularis* in der Nordsee. Natur und Museum 96: 209–20.

Tomlinson, J. T. 1953. A burrowing barnacle of the genus *Trypetesa* (order Acrothoracica). J. Wash. Acad. Sci. 43: 373–81.

————. 1955. The morphology of an acrothoracican barnacle, *Trypetesa lateralis*. J. Morphol. 96: 97–122.

————. 1960. Low hermit crab migration rates. Veliger 2: 61.

————. 1966. The advantages of hermaphroditism and parthenogenesis. J. Theoret. Biol. 11: 54–58.

————. 1969a. The burrowing barnacles (Cirripedia: Order Acrothoracica). Bull. U.S. Nat. Mus. 296: 1–162.

————. 1969b. Shell-burrowing barnacles. Amer. Zool. 9: 837–40.

Törnävä, S. R. 1948. The alimentary canal of *Balanus improvisus* Darwin. Acta Zool. Fennica 52: 1–52.

Turquier, Y. 1971. Recherches sur la biologie des Cirripèdes Acrothoraciques. IV. La métamorphose des cypris mâles de *Trypetesa nassarioides* Turquier et de *Trypetesa lampas* (Hancock). Arch. Zool. Expér. Gén. 112: 301–48.

Utinomi, H. 1958. Studies on the cirripedian fauna of Japan. VII. Cirripeds from Sagami Bay. Publ. Seto Mar. Biol. Lab. 6: 281–311.

————. 1964. Studies on the Cirripedia Acrothoracica. V. Morphology of *Trypetesa habei* Utinomi. Publ. Seto Mar. Biol. Lab. 12: 117–32.

————. 1970. Studies on the cirripedian fauna of Japan. IX. Distributional survey of thoracic cirripeds in the southeastern part of the Japan sea. Publ. Seto Mar. Biol. Lab. 17: 339–72.

Valentine, J. W. 1966. Numerical analysis of marine molluscan ranges on the extratropical northeastern Pacific shelf. Limnol. Oceanogr. 11: 198–211.

Visscher, J. P. 1928. Nature and extent of fouling of ships' bottoms. Bull. U.S. Bur. Fish. 43: 193–252.

Wagin, W. L. 1946. *Ascothorax ophioctenis* and the position of Ascothoracica Wagin in the system of the Entomostraca. Acta Zool. 27: 155–267.

Walker, G. 1972. The biochemical composition of the cement of two barnacle species, *Balanus hameri* and *Balanus crenatus*. J. Mar. Biol. Assoc. U.K. 52: 429–36.

————. 1974. The occurrence, distribution and attachment of the pedunculate barnacle *Octolasmis mulleri* (Coker) on the gills of crabs, particularly the blue crab *Callinectes sapidus* Rathbun. Biol. Bull. 147: 678–89.

Walker, G., P. S. Rainbow, P. Foster, and D. L. Holland. 1975. Zinc

phosphate granules in tissue surrounding the midgut of the barnacle *Balanus balanoides.* Mar. Biol. 33: 161–66.

Werner, W. E. 1967. The distribution and ecology of the barnacle
Balanus trigonus. Bull. Mar. Sci. 17: 64–84.

White, F. 1970. The chromosomes of *Trypetesa lampas* (Cirripedia,
Acrothoracica). Mar. Biol. 5: 29–34.

Whitney, J. O. 1970. Absence of sterol biosynthesis in the blue
crab *Callinectes sapidus* Rathbun and in the barnacle *Balanus nubilus* Darwin. J. Exper. Mar. Biol. Ecol. 4: 229–37.

Willett, G. 1937. An Upper Pleistocene fauna from the Baldwin
Hills, Los Angeles County, California. Trans. San Diego Soc.
Natur. Hist. 8: 379–406.

Worley, L. G. 1939. Correlation between salinity, size and abundance of intertidal barnacles. Publ. Puget Sound Biol. Sta. 7:
233–40.

Yanagimachi, R. 1961a. Studies on the sexual organization of the
Rhizocephala. III. The mode of sex-determination in *Peltogasterella.* Biol. Bull. 120: 272–83.

———. 1961b. The life cycle of *Peltogasterella* (Cirripedia, Rhizocephala). Crustaceana 2: 183–86.

Zullo, V. A. 1963. A review of the subgenus *Armatobalanus* Hoek
(Cirripedia: Thoracica) with the description of a new species
from the California coast. Ann. Mag. Natur. Hist. (13) 6: 587–94.

———. 1966. Zoogeographic affinities of the Balanomorpha (Cirripedia: Thoracica) of the eastern Pacific, pp. 139–44, *in* R. I. Bowman, ed., The Galápagos. Berkeley and Los Angeles: University
of California Press. 318 pp.

———. 1968. Extension of range for *Balanus tintinnabulum californicus* Pilsbry, 1916 (Cirripedia, Thoracica). Occas. Pap. Calif.
Acad. Sci. 70: 1–3.

———. 1969a. A late Pleistocene marine invertebrate fauna from
Bandon, Oregon. Proc. Calif. Acad. Sci. 36: 347–61.

———. 1969b. Thoracic Cirripedia of the San Diego Formation,
San Diego County, California. Los Angeles Co. Mus. Contr. Sci.
159: 1–25.

Zullo, V. A., D. B. Beach, and J. T. Carlton. 1972. New barnacle
records (Cirripedia, Thoracica). Proc. Calif. Acad. Sci. 39: 65–74.

Chapter 21

Isopoda and Tanaidacea: *The Isopods and Allies*

Welton L. Lee and Milton A. Miller

The Isopoda (about 4,000 species) and Tanaidacea (some 400 species) are closely related orders of the crustacean subclass Malacostraca. The isopods include the familiar terrestrial pill bugs, sow bugs, and wood lice and their freshwater and marine relatives. The tanaidaceans, sometimes mistaken for isopods, are entirely aquatic and mostly marine.

The following characteristics are useful in distinguishing isopods from other crustaceans. (1) The body, ovoid or elongate as seen from above, is often dorsoventrally depressed, though the dorsal surface may be either flattened or convex. (2) The three divisions of the body (cephalon, peraeon, and pleon) are often not sharply marked off from one another, but may be readily distinguished by their appendages. (3) The cephalon, consisting of the head fused together with the first thoracic segment, bears a pair of sessile compound eyes, fused immovably with the head shield, and six pairs of appendages. The latter include two pairs of antennae, the first pair smaller than the second and rudimentary in land isopods, and four pairs of mouthparts—a pair of mandibles, two pairs of maxillae, and one pair of maxillipeds. (4) The peraeon consists of the seven free (clearly demarcated) thoracic segments (peraeonites), each bearing a pair of legs (peraeopods), all uniramous (unbranched, I-shaped) and nearly alike ("iso-poda"="equal legs"). (5) The pleon consists of the six abdominal segments (pleonites) and a terminal anal plate (telson), mutually distinct or variously fused to one another. Each of the first five segments usually bears a pair of bira-

mous (branched, Y-shaped) pleopods used in respiration and, in some forms, for swimming as well. The sixth abdominal segment is coalesced with the telson and bears a pair of biramous (occasionally uniramous) uropods (the term "pleotelson" refers to the telson plus all pleonites fused to or consolidated with it).

The uropods are variously adapted and hence of prime importance in distinguishing isopod suborders. They may be hinged laterally to the pleotelson, folding together like cabinet doors beneath the pleopods (suborder Valvifera and family Tylidae of the terrestrial suborder Oniscoidea). They may hinge laterally and bear flattened branches that extend alongside the telson (or arch above it) to form with it a caudal fan (suborders Flabellifera, Gnathiidea, and Anthuridea). Or they may be terminal and bear slender branches that usually extend beyond the telson (suborders Asellota and Oniscoidea; also the order Tanaidacea). For further diagnoses of the isopod suborders, see the keys in Miller (1975).

Tanaidaceans (cheliferans) differ from isopods in several respects. They have pincerlike (chelate) first legs; in isopods the first legs may be subchelate like jackknives, but never truly chelate. Tanaidaceans have a short carapace (headfold) that is fused with the first two thoracic segments, leaving six free peraeonites, rather than seven as in isopods; and tanaidaceans have stalked (rather than sessile) eyes. Tanaidacean pleopods, unlike those of isopods, are not modified for respiration, that function being served by the lining of the carapace. Finally, tanaidaceans bear uropods in which the inner branch and sometimes the outer branch, is jointed, rather than being without joints as in isopods.

The Tanaidacea are divided into two suborders, Monokonophora and Dikonophora, primarily on the basis that monokonophoran

Welton L. Lee is Chairman of the Department of Invertebrate Zoology at the California Academy of Sciences. Milton A. Miller is Professor Emeritus of Zoology at the University of California, Davis.

males bear a single, rather than a double, penial or genital process. This is a small conical elevation between the last pair of peraeopods, either on the midventral line (if single) or on each side of it (if paired). In female monokonophorans, the brood pouch is always formed by four pairs of plates (oostegites) from the bases of peraeopods 2–5; whereas in ovigerous dikonophorans, it is formed by either one pair or four pairs of oostegites (depending on the species). Regardless of sex, members of the two suborders may be distinguished by the flagellum of the first antenna, which is biramous in Monokonophora but uniramous in Dikonophora. It should be noted that protogyny (sex reversal, with female characteristics developed first) is common within the Dikonophora, in which life histories are better understood than in the Monokonophora (Gardiner, 1973).

Both isopods and tanaidaceans are predominantly benthic animals, although many can swim and some isopods are planktonic (Dow & Menzies, 1957). Many occur intertidally, especially on reefs where they hide under rocks, in crevices, empty shells, or worm tubes, or among seaweeds and sessile animals such as tunicates, bryozoans, hydroids, and sponges; some burrow in driftwood, sand, or mud (see, for example, Lang 1960a). Many are nocturnal, like the beach pill bugs that emerge at night to forage. Most are inconspicuous because of their small size, camouflaging shape and coloration, and secretive habits; some can even change their color or pattern to match their background. Few are large enough or common enough to be seen by the casual observer, hence the exclusion of many smaller species from this chapter.

Salinity and temperature are the major physical factors limiting the distribution of aquatic isopod species. Along the Pacific coast, one finds many strictly marine forms, others that are confined to bays and estuaries and tolerant of the lower and fluctuating salinities there (e.g., *Gnorimosphaeroma luteum*, 21.6, which occurs in waters ranging from fresh to nearly 100 percent seawater), and a very few strictly freshwater species (notably, *Asellus tomalensis* of coastal streams and lakes in California and *A. occidentalis* on the Pacific coast from British Columbia to Oregon; see Bowman, 1974.) The degree to which the animals can osmoregulate, especially during early development, surely helps set limits to the distribution and dispersion of aquatic species into brackish or freshwater habitats.

The latitudinal distribution of species along the Pacific coast suggests that temperature also plays a role in their distribution. Point Conception (Santa Barbara Co.) is a major landmark separating warm-water from temperate and cold-water faunas. Some isopods, like *Idotea* (*Pentidotea*) *resecata* (21.16), which ranges from Alaskan waters to the Gulf of California, can tolerate a wide range of ambient temperatures. Other species are more restricted in both thermal tolerance and geographical distribution.

A third factor, humidity, is important in the ecology of terrestrial isopods (suborder Oniscoidea). Maritime isopods, those of the upper shore, like *Ligia* (e.g., 21.12, 21.13), are imperfectly adapted for a truly terrestrial existence. The need to prevent lethal drying of their gill-like pleopods and excessive evaporation through their permeable exoskeletons confines them to damp habitats. They probably represent a transitional stage in the evolutionary emigration from the ancestral sea to land.

Isopods occupy various ecological niches. They form important intermediate links in food chains as herbivores, as predators, and parasites, and as scavengers and detritus feeders that help reduce and recycle dead and decaying organic matter. Some are ectocommensal on other invertebrates, notably crustaceans, echinoderms, and mollusks. Members of the suborder Epicaridea parasitize crustaceans exclusively, commonly decapods; large members of the family Cymothoidae (suborder Flabellifera) are conspicuous parasites in the mouths and on the gills of fishes. Woodborers, notably the "gribbles" (*Limnoria*, 21.8, and allies), become economic pests when they riddle pilings, wooden hulls, and other submerged structures.

Female isopods and tanaidaceans brood their young in a pouch (marsupium) formed beneath the thorax by thin, overlapping plates (oostegites) that project medially from the basal joint of one or more pairs of legs. Development of free-living forms is direct, and the young that emerge from the brood pouch are essentially miniatures of their parents (lacking only the last pair of thoracic legs, which are acquired later). Parasitic isopods, however, have larval stages rather different from the adult form. The molting process is unique in isopods: most shed half their exoskeletons at a time, rear half first. Since considerable time (perhaps days) elapses between anterior and posterior molts, the new exoskeleton has sufficient time to harden, so that the animals can use about one-half of their appendages during the process, and motility is not completely lost as it is in arthropods that shed their entire exoskeleton at one time.

The sexes generally look alike, but marked sexual dimorphism is exhibited by some species, for example, *Idotea* (*Pentidotea*) *montereyensis* (21.19), *Paracerceis cordata* (21.5), and *Phyllodurus abdominalis* (21.9). In *Phyllodurus* (as in other members of the parasitic suborder Epicaridea) the males are diminutive and simple compared to the relatively huge and often grotesquely modified females.

The wide distribution and great diversity of the Isopoda may be partly a function of the age of the group. Fossils first appear in the Middle Pennsylvanian period; they represent the earliest certain members of the superorder Peracarida of which we have record. The abundance of isopods both at the seashore and on land, their convenient size, and the ease with which they are collected, maintained, and handled, have led to their use in a variety of studies.

The most useful references for identifying west coast isopods and tanaidaceans are Miller (1975, central California only), Schultz (1969, marine isopods only), Hatch (1947, limited to Washington and vicinity), Richardson (1905, 1909, classical monographs now greatly outdated), and Sieg & Winn (1978, tanaidacean suborders and families only). Miller, Schultz, and Sieg & Winn give only keys and figures, whereas the other sources also provide descriptions. A comprehensive monograph on Pacific coast isopods and tanaidaceans is long overdue. Additional taxonomic literature and other references are cited in the text.

For identification of land isopods, see especially Miller (1975), Hatch (1947), and Van Name (1936, 1940, 1942, no keys); see also papers by Menzies (1950a), Schultz (1970, 1972), and Vandel (1951). The substantial literature on the economically important woodborers is cited under *Limnoria* spp. (21.8).

Phylum Arthropoda / Class Crustacea / Subclass Malacostraca
Superorder Peracarida / Order Isopoda / Suborder Flabellifera
Family Cirolanidae

21.1 **Cirolana harfordi** (Lockington, 1877)

Common under stones resting in sand and mussel beds, middle intertidal zone; British Columbia to Baja California; Channel Islands.

Body to 20 mm long; telson rounded posteriorly, its border armed with about 26 stout, non-plumose setae; color variable, pale gray to nearly black; "two-tone" individuals which have just molted half the exoskeleton are not uncommon.

These isopods sometimes occur in enormous numbers, up to 12,600 per m² in mussel beds at Pacific Grove, Monterey Co. (Hewatt, 1937). Females live up to 18 months, each producing one or two broods of 18–68 young, with an incubation period of about 3–4 months. Ovigerous females have been reported from central California in April. These isopods are voracious scavengers on dead animal matter and may also be predators. They will swarm over dead fish and strip the bones in a short time. Ichthyologists sometimes use them to clean fish skeletons for study. They will eat any meat (which they locate by olfaction) and they bolt it down in huge bites, but digestion is slow—a gorged gut may take a month to empty. Examination of the gut contents of freshly captured isopods shows that the diet consists mainly of polychaete worms, with amphipods second in importance, and copepods last. It is estimated that 1,000 *Cirolana* ingest annually 300 kcal, of which 262 are assimilated (Johnson, 1974).

Because of their relatively large size and numbers, these isopods are an important link in intertidal food chains. They are eaten by several inshore fishes, including the kelpfish (*Gibbonsia elegans*), the rockweed gunnel (*Xererpes fucorum*), the dwarf perch (*Micrometrus minimus*), and probably others.

Two other species of *Cirolana*, **C. californiensis** Schultz and **C. joanneae** Schultz, have been described from the depths of submarine canyons off southern California (Schultz, 1966).

Three species in the related genus *Excirolana* occur in sandy beaches along the Pacific coast. **E. linguifrons** (Richardson) is commonest in central California in beaches containing beds of the sand crab *Emerita analoga* (24.4), where it feeds on dead sand crabs, sand crab eggs, and such other animal matter as comes along. MacGinitie (1939) noted "it is almost impossible for one to stand still barefooted in a bed of *Emeritas* because these little scavengers begin eating on one's foot, and each bite is like the prick of a needle." Another form, **E. chiltoni** (Richardson), which ranges from Oregon to southern California, has been studied by Enright (1965). These isopods remain burrowed most of the time, but many come out and swim about for a few hours as a high tide recedes. This rhythmic pattern of activity is evidently inherent, for it persists for several days when animals are kept in laboratory aquaria. Experiments show that it is probably the turbulence of the waves on the beach at high tide that serves to keep the rhythm synchronous in the isopod population. A third species, **E. kincaidi** (Hatch), ranges from Washington and Oregon to Half Moon Bay, San Mateo Co. (J. Carlton, pers. comm.).

See especially W. Johnson (1974, 1976a, b); see also Brusca (1966), Enright (1965), Hatch (1947), Heusner & Enright (1966), Hewatt (1935, 1937, 1938), Johnson & Snook (1927), MacGinitie (1939), MacGinitie & MacGinitie (1968), Miller (1975), Mitchell (1953), Richardson (1905), and Schultz (1966, 1969).

*Phylum Arthropoda / Class Crustacea / Subclass Malacostraca
Superorder Peracarida / Order Isopoda / Suborder Flabellifera
Family Sphaeromatidae*

21.2 **Tecticeps convexus** Richardson, 1899

Present in lowest intertidal zone, but more commonly subtidal on sandy, rocky bottoms; Crescent City (Del Norte Co.) to Point Conception (Santa Barbara Co.).

Body to 15 mm long, oval, flattened; first pair of legs in females, and first and second pairs in males, subchelate and prehensile; telson entire, posteriorly rounded; uropods with two branches pointed, subequal in length.

These animals are good swimmers. When disturbed they fold the body double to form a flattened ball and extend laterally a pair of sharp daggerlike processes borne on the uropods. They also burrow into the sand or gravel, where they are difficult to detect because their color pattern closely matches the substratum. The neurosecretory system of the head has been studied in a related Japanese species.

See especially Richardson (1905); see also Hansen (1905), Miller (1975), Oguro (1960), and Schultz (1969).

*Phylum Arthropoda / Class Crustacea / Subclass Malacostraca
Superorder Peracarida / Order Isopoda / Suborder Flabellifera
Family Sphaeromatidae*

21.3 **Sphaeroma quoyana** H. Milne-Edwards, 1840 (=S. pentodon)

Often abundant, burrowing in hard clay, mud, soft sandstone, wood, and the foam plastic of mooring flats, low intertidal and adjacent subtidal shallows in protected bays and estuaries; Humboldt Bay to Bahía de San Quintín (Baja California); Australia, Tasmania, New Zealand.

Body to about 11 mm long; telson with two longitudinal rows of four tubercles each; outer branch of uropod bearing five teeth.

Until the studies of Rotramel (1972), North American specimens of *S. quoyana* were misidentified as *S. pentodon* Richardson. These animals seek protection by boring holes up to 9 mm in diameter and 35 mm deep in a variety of substrata. Boring is done with the powerful mandibles. The animals brace themselves, bite off chunks of material, and wash away the fragments in a current created by the pleopods. They often sit at the mouths of their burrows, retreating and rolling up into a ball when disturbed. They feed mainly on algae; when boring in wood, they do not eat the wood itself as does *Limnoria*, a much more destructive boring isopod. *S. quoyana* tolerates brackish conditions well, and in laboratory experiments half the animals lived for 11 days immersed in fresh water. The salinity of the blood fluctuates along with changes in the salinity of the surrounding environment.

Breeding probably occurs the year around. As in other sphaeromatid isopods, the young are brooded in deep paired pits ("uteri") in the ventral body wall, and the plates (oostegites) forming the brood pouch serve merely to cover the openings to the uteri.

The tiny commensal isopod **Iais californica** (Richardson) is a common companion of *S. quoyana* in central California; it lives on the body of its larger host, not physically attached but usually crawling about the ventral side or nestling between the pleopods. An east coast *Sphaeroma*, *S. quadridentatum*, is host to an intestinal fungus (*Palavascia*, a trichomycete), which lives attached to the exoskeletal lining of the hindgut.

This lining, along with the fungus, is eliminated each time the isopod molts, yet nearly 100 percent of the isopods bear the fungi. Reinfection occurs when isopods ingest fecal pellets containing the thick-walled fungus spores; the spores germinate only after ingestion by an isopod. Ingestion of fecal pellets occurs in other isopods, and indeed for terrestrial forms it has been suggested that this may be essential; one passage through the gut concentrates some materials (e.g., copper) to the degree that they can be absorbed effectively from the food the second time around.

For *S. quoyana* (usually as *S. pentodon*), see Barrows (1919), Johnson & Snook (1927), Menzies (1962), Menzies & Barnard (1951), R. Miller (1926), Richardson (1905), Ricketts & Calvin (1968), Riegel (1959b), Rotramel (1972), and Schultz (1969). For studies of related species, see Audoinoit (1956), Bishop (1969), Bouquet, Levi & Teissier (1951), Hansen (1905), Lichtwardt (1961), Menzies (1954b), Okay (1943, 1945), Potier (1951), West (1964), and Wieser (1965, 1966).

Phylum Arthropoda / Class Crustacea / Subclass Malacostraca
Superorder Peracarida / Order Isopoda / Suborder Flabellifera
Family Sphaeromatidae

21.4 **Exosphaeroma amplicauda** (Stimpson, 1857)

Under stones, low intertidal zone on rocky shores and in bays; Aleutian Islands (Alaska) to San Diego and Santa Catalina Island (Channel Islands).

Body to 8 mm long, increasing in width posteriorly with relatively huge, expanded uropods.

Little is known of the biology of this species. The function of the gigantic uropods is certainly worth investigation. Two related species, **E. octoncum** (Richardson) and **E. rhomburum** (Richardson), occur intertidally in California waters, and a third form, **E. inornata** Dow, has been described from holdfasts of the kelp *Macrocystis* off southern California.

See Dow (1958), Given & Lees (1967), Hansen (1905), Hatch (1947), Johnson & Snook (1927), Menzies (1954a), Miller (1975), Richardson (1905), and Schultz (1969).

Phylum Arthropoda / Class Crustacea / Subclass Malacostraca
Superorder Peracarida / Order Isopoda / Suborder Flabellifera
Family Sphaeromatidae

21.5 **Paracerceis cordata** (Richardson, 1899)

On pink coralline algae and kelp holdfasts on exposed rocky shores, on sandy mud in bays, low intertidal zone to 55 m depth; Aleutian Islands (Alaska) to southern California.

Body about 7 mm long; pleotelson with terminal notch formed by three parallel sinuses connected in series in older males, with a simple notch in females and young males.

Sexual dimorphism is so marked in this species that males and females were originally placed, not just in separate species, but in separate genera. Up to five individuals per square meter have been found in holdfasts of the kelp *Macrocystis* in Monterey Bay. Two other morphologically similar species of *Paracerceis* occur in southern California, **P. sculpta** (Holmes) and **P. gilliana** (Richardson).

For *P. cordata*, see Andrews (1945, as *Cilicaea cordata*), Hansen (1905), Miller (1975), Richardson (1905, as *Cilicaea cordata*), and Schultz (1969). For other species of *Paracerceis*, see Miller (1968) and Schultz (1969).

Phylum Arthropoda / Class Crustacea / Subclass Malacostraca
Superorder Peracarida / Order Isopoda / Suborder Flabellifera
Family Sphaeromatidae

21.6 **Gnorimosphaeroma luteum** Menzies, 1954
(=G. oregonensis lutea, G. lutea)

Sometimes common, sheltered in mud under logs and stones in brackish bays and coastal streams (e.g., mouth of Salinas River, Monterey Co.), low intertidal and shallow subtidal zones; Alaska to central California.

Body to 8.5 mm long, smooth; legs all ambulatory; telson broadly rounded and entire.

Gnorimosphaeroma is exceptional among isopod genera in containing both freshwater and marine species. *G. luteum* can meet and withstand salinities of 0.6–135 percent seawater; the salinity of the blood shifts to some extent along with that of the external environment. The closely related form **G. oregon-**

ensis (Dana) lives on the open coast, on rocky or shell substrata, and is much less tolerant of salinity changes. A third species, **G. noblei** Menzies, described from Tomales Bay (Marin Co.) is now known from several localities along the California coast between Humboldt Bay and Palos Verdes Peninsula (Los Angeles Co.), and also from the Kurile Islands. A fourth species, **G. rayi** Hoestlandt, is known only from Tomales Bay.

See Eriksen (1968), Filice (1958), Hoestlandt (1964, 1973), Lockwood (1962), MacGinitie & MacGinitie (1968), Menzies (1954a), Miller (1975), Ricketts & Calvin (1968), Riegel (1959a,b), and Schultz (1969).

Phylum Arthropoda / Class Crustacea / Subclass Malacostraca
Superorder Peracarida / Order Isopoda / Suborder Flabellifera
Family Cymothoidae

21.7 Lironeca vulgaris Stimpson, 1857
(=Livoneca vulgaris)   Fish Louse

Often common, on gills of the olive rockfish, starry flounder, jack mackerel, sand bass, lingcod, and many other west coast fishes, surface to at least 550 m depth in coastal waters; Washington to Baja California.

Adult females to 43 mm long, the body broad, the head and abdomen more or less buried in thorax, the telson nearly twice as broad as long; males to 31 mm long, the body smaller and narrower, cream-colored.

This common parasite of fishes, though not a resident of the intertidal zone, is often noted in fish markets. Juveniles are free-swimming; adults are generally attached in the gill chamber of the host. After settlement, the animals first become functional males, then with further growth transform into females. Ovigerous adults have been observed from August through April on kelp-bed fishes off Santa Monica (Los Angeles Co.); females bearing juveniles ready to hatch were commonest in January and February.

Studies of the related species *L. convexa* from the bait fish *Chloroscombrus orqueta* show that this isopod, too, functions first as a male, later as a female. Males occur in the gill chamber, with their heads directed anteriorly; only the male

stage destroys fish tissue, but whether or not it eats it is uncertain. The female stage of *L. convexa* lives on the floor of the host's mouth, close to the gill rakers, and faces caudally. Females appear to cause no damage to the fish, but they are in a position to pick up food filtered from the water by the gill rakers. Ovaries are already present in the male in rudimentary form, enlarging as the testes atrophy during the change in sex. The plates that form the female brood pouch (oostegites) apparently develop wholly in connection with a single molt.

See Brusca (1978), Calman (1898), Hatch (1947), Menzies, Bowman & Alverson (1955), Miller (1975), Nicola (1949), Orcutt (1950), Richardson (1905), Schultz (1969), and Turner, Ebert & Given (1969).

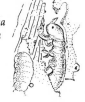

Phylum Arthropoda / Class Crustacea / Subclass Malacostraca
Superorder Peracarida / Order Isopoda / Suborder Flabellifera
Family Limnoriidae

21.8 Limnoria spp.
Gribble

Common, boring into wooden pilings, floats, ships, and the holdfasts of larger brown algae; genus distributed worldwide except in high Arctic and Antarctic waters.

These small isopods are mostly under 4 mm long. The species fall into two subgenera, distinguished as follows: *Limnoria* (*Limnoria*) species bore in wood, and the incisors of the right and left mandibles are equipped, respectively, for filing and rasping; *Limnoria* (*Phycolimnoria*) species bore in holdfasts of larger kelps, and the incisors of the mandibles are not equipped for rasping or filing.

Four species of the genus *Limnoria* occur in California waters:

L. (Limnoria) lignorum Rathke, 1799; in wood; Aleutian Islands (Alaska) and southern Alaska to Point Arena (Mendocino Co.), Newfoundland to Rhode Island, Iceland and northern Europe; a boreal species and the only *Limnoria* on the west coast north of Crescent City (Del Norte Co.); no tubercles on telson.

L. (Limnoria) quadripunctata Holthuis, 1949; in wood;

Crescent City (Del Norte Co.) to La Jolla (San Diego Co.), New Zealand, southern Europe, Mediterranean Sea, South Africa; a temperate species limited to regions where ocean temperatures average between 11.4 and 16.2°C for at least 5 consecutive months during the year; four tubercles on telson. (Fig. 21.8.)

L. (Limnoria) tripunctata Menzies, 1951; in wood; San Francisco Bay to Mazatlán (Mexico); a temperate-tropical species distributed worldwide in warmer waters; three tubercles on telson.

L. (Phycolimnoria) algarum Menzies, 1956; in holdfasts of kelps such as *Egregia*, *Macrocystis*, *Nereocystis*, *Laminaria*, *Eisenia*, and *Postelsia* (brown algae of the order Laminariales); Cape Arago (Oregon) to Bahía de Tortuga (Baja California); telson lacking tubercles but with two small longitudinal ridges medially. This large (up to 6 mm long) species can seriously weaken algal holdfasts by its tunnels in the rhizoids.

Along the California coast, *L. (L.) quadripunctata* is found in cooler open coastal situations, and is the only gribble found to date in Tomales Bay (Marin Co.), Monterey Bay, Morro Bay (San Luis Obispo Co.), and Santa Barbara. *L. (L.) tripunctata* is the only gribble in the warmer and less-saline parts of San Francisco Bay and the Los Angeles–Long Beach harbor area. The two species occur together in intermediate environments in both San Francisco Bay and the Los Angeles area. At the northern end of the state, mixed infestations of *L. (L.) quadripunctata* and *L. (L.) lignorum* have been reported at Eureka.

Limnoria species bore for both protection and food. The wood borers, with their asymmetrical rasp-and-file mandibles, bite off and swallow bits of wood. A gastric mill in the foregut, a sort of internal set of jaws, further grinds the food. Investigators report that up to 30 percent (in some experiments up to 60 percent) of the wood is utilized by the isopods. The isopod midgut secretes a battery of enzymes capable of digesting cellulose, the main component used, as well as other materials. The lignins are not digested; they preserve much of the structure of the wood (even as seen under the scanning electron microscope), so that ingested fragments may show little morphological change in their passage through the mid- and hindguts.

The cellulose-splitting enzymes (cellulases) definitely come from the isopod, and not from bacteria, though the marine environment abounds with bacteria that digest cellulose. Scanning electron microscope studies of the entire lining of the gut, including digestive glands, of *Limnoria tripunctata* and *L. lignorum* (and of the marine wood-inhabiting amphipod *Chelura terebrans* and the land isopod *Oniscus asellus*) demonstrate the remarkable fact that the gut is sterile, completely lacking resident microorganisms. These species seem to be the first multicellular animals of any sort that are *known* not to contain microbial residents in the gut; even the fecal pellets after ejection have few associated bacteria. The causes are not clear. Are the microorganisms all swiftly digested? Does the gut contain powerful antibiotics?

In any event, *Limnoria* individuals get their basic energy requirements from the wood they eat. They can live for a while on a diet of cellulose filter paper, and longer on one of whole wood, but even wood does not provide a balanced diet. Other nitrogen sources for *Limnoria* have not been pinpointed, but it seems likely that fungi, universally present in waterlogged wood, may contribute important elements to the diet. So, too, may the microorganisms of various sorts that abound on the outer surface of the body and appendages, especially the pleopods, and that may be ingested when the animal grooms itself.

Food reserves, stored mainly as glycogen, are rapidly depleted after a few days of starvation and do not appear sufficient to support life under anaerobic conditions. The need to keep the burrow well ventilated perhaps explains why *Limnoria* species bore only in the superficial layers of wood (10–15 mm below the surface) and periodically construct small tunnels to the surface as well as interconnections between adjacent burrows.

Limnoria (L.) lignorum breeds all year with a peak in April and May and a lull in January and February at Friday Harbor (Washington). Ovigerous females carry an average of 22 eggs. *L. (L.) tripunctata* at La Jolla (San Diego Co.) likewise breeds all year, with a peak in September and a midwinter decline. In laboratory experiments, breeding occurred between 15 and 30°C, and the fastest population increase took place at 25°C. Brood size in natural populations at La Jolla averages four to ten per female, the broods being smallest under situations of highest population density (reminiscent of results observed in mice under crowded conditions). The breeding season of *L.(L.) quadripunctata* is uncertain; brood size averages 9.5. The mean number of broods per year is still unknown for any species, but is probably more than one. In all species examined, burrows large enough to hold two animals are usually occupied by a sexual pair. When young are released, they generally remain in the same wood mass as their parents, boring their own tunnels. Animals leaving home and swimming to new and separate boring sites are almost invariably sexually mature males and females (but not ovigerous females). Migration is greatest in the spring or summer months. Arrivals at a new site tend to settle and bore near burrows already established, if any are present.

Digging *Limnoria* out of infested wood for laboratory examination used to be a laborious process, but no longer (Johnson & Ray, 1962). A hole is bored in the center of the chunk of wood to be examined, filled with a supersaturated salt solution, and corked, and the block is then placed in a bowl of seawater. As the salt diffuses outward, the *Limnoria* migrate to the surface of the block.

The annual cost of damage caused by gribbles to floats and pilings is high, and even creosote treatment of the wood before immersion is no longer more than a temporary deterrent to some forms, especially *L. (L.) tripunctata*. Perhaps here, as in the case of insect pests and pesticides, we are aiding nature in the development of resistant strains.

See especially the excellent summary papers of Menzies (1954b, 1957, 1958, 1959); see also Becker (1959), Beckman & Menzies (1960), Boyle & Mitchell (1978), Coker (1923), Fahrenbach, (1959), George (1966), Haderlie (1969, 1971, 1976), Henderson (1924), F. Johnson & Ray (1962), M. W. Johnson (1935), M. W. Johnson & Menzies (1956), Kühne (1973), Lane (1959), Menzies (1951, 1955), Menzies, Mohr & Wakeman (1963), Meyers & Reynolds (1957), R. Miller (1926), Mohr (1959), North (1964, with observations on *L. (P.) algarum*), Ray (1956, 1959a,b,c), Ray & Julian (1952), Ray & Stuntz (1959), Reish & Hetherington (1969), Reynolds & Meyers (1959), Schultz (1969), Sleeter et al. (1978), and Strunk (1959).

Phylum Arthropoda / Class Crustacea / Subclass Malacostraca
Superorder Peracarida / Order Isopoda / Suborder Gnathiidea / Family Bopyridae

21.9 **Phyllodurus abdominalis** Stimpson, 1857

External parasite on the abdomen of the burrowing blue mud shrimp, *Upogebia pugettensis* (24.1), low intertidal mud flats in protected bays; British Columbia to central California.

Eyes absent; female body to 14 mm long, broad, ovate, with leaflike pleopods projecting laterally from abdomen; males small, to 6 mm long, less modified.

In this fairly common ectoparasitic species, the isopods usually occur in pairs. The female clings firmly with pointed recurved legs, usually placed across the host's abdomen just anterior to the first pair of pleopods; males apparently move about more freely on the host. Details on the method of feeding and degree of damage caused to the host are lacking.

See Danforth (1970), Fee (1926), Hatch (1947), MacGinitie & MacGinitie (1968), Richardson (1905), and Schultz (1969).

Phylum Arthropoda / Class Crustacea / Subclass Malacostraca
Superorder Peracarida / Order Isopoda / Suborder Gnathiidea / Family Bopyridae

21.10 **Ione** sp.

Externally parasitic on the gills of the bay ghost shrimp, *Callianasssa californiensis* (24.2), in burrows on low intertidal mud flats in protected bays; British Columbia to California.

Females to 25 mm long, the lateral plates of abdomen fingerlike, as long as uropods, bearing a featherlike fringe; males smaller, the lateral plates of abdomen not fringed.

These parasites occur in the branchial chamber of *Callianassa* along with another crustacean invader, the copepod *Hemicyclops thysanotus*. Little is known of the biology or host relations of either species.

See Danforth (1970), Fee (1926), Gooding (1960), Hatch (1947), Hiraiwa (1936), Richardson (1905), Schultz (1969), and Shiino (1964, 1965).

Phylum Arthropoda / Class Crustacea / Subclass Malacostraca
Superorder Peracarida / Order Isopoda / Suborder Gnathiidea / Family Bopyridae

21.11 **Aporobopyrus muguensis** Shiino, 1964

Parasitic in the gill chamber of the crablike anomuran *Pachycheles rudis* (24.20); taken sporadically at Dillon Beach (Marin Co.), Pacific Grove (Monterey Co.), and Point Mugu (Ventura Co.).

Females to 6 mm long, with lateral plates of abdomen flat but not elongated, the uropods uniramous; males to 3.5 mm long, the last abdominal segment triangulate but posterolateral angles not produced into long processes.

Parasites on the gills of crabs, concealed in the gill chamber, are easily overlooked, and the present species is doubtless more widespread than indicated here. Internal parasites of crabs, too, are easily missed by the casual observer. A case in point is the endoparasitic isopod **Portunion conformis** Muscatine found in the visceral cavity of the crab *Hemigrapsus oregonensis* (25.45) curled below and around the gut. There are no external symptoms of infection, but in the San Francisco Bay area an examination of 372 crabs revealed 85 female and 6 male parasites.

See especially Shiino (1964); see also Danforth (1970), Hiraiwa (1936), Muscatine (1956), Schultz (1969), and Shiino (1965).

Phylum Arthropoda / Class Crustacea / Subclass Malacostraca
Superorder Peracarida / Order Isopoda / Suborder Oniscoidea / Family Ligiidae

21.12 **Ligia (Megaligia) occidentalis** Dana, 1853

Often fairly common under stones and in crevices, high intertidal zone and just above the highest tide marks on rocky beaches; Sonoma Co. to Central America.

Body length to 25 mm; eyes separated by a distance equal to the greatest diameter of an eye; uropods long, with basal segment several times longer than broad.

These fast-moving isopods sometimes scurry over rocks in full sunlight, but during daylight hours they are more commonly hidden under rocks or stranded *Macrocystis*. Activity commences in the late afternoon and continues through the night until early morning. The animals are scavengers and scrapers, feeding on dead plant and animal material and also on the algal films present on intertidal rocks.

L. (M.) occidentalis, although essentially terrestrial, represents a transitional stage in the evolution from aquatic habitats to life on land. It avoids entering the water and dies if kept continuously submerged. However, it is confined to the shore by the need to keep its pleopod gills moist and the need to replace water lost by evaporation from the body. When dry, the animals locate tidepools, possibly by following humidity gradients. Once water is found it is usually first tested with the antennae, then the animal turns and dips its uropods and tail. Water is then drawn up to the pleopods by capillary action. It is possible that at this time the animals engage in "anal drinking" as the major means of replacing body water stores, a habit common among terrestrial isopods. *L. (M.) occidentalis* is actually able to survive longer in experiments with constant lowered humidities than most California terrestrial isopods (although none can withstand suboptimal dryness for long). Its longer survival is likely associated with the fact that its body holds a larger store of water, and its ratio of body surface to body volume, an important factor in evaporation, is lower than in most of the terrestrial forms that are smaller in overall size. With water loss, the body fluids become more concentrated. Recent work shows that *L. (M.) occidentalis* has a relatively high level of free amino acids in the tissues, and it is possible that these play a role in regulating the osmotic pressures of body cells to compensate for changes in salinity of the blood.

Like some other intertidal isopods, this species shows a diurnal rhythm of body-color change; the animals are pale at night and darker during the day, due to clustering and dispersal, respectively, of pigment granules within the body pigment cells (chromatophores). Maximum pigment dispersal occurs by 8:00 a.m., and maximum concentration by 10:00 p.m. Under laboratory conditions of constant light, the rhythm is quickly lost, and animals on dark backgrounds remain dark while those on light backgrounds remain light. In animals held in constant darkness, the rhythm is maintained for a few days but is gradually lost. The pigmentary change is probably related to protective coloration.

Little is known of the reproductive cycle of *L. (M.) occidentalis*, but ovigerous females have been noted at Pacific Grove (Monterey Co.) in March, May, and June.

See Abbott (1939), Armitage (1960), Brusca (1966), Fingerman (1963), Given & Lees (1967), Hewatt (1938), Hewitt (1907), Hilton (1915), Jackson (1922), Johnson & Snook (1927), Kittredge et al. (1962), Miller (1938, 1975), Richardson (1905), Ricketts & Calvin (1968), Ruck & Jahn (1954), Van Name (1936), and Wilson (1970), and the account and references cited under *L. (L.) pallasii* (21.13). Numerous studies relating to structure, behavior, ecology, and physiology have been carried out on other species of *Ligia*; for a sampling of these, see Alexandrowicz (1952), Barnes (1935, 1940), Carlisle (1956), Dresel & Moyle (1950), Edney (1954, 1957, 1958), Ellenby (1951), Ellenby & Evans (1956), Enami (1941), Jackson (1926), Kerkut & Krishnaswamy (1960), Kleinholz (1937), Kleitman (1940), Lagerspetz & Lehtonen (1961), Lockwood (1962), Nicholls (1931), Numanoi (1933, 1934, 1937), Ondo & Mori (1956), Parry (1953), Stewart (1913), Suzuki (1934), Tait (1910, 1917, 1927), Tinturier-Hamelin (1960, 1961), and Todd (1963).

Phylum Arthropoda / Class Crustacea / Subclass Malacostraca
Superorder Peracarida / Order Isopoda / Suborder Oniscoidea / Family Ligiidae

21.13 **Ligia (Ligia) pallasii** Brandt, 1833

Sometimes common in caves and crevices on rocky cliffs, especially those with freshwater seep, often with green alga *Enteromorpha*, high intertidal zone and above; western Aleutian Islands and Alaskan mainland to Santa Cruz Co.

Body to 35 mm long; eyes separated by a distance equal to twice the greatest diameter of an eye; uropods short, with basal segment as long as broad; adult male with greatly expanded side plates (epimera).

Sexual dimorphism in body proportions is marked in *L. (L.) pallasii*. Males, with their large, laterally expanded epimeral plates, have a length/width ratio of about 1.6; females and immature males are much narrower, with a corresponding ratio of about 2.1. Half-molted males are occasionally found in which the rear half of the body (from the fourth thoracic segment back) has molted and is abruptly and significantly wider than the as-yet-unmolted front half—an odd sight. The normal aspect is restored when the anterior molt occurs, usually several days after the posterior molt. On the walls of

cliffs and sea caves, the larger and broader males often cover and shield the females and juveniles.

The distribution of this species overlaps that of its relative, *L. (M.) occidentalis* (21.12), in central California. Northward, *L. (L.) pallasii* replaces *L. (M.) occidentalis*, whereas the converse is true southward. Where their distributions overlap geographically, the two species are generally ecologically segregated, in that *L. (L.) pallasii* prefers sea cliffs and *L. (M.) occidentalis* is commoner on rocky beaches. Wilson (1970) reports differences in the osmoregulatory responses shown by these species associated with their behavior and ecology. He found that the slower-moving *L. (L.) pallasii* lives permanently in cool, moist habitats characterized by fluctuating hyposaline conditions, whereas the faster-moving *L. (M.) occidentalis* alternates its activity between periods of foraging in drier, more terrestrial situations and periods of exposure to wet conditions of variable salinity where it can replace water previously lost by evaporation. Both species are air breathers with gill-like pleopods not equipped with tracheal trees (pseudotracheae). The respiratory pleopods must be kept moist to function properly. This is done by immersion or by dipping the tail in water in such a way that the uropods serve as capillary siphons.

Ligia species are scavengers on dead plants and animals, and also feed by scraping the algal film from upper intertidal rocks. In turn, they are fed upon by birds, especially gulls, and by the intertidal crab *Pachygrapsus crassipes* (25.43). Carefoot (1973) found that the life span of *L. (L.) pallasii* is 1.5–2 years, that breeding occurs in the spring and early summer (although some females carry winter broods), and that the average brood size is 48 ± 11 young. The overall sex ratio is 1:1.

See especially Abbott (1939), Carefoot (1973), Fee (1926), Hatch (1947), Johnson & Snook (1927), Miller (1938, 1975), Richardson (1905), Ricketts & Calvin (1968), Van Name (1936), and Wilson (1970); see also the references under *L. (M.) occidentalis* (21.12).

21.14 Alloniscus perconvexus Dana, 1856

Fairly common, burrowing in damp sand above high-tide mark, or under stranded seaweed and debris on sandy beaches; British Columbia to southern California.

Body to about 16 mm long, convex, able to roll into a ball; second antenna with three-jointed flagellum, the distal joints small; head with hornlike tubercles; five pairs of pleopods present; color cream, brown, or grayish, with white patches.

At the southern end of its range the distribution of this species is overlapped by that of another sandy-beach isopod, **Tylos** sp., which superficially resembles it and has a similar mode of life. *Tylos* is distinguished by having a one- or two-jointed flagellum on the second antenna and only four pairs of pleopods.

Alloniscus can often be located by the molelike ridges and humps it makes in its shallow burrowing. Buried or in hiding during the day, it emerges at night and wanders over the beach, feeding on beach wrack. It is an air breather, and can drown on prolonged submersion in salt water. It respires through its pleopods, like the terrestrial pill bugs *Oniscus* and *Philoscia*; it has special air chambers at the lateral margins of the pleopods but no tracheal trees (pseudotracheae) such as are found in *Porcellio* and *Armadillidium*.

See Arcangeli (1958–60), Brusca (1966), Hamner, Smyth & Mulford (1969), Hatch (1947), Hayes (1970a,b), Heeley (1941), Miller (1938), Richardson (1905), Ricketts & Calvin (1968), and Van Name (1936, 1940). For the behavior and biology of beach forms like *Alloniscus*, the literature on land isopods may be more useful than that on marine forms; see Abbott (1918), Allee (1931), Beerstecher et al. (1954), Bursell (1955), Cloudsley-Thompson (1951, 1952, 1956, 1961), Gunn (1937), Henke (1930), Howard (1940), Lemos de Castro (1965), Miller (1936), Silén (1954), Thompson (1957), Vandel (1943, 1945, 1947), Waloff (1941), Warburg (1964), Wieser (1966), and the references under *Ligia (M.) occidentalis* (21.12) and *L. (L.) pallasii* (21.13).

21.15 Idotea (Pentidotea) stenops (Benedict, 1898)

On brown algae, especially *Egregia*, less often on surfgrass (*Phyllospadix*) or under rocks and boulders, middle intertidal zone to subtidal waters on exposed coasts; Alaska to Point Conception (Santa Barbara Co.).

Body to about 40 mm long; eyes narrow; telson slender, pointed; color always olive-green to brown.

This large and bulky idoteid isopod may be difficult to find, since it so closely matches the color of the *Egregia* on which it is commonly found and to which it clings tenaciously. In some areas where it is relatively rare intertidally, it is consistently found in the stomachs of sea trout (*Hexogrammos* sp.), suggesting that it is more abundant subtidally. Ovigerous females have been taken in June in central California.

See Hatch (1947), Menzies (1950b), Miller (1968, 1975), Miller & Lee (1970), Richardson (1905), Ricketts & Calvin (1968), and Schultz (1969).

21.16 Idotea (Pentidotea) resecata Stimpson, 1857

On the eelgrass *Zostera* in bay localities, and on the kelps *Macrocystis* and *Pelagophycus*, surface to 18 m depth; Alaska to tip of Baja California and Mazatlán (Mexico).

Body to 39 mm long; posterior border of telson in adult markedly concave, the terminal excavation bounded laterally by pointed projections; color green when on *Zostera*, brown when on kelp.

With its elongate body and recurved clinging legs this animal is very well adapted to life on the slender stems of *Macrocystis* and the blades of *Zostera*. It always orients its body along the long axis of its substratum, and if placed crosswise it immediately reorients itself.

Like its relative *Idotea (P.) montereyensis* (21.19), this species feeds on the plants to which it clings. On *Macrocystis* the animal often grazes on the basal portion of the blade near the float, causing the blade to separate from the plant before it is completely eaten. However, the isopod does far less damage to kelp than do the bottom-grazing sea urchins. In feeding experiments conducted in southern California, *I. (P.) resecata* preferred the kelps *Macrocystis, Pelagophycus porra, Eisenia arborea, Pterygophora californica,* and *Egregia laevigata* to other foods. It hardly touched *Phyllospadix* and *Zostera,* and eventually died on these substrata. This accords with its distribution in southern California, where it is most abundant on kelp and is not found on *Zostera.* However, it is an important member of the *Zostera* community in central and northern California, and the inability of southern forms to live when fed only on *Zostera* is puzzling.

Studies on kelp-bed ecology have shown that *I. (P.) resecata* is a common food of more than 20 species of marine fishes. For the young of the kelp bass, *Paralabrax clathratus,* it is more important in the diet than all other isopods taken together.

Reproductive data are lacking, but ovigerous females have been taken in July in central California. Studies of development show that the general body form changes little during growth, but some features (such as the number of segments in the flagellum of the second antenna, the number of segments in the palp of the maxilliped, and the number of setae on the maxilliped and walking legs) vary directly with size, each molt providing further change.

Body coloration in *I. resecata* involves the same system of pigments as that reported for *I. (P.) montereyensis.* However, it has been impossible to demonstrate a capability for color change in *I. resecata.* The red and green color varieties may well represent distinct physiological races. Like *I. (P.) montereyensis,* this animal deposits its pigments in the cuticle, and the green color variety here is in some respects virtually identical to the green color variety of *I. (P.) montereyensis.* However, the brown variety is quite different. In *I. (P.) resecata* a green pigment is deposited in the endocuticle and a red pigment in the exocuticle, resulting in an over-all brown coloration.

Idotea (P.) resecata dies if environmental temperatures reach 29–31°C. The animals appear to become scarce at the surface of kelp beds when the surface temperature exceeds 18°C, a temperature which also adversely affects *Macrocystis.* Respira-

tion rates for the isopod show an optimum at 15°C and a distinct drop at 25°C, a temperature which is lethal to some. *I. (P.) resecata* is less tolerant of salinity changes than its intertidal relative *Idotea (P.) wosnesenskii* (21.18); test animals survived 60–83 minutes in fresh water, 79 minutes in 200 percent seawater, 445 minutes in 150 percent seawater, and 33 hours in 125 percent seawater (experiments presumably conducted at room temperature).

See especially Abbott (1939), Lee & Gilchrist (1972, 1974), Menzies (1950b), Menzies & Waidzunas (1948), North (1964), and Quast (1968a, 1968b); see also Andrews (1945), Fee (1926), MacGinitie & MacGinitie (1968), Miller (1968, 1975), Miller & Lee (1970), Richardson (1905), Ricketts & Calvin (1968), and Schultz (1969).

Phylum Arthropoda / Class Crustacea / Subclass Malacostraca
Superorder Peracarida / Order Isopoda / Suborder Valvifera / Family Idoteidae

21.17 **Idotea (Pentidotea) aculeata** Stafford, 1913

On the surfgrass *Phyllospadix,* the pink coralline alga *Bossea,* and other intertidal seaweeds, low intertidal zone along exposed coastal area; Sonoma Co. to La Jolla (San Diego Co.); common in southern California, rare in central California.

Body to 23.5 mm long; telson bearing distinct projection on terminal end; anterior three thoracic segments with margins incised; color pink to red or mahogany, often with patches of white spots; species easily confused with *Idotea (P.) kirchanskii* (21.20), from which it is distinguished primarily by its color and its prominent terminal projection.

Little is known of the biology of this isopod except that, like many other idoteids, it is eaten by intertidal fishes; *I. (P.) aculeata* has been reported from the stomachs of such tidepool fishes as the striped kelpfish, *Gibbonsia metzi,* and the dwarf perch, *Micrometrus minimus.*

See Johnson & Snook (1927), Menzies (1950b), Miller (1975), Miller & Lee (1970), Mitchell (1953), and Schultz (1969).

Phylum Arthropoda / Class Crustacea / Subclass Malacostraca
Superorder Peracarida / Order Isopoda / Suborder Valvifera / Family Idoteidae

21.18 **Idotea (Pentidotea) wosnesenskii** (Brandt, 1851)

Common in mussel beds, also present on the seaweeds *Ulva*, *Porphyra*, and various brown algae, and under rocks in bays, upper middle intertidal zone to 16 m depth on exposed and protected rocky coasts; Sea of Okhotsk (U.S.S.R.) and Alaska to Estero Bay (San Luis Obispo Co.).

Body to 35 mm long, robust, not tapered; eyes kidney-shaped; telson broad, evenly rounded, ending in a small tooth; first segment of pleon with acute lateral borders; color usually very dark, almost black, but light-brown, red, and green color varieties do occur.

This species was named for the Russian zoologist Ilya Gavrilovich Vosnesensky, who collected extensively in Siberia, Alaska, and California in the period 1839–48.

Male individuals tend to be larger and paler than females, and have somewhat heavier legs. Nothing is known of the reproductive cycle except that ovigerous females have been found in July. The species has been taken from the stomachs of such intertidal fishes as the spotted kelpfish, *Gibbonsia elegans*, and the dwarf perch, *Micrometrus minimus*.

See Brusca (1966), Fee (1926), Hatch (1947), Johnson & Snook (1927), Menzies (1950b), Miller (1968, 1975), Miller & Lee (1970), Mitchell (1953), Richardson (1905), Ricketts & Calvin (1968), and Schultz (1969).

Phylum Arthropoda / Class Crustacea / Subclass Malacostraca
Superorder Peracarida / Order Isopoda / Suborder Valvifera / Family Idoteidae

21.19 **Idotea (Pentidotea) montereyensis** Maloney, 1933

Abundant to rare on *Phyllospadix* in low intertidal zone, and on various red algae including *Farlowia*, *Endocladia*, *Leptocladia*, and *Plocamium* in middle intertidal zone, on exposed rocky shores; Boundary Bay (British Columbia) to Point Conception (Santa Barbara Co.), also reported from Pismo Beach (San Luis Obispo Co.), on drifting brown algae, and from San Diego.

Body averaging about 16 mm long; telson broadly rounded, with a small median tooth more prominent in females than males; females broader and less linear in outline than males; color red, green, or brown, often with black-and-white patterns as well.

This remarkable species feeds on the plants on which it lives, and in turn is prey to a variety of marine fishes. Populations can be found both on *Phyllospadix* at about the zero tide level and on various red algae in deeper pools inshore of the *Phyllospadix* beds. Individuals in each of these populations match their plant substrata, those on *Phyllospadix* being bright green and those inshore matching the exact shade of red algae to which they cling. These animals are capable of changing color in response to a change in the color of the substratum.

The animal's overall color is a result of the pigments deposited in the cuticle. Red animals deposit the free, red carotenoid pigment canthaxanthin in their cuticles, whereas green animals deposit a green canthaxanthin-protein complex. Since pigmentation is primarily a result of this cuticular color, molting always accompanies color change. This ability to change color allows the species to use for food and habitat a large variety of intertidal plants, without young and adults competing directly with each other for the same resources and without directly sacrificing their protective color camouflage. The story is a remarkable one.

Although some breeding occurs all year, it is in the early summer that the largest number of young is released. The youngsters cannot effectively cling to the wave-swept surf-grass, and they are swept inshore onto a variety of red algae in more protected locations. Here they become red in color, matching the plants. As the population density inshore increases, the previous year's population, now red adults, begins to move from the red algae and migrate back to the *Phyllospadix* beds. There the animals soon molt and turn green. The result is the separation of adults from young, with the adults (which can hold more effectively in the face of heavy surf) occupying the wave-swept *Phyllospadix*, and the less-adept young placed on the more protected middle intertidal red algae. Speed of molting, color change, and chromatophore responses are geared to maximize both the separation of young and adults and the protection of both populations from predators.

Studies of a series of closely related intertidal idoteid isopods have shown that all have the same basic pigments and all synthesize them in the same way, but each species has its own pattern of pigment deposition and a color change capability related to its own particular ecological needs.

See especially Lee (1966a,c, 1972); see also Amar (1951), Fingerman (1956), Lee (1966b), Lee & Gilchrist (1972, 1974), Maloney (1933), Menzies (1950b), Miller & Lee (1970), Miyawaki (1958), Nagano (1949), Naylor (1955a,b,c), North (1964), Oguro (1959a,b,c, 1962), Peabody (1939), Ricketts & Calvin (1968), Schultz (1969), Smith (1938), Sunesone (1947), and Tinturier-Hamelin (1961).

Phylum Arthropoda / Class Crustacea / Subclass Malacostraca
Superorder Peracarida / Order Isopoda / Suborder Valvifera / Family Idoteidae

21.20 **Idotea (Pentidotea) kirchanskii** Miller & Lee, 1970

On the surfgrass *Phyllospadix scouleri*, low intertidal zone on rocky shores; central California.

Body to 15 mm long, linear, compact; antennae short; telson broadly rounded; color bright green, often mottled with pink, and often red on tips of appendages; species easily confused with *Idotea* (*P.*) *aculeata* (21.17), but differing in the bright-green color and rounded telson.

The bright green color of this species closely matches that of *Phyllospadix*, the only substratum on which it has been found. The pink mottling often seen matches the color of the epiphyte *Melobesia*, so common on *Phyllospadix*. This isopod is remarkably well adapted for holding on to the wave-swept *Phyllospadix* and often defies the collector's attempts to pull it off.

Idotea (*P.*) *kirchanskii* is frequently found together with *Idotea* (*P.*) *montereyensis* (21.19), but never reaches the high population densities reported for the latter. It appears to have the same pigmentary system as that in other intertidal idoteids so far investigated, but apparently does not change its color in response to a change in the color of its substratum as does *I.* (*P.*) *montereyensis*.

See Lee (1966b), Lee & Gilchrist (1972), Miller (1975), and Miller & Lee (1970).

Phylum Arthropoda / Class Crustacea / Subclass Malacostraca
Superorder Peracarida / Order Isopoda / Suborder Valvifera / Family Idoteidae

21.21 **Idotea (Idotea) urotoma** Stimpson, 1864

Rare to locally abundant under rocks encrusted with bryozoans and on kelp holdfasts, middle intertidal zone to shallow subtidal waters on rocky shores, occasionally in bays; Puget Sound (Washington) to Baja California; Channel Islands.

Body to 16 mm long, slender, telson broad, bluntly pointed, spade-shaped; color extremely variable.

Almost nothing is known of the biology of this attractive isopod. The animals occur in patches, but within these, population densities may be relatively high. One wonders why this species occurs in so many different color forms. It seems improbable that the coloration is protective, at least during low tide, for then the animals are usually under rocks, but (as with other isopods) we know very little of their behavior when the tide is high.

See Hatch (1947), Menzies (1950b), Miller (1975), North (1964), Richardson (1905), Ricketts & Calvin (1968), and Schultz (1969).

Phylum Arthropoda / Class Crustacea / Subclass Malacostraca
Superorder Peracarida / Order Isopoda / Suborder Valvifera / Family Idoteidae

21.22 **Idotea (Idotea) fewkesi** Richardson, 1905

Rare, on algae on exposed rocky coasts, occasionally intertidal, but more commonly subtidal to depths of 8–12 m; Alaska to central California.

Body to 42 mm in length, narrow and elongate; telson with right lateral angles and an acute median spine; color variable.

Ovigerous females of this species have been reported to occur in July and October.

See Fee (1926), Hatch (1947), Menzies (1950b), Miller (1975), Miller & Lee (1970), Richardson (1905), and Schultz (1969).

Phylum Arthropoda / Class Crustacea / Subclass Malacostraca
Superorder Peracarida / Order Isopoda / Suborder Valvifera / Family Idoteidae

21.23 **Colidotea rostrata** (Benedict, 1898)

On spines of the sea urchins *Strongylocentrotus purpuratus* (11.5) and *S. franciscanus* (11.6), low intertidal and shallow subtidal zones; southern California.

Body to 12 mm long; anterior dorsal projection of the head (rostrum) pointed; abdominal segments solidly fused dorsally, with only one partial suture visible on either side.

These animals match almost perfectly the color of the spines to which they cling. Beyond that aspect of coloration, almost nothing is known of their biology, and a study of the species should prove most interesting.

See Johnson & Snook (1927), Richardson (1905), and Schultz (1969).

Phylum Arthropoda / Class Crustacea / Subclass Malacostraca
Superorder Peracarida / Order Isopoda / Suborder Valvifera / Family Idoteidae

21.24 **Synidotea bicuspida** (Owen, 1839)

Relatively rare, among hydroids and bryozoans, low intertidal zone and adjacent subtidal waters along rocky shores on open coast; Alaska to central California; circumpolar Arctic.

Body to 12 mm long, lacking conspicuous tubercles; telson notched; head indented anteriorly, lacking tubercles.

Synidotea bicuspida is evidently a cold-water form; its southern limit of distribution is central California, north of Point Conception (Santa Barbara Co.).

See Hatch (1947), Menzies & Miller (1972), Miller (1975), Richardson (1905), and Schultz (1969).

Phylum Arthropoda / Class Crustacea / Subclass Malacostraca
Superorder Peracarida / Order Isopoda / Suborder Valvifera / Family Idoteidae

21.25 **Synidotea ritteri** Richardson, 1904

Often found on the hydroid *Aglaophenia*, low intertidal zone on rocky shores; Coos Bay (Oregon) to San Francisco Bay.

Body to about 8 mm long; head bearing conspicuous horn-like projections; telson notched; thorax bearing three rows of prominent tubercles.

Little is known of the biology of this and several other *Synidotea* species reported from California shore waters. **S. laticauda** Benedict is common in San Francisco Bay, especially in the shallow subtidal zone; it has been found in areas where the salinity averages about 17 percent that of seawater, and is an important food for young striped bass, starry flounder, steelhead rainbow trout, king salmon, white sturgeon, and other fishes in the Bay.

See Filice (1958), Ganssle (1966), Hatch (1947), Maloney (1933), Menzies & Miller (1972), Miller (1975), Painter (1966), Richardson (1905), and Schultz (1969).

Phylum Arthropoda / Class Crustacea / Subclass Malacostraca
Superorder Peracarida / Order Tanaidacea / Suborder Monokonophora
Family Metapseudidae

21.26 **Synapseudes intumescens** Menzies, 1949

Not often common, in algal holdfasts, on bryozoans, ascidians, sea stars, and abalones, occasionally in mussel beds, especially in regions dominated by larger brown algae (order Laminariales), very low intertidal zone on rocky, wave-swept open coast; Sonoma Co. and Farallon Islands (San Francisco Co.) to Del Mar (San Diego Co.).

Only a few millimeters long, but under magnification recognizable by its greatly reduced pleon consisting of only two narrow pleonites plus a triangular pleotelson bearing dorsally three setiferous swellings (intumescences); pleopods lacking; basal joint of first antenna with conspicuous spines on inner border; peraeon with prominent dorsolateral ridges.

Little is known of the ecology of this species. Because of their small size and cryptic habits, the creatures are seldom seen in the field, although they are fairly common in benthic samples from the rocky intertidal zone. Lacking pleopods, they are unable to swim. They are amusing to watch as they crawl awkwardly on the substratum and over the sessile or slow-moving organisms with which they are associated.

See Gardiner (1973), Menzies (1949), and Miller (1975).

Phylum Arthropoda / Class Crustacea / Subclass Malacostraca
Superorder Peracarida / Order Tanaidacea / Suborder Monokonophora
Family Pagurapseudidae

21.27 **Pagurapseudes** sp.

Occupying small snail shells among holdfasts of red algae, low intertidal zone on rocky shores protected from strong surf; so far known only from Pacific Grove (Monterey Co.) but probably present in similar habitats elsewhere.

About 3 mm long, resembling small hermit crabs but with thoracic region clearly segmented posterior to carapace.

Because of their striking similarity to tiny hermit crabs, these small cheliferans are easily overlooked by collectors. The Pacific Grove forms are not conspecific with any of the three known species of *Pagurapseudes*. They show closest affinity to **P. laevis** Menzies, which has been taken 1.6 km northwest of White Cove, Santa Catalina Island, Channel Islands (type locality) and at about 90 m depth in Melpomene Cove, Isla Guadalupe, Mexico. The other two described species are *P. bouryi* from Cuba and *P. spinipes* from Australia.

Pagurapseudes is aptly named after *Pagurus*, a genus of hermit crabs that it mimics by living in gastropod shells, and *Apseudes*, a related tanaidacean. Besides adopting the paguroid habit, species of *Pagurapseudes* show certain structural adaptations in common with hermit crabs for that mode of life—a remarkable case of convergent evolution! Foremost among these is a coiled abdomen, which fits nicely into the spiral snail shell. Unlike that of the hermit crab, however, the pleon in pagurapseudans is segmented (six somites including the pleotelson) and, though twisted, seems to be symmetrical. Another convergent feature is that the pleopods are reduced in number (zero to three pairs in *Pagurapseudes*, depending on the species and sex). Pleopods are completely lacking in females of California species (including *P. laevis*); males of Pacific Grove forms have not yet been observed, but presumably they would possess the first pair as do males of their southern relatives. The reduction or absence of pleopods is not necessarily associated with living in shells (though that

might contribute to a better fit), as some free-living tanaidaceans also lack pleopods (e.g., *Synapseudes*, 21.26). A rather curious difference is that the uropods in hermit crabs are greatly reduced, whereas those of pagurapseudans appear quite normal.

Although pagurapseudans removed from their shells can readily be distinguished from hermit crabs, this is more difficult to do when they are in their shells and only the front part of the body is visible. Nevertheless, especially using magnification, one can observe several distinct differences. For instance, in *Pagurapseudes* the first antennae are much shorter, having a double flagellum composed of only a few joints; the eyestalks are much more compact, conical, immovable, and bear a few relatively large ocelli; and the chelae are of equal size.

It is not known how and when pagurapseudans come to occupy snail shells and whether they move into larger shells, as do hermit crabs, when they outgrow them. Studies of this and other features of the life history and ecology of these unusual creatures should be most interesting.

The literature on *Pagurapseudes* is sparse and mainly taxonomic. See Bouvier (1918), Lang (1949), Menzies (1953), Miller (1975), Sieg & Winn (1978), and Whitelegge (1901).

Phylum Arthropoda / Class Crustacea / Subclass Malacostraca
Superorder Peracarida / Order Tanaidacea / Suborder Dikonophora
Family Tanaidae

21.28 **Pancolus californiensis** Richardson, 1905

Often among hydroids, low intertidal zone on rocky shores; reported only from the Monterey area, but probably more widespread.

Body about 5.5 mm long; pleon with three distinct somites plus lateral incisions of a fourth; two distinct pairs of pleopods, and a very tiny, obscure third pair; uropods uniramous, with four short joints.

Tanaids, like isopods and cumaceans (and unlike amphipods and mysids) leave the brood pouch at the first "manca" stage of development, in which the last pair of thoracic legs is not yet formed. In *Pancolus* there are probably at least two and perhaps as many as four manca instars, over the course of which the last legs develop. At sexual maturity, females produce only one pair of brood pouch plates (oostegites). These plates, arising from the bases of the fifth pair of walking legs, are very large, thin, and leaflike, and roll up about the eggs as they are laid. Usually they roll about one another, as well, so that a single brood sac is formed, but occasionally they roll separately and produce a pair of brood sacs, one on each side. The brood pouch is so different in appearance from that of a normal isopod or tanaid that Richardson originally described the pouch as a parasite, and suggested it be called "Oosaccus."

The life history of *Pancolus* has not been studied in detail, but in related forms the sexually mature females continue to molt and produce several successive broods of young, with only one non-breeding or preparatory instar between successive broods. Females in the preparatory stage have rudimentary oostegites, and in this stage one cannot tell whether the animals have produced a previous brood or not. With males it is virtually impossible to tell whether or not one has a fully developed animal without a sizable series of specimens for comparison, and unfortunately for the tanaid systematist it appears that numerous species have been described on the basis of immature or subadult animals.

See Lang (1949, 1950, 1953, 1956, 1958, 1960, 1961, 1970), Miller (1975), Richardson (1905, as *P. californiensis* and *Oosaccus*); other useful papers on tanaidaceans include Menzies (1949, 1953) and Roubault (1937).

Phylum Arthropoda / Class Crustacea / Subclass Malacostraca
Superorder Peracarida / Order Tanaidacea / Suborder Dikonophora
Family Paratanaidae

21.29 **Leptochelia dubia** (Krøyer, 1842)

Sometimes common in tidepools, on mud flats (usually on green algae), among fouling growth on submerged structures in harbors and bays; cosmopolitan, widely distributed along Pacific coast.

Tiny, but easily recognizable by its slender, subcylindrical, smooth, whitish body comprising a shieldlike head, six

peraeonites, and six pleonal divisions (including pleotelson), all clearly visible under dissecting microscope; males differing markedly from females in having attenuated chelipeds with two teeth on immovable finger of pincer, longer antennae, larger eyes, and metamorphosed (non-functional?) mouthparts.

Conflicting views on the synonomy of this dikonophoran tanaidacean make it difficult to state with certainty its geographical and ecological distribution (for a review, see Miller, 1968). Some authors (notably Monod, 1933) regard it as a tropical-subtropical cosmopolite. Most writers, however, consider several other described species (notably *L. savignyi*) to be junior synonyms of *L. dubia*, and thus extend its distribution into temperate waters, too. Until the taxonomic difficulties are resolved, the present authors will use the name *L. dubia* for the Pacific coast representatives of this form.

It is remarkable that this benthic species, lacking pelagic larvae or appreciable ability to swim, should enjoy worldwide distribution. The most plausible explanation is that *L. dubia* has been transported passively on ship bottoms and on natural rafts carried considerable distances by currents. Fouling growths on hulls or rafts would favor such transport, serving the "hitchhikers" for food and shelter en route. The immigrants would have to adapt to changed environments and to establish viable populations in new localities.

Little else is known about the biology of *L. dubia*. One infers from their small size and narrow shape that these creatures crawl easily into cracks and crevices for protection and in search of food. Their abundance suggests that they may be important links in benthic food chains.

For further information (mostly taxonomic), see Miller (1968, 1975), Monod (1933), and Richardson (1905).

Literature Cited

Abbott, C. H. 1918. Reactions of land isopods to light. J. Exper. Zool. 27: 193–246.

———. 1939. Shore isopods; niches occupied and degrees of transition toward land life with special reference to the family Ligydae. Proc. 7th Pacific Sci. Congr. 3: 505–11.

Alexandrowicz, J. S. 1952. Innervation of the heart of *Ligia oceanica*. J. Mar. Biol. Assoc. U.K. 31: 85–96.

Allee, W. C. 1931. Animal aggregations. A study in general sociology. Chicago: University of Chicago Press. 431 pp.

Amar, R. 1951. Formations endocrines cérébrales des isopodes marins et comportement chromatique d'*Idotea*. Ann. Fac. Sci. Marseille (2) 20: 167–305.

Andrews, H. L. 1945. The kelp beds of the Monterey region. Ecology 26: 24–37.

Arcangeli, A. 1958–60. Revisione del genere *Alloniscus* Dana. Il sistema respiratorio speciale agli exopoditi dei pleopodi delle specie appartenenti allo stesso genere. Boll. Inst. Zool. Univ. Torino 6: 17–79.

Armitage, K. B. 1960. Chromatophore behavior of the isopod *Ligia occidentalis* Dana, 1853. Crustaceana 1: 193–207.

Audoinoit, R. 1956. Sur l'existence de territoires de régénération des péréiopodes de l'isopode *Sphaeroma serratum* (Fab.). C. R. Acad. Sci., Paris, 242: 1239–40.

Barnes, T. C. 1935. Salt requirements and orientation of *Ligia* in Bermuda. III. Biol. Bull. 69: 259–68.

———. 1940. Experiments on *Ligia* in Bermuda. VII. Further effects of sodium, ammonium, and magnesium. Biol. Bull. 78: 35–41.

Barrows, A. L. 1919. The occurrence of a rock-boring isopod along the shore of San Francisco Bay, California. Univ. Calif. Publ. Zool. 19: 299–316.

Becker, G. 1959. Biological investigations on marine borers in Berlin-Dahlem, pp. 62–83, *in* Ray (1959a).

Beckman, C., and R. J. Menzies. 1960. The relationship of reproductive temperature and the geographical range of the marine woodborer *Limnoria tripunctata*. Biol. Bull. 118: 9–16.

Beerstecher, E., Jr., J. Cornyn, C. Volkmann, L. Cardo, and R. Harper. 1954. Invertebrate nutrition. I. A preliminary survey of the nutritional requirements of an isopod: *Oniscus asellus*. Texas Rep. Biol. Med. 12: 207–11.

Bishop, J. A. 1969. Changes in genetic constitution of a population of *Sphaeroma rugicauda* (Crustacea: Isopoda). Evolution 23: 589–601.

Bocquet, C., C. Levi, and G. Teissier. 1951. Recherches sur le polychromatisme de *Sphaeroma serratum* (Fab.). Arch. Zool. Expér. Gén. 87: 245–97.

Bouvier, E. L. 1918. Sur une petite collection de Crustacés de Cuba offerte au Muséum par M. de Boury. Bull. Mus. Nat. Hist. Natur. 24: 6–15.

Bowman, T. E. 1974. The California freshwater isopod, *Asellus tomalensis*, rediscovered and compared with *Asellus occidentalis*. Hydrobiologia 44: 431–41.

Boyle, P. J., and R. Mitchell. 1978. Absence of microorganisms in crustacean digestive tracts. Science 200: 1157–59.

Brusca, G. J. 1966. Studies on the salinity and humidity tolerances of five species of isopods in a transition from marine to terrestrial life. Bull. So. Calif. Acad. Sci. 65: 146–54.

Brusca, R. C. 1978. Studies on the cymothoid fish symbionts of the eastern Pacific (Crustacea: Cymothoidae). II. Systematics and biology of *Lironeca vulgaris* Stimpson, 1857. Allan Hancock Found. Occas. Pap. (n.s.) 2: 1–19.

Bursell, E. 1955. The transpiration of terrestrial isopods. J. Exper. Biol. 32: 238–55.

Calman, W. T. 1898. On a collection of Crustacea from Puget Sound. Ann. N.Y. Acad. Sci. 11: 259–92.

Carefoot, T. H. 1973. Studies on the growth, reproduction, and life cycle of the supralittoral isopod *Ligia pallasii*. Mar. Biol. 18: 302–11.

Carlisle, D. B. 1956. Studies on the endocrinology of isopod crustaceans. Moulting in *Ligia oceanica* (L.). J. Mar. Biol. Assoc. U.K. 35: 515–20.

Cloudsley-Thompson, J. L. 1951. Rhythmicity in the woodlouse, *Armadillidium vulgare*. Entomol. Monthly Mag. 87: 275–78.

———. 1952. Studies in diurnal rhythms. II. Changes in the physiological responses of the woodlouse *Oniscus asellus* to environmental stimuli. J. Exper. Biol. 29: 295–303.

———. 1956. Studies in diurnal rhythms. VII. Humidity responses and nocturnal activity in woodlice (Isopoda). J. Exper. Biol. 33: 576–82.

———. 1961. Rhythmic activity in animal physiology and behaviour. New York: Academic Press. 236 pp.

Coker, R. E. 1923. Breeding habits of *Limnoria* at Beaufort, N.C. J. Elisha Mitchell Sci. Soc. 39: 95–100.

Danforth, C. G. 1970. Epicaridea (Crustacea: Isopoda) of North America. Ann Arbor, Mich.: University Microfilms. 190 pp.

Dow, T. C. 1958. Description of a new isopod from California, *Exosphaeroma inornata*. Bull. So. Calif. Acad. Sci. 57: 93–97.

Dow, T. C., and R. J. Menzies. 1957. The pelagic isopod *Idotea metallica* in the Mediterranean. Pubbl. Staz. Zool. Napoli 30: 330–36.

Dresel, E. I. B., and V. Moyle. 1950. Nitrogenous excretion of amphipods and isopods. J. Exper. Biol. 27: 210–25.

Edney, E. B. 1954. Woodlice and the land habitat. Biol. Rev. 29: 185–219.

———. 1957. The water relations of terrestrial arthropods. London: Cambridge University Press. 109 pp.

———. 1956 (publ. 1958). (1) A new interpretation of the relation between temperature and transpiration in arthropods. (2) The microclimate in which woodlice live. Proc. Internat. Congr. Entomol. 2: 329–32, 709–12.

Ellenby, C. 1951. Body size in relation to oxygen consumption and pleopod beat in *Ligia oceanica* (L.). J. Exper. Biol. 28: 492–507.

Ellenby, C., and D. A. Evans. 1956. On the relative importance of body weight and surface area measurements for the prediction of the level of oxygen consumption of *Ligia oceanica* (L.) and prepupae of *Drosophila melanogaster* Meig. J. Exper. Biol. 33: 134–41.

Enami, M. 1941. Melanophore responses in an isopod crustacean, *Ligia exotica*. I. General responses. II. Hormonal control of melanophores. Japon. J. Zool. 9: 497–531.

Enright, J. T. 1965. Entrainment of a tidal rhythm. Science 147: 864–66.

Eriksen, C. H. 1968. Aspects of the limno-ecology of *Corophium spinicorne* Stimpson (Amphipoda) and *Gnorimosphaeroma oregonensis* (Dana) (Isopoda). Crustaceana 14: 1–12.

Fahrenbach, W. H. 1959. Studies on the histology and cytology of midgut diverticula of *Limnoria lignorum*, pp. 96–107, *in* Ray (1959a).

Fee, A. R. 1926. The Isopoda of Departure Bay and vicinity with descriptions of new species, variations and colour notes. Contr. Canad. Biol. & Fish. 3: 13–47.

Filice, F. P. 1958. Invertebrates from the estuarine portion of San Francisco Bay and some factors influencing their distribution. Wasmann J. Biol. 16: 159–211.

Fingerman, M. 1956. The physiology of the melanophores of the isopod, *Ligia exotica*. Tulane Stud. Zool. 3: 139–48.

———. 1963. The control of chromatophores. New York: Macmillan. 184 pp.

Ganssle, D. 1966. Fishes and decapods of San Pablo and Suisun Bays, pp. 64–94, *in* D. W. Kelley, ed., Ecological studies of the Sacramento–San Joaquin estuary, part I. Calif. Dept. Fish & Game, Fish Bull. 133: 1–133.

Gardiner, L. F. 1973. New species of the genera *Synapseudes* and *Cycloapseudes* with notes on morphological variation, postmarsupial development, and phylogenetic relationships within the family Metapseudidae (Crustacea: Tanaidacea). Zool. J. Linn. Soc. 53: 25–58.

George, R. Y. 1966. Glycogen content in the wood-boring isopod, *Limnoria lignorum*. Science 153: 1262–64.

Given, R. R., and D. C. Lees. 1967. Santa Catalina Island biological survey. Survey Rep. 1. Allan Hancock Found., Univ. So. Calif. 126 pp.

Gooding, R. U. 1960. North and South American copepods of the genus *Hemicyclops* (Cyclopoida: Clausiidae). Proc. U.S. Nat. Mus. 112: 159–95.

Gunn, D. L. 1937. The humidity reactions of the wood-louse, *Porcellio scaber* (Latr.). J. Exper. Biol. 14: 178–86.

Haderlie, E. C. 1969. Marine fouling and boring organisms in Monterey harbor. II. Second year of investigation. Veliger 12: 182–92.

———. 1971. Marine fouling and boring organisms at 100 foot depth in open water of Monterey Bay. Veliger 13: 249–60.

———. 1976. Destructive marine wood and stone borers in Monterey Bay, pp. 947–53, *in* J. M. Sharpley and A. M. Kaplan, eds., Proc. 3rd Internat. Symp. on Biodeterioration. London: Applied Science Publ. 1,138 pp.

Hamner, W. M., M. Smyth, and E. D. Mulford, Jr. 1969. The behavior and life history of a sand-beach isopod, *Tylos punctatus*. Ecology 50: 442–53.

Hansen, H. J. 1905. On the propagation, structure, and classification of the family Sphaeromidae. Quart. J. Microscop. Sci. 49: 69–135.

Hatch, M. H. 1947. The Chelifera and Isopoda of Washington and adjacent regions. Univ. Washington Publ. Biol. 10: 155–274.

Hayes, W. B. 1970a. The accuracy of pitfall trapping for the sand-beach isopod *Tylos punctatus*. Ecology 51: 514–16.

———. 1970b. Copper concentrations in the high-beach isopod *Tylos punctatus*. Ecology 51: 721–23.

Heeley, W. 1941. Observations on the life-histories of some terrestrial isopods. Proc. Zool. Soc. London 111: 79–149.

Henderson, J. T. 1924. The gribble: A study of the distribution factors and life history of *Limnoria lignorum* at St. Andrews, N.B. Contr. Canad. Biol. (n.s.) 2: 309–25.

Henke, K. 1930. Die lichtorientierung und die Bedingungen der Lichtstimmung bei der Rollassel *Armadillidium cinereum* Zenker. Z. Vergl. Physiol. 13: 534–625.

Heusner, A. A. and J. T. Enright. 1966. Long-term activity recording in small aquatic animals. Science 154: 532–33.

Hewatt, W. G. 1935. Ecological succession in the *Mytilus californianus* habitat as observed in Monterey Bay, California. Ecology 16: 244–51.

———. 1937. Ecological studies on selected marine intertidal communities of Monterey Bay, California. Amer. Midl. Natur. 18: 161–206.

———. 1938. Notes on the breeding seasons of the rocky beach fauna of Monterey Bay, California. Proc. Calif. Acad. Sci. 23: 283–88.

Hewitt, C. G. 1907. *Ligia*. Liverpool Mar. Biol. Comm. [LMBC] Memoir 14: 1–37.

Hilton, W. A. 1915. Early development of *Ligia*. Pomona J. Entomol. Zool. 7: 211–27.

Hiraiwa, Y. K. 1936. Studies on a bopyrid *Epipenaeon japonica* Thieleman. III. Development and life cycles with special reference to the sex differentiation in the bopyrid. J. Sci. Hiroshima Univ. (B) 4: 101–41.

Hoestlandt, H. 1964. Examen comparé de races polychromatiques de Sphéromes (Crustaces: Isopodes) des côtes atlantique européenne et pacifique américaine. Verh. Internat. Verein. Limnol. 15: 871–78.

———. 1973. Étude systématique et génétique de trois espèces Pacifiques Nord-Américaines du genre *Gnorimosphaeroma* Menzies (Isopodes: Flabellifères). I. Considérations générales et systématique. Arch. Zool. Expér. Gén. 114: 349–95.

Howard, A. D. 1952. Molluscan shells occupied by tanaids. Nautilus 65: 75–76.

Howard, H. W. 1940. The genetics of *Armadillidium vulgare* Latr. J. Genet. 40: 83–108.

Jackson, H. G. 1922. A revision of the isopod genus *Ligia* (Fab.). Proc. Zool. Soc. London 1922: 683–703.

———. 1926. The morphology of the isopod head. Part 1. Proc. Zool. Soc. London 1928: 885–911.

Johnson, F. H., and D. L. Ray. 1962. Simple method of harvesting *Limnoria* from nature. Science 135: 795.

Johnson, M. E., and H. J. Snook. 1927. Seashore animals of the Pacific coast. New York: Macmillan. 659 pp.

Johnson, M. W. 1935. Seasonal migrations of the wood-borer *Limnoria lignorum* (Rathke) at Friday Harbor, Washington. Biol. Bull. 69: 427–38.

Johnson, M. W., and R. J. Menzies. 1956. The migratory habits of the marine gribble *Limnoria tripunctata* Menzies in San Diego harbor, California. Biol. Bull. 110: 54–68.

Johnson, W. S. 1974. Population dynamics, energetics and biology of the marine isopod *Cirolana harfordi* (Lockington) in Monterey Bay, California. Doctoral thesis, Biological Sciences, Stanford University, Stanford, Calif. 99 pp.

———. 1976a. Biology and population dynamics of the intertidal isopod *Cirolana harfordi*. Mar. Biol. 36: 343–50.

———. 1976b. Population energetics of the intertidal isopod *Cirolana harfordi*. Mar. Biol. 36: 351–57.

Kerkut, G. A., and S. Krishnaswamy. 1960. Electrical potentials in the Isopoda. Comp. Biochem. Physiol. 1: 293–301.

Kittredge, J. S., D. J. Simonsen, E. Roberts, and B. Jelinek. 1962. Free amino acids of marine invertebrates, pp. 176–86, *in* J. T. Holden, ed., Amino-acid pools: Distribution, formation, and function of free amino acids. Amsterdam: Elsevier. 815 pp.

Kleinholz, L. H. 1937. Studies in the pigmentary system of crustacea. I. Color changes and diurnal rhythm in *Ligia baudiniana*. Biol. Bull. 72: 24–36.

Kleitman, N. 1940. The modifiability of the diurnal pigmentary rhythm in isopods. Biol. Bull. 78: 403–6.

Kühne, H. 1973. On the nutritional requirements of wood-boring Crustacea, pp. 814–21, *in* R. F. Acker, B. F. Brown, J. R. DePalma, and W. P. Iverson, eds., Proc. 3rd Internat. Congr. on Marine Corrosion and Fouling. Evanston, Ill.: Northwestern University Press. 1,031 pp.

Lagerspetz, K., and A. Lehtonen. 1961. Humidity reactions of some aquatic isopods in the air. Biol. Bull. 120: 38–43.

Lane, C. E. 1959. The general histology and nutrition of *Limnoria*, pp. 34–45, *in* Ray (1959a).

Lang, K. 1949. Contribution to the systematics and synonymics of the Tanaidacea. Ark. Zool. 42A: 1–14.

———. 1950. The genus *Pancolus* Richardson, and some remarks on *Paratanais euelpis* Barnard (Tanaidacea). Ark. Zool. (2) 1: 357–60.

———. 1953. The postmarsupial development of the Tanaidacea. Ark. Zool. (2) 4: 409–22.

———. 1956. Neotanaidae nov. fam., with some remarks on the phylogeny of the Tanaidacea. Ark. Zool. (2) 9: 469–75.

———. 1957. Tanaidacea from Canada and Alaska. Contr. Dépt. Pêche. Québec No. 52: 1–53.

———. 1958. Protogynie bei zwei Tanaidaceen–Arten. Ark. Zool. (2) 11: 535–40.

———. 1960a. Contributions to the knowledge of the genus *Microcerberus* Karaman (Crustacea: Isopoda) with a description of a new species from the central Californian coast. Ark. Zool. (2) 13: 493–510.

———. 1960b. The genus *Oösaccus* Richardson and the brood pouch of some tanaids. Ark. Zool. (2) 13: 77–79.

———. 1961. Further notes on *Pancolus californiensis* Richardson. Ark. Zool. (2) 13: 573–77.

———. 1970. Taxonomische und phylogenetische Untersuchungen über die Tanaidaceen. Ark. Zool. (2) 22: 595–626.

Lee, W. L. 1966a. Pigmentation of the marine isopod *Idothea montereyensis*. Comp. Biochem. Physiol. 18: 17–36.

———. 1966b. Pigmentation of the marine isopod *Idothea granulosa* (Rathke). Comp. Biochem. Physiol. 19: 13–27.

———. 1966c. Color change and the ecology of the marine isopod *Idotea* (*Pentidotea*) *montereyensis* Maloney, 1933. Ecology 47: 930–41.

———. 1972. Chromatophores and their role in color change in the marine isopod *Idotea montereyensis* (Maloney). J. Exper. Mar. Biol. Ecol. 8: 201–15.

Lee, W. L., and B. M. Gilchrist. 1972. Pigmentation, color change, and the ecology of the marine isopod *Idotea resecata* (Stimpson, 1857). J. Exper. Mar. Biol. Ecol. 10: 1–27.

———. 1974. Monohydroxy carotenoids in *Idotea* (Crustacea: Isopoda). Comp. Biochem. Physiol. 51B: 247–53.

Lemos de Castro, A. 1965. On the systematics of the genus *Littorophiloscia* Hatch (Isopoda, Oniscidae). Arquiros do Museu Nacional 53: 85–98.

Lichtwardt, R. W. 1961. A *Palavascia* (Eccrinales) from the marine isopod *Sphaeroma quadridentatum* Say. J. Elisha Mitchell Sci. Soc. 77: 242–49.

Lockwood, A. P. M. 1962. The osmoregulation of Crustacea. Biol. Rev. 37: 257–305.

MacGinitie, G. E. 1939. Littoral marine communities. Amer. Midl. Natur. 21: 28–55.

MacGinitie, G. E., and N. MacGinitie. 1968. Natural history of marine animals. 2nd ed. New York: McGraw-Hill. 523 pp.

Maloney, J. O. 1933. Two new species of isopod crustaceans from California. J. Wash. Acad. Sci. 23: 144–47.

Menzies, R. J. 1949. A new species of apseudid crustacean of the genus *Synapseudes* from northern California (Tanaidacea). Proc. U.S. Nat. Mus. 99: 509–15.

————. 1950a. Notes on California isopods of the genus *Armadillon-iscus*, with the description of *Armadilloniscus coronacapitalis* n. sp. Proc. Calif. Acad. Sci. 26: 467–81.

————. 1950b. The taxonomy, ecology and distribution of northern California isopods of the genus *Idotea* with the description of a new species. Wasmann J. Biol. 8: 155–95.

————. 1951. A new species of *Limnoria* (Crustacea: Isopoda) from southern California. Bull. So. Calif. Acad. Sci. 50: 86–88.

————. 1953. The apseudid Chelifera of the Eastern Tropical and North Temperate Pacific Ocean. Bull. Mus. Compar. Zool., Harvard, 107: 443–96.

————. 1954a. A review of the systematics and ecology of the genus "Exosphaeroma" with the description of a new genus, and a new species, and a new subspecies (Crustacea, Isopoda, Sphaeromidae). Amer. Mus. Novitates 1683: 1–24.

————. 1954b. The comparative biology of reproduction in the wood-boring isopod crustacean *Limnoria*. Bull. Mus. Compar. Zool., Harvard, 112: 364–88.

————. 1955. Aggregation by the marine wood-boring isopod, *Limnoria*. Oikos 6: 149–52.

————. 1957. The marine borer family Limnoriidae (Crustacea, Isopoda). Bull. Mar. Sci. Gulf Caribb. 7: 101–200.

————. 1958. The distribution of wood-boring *Limnoria* in California. Proc. Calif. Acad. Sci. 29: 267–72.

————. 1959. The identification and distribution of the species of *Limnoria*, pp. 10–33d, *in* Ray (1959a).

————. 1962. The marine isopod fauna of Bahía de San Quintín, Baja California, Mexico. Pacific Natur. 3: 337–48.

Menzies, R. J., and J. L. Barnard. 1951. The isopodan genus *Iais* (Crustacea). Bull. So. Calif. Acad. Sci. 50: 136–51.

Menzies, R. J., T. E. Bowman, and F. G. Alverson. 1955. Studies of the biology of the fish parasite *Livoneca convexa* Richardson (Crustacea, Isopoda, Cymothoidae). Wasmann J. Biol. 13: 277–95.

Menzies, R. J., and M. A. Miller. 1972. Systematics and zoogeography of the genus *Synidotea* (Crustacea: Isopoda) with an account of Californian species. Smithsonian Contr. Zool. 102: 1–33.

Menzies, R. J., J. Mohr, and C. M. Wakeman. 1963. The seasonal settlement of wood-borers in Los Angeles–Long Beach harbors. Wasmann J. Biol. 21: 97–120.

Menzies, R. J., and R. J. Waidzunas. 1948. Post-embryonic growth changes in the isopod *Pentidotea resecata* (Stimpson), with remarks on their taxonomic significance. Biol. Bull. 95: 107–13.

Meyers, S. P., and E. S. Reynolds. 1957. Incidence of marine fungi in relation to wood-borer attack. Science 126: 969.

Miller, M. A. 1936. California isopods of the genus *Porcellio* with descriptions of a new species and a new subspecies. Univ. Calif. Publ. Zool. 41: 165–72.

————. 1938. Comparative ecological studies on the terrestrial isopod Crustacea of the San Francisco Bay region. Univ. Calif. Publ. Zool. 43: 113–42.

————. 1968. Isopoda and Tanaidacea from buoys in coastal waters of the continental United States, Hawaii, and the Bahamas. Proc. U.S. Nat. Mus. 125: 1–53.

————. 1975. Phylum Arthropoda: Crustacea, Tanaidacea and Isopoda, pp. 277–312, *in* R. I. Smith and J. T. Carlton, eds., Light's manual: Intertidal invertebrates of the central California coast. 3rd ed. Berkeley and Los Angeles: University of California Press. 716 pp.

Miller, M. A., and W. L. Lee. 1970. *Idotea kirchanskii*. A new idoteid isopod from central California with notes on its distribution and ecology. Proc. Biol. Soc. Washington 82: 789–98.

Miller, R. C. 1926. Ecologic relations of marine wood-boring organisms in San Francisco Bay. Ecology 7: 247–52.

Mitchell, D. F. 1953. An analysis of stomach contents of California tide pool fishes. Amer. Midl. Natur. 49: 862–71.

Miyawaki, M. 1958. On the neurosecretory system of the isopod, *Idotea japonica*. Annot. Zool. Japon. 31: 216–21.

Mohr, J. L. 1959. On the protozoan association of *Limnoria*, pp. 84–95, *in* Ray (1959a).

Monod, T. 1933. Mission Robert-Ph. Dollfus en Égypte. Tanaidacea et Isopoda. Mém. Inst. Égypte 21: 162–264.

Muscatine, L. 1956. A new entoniscid (Crustacea: Isopoda) from the Pacific coast. J. Wash. Acad. Sci. 46: 122–26.

Nagano, T. 1949. Physiological studies on the pigmentary system of Crustacea. III. The color change of an isopod *Ligia exotica* (Roux). Sci. Rep. Tôhôku Univ. (4) 18: 167–75.

Naylor, E. 1955a. The ecological distribution of British species of *Idotea*. J. Anim. Ecol. 24: 255–69.

————. 1955b. The life cycle of *Idotea emarginata* (Fabricius). J. Anim. Ecol. 24: 270–81.

————. 1955c. The diet and feeding mechanism of *Idotea*. J. Mar. Biol. Assoc. U.K. 34: 347–55.

Nicholls, A. G. 1931. Studies on *Ligia oceanica*. I.A. Habitat and effect of change of environment on respiration. B. Observations on moulting and breeding. II. The processes of feeding, digestion, and absorption with a description of the structure of the foregut. J. Mar. Biol. Assoc. U.K. 17: 655–708.

Nicola, M. de. 1949. Alkaline phosphatases and the cycle of nucleic acids in the gonads of some isopod crustaceans. Quart. J. Microscop. Sci. 90: 391–99.

North, W. J., ed. 1964. An investigation of the effects of discharged wastes on kelp. Resources Agency of Calif., State Water Qual. Control Bd., Publ. 26: 1–124.

Numanoi, H. 1933. (1) Effects of temperature on the frequency of the rhythmical beatings of the pleopods of *Ligia exotica*. (2) Temperature characteristics of the pleopodal beatings in *Asellus nipponensis*. J. Fac. Sci., Univ. Tokyo, Zool. 3: 217–31.

————. 1934. (1) Relation between atmospheric humidity and evaporation of water in *Ligia exotica*. (2) Calcium in the blood of *Ligia exotica* during molting and non-molting phases. (3) Calcium contents of the carapace and other organs of *Ligia exotica* during molting and non-molting phases. J. Fac. Sci., Univ. Tokyo, Zool. 3: 343–64.

———. 1937. Migration of calcium through the blood in *Ligia exotica* during its molting. Japon. J. Zool. 7: 241–49.

Oguro, C. 1959a. On the sinus glands in four species belonging to the Idoteidae (Crustacea, Isopoda). J. Fac. Sci., Hokkaido Imp. Univ., (6) Zool. 14: 261–64.

———. 1959b. On the physiology of melanophores in the marine isopod *Idotea japonica*. J. Endocrinol. Japon. 6: 246–52.

———. 1959c. Occurrence of accessory sinus gland in the isopod *Idotea japonica*. Annot. Zool. Japon. 32: 71.

———. 1960. On the neurosecretory system in the cephalic region of the isopod *Tecticeps japonica*. J. Endocrinol. Japon. 7: 137–45.

———. 1962. On the physiology of melanophores in the marine isopod *Idotea japonica*. III. The role of the eyes in background response. Crustaceana 4: 85–92.

Okay, S. 1943. Changement de coloration chez *Sphaeroma serratum* (Fab.). Rev. Fac. Sci. Univ., Istanbul, (B) 9: 204–25.

———. 1945. Sur l'excitabilité directe des chromatophores, les changements périodiques de coloration et le centre chromatophorotropique chez *Sphaeroma serratum* (Fab.). Rev. Fac. Sci. Univ., Istanbul, (B) 9: 366–86.

Ondo, Y., and S. Mori. 1956. Periodic behavior of the shore isopod *Megaligia exotica* (Roux). I. Observations under natural conditions. Japon. J. Ecol. 5: 161–67.

Orcutt, H. G. 1950. The life history of the starry flounder *Platichthys stellatus* (Pallas). Calif. Dept. Fish & Game, Fish Bull. 78: 1–64.

Painter, R. E. 1966. Zoobenthos of San Pablo and Suisun Bays, pp. 40–56, *in* D. W. Kelley, ed., Ecological studies of the Sacramento–San Joaquin estuary, part I. Calif. Dept. Fish & Game, Fish Bull. 133: 1–133.

Parry, G. 1953. Osmotic and ionic regulation in the isopod crustacean, *Ligia oceanica*. J. Exper. Biol. 30: 567–74.

Peabody, E. B. 1939. Pigmentary responses in the isopod, *Idothea*. J. Exper. Zool. 82: 47–83.

Potier, L. 1951. Croissance relative et profils de croissance des péréiopodes du Crustacé Isopode *Sphaeroma serratum* (Fab.). C. R. Acad. Sci., Paris, 232: 2041–43.

Quast, J. C. 1968a. Observations on the food and biology of the kelp bass, *Paralabrax clathratus*, with notes on its sportfishery at San Diego, California, pp. 81–108, *in* W. J. North and C. L. Hubbs, eds., Utilization of kelp-bed resources in southern California. Calif. Dept. Fish & Game, Fish Bull. 139: 1–264.

———. 1968b. Observations on the food of the kelp-bed fishes, pp. 109–42, *in* W. J. North and C. L. Hubbs, eds., Utilization of kelp-bed resources in southern California. Calif. Dept. Fish & Game, Fish Bull. 139: 1–264.

Ray, D. L. 1956. Some marine invertebrates useful for genetic research, pp. 497–512, *in* A. A. Buzzati-Traverso, ed., Perspectives in marine biology. Berkeley and Los Angeles: University of California Press. 621 pp.

———, ed. 1959a. Marine boring and fouling organisms. Seattle: University of Washington Press. 543 pp.

———. 1959b. Nutritional physiology of *Limnoria*, pp. 46–61; Some properties of cellulase from *Limnoria*, pp. 372–96; *in* Ray (1959a).

———. 1959c. An integrated approach to some problems of marine biological deterioration: Destruction of wood in sea water, pp. 70–87, *in* I. Pratt and J. E. McCauley, eds., Marine biology, 20th Ann. Biol. Colloq., Oregon State College, Corvallis. 87 pp.

Ray, D. L., and J. R. Julian. 1952. Occurrence of cellulase in *Limnoria*. Nature 169: 32–33.

Ray, D. L., and D. E. Stuntz. 1959. Possible relation between marine fungi and *Limnoria* attack on submerged wood. Science 129: 93–94.

Reish, D. J., and W. M. Hetherington. 1969. The effects of hyper- and hypo-chlorinities on members of the wood-boring genus *Limnoria*. Mar. Biol. 2: 137–39.

Reynolds, E. S., and S. P. Meyers. 1959. Marine fungi and *Limnoria*. Science 130: 4.

Richardson, H. 1905. A monograph of the isopods of North America. Bull. U.S. Nat. Mus. 54: 1–727.

———. 1909. Isopods collected in the northwest Pacific by the U.S. Bureau of Fisheries steamer "Albatross" in 1906. Proc. U.S. Nat. Mus. 37: 75–129.

Ricketts, E. F., and J. Calvin. 1968. Between Pacific tides. 4th ed. Revised by J. W. Hedgpeth. Stanford, Calif.: Stanford University Press. 614 pp.

Riegel, J. A. 1959a. A revision in the sphaeromid genus *Gnorimosphaeroma* Menzies (Crustacea: Isopoda) on the basis of morphological, physiological and ecological studies on two of its "subspecies." Biol. Bull. 117: 154–62.

———. 1959b. Some aspects of osmoregulation in two species of sphaeromid isopod Crustacea. Biol. Bull. 116: 272–84.

Rotramel, G. 1972. *Iais californica* and *Sphaeroma quoyanum*, two symbiotic isopods introduced to California (Isopoda, Janiridae and Sphaeromatidae). Crustaceana 3 (Suppl.): 193–97.

Roubault, A. 1937. Dimorphisme et croissance chez un Tanaidacé. Trav. Sta. Biol. Roscoff 15: 135–52.

Ruck, P., and T. L. Jahn. 1954. Electrical studies on the compound eye of *Ligia occidentalis* Dana (Crustacea: Isopoda). J. Gen. Physiol. 37: 825–49.

Schultz, G. A. 1966. Submarine canyons of southern California. 4. Systematics: Isopoda. Allan Hancock Pacific Exped. 27: 1–56.

———. 1969. The marine isopod crustaceans. Dubuque, Iowa: Brown. 359 pp.

———. 1970. A review of the species of the genus *Tylos* Latreille from the New World (Isopoda, Oniscoidea). Crustaceana 19: 297–305.

———. 1972. A review of species of the family Scyphacidae in the New World (Crustacea, Isopoda, Oniscoidea). Proc. Biol. Soc. Washington 84: 477–88.

Shiino, S. M. 1964. On three bopyrid isopods from California. Rep. Fac. Fish., Univ. Mie 5: 19–25.

———. 1965. Phylogeny of the genera within the family Bopyridae. Bull. Mus. Nat. Hist. Natur. (2) 37: 462–65.

Sieg, J., and R. Winn. 1978. Keys to suborders and families of Tanaidacea (Crustacea). Proc. Biol. Soc. Washington 91: 840–46.

Silén, L. 1954. On the circulatory system of the Isopoda Oniscoidea. Acta Zool. 35: 11–70.

Sleeter, T. D., P. J. Boyle, A. M. Cundell, and R. Mitchell. 1978. Relationship between marine microorganisms and the wood-boring isopod *Limnoria tripunctata*. Mar. Biol. 45: 329–36.

Smith, H. G. 1938. The receptive mechanism of the background response in chromatic behaviour of Crustacea. Proc. Roy. Soc. London B125: 250–63.

Stewart, D. A. 1913. Changes in the branchial lamellae of *Ligia oceanica* after prolonged immersion in fresh and salt water. Mem. Proc. Manchester Lit. Phil. Soc. 58: 1–12.

Strunk, S. W. 1959. The formation of intracellular crystals in the midgut diverticula of *Limnoria lignorum*, pp. 108–19, *in* Ray (1959a).

Suneson, S. 1947. Color change and chromatophore activators in *Idothea*. Kungl. Fysiogr. Sällskap. Lund Handl. (N.F.) 58: 1–34.

Suzuki, S. 1934. Ganglion cells in the heart of *Ligia exotica* (Roux). Sci. Rep. Tôhôku Imp. Univ. (4) 9: 214–18.

Tait, J. 1910. Colour change in the isopod *Ligia oceanica*. J. Physiol. 40: 40–41.

———. 1917. Experiments and observations on Crustacea. I. Immersion experiments on *Ligia*. II. Moulting of isopods. Proc. Roy. Soc. Edinburgh 37: 50–68.

———. 1927. Experiments and observations on Crustacea. VII. Some structural and physiological features of the valviferous isopod *Chiridotea*. Proc. Roy. Soc. Edinburgh 46: 334–48.

Thompson, R. 1957. Successive reversal of a position habit in an invertebrate. Science 126: 163–64.

Tinturier-Hamelin, E. 1960. Recherches sur le polytypisme d'*Idotea baltica*. C. R. Acad. Sci., Paris, 251: 2408–10.

———. 1961. Influence d'une ovariectomie totale sur la formation du marsupium chez *Idotea baltica* (Pallas) (Isopode Valvifere). Recherches Preliminaires. Bull. Soc. Linn. Normandie (10) 2: 65–66.

Todd, M. E. 1963. Osmoregulation in *Ligia oceanica* and *Idotea granulosa*. J. Exper. Biol. 40: 381–92.

Turner, C. H., E. E. Ebert, and R. R. Given. 1969. Man-made reef ecology. Calif. Dept. Fish & Game, Fish Bull. 146: 1–221, App. 1.

Vandel, A. 1943. Essai sur l'origine, l'évolution et la classification des Oniscoidea (Isopodes terrestres). Bull. Biol. France Belg. 30 (Suppl.): 1–136.

———. 1945. La répartition géographique des Oniscoidea (Crustacés Isopodes terrestres). Bull. Biol. France Belg. 79: 221–72.

———. 1947. Recherches sur la génétique et la sexualité des Isopodes terrestres X. Étude des garnitures chromosomiques de quelques espèces d'Isopodes marins, dulçaquicoles et terrestres. Bull. Biol. France Belg. 81: 154–76.

———. 1951. Le genre *"Porcellio"* (Crustacés: Isopodes, Oniscoidea) évolution et systématique. Mém. Mus. Hist. Natur. Paris (n.s.) 3: 81–192.

Van Name, W. G. 1936. The American land and fresh-water isopod Crustacea. Bull. Amer. Mus. Natur. Hist. 71: 1–535.

———. 1940. A supplement to the American land and freshwater isopod Crustacea. Bull. Amer. Mus. Natur. Hist. 77: 109–42.

———. 1942. A second supplement to the American land and freshwater isopod Crustacea. Bull. Amer. Mus. Natur. Hist. 80: 299–329.

Waloff, N. 1941. The mechanism of humidity reactions of terrestrial isopods. J. Exper. Biol. 18: 115–35.

Warburg, M. R. 1964. The response of isopods towards temperature, humidity and light. Anim. Behav. 12: 175–83.

West, D. A. 1964. Polymorphism in the isopod *Sphaeroma rugicauda*. Evolution 18: 671–84.

Whitelegge, T. 1901. Crustacea, part 2. Isopoda, part 1. Sci. Res. Trawling Exped. H.M.C.S. "Thetis." Mem. Australian Mus. 4: 203–46.

Wieser, W. 1965. Electrophoretic studies in blood proteins in an ecological series of isopod and amphipod species. J. Mar. Biol. Assoc. U.K. 45: 507–23.

———. 1966. Copper and the role of isopods in degradation of organic matter. Science 153: 67–69.

Wilson, W. J. 1970. Osmoregulatory capabilities in isopods: *Ligia occidentalis* and *Ligia pallasii*. Biol. Bull. 138: 96–108.

Chapter 22

Amphipoda: *The Amphipods and Allies*

J. Laurens Barnard, Darl E. Bowers, and Eugene C. Haderlie

Members of the order Amphipoda are among the most abundant crustaceans in the intertidal zone of California. More than 150 species occur in southern California alone. Because they are often hard for students to identify, and difficult to photograph, this chapter considers only a few of the common species. Some 5,500 amphipods have been described worldwide. Most of them are marine, but more than 1,200 species live in fresh water, and nearly 100 species are terrestrial. Terrestrial amphipods live in damp places and may be found in litter on forest floors high on mountains in Japan, India, and Australia (Bousfield, 1958; Hurley, 1968). Several terrestrial species have been introduced into California and live in gardens and greenhouses; one species is common in fuchsia gardens in Golden Gate Park, San Francisco (Bousfield & Carlton, 1967). The majority of described species of amphipods are benthic or bottom-dwelling marine animals, but some are pelagic and either swim freely or make homes under the bells of jellyfishes or in abandoned swimming bells of siphonophores or old pelagic tunicate tests. Many pelagic forms have transparent bodies and possess enormous eyes covering most of the head. Amphipods generally are small, usually less than 1 cm long, but some giant forms 28 cm long have been photographed at depths of over 5,300 m in the Pacific Ocean (Hessler, Isaacs & Mills, 1972). In certain areas in the intertidal zone, such as in tidepools, among fouling growth, and under decaying kelp in the strand zone on beaches, amphipods are often exceptionally abundant.

J. Laurens Barnard is Curator of Crustacea at the National Museum of Natural History. Darl E. Bowers is Professor of Biology at Mills College, Oakland. Eugene C. Haderlie is Professor of Oceanography at the Naval Postgraduate School, Monterey.

The order Amphipoda is commonly divided into three suborders: the Gammaridea, containing most of the common amphipods seen along the seashore, the Hyperiidea, a pelagic group, and the Caprellidea, the skeleton shrimps and whale lice.

The bodies of most amphipods tend to be laterally compressed. The animals have compound eyes that are sessile (not stalked) and lateral in position on the head. The first thoracic segment (and sometimes the second as well) is fused with the head, and the abdominal segments are not distinctly demarcated from those in the thoracic region. The appendages show great diversity in form and function: the first pair of thoracic appendages is modified to form maxillipeds, mouthparts that aid in manipulating food; the second and third are often modified for grasping and are called gnathopods; and the remaining thoracic appendages are "walking" legs. From two to six pairs of gills are located on the posterior thoracic legs. In females the bases of thoracic legs 3–6 usually bear leaflike projections that overlap below the body to form a brood pouch or marsupium, in which the young are carried until able to fend for themselves. Thoracic appendages 2–8 have basal articles (coxae) consisting of long flattened plates that create a lateral shield; this shield protects the brood pouch in the female and the gills in both sexes. The abdomen bears six pairs of appendages: the anterior three pairs are pleopods, used for swimming and creating a respiratory current over the gills: the remaining three pairs are all uropods, which point backward and are used, along with the terminal part of the abdomen, in jumping.

Most common amphipods seen in the intertidal zone can swim using the pleopods and sometimes the thoracic legs. They can also crawl about on the bottom using the legs. When exposed on a wet surface or on a sandy beach, they lean over on one side and scull along rapidly, using the thoracic legs and gaining additional power by periodically pushing with the terminal end of the abdomen. Some can jump vertically, using the posterior part of the abdomen and the uropods as a spring. One species of beach amphipod 2 cm long has been known to leap forward through the air for a distance of up to a meter.

Amphipods living on sandy beaches are active burrowers. Others living on more solid substrata in the intertidal zone build more permanent tubes of mud, debris, or material secreted by the first two pairs of walking legs (thoracic appendages 4 and 5). The animals live in the tubes but venture forth to feed. Still other species live in natural depressions in rocks or mollusk shells. Some have taken up a semi-parasitic mode of life and burrow into sponges or compound ascidians. Others make burrows in living or dead kelp, and one worldwide species (*Chelura terebrans*) is found in burrows in submerged wood, usually along with the wood-boring isopod *Limnoria* (21.8; see Barnard, 1950).

Most amphipods appear to be scavengers or detritus feeders, but some consume tiny plants growing on rockweeds and kelp, and a few capture and eat small animals such as copepods and bryozoans. A few live in the branchial cavities of ascidians and may pierce and feed on host tissue as true parasites. One species (*Dulichia rhabdoplastis*) has a remarkable symbiotic relationship with the large red sea urchin *Strongylocentrotus franciscanus* (11.6). The amphipod fastens detritus, composed of fecal pellets and rejected food, to the ends of the urchin's spines and lengthens these to form a compact strand or rod that trails off from the spine tip. Up to 30 strands may be found on one urchin, and each is occupied by one or more amphipods that live and reproduce there. Adult and young amphipods feed on the benthic diatoms that flourish on the strands, and they keep other organisms from invading the diatom garden. This is the only recorded example of "farming" in a marine crustacean (McCloskey, 1970, 1971).

Amphipods, like the Isopoda and Tanaidacea, differ from the majority of marine invertebrates by brooding eggs that hatch into juveniles very similar in general appearance to the adults. Sexes are separate. The gonads are paired and tubular; the male ducts open on the ventral part of the last thoracic segment, the oviducts on the coxae of the sixth thoracic segment. The mating female is often seen carrying the male about on her back for days prior to her last molt. During copulation the male often twists his abdomen around and discharges sperm into the marsupium of the female where the discharged eggs are fertilized. Marine amphipods brood from a few to several hundred eggs at a time and may have several broods each year.

Among the most bizarre and interesting amphipods are the caprellids or skeleton shrimps. These elongate animals bear clinging feet at both the forward and rear ends. Most of the body consists of elongate thoracic segments; the abdomen is much reduced. These animals have gills on the fourth and fifth pairs of thoracic segments only. Skeleton shrimps are abundant on hydroid colonies and among small algal clumps.

Related to the caprellids, but with a much shorter and stouter, dorsoventrally flattened body, are the "whale lice," which are highly specialized ectoparasites of cetaceans. These can be seen on the skin of stranded whales and dolphins (Leung, 1967).

For general accounts of west coast amphipods, see especially Barnard (1952, 1954, 1958, 1969a,b); see also Alderman (1936), Barnard & Reish (1959), Hazel & Kelley (1966), Kelley (1967), and Shoemaker (1949). Anatomical studies have been made by Hilton (1917) and Schmitz (1967), and aspects of development have been investigated by Rappaport (1960). Studies on pressure sensitivity and temperature compensation are found in Enright (1961b, 1967); and on orientation behavior, in Craig (1971, 1973) and Menaker (1958). Studies on east coast species, of niche diversity and life history (see especially Croker, 1967a,b and Dexter, 1971), suggest potential research projects on California sandy-beach amphipods. Aids for identification of the numerous gammarid amphipods of the west coast, including extensive keys, are provided in Smith & Carlton (1975); see especially Barnard (1975), Bousfield (1975), and Bowers (1963, 1975), and the specialized systematic references cited there. McCain (1975) and Martin (1977) provide keys for the identification of the common caprellids of central and northern California.

Phylum Arthropoda / Class Crustacea / Subclass Malacostraca
Superorder Peracarida / Order Amphipoda / Suborder Gammaridea
Family Talitridae

22.1 **Orchestoidea californiana** (Brandt, 1851)

Common on wide exposed beaches of fine sand, high intertidal zone; Vancouver Island (British Columbia) to Laguna Beach (Orange Co.)

Mature animals to 28 mm long, distinguished from *O. corniculata* (22.2) by having long, thin, orange or rosy-red second antennae; immature animals pigmented dorsally with a dark "butterfly" pattern and a dark middorsal line, but with sides of body relatively free of markings.

This relatively large amphipod, sometimes called the long-horned beach hopper, survives best on exposed beaches of fine sand backed by dunes. The backshore areas provide a refuge for the hoppers during storms, when cutting surf erodes the beach sand. It is sometimes difficult to locate the large breeding adults in their isolated burrows in the high intertidal zone during the reproductive season, but smaller animals are easily found in and under washed-up clumps of seaweed, where they find food and daytime shelter from high temperatures, drying winds, and diurnal predators. On a summer night at an infested beach, one can see hundreds of hoppers feeding on washed-up seaweeds. As dawn approaches, one sees the predawn skirmishes of large males fighting over an open burrow.

When digging a new burrow high on the beach, a mature animal actively kicks sand outward in two opposite elongate rays. The burrows are elliptical in cross section and may be as deep as 30 cm in the summer, less in the winter, with a plug of sand at the mouth of the burrow.

Molting in this species is easy to observe. Just prior to molting, the dorsal surface of the posterior abdomen becomes opaque white; the change in color rapidly spreads over the entire body. The animal then twitches violently and splits the old exoskeleton transversely on the dorsal side between the first and second thoracic segments, and longitudinally along each side of the thorax above the legs. The dorsal plates then lift up and the animal backs out of the slit. The process requires only a few minutes, but the animal cannot jump nor-

mally for 24 hours or more. The molting process normally occurs deep in the burrow. Techniques have been developed for recognizing various instars between molts.

Mating occurs in the burrows from June until November in central California. The dark-blue eggs of the female can be seen through the body wall. After being in contact with the female for up to 48 hours, the male deposits sperm in a gelatinous mass on the ventral surface of the female, then leaves the burrow. The female discharges from 10–100 eggs into the marsupium, where they are fertilized and retained until hatched.

Orchestoidea californiana is fed upon by many shorebirds, by staphylinid beetles, and by racoons patrolling the beach at night. The adults are often infested by at least two species of parasitic mites, which cling to the ventral surface of the body (see the account of *O. corniculata*, 22.2).

See Bousfield (1957, 1961, 1975), Bowers (1963, 1964, 1975), Canaris (1962), McClurkin (1953), and Scurlock (1975).

Phylum Arthropoda / Class Crustacea / Subclass Malacostraca
Superorder Peracarida / Order Amphipoda / Suborder Gammaridea
Family Talitridae

22.2 **Orchestoidea corniculata** Stout, 1913

Common on steep protected beaches of coarse sand, high to middle intertidal zones; Humboldt Co. to southern California.

Mature animals to 25 mm long, distinguished from *O. californiana* (22.1) by having second antenna salmon-pink or orange in color, and with a shorter, thicker flagellum; immature animals pigmented with dark T-shaped designs dorsally, and with two diffuse dark spots on each side laterally.

This amphipod, often called the short-horned beach hopper, may sometimes be found on the same beach as *O. californiana*, but normally occurs in more sheltered locations and in coarser, damper sand lower in the intertidal zone. Its burrows are circular in cross section and normally about 15 cm deep. In excavating the sand, the beach hopper forms rounded mounds much like those of a pocket gopher on land.

The animals display a circadian rhythm in their activities. During daylight hours they remain in their burrows or under beach drift. During low tide at night they emerge and migrate down the beach to feed primarily on drift algae. At night the animals may be active over a band of beach 15–25 m wide (the larger animals foraging at the upper edge of the band), but at dawn they retreat to burrows arranged in a band 5–7 m wide in the upper intertidal zone along the beach. Sand penetrability appears to be important in determining the location of the burrows.

Early observations suggested that the animals oriented their movements by the position of the moon, but recent studies have not confirmed this. During the day, when unearthed, the amphipods orient in a landward direction despite overcast conditions, moisture gradients, beach slope, or displacement to a new beach. They must visually orient to some landmark or the backshore.

Like many sandy-beach amphipods, *O. corniculata* exhibits an activity pattern in rhythm with the ebb and flow of the tide, as well as with the alternation of day and night. Since each tidal cycle lasts slightly more than a day (about 24 hours and 50 minutes), the relation between time of day and high tide is constantly changing. In calm weather, the part of the beach inhabited by these amphipods is covered only at higher high water, at which time the animals always burrow below the surface, regardless of the time of day. When higher high water occurs during daylight hours, there is no noticeable effect, since the animals are already buried. When higher high water occurs at night, the animals dig in until the tide ebbs again, then re-emerge to feed. Adults and juveniles show slightly different periods of peak activity. Experiments show that the circadian (24-hour) and tidal (nearly 25-hour) rhythms are governed by biological clocks. Animals isolated from the sea and kept in boxes of damp sand in the laboratory in constant darkness continue for several days to show a normal diurnal rhythm of behavior, and—50 minutes later each day—they dig down as though anticipating higher high water.

This species is fed upon by many shore beetles and by birds. It serves as an intermediate host for larvae of a parasitic fluke (*Maritrema pacifica*), the mature adult of which lives in the intestine of several species of gulls. Both *O. corniculata* and *O. californiana* are infested on the ventral surface with small ectoparasitic mites (*Gammaridacarus brevisternalis*), which occur on up to 83 percent of the larger amphipods. After death of a host, the mites crawl over the sand and can locate new hosts at a distance of several centimeters.

See Baker & Yip (1978), Bousfield (1957, 1961, 1975), Bowers (1963, 1964, 1975), Ching (1974), Craig (1971, 1973), Enright (1961a), Ercolini & Scapini (1974), McClurkin (1953), McGinnis (1972), Osbeck (1970), and Scurlock (1975).

Phylum Arthopoda / Class Crustacea / Subclass Malacostraca
Superorder Peracarida / Order Amphipoda / Suborder Gammaridea
Family Pleustidae

22.3 **Pleustes platypa** Barnard & Given, 1960

Encountered on the kelp *Macrocystis pyrifera*; sometimes washed ashore on drift kelp; Gaviota Beach (Santa Barbara Co.) to Point Loma (San Diego Co.)

Length to 8.5 mm; head with long, broad, duckbill-shaped rostrum, longer than remainder of head; color pattern highly variable, but normally with yellow or brownish bands around the body.

This amphipod has been found only in shallow water offshore, living on the giant kelp *Macrocystis* in close association with the small snail *Mitrella carinata* (13.91). When on the kelp, the amphipod tucks its abdomen under the thorax and rests at about a 30° angle. In size, shape, color pattern, and posture it very closely resembles *Mitrella*, for which it may be mistaken. This is one of the few recorded instances of amphipod-molluscan mimicry (see also *Lacuna marmorata*, 13.41).

See Barnard & Given (1960) and Crane (1969).

Phylum Arthropoda / Class Crustacea / Subclass Malacostraca
Superorder Peracarida / Order Amphipoda / Suborder Gammaridea
Family Ampithoidae

22.4 **Ampithoe plumulosa** Shoemaker, 1938

Common in tubes on wet, low intertidal algae and surfgrass; also floating docks in harbors; British Columbia to Ecuador.

Length to 13 mm; telson thick, puffy, uncleft; third uropod large, visible, with stout outer ramus showing two hooked spines.

This species is often abundant on floating docks in southern California. It apparently requires virtually continuous immersion, but nothing is known of its biology.

See Barnard (1965, 1969a,b, 1975).

Phylum Arthropoda / Class Crustacea / Subclass Malacostraca
Superorder Peracarida / Order Amphipoda / Suborder Gammaridea
Family Ampithoidae

22.5 **Cymadusa uncinata** (Stout, 1912)

Common in tubes on *Macrocystis*; San Juan Islands (Washington) to Laguna Beach (Orange Co.).

Length to 35 mm, larger than *Ampithoe plumulosa* (22.4), otherwise similar.

This is the largest species of amphipod reported from shallow inshore waters in California. It has been reported to utilize a sticky web to cement rolled-over edges of blades of the giant kelp *Macrocystis* to form cigar-shaped tunnels for habitation. Its diet is uncertain.

See Barnard (1965, 1969a,b, 1975).

Phylum Arthropoda / Class Crustacea / Subclass Malacostraca
Superorder Peracarida / Order Amphipoda / Suborder Gammaridea
Family Dexaminidae

22.6 **Polycheria osborni** Calman, 1898

Rare to abundant, burrowed into tests of tunicates, especially *Aplidium californicum* (12.1), low intertidal zone and subtidal waters; known from Oregon to southern California, probably more widespread.

Body 3–8 mm long, semitransparent with dark markings; eyes dark purple.

These animals form shallow pits in the tests of compound ascidians, where they lie on their backs with their legs projecting toward the opening. In this position they use the pleo-

pods to propel water forward along the ventral body surface. Diatoms and detritus are filtered out of the stream for food. The animal holds the tunicate tissue at the sides of the burrow with the distal ends of the anterior legs, and can open and close the burrow by moving its legs. When closed, the aperture looks like a zigzag seam. *Polycheria* normally does not leave the burrow, but if forced to do so can swim sluggishly. When an animal finds a suitable ascidian, it forms a burrow not by digging but by lying on its back, grasping the ascidian test with the anterior legs, and pulling slowly, thereby pushing the dorsal surface down into the test; the operation may require several hours. The ascidian seems to be neither harmed nor benefited, and *Polycheria* is provided with a home. In its method of orientation and feeding, this amphipod shows adaptations analogous to those of barnacles, but has undergone much less modification.

Breeding occurs in early summer, and females brood up to 80 young in a brood pouch throughout the summer months, releasing a few young at a time. The young immediately search out depressions in the tunicate where new burrows can be formed.

See Alderman (1936), Barnard (1969a,b, 1975), Ricketts & Calvin (1968), and Skogsberg & Vansell (1928).

Phylum Arthropoda / Class Crustacea / Subclass Malacostraca
Superorder Peracarida / Order Amphipoda / Suborder Gammaridea
Family Calliopidae

22.7 **Oligochinus lighti** Barnard, 1969

Abundant to scarce, associated with algae and surfgrass (particularly the red algae *Gigartina papillata* and *Endocladia muricata* at Monterey), high and middle intertidal zones in rocky areas; Mendocino Co. to Goleta (Santa Barbara Co.)

Body up to 11.5 mm long, reddish brown, differing from other members of the family in having a shortened cleft telson, quite long and oarlike third uropods, and a small, scale-like accessory flagellum on the first antenna.

The ecology of this species has been investigated at Pacific Grove (Monterey Co.) by Johnson (1973). It is the most common amphipod in the *Gigartina-Endocladia* association. Animals found between 102 and 57 cm above zero tide level showed highest populations in the fall of each year and lowest populations in the spring. Population densities appear to be controlled mainly by physical, rather than biological, factors.

The animals feed on diatoms or epiphytic algae they remove from benthic algae, and on the surface layers of cells from *Gigartina* and *Endocladia*. Because of their cryptic coloration and habit of secreting themselves among the plants, they are commonly overlooked. They avoid direct light and come out of the algal clump only during darkness, regardless of the tidal condition.

In any one population there are always gravid females and juveniles, indicating no strong seasonal reproductive cycle. Differences in population density between fall and spring are correlated with seasonal climatic factors, such as aerial temperature, amount of exposure, and desiccation, and with the tolerances of the animals for various combinations of these variables.

See Barnard (1969a,b, 1975) and Johnson (1973, 1975).

*Phylum Arthropoda / Class Crustacea / Subclass Malacostraca
Superorder Peracarida / Order Amphipoda / Suborder Caprellidea
Family Caprellidae*

22.8 **Caprella californica** Stimpson, 1857
(=Caprella scaura var. californica)   Skeleton Shrimp

Common on eelgrass (*Zostera*), low intertidal zone and subtidal waters in bays; central and southern California.

Body to 35 mm long, distinguished from other common caprellids by its habitat and color, by its large size, and by the long, dorsal, anteriorly directed spine on head; female abdomen tiny, bearing a pair of lobes but no appendages.

Caprellid amphipods are the praying mantises of the sea (though no relation to mantis shrimps). Skeleton shrimps apparently have evolved from more conventional gammarid amphipods through reduction of the abdomen, elongation of the free thoracic segments (segments 1 and often 2 are more or less fused to the head), and loss of the legs on thoracic segments 4 and 5. The animals cling primarily to hydroids, bryo-

zoans, and plants by means of the last three pairs of thoracic legs, leaving the large subchelate appendages of thoracic segments 2 and 3 for use as "arms" for capturing food, for defense, and for cleaning. While clinging to the substratum, they exhibit bowing-and-scraping motions somewhat similar to those of the terrestrial mantis. The arms are also important to locomotion; caprellids move very rapidly in inchworm fashion, grasping the substratum alternately with "arms" and "legs." When free in the water, caprellids can swim by rapidly flexing and straightening the body.

Caprella californica is an omnivrous, opportunistic feeder, taking organic detritus from the water and scraping diatoms, dinoflagellates, and suctorians off the substratum. It has been observed to feed on small dead crustaceans. It also captures small swimming amphipods, and while holding them in a claw, devours them piecemeal.

Both sexes bear short gills on thoracic appendages 4 and 5, and the females carry their young in a ventral brood pouch formed of leaflike projections from these body segments.

See Caine (1976, 1977), Dougherty & Steinberg (1953), Keith (1969, 1971), Martin (1977), McCain (1968, 1975), and Ricketts & Calvin (1968).

*Phylum Arthropoda / Class Crustacea / Subclass Malacostraca
Superorder Peracarida / Order Amphipoda / Suborder Caprellidea
Family Caprellidae*

22.9 **Caprella** sp.

This unidentified caprellid is an example of skeleton shrimps often found in great abundance on hydroids in the California intertidal zone. They are hardy and live well in aquaria if collected with their natural cnidarian hosts. Their behavior is interesting to watch under a low-power microscope.

Food habits and methods of feeding vary with the species; there are, in California waters, different forms that filter feed, browse on plants, scrape the substratum for diatoms, scavenge on dead material, and capture living prey.

Small caprellids are often rather transparent, and live animals mounted in seawater on a glass slide under a coverslip are excellent for study under the compound microscope. Action of the elongate heart, muscles of the appendages and body, circulation of blood, and movement of the digestive tract can be seen very clearly.

See Caine (1977) and McCain (1975).

Literature Cited

Alderman, A. L. 1936. Some new and little known amphipods of California. Univ. Calif. Publ. Zool. 41: 53–74.

Baker, A., and L. Yip. 1978. Circadian and circatidal rhythms of population movement in adult and juvenile *Orchestoidea corniculata* (Crustacea: Amphipoda) on a central California sandy beach. Research paper, Biol. 175h, library, Hopkins Marine Station of Stanford University, Pacific Grove, Calif. 27 pp., 9 pls.

Barnard, J. L. 1950. The occurrence of *Chelura terebrans* Philippi in Los Angeles and San Francisco harbors. Bull. So. Calif. Acad. Sci. 49: 90–97.

———. 1952. Some Amphipoda from central California. Wasmann J. Biol. 10: 9–36.

———. 1954. Marine Amphipoda of Oregon. Oregon State Monogr., Stud. Zool. 8: 1–103.

———. 1958. Amphipod crustaceans as fouling organisms in Los Angeles–Long Beach harbors, with reference to the influence of seawater turbidity. Calif. Fish & Game 44: 161–70.

———. 1965. Marine amphipods of the family Ampithoidae from southern California. Proc. U.S. Nat. Mus. 118: 1–46.

———. 1969a. Gammaridean Amphipoda of the rocky intertidal of California: Monterey Bay to La Jolla. Bull. U.S. Nat. Mus. 258: 1–230.

———. 1969b. The families and genera of marine gammaridean Amphipoda. Bull. U.S. Nat. Mus. 271: 1–535.

———. 1975. Identification of gammaridean amphipods, pp. 314–52, *in* Smith & Carlton (1975).

Barnard, J. L., and R. R. Given. 1960. Common pleustid amphipods of southern California, with a projected revision of the family. Pacific Natur. 1: 41–42.

Barnard, J. L., and D. J. Reish. 1959. Ecology of Amphipoda and Polychaeta of Newport Bay, California. Allan Hancock Found. Occas. Pap. 21: 1–106.

Bousfield, E. L. 1957. Notes on the amphipod genus *Orchestoidea* on the Pacific coast of North America. Bull. So. Calif. Acad. Sci. 56: 119–29.

———. 1958. Distributional ecology of terrestrial Talitridae (Crustacea; Amphipoda) of Canada. Proc. 10th Internat. Congr. Entomol. 1: 883–98.

———. 1961. New records of beach hoppers (Crustacea: Amphipoda) from the coast of California. Nat. Mus. Canada Contr. Zool., Bull. 172: 1–12.

———. 1975. Morphological key to Talitridae, pp. 352–55, *in* Smith & Carlton (1975).

Bousfield, E. L., and J. T. Carlton. 1967. New records of Talitridae (Crustacea: Amphipoda) from the central California coast. Bull. So. Calif. Acad. Sci. 66: 277–84.

Bowers, D. E. 1963. Field identification of five species of California beach hoppers (Crustacea: Amphipoda). Pacific Sci. 17: 315–20.

———. 1964. Natural history of two beach hoppers of the genus *Orchestoidea* (Crustacea: Amphipoda) with reference to their complemental distribution. Ecology 45: 677–96.

———. 1975. A field (color pattern) key to *Orchestoidea*, pp. 355–57, *in* Smith & Carlton (1975).

Caine, E. A. 1976. Cleansing mechanisms of caprellid amphipods (Crustacea) from North America. Mar. Behav. Physiol. 4: 161–69.

———. 1977. Feeding mechanisms and possible resource partitioning of the Caprellidae (Crustacea: Amphipoda) from Puget Sound. U.S.A. Mar. Biol. 42: 331–36.

Canaris, A. G. 1962. A new genus and species of mite (Laelaptidae) from *Orchestoidea californiana* (Gammaridea). J. Parasitol. 48: 467–69.

Ching, H. L. 1974. Two new species of *Maritrema* (Trematoda; Microphallidae) from the Pacific coast of North America. Canad. J. Zool. 52: 865–69.

Craig, P. C. 1971. An analysis of the concept of lunar orientation in *Orchestoidea corniculata* (Amphipoda). Anim. Behav. 19: 368–74.

———. 1973. Behaviour and distribution of the sand-beach amphipod *Orchestoidea corniculata*. Mar. Biol. 23: 101–9.

Crane, J. M., Jr. 1969. Mimicry of the gastropod *Mitrella carinata* by the amphipod *Pleustes platypa*. Veliger 12: 200.

Croker, R. A. 1967a. Niche specificity of *Neohaustorius schmitzi* and *Haustorius* sp. (Crustacea: Amphipoda) in North Carolina. Ecology 48: 971–75.

———. 1967b. Niche diversity in five sympatric species of intertidal amphipods (Crustacea: Haustoriidae). Ecol. Monogr. 37: 173–200.

Dexter, D. M. 1971. Life history of the sandy-beach amphipod *Neohaustorius schmitzi* (Crustacea: Haustoriidae). Mar. Biol. 8: 232–37.

Dougherty, E. C., and J. E. Steinberg. 1953. Notes on the skeleton shrimps (Crustacea, Caprellidae) of California. Proc. Biol. Soc. Washington 66: 39–50.

Enright, J. T. 1961a. Lunar orientation of *Orchestoidea corniculata* Stout (Amphipoda). Biol. Bull. 120: 148–56.

———. 1961b. Pressure sensitivity of an amphipod. Science 133: 758–60.

———. 1967. Temperature compensation in short duration time measurement by an intertidal amphipod. Science 156: 1510–12.

Ercolini, A., and F. Scapini. 1974. Sun compass and shore slope in the orientation of littoral amphipods (*Talitrus saltator* Montagu). Monitore Zool. Ital. 8: 85–115.

Hazel, C. R., and D. W. Kelley. 1966. Zoobenthos of the Sacramento–San Joaquin delta, pp. 113–33, *in* D. W. Kelley, comp., Ecological studies of the Sacramento–San Joaquin estuary, part 1. Calif. Dept. Fish & Game, Fish Bull. 133: 1–133.

Hessler, R. R., J. D. Isaacs, and E. L. Mills. 1972. Giant amphipods from the abyssal Pacific Ocean. Science 175: 636–37.

Hilton, W. A. 1917. The central nervous system of the amphipod *Orchestia*. Pomona Coll. J. Entomol. Zool. 9: 88–91.

Hurley, D. E. 1968. Transition from water to land in amphipod crustaceans. Amer. Zool. 8: 327–53.

Johnson, S. E. 1973. The ecology of *Oligochinus lighti* J. L. Barnard, 1969, a gammarid amphipod from the high rocky intertidal region of Monterey Bay, California. Doctoral thesis, Biological Sciences, Stanford University, Stanford, Calif. 267 pp.

_____. 1975. Microclimate and energy flow in the marine rocky intertidal, pp. 559–87, *in* D. W. Gates and R. B. Schmerl, eds., Perspectives of biophysical ecology. New York: Springer-Verlag.

Keith, D. E. 1969. Aspects of feeding in *Caprella californica* Stimpson and *Caprella equilibra* Say (Amphipoda). Crustaceana 16: 119–24.

_____. 1971. Substrate selection in caprellid amphipods of southern California, with emphasis on *Caprella californica* Stimpson and *Caprella equilibra* Say (Amphipoda). Pacific Sci. 25: 387–94.

Kelley, D. W. 1967. Identification of *Corophium* from the Sacramento–San Joaquin delta. Calif. Fish & Game 53: 295–96.

Leung, Y.-M. 1967. An illustrated key to the species of whale-lice (Amphipoda, Cyamidae), ectoparasites of Cetacea, with a guide to the literature. Crustaceana 12: 279–91.

Martin, D. M. 1977. A survey of the family Caprellidae (Crustacea, Amphipoda) from selected sites along the northern California coast. Bull. So. Calif. Acad. Sci. 76: 146–67.

McCain, J. C. 1968. The Caprellidae (Crustacea: Amphipoda) of the western North Atlantic. Bull. U.S. Nat. Mus. 278: 1–147.

_____. 1975. Phylum Arthropoda: Crustacea, Amphipoda: Caprellidea, pp. 367–76, *in* Smith & Carlton (1975).

McCloskey, L. R. 1970. A new species of *Dulichia* (Amphipoda, Podoceridae) commensal with a sea urchin. Pacific Sci. 24: 90–98.

_____. 1971. A marine farmer: The rod-building amphipod. Fauna 1: 20–25.

McClurkin, J. I. 1953. Studies on the genus *Orchestoidea* (Crustacea: Amphipoda) in California. Doctoral thesis, Biological Sciences, Stanford University, Stanford, Calif. 207 pp.

McGinnis, J. W. 1972. A tidal rhythm in the terrestrial sand beach amphipod *Orchestoidea corniculata*. Research paper, Biol. 175h, library, Hopkins Marine Station of Stanford University, Pacific Grove, Calif. 23 pp.

Menaker, M. 1958. Celestial time compensated orientation of east coast amphipods. Anat. Rec. 132: 476.

Osbeck, B. L. 1970. Circadian and tidal rhythms in the sand beach amphipod genus *Orchestoidea* (Talitridae). Master's thesis, University of California, Santa Barbara. 47 pp., 26 pls.

Rappaport, R. 1960. The origin and formation of blastoderm cells of gammarid Crustacea. J. Exper. Zool. 144: 43–59.

Ricketts, E. F., and J. Calvin. 1968. Between Pacific tides. 4th ed. Revised by J. W. Hedgpeth. Stanford, Calif.: Stanford University Press. 614 pp.

Schmitz, E. H. 1967. Visceral anatomy of *Gammarus lacustris lacustris* Sars (Crustacea: Amphipoda). Amer. Midl. Natur. 78: 1–54.

Scurlock, D. 1975. Infestation of the sandy beach amphipod *Orchestoidea corniculata* by *Gammaridacarus brevisternalis* (Acari: Laelaptidae). Bull. So. Calif. Acad. Sci. 74: 5–9.

Shoemaker, C. R. 1949. The amphipod genus *Corophium* on the west coast of America. J. Wash. Acad. Sci. 39: 66–82.

Skogsberg, T., and G. H. Vansell. 1928. Structure and behavior of the amphipod *Polycheria osborni*. Proc. Calif. Acad. Sci. (4) 17: 267–95.

Smith, R. I., and J. T. Carlton, eds. 1975. Light's manual: Intertidal invertebrates of the central California coast. 3rd ed. Berkeley and Los Angeles: University of California Press. 716 pp.

Chapter 23
Caridea: *The Shrimps*

Fenner A. Chace, Jr., and Donald P. Abbott

Shrimps, as the word is used in America, include the presumably primitive members of the crustacean order Decapoda—the diverse assemblage that also contains the lobsters, hermit crabs, true crabs, and lesser-known groups all characterized by a complete head shield, or carapace, and by five pairs of usually prominent "legs" adapted for walking or grasping. Until rather recently, all of the shrimps were assigned to the Natantia, a suborder characterized by having the anterior somite of the abdomen not much smaller than the succeeding somites, and five pairs of abdominal appendages (pleopods) well developed and adapted for swimming. This convenient concept has now been abandoned by some, but not all, decapod specialists in the belief that one of the sections comprising the Natantia (the Penaeidea) differs more significantly from the other two sections (the Caridea and the Stenopodidea) than all three do from the remaining decapod subdivisions (see Glaessner, 1969, pp. R443–46). Only one of these sections, the Caridea, is known to be represented in the intertidal zone of the Pacific coast of California. In addition to the natantian characters mentioned above, the carideans are distinguished by having the lateral flap of the second abdominal somite overlapping the flaps of both the first and third somites, and none of the three posterior pairs of thoracic "legs" provided with true pincers.

The earliest known carideans were found in Upper Permian or Lower Triassic rocks, but the fossil record is far too incomplete to afford reliable information about the evolutionary origins and an-

cestors of the group. The approximately 1,700 living species currently recognized occur in all the world oceans to a depth of at least 5,000 m and in tropical to warm-temperate fresh waters around the world; one caridean even lives arboreally in mangrove thickets. They inhabit virtually every kind of aquatic environment: the open sea, the mud and sand beneath, the alga-draped rocks and rock pools of the low intertidal zone, and even sanctuaries furnished by other animals such as mollusks and sponges, where security may mean imprisonment. Whatever the habitat, most carideans have not become so peculiarly altered as to lose their "shrimplike" appearance, whether the body is robust and somewhat depressed as in *Crangon* (the true "shrimps" according to British usage) or slender and compressed laterally, with a prominent head spine, or rostrum, as in *Palaemon* ("prawns" by British standards). They range in size from tiny parasitic and commensal species scarcely 5 mm long to the giant *Macrobrachium* of tropical fresh waters whose heavy bodies may measure well over 30 cm in length and whose massive claws may span nearly 120 cm. Most California intertidal shrimps are relatively small, ranging from a few millimeters to a few centimeters in length.

Like all decapods, caridean shrimps have stalked eyes, but the eyes are sometimes covered by the head shield, or carapace, as in *Alpheus* and *Betaeus*, so that vision is greatly restricted and tactile organs must assume much of the function usually performed by the eyes. The central nervous system consists of a supraesophageal ganglion and a ventral series of ganglia extending through the length of the body, only those associated with the mouthparts being fused. The principal organs of respiration are the gills, attached to and above the bases of the thoracic legs and well pro-

Fenner A. Chace, Jr., is Zoologist Emeritus, Department of Invertebrate Zoology, National Museum of Natural History, Smithsonian Institution. Donald P. Abbott is Professor, Department of Biological Sciences and Hopkins Marine Station, Stanford University.

tected under the lateral folds of the carapace. Water is circulated about the gills by the pumping action of a small "bailer" on each side, attached to the base of one of the mouthparts (second maxilla). Excretion in shrimps is accomplished chiefly by the paired antennal glands, opening at the base (peduncle) of the second antenna. The excretory canal in many of the carideans is shorter and less complicated than in some other decapods. The caridean heart lies dorsally in the thorax, and is provided with five pairs of ostia (for the entrance of blood), in contradistinction to only three pairs of ostia in most other decapods. The rapid heartbeat is easily seen through the transparent carapace in small living tidepool shrimps.

In most marine carideans, the males are somewhat smaller than the females, and they are usually distinguishable by the presence of two pendants, or processes, on the mesial margin of the mesial branch of the second abdominal limb; in females there is but one pendant in this location. After mating, the fertilized eggs are carried about by the females, attached to the pleopods, until hatching occurs. The young usually hatch as zoea larvae and pass through a variable number of zoeal and mysis stages before attaining the adult form in a few weeks to a few months. Little is known about the life span of most carideans, but some have survived for more than five years under laboratory conditions.

Probably the majority of caridean shrimps are omnivorous, catching or scavenging whatever may be encountered. Pelagic species often have extensively setose "legs" that can be positioned to form a basket for filtering the water through which the animal swims. Others have pincers especially adapted for scraping algae off rocks or, in commensal forms, for scraping mucus from the gills of mollusk hosts. Snapping shrimps (*Alpheus*, e.g., 23.2) may be able to stun their piscine prey with their explosive claws, and the cleaner shrimps (see, for example, Limbaugh et al., 1961) pick parasites and necrotic tissue from cooperating fishes.

The foregut of the digestive system, in shrimps as in other familiar decapod crustaceans, contains a masticating apparatus or gastric mill capable of grinding up food that has been ingested in larger pieces. This permits the animals to feed quickly, then retire to sheltered situations to masticate at leisure.

Shrimps, in their turn, provide sustenance to a variety of parasites and predators. Carideans are rather frequent victims of bopyrid isopod parasites, especially those that invade the gill chamber of the host. These infestations seem rarely to be fatal, and represent a minor hazard. Probably the chief source of danger to shrimps comes from fishes (see, for example, Quast, 1968), followed by diving birds and man. And, as fishermen and shrimpers know, both fishes and men can be lured to an untimely end by the promise of shrimps.

Shrimps are of considerable economic importance to man, both directly as human food, and indirectly, as sustenance for other marine animals important to man. Of greatest direct commercial value are the species of *Penaeus* (penaeid shrimps, not covered in this book), which are most abundant in the warmer seas of the world (see, for example, Edwards, 1978). However, many inhabitants of northern Europe are beyond convincing that any shrimps surpass in flavor the little *Crangon* of more northern seas and estuaries. In California, species of *Crangon* (bay shrimps) are taken commercially only in San Francisco Bay, where they are used mainly for fish bait.

The main commercial shrimp fished in California waters since 1952 is the Pacific ocean shrimp, *Pandalus jordani* Rathbun, a caridean found in deeper waters (mainly 45–370 m depth) ranging from Unalaska (Aleutian Islands, Alaska) south at least to San Diego. Averaging about 10 cm in length (though reaching 14 cm), the shrimp are carnivores with a life span of probably 3–4 years. They change sex during life, breeding first as males, then the following year or two as females. The bulk of the catch in this strictly regulated fishery occurs at the northern end of the state, between Eureka (Humboldt Co.) and the Oregon border. Catches for 1971–74 were 0.56–1.4 million kg, worth $220,000–$602,000 (McAllister, 1975, 1976; Pinkas, 1974).

Most species of California intertidal shrimps have received little attention except from systematists, and much remains to be learned of their biology. The smaller species, particularly those from pools in the middle intertidal zone, thrive in seawater aquaria and feed readily on bits of meat or living brine shrimp. They make excellent subjects for the study of behavior and some aspects of physiology. Much of the classical work on crustacean hormones and color change was carried out on shrimps.

The following characters are significant to the identification of California intertidal shrimps: the form of the rostrum (head spine); the configuration of the frontal margin of the carapace and the degree of eye concealment; the development and symmetry of the anterior pair of thoracic legs; the integrity of, or the number of articles in, the antepenultimate segment of the second leg; the form of the terminal segment of the fourth and fifth legs; color pattern; and size. Because many of the shrimps most likely to be found in this habitat are not easily recognized from photographs, only a few representative species are illustrated and discussed here. To supplement this incomplete coverage, there appears below a list of the shrimps occurring naturally on California shores. Species followed ·by a decimal number in parentheses are covered in the text.

Section Caridea
  Family Palaemonidae
    Palaemon (Palaemon) macrodactylus (23.1). Oriental shrimp.
    Palaemon (Palaemon) ritteri Holmes, 1895. San Diego to Peru. Along rocky and sandy shores and in tidepools.
    Palaemonetes (Palaemonetes) hiltoni Schmitt, 1921. San Pedro (Los Angeles Co.) to Sinaloa (Mexico). Estuarine?
    Palaemonella holmesi (Nobili, 1907). Santa Cruz and Santa Catalina Islands (Channel Islands) to Ecuador. Among algae.
    Periclimenes (Periclimenes) infraspinis (Rathbun, 1902). San Diego to Costa Rica. Subtidal on sand and among algae on rocks.
  Family Alpheidae
    Alpheus californiensis Holmes, 1900. San Pedro (Los Angeles Co.) to Bahía Magdalena (Baja California). Lowest tide level and below.
    Alpheus clamator (23.2). Snapping shrimp, Pistol shrimp.
    Betaeus ensenadensis Glassell, 1938. San Diego to Ensenada (Baja California). In burrows of the bay ghost shrimp, *Callianassa californiensis* (24.2), and the blue mud shrimp, *Upogebia pugettensis* (24.1), in mud flats.
    Betaeus gracilis Hart, 1964. Monterey to Laguna Beach (Orange Co.). Along shore and in kelp holdfasts.
    Betaeus harfordi (23.3).
    Betaeus harrimani (23.4).
    Betaeus longidactylus (23.5).
    Betaeus macginitieae Hart, 1964. Monterey to Santa Catalina Island (Channel Islands). Living in pairs in association with the sea urchins *Strongylocentrotus franciscanus* (11.6) and *S. purpuratus* (11.5).
    Betaeus setosus Hart, 1964. Queen Charlotte Islands (British Columbia) to Morro Bay (San Luis Obispo Co.). Among rocks and algae in tidepools.

Synalpheus lockingtoni Coutière, 1909. Santa Monica Bay (Los Angeles Co.) to Gulf of California. Littoral to 550 m.
  Family Hippolytidae
    Heptacarpus brevirostris (23.6).
    Heptacarpus carinatus Holmes, 1900. Monterey Bay to Point Loma (San Diego Co.) Among algae.
    Heptacarpus [or Eualus] layi (Owen, 1839). Esquimalt harbor (Vancouver Island, British Columbia) to Monterey. Habitat unknown.
    Heptacarpus palpator (Owen, 1839). San Francisco to Bahía Magdalena (Baja California). Among low intertidal rocks and algae.
    Heptacarpus paludicola Holmes, 1900. British Columbia to San Diego. Among algae in tidepools.
    Heptacarpus pictus (23.7). Red-banded transparent shrimp.
    Heptacarpus taylori (Stimpson, 1857). San Francisco to Bahía Magdalena (Baja California). Subtidal among rocks and algae and on pilings.
    Hippolyte californiensis (23.9). Slender green shrimp, Grass shrimp.
    Hippolyte clarki Chace, 1951. Sitka (Alaska) to Palos Verdes Peninsula (Los Angeles Co.) and Santa Catalina Island (Channel Islands).
    Lysmata californica (23.8). Red rock shrimp.
    Spirontocaris affinis (Owen, 1839). Known only from type specimen, Monterey.
    Spirontocaris prionota (23.10). Broken-back shrimp.
  Family Crangonidae
    Crangon franciscorum Stimpson, 1856. Franciscan bay shrimp. Southeastern Alaska to San Diego. Subtidal to 55 m on sand or mud.
    Crangon nigricauda Stimpson, 1856. Black-tailed bay shrimp. British Columbia to Baja California. Littoral to 57 m.
    Crangon stylirostris (23.11). Bay shrimp.

Carlton & Kuris (1975) and Word & Charwat (1976) present good keys to California shore shrimps, including nearly all of the species listed above. Full descriptions of most of the species are available in Schmitt (1921). Common names used are those adopted by the California Department of Fish and Game (Gates & Frey, 1974). Aspects of commercial shrimp fishing in California are reviewed in Bonnot

(1932), Dahlstrom (1961, 1967, 1972), Frey (1971), Geibel & Heimann (1976), Gotshall (1967), and Young & Withycombe (1949). For further information on the biology and physiology of crustaceans in general, including shrimps, see Gurney (1939, 1942), Kaestner (1970), Lockwood (1967), Schmitt (1965), and Watermann (1960, 1961).

Phylum Arthropoda / Class Crustacea / Subclass Malacostraca
Superorder Eucarida / Order Decapoda / [Suborder Natantia] / Section Caridea
Family Palaemonidae

23.1 **Palaemon (Palaemon) macrodactylus** Rathbun, 1902
Oriental Shrimp

Found subtidally among rocks and debris and on pilings and sea walls; north China, Korea, Japan; introduced into San Francisco Bay.

Length to about 58 mm; dorsal margin of rostrum (head spine) with 9–15 teeth, the proximal three teeth lying posterior to level of margin of orbit of eye, the most distal tooth usually arising subapically; ventral of two lateral spines near anterior margin of carapace originating behind margin; living shrimps usually translucent or nearly transparent, lacking conspicuous markings or color pattern, occasionally dark green or olive drab.

This species was accidentally introduced into San Francisco Bay, probably about 1954, and has since become abundant in many tidal creeks where brackish water conditions exist. In Suisun Bay and San Pablo Bay, populations extend into the open-bay waters, where they mix with schools of the native bay shrimps (*Crangon* spp.) and thus become a significant part of the commercial bay shrimp catch.

The species survives well in the laboratory at temperatures of 14–26°C, and in both normal and dilute seawater. Females bearing eggs are found at least from the middle of April through October. Hatched larvae thrive in aquaria. Development, from hatching to the time of metamorphosis into juvenile shrimp resembling adults, generally takes 12–18 days; during this period the animals pass through six (occasionally five or seven) larval stages, or instars. The larvae have been the subject of physiological studies, including one on the effects of eyestalk extirpation on development.

See Carlton & Kuris (1975), Ganssle (1966), Little (1969), Newman (1963), and Ricketts & Calvin (1968).

Phylum Arthropoda / Class Crustacea / Subclass Malacostraca
Superorder Eucarida / Order Decapoda / [Suborder Natantia] / Section Caridea
Family Alpheidae

23.2 **Alpheus clamator** Lockington, 1877
(formerly called Crangon dentipes)
Snapping Shrimp, Pistol Shrimp

Common, low intertidal pools and crevices on rocky shores; Farallon Islands (San Francisco Co.) to Bahía San Bartolomé (Baja California).

Length to about 37 mm; first pair of legs chelate, with one chela powerfully developed into snapping claw; hood over each eye armed with small tooth on anterior margin; fourth (largest) joint of walking legs with prominent tooth near distal end of lower margin; body tan with some dark-brown spots.

Several species of snapping shrimp occur in California. They are among the noisiest of the animals producing underwater sounds in the sea, and their metallic clicking is familiar to those visiting tidepool areas at ebb tide. The sound is produced by the enlarged claw, which may be at least half the size of the body. The movable finger of the claw can be elevated through an angle of 90°, then "cocked" in this position. Studies of the related species **A. californiensis** (an inhabitant of mud flats in bays) show that both the movable finger and the "hand" of the claw bear disks with microscopically smooth surfaces that are brought into contact when the claw is fully opened. In the cocked position the disks fit together perfectly, trapping a thin layer of water between them. It appears that when the closing muscles of the claw contract, the cohesive force of water between the disks holds the claw open until the closing force approaches 2×10^5 dynes. This exceeds the tensile strength of the water layer between the disks; they separate, and the claw snaps shut with great force (in some snapping shrimps, squirting a jet of water outward from the tip of the chela).

At least a good part of the snapping noise appears to accompany the separation of the two disks, rather than the clashing together of the two fingers of the claw a fraction of a second later. Placing cotton or a layer of cloth between the fingers dulls the snap a bit, but a person handling an excised claw can produce a fairly respectable snap over and over again by repeatedly cocking the claw and then separating the disks, without ever allowing the claw to close fully (S. Davis, pers. comm.). On the other hand, a tiny scratch on one of the disks destroys the claw's ability to cock and snap, though not its ability to open and close. Other effects accompany the sound. Snapping *Alpheus* have been known to crack the glass vessels containing them if the vessels were previously scratched. A snapping alpheid held within the hand causes a stinging sensation, and in natural situations small animals near an active snapper are stunned or killed. Both the mechanism of snapping and the nature of the effects of the pistol claw on other animals a short distance away need further investigation.

Observations on *A. californiensis* show the claw is used both offensively and defensively. Prey, including the small goby fishes *Clevelandia ios* and juvenile *Gillichtys mirabilis*, are stunned while still beyond reach of the shrimp, and are then easily captured and drawn into the shrimp's burrow. Sometimes prey is held with the smaller cheliped and given the quietus by whacks from the pistol claw.

Both *A. clamator* and *A. californiensis* are often found in sexual pairs within their burrows.

See Carlton & Kuris (1975), Hazlett & Winn (1962), Johnson, Everest & Young (1947), Knowlton & Moulton (1963), MacGinitie (1937), MacGinitie & MacGinitie (1968), Ricketts & Calvin (1968), Ritzmann (1973, 1974), Schein (1975), and Schmitt (1921).

eyes shallowly notched; fingers and palm of large claws subequal in length; terminal joint of walking legs stout, bifid; body purple, blue, or blue-black.

There appears to be a definite correlation in size between the abalone and the shrimp associated with it. The shrimp rarely leaves the protection of its host; when it does venture a short distance onto the outer surface of the shell, the slightest disturbance causes it to dart back beneath the mantle and the abalone to close down over it. Usually a host bears only one shrimp, but occasionally as many as four are found. On the host, the shrimp resides in the pallial groove or mantle cavity, generally remaining with its head near the abalone's mouth. The shrimp's activities seem not to disturb the host.

Betaeus harfordi, experimentally removed from its host, relocates the abalone by scent, that is, by detecting water-borne materials emitted by the host. Shrimps respond to the presence of "host factor" in the water by moving upstream. The distance chemoreceptors involved are located on the flagella of the first antennae. The eyes are little used in locating the host. Ovigerous females of *B. harfordi* have been noted in March, April, May, September, and December in southern California.

In a related species, **B. macginitieae**, which lives associated with the sea urchins *Strongylocentrotus purpuratus* (11.5) and *S. franciscanus* (11.6), both distance chemoreception (scent) and vision appear to play roles in finding a host. The chemoreceptors are located as in *B. harfordi*.

See Ache (1975), Ache & Case (1969), Ache & Davenport (1972), Carlton & Kuris (1975), Hart (1964), MacGinitie & MacGinitie (1968), and Ricketts & Calvin (1968).

Phylum Arthropoda / Class Crustacea / Subclass Malacostraca
Superorder Eucarida / Order Decapoda / [Suborder Natantia] / Section Caridea
Family Alpheidae

23.3 **Betaeus harfordi** (Kingsley, 1878)

Intertidal zone to 25 m depth; commensal with all California species of abalone (*Haliotis*, e.g., 13.1–4); Fort Bragg (Mendocino Co.) to Bahía Magdalena (Baja California).

Length to about 18 mm; frontal margin of carapace between

Phylum Arthropoda / Class Crustacea / Subclass Malacostraca
Superorder Eucarida / Order Decapoda / [Suborder Natantia] / Section Caridea
Family Alpheidae

23.4 **Betaeus harrimani** Rathbun, 1904

Under low intertidal rocks, logs, and debris on protected shores, and sometimes in tunnels of the blue mud shrimp, *Upogebia pugettensis* (24.1) and the bay ghost shrimp, *Callianassa californiensis* (24.2), on mud flats in bays; Sitka (Alaska) to Dana Point (Orange Co.).

Length to about 30 mm; frontal margin of carapace between eyes evenly convex; fingers of large claws not longer than palm; terminal joint of walking legs slender, not bifid; body transparent except where pigmented.

This is an active and agile shrimp, not easily seen in the field because of its transparency and protective coloration, and rarely taken in any numbers. The body bears numerous red pigment cells (chromatophores) typically surrounded by dark-blue spots. The shrimps appear reddish by day, when the pigment is dispersed in chromatophores, and bluish at night, when chromatophores are "contracted."

Egg-bearing females have been observed in the field from June to September. Newly laid eggs are bright green.

See Carlton & Kuris (1975) and Hart (1964).

Phylum Arthropoda / Class Crustacea / Subclass Malacostraca Superorder Eucarida / Order Decapoda / [Suborder Natantia] / Section Caridea Family Alpheidae

23.5 **Betaeus longidactylus** Lockington, 1877

Among rocks and algae in intertidal pools on rocky shores, in burrows of the blue mud shrimp, *Upogebia pugettensis* (24.1), and the echiuroid worm *Urechis caupo* (19.7) in intertidal mud flats in bays; subtidal in kelp holdfasts and fouling growth on boats; Monterey Bay to Gulf of California.

Length to about 40 mm; frontal margin of carapace between eyes evenly convex; fingers of large claws longer than palm; terminal joint of walking legs slender, not bifid; living animal olive-green, red-brown, or blue-green, with a light mid-dorsal stripe, the legs reddish with clear tips, the tail fan dark with yellow setae.

Common in southern California along the outer coast. In the north, this species is restricted to quiet water, such as Elkhorn Slough (Monterey Co.). The animals are often found as sexual pairs. Egg-bearing females have been collected in January, June, August, and September.

See Carlton & Kuris (1975), Hart (1964), and Ricketts & Calvin (1968).

Phylum Arthropoda / Class Crustacea / Subclass Malacostraca Superorder Eucarida / Order Decapoda / [Suborder Natantia] / Section Caridea Family Hippolytidae

23.6 **Heptacarpus brevirostris** (Dana, 1852)

In harbors on floats and pilings, along open coast among rocks and algae in low intertidal pools and subtidal waters; Attu (Aleutian Islands, Alaska) to south of Carmel (Monterey Co.).

Length to about 50 mm; rostrum (head spine) very short, reaching little if at all beyond eyes, directed obliquely downward; no spines above eyes on each side near base of rostrum; body color variable.

Nothing is known about the biology of this species.

See Carlton & Kuris (1975) and Schmitt (1921).

Phylum Arthropoda / Class Crustacea / Subclass Malacostraca Superorder Eucarida / Order Decapoda / [Suborder Natantia] / Section Caridea Family Hippolytidae

23.7 **Heptacarpus pictus** (Stimpson, 1871)
Red-Banded Transparent Shrimp

Abundant among rocks and algae in tidepools on rocky shores, also in bays in beds of eelgrass *Zostera* and among fouling growth on floats; San Francisco to San Diego.

Length to about 24 mm; rostrum (head spine) straight, shorter than portion of carapace lying posterior to orbits of eyes but reaching beyond middle of antennal scales (blade-like branch of second antenna); no spines above eyes on each side near base of rostrum.

Color in this species is highly variable, as is that of the habitat. Four major body-color patterns are recognized, and disruptive coloration involving bands and blotches is common. The various pigments are all borne in the living tissues, and not in the overlying exoskeleton.

Females of *H. pictus* produce more than one brood of young, molting between successive broods. Newly molted females are receptive, and males apparently recognize them as such only upon contact. Mating is rapid, and the male simply attaches packets of sperm to the ventral surface of the female near the oviducal apertures, probably using his first pair of abdominal appendages to aid in placement.

See Bauer (1975, 1977), Carlton & Kuris (1975), and Schmitt (1921).

Phylum Arthropoda / Class Crustacea / Subclass Malacostraca
Superorder Eucarida / Order Decapoda / [Suborder Natantia] / Section Caridea
Family Hippolytidae

23.8 Lysmata californica (Stimpson, 1866)
(=Hippolysmata californica)   Red Rock Shrimp

Often common among rocks and algae in low intertidal pools and crevices; subtidal to over 60 m; Santa Barbara to Bahía Sebastián Vizcaíno (Baja California).

Length to about 70 mm; two longest joints of second legs multiarticulate (about 50 annulations); color pattern conspicuous and relatively constant; eggs of ovigerous female light pea green.

This is one of the more robust and least specialized of the "cleaning" shrimps, so-called because they pick parasites and other materials from the bodies of other animals, especially fishes. The red rock shrimps often occur in groups of several hundred, tending to cluster in sheltered spots by day and to disperse at night. When crowded, they are aggressive toward one another and often fight over food. Cleaning other animals is only a part-time occupation, and does not provide the only source of food. The shrimps seem not to seek out hosts, but they tend to frequent areas that are visited from time to time by larger animals with thick integuments, particularly the California moray (*Gymnothorax mordax*), the bright orange garibaldi (*Hypsypops rubicunda*), and the California spiny lobster (*Panulirus interruptus*). When one of these forms comes around, the shrimp goes to it and moves all over the surface of its body, picking off almost anything removable, including parasites and decaying tissue. The shrimp may even enter the host's mouth, usually without exciting the host. Scuba divers observing the process find that even a diver's hand extended into the area is carefully cleaned by the shrimp, especially the areas around the fingernails.

The cleaner-host relationship is not a highly evolved or fixed one. In a closed situation, such as an aquarium, a group of shrimp may "clean" a fish down to the bones. And in nature the shrimp themselves get eaten; they are often found in the stomachs of the California moray, the garibaldi, and the black croaker (*Cheilotrema saturnum*). Small numbers of the shrimp are taken by man as fish bait for any fishes living near bottom rocks or pilings, such as rockfishes.

Female shrimp bearing eggs have been noted in May and June.

See Feder, Turner & Limbaugh (1974), Limbaugh (1961), Limbaugh, Pederson & Chace (1961), MacGinitie & MacGinitie (1968), Ricketts & Calvin (1968), and Turner & Sexsmith (1964).

Phylum Arthropoda / Class Crustacea / Subclass Malacostraca
Superorder Eucarida / Order Decapoda / [Suborder Natantia] / Section Caridea
Family Hippolytidae

23.9 Hippolyte californiensis Holmes, 1895
Slender Green Shrimp, Grass Shrimp

Common on eelgrass *Zostera* in quiet bays, low intertidal and shallow subtidal zones; Bodega Bay harbor (Sonoma Co.) to Bahía Santa Inez (Gulf of California).

Length to about 40 mm, usually less; rostrum (head spine) nearly horizontal (not turning up), the tip usually bifid (not trifid); terminal joint of walking legs of females slender, nearly half as long as penultimate joint; color uniformly green.

In *Hippolyte* the males are considerably smaller than the females, and have a much less conspicuous rostrum and markedly prehensile legs. During daylight hours *H. californiensis* is difficult to detect, for it seeks shelter or orients itself lengthwise on *Zostera* blades, which match its body color. At night it swims in swarms in the eelgrass beds. In the Gulf of California many specimens were found in the stomachs of the eared grebe, an aquatic diving bird. Occasional grass shrimps bear a large parasitic isopod, *Bopyrina striata*, under the carapace.

Hippolyte clarki, which has a proportionately longer and more upturned rostrum usually terminating in three, rather than two, spines, also occurs in southern California but replaces *H. californiensis* in the Pacific northwest and ranges as far north as Sitka (Alaska). The larva shown in Needler (1934, pp. 4–5) labeled *H. californiensis* probably belongs to *H. clarki*. Larvae of related species and a key to larvae of genera in the family Hippolytidae are presented in Williamson (1957).

See Carlton & Kuris (1975), Chace (1951), MacGinitie & MacGinitie (1968), Needler (1934), Ricketts & Calvin (1968), Schmitt (1921, *H. californiensis* in part), and Williamson (1957).

Phylum Arthropoda / Class Crustacea / Subclass Malacostraca
Superorder Eucarida / Order Decapoda / [Suborder Natantia] / Section Caridea
Family Hippolytidae

23.10 **Spirontocaris prionota** (Stimpson, 1864)
Broken-Back Shrimp

Under stones in tidepools on rocky shores, low intertidal zone; subtidal to 150 m; Bering Sea to Monterey Bay; Japan.

Length to about 35 mm; body relatively massive, opaque; rostrum (head spine) strongly compressed, leaflike, very deep; dorsal margin of carapace grossly serrate; usually three strong spines above eyes on each side near base of rostrum; color variable.

Little is known of the biology of this animal. In Europe the larval stages of related species have been studied in detail.

See Pike & Williamson (1961), Ricketts & Calvin (1968), and Schmitt (1921).

Phylum Arthropoda / Class Crustacea / Subclass Malacostraca
Superorder Eucarida / Order Decapoda / [Suborder Natantia] / Section Caridea
Family Crangonidae

23.11 **Crangon stylirostris** Holmes, 1900
(=Crago stylirostris)   Bay Shrimp

Surf zone on semiprotected beaches; most commonly subtidal to 50 m on hard sandy bottoms; Chirikof Island (Alaska) to San Luis Obispo Bay.

Length to about 55 mm; rostrum (head spine) short, dorsally flattened, lacking dorsal teeth, curving strongly down-

ward, tapering to acute tip; first pair of legs subchelate; carapace lacking a middorsal spine or tooth in gastric region (gastric spine present in all 15 other *Crangon* species found in California waters).

This species is mainly a sublittoral form and is the least abundant of the four species of bay shrimps commonly caught by shrimp trawlers in San Francisco Bay (all *Crangon* spp., called *Crago* in most older references). *C. stylirostris* has been taken there at water temperatures ranging from 8.7° to 16°C, and at salinities of 52–100 percent seawater. Little is known of its biology, though the other species of *Crangon* on this coast have been studied and the literature on European species is extensive.

In California, bay shrimps are trawled commercially only in San Francisco Bay, but the fishery here is more than 100 years old. Three species are taken in significant numbers: the Franciscan bay shrimp, **C. franciscorum** (Alaska to San Diego at depths to 55 m), always forms the bulk of the catch; the black-tailed bay shrimp, **C. nigricauda**, is second in abundance; and the spotted bay shrimp, **C. nigromaculata**, is only occasionally taken in numbers. All three species are now fished mainly for bait (annual catch 2,300–25,000 kg in recent years), but some are taken for human consumption, as well (averaging 5,000 kg per year in recent years). A generation ago the San Francisco Bay fishery was more active; in the 1940–57 period, bay shrimp catches averaged 320,000 kg per year, with peaks as high as 1,360,000 kg. For bay shrimping in still earlier times, see Bonnot's (1932) fascinating account of the former Chinese shrimp fishery in San Francisco Bay.

For *C. stylirostris*, see Carlton & Kuris (1975), Ricketts & Calvin (1968), Schmitt (1921), and Wicksten (1977). For other bay shrimps, see Bonnot (1932), Frey (1971), Ganssle (1966), and Israel (1936).

Literature Cited

Ache, B. W. 1975. Antennular mediated host location by symbiotic crustaceans. Mar. Behav. Physiol. 3: 125–30.

Ache, B. W., and J. Case. 1969. An analysis of antennular chemoreception in two commensal shrimps of the genus *Betaeus*. Physiol. Zool. 42: 361–71.

Ache, B. W., and D. Davenport. 1972. The sensory basis of host recognition by symbiotic shrimps, genus *Betaeus*. Biol. Bull. 143: 94–111.

Bauer, R. 1975. Mating behavior and spermatophore transfer in tidepool shrimp *Heptacarpus* (Decapoda: Caridea). 56th Ann. Meet., West. Soc. Natur., Abs. Symp. & Contr. Pap., pp. 1–2 (abstract).

———. 1977. Polymorphism of camouflage color patterns in some Pacific coast shrimps. 58th Ann. Meet., West. Soc. Natur., Abs. Symp. & Contr. Pap., p. 40 (abstract).

Bonnot, P. 1932. The California shrimp industry. Calif. Div. Fish & Game, Fish Bull. 38: 1–20.

Carlton, J. T., and A. M. Kuris. 1975. Keys to decapod Crustacea, pp. 385–412, *in* R. I. Smith and J. T. Carlton, eds., Light's manual: Intertidal invertebrates of the central California coast. 3rd ed. Berkeley and Los Angeles: University of California Press. 716 pp.

Chace, F. A., Jr. 1951. The grass shrimps of the genus *Hippolyte* from the west coast of North America. J. Wash. Acad. Sci. 41: 35–39.

Dahlstrom, W. A. 1961. The California ocean shrimp fishery. Bull. Pacific Mar. Fish. Comm. 5: 17–23.

———. 1967. Synopsis of biological data on the ocean shrimp *Pandalus jordani* Rathbun. F.A.O. world scientific conference on biology and culture of shrimps and prawns, Mexico, 1967. F.A.O. Species Synop. 9: 1–42.

———. 1972. Status of the California ocean shrimp resource and its management. Mar. Fish. Rev. 35: 55–59.

Edwards, R. R. C. 1978. The fishery and fisheries biology of penaeid shrimp on the Pacific coast of Mexico. Oceanogr. Mar. Biol. Ann. Rev. 16: 145–80.

Feder, H. M., C. H. Turner, and C. Limbaugh. 1974. Observations on fishes associated with kelp beds in southern California. Calif. Dept. Fish & Game, Fish Bull. 160: 1–138.

Frey, H. W., ed. 1971. California's living marine resources and their utilization. Calif. Dept. Fish & Game. 148 pp. (shrimps pp. 11–16, 137, 148).

Ganssle, D. 1966. Fishes and decapods of San Pablo and Suisun Bays, pp. 64–94, *in* D. W. Kelley, comp., Ecological studies of the Sacramento–San Joaquin estuary. Calif. Dept. Fish & Game, Fish Bull. 133: 1–133.

Gates, D. E., and H. W. Frey. 1974. Designated common names of certain marine organisms in California. Calif. Dept. Fish & Game, Fish Bull. 161: 55–90.

Geibel, J. J., and R. F. G. Heimann. 1976. Assessment of ocean shrimp management in California resulting from widely fluctuating recruitment. Calif. Fish & Game 62: 255–73.

Glaessner, M. F. 1969. Decapoda, pp. R399–533, *in* R. C. Moore, ed., Treatise on invertebrate paleontology, part R, Arthropoda 4, vol. 2. New York: Geol. Soc. Amer.; Lawrence: University of Kansas Press.

Gotshall, D. W. 1967. The use of predator food habits in estimating relative abundance of the ocean shrimp, *Pandalus jordani* Rathbun. F.A.O. world scientific conference on biology and culture of shrimps and prawns, Mexico, 1967. F.A.O. Experience Pap. 35: 1–20.

Gurney, R. 1939. Bibliography of the larvae of decapod Crustacea. London: Ray Society. 123 pp.

———. 1942. Larvae of decapod Crustacea. London: Ray Society. 306 pp.

Hart, J. F. L. 1964. Shrimps of the genus *Betaeus* on the Pacific coast of North America with descriptions of three new species. Proc. U.S. Nat. Mus. 115: 431–66.

Hazlett, B. A., and H. E. Winn. 1962. Sound production and associated behavior of Bermuda crustaceans (*Panulirus*, *Gonodactylus*, *Alpheus*, *Synalpheus*). Crustaceana 4: 25–38, pl. 1.

Israel, H. R. 1936. A contribution toward the life histories of two California shrimps, *Crago franciscorum* (Stimpson) and *Crago nigricauda* (Stimpson). Calif. Dept. Fish & Game, Fish Bull. 46: 1–28.

Johnson, M. W., F. A. Everest, and R. W. Young. 1947. The role of snapping shrimp (*Crangon* and *Synalpheus*) in the production of underwater noise in the sea. Biol. Bull. 93: 122–38.

Kaestner, A. 1970. Invertebrate zoology. 3. Crustacea. Translation from 2nd German edition by H. W. Levi and L. R. Levi. New York: John Wiley. 512 pp.

Knowlton, R. E., and Moulton, J. M. 1963. Sound production in the snapping shrimp *Alpheus* (*Crangon*) and *Synalpheus*. Biol. Bull. 125: 311–31.

Limbaugh, C. 1961. Life-history and ecologic notes on the black croaker. Calif. Fish & Game 47: 163–74.

Limbaugh, C., H. Pederson, and F. A. Chace, Jr. 1961. Shrimps that clean fishes. Bull. Mar. Sci. Gulf Caribb. 11: 237–57.

Little, G. 1969. The larval development of the shrimp, *Palaemon macrodactylus* Rathbun, reared in the laboratory, and the effect of eyestalk extirpation on development. Crustaceana 17: 69–87.

Lockwood, A. P. M. 1967. Aspects of the physiology of Crustacea. San Francisco: Freeman. 328 pp.

MacGinitie, G. E. 1937. Notes on the natural history of several marine Crustacea. Amer. Midl. Natur. 18: 1031–37.

MacGinitie, G. E., and N. MacGinitie. 1968. Natural history of marine animals. 2nd ed. New York: McGraw-Hill. 523 pp.

McAllister, R. 1975. California marine fish landings for 1973. Calif. Dept. Fish & Game, Fish Bull. 163: 1–53.

———. 1976. California marine fish landings for 1974. Calif. Dept. Fish & Game, Fish Bull. 166: 1–53.

Needler, A. B. 1934. Larvae of some British Columbia Hippolytidae. Contr. Canad. Biol. & Fish. 8: 239–42.

Newman, W. A. 1963. On the introduction of an edible oriental shrimp (Caridea, Palaemonidae) to San Francisco Bay. Crustaceana 5: 119–32.

Pike, R. B., and D. I. Williamson. 1961. The larvae of *Spirontocaris* and related genera (Decapoda: Hippolytidae). Crustaceana 2: 187–208.

Pinkas, L. 1974. California marine fish landings for 1972. Calif. Dept. Fish & Game, Fish Bull. 161: 1–53.

Quast, J. C. 1968. Observations on the food and biology of the kelp bass, *Paralabrax clathratus*, with notes on its sportfishery at San Diego, California, pp. 81–108; Observations on the food of the kelp-bed fishes, pp. 109–42; *in* W. J. North and C. L. Hubbs, eds., Utilization of kelp-bed resources in southern California. Calif. Dept. Fish & Game, Fish Bull. 139: 1–264.

Ricketts, E. F., and J. Calvin. 1968. Between Pacific tides. 4th ed. Revised by J. W. Hedgpeth. Stanford, Calif.: Stanford University Press. 614 pp.

Ritzmann, R. E. 1973. Snapping behavior of the shrimp *Alpheus californiensis*. Science 181: 459–60.

———. 1974. Mechanisms for the snapping behavior of two alpheid shrimp, *Alpheus californiensis* and *Alpheus heterochelis*. J. Compar. Physiol. 95: 217–36.

Schmitt, W. L. 1921. The marine decapod Crustacea of California. Univ. Calif. Publ. Zool. 23: 1–359.

———. 1965. Crustaceans. Ann Arbor: University of Michigan Press. 204 pp.

Turner, C. H., and J. C. Sexsmith. 1964. Marine baits of California. Calif. Dept. Fish & Game. 71 pp.

Waterman, T. H., ed. 1960. The physiology of Crustacea. 1. Metabolism and growth. New York: Academic Press. 670 pp.

———, ed. 1961. The physiology of Crustacea. 2. Sense organs, integration, and behavior. New York: Academic Press. 681 pp.

Wicksten, M. K. 1977. Range extensions of four species of crangonid shrimps from California and Baja California, with a key to the genera (Natantia: Crangonidae). Proc. Biol. Soc. Washington 90: 963–67.

Williamson, D. I. 1957. Crustacea, Decapoda: Larvae. V. Caridea, Family Hippolytidae. Conseil. Internat. l'Explor. Mer, Zooplankton Sheet 68: 1–5.

Word, J. Q., and D. K. Charwat. 1976. Natantia. Invertebrates of southern California coastal waters. Vol. 2. Southern Calif. Coastal Water Research Project, El Segundo, Calif. 238 pp.

Young, P. H., and J. W. Withycombe. 1949. Shrimp and prawn, pp. 157–60, *in* Bureau of Marine Fisheries, The commercial fish catch of California for the year 1947 with an historical review 1916–1947. Calif. Dept. Fish & Game, Fish Bull. 74: 1–267.

Chapter 24

Macrura and Anomura:
The Ghost Shrimps, Hermit Crabs, and Allies

Janet Haig and Donald P. Abbott

The decapod or "ten-legged" Crustacea are subdivided by some authorities into two large groups, Natantia and Reptantia. In the natant ("swimming") decapods, the shrimps and prawns, the body is laterally compressed, and there are five pairs of well-developed abdominal appendages used for swimming. In the reptant ("crawling") decapods, the body is often strongly depressed; the abdominal appendages are not used for swimming, and some or all of them may be reduced or absent.

The Reptantia, in turn, are divided into the Macrura, Anomura, and Brachyura. The brachyurans (Chapter 25) are the "true crabs," characterized by having, among other things, a flattened carapace and a small abdomen that folds tightly under the body and lacks a tail fan. The macrurans, which include such forms as mud shrimps and ghost shrimps, spiny lobsters, true lobsters, and freshwater crayfishes, have a long, straight, symmetrical abdomen with a tail fan; the carapace is cylindrical or depressed. They are distributed world wide, and roughly 700 living species are known. Fossil macrurans first appear near the end of the Paleozoic.

The Anomura, standing between the Macrura and the Brachyura, contains an assemblage of dissimilar forms, and many authorities consider the group to be an unnatural one. It includes the sand crabs, stone crabs, hermit crabs, galatheid crabs, and porcelain crabs. In some classifications the mud shrimps, ghost shrimps, and their relatives are placed here instead of with the Macrura. With these omitted, the Anomura may be defined as follows: the abdomen is well developed; it may be folded under the body as in the brachyuran crabs, or bent upon itself, or soft and asymmetrical; the tail fan may be either present or absent; the fifth pair of legs (and sometimes the fourth as well) is much reduced in size. Anomurans occur in all seas, and roughly 1,300 living species are known. Fossil forms first appear in the Jurassic period of the Mesozoic.

As in all decapod crustaceans, the body of macrurans and anomurans is divided into the cephalothorax and abdomen. The cephalothorax is covered dorsally by the carapace, which is often produced forward as a rostrum between the eyes. In hermit crabs the posterior part of the carapace is normally soft, and the hard anterior portion is sometimes referred to as the shield. The cephalothorax bears the following structures: the compound eyes; six pairs of mouthparts, the outermost being the third maxillipeds; the antennules (or first antennae) and antennae (or second antennae), each terminating in a flagellum; and five pairs of legs. Each leg is composed of several segments or articles separated by flexible joints; the outermost four segments, counting from the body toward the limb tip, are called the merus, carpus, propodus, and dactylus. Often the legs of the first pair are larger than the others, and the propodus is distally produced into a "fixed finger" that, together with the dactylus, or "movable finger," forms a pincer; the propodus and dactylus together are called the chela, and the whole

Janet Haig is Associate Curator of Crustacea at the Allan Hancock Foundation, University of Southern California. Donald P. Abbott is Professor, Department of Biological Sciences and Hopkins Marine Station, Stanford University.

leg is the cheliped. The remaining four pairs of thoracic limbs, where rather similar and used for locomotion or clinging, are often termed walking legs. In anomurans the last pair of thoracic limbs is typically specialized in connection with such uses as cleaning the gills and brooding eggs, and is carried tucked under the carapace. In hermit crabs the next to last pair of legs, also small, helps the crab move out of the shell.

The abdomen is divided into six true segments and a terminal unit, the telson; in some forms, like the hermit crabs, all but traces of body segmentation are lost. Some or all of the anterior five abdominal segments may bear appendages (pleopods) serving various functions. In Macrura and some Anomura, the sixth abdominal segment has an enlarged pair of appendages (uropods), which together with the telson form a tail fan.

Only the mud shrimps (Upogebiidae) and ghost shrimps (Callianassidae) among the Macrura are truly members of the California intertidal fauna. They are adapted for permanent life in burrows that they dig in muddy or gravelly sand. The spiny lobsters (family Palinuridae) have one representative, **Panulirus interruptus** (Randall), in California waters, but it is typically an offshore form and is only occasionally found in lower tidepools.

Among the Anomura, the sand crabs (Hippidae and Albuneidae) are highly adapted to a life in shifting sand. The abdomen is somewhat reduced in length (but, in the Hippidae, with a greatly elongated telson), and is held forward below the cephalothorax. These forms are effective burrowers, but do not form tunnels as do the mud and ghost shrimps.

The stone crabs (Lithodidae) are still more crablike in appearance. The abdomen is short, fleshy, and folded forward below the body, and uropods are lacking. Most stone crabs live offshore in deeper water. Some of them, notably the Alaska king crab, are of economic importance, but no fishery is developed for any of the large California species. One small subfamily contains mostly littoral forms, some of which are likely to be encountered.

The hermit crabs are represented in the California intertidal by two families, Diogenidae and Paguridae. They live in the abandoned shells of gastropod mollusks and sometimes in other hollow objects, thereby protecting their soft abdomens. The littoral *Pagurus* species (24.10–13) scramble about in tidepools, and are the most noticeable of the anomurans to the casual observer; *Isocheles pilosus* (24.9) is more sluggish and often conceals itself under the sand in bays and estuaries. For the identification of hermit crabs, more than for other anomurans, knowledge of the color of living animals is useful and sometimes necessary.

The porcelain crabs or rocksliders (Porcellanidae) have a crablike structure and are secretive in their habits, living under or in crevices of rocks, in kelp holdfasts, among mussels, and in other sheltered situations. In the tropics they often hide in or around corals, and some live commensally with hermit crabs, sea anemones, sea stars, or other animals. Only one California species, **Polyonyx quadriungulatus** Glassell, is adapted for commensal life, occupying the tubes of the polychaete worm *Chaetopterus variopedatus* (18.20); since it rarely occurs intertidally in California, it is not treated in this chapter.

A family related to the porcelain crabs, the Galatheidae, is represented in California by several species, nearly all from deep water. The one exception is the pelagic red crab **Pleuroncodes planipes** Stimpson, an inhabitant of Mexico that drifts into California waters during years of unusually warm ocean temperatures. At such times it attracts a good deal of attention because large numbers of these animals may be washed ashore and stranded on California beaches (see Boyd, 1960, 1967; Glynn, 1961; Longhurst, Lorenzen & Thomas, 1967; Longhurst & Seibert, 1971).

For the most frequently followed classifications of Macrura and Anomura, see Balss (1957), Borradaile (1903, 1907), Glaessner (1969), and Kaestner (1970). Balss and Glaessner review the history of other systems of classification, and Glaessner discusses the known fossil macrurans and anomurans. Three recent books giving much general information on the biology of these groups are Kaestner (1970), Schmitt (1965), and Warner (1977). The works best suited for identification of California species as a whole (including non-intertidal forms) are Holmes (1900), Johnson & Snook (1927), and Schmitt (1921), though they are out of date, and more recent information is widely scattered through the literature. Useful keys to common intertidal forms are presented in Carlton & Kuris (1975).

Color notes on several species of hermit crabs and other anomurans were provided by Dr. J. F. L. Hart and by Dr. Mary K. Wicksten, and are used in this chapter with their permission. Their help is gratefully acknowledged.

Phylum Arthropoda / Class Crustacea / Subclass Malacostraca
Superorder Eucarida / Order Decapoda / Suborder Reptantia / Section Macrura
Superfamily Thalassinidea / Family Upogebiidae

24.1 **Upogebia pugettensis** (Dana, 1852)
Blue Mud Shrimp

Burrowing in mud or sandy mud flats of estuaries, low intertidal zone in northern part of range, middle intertidal zone from Monterey southward; southern Alaska to Bahía de San Quintín (Baja California).

Total length to about 150 mm, Monterey Bay and northward, to about 65 mm in southern part of range; rostrum broad, both rostrum and anterior part of carapace rough and hairy; first pair of legs subequal in form and size.

These animals build permanent burrows in which they live in pairs. A current of water is fanned through the burrow by beating of the pleopods. Detritus and plankton, strained from the current by hairs on the first two pairs of legs, provide food. *Upogebia* has a higher respiratory rate, and survives anoxia less well, than *Callianassa californiensis* (24.2). It can tolerate brackish conditions down to 10 percent the salinity of seawater, and regulates osmotically when the salinity drops below 75 percent that of seawater. There is a definite breeding season, from winter to early summer: females with eggs have been noted from December through April. Three swimming larval stages precede the post-larva.

Numerous commensals are reported to occur with *Upogebia*, and several, though not all, live together in a single burrow. Among them are the pea crabs *Scleroplax granulata* (25.41; the most prevalent commensal), *Pinnixa schmitti*, and *P. franciscana* (25.38) and the snapping shrimps *Betaeus ensenadensis, B. longidactylus* (23.5), and *B. harrimani* (23.4); *B. ensenadensis*, and probably also *B. longidactylus* occur in pairs. Three bivalves may be associated with *Upogebia*: *Pseudopythina rugifera, P. compressa* (actually attached to the mud shrimp), and *Cryptomya californica* (15.66), which lives in the adjacent mud and extends its siphons into the burrow. A scale worm, *Hesperonoe* (probably *H. complanata*; see under 18.6), lives with *Upogebia*, and the goby fish *Clevelandia ios* occasionally uses the burrows for refuge. The copepod *Clausidium vancouverense*, a small crustacean, lives on the outside of the body, and a pair of parasitic iso-

pods, *Phyllodurus abdominalis* (21.9), is often found lying across the abdomen and attached to the pleopods (one is shown attached to a shrimp in Fig. 21.9a). The phoronid *Phoronis pallida* may extend its tentacles into *Upogebia* burrows.

In the Puget Sound area (Washington), *Upogebia* is a destructive element in the oyster industry; it covers young oysters with mud and debris, smothering them, and undermines the dykes built to keep the oysters covered with water at low tide.

See especially MacGinitie (1930); see also Fingerman & Oguro (1963), Frey (1971), Gooding (1957), Gross (1957), Hart (1937), M. E. Johnson & Snook (1927), Light & Hartman (1937), MacGinitie (1935, 1937), MacGinitie & MacGinitie (1968), Powell (1974), Ricketts & Calvin (1968), Stevens (1928), L. Thompson & Pritchard (1969), R. Thompson (1972), and R. Thompson & Pritchard (1969).

Phylum Arthropoda / Class Crustacea / Subclass Malacostraca
Superorder Eucarida / Order Decapoda / Suborder Reptantia / Section Macrura
Superfamily Thalassinidea / Family Callianassidae

24.2 **Callianassa californiensis** Dana, 1854
Bay Ghost Shrimp

Abundant, burrowing in mixed sand and mud, middle intertidal zone on flats of bays and estuaries; southern Alaska to Estero de Punta Banda (Baja California).

Total length to about 115 mm; carapace smooth; chelipeds unequal, the larger one in adult males with a short, broad palm and with a distinct gap between the fingers when they are closed; body whitish.

Callianassa gigas Dana is similar in habits, habitat, and distribution, but is larger (125–150 mm long). The large cheliped of adult males is proportionally longer and narrower, and there is practically no gap between the fingers when they are closed; females and young are very similar to those of *C. californiensis*.

Ghost shrimps burrow continuously in loose sand, forming impermanent tunnels with many branches that may extend to a depth of 75 cm. They were believed to feed exclusively on organic detritus sifted from the soft substratum by hairs on the second and third pairs of legs; recent evidence indicates that they also ingest surface detritus and plankton from water moving through the burrow, as does the blue mud shrimp, *Upogebia pugettensis* (24.1). Water movement, needed for respiration as well as feeding, is aided by beating of the broad, paddle-like pleopods on the abdomen; multiple openings to the burrows also facilitate circulation. Respiratory gas exchange occurs very largely through the gills (not the pleopods, as often suggested), which are unusually small for an aquatic crustacean. The respiratory rate is low, even when oxygen is plentiful. The blood contains a hemocyanin respiratory pigment that functions not only in oxygen uptake and transport but liberates more bound oxygen to the tissues of the shrimp under conditions of very low oxygen tension. These features, and the ability to switch to anaerobic metabolism, allow the animals to survive anoxia for nearly 6 days.

The bay ghost shrimp tolerates salinities ranging from 25 or 30 percent to 125 percent the salinity of normal seawater. The salinity of the blood changes along with that of the environment and the animals are less resistant to brackish conditions than *Upogebia*. Some breeding goes on all year, but more females carry eggs in June and July, at least in the Monterey area.

Numerous commensal species are reported to live with *C. californiensis*. Occupying the burrows may be the scale worm *Hesperonoe complanata* (see under 18.6; one worm to a burrow), the snapping shrimps *Betaeus harrimani* (23.4) and *B. ensenadensis* (the latter, one pair to a burrow), and the pea crabs *Scleroplax granulata* (25.41), *Pinnixa franciscana* (25.38), and (rarely) *P. schmitti*. The burrowing clam *Cryptomya californica* (15.66) extends its siphons into the burrows, and the goby fish *Clevelandia ios* occasionally uses the passageways for refuge. Two species of copepods live on the ghost shrimp itself: small, bright-red *Clausidium vancouverense* occurs on the body surface and in the gill chambers; *Hemicyclops thysanotus* (formerly called *H. callianassae*) occurs in the gill chambers and also on the egg masses of ovigerous females. Parasitic isopods (*Ione* sp., 21.10) may also be present on the gills (see Fig. 21.10).

Fishermen use ghost shrimps as bait for such species as the spotfin croaker and the diamond turbot; an average of about 2,700 kg is taken yearly for this purpose in the San Diego and Los Angeles areas alone. At the same time, *Callianassa* may be a pest in oyster beds (as in Puget Sound, Washington), where their burrowing activities smother young oysters in sand and silt.

Fossils of the genus *Callianassa* appear from as far back as the Mesozoic. A number of fossil species are known from Pacific North America, and the Recent species *C. gigas* is reported (as *C. longimana*, a synonym) from the Pleistocene.

See especially MacGinitie (1934); see also Bybee (1969), Farley & Case (1968), Frey (1971), Gooding (1957, 1960), M. E. Johnson & Snook (1927), Light & Hartman (1937), Lindsay (1963), MacGinitie (1935), MacGinitie & MacGinitie (1968), K. Miller & Van Holde (1974), Powell (1974), Rathbun (1926), Ricketts & Calvin (1968), Roxby et al. (1974), Stevens (1928), L. Thompson & Pritchard (1969), R. Thompson & Pritchard (1969), Torres, Gluck & Childress (1977), and Turner & Sexsmith (1964).

Phylum Arthropoda / Class Crustacea / Subclass Malacostraca
Superorder Eucarida / Order Decapoda / Suborder Reptantia / Section Macrura
Superfamily Thalassinidea / Family Callianassidae

24.3 **Callianassa affinis** Holmes, 1900
Beach Ghost Shrimp

Burrowing in sandy gravel between and beneath boulders, middle intertidal zone on protected outer coasts; Goleta (Santa Barbara Co.) to Bahía de San Quintín (Baja California).

Total length to about 65 mm; carapace smooth; chelipeds unequal, the large one in males proportionately longer and narrower than in *C. californiensis* (24.2); species most easily distinguished by its different habitat and habits; body white.

This is the smallest of the California species of *Callianassa*, and in some respects its natural history resembles that of *Upogebia pugettensis* (24.1). The animals build permanent burrows, in which they live in pairs, and they feed on suspended plankton and detritus that is carried in a current of water pumped through the burrow by the fanning of the pleopods. They cannot regulate osmotically.

A pair of blind gobies, *Typhlogobius californiensis*, lives in each burrow with the callianassids; the bodies of the fishes are always about the same size as the bodies of the hosts (see the photograph). No other commensals are present.

See Farley & Case (1968), Gross (1957), Holmes (1900), MacGinitie (1937), MacGinitie & MacGinitie (1968), and Ricketts & Calvin (1968).

Phylum Arthropoda / Class Crustacea / Subclass Malacostraca
Superorder Eucarida / Order Decapoda / Suborder Reptantia / Section Anomura
Superfamily Hippidea / Family Hippidae

24.4 **Emerita analoga** (Stimpson, 1857)
Sand Crab, Mole Crab

Usually abundant, burrowed in sand between tide marks on surf-swept sandy beaches; Kodiak Island (Alaska) to Bahía Magdalena (Baja California); Peru and Chile; isolated populations reported from head of Gulf of California and from Caleta Falso (Argentina).

Carapace oval, to about 35 mm long, with a strongly convex surface; no lateral spines; first pair of legs lacking claws, the terminal segments oval and flattened; body grayish or sand-colored.

This is one of the most widely studied anomurans in the Pacific. The crabs occur in large aggregations in areas in or near the wash of the waves, and the aggregations move along the beach with the long-shore movement of sand. Part of the population also moves up and down the beach with the rise and fall of the tide. While feeding, the animals lie just under the surface, facing seaward, with only their eyes and first antennae visible. As the outgoing backwash of the wave flows over them, they extend their long, feathery second antennae and strain from the water food particles 4 μm to 2 mm in diameter. The crabs must bury themselves rapidly and repeatedly to maintain their position in the shifting sand. If dislodged, they can swim or tread water in a manner most unusual for crustaceans, by beating the uropods, the most posterior of the body appendages. The crabs flourish best on beaches exposed to considerable surf. They are not able to regulate osmotically, and are intolerant of brackish conditions.

Mating occurs in the spring and summer, usually at night. A few ovigerous females are found at all seasons, but the main breeding period is from March to November. Both temperature and food supply control breeding and the number of eggs produced. Most females probably breed in two successive summers, and the life span may be 2–3 years. The bright-orange eggs are carried attached to the pleopods. On hatching they release pelagic zoea larvae that have a rounded carapace, large eyes, and a short anterior (rostral) spine. The zoea larvae molt at least five times, the rostral spine becoming longer and more prominent at each instar, then undergo metamorphosis to a megalops stage resembling a juvenile *Emerita*. The megalops stage animals come ashore; their time of arrival on the beach varies greatly in different areas, for reasons not yet understood. Laboratory-reared *E. analoga* may pass through 8–11 molts before reaching the megalops stage; similar differences in molting frequency in planktonic, as opposed to cultured, larvae have been observed for *E. rathbunae* of the west American tropics and *E. talpoida* of the east coast of the United States.

Adult male *Emerita analoga* (carapace length 10–22 mm, no pleopods on abdomen) are smaller than adult females (carapace length 14–35 mm, three pairs of pleopods). In some areas the ratio of males to females in the population appears to shift with season, the population being 75 percent males in June and 25 percent males in October. These and other observations suggest (but do not prove) that a sex change from male to female is widespread. However, other factors complicate the picture. At least in southern California the population tends to migrate to subtidal waters offshore during the winter months; some larger adults reappear on the beaches in the spring, but the population is augmented by the settlement of megalops stages, which are predominantly males. Growth rates may differ markedly in different areas, but females appear to grow faster than males, bringing about a change in the ratios of males to females within a given size class. The possibility of a sex change from male to female in at least a part of the population remains, but it is also possible that changes in the ratios of males to females in whole intertidal populations might result from sex differences in recruitment rates, migration rates, and/or mortality.

Emerita analoga are eaten by many shorebirds (e.g., semipalmated plover, snowy plover, western sandpiper, sanderling), by several fishes (barred surfperch, California corbina, and young black croaker), and by the swimming crab *Portunus xan-*

tusii xantusii (25.14). Dead *Emerita* form the main diet of the large sand crab *Blepharipoda occidentalis* (24.5) and of the beach isopod *Excirolana linguifrons*. About 4,100 kg (rougly two million *Emerita*) were taken for bait in southern California in 1967.

The biology of the species has been studied in Chile, and investigations have been made on its role in the food web on Peruvian beaches.

Emerita populations have served as indicators of DDT residues in California coastal waters. They have also provided excellent material for laboratory studies in neurobiology. Swimming activity has been analyzed in detail in terms of anatomy, behavior, and neurophysiology. The tail contains the largest sensory neurons yet described in *any* animal.

For *E. analoga*, see Barnes & Wenner (1968), Boolootian et al. (1959), Burnett (1971), Carlisle, Schott & Abramson (1960), Cox & Dudley (1968), Cubit (1969), Dillery & Knapp (1970), Efford (1965, 1966, 1967, 1969, 1970), Eickstaedt (1969), Frey (1971), Fusaro (1977, 1978), Gilchrist & Lee (1972), Gross (1957), M. E. Johnson & Snook (1927), M. W. Johnson (1940), M. W. Johnson & Lewis (1942), Joseph (1962), Knox & Boolootian (1963), Koepcke & Koepcke (1952), MacGinitie (1938, 1939), MacGinitie & MacGinitie (1968), Osorio, Bahamonde & López (1967), Paul (1971, 1972, 1976a,b, 1979), Reeder (1951), Ricketts & Calvin (1968), Schmitt (1921), Turner & Sexsmith (1964), Wenner (1972), and Weymouth & Richardson (1912). For other species, see Costello & Henley (1971), Edwards & Irving (1943), Haley (1979), Knight (1967), Rees (1959), Snodgrass (1952), Subramoniam (1977), and Wenner (1977).

Phylum Arthropoda / Class Crustacea / Subclass Malacostraca Superorder Eucarida / Order Decapoda / Suborder Reptantia / Section Anomura Superfamily Hippidea / Family Albuneidae

24.5 **Blepharipoda occidentalis** Randall, 1839
Spiny Mole Crab

Burrowing in sand on surf-swept beaches, low intertidal zone and subtidal to 9 m depth; Stinson Beach (Marin Co.) to Bahía Santa Rosalía (Baja California); larvae found in plankton near shore as far south as Bahía Ballenas (Baja California).

Carapace to about 60 mm long, bearing three frontal and four lateral spines; claws spiny; carapace dark gray, legs cream colored.

Blepharipoda occur with *Emerita analoga* (24.4) but in much smaller numbers; they do not migrate up and down the beach with the tide. Juveniles feed on plankton and suspended particles, snaring them from the water with the enlarged feathery second antennae; in the laboratory they can be reared on brine shrimp larvae. The adults are scavengers and depend to a great extent on dead *Emerita* for their food. The *Blepharipoda*, in turn, are preyed upon by fishes like the barred surfperch, *Amphistichus argenteus*, which also eats *Emerita*. Females bear bright-red eggs that hatch mainly at night; the larvae pass through five zoeal stages and one megalops stage before becoming juvenile sand crabs. The zoeae swim backward, with the abdomen extended; the megalops stage can swim, but usually settles and burrows as do juveniles and adults.

Lepidopa californica Efford, a related form, occurs on sheltered sandy beaches but is more common in subtidal waters offshore. The main population is distributed from San Pedro (Los Angeles Co.) southward, but individuals are taken occasionally in Monterey Bay. The animals differ from *Blepharipoda* in several respects: they are smaller in size, the carapace has only one lateral spine on each side, the claws are not spiny, and the body is white with an iridescent tinge.

See Carlisle, Schott & Abramson (1960), Holmes (1900), M. E. Johnson & Snook (1927), M. W. Johnson & Lewis (1942), Knight (1968), MacGinitie & MacGinitie (1968), Ricketts & Calvin (1968), Schmitt (1921), and Turner & Sexsmith (1964).

Phylum Arthropoda / Class Crustacea / Subclass Malacostraca Superorder Eucarida / Order Decapoda / Suborder Reptantia / Section Anomura Superfamily Paguridea / Family Lithodidae

24.6 **Hapalogaster cavicauda** Stimpson, 1859

Uncommon under rocks on protected rocky shores, low intertidal zone and subtidal to at least 15 m; Cape Mendocino to Isla San Gerónimo (Baja California); reported once from Guaymas (Mexico).

Carapace to about 19 mm in length; carapace and legs nearly smooth, densely covered with brownish short hair.

This attractive, flattened, and fuzzy crustacean has habits similar to those of the porcelain crabs (*Petrolisthes*, 24.14 –18, and *Pachycheles* 24.19–21). Food is obtained by straining the water for particles with brushes of long hairs on the outer maxillipeds. Sparse data indicate that breeding takes place in the colder months.

A related northern species, *H. mertensii*, has been studied in Puget Sound (Washington). It is omniverous, fanning the water for plankton, gleaning scraps of brown, red, and green algae, and crushing barnacles with the chelae. Nearly all females carried eggs in November, December, January, and April, but none in June or July. Sizes of the broods varied from 600 to 2,076 eggs in animals with a carapace width of 15.6–19.1 mm. The eggs hatch as prezoea larvae, a stage lasting 5–6 hours, then go through four swimming zoeal stages lasting altogether about 6 weeks (44 days) in the laboratory. The zoeae swam actively and fed on brine shrimp nauplius larvae when these were presented. The zoeal phase is followed by a glaucothoe larval stage, which can swim with its pleopods but generally remains on the bottom and feeds on small algae and other materials scraped from stones. The first juvenile stage, still very small (carapace 2 mm long, 1.3 mm wide), does not swim; it folds its abdomen beneath the thorax like an adult, and seeks shelter under pebbles.

See Bouvier (1895, 1896), Hewatt (1937, 1938), Holmes (1900), M. E. Johnson & Snook (1927), Knudsen (1964), McLean (1962), P. Miller & Coffin (1961), Ricketts & Calvin (1968), and Schmitt (1921).

Phylum Arthropoda / Class Crustacea / Subclass Malacostraca
Superorder Eucarida / Order Decapoda / Suborder Reptantia / Section Anomura
Superfamily Paguridea / Family Lithodidae

24.7 **Oedignathus inermis** (Stimpson, 1860)

Scattered, under encrusting coralline algae, among the bases of mussels in beds of *Mytilus californiensis* (15.9), and in similar sheltered situations, middle to low intertidal zones on rocky shores; Unalaska (Aleutian Islands, Alaska) to Monterey Peninsula; Peter the Great Bay (U.S.S.R.), northern Japan, and Korea.

Carapace to about 30 mm long, dull brown, bearing flat, scalelike plates; body hair sparse or lacking; chelipeds unequal, with wartlike tubercles.

These crabs, normally found in pairs, often appear trapped, with little opportunity to move about in the secluded spaces they inhabit. They obtain food by sweeping plankton from the water with their outer maxillipeds. Almost nothing is known of their biology.

See Bouvier (1896), Holmes (1900), M. E. Johnson & Snook (1927), MacGinitie (1937), Makarov (1938), Ricketts & Calvin (1968), and Schmitt (1921).

Phylum Arthropoda / Class Crustacea / Subclass Malacostraca
Superorder Eucarida / Order Decapoda / Suborder Reptantia / Section Anomura
Superfamily Paguridea / Family Lithodidae

24.8 **Cryptolithodes sitchensis** Brandt, 1853
Turtle Crab, Sitka Crab, Umbrella Crab

Scattered, on and under rocks, low intertidal zone on protected rocky shores; subtidal on rocks to depths of at least 15 m; Sitka (Alaska) to Point Loma (San Diego Co.).

Carapace to 50 mm long, to 70 mm wide; rostrum widened toward distal end; lateral expansions of carapace produced forward, forming deep indentations above the eyes; chelae nearly smooth, with a longitudinal ridge along middle of outer surface; color extremely variable.

The butterfly crab, **C. typicus** Brandt, reported from Unalaska (Aleutian Islands, Alaska) to Santa Rosa Island (Channel Islands), is similar in habits and habitat to *C. sitchensis* and shows equally remarkable color variation. The rostrum narrows toward the distal end, the carapace does not come as far forward, the orbits are shallower, and the chelae are tuberculate.

Cryptolithodes sitchensis is slow-moving and clings tightly to rocks. Because of its form and color, it is often well camouflaged and difficult to detect. Little is known of its biology. Larval development of *C. typicus*, followed in the laboratory, shows a very short (30-minute) prezoeal stage, four zoeal instars, and a megalops stage, which settles from the plankton. Time from hatching to the first juvenile stage was 24 days at a temperature of 12–20°C.

See Hart (1930, 1965), Holmes (1900), M. E. Johnson & Snook (1927), McLean (1962), Odenweller (1972), Ricketts & Calvin (1968), and Schmitt (1921).

Phylum Arthropoda / Class Crustacea / Subclass Malacostraca
Superorder Eucarida / Order Decapoda / Suborder Reptantia / Section Anomura
Superfamily Paguridea / Family Diogenidae

24.9 Isocheles pilosus (Holmes, 1900)
(=Holopagurus pilosus)   Hermit Crab

Abundant on sand and mud flats in bays, estuaries, and other sheltered situations, low intertidal zone and offshore to depths of about 55 m; Bodega Bay harbor (Sonoma Co.) south at least to Estero de Punta Banda (Baja California).

Carapace to about 30 mm long; species distinctive as the only common California intertidal hermit crab with very hairy antennae and with the left chela larger than the right; carapace and antennae bluish, the chelipeds creamy orange tinged with blue, the walking legs creamy orange.

These hermit crabs bury themselves in sand with only their eyes, antennae, and a part of the mouth region exposed. They feed by picking up particles from the surface with the chelipeds. They are occasionally found living together in dense aggregations of up to 200 individuals, most of them completely buried in the substratum. The majority of large animals inhabit shells of moon snails (*Polinices,* 13.62, 13.63), but a variety of shells are used by smaller individuals.

See Fager (1968), Holmes (1900), Ricketts & Calvin (1968), and Schmitt (1921).

Phylum Arthropoda / Class Crustacea / Subclass Malacostraca
Superorder Eucarida / Order Decapoda / Suborder Reptantia / Section Anomura
Superfamily Paguridea / Family Paguridae

24.10 Pagurus samuelis (Stimpson, 1857)
Hermit Crab

Abundant in tidepools on protected rocky shores, mainly upper and middle intertidal zones but extending subtidally to *Macrocystis* holdfasts; Vancouver Island (British Columbia) to Punta Eugenia (Baja California); not Japan, as sometimes reported.

Carapace to about 19 mm long, bearing a prominent and sharp median frontal tooth; carapace shield longer than wide; carapace and legs hairy; antennal flagella scarlet; chelae khaki with blue or white fingers, left fingertips often red; walking legs with distal end of propodus bright blue, dactylus bright blue with a red longitudinal stripe and terminal patch on either side (not visible in photograph).

This is the most abundant tidepool hermit crab in southern and central California. It occurs higher in the intertidal zone, and is more resistant to the effects of exposure to air and sunlight, than any other California species. At Pacific Grove (Monterey Co.) more than 80 percent of the animals of reproductive size live in the shells of species of *Tegula*, especially *T. funebralis* (13.32), the remainder in shells of *Littorina planaxis* (13.42), *Nucella emarginata* (13.83), *Acanthina punctulata* (13.80), and other snails. Laboratory experiments also indicate a marked preference for *Tegula* shells. Shell selection , studied in detail in this species, involves an elaborate pattern of behavior; learning appears not to be involved, for crabs lacking previous experience with shells exhibit the full range of behavior seen in "experienced" crabs. Crabs often steal shells from each other, but do not attempt to take shells still occupied by snails.

The adult population is rather inactive during the day; activity picks up in late afternoon and continues through the night until dawn. The activity pattern appears to be a direct response to light levels, not an inherent rhythm of behavior. The compound eyes adapt to day and night conditions by shifts in the position of pigments, only as a direct response to light levels.

The diet is varied; adults scavenge both plant materials (especially pieces of the large brown algae *Macrocystis* and *Pelvetia*) and dead animal matter. They have been kept indefinitely in the laboratory on a diet of *Pelvetia*. In California, hermit crabs are an important item in the diet of the pile perch and are eaten by other fishes such as the California sheephead and spotted kelpfish; *P. samuelis* may well be included among the species preyed upon, since specimens have been found in the stomachs of the striped kelpfish, *Gibbonsia metzi*.

Females bearing eggs are found through much of the year in southern California, and in Monterey Bay are noted at least from February through August. In courtship, the male grips the female by the edge of her shell and may carry her about for a day or more, pausing periodically to knock his shell repeatedly against hers. Mating itself, for which the animals must extend mostly out of their shells, lasts less than a second. Sexual maturity occurs early in life: the smallest ovigerous females, measuring only 1.8 mm across the carapace shield, carry less than 100 eggs; large females may carry more than 2,000 eggs.

The eggs hatch as zoea larvae. There are four zoeal instars, lasting a total of 26–30 days at 15–17°C. Under laboratory conditions the larvae thrive on a diet of newly hatched brine shrimp larvae. The fourth zoeal stage molts to a glaucothoe larval stage, which settles from the plankton, finds a tiny empty shell, and begins to feed on organic materials scraped from stones. After several days it molts to become a very small juvenile hermit crab.

Studies were made on the embryology of a closely related Japanese species, then believed to be a Japanese population of *P. samuelis*. Careful comparisons of both adults and larvae of Japanese and American forms show that they represent different species.

See Andrews (1945), Ball (1968), Bollay (1964), Coffin (1954, 1958, 1960), Hart (1971), Hewatt (1937, 1938), Holmes (1900), MacGinitie (1935), MacMillan (1972b), McLaughlin (1974, 1976), Mitchell (1953), Ōishi (1960), Orians & King (1964), Quast (1968), Reese (1962, 1963, 1969), Ricketts & Calvin (1968), and Schmitt (1921).

Phylum Arthropoda / Class Crustacea / Subclass Malacostraca
Superorder Eucarida / Order Decapoda / Suborder Reptantia / Section Anomura
Superfamily Paguridea / Family Paguridae

24.11 **Pagurus hirsutiusculus** (Dana, 1851)
Hermit Crab

In tidepools with sandy or rocky bottoms, middle and low intertidal zones on protected rocky shores; subtidal to 110 m; Pribilof Islands (Alaska) to San Diego; Bering Strait to northern Japan.

Carapace to 19 mm long, bearing a prominent and sharp median frontal tooth; antennal flagella greenish brown with yellow spots; chelae mud-colored, the fingertips white or blue-white; propodi of walking legs white at distal end; dactyli of walking legs with a blue patch proximally and with four longitudinal red stripes (not shown in photograph); northern specimens (**P. hirsutiusculus hirsutiusculus**) with carapace shield wider than long, and carapace and legs hairy; specimens from about the Monterey Peninsula south (**P. hirsutiusculus venturensis** Coffin) less hairy, with carapace shield often a little longer than wide.

This species lives lower down in the intertidal zone than *P. samuelis* (24.10), though the populations overlap, and is less resistant to the effects of exposure out of water than either *P. samuelis* or *P. granosimanus* (24.13). However, it tolerates brackish conditions and occurs in San Francisco Bay in areas where the salinity averages 75 percent that of normal seawater.

Laboratory tests of these hermit crabs in southern California show that they can discriminate between, and have preferences among, certain kinds of shells. *Acanthina* (e.g., 13.80) and *Olivella* (e.g., 13.99) shells are commonly used in southern California. In the Monterey Bay area less than 15 percent of the adults of *P. hirsutiusculus* occupy the large *Tegula* shells favored by *P. samuelis*; instead, the great majority live in the smaller shells of *Nassa*, *Olivella biplicata* (13.99), *Nucella emarginata* (13.83), *Littorina planaxis* (13.42), *Epitonium tinctum* (13.50), *Mitrella carinata* (13.91), *Homalopoma luridum* (13.40), etc. At Coyote Point (San Mateo Co.) in San Francisco Bay, *P. hirsutiusculus* is usually found inhabiting the shells of *Nassarius obsoletus* (13.95) and *Urosalpinx cinerea* (13.78), and occasionally *Busycotypus canaliculatus* (13.85), all introduced gastropods whose shells were not available there a century ago.

Larvae have been reared in the same manner as those of *P. samuelis*. Adult females carrying eggs are reported from the Monterey area in July and August, and in other parts of California in every month from December through April. A parasitic isopod, *Pseudione giardi*, has been found with *Pagurus hirsutiusculus* in the San Juan Archipelago (Washington).

The fine structure of the sensory (aesthetasc) hairs on the antennules has been investigated; highly innervated and permeable, the hairs appear to contain chemoreceptors.

See Bollay (1964), Coffin (1954, 1957, 1958), Filice (1958), Forss & Coffin (1960), George & Strömberg (1968), Ghiradella, Case & Cronshaw (1968), Ghiradella, Cronshaw & Case (1968), Holmes (1900), MacGinitie (1935), McLaughlin (1974), Orians & King (1964), Reese (1962, 1969), Ricketts & Calvin (1968), Schmitt (1921), Spight (1977), Vance (1972a,b), and Wicksten (1977).

Phylum Arthropoda / Class Crustacea / Subclass Malacostraca
Superorder Eucarida / Order Decapoda / Suborder Reptantia / Section Anomura
Superfamily Paguridea / Family Paguridae

24.12 **Pagurus hemphilli** (Benedict, 1892)
Hermit Crab

In tidepools, middle and low intertidal zones on protected rocky shores; on rock and gravel bottoms to at least 15 m depth; Queen Charlotte Islands (British Columbia) to San Miguel Island (Channel Islands).

Carapace less than 19 mm long, bearing a prominent and sharp median frontal tooth; carapace shield distinctly longer than wide; antennal flagella orange-red with yellow spots; chelae mahogany-red with bluish granules, the fingertips white on right claw, orange on left; walking legs not hairy, with the propodi mahogany-red with white granules, the dactyli same but with a white patch distally.

In California these crabs occur lower in the intertidal than other littoral *Pagurus* species, and the main population is probably subtidal. Little is known of the biology of this species, but ovigerous females have been noted in February and March in the Monterey area.

See Hewatt (1937), Holmes (1900), McLaughlin (1974), McLean (1962), Ricketts & Calvin (1968), and Schmitt (1921).

Phylum Arthropoda / Class Crustacea / Subclass Malacostraca
Superorder Eucarida / Order Decapoda / Suborder Reptantia / Section Anomura
Superfamily Paguridea / Family Paguridae

24.13 **Pagurus granosimanus** (Stimpson, 1859)
Hermit Crab

Common in tidepools, lower middle and low intertidal zones on protected rocky shores; subtidal to 15 m or more on gravel bottoms and in kelp holdfasts; Unalaska (Aleutian Islands, Alaska) to Bahía de Todos Santos (Baja California).

Carapace to 19 mm long, bearing a broad and rounded median frontal tooth; antennal flagella scarlet; chelae olive-brown with blue or white granules, the larger one evenly and finely granulated on both upper and lower surfaces; walking legs with propodi olive-brown with light granules, dactyli colored similarly, lacking patches or stripes.

Pagurus beringanus (Benedict), also with the median frontal tooth of the carapace broad and rounded, rarely occurs intertidally in California waters. In that species the propodi of the walking legs have red spots and a distal red band, and the large chela has orange tubercles on the upper surface only.

Pagurus granosimanus lives lower in the intertidal, and is less resistant to the effects of exposure above water, than *P. samuelis* (24.10). At Pacific Grove (Monterey Co.) more than 95 percent of the animals of breeding size occupy *Tegula* shells, especially those of *T. funebralis* (13.32); smaller crabs show no marked preference for shells of particular snails here or elsewhere in California. The diurnal pattern of activity is like that of *P. samuelis*, but the animals do not become active until later in the afternoon. Retinal pigments in the eyes shift position with day and night, apparently only in direct response to the ambient light, again as in *P. samuelis*.

Details on reproduction are lacking, but at Pacific Grove few females with eggs are seen in February, many in April and May. Adults have been maintained for long periods in aquaria on a diet of brine shrimp and the brown alga *Pelvetia*.

Experimental neurophysiological investigations have been made of the abdominal nervous system, which serves the parts of the body normally held deep within the shell. The hard parts of the abdomen have sensory hairs much as in the crayfish. The softer parts have epidermal receptors more similar to those of the earthworm's skin, and not unlike those in the paws of a cat. The right side of the abdomen, which curls around the central axis of the snail shell, is less sensitive than the left side, which bears the pleopods (lacking on the right in *Pagurus*).

Fossil remains of *P. granosimanus* have been reported from Pleistocene deposits.

See Andrews (1945), Ball (1968), Bollay (1964), Chapple (1966), Hewatt (1937, 1938), Holmes (1900), McLaughlin (1974), McLean (1962), Orians & King (1964), Rathbun (1926), Reese (1962), Ricketts & Calvin (1968), Schmitt (1921), Spight (1977), and Vance (1972a,b).

Phylum Arthropoda / Class Crustacea / Subclass Malacostraca
Superorder Eucarida / Order Decapoda / Suborder Reptantia / Section Anomura
Superfamily Galatheidea / Family Porcellanidae

24.14 **Petrolisthes rathbunae** Schmitt, 1921
Porcelain Crab

Under stones, low intertidal zone on rocky shores; Monterey Bay to Laguna Beach (Orange Co.), also Santa Barbara Islands (Channel Islands) and Isla Guadalupe (Mexico).

Carapace to about 19 mm long, covered with short, scale-like striations, their anterior edges hairy; carpus of chelipeds about 2.5 times as long as wide, with parallel margins; body in life greenish with reddish-purple spots (photograph shows faded, preserved specimen); red bands on legs, red also on distal segments of maxillipeds and on tips and inner edges of fingers.

Practically nothing is known of the biology of this species. In California, females bearing eggs have been observed in March, April, June, and July. Specimens have not been collected often, and it is probable that the larger part of the population is subtidal.

See Haig (1960) and Schmitt (1921).

Phylum Arthropoda / Class Crustacea / Subclass Malacostraca
Superorder Eucarida / Order Decapoda / Suborder Reptantia / Section Anomura
Superfamily Galatheidea / Family Porcellanidae

24.15 **Petrolisthes eriomerus** Stimpson, 1871
Porcelain Crab

Common under stones, on submerged jetties, in mussel beds and kelp holdfasts, low intertidal zone on protected rocky shores in northern part of range, offshore to depths of about 90 m from San Luis Obispo Co. southward; Chicagof Island (Alaska) to La Jolla (San Diego Co.).

Carapace to about 19 mm long; anterior portions ornamented with rough granules and small tubercles; carpus of chelipeds about twice as long as wide, with parallel margins, and covered with large rough granules and small tubercles; legs banded; body in life reddish brown, often mottled with blue; blue also on distal segments of maxillipeds and inner side of base of movable fingers of chelae.

Natural history studies of this species have been made on populations in Puget Sound (Washington). The animals are commonest in regions of swift currents. The main food is diatoms, strained from the water by the hairy maxillipeds, though stomach analyses show that some sessile diatoms and scraps of green algae are also eaten.

Females with eggs occur in Puget Sound from February through October. About 80 percent of the females produce two broods of young a year, the first batch hatching from May to August, the second from August to October. The eggs are very large (0.75–0.8 mm in diameter), and the number carried (10–1,580, mean 621) varies with the size of the female.

Laboratory-reared larvae hatch as prezoeae; these molt to become the first zoeal stage in an hour or less. As in other porcelain crabs, the larva passes through two zoeal instars and a megalops stage. Zoeae accept brine shrimp nauplius larvae as food, but the megalops is a filter feeder like the adult.

See Forss & Coffin (1960), Gonor & Gonor (1973a, 1973b), Haig (1960), Knudsen (1964), Schmitt (1921), and Turner, Ebert & Given (1969).

Phylum Arthropoda / Class Crustacea / Subclass Malacostraca
Superorder Eucarida / Order Decapoda / Suborder Reptantia / Section Anomura
Superfamily Galatheidea / Family Porcellanidae

24.16 **Petrolisthes manimaculis** Glassell, 1945
Porcelain Crab

Common under stones, low intertidal zone on rocky shores; Bodega Bay harbor (Sonoma Co.) to Punta Eugenia (Baja California).

Carapace to about 19 mm long, smooth to granular anteriorly, nearly smooth posteriorly; carpus of chelipeds from a little over two times to nearly three times as long as wide, with parallel margins; body in life brown with a transverse series of blue dots on carapace, distal segments of maxillipeds blue, chela often with a longitudinal row of blue spots on palm, and with movable finger orange on inner side of base (photograph shows faded, preserved specimen).

Practically nothing is known of the biology of the species. Females carrying eggs have been found in every month from October through March and also in June.

See Glassell (1945), Haig (1960), and Schmitt (1921, as *P. gracilis*).

Phylum Arthropoda / Class Crustacea / Subclass Malacostraca
Superorder Eucarida / Order Decapoda / Suborder Reptantia / Section Anomura
Superfamily Galatheidea / Family Porcellanidae

24.17 **Petrolisthes cabrilloi** Glassell, 1945
Porcelain Crab

Abundant under stones, in mussel beds, and on wharf pilings, middle intertidal zone on rocky shores; Morro Bay (San Luis Obispo Co.) to Bahía Magdalena (Baja California).

Carapace to about 16 mm long, ornamented with fine granules; carpus of chelipeds about twice as long as wide, its margins parallel except for a small, sometimes indistinct lobe on proximal fourth of inner margin; body in life usually orange with many pale spots and transverse lines, distal segments of maxillipeds orange, walking legs banded; ventral view well shown in Fig. 20.36.

This species is common along the shore in southern California (where it has frequently been confused with *Petrolisthes cinctipes*, 24.18), but its life history has never been studied. Egg-bearing females have been recorded in February, April, May, October, and November. Occasional individuals are found serving as host to the parasitic barnacle *Lernaeodiscus porcellanae* (20.36).

See Glassell (1945) and Haig (1960).

Phylum Arthropoda / Class Crustacea / Subclass Malacostraca
Superorder Eucarida / Order Decapoda / Suborder Reptantia / Section Anomura
Superfamily Galatheidea / Family Porcellanidae

24.18 **Petrolisthes cinctipes** (Randall, 1839)
Flat Porcelain Crab

Common under stones and in beds of mussels, sometimes among sponges and tunicates, middle and low intertidal zones on exposed rocky coast; Queen Charlotte Islands (British Columbia) to Point Conception (Santa Barbara Co.) and northern Channel Islands (San Miguel, Santa Rosa, and Santa Cruz Islands).

Carapace to about 24 mm long, ornamented with fine granules; carpus of chelipeds 1.5–2 times as long as wide, the margins not parallel, the inner margin with a distinct proximal lobe; color in life reddish brown, distal segments of maxillipeds reddish, walking legs banded.

This is one of the most abundant macroscopic animals living in the interstices of well-developed beds of the mussel *Mytilus californianus* (15.9); up to 860 specimens per m² were found in mussel beds at Pacific Grove (Monterey Co.). It feeds on plankton and suspended organic detritus (such as diatoms) swept from the water by long feathery hairs on the outer maxillipeds; rarely it ingests algae and dead animal tissue.

Reproduction occurs all or nearly all year; substantial numbers of egg-bearing females been found at Monterey in all months of the year except November, and at no time are the gonads wholly spawned out. The eggs, 0.8–0.84 mm in diameter, are deep scarlet to maroon when newly extruded, gradually changing to brownish red. The two zoea larval stages are carnivores. The megalops stage is a filter feeder like the adult.

Studies have been made of the gross anatomy of the central nervous system, the organic constituents of the body, the intermolt cycle, and the endocrine organs of the eyestalk and brain.

See Boolootian et al. (1959), Giese (1966), Gonor & Gonor (1973a,b), Haig (1960), Hewatt (1935, 1937, 1938), Kurup (1964a,b), MacGinitie (1935, 1939), MacGinitie & MacGinitie (1968), Ricketts & Calvin (1968), Sayed (1963), Schmitt (1921), and Wicksten (1973).

Phylum Arthropoda / Class Crustacea / Subclass Malacostraca
Superorder Eucarida / Order Decapoda / Suborder Reptantia / Section Anomura
Superfamily Galatheidea / Family Porcellanidae

24.19 **Pachycheles pubescens** Holmes, 1900
Porcelain Crab

Under stones and in rock crevices, low intertidal zone on
rocky shores in northern part of range, offshore to depths of
about 55 m from San Luis Obispo Co. southward; Goose Island (British Columbia) to Cabo Thurloe (Baja California).

Carapace to about 18 mm long; chelae densely covered
with short soft hairs which conceal large granules, and bearing also scattered tufts of long stiff hairs; body dull brown;
species differing from *P. rudis* (24.20) and *P. holosericus* (24.21)
in having seven plates on the telson instead of five.

Females bearing eggs have been found in nearly every
month of the year. The eggs are 0.5–0.58 mm in diameter.
They are brilliant yellow-orange when newly extruded, this
color gradually changing to amber. They may be carried for
as long as 49 days. The larvae hatch as prezoeae; in the laboratory the first zoeal stage is reached in 10 minutes to an hour.
Animals cultured at 14°C went through the two zoeal stages
in 34–40 days. Zoeae have been reared on a diet of brine
shrimp nauplius larvae; the crab megalops stage is exclusively
a filter feeder.

A parasitic isopod, *Aporobopyrus oviformis*, sometimes occurs
in the gill chamber.

See Gonor & Gonor (1973a,b), Gore (1973), Haig (1960), MacMillan (1972a), Schmitt (1921), Shiino (1964), and Turner, Ebert &
Given (1969).

Phylum Arthropoda / Class Crustacea / Subclass Malacostraca
Superorder Eucarida / Order Decapoda / Suborder Reptantia / Section Anomura
Superfamily Galatheidea / Family Porcellanidae

24.20 **Pachycheles rudis** Stimpson, 1859
Thick-Clawed Porcelain Crab

Common in various sheltered situations on rocky shores:
under stones, in kelp holdfasts, beneath eelgrass root mats, in

sponge cavities, in empty burrows of rock-boring clams,
among mussels and rock oysters, and on pilings, low intertidal zone to depths of at least 18 m; Kodiak Island (Alaska) to
Bahía Magdalena (Baja California).

Carapace to about 18 mm long; chelae with scattered long
hairs that usually do not conceal the granular surface; carpus
of chelipeds with a broad triangular lobe on inner margin, its
edge not strongly toothed; telson of abdomen with five plates;
body dull brown.

Crabs of this species normally live in pairs in secluded
nooks where they may reach a size too large to escape. They
feed on plankton and particles of organic detritus combed
from the water by setae on the outer maxillipeds. In Puget
Sound (Washington), females bearing eggs were found to
occur from December at least through August (when observations ended), and in California they have been noted in every
month of the year except September. Females probably produce two or three broods of young a year. In the study at
Puget Sound, numbers of eggs carried by ovigerous females
of carapace width 12–16.8 mm ranged from 210 to 2,130.

Zoea larvae have been reared in the laboratory on a diet of
newly hatched brine shrimp larva. The first zoeal stage lasts
10–11 days, the second 14 days, at 15–18°C. The megalops
stage, like the adult crab, is exclusively a filter feeder.

A parasitic isopod, *Aporobopyrus muguensis* (21.11), sometimes occurs in the gill chamber of the adult (see Fig. 21.11).

See Andrews (1945), Gonor & Gonor (1973a,b), Gore (1973), Haig
(1960), Hewatt (1937, 1938), Knight (1966), Knudsen (1964), Mac-
Ginitie (1935, 1937), MacGinitie & MacGinitie (1968), MacMillan
(1972a), Ricketts & Calvin (1968), Schmitt (1921), Shiino (1964), and
Turner, Ebert & Given (1969).

Phylum Arthropoda / Class Crustacea / Subclass Malacostraca
Superorder Eucarida / Order Decapoda / Suborder Reptantia / Section Anomura
Superfamily Galatheidea / Family Porcellanidae

24.21 **Pachycheles holosericus** Schmitt, 1921
Porcelain Crab

Frequently under stones, sometimes in other sheltered situations such as sponge cavities and kelp holdfasts, low interti-

dal zone on rocky shores; subtidal to at least 18 m, under rocks, in heavy undergrowth on stones, and among mussels; Santa Barbara to Bahía Magdalena (Baja California).

Carapace to about 18 mm long; chelae densely covered with short soft hairs, through which some granules or tubercles protrude; carpus of chelipeds with a toothed lobe on inner margin; telson of abdomen with five plates; body dull brown.

These crabs are practically always found in pairs; more than one pair may occupy a single sponge. In California, egg-bearing females have been collected from November to March and in June and July.

See Haig (1960), Schmitt (1921), and Turner, Ebert & Given (1969).

Literature Cited

Andrews, H. L. 1945. The kelp beds of the Monterey region. Ecology 26: 24–37.

Ball, E. E., Jr. 1968. Activity patterns and retinal pigment migration in *Pagurus* (Decapoda, Paguridea). Crustaceana 14: 302–6.

Balss, H. 1957. Systematik, pp. 1505–1672, *in* Dr. H. G. Bronns Klassen und Ordnungen des Tierreichs, Bd. 5, Abt. 1, Buch 7, Lief. 12. Leipzig: Winter.

Barnes, N. B., and A. M. Wenner. 1968. Seasonal variation in the sand crab *Emerita analoga* (Decapoda, Hippidae) in the Santa Barbara area of California. Limnol. Oceanogr. 13: 465–75.

Bollay, M. 1964. Distribution and utilization of gastropod shells by the hermit crabs *Pagurus samuelis, Pagurus granosimanus,* and *Pagurus hirsutiusculus* at Pacific Grove, California. Veliger 6 (Suppl.): 71–76.

Boolootian, R. A., A. C. Giese, A. Farmanfarmaian, and J. Tucker. 1959. Reproductive cycles of five west coast crabs. Physiol. Zool. 32: 213–20.

Borradaile, L. A. 1903. On the classification and genealogy of the reptant decapods, pp. 690–98, *in* J. S. Gardiner, ed., The fauna and geography of the Maldive and Laccadive archipelagoes, part 2. Cambridge, Eng.: Cambridge University Press.

———. 1907. On the classification of the decapod crustaceans. Ann. Mag. Natur. Hist. (7) 19: 457–86.

Bouvier, E.-L. 1895. Recherches sur les affinités des *Lithodes* et des *Lomis* avec les Paguridés. Ann. Sci. Natur., Zool. (7) 18: 157–213.

———. 1896. Sur la classification des Lithodinés et sur leur distribution dans les océans. Ann. Sci. Natur., Zool. (8) 1: 1–46.

Boyd, C. M. 1960. The larval stages of *Pleuroncodes planipes* Stimpson (Crustacea, Decapoda, Galatheidae). Biol. Bull. 118: 17–30.

———. 1967. The benthic and pelagic habitats of the red crab, *Pleuroncodes planipes.* Pacific Sci. 21: 394–403.

Burnett, R. 1971. DDT residues: Distribution of concentrations in *Emerita analoga* (Stimpson) along coastal California. Science 174: 606–8.

Bybee, J. R. 1969. Effects of hydraulic pumping operations on the fauna of Tijuana Slough. Calif. Fish & Game 55: 213–20.

Carlisle, J. G., Jr., J. W. Schott, and N. J. Abramson. 1960. The barred surfperch (*Amphistichus argenteus* Agassiz) in southern California. Calif. Dept. Fish & Game, Fish Bull. 109: 1–79.

Carlton, J. T., and A. M. Kuris. 1975. Keys to decapod Crustacea, pp. 385–412, *in* R. I. Smith and J. T. Carlton, eds., Light's manual: Intertidal invertebrates of the central California coast. 3rd ed. Berkeley and Los Angeles: University of California Press. 716 pp.

Chapple, W. D. 1966. Sensory modalities and receptive fields in the abdominal nervous system of the hermit crab, *Pagurus granosimanus* (Stimpson). J. Exper. Biol. 44: 209–23.

Coffin, H. G. 1954. The biology of *Pagurus samuelis* (Stimpson). Doctoral thesis, Biological Sciences, University of Southern California. 212 pp.

———. 1957. A new southern form of *"Pagurus hirsutiusculus"* (Dana) (Crustacea, Decapoda). Publ. Dept. Biol. Sci., Walla Walla College 21: 1–6.

———. 1958. The laboratory culture of *Pagurus samuelis* (Stimpson) (Crustacea, Decapoda). Publ. Dept. Biol. Sci., Walla Walla College 22: 1–5.

———. 1960. The ovulation, embryology and developmental stages of the hermit crab *Pagurus samuelis* (Stimpson). Publ. Dept. Biol. Sci., Walla Walla College 25: 1–28.

Costello, D. P., and C. Henley. 1971. Methods for obtaining and handling marine eggs and embryos. 2nd ed. Marine Biological Laboratory, Woods Hole, Mass. 247 pp.

Cox, G. W., and G. H. Dudley. 1968. Seasonal pattern of reproduction of the sand crab, *Emerita analoga,* in southern California. Ecology 49: 746–51.

Cubit, J. 1969. Behavior and physical factors causing migration and aggregation of the sand crab *Emerita analoga* (Stimpson). Ecology 50: 118–23.

Dillery, D. G., and L. V. Knapp. 1970. Longshore movements of the sand crab, *Emerita analoga* (Decapoda, Hippidae). Crustaceana 18: 233–40.

Edwards, G. A., and L. Irving. 1943. The influence of temperature and season upon the oxygen consumption of the sand crab, *Emerita talpoida* (Say). J. Cell. Compar. Physiol. 21: 169–82.

Efford, I. E. 1965. Aggregation in the sand crab, *Emerita analoga* (Stimpson). J. Anim. Ecol. 34: 63–75.

_____. 1966. Feeding in the sand crab, *Emerita analoga* (Stimpson) (Decapoda, Anomura). Crustaceana 10: 167–82.

_____. 1967. Neoteny in sand crabs of the genus *Emerita* (Anomura, Hippidae). Crustaceana 13: 81–93.

_____. 1969. Egg size in the sand crab, *Emerita analoga* (Decapoda, Hippidae). Crustaceana 16: 15–26.

_____. 1970. Recruitment to sedentary marine populations as exemplified by the sand crab, *Emerita analoga* (Decapoda, Hippidae). Crustaceana 18: 293–308.

Eickstaedt, L. L. 1969. The reproductive biology of the sand crab *Emerita analoga* (Stimpson). Doctoral thesis, Biological Sciences, Stanford University, Stanford, Calif. 100 pp.

Fager, E. W. 1968. A sand-bottom epifaunal community of invertebrates in shallow water. Limnol. Oceanogr. 13: 448–64.

Farley, R. D., and J. F. Case. 1968. Perception of external oxygen by the burrowing shrimp, *Callianassa californiensis* Dana and *C. affinis* Dana. Biol. Bull. 134: 261–65.

Filice, F. P. 1958. Invertebrates from the estuarine portion of San Francisco Bay and some factors influencing their distribution. Wasmann J. Biol. 16: 159–211.

Fingerman, M., and C. Oguro. 1963. Chromatophore control and neurosecretion in the mud shrimp, *Upogebia affinis*. Biol. Bull. 124: 24–30.

Forss, C. A., and H. G. Coffin. 1960. The use of the brine shrimp nauplii, *Artemia salina*, as food for the laboratory culture of decapods. Publ. Dept. Biol. Sci., Walla Walla College 26: 1–15.

Frey, H. W., ed. 1971. California's living marine resources and their utilization. Calif. Dept. Fish & Game. 148 pp.

Fusaro, C. 1977. Population structure, growth rate and egg production of the sand crab, *Emerita analoga* (Hippidae): A comparative analysis. Doctoral thesis, Biological Sciences, University of California, Santa Barbara. 182 pp.

_____. 1978. Growth rate of the sand crab, *Emerita analoga* (Hippidae), in two different environments. Fish. Bull. 76: 369–75.

George, R. Y., and J.-O. Strömberg. 1968. Some new species and new records of marine isopods from San Juan Archipelago, Washington, U.S.A. Crustaceana 14: 225–54.

Ghiradella, H. T., J. F. Case, and J. Cronshaw. 1968. Structure of aesthetascs in selected marine and terrestrial decapods: Chemoreceptor morphology and environment. Amer. Zool. 8: 603–21.

Ghiradella, H. T., J. Cronshaw, and J. Case. 1968. Fine structure of the aesthetasc hairs of *Pagurus hirsutiusculus* Dana. Protoplasma 66: 1–20.

Giese, A. C. 1966. Lipids in the economy of marine invertebrates. Physiol. Rev. 46: 244–98.

Gilchrist, B. M., and W. L. Lee. 1972. Carotenoid pigments and their possible role in reproduction in the sand crab, *Emerita analoga* (Stimpson, 1857). Comp. Biochem. Physiol. 42B: 263–94.

Glaessner, M. F. 1969. Decapoda, pp. R399–533, *in* R. C. Moore, ed., Treatise on invertebrate paleontology, part R, Arthropoda 4, vol. 2. New York: Geol. Soc. Amer.; Lawrence: University of Kansas Press.

Glassell, S. A. 1945. Four new species of North American crabs of the genus *Petrolisthes*. J. Wash. Acad. Sci. 35: 223–29.

Glynn, P. W. 1961. The first recorded mass stranding of pelagic red crabs, *Pleuroncodes planipes*, at Monterey Bay, California, since 1859, with notes on their biology. Calif. Fish & Game 47: 97–101.

Gonor, S. L., and J. J. Gonor. 1973a. Descriptions of the larvae of four North Pacific Porcellanidae (Crustacea: Anomura). Fish. Bull. 71: 189–223.

_____. 1973b. Feeding, cleaning, and swimming behavior in larval stages of porcellanid crabs (Crustacea: Anomura). Fish. Bull. 71: 225–34.

Gooding, R. U. 1957. "*Callianassa pugettensis*" (Decapoda, Anomura), type host of the copepod *Clausidium vancouverense* (Haddon). With a note on *Hemicyclops pugettensis* Light & Hartman, another copepod associated with callianassids. Ann. Mag. Natur. Hist. (12) 10: 695–700.

_____. 1960. North and South American copepods of the genus *Hemicyclops* (Cyclopoida: Clausiidae). Proc. U.S. Nat. Mus. 112: 159–95.

Gore, R. H. 1973. *Pachycheles monilifer* (Dana, 1852): The development in the laboratory of larvae from an Atlantic specimen with a discussion of some larval characters in the genus (Crustacea: Decapoda: Anomura). Biol. Bull. 144: 132–50.

Gross, W. J. 1957. An analysis of response to osmotic stress in selected decapod Crustacea. Biol. Bull. 112: 43–62.

Haig, J. 1960. The Porcellanidae (Crustacea Anomura) of the eastern Pacific. Allan Hancock Pacific Exped. 24: 1–440.

Haley, S. R. 1979. Sex ratio as a function of size in *Hippa pacifica* Dana (Crustacea, Anomura, Hippidae): A test of the sex reversal and differential growth rate hypotheses. Amer. Natur. 113: 391–97.

Hart, J. F. L. 1930. Some decapods from the south-eastern shores of Vancouver Island. Canad. Field Natur. 44: 101–9.

_____. 1937. Larval and adult stages of British Columbia Anomura. Canad. J. Res. D15: 179–220.

_____. 1965. Life history and larval development of *Cryptolithodes typicus* Brandt (Decapoda, Anomura) from British Columbia. Crustaceana 8: 255–76.

_____. 1971. New distribution records of reptant decapod Crustacea, including descriptions of three new species of *Pagurus*, from the waters adjacent to British Columbia. J. Fish. Res. Bd. Canada 28: 1527–44.

Hewatt, W. G. 1935. Ecological succession in the *Mytilus californianus* habitat as observed in Monterey Bay, California. Ecology 16: 244–51.

_____. 1937. Ecological studies on selected marine intertidal communities of Monterey Bay, California. Amer. Midl. Natur. 18: 161–206.

_____. 1938. Notes on the breeding seasons of the rocky beach fauna of Monterey Bay, California. Proc. Calif. Acad. Sci. (4) 23: 283–88.

Holmes, S. J. 1900. Synopsis of California stalk-eyed Crustacea. Occas. Pap. Calif. Acad. Sci. 7: 1–262.

Johnson, M. E., and H. J. Snook. 1927. Seashore animals of the Pacific coast. New York: Macmillan. 659 pp. (Paperback reprint, with original pagination, Dover, New York, 1967.)

Johnson, M. W. 1940. The correlation of water movements and dispersal of pelagic larval stages of certain littoral animals, especially the sand crab, *Emerita*. J. Mar. Res. 2: 236–45.

Johnson, M. W., and W. M. Lewis. 1942. Pelagic larval stages of the sand crabs *Emerita analoga* (Stimpson), *Blepharipoda occidentalis* Randall, and *Lepidopa myops* Stimpson. Biol. Bull. 83: 67–87.

Joseph, D. C. 1962. Growth characteristics of two southern California surffishes, the California corbina and spotfin croaker, family Sciaenidae. Calif. Dept. Fish & Game, Fish Bull. 119: 1–54.

Kaestner, A. 1970. Invertebrate zoology. 3. Crustacea. Translated and adapted from 2nd German ed. by H. W. Levi & L. R. Levi. New York: Interscience. 523 pp.

Knight, M. D. 1966. The larval development of *Polyonyx quadriungulatus* Glassell and *Pachycheles rudis* Stimpson (Decapoda, Porcellanidae) cultured in the laboratory. Crustaceana 10: 75–97.

_____. 1967. The larval development of the sand crab *Emerita rathbunae* Schmitt (Decapoda, Hippidae). Pacific Sci. 21: 58–76.

_____. 1968. The larval development of *Blepharipoda occidentalis* Randall and *B. spinimana* (Philippi) (Decapoda, Albuneidae). Proc. Calif. Acad. Sci. (4) 35: 337–70.

Knox, C., and R. A. Boolootian. 1963. Functional morphology of the external appendages of *Emerita analoga*. Bull. So. Calif. Acad. Sci. 62: 45–68.

Knudsen, J. W. 1964. Observations of the reproductive cycles and ecology of the common Brachyura and crablike Anomura of Puget Sound, Washington. Pacific Sci. 18: 3–33.

Koepcke, H.-W., and M. Koepcke. 1952. Sobre el proceso de transformación de la materia orgánica en las playas arenosas del Perú. Publ. Mus. Hist. Natur. "Javier Prado" (A) Zool. 8: 1–25.

Kurup, N. G. 1964a. The incretory organs of the eyestalk and brain of the porcelain crab, *Petrolisthes cinctipes* (Reptantia–Anomura). Gen. Comp. Endocrinol. 4: 99–112.

_____. 1964b. The intermolt cycle of an anomuran, *Petrolisthes cinctipes* Randall (Crustacea–Decapoda). Biol. Bull. 127: 97–107.

Light, S. F., and O. Hartman. 1937. A review of the genera *Clausidium* Kossmann and *Hemicyclops* Boeck (Copepoda, Cyclopoida), with the description of a new species from the northeast Pacific. Univ. Calif. Publ. Zool. 41: 173–88.

Lindsay, C. E. 1963. Pesticide tests in the marine environment in the State of Washington. Proc. Nat. Shellfish Assoc. 52: 87–98.

Longhurst, A. R., C. J. Lorenzen, and W. H. Thomas. 1967. The role of pelagic crabs in the grazing of phytoplankton off Baja California. Ecology 48: 190–200.

Longhurst, A. R., and D. L. R. Seibert. 1971. Breeding in an oceanic population of *Pleuroncodes planipes* (Crustacea, Galatheidae). Pacific Sci. 25: 426–28.

MacGinitie, G. E. 1930. The natural history of the mud shrimp *Upogebia pugettensis* (Dana). Ann. Mag. Natur. Hist. (10) 6: 36–44.

_____. 1934. The natural history of *Callianassa californiensis* Dana. Amer. Midl. Natur. 15: 166–77.

_____. 1935. Ecological aspects of a California marine estuary. Amer. Midl. Natur. 16: 629–765.

_____. 1937. Notes on the natural history of several marine Crustacea. Amer. Midl. Natur. 18: 1031–37.

_____. 1938. Movements and mating habits of the sand crab, *Emerita analoga*. Amer. Midl. Natur. 19: 471–81.

_____. 1939. Littoral marine communities. Amer. Midl. Natur. 21: 28–55.

MacGinitie, G. E., and N. MacGinitie. 1968. Natural history of marine animals. 2nd ed., rev. New York: McGraw-Hill. 523 pp.

MacMillan, F. E. 1972a. The larval development of northern California Porcellanidae (Decapoda, Anomura). I. *Pachycheles pubescens* Holmes in comparison to *Pachycheles rudis* Stimpson. Biol. Bull. 142: 57–70.

_____. 1972b. The larvae of *Pagurus samuelis* (Decapoda: Anomura) reared in the laboratory. Bull. So. Calif. Acad. Sci. 70: 58–68.

Makarov, V. V. 1938. Anomura. Fauna S.S.S.R., Rakoobraznye [Crustacea] 10(3). Zool. Inst. Acad. Sci. U.S.S.R., n.s. 16. 324 pp. English transl., 1962. Jerusalem: Israel Program of Scientific Translation (I.P.S.T. Press), available through U.S. Dept. Commerce, Washington, D.C. 283 pp.

McLaughlin, P. A. 1974. The hermit crabs (Crustacea Decapoda, Paguridea) of northwestern North America. Zool. Verh. Leiden 130: 1–396.

_____. 1976. A new Japanese hermit crab (Decapoda, Paguridae) resembling *Pagurus samuelis* (Stimpson). Crustaceana 30: 13–26.

McLean, J. H. 1962. Sublittoral ecology of kelp beds of the open coast area near Carmel, California. Biol. Bull. 122: 95–114.

Miller, K., and K. E. Van Holde. 1974. Oxygen binding by *Callianassa californiensis* hemocyanin. Biochemistry 13: 1668–74.

Miller, P. E., and H. G. Coffin. 1961. A laboratory study of the developmental stages of *Hapalogaster mertensii* (Brandt) (Crustacea, Decapoda). Publ. Dept. Biol. Sci., Walla Walla College 30: 1–18.

Mitchell, D. F. 1953. An analysis of stomach contents of California tide pool fishes. Amer. Midl. Natur. 49: 862–71.

Odenweller, D. B. 1972. A new range record for the umbrella crab, *Cryptolithodes sitchensis* Brandt. Calif. Fish & Game 58: 240–43.

Ōishi, S. 1960. Studies on the teloblasts in the decapod embryo. II. Origin of teloblasts in *Pagurus samuelis* (Stimpson) and *Hemigrapsus sanguineus* (De Haan). Embryologia 5: 270–82.

Orians, G. H., and C. E. King. 1964. Shell selection and invasion rates of some Pacific hermit crabs. Pacific Sci. 18: 297–306.

Osorio, C., N. Bahamonde, and M. T. López. 1967. El limanche [*Emerita analoga* (Stimpson)] en Chile (Crustacea, Decapoda, Anomura). Bol. Mus. Nac. Hist. Natur. Santiago 29: 61–116.

Paul, D. H. 1971. Swimming behavior of the sand crab, *Emerita analoga* (Crustacea, Anomura). I. Analysis of the uropod stroke. II. Morphology and physiology of the uropod neuromuscular system. III. Neuronal organization of uropod beating. Z. Vergl. Physiol. 75: 233–302.

_____. 1972. Decremental conduction over "giant" afferent processes in an arthropod. Science 176: 680–82.

_____. 1976a. Proprioception from nonspiking sensory cells in a swimming behavior of the sand crab *Emerita analoga*, pp. 785–87, *in* R. M. Herman, S. Grillner, P. S. G. Stein, and D. G. Stuart, eds., Neural control and locomotion. New York: Plenum.

_____. 1976b. Role of proprioceptive feedback from nonspiking mechanosensory cells in the sand crab, *Emerita analoga*. J. Exper. Biol. 65: 243–58.

_____. 1979. An endogenous motor program for sand crab uropods. J. Neurobiol. 10: 273–89.

Powell, R. R. 1974. The functional morphology of the fore-guts of the thalassinid crustaceans, *Callianassa californiensis* and *Upogebia pugettensis*. Univ. Calif. Publ. Zool. 102: 1–41, 39 figs.

Quast, J. C. 1968. Observations on the food of the kelp-bed fishes, pp. 109–42, *in* W. J. North and C. L. Hubbs, Utilization of kelp-bed resources in southern California. Calif. Dept. Fish & Game, Fish Bull. 139: 1–264.

Rathbun, M. J. 1926. The fossil stalk-eyed Crustacea of the Pacific slope of North America. U.S. Nat. Mus., Bull. 138: 1–155.

Reeder, W. G. 1951. Stomach analysis of a group of shorebirds. Condor 53: 43–45.

Rees, G. H. 1959. Larval development of the sand crab *Emerita talpoida* (Say) in the laboratory. Biol. Bull. 117: 356–70.

Reese, E. S. 1962. Shell selection behaviour of hermit crabs. Anim. Behav. 10: 347–60.

_____. 1963. The behavioral mechanisms underlying shell selection by hermit crabs. Behaviour 21: 78–126.

_____. 1969. Behavioral adaptations of intertidal hermit crabs. Amer. Zool. 9: 343–55.

Ricketts, E. F., and J. Calvin. 1968. Between Pacific tides. 4th ed. Revised by J. W. Hedgpeth. Stanford, Calif.: Stanford University Press. 614 pp.

Roxby, R., K. Miller, D. P. Blair, and K. E. Van Holde. 1974. Subunits and association equilibria of *Callianassa californiensis* hemocyanin. Biochemistry 13: 1662–74.

Sayed, S. M. 1963. The central nervous system of *Petrolisthes cinctipes* (Randall). Crustaceana 5: 251–56.

Schmitt, W. L. 1921. The marine decapod Crustacea of California. Univ. Calif. Publ. Zool. 23: 1–470.

_____. 1965. Crustaceans. Ann Arbor: University of Michigan Press. 204 pp.

Shiino, S. M. 1964. On three bopyrid isopods from California. Rep. Fac. Fish., Univ. Mie 5: 19–25.

Snodgrass, R. E. 1952. The sand crab *Emerita talpoida* (Say) and some of its relatives. Smithsonian Misc. Collec. 117 (8): 1–34.

Spight, T. M. 1977. Availability and use of shells by intertidal hermit crabs. Biol. Bull. 152: 120–33.

Stevens, B. A. 1928. Callianassidae from the west coast of North America. Publ. Puget Sound Biol. Sta. 6: 315–69.

Subramoniam, T. 1977. Aspects of sexual biology of the anomuran crab *Emerita asiatica*. Mar. Biol. 43: 369–77.

Thompson, L. C., and A. W. Pritchard. 1969. Osmoregulatory capacities of *Callianassa* and *Upogebia* (Crustacea: Thalassinidea). Biol. Bull. 136: 114–29.

Thompson, R. K. 1972. Functional morphology of the hind-gut gland of *Upogebia pugettensis* (Crustacea, Thalassinidea) and its role in burrow construction. Doctoral thesis, Zoology, University of California, Berkeley. 202 pp.

Thompson, R. K., and A. W. Pritchard. 1969. Respiratory adaptations of two burrowing crustaceans, *Callianassa californiensis* and *Upogebia pugettensis* (Decapoda, Thalassinidea). Biol. Bull. 136: 274–87.

Torres, J. J., D. L. Gluck, and J. J. Childress. 1977. Activity and physiological significance of the pleopods in the respiration of *Callianassa californiensis* (Dana) (Crustacea: Thalassinidea). Biol. Bull. 152: 134–46.

Turner, C. H., E. E. Ebert, and R. R. Given. 1969. Man-made reef ecology. Calif. Dept. Fish & Game, Fish Bull. 146: 1–221.

Turner, C. H., and J. C. Sexsmith. 1964. Marine baits of California. Calif. Dept. Fish & Game. 71 pp.

Vance, R. R. 1972a. Competition and mechanism of coexistence in three sympatric species of intertidal hermit crabs. Ecology 53: 1062–74.

_____. 1972b. The role of shell adequacy in behavioral interactions involving hermit crabs. Ecology 53: 1075–83.

Warner, G. F. 1977. The biology of crabs. New York: Van Nostrand Reinhold. 202 pp.

Wenner, A. M. 1972. Sex ratio as a function of size in marine Crustacea. Amer. Natur. 106: 321–50.

_____. 1977. Food supply, feeding habits, and egg production in Pacific mole crabs (*Hippa pacifica* Dana). Pacific Sci. 31: 39–47.

Weymouth, F. W., and C. H. Richardson, Jr. 1912. Observations on the habits of the crustacean *Emerita analoga*. Smithsonian Misc. Collec. 59(7): 1–13.

Wicksten, M. K. 1973. Feeding in the porcelain crab, *Petrolisthes cinctipes* (Randall) (Anomura: Porcellanidae). Bull. So. Calif. Acad. Sci. 72: 161–63.

_____. 1977. Shells inhabited by *Pagurus hirsutiusculus* (Dana) at Coyote Point Park, San Francisco Bay, California. Veliger 19: 445–46.

Chapter 25

Brachyura: *The True Crabs*

John S. Garth and Donald P. Abbott

The Brachyura represent the highest development attained by articulated animals in the sea. Their hormonally controlled molting cycle, autotomy reflex, and ability to regenerate lost limbs excite the physiologist; their highly organized nervous systems, complex organs of sight and sound production, and incipient social organization beguile the animal behaviorist as well. Their commensal and mutualistic relationships with other invertebrates intrigue the marine ecologist, and their role as hosts to invading arthropods engages the parasitologist's attention.

Crabs first appear in the fossil record early in the Jurassic period of the Mesozoic, nearly 200 million years ago. As a group they show a continuation of the trend toward shortening the body and reducing the abdomen expressed in various anomuran groups (Chapter 24). The crab cephalothorax, formed by fusion of head and thorax and covered by the carapace, is short and broad, and forms virtually the whole body. The crab abdomen (corresponding to what gourmets call the "tail" of a lobster) is reduced to a thin, flat plate, tucked forward out of sight below the cephalothorax, hence the name "Brachyura," or "short-tailed" crabs.

The original metamerism, or serial segmentation, of the cephalothorax is largely obscured except as represented by the appendages. The five pairs of head appendages include the first and second antennae and the innermost three pairs of mouthparts (the mandibles and the first and second maxillae). The eight pairs of thoracic appendages include the outermost three pairs of mouthparts (the first, second, and third maxillipeds) and the five pairs of walking legs, the first of which are modified as chelipeds, or pincers.

The abdomen in crabs usually retains some external indications of serial segmentation, and some, but not all, of the original abdominal limbs. Female crabs possess four pairs of pleopods, which are used to carry eggs during the reproductive season. The males bear two pairs of pleopods, used as copulatory organs. Uropods are lacking in the adult stages of all but the most primitive crabs, and in most crabs they are absent even in the hatched larvae.

Crabs show some very effective locomotory innovations. Below their short, wide bodies the bases of the right and left members of each pair of walking legs are well separated; moreover, the legs themselves are relatively long and are hinged and suspended for the sideways gait that is so characteristic of true crabs. Many crabs can take relatively large steps, some species are capable of very rapid running, and nearly all can cling firmly to hard substrata when at rest. The limbs consist of non-repetitive segments or articles, separated by joints. The movable segments, starting at the body and moving outward on a leg, are called the coxa, basi-ischium, merus, carpus, propodus, and dactylus. Analogy of the cheliped and the human forelimb equates the merus with the forearm, the carpus with the wrist, the propodus with the hand or manus, the movable dactylus with the forefinger, and an opposable projection of the propodus with the pollex or thumb. The pollex is sometimes referred to as the fixed or immovable finger. Carcinologists (those who study crabs) often use the crustacean and human anatomical terms interchangeably.

A crab whose leg becomes trapped or is gripped by a predator

John S. Garth is Professor Emeritus of the Department of Biological Sciences, and Curator Emeritus of the Allan Hancock Foundation, University of Southern California. Donald P. Abbott is Professor, Department of Biological Sciences and Hopkins Marine Station, Stanford University.

can escape by quickly autotomizing (shedding) the limb and running away. Locomotory effectiveness is scarcely affected by the loss. Detachment of the limb is brought about by special muscles, and occurs at a preformed fracture plane in the basi-ischium.

The crab body is covered by a hard exoskeleton that offers protection but limits growth. Periodically, a crab seeks a sheltered hiding place, resorbs some materials from its exoskeleton, then splits the remaining shell and crawls out of it. In a short time the new exoskeleton, already partly formed inside the old shell but still thin and soft, thickens and hardens, but before it does the crab body absorbs considerable water and expands to a size larger than before the molt. Missing limbs, if any, are regenerated along with molting, provided the loss has occurred sufficiently early in the intermolt period to permit their internal preformation.

In crabs, as in other arthropods, increase in size must always be accomplished by molting, from the larval stages onward. The fertilized eggs, carried on special hairs on the pleopods of females, hatch as protozoea larvae and immediately molt to zoea larvae. These larvae are much more shrimplike in appearance than the adults. During the zoeal period the young exist as part of the plankton. They swim actively by means of appendages on the cephalothorax that become the antennae and mandibles of the adult and feed on metazoans, larval forms, and protozoa even smaller than themselves. Periodically they molt their larval exoskeletons, the body enlarging and undergoing some change in form at each molt. The stages between molts are called instars, and different species of crabs exhibit one to five zoeal instars. The last zoeal instar is followed by a molt and metamorphosis to the megalops stage. The megalops looks a bit like a small crab. The abdomen is somewhat flattened and can be folded forward below the cephalothorax, allowing the animal to sit on the substratum like an adult. The abdomen can also be extended posteriorly, allowing the megalops to swim swiftly by the beating of its strong pleopods. The megalops instar represents a transitional stage in the life of a crab, during which the animal swims about but settles from time to time, testing the bottom. The megalops molts to produce the first juvenile crab instar, which in nearly all crabs remains on the bottom and commences an adult type of existence. Molting occurs less frequently as crabs get larger and older. Some species cease molting on becoming sexually mature; others continue to molt throughout life.

Nearly 4,500 living species of crabs have been described. Some occur in brackish or fresh water, or in damp habitats on land; some occur in the ocean deeps, and others are pelagic in the open sea; still others live on or in the bodies of other organisms. However, the great majority of species are free-living and benthic, and inhabit the sea floor on continental shelves and slopes in comparatively shallow water. In terms of ecological diversity, they occupy a great variety of specific habitats and consume a great variety of foods. Body form varies along with habitat and way of life.

The intertidal Brachyura fall into three large groups: the spider crabs, the cancroid crabs, and the grapsoid crabs (Rathbun, 1918, 1925, 1930). Additional groups occur in deeper water off our coast (Rathbun, 1937). The carapace tends to be longer than broad in spider crabs, its shape pyriform (pear-shaped), lyrate (broadly triangular), or alate (winged). In other crabs the carapace tends to be broader than long: in cancroid crabs like *Cancer* it is broadly oval; in grapsoid crabs like *Pachygrapsus* it is rectangular or squarish. In spider crabs the frontal portion of the carapace develops into a prominent rostrum (beak) consisting of a single or a double horn; in other crabs the front is either straight and flat or many-toothed. Hooked hairs are present in spider crabs for the attachment of foreign objects; body hairs (setae) in other crabs, when present, are never of the hooked kind.

For both the serious student of crabs and the amateur beachcomber who enjoys the pastime of crab watching, California is ideally situated. Between Crescent City (Del Norte Co.) and Monterey one may find cold-water species that occur north to Alaska; between Santa Barbara and San Diego one may discover warm-water species that occur south to Bahía Magdalena (Baja California). Some crabs collectible at low tide in northern California are obtainable only by dredging in southern California, whereas crabs that occupy middle intertidal levels in California may dominate high intertidal levels in Puget Sound (Washington). A roll call of crab families illustrates the diversity to be found in the California intertidal region.

The spider crabs, family Majidae, include the decorators, such as *Loxorhynchus* (25.10, 25.11) and *Scyra* (25.9), which disguise themselves with bits of hydroids, sponges, and algae, and the kelp crabs *Pugettia* (25.4–7) and *Taliepus* (25.3), which are usually free of adornment.

Among cancroid crabs are the Portunidae, a tropical family ranging as far north as Santa Barbara. These swimming crabs may be recognized by the paddle-shaped last pair of legs. The family Cancridae is well represented, with nine species. A commercially important fishery is based on the Dungeness crab, *Cancer magister* (25.20), and several species of *Cancer* have been used in significant experimental studies. The pebble crabs, family Xanthidae, are abundant in the tropics; *Cycloxanthops* (25.23) and *Paraxanthias* (25.28) extend northward to Monterey, and some *Lophopanopeus* (e.g., 25.24) reach Alaska. An eastern xanthid, *Rhithropanopeus harrisii* (25.27), recently introduced into San Francisco Bay, has spread into even severely polluted areas.

Among grapsoid crabs, the family Grapsidae includes the shore crabs *Pachygrapsus* (25.43) and *Hemigrapsus* (25.44, 25.45), which dominate the higher intertidal region of rocky or muddy beaches. Semi-terrestrial and relatively hardy, they have been favorite experimental subjects for a variety of physiological studies (e.g., Bollenbacher, Borst & O'Conner, 1972; R. I. Smith & Rudy, 1972). The family Ocypodidae includes the fiddler crabs of the genus *Uca* (e.g., 25.46), whose ritualistic courtship patterns have been studied by animal behaviorists (Altevogt, 1957; Crane, 1957, 1975; Dembrowski, 1926), and whose color changes, sexual dimorphism, and acclimation are of interest to physiologists and ecologists (e.g., Fingerman, 1966; Valiela et al., 1974; Vernberg, 1969). The family Goneplacidae is represented by *Malacoplax* (25.30), usually found burrowing with *Uca* on intertidal mud flats, but ranging into the subtidal as well. Finally, the family Pinnotheridae embraces the pea crabs that usually live as commensals with other invertebrates (Cheng, 1967; Clark, 1956; Quayle, 1960). Free-living males of some genera bear little resemblance to females, which are animated egg sacs imprisoned within their unresisting hosts. Invertebrate species with which pinnotherids have been found as commensals are listed in Schmitt, McCain & Davidson (1973).

Crabs contribute to human welfare both directly as food for man (Frey, 1971) and indirectly as food for fishes and other animals (Quast, 1968). Crab claws excavated from kitchen middens testify to the importance of this food source to early California Indians (Hubbs, 1967). A single species of crab, *Cancer magister*, now accounts for 99 percent of the commercial catch in California waters, annually averaging 4,850 metric tons for the state, and over 16,000 metric tons for the west coast. The planktonic larvae of this species are consumed in vast quantities by such pelagic fishes as the herring, pilchard, and salmon; these in turn are fed upon by the tuna,

fur seal, and bear, respectively. Juvenile *Cancer productus* (25.22) are eaten in large numbers by the sculpin *Scorpaena guttata,* the kelp bass *Paralabrax clathratus,* and the sand bass *P. nebulifer.* The rock crab *Cancer antennarius* (25.16) is a delicacy esteemed by the sea otter of the Monterey coast, and adults of *Pachygrapsus crassipes* (25.43) are devoured by rats, raccoons, and sea gulls. Mink, ducks, and octopuses are crab predators in British Columbia. *Hemigrapsus oregonensis* (25.45) is stalked by the willet and other shorebirds, and is used by fishermen as bait for the pile perch in the Monterey area. *Loxorhynchus crispatus* (25.11) and *Paraxanthias taylori* (25.28) are important food sources for some fishes, notably the black croaker and the cabezon; *P. taylori* is eaten by the scorpion fish as well. *Lophopanopeus frontalis* (25.26), *Mimulus foliatus* (25.8), and *Pugettia* spp. have been found among stomach contents of the tidepool fish *Gibbonsia elegans* in the Los Angeles area.

Rathbun (1926), Menzies (1951), and more recently Zullo & Chivers (1969) and Nations (1970, 1975) have contributed to our knowledge of fossil crabs of the Pacific coast. Of these the Cancridae are best known because their heavy claws fossilize readily. Of living species *Cancer magister* (25.20) appeared in the Lower Pliocene, *C. antennarius* (25.16), *C. anthonyi* (25.17), *C. gracilis* (25.18), *C. oregonensis* (25.21), and *C. productus* (25.22) in the Middle Pliocene, *C. branneri* in the Upper Pliocene, and *C. jordani* (25.19) in the Lower Pleistocene. When these and other living species are found as fossils either north or south of their present range, we have the best possible evidence of climatic change, assuming that past and present ecological requirements are similar.

Warner (1977) provides an excellent modern introduction to the biology of crabs. For further general reading and reference the following are also recommended: for crustacean biology, Green (1961), Kaestner (1970), and Schmitt (1965); for crustacean physiology, Lockwood (1967) and Waterman (1960, 1961); for life history studies, Costlow & Bookhout (1960), Gurney (1960), Lebour (1928), and Wear (1970); for economic importance, Frey (1971) and Scheuerman (1958); for fossil crustaceans, Glaessner (1969); and for phylogeny and evolution, Whittington & Rolfe (1963), and Williamson (1974). Carlton & Kuris (1975) offers a key to species common in the intertidal zone in central California. Planktonic larvae found off the California coast may be identified to the family level using the key in Hart (1971a).

Phylum Arthropoda / Class Crustacea / Subclass Malacostraca
Superorder Eucarida / Order Decapoda / Suborder Reptantia
Section Brachyura / Family Majidae

25.1 **Pyromaia tuberculata tuberculata** (Lockington, 1877)
(=Inachoides tuberculatus)

Common under rocks in bays and on protected wharf pilings, low intertidal zone; subtidal to 411 m; Tomales Bay (Marin Co.) to Cabo Corrientes (Colombia); Japan; Pleistocene fossils from San Pedro (Los Angeles Co.) and San Diego Bay.

Carapace pyriform, tuberculate, to 17.7 mm wide in males and 15.1 mm in females; rostrum bearing a single spine; postorbital spine curving around eye; manus of cheliped inflated in male; walking legs long and slender; body and appendages often heavily overgrown with sponges and seaweeds, providing excellent camouflage; movements sluggish.

Although this crab may be collected intertidally in the southern part of the state, it is more properly considered a subtidal species; northern and central California collections have been made with a dredge or trawl, as in Tomales Bay (Marin Co.) and Monterey Bay.

See Garth (1958), Rathbun (1925, 1926), Ricketts & Calvin (1968), Sakai (1976), Schmitt (1921), and Willett (1937); the figure used in Schmitt and in Rathbun (1925) is of the Gulf of California subspecies, *P. tuberculata mexicana* (Rathbun).

Phylum Arthropoda / Class Crustacea / Subclass Malacostraca
Superorder Eucarida / Order Decapoda / Suborder Reptantia
Section Brachyura / Family Majidae

25.2 **Epialtoides hiltoni** (Rathbun, 1923)
(=Epialtus hiltoni)   Small Kelp Crab

Among holdfasts of eelgrass (*Zostera*) and surfgrass (*Phyllospadix*), low intertidal zone; Santa Catalina Island (Channel Islands) and Laguna Beach (Orange Co.) to Bahía Magdalena, also Isla Guadalupe, Islas San Benito, and Isla Cedros (Baja California).

Carapace width to 15.7 mm in males, to 9.6 mm in females; species distinguished from the young of *Pugettia* (25.4–7) or *Taliepus* (25.3) by the oblong rostrum with bilobed tip, the broad and advancing anterior marginal lobes, the prominent preorbital tooth, and the elongated manus of the cheliped in the male.

The discovery of this small kelp crab in an eelgrass holdfast in a cave at La Jolla (San Diego Co.) provided the clue that enabled others to obtain it in goodly numbers and to extend its known range to Mexico and offshore islands.

See Garth (1958) and Rathbun (1925).

Phylum Arthropoda / Class Crustacea / Subclass Malacostraca
Superorder Eucarida / Order Decapoda / Suborder Reptantia
Section Brachyura / Family Majidae

25.3 **Taliepus nuttallii** (Randall, 1839)
(=Epialtus nuttallii)   Southern Kelp Crab, Globose Kelp Crab

Clinging to larger brown algae, low intertidal zone on rocky shores, and among kelp beds offshore; subtidal to 93 m; Santa Barbara to Bahía Magdalena (Baja California); less abundant than formerly, owing to absence of large kelps; Pleistocene fossils from Playa del Rey (Los Angeles Co.).

Carapace to 92 mm wide in males, to half that in females; species differing from *Pugettia producta* (25.4) in having a more convex carapace and a more prominent rostrum with a small triangular notch at the tip, and in lacking the small tooth in front of the eye; color dark red-brown.

Rather than a replacement of *Pugettia producta* (25.4) south of Point Conception (Santa Barbara Co.), as suggested in Ricketts & Calvin (1968), *T. nuttallii* is the northern hemispheric analogue of the south-temperate *T. marginatus* (Bell, 1835) from north Chile and Peru, and as such defines a north-temperate region.

This herbivorous crab shows a marked preference for large brown algae, especially *Macrocystis* and *Egregia*.

See Garth (1955, 1958), Hinton (1969), Johnson & Snook (1927), Leighton (1966), North (1964), Rathbun (1925), Ricketts & Calvin (1968), Schmitt (1921), and Willett (1937).

Phylum Arthropoda / Class Crustacea / Subclass Malacostraca
Superorder Eucarida / Order Decapoda / Suborder Reptantia
Section Brachyura / Family Majidae

25.4 Pugettia producta (Randall, 1839)
(=*Epialtus productus*)   Shield-Backed Kelp Crab, Kelp Crab

Young animals common among rocks or on the brown alga *Egregia*, low intertidal zone on rocky shores of protected outer coast in winter, migrating to floating kelp (*Macrocystis*) in summer as they grow older; subtidal to 73 m; Prince of Wales Island (Alaska) to Punta Asunción (Baja California); Pleistocene fossils from Santa Monica and San Pedro (both Los Angeles Co.).

Carapace width to 93 mm in males, 78 mm in females; color reddish or olive-brown, mottled with small round spots of darker shade, frequently lighter in young; species distinguished from *Taliepus nuttallii* (25.3) by its depressed and alate margins.

Both this species and *T. nuttallii*, the other large kelp crab of the California coast, were once placed in the genus *Epialtus*, now restricted to the smaller kelp crabs.

Pugettia producta is mainly an herbivore on larger brown algae, aggregating in densities up to 27 per m² of kelp holdfast in the Monterey region. In California waters the preferred foods are *Macrocystis*, *Egregia*, and *Pterygophora*. In Puget Sound (Washington) the main summer diet is brown algae (*Fucus*, *Sargassum*, and *Nereocystis*); in winter, on pilings where the normal plant foods were not available, individuals changed to a carnivorous diet including barnacles, hydroids, and bryozoans. Brackish conditions are not tolerated; the crabs do not osmoregulate, and the body wall is several times more permeable than that of the common shore crabs *Pachygrapsus crassipes* (25.43) and *Hemigrapsus nudus* (25.44).

Breeding in California occurs all year. Females mate while hard-shelled, and at least half the females are found carrying eggs at all months of the year. Pairs have been observed to mate in a male-under-female position, while eggs were still attached to the pleopods of the female. In Monterey Bay, eggs are carried for 28–31 days before hatching. Females held in the laboratory typically deposited a new brood of eggs within 2 days after the hatching of the previous brood, and appear

capable of producing a new crop of offspring approximately every 30 days. In Puget Sound populations, brood size for females 41–56 mm in carapace width averaged about 61,000 (range 34,000–84,000). In Monterey Bay, 66 percent of the ovigerous females examined had egg masses infested with the small (1–2 mm long) nemertean worm *Carcinonemertes epialti*. About 3 percent of all crabs were parasitized by the highly modified barnacle *Heterosaccus californicus* (20.34), which is visible beneath the abdomen (see Fig. 20.34) and interferes with normal reproductive activity.

Additional studies have been conducted on the structure of the reproductive and nervous systems, the composition of the exoskeleton, and the physiology of respiration.

See Andrews (1945), Boolootian et al. (1959), Fasten (1915), Garth (1958), Gross (1957a), Hart (1940, 1968), Heath (1941), Hewatt (1938), Hines (1978), Hinton (1969), Huang & Giese (1958), Johnson & Snook (1927), Knudsen (1964a,b), Leighton (1966), MacGinitie (1935), MacGinitie & MacGinitie (1968), McLean (1962), North (1964), Rathbun (1925, 1926), Ricketts & Calvin (1968), Schmitt (1921), Thompson & Chow (1955), Weymouth et al. (1944), and Willett (1937).

Phylum Arthropoda / Class Crustacea / Subclass Malacostraca
Superorder Eucarida / Order Decapoda / Suborder Reptantia
Section Brachyura / Family Majidae

25.5 Pugettia gracilis Dana, 1851
Graceful Kelp Crab

In eelgrass and kelp, low intertidal zone on rocky shores; subtidal to at least 73 m depth; Attu Island (Aleutian Islands, Alaska) to Monterey Bay.

Carapace width to 39.2 mm in males, to 28.0 mm in females; species differing from *Pugettia richii* (25.6) in having the anterior marginal tooth broad and completely joined to the postorbital, in having the merus of the male cheliped with a superior (dorsal) crest, and in having the legs stout rather than slender.

Like the larger kelp crab *P. producta* (25.4), the graceful kelp crab keeps its carapace naked and smooth. It is active and has hooked and clawed legs, and does not disguise itself with bits of sponge, bryozoan, or algae, as do the more sluggish of the spider crabs. In places it is quite numerous, and up to 92 per m² of kelp holdfast have been reported in Monterey Bay.

Mating occurs as in *Pachygrapsus* (25.43) and *Hemigrapsus* (25.44, 25.45), with the male supine on his back and the female above, facing him. Good reproductive data are lacking, but ovigerous females have been taken almost throughout the year in Puget Sound (Washington) and through much of the year in Oregon and California. Number of eggs per brood, for five females of carapace width 20–25 mm, ranged from 6,200 to 13,300, and averaged 10,500.

Small crabs found in the stomachs of the tidepool fishes *Gibbonsia metzi* and *Clinocottus analis* in the Los Angeles area were probably *P. richii* (25.6), rather than this species as reported.

See Andrews (1945), Fasten (1915), Garth (1958), Hart (1940), Johnson & Snook (1927), Kundsen (1964b), Mitchell (1953), Rathbun (1925), Ricketts & Calvin (1968), and Schmitt (1921).

Phylum Arthropoda / Class Crustacea / Subclass Malacostraca
Superorder Eucarida / Order Decapoda / Suborder Reptantia
Section Brachyura / Family Majidae

25.6 **Pugettia richii** Dana, 1851

Under stones, low intertidal zone on rocky shores; subtidal to 97 m; Prince of Wales Island (Alaska) to Isla San Gerónimo (Baja California); Pleistocene fossils from Playa del Rey (Los Angeles Co.) and San Diego.

Carapace to 36.0 mm wide in males, to 26.5 mm wide in females, constricted at base of anterior marginal tooth, this tooth narrow and incompletely joined to postorbital tooth; merus of cheliped with a few tubercles but lacking the superior crest present in *P. gracilis* (25.5); color variable, red to brown, the slender legs usually with light and dark bands.

This is another species in which the body is often effectively masked by an overgrowth of hydroids, bryozoans, and algae. In places the crabs are very abundant: up to 245 per m² of kelp holdfast were reported from Monterey Bay. Reproductive data are sketchy, but ovigerous females have been noted in Monterey Bay throughout the year, and in the Los Angeles area in September.

Small individuals found in the stomachs of the tidepool fishes *Gibbonsia metzi* and *Clinocottus analis* in the Los Angeles area were probably this species, rather than *P. gracilis* (25.5) as reported.

See Andrews (1945), Garth (1958), Hart (1940, 1953, 1962, 1968), Hewatt (1938), Hines (1978), Johnson & Snook (1927), McLean (1962), Mitchell (1953), Rathbun (1925, 1926), Schmitt (1921), Turner, Ebert & Given (1969), and Willett (1937).

Phylum Arthropoda / Class Crustacea / Subclass Malacostraca
Superorder Eucarida / Order Decapoda / Suborder Reptantia
Section Brachyura / Family Majidae

25.7 **Pugettia dalli** Rathbun, 1893

Among holdfasts of various marine plants and sessile animals, low intertidal zone on rocky shores, and on harbor floats; subtidal to 117 m; San Miguel Island (Channel Islands) to Cabo Thurloe (Baja California).

Carapace to 13.8 mm wide in males, to 10.3 mm wide in females, covered with minute vesicles; anterior marginal projection a transverse spine, the postorbital projection an ovate lobe inclined at a right angle to it; propodus of male cheliped crested, carpus doubly so; walking legs slender; color varying with algal surroundings.

Within its range this crab is distinguished from *P. richii* (25.6) by its slender walking legs and transverse anterior marginal spine. The body is frequently clothed with calcareous algae and the bryozoan *Holoporella*. The species is known from all of the Channel Islands and from numerous mainland localities from Los Angeles Co. southward. At Santa Catalina Island it is found in holdfasts of the kelp *Eisenia* and in clumps of the red algae *Lithothrix* and *Liagora*. At Redondo Beach it is common among hydroids on floating docks and on pilings, and at Corona del Mar it occurs on the surfgrass *Phyllospadix*.

Ovigerous females may be found in any month, but most abundantly in August. The young have proportionately longer rostral spines than do the adults.

See Garth (1958), Johnson & Snook (1927), Rathbun (1925), and Schmitt (1921).

Phylum Arthropoda / Class Crustacea / Subclass Malacostraca
Superorder Eucarida / Order Decapoda / Suborder Reptantia
Section Brachyura / Family Majidae

25.8 **Mimulus foliatus** Stimpson, 1860

Uncommon under rocks, low intertidal zone on protected outer coast; common on kelp (*Macrocystis*) holdfasts in subtidal waters offshore and ranging to depths of 128 m; Captain's Bay (Unalaska, Alaska) to Point Arguello (Santa Barbara Co.) and Santa Cruz Island (Channel Islands), and occasionally San Diego.

Carapace to 39 mm wide in males, 32.4 mm wide in females, the lateral expansions leaflike and overlapping at lateral fissures, the two rostral horns short, narrowly separated; predominant color red or red-brown, lighter on chelae than on carapace, with a white "V" and red-and-white striped legs (neither shown in photograph).

Adults may have a covering growth of bryozoans or sponges. In the kelp forests off Pacific Grove in Monterey Bay, populations of up to 73 crabs per m² of kelp holdfast have been reported. Individuals are more abundant among the kelp in summer and fall than in late winter, but ovigerous females occur in the population throughout the year. *Macrocystis* is a major element in the diet.

Specimens of *M. foliatus* taken from the stomach of the tidepool fish *Gibbonsia metzi* in the Los Angeles area and from the surface of a rock cod in 9 m of water near Point Loma (San Diego Co.; unpublished record) confirm the presence of this cold-water species in regions of upwelling well south of its normal range.

See Andrews (1945), Garth (1958), Hart (1940), Hewatt (1938), Hines (1978), Johnson & Snook (1927), McLean (1962), Mitchell (1953), Rathbun (1925), Ricketts & Calvin (1968), and Schmitt (1921).

Phylum Arthropoda / Class Crustacea / Subclass Malacostraca
Superorder Eucarida / Order Decapoda / Suborder Reptantia
Section Brachyura / Family Majidae

25.9 **Scyra acutifrons** Dana, 1851
Sharp-Nosed Crab

Among algae and associated sessile animals, on and under rocks, low intertidal zone on protected outer coast; also among fouling growth on pilings; subtidal to 91 m; Kachemak Bay (Cook Inlet, Alaska) to Punta San Carlos (Baja California); Pleistocene fossils from San Pedro (Los Angeles Co.).

Carapace to 37.7 mm wide in males, 30 mm in females, pyriform, tuberculate but not spinose, the rostrum composed of two flattened horns; chelipeds of male enlarged, the propodus compressed and crested.

The carapace of larger specimens is usually covered with an encrustation of sponges, tunicates, barnacles, bryozoans, and hydroids. Although not so efficient a masker as *Loxorhynchus crispatus* (25.11), *S. acutifrons* is more truly an intertidal crab than the larger species. Like other spider crabs, it usually sits with the anterior end pointed down.

Ovigerous females may be found almost any month of the year. In females 19.5–30 mm across the carapace, the number of eggs carried ranged from 2,700 to 16,300, averaging 8,600. A second batch of eggs may be produced shortly after the first has hatched.

See Garth (1958), Hart (1940), Hines (1978), Johnson & Snook (1927), Knudsen (1964b), McLean (1962), Rathbun (1925, 1926), Ricketts & Calvin (1968), Schmitt (1921), and Turner, Ebert & Given (1969).

Phylum Arthropoda / Class Crustacea / Subclass Malacostraca
Superorder Eucarida / Order Decapoda / Suborder Reptantia
Section Brachyura / Family Majidae

25.10 **Loxorhynchus grandis** Stimpson, 1857
Sheep Crab

Occasional, low intertidal zone (as at Hedionda Lagoon, San Diego Co.); characteristically subtidal to depth of 124 m; Cordell Bank (Marin Co.) to Punta San Bartolomé (Baja California); Upper Oligocene fossils from Washington; Pliocene fossils from central California; Pleistocene fossils from Santa Monica and Playa del Rey (both Los Angeles Co.).

Carapace to 159 mm wide in males, 115 mm wide in females, inflated, ovate, sparsely tuberculate but with two anterior marginal tubercles; rostrum strongly deflexed; chelipeds of males greatly enlarged, especially in old individuals.

Because of its large size, peculiar shape, and deliberate movements, the sheep crab presents a ludicrous appearance. The young mask themselves with living barnacles, bryozoans, hydroids, and algae and tend to remain in environments where they match the background. The adults have less need for concealment; they lose the instinct to decorate themselves, and divers report seeing them walking on open, sandy, subtidal bottoms. The animals are carnivores and scavengers; hungry crabs in aquaria have been observed to take apart and feed on starfishes, clams, and even octopuses.

Mating has been noted in the spring and early summer off southern California.

The polychaete *Iphitime loxorhynchi* is found in the branchial cavity.

See Garth (1958), Hartman (1952), Johnson & Snook (1927), MacGinitie (1937), MacGinitie & MacGinitie (1968), Nations (1975), Rathbun (1925, 1926), Schmitt (1921), Turner, Ebert & Given (1969), Wicksten (1979b), and Willett (1937).

Phylum Arthropoda / Class Crustacea / Subclass Malacostraca
Superorder Eucarida / Order Decapoda / Suborder Reptantia
Section Brachyura / Family Majidae

25.11 Loxorhynchus crispatus Stimpson, 1857
Masking Crab, Moss Crab

Scarce in low intertidal zone on rocky shores of protected outer coast, much more common on subtidal pilings, kelp holdfasts, and rocks to 183 m depth; Redding Rock (Humboldt Co.) to Isla Natividad (Baja California)—unpublished records; Pleistocene fossils from Santa Monica and San Pedro (both Los Angeles Co.).

Carapace width to 88 mm in males, 68 mm in females; spe-

cies distinguished from *L. grandis* (25.10) by: (1) smaller size, (2) a carapace with fewer tubercles and only one anterior marginal tubercle, (3) a rostrum not strongly deflexed, and (4) a plushlike coat of short, thick hairs; chelipeds much longer in males than in females, to 335 mm long in old males; walking legs short for a spider crab.

These slow-moving crabs are generally found so thickly covered with foreign growth, such as hydroids, anemones, seaweeds, bryozoans, and sponges, that in their natural environment they are scarcely recognizable as crabs at all. Crabs experimentally denuded quickly redecorate themselves, using scraps of whatever is available. Masking materials are attached by wedging them among strong hooked setae provided with spines or barbs. Crabs with these setae removed are unable to decorate themselves until the setae are regenerated, at the next molt. Secretions from glands on the mouthparts play no role in attachment of masking materials.

Large individuals are often found clinging, head down, on vertical walls and pilings. They eat a wide variety of materials including algae, sponges, small crustaceans, erect bryozoan colonies, and several other invertebrates. Under aquarium conditions the purple sea urchin, *Strongylocentrotus purpuratus* (11.5), is torn apart and consumed. Sessile or sedentary prey are preferred. The crabs, in turn, form an important food for some fishes, including the black croaker and the cabezon.

See Garth (1958), Haderlie (1968, 1969), Haig & Wicksten (1975), Hines (1978), Holmes (1900), Johnson & Snook (1927), Limbaugh (1961), McLean (1962), O'Connell (1953), Rathbun (1925, 1926), Schmitt (1921), Turner, Ebert & Given (1969), and Wicksten (1975, 1977a,b, 1978, 1979b).

Phylum Arthropoda / Class Crustacea / Subclass Malacostraca
Suborder Eucarida / Order Decapoda / Suborder Reptantia
Section Brachyura / Family Majidae

25.12 Pelia tumida (Lockington, 1877)
Dwarf Crab

Under stones, low intertidal zone on rocky shores; subtidal to 100 m; Monterey Bay to Bahía de Petatlán (Mexico), including Gulf of California.

Carapace small, to 14.5 mm wide in males, 13 mm in females, pear-shaped, covered with pubescence to which sponges and other foreign particles adhere; rostrum bifid; walking legs compressed, bearing stiff setae on margins.

When turned on its back, the dwarf crab usually remains motionless with legs upcurved, almost indistinguishable from its surroundings. When placed in an aquarium and deprived of its normal covering, it decorates itself with any material available. A common inhabitant of kelp holdfasts, up to 65 per m² of *Macrocystis* holdfasts were reported in Monterey Bay.

See Andrews (1945), Garth (1958), Johnson & Snook (1927), Rathbun (1925), and Schmitt (1921).

Phylum Arthropoda / Class Crustacea / Subclass Malacostraca Superorder Eucarida / Order Decapoda / Suborder Reptantia Section Brachyura / Family Majidae

25.13 **Herbstia parvifrons** Randall, 1839
Spider Crab

Infrequent, under stones, low intertidal zone; occasionally dredged to 73 m; Monterey Bay to Bahía Magdalena (Baja California).

Carapace to 41.6 mm wide in males, 17.1 mm wide in females, flattened, ovate, usually overgrown with sponges; rostrum short; legs long, spiny; color (not shown well in the photograph) light tan to mottled brown, the chelipeds a more pronounced red with tips of fingers white, the walking legs banded with reddish brown.

This species, not likely to be confused with any other spider crab within its range, is distinguished from *H. camptacantha* of the Gulf of California by having two teeth between the inner and outer orbital teeth, instead of only one.

See Garth (1958), Rathbun (1925), Schmitt (1921), and Weymouth (1910).

Phylum Arthropoda / Class Crustacea / Subclass Malacostraca Superorder Eucarida / Order Decapoda / Suborder Reptantia Section Brachyura / Family Portunidae

25.14 **Portunus xantusii xantusii** (Stimpson, 1860)
Swimming Crab

On sand flats, low intertidal zone; in beds of eelgrass (*Zostera*) in Mission Bay (San Diego Co.); subtidal to 179 m; occasionally found swimming at water surface, attracted to ship's lights at night; Santa Barbara to Topolobampo (Mexico); not likely to be found intertidally north of Newport Bay (Orange Co.); Pleistocene fossils from Playa del Rey (Los Angeles Co.) and San Diego.

Carapace broad, flat, finely pubescent, each lateral angle armed with a long spine; body width (including spines) to 70.7 mm in males, 73.1 mm in females; last pair of legs paddlelike, each merus armed with spinules but no spine.

These crabs swim by sculling with the flattened last pair of legs. Swimming, like walking, is sideways, with one "elbow" leading and the cheliped and legs of the opposite side trailing. Predation of *Portunus* on the mole crab *Emerita analoga* (24.4) has been described by the MacGinities. A portunid moving across a bed of *Emerita* pauses frequently to push one of the large claws straight down into the sand. If an *Emerita* is found and grasped, *Portunus* starts running in a circle about the submerged claw while pulling it upward till the prey breaks the sand surface.

Ovigerous females of *Portunus* are found from May through September.

This crab has close relatives to the south. A subspecies, *P. xantusii minimus* Rathbun, occurs in the Gulf of California, and another subspecies, *P. xantusii affinis* (Faxon), extends south to Ecuador and Peru.

See Ally (1974), Garth & Stephenson (1966), Glassell (1935), Johnson & Snook (1927), MacGinitie & MacGinitie (1968), Rathbun (1926, 1930), Ricketts & Calvin (1968), Schmitt (1921), and Willett (1937).

Phylum Arthropoda / Class Crustacea / Subclass Malacostraca
Superorder Eucarida / Order Decapoda / Suborder Reptantia
Section Brachyura / Family Portunidae

25.15 Callinectes arcuatus Ordway, 1863
Swimming Crab

On sand and mud bottoms of lagoons and estuaries, often seined at mouths of shallow sloughs such as Anaheim Slough (Orange Co.), low intertidal zone to 27 m depth; Los Angeles harbor to Peru, including Gulf of California.

Carapace width, including lateral spines, to 120.5 mm in males, 98.5 mm in females; species differing from *Portunus xantusii* (25.14) in having the male abdomen in the shape of an inverted letter "T" and in lacking the inner spine on the wrist of the cheliped.

Along the California coast, *C. arcuatus* could be confused only with its congener **C. bellicosus** (Stimpson), which has been reported from San Diego and Point Loma (San Diego Co.) and as a Pleistocene fossil from Playa del Rey and Long Beach (both Los Angeles Co.). Both species are related to the edible blue crab of the east coast, *C. sapidus*, which is marketed in the soft-shelled condition, and they are said to be as savory.

Ovigerous females have been found from March to September.

See Garth & Stephenson (1966), Rathbun (1926, 1930), and Willett (1937).

Phylum Arthropoda / Class Crustacea / Subclass Malacostraca
Superorder Eucarida / Order Decapoda / Suborder Reptantia
Section Brachyura / Family Cancridae

25.16 Cancer antennarius Stimpson, 1856
Rock Crab

Common in some areas, low intertidal region on rocky shores; subtidal around bases of kelp and on gravel bottoms to 40 m depth; Coos Bay, Oregon (unpublished record) to Baja California, including Islas de Todos Santos; probably not British Columbia as reported; Pliocene fossils from central and southern California; Pleistocene fossils from Playa del Rey and San Pedro (both Los Angeles Co.).

Carapace to 151 mm wide in males, to 144 mm wide in females (J. C. Carroll, pers. comm.); juveniles pubescent, adults with only a fringe of hair on the lower surface, abdomen, and walking legs; species readily recognized by the red speckling on the lower surface, especially anteriorly.

Considered by some to be as tasty as the Dungeness crab, *C. antennarius* supports only a very limited sport fishery. The animals are caught by hand or with a baited hoop net. Crabs 4 inches (10 cm) or more across the carapace may be taken at any season; the bag limit for this and all other members of the genus *Cancer* except the Dungeness crab (*C. magister*, 25.20) is 35 crabs of all species combined (1979 State Department of Fish and Game regulations, subject to change).

The rock crab is mainly an inhabitant of the outer coast. Although it may invade sloughs, it does not tolerate brackish conditions and cannot osmoregulate. It is both a scavenger and an active predator, eating a wide variety of other animals. It captures hermit crabs by walking over them and sitting on them, or several may be flicked under the body by the walking legs and held there in a cage formed by the body and appendages. A crab in its shell is then removed from the cage, held in both chelae, and inspected. The fingers of both large claws are then inserted into the aperture of the hermit crab's shell, and the shell cracked away a bit at a time until the hermit crab can withdraw no further and is removed and eaten. The rock crab itself is often eaten by the California sea otter in the Monterey area. The polychaete *Iphitime holobranchiata* lives in the branchial cavity.

Mating occurs after female crabs molt and while they are still soft-shelled. Ripe females ready to molt release a substance (pheromone) in the urine, which attracts the male and stimulates courtship behavior. Ovigerous females are most common from November to January, but a few may be found at other seasons in southern California. The first zoeal stage has been described.

See Case (1964), Cook (1965), Frey (1971), Ghiradella, Cronshaw & Case (1970), Giese (1966), Gross (1957a), Johnson & Snook (1927), Jones (1941), Kittredge, Terry & Takahashi (1971), MacGinitie (1935), McLean (1962), Mir (1961), Nations (1975), Phillips (1939), Pilger (1971), Rathbun (1930), Ricketts & Calvin (1968), Schmitt (1921), Thompson & Chow (1955), Turner, Ebert & Given (1969), and Willett (1937).

Phylum Arthropoda / Class Crustacea / Subclass Malacostraca
Superorder Eucarida / Order Decapoda / Suborder Reptantia
Section Brachyura / Family Cancridae

25.17 **Cancer anthonyi** Rathbun, 1897
Yellow Crab

In rocky areas of bays and estuaries, low intertidal zone; subtidal to 132 m; Humboldt Bay to Bahía Magdalena (Baja California); uncommon north of San Pedro (Los Angeles Co.); Pliocene and Pleistocene fossils from central California; Pleistocene fossils also from San Pedro and San Diego.

Carapace to 152 mm wide in males, to 98.7 mm wide in females; carpus of cheliped bearing only one spine, no spines on hand; black color of fingers extending less than half the length of outer margins; underparts uniformly light, not red-spotted as in *C. antennarius* (25.16).

This is an edible crab frequently taken in southern California by sport fishermen using baited hoop nets; size, season, and bag limits are the same as those listed under *C. antennarius*. The yellow crab is especially abundant on artificial reefs off southern California in April and May, and in Elkhorn Slough (Monterey Co.) in June and July.

Mating, which occurs immediately after the females molt, has been observed in June. Females ripe and ready to molt emit substances (pheromones) that attract males and stimulate courtship behavior. The first zoeal stage has been described.

Divers off southern California noted a yellow crab actively removing ectoparasites from a sand bass 46 cm long. Juvenile crabs are eaten by a variety of reef fishes.

A smaller species, **C. amphioetus** Rathbun, has been taken intertidally at Newport Bay and Anaheim Landing (both Orange Co.). It occurs on sandy and muddy bottoms of shallow bays and in deeper water on bottoms of rock, pebbles, sand, and broken shells. Commonest subtidally, it ranges to depths of 148 m in American waters and to 309 m in Japanese waters. It is known from El Segundo, Los Angeles Co. (unpublished record) to Bahía Magdalena (Baja California) and the Gulf of California, and from Japan, Korea, and northern China; Pleistocene fossils are found at San Pedro (Los Angeles Co.) It is distinguished by having a carapace that is relatively small (width to 29.5 mm in American specimens, to 41.4 mm in Japanese specimens), not hairy, strongly areolated (especially in the female), with marginal teeth broadly triangular and moderately produced. It differs from *Cancer anthonyi* in having the edges of the lateral teeth non-serrated and the median tooth of the front blunted.

For *C. anthonyi*, see Carlisle (1969), Frey (1971), Johnson & Snook (1927), Kittredge, Terry & Takahashi (1971), MacGinitie (1935), Mir (1961), Nations (1975), Phillips (1939), Rathbun (1926, 1930), Schmitt (1921), Turner, Ebert & Given (1969), and Willis (1968). For *C. amphioetus*, see Johnson & Snook (1927), Kim (1973), Nations (1975), Rathbun (1904, 1930), Sakai (1965, 1976), Schmitt (1921), and Shen (1932).

Phylum Arthropoda / Class Crustacea / Subclass Malacostraca
Superorder Eucarida / Order Decapoda / Suborder Reptantia
Section Brachyura / Family Cancridae

25.18 **Cancer gracilis** Dana, 1852
Slender Crab, Graceful Crab

Sometimes common on mud flats and in beds of eelgrass (*Zostera*), low intertidal zone in bays; subtidal to 174 m offshore; Prince William Sound (Alaska) to Bahía Playa María (Baja California); usually not taken intertidally south of central California; Pliocene fossils from Los Angeles; Pleistocene fossils from Santa Monica (Los Angeles Co.) to San Diego.

Carapace to 91 mm wide in males, to 64 mm wide in females; species distinguished from *C. magister* (25.20) by its smaller size, more convex carapace, and absence of tubercles upon its surface; walking legs slender and graceful, suggesting the specific name.

Although seasonally found in bays, the slender crab does not tolerate brackish conditions; its body wall is relatively permeable to water and salts, and it does not osmoregulate. It feeds mainly on animal remains, and on barnacles when available. Young crabs (carapace 8–35 mm wide) are a major food of the starry flounder, *Platichthys stellatus*, in Monterey Bay. Some adults are taken for human consumption. California sport fishing regulations (1979, and subject to change) are the same as for *C. antennarius* (25.16).

In Elkhorn Slough (Monterey Co.), mating is common in November, and ovigerous females were noted in July and August. Farther north, in Puget Sound (Washington), animals held in the laboratory bore eggs from December to April, and a few females produced a small second brood. The males remain with the females after mating, and appear to protect them. Larval stages are planktonic, as in other crabs, and the megalops larvae and juvenile crabs are frequently found crawling unharmed on and under the bells, and even in the stomachs, of the larger jellyfishes, especially *Pelagia colorata* (3.21).

See Gross (1957a), Hart (1968), Johnson & Snook (1927), Knudsen (1964b), MacGinitie (1935), Nations (1975), Orcutt (1950), Phillips (1939), Rathbun (1926, 1930), Schmitt (1921), Turner, Ebert & Given (1969), and Weymouth (1910).

Phylum Arthropoda / Class Crustacea / Subclass Malacostraca Superorder Eucarida / Order Decapoda / Suborder Reptantia Section Brachyura / Family Cancridae

25.19 Cancer jordani Rathbun, 1900
Hairy Cancer Crab

Under rocks, low intertidal zone; subtidal to 104 m; Coos Bay (Oregon; unpublished record) to Cabo Thurloe (Baja California); Pleistocene fossils from San Pedro (Los Angeles Co.).

A small species of *Cancer* with a hairy carapace to 33.4 mm wide in males and 19.5 mm wide in females (the only small hairy *Cancer* besides *C. branneri*, see below); anterolateral carapace margin with sharp teeth, alternately large and small, the tenth (most posterior) tooth inconspicuous; carpus of cheliped bearing two spines at distal end.

Up to 78 *C. jordani* were recorded per m² of kelp holdfast in Monterey Bay. The natural history of this crab is poorly known, but ovigerous females have been noted in Monterey Bay in October and November.

A similar species, **C. branneri** Rathbun, ranges from Alaska to Isla Cedros, Baja California (unpublished record), and is known from Pliocene fossils at Santa Barbara and Los Angeles and from Pleistocene fossils at Tomales Bay (Marin Co.) and San Pedro (Los Angeles Co.). *C. branneri* differs from *C.*

jordani in having the tenth (most posterior) of the teeth on the anterolateral carapace margin spiniform, in having the middle orbital tooth resembling the two outer teeth but less advanced, and in having a markedly areolated carapace.

See Andrews (1945), Garth (1961), Hewatt (1938), Johnson & Snook (1927), MacGinitie (1935), McLean (1962), Nations (1975), Rathbun (1930), Reish & Winter (1954), and Schmitt (1921).

Phylum Arthropoda / Class Crustacea / Subclass Malacostraca Superorder Eucarida / Order Decapoda / Suborder Reptantia Section Brachyura / Family Cancridae

25.20 Cancer magister Dana, 1852
Dungeness Crab, Market Crab, Common Edible Crab

Sometimes in sandy pools between low intertidal rock outcroppings, but more commonly from deeper sandy bottoms; subtidal to 230 m, but not abundant below 90 m; Tanaga Island (Aleutian Islands, Alaska) to Pismo Beach, San Luis Obispo Co. (unpublished record) and rarely to Santa Barbara (unpublished record), but probably not Bahía Magdalena (Baja California) as reported; Pliocene fossils from Coos Bay (Oregon) to Ventura and Pleistocene fossils from Cape Blanco (Oregon) to San Pedro (Los Angeles Co.).

Carapace to 230 mm wide in males, to about 165 mm wide in females, granular, its anterolateral margins each bearing ten small teeth or denticles, the posterior one the largest.

This crab, the largest edible true crab on the west coast, supports a large fishery from California to British Columbia. Annual catch for the whole coast averages over 16 million kg (35 million lbs), with an annual price to the fishermen of about $5.5 million. This species alone accounts for more than 99 percent of all crab species taken commercially in California. In California waters there appear to be five subpopulations, which show little or no mixing, centering in the following areas: (1) Avila/Morro Bay (San Luis Obispo Co.), (2) Monterey Bay, (3) San Francisco, (4) the region from Fort Bragg (Mendocino Co.) to Cape Mendocino (Humboldt Co.), and (5) the Eureka/Crescent City region, north of Cape Mendocino (Humboldt and Del Norte Cos.). The three southern areas now yield few crabs, and the bulk of California's catch comes from Fort Bragg northward to the Oregon border. Crabs are taken in traps, mainly at 10–45 m depth; the California catch alone has averaged over 4.8 million kg (about 10.7 million lbs) annually over the past 25 years, but the catch fluctuates greatly from year to year. Both commercial and sport fishing for the species is strictly controlled. According to the 1979 sport fishing regulations of the California Department of Fish and Game (subject to change), the bag limit is 10 crabs, and only male animals measuring 6.25 inches (16 cm) or more across the carapace may be taken; south of Mendocino Co. the fishing season is from the second Tuesday in November through June 30, and in Mendocino Co. and northward it is from December 1 through July 30. No crabs may be taken in San Francisco and San Pablo Bays.

The crabs are carnivores, and 40 different identifiable food items have been found in their stomachs. Smaller crustaceans of all sorts are favored foods; small clams and oysters are also eaten, the shells being crushed or opened by chipping away the margins with the strong claws. Some worms are taken, and even some fishes caught with the chelipeds. Dead animal flesh is eaten only if fresh and unspoiled. On a soft bottom the crabs often bury themselves with only their eyes and antennae exposed. Water is circulated through the gill chamber even while the crabs are buried; normally water enters from below and passes out anteriorly, but currents can be reversed to flush the gill chamber. Gill surface area is large, about 877 cm² in a crab 11.8 cm across the carapace.

Mating occurs from April to September in British Columbia, and may begin earlier in California. Female crabs may store viable sperm for several months before use. California females spawn from September to December, and most adult females are found with eggs from November through February. The smallest females with eggs are about 10 cm across the carapace. The number of eggs carried in a brood varies with the size of the female, ranging from 700,000 to about 2.5 million. Hatching starts in December and reaches a peak in March. Molting of the adults follows reproduction, from May to September, with a peak in the summer.

The eggs hatch as protozoea larvae, a stage lasting only a few minutes, then pass through five successive zoeal stages to a megalops stage. The larval period, from hatching through the megalops phase, lasts 3–5 months, depending on temperature, salinity, and other factors. These larval stages are pelagic, and tend to stay in the upper 20 m of the sea; enormous numbers are eaten, especially by such fishes as the herring, pilchard, and salmon.

Metamorphosis of the megalops larvae into juvenile crabs occurs mainly from April through June off California. The young crabs take up residence on the bottom, where many more are consumed by bottom predators. At this stage males and females can be told apart by differences in primary sex organs. Thereafter it takes about 1.5 years and about 11 molts to reach sexual maturity (at a carapace width of about 10 cm). Males reach legal size at an age of 3.5–4 years in California waters, and after sexual maturity both males and females molt only once a year. The largest animals, about 23 cm across, are probably about 6 years old in California, and perhaps 1–2 years older in more northerly waters. Hermaphroditism has been reported only once.

The megalops larvae of *Cancer magister* often attach to the bells of pelagic jellyfishes and to the tentacles of the chondrophoran *Velella velella* (3.18). Off Bodega Bay (Sonoma Co.) and outside San Francisco Bay, up to 88 percent of the *Velella* carried from one to three megalopa, and the guts of the crab larvae were full of undischarged *Velella* nematocysts.

Recent studies on *C. magister*, largely carried out or sponsored by the California Department of Fish and Game, include investigations of the yearly migrations of tagged crabs, the life history of the crab under laboratory and field conditions, the distribution and behavior of the pelagic larval stages, and the possible causes of recent declines in commercial catches. It has been discovered that in general, from Fort Bragg northward, male crabs in most areas move farther offshore from November to March and closer inshore from March through June. North of Cape Mendocino there are population shifts northward or southward with season, but the movements differ in different areas. Causes of fluctuations in the crab populations in nature are difficult to pinpoint, but among those suggested and investigated are (1) the effects of pesticides, sewage, and industrial pollutants on the mortality of larval and adult crabs and on crab reproduction, (2) predation by a nemertean worm, *Carcinonemertes errantia*, on the fertilized eggs and embryos carried on the pleopods by brooding female crabs, and (3) predation on the pelagic larval stages of the crab by such fishes as the coho salmon, recently successfully introduced into rivers not far from the fishery area. *Velella* may serve to transport crab larvae to the near-shore habitats of juvenile Dungeness crabs, and the presence of the hydrozoan in coastal waters may also contribute to the fluctuations in abundance.

Fossil remains of species of *Cancer* first appear in the Eocene epoch, about 60 million years ago. The species known today are confined mainly to temperate seas where annual mean temperatures range between 4° and 24°C (40–75°F).

See Butler (1967) for a bibliography of 131 papers dealing with various aspects of this crab; sources for the present account, some not listed in Butler's bibliography, include Buchanan & Millemann (1969), Fasten (1915), Frey (1971), Giese (1966), Gotshall (1977, 1978a,b,c), Johnson & Snook (1927), Jones (1941), MacGinitie (1935), MacKay (1934, 1937, 1942, 1943a,b), Mir (1961), Nations (1975), Phillips (1935, 1939), Poole (1966a, 1967), Rathbun (1926, 1930), Reed (1969), Ricketts & Calvin (1968), Scheer & Meenakshi (1961), Scheffer (1959), Schmitt (1921), M. Smith (1963), Snow & Nielsen (1966), Snow & Wagner (1965), Thompson & Chow (1955), Waldron (1958), Welsh (1974), Weymouth (1915, 1916), Weymouth & MacKay (1934), Wickham (1977, 1979), Wickham, Schleser & Schuur (1976), and Willis (1971).

Phylum Arthropoda / Class Crustacea / Subclass Malacostraca
Superorder Eucarida / Order Decapoda / Suborder Reptantia
Section Brachyura / Family Cancridae

25.21 **Cancer oregonensis** (Dana, 1852)
Oregon Cancer Crab

Occasionally found under rocks or wedged in crevices and holes, low intertidal zone on rocky shores, also on pilings heavily encrusted with barnacles and mussels; more abundant subtidally, to 435 m depth; St. George Island (Pribilof Islands, Alaska) to Palos Verdes (Los Angeles Co.): probably not Bahía Magdalena (Baja California) as reported; Pliocene fossils from Trinidad (Humboldt Co.) to San Diego; Pleistocene fossils from Oregon to San Pedro (Los Angeles Co.).

Carapace to 31.9 mm wide in males, 47.1 mm wide in females, widest opposite the seventh or eighth tooth, differing

from the carapace in all other *Cancer* species in not having the anterolateral and posterolateral margins meeting at an angle, thus body appears more rounded when viewed from above; walking legs hairy; body dark red above, lighter below; chelipeds with dark-colored claws.

Biological studies have been carried out mainly on populations from Puget Sound (Washington). The crabs are primarily carnivores. Barnacles, the staple food when available, are crushed with the chelipeds. Polychaete worms and smaller crustaceans are also taken, along with a few scraps (possibly incidental) of smaller green algae.

Courtship, molting, and mating occur in late spring and summer, mainly from April to June. The male grasps and carries a female for several days before she molts, and mating occurs after the molting. Courting pairs conceal themselves well, and the male remains with his mate until her shell has hardened enough for normal activity. Sperm are stored by the female until the eggs are laid, and ovigerous females are seen mainly from November to February. One brood a year seems normal, but occasional ovigerous females found from April to June are probably carrying a second brood. The number of eggs borne by ovigerous females of carapace width 17–26 mm ranged from 10,000 to 33,000, averaging 20,540.

See especially Knudsen (1964a); see also Hart (1940), Johnson & Snook (1927), Nations (1975), Rathbun (1930), Ricketts & Calvin (1968), Schmitt (1921), and Wicksten (1979a).

Phylum Arthropoda / Class Crustacea / Subclass Malacostraca
Superorder Eucarida / Order Decapoda / Suborder Reptantia
Section Brachyura / Family Cancridae

25.22 **Cancer productus** Randall, 1839
Red Crab

Often half buried in sandy substratum under rocks, low intertidal zone, primarily in bays and estuaries, but also on protected rocky coasts; subtidal to 79 m; Kodiak Island (Alaska) to San Diego; probably not Bahía Magdalena (Baja California) as reported; Pliocene fossils from Trinidad (Humboldt Co.) to San Ysidro (Baja California); Pleistocene fossils from Cape Blanco (Oregon) to Punta Descanso (Baja California).

Carapace to 158 mm wide in both sexes, smooth, the margins serrate, the anterolateral teeth closely spaced; front produced beyond orbital angles as five equally spaced teeth; adults usually with upper surface dark red, lower surface yellowish white; young pure white or exhibiting a striking variety of color patterns including bands of brown and white, stripes of red and white, and brown stripes (as shown in Figs. 25.22c–h).

According to Rathbun (1930), many specimens formerly referred to the young of the common species *C. magister* (25.20), *C. productus*, and *C. antennarius* (25.16) were later found to belong to the more recently defined species *C. anthonyi* (25.17), *C. jordani* (25.19), and *C. branneri*. For this reason, reported occurrences prior to 1900 are questionable unless specimens are available for verification.

Like other *Cancer* species, *C. productus* is a carnivore, and any animal matter available is eaten. In Puget Sound (Washington), barnacles are crushed with the large claw and passed to the mouth. Small living crabs and dead fishes are also eaten.

Mating occurs when females are soft-shelled, especially in the summer months. Ovigerous females occur from October to June (especially December to March) in Puget Sound, and from January to August (especially April to June) in southern California. The larvae have been reared in the laboratory and are similar to those of *C. magister*. Juvenile crabs, common in southern California in the fall and winter, are eaten in large numbers by the sculpin *Scorpaena guttata*, the sand bass *Paralabrax nebulifer*, and the kelp bass *P. clathratus*. Adults are taken in small numbers by fishermen for food. Under 1979 sport fishing regulations there is no closed season; size and bag limits are the same as those listed under *C. antennarius* (25.16).

See Case (1964), Fasten (1915), Frey (1971), Ghiradella, Cronshaw & Case (1970), Hart (1940), Johnson & Snook (1927), Knudsen (1964b), MacGinitie (1935), McLean (1962), Nations (1975), Phillips (1939), Poole (1966b), Rathbun (1926, 1930), Schmitt (1921), Thompson & Chow (1955), Trask (1970), and Turner, Ebert & Given (1969).

Phylum Arthropoda / Class Crustacea / Subclass Malacostraca
Superorder Eucarida / Order Decapoda / Suborder Reptantia
Section Brachyura / Family Xanthidae

25.23 Cycloxanthops novemdentatus (Lockington, 1877)
Large Pebble Crab

Common in pools and under rocks, low intertidal zone; young in crevices or in spaces between snail tubes; subtidal to 73 m; Monterey Bay to Punta Abreojos (Baja California), including offshore islands; not common in northern part of range, but conspicuous in the intertidal zone at La Jolla (San Diego Co.); Pleistocene fossils from Santa Monica and Playa del Rey (both Los Angeles Co.).

Carapace to 94 mm wide in males and 52.4 mm wide in females, flattened, granular and rugose anteriorly, smoother posteriorly, its anterolateral border bearing nine small teeth; color brown, tinged with purple and occasionally red, the chelae black.

This is the largest California pebble crab and the subject of extensive studies by Knudsen. The animals feed mainly on coralline algae and other red algae, but some green algae and surfgrass (*Phyllospadix*) are taken, and some animals, such as crustaceans and worms. Crabs in the field have been observed to break open and eat purple sea urchins (*Strongylocentrotus purpuratus*, 11.5) and to attempt to dislodge black abalones (*Haliotis cracherodii*, 13.4). In turn, they are preyed upon by larger crabs, fishes, and probably octopuses.

The crabs have cryptic habits, seek shelter if exposed, and tend to run toward dark places; they can bury themselves rapidly in sand if no other cover is available. When running, the last pair of legs is often held above the ground, serving as rear "feelers." Crabs backed against a wall use the back legs to walk along, or to grip, the wall, and crabs in low tunnels use the rear legs to gauge the height of the ceiling. Suddenly disturbed crabs may freeze motionless, some even exhibiting a cataleptic rigidity if roughly handled.

Females are sexually mature at a carapace width of 33 mm or more. There is no courtship prior to mating, and females mate while hard-shelled. Eggs are carried from mid-June to September in southern California, and ovigerous females move 1–2 m below the low tide level before extruding their egg masses. An average female carries about 45,000 eggs, which hatch in 25–30 days. Larval stages include a prezoea, four zoeal instars, and a megalops stage, lasting altogether about 5 weeks under laboratory conditions. The larvae are active swimmers; the zoeae can cover a meter in 45–60 seconds, and the megalops nearly twice that.

See especially Knudsen (1960a,b,c); see also Johnson & Snook (1927), Rathbun (1926, 1930), Ricketts & Calvin (1968), Schmitt (1921), and Willett (1937).

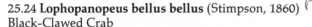

Phylum Arthropoda / Class Crustacea / Subclass Malacostraca
Superorder Eucarida / Order Decapoda / Suborder Reptantia
Section Brachyura / Family Xanthidae

25.24 **Lophopanopeus bellus bellus** (Stimpson, 1860)
Black-Clawed Crab

Commonly burrowed in sand under rocks, low intertidal zone, especially in quiet waters of bays and estuaries where a single layer of rocks covers fine sand or mud; also in tidepools and kelp holdfasts on rocky shores; subtidal to 73 m; Resurrection Bay (Alaska) to Cayucos (San Luis Obispo Co.); Pleistocene fossils from Playa del Rey (Los Angeles Co.).

Carapace to 34.2 mm wide in males, 24.1 mm wide in females, pubescent; cheliped with the carpus smooth or rugose and bearing a deep dorsal groove, the propodus with a small lobe on upper margin, the dactylus with a large tooth at base; walking legs pubescent, the carpus slightly bilobed.

This crab has been studied in the Puget Sound area (Washington). The stomachs of the animals examined there contained brown algae, green plant matter, coralline algae, and some remains of mussels, barnacles, and other crustaceans. Females mate while hard-shelled, and most (60–70 percent) produce two broods of young a year. Brood 1 eggs are laid from December to April, and by early March over 90 percent of the females are ovigerous. Hatching occurs from April to August. Brood 2 eggs are deposited from May to mid-August, and hatching begins in August. Egg number per brood ranged from 6,000 to 36,000, averaging 15,640. For the population, egg production per female per year is about 25,000. The diploid chromosome number is 124.

A subspecies, **L. bellus diegensis** Rathbun, occurs from Monterey Bay to San Diego. It differs from *L. bellus bellus* in having the carpus of the cheliped tuberculated and the carpus of the walking legs markedly bilobed. The carapace of the males may be 21.8 mm wide, that of the females 19.3 mm wide. Studied extensively in southern California, this crab is similar in many ways to *L. bellus bellus*, but shows some differences. Ovigerous females occur from February through October. They carry fewer eggs (an average of 3,500), and there is evidence that some females may produce more than two broods per year. The crabs are mainly herbivores, but some animal matter is taken as well. Parasites found on the crabs include two species of rhizocephalan barnacles.

See especially Knudsen (1959a,b, 1960a,c, 1964b); see also Andrews (1945), Fasten (1921, 1926), Hart (1935, 1968), Johnson & Snook (1927), Menzies (1948), Rathbun (1930), Ricketts & Calvin (1968), Schmitt (1921), and Willett (1937).

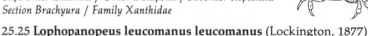

Phylum Arthropoda / Class Crustacea / Subclass Malacostraca
Superorder Eucarida / Order Decapoda / Suborder Reptantia
Section Brachyura / Family Xanthidae

25.25 **Lophopanopeus leucomanus leucomanus** (Lockington, 1877)

Rather plentiful under rocks, buried in sand under rocks, among snail tubes, and in piles of drifted algae, low intertidal zone on protected outer rocky coast; subtidal to 200 m; Carmel (Monterey Co.) to beach at Rosarito (Baja California); Pleistocene fossils from San Pedro and Long Beach (both Los Angeles Co.).

Carapace to 20.8 mm wide in males and 17.3 mm wide in females, not pubescent; cheliped with the carpus bearing reticulations but no deep dorsal groove, the propodus with a large lobe on upper margin, the dactylus with a large tooth at base; walking legs not pubescent, the carpus not bilobed.

Lophopanopeus leucomanus leucomanus, investigated off southern California, feeds mainly on algae, especially encrusting species of coralline algae, but some animal material is also taken. Ovigerous females are found from February to October, and carry an average of 3,500 eggs (ranging from 1,000 to 6,400); they probably produce more than two broods of young a year. Although the species extends into depths of 200 m, animals taken in deeper water are always small, suggesting that conditions there are less favorable.

Another subspecies, **L. leucomanus heathii** Rathbun (Fig. 25.25b–f), ranges from Moss Beach (San Mateo Co.) to La Jolla (San Diego Co.). It differs in that the carpus of the cheliped lacks reticulating ridges and the carpus of the walking legs is slightly bilobed. The carapace is up to 24.8 mm wide in males and 17.3 mm in females. Johnson & Snook (1927) first recognized that *Lophopanopeus heathii* was a northern replacement for *L. leucomanus*, a relationship formalized by Menzies (1948), who made the former a subspecies of the latter.

See especially Knudsen (1958, 1960a,c); see also Andrews (1945), Garth (1961), Hewatt (1938), Johnson & Snook (1927), Jones (1941), McLean (1962), Menzies (1948), Rathbun (1926, 1930), and Schmitt (1921).

Phylum Arthropoda / Class Crustacea / Subclass Malacostraca
Superorder Eucarida / Order Decapoda / Suborder Reptantia
Section Brachyura / Family Xanthidae

25.26 **Lophopanopeus frontalis** (Rathbun, 1893)

Under rocks in bays and in clusters of mussels on pilings, low intertidal zone; subtidal to 37 m; Santa Monica Bay (Los Angeles Co.) to Bahía Magdalena (Baja California), and Gulf of California; now rare in northern part of its range; Pleistocene fossils from Playa del Rey and San Pedro (both Los Angeles Co.).

Carapace to 23.7 mm wide in males, 13 mm wide in females, lacking pubescence dorsally; cheliped with the carpus smooth or reticulate, with a deep dorsal groove, the propodus (hand) with a large lobe on upper margin, the dactylus (of major cheliped) lacking a large tooth; dark band on pollex (thumb) of chela extending far backward and upward on propodus.

The involved synonymy of this species includes *Xanthodes leucomanus* Lockington, 1877 (p. 100, not p. 32), which became *Lophopanopeus lockingtoni* Rathbun, 1900, which in turn became a synonym of *L. frontalis* (Rathbun, 1893). The distribution given above includes that recorded under these synonyms.

This species is recorded among the stomach contents of the tidepool fish *Gibbonsia elegans* in the Los Angeles area, but little is known of its natural history.

See Johnson & Snook (1927), Menzies (1948), Mitchell (1953), Rathbun (1926, 1930), Ricketts & Calvin (1968), Schmitt (1921), and Willett (1937).

Phylum Arthropoda / Class Crustacea / Subclass Malacostraca
Superorder Eucarida / Order Decapoda / Suborder Reptantia
Section Brachyura / Family Xanthidae

25.27 **Rhithropanopeus harrisii** (Gould, 1841)
Brackish-Water Crab

Locally common on muddy bottoms of sloughs and estuaries, and even into fresh water; subtidal to 10.5 m in San Francisco Bay (on the east coast to 36.5 m); sloughs south of Coos Bay (Oregon); San Francisco Bay and Delta areas as far upriver as Stockton (San Joaquin Co.); Atlantic coast from New Brunswick (Canada) to Tampico (Mexico); Europe.

Carapace to 19.1 mm wide in males, 10.6 mm wide in females, the first two anterolateral teeth on margin fused, the last three dentiform; front (rostrum) truncate, dorsal ridges prominent; walking legs slender.

This little crab has been introduced to both Europe and the Pacific coast from the Atlantic shores of North America. It was first reported from the west coast (San Francisco Bay), where it is now abundant, by Jones (1940). The crab tolerates brackish and even fresh water for prolonged periods; it osmoregulates efficiently at low salinities, and in dilute media may excrete a fourth of its body weight in urine each day. The permeability of the body wall appears to *decrease* at low salinities, a response found in a few other brackish-water crustaceans and in the annelid *Nereis*. In San Francisco Bay not only is it common near domestic sewage outfalls, but it is almost the only larger invertebrate that seems undisturbed by effluent from industrial outfalls that is toxic to most other macroscopic animal life. Development in Atlantic populations of this species has been studied from the standpoints of morphology, physiology, biochemistry, and behavior.

See Buitendijk & Holthuis (1949), Chamberlain (1962), Connolly (1925), Costlow (1966), Costlow & Bookhout (1971), Costlow, Bookhout & Monroe (1966), Filice (1958, 1959), Forward & Costlow (1974), Frank, Sulkin & Morgan (1975), Hood (1962), Jones (1940, 1941), Kalber (1970), Kalber & Costlow (1966), Kinne & Rotthauwe (1952), Morgan, Kramarsky & Sulkin (1978), Ott & Forward (1976), Pautsch, Lawinski & Turoboyski (1969), Payen, Costlow & Charniaux-Cotton (1971), Rathbun (1930), Ricketts & Calvin (1968), R. I. Smith (1967), Skorkowski (1972), Sulkin (1973, 1975), Via & Forward (1975), and Whitney (1969).

Phylum Arthropoda / Class Crustacea / Subclass Malacostraca
Superorder Eucarida / Order Decapoda / Suborder Reptantia
Section Brachyura / Family Xanthidae

25.28 Paraxanthias taylori (Stimpson, 1860)
(=Xanthias taylori)   Lumpy Crab

Common, particularly in holes and crevices in region of algal turf, middle and low intertidal zones on rocky shores; subtidal to 100 m; Monterey Bay to Bahía Magdalena (Baja California).

Carapace width to 25.2 mm in males, 42 mm in females; carapace and chelipeds bearing blunt tubercles in adults and sharp curved teeth in juveniles; walking legs covered with stiff bristles; color dull red above, lower surface much lighter; chelae characteristically with dark-brown fingers.

A prominent feature of the intertidal region at La Jolla (San Diego Co.), these crabs are less common in the northern part of their range, though up to six per m² have been reported in kelp holdfasts in Monterey Bay. The diet is chiefly red and green algae, especially coralline algae. Some living and dead animal matter is eaten, and in aquaria nudibranch eggs may be taken if available. In turn, the crabs are a favorite food of the black croaker, and are also eaten by the cabezon and scorpionfish.

Out in the open on submerged rocks, individuals of *P. taylori* generally move only in the period between waves. They stop and cling firmly as a wave crest passes over them. Some stimulus, perhaps the "lift" exerted by the wave on the body or on the tactile hairs on the carapace or legs, signals the crabs to cling.

The abdomen of the female begins to acquire its feminine

characteristics when the carapace is only 6.5 mm wide, and at a width of 11.5–13.5 mm the first egg mass is produced. Mating occurs with the hard-shelled female on her back, the male astraddle above; mating may last up to 3 hours and is sometimes repeated several days in a row. Ovigerous females are abundant in southern California from April to September. The number of eggs per brood varies with the size of the female, from about 3,800 in a mother with a carapace 15 mm wide to about 40,000 in one 40 mm across the back; the average female carries about 21,000 eggs. More than two broods per year are probably produced. The larval stages have been described.

Individual crabs sometimes bear the rhizocephalan barnacle parasite *Thompsonia*, and occasional animals are overgrown with sponge or algae.

See especially Knudsen (1959b, 1960a,c); see also Andrews (1945), Johnson & Snook (1927), Limbaugh (1955), Rathbun (1930), Ricketts & Calvin (1968), and Schmitt (1921).

Phylum Arthropoda / Class Crustacea / Subclass Malacostraca
Superorder Eucarida / Order Decapoda / Suborder Reptantia
Section Brachyura / Family Xanthidae

25.29 Pilumnus spinohirsutus (Lockington, 1877)
Retiring Southerner, Hairy Crab

Under and among rocks, often hidden in sand, low intertidal zone; subtidal to 25 m; San Pedro (Los Angeles Co.) to Bahía Magdalena (Baja California); erroneously reported from tropical western Atlantic.

Carapace to 30.8 mm wide in males and 34.6 mm wide in females, easily recognizable by its hairs and spines.

This species is distinguished from *P. townsendi* Rathbun, which occurs in Bahía Magdalena and the Gulf of California, by having five anterolateral spines instead of four.

The retiring habits of *P. spinohirsutus* have earned it one common name. It hides in the sand under and among rocks, where its light-brown color makes it difficult to spot. The crabs *Cycloxanthops novemdentatus* (25.23) and *Paraxanthias taylori* (25.28) occur in the same habitat, sometimes even under the same rock. Formerly obtainable at San Pedro (Los Angeles

Co.) and at the entrance to Newport Bay (Orange Co.), this warm-temperate species is disappearing as harbors are dredged and deepened. It persists at outfalls of hydroelectric and nuclear power plants, where discharge water has been heated.

See Fabbiani (1972), Johnson & Snook (1927), Rathbun (1930), Ricketts & Calvin (1968), and Schmitt (1921).

Phylum Arthropoda / Class Crustacea / Subclass Malacostraca
Superorder Eucarida / Order Decapoda / Suborder Reptantia
Section Brachyura / Family Goneplacidae

25.30 **Malacoplax californiensis** (Lockington, 1877)
(= Speocarcinus californiensis)   Burrowing Crab

Sometimes common, burrowing in mud flats of bays and estuaries, middle and low intertidal zones; subtidal to 33 m; Mugu Lagoon (Ventura Co.) to Bahía Magdalena (Baja California); common in suitable situations.

Carapace width to 22.6 mm in males, the dorsal surface convex longitudinally, almost straight transversely, the margin bearing three anterolateral teeth; front and sides of carapace, tapering eyestalks, and legs all fringed with light-colored hairs; body brownish to white; chelae with tips of fingers black.

Like the fiddler crab, the burrowing crab is limited to the southern part of the state, where it may form deep burrows. It, too, threatens the intruder from the mouth of its burrow but retreats to safety when danger approaches. Unlike the fiddler, however, it ranges throughout the intertidal region and may be dredged from the bottom of the bay, as at Newport (Orange Co.) or San Diego.

See Guinot (1969) for name change; see also Johnson & Snook (1927), Rathbun (1918), Ricketts & Calvin (1968), and Schmitt (1921).

Phylum Arthropoda / Class Crustacea / Subclass Malacostraca
Superorder Eucarida / Order Decapoda / Suborder Reptantia
Section Brachyura / Family Pinnotheridae

25.31 **Fabia subquadrata** Dana, 1851
Mussel Crab, Pea Crab

Commensal in the mantle cavities of bivalve mollusks, especially *Mya arenaria* (15.67) and the mussels *Mytilus edulis* (15.10) and *M. californianus* (15.9); adults occasionally reported in the clams *Cyclocardia borealis*, *C. ventricosa*, *Tivela stultorum* (15.32), *Protothaca staminea* (15.41), *Saxidomus giganteus* (15.33), *Tresus capax* (15.47), *T. nuttallii* (15.48), and *Tapes* sp., the sea urchin *Strongylocentrotus purpuratus* (11.5), and the tunicate *Styela gibbsii*; low intertidal zone to 88 m depth; Akutan Pass (Aleutian Islands, Alaska) to La Jolla (San Diego Co.); uncommon south of San Pedro (Los Angeles Co.), where the related species *F. concharum* (25.32) predominates.

Carapace width in males to 7.3 mm, in females to 16.2 mm; distinguished from *F. concharum* by the transverse groove across the front, the distally widened palm of the chela, and an outer maxilliped with the terminal segment reaching the end of the subterminal segment.

The strange life history of *F. subquadrata* has been studied in the Puget Sound area (Washington) by Pearce (1966a), and although some of the details are unknown, the whole is a remarkable story. The larger adult females have only a membranous outer shell, and their soft bodies are quite unsuited to life outside the protecting body of the host. The clams in which they live subsist by filtering the water for plankton, trapping this in mucus on the gills, and passing strings of mucus laden with food particles along food grooves to the mouth. The crab, huddled within the mantle cavity, plucks the food and mucus strings from *one* of the clam's two gills, often damaging that gill severely over a period of time; the other gill, relatively unharmed, provides food for the clam. Two such crabs in a single host might well destroy it, but two are almost never found together; females are hostile toward each other when confronted outside a clam, and if a second is inserted into an already occupied clam, one or both crabs shortly leave. In California the main host is *Mytilus californianus* (*Modiolus modiolus* is preferred in Puget Sound), and about 3 percent of the mussels contain pea crabs.

Some ovigerous females are found in mussels nearly all year; females carrying eggs are rare in August in the Puget Sound area, but increase in number to a peak from November to January and then decline steadily until the next August. One female examined bore about 1,200 eggs. Major hatching begins in February, and the liberated young pass through four zoeal stages and a megalops stage in the plankton before reaching the first crab instar. Development time from hatching to the first crab stage was about 52 days for animals reared in the laboratory at 11.6–13.6°C. The first crab stage, 0.76–0.9 mm across the carapace, has hairy legs and is adapted for swimming. It appears to be the stage that seeks a clam host, for the smallest crab found in a clam had a carapace width of 0.86 mm. Both males and females seek clam hosts, often such smaller species as *Astarte compacta*, *Cyclocardia ventricosa*, *Crenella columbiana*, and *Kellia* sp., in addition to *Modiolus*. Within the host's mantle cavity the crabs grow in April and May, probably molting at least seven times. The exoskeleton during this period is thin and soft.

Then, in late May, the crabs molt again, and this time both sexes produce a hard and calcified outer shell. Hard-shelled males average 4.1 mm in carapace width (range 1.3–6.8 mm), and females average 3.5 mm (range 1.5–6.2 mm). These tiny, hard-shelled crabs then leave their molluscan hosts and swarm into the plankton, where males and females meet and mate for the first and last time. This is the only period of their lives when the two sexes are found together. Females then return to a clam host, but never to the smaller species of clams they occupied as juveniles. Safe in the mantle cavities of mussels, the fertilized females molt and become soft-shelled again, then grow and start to produce eggs. Ovigerous females range in size from 5 to 13.4 mm across the carapace. The females do not molt the lining of their seminal receptacles (spermathecae); thus they appear to retain for a lifetime the store of sperm from their one mating in the plankton. Most females live only a year, but some may persist and produce broods a second year, still using the initial store of sperm.

The fate of males is less certain; a few appear to return to hosts, for a few males are found throughout the year (always alone) in mussels, where they remain small and hard-shelled.

The vast majority are never seen again, and it seems likely that, having done their essential job, they perish. There is no evidence that they ever visit clams containing the females, which shortly become so much larger than the males that further mating appears anatomically impossible.

Mussels more than 85 mm in length rarely contain pea crabs. Juvenile crabs and newly mated females usually seek mussels less than 15 mm long, and the larger *Modiolus* either were never infected or had outlived their symbionts.

See especially Pearce (1966a) and Schmitt, McCain, & Davidson (1973); see also Davidson (1968) and Irvine & Coffin (1960). Because of the confusion between *F. subquadrata* and *F. concharum*, references to older works are omitted.

Phylum Arthropoda / Class Crustacea / Subclass Malacostraca
Superorder Eucarida / Order Decapoda / Suborder Reptantia
Section Brachyura / Family Pinnotheridae

25.32 **Fabia concharum** (Rathbun, 1893)
(=F. lowei, Pinnotheres concharum)

Commensal in the mantle cavities of bivalve mollusks, including *Barnea subtruncata* (15.71), *Donax gouldii* (15.54), *Mya arenaria* (15.67), *Cryptomya californica* (15.66), *Modiolus capax* (15.5), *M. modiolus*, *Tivela stultorum* (15.32), *Parapholas californica* (15.75), *Protothaca staminea* (15.41), and *Tapes* sp.; San Pedro (Los Angeles Co.) to Bahía de Tortuga (Baja California).

Distinguished from *F. subquadrata* (25.31) in having the front smooth and lacking a transverse groove, the palm of the chela not distally widened, and the terminal segment of the outer maxilliped not reaching the end of the subterminal segment.

This small commensal crab occupies some of the same bivalve hosts as the related species *F. subquadrata*. *F. concharum* predominates in southern California, where the two species overlap in distribution.

Although *F. concharum* is not known to inhabit ascidians, and *F. subquadrata* only occasionally does so, a somewhat similar pea crab, **Pinnotheres pugettensis** Holmes, typically occurs as a commensal in large tunicates. At Departure Bay (British Columbia) and Puget Sound (Washington) it is found in *Halocynthia* (= *Tethyum*) *aurantium*, *H. hilgendorfi igaboja* (12.40), and *Ascidia paratropa* (12.26); in Monterey harbor it occurs in *Styela montereyensis* (12.33). The carapace is subpentagonal, and widest anteriorly (width 10.5 mm in females, unknown in males); the chela in females is widest behind the fingers, and the outer surface of the manus is brownish with light reticulations; the walking legs increase in length from first to last, and the dactylus of the fourth leg is the longest.

The type specimen, from Puget Sound, was found in the branchial cavity of *Cynthia* (= *Halocynthia*). Wells (1928) illustrates the enlargement of the atrial cavity of *H. aurantium* caused by the presence of this crab. The discovery of two specimens of *P. pugettensis* in *Styela montereyensis* at Monterey in 1978 (C. C. Lambert, pers. comm.) not only extends the range to central California, but adds a new host.

For *F. concharum*, see Davidson (1968) and the account and references for *F. subquadrata* (25.31). For *P. pugettensis*, see Holmes (1900), Pearce (1966a), Rathbun (1918), Schmitt, McCain & Davidson (1973), and Wells (1928).

Phylum Arthropoda / Class Crustacea / Subclass Malacostraca
Superorder Eucarida / Order Decapoda / Suborder Reptantia
Section Brachyura / Family Pinnotheridae

25.33 Parapinnixa affinis Holmes, 1900
Pea Crab

Commensal in tubes of the polychaete worms *Terebella californica* and *Loimia medusa* (= *L. montagui*) among clumps of shells and marine plants, low intertidal zone in protected situations; San Pedro (Los Angeles Co.), Anaheim Landing and Newport Bay (Orange Co.), and San Diego; Sakhalin and Kurile Islands (U.S.S.R.).

Carapace to 4.1 mm wide in males, 6 mm wide in females, less than twice as wide as long, smooth and shining; pollex (thumb) of cheliped with two teeth at tip and a large tooth at center; first pair of walking legs larger than others, the dactyli short and stout; carapace light amber mottled with ochre, legs pale ochre with greenish tinge (not shown in the photograph), dactyli yellow with white tips.

The worm that serves as host for this minute commensal

crab builds a mucus tube with embedded mud and sand. Generally one finds only one crab to a worm tube, but occasionally a male and female are found together. In one locality, females outnumbered males by three to one; in another spot males formed a two-to-one majority. Of 100 specimens examined, five were albino crabs.

See Berkeley & Berkeley (1941), Glassell (1933), Holmes (1900), and Kobjakova (1967).

Phylum Arthropoda / Class Crustacea / Subclass Malacostraca
Superorder Eucarida / Order Decapoda / Suborder Reptantia
Section Brachyura / Family Pinnotheridae

25.34 Pinnixa barnharti Rathbun, 1918
Pea Crab

Commensal in the cloaca of the sea cucumber *Caudina arenicola*, low intertidal and subtidal sand flats of bays and estuaries; Venice (Los Angeles Co.) to Bahía Ballenas (Baja California); reported from Puget Sound (Washington), presumably in *Caudina intermedia*, and from Zihuatanejo (Mexico), in *Holothuria* (*Paraholothuria*) *riojai*.

Carapace to 15 mm wide in males, 16.2 mm wide in females, convex, laterally truncate; chelipeds massive, the fingers widely gaping, the dactylus with a strong tooth at the middle, the pollex (thumb) strap-shaped; walking legs with the dactyli straight.

The crab inhabiting the sweet potato sea cucumber was formerly thought to be invariably this species. However, 25 percent of the *Caudina* (= *Molpadia*) examined in the San Diego area were found to contain *Opisthopus transversus* (25.42) instead. Ricketts noted that the crabs often crawled out of the holothurian while the host was being relaxed with epsom salts prior to preservation.

See Caso (1963), Hopkins & Scanland (1964), Lie (1968), MacGinitie & MacGinitie (1968), Rathbun (1918), Ricketts & Calvin (1968), and Schmitt (1921).

Phylum Arthropoda / Class Crustacea / Subclass Malacostraca
Superorder Eucarida / Order Decapoda / Suborder Reptantia
Section Brachyura / Family Pinnotheridae

25.35 **Pinnixa faba** (Dana, 1851)
Large Pea Crab

Adults encountered intertidally as commensals in the mantle cavities of the burrowing bivalves *Tresus capax* (15.47; Puget Sound, Washington; Humboldt Bay) and *T. nuttallii* (15.48; Bodega Bay, Sonoma Co., and southward), in the mantle folds of the sea hare *Aplysia vaccaria* (14.9), in the atrial cavity of the tunicate *Styela gibbsii*, and in the cloaca of the holothurians *Caudina arenata* and *C. arenicola*; juveniles commensal in an assortment of clams (see below); Prince of Wales Island (Alaska) to Newport Bay (Orange Co.); Pleistocene fossils from Cape Blanco (Oregon).

Carapace width to 15 mm in males, 22.8 in females (see also Fig. 18.26); species distinguished from *P. littoralis* (25.36) by (1) carapace oblong, with orbits oval rather than pointed, (2) chelae with pollex (thumb) horizontal rather than deflexed in males, and fingers closed rather than gaping in females, and (3) merus of third leg in males more than twice as long as wide rather than only twice as long as wide.

Biologically, *P. faba* and its close relative *P. littoralis* show many similarities. Both species are best known from studies made in the Puget Sound area, where the adults occur only in *Tresus capax*. The two crabs occur in the same clam beds, but only one species of adult crab occupies a given clam host. Adult crabs occur in sexual pairs, often along with a few juveniles whose further development seems inhibited by the presence of adults of the same sex. Mature females hide under the visceral fold of *T. capax*, feeding on plankton and other particulate matter brought in with the clam's feeding current and trapped in its mucus. Food is scraped from the clam's visceral fold, where feeding activities cause some lesions but no damage to the clam's most important feeding organs, its gills and palps. The much smaller males and juvenile crabs range freely through the host's mantle cavity.

In Washington waters some females with eggs occur nearly all year. Breeding peaks for *P. faba* occur from January to April or May, and again in June and July, representing two broods. Breeding peaks of *P. littoralis* are similar, but come about a month earlier. Average females of both species carry 7,000–8,000 eggs, though a large California specimen of *P. faba* carried more than 54,000 eggs. A hiatus in breeding occurs when the females molt, from late August through October. Hatched larvae of both crab species are planktonic, and take about 47 days (under laboratory conditions) to reach the last larval or megalops stage, which seeks a bivalve host.

Males of both crabs subsequently pass through about 14 post-larval molts to reach terminal anecdysis (a final adult stage at which molting ceases), with an average carapace width of 13.1 mm. Females first become ovigerous at a carapace width of 12.8–14 mm (15 or 16 instars), but continue to grow to an average carapace width of 19.7 mm (23 or 24 instars). If an adult dies, it is replaced by the growth of one of the juvenile crabs, if any of that sex are present in the host clam.

California records are sparse, but both species of crabs, and both species of the host clam *Tresus*, occur in Humboldt Bay; today the crabs apparently inhabit only *T. capax*, but an earlier report put *P. faba* in *T. nuttallii*. *P. faba* has also been reported from *T. nuttallii* in Tomales Bay (Marin Co.), where it was collected by R. H. Morris, in Elkhorn Slough (Monterey Co.), Morro Bay (San Luis Obispo Co.), and Newport Bay (Orange Co.). *T. nuttallii* lacks the visceral fold present in *T. capax*, from which the *Pinnixa* females typically feed. Moreover, in California waters, *P. faba* has been reported from the clams *Macoma nasuta* (15.50), *Mya arenaria* (15.67), *Saxidomus giganteus* (15.33), *Siliqua gibbsi*, and *Protothaca staminea* (15.41), only one to a host under most circumstances. Further investigation of the biology of both *P. faba* and *P. littoralis* seems desirable, since older records need further verification.

See especially Pearce (1966b); see also Hart (1930), Johnson & Snook (1927), MacGinitie (1935), MacGinitie & MacGinitie (1968), Pearce (1962, 1965), Rathbun (1918), Ricketts & Calvin (1968), Schmitt (1921), Schmitt, McCain & Davidson (1973), Taylor (1912), Wells (1928, 1940), and Zullo & Chivers (1969).

Phylum Arthropoda / Class Crustacea / Subclass Malacostraca
Superorder Eucarida / Order Decapoda / Suborder Reptantia
Section Brachyura / Family Pinnotheridae

25.36 **Pinnixa littoralis** Holmes, 1894
Pea Crab

Found as a commensal in the mantle cavities of burrowing clams; in California, adults are reported in the gaper clams *Tresus capax* (15.47) and *T. nuttallii* (15.48) and the Washington clam *Saxidomus nuttallii* (15.34) (in Puget Sound they occur only in *T. capax*); juvenile crabs in a wide variety of bivalve hosts (see below); Prince William Sound (Alaska) to San Diego.

Carapace width to 16 mm in males, 26 mm in females; species differing from *P. faba* (25.35) in having (1) side walls of carapace less steep, the outline as seen from above not longitudinally truncate but inclined obliquely backward and outward, (2) orbits continuing laterally in an acute angle, not oval, (3) chelae with pollex (thumb) deflexed rather than horizontal in males, and fingers gaping rather than closed in females, and (4) merus of third leg in males no more than twice as long as wide; color gray or greenish white, with brown bands on walking legs; yellow liver and orange ovary ordinarily visible through transparent carapace (though not seen in the photograph).

Juvenile crabs appear to be far less specific than adults in their choice of hosts. In California, juveniles have been found in the clams *Mya arenaria* (15.67), *Clinocardium nuttallii* (15.29), and *Macoma secta* (15.51), as well as in the definitive hosts; in Puget Sound they inhabit these species and in addition the clams *Entodesma saxicola* (15.81), *Saxidomus giganteus* (15.33), *Macoma nasuta* (15.50), *M. inquinata*, *M. indentata*, *M. secta* (15.51), *Serripes groenlandicus*, *Siliqua patula* (15.65), *Protothaca staminea* (15.41), and *Tapes*, and are sometimes found in the pallial cavities of the larger limpets *Notoacmea* and *Collisella* spp. Further details of biology are included under *P. faba*. Because of the similarity of these two crabs, the close resemblance of the two species of *Tresus* that serve as their definitive hosts, and the fact that reports seldom distinguished the crabs as sexually mature or immature, there is considerable doubt about the validity of early California records. This is particularly true south of Monterey, where some reports of *P. faba* may actually refer to *P. littoralis*.

See especially Pearce (1966b); see also Hart (1930, 1940, 1968), Johnson & Snook (1927), Rathbun (1918), Ricketts & Calvin (1968), Schmitt (1921), Schmitt, McCain and Davidson (1973), and Wells (1928, 1940).

Phylum Arthropoda / Class Crustacea / Subclass Malacostraca
Superorder Eucarida / Order Decapoda / Suborder Reptantia
Section Brachyura / Family Pinnotheridae

25.37 **Pinnixa longipes** (Lockington, 1877)
Pea Crab

Commensal in tubes of the annelid worm *Axiothella* (= *Clymenella*) *rubrocincta*, abundant on sand bars exposed at low tide; also in tubes of the polychaete worms *Pectinaria californiensis* (18.32) and *Pista elongata* (18.33), and occasionally in burrows of the echiuroid worm *Urechis caupo* (19.7); subtidal to 128 m; Bodega Bay harbor (Sonoma Co.) to Ensenada (Baja California); reported erroneously from the east coast of North America.

Carapace to 6 mm wide in males, to 6.3 mm wide in females, nearly three times as wide as long; third leg greatly enlarged, the merus bearing a flange on posterior margin; fourth leg small, not reaching beyond merus of third.

The transverse elongation of the body expressly adapts this little crab for life in long, slender worm tubes. The crab insinuates itself through the narrow opening by inserting one of the enormously elongated third legs first, then carefully edging itself in sideways.

Ovigerous females were found in July and August in Elkhorn Slough (Monterey Co.).

See Carlisle (1969), Glassell (1935), Johnson & Snook (1927), MacGinitie (1935), MacGinitie & MacGinitie (1968), Rathbun (1918), Ricketts & Calvin (1968), Schmitt (1921), and Schmitt, McCain & Davidson (1973).

Phylum Arthropoda / Class Crustacea / Subclass Malacostraca
Superorder Eucarida / Order Decapoda / Suborder Reptantia
Section Brachyura / Family Pinnotheridae

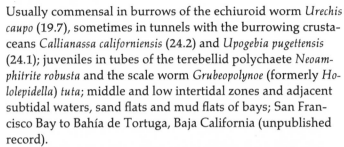

25.38 **Pinnixa franciscana** Rathbun, 1918
Pea Crab

Phylum Arthropoda / Class Crustacea / Subclass Malacostraca
Superorder Eucarida / Order Decapoda / Suborder Reptantia
Section Brachyura / Family Pinnotheridae

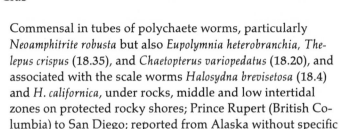

25.39 **Pinnixa tubicola** Holmes, 1894
Pea Crab

Usually commensal in burrows of the echiuroid worm *Urechis caupo* (19.7), sometimes in tunnels with the burrowing crustaceans *Callianassa californiensis* (24.2) and *Upogebia pugettensis* (24.1); juveniles in tubes of the terebellid polychaete *Neoamphitrite robusta* and the scale worm *Grubeopolynoe* (formerly *Hololepidella*) *tuta*; middle and low intertidal zones and adjacent subtidal waters, sand flats and mud flats of bays; San Francisco Bay to Bahía de Tortuga, Baja California (unpublished record).

Carapace to 22.7 mm wide in males, to 11 mm wide in females, the sides pointed; cheliped with the dactylus (thumb) horizontal, the propodus bearing a sharp, raised line of granules near lower edge; merus of the third leg broadened (about 1.7 times as long as wide).

Pinnixa schmitti Rathbun, which occupies tunnels of *Callianassa californiensis* and *Upogebia pugettensis* from Alaska to San Francisco Bay, differs from this species by having a more oblong carapace, the cheliped with a less prominent line of granules on the propodus extending onto the distally inclined pollex (thumb), and the merus of the third leg narrower (twice as long as wide).

The crab *Scleroplax granulata* (25.41), which also occurs in *Urechis* burrows, is smaller than *P. franciscana*. Both crabs filter particles from the water using the second pair of maxillipeds, and may also eat bits of detritus. The males migrate from burrow to burrow.

Ovigerous females of *P. franciscana* have been found in March at Elkhorn Slough (Monterey Co.) and Morro Bay (San Luis Obispo Co.), and in July at Morro Bay.

See MacGinitie (1935), MacGinitie & MacGinitie (1968), Rathbun (1918), Reish (1961b), Ricketts & Calvin (1968), Schmitt (1921), and Schmitt, McCain & Davidson (1973).

Commensal in tubes of polychaete worms, particularly *Neoamphitrite robusta* but also *Eupolymnia heterobranchia*, *Thelepus crispus* (18.35), and *Chaetopterus variopedatus* (18.20), and associated with the scale worms *Halosydna brevisetosa* (18.4) and *H. californica*, under rocks, middle and low intertidal zones on protected rocky shores; Prince Rupert (British Columbia) to San Diego; reported from Alaska without specific locality.

Carapace to 9 mm wide in males, 13 mm wide in females, smooth.

Pinnixa tubicola, *P. schmitti* (see under 25.38), and *P. weymouthi* (25.40) all have smooth carapaces, swollen propodi on the chelae, and nearly straight dactyli on the legs. The male abdomen of *P. tubicola* is more triangular, that of *P. schmitti* more convexly margined, and that of *P. weymouthi* more slender than the other two. **Pinnixa tomentosa** Lockington (San Clemente Island, Channel Islands; Mission Bay, San Diego Co.; San Felipe and Bahía de Los Angeles, Baja California) is morphologically similar to *P. tubicola* but differs in having the propodus of the third walking leg terminally spinulous, the palm of the chelae in males smooth. *P. tomentosa* shares the tubes of the polychaete *Chaetopterus variopedatus* with *P. tubicola* where their ranges overlap at San Diego.

Terebellid worms, the most frequent hosts of *P. tubicola*, extend their feeding tentacles into the sediments surrounding their tubes and use cilia and mucus to transport back to the tube a supply of detritus particles. The filter feeder *Chaetopterus* traps plankton and suspended detritus from the water. In both cases the food intended for the worm consists of organic particles in mucus. Possibly the commensal crabs take a share for themselves. The crabs usually occur in the host tube in pairs. Females have not been reported outside of worm tubes, and males have been noted free only once, in Elkhorn Slough (Monterey Co.).

Ovigerous females were reported in Monterey Bay in June and August.

See Hart (1930), Johnson & Snook (1927), MacGinitie (1935), MacKay (1931), Pearce (1966b), Pearse (1913), Rathbun (1918), Ricketts & Calvin (1968), Scanland & Hopkins (1978), Schmitt (1921), and Schmitt, McCain and Davidson (1973).

Phylum Arthropoda / Class Crustacea / Subclass Malacostraca
Superorder Eucarida / Order Decapoda / Suborder Reptantia
Section Brachyura / Family Pinnotheridae

25.40 **Pinnixa weymouthi** Rathbun, 1918
Pea Crab

Reported in unoccupied tubes of annelids on undersides of rocks, low intertidal zone; Monterey Bay and Peninsula.

Carapace about 5.3 mm wide in males, 6.2 mm wide in females; species distinguished from *P. tubicola* (25.39) by its narrower body, truncated sides, and more prominent anterolateral angle.

Almost nothing is known of the natural history of this small crab.

See Rathbun (1918), Schmitt (1921), and Schmitt, McCain & Davidson (1973).

Phylum Arthropoda / Class Crustacea / Subclass Malacostraca
Superorder Eucarida / Order Decapoda / Suborder Reptantia
Section Brachyura / Family Pinnotheridae

25.41 **Scleroplax granulata** Rathbun, 1893
Pea Crab

A common commensal in burrows of the echiuroid worm *Urechis caupo* (19.7) and tunnels of the burrowing shrimps *Callianassa californiensis* (24.2) and *Upogebia pugettensis* (24.1), middle and low intertidal zones on sandy mud and mud flats in protected bays; Roller Bay (Hope Island, off northern tip of Vancouver Island, British Columbia) to Ensenada (Baja California).

Carapace to 11.5 mm wide in males, 12.9 mm wide in females, hard, highly convex, the anterolateral margins not forming an angle with posterolateral margins (as they do in *Pinnixa*, e.g., 25.34–40) but curving gradually into them; chela

in male much larger than in female, the short pollex (thumb) bearing a large tooth filling gape caused by curved dactylus (finger); walking legs similar, fourth leg not noticeably smaller than others.

Up to six crabs have been found in a single *Urechis* burrow in Elkhorn Slough (Monterey Co.), but the animals often occur singly. They feed on particles of flesh and detritus and can also strain plankton from the water with their outer mouthparts. The bryozoan *Triticella elongata* is often found growing on the carapace and appendages and in the gill chamber of this crab (see Fig. 6.19).

Ovigerous females (some only 4.9 mm across the carapace) are most abundant in Elkhorn Slough in February, March and April, but a few were noted from June to August. Females with eggs were seen in Humboldt Bay in January and in Morro Bay (San Luis Obispo Co.) in February. Males may migrate from burrow to burrow. Some crabs "play possum" for a few seconds or up to 2 minutes when disturbed; crabs in *Urechis* burrows may move to the host worm and cling to its body.

See Hart (1930, 1935, 1937, 1940, 1971b), Johnson & Snook (1927), MacGinitie (1935), MacGinitie & MacGinitie (1968), Rathbun (1918), Ricketts & Calvin (1968), Schmitt (1921), Schmitt, McCain & Davidson (1973), and Wells (1928).

Phylum Arthropoda / Class Crustacea / Subclass Malacostraca
Superorder Eucarida / Order Decapoda / Suborder Reptantia
Section Brachyura / Family Pinnotheridae

25.42 **Opisthopus transversus** Rathbun, 1893
Mottled Pea Crab

A commensal crab found in the cloaca of the sea cucumbers *Parastichopus californicus* (9.7), *P. parvimensis* (9.8), and *Caudina* (formerly *Molpadia*) *arenicola*, in the tubes of the polychaete *Chaetopterus variopedatus* (18.20), on the gills of the amphineuran *Cryptochiton stelleri* (16.25), and in the mantle cavities or pallial grooves of a variety of clams and gastropods (see below); low intertidal zone and adjacent subtidal waters on rocky shores and in mud flats of protected bays; Monterey to Laguna San Ignacio, Baja California (unpublished record); San Felipe (Baja California).

Carapace to 11.8 mm wide in males, 21 mm wide in females, squarish, a little wider than long, firm, harder in male than in female; chelipeds pubescent; walking legs pubescent, similar in form, the second pair slightly longer than others; carapace richly spotted with vermilion to deep red, polished to almost pearly smoothness.

Unlike most commensal crabs, this little species is found living with a wide variety of hosts, always in a protected place. In addition to the above hosts, *Opisthopus* has been found in the mantle cavities or mantle folds of the giant keyhole limpet *Megathura crenulata* (13.8), the bubble snail *Bulla gouldiana* (14.6), the sea hare *Aplysia vaccaria* (14.9), the slug *Navanax inermis* (14.4), the cone snail *Conus californicus* (13.103), and the large snails *Polinices lewisii* (13.62) and *Astraea undosa* (13.39). Clams likewise may serve as hosts, though less commonly, and the crab has been found in the mantle cavities or siphon tubes of the gaper *Tresus nuttallii* (15.48), the clay borer *Zirfaea pilsbryi* (15.72), the rock borer *Parapholas californica* (15.75), the mussels *Modiolus modiolus*, *Mytilus edulis* (15.10), and *Megapitaria squalida*, the scallop *Hinnites multirugosus*, and the bivalves *Sanguinolaria nuttallii*, *Platyodon cancellatus* (15.68), and *Dinocardium* (= *Trachycardium*) *robustum*.

In the San Diego area alone Hopkins and Scanland (1964) found *O. transversus* in association with 13 different species of hosts distributed among three phyla. They found the largest and most mature crabs (ovigerous females) in the mantle cavity of *Hinnites*, the intestine of *Caudina*, and the siphon of *Zirfaea*. Crabs inhabiting the gastropods *Astraea* and *Megathura* and the sea cucumber *Parastichopus* were somewhat smaller, and those inhabiting the gastropod *Bulla* and the tubes of *Chaetopterus* were smaller still, suggesting that the size of the cavity available may limit the size of the resident crab. Regarding color, they found that the bright-red mottling of the young crabs persists in individuals having a food supply independent of that of the host, such as the crabs inhabiting hosts like *Astraea*, *Bulla*, and *Megathura*; however, the color is lost in crabs that use food already gathered by filter-feeding hosts such as *Hinnites*, *Dinocardium*, and *Zirfaea*. This suggests that, indirectly, the host may influence the color of crabs by limiting the amount of carotenoid pigments available to them. In sea cucumber hosts, which ingest large quantities of

mud rich in carotenoids, crabs maintain a color pattern of reddish brown and white (in *Caudina*) or a pale red mottling (in *Parastichopus*). The similarity of pattern in the crabs inhabiting *Caudina* to that in *Pinnixa barnharti* (25.34), at one time thought to be the only pinnotherid commensal with this holothurian, suggests convergence, since this pattern was not found among *O. transversus* specimens occurring in any other host. Of 53 *Caudina* from the San Diego area, 52 contained a commensal crab, and of this number, 13 (or 25 percent) were *O. transversus*.

See especially Hopkins & Scanland (1964); see also Beondé (1968), Glassell (1935), Johnson & Snook (1927), MacGinitie (1935), MacGinitie & MacGinitie (1968), Rathbun (1904, 1918), Ricketts & Calvin (1968), Schmitt (1921), Schmitt, McCain and Davidson (1973), Webster (1968), and Wolfson (1974).

*Phylum Arthropoda / Class Crustacea / Subclass Malacostraca
Superorder Eucarida / Order Decapoda / Suborder Reptantia
Section Brachyura / Family Grapsidae*

25.43 **Pachygrapsus crassipes** Randall, 1839
Striped Shore Crab, Lined Shore Crab

Abundant in crevices, tidepools, and mussel beds, high and middle intertidal zones on rocky shores; occasionally on pilings in harbors; sometimes on hard muddy shores of bays and estuaries, at the mouths of burrows; Charleston (Oregon) to Isla de Santa Margarita (Baja California), and Gulf of California; Japan and Korea.

Carapace to 47.8 mm wide in males, 40.8 mm wide in females, squarish, broader than long, the sides converging posteriorly, the upper surface transversely striated anteriorly; distinguished from *Hemigrapsus* species (25.44, 25.45) by its darker coloration, marked with shades of red, purple, or green, and by having only one lateral tooth, and from *P. transversus* (Gibbes) of the Panamic region by the smooth merus of the last walking leg.

This abundant crab, a familiar resident of higher pools and crevices, has been the object of many investigations. Native to the west coast of North America, it was not reported from Asia until 1890; it may have been introduced to the Orient by ships carrying zoea larvae in water ballast.

More than any other north-temperate, west coast crab, *P. crassipes* is adapted to a semi-terrestrial life. In the field the animals spend at least half their time out of water, but visit pools at intervals to feed and to moisten the gills. Water (about 3 percent of the body volume) is retained in the gill chamber when the crabs are above the surface, and in experiments some animals have remained alive high and dry in shaded enclosures for up to 70 hours. Respiratory rate decreases as the crabs get drier. Partly desiccated crabs take to water quickly when immersed, and they can even absorb water from saline media against a diffusion gradient. The crabs regulate osmotically in both brackish and hypersaline media; given a choice of waters of different salinities, individuals choose normal seawater, even if found living under other conditions. *P. crassipes* tolerates a wide range of environmental temperatures. The composition of the pool of free amino acids found in the tissues is intermediate between that of typical marine crustaceans, like the lobsters and shrimps, and that of such semi-terrestrial forms as the isopod *Ligia* and the beachhopper *Orchestoidea*.

The main food of *Pachygrapsus* on rocky shores is the film of algae and diatoms growing on rocks in higher pools and crevices, which the animal scrapes off by means of the small spoon-shaped cups at the tips of the chelae. Some larger algae are also eaten, especially the green algae *Ulva* and *Enteromorpha*, the brown seaweed *Fucus*, and the red algae *Endocladia*, *Rhodoglossum*, and *Grateloupia*. Dead animal matter and detritus are also eaten when available. Living prey, such as limpets, the snails *Littorina* and *Tegula*, hermit crabs, and isopods, are occasionally taken. The limpets *Notoacmea scutum* (13.23) and *Collisella limatula* (13.14) are attacked; the crab presses the fingertips of its open chela straight down on top of the shell and pinches, popping the top off the limpet shell with the exertion of relatively little force. The crabs have also been observed catching flies that settle nearby; a quick lunge and sudden snap of the chela often results in a successful catch. In turn, *P. crassipes* adults are eaten by such air-breathing predators as sea gulls, rats, raccoons, and even man. Juvenile crabs and larvae are consumed by sea anemones and fishes. There is even some cannibalism, especially on crabs that have just molted and have soft shells. Out of water, food is located mainly by vision and by tactile stimuli, and the crab's excellent eyesight plays an important role in much of its behavior. Regular shifts in position of retinal pigments adapt the eyes for both day and night vision. Crabs held in constant darkness (but not in constant light) show a similar shift in retinal pigments, and the daily changes appear to be regulated by a "biological clock."

The crabs are rather aggressive toward one another. They do not defend feeding territories, but they may defend a food item, and often defend a crevice refuge against intruders when under water. Two crabs in confrontation often engage in a "boxing match," their chelae held outspread and gaping. The larger animal almost always dominates the smaller, and males tend to dominate females, but the major weapon here is bluff; crabs rarely injure each other in such bouts in the field.

Fertile females that are ready to molt emit a chemical signal, or pheromone, called crustecdysone, which attracts and stimulates males. Mating occurs after the female molts, while she is still soft-shelled. A courtship dance precedes the act, after which the male rolls over on his back and the female walks above him. Ovigerous females are found from March or April to August or September in central California, and from late February to October in southern California; peak reproduction occurs in June and July in both regions. A medium-sized female carries about 50,000 eggs and produces one or two broods a year. Larval stages occur in the plankton, and juvenile crabs are found in tufts of the alga *Endocladia*, well up in the intertidal zone, from January to July in the Monterey area. It takes at least 3 years after hatching for adults to reach full size; this involves 21 molts for females and 18 molts for males. Males reach sexual maturity 7 months after hatching; females require 11–12 months. The smallest ovigerous females reported were only 19 mm in carapace width. Old crabs are sometimes seen encrusted with bryozoans and barnacles, suggesting that they may have ceased to molt.

See especially Hiatt (1948); see also Baker (1912), Bliss (1968), Bollenbacher, Borst & O'Conner (1972), Boolootian (1965), Boolootian et al. (1959), Bovbjerg (1960a,b), Chapin (1968), Glynn (1965), Gross (1955, 1957a,b, 1958, 1959, 1961), Gross & Marshall (1960),

Hewatt (1935, 1937, 1938), Johnson & Snook (1927), Jones (1941), Kittredge, Terry & Takahashi (1971), Kittredge et al. (1962), Mac-Ginitie (1935), MacGinitie & MacGinitie (1968), Prosser, Green & Chow (1955), Rathbun (1918), Ricketts & Calvin (1968), Roberts (1957), Sakai (1965, 1976), Schmitt (1921), and R. I. Smith (1948).

Phylum Arthropoda / Class Crustacea / Subclass Malacostraca
Superorder Eucarida / Order Decapoda / Suborder Reptantia
Section Brachyura / Family Grapsidae

25.44 **Hemigrapsus nudus** (Dana, 1851)
Purple Shore Crab

Common in some areas under stones and among seaweeds, middle and low intertidal zones on rocky shores; less commonly along clay banks in sloughs and estuaries, but never in burrows; Yakobi Island (Alaska) to Bahía de Tortuga (Baja California); uncommon in southern California and southward; Pleistocene fossils from San Pedro (Los Angeles Co.) to San Diego Bay.

Carapace to 56.2 mm wide in males, 34 mm wide in females, smooth and convex anteriorly, flat and punctate posteriorly, the anterolateral margins rounded and bearing two teeth; chelae of males with a patch of hair inside; color usually purple, sometimes greenish yellow or reddish brown; easily distinguished from *H. oregonensis* (25.45) and from the slimmer *Pachygrapsus crassipes* (25.43) by the red or purplish spots on the chelae.

In the southern part of its range, which it shares with *Pachygrapsus crassipes*, *H. nudus* remains at middle intertidal levels and below, generally lower down than *Pachygrapsus*. In the Puget Sound region (Washington), however, it inhabits also the higher intertidal levels on rocky shores and invades the gravel-shore habitat of *H. oregonensis*. The purple shore crab tolerates both brackish and hypersaline conditions, regulating osmotically in both media. The body wall is slightly more permeable to water than in *Pachygrapsus*, but less so than in *H. oregonensis*. Osmotic activity increases with degree of osmotic stress.

The diet in Puget Sound, where studies have been made, consists mainly of diatoms, desmids, and small green algae scraped from rocks with the cup-shaped tips of the chelae, as in *Pachygrapsus*. A small amount of animal material is also found in the gut. Vision appears to be an important sense, and the retinal pigments shift in position regularly to adapt the eye for day and night seeing. These shifts continue under experimental conditions of constant darkness, but not under constant illumination, and appear to be controlled by a "biological clock."

In the Monterey Bay area, breeding occurs in winter; ovigerous females are found only from November to April. In Puget Sound, where the season is a bit later, ovigerous females are present from early January to mid-July, with a seasonal peak in April. Copulation is much as in *Pachygrapsus*, but without preliminary courtship. Size of the brood varies with the size of the female, from a mere 441 eggs in a female 11.9 mm in carapace width to over 36,400 in a female 34 mm across the back; the average for 51 ovigerous females was about 13,000 eggs. One brood per year is the rule, and second broods appear to be rare. Hatched larvae enter the plankton and pass through five zoeal stages and a megalops stage before the first juvenile crab instar. Larvae have been raised in the laboratory and the instars described.

The parasitic isopod *Portunion conformis* occurs in the perivisceral cavity in some individuals of *H. nudis*.

See Boolootian et al. (1959), Dehnel (1958, 1960, 1962, 1966), Dehnel & Carefoot (1965), Dehnel & McCaughran (1964), Dehnel & Stone (1964), Fasten (1915), Gross (1957a), Hart (1935, 1940, 1968), Hu (1958), Johnson & Snook (1927), Jones (1941), Knudsen (1964b), MacGinitie (1935), MacGinitie & MacGinitie (1968), McWhinnie & Scheer (1958), Meenakshi & Scheer (1959), Piltz (1969), Rathbun (1918, 1926), Ricketts & Calvin (1968), Scheer & Meenakshi (1961), Schmitt (1921), R. I. Smith (1948), R. I. Smith & Rudy (1972), Thompson & Chow (1955), and Todd & Dehnel (1960).

Phylum Arthropoda / Class Crustacea / Subclass Malacostraca
Superorder Eucarida / Order Decapoda / Suborder Reptantia
Section Brachyura / Family Grapsidae

25.45 **Hemigrapsus oregonensis** (Dana, 1851)
Yellow Shore Crab, Mud-Flat Crab

On open mud flats, in mats of the green alga *Enteromorpha*, and in beds of the eelgrass *Zostera*, high to low intertidal zones in bays and estuaries: Resurrection Bay (Alaska) to

Bahía de Todos Santos (Baja California); not Gulf of California (specimens reported from there are now recognized as *Goetice americanus* Rathbun); Pleistocene fossils from San Pedro (Los Angeles Co.).

Carapace width to 34.7 mm in males and 29.1 mm in females; color yellow or gray, the carapace and legs mottled with brownish purple or black, the claw tips light yellow or white; species distinguished from *H. nudus* (25.44) by its dull color, hairy legs, and four-lobed anterior margin of carapace, and by absence of red spots on chelipeds.

This species tends to replace *Pachygrapsus crassipes* (25.43) and *Hemigrapsus nudus* in bays, on mud flats, and on gravel shores. Though its body is more permeable to water than those of the other two species, it osmoregulates effectively in both brackish and hypersaline media. The crabs are common in parts of San Francisco Bay where the salinity averages 73 percent that of seawater, and a population in Los Peñasquitos Lagoon (San Diego Co.) was found living at a salinity of 175 percent that of seawater, far beyond the tolerance limits of most members of the species. The crabs endure moderately polluted conditions in Los Angeles harbor.

The mud-flat crab feeds mainly at night, primarily on diatoms and green algae. Although largely an herbivore, it takes meat when available and can locate it at a distance under water by means of chemoreceptors on the first antennae. The eyes are not essential in finding food, but they show adaptive changes for day and night vision; as in other grapsoid crabs, these changes continue to occur when the crabs are held in constant darkness, but not when they are constantly illuminated. The crabs are good diggers and can bury themselves rapidly, providing some protection against such important predators as the western willet, a marsh bird.

Reproduction in the mud-flat crab is best known in populations in Washington and adjacent British Columbia. Mating occurs much as in *Pachygrapsus crassipes* and *H. nudus*. Under laboratory conditions a female may mate with several males, and observations suggest that males may be stimulated to mate by substances secreted by the female. In northern waters ovigerous females are seen from February to September, and by May some 90 percent are carrying eggs. Hatching occurs from May to July, and in August some fe-

males produce a second brood. The number of eggs carried by individual females ranges from 800 to 11,000 (averaging 4,500). The larval stages have been described and illustrated by Hart (1935). Detailed studies of reproduction in California waters are lacking, but females with eggs have been noted in Monterey Bay from November through February and in May. Young crabs are abundant from May to August and are sometimes used by fishermen as bait for the pile perch, *Rhacochilus vacca*.

The parasitic isopod *Portunion conformis* occurs in the perivisceral cavity of *H. oregonensis*. Infected hosts appear normal, and the presence of the parasites (one to four per infected host) can be determined only by dissection. Infection was present in 21.2 percent of the 372 crabs sampled from San Francisco Bay, and 12.3 percent of the 122 crabs from Bodega Bay.

See Bowman (1949), Dehnel (1958, 1960, 1962), Dehnel & Carefoot (1965), Dehnel & McCaughran (1964), Dehnel & Stone (1964), Easton (1972), Filice (1958), Gross (1957a, 1961), Hart (1935, 1940, 1968), Holmes (1900), Johnson & Snook (1927), Jones (1941), Knudsen (1964b), MacGinitie (1935), MacGinitie & MacGinitie (1968), MacKay (1943a), Muscatine (1956), Piltz (1969), Rathbun (1918, 1926), Reeder (1951), Reish (1961a), Reish & Winter (1954), Ricketts & Calvin (1968), Schmitt (1921), R. I. Smith (1948), Symons (1964), Todd & Dehnel (1960), Turner & Sexsmith (1964), and Vassallo (1969).

Phylum Arthropoda / Class Crustacea / Subclass Malacostraca
Superorder Eucarida / Order Decapoda / Suborder Reptantia
Section Brachyura / Family Ocypodidae

25.46 Uca crenulata crenulata (Lockington, 1877)
Fiddler Crab

On sand and mud flats, high and middle intertidal zones in protected bays and estuaries, forming permanent burrows marked by presence of sand and mud pellets around openings; Playa del Rey, Los Angeles Co. (existence of small colony recently verified) to Isla de los Mangles (Baja California); former large colonies at Newport Bay (Orange Co.) and Mission Bay (San Diego Co.) now facing extinction.

Carapace up to 19.7 mm wide in males, 17 mm wide in females, rectangular, wider than long, convex, smooth; females with chelae small and of equal size; males with one chela small, the other greatly enlarged, the large one with an oblique ridge inside propodus, bending at an obtuse angle and continuing to upper margin.

Another subspecies, *U. crenulata coloradensis* (Rathbun), occurs in the Gulf of California (Bahía San Felipe to Bahía de Tenacatita).

Fiddler crabs dig burrows at the levels of the highest tides and feed on minute plants and animals contained in sandy mud. Females feed with both chelae, but males use only the small claw in taking food. Feeding is selective, and the rejected mud remains as pellets about the burrows. Males use the large claw in ritualistic encounters with other males (which seldom result in injury), and in beckoning to the females, which are lured into the male's burrow for mating. Motions of the large and small claw in males were probably responsible for the name "fiddler crab," although some species stridulate, producing a rasping sound; one of the more common of these is *Uca musica* Rathbun, which occurs from Baja California and the Gulf of California to Puerto Pizarro, Peru, and perhaps formerly near San Diego as well.

Uca crenulata tolerates brackish conditions well and also occurs in lagoons where salinities are high (up to 150 percent seawater). The animals osmoregulate in both dilute and hypersaline media. In other respects *U. crenulata* has not been the object of much study, but a great deal is known of the behavior and physiology of other species of fiddler crabs.

Although subgenera are not used in this account, it should be noted that the correct subgeneric name for both *Uca crenulata* and *U. musica* is *Leptuca* Bott, 1973, rather than *Celuca* Crane, 1975 (see von Hagen, 1976).

See especially Crane (1975); see also Altevogt (1957), Bliss & Mantel (1968), Bott (1973), Crane (1941, 1957), Dembrowski (1926), Fingerman (1966), Gross (1961), Johnson & Snook (1927), Jones (1941), Peters (1955), Rathbun (1918), Ricketts & Calvin (1968), Schmitt (1921), Valiela et al. (1974), Vernberg (1969), von Hagen (1976), and Waterman (1960, 1961, indexes).

Literature Cited

Ally, J. R. R. 1974. A description of the laboratory-reared first and second zoeae of *Portunus xantusii* (Stimpson) (Brachyura, Decapoda). Calif. Fish & Game 60: 74–78.

Altevogt, R. 1957. Untersuchungen zur Biologie, Ökologie, und Physiologie indischer Winkerkrabben. Z. Morphol. Ökol. Tiere 46: 1–110.

Anderson, G. L. 1975. The effect of intertidal height and the parasitic crustacean *Fabia subquadrata* Dana on the nutrition and reproductive capacity of the California sea mussel *Mytilus californianus* Conrad. Veliger 17: 299–306.

Andrews, H. L. 1945. The kelp beds of the Monterey region. Ecology 26: 24–37.

Baker, C. F. 1912. Notes on the Crustacea of Laguna Beach. Rep. Laguna Mar. Lab. 1: 100–117.

Beondé, A. C. 1968. *Aplysia vaccaria*, a new host for the pinnotherid crab *Opisthopus transversus*. Veliger 10: 375–78.

Berkeley, E., and C. Berkeley. 1941. On a collection of Polychaeta from southern California. Bull. So. Calif. Acad. Sci. 40: 16–60.

Bliss, D. E. 1968. Transition from water to land in decapod crustaceans. Amer. Zool. 8: 355–92.

Bliss, D. E., and L. H. Mantel, eds. 1968. Terrestrial adaptations in Crustacea. Amer. Zool. 8: 307–685.

Bollenbacher, W. E., D. W. Borst, and J. D. O'Conner. 1972. Endocrine regulation of lipid synthesis in decapod crustaceans. Amer. Zool. 12: 381–84.

Boolootian, R. A. 1965. Aspects of reproductive biology in the striped shore crab *Pachygrapsus crassipes*. Bull. So. Calif. Acad. Sci. 64: 43–49.

Boolootian, R. A., A. C. Giese, A. Farmanfarmaian, and J. Tucker. 1959. Reproductive cycles of five west coast crabs. Physiol. Zool. 32: 213–20.

Bott, R. 1973. Die verwandtschaftlichen Beziehungen der *Uca*-Arten (Decapoda: Ocypodidae). Senckenbergiana Biol. (Frankfurt) 54: 315–25.

Bovbjerg, R. V. 1960a. Behavioral ecology of the crab, *Pachygrapsus crassipes*. Ecology 41: 668–72.

———. 1960b. Courtship behavior of the lined shore crab, *Pachygrapsus crassipes* Randall. Pacific Sci. 14: 421–22.

Bowman, T. E. 1949. Chromatophorotropins in the central nervous system of the crab, *Hemigrapsus oregonensis*. Biol. Bull. 96: 238–45.

Buchanan, D. V., and R. E. Millemann. 1969. The prezoeal stage of the Dungeness crab, *Cancer magister* Dana. Biol. Bull. 137: 250–55.

Buitendijk, A. M., and L. B. Holthuis. 1949. Note on the Zuiderzee crab, *Rhithropanopeus harrisii* (Gould) subspecies *tridentatus* (Maitland). Zool. Meded. 30: 95–106.

Butler, T. H. 1967. A bibliography of the Dungeness crab, *Cancer magister* Dana. Fish. Res. Bd. Canada, Tech. Rep. 1: 1–12.

Carlisle, J. G., Jr. 1969. Invertebrates taken in six year trawl study in Santa Monica Bay. Veliger 11: 237–42.

Carlton, J. T., and A. M. Kuris. 1975. Keys to decapod Crustacea, pp. 385–412, *in* R. I. Smith and J. T. Carlton, eds. Light's manual: Intertidal invertebrates of the central California coast. 3rd ed. Berkeley and Los Angeles: University of California Press. 716 pp.

Case, J. 1964. Properties of the dactyl chemoreceptors of *Cancer antennarius* Stimpson and *C. productus* Randall. Biol. Bull. 127: 428–46.

Caso, M. E. 1963. Estudios sobre equinodermos de México. Contribución al conocimiento de los holoturoideos de Zijuatanejo y de la Isla de Ixtapa (primera parte). Anales Inst. Biol. Univ. México, 36: 253–91.

Chamberlain, N. A. 1962. Ecological studies of the larval development of *Rhithropanopeus harrisii* (Xanthidae, Brachyura). Tech. Rep. Chesapeake Bay Inst. 28: 1–47.

Chapin, D. 1968. Some observations of predation on *Acmaea* species by the crab *Pachygrapsus crassipes*. Veliger 11 (Suppl): 67–68.

Cheng, T. C. 1967. Advances in marine biology. 5. Marine mollusks as hosts for symbioses, with a review of known parasites of commercially important species. New York: Academic Press. 424 pp.

Clark, R. B. 1956. *Capitella capitata* as a commensal, with a bibliography of parasites and commensalism in the polychaetes. Ann. Mag. Natur. Hist. (12) 9: 433–48 (see especially pp. 442–44).

Connolly, C. J. 1925. The larval stages and megalops of *Rhithropanopeus harrisii* (Gould). Contr. Canad. Biol. & Fish. (n.s.) 2: 329–34.

Cook, R. T. 1965. Predators of *Pagurus*. Research paper, Biology 175h, library, Hopkins Marine Station of Stanford University, Pacific Grove, Calif. 24 pp.

Costlow, J. D., Jr. 1966. The effect of eyestalk extirpation on larval development of the mud crab, *Rhithropanopeus harrisii* (Gould). Gen. Comp. Endocrinol. 7: 255–74.

Costlow, J. D., Jr., and C. G. Bookhout. 1960. A method for developing brachyuran eggs *in vitro*. Limnol. Oceanogr. 5: 212–15.

———. 1971. The effect of cyclic temperatures on larval development in the mud-crab *Rhithropanopeus harrisii*, pp. 211–20, *in* D. J. Crisp, ed., 4th European Mar. Biol. Sympos. Cambridge, Eng.: Cambridge University Press.

Costlow, J. D., Jr., C. G. Bookhout, and R. J. Monroe. 1966. Studies on the larval development of the crab *Rhithropanopeus harrisii* (Gould). I. Effect of salinity and temperature on larval development. Physiol. Zool. 39: 81–100.

Crane, J. 1941. Crabs of the genus *Uca* from the west coast of Central America. Zoologica 26: 145–208.

———. 1957. Basic patterns of display in fiddler crabs (Ocypodidae, genus *Uca*). Zoologica 42: 69–82.

———. 1975. Fiddler crabs of the world (Ocypodidae: genus *Uca*). Princeton, N.J.: Princeton University Press. 736 pp.

Davidson, E. S. 1968. The *Pinnotheres concharum* complex (Crustacea, Decapoda, family Pinnotheridae). Bull. So. Calif. Acad. Sci. 67: 85–88.

Dehnel, P. A. 1958. Effect of photoperiod on the oxygen consumption of two species of intertidal crabs. Nature 181: 1415–17.

———. 1960. Effect of temperature and salinity on the oxygen consumption of two intertidal crabs. Biol. Bull. 118: 215–49.

———. 1962. Aspects of osmoregulation in two species of intertidal crabs. Biol. Bull. 122: 208–27.

———. 1966. Chloride regulation in the crab *Hemigrapsus nudus*. Physiol. Zool. 39: 259–65.

Dehnel, P. A., and T. H. Carefoot. 1965. Ion regulation in two species of intertidal crabs. Comp. Biochem. Physiol. 15: 377–97.

Dehnel, P. A., and D. A. McCaughran. 1964. Gill tissue respiration in two species of estuarine crabs. Comp. Biochem. Physiol. 13: 233–59.

Dehnel, P. A., and D. Stone. 1964. Osmoregulatory role of the antenary gland in two species of estuarine crabs. Biol. Bull. 126: 354–72.

Dembrowski, J. B. 1926. Notes on the behavior of the fiddler crab. Biol. Bull. 50: 179–201.

Easton, D. M. 1972. Autotomy of walking legs in the Pacific shore crab *Hemigrapsus oregonensis*. Mar. Behav. Physiol. 1: 209–17.

Fabbiani, L. A. 1972. Dos nuevas citas de xánthidos (Decápodos braquiuros) para las costas del Atlántico, *Eriphia squamata* y *Pilumnus spinohirsutus*. Mem. Soc. Cien. Nat. La Salle 32: 47–54.

Fasten, N. 1915. The male reproductive organs of some common crabs of Puget Sound. Puget Sound Mar. Sta. Publ. 1: 35–41.

———. 1921. The explosion of the spermatozoa of the crab *Lophopanopeus bellus* (Stimpson) Rathbun. Biol. Bull. 41: 288–301.

———. 1926. Spermatogenesis of the black-clawed crab, *Lophopanopeus bellus* (Stimpson) Rathbun. Biol. Bull. 50: 277–93.

Filice, F. P. 1958. Invertebrates from the estuarine portion of San Francisco Bay and some factors influencing their distribution. Wasmann J. Biol. 16: 159–211.

———. 1959. The effect of wastes on the distribution of bottom invertebrates in the San Francisco Bay estuary. Wasmann J. Biol. 17: 1–17.

Fingerman, M. 1966. Neurosecretory control of pigmentary effectors in crustaceans. Amer. Zool. 6: 169–79.

Forward, R. B., Jr., and J. D. Costlow, Jr. 1974. The ontogeny of phototaxis by larvae of the crab *Rhithropanopeus harrisii* (Gould). Mar. Biol. 26: 27–33.

Frank, J. R., S. D. Sulkin, and R. P. Morgan II. 1975. Biochemical changes during larval development of the xanthid crab *Rhithropanopeus harrisii*. I. Protein, total lipid, alkaline phosphatase, and glutamic oxaloacetic transaminase. Mar. Biol. 32: 105–11.

Frey, H. W., ed. 1971. California's living marine resources and their utilization. Calif. Dept. Fish & Game. 148 pp.

Garth, J. S. 1955. The case for a warm-temperate marine fauna on the west coast of North America, pp. 19–27, *in* Essays in the natural sciences in honor of Captain Allan Hancock. Los Angeles: University of Southern California Press.

———. 1958. Brachyura of the Pacific coast of America. Oxyrhyncha. Allan Hancock Pacific Exped. 21: 1–854.

———. 1961. Distribution and affinities of the brachyuran Crustacea. Syst. Zool. 9: 105–23.

Garth, J. S., and W. Stephenson. 1966. Brachyura of the Pacific coast of America. Brachyrhyncha: Portunidae. Allan Hancock Monogr. Mar. Biol. 1: 1–154.

Ghiradella, H., J. Cronshaw, and J. Case. 1970. Surface of the cuticle on the aesthetascs of *Cancer*. Protoplasma 69: 145–50.

Giese, A. C. 1966. Lipids in the economy of marine invertebrates. Physiol. Rev. 46: 244–98.

Glaessner, M. F. 1969. Decapoda, pp. 399–651, *in* R. C. Moore et al., eds., Treatise on invertebrate paleontology, part R, Arthropoda 4, vol. 2. New York: Geol. Soc. Amer.; Lawrence: University of Kansas Press.

Glassell, S. A. 1933. Notes on *Parapinnixa affinis* Holmes and its allies. Trans. San Diego Soc. Natur. Hist. 7: 319–30.

———. 1935. New or little known crabs from the Pacific coast of northern Mexico. Trans. San Diego Soc. Natur. Hist. 8: 93–105.

Glynn, P. W. 1965. Community composition, structure, and interrelationships in the marine intertidal *Endocladia muricata–Balanus glandula* association in Monterey Bay, California. Beaufortia 12: 1–198.

Gotshall, D. W. 1977. Stomach contents of northern California Dungeness crabs, *Cancer magister*. Calif. Fish & Game 63: 43–51.

———. 1978a. Relative abundance studies of Dungeness crabs, *Cancer magister*, in northern California. Calif. Fish & Game 64: 24–37.

———. 1978b. Catch-per-unit-of-effort studies of northern California Dungeness crabs, *Cancer magister*. Calif. Fish & Game 64: 189–99.

———. 1978c. Northern California Dungeness crab, *Cancer magister*, movements as shown by tagging. Calif. Fish & Game 64: 234–54.

Green, J. 1961. A biology of Crustacea. London: Witherby. 180 pp.

Gross, W. J. 1955. Aspects of osmotic regulation in crabs showing the terrestrial habit. Amer. Natur. 89: 205–22.

———. 1957a. An analysis of response to osmotic stress in selected decapod Crustacea. Biol. Bull. 112: 43–62.

———. 1957b. A behavioral mechanism for osmotic regulation in a semi-terrestrial crab. Biol. Bull. 113: 268–74.

———. 1958. Potassium and sodium regulation in an intertidal crab. Biol. Bull. 114: 334–47.

———. 1959. The effect of osmotic stress on the ionic exchange of a shore crab. Biol. Bull. 116: 248–57.

———. 1961. Osmotic tolerance and regulation in crabs from a hypersaline lagoon. Biol. Bull. 121: 290–301.

Gross, W. J., and L. A. Marshall. 1960. The influence of salinity on the magnesium and water fluxes of a crab. Biol. Bull. 119: 440–53.

Guinot, D. 1969. Recherches préliminaires sur les groupements naturels chez les Crustacés Décapodes Brachyoures. VII. Les Goneplacidae. Bull. Mus. Nat. Hist. Natur. Paris (2) 41: 241–65.

Gurney, R. 1960. Bibliography of the larvae of decapod Crustacea *and* Larvae of decapod Crustacea. Codicote, Herts, Eng.: Wheldon & Wesley. 429 pp. (Reprinting, in one volume, of Ray Society publications 125 and 129, originally issued in 1939 and 1942.)

Haderlie, E. C. 1968. Marine fouling organisms in Monterey harbor. Veliger 10: 327–41.

———. 1969. Marine fouling and boring organisms in Monterey harbor. II. Second year of investigation. Veliger 12: 182–92.

Haig, J., and M. K. Wicksten. 1975. First records and range extensions of crabs in California waters. Bull. So. Calif. Acad. Sci. 74: 100–104.

Hart, J. F. L. 1930. Some decapods from the south-eastern shores of Vancouver Island. Canad. Field Natur. 44: 101–9.

———. 1935. The larval development of British Columbia Brachyura. I. Xanthidae, Pinnotheridae (in part), and Grapsidae. Canad. J. Res. D12: 411–32.

———. 1937. Culture methods for Brachyura and Anomura, pp. 237–38, *in* P. S. Galtsoff, F. E. Lutz, P. S. Welch, and J. G. Needham, eds., Culture methods for invertebrate animals. Ithaca, N.Y.: Comstock. 590 pp. (Paperback ed., Dover, New York, 1959.)

———. 1940. Reptant decapod Crustacea of the west coasts of Vancouver and Queen Charlotte Islands, British Columbia. Canad. J. Res. D18: 86–105.

———. 1953. Northern extensions of range of some reptant decapod Crustacea of British Columbia. Canad. Field Natur. 67: 139–40.

———. 1962. Records of distribution of some Crustacea in British Columbia. Rep. Prov. Mus. Natur. Hist. Anthropol. for 1961: W17–19.

———. 1968. Crab-like Anomura and Brachyura (Crustacea: Decapoda) from southeastern Alaska and Prince William Sound. Nat. Mus. Canada Natur. Hist. Pap. 38: 1–6.

———. 1971a. Key to planktonic larvae of families of decapod Crustacea of British Columbia. Syesis 4: 227–34.

———. 1971b. New distribution records of reptant decapod Crustacea, including descriptions of three new species of *Pagurus*, from the waters adjacent to British Columbia. J. Fish. Res. Bd. Canada 28: 1527–44.

Hartman, O. 1952. *Iphitime* and *Ceratocephala* (polychaetous annelids) from California. Bull. So. Calif. Acad. Sci. 51: 9–20.

Heath, J. P. 1941. The nervous system of the kelp crab, *Pugettia producta*. J. Morphol. 69: 481–92.

Hewatt, W. G. 1935. Ecological succession in the *Mytilus californianus* habitat as observed in Monterey Bay, California. Ecology 16: 244–51.

————. 1937. Ecological studies on selected marine intertidal communities of Monterey Bay, California. Amer. Midl. Natur. 18: 161–206.

————. 1938. Notes on the breeding seasons of the rocky beach fauna of Monterey Bay, California. Proc. Calif. Acad. Sci. (4) 23: 283–88.

Hiatt, R. W. 1948. The biology of the lined shore crab, *Pachygrapsus crassipes* Randall. Pacific Sci. 2: 135–213.

Hines, A. H. 1978. Population dynamics and niche analysis of five species of spider crabs (Brachyura, Majidae) sympatric in a kelp forest. 59th Ann. Meet., West. Soc. Natur., Abs. Symp. Contr. Pap., p. 6 (abstract).

Hinton, S. 1969. Seashore life of southern California. Calif. Natur. Hist. Guides 26: 1–181.

Holmes, S. J. 1900. Synopsis of the California stalk-eyed Crustacea. Occas. Pap. Calif. Acad. Sci. 7: 1–262 (including complete bibliography to 1900).

Hood, M. R. 1962. Studies on the larval development of *Rhithropanopeus harrisii* (Gould) of the family Xanthidae (Brachyura). Gulf Res. Rep. 1: 122–30.

Hopkins, T. S., and T. B. Scanland. 1964. The host relations of a pinnotherid crab, *Opisthopus transversus* Rathbun (Crustacea: Decapoda). Bull. So. Calif. Acad. Sci. 63: 175–80.

Hu, A. S. L. 1958. Glucose metabolism in the crab *Hemigrapsus nudus*. Arch. Biochem. Biophys. 75: 387–95.

Huang, H., and A. C. Giese. 1958. Tests for digestion of algal polysaccharides by some marine herbivores. Science 127: 475.

Hubbs, C. L. 1967. A discussion of the geochronology and archeology of the California islands, pp. 337–41, in Proc. Symp. Biol. Calif. Islands. Santa Barbara, Calif.: Santa Barbara Botanic Garden.

Irvine, J. A., and H. G. Coffin. 1960. Laboratory culture and early stages of *Fabia subquadrata* (Dana) (Crustacea, Decapoda). Publ. Dept. Biol. Sci., Walla Walla College 28: 1–24.

Johnson, M. E., and H. J. Snook. 1927. Seashore animals of the Pacific coast. New York: Macmillan. (Paperback ed., Dover, New York, 1967.)

Jones, L. L. 1940. An introduction of an Atlantic crab into San Francisco Bay. Proc. 6th Pacific Sci. Congr. 3: 485–86.

————. 1941. Osmotic regulation in several crabs of the Pacific coast of North America. J. Cell. Compar. Physiol. 18: 79–92.

Kaestner, A. 1970. Crustacea. 3. Invertebrate zoology. (Translated from the 2nd German ed. by H. W. Levi and L. R. Levi.) New York: Wiley. 512 pp.

Kalber, F. A. 1970. Osmoregulation in decapod larvae as a consideration in culture technique. Helgoländer Wiss. Meeresunters. 20: 697–706.

Kalber, F. A., Jr., and J. D. Costlow, Jr. 1966. The ontogeny of osmoregulation and its neurosecretory control in the decapod crustacean, *Rhithropanopeus harrisii*. Amer. Zool. 6: 221–29.

Kim, H. S. 1973. Anomura-Brachyura. Illustrated encyclopedia of fauna & flora of Korea (Seoul) 14: 1–694.

Kinne, O., and H. W. Rotthauwe. 1952. Biologische Beobachtungen über die Blutkonzentration an *Heteropanope tridentatus* (Maitland) (Dekap.). Kieler Meeresforsch. 8: 212–17.

Kittredge, J. S., and F. T. Takahashi. 1972. The evolution of sex pheromone communication in the Arthropoda. J. Theoret. Biol. 35: 467–71.

Kittredge, J. S., M. Terry, and F. T. Takahashi. 1971. Sex pheromone activity of the molting hormone, crustecdysone, on male crabs (*Pachygrapsus crassipes, Cancer antennarius,* and *C. anthonyi*). Fish. Bull. 69: 337–43.

Kittredge, J. S., D. G. Simonsen, H. Roberts, and B. Jelinek. 1962. Free amino acids of marine invertebrates, pp. 176–86, in J. T. Holden, ed., Amino acid pools: Distribution, formation and function of free amino acids. Amsterdam: Elsevier. 815 pp.

Knudsen, J. W. 1958. Life cycle studies of the Brachyura of western North America. I. General culture methods and the life cycle of *Lophopanopeus leucomanus leucomanus* (Lockington). Bull. So. Calif. Acad. Sci. 57: 51–59.

————. 1959a. Life cycle studies of the Brachyura of western North America. II. The life cycle of *Lophopanopeus bellus diegensis* Rathbun. Bull. So. Calif. Acad. Sci. 58: 57–64.

————. 1959b. Life cycle studies of the Brachyura of western North America. III. The life cycle of *Paraxanthias taylori* (Stimpson). Bull. So. Calif. Acad. Sci. 58: 138–45.

————. 1960a. Aspects of the ecology of the California pebble crabs (Crustacea: Xanthidae). Ecol. Monogr. 30: 165–85.

————. 1960b. Life cycle studies of the Brachyura of western North America. IV. The life cycle of *Cycloxanthops novemdentatus* (Stimpson) [error for Lockington]. Bull. So. Calif. Acad. Sci. 59: 1–8.

————. 1960c. Reproduction, life history, and larval ecology of the California Xanthidae, the pebble crabs. Pacific Sci. 14: 3–17.

————. 1964a. Observations of the mating process of the spider crab *Pugettia producta* (Majidae, Crustacea). Bull. So. Calif. Acad. Sci. 63: 38–41.

————. 1964b. Observations of the reproductive cycles and ecology of the common Brachyura and crablike Anomura of Puget Sound, Washington. Pacific Sci. 18: 3–33.

Kobjakova, Z. I. 1967. Decapoda (Crustacea, Decapoda) from the Possjet Bay (the Sea of Japan). [In Russian.] Zool. Inst. Acad. Sci. U.S.S.R., Issledovanija Fauny Morei 5: 230–47.

Lebour, M. V. 1928. The larval stages of the Plymouth Brachyura. Proc. Zool. Soc. London 1928: 473–560.

Leighton, D. L. 1966. Studies of food preference in algivorous invertebrates of southern California kelp beds. Pacific Sci. 20: 104–13.

Lie, U. 1968. A quantitative study of benthic infauna in Puget Sound, Washington, U.S.A., in 1963–1964. Fiskeridir. Skr. Ser. Havunders. 14: 229–556.

Limbaugh, C. 1955. Fish life in the kelp beds and the effects of kelp harvesting. Univ. Calif. Inst. Mar. Res., IMR Ref. 55-9: 1–158.

———. 1961. Life-history and ecologic notes on the black croaker. Calif. Fish & Game 47: 163–74.

Lockwood, A. P. M. 1967. Aspects of the physiology of Crustacea. San Francisco: Freeman. 328 pp.

MacGinitie, G. E. 1935. Ecological aspects of a California marine estuary [Elkhorn Slough]. Amer. Midl. Natur. 16: 629–765.

———. 1937. Notes on the natural history of several marine Crustacea. Amer. Midl. Natur. 18: 1031–37.

MacGinitie, G. E., and N. MacGinitie. 1968. Natural history of marine animals. 2nd ed., rev. New York: McGraw-Hill. 523 pp.

MacKay, D. C. G. 1931. Notes on brachyuran crabs of northern British Columbia. Canad. Field Natur. 45: 187–89.

———. 1934. The growth and life history of the Pacific edible crab, *Cancer magister* Dana. Doctoral thesis, Biological Sciences, Stanford University, Stanford, Calif. 253 pp. (Numerous publications 1931–43; see Butler, 1967.)

———. 1937. Notes on rearing the Pacific edible crab, *Cancer magister*, pp. 239–41, in P. S. Galtsoff, F. E. Lutz, P. S. Welch, and J. G. Needham, eds., Culture methods for invertebrate animals. Ithaca, N.Y.: Comstock. 590 pp. (Paperback ed., Dover, New York, 1959.)

———. 1942. The Pacific edible crab, *Cancer magister*. Fish. Res. Bd. Canada Bull. 62: 1–32.

———. 1943a. The behavior of the Pacific edible crab, *Cancer magister* Dana. J. Compar. Psychol. 36: 255–68.

———. 1943b. Temperature and world distribution of crabs of the genus *Cancer*. Ecology 24: 113–15.

McLean, J. H. 1962. Sublittoral ecology of kelp beds of the open coast area near Carmel, California. Biol. Bull. 122: 95–114.

McWhinnie, M. A., and B. T. Scheer. 1958. Blood glucose of the crab *Hemigrapsus nudus*. Science 128: 90.

Meenakshi, V. R., and B. T. Scheer. 1959. Acid mucopolysaccharide of the crustacean cuticle. Science 130: 1189–90.

Menzies, R. J. 1948. A revision of the brachyuran genus *Lophopanopeus*. Allan Hancock Found. Occas. Pap. 4: 1–45.

———. 1951. Pleistocene Brachyura from the Los Angeles area: Cancridae. J. Paleontol. 25: 165–70.

Mir, R. D. 1961. The external morphology of the first zoeal stages of the crabs, *Cancer magister* Dana, *Cancer antennarius* Stimpson, and *Cancer anthonyi* Rathbun. Calif. Fish & Game 47: 103–11.

Mitchell, D. F. 1953. An analysis of stomach contents of California tidepool fishes. Amer. Midl. Natur. 49: 862–71.

Morgan, R. P., E. Kramarsky, and S. D. Sulkin. 1978. Biochemical changes during larval development of the xanthid crab *Rhithropanopeus harrisii*. III. Isozyme changes during ontogeny. Mar. Biol. 48: 223–26.

Muscatine, L. 1956. A new entoniscid (Crustacea: Isopoda) from the Pacific coast. J. Wash. Acad. Sci. 46: 122–26.

Nations, J. D. 1970. The family Cancridae and its fossil record on the west coast of North America. Doctoral thesis, Paleontology, University of California, Berkeley. 252 pp.

———. 1975. The genus *Cancer* (Crustacea: Brachyura): Systematics, biogeography and fossil record. Los Angeles Co. Mus. Natur. Hist., Sci. Bull. 23: 1–104.

North, W. J., ed. 1964. An investigation of the effects of discharged wastes on kelp. Resources Agency of Calif., State Water Qual. Control Bd., Publ. 26: 1–126.

Orcutt, H. G. 1950. The life history of the starry flounder, *Platichthys stellatus* (Pallas). Calif. Dept. Fish & Game, Fish Bull. 78: 1–64.

Ott, F. S., and R. B. Forward, Jr. 1976. The effect of temperature on phototaxis and geotaxis by larvae of the crab *Rhithropanopeus harrisii* (Gould). J. Exper. Mar. Biol. Ecol. 23: 97–107.

Pautsch, F., L. Lawinski, and K. Turoboyski. 1969. Zur Ökologie der Krabbe *Rhithropanopeus harrisii* (Gould) (Xanthidae). Limnologica 7: 63–68.

Payen, G., J. D. Costlow, Jr., and H. Charniaux-Cotton. 1971. Étude comparative de l'ultrastructure des glandes androgènes de crabes normaux et pédonuclectomisés pendant la vie larvaire ou après la puberté chez les espèces: *Rhithropanopeus harrisii* (Gould) et *Callinectes sapidus* Rathbun. Gen. Comp. Endocrinol. 17: 526–42.

Pearce, J. B. 1962. Adaptation in symbiotic crabs of the family Pinnotheridae. Biologist 45: 11–15.

———. 1965. On the distribution of *Tresus capax* and *Tresus nuttalli* in the waters of Puget Sound and the San Juan Archipelago (Pelecypoda: Mactridae). Veliger 7: 166–70.

———. 1966a. The biology of the mussel crab, *Fabia subquadrata*, from the waters of the San Juan Archipelago, Washington. Pacific Sci. 20: 3–35.

———. 1966b. On *Pinnixa faba* and *Pinnixa littoralis* (Decapoda: Pinnotheridae) symbiotic with the clam, *Tresus capax* (Pelecypoda: Mactridae), pp. 565–89, in H. Barnes, ed., Some contemporary studies in marine science. London: Allen & Unwin. 716 pp.

Pearse, A. S. 1913. On the habits of Crustacea found in *Chaetopterus* tubes at Woods Hole, Massachusetts. Biol. Bull. 24: 102–14, pl. 1.

Peters, H. M. 1955. Die Winkgebärde von *Uca minuta* (Brachyura) in vergleichend-ethologischer, -ökologischer und -morphologischer-anatomischer Betrachtung. Z. Morphol. Ökol. Tiere 43: 425–500.

Phillips, J. B. 1935. The crab fishery of California. Calif. Fish & Game 21: 38–60.

———. 1939. The market crab of California and its close relatives. Calif. Fish & Game 25: 18–29.

Pilger, J. 1971. A new species of *Iphitime* (Polychaeta) from *Cancer antennarius* (Crustacea: Decapoda). Bull. So. Calif. Acad. Sci. 70: 84–87.

Piltz, F. M. 1969. A record of the entoniscid parasite, *Portunion conformis* Muscatine (Crustacea: Isopoda) infecting two species of *Hemigrapsus*. Bull. So. Calif. Acad. Sci. 68: 257–59.

Poole, R. L. 1966a. A description of laboratory-reared zoeae of *Cancer magister* Dana, and megalopae taken under natural conditions (Decapoda Brachyura). Crustaceana 11: 83–97.

———. 1966b. A sexually abnormal red crab, *Cancer productus* Randall. Calif. Fish & Game 52: 117.

———. 1967. Preliminary results of the age and growth study of the market crab (*Cancer magister*) in California: The age and growth of *Cancer magister* in Bodega Bay. Mar. Biol. Assoc. India, Symp. on Crustacea, Proc. Pt. II, pp. 553–67.

Prosser, C. L., J. W. Green, and T. J. Chow. 1955. Ionic and osmotic concentrations in blood and urine of *Pachygrapsus crassipes* acclimated to different salinities. Biol. Bull. 109: 99–107.

Quast, J. C. 1968. Observations on the food of the kelp-bed fishes, pp. 109–42, *in* W. J. North and C. L. Hubbs, eds., Utilization of kelp-bed resources in southern California. Calif. Dept. Fish & Game, Fish Bull. 139: 1–264.

Quayle, D. B. 1960. The intertidal bivalves of British Columbia. British Columbia Prov. Mus., Handbook 17: 1–104 (see especially p. 14).

Rathbun, M. J. 1904. Decapod crustaceans of the northwest coast of North America. Harriman Alaska Expedition 10: 1–210.

———. 1918. The grapsoid crabs of America. Bull. U.S. Nat. Mus. 97: 1–461.

———. 1925. The spider crabs of America. Bull. U.S. Nat. Mus. 129: 1–613.

———. 1926. The fossil stalk-eyed Crustacea of the Pacific slope of North America. Bull. U.S. Nat. Mus. 138: 1–155.

———. 1930. The cancroid crabs of America. Bull. U.S. Nat. Mus. 152: 1–609.

———. 1937. The oxystomatous and allied crabs of America. Bull. U.S. Nat. Mus. 166: 1–278.

Reed, P. H. 1969. Culture methods and effects of temperature and salinity on survival and growth of Dungeness crab (*Cancer magister*) larvae in the laboratory. J. Fish. Res. Bd. Canada 26: 389–97.

Reeder, W. G. 1951. Stomach analysis of a group of shorebirds. Condor 53: 43–45.

Reish, D. J. 1961a. A study of benthic fauna in a recently constructed boat harbor in southern California. Ecology 42: 84–91.

———. 1961b. The use of the sediment bottle collector for monitoring polluted marine waters. Calif. Fish & Game 47: 261–72.

Reish, D. J., and H. A. Winter. 1954. The ecology of Alamitos Bay, California, with special reference to pollution. Calif. Fish & Game 40: 105–21.

Ricketts, E. F., and J. Calvin. 1968. Between Pacific tides. 4th ed. Revised by J. W. Hedgpeth. Stanford, Calif.: Stanford University Press. 614 pp.

Roberts, J. L. 1957. Thermal acclimation of metabolism in the crab *Pachygrapsus crassipes* Randall. I. The influence of body size, starvation, and moulting. II. Mechanisms and the influence of season and latitude. Physiol. Zool. 30: 232–55.

Sakai, T. 1965. The crabs of Sagami Bay. Tokyo: Maruzen. 206 pp.

———. 1976. Crabs of Japan and adjacent seas. Tokyo: Kodansha. 773 pp.

Scanland, T. D., and T. S. Hopkins. 1978. A supplementary description of *Pinnixa tomentosa* and comparison with the geographically adjacent *Pinnixa tubicola* (Brachyura, Pinnotheridae). Proc. Biol. Soc. Washington 91: 636–41.

Scheer, B. T., and V. R. Meenakshi. 1961. The metabolism of carbohydrates in arthropods, pp. 65–83, *in* A. W. Martin, ed., Comparative physiology of carbohydrate metabolism in heterothermic animals. Seattle: University of Washington Press. 147 pp.

Scheffer, V. B. 1959. Invertebrates and fishes collected in the Aleutians, 1936–38. No. Amer. Fauna 61: 365–406.

Scheuermann, W. F. 1958. An annotated bibliography of research in economically important species of California fish and game. Calif. Legisl. Assembly Interim Comm. Rep. 5: 1–271.

Schmitt, W. L. 1921. The marine decapod Crustacea of California. Univ. Calif. Publ. Zool. 23: 1–470.

———. 1965. Crustaceans. Ann Arbor: University of Michigan Press. 204 pp.

Schmitt, W. L., J. C. McCain, and E. S. Davidson. 1973. Crustaceorum Catalogus editus a H.-E. Gruner et L. B. Holthius. 3. Decapoda I, Brachyura I, Fam. Pinnotheridae. The Hague: Junk. 160 pp.

Shen, C.-J. 1932. The brachyuran Crustacea of North China. Zoologica Sinica, ser. A, Invertebrates of China, 9 (1): 1–320.

Skorkowski, E. F. 1972. Separation of three chromatophorotropic hormones from the eyestalk of the crab *Rhithropanopeus harrisii* (Gould). Gen. Comp. Endocrinol. 18: 329–34.

Smith, M. 1963. Desoxyribonucleic acids in crabs of the genus *Cancer*. Biochem. Biophys. Res. Commun. 10: 67–72.

Smith, R. I. 1948. The role of the sinus glands in retinal pigment migration in grapsoid crabs. Biol. Bull. 95: 169–85.

———. 1967. Osmotic regulation and adaptive reduction of water-permeability in a brackish-water crab, *Rhithropanopeus harrisii* (Brachyura: Xanthidae). Biol. Bull. 133: 643–58.

Smith, R. I., and P. P. Rudy. 1972. Water-exchange in the crab *Hemigrapsus nudus* measured by use of deuterium and tritium oxides as tracers. Biol. Bull. 143: 234–46.

Snow, C. D., and J. R. Nielsen. 1966. Pre-mating behavior of the Dungeness crab (*Cancer magister* Dana). J. Fish. Res. Bd. Canada 23: 1319–23.

Snow, C. D., and E. J. Wagner. 1965. Tagging of Dungeness crabs with spaghetti and dart tags. Oregon Res. Briefs 11: 5–13.

Sulkin, S. D. 1973. Depth regulation of crab larvae in the absence of light. J. Exper. Mar. Biol. Ecol. 13: 73–82.

———. 1975. The influence of light in the depth regulation of crab larvae. Biol. Bull. 148: 333–43.

Symons, P. E. K. 1964. Behavioral responses of the crab *Hemigrapsus oregonensis* to temperature, diurnal light variation, and food stimuli. Ecology 45: 580–91.

Taylor, G. W. 1912. Preliminary list of one hundred and twenty-nine species of British Columbia decapod crustaceans. Contr. Canad. Biol. 1906–10: 187–214.

Thompson, T. G., and T. J. Chow. 1955. The strontium-calcium atom ratio in carbonate-secreting marine organisms. Pap. Mar. Biol. Oceanogr., Deep-Sea Res. 3 (Suppl.): 20–39.

Todd, M.-E., and P. A. Dehnel. 1960. Effect of temperature and salinity on heat tolerance of two grapsoid crabs, *Hemigrapsus nudus* and *Hemigrapsus oregonensis*. Biol. Bull. 118: 150–72.

Trask, T. 1970. A description of laboratory-reared larvae of *Cancer productus* Randall (Decapoda, Brachyura) and a comparison to larvae of *Cancer magister* Dana. Crustaceana 18: 133–46.

Turner, C. H., E. E. Ebert, and R. R. Given. 1969. Man-made reef ecology. Calif. Dept. Fish & Game, Fish Bull. 146: 1–221.

Turner, C. H., and J. C. Sexsmith. 1964. Marine baits of California. Calif. Dept. Fish & Game. 71 pp.

Valiela, I., D. F. Babiec, W. K. Atherton, S. Seitzinger, and C. Krebs. 1974. Some consequences of sexual dimorphism: Feeding in male and female fiddler crabs, *Uca pugnax* (Smith). Biol. Bull. 147: 652–60.

Vassallo, M. T. 1969. The ecology of *Macoma inconspicua* (Broderip & Sowerby, 1829) in central San Francisco Bay. I. The vertical distribution of the *Macoma* community. Veliger 11: 223–34.

Vernberg, F. J. 1969. Acclimation of intertidal crabs. Amer. Zool. 9: 333–41.

Via, S. E., and R. B. Forward, Jr. 1975. The ontogeny and spectral sensitivity of polarotaxis in larvae of the crab *Rhithropanopeus harrisii* (Gould). Biol Bull. 149: 251–66.

von Hagen, H. O. 1976. Review: Jocelyn Crane, Fiddler crabs of the world. Ocypodidae: genus *Uca*. Crustaceana 31: 221–24.

Waldron, K. D. 1958. The fishery and biology of the Dungeness crab (*Cancer magister* Dana) in Oregon waters. Fish Comm. Oregon, Contr. 24: 1–43.

Warner, G. F. 1977. The biology of crabs. New York: Van Nostrand Reinhold. 202 pp.

Waterman, T. H., ed. 1960. The physiology of Crustacea. 1. Metabolism and growth. New York: Academic Press. 670 pp.

————, ed. 1961. The physiology of Crustacea. 2. Sense organs, integration, and behavior. New York: Academic Press. 681 pp.

Wear, R. G. 1970. Notes and bibliography on the larvae of xanthid crabs. Pacific Sci. 24: 84–89.

Webster, S. K. 1968. An investigation of the commensals of *Cryptochiton stelleri* (Middendorff, 1846) in the Monterey Peninsula area, California. Veliger 11: 121–25.

Wells, W. W. 1928. Pinnotheridae of Puget Sound. Puget Sound Biol. Sta. Publ. 6: 283–314.

————. 1940. Ecological studies on the pinnotherid crabs of Puget Sound. Univ. Washington Publ. Oceanogr. 2: 19–50.

Welsh, J. P. 1974. Mariculture of the crab *Cancer magister* (Dana) utilizing fish and crustacean wastes as food. Arcata, Calif.: Humboldt State University. 76 pp.

Weymouth, F. W. 1910. Synopsis of the true crabs (Brachyura) of Monterey Bay, California. Stanford Univ. Publ., Univ. Ser. 4: 1–64.

————. 1915. Contributions to the life history of the Pacific coast edible crab (*Cancer magister*). Rep. Comm. Fish., British Columbia, for 1914, 1: 123–29.

————. 1916. Contributions to the life history of the Pacific coast edible crab (*Cancer magister*). Rep. Comm. Fish., British Columbia, for 1915, 2: 161–63.

Weymouth, F. W., and D. C. G. MacKay. 1934. Relative growth in the Pacific edible crab, *Cancer magister*. Proc. Soc. Exper. Biol. Med. 31: 1137–39.

Weymouth, F. W., J. M. Crismon, V. E. Hall, H. S. Belding, and J. Field II. 1944. Total and tissue respiration in relation to body weight. A comparison of the kelp crab with other crustaceans and with mammals. Physiol. Zool. 17: 50–71.

Whitney, J. O. 1969. Absence of sterol synthesis in larvae of the mud crab *Rhithropanopeus harrisii* and of the spider crab *Libinia emarginata*. Mar. Biol. 3: 134–35.

Whittington, H. B., and W. D. I. Rolfe. 1963. Phylogeny and evolution of Crustacea. Mus. Compar. Zool., Harvard Univ., Cambridge, Mass. 192 pp.

Wickham, D. E. 1977. Effects of the crab-egg predator, *Carcinonemertes errantia*, on the population dynamics of its host, the Dungeness crab. 58th Ann. Meet., West. Soc. Natur., Abs. Symp. Contr. Pap., pp. 49–50 (abstract).

————. 1979. The relationship between megalopae of the Dungeness crab, *Cancer magister*, and the hydroid, *Velella velella*, and its influence on abundance estimates of *C. magister* megalopae. Calif. Fish & Game 65: 184–86.

Wickham, D. E., R. Schleser, and A. Schuur. 1976. Observations on the inshore population of Dungeness crab in Bodega Bay. Calif. Fish & Game 62: 89–92.

Wicksten, M. K. 1975. Observations on decorating behavior following molting in *Loxorhynchus crispatus* Stimpson (Decapoda, Majidae). Crustaceana 29: 315–16.

————. 1977a. Decorating in the crab *Loxorhynchus crispatus* Stimpson (Brachyura: Majidae). Doctoral thesis, Biology, University of Southern California, Los Angeles. 80 pp.

————. 1977b. Feeding in the decorator crab, *Loxorhynchus crispatus* (Brachyura: Majidae). Calif. Fish & Game 63: 122–24.

————. 1978. Attachment of decorating materials in *Loxorhynchus crispatus* (Brachyura: Majidae). Trans. Amer. Microscop. Soc. 97: 217–20.

————. 1979a. Records of *Cancer oregonensis* in California (Brachyura: Cancridae). Calif. Fish & Game 65: 118–20.

_____. 1979b. Decorating behavior in *Loxorhynchus crispatus* Stimpson and *Loxorhynchus grandis* Stimpson (Brachyura, Majidae). Crustaceana 5 (Suppl.): 37–46.

Willett, G. 1937. An Upper Pleistocene fauna from the Baldwin Hills, Los Angeles County, California. Trans. San Diego Soc. Natur. Hist. 8: 379–406.

Williamson, D. I. 1974 (publ. 1976). Larval characters and the origin of crabs (Crustacea, Decapoda, Brachyura). Thalassia Jugoslavica 10: 401–14.

Willis, M. 1968. Northern range extension for the yellow crab, *Cancer anthonyi*. Calif. Fish & Game 54: 217.

_____. 1971. Occurrence of hermaphroditism in the market crab, *Cancer magister*. Calif. Fish & Game 57: 131–32.

Wolfson, F. H. 1974. Two symbioses of *Conus* (Mollusca: Gastropoda) with brachyuran crabs. Veliger 16: 427–29.

Zullo, V. A., and D. D. Chivers. 1969. Pleistocene symbiosis: Pinnotherid crabs and pelecypods from Cape Blanco, Oregon. Veliger 12: 72–73.

Chapter 26

Three Other Crustaceans:
A Copepod, a Leptostracan, and a Stomatopod

Eugene C. Haderlie, Donald P. Abbott, and Roy L. Caldwell

The great majority of the larger and more conspicuous crustacean species inhabiting the California intertidal zone belong to the groups considered in Chapters 20–25. Several other crustacean groups have been omitted, either because they are not well represented intertidally or because the individual animals are very small and inconspicuous. Three species, however, representing crustacean taxa not considered earlier, are included here, for they are often seen in the intertidal zone or in the shallow subtidal waters of California.

The subclass Copepoda is one of the largest crustacean groups. Copepods occur in vast numbers as free-swimming pelagic animals in the upper levels of the open seas, but in coastal regions many are bottom dwellers that live on soft substrata or crawl about among hydroids, bryozoans, and seaweeds. Many copepods are found in fresh water, and still others are highly modified parasites on or in fishes, marine mammals, crustaceans, cnidarians, echinoderms, tunicates, and other invertebrates. Among the most conspicuous of the intertidal copepods is the little orange species *Tigriopus californicus* (26.1), which lives in high splash pools along the shore.

The subclass Malacostraca includes many large and important groups of crustaceans such as the isopods (Chapter 21), the amphipods (Chapter 22), and the decapods (Chapters 23–25). Also included in the subclass Malacostraca is the order Leptostraca, a small relict group consisting of less than a dozen living species but related to fossil forms extending back into the early Paleozoic. An account is presented of one leptostracan, *Nebalia pugettensis* (26.2). We also include here another malacostracan group, the order Stomatopoda, which contains the highly specialized predaceous mantis shrimps. *Hemisquilla ensigera californiensis* (26.3), the form considered here, is one of three species of stomatopods found in shallow water on California shores.

Phylum Arthropoda / Class Crustacea / Subclass Copepoda
Order Harpacticoida / Family Harpacticidae

26.1 **Tigriopus californicus** (Baker, 1912)
(=T. triangulus)

Common in splash pools at or above high tide level on rocky shores; Torch Bay, Glacier Bay National Park, Alaska (unpublished record; M. Dethier, pers. comm.) to Baja California; possibly not distinct from species in Japan and Chile.

Body 1–1.4 mm long, divided into a broader anterior portion (prosome) and a narrower posterior portion (urosome), colored orange to brick red; first antennae large and conspicuous, hooked at tips in males; terminal segment of body bearing two long caudal processes (rami); females often with a single cluster of eggs ventrally.

Eugene C. Haderlie is Professor of Oceanography at the Naval Postgraduate School, Monterey. Donald P. Abbott is Professor, Department of Biological Sciences and Hopkins Marine Station, Stanford University. Roy L. Caldwell is Associate Professor of Zoology, University of California, Berkeley.

Tigriopus californicus is limited largely to tidepools that are only occasionally filled by splash, such as at high tides and during storms, but in these pools it often occurs in enormous numbers. The species is easily kept in the laboratory in bowls of seawater at room temperature. *Tigriopus* can filter feed to some extent, like most free-living copepods, but it is primarily a browser, and rasps the film of algae and detritus from rocks and gravel in pools, using specialized mouthparts. It can also catch small animals, and will even cannibalize its own young if the population density gets too high in the pool. In general, females are more predaceous than males.

At Pacific Grove (Monterey Co.), the reproduction and population biology of *Tigriopus* have been investigated (Egloff, 1966). As the eggs are laid by a mature female, they are fertilized by sperm stored in her body from an earlier mating. From 32 to 140 eggs are laid at one time and cemented together into a rounded mass that is held on the ventral surface of the female urosome. The time required for hatching is temperature-dependent: 8.2 days at 10°C, 4.8 days at 15°C, and 2.4 days at 23°C. Developing eggs are greenish, but turn light pink as the larvae become ready to hatch. Removal of a pink egg mass from a female stimulates hatching within a few minutes (V. Vacquier, pers. comm.). Following hatching, the free-swimming young pass through six naupliar stages and six copepodid (juvenile copepod) stages, molting between each, before becoming adults. Increased hydrostatic pressure applied experimentally during the naupliar stages results in a higher proportion of females. Development from hatching to the adult form requires 22.5 days at 15°C, but only 14 days at 23°C. Sexual differences in males and females appear in the last four copepodid stages; some females in copepodid stage 6 even have eggs in the body or on the ventral surface. Mature males may attempt to mate with any other individual regardless of sex and may even clasp animals as young as copepodid stage 3. The hooked antennae are used for attachment in the mating embrace. Only sexually mature females who have not previously mated accept the male for mating (R. Burnett, pers. comm.). Fertilized females lay up to 12 batches of fertilized eggs from one mating, and at 23°C can produce a batch every 4 days. With optimal levels of oxygen and light and a good food supply, a population of *Tigriopus* can double every 6.6 days at 15°C and every 3.9 days at 23°C.

The tolerance of *Tigriopus* for great fluctuations in salinity and temperature allow it to live in high splash pools devoid of most other animals. During summer months along the California coast, where there is little or no rainfall and few storm waves, evaporation increases the salinity of high pools to tenfold that of normal seawater, and salt crystals form at the edges of the pool. In pools of increasing salinity, populations of *Tigriopus* remain active up to a salinity of about 300 percent that of seawater, but become inactive and lie on the bottom at salinities over 600 percent. The animals can remain in brine at the saturation level for 12 days or more and still become active again if removed to normal seawater. The ripe eggs also tolerate high salinities, and remain capable of hatching if returned to normal seawater. *Tigriopus* also withstands temperatures up to 40°C (for short exposures) and can survive exposure to air for limited periods. Because of the ease with which it can be reared and maintained in the laboratory, *Tigriopus californicus* has been used in a variety of physiological, behavioral, and ecological investigations.

In the field the copepods are dispersed from pool to pool by wave splash during the highest tides. They also regularly leave tidepools on their own and migrate up and down rivulets of water, ending up in different pools. In one study using animals marked with a vital dye, over half the population in one pool changed within a few day's time, though the total population remained about the same size (R. Burnett, pers. comm.). In addition, *Tigriopus* may be carried from pool to pool, clinging to the fine bristles on the legs of the shore crab *Pachygrapsus crassipes* (25.43) as it moves about (M. Ghiselin, pers. comm., 1958; Egloff, 1966). Up to 73 adult and larval copepods have been found on a single crab.

The assistance of Robin Burnett in preparing this account is gratefully acknowledged.

For *T. californicus,* see Barnett & Kontogiannis (1975), Buzatti-Traverso (1958), Clogston (1965), Cooper (1977), Dethier (1980), Egloff (1966), Hopkins Marine Station (1977), Kontogiannis (1973), Kontogiannis & Barnett (1973), Lang (1948), Lear & Oppenheimer (1962), Mistakidis (1949), Monk (1941), Ohman (1977), Patterson (1968), Provasoli, Shiraishi & Lance (1959), Stoller (1977), Vacquier (1962), Vacquier & Gelser (1965), and Vittor (1971). For other species of *Tigriopus,* see Bozić (1960, 1975), Comita & Comita (1966), Fraser (1936a,b), Gilat (1967), Harris (1973), Igarashi (1963), Lang (1948), Mistakidis (1949), Provasoli, Conklin &

D'Agostino (1970), Provasoli, Shiraishi & Lance (1957), and Ranade (1957). For other California shore copepods, see Clogston (1965), Lang (1965), and the references cited in Illg (1975).

Phylum Arthropoda / Class Crustacea / Subclass Malacostraca
Superorder Phyllocarida / Order Leptostraca / Family Nebaliidae

26.2 **Nebalia pugettensis** (Clark, 1932)

In pockets of silt and organic mud among rocks, under mats of green alga (*Enteromorpha*) on estuarine mud flats in summer, low intertidal zone in protected areas; in black, sulfide-rich mud covered with decaying vegetation; Puget Sound (Washington) to Bahía de San Quintín (Baja California).

Body to 10 mm long; thorax, its appendages, and anterior part of abdomen enclosed by a laterally compressed, hoodlike carapace, the two sides held together by an adductor muscle passing through the head; hinged plate (rostrum) covering head anteriorly; eyes stalked; thoracic appendages fringed with long setae, bearing paddlelike projections laterally; first four abdominal appendages (pleopods) longer; abdomen terminating in two long bristles (caudal rami).

Nebalia pugettensis, along with only a few other species worldwide, belongs to the Leptostraca; the adults in this archaic order differ from those in most other groups in the subclass Malacostraca by having seven abdominal segments rather than the usual six and by retaining the caudal rami. The animals may occur in enormous numbers in areas of black sulfurous mud. They are filter feeders, using the bristle-laden thoracic appendages to create a current under the carapace and to strain from the water small organisms and fine particles of detritus. Collected food is then transferred to a ventral groove and moved forward to the mouth. The four anterior abdominal pleopods help create the feeding currents. The pleopods are also the main swimming organs. The large and powerful first antennae beat along with the pleopods, and are especially important in locomotion when the animals are moving through soft sediments or pushing their way through mats of the green alga *Enteromorpha*. The animals often swim on their sides, and can flex the abdomen to give a powerful escape response.

The eggs are carried under the carapace, attached to the thoracic legs of the female. Development is direct, and

the young hatch at a post-larval stage. After hatching they are brooded under the mother's carapace until they resemble miniature adults, except that their own carapaces are incompletely developed.

See Cannon (1927, 1960), Clark (1932), LaFollette (1914), and Menzies & Mohr (1952).

Phylum Arthropoda / Class Crustacea / Subclass Malacostraca
Superorder Hoplocarida / Order Stomatopoda / Family Gonodactylidae

26.3 **Hemisquilla ensigera californiensis** (Stephenson, 1967)
Mantis Shrimp

Moving about on open bottom or secluded in burrows in mud or sand-shell substrata, shallow subtidal waters to 90 m depth; Point Conception (Santa Barbara Co.) to Gulf of California; scattered localities south to Golfo de Chiriquí (Panama).

Adults to 300 mm long; post-larvae under 35 mm; overall color yellow-brown to orangish tan; telson and distal segments of raptorial appendages (see below) greenish yellow to bright yellow; distal segments of antennules, maxillipeds, walking legs, and pleopods blue; distal segment of uropod very deep blue with red setae fringe; antennal scales bluish at base, brownish yellow distally with pink setae; adult males with bright red lateral patches on carapace.

This is the largest and most brightly colored of the California stomatopods. Two other subspecies are known: *H. ensigera ensigera* is found off the coast of Peru and Chile, and *H. ensigera australiensis* occurs off Australia and New Zealand.

The biology of *H. ensigera* is poorly known. Although usually taken by dredge or trawl, it is occasionally seen by divers near its burrow or cruising in shallow channels near shore; it is also taken on hook and line. Each animal digs a large burrow 1–2 m deep and 5–8 cm in diameter. The animals apparently leave their burrows infrequently, except during the spring, when males are often caught in trawls. Mac-Ginitie and MacGinitie (1968) report a single haul of over 200 *H. ensigera* taken at 20 m off Ventura Co., samples of which were all males. As in many other stomatopods, the males probably leave their burrows during breeding season to seek out receptive females. Several divers have reported hearing low, "groaning" sounds emitted from *Hemisquilla* burrows. It is possible that they produce these sounds by stridulation, perhaps for mate attraction.

The most striking feature of *H. ensigera*, as in all stomatopods, is the greatly enlarged second maxillipeds, commonly called the raptorial appendages. These modified mouthparts are folded back on themselves, forming a collapsed Z. When the animal strikes, the two terminal segments (propodus and dactylus), swing out rapidly, delivering a powerful blow. Burrows (1969) reported that the strike of *H. ensigera* takes only 4 msec. Several fishermen have come to regret trying to dislodge a hook from a *Hemisquilla*, since the strike of a large individual is capable of severely damaging a finger. The power of the strike was demonstrated by a specimen kept in the laboratory; it struck and broke the side of a large aquarium constructed of double safety glass.

Caldwell and Dingle (1976) have divided stomatopods into two functional groups, the spearers and the smashers, which differ in the morphology of the raptorial appendages. Spearers have a series of spines on the dactylus used to impale soft-bodied prey. Smashers have a smooth, swollen dactylus used to crush hard-shelled organisms. *Hemisquilla* possesses an unspecialized smashing appendage, suggesting that it may feed on prey such as bivalves. Its natural diet is unknown, but in the laboratory it eats nereid worms, clams, oysters, and scallops. One large male was observed to crush and consume a 12 cm abalone.

Most stomatopods are tropical. However, three other mantis shrimps, all spearers, are occasionally reported from southern California. **Pseudosquillopsis marmorata** (Lockington, 1877), referred to as *Pseudosquilla lessonii* in Schmitt (1940) and Ricketts and Calvin (1968), is recorded from south of Point Conception to the Gulf of California and from the Galápagos Islands. It may be taken from burrows under rocks of the very low intertidal zone. Specimens reach a length of approximately 120 mm and bear three spines on the raptorial dactylus. Manning (1969) describes the post-larval and juvenile stages.

Nannosquilla anomala Manning, 1967, has been reported from San Clemente Island (Channel Islands). It occurs in burrows on sand bottoms at water depths of 5–23 m. It attains a total length of 34–41 mm (Manning, 1967) and bears 10–14

spines on the dactylus of the raptorial appendage.

Schmittius politus (Manning, 1972), formerly *Squilla polita*, occurs from Monterey Bay south to Punta Abreojos (Baja California) at depths of 42–168 m; it will probably not be found in the intertidal zone. Individuals are up to 60 mm in length and bear four spines on the raptorial dactylus.

See Caldwell & Dingle (1976) for a general account of stomatopods; see also Burrows (1969), MacGinitie & MacGinitie (1968), Manning (1967, 1968, 1969, 1971, 1972), Ricketts & Calvin (1968), Schmitt (1940), and Stephenson (1967).

Literature Cited

Barnett, C. J., and J. E. Kontogiannis. 1975. The effect of crude oil fractions on the survival of a tidepool copepod, *Tigriopus californicus*. Environ. Pollut. 8: 45–54.

Bozić, B. 1960. Le genre *Tigriopus* Norman (copépodes harpacticoides) et ses formes européenes; recherches morphologiques et expérimentales. Arch. Zool. Expér. Gén. 98: 167–269.

———. 1975. Détection actometrique d'un facteur d'interattraction chez *Tigriopus* (crustacés, copépodes, harpacticoides). Bull. Soc. Zool. France 100: 305–11.

Burrows, M. 1969. The mechanics and neural control of the prey capture strike in the mantid shrimps *Squilla* and *Hemisquilla*. Z. Vergl. Physiol. 62: 361–81.

Buzzati-Traverso, A. A. 1958. Genetische probleme der marine biologie. Rev. Suisse Zool. 65: 461–84.

Caldwell, R., and H. Dingle. 1976. Stomatopods. Sci. Amer. 234: 80–89.

Cannon, H. G. 1927. On the feeding mechanism of *Nebalia bipes*. Trans. Roy. Soc. Edinburgh 55: 355–69.

———. 1960. Leptostraca, *in* H. G. Bronns, Klassen und Ordnungen des Tierreichs, Bd. 5, Abt. 1, Buch 4. Leipzig: Winter. 81 pp.

Clark, A. E. 1932. *Nebaliella caboti* n. sp. with observations on other Nebaliacea. Trans. Roy. Soc. Canada (3) 26: 217–35.

Clogston, F. L. 1965. Postembryonic development of species of harpacticoid copepods from the Pacific coast of the United States and an application of developmental patterns to their systematics. Doctoral thesis, Zoology, University of Washington. 246 pp.

Comita, G. W., and J. J. Comita. 1966. Egg production in *Tigriopus brevicornis*, pp. 171–85, *in* H. Barnes, ed., Some contemporary studies in marine science. London: George Allen & Unwin. 716 pp.

Cooper, J. R. 1977. Migration behavior in *Tigriopus californicus*. Research paper, Biology 175h, library, Hopkins Marine Station of Stanford University, Pacific Grove, Calif. 25 pp.

Dethier, M. N. 1980. Tidepools as refuges: Predation and the limits of the harpacticoid copepod *Tigriopus californicus*. J. Exper. Mar. Biol. Ecol. 42 (in press).

Egloff, D. 1966. Ecological aspects of sex ratio and reproduction in experimental and field populations of the marine copepod *Tigriopus californicus*. Doctoral thesis, Biological Sciences, Stanford University, Stanford, Calif. 141 pp.

Fraser, J. H. 1936a. The occurrence, ecology and life history of *Tigriopus fulvus* (Fischer). J. Mar. Biol. Assoc. U.K. 20: 523–36.

———. 1936b. The distribution of rock pool copepods according to tidal level. Anim. Ecol. 5: 23–38.

Gilat, E. 1967. On the feeding of a benthonic copepod, *Tigriopus brevicornis* O. F. Müller. Bull. Sea Fish. Res. Sta. Israel (Haifa) 45: 79–95.

Harris, R. P. 1973. Feeding, growth, reproduction and nitrogen utilization by the harpacticoid copepod *Tigriopus brevicornis*. J. Mar. Biol. Assoc. U.K. 35: 785–800.

Hopkins Marine Station of Stanford University. 1977. The biology of high tidepools. Research papers, Biology 175h, library, Hopkins Marine Station of Stanford University, Pacific Grove, Calif. (Primarily papers on *Tigriopus californicus*.)

Igarashi, S. 1963. (3 papers) Developmental cycle of *Tigriopus japonicus* Mori. The primary sex ratio of a marine copepod, *Tigriopus japonicus*. On the stability of sex ratio and the prediction of its abnormality depending on the color of ovisac in *Tigriopus japonicus* Mori. Sci. Rep. Tôhôku Univ. (4) Biol. 29: 59–90.

Illg, P. L. 1975. Subclasses Copepoda and Branchiura, pp. 250–58, *in* R. I. Smith and J. T. Carlton, eds., Light's manual: Intertidal invertebrates of the central California coast. 3rd ed. Berkeley and Los Angeles: University of California Press. 716 pp.

Kontogiannis, J. E. 1973. Acquisition and loss of heat resistance in adult tide-pool copepod *Tigriopus californicus*. Physiol. Zool. 46: 50–54.

Kontogiannis, J. E., and C. J. Barnett. 1973. The effect of oil pollution on survival of the tidal pool copepod, *Tigriopus californicus*. Environ. Pollut. 4: 69–79.

LaFollette, R. 1914. A *Nebalia* from Laguna Beach. J. Entomol. Zool. Pomona 6: 204–6.

Lang, K. 1948. Monographie der Harpacticiden. Lund, Sweden: Ohlssons. 1,682 pp.

———. 1965. Copepoda Harpacticoidea from the California Pacific coast. K. Svenska Vetenskakad. Handl. vol. 10(2). 560 pp.

Lear, D. W., Jr., and C. H. Oppenheimer, Jr. 1962. Consumption of microorganisms by the copepod *Tigriopus californicus*. Limnol. Oceanogr. 7 (Suppl.): lxiii–lxv.

MacGinitie, G., and N. MacGinitie. 1968. Natural history of marine animals. 2nd ed. New York: McGraw-Hill. 523 pp.

Manning, R. 1967. *Nannosquilla anomala*, a new stomatopod crustacean from California. Proc. Biol. Soc. Washington 80: 147–50.

———. 1968. A revision of the family Squillidae (Crustacea, Stomatopoda), with the description of eight new genera. Bull. Mar. Sci. 18: 105–42.

———. 1969. The postlarvae and juvenile stages of two species of *Pseudosquillopsis* (Crustacea, Stomatopoda) from the Eastern Pacific region. Proc. Biol. Soc. Washington 82: 525–37.

———. 1971. Eastern Pacific expeditions of the New York Zoological Society. Stomatopod Crustacea. Zoologica 56: 95–113.

———. 1972. Notes on some stomatopod crustaceans from Peru. Proc. Biol. Soc. Washington 85: 297–308.

Menzies, R. J., and J. L. Mohr. 1952. The occurrence of the wood-boring crustacean *Limnoria* and the Nebaliacea in Morro Bay, California. Wasmann J. Biol. 10: 81–86.

Mistakidis, M. 1949. A new variety of *Tigriopus lillijeborgii* Norman. Dove Mar. Lab. Rep. (3) 10: 55–70 (compares known *Tigriopus* spp.).

Monk, C. R. 1941. Marine harpacticoid copepods from California. Trans. Amer. Microscop. Soc. 60: 75–99.

Ohman, M. D. 1977. Isozymes in *Tigriopus californicus* (Crustacea: Harpacticoida): Plasticity of individuals or of populations? Master's thesis, Biology, San Francisco State University. 86 pp.

Patterson, R. E. 1968. Physiological ecology of *Tigriopus californicus*, a high intertidal copepod. Master's thesis, Zoology, University of California, Berkeley. 22 pp.

Provasoli, L., D. E. Conklin, and A. S. D'Agostino. 1970. Factors inducing fertility in aseptic Crustacea. Helgoländer Wiss. Meeresunters. 20: 443–54.

Provasoli, L., K. Shiraishi, and J. R. Lance. 1959. Nutritional idiosyncrasies of *Artemia* and *Tigriopus* in monoxenic culture. Ann. N.Y. Acad. Sci. 77: 250–61.

Ranade, M. R. 1957. Observations on the resistance of *Tigriopus fulvus* (Fischer) to changes in the temperature and salinity. J. Mar. Biol. Assoc. U.K. 36: 115–19.

Ricketts, E. F., and J. Calvin. 1968. Between Pacific tides. 4th ed. Revised by J. W. Hedgpeth. Stanford, Calif.: Stanford University Press. 614 pp.

Schmitt, W. 1940. The stomatopods of the west coast of America based on collections made by the Allan Hancock Expeditions, 1933–38. Allan Hancock Pacific Exped. 5: 129–225.

Stephenson, W. 1967. A comparison of Australian and American specimens of *Hemisquilla ensigera* (Owen, 1832) (Crustacea: Stomatopoda). Proc. U.S. Nat. Mus. 120: 1–18.

Stoller, D. W. 1977. Tolerance of *Tigriopus californicus* (Baker, 1912) to slow increases in salinity produced by evaporation and hypersaline solutions. Research paper, Biology 175h, library, Hopkins Marine Station of Stanford University, Pacific Grove, Calif. 29 pp.

Vacquier, V. D. 1962. Hydrostatic pressure has a selective effect on the copepod *Tigriopus*. Science 135: 724–25.

Vacquier, V. D., and W. L. Gelser. 1965. Sex conversion induced by hydrostatic pressure in the marine copepod *Tigriopus californicus*. Science 150: 1619–21.

Vittor, B. A. 1971. Effects of the environment on fitness-related life history characters in *Tigriopus californicus*. Doctoral thesis, Biology, University of Oregon. 115 pp.

Chapter 27

Pycnogonida: *The Sea Spiders*

Joel W. Hedgpeth and Eugene C. Haderlie

The pycnogonids, or sea spiders, form a class of exclusively marine arthropods allied to the arachnids, the horseshoe crabs, and other stocks sometimes grouped together in a subphylum or phylum Chelicerata. Only three fossil species of pycnogonids are known, but *Paleopantopus* from the Lower Devonian period of the Paleozoic is not very different from modern forms, suggesting that the group is an ancient one. About 600 living species have been described; they occur from polar seas to the tropics and from the intertidal zone to depths of over 6,000 m. Pycnogonids are especially well represented in the benthic fauna around Antarctica, and in deep water many specimens have a leg span of 15 cm or more and a body length of 3–5 cm. In the temperate intertidal zones they are relatively small and are rarely present in large numbers, except on fouling growth on pilings and on buoys, where they are occasionally abundant. Pycnogonids are often overlooked because they are inconspicuous animals entangled among hydroids and bryozoans. Most live as bottom-dwelling animals throughout life, and move about sluggishly, but a few can swim briefly and are occasionally caught in plankton tows.

The segmented trunk of the body of most pycnogonids is much reduced (an older English name for the group was the "no body crabs"). Some species have long gangling legs. At the anterior end of the body is the head, or cephalon, which bears a conspicuous proboscis terminating in the mouth. The proboscis may be less than half to three or four times the length of the body. Its complex internal structures include denticles, a sievelike arrangement of

Joel W. Hedgpeth is Emeritus Professor of Oceanography, Oregon State University. Eugene C. Haderlie is Professor of Oceanography at the Naval Postgraduate School, Monterey.

straining setae, and associated muscles, used together for macerating and straining food. A tubercle on the dorsal surface of the head bears four or two simple eyes (or none, in some deep-sea species). The first trunk segment is fused with the head. The abdomen, terminating in the anus, forms merely a small, unsegmented posterior protuberance.

Some pycnogonids have a full complement of appendages: chelifores (chelate or pinching appendages used in food capture and handling), a pair of sensory palps, a pair of ovigerous (egg-bearing) legs, and usually four pairs of ambulatory legs borne on lateral extensions of the trunk. Some Antarctic species have five pairs of walking legs, and a few have six pairs. An ambulatory leg is normally composed of nine movable articles or segments. Starting from the base of the leg, these segments are the first, second, and third coxa (sometimes called coxa and first and second trochanter), the femur, the first and second tibia (sometimes termed patella and tibia), the first and second tarsus, and the terminal claw.

In some species the chelifores and palps are reduced or absent, and the ovigerous legs are often present only in the males. When present, the ovigerous legs are on the ventral surface of the body posterior to the palps, and consist of up to ten articles. In many species they are the same in both sexes, except for somewhat different proportions in some of the articles, but they function as egg bearers only in the males. The function of the ovigerous legs in the females of many of the small intertidal species is unknown. However, in the large Antarctic and deep-sea species the ovigerous legs serve as cleaning appendages in both sexes, and the terminal segments form a structure like a shepherd's crook that is worked along the surface of the long leg joints. In these species, where leg cleaning has been observed, there is no morphological difference between the ovigers of the two sexes. In the deep-sea genus *Colossendeis*, cleaning seems to be the only function, for the males have never been observed bearing egg masses.

The trunk in many pycnogonids is relatively small, and some of the viscera extend well down into the legs. The digestive system of pycnogonids, centered in the trunk, has lateral branches (caeca) in the legs that reach almost to the terminal claws. There are no discrete excretory or respiratory systems; exchanges with the seawater appear to take place through thin parts of the cuticle.

The sexes are separate in all but one species of sea spider. The ovaries or testes lie above the gut on either side of the heart, and like the digestive system, each gonad sends lateral branches into the ambulatory legs, often reaching the end of the tibial joints. The gonopores are located on the second coxal joint of some or all of the legs. Early development of the eggs has been studied in a European species; the yolk is synthesized within the enlarging ova (in a pattern similar to that in another aquatic chelicerate, the horseshoe crab) with little contribution from the outside (King & Jarvis, 1970).

In those pycnogonids in which mating has been observed, the male clings to the ventral surface of the female in such a way that the gonopores of each sex are closely aligned in a sort of pseudocopulation. As the eggs are released, they are fertilized; in many pycnogonids, including *Pycnogonum*, the sperm are not motile. Studies of the ultrastructure (Van Deurs, 1973, 1974a,b) reveal an arrangement of the microtubules of the flagellum in 8 to 12 doublets without a central placement. The variation represented by the reduction of the microtubules indicates an evolutionary trend toward reduction of movement, until, as in *Pycnogonum*, there are only isolated microtubules without any other organelle. This trend toward reduction of movement, with more stages, is encountered in the Protura, an order of insects (Baccetti, 1979).

As the eggs are laid and fertilized, the male gathers them, forming them into a ball that he carries on the ovigerous legs until the young hatch. Some males have been observed carrying up to 14 eggs balls, some eggs newly collected and others in the process of hatching.

In some species the young hatch out as post-larvae with three or four pairs of appendages. In others, the young hatch as protonymphon larvae with only three pairs of appendages—chelifores, palps, and ovigers. These larvae are generally oval in shape and superficially resemble barnacle nauplii, but they cannot swim. Further segments and additional legs are added at the posterior end as the protonymphon grows into an adult. During development after hatching, juveniles grow by a series of molts, but adults may grow by a process not involving molting (Jarvis & King, 1972).

Pycnogonids feed mainly on hydroids and bryozoans, but some reportedly attack sponges, small polychaetes, or even the shoots of red algae. Among our intertidal pycnogonids, two species are parasitic on sea anemones, and another is parasitic on sea mussels (Benson & Chivers, 1960). The young of some sea spiders are parasites in clams; others develop in close association with hydromedusae, nudibranchs, holothurians, ophiurans, and hydroids such as *Obelia*, *Tubularia*, *Aglaophenia*, and *Coryne* (see King, 1973). Pycnogonids in turn are fed upon by isopods, anemones, and fishes.

Some large (tarantula-sized) sea spiders from Antarctic seas are hardy animals and live well in oxygenated aquaria at temperatures below 11°C. They have been used for physiological studies on oxygen uptake and consumption (Douglas, Hedgpeth & Hemmingsen, 1970; Redmond & Swanson, 1968) and digestion (Richards & Fry, 1978). In Europe some intertidal pycnogonids show a rhythm in activity correlated with tidal level, becoming most active at or just after high tide (Isaac & Jarvis, 1973).

The swimming mechanics of several species of pycnogonids have been studied; in some the legs beat ventrally so that the animal swims dorsal side foremost (Morgan, 1971).

For modern reviews of the Pycnogonida, see Hedgpeth (1955) and King (1973, 1974). For a discussion of the problems of pycnogonid phylogeny, see Hedgpeth (1954). An old but still useful account of pycnogonids of the entire west coast is that of Cole (1904). Hedgpeth (1941) provides a key for Pacific coast forms. Giltay (1934) lists sea spiders from British Columbia. Exline (1936) considers species occurring in Puget Sound (Washington), but many errors in identification are included. California pycnogonids are treated in Cole (1904), Hall (1913), Hedgpeth (1939, 1951), and Ziegler (1960). A key to the intertidal species of central California is given in Hedgpeth (1975).

Phylum Arthropoda / Class Pycnogonida / Order Pantopoda
Family Pycnogonidae

27.1 **Pycnogonum stearnsi** Ives, 1892

Common under rocks or associated with anemones, hydroids, or ascidians, sometimes in root masses of the surfgrass *Phyllospadix*, middle to low intertidal zones; British Columbia to central California; Japan.

Body to 13 mm long in females, 10 mm long in males, oval in outline; surface smooth but granular, dorsal tubercles inconspicuous or absent; trunk segmentation distinct; legs short and stout; ovigerous legs present only in males; overall color pale salmon-pink, ivory, or white.

This species is one of the most conspicuous and commonly collected pycnogonids on the California coast. It is typically found at the base of the solitary green sea anemone, *Anthopleura xanthogrammica* (3.30), but it is also associated with the white anemone *Metridium senile* (3.37), the hydroid *Aglaophenia* (e.g., 3.4, 3.5), and the social ascidian *Clavelina huntsmani* (12.20). *P. stearnsi* is parasitic on the anemones, and possibly other hosts, and feeds by inserting its proboscis into the column of the anemone and withdrawing fluid and cellular food. Investigation of the proboscis shows that it has a wide aperture for taking in particulate food. Nematocysts have been found in the gut. The circulatory system of related species (*P. rhinoceros* and *P. littorale*) has been studied by Firstman (1973), who was unable to find a heart in the specimens. Apparently peristaltic contractions of the extensive digestive caeca (blind diverticula of the gut) may be sufficient to effect blood movement.

The males are sometimes found with masses of eggs held ventrally by their ovigerous legs.

See Firstman (1973), Fry (1965), Hedgpeth (1941, 1949, 1951, 1975), Ricketts & Calvin (1968), and Ziegler (1960).

Phylum Arthropoda / Class Pycnogonida / Order Pantopoda
Family Pycnogonidae

27.2 **Pycnogonum rickettsi** Schmitt, 1934

Uncommon, associated with the anemones *Metridium senile* (3.37) on wharf pilings and *Anthopleura xanthogrammica* (3.30) on rocky shores and with the hydroid *Aglaophenia* (e.g., 3.14, 3.15), middle to low intertidal zones and subtidal waters; Friday Harbor (Washington) to Monterey.

Body 7–10 mm long; species resembling *P. stearnsi* (27.1) in size and shape but distinguished by presence of prominent dorsal tubercles on trunk and legs; ovigerous legs present only in males; color usually light brown or tan with darker lines or clear areas giving reticulate pattern, whitish or pale flesh-colored in young.

On rocky shores when the tide is out the pycnogonid is normally found near the base of the anemone *Anthopleura xanthogrammica* (3.30), but in pools and subtidal waters it occurs along the column up to the crown of the tentacles. *P. rickettsi* may be found in the same association as *P. stearnsi*, but as far as is known not on the same individual anemone.

As in *P. stearnsi*, the males often carry fertilized eggs, which are held to the ventral surface by the ovigerous legs.

The species was named for biologist E. F. Ricketts, senior author of *Between Pacific tides* (1968; first published 1939) and the model for Doc in John Steinbeck's novel *Cannery Row*.

See Hedgpeth (1941, 1951, 1975), Ricketts & Calvin (1968), and Ziegler (1960).

Phylum Arthropoda / Class Pycnogonida / Order Pantopoda
Family Ammotheidae

27.3 **Ammothea hilgendorfi** (Böhm, 1879)
(=Lecythorhynchus marginatus, L. hilgendorfi)

Common on hydroids in crevices along rocky shores and on undersides of rocks on breakwaters, low intertidal zone and subtidal waters; North Pacific to Isla Cedros (Baja California); Japan.

Body to 25 mm long; legs long, gangling, with brownish or purplish banding; trunk elongate, sometimes with fine red spotting along dorsal midline; proboscis cylindrical with one joint; anterodorsal appendages (chelifores) rudimentary, consisting of one-jointed stumps; ovigerous legs fully developed in both sexes.

This is one of the characteristic pycnogonids of central California. Although it is large, it may be overlooked, for it blends in well with the hydroid *Aglaophenia* (e.g., 3.14, 3.15) on which it commonly lives. The larval stages live within the tissues of the hydroid, entering by means of sharply pointed appendages. The gross anatomy of the nervous system has been illustrated, but little is known of the biology of the species.

Ammothea hilgendorfi has long been known on this coast as *Lecythorhynchus marginatus* Cole, described in the Harriman Alaska Reports. It is the only one of eleven species considered as "new" by Cole (1904) that has fallen to synonymy, a remarkable record for one who had apparently not made a critical observation of these animals before 1899. The genus *Ammothea* is for the most part a group of the southern hemisphere, and is represented by a complex of large, shallow-water species in the Antarctic Ocean.

See Cole (1904), Hall (1913), Hedgpeth (1941, 1949, 1951, 1975), Hilton (1915), Ricketts & Calvin (1968), and Ziegler (1960).

Phylum Arthropoda / Class Pycnogonida / Order Pantopoda Family Ammotheidae

27.4 **Nymphopsis spinosissima** (Hall, 1912)

Found in sheltered crevices among stones and hydroids, low intertidal zone of rocky shores; also subtidal; Alaska to Laguna Beach (Orange Co.).

Body 13–15 mm long, compact; tubercles tall, spiny, numbering three; trunk segmentation indistinct; proboscis broad and bulbous, nearly as long as body, sometimes directed ventrally; legs relatively short and stout, the longer joints with two distinct rows of spines; ovigerous legs present in both sexes.

Nothing is known about the biology of this species.

See Hedgpeth (1939, 1941, 1951, 1975) and Hilton (1942).

Literature Cited

Baccetti, B. 1979. Ultrastructure of sperm and its bearing on arthropod phylogeny, pp. 609–44, *in* A. P. Gupta, ed., Arthropod phylogeny. New York: Van Nostrand Reinhold. 762 pp.

Benson, P. H., and D. C. Chivers. 1960. A pycnogonid infestation of *Mytilus californianus*. Veliger 3: 16–18.

Cole, L. J. 1904. Pycnogonids of the west coast of North America. Harriman Alaska Expedition 10: 249–98.

Douglas, E. L., J. W. Hedgpeth, and E. A. Hemmingsen. 1970. Oxygen consumption of some Antarctic pycnogonids. Antarc. J., U.S., 4: 109.

Exline, H. I. 1936. Pycnogonida from Puget Sound. Proc. U.S. Nat. Mus. 83: 413–22.

Firstman, B. L. 1973. The relationship of the chelicerate arterial system to the evolution of the endosternite. J. Arachnol. 1: 1–54.

Fry, W. G. 1965. The feeding mechanisms and preferred foods of three species of Pycnogonida. Bull. Brit. Mus. (Natur. Hist.), Zool. 12: 195–233.

_____, ed. 1978. Sea spiders (Pycnogonida). Proceedings of a meeting held in honor of Joel W. Hedgpeth on 7 October 1976 in the rooms of the Linnean Society of London. Zool. J. Linn. Soc. 63 (1 & 2): 1–238.

Giltay, L. 1934. Pycnogonids from the coast of British Columbia. Canad. Field Natur. 48: 49–50.

Hall, H. V. M. 1913. Pycnogonida from the coast of California, with descriptions of two new species. Univ. Calif. Publ. Zool. 11: 127–42.

Hedgpeth, J. W. 1939. Some pycnogonids found off the coast of southern California. Amer. Midl. Natur. 22: 458–65.

_____. 1941. A key to the Pycnogonida of the Pacific coast of North America. Trans. San Diego Soc. Natur. Hist. 9: 253–64.

_____. 1949. Report on the Pycnogonida collected by the Albatross in Japanese waters in 1900 and 1906. Proc. U.S. Nat. Mus. 98: 233–321.

_____. 1951. Pycnogonids from Dillon Beach and vicinity, California, with descriptions of two new species. Wasmann J. Biol. 9: 105–17.

_____. 1954. On the phylogeny of the Pycnogonida. Acta Zool. 35: 193–213.

_____. 1955. Pycnogonida, pp. 163–73, *in* R. C. Moore, ed., Treatise on invertebrate paleontology, part P, Arthropoda 2. New York: Geol. Soc. Amer.; Lawrence: University of Kansas Press.

_____. 1975. Pycnogonida, pp. 413–24, *in* R. I. Smith and J. T. Carlton, eds., Light's manual: Intertidal invertebrates of the central California coast. 3rd ed. Berkeley and Los Angeles: University of California Press. 716 pp.

_____. 1978. A reappraisal of the Palaeopantopoda with description of a species from the Jurassic. Zool. J. Linn. Soc. London 63: 23–34.

Hilton, W. A. 1915. The central nervous system of the pycnogonid *Lecythorhynchus*. Pomona Coll. J. Entomol. Zool. 6: 134–36.

———. 1942. Pycnogonids from the Allan Hancock Expeditions. Allan Hancock Pacific Exped. 5: 277–339.

Isaac, M. J., and J. H. Jarvis. 1973. Endogenous tidal rhythmicity in the littoral pycnogonid *Nymphon gracile* (Leach). J. Exper. Mar. Biol. Ecol. 13: 83–90.

Jarvis, J. H., and P. E. King. 1972. Reproduction and development in the pycnogonid *Pycnogonum littorale*. Mar. Biol. 13: 146–54.

King, P. E. 1973. Pycnogonids. New York: St. Martins Press. 144 pp.

———. 1974. British sea spiders (Arthropoda: Pycnogonida). Keys and notes for the identification of the species. Synopses British Fauna 5: 1–68.

King, P. E., and J. H. Jarvis. 1970. Egg development in a littoral pycnogonid *Nymphon gracile*. Mar. Biol. 7: 294–304.

Morgan, E. 1971. The swimming of *Nymphon gracile* (Pycnogonida). J. Exper. Biol. 55: 273–87.

Redmond, J. R., and C. D. Swanson. 1968. Preliminary studies on the physiology of the Pycnogonida. Antarc. J., U.S., 3: 130–31.

Richards, P. R., and W. G. Fry. 1978. Digestion in pycnogonids: A study of some polar forms. Zool. J. Linn. Soc. 63: 75–97.

Ricketts, E. F., and J. Calvin. 1968. Between Pacific tides. 4th ed. Revised by J. W. Hedgpeth. Stanford, Calif.: Stanford University Press. 614 pp.

Van Deurs, B. 1973. Axonemal 12 + 0 pattern in the flagellum of motile spermatozoa of *Nymphon leptocheles*. J. Ultrastruc. Res. 42: 594–98.

———. 1974a. Pycnogonid sperm. An example of inter- and intraspecific axonemal variation. Cell Tissue Res. 149: 105–11.

———. 1974b. Spermatology of some Pycnogonida (Arthropoda), with special reference to a microtubule-nuclear-envelope complex. Acta Zool. 55: 151–62.

Ziegler, A. C. 1960. Annotated list of Pycnogonida collected near Bolinas, California. Veliger 3: 19–22.

Chapter 28

Insecta, Chilopoda, and Arachnida: *Insects and Allies*

William G. Evans

Insects are found in all the major marine littoral habitats of California, ranging from wave-swept rocky shores and sand beaches to quiet mud flats and salt marshes in protected bays. Within these habitats insects are represented in the intertidal region from the splash zone to the lowest tidemarks. Considering the great number of species in the class Insecta, as well as their diverse habits, it is not surprising that some have become adapted to living in maritime environments. Terrestrial habitats with features similar to those found in the intertidal zone, such as sand dunes, marshes, and mud flats, and the littoral zones of streams, rivers, ponds, and freshwater and salt-water lakes, are characterized by a conspicuous insect fauna, and the same is true for the marine littoral.

Intertidal insects occur all over the world, especially in those areas that have abundant intertidal algae. It is highly likely that the ancestors of these insects did not require much of a transition from terrestrial habitats to intertidal ones, for the distinguishing feature of intertidal insects is their terrestriality. They share adaptive characters with their relatives in closely similar terrestrial habitats, enabling them to live in a tidally dominated environment but not in the sea itself.

Clearly, some physiological adaptations (sensory, behavioral, osmoregulatory) have developed, so that these insects have become an integral component of this ecosystem. Unusual adaptive behavior, however, has been demonstrated in only two cases: the development of a lunar (tidal) rhythm of adult emergence from pupal cases in the European intertidal chironomid fly *Clunio marinus* (Caspers, 1951), and a tidal rhythm of locomotory activity in a Cali-

fornia intertidal beetle (Evans, 1976). Even these behavioral changes may have involved only minor genetic changes. Unfortunately, little is known of the biology of most intertidal insects, including the California species, and much work is needed before an understanding of any of these adaptations can be gained.

No attempt is made here to cover all of the insect groups found at the seashore. Of the six orders of intertidal insects known to be represented in California, only three are taken up here: the bristletails (Thysanura), the flies (Diptera), and the beetles (Coleoptera); of these, the flies and the beetles are the most noticeable and are represented by the most species. Some arthropod relatives of the insects are also treated: a representative mite, a spider, and a pseudoscorpion (all Arachnida), and a centipede (Chilopoda).

The three insect orders found in the intertidal zone but not illustrated or dealt with in separate species accounts in this chapter are the springtails (Collembola), the earwigs (Dermaptera), and the true bugs (Hemiptera). Several kinds of springtails inhabit the interstitial spaces of sand beaches, and some are found in decaying wrack (remains of algae cast ashore by waves) or in rock crevices. Because of their small size they are usually overlooked, though they are active and at certain times and places enormously abundant (for example, a circular trap 12.5 cm across exposed from 5 a.m. to 7 a.m. in the intertidal zone on a sandy beach at Pacific Grove (Monterey Co.) captured nearly a quarter of a million individuals; D. P. Abbott, pers. comm.). Springtails sometimes need special preparation, such as clearing and mounting on microscope slides, before they can be identified. It is usually best to send them to specialists, since many of the marine springtails are undescribed. California intertidal springtails are discussed by Scott (1956). The

William G. Evans is Professor of Entomology at the University of Alberta.

order Dermaptera is represented by one species in the California marine littoral, the maritime earwig *Anisolabis maritima*. This subsocial insect is distributed mainly on the Atlantic and Gulf coasts, but it also occurs in San Francisco Bay and Orange Co. Langston (1974) describes its biology and distribution in California; information on nesting and egg care is provided by Kramer (1962). The order Hemiptera contains several groups, such as shore bugs (Saldidae) and water boatmen (Corixidae), that inhabit salt marshes and brine pools; California representatives are treated by Daly (1975), and Usinger (1956).

The bristletails (Thysanura), covered in more detail in the chapter, are fast-moving, primitive insects that closely resemble the familiar silverfish found in buildings. Members of this order are sometimes considered to be the only true wingless group of insects. They are usually clothed in scales and have biting mouthparts, well-developed compound eyes, long cerci, and a long median caudal filament. Since these insects undergo a slight metamorphosis, the young, called nymphs, resemble adults except for size. Sexual maturity is reached in the eighth or ninth instar in *Petrobius maritimus*, a thysanuran found in the marine supralittoral in Europe. The length of the life cycle is variable depending on the species but extends from about 2 months to 2 years. There are probably several species that inhabit splash-zone crevices in California; like the single species treated, *Neomachilis halophila* (28.5), they are active at night when the tide is low.

Aside from the springtails, flies (Diptera) are by far the most abundant insects, in terms of the numbers of individuals, on the shore. Swarms of kelp flies are very noticeable during most of the year, and other flies are often seen flying around rock surfaces at low tide. Dipterous larvae are, of course, just as abundant as the adults, but less readily seen. A close examination of cast-up kelp will usually reveal masses of maggots, and small pieces of growing algae from intertidal rocks will, at certain times of the year, yield crane fly larvae or midge larvae.

Flies go through a complete metamorphosis with the usual stages of development being the egg, larva, pupa, and adult. In some cases, however, living young are produced by the female, or the egg hatches immediately after being deposited in or on the substratum. Some dipterous larvae of the more-advanced families are legless and maggotlike, with no head capsule and with mouthparts consisting only of retractile, sclerotized mouthhooks. Other larvae are also legless, though prolegs (false legs) are often present as well as a distinct head capsule and chewing mouthparts. The usual number of larval instars is four. The pupal stage of some flies is enclosed in a tough, hardened case (puparium), and in order to emerge, the adult has to force its way through a removable cap on the anterior end. In other flies the outer skin of the pupa is hardened, and this splits on the thorax when the adult is ready to emerge. The unhardened, delicate pupae of others are found in cocoons constructed by the last larval instar from substratum materials such as plant debris. The length of the life cycle varies considerably. *Coelopa vanduzeei* (28.10), for instance, can complete a generation in a few weeks, but most flies have longer life cycles, such as up to a year or more.

As is evident, most flies are scavengers, but some, like the dolichopodids, are predators and others are phytophagous, feeding on algae. The mouthparts of adult flies are adapted for sucking, so the general form is of an elongate proboscis. Maxillary palps are usually present, but the labial palps in many flies form a fused or sometimes separated lobe (labellum).

The intertidal Diptera of California are represented by at least seven families, and this chapter does not really do them justice. The crane flies (Tipulidae), midges (Chironomidae), long-legged flies (Dolichopodidae), kelp flies (Coelopidae), and beach flies (Canaceidae) are included, but the brine flies of the family Ephydridae and kelp flies in the family Anthomyiidae are not covered. Descriptions of the larvae and pupae of several California wrack dipterans, with observations on their biology, are given by Kompfner (1974) and Wirth & Stone (1956). For flies in general, Cole (1969) should be consulted.

Many different kinds of beetles (Coleoptera) are found in maritime situations in California. On rocky shores, species are distributed at all levels between high and low tidemarks and in the splash zone; on sandy beaches they are most abundant in the high and middle intertidal zones, where some forms live near the surface while others may be buried to a depth of 30 cm or more. At least ten families of beetles are represented. The species show considerable diversity in form and habit, though most forms are predators and/or scavengers.

The primary source of food for the scavengers is algae, especially the brown algae *Macrocystis* and *Nereocystis*, which have been detached by wave action from offshore kelp beds and cast up on the shore. These large masses of dead and decaying algae provide the energy for the distinctive wrack fauna, with its scavengers and attendant predators. Some beetles, however, rely on less-conspicuous food, such as the remains of stranded planktonic animals.

Beetles go through egg, larval, pupal, and adult stages in their development, but because of the sheer size of the order (approximately 350,000 named species; see Arnett, 1968) and because of their varied habits and ways of life, there is great diversity in larval form. Some are completely legless, such as weevil larvae; others have reduced legs; still others have highly developed legs that are used, for instance, for running or for digging into soil or other substrata. The usual number of larval instars is three or four, and the pupae do not have protective cases or cocoons; thus the appendages are free and the pupa is often soft-skinned, its only protection being the medium in which it is found. The length of the life cycle is quite variable, but most species take about a year to develop, with the immature stages occurring at various times of the year depending on the species. The mouthparts of adult and larval beetles are adapted for chewing and consist of structures found in generalized chewing mouthparts of insects, namely, an upper lip (labrum), paired mandibles, paired maxillae with maxillary palps, and the labium with labial palps.

In the intertidal zone, rove beetles (family Staphylinidae) with their characteristic, shortened wing covers (elytra) are very common and, along with the colorful, soft-winged flower beetles (Melyridae), hister beetles (Histeridae), and ground beetles (Carabidae), are mainly predators on insects and crustaceans. Some, however, feed on freshly killed or dying animals. Some staphylinids, as well as slow-moving darkling beetles (Tenebrionidae) and weevils or snout beetles (Curculionidae), are mainly scavengers in the decaying kelp community. By contrast, members of the family Salpingidae are probably phytophagous; and the water-scavenging beetles (Hydrophilidae) are predaceous as larvae but, like the minute, feather-winged beetles (Ptiliidae), feed on fungi during the adult stage. Arnett (1968), Blackwelder (1957), Doyen (1975), Fall (1901), Hatch (1957a,b, 1961, 1965, 1971), Leech & Chandler (1956), Moore (1956a,b, 1964a) and Moore & Legner (1974, 1975) are particularly useful works for serious studies of the beetles, either for a general view of the whole order or for specific information on such items as distribution or identification of marine littoral forms.

Other related arthropods, usually associated with the terrestrial environment, such as spiders, mites, pseudoscorpions, and centipedes, also inhabit the marine littoral, although little is known about them, particularly of their biology. Extensive taxonomic studies have been made, however, on a unique family of mites, the Halacaridae (Newell, 1947, 1949, 1951, 1975). This ecologically diverse family, though represented in the fauna of freshwater streams and lakes as well as that of caves, is the most successful of the primarily terrestrial groups of arthropods in adapting to the marine environment, and exhibits a wide range of habit (herbivorous, carnivorous, scavenging) and habitat. Its members are found intertidally on most coastlines and subtidally to depths of 4,100 m in the deep trenches of the Pacific Ocean (Newell, 1967). Unlike marine insects, which rely (with one or two exceptions) on atmospheric oxygen for respiratory purposes, these mites can use dissolved oxygen, and thus can inhabit the subtidal zone. This is an evolutionary step that the insects and other arachnids have not successfully taken.

Pacific coast intertidal insects are discussed in Evans (1968), Moore (1975), Saunders (1928), Schlinger (1975), Smith & Carlton (1975), and Usinger (1954, 1956), and accounts of a marine tipulid and chironomid from Japan are given by Tokunaga (1930, 1932). Baduoin (1946) represents a class of European works on intertidal insects that emphasizes physiological adaptation to submergence in these animals. Marine insects in general are treated by Cheng (1976), MacKerras (1950), Thorpe (1932), and Usinger (1957).

Several persons helped me identify specimens or gave me encouragement and advice in the preparation of this chapter. I would like to extend my thanks to D. P. Abbott, G. E. Ball, R. E. Crabill, Jr., J. E. Evans, H. B. Leech, R. E. Leech, and P. Wygodzinsky.

Phylum Arthropoda / Class Chilopoda / Order Geophilomorpha
Family Schendylidae

28.1 **Nyctunguis heathii** (Chamberlin, 1909)
Centipede

Fairly common in some areas in crevices during day or on surfaces of rocks at night during low tide, high intertidal zone on rocky shores; California.

Length 45–50 mm; color bright reddish yellow; for nonspecialists, very difficult to separate from other littoral centipedes.

Centipedes are primarily carnivorous, using the poison claws on the first segment behind the head to capture and kill prey, and the littoral forms are no exception. The European marine centipede *Scolioplanes maritimus* has been observed feeding on the barnacle *Balanus balanoides* and on the snail *Littorina saxatilis*, but *Nyctunguis* probably feeds on the extensive insect and mite fauna of the intertidal zone.

See Blower (1957), R. Chamberlin (1960), and Cloudsley-Thompson (1968).

Phylum Arthropoda / Class Arachnida / Order Pseudoscorpionida
Family Garypidae

28.2 **Garypus californicus** Banks, 1909
Pseudoscorpion

Sometimes common, high intertidal zone under stones or debris, on sandy beaches, in crevices, or under cobbles on rocky shores; California.

Length 4.5 mm including the small appendages flanking the mouth (chelicerae); cephalothorax dark brown, subtriangular, with four eyes; dorsal plates of segments in posterior body region pale, yellowish, each plate divided laterally into two brownish-edged rectangles each with a central dark spot.

Like other pseudoscorpions, this species is predaceous, and feeds on intertidal arthropods such as flies, springtails, beetle larvae, and probably small crustaceans. *Garypus* are often found among beach-fly puparia washed by high tides into rock crevices. Here the pseudoscorpions grasp the emerging adults before they have a chance to fly away, then commence feeding. Often, *Garypus* are abundant under stones and debris at the bases of low banks or cliffs bordering sandy beaches. The silken chamber in which the female broods her eggs is sometimes the first indication of the presence of these animals. They are gregarious and move surprisingly quickly when disturbed.

Pseudoscorpions are recognized by the characteristic scorpionlike pedipalps (second pair of appendages) that bear the

large chelae, or pincers, that are used for prey capture and manipulation, for nestbuilding, and also in social interactions such as fighting, courting, and mating. Each chela consists of a fixed finger and a movable finger. One or both of these types of fingers bear an inwardly curving poison tooth that is used for immobilizing prey. Once the prey has stopped moving, it is passed from the chelae to the chelicerae, and digestive fluid is injected into its body. The digested body contents are then ingested.

The majority of pseudoscorpions are small, seldom exceeding 3–4 mm in length, but some members of the genus *Garypus* are larger; for instance, *G. giganteus* from Baja California is 7.5 mm long. Along with some other pseudoscorpion genera, *Garypus* are associated with the seashore, most commonly on tropical or subtropical shores. Species of the genus are found along the coasts of the Mediterranean Sea, Florida, California, and Baja California.

See J. Chamberlin (1921), Lee (1979), Muchmore (1972), and Weygoldt (1969).

Phylum Arthropoda / Class Arachnida / Order Acarina / Family Bdellidae

28.3 **Neomolgus littoralis** (Linnaeus, 1758)
Mite

Fairly abundant under stones and wrack, and on rock surfaces, high intertidal zone on rocky shores and sandy beaches; Alaska to California; Japan; Connecticut, Europe, Mediterranean Sea; circumpolar in the Arctic.

Length 2.5–3.5 mm; color bright red; mouthparts long, pointed, snoutlike.

Highly conspicuous because of their large size and scarlet color, these mites are diurnal predator-scavengers. In Japan, they were observed actively searching for food even in direct sunlight; they readily attacked crane flies (*Limonia monostromia*) emerging from pupal cases.

See Halbert (1920), King (1914), Thor (1931), and Tokunaga (1930).

Phylum Arthropoda / Class Arachnida / Order Araneae / Family Linyphiidae

28.4 Spirembolus mundus Chamberlin & Ivie, 1933
Spider

Locally abundant on rock surfaces in the green alga *Enteromorpha* and under debris on sand, high intertidal zone on rocky shores and beaches; Oregon to California.

Body 2.5 mm in length, antlike in outline, black with pale-brown legs; male readily distinguished from female by large palpal organ (see the photograph).

Although *S. mundus* has been found next to freshwater creeks in Oregon, it also occurs in maritime habitats. This species is gregarious and must be an active hunter of prey, since webs are not found in the spider's usual habitat.

The only other record of an intertidal linyphiid spider is that of *Mynoglenes marrineri*, which is found in the subantarctic islands of New Zealand. It is interesting to note that this spider is also eurytopic, being found in areas away from the beach as well as intertidally.

See R. Chamberlin & Ivie (1933) and Forster (1964).

Phylum Arthropoda / Class Insecta / Order Thysanura / Family Machilidae

28.5 Neomachilis halophila Silvestri, 1911
Bristletail

Common under stones and in rock crevices, high intertidal zone on rocky shores; also on sandy beaches; California.

Length 13–16 mm including long caudal filament and cerci; color dark, silvery gray; pointed processes (styli) on middle and hind legs; compound eyes large, contiguous; species superficially similar to common household silverfish though in a different family and darker in color.

Secluded by day, *N. halophila* emerges to feed after dark, its numbers out in the open increasing steadily until dawn. The animals are scavengers and grazers. In the spring at Pacific Grove (Monterey Co.), analyses of stomach contents show that they consume unicellular green algae (derived from lichens growing on the rocks under which the insects hide during the day), pollen grains from the Monterey pine *Pinus*

radiata (dispersed by the wind in great quantities at that season), yeasts, and plant detritus left on the beach by the receding tide.

See Benedetti (1973).

Phylum Arthropoda / Class Insecta / Order Diptera / Family Tipulidae

28.6 Limonia (Idioglochina) marmorata (Osten-Sacken, 1861)
(=Dicranomyia signipennis)   Crane Fly

Swarming in low numbers over algae or resting on vertical or overhanging sides of rocks, high to middle intertidal zone on rocky shores; British Columbia to San Luis Obispo Co. and San Miguel Island (Channel Islands).

Adult 5–9 mm long, a typical crane fly. Larva 12–15 mm long when full grown; abdomen with toothed ridges on all segments dorsally and on all but the first segment ventrally; color varying from dull olive-green to dark reddish brown and mottled, owing entirely to blood, gut contents, and fat body.

This is the only marine crane fly on the Pacific coast, and the adults are readily recognized as they swarm, usually in the evening, or rest on rocks. The females place their eggs singly or in small groups in algae. The eggs are subspherical, about 0.4 mm by 0.3 mm in size, and finely reticulate in texture. They can be very light brown or orange-yellow when the embryo is well developed.

L. marmorata passes the winter in the larval stage in British Columbia. Adults are more numerous in May, becoming scarcer during the summer months.

The larvae are found mainly in *Enteromorpha*, but *Porphyra* and *Pelvetia* also serve as food. They feed on the algae that can be reached from the loose silken tubes in which they reside, and which are usually situated deep in the algae or in wide crevices or between barnacles. The pupae remain in the larval tubes until the adult is ready to emerge. At low tide, pupal exuviae may be found in great numbers during the spring.

Adult crane flies usually feed by sucking the nectar of flowers, or they may only drink water, but *L. marmorata* appears to be a scavenger, since it feeds indiscriminately on the body fluids of dead animals such as the barnacle *Balanus glandula* (20.24) and the limpet *Collisella digitalis* (13.12). The adult stage is probably short lived; in the related species *L. monostromia* (found in Japan) it lasts only 2–3 days. This latter species is apparently preyed on by robber flies, mites (*Neomolgus littoralis*), and bats; except for the robber flies, which have not been observed in the intertidal zone in California, *L. marmorata* probably has similar predators.

See Alexander (1967), Glynn (1965), Saunders (1928), and Tokunaga (1930).

Phylum Arthropoda / Class Insecta / Order Diptera / Family Chironomidae

28.7 **Paraclunio trilobatus** Kieffer, 1911
Midge

On rocks and in crevices, low intertidal zone on rocky shores exposed to heavy surf; Oregon to Monterey Co.

Adult 5–7 mm in length, long-legged, blackish; setae with flattened scalelike modification in males and to a certain extent in females.

The larvae of this midge have been reported in the same zone as abalones (*Haliotis*, e.g., 13.1, 13.4), but they are difficult to find. The adults can be seen skittering over rocks and algae at low tide at night, particularly in the spring and summer.

The related *P. alaskensis* is very abundant along the southern shore of Vancouver Island (British Columbia). Population densities of the larvae averaged 1.7 per cm², though the distributions were clumped because of environmental factors. The larvae feed on green algae and diatoms and apparently can select certain groups of these plants from the substratum. Development time for *P. alaskensis* at 10°C from egg laying to the emergence of adults from pupal cases is 204 days.

See Morley & Ring (1972a,b) and Wirth (1949).

Phylum Arthropoda / Class Insecta / Order Diptera / Family Chironomidae

28.8 **Tethymyia aptena** Wirth, 1949
Midge

Seasonally abundant, high intertidal zone on rocky shores; larvae sometimes in algae; Mendocino Co. to Monterey.

Adults 1–2 mm long, grayish; antennae seven-segmented; maxillary palps one-segmented; wings vestigial. Larva to 4 mm long; body whitish, mottled with bluish to violet pigmentation; head dark brown, almost blackish; hooks of anterior pseudopods (footlike appendages) light amber; hooks of posterior pseudopods black.

The larvae of this midge are easier to find than the adults. Large numbers of the larvae sometimes occur in algae such as *Enteromorpha* and at certain times of the year form the major food item of the intertidal ground beetle *Thalassotrechus barbarae* (28.14). Both the adults and larvae are very abundant during the winter months and rare during the summer.

The nearest relative of *T. aptena* is *Belgica antarctica*, one of the few insects that is truly antarctic, being found at about 65° South latitude. Like all members of the subfamily Clunioninae (for instance, *Paraclunio trilobatus*, 28.7), *B. antarctica* is intertidal.

See Wirth (1949).

Phylum Arthropoda / Class Insecta / Order Diptera / Family Dolichopodidae

28.9 **Aphrosylus praedator** Wheeler, 1897
Long-Legged Fly

On algae, high and middle intertidal zones on rocky shores; British Columbia to San Luis Obispo Co.

Adult 2.5–3 mm long; wings somewhat smoky; body dull black with light-colored halteres; antennae with the third segment enlarged, the apical segment terminating as a long flagellum; head held in characteristic posterior-ventral position.

In British Columbia the adults have been seen flying about or resting between tides on algae-covered rocks on steep rocky shores. They are predaceous (as indicated by their name) and feed on both larval and adult chironomid flies, particularly *Smittia*, as well as other small insects, and possibly on small crustaceans found in the same habitat. The larvae are also predaceous and are found singly in the matting of filamentous algae covering the rocks. The pupae are encased in rounded, elliptical cocoons loosely spun in the filamentous algae. The short, stout pupa thrusts its long, black, straplike prothoracic horns out through an opening in the cocoon to a distance equal to the length of the thorax. The head bears a short, pointed tubercle on the face, which probably aids the adult in working its way out of the cocoon.

During early summer in British Columbia, exuviae may be picked off the algae-covered rocks in large numbers at low tide, but it is not easy to find the pupae hidden in the cocoons.

See Saunders (1928).

Phylum Arthropoda / Class Insecta / Order Diptera / Family Coelopidae

28.10 **Coelopa (Neocoelopa) vanduzeei** Cresson, 1914
Kelp Fly

On wrack banks, middle and low intertidal zones on sandy beaches; southern Alaska to Baja California.

Adult 5–7 mm long; body dorsoventrally flattened; thorax and head dark brown; abdomen and legs brown; eyes light reddish tan; head, maxillary palps, and legs strongly bristled; abdomen hairy; wings held flat over abdomen. Third instar larva 9–12 mm long; narrowing anteriorly; greatest width and height 1.5 mm at caudal end; color white with gray ventral spines on abdominal segments; caudal plates yellow with black borders, surrounded by pale hairs; spiracles three in number; anus four-lobed, located ventroposteriorly on segment 12; anterior prothoracic spiracles bilobed, each lobe marginally scalloped to form ten petal-like processes. Puparium 3.5–6 mm long; ventral surface and dorsal surface of anterior three segments flattened, dorsum elsewhere arched; lateral edges of each segment indented; color yellow-brown to brown-black.

Members of the family Coelopidae are restricted to the wrack habitat of beaches in temperate and high-latitude regions of both the Northern and Southern Hemispheres. The larvae are saprophagous, feeding on many kinds of cast-up algae. *C. vanduzeei* is probably the most abundant fly associated with wrack on the Pacific coast. Swarms of adults are often seen flying over the wrack banks, and large numbers can be found resting within the wrack itself. Sometimes the adults are found with adults of **Fucellia costalis** Stein (Anthomyiidae), a larger, dark fly that has bristles on the anterior margins of the front wings. *C. vanduzeei* larvae are found mainly in the lower-beach wrack banks and usually in the upper 10 cm of the bank. They feed on algae that have started to decompose and prefer *Macrocystis* to other algae in the wrack. These flies are adapted for rapid population growth in temporary habitats, and throughout the year their numbers may vary from very low to very high. Large numbers of the pupae are often parasitized by small wasps (*Eupteromalus*, family Pteromalidae), further reducing populations.

Techniques for the laboratory culture of *C. vanduzeei*, using rehydrated powdered *Macrocystis pyrifera*, have been developed at the Hopkins Marine Station (Pacific Grove, Monterey Co.). At temperatures of 22–29°C, a life cycle from adults of one generation to those of the next takes 13–16 days; eggs hatch in 48 hours, the three larval instars occupy 8–9 days, and pupae are present for 3–6 days before adults emerge (S. L. Spencer, pers. comm.).

See Cole (1969), Egglishaw (1960), and Kompfner (1974).

Phylum Arthropoda / Class Insecta / Order Diptera / Family Canaceidae

28.11 **Canaceoides nudatus** (Cresson, 1926)
Beach Fly

High and middle intertidal zones on rocky shores; Washington to Isla San Gerónimo (Baja California).

Adult wing 3–3.5 mm long, blackish. Larva 4–5 mm long, covered with fine spicules interspersed with a sparse assortment of large spines; anterior respiratory processes well developed; posterior spiracles borne on tubelike extension of body.

Canaceoides nudatus feeds on algae such as *Enteromorpha* and is only one of several species of seashore-inhabiting canaceids found on the Pacific coast of North America.

Hawaiian records of this species are apparently in error. Two species of the genus do occur in Hawaii, the widely distributed *C. angulatus* and the Hawaiian *C. hawaiiensis*.

See Williams (1938) and Wirth (1951, 1969).

Phylum Arthropoda / Class Insecta / Order Coleoptera / Family Carabidae

28.12 Dyschirius marinus (LeConte, 1880)
Ground Beetle

High intertidal zone on sandy beaches; central and southern California.

Adult 6.5 mm long; body narrowly constricted between thorax and abdomen, light reddish brown; head strongly prognathous (mouthparts anterior), the mouthparts with stout mandibles; eyes black; elytra fused together along suture; wings absent. Larva 9.5 mm long, elongate, pale with yellow triangular markings dorsally on abdominal segments 2 to 8; head strongly prognathous, brown on dorsal-posterior part; mandibles sickle-shaped; maxillary palps, labial palps, and antennae elongate.

The larvae and adults of this beetle are predaceous on other high-beach insects and crustaceans associated with decaying kelp. Generally populations of *Dyschirius* are low and specimens are difficult to find. Unlike rock-dwelling carabids, which tend to be flattened, *Dyschirius* adults are somewhat cylindrical and adapted for burrowing in sand much like the staphylinid beetle *Bledius monstratus* (28.18), with which it is usually associated.

See Lindroth (1961).

Phylum Arthropoda / Class Insecta / Order Coleoptera / Family Carabidae

28.13 Trechus ovipennis Motschulsky, 1845
Ground Beetle

Uncommon, in crevices and under loose stones, high intertidal zone on rocky shores; southern Alaska to central California.

Adult 4–5 mm long, dark rufous to grayish black; appen-

dages paler; head very narrow; prothorax strikingly small and narrow; elytra very broad.

Like most other ground beetles, this species is probably a predator and/or scavenger. It is usually found hidden under stones and in crevices during the day but is active at night.

See Jeannel (1927).

Phylum Arthropoda / Class Insecta / Order Coleoptera / Family Carabidae

28.14 Thalassotrechus barbarae (Horn, 1892)
Ground Beetle

In crevices, high intertidal zone on rocky shores; northern California to Baja California.

Adult 5–7 mm long; head, pronotum (dorsal part of prothorax), and legs light reddish brown; head prognathous (with anterior mouthparts); elytra black. Larva 6–8 mm long, slim, cylindrical; head and prothorax heavily sclerotized; mouthparts long, anterior on head; each abdominal segment bearing yellowish-brown dorsal plates, the posterior segment with two long sclerotized cerci.

The adults vary in size and color along a north-south cline, those from Baja California being about 5 mm long and light-colored, those from northern California about 7 mm long and with black elytra.

When tidal conditions permit, the adults emerge from crevices at dusk and walk over adjacent rock surfaces and algae, feeding on stranded zooplankton, algal-feeding dipterous larvae such as *Tethymyia aptena* (28.8), and small worms (oligochaetes). Foraging takes place over a broad zone extending from the splash zone to about 0.6 m above zero tide level. Mating also takes place when the adults emerge from their crevices at night, but aside from this interaction, adults out in the open are always alone, in contrast to the aggregations that exist in crevices during the day. The beetles retreat into crevices during a full moon, at dawn, or when the tide floods, but come out again on an ebbing tide if it occurs 2–3 hours before dawn. Activity is regulated by a combination of circadian and circatidal locomotory rhythms. These insects are active throughout the year, and all stages of development can be found during any month. The eggs, larvae, and pupae are restricted to crevices.

During 1969, this insect could not be found near Santa Barbara, after which it was named, presumably because of the massive oil spill that occurred earlier that year.

See Evans (1970, 1976, 1977) and Van Dyke (1918).

Phylum Arthropoda / Class Insecta / Order Coleoptera / Family Hydrophilidae

28.15 **Cercyon fimbriatus** Mannerheim, 1852
Water-Scavenger Beetle

Adults sometimes very numerous, high intertidal zone on rocky shores and sandy beaches; Alaska to Baja California.

Adult 2.5–3.5 mm long; body oval, convex; head held obliquely; mouthparts ventral; elytra deeply striated, the color variable, usually brownish to black.

The larvae of other species of *Cercyon* (*C. litoralis* and *C. depressus*) in Europe prey on slow-moving, soft-skinned animals, such as annelid (enchytraeid) worms and dipterous larvae associated with decaying kelp. The adults, however, feed on saprophytic fungi and bacteria.

See Backlund (1945) and Leech & Chandler (1956).

Phylum Arthropoda / Class Insecta / Order Coleoptera / Family Hydrophilidae

28.16 **Cercyon luniger** Mannerheim, 1853
Water-Scavenger Beetle

High intertidal zone on rocky shores and sandy beaches; Alaska to Baja California.

Adult 2.5–4 mm long; body oval, convex; head held obliquely; mouthparts ventral; elytra with faint longitudinal striations; color variable, generally reddish brown with black scalloped pattern on apical quarter of elytra.

The larvae of *C. luniger*, like those of *C. fimbriatus* (28.15), are probably predaceous on dipterous larvae and other slow-moving, soft-skinned animals associated with decaying kelp. The adults probably feed on saprophytic fungi.

Cercyon luniger adults have frequently been observed swarming on beaches in large numbers with other beetles of the decaying kelp community, such as *Cafius luteipennis* (28.19).

See Backlund (1945), Leech (1971), and Leech & Chandler (1956).

Phylum Arthropoda / Class Insecta / Order Coleoptera / Family Limnebiidae

28.17 **Ochthebius vandykei** Knisch, 1924
Limnebiid Beetle

Common in crevices, high intertidal zone on rocky shores; British Columbia to San Luis Obispo, including San Francisco Bay.

Adult 1–1.5 mm long, highly sculptured; color black; antennae-like maxillary palps projecting forward; antennae extended posterolaterally.

The adults, which often occur in large numbers, are probably detritus feeders; the larvae, like other limnebiid larvae, are carnivores.

Beetles in the family Limnebiidae (=Hydraenidae) are small (1–2.5 mm long), and are always associated with wet or damp habitats. Some occur among the stones and sand adjacent to fast-running water, some along the margins of ponds and lakes, and others in brackish waters and in the intertidal zone. The adult antennae have lost their function as sensory appendages (a role which has been taken over by the elongated maxillary palps) and serve instead to break through the surface film of water and allow a bubble of air to travel to the ventral surface of the thorax and abdomen. The bubble is held there for respiratory purposes by an unwettable pile of short, dense hairs (plastron).

The genus *Ochthebius* is widely distributed, being common in Europe and North America and also represented in Africa, India, Siberia, and Australia. In California, species are found in hot springs with a temperature of 38.8°C, in saline pools in Death Valley, and in a variety of other aquatic habitats. *O. vandykei* is related to species found in the intertidal zone in Europe, such as *O. marinus* and *O. lejolisi*, which occur in Britain, and the Adriatic species *O. steinbuhleri* and *O. adriaticus*.

See Leech & Chandler (1956) and Riedl (1963).

Phylum Arthropoda / Class Insecta / Order Coleoptera / Family Staphylinidae

28.18 Bledius monstratus Casey, 1889
(=B. monstrosus)   Rove Beetle

Relatively uncommon, burrowed in sand near wrack, high intertidal zone on sandy beaches; Washington to California.

Adult 4.5–5 mm long; head and abdomen usually dark brown to black, sometimes light brown throughout; elytra yellowish brown owing to dense pubescence; legs brownish; front femur and tibia expanded; front tibia with two longitudinal series of short blunt spines; middle tibia bearing two rows of sharp spines; prothorax coarsely pitted.

The adult of this species appears to be predaceous, though many staphylinids are scavengers. Members of this large genus are found throughout the world, particularly in temperate regions. The adults and larvae inhabit burrows in the sand or mud near rivers, streams, ponds, and lakes, on beaches and salt flats, and in salt marshes. The eggs are deposited in chambers next to the main burrow in the substratum. The beach forms are associated with the decaying kelp community.

See Casey (1889), Hatch (1957b), Herman (1972), and Moore (1964a).

Phylum Arthropoda / Class Insecta / Order Coleoptera / Family Staphylinidae

28.19 Cafius luteipennis Horn, 1884
Rove Beetle

On or under decaying kelp, high to middle intertidal zone on sandy beaches; British Columbia to Baja California.

Adult 6–8 mm long; head coarsely punctate above; head, thorax, abdomen, and legs shiny black; elytra yellowish brown; apex of abdomen and mouthparts dark brown; tarsi of front legs spiny, hairy, enlarged for burrowing.

This species may be found under freshly cast-up kelp in company with fly maggots (for example, *Fucellia*, family Anthomyiidae) or as a member of the decomposing kelp community at a higher level on the beach. Both adults and larvae prey on *Fucellia* larvae and pupae, amphipods, and other species of *Cafius*.

Six other species of *Cafius* occur with *C. luteipennis* in southern California as part of the decaying kelp community. Occasionally, spectacular mass flights of various species of beetles associated with this community can be seen on beaches, the swarms generally flying in a direction parallel with the beach and lasting for an hour or more. These swarms are usually mistaken for swarms of kelp flies (Anthomyiidae and Coelopidae). *C. luteipennis* has been captured from such swarms in Baja California, and in Monterey and Marin Cos.

Cafius is a large genus restricted to coastal habitats. There are at least eight intertidal species on the Pacific coast of North America. Other species occur on the Atlantic coasts of Europe (*C. xantholoma* and *C. sericeus*, for example) and North America (*C. bistriatus*), and in the eastern Pacific (*C. nauticus*).

See Hatch (1957b), James, Moore & Legner (1971), Keen (1895), and Leech (1971).

Phylum Arthropoda / Class Insecta / Order Coleoptera / Family Staphylinidae

28.20 Cafius seminitens (Horn, 1884)
Rove Beetle

Common on decomposing kelp, high intertidal zone on sandy beaches; British Columbia to Baja California.

Adult 9–12 mm long, black, shiny; head coarsely pitted; prothorax pitted dorsally with two longitudinal rows of seven to nine pits; dorsal plates of abdomen with distinct wavy pattern of pubescence; tarsi of front legs spiny, modified for burrowing.

Larvae and adults of this species are predators in the decomposing kelp community of sandy beaches. They feed voraciously on larvae and pupae of the anthomyiid kelp fly *Fucellia* and on amphipods, other scavengers of cast-up kelp, and other species of *Cafius*. Occasionally they scavenge on dead fish, but they appear to subsist mainly on *Fucellia*. As soon as a maggot is attacked and the body wall ruptured, the oozing body fluids attract other staphylinids in the immediate area and as many as seven beetles have been seen feeding on one maggot. Six other species of *Cafius* as well as other species of beetles occur with *C. seminitens* in the decaying kelp community, but it is not known how the different ecological niches are partitioned.

See Hatch (1957b), James, Moore & Legner (1971), and Keen (1895).

Phylum Arthropoda / Class Insecta / Order Coleoptera / Family Staphylinidae

28.21 Hadrotes crassus (Mannerheim, 1846)
Rove Beetle

Under decaying kelp or buried in sand by day, on beach surface at night, high and middle intertidal zones on sandy beaches; Alaska to Baja California.

Adult 11–17 mm long, the largest black staphylinid beetle on the beach, elongate, shiny black with brownish abdomen. Larva 14–16 mm long; head and thorax reddish brown; abdomen light reddish brown above; legs yellowish.

Both the larvae and adults of this species are nocturnal predators on crustaceans and other beach-inhabiting insects.

See Hatch (1957b) and Moore (1964b).

Phylum Arthropoda / Class Insecta / Order Coleoptera / Family Staphylinidae

28.22 Thinopinus pictus LeConte, 1852
Rove Beetle

Burrowed in damp or wet sand, on surface at night, high and middle intertidal zones on sandy beaches; Alaska to Baja California.

Adult 14–22 mm long; mandibles large; elytra short; wings absent; front legs adapted for burrowing in sand and grasping prey; color pale yellowish brown, inconspicuous against sand (specimens found on dark sand bear black markings of varying shape and amount on pronotum, elytra, and abdominal tergites); ventral cleft in last abdominal segment of adult males, lacking in females. Larva 12–20 mm long, widest at head, tapering slightly posteriorly, blunter than larva of *Hadrotes crassus* (28.21); mandibles large, sickle-shaped; legs adapted for digging and catching prey.

These remarkable animals are among the largest of intertidal beetles. Burrowed in sand not too firmly packed during the day and at high tide, they occupy a belt of beach lying between the upper limits of higher high water and lower high water. The population moves gradually up and down the beach with shifts between spring and neap tides. The animals emerge from the sand when uncovered by the sea at night. At the surface they usually remain immobile, but lunge out and grasp with their mandibles prey coming within about 5 mm of them. On some beaches the populations may reach 1 or 2 per m². They feed mainly on the beach hoppers *Orchestoidea* (22.1, 22.2) and *Orchestia*, but they also eat the sand isopod *Alloniscus perconvexus* (21.14), beach flies, and other *T. pictus*. Unable to swim, they can nevertheless survive up to a few days floating at the surface of turbulent water. Numerous mites may occur beneath the elytra of adults.

Breeding probably occurs mainly from August through October in the Santa Barbara area. The eggs are large (3 mm by 2.2 mm), and are laid individually below the sand. They hatch in about 2 weeks.

See especially Craig (1970); see also Hamilton (1894) and Malkin (1958).

Phylum Arthropoda / Class Insecta / Order Coleoptera / Family Staphylinidae

28.23 Bryobiota bicolor (Casey, 1885)
Rove Beetle

On decaying kelp, high intertidal zone on sandy beaches; San Mateo Co. to San Diego.

Adult 2.5 mm long; head, thorax, elytra, and legs reddish brown; abdominal segments black; eyes small; each elytron much longer than wide.

Bryobiota bicolor is a member of the decaying kelp community on the upper beach. In San Diego, the type locality of this species, adults were originally found in great numbers under densely packed algae cast up on the shores of the inner harbor in the spring. Occurring with it and also in great abundance were three other beetles, *Cafius decipiens* (Staphylinidae), *Motschulskium sinuatocolle* (28.28) and *Phycocoetes testa-*

ceus (28.35). Another staphylinid, *Cafius opacus*, occurred in smaller numbers. The conditions in San Diego Bay have changed so much since Casey described this species in 1885 that it is unlikely that high densities of these insects will occur there again.

See Casey (1885) and Moore (1956a).

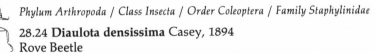

Phylum Arthropoda / Class Insecta / Order Coleoptera / Family Staphylinidae

28.24 **Diaulota densissima** Casey, 1894
Rove Beetle

In protected crevices or among algae and barnacles, high and middle intertidal zones on rocky shores; Alaska to Coronado (San Diego Co.).

Adult 2.5–3 mm long; head, thorax, and abdomen black; each elytron as long as wide; body covered with short, dense pubescence.

Diaulota densissima adults and larvae occur in a variety of habitats, such as in crevices, or with *D. vandykei* (28.25) and *Liparocephalus cordicollis* (28.26) among algae, or in the association dominated by the red alga *Endocladia* and the barnacle *Balanus glandula* (20.24). At high tide they are sometimes found in the air pockets trapped below the basal plates of barnacles. Population densities averaged 107 per 1,000 cm² in favorable *Endocladia* beds at Pacific Grove, Monterey Co. (8.39 mg dry weight biomass per 1,000 cm²), and the numbers sometimes exceeded twice that. Both the adults and larvae move about actively at low tide, exploring the crevices between barnacles and similar sites. Stomachs of some adults examined were full of green and blue-green algae and some contained diatoms. Both the adults and larvae also prey and scavenge on animal food as well.

Breeding probably occurs in the fall, winter, and spring at Pacific Grove. The large yolky eggs are laid in narrow crevices, where they remain till hatching.

See Casey (1893), Glynn (1965), Jones (1968), Moore (1956a,b) and Saunders (1928).

Phylum Arthropoda / Class Insecta / Order Coleoptera / Family Staphylinidae

28.25 **Diaulota vandykei** (Moore, 1956)
Rove Beetle

Among algae, middle intertidal region on rocky shores; San Mateo and Monterey Cos.

Adult 2.8 mm long; head, thorax, elytra, and legs dark reddish brown; abdomen dark brown.

Both adults and larvae of this species are predators and can generally be found in association with *D. densissima* (28.24) and *Liparocephalus cordicollis* (28.26), but they are more numerous from zero tide level to 1.5 m above it. At Pacific Grove (Monterey Co.), the larvae appear to be more abundant than the adults during the months of June, July, and August, perhaps indicating that there is one generation of this beetle each year. The larvae are most readily found in the debris trapped in densely growing stands of the green alga *Ulva californica*.

See J. Chamberlin & Ferris (1929), Jones (1968), and Moore (1956a).

Phylum Arthropoda / Class Insecta / Order Coleoptera / Family Staphylinidae

28.26 **Liparocephalus cordicollis** LeConte, 1880
Rove Beetle

Among algae, middle and low intertidal zones on rocky shores; Alaska to central California.

Adult 4 mm long, dark reddish brown, covered with short, dense pubescence; head broader than prothorax, large compared with that of other intertidal staphylinids of similar size; mouthparts clearly ventral.

This insect is a predator on small crustaceans and other animals found in holdfasts of algae such as *Egregia*, and in crevices and on rock surfaces among algae.

Liparocephalus cordicollis is usually found in association with two other staphylinids, *Diaulota densissima* (28.24) and *D. vandykei* (28.25), though it is most abundant 0.15 to 0.6 m above zero tide level, whereas the other two species are found at higher tide levels. At Pacific Grove (Monterey Co.) the larvae prefer algal holdfasts such as *Egregia*, in contrast to the adults, which seem to prefer open rock surfaces under thalli of the red alga *Iridaea flaccida*. In such areas adults can be seen walking over the rock surfaces when they dry out at ebb tide. At low spring tides the insects frequently move down into the −0.3 to +0.4 m tide level.

See J. Chamberlin & Ferris (1929), Jones (1968), Keen (1897), Le-Conte (1880), and Saunders (1928).

Phylum Arthropoda / Class Insecta / Order Coleoptera / Family Staphylinidae

28.27 Pontomalota nigriceps Casey, 1885
Rove Beetle

High and middle intertidal zones on sandy beaches; California.

Adult 3 mm long, elongate; head and abdomen brownish black; elytra pale yellowish; antennae and legs yellowish brown; abdominal tergites pubescent.

Pontomalota often occurs in large numbers and, being nocturnal, is easily found at night when it forages on the sand surface. Dead crustaceans such as the amphipod *Orchestoidea* (22.1, 22.2) serve as food, and this scavenging habit has led to an association with *Thinopinus pictus* (28.22), the large staphylinid predator of the sandy beach. *Pontomalota* can often be found feeding on the remains of the prey of *Thinopinus*.

Pontomalota nigriceps may be a synonym for *P. opaca* (Le-Conte), which occurs on southern California beaches along with two other staphylinids, *Thinopinus pictus* (28.22) and **Thinusa maritima** (Casey).

See Kincaid (1961) and Moore (1975).

Phylum Arthropoda / Class Insecta / Order Coleoptera / Family Ptiliidae

28.28 Motschulskium sinuatocolle Matthews, 1872
Feather-Winged Beetle

On wrack, high intertidal zone on sandy beaches; British Columbia to California.

Adult 1.3 mm long; body broad and highly convex; edges of wings fringed with long hairs; antennae pubescent.

Motschulskium sinuatocolle, the only species in the genus, is associated with decaying kelp, where it probably feeds chiefly on the spores of fungi, as do other ptiliid beetles.

The feather-winged beetles, formerly known as the family Trichopterygidae, are the smallest of all beetles, ranging in size from about 0.25 to 1 mm. They are usually found in decaying vegetation, rotting logs, or ant nests, and some, like *Motschulskium*, live in decaying seaweed on beaches. *Actinopteryx fucicola*, for instance, is widely distributed on seacoasts throughout the world; *Ptenidium punctatum* and *Actinidium coarctatum* are European intertidal species.

See Marine Biological Association (1957) and Riedl (1963).

Phylum Arthropoda / Class Insecta / Order Coleoptera / Family Histeridae

28.29 Neopachylopus sulcifrons (Mannerheim, 1843)
Hister Beetle

In wrack, high intertidal zone on sandy beaches; British Columbia to southern California.

Adult 3.5–5 mm long, shiny, dark brown to blackish; head with mouthparts clearly ventral (hypognathous); antennae clubbed; legs heavily spined, adapted for burrowing; tarsi flattened, platelike; elytra striated, extending posteriorly two-thirds the length of the body.

The adults and larvae of most histerids are carnivorous on insects associated with such decaying organic matter as carrion, excrement, and vegetation. *Neopachylopus* probably feeds on very small animals, such as collembolans, associated with decaying kelp on the upper beach. European species with similar habits, *Acritus punctum* and *Pachylopus maritimus*, are also members of the wrack community.

See Marine Biological Association (1957) and Riedl (1963).

Phylum Arthropoda / Class Insecta / Order Coleoptera / Family Melyridae

28.30 **Endeodes collaris** (LeConte, 1853)
Soft-Winged Flower Beetle

High intertidal zone on rocky shores and sandy beaches; Vancouver (British Columbia) to Monterey.

Adult 5–7 mm long; prothorax reddish brown, rest of body bluish black; wings absent; elytra abbreviated; protrusible membranous vesicles on prothorax and between metathorax and abdomen. Larva 6 mm long; head reddish brown; abdomen and thorax yellowish brown with a dorsal pair of pale circular spots on each segment; processes on terminal segments (urogomphi) black, pointed.

Predaceous *Endeodes* adults can be found on the surfaces of rocks in the daytime when the tide is low, whereas the larvae are confined to crevices near the highest tide mark. In Pacific Grove (Monterey Co.), larvae have been found in crevices just above the high tide mark in the same area in which the staphylinids *Liparocephalus cordicollis* (28.26) and *Diaulota* (28.24, 28.25) were found. The rock surfaces were heavily coated with bird droppings, and against such a background the colorful adults are easily seen. They resemble staphylinid beetles because of the reduced elytra, but their conspicuous color and the bladderlike membranous vesicles on the thorax and abdomen readily identify them as melyrids. Both the vesicles and the coloration are probably elements of a warning display directed toward predators.

Other Pacific coast species of *Endeodes* are also intertidal, all adults being found in the upper intertidal zone, often under debris on beaches and on rock surfaces; the larvae are crevice dwellers. **E. insularis** Blackwelder, **E. blaisdelli** Moore, and **E. terminalis** Marshall are found in the warm temperate zone extending from Point Conception (Santa Barbara Co.) to Baja California. **E. basalis** (LeConte) and **E. rugiceps** Blackwelder are found north and south of Point Conception, but *E. collaris* is the only species that is found exclusively in the cold temperate zone. These species are readily separated by color and by external characters.

See Blackwelder (1932), Leech & Chandler (1956), and Moore (1954).

Phylum Arthropoda / Class Insecta / Order Coleoptera / Family Tenebrionidae

28.31 **Epantius obscurus** (LeConte, 1851)
Darkling Beetle

Under wrack, high intertidal zone, and occasionally above high tide mark on sandy beaches; Marin Co. to Baja California (both coasts).

Adult 7–8 mm long, dull black; elytra with longitudinal striations; lateral edges of thorax sinuate.

The slow-moving adults of *Epantius* can be found under dried kelp on the high beach. Both the adults and the larvae are scavengers, feeding for the most part on dried kelp.

See Blaisdell (1932, 1943).

Phylum Arthropoda / Class Insecta / Order Coleoptera / Family Tenebrionidae

28.32 **Phaleria rotundata** LeConte, 1851
Darkling Beetle

Adults often abundant under dried kelp by day, high intertidal zone on sandy beaches; central California to northern Baja California.

Adult 5 mm long, oval, reddish brown; elytra striated, covering abdomen except for apical segment; front tibiae flattened, each enlarged at apex, forming semilobed structure with spine or inner reduced lobe.

This primarily nocturnal beetle is a scavenger, feeding on kelp washed up on sandy beaches.

See Blaisdell (1943).

Phylum Arthropoda / Class Insecta / Order Coleoptera / Family Salpingidae

28.33 **Aegialites subopacus** (Van Dyke, 1918)
Salpingid Beetle

In crevices, high intertidal zone on rocky shores; Sonoma, Marin, and San Mateo Cos.

Adult 3.5–3.8 mm long, bluish black above (metallic green when wet), brown below, wingless.

Adults and larvae of this insect are gregarious. The adults, and perhaps the larvae, are thought to be phytophagous, since adult specimens from Bodega Bay harbor (Sonoma Co.) had diatoms in their midguts. They could, however, be scavengers. Aside from these observations, nothing else is known of the biology of this member of a unique group of exclusively marine insects. Other intertidal insects have terrestrial relatives from which the marine forms are, in most cases, obviously derived. This is not the case with *Aegialites*, which is not clearly related to any terrestrial group. It has been placed in five different families by various authors, and its inclusion in a subfamily of the Salpingidae, on the basis of characteristics of the male genitalia, does not imply that it is very close to any group in this family.

The genus *Aegialites*, with four intertidal species, is one of two genera in the subfamily Aegialitinae that were formerly placed in the well-known family Eurystethidae. A. **fuchsii** Horn, with striated, shiny elytra, is distributed from Mendocino Co. south to the Farallon Islands (San Francisco Co.) *A. californicus*, the type locality of which was mistakenly thought to be California, ranges from the Aleutian Islands (Alaska) to Oregon. Farther west, *A. stejnegeri* Linnell, with two subspecies, ranges from the Sea of Okhotsk, U.S.S.R. (*A. stejnegeri stejnegeri*) to the Kurile Islands, U.S.S.R. (*A. stejnegeri sugiharai*). Like *A. subopacus*, the latter is phytophagous, but immature oribatid mites have been dissected from the fore- and midguts of *A. stejnegeri stejnegeri*.

The other genus in the Aegialitinae is the intertidal *Antarcticodomas*, with one species, *A. fallai*, occurring in the subantarctic islands south of New Zealand.

See especially Spilman (1967); see also Leech & Chandler (1956).

Phylum Arthropoda / Class Insecta / Order Coleoptera / Family Curculionidae

28.34 **Emphyastes fucicola** Mannerheim, 1852
Weevil

Under dried kelp, high and upper middle intertidal zones on sandy beaches; Alaska to southern California.

Adult 6–8 mm long (the largest intertidal weevil in California); color varying from black through light brown to pale yellowish brown; apices of hind tibiae strongly expanded; apices of front tibiae paddle-shaped. Larva 7–8 mm long; lacking legs; body creamy-white, the head light brownish yellow.

The adults of this weevil, which are probably saprophagous, can generally be found in the daytime under dried kelp, but at night they move around on the surface of the sand. The habits of the adult can readily be inferred by observing the peculiarly shaped fore and hind tibiae, which are adapted for digging into sand. Active adults have been found buried in sand to a depth of 30 cm in beaches at Pacific Grove (Monterey Co.). The larvae feed on decaying kelp below the surface of the sand and consequently are not readily seen unless the subsurface material is sifted.

See Hatch (1971) and LeConte & Horn (1883).

Phylum Arthropoda / Class Insecta / Order Coleoptera / Family Curculionidae

28.35 **Phycocoetes testaceus** LeConte, 1876
Weevil

High intertidal zone on sandy beaches and rocky shores; British Columbia and probably Alaska to southern California.

Adult 2–3 mm long, light to dark brown; tibiae normal (not expanded as in *Emphyastes fucicola*, 28.34).

The adults of *Phycocoetes* are probably detritus feeders. They are commonly found in large groups in crevices in high intertidal rocks, and occasionally on sand under washed-up kelp.

See Hatch (1971), Keen (1895), and LeConte & Horn (1883).

Literature Cited

Alexander, C. P. 1967. The crane flies of California. Bull. Calif. Insect Survey 8: 1–269.

Arnett, R. H., Jr. 1968. The beetles of the United States. Washington, D.C.: Catholic University Press. 1,112 pp.

Backlund, H. O. 1945. Wrack fauna of Sweden and Finland, ecology and chorology. Opuscula Entomol. Suppl. 5: 1–238.

Baduoin, R. 1946. Contribution a l'éthologie d'*Aepophilus bonnairei*.

Signoret et à celle de quelques autres arthropodes à respiration aérienne de la zone intercotidale. Bull. Soc. Zool. France 71: 109–13.

Benedetti, R. 1973. Notes on the biology of *Neomachilis halophila* on a California sandy beach (Thysanura: Machilidae). Pan-Pac. Entomol. 49: 246–49.

Blackwelder, R. E. 1932. The genus *Endeodes* LeConte (Coleoptera, Melyridae). Pan-Pac. Entomol. 8: 128–36.

———. 1944–57. Checklist of the coleopterous insects of Mexico, Central America and West Indies, and South America. U.S. Natur. Mus., Bull. 185. Washington, D.C. (issued in separate parts, 1944–57). 1,492 pp.

Blaisdell, F. E., Sr. 1932. Studies in the tenebrionid tribe Scaurini: A monographic revision of the Eulabes (Coleoptera). Trans. Amer. Entomol. Soc. 58: 35–101.

———. 1943. Contributions toward a knowledge of the insect fauna of Lower California. 7. Coleoptera: Tenebrionidae. Proc. Calif. Acad. Sci. (4) 24: 171–288.

Blower, J. G. 1957. Feeding habits of a marine centipede. Nature 180: 560.

Casey, T. L. 1885. New genera and species of California Coleoptera. Bull. Calif. Acad. Sci. 1: 285–336.

———. 1889. Coleopterological notices I. Ann. N.Y. Acad. Sci. 5: 39–198.

———. 1893. Coleopterological notices V. Ann. N.Y. Acad. Sci. 7: 281–606.

Caspers, H. 1951. Rhythmische Erscheinungen in der Fortpflanzung von *Clunio marinus* (Dipt. Chiron.) und das Problem der lunaren Periodizität bei Organismen. Archiv f. Hydrobiologie 18 (Suppl.): 415–594.

Chamberlin, J. C. 1921. Notes on the genus *Garypus* in North America (Pseudoscorpionida-Cheliferidae). Canad. Entomol. 53: 186–91.

Chamberlin, J. C., and G. F. Ferris. 1929. On *Liparocephalus* and allied genera. Pan-Pac. Entomol. 5: 137–43, 153–62.

Chamberlin, R. V. 1960. A new marine centipede from the California littoral. Proc. Biol. Soc. Washington 73: 99–102.

Chamberlin, R. V., and W. Ivie. 1933. Spiders of the Raft River Mountains of Utah. Bull. Univ. Utah 23: 1–79.

Cheng, L., ed. 1976. Marine insects. New York: Elsevier/North-Holland. 581 pp.

Cloudsley-Thompson, J. L. 1968. Spiders, scorpions, centipedes and mites. Oxford: Pergamon. 288 pp.

Cole, F. R. 1969. The flies of western North America. Berkeley and Los Angeles: University of California Press. 693 pp.

Craig, P. C. 1970. The behavior and distribution of the intertidal sand beetle, *Thinopinus pictus* (Coleoptera: Staphylinidae). Ecology 51: 1012–17.

Daly, H. V. 1975. Orders of intertidal insects; Collembola, Hemiptera, pp. 432–35, *in* Smith & Carlton (1975).

Doyen, J. T. 1975. Intertidal insects: Order Coleoptera, pp. 446–52, *in* Smith & Carlton (1975).

Egglishaw, H. J. 1960. Studies on the family Coelopidae (Diptera). Trans. Roy. Entomol. Soc. London 112: 109–40.

Evans, W. G. 1968. Some intertidal insects from western Mexico. Pan-Pac. Entomol. 44: 236–41.

———. 1970. *Thalassotrechus barbarae* and the Santa Barbara oil spill. Pan-Pac. Entomol. 46: 233–37.

———. 1976. Circadian and circatidal locomotory rhythms in the intertidal beetle *Thalassotrechus barbarae* (Horn): Carabidae. J. Exper. Mar. Biol. Ecol. 22: 79–90.

———. 1977. Geographic variation, distribution and taxonomic status of the intertidal insect *Thalassotrechus barbarae* (Horn) (Coleoptera: Carabidae). Quaest. Entomol. 13: 83–90.

Fall, H. C. 1901. List of the Coleoptera of southern California, with notes on habits and distribution and descriptions of new species. Occas. Pap. Calif. Acad. Sci. 8: 1–282.

Forster, R. R. 1964. The Araneae and Opiliones of the subantarctic islands of New Zealand. Pacific Insects Monogr. 7: 58–115.

Glynn, P. W. 1965. Community composition, structure, and interrelationships in the marine intertidal *Endocladia muricata–Balanus glandula* association in Monterey Bay, California. Beaufortia 12: 1–198.

Halbert, J. N. 1920. The Acarina of the seashore. Proc. Roy. Irish Acad. 35: 106–52.

Hamilton, J. 1894. Catalogue of the Coleoptera of Alaska, with synonymy and distribution. Trans. Amer. Entomol. Soc. 21: 1–38.

Hatch, H. 1957a. The beetles of the Pacific northwest. 1. Introduction and Adephaga. Univ. Washington Publ. Biol. 16: 1–340.

———. 1957b. The beetles of the Pacific northwest. 2. Staphyliniformia. Univ. Washington Publ. Biol. 16: 1–384.

———. 1961. The beetles of the Pacific northwest. 3. Pselaphidae and Diversicornia I. Univ. Washington Publ. Biol. 16: 1–503.

———. 1965. The beetles of the Pacific northwest. 4. Macrodactyles, Palpicornes and Heteromera. Univ. Washington Publ. Biol. 16: 1–268.

———. 1971. The beetles of the Pacific northwest. 5. Rhipiceroidea, Sternoxi, Phytophaga, Rhynchophora and Lamellicornia. Univ. Washington Publ. Biol. 16: 1–662.

Herman, L. H., Jr. 1972. Revision of *Bledius* and related genera. 1. The *aequatorialis, mandibularis,* and *semiferrugineus* groups and two new genera (Coleoptera, Staphylinidae, Oxytelinae). Bull. Amer. Mus. Natur. Hist. 149 (art. 2): 111–254.

James, G. J., I. Moore, and E. F. Legner. 1971. The larval and pupal stages of four species of *Cafius* (Coleoptera: Staphylinidae) with notes on their biology and ecology. Trans. San Diego Soc. Natur. Hist. 16: 279–90.

Jeannel, R. 1927. Monographie des Trechinae. Morphologie comparée et distribution geographique d'un groupe de Coleopteres, Part 2. L'Abeille 33: 1–592.

Jones, T. W. 1968. The zonal distribution of three species of Staphylinidae in the rocky intertidal zone of California. Pan-Pac. Entomol. 44: 203–10.

Keen, J. H. 1895. List of Coleoptera collected at Massett, Queen Charlotte Islands, B.C. Canad. Entomol. 27: 165–72, 217–20.

———.1897. Three interesting Staphylinidae from Queen Charlotte Islands, B.C. Canad. Entomol. 29: 285–87.

Kincaid, T. 1961. The staphylinid genera *Pontomalota* and *Thinusa*. Privately published. The Calliostoma Co., 1904 N. East 52nd, Seattle, Wash. 10 pp.

King, L. A. L. 1914. Notes on the habits and characteristics of some littoral mites of Millport. Proc. Roy. Phys. Soc. Edinburgh 19: 129–41.

Kompfner, H. 1974. Larvae and pupae of some wrack dipterans on a California beach (Diptera: Coelopidae, Anthomyiidae, Sphaeroceridae). Pan-Pac. Entomol. 50: 44–52.

Kramer, S. 1962. Nesting and egg care in the seaside earwig, *Anisolabis maritima* (Insecta, Dermaptera). Amer. Zool. 2: 421 (abstract).

Langston, L. 1974. The maritime earwig in California (Dermaptera: Carcinophoridae). Pan-Pac. Entomol. 50: 28–34.

LeConte, J. L. 1880. Short studies of North American Coleoptera. Trans. Amer. Entomol. Soc. 8: 163–218.

LeConte, J. L., and G. H. Horn. 1883. Classification of the Coleoptera of North America. Smithsonian Misc. Coll. 507. 567 pp.

Lee, V. F. 1979. The maritime pseudoscorpions of Baja California, Mexico (Arachnida: Pseudoscorpionida). Occas. Pap. Calif. Acad. Sci. 131: 1–38.

Leech, H. B. 1971. Nearctic records of flights of *Cafius* and some related beetles at the seashore (Coleoptera: Staphylinidae and Hydrophilidae). Wasmann J. Biol. 29: 65–70.

Leech, H. B., and H. P. Chandler. 1956. Aquatic Coleoptera, pp. 293–371, *in* Usinger (1956).

Lindroth, C. H. 1961. The ground-beetles (Carabidae, excl. Cicindelinae) of Canada and Alaska. 1. Opuscula Entomol. Suppl. 20: 1–200.

MacKerras, I. M. 1950. Marine insects. Proc. Roy. Soc. Queensland 61: 19–29.

Malkin, B. 1958. Protective coloration in *Thinopinus pictus*. Coleop. Bull. 12: 20.

Marine Biological Association, U.K. 1957. Plymouth marine fauna. 3rd ed. Plymouth: Marine Biological Association of the U.K. 457 pp.

Moore, I. 1954. Notes on *Endeodes* LeConte with a description of a new species from Baja California (Coleoptera: Malachiidae). Pan-Pac. Entomol. 30: 195–98.

———. 1956a. A revision of the Pacific coast Phytosi with a review of the foreign genera (Coleoptera: Staphylinidae). Trans. San Diego Soc. Natur. Hist. 12: 103–52.

———. 1956b. Notes on some intertidal Coleoptera with description of the early stages (Carabidae, Staphylinidae, Malachiidae). Trans. San Diego Soc. Natur. Hist. 12: 207–30.

———. 1964a. The Staphylinidae of the marine mud flats of southern California and northwestern Baja California (Coleoptera). Trans. San Diego Soc. Natur. Hist. 13: 269–84.

———. 1964b. The larva of *Hadrotes crassus* (Mannerheim) (Coleoptera: Staphylinidae). Trans. San Diego Soc. Natur. Hist. 13: 309–12.

———. 1975. Nocturnal Staphylinidae of the southern California sea beaches. Entomol. News 86: 91–93.

Moore, I., and E. F. Legner. 1974. Bibliography (1758 to 1972) to the Staphylinidae of America north of Mexico (Coleoptera). Hilgardia 42: 511–47.

———. 1975. A catalogue of the Staphylinidae of America north of Mexico (Coleoptera). Univ. Calif. Spec. Publ. 3015. 514 pp.

Morley, R. L., and R. A. Ring. 1972a. The intertidal Chironomidae (Diptera) of British Columbia. 1. Keys to their life stages. Canad. Entomol. 104: 1093–98.

———. 1972b. The intertidal Chironomidae (Diptera) of British Columbia. 2. Life history and population dynamics. Canad. Entomol. 104: 1099–1121.

Muchmore, B. 1972. The pseudoscorpion genus *Paraliochthonius* (Arachnida, Pseudoscorpionida, Chthioniidae). Entomol. News 83: 248–56.

Newell, I. M. 1947. A systematic and ecological study of the Halacaridae of eastern North America. Bingham Oceanogr. Coll. Bull. 10: 1–232.

———. 1949. New genera and species of Halacaridae (Acari). Amer. Mus. Novitates 1411: 1–22.

———. 1951. *Copidognathus curtus* Hall, 1912, and other species of *Copidognathus* from western North America (Acari, Halacaridae). Amer. Mus. Novitates 1499: 1–23.

———. 1967. Abyssal Halacaridae (Acari) from the southeast Pacific. Pacific Insects 9: 693–708.

———. 1975. Marine mites (Halacaridae), pp. 425–31, *in* Smith & Carlton (1975).

Riedl, R., ed. 1963. Fauna und Flora der Adria. Hamburg and Berlin: Verlag Paul Parey. 640 pp.

Saunders, L. G. 1928. Some marine insects of the Pacific coast of Canada. Ann. Entomol. Soc. Amer. 21: 521–45.

Schlinger, E. I. 1975. Intertidal insects: Order Diptera, pp. 436–46, *in* Smith & Carlton (1975).

Scott, D. B., Jr. 1956. Aquatic Collembola, pp. 74–78, *in* Usinger (1956).

Smith, R. I., and J. T. Carlton, eds. 1975. Light's manual: Intertidal invertebrates of the central California coast. 3rd ed. Berkeley and Los Angeles: University of California Press. 716 pp.

Spilman, T. J. 1967. The heteromerous intertidal beetles (Coleoptera: Salpingidae: Aegialitinae). Pacific Insects 9: 1–21.

Thor, Sig. 1931. Acarina. Bdellidae, Nicoletiellidae, Cryptognathidae. Tierreich 56: 1–87.

Thorpe, W. H. 1932. Colonization of the sea by insects. Nature 130: 629–30.

Tokunaga, M. 1930. The morphological and biological studies on a new marine cranefly, *Limonia (Dicranomyia) monostromia*, from Japan. Mem. Coll. Agr., Kyoto Imp. Univ., 10: 1–127.

———. 1932. Morphological and biological studies on a new marine chironomid fly. Part I. Mem. Coll. Agr., Kyoto Imper. Univ., 19: 1–56.

Usinger, R. L., ed. 1954. Class Insecta, pp. 189–94, *in* S. F. Light et al., Intertidal invertebrates of the central California coast. Berkeley and Los Angeles: University of California Press. 446 pp.

———, ed. 1956. Aquatic insects of California. Berkeley and Los Angeles: University of California Press. 508 pp.

———. 1957. Marine insects, pp. 1177–82, *in* J. W. Hedgpeth, ed., Treatise on marine ecology and paleoecology, vol. 1. Mem. Geol. Soc. Amer. 67. 1,296 pp.

Van Dyke, E. C. 1918. New intertidal rock-dwelling Coleoptera from California. Entomol. News 29: 303–8.

Weygoldt, P. 1969. The biology of pseudoscorpions. Cambridge, Mass.: Harvard University Press. 145 pp.

Williams, F. X. 1938. Biological studies in water-loving insects. 3. Diptera or flies. A. Ephydridae and Anthomyiidae. Proc. Hawaii Entomol. Soc. 10: 85–119.

Wirth, W. W. 1949. A revision of the clunionine midges with descriptions of a new genus and four new species (Diptera: Tendipedidae). Univ. Calif. Publ. Entomol. 8: 151–82.

———. 1951. A revision of the dipterous family Canaceidae. Occas. Pap. Bernice P. Bishop Mus. 20: 245–75.

———. 1969. The shore flies of the genus *Canaceoides* Cresson (Diptera: Canaceidae). Proc. Calif. Acad. Sci. (4) 36: 551–70.

Wirth, W. W., and A. Stone. 1956. Aquatic Diptera, pp. 372–482, *in* Usinger (1956).

Maps

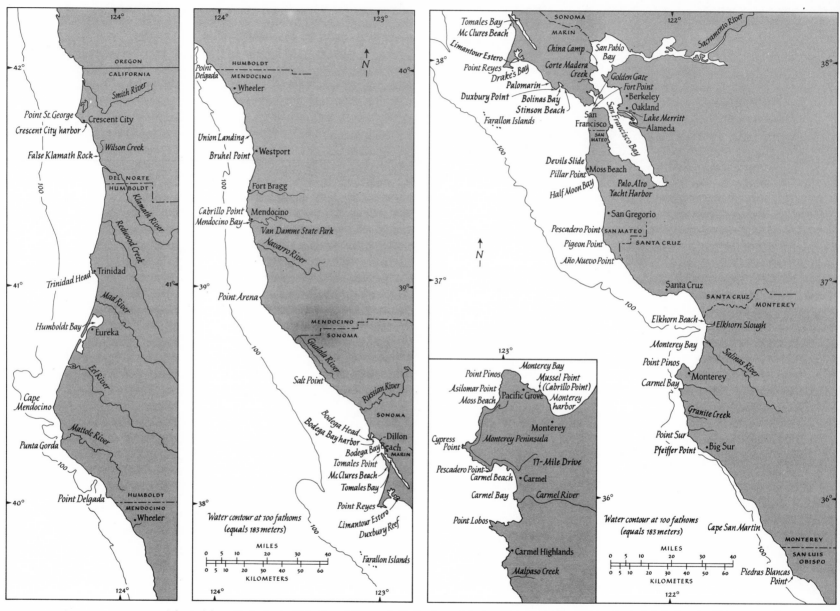

Adapted from Isabella A. Abbott and George J. Hollenberg, *Marine Algae of California* (Stanford, Calif.: Stanford University Press, 1976).

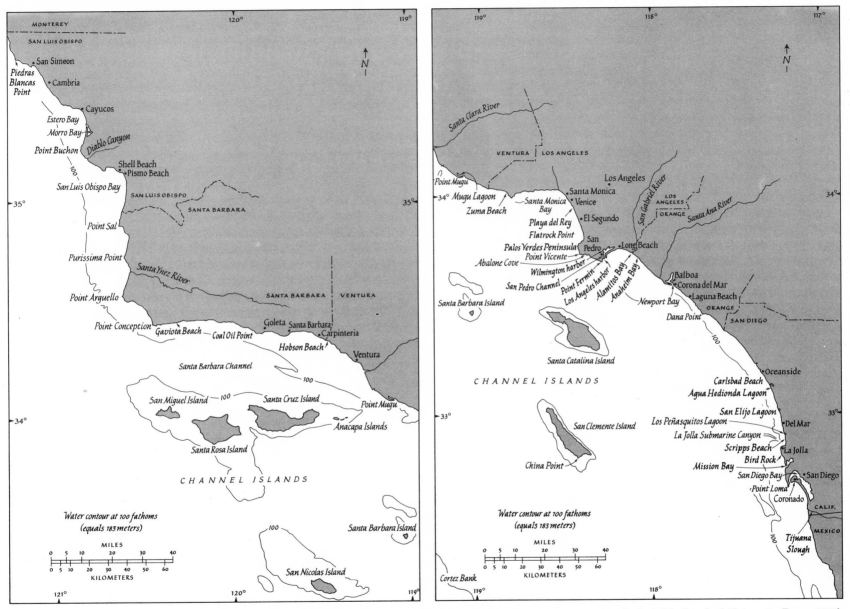

Left map:

MONTEREY
SAN LUIS OBISPO
120°
119°
N

• San Simeon
Piedras Blancas Point
• Cambria
• Cayucos
Estero Bay
Morro Bay
Diablo Canyon
Point Buchon
Shell Beach
Pismo Beach
San Luis Obispo Bay
SAN LUIS OBISPO
SANTA BARBARA
35°
100
Point Sal
Purissima Point
Santa Ynez River
Point Arguello
SANTA BARBARA | VENTURA
Point Conception
Gaviota Beach
Coal Oil Point
• Goleta
Santa Barbara
Carpinteria
Hobson Beach
Ventura
Santa Barbara Channel
100
San Miguel Island — 100
Santa Cruz Island
Point Mugu
Anacapa Islands
34°
Santa Rosa Island
100
CHANNEL ISLANDS
Santa Barbara Island
100
121°
120°
119°

Water contour at 100 fathoms
(equals 183 meters)

MILES
0 5 10 20 30 40
0 5 10 20 30 40 50 60
KILOMETERS

San Nicolas Island

Right map:

119°
118°
117°
N
Santa Clara River
VENTURA | LOS ANGELES
Los Angeles
Point Mugu
San Gabriel River
34°
Mugu Lagoon
Santa Monica
LOS ANGELES
Santa Ana River
Zuma Beach
Santa Monica Bay
Venice
ORANGE
• El Segundo
Playa del Rey
Flatrock Point
San Pedro
Palos Verdes Peninsula
Long Beach
Point Vicente
Abalone Cove
Wilmington harbor
San Pedro Channel
Point Fermin
Los Angeles harbor
Alamitos Bay
Anaheim Bay
Balboa
Santa Barbara Island
Corona del Mar
Newport Bay
Laguna Beach
ORANGE
Dana Point
SAN DIEGO
100
Santa Catalina Island
CHANNEL ISLANDS
• Oceanside
Carlsbad Beach
33°
Agua Hedionda Lagoon
San Clemente Island
San Elijo Lagoon
• Del Mar
Los Peñasquitos Lagoon
La Jolla Submarine Canyon
• La Jolla
Scripps Beach
China Point
Bird Rock
Mission Bay
San Diego Bay
• San Diego
Point Loma
Coronado
CALIF.
Water contour at 100 fathoms
(equals 183 meters)
MEXICO
Tijuana Slough
MILES
0 5 10 20 30 40
0 5 10 20 30 40 50 60
KILOMETERS
Cortez Bank
119°
118°
100

Adapted from Isabella A. Abbott and George J. Hollenberg, *Marine Algae of California* (Stanford, Calif.: Stanford University Press, 1976).

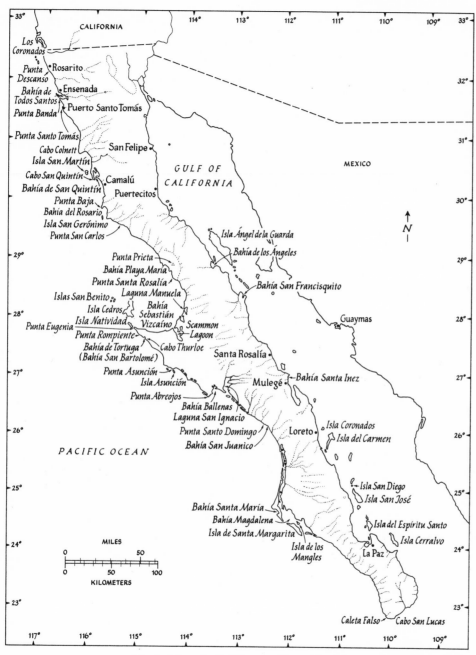

CALIFORNIA

Los Coronados
Punta Descanso
•Rosarito
Bahía de Todos Santos •Ensenada
Punta Banda •Puerto Santo Tomás
Punta Santo Tomás
Cabo Colnett
Isla San Martín
Cabo San Quintín
Bahía de San Quintín •Camalú
Punta Baja •Puertecitos
Bahía del Rosario
Isla San Gerónimo
Punta San Carlos
Punta Prieta
Bahía Playa María
Punta Santa Rosalía
Laguna Manuela
Islas San Benito
Isla Cedros
Isla Natividad
Punta Eugenia
Punta Rompiente
Bahía de Tortuga
(Bahía San Bartolomé)
Cabo Thurloe
Bahía Sebastián Vizcaíno
Scammon Lagoon
Punta Asunción
Isla Asunción
Punta Abreojos
Bahía Ballenas
Laguna San Ignacio
Punta Santo Domingo
Bahía San Juanico

San Felipe

GULF OF CALIFORNIA

MEXICO

Isla Ángel de la Guarda
Bahía de los Ángeles

Bahía San Francisquito

Guaymas

Santa Rosalía

Mulegé •← Bahía Santa Inez

Loreto
Isla Coronados
Isla del Carmen

Isla San Diego
Isla San José

Bahía Santa María
Bahía Magdalena
Isla de Santa Margarita
Isla de los Mangles

Isla del Espíritu Santo
Isla Cerralvo

•La Paz

PACIFIC OCEAN

N

Caleta Falso— Cabo San Lucas

MILES
0   50

0   50   100
KILOMETERS

Adapted from Ira L. Wiggins, *Flora of Baja California* (Stanford, Calif.: Stanford University Press, 1980).

Photographs

Photographic Techniques and Credits

Robert H. Morris

I took most of the photographs in this book using an Exacta 35 mm single-lens reflex camera. Producing the photographs was a long and difficult job, but also an interesting and challenging one. Many trips at low tides were required, to more than 60 different locations along the coast from the Oregon border to the Mexican border. Approximately 900 identifiable specimens were collected over a five-year period, and over 2,000 photographs were taken. All of the types of habitats found between the limits of the lowest low tide zone and the highest high tide zone were covered. Kodacolor film was used exclusively, to obtain the best possible prints for publication.

Ideally, a project of this type would have involved a two- or three-man team using a mobile laboratory equipped with both photographic and biological equipment. Instead, it was completed as a one-man effort using a second-hand Exacta, an unscientific assortment of bottles and trays, and always a large jug of fresh seawater. To obtain the desired range of enlargements, 50 mm or 100 mm lenses were used in combination with bellows, extension tubes, and/or close-up lenses. This equipment seemed to function more effectively when it was supplemented by ample amounts of sweating and cursing.

Approximately 75 percent of my photographs were taken using controlled floodlighting, because this gave more consistently good results. The closest thing to a laboratory was the kitchen sink and refrigerator in my apartment or the nearest motel. Time was always a critical factor in collecting, transporting, preparing, and photo-

Robert H. Morris is a biological illustrator-photographer, landscape architect, and recreation planner.

graphing the specimens before they lost their natural forms and colors. To enhance a natural color, specimens were often photographed on backgrounds of complementary color. For example, to emphasize the rose-pink color of *Hopkinsia rosacea*, it was photographed on a green background. Later, to give the book greater uniformity and to further emphasize the form, color, and texture of individual species, the backgrounds of most prints were blacked out using an acrylic-base paint that would adhere to the glossy photographs.

Once I had developed a system that could be set up quickly to provide proper lighting conditions for all specimens of a particular size, the two major problems remaining were time and avoiding surface reflections (where the specimen was photographed under water). In most cases, the first try had to be right, because very few specimens survived long enough to provide a second chance if the first print was not satisfactory. This was especially a problem when the specimen involved was a comparatively rare one that was not easy to find again.

The remaining 25 percent of my photographs were taken in the field using natural lighting. Some of these came out very well, but generally the quality of photographs done with natural lighting was not as consistently good as that done under controlled floodlighting conditions. Turbid water, too little or too much light, reflections, and a misty lens were some of the major causes of the reduced quality.

Wherever possible, specimens were photographed out of water, because in most cases this produced a more three-dimensional picture with highlights. However, the form and structure of many animals, for example, all of the nudibranchs, required that they be photographed under water. Many of the specimens collected and photographed were preserved for future identification. All others that survived the collecting, transporting, and photographing process were returned to their original natural habitats.

All of the photographs of specimens collected by me were made from either live or freshly killed animals. These photographs most effectively show the natural colors and forms of the different species. In a few cases, additional specimens were provided by some of the chapter authors. Unfortunately, most of these specimens were preserved animals, and the photographs made from these specimens generally do not portray the species in their best colors and forms.

The only other major difficulty in taking the photographs occurred with animals having a very strong negative reaction to light. For example, several flatworm species fragmented or rolled into a tight knot when exposed to a controlled floodlight setup. The problem became so acute in three cases, *Thysanozoon sandiegense, Eurylepta aurantiaca*, and *Eurylepta* sp., that I decided it would be more effective to make carefully painted illustrations of the three species from living specimens and use these in place of photographs. In addition to these three painted illustrations, a very selective amount of retouching was done on several of the photographs to emphasize colors or features that were particularly important to portray. This limited amount of retouching was done in a manner that enhanced but maintained the authenticity of the original photographs.

Assembling all of these photographs has been demanding work, but also a great joy, and I welcome inquiries.

The photographs by other contributors were taken primarily with 35 mm cameras using Ektachrome or Kodachrome film. Prints for publication were made from the slides. An alphabetical listing of contributors is given below; for all photographs not taken by me, appropriate credit is given with each photograph as it appears in the book.

Donald P. Abbott
Robert Ames
Zach M. Arnold
William C. Austin
Eldon E. Ball
Charles H. Baxter
Robert D. Beeman

Charles Birkeland
Ralph Buchsbaum
Nancy Burnett
Roy L. Caldwell
James E. Carlson
Fred L. Clogston
Clinton L. Collier

Robert Evans
William G. Evans
Wesley M. Farmer
Veronica A. Gauley
William B. Gladfelter
Thomas Gore
Eugene C. Haderlie
Janine E. Haderlie
Michael G. Hadfield
Walter Eden Harvey
Frederick G. Hochberg
Anne Hurst
Robert Emil King
Gretchen Lambert
James R. Lance
Welton L. Lee
Kenneth E. Lucas

Richard N. Mariscal
Gary R. McDonald
Todd Newberry
William A. Newman
Chris William Patton
George Reeves
Peter A. Roy
Paul C. Schroeder
DeBoyd L. Smith
Robert H. Smith
Joan Emily Steinberg
Steven K. Webster
Lani A. West
Donald Wobber
Ronald Wolff
Russel L. Zimmer

Typical intertidal habitats of the California coast

(next four pages)

1. Rocky shore on open coast, upper and middle intertidal zones exposed. Dana Point (Orange Co.).

2. Rocky shore on open coast, by "Great Tide Pool," upper and middle intertidal zones exposed. Pacific Grove (Monterey Co.).

3. Tidal channel on rocky shore, open coast, exposed at low tide. Malpaso Creek area (Monterey Co.).

4. Rocky shore exposed to strong surf, upper and middle intertidal zones exposed, showing mainly clusters of the sea mussel *Mytilus californianus* (15.9) and the leaf barnacle *Pollicipes polymerus* (20.11). Mussel Point (Pacific Grove, Monterey Co.). (C. Baxter)

5. Rocky shore on open coast, showing beds of surfgrass (*Phyllospadix*) and pink coralline algae, exposed at very low tide. Dana Point (Orange Co.).

6. Kelp bed (*Macrocystis*) adjacent to rocky shore. Pacific Grove (Monterey Co.).

7. Small barnacles (*Chthamalus*, e.g., 20.12) and limpets (*Collisella scabra*, 13.13), upper intertidal zone on granite rock. Pacific Grove (Monterey Co.). (C. Baxter)

8. Cluster of the sea mussel *Mytilus californianus* (15.9) surrounded by dark tufts of red algae (*Endocladia muricata*), upper middle intertidal zone on rocky shore. Pacific Grove (Monterey Co.). (C. Baxter)

9. The predatory snail *Nucella emarginata* (13.83) and the barnacle *Balanus glandula* (20.24), among tufts of red algae, upper middle intertidal zone on rocky shore. Pacific Grove (Monterey Co.). (C. Baxter)

10. The sea anemone *Anthopleura elegantissima* (3.31), partly concealed by adhering gravel and shell fragments, forming a dense bed on middle intertidal rocks that also bear brown algae (*Pelvetia fastigiata* and *Fucus distichus*) and red algae (*Gigartina papillata*). Pacific Grove (Monterey Co.). (C. Baxter)

11. The black turban snail *Tegula funebralis* (13.32), the chiton *Mopalia* (center), the sea anemone *Anthopleura elegantissima* (3.31; contracted specimens partly concealed by attached shell fragments), and a colony of the tube worm *Dodecaceria fewkesi* (18.25; bottom), middle intertidal zone on protected rocky shore. Pacific Grove (Monterey Co.). (C. Baxter)

12. Bed of the sea mussel *Mytilus californianus* (15.9), with an unusually large population of the predatory ochre sea star *Pisaster ochraceus* (8.13), middle intertidal zone on rocky shore. Westport (Mendocino Co.).

13. Clusters of the leaf barnacle *Pollicipes polymerus* (20.11) on wave-swept rocky shore, middle intertidal zone. Pacific Grove (Monterey Co.). (C. Baxter)

14. The sea palm (*Postelsia palmaeformis*, a brown alga), exposed in middle intertidal zone at very low tide, outer coastal rocks exposed to heavy surf. Sea Ranch (Sonoma Co.). (S. Webster)

15. The brown alga *Laminaria* exposed at very low tide, low intertidal zone on wave-swept rocky shore. Pescadero State Beach (San Mateo Co.). (S. Webster)

16. Pink crustose coralline algae bearing the chiton *Tonicella lineata* (16.11; lower left), a very small specimen of the red abalone *Haliotis rufescens* (13.1; center), and a cluster of sea grapes (*Botryocladia pseudodichotoma*, a red alga; center right), very low intertidal zone on rocky shore. Pacific Grove (Monterey Co.). (C. Baxter)

17. Sandy beach on open coast. Stinson Beach (Marin Co.).

18. Sand and mud flats of inner bay, exposed at low tide, showing mat of the green alga *Enteromorpha*. Newport Bay (Orange Co.).

19. Mud bank of inner bay, showing crab burrows and pickleweed (*Salicornia*; above). Mission Bay (San Diego Co.).

20. Concrete-sheathed pilings, showing clusters of fouling organisms exposed at very low tide. Municipal Wharf, Monterey harbor.

21. Close-up of piling exposed at very low tide, showing fouling community dominated by the sea anemones *Metridium senile* (3.37; center and below) and *Anthopleura elegantissima* (3.31; top), covering giant acorn barnacles (*Balanus nubilus*, 20.32). Monterey. (J. Haderlie)

22. Mud flats at low tide. Elkhorn Slough (Moss Landing, Monterey Co.). (N. Burnett)

1

2

3

4

5

6

P1

7

8

9

10

11

12

13

14

15

16

17

18

19

20

21

22

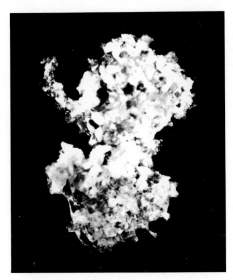

1.1 **Iridia serialis** Le Calvez. 4.5 mm long. Dredged, Monterey Bay. (Z. Arnold)

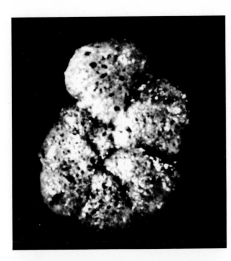

1.2 **Haplophragmoides columbiensis** Cushman var. **evolutum** Cushman & Mc-Culloch. 0.85 mm diam. Point Pinos (Pacific Grove, Monterey Co.). (Z. Arnold)

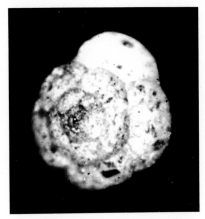

1.3 **Trochammina pacifica** Cushman. 0.33 mm diam. From washings, Moss Beach (San Mateo Co.). (Z. Arnold)

1.4 **Textularia schencki** Cushman & Valentine. 0.75 mm long. Point Pinos (Pacific Grove, Monterey Co.). (Z. Arnold)

1.5 **Cyclogyra lajollaensis** (Uchio). 0.35 mm diam. Cannery Row (Monterey Co.). (Z Arnold)

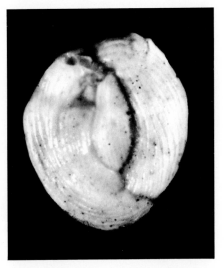

1.6 **Massilina pulchra** Cushman & Gray. 0.87 mm diam. From washings, Moss Beach (San Mateo Co.). (Z. Arnold)

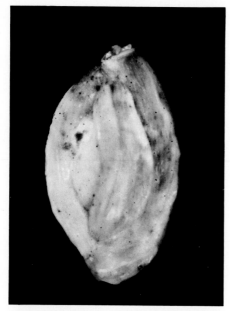

1.9 Elphidium crispum (Linnaeus). 0.95 mm diam. Pillar Point (San Mateo Co.). (Z. Arnold)

1.8 Quinqueloculina vulgaris d'Orbigny. 0.85 mm long. From washings, Moss Beach (San Mateo Co.). (Z. Arnold)

1.7 Quinqueloculina angulostriata Cushman & Valentine. 1.41 mm long. Moss Beach (San Mateo Co.). (Z. Arnold)

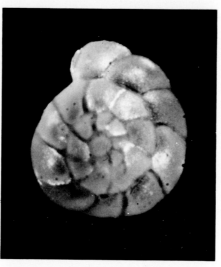

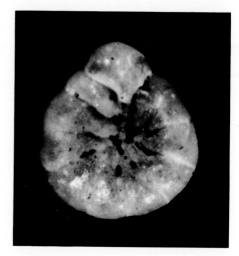

1.11b Ammonia beccarii (Linnaeus). 0.7 mm diam; ventral view. Strandline deposit, San Francisco Bay. (Z. Arnold)

1.10 Nonionella basispinata (Cushman & Moyer). 0.85 mm long. Dredged, Monterey Bay. (Z. Arnold)

1.11a Ammonia beccarii (Linnaeus). 0.7 mm diam; dorsal view. Strandline deposit, San Francisco Bay. (Z. Arnold)

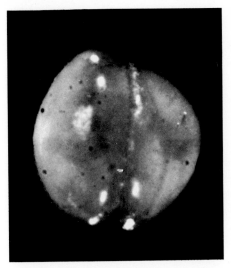

1.12 **Glabratella ornatissima** (Cushman). 0.5 mm long. Shell Beach (San Luis Obispo Co.). (Z. Arnold)

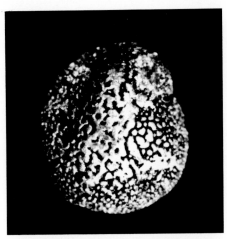

1.13a **Eponides columbiensis** (Cushman). 0.9 mm diam; ventral view. Shell Beach (San Luis Obispo Co.). (Z. Arnold)

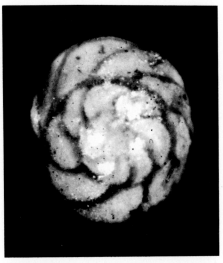

1.13b **Eponides columbiensis** (Cushman). 1.0 mm diam; dorsal view. Shell Beach (San Luis Obispo Co.). (Z. Arnold)

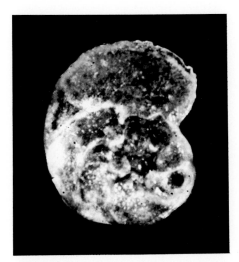

1.14a **Rosalina columbiensis** (Cushman). 0.5 mm diam; ventral view. Cabrillo Point (Pacific Grove, Monterey Co.). (Z. Arnold)

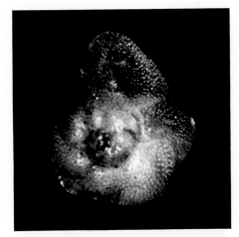

1.14b **Rosalina columbiensis** (Cushman). 0.5 mm diam; dorsal view. Cabrillo Point (Pacific Grove, Monterey Co.). (Z. Arnold)

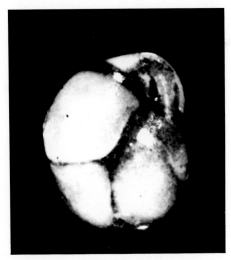

1.15 **Cassidulina subglobosa** Brady. 0.6 mm diam. San Pedro Channel (Los Angeles Co.). (Z. Arnold)

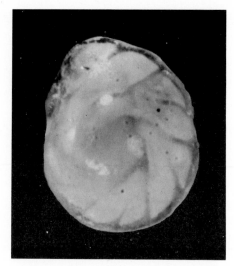

1.16 **Cassidulina limbata** Cushman & Hughes. 0.6 mm diam. San Pedro Channel (Los Angeles Co.). (Z. Arnold)

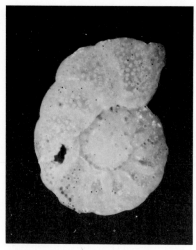

1.17 **Cibicides lobatulus** (Walker & Jacob). 0.55 mm diam. Cabrillo Point (Pacific Grove, Monterey Co.). (Z. Arnold)

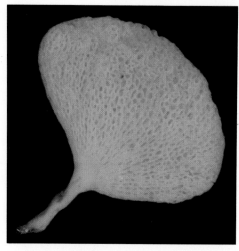

2.1 **Clathrina blanca** (Miklucho-Maclay). 2 cm high. La Jolla (San Diego Co.).

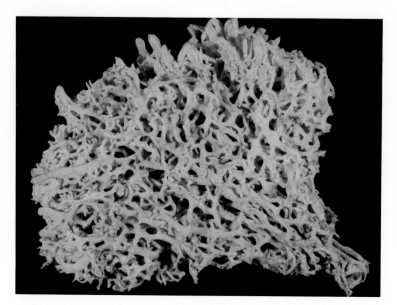

2.2 **Leucosolenia eleanor** Urban. 5 cm diam. Pacific Grove (Monterey Co.).

2.3 **Leucilla nuttingi** (Urban). Largest individuals 2 cm long. Color normally off-white. Pacific Grove (Monterey Co.).

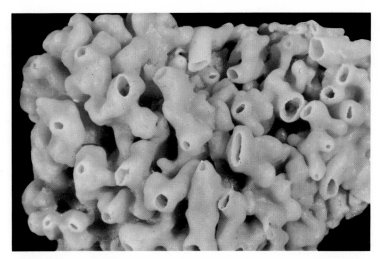

2.4 **Leucetta losangelensis** (De Laubenfels). Portion of colony shown 10 cm long. Corona del Mar (Orange Co.).

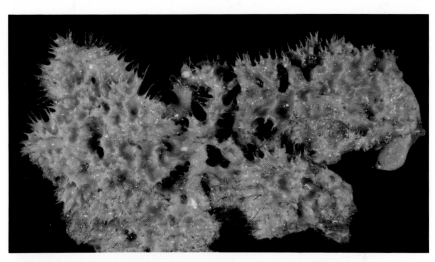

2.5 **Aplysilla glacialis** (Merejkowsky). Keratose Sponge. Greatest dimension shown 5 cm long. Pacific Grove (Monterey Co.).

2.6 **Dysidea fragilis** (Montagu). Keratose Sponge. Greatest dimension shown 8 cm. Corona del Mar (Orange Co.).

2.7 **Aplysina fistularis** (Pallas). Yellow Sponge, Keratose Sponge. Greatest dimension shown 10 cm. La Jolla (San Diego Co.).

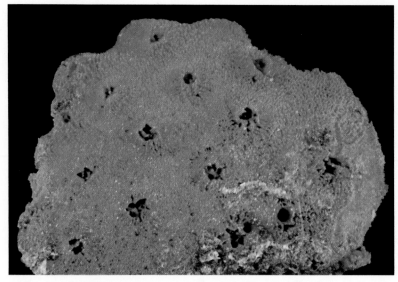

2.8 **Acarnus erithacus** De Laubenfels. Greatest dimension shown 5 cm. Pacific Grove (Monterey Co.).

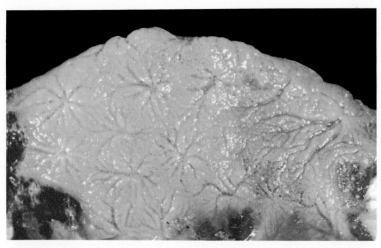

2.9 **Axocielita originalis** (De Laubenfels). Greatest dimension shown 5 cm. Dana Point (Orange Co.).

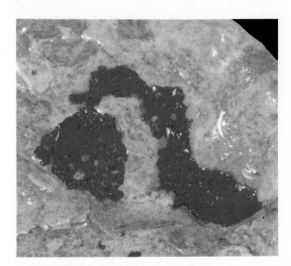

2.10 **Hymenamphiastra cyanocrypta** De Laubenfels. Greatest dimension shown about 3 cm. Point Vicente (Palos Verdes, Los Angeles Co.).

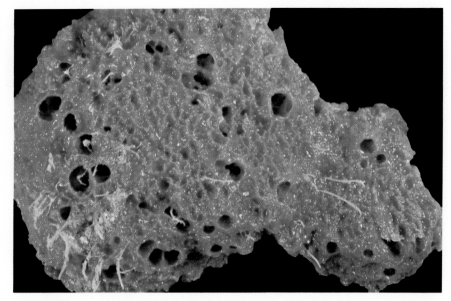

2.11 **Lissodendoryx topsenti** (De Laubenfels). Greatest dimension shown 6 cm. Pacific Grove (Monterey Co.).

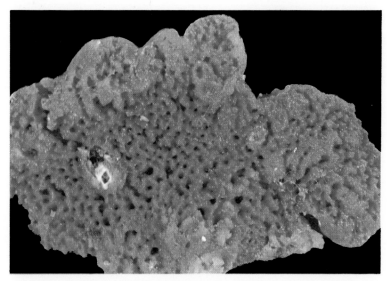

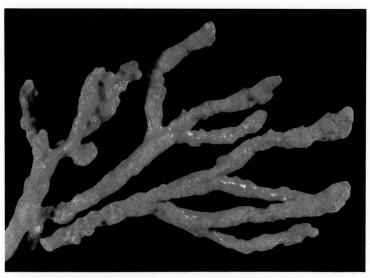

2.12 **Microciona microjoanna** De Laubenfels. Greatest dimension shown 7 cm. Pacific Grove (Monterey Co.).

2.13 **Microciona prolifera** (Ellis & Solander). Redbeard Sponge. Longest portion shown 12 cm. San Francisco Bay.

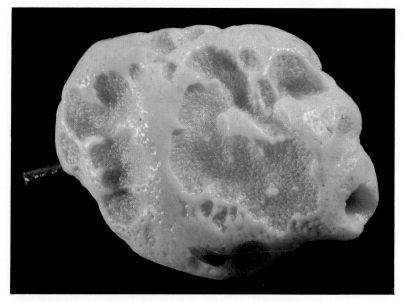

2.14 **Ophlitaspongia pennata** (Lambe) var. **californiana** De Laubenfels. Greatest dimension shown 15 cm, growing with nudibranch *Rostanga pulchra* (14.26). Pacific Grove (Monterey Co.).

2.15 **Halichondria panicea** (Pallas). Crumb-of-Bread Sponge. Greatest dimension shown 7 cm diam. Dillon Beach (Marin Co.).

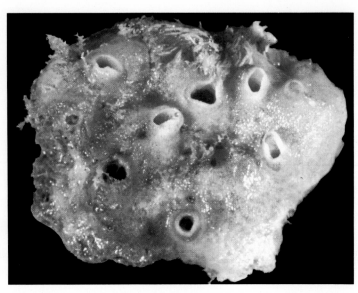

2.16 **Haliclona** sp. Greatest dimension shown 7.5 cm. Pacific Grove (Monterey Co.).

2.17 **Haliclona** sp. 4 cm diam. Pacific Grove (Monterey Co.).

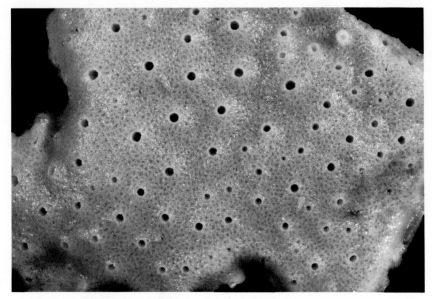

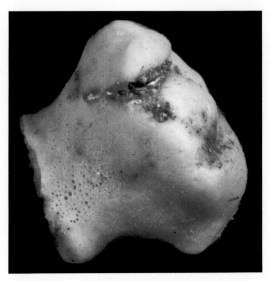

2.18 **Haliclona** sp. Greatest dimension shown 5 cm. Pacific Grove (Monterey Co.).

2.19 **Geodia mesotriaena** Lendenfeld. 4 cm diam. Pacific Grove (Monterey Co.).

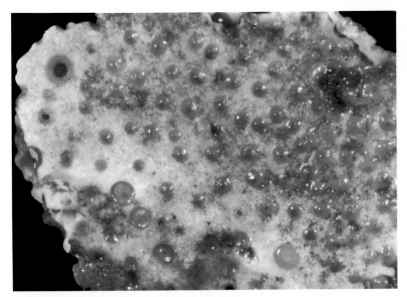

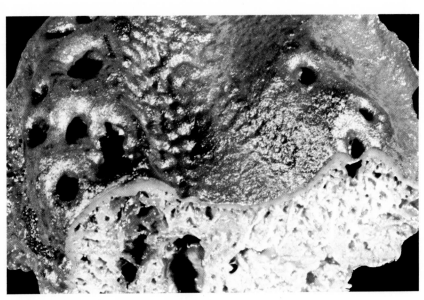

2.20 Cliona celata Grant var. **californiana** De Laubenfels. Boring Sponge, Sulfur Sponge. Yellow lobes, projecting from abalone shell, to 3 mm diam. Dillon Beach (Marin Co.).

2.21 Spheciospongia confoederata De Laubenfels. 7.5 cm across cut. Pacific Grove (Monterey Co.).

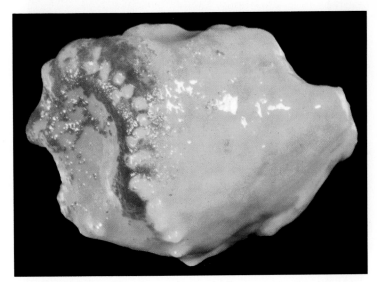

2.22 Suberites ficus (Johnson). Greatest dimension 6 cm. Point Buchon (San Luis Obispo Co.).

2.23 Tethya aurantia (Pallas). 4 cm diam. 17-Mile Drive (Monterey Co.).

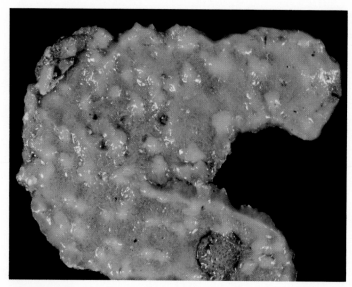

3.1 **Sarsia** sp. Colony 5 cm tall. Monterey Bay. (W. Gladfelter)

2.24 **Higginsia** sp. Greatest dimension 5 cm. Pacific Grove (Monterey Co.).

3.2 **Tubularia marina** (Torrey). Feeding polyps bearing pink medusoids; polyp and stalk together 4 cm long. Monterey Bay. (D. Wobber)

3.3 **Tubularia crocea** (Agassiz). Polyps and stalks together 8 cm long. Morro Bay (San Luis Obispo Co.).

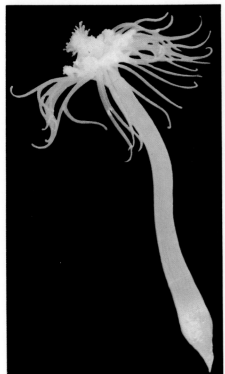

3.5 **Hydractinia milleri** Torrey. Expanded polyps to 1 cm tall. Monterey Bay. (D. Wobber)

3.4 (*left*) **Corymorpha palma** Torrey. 7.5 cm tall. Mission Bay (San Diego Co.).

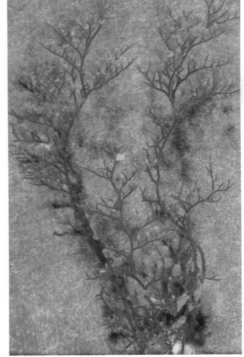

3.7 (*right*) **Eudendrium californicum** Torrey. Colony 8 cm tall. Dillon Beach (Marin Co.).

3.6 (*left*) **Garveia annulata** Nutting. Colony 3 cm tall, growing on branches of coralline algae. Carmel Bay (Monterey Co.). (D. Abbott)

3.8 **Polyorchis montereyensis** Skogsberg. Bell 4 cm tall. Monterey Bay. (W. Gladfelter)

3.9 **Clytia bakeri** Torrey. Colony 12 cm tall, growing on clam *Tivela stultorum* (15.32). Elkhorn Beach (Monterey Co.).

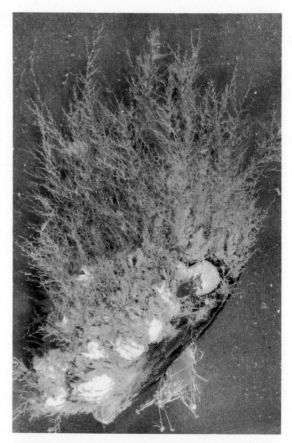

3.10 **Obelia** sp. Colony 7.5 cm tall, growing on mussel *Mytilus californianus* (15.9) and barnacles. Morro Bay (San Luis Obispo Co.).

3.11 **Sertularia furcata** Trask. Stalks to 1 cm tall, growing on surfgrass *Phyllospadix*. Santa Barbara.

3.12 **Sertularella turgida** (Trask). Colony 4 cm tall. Trinidad Head (Humboldt Co.).

3.13 **Abietinaria** sp. Colony 4.5 cm tall. Pacific Grove (Monterey Co.).

3.16 **Plumularia** sp. Colony to 2 cm tall. Duxbury Point (Marin Co.).

3.14 **Aglaophenia struthionides** (Murray). Ostrich-Plume Hydroid. Plume 8.5 cm tall. Point Pinos (Pacific Grove, Monterey Co.).

3.15 **Aglaophenia latirostris** Nutting. Ostrich-Plume Hydroid. Plume 6 cm tall. Pacific Grove (Monterey Co.).

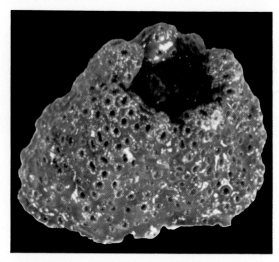

3.17 **Allopora porphyra** (Fisher). Colony 4 cm across, encrusting dead barnacle. Pigeon Point (San Mateo Co.).

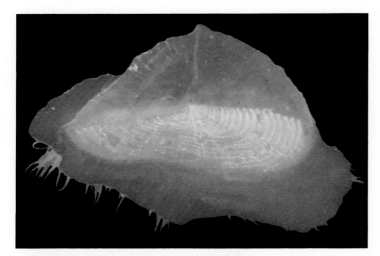

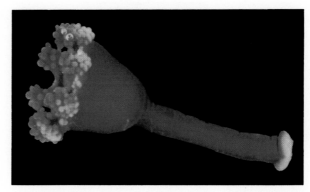

3.19 **Manania** sp. 1.5 cm long. Malpaso Creek (Monterey Co.).

3.18 **Velella velella** (Linnaeus). By-the-Wind-Sailor. 7.5 cm long.
Point Arena (Mendocino Co.).

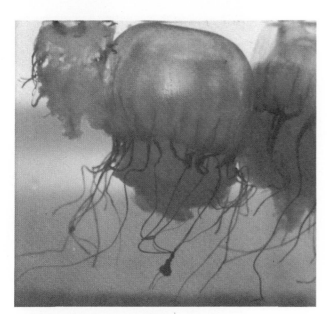

3.20 **Chrysaora melanaster** Brandt. 20 cm
diam. Monterey Bay. (D. Smith)

3.21 **Pelagia colorata** Russell. 30 cm diam. Monterey Bay. (W. Gladfelter)

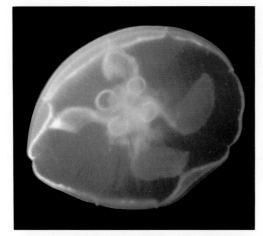

3.22 **Aurelia aurita** (Linnaeus). Moon Jelly. 15 cm diam. Monterey Bay. (W. Gladfelter)

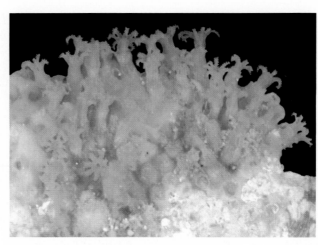

3.23 **Clavularia** sp. Colony 4 cm across. Flatrock Point (Palos Verdes, Los Angeles Co.).

3.24 **Muricea appressa** (Verrill). Sea Fan. Colony 15 cm tall. Agua Hedionda Lagoon (San Diego Co.).

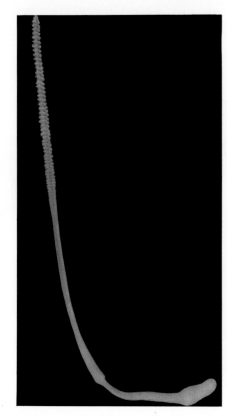

3.26 **Renilla köllikeri** Pfeffer. Sea Pansy. 7 cm diam. Coal Oil Point (Santa Barbara Co.). (E. Ball)

3.25 **Stylatula elongata** (Gabb). Sea Pen. 35 cm tall; base normally straight, not curved. Newport Bay (Orange Co.).

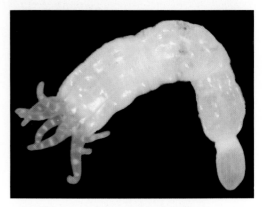

3.27a **Halcampa decemtentaculata** Hand. Whole animal, with basal physa, 2.5 cm long. Bruhel Point (Mendocino Co.).

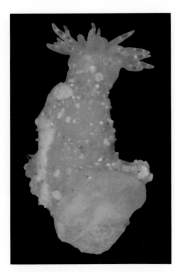

3.27b **Halcampa decmentaculata** Hand. Two individuals, with tentacles expanded, growing alongside red tunicate *Ritterella rubra* (12.8). Monterey Bay. (N. Burnett)

3.28 **Cactosoma arenaria** Carlgren. Detached specimen 2 cm tall, showing pedal disk. Pacific Grove (Monterey Co.).

3.29 **Epiactis prolifera** Verrill. Proliferating Anemone. Tentacular crown 3 cm diam. Juveniles on column. Westport (Mendocino Co.).

3.30 **Anthopleura xanthogrammica** (Brandt). Tentacular crown 18 cm diam. False Klamath Rock (Del Norte Co.).

3.31 **Anthopleura elegantissima** (Brandt). Aggregating Anemone. Tentacular crown 5 cm diam. Moss Beach (San Mateo Co.).

3.32 Anthopleura artemisia (Pickering). Tentacular crown 4 cm diam. Point Saint George (Del Norte Co.).

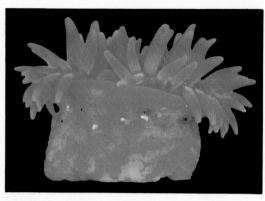

3.33 Tealia crassicornis (Müller). Column 5 cm diam. Westport (Mendocino Co.).

3.34 Tealia coriacea (Cuvier). Column 7 cm diam. Bruhel Point (Mendocino Co.).

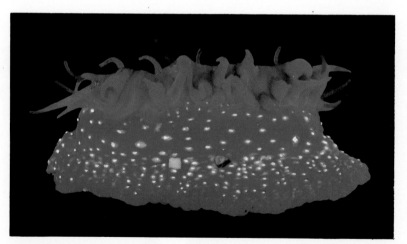

3.35 Tealia lofotensis (Danielssen). Column 7 cm diam. Cabrillo Point (Mendocino Co.).

3.36 Zaolutus actius Hand. 14 cm tall. San Diego Bay.

3.37a **Metridium senile** (Linnaeus). Plumose Anemone. Tentacular crown 4 cm diam; brown and tan forms. Monterey harbor.

3.37b **Metridium senile** (Linnaeus). Tentacular crown 6 cm diam; gray-and-white form. Monterey harbor. (W. Lee)

3.38 **Metridium exilis** Hand. Tentacular crown 1.4 cm diam. Pacific Grove (Monterey Co.).

3.39 **Haliplanella luciae** (Verrill). Tentacular crown 2.5 cm diam. Crescent City (Del Norte Co.).

3.40 **Harenactis attenuata** Torrey. 12 cm long; expanded base shown, used for anchoring in soft substrata. San Diego Bay.

3.41 **Corynactis californica** Carlgren. Column 1 cm diam. Monterey Bay. (W. Lee)

3.42a **Balanophyllia elegans** Verrill. Orange Cup Coral. Expanded animal; column 1 cm diam, growing with blue sponge *Hymenamphiastra cyanocrypta* (2.10) and pink coralline alga. Monterey Bay. (D. Wobber)

3.42b **Balanophyllia elegans** Verrill. Contracted animals; column 1 cm diam. Monterey Bay. (W. Lee)

3.42c Balanophyllia elegans Verrill. Calcareous skeleton; column 1 cm diam. Pacific Grove (Monterey Co.).

3.43 Pachycerianthus fimbriatus McMurrich. Tentacular crown 5 cm diam. Subtidal, Monterey Bay. (D. Wobber)

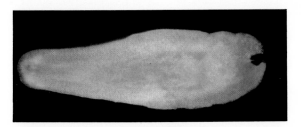

4.1 Polychoerus carmelensis Costello & Costello. 6 mm long. Pacific Grove (Monterey Co.). (R. Buchsbaum)

4.2 Stylochus tripartitus Hyman. 30 mm long. Moss Beach (San Mateo Co.).

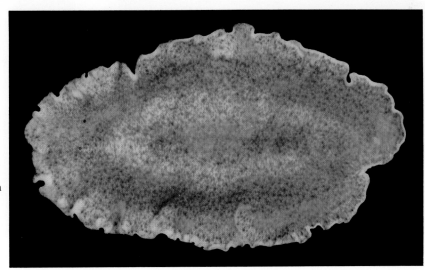

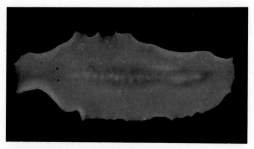

4.4 Notoplana acticola (Boone). 30 mm long. Pacific Grove (Monterey Co.).

4.3 Kaburakia excelsa Bock. 60 mm long. Agua Hedionda Lagoon (San Diego Co.).

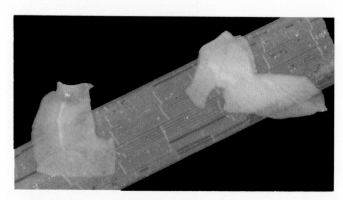

4.5 Phylloplana viridis (Freeman). 20 mm long, growing on eelgrass *Zostera*. Humboldt Bay.

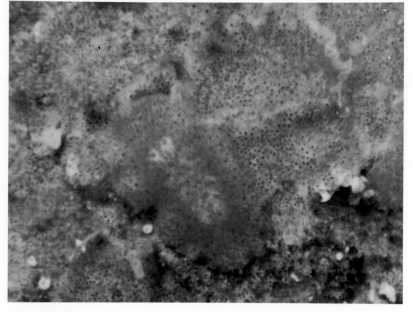

4.6a Hoploplana californica Hyman. 10 mm long, growing on encrusting bryozoan *Celleporaria brunnea* (6.12); dorsal view. Monterey harbor. (E. Haderlie)

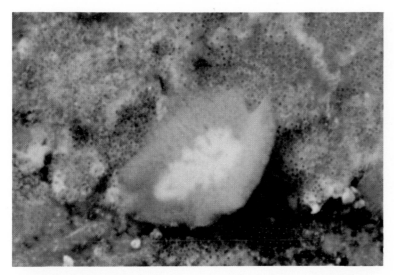

4.6b **Hoploplana californica** Hyman. Ventral view of same specimen.

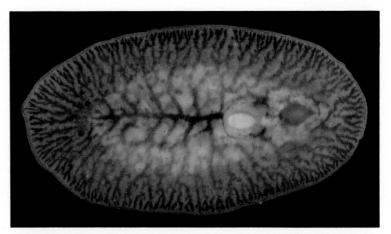

4.7 **Alloioplana californica** (Heath & McGregor). 18 mm long. Pacific Grove (Monterey Co.).

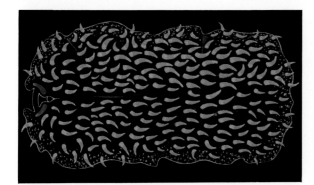

4.8 **Thysanozoon sandiegense** Hyman. 28 mm long. La Jolla (San Diego Co.). (painting, Morris)

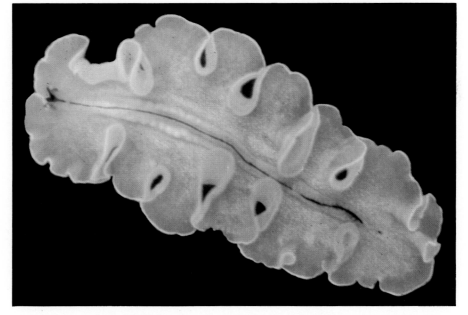

4.9 **Pseudoceros luteus** (Plehn). 40 mm long. Pacific Grove (Monterey Co.). (D. Smith)

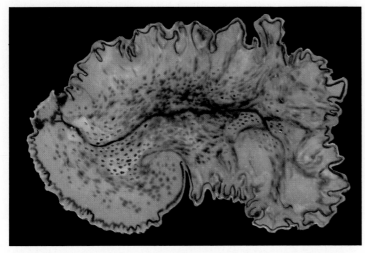

4.10 **Pseudoceros montereyensis** Hyman. 35 mm long. Pacific Grove (Monterey Co.).

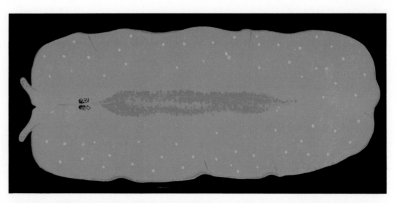

4.11 **Eurylepta aurantiaca** (Heath & McGregor). 25 mm long. Pacific Grove (Monterey Co.). (painting, Morris)

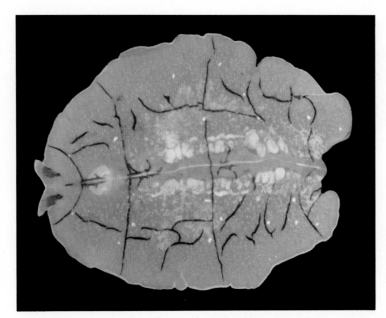

4.12 **Eurylepta californica** Hyman. 30 mm long. Pacific Grove (Monterey Co.).

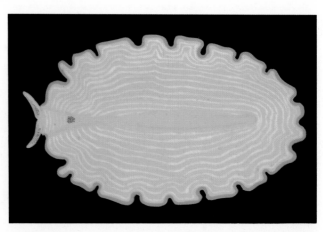

4.13 **Eurylepta** sp. 18 mm long. Pacific Grove (Monterey Co.). (painting, Morris)

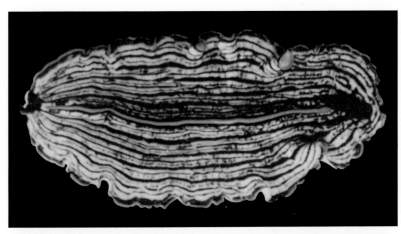

4.14 Prostheceraeus bellostriatus Hyman. 35 mm long. Monterey harbor. (E. Haderlie)

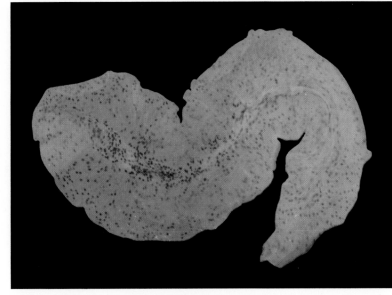

4.15 Enchiridium punctatum Hyman. 36 mm long. Dana Point (Orange Co.).

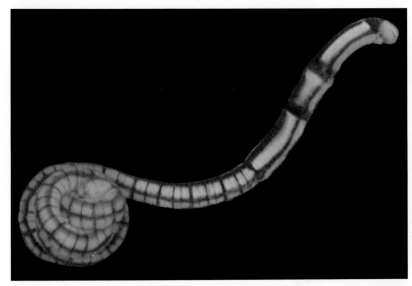

5.1 Tubulanus frenatus (Coe). 20 cm long. Newport Bay (Orange Co.).

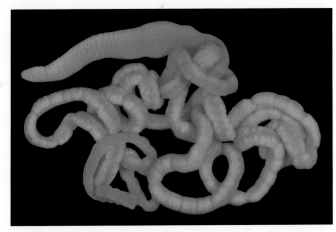

5.2 Tubulanus polymorphus Renier. 45 cm long. Point Saint George (Del Norte Co.).

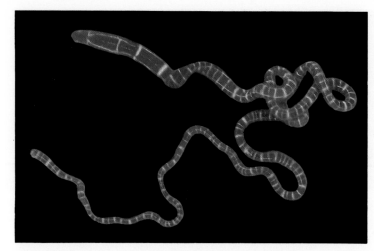

5.3 **Tubulanus sexlineatus** (Griffin). 18 cm long. Pacific Grove (Monterey Co.).

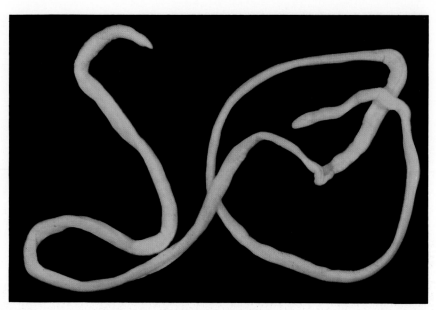

5.4 **Carinoma mutabilis** Griffin. 20 cm long. Humboldt Bay.

5.5 **Baseodiscus punnetti** (Coe). 20 cm long. Cabrillo Point (Mendocino Co.).

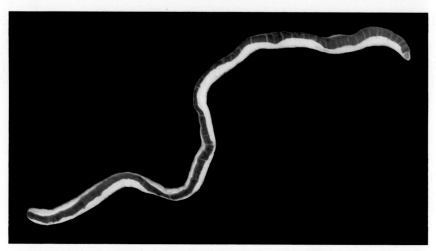

5.6 **Micrura verrilli** Coe. 15 cm long. Pacific Grove (Monterey Co.).

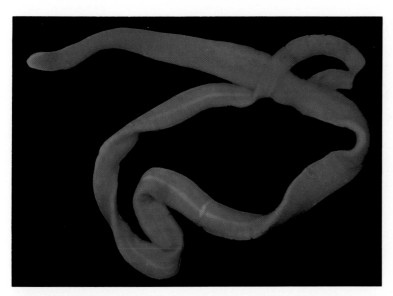

5.7 **Cerebratulus californiensis** Coe. 15 cm long. Elkhorn Slough (Monterey Co.).

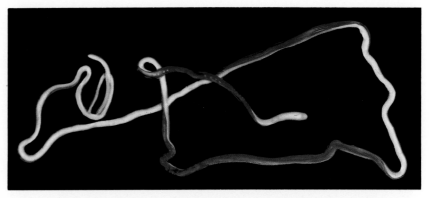

5.8 **Emplectonema gracile** (Johnson). 20 cm long. Tomales Bay (Marin Co.).

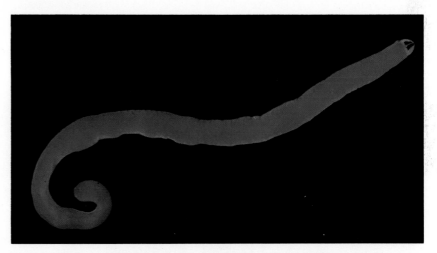

5.10 **Amphiporus bimaculatus** Coe. 10 cm long. Pacific Grove (Monterey Co.).

5.9 **Paranemertes peregrina** Coe. 10 cm long. Pacific Grove (Monterey Co.).

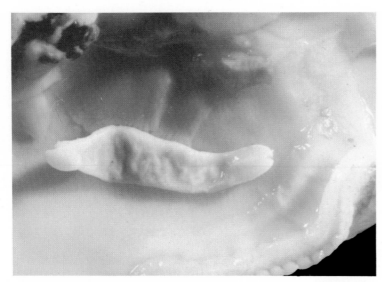

5.11 **Malacobdella grossa** (O. F. Müller). 2.5 cm long, living in mantle cavity of *Siliqua*. Point Saint George (Del Norte Co.).

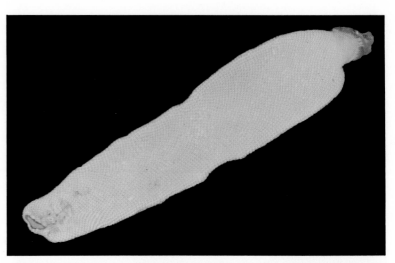

6.1 **Membranipora tuberculata** (Bosc). Colony 5 cm long, encrusted on algae. Pacific Grove (Monterey Co.).

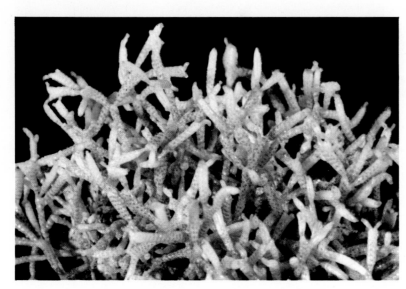

6.2a **Thalamoporella californica** (Levinsen). Portion of colony shown 8 cm across. Agua Hedionda Lagoon (San Diego Co.).

6.2b **Thalamoporella californica** (Levinsen). Zooids extended. Monterey Bay. (N. Burnett)

6.3 **Tricellaria occidentalis** (Trask). Colony 4 cm across. Moss Beach (San Mateo Co.).

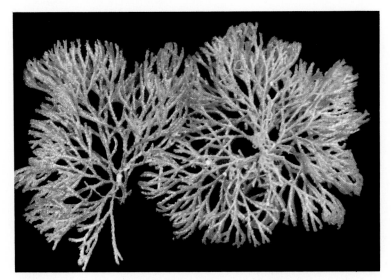

6.4a **Bugula californica** (Robertson). Colony 6 cm high. Dillon Beach (Marin Co.).

6.4b **Bugula californica** (Robertson). Individual frond. Monterey Peninsula. (N. Burnett)

6.5 **Bugula neritina** (Linnaeus). Two colonies, each about 8 cm across. Mission Bay (San Diego Co.).

6.7a **Schizoporella unicornis** (Johnston). Colonies 4 mm and 6 mm across, the right one showing globose ovicells. Color altered by preservation. Southern California.

6.6 **Dendrobeania laxa** (Robertson). Colony 6 cm across. Dillon Beach (Marin Co.).

6.7b **Schizoporella unicornis** (Johnston). Orange opercula of zooids. Dillon Beach (Marin Co.). (C. Patton)

6.8a **Hippodiplosia insculpta** (Hincks). Colony on left 4 cm across. Monterey Peninsula. (N. Burnett)

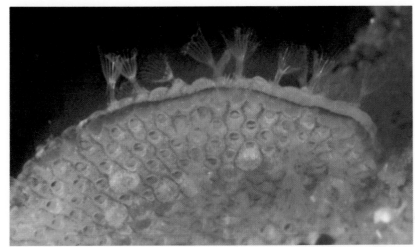

6.8b **Hippodiplosia insculpta** (Hincks). Zooids extended. Monterey Peninsula. (N. Burnett)

6.9 **Eurystomella bilabiata** (Hincks). Colony 4.5 cm long, encrusted on shell of prosobranch *Ceratostoma foliatum* (13.70). Pacific Grove (Monterey Co.).

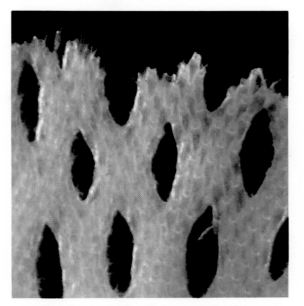

6.10a **Phidolopora pacifica** (Robertson). Colony 4 cm across. Monterey Peninsula. (N. Burnett)

6.10b **Phidolopora pacifica** (Robertson). Zooids extended. Monterey Peninsula. (N. Burnett)

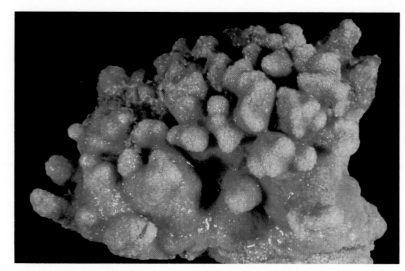

6.11 Costazia robertsoniae Canu & Bassler. Colony 50 mm high. Point Sal (Santa Barbara Co.).

6.12 **Celleporaria brunnea** (Hincks). Colony 7 cm across. Agua Hedionda Lagoon (San Diego Co.).

6.13 Diaperoecia californica (d'Orbigny). Colony 6 cm across. Point Arena (Mendocino Co.).

6.14 **Crisulipora occidentalis** Robertson. Colony 30 mm high, growing on mussel *Mytilus edulis* (15.10). Anaheim Bay (Orange Co.).

6.15 **Filicrisia franciscana** (Robertson). Zooids about 0.7 mm long. Subtidal, southern California. (C. Patton)

6.16 **Flustrellidra corniculata** (Smitt). Colony 6 cm long, growing on calcareous algae. Pacific Grove (Monterey Co.).

6.17 **Zoobotryon verticillatum** (Delle Chiaje). Colony 6 cm long. Mission Bay (San Diego Co.).

6.18 **Bowerbankia gracilis** (O'Donoghue). Zooecia 1–1.5 mm long. (C. Patton)

6.19 **Triticella elongata** (Osburn). Colony 15 mm across, growing on pea crab *Scleroplax granulata* (25.41). Elkhorn Slough (Monterey Co.).

6.20a **Barentsia benedeni** (Foettinger). Monterey Peninsula. (N. Burnett)

6.20b **Barentsia benedeni** (Foettinger). Stalk and calyx. Monterey Peninsula. (N. Burnett)

6.20c **Barentsia benedeni** (Foettinger). Calyces 0.2–0.3 mm deep, with extended tentacles. San Francisco Bay. (R. Mariscal)

7.1a Terebratalia transversa Sowerby. 38 mm wide. Trinidad Head (Humboldt Co.).

7.1b Terebratalia transversa Sowerby. About 35 mm wide. Friday Harbor (Washington). (R. Zimmer)

7.2 Phoronis vancouverensis Pixell. Portion of cluster shown 70 mm wide. Monterey. (N. Burnett)

7.3a (*see also overleaf*) **Phoronopsis viridis** Hilton. Lophophores of intact animals projecting above sandy mud. Elkhorn Slough (Monterey Co.). (R. Buchsbaum)

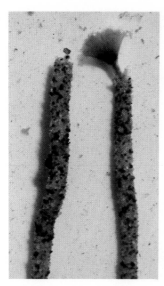

7.3b Phoronopsis viridis Hilton. Excavated tubes about 2 mm diam. Elkhorn Slough (Monterey Co.). (R. Buchsbaum)

7.3c Phoronopsis viridis Hilton. Lophophore of animal removed from tube. Elkhorn Slough (Monterey Co.). (R. Buchsbaum)

8.1 Astropecten armatus Gray. Arm radius 8 cm. Mission Bay (San Diego Co.).

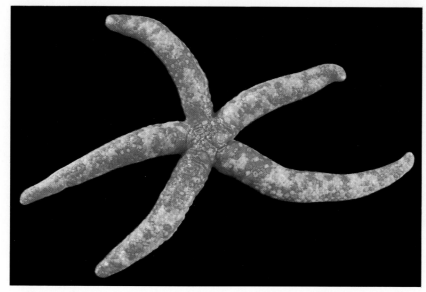

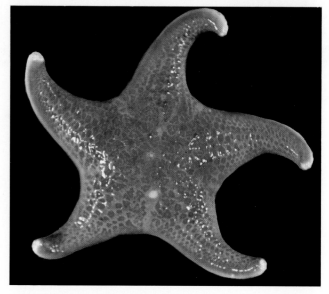

8.2 Linckia columbiae Gray. Arm radius 4 cm. La Jolla (San Diego Co.).

8.3 Dermasterias imbricata (Grube). Leather Star. Arm radius 9 cm. Malpaso Creek (Monterey Co.).

8.4a **Patiria miniata** (Brandt). Bat Star, Sea Bat. Arm radius 7 cm; color variant. Pacific Grove (Monterey Co.).

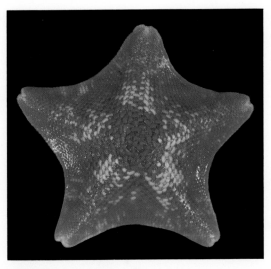

8.4b **Patiria miniata** (Brandt). Arm radius 4 cm; color variant. Pacific Grove (Monterey Co.).

8.4c **Patiria miniata** (Brandt). Arm radius 2.5 cm; color variant. Pacific Grove (Monterey Co.).

8.4d **Patiria miniata** (Brandt). Arm radius 8 cm; color variant. Pfeiffer Point (Monterey Co.).

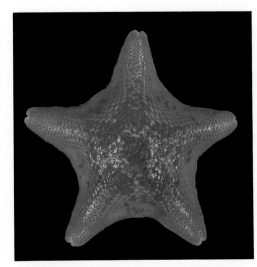

8.4e **Patiria miniata** (Brandt). Arm radius 5 cm; color variant. Pacific Grove (Monterey Co.).

8.4f **Patiria miniata** (Brandt). Arm radius 5 cm; color variant. Pigeon Point (San Mateo Co.).

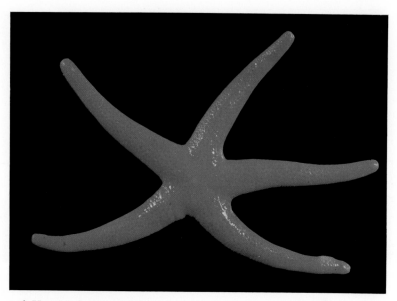

8.5a **Henricia leviuscula** (Stimpson). Arm radius 2 cm. Pacific Grove (Monterey Co.).

8.5b **Henricia leviuscula** (Stimpson). Arm radius 5 cm. Pigeon Point (San Mateo Co.).

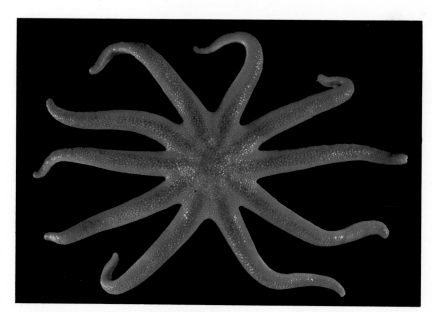

8.6 **Solaster stimpsoni** Verrill. Stimpson's Sun Star. Arm radius 8 cm. Trinidad Head (Humboldt Co.).

8.7 **Solaster dawsoni** Verrill. Dawson's Sun Star. Arm radius 16 cm. Puget Sound (Washington). (C. Birkeland)

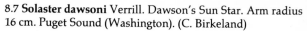

8.8a **Astrometis sertulifera** (Xantus). Adult; arm radius 8 cm. Corona del Mar (Orange Co.).

8.8b **Astrometis sertulifera** (Xantus). Juvenile; arm radius 5 cm. Dana Point (Orange Co.).

8.9 **Orthasterias koehleri** (de Loriol). Arm radius 10 cm. Monterey Bay.

8.10 **Leptasterias pusilla** (Fisher). Arm radius 1.5 cm. Monterey Peninsula. (R. Smith)

8.11 **Leptasterias hexactis** (Stimpson). Arm radius 2 cm. Pacific Grove (Monterey Co.).

8.12 **Evasterias troschelii** (Stimpson). Arm radius 11 cm. Monterey. (D. Abbott)

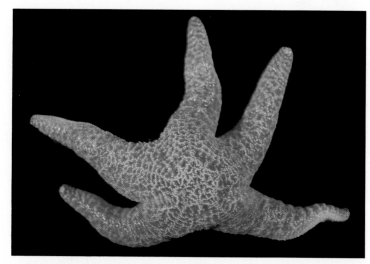

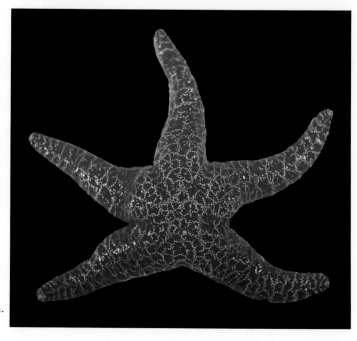

8.13a **Pisaster ochraceus** (Brandt). Ochre Starfish. Arm radius 13 cm. Pacific Grove (Monterey Co.).

8.13b **Pisaster ochraceus** (Brandt). Arm radius 12 cm. Cape Mendocino (Humboldt Co.).

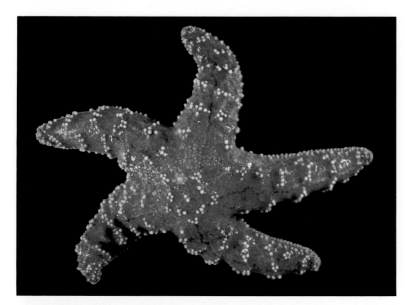

8.13c **Pisaster ochraceus** (Brandt). Juvenile; arm radius 8 cm. Pacific Grove (Monterey Co.).

8.14a **Pisaster giganteus** (Stimpson). Adult; arm radius 11 cm. Dana Point (Orange Co.).

8.14b **Pisaster giganteus** (Stimpson). Juvenile; arm radius 8 cm. Pacific Grove (Monterey Co.).

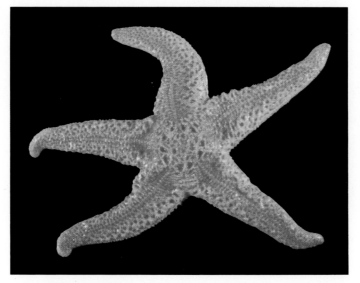

8.15 **Pisaster brevispinus** (Stimpson). Arm radius 14 cm. Moss Beach (San Mateo Co.).

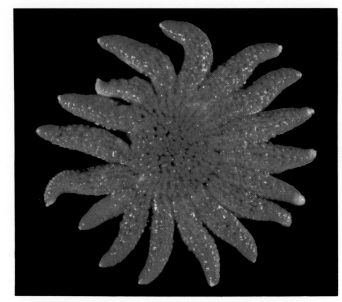

8.16 Pycnopodia helianthoides (Brandt). Sunflower Star, Twenty-Rayed Star. Arm radius 16 cm. Malpaso Creek (Monterey Co.).

9.1 Cucumaria pseudocurata Deichmann. Tar Spot. 2.5 cm long. 17-Mile Drive (Monterey Co.). (D. Abbott)

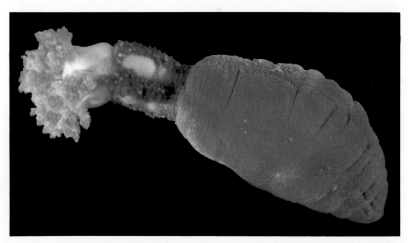

9.3 Cucumaria miniata Brandt. 15 cm long. Cabrillo Point (Mendocino Co.).

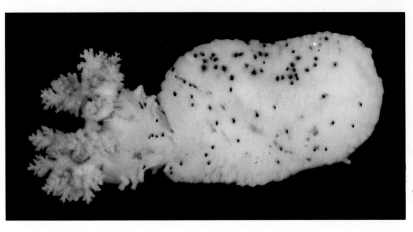

9.2 Cucumaria piperata (Stimpson). 5 cm long. Pacific Grove (Monterey Co.).

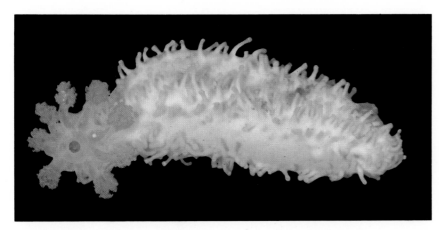

9.4 **Eupentacta quinquesemita** (Selenka). 5 cm long. Point Saint George (Del Norte Co.).

9.5 **Pachythyone rubra** (Clark). 2.5 cm long. Pacific Grove (Monterey Co.). (D. Abbott)

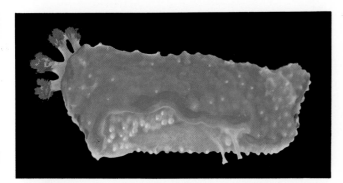

9.6 **Lissothuria nutriens** (Clark). 1.5 cm long. Pacific Grove (Monterey Co.).

9.7 **Parastichopus californicus** (Stimpson). About 20 cm long. Pacific Grove (Monterey Co.). (D. Abbott)

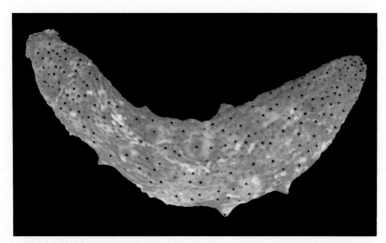

9.8 Parastichopus parvimensis (Clark). 18 cm long. Agua Hedionda Lagoon (San Diego Co.).

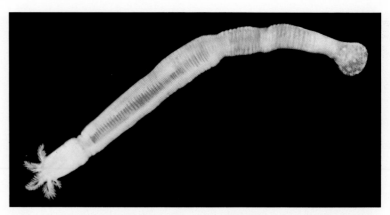

9.9 **Leptosynapta albicans** (Selenka). 12 cm long. Pacific Grove (Monterey Co.).

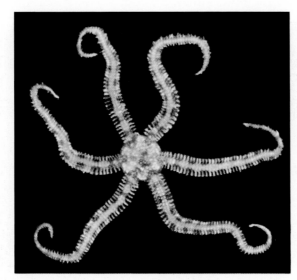

10.1 **Ophiactis simplex** (Le Conte). Arm radius 15 mm. Newport Bay (Orange Co.).

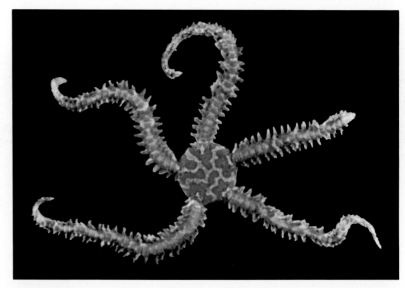

10.2 **Ophiopholis aculeata** (Linnaeus) var. **kennerlyi** Lyman. Daisy Brittle Star. Arm radius 35 mm. Pacific Grove (Monterey Co.).

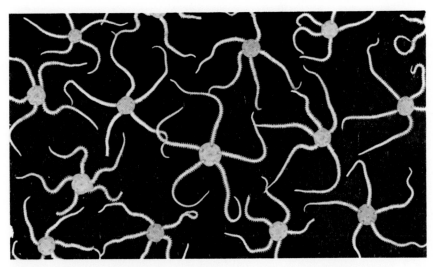

10.3 **Amphipholis squamata** (Delle Chiaje). Arm radius 9 mm. Pacific Grove (Monterey Co.).

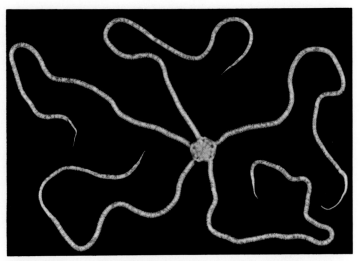

10.4 **Amphiodia occidentalis** (Lyman). Arm radius 110 mm. Pacific Grove (Monterey Co.).

10.5 **Amphiodia urtica** (Lyman). Arm radius 40 mm; amputated arms not yet regenerated. British Columbia. (W. Austin)

10.6 **Ophiothrix spiculata** Le Conte. Arm radius 55 mm. Corona del Mar (Orange Co.).

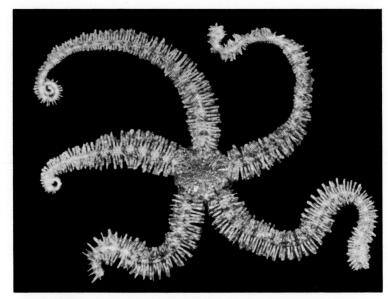

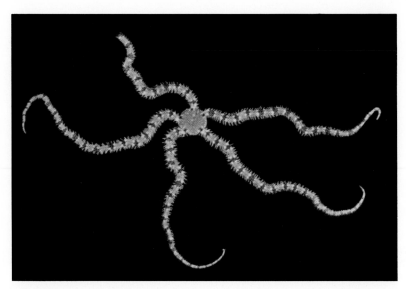

10.7 **Ophiothrix rudis** Lyman. Arm radius 50 mm. Dana Point (Orange Co.).

10.8 **Ophionereis annulata** (Le Conte). Arm radius 65 mm. Corona del Mar (Orange Co.).

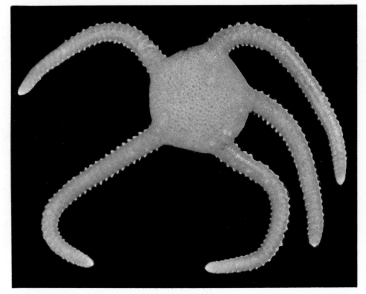

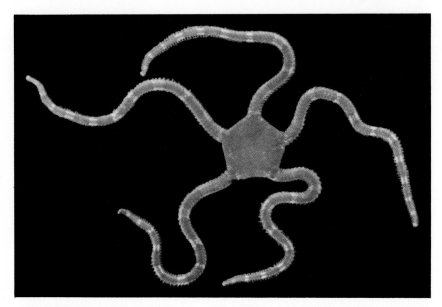

10.9 **Ophioplocus esmarki** Lyman. Arm radius 50 mm. Pacific Grove (Monterey Co.).

10.10 **Ophioderma panamense** Lütken. Arm radius 60 mm. Dana Point (Orange Co.).

10.11 Ophiopteris papillosa (Lyman). Arm radius 50 mm. Dredged, offshore Los Angeles.

10.12 Gorgonocephalus eucnemis (Müller & Troschel). Basket Star. 45 cm total diam. Monterey Bay. (J. Haderlie)

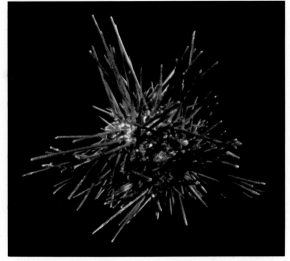

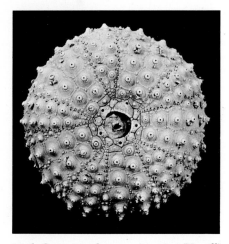

11.1 Eucidaris thouarsii (Valenciennes). 100 mm total diam. Gulf of California. (D. Abbott)

11.2a Centrostephanus coronatus (Verrill). 70 mm total diam. Spine arrangement altered by preservation. San Diego. (R. Wolff)

11.2b Centrostephanus coronatus (Verrill). Test 39 mm diam; aboral view. San Diego. (R. Wolff)

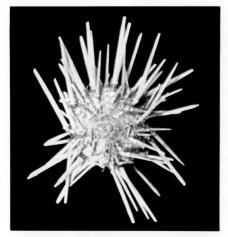

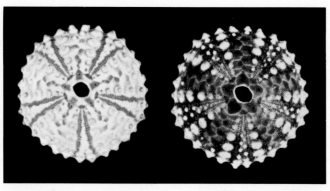

11.3b **Arbacia stellata** Gmelin. Test of light specimen 25 mm diam; of dark specimen, 26 mm. Gulf of California. (R. Wolff)

11.3a **Arbacia stellata** Gmelin. 42 mm total diam. Spine arrangement altered by preservation. Gulf of California. (R. Wolff)

11.4a **Lytechinus anamesus** H. L. Clark. 50–70 mm total diam. La Jolla (San Diego Co.). (D. Abbott)

11.4b **Lytechinus anamesus** H. L. Clark. Test 37 mm diam. San Pedro (Los Angeles Co.). (R. Wolff)

11.5 **Strongylocentrotus purpuratus** (Stimpson). Purple Sea Urchin. 60 mm total diam. Pacific Grove (Monterey Co.).

11.7a **Dendraster excentricus** (Eschscholtz). Sand Dollar. 75 mm diam; aboral view, bay form. Coos Bay (Oregon). (R. Wolff)

11.6 **Strongylocentrotus franciscanus** (A. Agassiz). Red Sea Urchin. 150 mm total diam. Coos Bay (Oregon). (R. Wolff)

11.7b (*see also overleaf*) **Dendraster excentricus** (Eschscholtz). Test 63.3 mm diam; aboral view, bay form. Mission Bay (San Diego Co.). (R. Wolff)

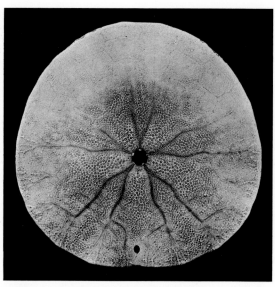

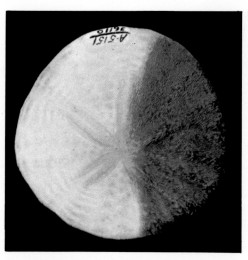

11.8a Dendraster vizcainoensis Grant &
Hertlein. Sand Dollar. 55 mm diam; aboral
view. Punta Prieta (Baja California). (R. Wolff)

11.7c Dendraster excentricus (Eschscholtz).
Test 60.9 mm diam; aboral view, outer coast
form. Stinson Beach (Marin Co.). (R. Wolff)

11.7d Dendraster excentricus (Eschscholtz).
Test 60.9 mm diam; oral view, outer coast
form. Stinson Beach (Marin Co.). (R. Wolff)

11.8c Dendraster vizcainoensis Grant & Hertlein. Test
58 mm diam; lateral profile. Bahía Sebastián Vizcaíno
(Baja California). (R. Wolff)

11.8b Dendraster vizcainoensis Grant &
Hertlein. Test 58 mm diam; aboral view.
Punta Prieta (Baja California). (R. Wolff)

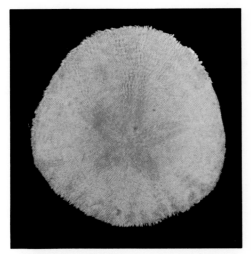

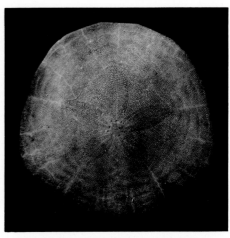

11.9a **Dendraster laevis** H. L. Clark. Sand Dollar. 44 mm diam; aboral view. La Jolla (San Diego Co.). (R. Wolff)

11.9b **Dendraster laevis** H. L. Clark. 44 mm diam; oral view. La Jolla (San Diego Co.). (R. Wolff)

11.9c **Dendraster laevis** H. L. Clark. Test 49.6 mm diam; aboral view. La Jolla (San Diego Co.). (R. Wolff)

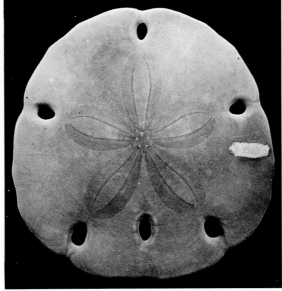

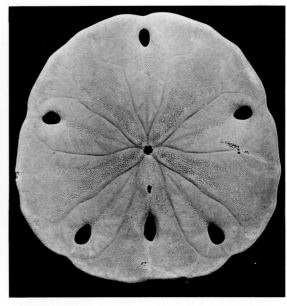

11.9d **Dendraster laevis** H. L. Clark. Test 51.4 mm diam; lateral profile. La Jolla (San Diego Co.). (R. Wolff)

11.10a **Encope micropora** L. Agassiz. Keyhole Urchin. Test 92 mm diam; aboral view. Baja California. (R. Wolff)

11.10b **Encope micropora** L. Agassiz. Test 92 mm diam; oral view. Baja California. (R. Wolff)

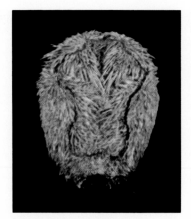

11.11a Brissopsis pacifica
(A. Agassiz). Heart Urchin.
28 mm long; aboral view. Baja
California. (R. Wolff)

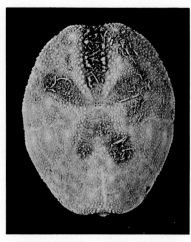

11.11b **Brissopsis pacifica** (A. Agassiz).
Test 24 mm long; aboral view. Baja
California. (R. Wolff)

11.12a **Lovenia cordiformis** (A. Agassiz).
Heart Urchin, Sea Porcupine. 40 mm long;
aboral view. Newport Bay (Orange Co.).
(R. Wolff)

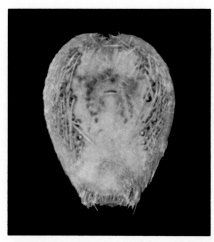

11.12b Lovenia cordiformis (A. Agassiz).
40 mm long; oral view. Newport Bay
(Orange Co.). (R. Wolff)

11.12c **Lovenia cordiformis**
(A. Agassiz). Test 32 mm long;
aboral view. Newport Bay
(Orange Co.). (R. Wolff)

11.13a **Nacospatangus laevis** (H. L. Clark).
Heart Urchin. 44 mm long; aboral view.
Spines partially lost. San Clemente Island
(Channel Islands). (R. Wolff)

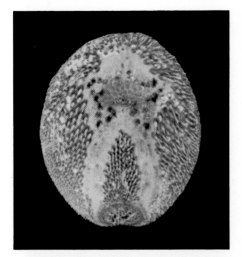

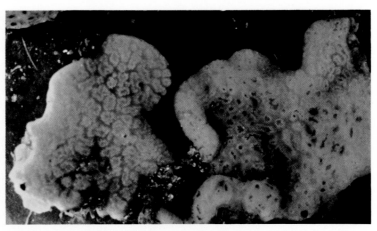

11.13b **Nacospatangus laevis** (H. L. Clark). 44 mm long; oral view. Spines partially lost. San Clemente Island (Channel Islands). (R. Wolff)

11.13c **Nacospatangus laevis** (H. L. Clark). Test 23 mm long; aboral view. San Clemente Island (Channel Islands). (R. Wolff)

12.1 **Aplidium californicum** (Ritter & Forsyth). 10 cm largest diam. Pacific Grove (Monterey Co.). (D. Abbott)

12.2 **Aplidium solidum** (Ritter & Forsyth). Portion of colony shown about 15 cm across. Monterey Bay. (R. King)

12.3 **Aplidium propinquum** (Van Name). Lobes 3 cm long. Pacific Grove (Monterey Co.).

12.4 **Aplidium arenatum** (Van Name). Lobes 1.5–2 cm long. Pacific Grove (Monterey Co.).

12.5 **Polyclinum planum** (Ritter & Forsyth). 10 cm across. Monterey Peninsula. (R. King)

12.6 **Synoicum parfustis** (Ritter & Forsyth). Lobes 5–8 cm long. Pacific Grove (Monterey Co.). (T. Newberry)

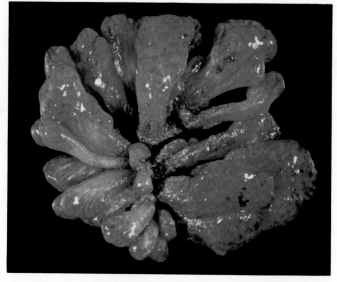

12.7 **Ritterella pulchra** (Ritter). Lobes 40 mm high. Pacific Grove (Monterey Co.).

12.9 Ritterella aequalisiphonis (Ritter & Forsyth). Lobes 3 cm long. Monterey Peninsula. (D. Abbott)

12.8 Ritterella rubra Abbott & Trason. Largest lobes 15 mm diam. Carmel Bay (Monterey Co.). (D. Abbott)

12.10 Euherdmania claviformis (Ritter). Cluster 8 cm across. Monterey Peninsula. (T. Newberry)

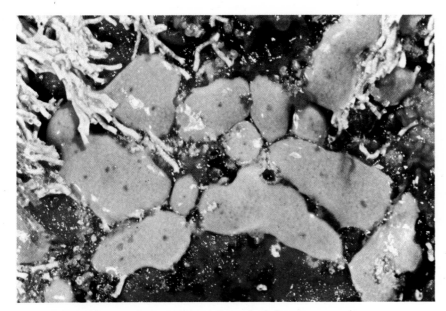

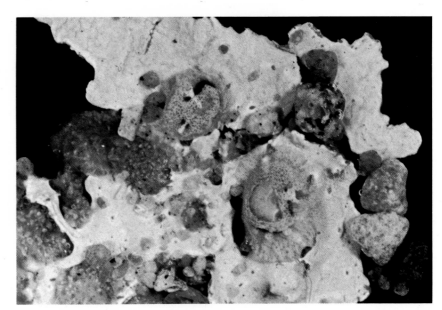

12.11 Didemnum carnulentum Ritter & Forsyth. Colonies 1–3 cm across. Corona del Mar (Orange Co.).

12.12 Trididemnum opacum (Ritter). Field about 12 cm wide. Two gastropods (*Lamellaria stearnsii*, 13.64) feed on the colony. 17-Mile Drive (Monterey Co.).

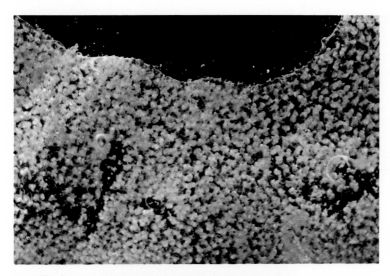

12.13 **Diplosoma macdonaldi** Herdman. Portion of colony shown 4.8 cm across. Pacific Grove (Monterey Co.). (D. Abbott)

12.14 **Pycnoclavella stanleyi** Berrill & Abbott. Gold thoraxes of zooids 1–2 mm long; growing with globular juvenile colony of tunicate *Archidistoma molle* (12.17). Pacific Grove (Monterey Co.). (N. Burnett)

12.15 **Archidistoma ritteri** (Van Name) Longest lobes about 4 cm. Monterey Peninsula. (W. Lee)

12.16 **Archidistoma psammion** (Ritter & Forsyth). Colony 12 cm across. Pacific Grove (Monterey Co.). (D. Abbott)

12.17 Archidistoma molle (Ritter). Colonies 5 cm across. Pacific Grove (Monterey Co.). (D. Abbott)

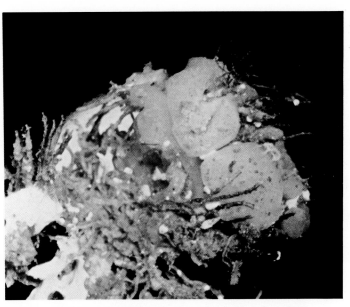

12.18 Archidistoma diaphanes (Ritter & Forsyth). Lobes about 3 cm diam; growing with dense white colonies of *Didemnum carnulentum* (12.11). Pacific Grove (Monterey Co.). (T. Newberry)

12.19 Cystodytes lobatus (Ritter). White and pink color variants. Oblong holes surrounding siphons of the clam *Mytilimeria nuttallii* (15.83). Monterey Bay. (G. Lambert)

12.20a (*see also overleaf*) **Clavelina huntsmani** Van Name. Light-Bulb Tunicate. Zooids 40 mm long. Pacific Grove (Monterey Co.). (D. Abbott)

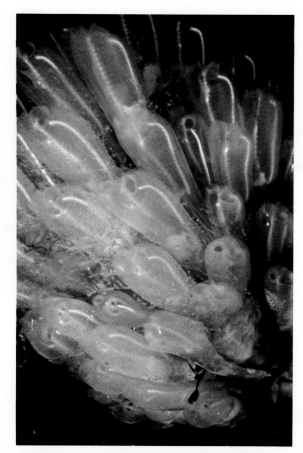

12.21 **Distaplia occidentalis** Bancroft. Larger colony 4 cm diam. Pacific Grove (Monterey Co.).

12.20b **Clavelina huntsmani** Van Name. Pink lines mark position of pharynx. Yellow-orange clumps in atrium represent developing tadpole larvae. Pacific Grove (Monterey Co.). (N. Burnett)

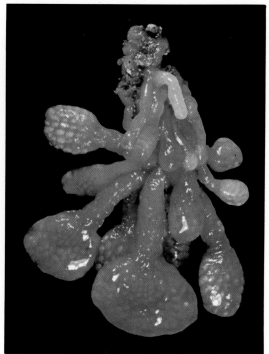

12.22 **Distaplia smithi** Abbott & Trason. Lobes 2–5 cm long. Pacific Grove (Monterey Co.).

12.23 **Ciona intestinalis** (Linnaeus). Cluster of animals, each 10 cm long. Center individual overgrown with *Botryllus* (12.29). Monterey harbor. (D. Abbott)

12.24 **Perophora annectens** Ritter. Zooids 2–3 mm diam. Pacific Grove (Monterey Co.). (N. Burnett)

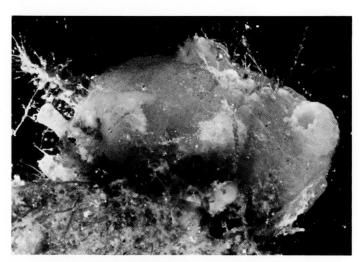

12.25 **Ascidia ceratodes** (Huntsman). 5 cm long; showing open oral siphon. Monterey harbor. (N. Burnett)

12.26 **Ascidia paratropa** (Huntsman). About 7 cm high. Carmel Bay (Monterey Co.). (W. Gladfelter)

12.27 **Corella willmeriana** Herdman. 3 cm long. Monterey Bay. (W. Gladfelter)

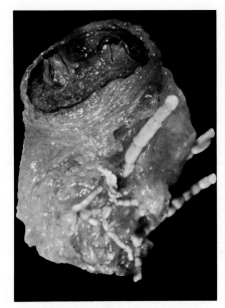

12.28 **Chelyosoma productum** Stimpson. 3 cm high; with adhering branches of coralline algae. Pacific Grove (Monterey Co.).

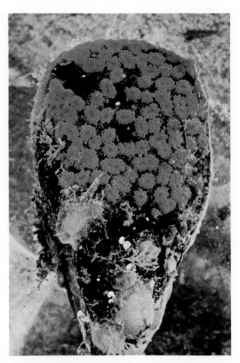

12.29 **Botryllus** sp. Zooid clusters 5 mm diam, growing on mussel *Mytilus edulis* (15.10). Monterey harbor. (D. Abbott)

12.30 **Botrylloides** sp. Zooids 1–2 mm long. (J. Carlson)

12.31 **Metandrocarpa taylori** Huntsman. Zooids 5 mm diam, growing on bryozoan *Eurystomella bilabiata* (6.9). Pacific Grove (Monterey Co.). (T. Newberry)

12.32 **Cnemidocarpa finmarkiensis** (Kiaer). About 3 cm across. Friday Harbor (Washington). (R. Zimmer)

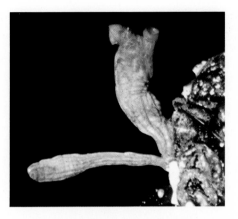

12.33 **Styela montereyensis** (Dall). 10 cm long. Monterey harbor. (T. Newberry)

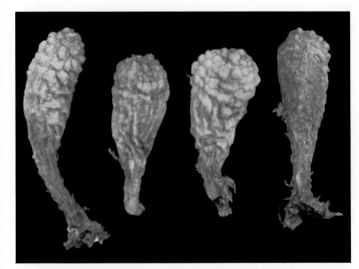

12.34 **Styela clava** Herdman. 6–8 cm long. San Diego.

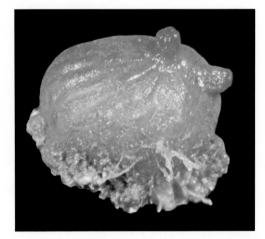

12.35 **Styela truncata** Ritter. 2 cm across. Dana Point (Orange Co.).

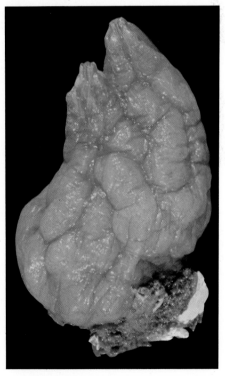

12.36 **Styela plicata** (Lesueur). 7 cm long. Newport Bay (Orange Co.).

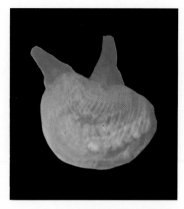

12.37b **Pyura haustor** (Stimpson). Animal removed from tunic 4 cm diam. 17-Mile Drive (Monterey Co.).

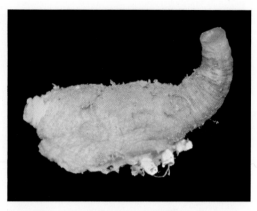

12.38 **Pyura mirabilis** (von Drasche). About 8 cm long (one siphon missing). Monterey Bay. (D. Abbott)

12.37a **Pyura haustor** (Stimpson). Two individuals with red siphons expanded, about 2 cm between siphon tips; growing on *Mytilus* shell along with serpulid worms forming whitish tubes. Monterey harbor. (T. Newberry)

12.39a **Boltenia villosa** (Stimpson). 5 cm long. Monterey Bay. (T. Newberry)

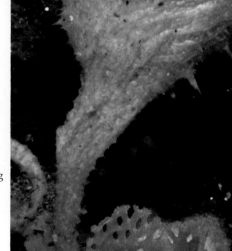

12.39b **Boltenia villosa** (Stimpson). About 3 cm long; growing with bryozoan *Phidolopora pacifica* (6.10). Monterey Bay. (N. Burnett)

12.40a **Halocynthia hilgendorfi igaboja** (Oka). Individual about 35 mm across. Pacific Grove (Monterey Co.).

12.40b **Halocynthia hilgendorfi igaboja** (Oka). Spines to 6 mm long. Pacific Grove (Monterey Co.).

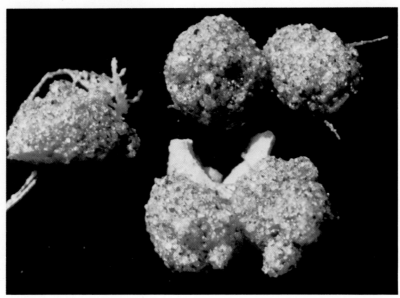

12.41 **Molgula manhattensis** (DeKay). Larger intact specimen about 2 cm diam. Top: intact; bottom: removed from tunic. San Francisco Bay. (T. Newberry)

12.42 **Molgula verrucifera** Ritter & Forsyth. Five specimens encrusted with sand and shell fragments, each about 8 mm diam. Palos Verdes (Los Angeles Co.). (J. Haderlie)

13.1 Haliotis rufescens Swainson. Red Abalone. 29 cm long. Bruhel Point (Mendocino Co.).

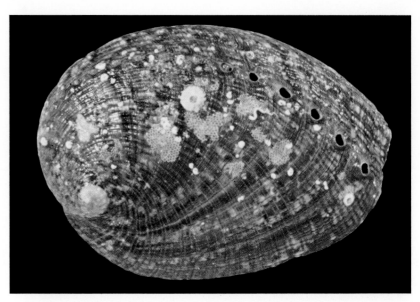

13.2 Haliotis fulgens Philippi. Green Abalone. 16 cm long. Dana Point (Orange Co.).

13.3 Haliotis corrugata Gray. Pink Abalone. 16 cm long. Dana Point (Orange Co.).

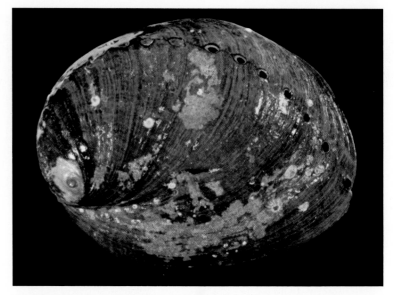

13.4 Haliotis cracherodii Leach. Black Abalone. 18 cm long. Pacific Grove (Monterey Co.).

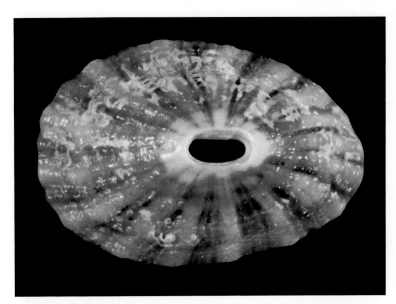

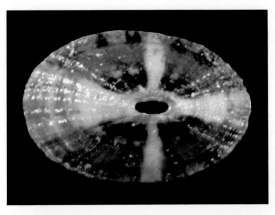

13.5b **Fissurella volcano crucifera** Dall. Keyhole Limpet. 20 mm long. Dana Point (Orange Co.).

13.5a **Fissurella volcano** Reeve. Keyhole Limpet. 30 mm long. Malpaso Creek (Monterey Co.).

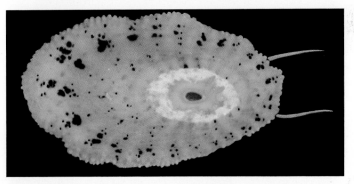

13.7a **Megatebennus bimaculatus** (Dall). Two-Spotted Keyhole Limpet. Whole animal 30 mm long. Malpaso Creek (Monterey Co.).

13.6 **Lucapinella callomarginata** (Dall). Southern Keyhole Limpet, Fleshy Keyhole Limpet. Whole animal 35 mm long. Newport Bay (Orange Co.).

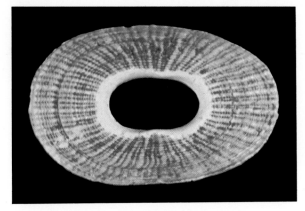

13.7b **Megatebennus bimaculatus** (Dall). Shell, 15 mm long. Malpaso Creek (Monterey Co.).

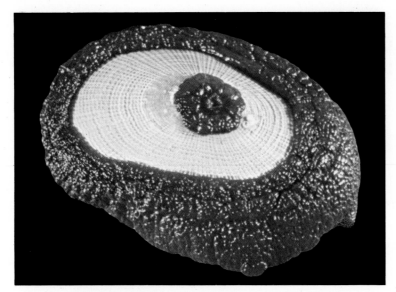

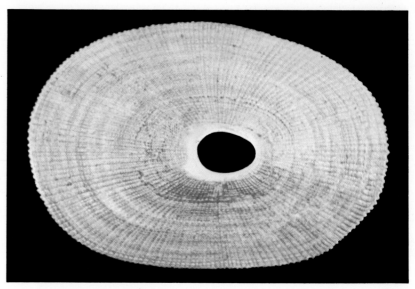

13.8a Megathura crenulata (Sowerby). Giant Keyhole Limpet. Whole animal 12 cm long. Corona del Mar (Orange Co.).

13.8b Megathura crenulata (Sowerby). Shell, 8 cm long. Flatrock Point (Palos Verdes, Los Angeles Co.).

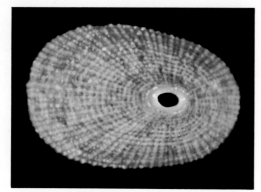

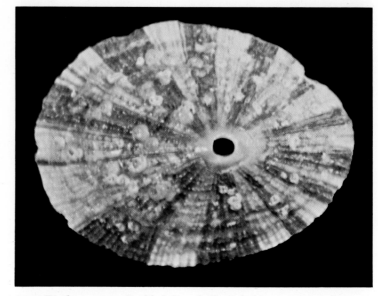

13.9 Diodora arnoldi McLean. 18 mm long. 17-Mile Drive (Monterey Co.).

13.10 Diodora aspera (Rathke). Rough Keyhole Limpet. 50 mm long. Point Arena (Mendocino Co.).

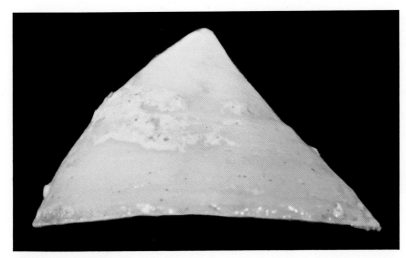

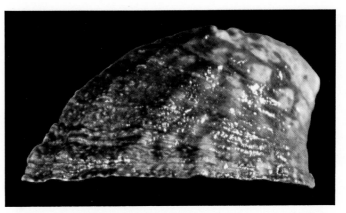

13.12 **Collisella digitalis** (Rathke). Ribbed Limpet. 25 mm long. McClures Beach (Marin Co.).

13.11 **Acmaea mitra** Rathke. White-Cap Limpet. 30 mm long. Pigeon Point (San Mateo Co.).

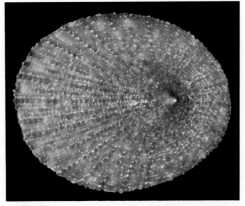

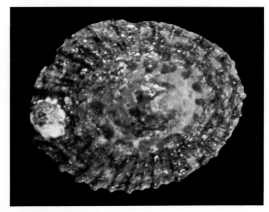

13.14a **Collisella limatula** (Carpenter). File Limpet. 30 mm long. Pacific Grove (Monterey Co.).

13.14b **Collisella limatula** (Carpenter). 28 mm long; bay form. Tomales Bay (Marin Co.).

13.13 **Collisella scabra** (Gould). Rough Limpet. 25 mm long. Point Sal (Santa Barbara Co.).

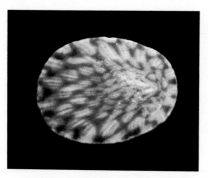

13.15 Collisella ochracea (Dall). 16 mm long. Hobson Beach (Ventura Co.).

13.16 Collisella pelta (Rathke). Shield Limpet. 32 mm long. Monterey.

13.17 Collisella asmi (Middendorff). Black Limpet. 10 mm long, growing on prosobranch *Tegula funebralis* (13.32). Monterey.

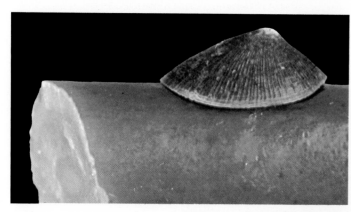

13.18 Collisella instabilis (Gould). Unstable Seaweed Limpet. 24 mm long, growing on stipe of alga *Laminaria*. Pacific Grove (Monterey Co.).

13.20 Notoacmea insessa (Hinds). Seaweed Limpet. 20 mm long, growing on stipe of alga *Egregia*. Moss Beach (San Mateo Co.).

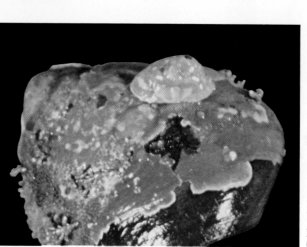

13.19 Collisella triangularis (Carpenter). Triangular Limpet. 7 mm long, growing on coralline-encrusted prosobranch *Tegula brunnea* (13.31). Cabrillo Point (Mendocino Co.).

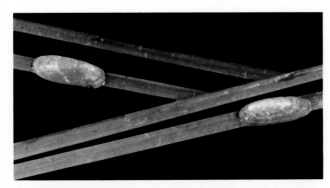

13.21 **Notoacmea paleacea** (Gould). Surfgrass Limpet. 8 mm long, growing on surfgrass *Phyllospadix*. Trinidad Head (Humboldt Co.).

13.22 **Notoacmea depicta** (Hinds). Painted Limpet. 12 mm long, growing on eelgrass *Zostera*. Agua Hedionda Lagoon (San Diego Co.).

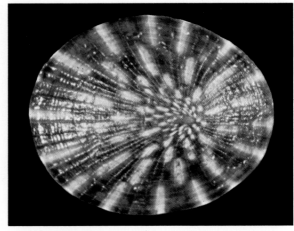

13.23 **Notoacmea scutum** (Rathke). Plate Limpet. 24 mm long. Point Buchon (San Luis Obispo Co.).

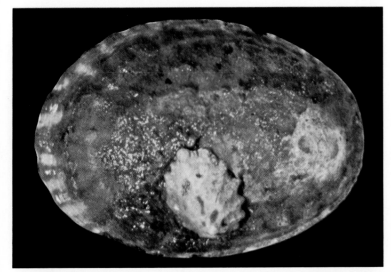

13.24 **Lottia gigantea** Sowerby. Owl Limpet. 50 mm long. Young proso-branch *Collisella scabra* (13.13) on back. Point Fermin (San Pedro, Los Angeles Co.).

13.25 **Calliostoma annulatum** (Lightfoot). Purple-Ringed Top Snail. 26 mm diam. Cannery Row (Monterey Co.).

13.26 Calliostoma canaliculatum (Lightfoot). Channeled Top Snail. 30 mm diam. Pacific Grove (Monterey Co.).

13.27 Calliostoma ligatum (Gould). Blue Top Snail. 22 mm diam. Devils Slide (San Mateo Co.).

13.28 Calliostoma gemmulatum Carpenter. Gem Top Snail. 14 mm diam. Hobson Beach (Ventura Co.).

13.29 Margarites pupillus (Gould). Little Margarite. 10 mm diam. Point Saint George (Del Norte Co.).

13.30 Norrisia norrisi (Sowerby). Norris's Top Snail. 30 mm shell diam. Laguna Beach (Orange Co.).

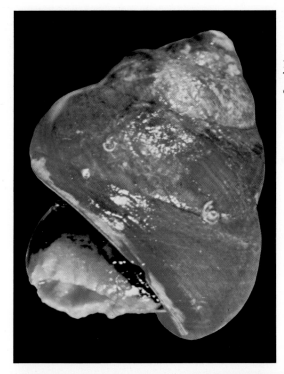

13.31 **Tegula brunnea** (Philippi). Brown Turban Snail, Brown Tegula. 28 mm diam. Duxbury Point (Marin Co.).

13.32 **Tegula funebralis** (A. Adams). Black Turban Snail, Black Tegula. 28 mm diam. Point Sal (Santa Barbara Co.).

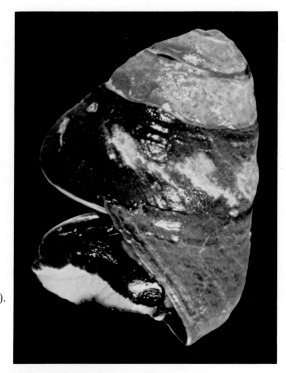

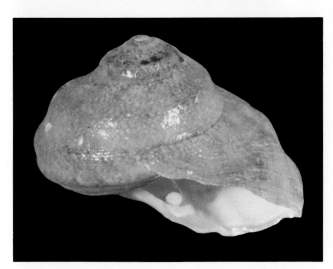

13.33 **Tegula aureotincta** (Forbes). Gilded Turban Snail, Gilded Tegula. 25 mm diam. Dana Point (Orange Co.).

13.34 **Tegula eiseni** Jordan. Banded Turban Snail, Banded Tegula. 20 mm diam. Newport Bay (Orange Co.).

13.35 Tegula gallina (Forbes). Speckled
Turban Snail, Speckled Tegula. 22 mm
diam. Corona del Mar (Orange Co.).

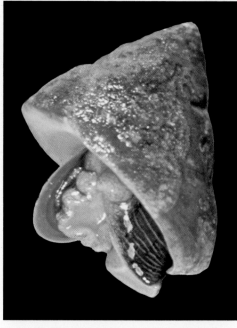

13.36a Tegula montereyi (Kiener). Monterey
Turban Snail, Monterey Tegula. 26 mm diam.
Pacific Grove (Monterey Co.).

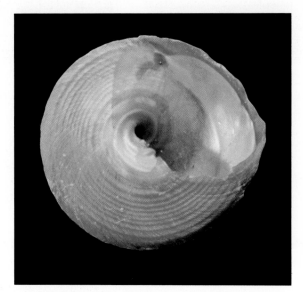

13.36b Tegula montereyi (Kiener). 24 mm diam, aperture and umbilicus. Pacific Grove (Monterey Co.).

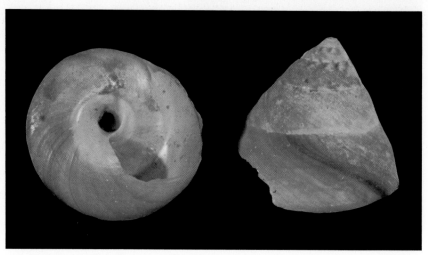

13.37 Tegula pulligo (Gmelin). Dusky Turban Snail, Dusky Tegula.
25 mm diam. Pacific Grove (Monterey Co.).

13.38 **Astraea gibberosa** (Dillwyn). Red Top Snail, Red Turban Snail. 35 mm diam. Pacific Grove (Monterey Co.).

13.39 **Astraea undosa** (Wood). Wavy Top Snail, Wavy Turban Snail. 75 mm diam. La Jolla (San Diego Co.).

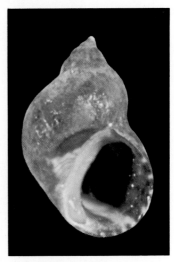

13.40 **Homalopoma luridum** (Dall). 5–9 mm diam; color variants. Pacific Grove (Monterey Co.). (N. Burnett)

13.41 **Lacuna marmorata** Dall. Chink Snail. 7 mm high, on surfgrass *Phyllospadix*. Bruhel Point (Mendocino Co.).

13.42 **Littorina planaxis** Philippi. Eroded Periwinkle. 13 mm high. Monterey.

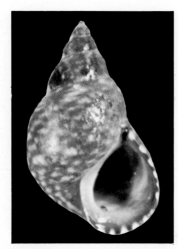

13.43 Littorina scutulata
Gould. Checkered Periwinkle.
13 mm high. Pacific Grove
(Monterey Co.).

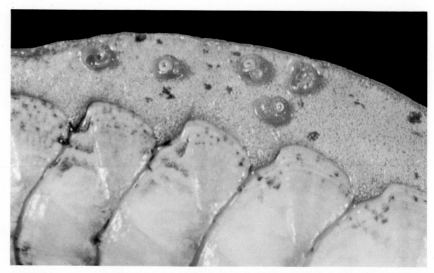

13.44 Vitrinella oldroydi Bartsch. 2 mm diam, growing on chiton *Stenoplax heathiana* (16.8). Moss Beach (San Mateo Co.).

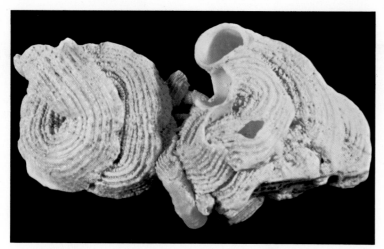

13.45a Serpulorbis squamigerus (Carpenter). Scaled Worm Snail.
Tube diam 12 mm. Agua Hedionda Lagoon (San Diego Co.).

13.45b Serpulorbis squamigerus (Carpenter). Snail, removed from shell, 72 mm long. Agua Hedionda Lagoon (San Diego Co.).

13.46 **Petaloconchus montereyensis** Dall.
Cluster of specimens; tube diam to 2 mm.
Pacific Grove (Monterey Co.). (M. Hadfield)

13.47 **Cerithidea californica** (Haldeman).
California Horn Snail. 40 mm high. Anaheim Bay (Orange Co.).

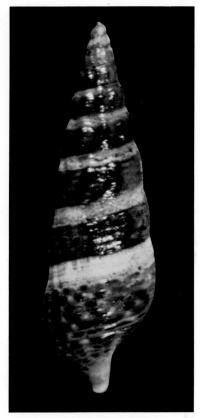

13.48 **Batillaria attramentaria**
(Sowerby). 21 mm high. Tomales
Bay (Marin Co.).

13.49 **Bittium eschrichtii** (Middendorff). Threaded Bittium.
10 mm high. Dana Point (Orange Co.).

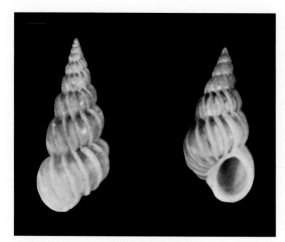

13.50 Epitonium tinctum (Carpenter).
Tinted Wentletrap. 13 mm high. Monterey.

13.51 Opalia chacei Strong. Chace's
Wentletrap. 30 mm high. 17-Mile Drive
(Monterey Co.).

13.52 Opalia funiculata (Carpenter).
Sculptured Wentletrap. 16 mm high. Flatrock Point (San Pedro, Los Angeles Co.).

13.53 Hipponix cranioides Carpenter. Hoof Snail. Side view, 18 mm
diam; dorsal view, 15 mm diam. Point Buchon (San Luis Obispo Co.).

13.54a Hipponix tumens Carpenter. Ribbed
Hoof Snail. 12 mm diam. La Jolla (San Diego Co.).

13.54b **Hipponix tumens** Carpenter. Side view of same specimen.

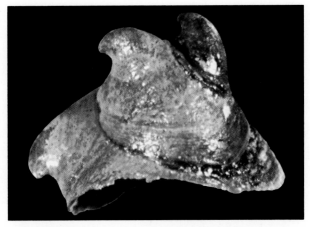

13.55 **Crepidula adunca** Sowerby. Hooked Slipper Snail. Three specimens: lower, female, 20 mm long; middle, probably intersex; top, male. Pacific Grove (Monterey Co.).

13.56a **Crepidula aculeata** (Gmelin). Spiny Slipper Snail. 10 mm long; dorsal view. Piedras Blancas Point (San Luis Obispo Co.).

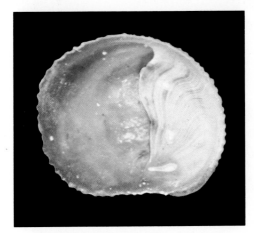

13.56b **Crepidula aculeata** (Gmelin). Ventral view of same specimen.

13.57a **Crepidula nummaria** Gould. White Slipper Snail. 20 mm long; dorsal view. Moss Beach (San Mateo Co.).

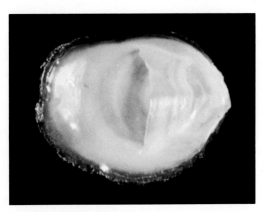

13.57b **Crepidula nummaria** Gould. Ventral view of same specimen.

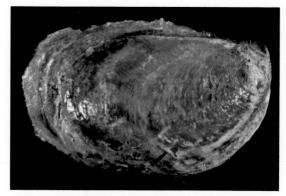

13.58a Crepidula onyx Sowerby. Onyx Slipper Snail. 38 mm long. Mission Bay (San Diego Co.).

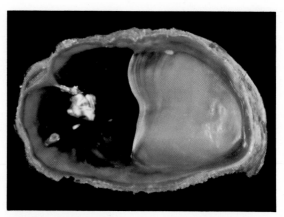

13.58b Crepidula onyx Sowerby. Ventral view of same specimen.

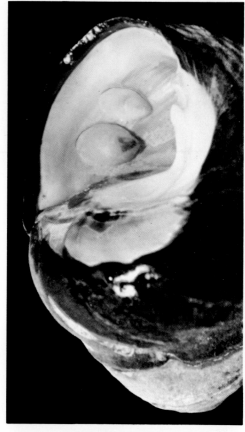

13.59 Crepidula perforans (Valenciennes). Western White Slipper Snail. 4 mm and 6 mm long, attached to aperture of prosobranch *Tegula funebralis* (13.32). Flatrock Point (San Pedro, Los Angeles Co.).

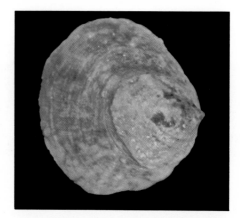

13.60a Crepipatella lingulata (Gould). Half-Slipper Snail. 15 mm long. Mission Bay (San Diego Co.).

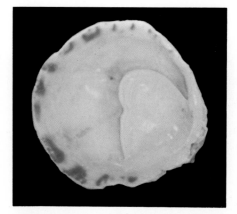

13.60b Crepipatella lingulata (Gould). Ventral view.

13.61 Crucibulum spinosum Sowerby. Cup-and-Saucer Limpet. 7–20 mm diam, growing on shell of mussel *Mytilus edulis* (15.10). Newport Bay (Orange Co.).

13.62a Polinices lewisii (Gould). Lewis's Moon Snail. 105 mm diam. Bodega Bay harbor (Sonoma Co.).

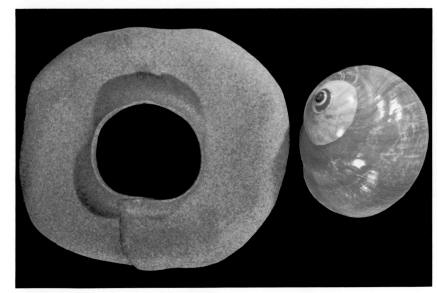

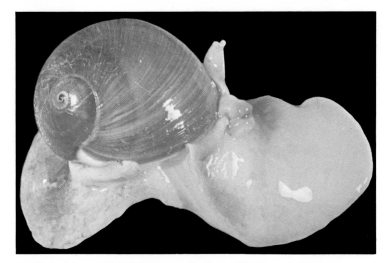

13.63a (*see also overleaf*) **Polinices reclusianus** (Deshayes). Southern Moon Snail. Shell diam 47 mm; with expanded foot. Mission Bay (San Diego Co.).

13.62b Polinices lewisii (Gould). Snail with sand collar (egg mass). Shell diam 100 mm. Elkhorn Slough (Monterey Co.).

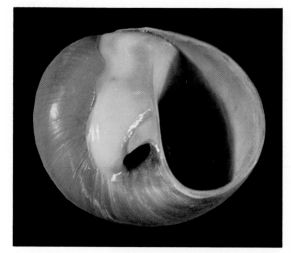

13.63b Polinices reclusianus (Deshayes). Shell diam 50 mm; aperture and umbilicus shown. Agua Hedionda Lagoon (San Diego Co.).

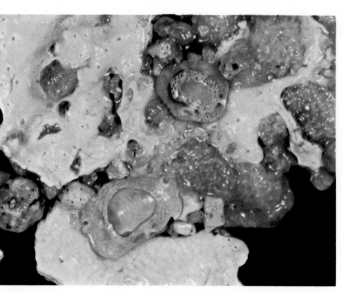

13.64 Lamellaria stearnsii Dall. Shell 15 mm long, on whitish compound ascidian *Trididemnum opacum* (12.12). Pacific Grove (Monterey Co.).

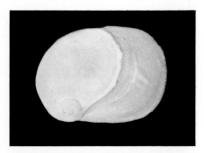

13.65a Lamellaria rhombica Dall. Shell, 15 mm long. Monterey.

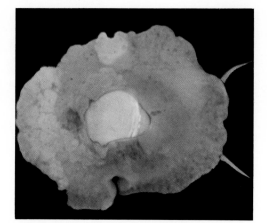

13.65b Lamellaria rhombica Dall. Whole animal; shell 15 mm long. Monterey.

13.66 Velutina sp. Smooth Velutina. 8 mm long. Pigeon Point (San Mateo Co.).

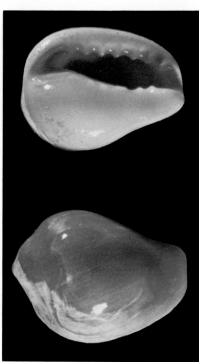

13.67a Erato vitellina Hinds. Apple-Seed Erato. 14 mm long. Pacific Grove (Monterey Co.).

13.67b **Erato vitellina** Hinds. Whole animal; shell 14 mm long. Pacific Grove (Monterey Co.).

13.68 **Trivia solandri** (Sowerby). Solander's Trivia. 16 mm long. Dana Point (Orange Co.).

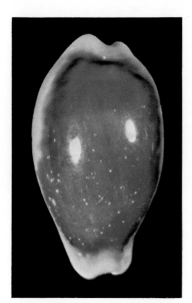

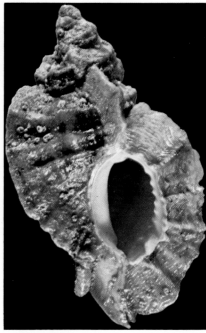

13.70 **Ceratostoma foliatum** (Gmelin). Leafy Hornmouth. 5.2 cm high. Cape Mendocino (Humboldt Co.).

13.69a **Cypraea spadicea** Swainson. Chestnut Cowrie. 40 mm long. Agua Hedionda Lagoon (San Diego Co.).

13.69b **Cypraea spadicea** Swainson. Aperture of same specimen.

13.71 Ceratostoma nuttalli (Conrad). Nuttall's Hornmouth. 37–45 mm high; color variants. Point Fermin (San Pedro, Los Angeles Co.).

13.72a Pteropurpura trialata (Sowerby). Three-Winged Murex. 60 mm high. Dana Point (Orange Co.).

13.72b Pteropurpura trialata (Sowerby). 80 mm high, depositing egg capsules on mussel *Mytilus edulis* (15.10). Dana Point (Orange Co.).

13.74a Ocenebra circumtexta
Stearns. Circled Rock Snail.
17 mm high. Piedras Blancas
Point (San Luis Obispo Co.).

13.73b Maxwellia gemma (Sowerby).
Immature specimen, 28 mm high. Dana
Point (Orange Co.).

13.73a Maxwellia gemma (Sowerby). Gem
Murex. Mature specimen, 50 mm high. Agua
Hedionda Lagoon (San Diego Co.).

13.74b Ocenebra circumtexta
Stearns. Aperture view.

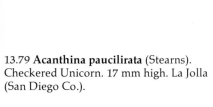

13.75 **Ocenebra interfossa** Carpenter. Sculptured Rock Snail. 15 mm high. Cape Mendocino (Humboldt Co.).

13.76 **Ocenebra lurida** (Middendorff). Lurid Rock Snail. 20 mm high. Cabrillo Point (Mendocino Co.).

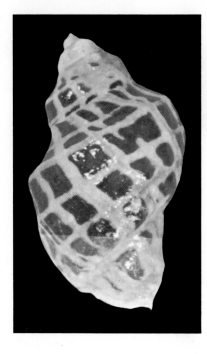

13.77 **Roperia poulsoni** (Carpenter). Poulson's Rock Snail. 50 mm high. Anaheim Bay (Orange Co.).

13.78 **Urosalpinx cinerea** (Say). Atlantic Oyster Drill. 35 mm high. San Francisco Bay.

13.79 **Acanthina paucilirata** (Stearns). Checkered Unicorn. 17 mm high. La Jolla (San Diego Co.).

13.80 Acanthina punctulata (Sowerby). Spotted Unicorn. Left: immature specimen, 24 mm high; right: adult specimen, 30 mm high. Tomales Bay (Marin Co.).

13.81 Acanthina spirata (Blainville). Angular Unicorn. 30 mm high. Monterey harbor.

13.82 Nucella canaliculata (Duclos). Channeled Dogwinkle. 32 mm high. Malpaso Creek (Monterey Co.).

13.83a Nucella emarginata (Deshayes). Emarginate Dogwinkle. 35 mm high. Pacific Grove (Monterey Co.).

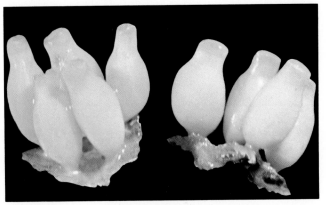

13.83b Nucella emarginata (Deshayes). Egg capsules 6 mm high. Pacific Grove (Monterey Co.).

13.84 **Nucella lamellosa** (Gmelin). Wrinkled Purple, Frilled Dogwinkle. 47 mm high. Bodega Bay harbor (Sonoma Co.).

13.85 **Busycotypus canaliculatus** (Linnaeus). Channeled Whelk. 60 mm high. San Francisco Bay.

13.86 **Searlesia dira** (Reeve). Dire Whelk. 30 mm high. Duxbury Point (Marin Co.).

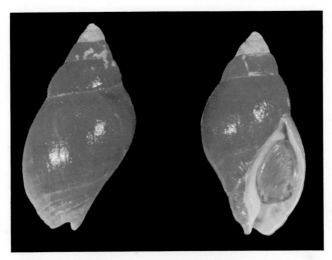

13.88 **Macron lividus** (A. Adams). Livid Macron. 20 mm high. Dana Point (Orange Co.).

13.89 **Amphissa columbiana** Dall. Wrinkled Dove Snail. 18 mm high. Point Saint George (Del Norte Co.).

13.87 **Kelletia kelletii** (Forbes). Kellet's Whelk. 58 mm high. Corona del Mar (Orange Co.).

13.90 **Amphissa versicolor** Dall. Variegated Amphissa. 10–13 mm high; color variants. Pacific Grove (Monterey Co.).

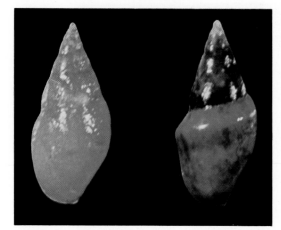

13.91 Mitrella carinata (Hinds). Carinated Dove Snail. 10 mm high; form and color variants. Monterey.

13.92a Nassarius fossatus (Gould). Channeled Nassa. Shell 40 mm high. San Diego Bay.

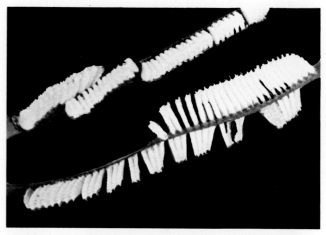

13.92b Nassarius fossatus (Gould). Egg capsules, each 6 mm high, on eelgrass *Zostera*. Mission Bay (San Diego Co.).

13.93 Nassarius mendicus (Gould). Lean Nassa. 19 mm high. Humboldt Bay.

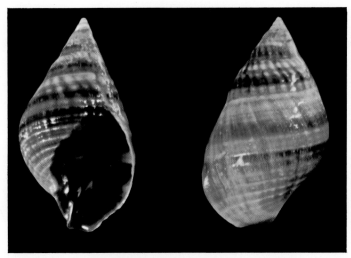

13.94 **Nassarius mendicus cooperi** (Forbes).
Cooper's Lean Nassa. 18 mm high. Bolinas Lagoon
(Marin Co.).

13.95 **Nassarius obsoletus** (Say). Eastern Mud Snail. 20 mm
high. San Francisco Bay.

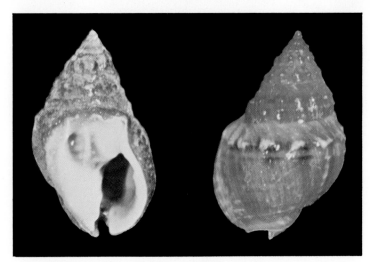

13.98 **Olivella baetica**
Carpenter. Little Olive,
Beatic Olivella. 12 mm
high. San Diego Bay.

13.96 **Nassarius tegula** (Reeve). Covered-Lip Nassa. 16 mm high.
San Diego Bay.

13.97 **Fusinus luteopictus** (Dall). Painted
Spindle. 20 mm high. Pacific Grove
(Monterey Co.).

13.99 Olivella biplicata (Sowerby). Purple Olive, Purple Olivella. 19 mm high. Humboldt Bay.

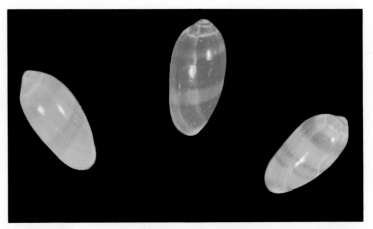

13.101 Volvarina taeniolata Mörch. Banded California Marginella. 9 mm high. La Jolla (San Diego Co.).

13.100 Mitra idae Melville. Ida's Miter. 21 mm high. Dana Point (Orange Co.).

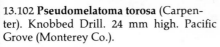

13.102 Pseudomelatoma torosa (Carpenter). Knobbed Drill. 24 mm high. Pacific Grove (Monterey Co.).

13.103 Conus californicus Hinds. California Cone. 26 mm high. Mission Bay (San Diego Co.).

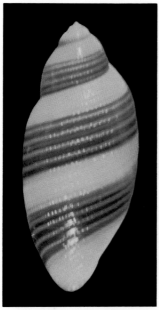

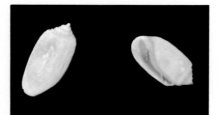

14.2 Cylichnella inculta (Gould).
Barrel Bubble Snail. 5 mm long.
Morro Bay (San Luis Obispo Co.).

14.3 Aglaja ocelligera (Bergh). Yellow-Spotted Aglaja.
26 mm long. Tomales Bay (Marin Co.).

14.1 Rictaxis punctocaelatus (Carpenter).
Striped Barrel Snail. 12 mm long.
Humboldt Bay.

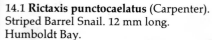

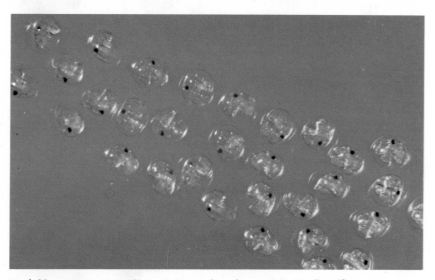

14.4b Navanax inermis (Cooper). Live veliger larvae. Mission Bay (San
Diego Co.). (J. Lance)

14.4a Navanax inermis (Cooper). Navanax. 100 mm long, with egg strings.
Santa Catalina Island (Channel Islands). (R. Ames)

14.5 **Gastropteron pacificum** Bergh. 13 mm long. Tomales Bay (Marin Co.).

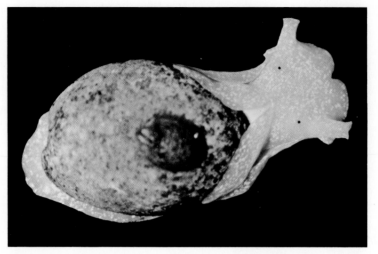

14.6a **Bulla gouldiana** Pilsbry. Cloudy Bubble Snail. Animal 45 mm long, with small commensal snail *Crepidula* on top. San Diego Bay. (W. Farmer)

14.6b **Bulla gouldiana** Pilsbry. 50 mm long. Newport Bay (Orange Co.).

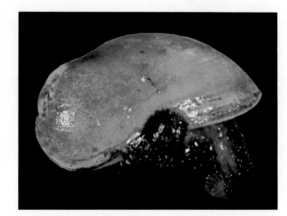

14.7 **Haminoea vesicula** (Gould). White Bubble Snail. 14 mm long. Newport Bay (Orange Co.).

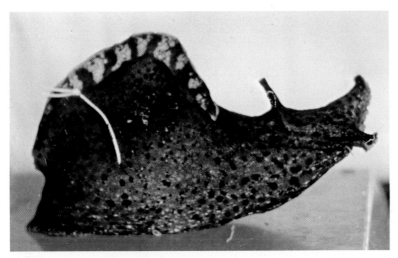

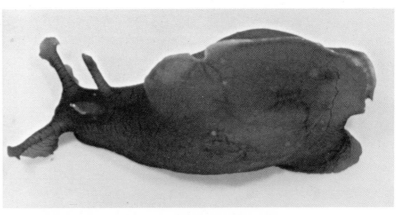

14.8b **Aplysia californica** Cooper. Juvenile, 125 mm long. Elkhorn Slough (Monterey Co.). (R. Beeman)

14.8a **Aplysia californica** Cooper. California Brown Sea Hare. Adult (with yellow marking tag), 350 mm long. Elkhorn Slough (Monterey Co.). (R. Beeman)

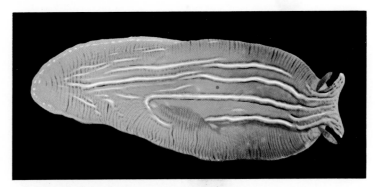

14.10 **Phyllaplysia taylori** Dall. Taylor's Sea Hare. 40 mm long. Tomales Bay (Marin Co.).

14.9 **Aplysia vaccaria** Winkler. California Black Sea Hare. 290 mm long. Dana Point (Orange Co.). (R. Beeman)

14.11 Stiliger fuscovittatus Lance. Streaked Stiliger. 10 mm long, creeping on red algae. Mission Bay (San Diego Co.). (J. Lance)

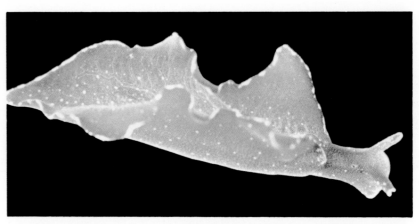

14.12 Elysia hedgpethi Marcus. Hedgpeth's Elysia. 25 mm long. Elkhorn Slough (Monterey Co.). (G. McDonald)

14.13 Tylodina fungina Gabb. Yellow Sponge Tylodina. 34 mm long. La Jolla (San Diego Co.). (W. Harvey/J. Lance)

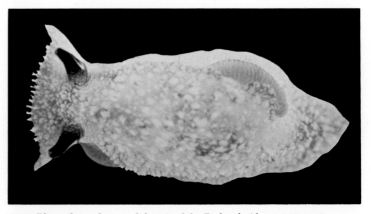

14.14 Pleurobranchaea californica MacFarland. About 200 mm long. Central California. (W. Austin)

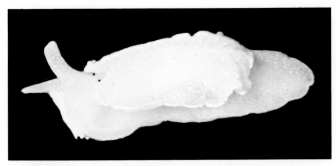

14.15a **Berthella californica** (Dall). White Berthella. Whole animal 20 mm long. Monterey Peninsula. (W. Lee)

14.15b **Berthella californica** (Dall). Internal shell 14 mm long. Pacific Grove (Monterey Co.).

14.16 **Conualevia alba** Collier & Farmer. Smooth-Horned Dorid. 22 mm long. Newport Bay (Orange Co.). (C. Collier)

14.17 **Cadlina limbaughi** Lance. Limbaugh's Cadlina. 31 mm long. La Jolla (San Diego Co.). (J. Lance)

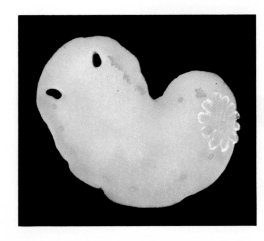

14.18 **Cadlina flavomaculata** MacFarland. Yellow-Spotted Cadlina. 20 mm long. Pacific Grove (Monterey Co.).

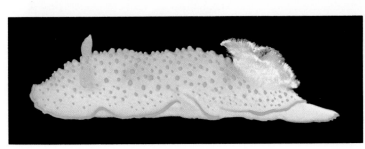

14.19 **Cadlina luteomarginata** MacFarland. Yellow-Edged Cadlina. 60 mm long. Pacific Grove (Monterey Co.).

14.20 **Cadlina modesta** MacFarland. Modest Cadlina. 20 mm long. Pacific Grove (Monterey Co.).

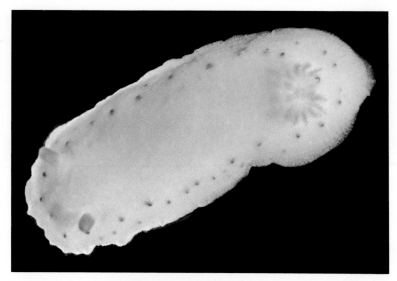

14.21 **Cadlina sparsa** (Odhner). Dark-Spotted Cadlina. 25 mm long. La Jolla (San Diego Co.). (J. Lance)

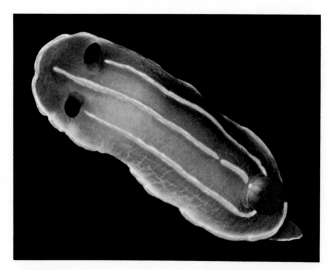

14.22 **Chromodoris macfarlandi** Cockerell. MacFarland's Chromodoris. 30 mm long. Anacapa Island (Channel Islands). (R. Ames)

14.23 **Chromodoris porterae** Cockerell. Porter's Nudibranch. 29 mm long. Point Loma (San Diego Co.). (J. Lance)

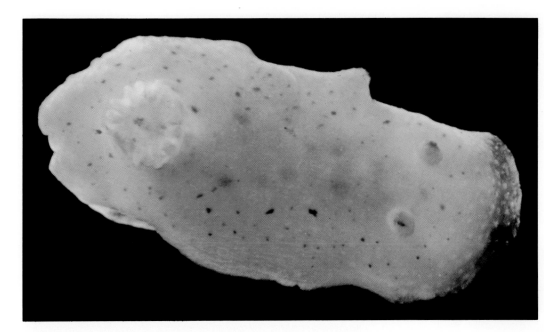

14.25 **Hallaxa chani** Gosliner & Williams. Chan's Nudibranch. 16 mm long. Duxbury Point (Marin Co.). (G. McDonald)

14.24 **Hypselodoris californiensis** (Bergh). Blue-and-Gold Nudibranch. 65 mm long. Bird Rock (San Diego Co.). (C. Collier)

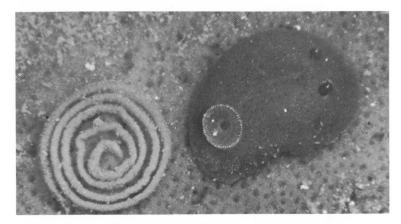

14.26 **Rostanga pulchra** MacFarland. Red Sponge Nudibranch. 20 mm long, with egg ribbon, attached to red sponge. Monterey. (R. Ames)

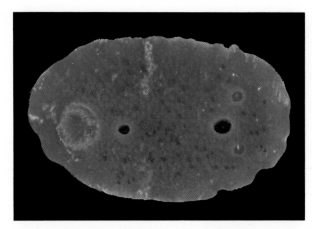

14.27 **Aldisa sanguinea** (Cooper). Blood Spot. 15 mm long. Pacific Grove (Monterey Co.).

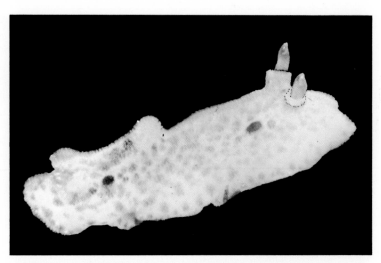

14.28 **Atagema quadrimaculata** Collier. Hunchback Nudibranch. 30 mm long. Point Loma (San Diego Co.). (C. Collier)

14.29 **Archidoris odhneri** (MacFarland). White-Knight Nudibranch. 80 mm long. Pacific Grove (Monterey Co.). (R. Ames)

14.30 **Archidoris montereyensis** (Cooper). Monterey Dorid. 45 mm long. Monterey. (R. Ames)

14.31a **Anisodoris nobilis** (MacFarland). Sea Lemon. 80 mm long. Pacific Grove (Monterey Co.).

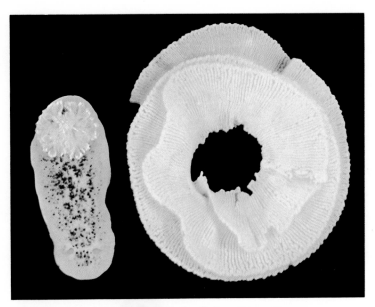

14.31b **Anisodoris nobilis** (MacFarland). Egg ribbon 95 mm diam. Pacific Grove (Monterey Co.).

14.32a **Diaulula sandiegensis** (Cooper). Ring-Spotted Dorid. 60 mm long. Central California. (J. Steinberg)

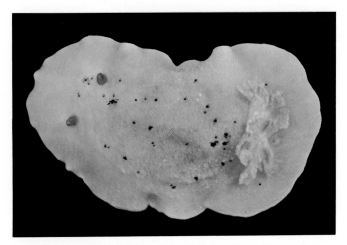

14.33 **Discodoris heathi** MacFarland. Gritty Dorid. 35 mm long. Pacific Grove (Monterey Co.).

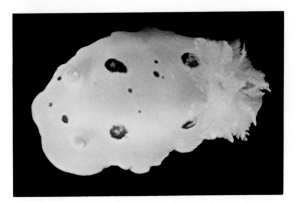

14.32b **Diaulula sandiegensis** (Cooper). 50 mm long; marking variant. Northern California. (A. Hurst)

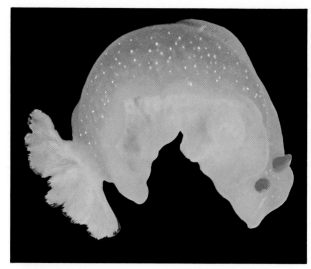

14.34 **Doriopsilla albopunctata** (Cooper). Salted Dorid. 55 mm long. Pacific Grove (Monterey Co.).

14.35 **Corambe pacifica** MacFarland & O'Donoghue. Frost Spot. 5 mm long, attached to encrusting bryozoan *Membranipora*. Anacapa Island (Channel Islands). (C. Collier)

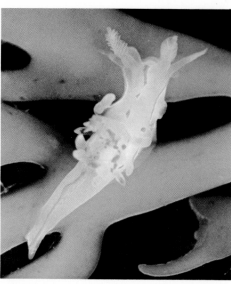

14.36 **Ancula pacifica** MacFarland. Pacific Ancula. 14 mm long, sitting on brown algae. La Jolla (San Diego Co.). (J. Lance)

14.37 **Hopkinsia rosacea** MacFarland. Hopkins's Rose. 28 mm long. Pacific Grove (Monterey Co.).

14.38 **Acanthodoris lutea** MacFarland. Orange-Peel Nudibranch. 22 mm long. Monterey. (R. Ames)

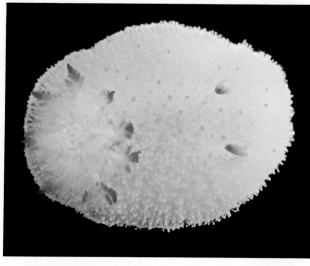

14.39 **Acanthodoris nanaimoensis** O'Donoghue. Wine-Plume Dorid. 30 mm long. Moss Beach (San Mateo Co.). (J. Steinberg)

14.40 Acanthodoris rhodoceras Cockerell &
Eliot. Black-and-White Dorid. 14 mm long.
Monterey harbor. (R. Ames)

14.41 Onchidoris bilamellata (Linnaeus). Many-
Gilled Onchidoris. 16 mm long. Monterey harbor.
(R. Ames)

14.42 Onchidoris hystricina (Bergh). White
Onchidoris. 11 mm long. Point Loma (San Diego
Co.). (J. Lance)

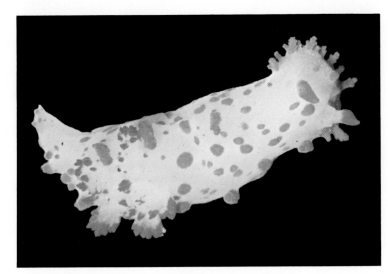

14.43 Triopha catalinae (Cooper). Sea-Clown Nudibranch. About 40
mm long. Monterey Peninsula. (W. Lee)

14.44a (*see also overleaf*) **Triopha maculata** MacFarland. Spotted Triopha.
40 mm long. Pacific Grove (Monterey Co.).

14.44b **Triopha maculata** MacFarland. 76 mm long, sitting on kelp blade covered with encrusting bryozoans. La Jolla (San Diego Co.). (W. Farmer)

14.45 **Crimora coneja** Marcus. Rabbit Nudibranch. 19 mm long. Point Loma (San Diego Co.). (J. Lance)

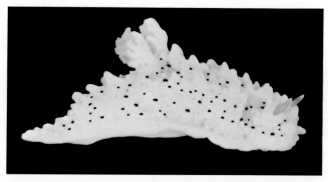

14.46 **Aegires albopunctatus** MacFarland. Salt-and-Pepper Nudibranch. 15 mm long. Pacific Grove (Monterey Co.).

14.47 **Polycera atra** MacFarland. Sorcerer's Nudibranch. 14 mm long. Mission Bay (San Diego Co.). (W. Farmer)

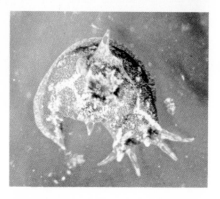

14.48 **Polycera hedgpethi** Marcus. Hedgpeth's Polycera. 25 mm long. San Francisco Bay. (W. Lee)

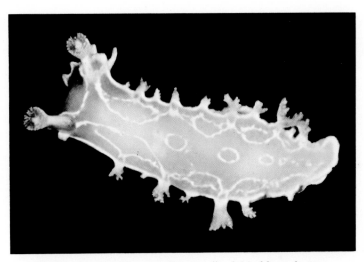

14.50 **Tritonia festiva** (Stearns). Diamondback Nudibranch. 30 mm long. Salt Point (Sonoma Co.). (R. Ames)

14.49 **Laila cockerelli** MacFarland. Cockerell's Nudibranch. 28 mm long. Monterey. (R. Ames)

14.52 **Dendronotus albus** MacFarland. White Dendronotus. 30 mm long. Salt Point (Sonoma Co.). (R. Ames)

14.51 **Dendronotus subramosus** MacFarland. Stubby Dendronotus. 22 mm long. Pacific Grove (Monterey Co.).

14.53 **Dendronotus iris** Cooper. Giant Dendronotus. 9 cm long, with white egg strings. Salt Point (Sonoma Co.). (R. Ames)

14.55 **Doto amyra** Marcus. Hammerhead Doto. 10 mm long. Mission Bay (San Diego Co.). (W. Farmer)

14.54 **Melibe leonina** (Gould). Melibe. About 80 mm long. Mission Bay (San Diego Co.). (W. Farmer)

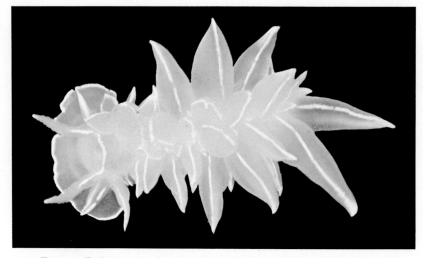

14.56 **Armina californica** (Cooper). Striped Armina. 50 mm long. Southern California. (P. Schroeder)

14.57 **Dirona albolineata** Cockerell & Eliot. Chalk-Lined Dirona. 35 mm long. Pacific Grove (Monterey Co.).

14.58 **Dirona picta** MacFarland. Spotted Dirona. 35 mm long. Pacific Grove (Monterey Co.).

14.59 **Antiopella barbarensis** (Cooper). Cockscomb Nudibranch. 30 mm long. 17-Mile Drive (Monterey Co.).

14.60 **Coryphella trilineata** O'Donoghue. Three-Lined Nudibranch. 23 mm long. Dillon Beach (Marin Co.). (W. Lee)

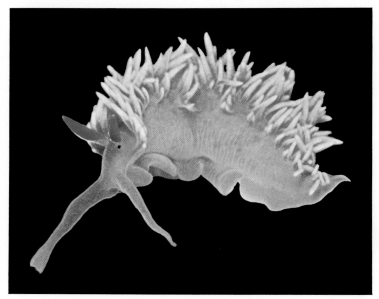

14.61 **Flabellinopsis iodinea** (Cooper). Elegant Eolid. 34 mm long. Santa Cruz Island (Channel Islands). (R. Ames)

14.62 **Cuthona lagunae** (O'Donoghue). Orange-Faced Nudibranch. 10 mm long. Rosarito (Baja California). (W. Farmer)

14.64 **Fiona pinnata** (Eschscholtz). Fiona. 30 mm long. Mission Bay (San Diego Co.). (W. Farmer)

14.63 **Cuthona divae** (Marcus). Rose-Pink Cuthona. 22 mm long. Moss Beach (San Mateo Co.). (J. Lance)

14.66 **Hermissenda crassicornis** (Eschscholtz). Opalescent Nudibranch. 45 mm long. Monterey Peninsula. (W. Lee)

14.65 **Phidiana pugnax** Lance. Fighting Phidiana. 34 mm long. Pacific Grove (Monterey Co.). (R. Ames)

14.67 **Aeolidia papillosa** (Linnaeus). Shag-Rug Nudibranch. 38 mm long. Monterey. (R. Ames)

14.68 **Spurilla chromosoma** Cockerell & Eliot. Frosted Spurilla. 28 mm long. La Jolla (San Diego Co.). (J. Lance)

14.69 **Spurilla oliviae** (MacFarland). Olive's Spurilla. 31 mm long. La Jolla (San Diego Co.). (W. Harvey/J. Lance)

14.70 **Trimusculus reticulatus** (Sowerby). Reticulate Button Snail. 16 mm and 10 mm diam. Smaller shell bearing tubes of spirorbid polychaetes (18.44). Pacific Grove (Monterey Co.).

14.71 **Melampus olivaceus** Carpenter. Olive Ear Snail. 13 mm long. Newport Bay (Orange Co.).

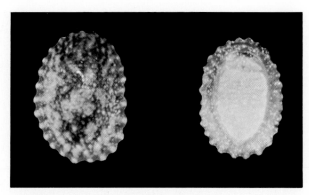

15.1 **Barbatia bailyi** (Bartsch). Baily's Ark. Right and left valves, 7 mm long; anterior ends toward center. Dana Point (Orange Co.).

14.72 **Onchidella borealis** Dall. Leather Limpet. 12 mm long; dorsal and ventral views. Pacific Grove (Monterey Co.).

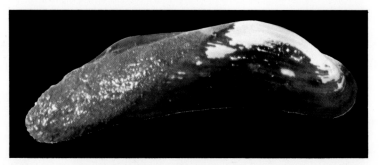

15.3 **Adula californiensis** (Philippi). California Pea-Pod Borer. Right valve, 36 mm long. Moss Beach (San Mateo Co.).

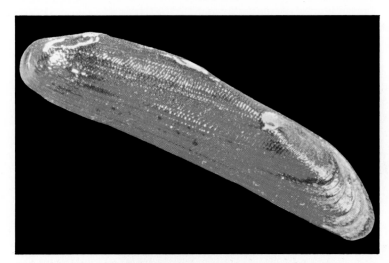

15.2 **Adula falcata** (Gould). Rough Pea-Pod Borer. Left valve, 60 mm long. Moss Beach (San Mateo Co.).

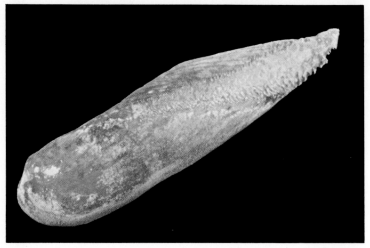

15.4 **Lithophaga plumula** (Hanley). Date Mussel. Left valve, 55 mm long. Newport Bay (Orange Co.).

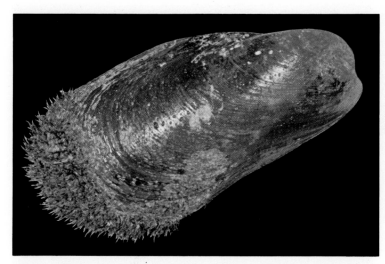

15.5 Modiolus capax (Conrad). Fat Horse Mussel. Right valve, 90 mm long. Newport Bay (Orange Co.).

15.6 Modiolus rectus (Conrad). Straight Horse Mussel. Left valve, 88 mm long. Tijuana Slough (San Diego Co.).

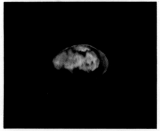

15.8 Musculus pygmaeus Glynn. Left valve, 3 mm long. Pacific Grove (Monterey Co.).

15.7 Ischadium demissum (Dillwyn). Ribbed Horse Mussel. Right valve, 92 mm long. San Francisco Bay.

15.9 Mytilus californianus Conrad. California Mussel, California Sea Mussel. Left valve, 100 mm long. Point Buchon (San Luis Obispo Co.).

15.10 **Mytilus edulis** Linnaeus. Bay Mussel. Left valve, 78 mm long. Bodega Bay harbor (Sonoma Co.).

15.11 **Septifer bifurcatus** (Conrad). Branch-Ribbed Mussel. Right valve, 38 mm long. Piedras Blancas Point (San Luis Obispo Co.).

15.12 **Ostrea lurida** Carpenter. Native Oyster, Olympia Oyster. Right valve, 35 mm long. San Diego Bay.

15.13 **Crassostrea gigas** (Thunberg). Pacific Oyster, Japanese Oyster. Right valve, 150 mm long. Tomales Bay (Marin Co.).

15.14 **Hinnites giganteus** (Gray). Rock Scallop. Right and left valves, 45 mm diam; left valve darker. Pacific Grove (Monterey Co.).

15.15 Argopecten aequisulcatus (Carpenter).
Speckled Scallop. Right valve, 65 mm diam.
Agua Hedionda Lagoon (San Diego Co.).

15.16 Leptopecten latiauratus (Conrad).
Broad-Eared Pecten. Left valve, 20 mm diam.
Anaheim Bay (Orange Co.).

15.17a Lima hemphilli Hertlein & Strong. File Shell.
Whole animal with expanded marginal tentacles;
shell 23 mm long. Pacific Grove (Monterey Co.).

15.17b Lima hemphilli Hertlein &
Strong. Left valve, 23 mm long.
Pacific Grove (Monterey Co.).

15.18a Pododesmus cepio (Gray). Abalone Jingle. Adult right
and left valves; left valve darker, 50 mm diam. Bolinas
Lagoon (Marin Co.).

15.18b Pododesmus cepio (Gray). Juvenile left
valve, 20 mm diam. Agua Hedionda Lagoon
(San Diego Co.).

15.19 Chama arcana Bernard. Left and right valves; right valve larger, 35 mm diam. Dana Point (Orange Co.).

15.20 Pseudochama exogyra (Conrad). Reversed Chama. Left valve, 36 mm diam. La Jolla (San Diego Co.).

15.21 Lasaea cistula Keen. Right valves, 2 mm long. Goleta Point (Santa Barbara Co.).

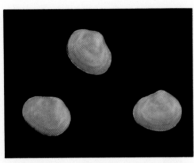

15.22 Lasaea subviridis Dall. 3.5 mm long. Upper: right valve; lower: left valves. San Diego.

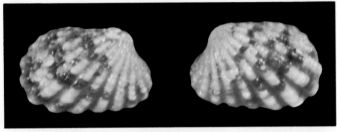

15.23 Glans carpenteri (Lamy). Little Heart Clam. Right and left valves, 10 mm long; anterior ends toward center. Monterey.

15.24 Kellia laperousii (Deshayes). Smooth Kelly Clam. Left valve, 20 mm long. Moss Beach (San Mateo Co.).

15.25 Epilucina californica (Conrad). California Lucine. Left valve, 35 mm diam. Malpaso Creek (Monterey Co.).

15.26 **Diplodonta orbella** (Gould). Round Diplodonta. Left valve, in "nest," 17 mm diam. Santa Barbara.

15.27 **Trachycardium quadragenarium** (Conrad). Spiny Cockle. Right valve, 60 mm diam. Newport Bay (Orange Co.).

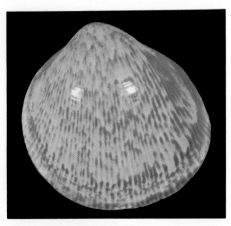

15.28 **Laevicardium substriatum** (Conrad). Egg Cockle. Left valve, 17 mm diam. San Diego Bay.

15.29 **Clinocardium nuttallii** (Conrad). Basket Cockle, Heart Cockle. Left valve; shell 60 mm diam. Bolinas Lagoon (Marin Co.).

15.30 **Gemma gemma** (Totten). Gem Clam. Left valve, 4 mm long. Bolinas Lagoon (Marin Co.).

15.31 **Mercenaria mercenaria** (Linnaeus). Quahog. Left valve, 58 mm long. Humboldt Bay.

15.32 **Tivela stultorum** (Mawe). Pismo Clam. Right valve, 100 mm long. Elkhorn Beach (Monterey Co.).

15.33 **Saxidomus giganteus** (Deshayes). Butter Clam, Smooth Washington Clam. Right valve, 85 mm long. Humboldt Bay.

15.34 **Saxidomus nuttalli** Conrad. Washington Clam. Left valve, 110 mm long. Elkhorn Slough (Monterey Co.).

15.35 **Irus lamellifer** (Conrad). Rock Venus. Right valve, 50 mm long. Monterey. (E. Haderlie)

15.36 **Tapes japonica** Deshayes. Japanese Littleneck Clam. Right valve, 47 mm long. San Francisco Bay.

15.37 **Chione californiensis** (Broderip). California Chione. Right valve, 52 mm long. Mission Bay (San Diego Co.).

15.38 **Chione fluctifraga** (Sowerby). Smooth Chione. Right valve, 34 mm long. Mission Bay (San Diego Co.).

15.39 **Chione undatella** (Sowerby). Wavy Chione. Left valve, 60 mm long. Tijuana Slough (San Diego Co.).

15.40 Protothaca laciniata (Carpenter). Rough-Sided Littleneck Clam. Right valve, 58 mm long. Agua Hedionda Lagoon (San Diego Co.).

15.41 Protothaca staminea (Conrad). Common Littleneck Clam. Right valve, 62 mm long. Humboldt Bay.

15.42 Protothaca tenerrima (Carpenter). Thin-Shelled Littleneck Clam. Left valve; shell 85 mm long. Elkhorn Slough (Monterey Co.).

15.43 Petricola pholadiformis Lamarck. False Angel Wing. Left valve, 50 mm long. San Francisco Bay.

15.44 **Petricola carditoides** (Conrad). Left valves (exterior view) and right valves (interior view); top, 30 mm long; bottom: 32 mm long. Agua Hedionda Lagoon (San Diego Co.).

15.45 **Petricola californiensis** Pilsbry & Lowe. Left valve; shell 26 mm long. San Diego Bay.

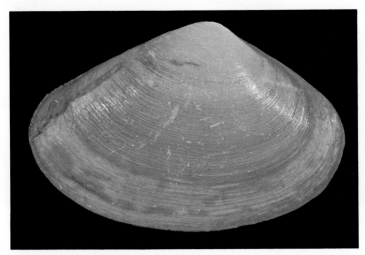

15.46 **Mactra californica** Conrad. California Mactra. Right valve, 35 mm long. San Diego Bay.

15.47 **Tresus capax** (Gould). Fat Gaper, Horse Clam. Left valve, 96 mm long. Humboldt Bay.

15.48 Tresus nuttallii (Conrad). Gaper. Right valve; shell 150 mm long. Morro Bay (San Luis Obispo Co.).

15.49 Tellina bodegensis Hinds. Bodega Tellen. Left valve, 55 mm long. Elkhorn Beach (Monterey Co.).

15.50 Macoma nasuta (Conrad). Bent-Nosed Clam. Left valve; shell 50 mm long. Tomales Bay (Marin Co.).

15.51 Macoma secta (Conrad). White Sand Clam. Right valve, 65 mm long. Bodega Bay harbor (Sonoma Co.).

15.52 Macoma balthica (Linnaeus). Baltic Macoma. Right valve, 23 mm long. San Francisco Bay.

15.53 Leporimetis obesa (Deshayes). Yellow Metis. Right valve, 80 mm long. Newport Bay (Orange Co.).

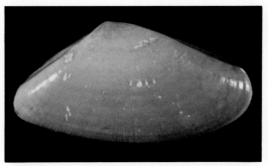

15.54 **Donax gouldii** Dall. Bean Clam. 18–25 mm long. Upper right: left valve; lower left and right: right valves. Corona del Mar (Orange Co.).

15.55 **Donax californicus** (Conrad). Wedge Clam. Right valve, 24 mm long. San Pedro (Los Angeles Co.).

15.56 **Gari californica** (Conrad). Sunset Clam. Right valve, 85 mm long. Pacific Grove (Monterey Co.).

15.58 **Tagelus californianus** (Conrad). California Jackknife Clam. Left valve, 60 mm long. Newport Bay (Orange Co.).

15.57 **Nuttallia nuttallii** (Conrad). Purple Clam. Right valve, 90 mm long. Tijuana Slough (San Diego Co.).

15.59 **Tagelus subteres** (Conrad). Right valve, 45 mm long. Mission Bay (San Diego Co.).

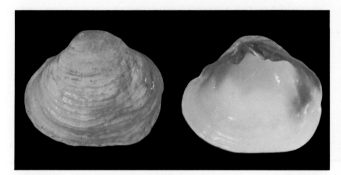

15.60 Semele rupicola Dall. Semele-of-the-Rocks. Left valve (exterior view) and right valve (interior view), 35 mm long. Dana Point (Orange Co.).

15.61 Semele decisa (Conrad). Clipped Semele. Left valve, 60 mm long. Santa Barbara.

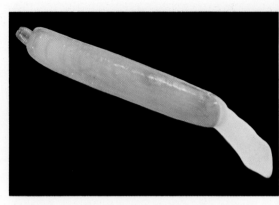

15.62 Solen rosaceus Carpenter. Rosy Razor Clam. Right valve; shell 52 mm long. Newport Bay (Orange Co.).

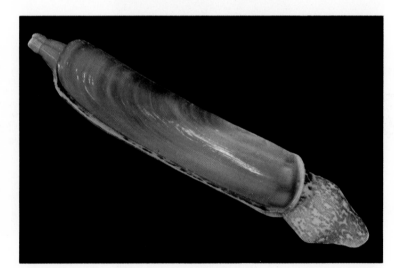

15.63 Solen sicarius Gould. Sickle Razor Clam. Right valve; shell 73 mm long. Tomales Bay (Marin Co.).

15.64 Siliqua lucida (Conrad). Bright Razor Clam. 22 mm long. Upper: left valve; lower: right valve. Tomales Bay (Marin Co.).

15.66 **Cryptomya californica** (Conrad). False Mya. Left valve, 25 mm long. Bodega Bay harbor (Sonoma Co.).

15.67 **Mya arenaria** Linnaeus. Soft-Shelled Clam. Left valve; shell 80 mm long. San Francisco Bay.

15.65 **Siliqua patula** (Dixon). Northern Razor Clam. Right valve; shell 117 mm long. Crescent City (Del Norte Co.).

15.68 **Platyodon cancellatus** (Conrad). Checked Borer. Left valve; shell 55 mm long. Duxbury Point (Marin Co.).

15.70 **Panopea generosa** (Gould). Geoduck. Left valve; shell 135 mm long. Humboldt Bay.

15.69 **Hiatella arctica** (Linnaeus). Little Gaper. Left valve; shell 40 mm long. Point Sal (Santa Barbara Co.).

15.71 **Barnea subtruncata** (Sowerby). Mud Piddock. Right valve, 63 mm long. Anaheim Landing (Orange Co.).

15.73 **Netastoma rostrata** (Valenciennes). Beaked Piddock. Right valve; 18 mm long overall. Monterey.

15.72 **Zirfaea pilsbryi** Lowe. Rough Piddock. Right valve; shell 80 mm long. Agua Hedionda Lagoon (San Diego Co.).

15.75 **Parapholas californica** (Conrad). Scale-Sided Piddock. Left valve, 105 mm long. Point Fermin (San Pedro, Los Angeles Co.).

15.74 **Chaceia ovoidea** (Gould). Wart-Necked Piddock. Right valve; shell 96 mm long. Agua Hedionda Lagoon (San Diego Co.).

15.76a **Penitella conradi** Valenciennes. Little Piddock, Abalone Piddock. 30 mm long, in chert bore; dorsal view, anterior end pointing left. Monterey. (E. Haderlie)

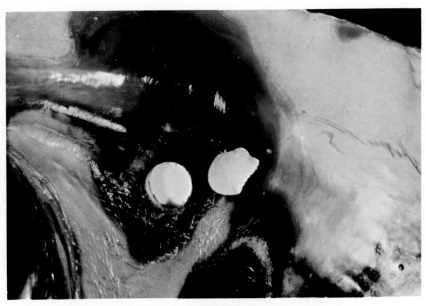

15.76b **Penitella conradi** Valenciennes. White spot, center right: right valve, removed from abalone shell, 6 mm long; white spot, center left: hole bored through abalone shell. Dana Point (Orange Co.).

15.77 **Penitella gabbii** (Tryon). Left valve, in rock bore; shell 35 mm long. Moss Beach (San Mateo Co.).

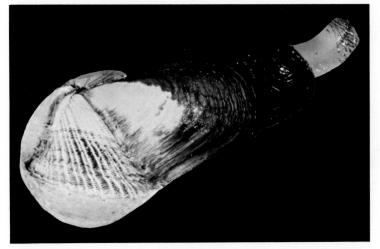

15.78 **Penitella penita** (Conrad). Flap-Tipped Piddock. Left valve; shell 45 mm long. Point Buchon (San Luis Obispo Co.).

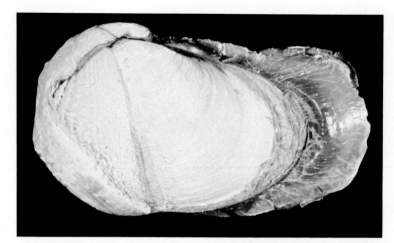

15.79 **Penitella fitchi** Turner. Left valve, 50 mm long. Monterey. (E. Haderlie)

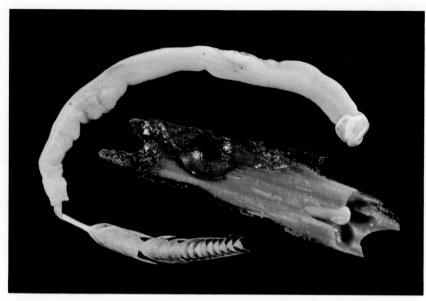

15.80 **Bankia setacea** (Tryon). Shipworm. Animal, removed from burrow, 160 mm long overall, the shell 7 mm diam; wood fragment showing burrows. Humboldt Bay.

15.81 **Entodesma saxicola** (Baird). Rock Entodesma. Right valve, 42 mm long. Point Arena (Mendocino Co.).

15.82 **Lyonsia californica** Conrad. Left valve, 28 mm long. Morro Bay (San Luis Obispo Co.).

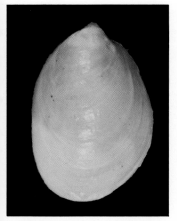

15.83a **Mytilimeria nuttallii** Conrad. Sea-Bottle Clam. Animal removed from burrow; left valve, 25 mm long. Pacific Grove (Monterey Co.).

15.83b **Mytilimeria nuttallii** Conrad. Slitlike apertures of burrow in colonial tunicate *Archidistoma psammion* (12.16). Pacific Grove (Monterey Co.).

16.1 **Leptochiton rugatus** (Pilsbry). 1 cm long. Point Pinos (Pacific Grove, Monterey Co.).

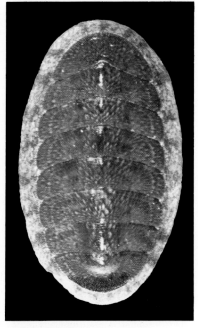

16.2 **Ischnochiton interstinctus** (Gould). 2.5 cm long. Piedras Blancas Point (San Luis Obispo Co.).

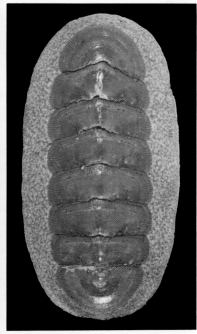

16.3 **Ischnochiton regularis** (Carpenter). 3 cm long. Malpaso Creek (Monterey Co.).

16.4 **Lepidozona mertensii** (Middendorff). 3 cm long. Devils Slide (San Mateo Co.).

16.5 **Lepidozona cooperi** (Pilsbry). 3.2 cm long. Trinidad Head (Humboldt Co.).

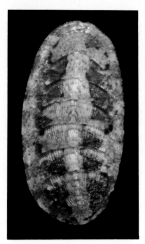

16.6 **Lepidozona sinudentata** (Carpenter). 2 cm long. Pacific Grove (Monterey Co.).

16.7 Lepidozona californiensis (Berry). 2.5 cm long. Dana Point (Orange Co.).

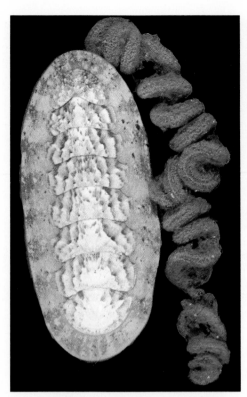

16.8 Stenoplax heathiana Berry. 8 cm long, with coiled egg strings. Pigeon Point (San Mateo Co.).

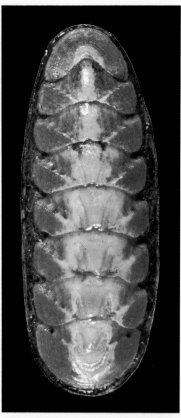

16.9 Stenoplax fallax (Pilsbry). 9 cm long. Pacific Grove (Monterey Co.).

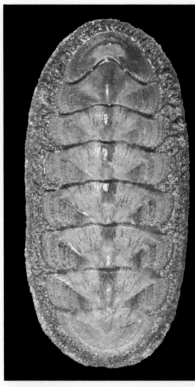

16.10 Stenoplax conspicua (Pilsbry). 9 c long. La Jolla (San Diego Co.).

16.11 Tonicella lineata (Wood). Lined Chiton. 3 cm long; anterior end on right. Point Buchon (San Luis Obispo Co.).

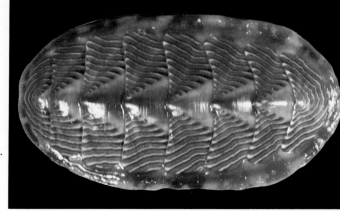

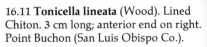

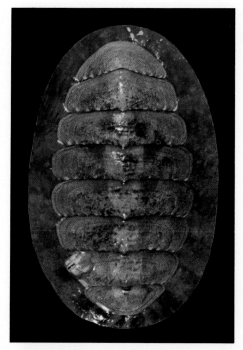

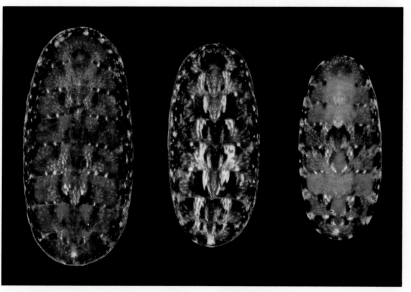

16.14 **Basiliochiton heathii**
(Pilsbry). 1.2 cm long. Pacific
Grove (Monterey Co.).

16.13 **Cyanoplax dentiens** (Gould). 1–1.3 cm long; color variants.
Moss Beach (San Mateo Co.).

16.12 **Cyanoplax hartwegii** (Carpenter). 2.8
cm long. Point Fermin (San Pedro, Los
Angeles Co.).

16.15a **Callistochiton crassicostatus** Pilsbry. 2 cm long; anterior end
on right. Point Pinos (Pacific Grove, Monterey Co.).

16.15b **Callistochiton crassicostatus** Pilsbry. 2.7 cm long. 17-
Mile Drive (Monterey Co.).

16.16 Callistochiton palmulatus Carpenter. 1.2 cm long. Pacific Grove (Monterey Co.).

16.17 Nuttallina californica (Reeve). 4 cm long. McClures Beach (Marin Co.).

16.18 Chaetopleura gemma Dall. 1.6 cm long. Pacific Grove (Monterey Co.).

16.19 Mopalia muscosa (Gould). Mossy Chiton. 5.5 cm long. Malpaso Creek (Monterey Co.).

16.20 Mopalia ciliata (Sowerby). 4.5 cm long; anterior end on right. Malpaso Creek (Monterey Co.).

16.23 Placiphorella velata Dall. Veiled Chiton.
3.3 cm long. Malpaso Creek (Monterey Co.).

16.21 Mopalia hindsii (Reeve). 4.3 cm long.
Point Saint George (Del Norte Co.).

16.22 Mopalia lignosa (Gould). 4 cm long.
Bruhel Point (Mendocino Co.).

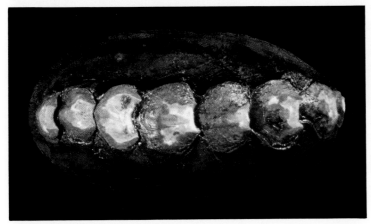

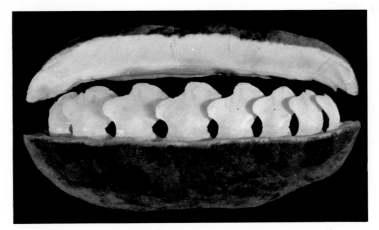

16.24 Katharina tunicata (Wood). 8 cm long; anterior end on right.
Point Arena (Mendocino Co.).

16.25 Cryptochiton stelleri (Middendorff). Gumboot Chiton. 30 cm
long; dissected specimen showing internal valves, anterior end on
right. Cabrillo Point (Mendocino Co.).

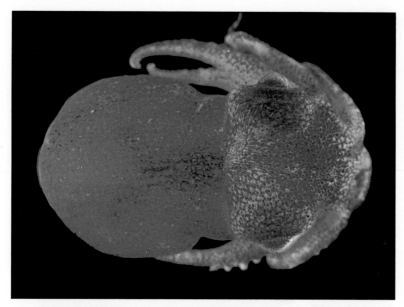

17.1a Rossia pacifica Berry. Stubby Squid, Short Squid. Common color phase, showing expanded red chromatophores. Victoria (British Columbia). (V. Gauley)

17.1b Rossia pacifica Berry. Chromatophores contracted except in mid-dorsal spot. Victoria (British Columbia). (T. Gore)

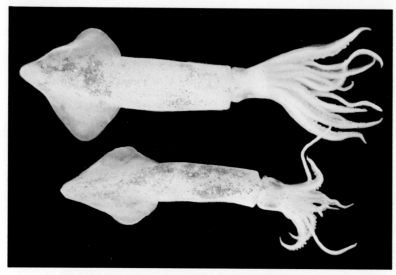

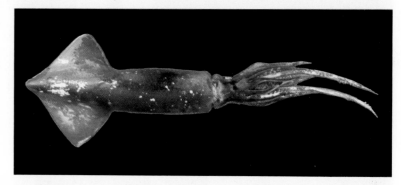

17.3 Dosidicus gigas (d'Orbigny). Jumbo Squid, Humboldt Squid. Dorsal mantle length 30 cm. Santa Barbara. (F. Hochberg)

17.2 Loligo opalescens Berry. Market Squid, Common Squid, Opalescent Squid, Sea Arrow, Calamary, Calamari. Large: male; small: female. Monterey Bay. (D. Smith)

17.4 Moroteuthis robusta (Verrill). North Pacific Giant Squid. Dorsal mantle length 75 cm. Santa Barbara Channel. (F. Hochberg)

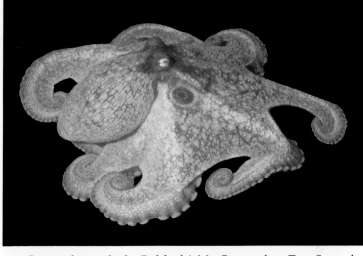

17.5 Octopus bimaculoides Pickford & MacConnaughey. Two-Spotted Octopus, Mud-Flat Octopus. Dorsal mantle length 12 cm. Santa Cruz Island (Channel Islands). (R. Evans)

17.6 Octopus rubescens Berry. Dorsal mantle length 7.5 cm. Santa Monica Bay (Los Angeles Co.). (R. Evans)

17.8 Octopus micropyrsus Berry. Dorsal mantle length 2 cm. Color altered by preservation. Santa Catalina Island (Channel Islands).

17.7 Octopus dofleini (Wülker). North Pacific Giant Octopus, Giant Pacific Octopus. Northern California. (K. Lucas)

18.1 **Arctonoe fragilis** (Baird). Scale Worm. 37 mm long, on sea star *Evasterias troschelii* (8.12). Point Saint George (Del Norte Co.).

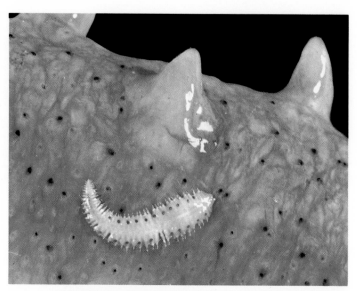

18.2 **Arctonoe pulchra** (Johnson). Scale Worm. 30 mm long, on sea cucumber *Parastichopus.* Corona Del Mar (Orange Co.).

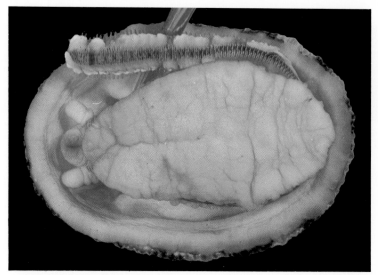

18.3 **Arctonoe vittata** (Grube). Scale Worm. 90 mm long, in pallial groove of keyhole limpet *Diodora aspera* (13.10). Malpaso Creek (Monterey Co.).

18.4 **Halosydna brevisetosa** Kinberg. Scale Worm. 55 mm long. Pacific Grove (Monterey Co.).

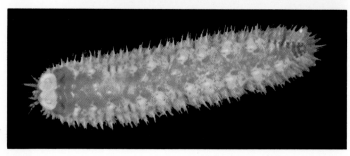

18.5 **Harmothoe imbricata** (Linnaeus). Scale Worm. 28 mm long. Pacific Grove (Monterey Co.).

18.6 **Hesperonoe adventor** (Skogsberg). Scale Worm. 45 mm long. Bodega Bay harbor (Sonoma Co.).

18.7 **Lepidasthenia longicirrata** Berkeley. Scale Worm. 80 mm long. Westport (Mendocino Co.).

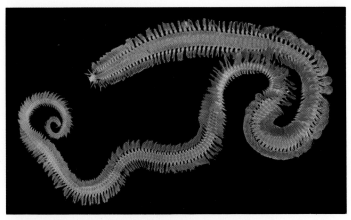

18.8 **Anaitides medipapillata** (Moore). 120 mm long. Dana Point (Orange Co.).

18.9 **Ophiodromus pugettensis** (Johnson). 33 mm long, in ambulacral groove of sea star. Agua Hedionda Lagoon (San Diego Co.).

18.10 **Nereis grubei** (Kinberg). 80 mm long. 17-Mile Drive (Monterey Co.).

18.11 **Neanthes brandti** (Malmgren). 40 cm long. Humboldt Bay.

18.12 **Platynereis bicanaliculata** (Baird). Male in spawning condition, 45 mm long. Agua Hedionda Lagoon (San Diego Co.).

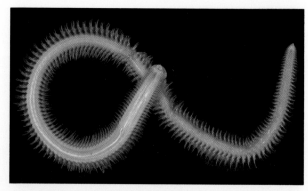

18.14 **Nephtys californiensis** Hartman. 140 mm long. Elkhorn Beach (Monterey Co.).

18.13 **Glycera americana** Leidy. 220 mm long, with proboscis fully extended, showing terminal teeth. Monterey.

18.15 **Pareurythoe californica** (Johnson). 110 mm long. Tijuana Slough (San Diego Co.).

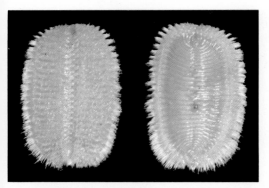

18.16 Euphrosine hortensis Moore. 30 mm long; dorsal and ventral views. Agua Hedionda Lagoon (San Diego Co.).

18.17 Diopatra splendidissima Kinberg. 150 mm long, shown emerging from parchment tube. Agua Hedionda Lagoon (San Diego Co.).

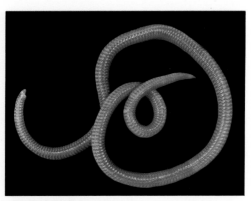

18.18 Arabella iricolor (Montagu). 170 mm long. Humboldt Bay.

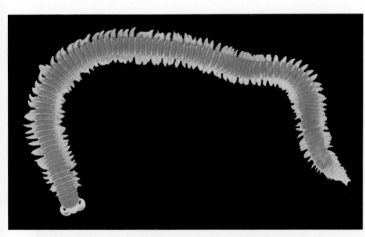

18.19 Dorvillea moniloceras (Moore). Candy-Striped Worm. 65 mm long. Monterey harbor.

18.20a Chaetopterus variopedatus (Renier). Parchment-Tube Worm. Worm, removed from tube, 100 mm long. Newport Bay (Orange Co.).

18.20b Chaetopterus variopedatus (Renier). Parchment tube, 280 mm long. Newport Bay (Orange Co.).

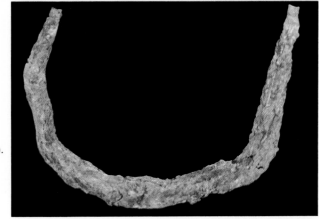

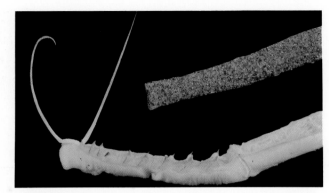

18.21 **Mesochaetopterus taylori** Potts. Worm 300 mm long, with tube. Crescent City harbor (Del Norte Co.).

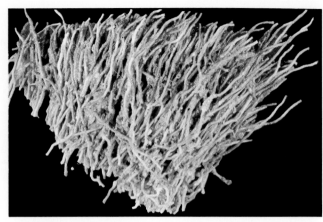

18.22 **Phyllochaetopterus prolifica** Potts. Cluster of tubes, each 60–120 mm long. Monterey harbor.

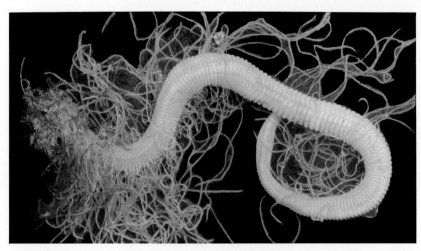

18.23 **Cirriformia luxuriosa** (Moore). 120 mm long. Piedras Blancas Point (San Luis Obispo Co.).

18.24 **Cirriformia spirabrancha** (Moore). About 60 mm long. Pacific Grove (Monterey Co.). (C. Patton)

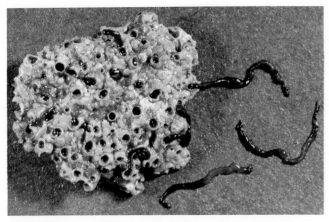

18.25 **Dodecaceria fewkesi** Berkeley & Berkeley. Calcareous tube cluster; worms 40 mm long. Cannery Row (Monterey Co.).

18.26 **Capitella capitata** (Fabricius). Worm 35 mm long, on pea crab *Pinnixa faba* (25.35). Humboldt Bay.

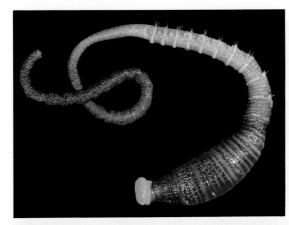

18.27 **Arenicola brasiliensis** Nonato. Lugworm. 145 mm long. San Francisco Bay.

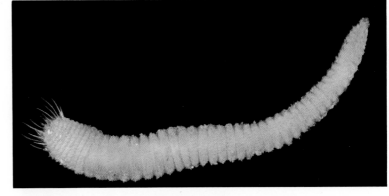

18.29 **Pherusa papillata** (Johnson). 70 mm long. Monterey.

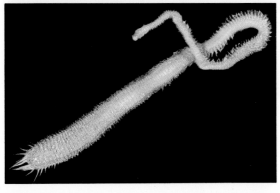

18.28 **Pherusa capulata** (Moore). 95 mm long. Newport Bay (Orange Co.).

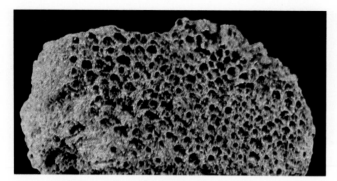

18.30a **Phragmatopoma californica** (Fewkes). Tube cluster 200 mm across. Pfeiffer Point (Monterey Co.).

18.30b **Phragmatopoma californica** (Fewkes). Expanded tentacles. Monterey Bay. (D. Wobber)

18.30c **Phragmatopoma californica** (Fewkes). Worm in tube. Pacific Grove (Monterey Co.). (P. Roy)

18.31a **Sabellaria cementarium** Moore. Honeycomb Worm. Sand tube. Cannery Row (Monterey Co.).

18.31b **Sabellaria cementarium** Moore. Worm, removed from tube, 55 mm long.

18.33 Pista elongata Moore. Tube, showing large terminal hood, 80 mm long. Cape Mendocino (Humboldt Co.).

18.32 Pectinaria californiensis Hartman. Worm 55 mm long, with conical sand tube. Tomales Bay (Marin Co.).

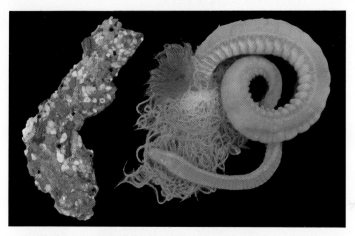

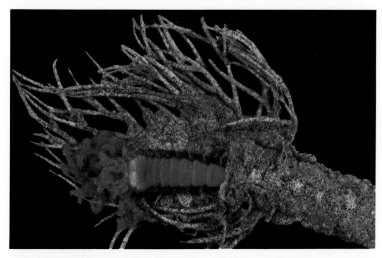

18.35 Thelepus crispus Johnson. Worm 115 mm long, with tube. Bruhel Point (Mendocino Co.).

18.34 Pista pacifica Berkeley & Berkeley. Worm 300 mm long, in tube. Bodega Bay harbor (Sonoma Co.).

18.36 Thelepus setosus (Quatrefages). Worm 120 mm long, with tube. Bruhel Point (Mendocino Co.).

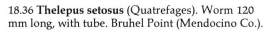

18.37 **Eudistylia polymorpha** (Johnson). Plume Worm, Feather-Duster Worm. Worms removed from tube. Top: 100 mm long; bottom: 170 mm long. Pacific Grove (Monterey Co.).

18.38 **Sabella crassicornis** Sars. Plume Worm, Feather-Duster Worm. Worm, removed from tube, 45 mm long. 17-Mile Drive (Monterey Co.).

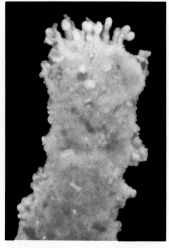

18.39a **Potamilla occelata** (Moore). Feather-Duster Worm. Worms in tubes. Pacific Grove (Monterey Co.). (N. Burnett)

18.39b **Potamilla occelata** (Moore). Tube with colony of symbiotic hydroid *Proboscidactyla circumsabella.* Monterey harbor. (N. Burnett)

18.40 **Myxicola infundibulum** (Renier). About 50 mm long. Pacific Grove (Monterey Co.). (C. Patton)

18.41 **Mercierella enigmatica** Fauvel. Tubes 2–3 mm diam. Berkeley Yacht Harbor (Alameda Co.). (E. Haderlie)

18.42a **Serpula vermicularis** Linnaeus. Worm in tube. Point Arena (Mendocino Co.). (W. Lee)

18.42b **Serpula vermicularis** Linnaeus. Tube, 50 mm across coil. Point Arena (Mendocino Co.).

18.43a **Salmacina tribranchiata** (Moore). Animals extended from tubes. Monterey Bay. (D. Wobber)

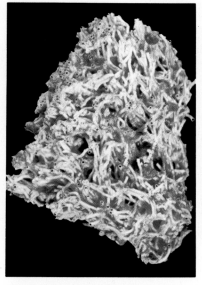

18.43b **Salmacina tribranchiata** (Moore). Tube cluster 35 mm tall. Pacific Grove (Monterey Co.).

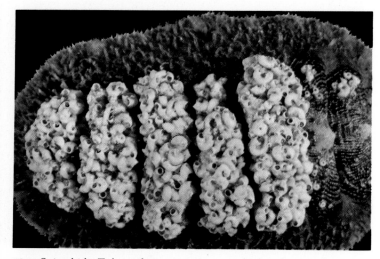

18.44 Spirorbids. Tube coils 3 mm across, attached to chiton valves. Pacific Grove (Monterey Co.).

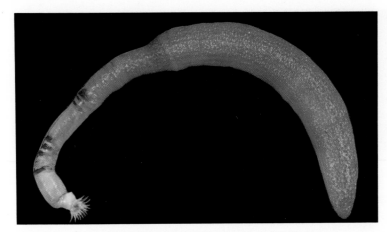

19.1 **Phascolosoma agassizii** Keferstein. 8 cm long. Pacific Grove (Monterey Co.).

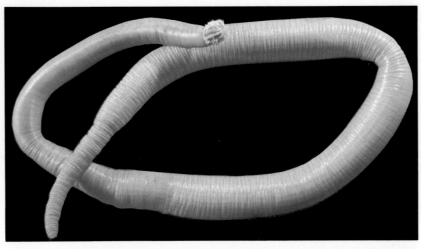

19.2 **Siphonosoma ingens** (Fisher). 38 cm long. Bodega Bay harbor (Sonoma Co.).

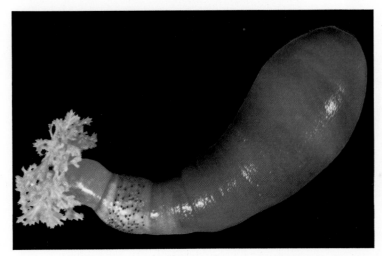

19.3 **Themiste pyroides** (Chamberlin). 9 cm long. 17-Mile Drive (Monterey Co.).

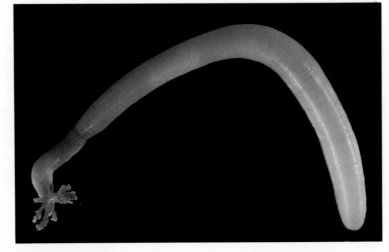

19.4 **Themiste zostericola** (Chamberlin). 12 cm long. Bodega Bay harbor (Sonoma Co.).

19.5 **Themiste dyscrita** (Fisher). 8 cm long. Color faded by preservation. Moss Beach (San Mateo Co.).

19.6 **Golfingia margaritacea californiensis** Fisher. 22 mm long. Pacific Grove (Monterey Co.).

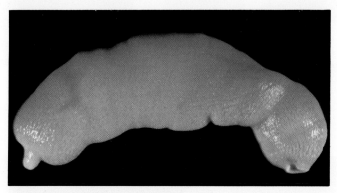

19.7a **Urechis caupo** Fisher & MacGinitie. Innkeeper Worm Fat Innkeeper. 20 cm long. Elkhorn Slough (Monterey Co.).

19.7b Commensal goby *Clevelandia*. Elkhorn Slough (Monterey Co.).

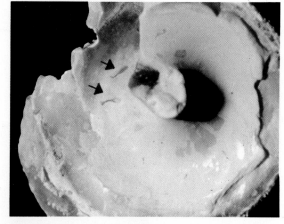

20.1a **Trypetesa lateralis** Tomlinson. Burrowed in interior of shell of prosobranch *Tegula funebralis* (13.32) inhabited by hermit crab; aperture of burrow about 2 mm long. Moss Beach (San Mateo Co.). (W. Newman)

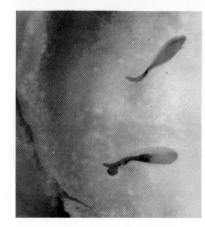

20.1b Close-up of burrow apertures.

20.2 Oxynaspis rossi Newman. 2.5 cm long. Santa Catalina Island (Channel Islands). (W. Newman)

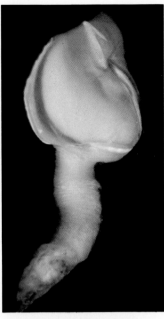

20.3 Octolasmis californiana Newman. 7 mm long. La Jolla (San Diego Co.). (W. Newman)

20.4 Lepas (Lepas) anatifera Linnaeus. Goose Barnacle. Cluster on floating algal stipe; largest barnacle 5 cm long. Piedras Blancas Point (San Luis Obispo Co.).

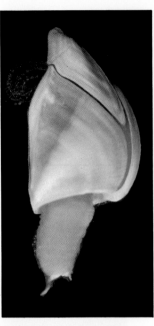

20.5 Lepas (Lepas) pacifica Henry. 22 mm long. Stranded on Scripps Beach (San Diego Co.). (W. Newman)

20.6 Lepas (Dosima) fascicularis Ellis & Solander. 4 cm long. Stranded near Crescent City (Del Norte Co.).

20.7 Conchoderma virgatum (Spengler). 4 cm long. Offshore from La Jolla (San Diego Co.). (W. Newman)

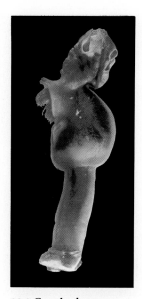

20.8 Conchoderma auritum (Linnaeus). Rabbit-Eared Barnacle. 7 cm long, attached to sessile barnacle *Coronula*. On whale stranded at San Francisco. (W. Newman)

20.9 Alepas pacifica Pilsbry. 2 cm long. Monterey Bay. (W. Newman)

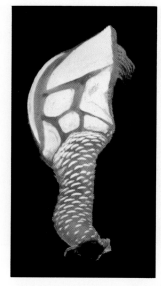

20.10 Arcoscalpellum californicum (Pilsbry). 2.4 cm long. Sumner Canyon (La Jolla, San Diego Co.). (W. Newman)

20.11 Pollicipes polymerus Sowerby. Leaf Barnacle. 6 cm long. Pacific Grove (Monterey Co.).

20.13 (*right*) **Chelonibia testudinaria** (Linnaeus). 45 mm diam. On turtle stranded at Del Mar (San Diego Co.). (W. Newman)

20.12 (*left*) **Chthamalus fissus** Darwin. 8 mm diam. San Diego Bay. (W. Newman)

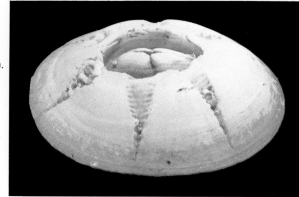

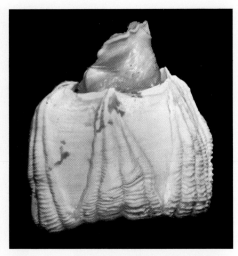

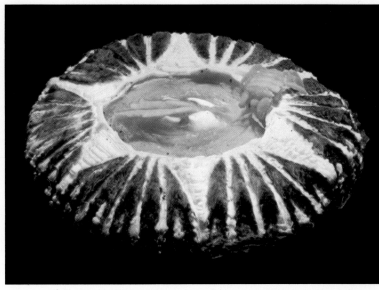

20.14 Platylepas hexastylos (Fabricius). 18 mm diam. On turtle stranded at Del Mar (San Diego Co.). (W. Newman)

20.15 Coronula diadema (Linnaeus). 50 mm diam. On stranded whale at San Francisco. (W. Newman)

20.16 Cryptolepas rachianecti Dall. 50 mm diam; whale louse (amphipod) attached. On whale stranded at Del Mar (San Diego Co.). (W. Newman)

20.17a Xenobalanus globicipitis Steenstrup. 40 mm long, growing on fin tip of pilot whale. Stranded at Del Mar (San Diego Co.). (W. Newman)

20.17b Xenobalanus globicipitis Steenstrup. Side view of same specimen.

20.18a Tetraclita rubescens Darwin. 32 mm diam. Malpaso Creek (Monterey Co.)

20.18b **Tetraclita rubescens elegans** Darwin. 20 mm diam. Moss Beach (San Mateo Co.). (W. Newman)

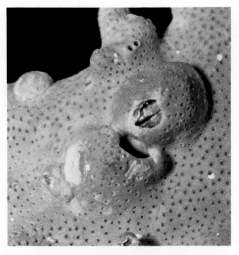

20.19 **Armatobalanus nefrens** (Zullo). 10 mm diam; obligate commensal of hydrocoral *Allopora californica.* Santa Catalina Island (Channel Islands). (W. Newman)

20.20 **Solidobalanus hesperius** (Pilsbry). 12 mm diam. Point Reyes (Marin Co.). (W. Newman)

20.21 **Membranobalanus orcutti** (Pilsbry). 10 mm diam; obligate commensal of sponges. Santa Barbara. (W. Newman)

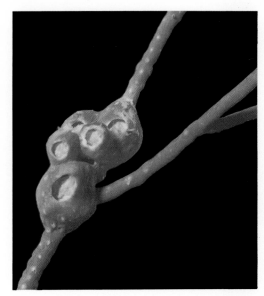

20.22 **Conopea galeata** (Linnaeus). 15 mm long; obligate commensal of gorgonians. La Jolla (San Diego Co.). (W. Newman)

20.23 **Semibalanus cariosus** (Pallas). 30 mm diam. McClures Beach (Marin Co.).

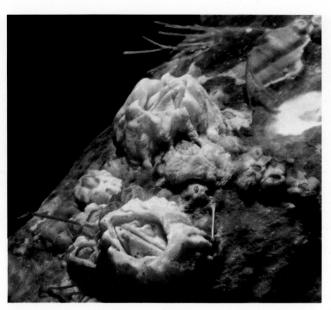

20.24 **Balanus glandula** Darwin. 15 mm diam; with small brown *Chthamalus dalli* (under 20.12). Monterey Bay. (W. Newman)

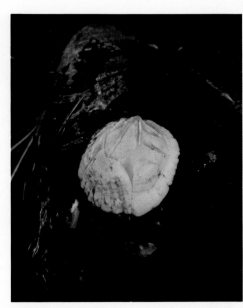

20.25 **Balanus crenatus** Bruguière. 17 mm diam. Monterey Bay. (W. Newman)

20.26 **Balanus trigonus** Darwin. 14 mm diam. La Jolla (San Diego Co.). (W. Newman)

20.27 **Balanus pacificus** Pilsbry. 18 mm diam. Mission Bay (San Diego Co.).

20.28 **Balanus amphitrite amphitrite** Darwin. 18 mm diam.
Newport Bay (Orange Co.).

20.29 **Balanus improvisus** Darwin. 10 mm diam.
San Francisco Bay. (W. Newman)

20.30 **Balanus regalis** Pilsbry. 25 mm
diam. La Jolla (San Diego Co.).
(W. Newman)

20.31 **Balanus aquila** Pilsbry.
80 mm diam. Monterey harbor.

20.32 **Balanus nubilus** Darwin. 70 mm diam.
Friday Harbor (Washington). (W. Newman)

20.33 **Megabalanus californicus** (Pilsbry). 30 mm diam.
Pfeiffer Point (Monterey Co.).

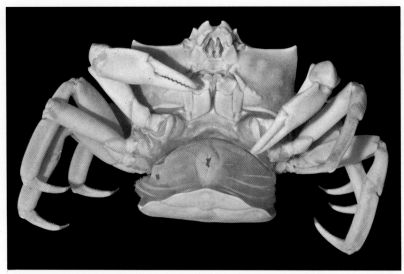

20.35 **Peltogasterella gracilis** (Boschma). Numerous cylindrical externae of parasites attached to abdomen of hermit crab *Pagurus*, each 5 mm long. Moss Beach (San Mateo Co.). (W. Newman)

20.34 **Heterosaccus californicus** Boschma. 40 mm long external portion of parasite protruding as an oval mound from ventral surface of abdomen of crab host, *Pugettia producta* (25.4). Monterey Bay. (W. Newman)

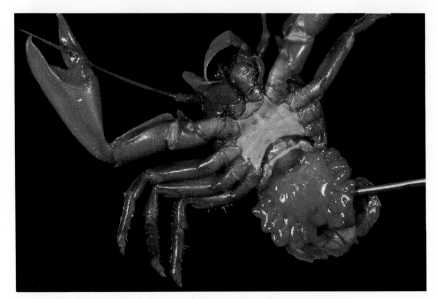

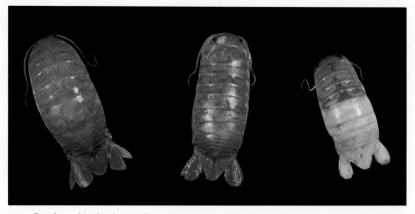

21.1 **Cirolana harfordi** (Lockington). 10–13 mm long; two-tone specimen on right has molted half of exoskeleton. Cape Mendocino (Humboldt Co.).

20.36 **Lernaeodiscus porcellanae** F. Müller. External portion of parasite protruding as a fleshy mass 6 mm long from beneath abdomen of anomuran *Petrolisthes cabrilloi* (24.17). Dana Point (Orange Co.).

21.2 Tecticeps convexus Richardson. 15 mm long. Bodega Bay harbor (Sonoma Co.).

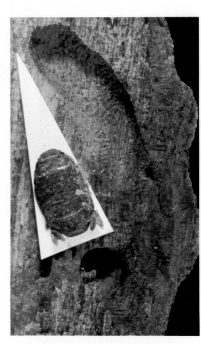

21.3 Sphaeroma quoyana H. Milne-Edwards. 10 mm long, shown adjacent to burrow in wood. Tomales Bay (Marin Co.).

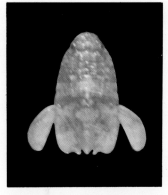

21.4 Exosphaeroma amplicauda (Stimpson). 7 mm long. Tomales Bay (Marin Co.).

21.5b Paracerceis cordata (Richardson). Female, 7 mm long. Pacific Grove (Monterey Co.).

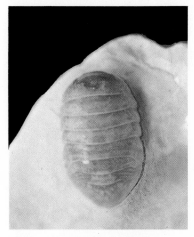

21.6 Gnorimosphaeroma luteum Menzies. Central California. (W. Lee)

21.5a Paracerceis cordata (Richardson). Male, 7 mm long, on pink coralline algae. Pillar Point (San Mateo Co.).

21.7 Lironeca vulgaris Stimpson. Fish Louse. 26 mm long. Pigeon Point (San Mateo Co.).

21.8a Limnoria (Limnoria) quadripunctata Holthuis. Gribble. 3.2 mm long, in wood burrow. Morro Bay (San Luis Obispo Co.). (E. Clogston)

21.8b Limnoria (Limnoria) quadripunctata Holthuis. 4 mm long, in wood burrow. Morro Bay (San Luis Obispo Co.). (E. Clogston)

21.9a Phyllodurus abdominalis Stimpson. Female, 14 mm long, on macruran *Upogebia pugettensis* (24.1). Tomales Bay (Marin Co.).

21.9b Phyllodurus abdominalis Stimpson. Left: female, 14 mm long; right: male, 6 mm long. Tomales Bay (Marin Co.).

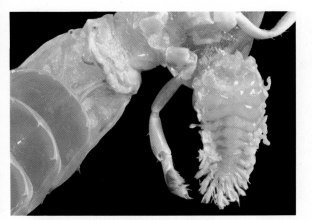

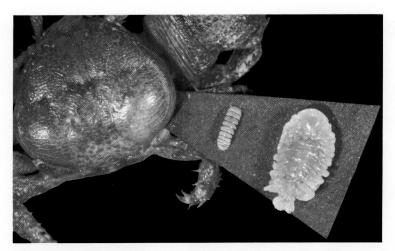

21.10 **Ione** sp. Female, 20 mm long, on macruran *Callianassa californiensis* (24.2). Elkhorn Slough (Monterey Co.).

21.11 **Aporobopyrus muguensis** Shiino. Left: male, 3 mm long; right: female, 6 mm long; taken from the gill chamber of anomuran *Pachycheles rudis* (24.20). Pacific Grove (Monterey Co.).

21.12 **Ligia (Megaligia) occidentalis** Dana. Central California. (W. Lee)

21.13a **Ligia (Ligia) pallasii** Brandt. Male. Santa Cruz Co. (W. Lee)

21.13b **Ligia (Ligia) pallasii** Brandt. Female. Central California. (W. Lee)

21.14 **Alloniscus perconvexus** Dana. Central California. (W. Lee)

21.15a Idotea (Pentidotea) stenops (Benedict).
40 mm long, shown on brown alga *Egregia*. Point
Pinos (Pacific Grove, Monterey Co.).

21.15b Idotea (Pentidotea) stenops
(Benedict). 40 mm long. Pacific Grove
(Monterey Co.).

21.16a Idotea (Pentidotea) resecata
Stimpson. 32 mm long; green
animal on eelgrass *Zostera*.
Tomales Bay (Marin Co.).

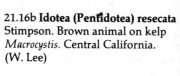

21.16b Idotea (Pentidotea) resecata
Stimpson. Brown animal on kelp
Macrocystis. Central California.
(W. Lee)

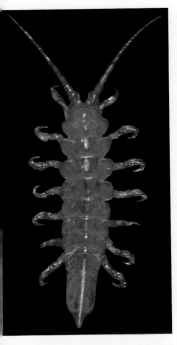

21.17 Idotea (Pentidotea) aculeata Stafford. 22 mm long. Dana Point (Orange Co.).

21.18 Idotea (Pentidotea) wosnesenskii (Brandt). 30 mm long. 17-Mile Drive (Monterey Co.).

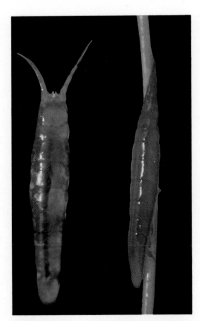

21.19 Idotea (Pentidotea) montereyensis Maloney. Color variants; green animal on surfgrass *Phyllospadix.* Central California. (W. Lee)

21.21 Idotea (Idotea) urotoma Stimpson. 16 mm long. Point Sur (Monterey Co.).

21.20 Idotea (Pentidotea) kirchanskii Miller & Lee. 15 mm long, on surfgrass *Phyllospadix.* Malpaso Creek (Monterey Co.).

21.22 Idotea (Idotea) fewkesi Richardson. Central California. (W. Lee)

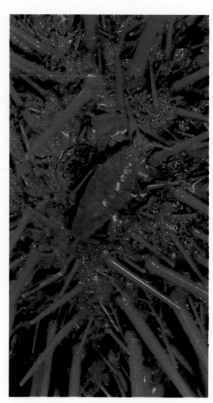

21.23 Colidotea rostrata (Benedict). 10 mm long, on sea urchin *Strongylocentrotus franciscanus* (11.6). Santa Barbara.

21.24 Synidotea bicuspida (Owen). 12 mm long. Crescent City (Del Norte Co.).

21.25 Synidotea ritteri Richardson. San Francisco Bay. (W. Lee)

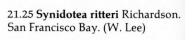

21.26 Synapseudes intumescens Menzies.
2 mm long. Pacific Grove (Monterey Co.).
(N. Burnett)

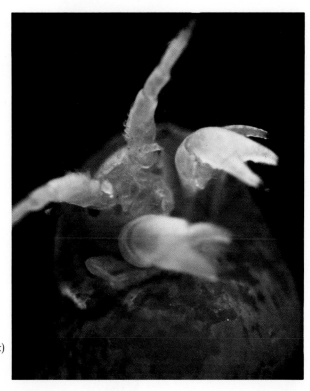

21.27 Pagurapseudes sp. Antennae 1 mm
long. Pacific Grove (Monterey Co.). (L. West)

21.28 **Pancolus californiensis** Richardson. 3 mm
long. Pacific Grove (Monterey Co.). (C. Patton)

22.1 **Orchestoidea californiana** (Brandt). 25 mm
long. Monterey Bay. (W. Lee)

21.29 **Leptochelia dubia** (Krøyer). 3 mm long. Pacific
Grove (Monterey Co.). (L. West)

22.2 Orchestoidea corniculata Stout. 22 mm long. Malpaso Creek (Monterey Co.).

22.3 Pleustes platypa Barnard & Given. Amphipod, on left, 8 mm long, closely mimicking color and form of snail *Mitrella carinata* (13.91) on right; from brown alga *Macrocystis.* Southern California. (G. Reeves)

22.4 Ampithoe plumulosa Shoemaker. 12 mm long, with tube. Mission Bay (San Diego Co.).

22.5 Cymadusa uncinata (Stout). 20 mm long. Pacific Grove (Monterey Co.).

22.6 Polycheria osborni Calman. 6 mm long. Point Pinos (Pacific Grove, Monterey Co.). (D. Abbott)

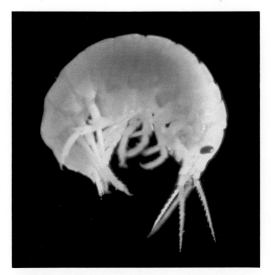

22.7 Oligochinus lighti Barnard. 5 mm long.
Pacific Grove (Monterey Co.). (E. Haderlie)

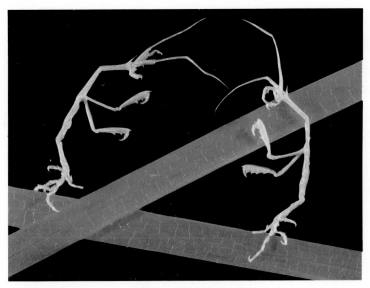

22.8 Caprella californica Stimpson. Skeleton Shrimp. 35 mm long,
on eelgrass *Zostera.* Tomales Bay (Marin Co.).

22.9 Caprella sp. 17 mm long, on hydroid *Abietinaria.* 17-Mile Drive
(Monterey Co.).

23.1 Palaemon (Palaemon) macrodactylus Rathbun. Oriental Shrimp.
Female, 50 mm long. San Francisco Bay.

23.3 Betaeus harfordi (Kingsley). Female, 18 mm long, taken from mantle cavity of abalone *Haliotis.* 17-Mile Drive (Monterey Co.).

23.2 Alpheus clamator Lockington. Snapping Shrimp, Pistol Shrimp. Male, 35 mm long. Pacific Grove (Monterey Co.).

23.4b Betaeus harrimani Rathbun. Female, 25 mm long. Elkhorn Slough (Monterey Co.).

23.4a Betaeus harrimani Rathbun. Male, 30 mm long. Laguna Beach (Orange Co.).

23.5 **Betaeus longidactylus** Lockington. 28 mm long. Dana Point (Orange Co.).

23.6 **Heptacarpus brevirostris** (Dana). 45 mm long. Color faded in dead specimen. Malpaso Creek (Monterey Co.).

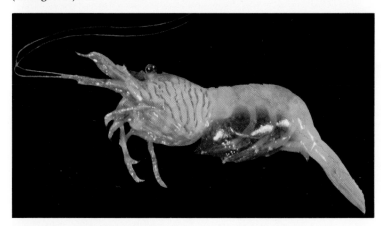

23.7 **Heptacarpus pictus** (Stimpson). Red-Banded Transparent Shrimp. 22 mm long. Pacific Grove (Monterey Co.).

23.8 **Lysmata californica** (Stimpson). Red Rock Shrimp. 45 mm long. Dana Point (Orange Co.).

23.9 **Hippolyte californiensis** Holmes. Slender Green Shrimp, Grass Shrimp. 28 mm long. Bodega Bay harbor (Sonoma Co.).

23.10 **Spirontocaris prionota** (Stimpson). Broken-Back Shrimp. 30 mm long. 17-Mile Drive (Monterey Co.).

23.11 **Crangon stylirostris** Holmes. Bay Shrimp. 48 mm long. Dillon Beach (Marin Co.).

24.1 **Upogebia pugettensis** (Dana). Blue Mud Shrimp. 95 mm long (overall, including appendages). Bolinas Lagoon (Marin Co.).

24.2 **Callianassa californiensis** Dana. Bay Ghost Shrimp. 100 mm long (overall). Elkhorn Slough (Monterey Co.).

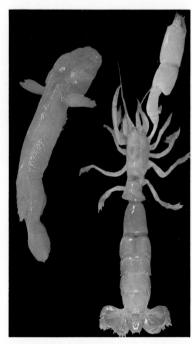

24.3 **Callianassa affinis** Holmes. Beach Ghost Shrimp. 62 mm long (overall), with blind goby *Typhlogobius*. Corona del Mar (Orange Co.).

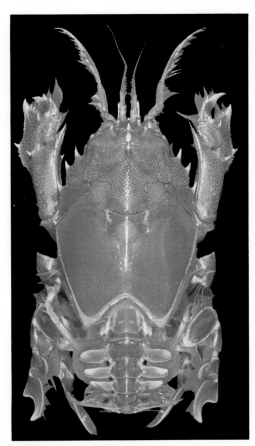

24.6 **Hapalogaster cavicauda** Stimpson. Carapace 18 mm long. Bruhel Point (Mendocino Co.).

24.4 **Emerita analoga** (Stimpson). Sand Crab, Mole Crab. Carapace 35 mm long. Half Moon Bay (San Mateo Co.).

24.5 **Blepharipoda occidentalis** Randall. Spiny Mole Crab. 80 mm long (overall). Elkhorn Beach (Monterey Co.).

24.7 **Oedignathus inermis** (Stimpson). Carapace 20 mm long. McClures Beach (Marin Co.).

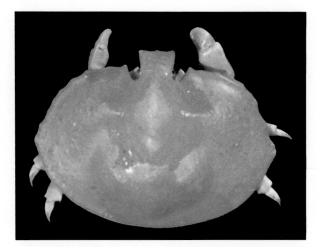

24.8a **Cryptolithodes sitchensis** Brandt. Turtle Crab, Sitka Crab, Umbrella Crab. 35 mm wide. Point Pinos (Pacific Grove, Monterey Co.).

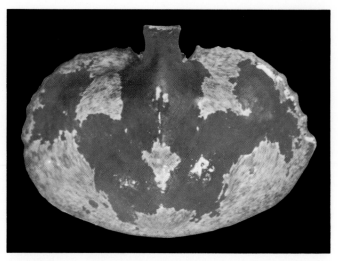

24.8b **Cryptolithodes sitchensis** Brandt. 50 mm wide; color variant. Malpaso Creek (Monterey Co.).

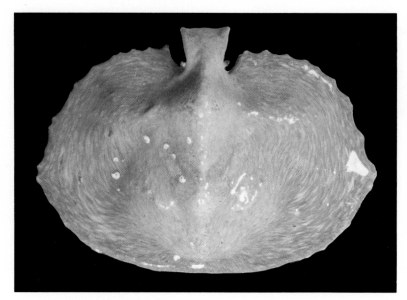

24.8c **Cryptolithodes sitchensis** Brandt. 52 mm wide; color variant. 17-Mile Drive (Monterey Co.).

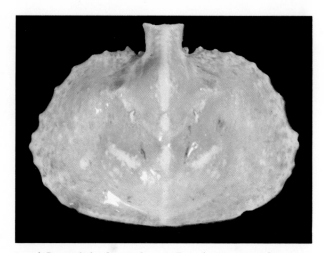

24.8d **Cryptolithodes sitchensis** Brandt. 36 mm wide; color variant. Bruhel Point (Mendocino Co.).

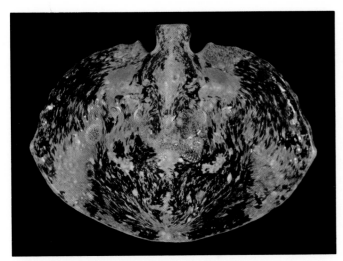

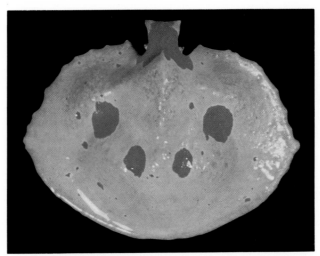

24.8e **Cryptolithodes sitchensis** Brandt. 40 mm wide; color
variant. Piedras Blancas Point (San Luis Obispo Co.).

24.8f **Cryptolithodes sitchensis** Brandt. 45 mm wide; color
variant. Pacific Grove (Monterey Co.).

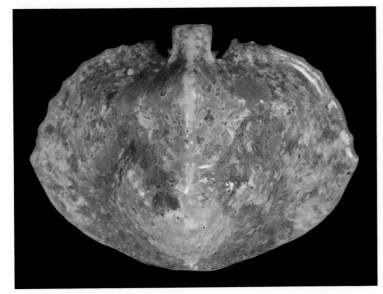

24.8g **Cryptolithodes sitchensis** Brandt. 40 mm wide; color variant.
Cabrillo Point (Mendocino Co.).

24.9 **Isocheles pilosus** (Holmes). Hermit Crab. Carapace 20 mm long;
inhabiting shell of moon snail *Polinices.* Mission Bay (San Diego Co.).

24.11 **Pagurus hirsutiusculus** (Dana). Hermit Crab. Carapace 14 mm long; inhabiting shell of snail *Nucella*. Pacific Grove (Monterey Co.).

24.10 **Pagurus samuelis** (Stimpson). Hermit Crab. Carapace 18 mm long; inhabiting shell of top snail *Calliostoma annulatum* (13.25). Pacific Grove (Monterey Co.).

24.12 **Pagurus hemphilli** (Benedict). Hermit Crab. Carapace 18 mm long; inhabiting shell of snail *Tegula brunnea* (13.31). Pacific Grove (Monterey Co.).

24.13 **Pagurus granosimanus** (Stimpson). Hermit Crab. Carapace 17 mm long; inhabiting shell of snail *Tegula brunnea* (13.31). Monterey Peninsula. (W. Lee)

24.14 **Petrolisthes rathbunae** Schmitt. Porcelain Crab.
Carapace 15 mm long. Color altered by preservation.
Isla Guadalupe (Mexico).

24.15 **Petrolisthes eriomerus** Stimpson. Porcelain Crab. Carapace
14 mm long. Cape Mendocino (Humboldt Co.).

24.16 **Petrolisthes manimaculis** Glassell. Porcelain Crab.
Carapace 14 mm long. Color altered by preservation.
Bodega Lagoon (Sonoma Co.).

24.17 **Petrolisthes cabrilloi** Glassell.
Porcelain Crab. Carapace 10 mm long.
Corona del Mar (Orange Co.).

24.18 **Petrolisthes cinctipes** (Randall). Flat Porcelain
Crab. Carapace 20 mm long. Point Saint George (Del
Norte Co.).

24.19 **Pachycheles pubescens** Holmes. Porcelain Crab. Carapace 12 mm long. Pacific Grove (Monterey Co.).

24.20 **Pachycheles rudis** Stimpson. Thick-Clawed Porcelain Crab. Carapace 16 mm long. Duxbury Point (Marin Co.).

24.21 **Pachycheles holosericus** Schmitt. Porcelain Crab. Carapace 14 mm long. Color altered by preservation. Abalone Cove (Palos Verdes, Los Angeles Co.).

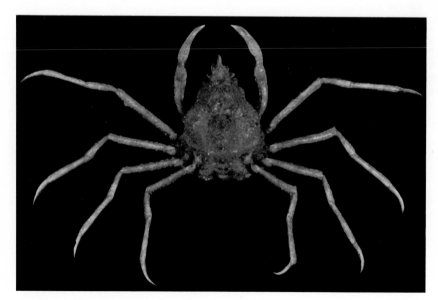

25.1 **Pyromaia tuberculata tuberculata** (Lockington). Carapace 15 mm wide. Newport Bay (Orange Co.).

25.2 **Epialtoides hiltoni** (Rathbun). Small Kelp Crab. Carapace 15 mm wide. Color altered by preservation. Bahía Magdalena (Baja California).

25.3 **Taliepus nuttallii** (Randall). Southern Kelp Crab, Globose Kelp Crab. Carapace 47 mm wide. Dana Point (Orange Co.).

25.4a **Pugettia producta** (Randall). Shield-Backed Kelp Crab, Kelp Crab. Adult, carapace 60 mm wide. Point Pinos (Pacific Grove, Monterey Co.).

25.4b **Pugettia producta** (Randall). Juvenile, carapace 20 mm wide, on brown alga *Macrocystis*. Pacific Grove (Monterey Co.).

25.5 **Pugettia gracilis** Dana. Graceful Kelp Crab. Carapace 20 mm wide. Cabrillo Point (Mendocino Co.).

25.6 **Pugettia richii** Dana. Carapace 30 mm wide. Malpaso Creek (Monterey Co.).

25.7 **Pugettia dalli** Rathbun. Male, carapace 13 mm wide. Color altered by preservation. Newport Channel (Newport Beach, Orange Co.).

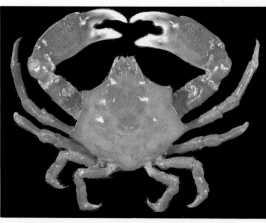

25.8 **Mimulus foliatus** Stimpson. Carapace 28 mm wide. Pacific Grove (Monterey Co.).

25.9a **Scyra acutifrons** Dana. Sharp-Nosed Crab. Carapace 26 mm wide. Pacific Grove (Monterey Co.).

25.9b **Scyra acutifrons** Dana. Carapace 13 mm wide, with attached *Gromia oviformis* (1.18). Pacific Grove (Monterey Co.).

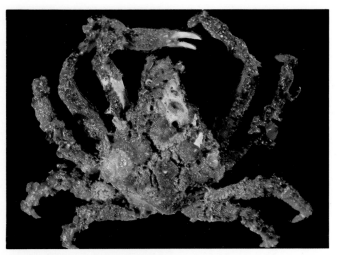

5.10a **Loxorhynchus grandis** Stimpson.
heep Crab. Adult, carapace 86 mm wide.
olor altered by preservation. Corona del
Mar (Orange Co.).

25.10b **Loxorhynchus grandis** Stimpson. Juvenile,
carapace 24 mm wide. Agua Hedionda Lagoon (San
Diego Co.).

25.11 **Loxorhynchus crispatus** Stimpson. Masking Crab, Moss
Crab. Carapace 64 mm wide. Pacific Grove (Monterey Co.).

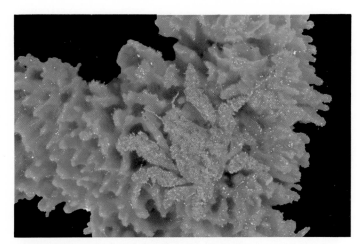

25.12b **Pelia tumida** (Lockington). Carapace 10 mm wide, on
encrusting sponge. La Jolla (San Diego Co.).

5.12a **Pelia tumida** (Lockington). Dwarf Crab. Carapace 12 mm wide.
ana Point (Orange Co.).

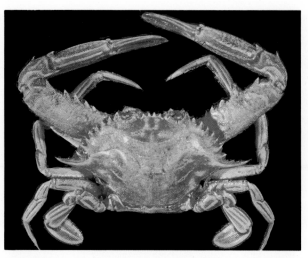

25.13 **Herbstia parvifrons** Randall. Carapace 30 mm wide. Color altered by preservation. Laguna Beach (Orange Co.).

25.14 **Portunus xantusii xantusii** (Stimpson). Swimming Crab. Carapace 37 mm wide. Mission Bay (San Diego Co.).

25.16 **Cancer antennarius** Stimpson. Rock Crab. Carapace 74 mm wide. Point Saint George (Del Norte Co.).

25.15 **Callinectes arcuatus** Ordway. Swimming Crab. Carapace 104 mm wide. Color altered by preservation. Los Angeles harbor.

25.17 **Cancer anthonyi** Rathbun. Yellow Crab. Carapace 110 mm wide. Agua Hedionda Lagoon (San Diego Co.).

25.18 **Cancer gracilis** Dana. Slender Crab, Graceful Crab. Carapace 60 mm wide. Tomales Bay (Marin Co.).

25.19 **Cancer jordani** Rathbun. Hairy Cancer Crab. Carapace 30 mm wide. Pacific Grove (Monterey Co.).

25.20 **Cancer magister** Dana. Dungeness Crab, Market Crab, Common Edible Crab. Carapace 70 mm wide. Humboldt Bay.

25.21 **Cancer oregonensis** (Dana). Oregon Cancer Crab. Carapace 36 mm wide. Cape Mendocino (Humboldt Co.).

25.22a Cancer productus Randall. Red Crab. Carapace 70 mm wide. Point Saint George (Del Norte Co.).

25.22b Cancer productus Randall. Carapace 60 mm wide; color variant. Humboldt Bay.

25.22c Cancer productus Randall. Carapace 36 mm wide; juvenile color variant. Pacific Grove (Monterey Co.).

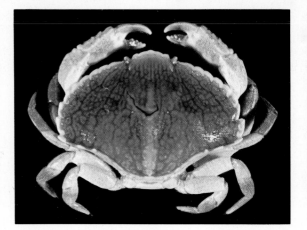

25.22d Cancer productus Randall. Carapace 45 mm wide; juvenile color variant. Elkhorn Slough (Monterey Co.).

25.22e Cancer productus Randall. Carapace 38 mm wide; juvenile color variant. San Francisco Bay.

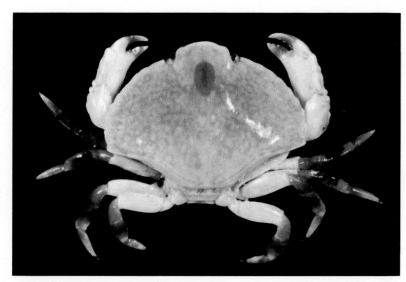

25.22f **Cancer productus** Randall. Carapace 26 mm wide; juvenile color variant. Bodega Bay harbor (Sonoma Co.).

25.22g **Cancer productus** Randall. Carapace 26 mm wide; juvenile color variant. Point Arena (Mendocino Co.).

25.22h **Cancer productus** Randall. Carapace 28 mm wide; juvenile color variant. Trinidad Head (Humboldt Co.).

25.23 **Cycloxanthops novemdentatus** (Lockington). Large Pebble Crab. Carapace 54 mm wide. Laguna Beach (Orange Co.).

25.24a **Lophopanopeus bellus bellus** (Stimpson). Black-Clawed Crab. Carapace 20 mm wide. Pacific Grove (Monterey Co.).

25.24b **Lophopanopeus bellus bellus** (Stimpson). Carapace 22 mm wide; color variant. False Klamath Rock (Del Norte Co.).

25.24c **Lophopanopeus bellus bellus** (Stimpson). Carapace 20 mm wide; color variant. Moss Beach (San Mateo Co.).

25.24d **Lophopanopeus bellus bellus** (Stimpson). Carapace 24 mm wide; color variant. Cabrillo Point (Mendocino Co.).

25.25a **Lophopanopeus leucomanus leucomanus** (Lockington). Carapace 16 mm wide. Point Fermin (San Pedro, Los Angeles Co.).

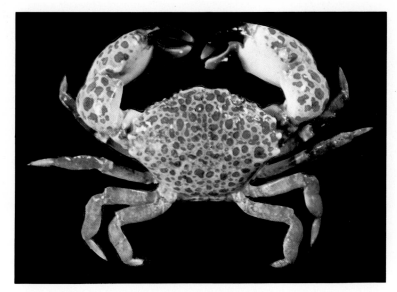

25.25b Lophopanopeus leucomanus heathii Rathbun. Carapace 24 mm wide. Pacific Grove (Monterey Co.).

25.25c Lophopanopeus leucomanus heathii Rathbun. Carapace 22 mm wide; color variant. Moss Beach (San Mateo Co.).

25.25d Lophopanopeus leucomanus heathii Rathbun. Carapace 23 mm wide; color variant. Pigeon Point (San Mateo Co.).

25.25e Lophopanopeus leucomanus heathii Rathbun. Carapace 22 mm wide; color variant. Corona del Mar (Orange Co.).

25.25f Lophopanopeus leucomanus heathii Rathbun. Carapace 23 mm wide; color variant. Pacific Grove (Monterey Co.).

25.26 **Lophopanopeus frontalis** (Rathbun). Carapace 20 mm wide. Newport Bay (Orange Co.).

25.27 **Rhithropanopeus harrisii** (Gould). Brackish-Water Crab. Carapace 17.5 mm wide. Color altered by preservation. Corte Madera Creek (San Francisco Bay).

25.28 **Paraxanthias taylori** (Stimpson). Lumpy Crab. Carapace 32 mm wide. La Jolla (San Diego Co.).

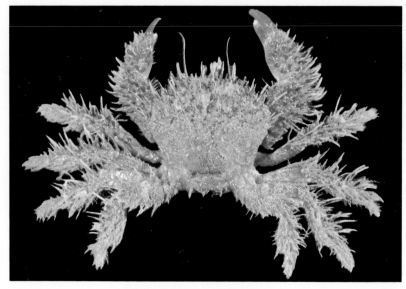

25.29 **Pilumnus spinohirsutus** (Lockington). Hairy Crab, Retiring Southerner. Carapace 25 mm wide. Dana Point (Orange Co.).

25.31 **Fabia subquadrata** Dana. Mussel Crab, Pea Crab. Female, carapace 15 mm wide; inhabiting mussel *Mytilus californianus* (15.9). Point Buchon (San Luis Obispo Co.).

25.30 **Malacoplax californiensis** (Lockington). Burrowing Crab. Carapace 26 mm wide. Newport Bay (Orange Co.).

25.32 **Fabia concharum** (Rathbun). Male, carapace 5 mm wide. Pacific Grove (Monterey Co.).

25.33 **Parapinnixa affinis** Holmes. Pea Crab. Carapace 5 mm wide. Color altered by preservation. Anaheim Bay (Orange Co.).

25.34 **Pinnixa barnharti** Rathbun. Pea Crab. Carapace 9.4 mm wide. Color altered by preservation. Newport Bay (Orange Co.).

25.35a Pinnixa faba (Dana). Large Pea Crab. Male, carapace 13 mm wide. Dillon Beach (Marin Co.).

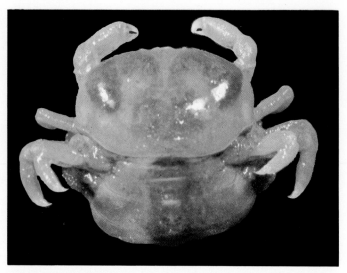

25.35b Pinnixa faba (Dana). Female, carapace 18 mm wide. Dillon Beach (Marin Co.).

25.36a Pinnixa littoralis Holmes. Pea Crab. Female, carapace 18 mm wide. Tomales Bay (Marin Co.).

25.36b Pinnixa littoralis Holmes. Juvenile, carapace 6 mm wide, growing in clam *Macoma secta* (15.51). Tomales Bay (Marin Co.).

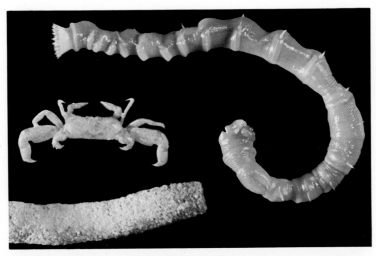

25.37 Pinnixa longipes (Lockington). Pea Crab. Carapace 6 mm wide; from tube (*below*) of annelid worm *Axiothella* (*above*). Bolinas Lagoon (Marin Co.).

25.38 Pinnixa franciscana Rathbun. Pea Crab. Carapace 16 mm wide. San Francisco Bay.

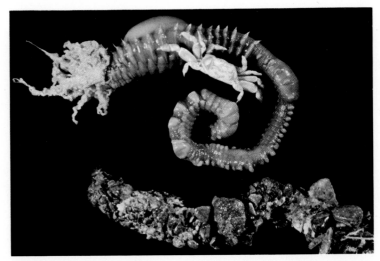

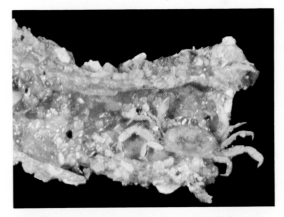

25.40 Pinnixa weymouthi Rathbun. Pea Crab. Carapace 5 mm wide; in unoccupied tube of annelid worm. Pacific Grove (Monterey Co.).

25.39 Pinnixa tubicola Holmes. Pea Crab. Carapace 9 mm wide; from tube (*below*) of terebellid worm (*above*). Bruhel Point (Mendocino Co.).

25.41 **Scleroplax granulata** Rathbun. Pea Crab. Carapace 11 mm wide. Bodega Bay harbor (Sonoma Co.).

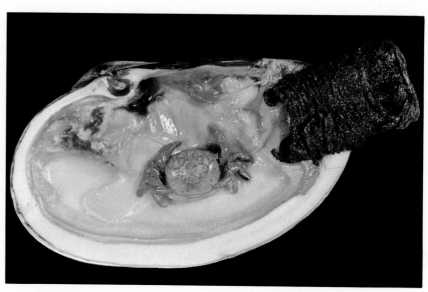

25.42 **Opisthopus transversus** Rathbun. Mottled Pea Crab. Female, carapace 18 mm wide; in mantle cavity of clam *Tresus nuttallii* (15.48). Tijuana Slough (San Diego Co.).

25.43a **Pachygrapsus crassipes** Randall. Striped Shore Crab, Lined Shore Crab. Carapace 38 mm wide. Pacific Grove (Monterey Co.).

25.43b **Pachygrapsus crassipes** Randall. Carapace about 40 mm wide; individual in tidepool, hiding under edge of carapace of rock crab *Cancer antennarius* (25.16). Pigeon Point (San Mateo Co.). (C. Baxter)

25.44 Hemigrapsus nudus (Dana). Purple Shore Crab. Male, carapace 45 mm wide. Pacific Grove (Monterey Co.).

25.45 Hemigrapsus oregonensis (Dana). Yellow Shore Crab, Mud-Flat Crab. Male, carapace 28 mm wide. San Diego Bay.

25.46 Uca crenulata crenulata (Lockington). Fiddler Crab. Male, carapace 18 mm wide. Newport Bay (Orange Co.).

26.1a Tigriopus californicus (Baker). Male, 1 mm long. Pacific Grove (Monterey Co.). (D. Smith)

26.1b Tigriopus californicus (Baker). Female, 1 mm long. Pacific Grove (Monterey Co.). (D. Smith)

26.2 **Nebalia pugettensis** (Clark). 6 mm long. Elkhorn Slough (Monterey Co.).

26.3a **Hemisquilla ensigera californiensis** (Stephenson). Mantis Shrimp. Male, 185 mm long. Dredged off Venice (Los Angeles Co.). (R. Caldwell)

26.3b **Hemisquilla ensigera californiensis** (Stephenson). Close-up of same specimen.

27.1 **Pycnogonum stearnsi** Ives. 10 mm long. 17-Mile Drive (Monterey Co.).

27.2 **Pycnogonum rickettsi** Schmitt. 9 mm long. Monterey.

27.3 **Ammothea hilgendorfi** (Böhm). 20 mm long. San Francisco Bay.

27.4 **Nymphopsis spinosissima** (Hall). 13 mm long. Pacific Grove (Monterey Co.).

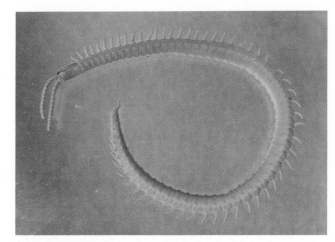

28.1 **Nyctunguis heathii** (Chamberlin). Centipede. 45 mm long. Monterey. (W. Evans)

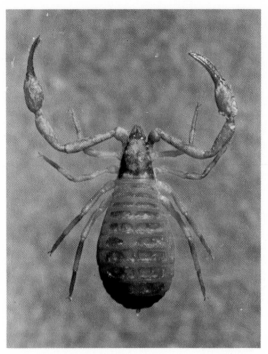

28.2 Garypus californicus Banks. Pseudoscorpion.
4.5 mm long. Monterey. (W. Evans)

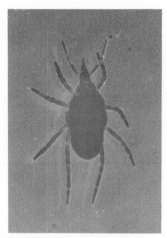

28.3 Neomolgus littoralis
(Linnaeus). Mite. 3 mm long.
Monterey. (W. Evans)

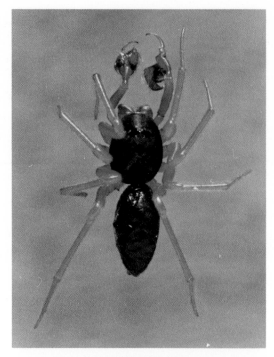

28.4 Spirembolus mundus Chamberlin & Ivie.
Spider. 2.5 mm long. Monterey. (W. Evans)

28.5 Neomachilis halophila Silvestri. Bristletail. 15 mm
long. Pacific Grove (Monterey Co.). (W. Evans)

28.6a Limonia (Idioglochina) marmorata (Osten-Sacken). Crane Fly. Adult, 9 mm long. Carmel (Monterey Co.). (W. Evans)

28.6b Limonia (Idioglochina) marmorata (Osten-Sacken). Left: pupa, 10 mm long; right: larva, 12 mm long. Carmel (Monterey Co.). (W. Evans)

28.7 Paraclunio trilobatus Kieffer. Midge. 6 mm long. Carmel (Monterey Co.). (W. Evans)

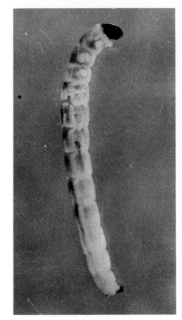

28.8 Tethymyia aptena Wirth. Midge. Larva, 2 mm long. Carmel (Monterey Co.). (W. Evans)

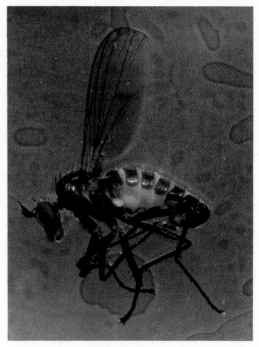

28.9 Aphrosylus praedator Wheeler. Long-Legged Fly. 3 mm long. Monterey. (W. Evans)

28.10a **Coelopa (Neocoelopa) vanduzeei** Cresson. Kelp Fly. Adult, 6.75 mm long; isolated wing showing venation. Pacific Grove (Monterey Co.). (W. Evans)

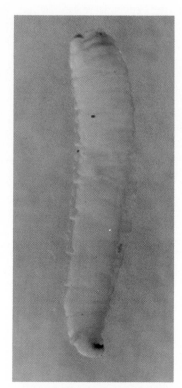

28.10b **Coelopa (Neocoelopa) vanduzeei** Cresson. Larva, 8.5 mm long. Pacific Grove (Monterey Co.). (W. Evans)

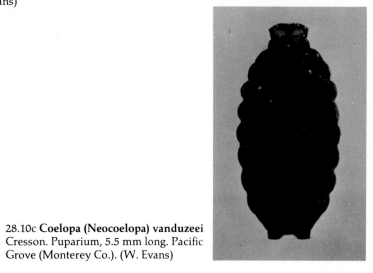

28.10c **Coelopa (Neocoelopa) vanduzeei** Cresson. Puparium, 5.5 mm long. Pacific Grove (Monterey Co.). (W. Evans)

28.11 Canaceoides nudatus (Cresson). Beach Fly. Larva, 5 mm long. Pacific Grove (Monterey Co.). (W. Evans)

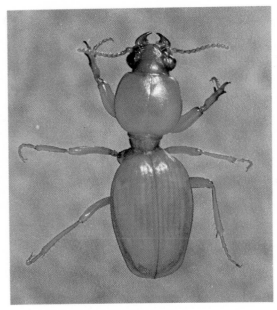

28.12a **Dyschirius marinus** (LeConte). Ground Beetle. Adult, 6.5 mm long. Pacific Grove (Monterey Co.). (W. Evans)

28.12b **Dyschirius marinus** (LeConte). Larva, 9.5 mm long. Pacific Grove (Monterey Co.). (W. Evans)

28.13 **Trechus ovipennis** Motschulsky. Ground Beetle. 5 mm long. Bodega Bay (Sonoma and Marin Cos.). (W. Evans)

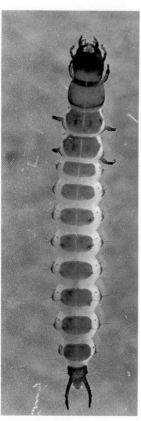

28.16 **Cercyon luniger** Mannerheim. Water-Scavenger Beetle. 3.5 mm long. Carmel (Monterey Co.). (W. Evans)

28.14a **Thalassotrechus barbarae** (Horn). Ground Beetle. Adult, 7 mm long. Pacific Grove (Monterey Co.). (W. Evans)

28.14b **Thalassotrechus barbarae** (Horn). Larva, 6 mm long. Pacific Grove (Monterey Co.). (W. Evans)

28.15 **Cercyon fimbriatus** Mannerheim. Water-Scavenger Beetle. 3 mm long. Carmel (Monterey Co.). (W. Evans)

28.17 **Ochthebius vandykei** Knisch. Limnebiid Beetle. 1.5 mm long. Monterey. (W. Evans)

28.18 **Bledius monstratus** Casey. Rove Beetle. 4.5 mm long. Pacific Grove (Monterey Co.). (W. Evans)

28.19 **Cafius luteipennis** Horn. Rove Beetle. 6 mm long. Pacific Grove (Monterey Co.). (W. Evans)

28.20 **Cafius seminitens** (Horn). Rove Beetle. 10 mm long. Pacific Grove (Monterey Co.). (W. Evans)

28.21a Hadrotes crassus
(Mannerheim). Rove Beetle.
Adult, 14 mm long. Carmel
(Monterey Co.). (W. Evans)

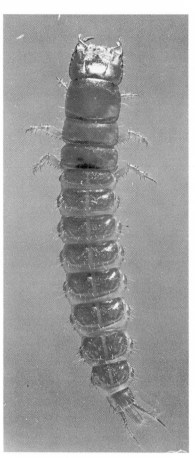

28.21b Hadrotes crassus (Mannerheim). Larva, 15 mm long. Carmel
(Monterey Co.). (W. Evans)

28.22a Thinopinus pictus LeConte. Rove
Beetle. Light adult, found on light-colored
sand, 16 mm long. Pacific Grove (Monterey
Co.). (W. Evans)

28.22b Thinopinus pictus LeConte. Dark
adult, found on dark-colored sand, 18 mm
long. Bodega Bay (Sonoma and Marin Cos.).
(W. Evans)

28.23 **Bryobiota bicolor** (Casey). Rove Beetle.
2.5 mm long. Monterey. (W. Evans)

28.24 **Diaulota densissima** Casey.
Rove Beetle. 2.5 mm long. Pacific
Grove (Monterey Co.). (W. Evans)

28.25 **Diaulota vandykei**
(Moore). Rove Beetle. 2.8 mm
long. Pacific Grove (Monterey
Co.). (W. Evans)

28.22c **Thinopinus pictus** LeConte. Larva,
15 mm long. Pacific Grove (Monterey Co.).
(W. Evans)

28.26 **Liparocephalus cordicollis** LeConte. Rove Beetle. 4 mm long. Pacific Grove (Monterey Co.). (W. Evans)

28.27 **Pontomalota nigriceps** Casey. Rove Beetle. 3 mm long. Carmel (Monterey Co.). (W. Evans)

28.28 **Motschulskium sinuatocolle** Matthews. Feather-Winged Beetle. 1.3 mm long. Pacific Grove (Monterey Co.). (W. Evans)

28.29 **Neopachylopus sulcifrons** (Mannerheim). Hister Beetle. 4 mm long. Pacific Grove (Monterey Co.). (W. Evans)

28.30a **Endeodes collaris** (LeConte). Soft-Winged Flower Beetle. Adult, 6 mm long. Monterey. (W. Evans)

28.30b **Endeodes collaris** (LeConte). Larva, 6 mm long. Monterey. (W. Evans)

28.31 **Epantius obscurus** (LeConte). Darkling Beetle. 8 mm long. Carlsbad Beach (San Diego Co.). (W. Evans)

28.32 **Phaleria rotundata** LeConte. Darkling Beetle. 5 mm long. Carmel (Monterey Co.). (W. Evans)

28.33 **Aegialites subopacus** (Van Dyke). Salpingid Beetle. 3.5 mm long. Bodega Bay (Sonoma and Marin Cos.). (W. Evans)

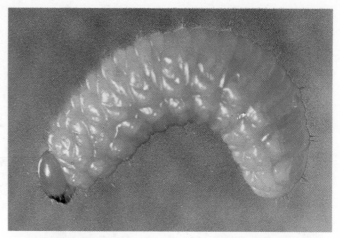

28.34a **Emphyastes fucicola** Mannerheim. Weevil. Adult, 8 mm long. Monterey. (W. Evans)

28.34b **Emphyastes fucicola** Mannerheim. Larva, 8 mm long. Monterey. (W. Evans)

28.35 **Phycocoetes testaceus** LeConte. Weevil. 3 mm long. Pacific Grove (Monterey Co.). (W. Evans)

Index

Index

For each species treated in the text, the scientific name is followed by its unique decimal serial number and the numbers of the pages where the species is mentioned; a boldface page number indicates the page on which the principal treatment begins. Names set in large and small capitals are of taxa at levels above genus. Names set in italic type are of synonyms or are names incorrectly applied. Numbers preceded by a P indicate pages in the photograph section.

Common species names of invertebrates are indexed for the principal account and for uses in the text not accompanied by the scientific name. Except for the principal account, in all cases where the scientific and common names of a particular invertebrate species occur together, only the scientific name is indexed. Common names of groups above the species level (e.g., barnacles, hydroids, octopuses, etc.) are indexed for uses in the text not accompanied by the scientific name. Someone desiring all references in the book to barnacles, for example, will need to consult the index references to "Cirripedia," to "barnacles," and to the scientific names of the barnacle species.

All references to species of marine plants and vertebrate animals (fishes, reptiles, birds, mammals) are indexed under their common or their scientific names as used in text.